BINA

Supplement D2:
The chemistry of
halides, pseudo-halides and azides

THE CHEMISTRY OF FUNCTIONAL GROUPS

A series of advanced treatises under the general editorship of
Professors Saul Patai and Zvi Rappoport

The chemistry of alkenes (2 volumes)
The chemistry of the carbonyl group (2 volumes)
The chemistry of the ether linkage
The chemistry of the amino group
The chemistry of the nitro and nitroso groups (2 parts)
The chemistry of carboxylic acids and esters
The chemistry of the carbon–nitrogen double bond
The chemistry of amides
The chemistry of the cyano group
The chemistry of the hydroxyl group (2 parts)
The chemistry of the azido group
The chemistry of acyl halides
The chemistry of the carbon–halogen bond (2 parts)
The chemistry of the quinonoid compounds (2 volumes, 4 parts)
The chemistry of the thiol group (2 parts)
The chemistry of the hydrazo, azo and azoxy groups (2 parts)
The chemistry of amidines and imidates (2 volumes)
The chemistry of cyanates and their thio derivatives (2 parts)
The chemistry of diazonium and diazo groups (2 parts)
The chemistry of the carbon–carbon triple bond (2 parts)
The chemistry of ketenes, allenes and related compounds (2 parts)
The chemistry of the sulphonium group (2 parts)
Supplement A: The chemistry of double-bonded functional groups (2 volumes, 4 parts)
Supplement B: The chemistry of acid derivatives (2 volumes, 4 parts)
Supplement C: The chemistry of triple-bonded functional groups (2 volumes, 3 parts)
Supplement D: The chemistry of halides, pseudo-halides and azides (3 volumes, 6 parts)
Supplement E: The chemistry of ethers, crown ethers, hydroxyl groups
and their sulphur analogues (2 volumes, 3 parts)
Supplement F: The chemistry of amino, nitroso and nitro compounds
and their derivatives (2 parts)
The chemistry of the metal–carbon bond (5 volumes)
The chemistry of peroxides
The chemistry of organic selenium and tellurium compounds (2 volumes)
The chemistry of the cyclopropyl group (2 parts)
The chemistry of sulphones and sulphoxides
The chemistry of organic silicon compounds (2 parts)
The chemistry of enones (2 parts)
The chemistry of sulphinic acids, esters and their derivatives
The chemistry of sulphenic acids and their derivatives
The chemistry of enols
The chemistry of organophosphorus compounds (3 volumes)
The chemistry of sulphonic acids, esters and their derivatives
The chemistry of alkanes and cycloalkanes
Supplement S: The chemistry of sulphur-containing functional groups
The chemistry of organic arsenic, antimony and bismuth compounds
The chemistry of enamines (2 parts)

UPDATES

The chemistry of α-haloketones, α-haloaldehydes and α-haloimines
Nitrones, nitronates and nitroxides
Crown ethers and analogs
Cyclopropanederived reactive intermediates
Synthesis of carboxylic acids, esters and their derivatives
The silicon–heteroatom bond
Syntheses of lactones and lactams
The syntheses of sulphones, sulphoxides and cyclic sulphides

Patai's 1992 guide to the chemistry of functional groups—*Saul Patai*

$$-\overset{|}{\underset{|}{C}}-X \; , \quad -\overset{|}{\underset{|}{C}}-N_3 \; , \quad -\overset{|}{\underset{|}{C}}-OCN$$

Supplement D2
The chemistry of
halides, pseudo-halides
and azides
Part 2

Edited by

SAUL PATAI

The Hebrew University, Jerusalem

and

ZVI RAPPOPORT

The Hebrew University, Jerusalem

1995

JOHN WILEY & SONS

CHICHESTER–NEW YORK– BRISBANE–TORONTO–SINGAPORE

An Interscience® Publication

Copyright © 1995 by John Wiley & Sons Ltd,
Baffins Lane, Chichester,
West Sussex PO19 1UD, England

Telephone: National   01243 779777
International (+44) 1243 779777

Other Wiley Editorial Offices

John Wiley & Sons, Inc., 605 Third Avenue,
New York, NY 10158-0012, USA

Jacaranda Wiley Ltd, 33 Park Road, Milton,
Queensland 4064, Australia

John Wiley & Sons (Canada) Ltd, 22 Worcester Road,
Rexdale, Ontario M9W 1L1, Canada

John Wiley & Sons (SEA) Pte Ltd, 37 Jalan Pemimpin #05-04,
Block B, Union Industrial Building, Singapore 2057

Library of Congress Cataloging-in-Publication Data

The Chemistry of halides, pseudo-halides, and azides / edited by Saul
Patai and Zvi Rappoport.—[Rev. ed.]
p.   cm.—(The Chemistry of functional groups. Supplement ;
D)
"An Interscience publication."
Includes bibliographical references and index.
ISBN 0–471–94209–X (v. 2)
1. Halides.   2. Azides.   I. Patai, Saul.   II. Rappoport, Zvi.
III. Series.
QD165.C48   1995
546′.73—dc20   94–18657
CIP

British Library Cataloguing in Publication Data

A catalogue record for this book is available from the British Library

ISBN   0 471 94209 X

Typeset in 9/10 Times by Thomson Press (India) Ltd, New Delhi, India.
Printed and bound in Great Britain by Biddles Ltd, Guildford, Surrey

To a good friend
Yuho Tsuno

Foreword

The present volume, *Supplement D2* retains the title of its direct predecessor, *Supplement D: The chemistry of halides, pseudo-halides and azides*, even though it deals only with derivatives of fluorine, chlorine, bromine, iodine and astatine. Hence, in order to keep the *Chemistry of the functional groups* series complete and up-to-date, it will be necessary to publish, in the not too distant future, an additional supplementary volume ('D3'?) discussing the recent advances in the chemistry of azides, cyanates, isocyanates and their thio derivatives.

Only four chapters which were planned and promised for the present volume did not materialize. These were the following: 'Acidity, basicity, H-bonding and complex formation'; 'Hypervalent halogen compounds'; 'Toxicology and pharmacology'; and 'Perfluorocarbons'.

We will be very grateful to readers who would call our attention to mistakes or omissions in the present volume, as well as in other volumes of the series.

Jerusalem
January 1995

SAUL PATAI
ZVI RAPPOPORT

The Chemistry of Functional Groups
Preface to the series

The series 'The Chemistry of Functional Groups' was originally planned to cover in each volume all aspects of the chemistry of one of the important functional groups in organic chemistry. The emphasis is laid on the preparation, properties and reactions of the functional group treated and on the effects which it exerts both in the immediate vicinity of the group in question and in the whole molecule.

A voluntary restriction on the treatment of the various functional groups in these volumes is that material included in easily and generally available secondary or tertiary sources, such as Chemical Reviews, Quarterly Reviews, Organic Reactions, various 'Advances' and 'Progress' series and in textbooks (i.e. in books which are usually found in the chemical libraries of most universities and research institutes), should not, as a rule, be repeated in detail, unless it is necessary for the balanced treatment of the topic. Therefore each of the authors is asked not to give an encyclopaedic coverage of his subject, but to concentrate on the most important recent developments and mainly on material that has not been adequately covered by reviews or other secondary sources by the time of writing of the chapter, and to address himself to a reader who is assumed to be at a fairly advanced postgraduate level.

It is realized that no plan can be devised for a volume that would give a complete coverage of the field with no overlap between chapters, while at the same time preserving the readability of the text. The Editors set themselves the goal of attaining reasonable coverage with moderate overlap, with a minimum of cross-references between the chapters. In this manner, sufficient freedom is given to the authors to produce readable quasimonographic chapters.

The general plan of each volume includes the following main sections:

(a) An introductory chapter deals with the general and theoretical aspects of the group.

(b) Chapters discuss the characterization and characteristics of the functional groups. i.e. qualitative and quantitative methods of determination including chemical and physical methods, MS, UV, IR, NMR, ESR and PES—as well as activating and directive effects exerted by the group, and its basicity, acidity and complex-forming ability.

(c) One or more chapters deal with the formation of the functional group in question, either from other groups already present in the molecule or by introducing the new group directly or indirectly. This is usually followed by a description of the synthetic uses of the group, including its reactions, transformations and rearrangements.

(d) Additional chapters deal with special topics such as electrochemistry. photochemistry, radiation chemistry, thermochemistry, syntheses and uses of isotopically labelled compounds, as well as with biochemistry, pharmacology and toxicology. Whenever applicable, unique chapters relevant only to single functional groups are also included (e.g. 'Polyethers', 'Tetraaminoethylenes' or 'Siloxanes').

This plan entails that the breadth, depth and thought-provoking nature of each chapter will differ with the views and inclinations of the authors and the presentation will necessarily be somewhat uneven. Moreover, a serious problem is caused by authors who deliver their manuscript late or not at all. In order to overcome this problem at least to some extent, some volumes may be published without giving consideration to the originally planned logical order of the chapters.

Since the beginning of the Series in 1964, two main developments have occurred. The first of these is the publication of supplementary volumes which contain material relating to several kindred functional groups (Supplements A, B, C, D, E, F and S). The second ramification was the publication of a series of 'Updates', which contained in each volume selected and related chapters, reprinted in the original form in which they were published, together with an extensive updating of the subjects, if possible, by the authors of the original chapters. Unfortunately the publication of these 'Updates' had to be discontinued.

A complete list of all mentioned volumes published to date will be found on the page opposite the inner title page of this book.

Advice or criticism regarding the plan and execution of this series will be welcomed by the Editors.

The application of this series would never have been started, let alone continued, without the support of many persons in Israel and overseas, including colleagues, friends and family. The efficient and patient co-operation of staff-members of the publisher also rendered us invaluable aid. Our sincere thanks are due to all of them.

The Hebrew University   SAUL PATAI
Jerusalem, Israel   ZVI RAPPOPORT

Contributing authors

Zeev B. Alfassi	Department of Nuclear Engineering, Ben Gurion University of the Negev, Beer Sheva 84102, Israel
Mark S. Baird	Department of Chemistry, University of Wales, Bangor, Gwynedd, LL57 2UW, Wales, UK
Klara Berei	KFKI Atomic Energy Research Institute, P. O. Box 49, H-1525 Budapest 114, Hungary
Ivan G. Bolesov	Department of Chemistry, Moscow State University, Moscow, Russia
James H. Brewster	Department of Chemistry, Purdue University, West Lafayette, Indiana 47907-3699, USA
Joseph Casanova	Department of Chemistry, California State University, Los Angeles, California 90032, USA
Gabriel Chuchani	Department of Chemistry, Instituto Venezolano de Investigaciones Científicas (IVIC), Apartado 21827, Caracas 1020-A, Venezuela
Jan Cornelisse	Leiden Institute of Chemistry, Gorlaeus Laboratories, Leiden University, P.O. Box 9502, 2300 RA Leiden, The Netherlands
Brian Everatt	Ciba Geigy plc, Hulley Road, Macclesfield, Cheshire, SK10 2NX, UK
Joseph S. Francisco	Department of Chemistry, Purdue University, 1393 Brown Building, West Lafayette, Indiana 47907-1393, USA
Michel Geoffroy	Departement de Chimie Physique, Université de Genève, Sciences II, 30 Quai Ernest-Ansermet, 1211 Genève 4, Switzerland
James R. Green	Department of Chemistry and Biochemistry, University of Windsor, Windsor, Ontario N9B 3P4, Canada
Jeff Hoyle	Department of Chemistry and Soil Science, Nova Scotia Agricultural College, P.O. Box 550, Truro, Nova Scotia B2N 5E3, Canada
Marianna Kańska	Department of Chemistry, Warsaw University, Warsaw, Poland
Kenneth L. Kirk	Department of Health & Human Services, National Institutes of Health, National Institute of Diabetes and Digestive and Kidney Diseases, Bethesda, Maryland 20892, USA
L. Klasinc	Institute Rugjer Bošković, Zagreb, Croatia
Gerald F. Koser	Department of Chemistry, Buchtel College of Arts and Sciences, The University of Akron, Akron, Ohio 44325-3601, USA

Joel F. Liebman   Department of Chemistry and Biochemistry, University of
                         Maryland, Baltimore County Campus, 5401 Wilkens Ave.,
                         Baltimore, Maryland 21228-5398, USA
Gerrit Lodder   Leiden Institute of Chemistry, Gorlaeus Laboratories, Leiden
                         University, P.O. Box 9502, 2300 RA Leiden, The Netherlands
Edwin A. C. Lucken   Departement de Chimie Physique, Université de Genève, Sciences
                         II, 30 Quai Ernest-Ansermet, 1211 Genève 4, Switzerland
Hisham G. Lutfi   Department of Chemistry and Biochemistry, Southern
                         Illinois University at Carbondale, Carbondale, Illinois 62901,
                         USA
W. Gary Mallard   Chemical Kinetics and Thermodynamics Division, National
                         Institute of Standards and Technology, Gaithersburg,
                         Maryland 20899-0001, USA
S. P. McGlynn   Department of Chemistry, Louisiana State University, Baton
                         Rouge, Louisiana 70803, USA
Cal Y. Meyers   Department of Chemistry and Biochemistry, Southern Illinois
                         University at Carbondale, Carbondale, Illinois 62901, USA
Jane S. Murray   Department of Chemistry, University of New Orleans, New
                         Orleans, Louisiana 70148, USA
I. Novak   Department of Chemistry, National University of Singapore,
                         Singapore
Heinz Oberhammer   Institut für Physikalische und Theoretische Chemie, Universität
                         Tübingen, 76072 Tübingen, Germany
Cristina Paradisi   Dipartimento di Chimica Organica dell'Università, Centro
                         Studio Meccanismi Reazioni del CNR, Via Marzolo 1, 35131
                         Padova, Italy
R. V. Parish   Department of Chemistry, UMIST, P.O. Box 88, Manchester,
                         M60 1QD, UK
Alicia B. Peñéñory   Departamento de Química Orgánica, Facultad de Ciencias
                         Químicas, Universidad Nacional de Córdoba, CC 61, Suc. 16,
                         5016 Córdoba, Argentina
Adriana B. Pierini   Departamento de Química Orgánica, Facultad de Ciencias
                         Químicas, Universidad Nacional de Córdoba, CC 61, Suc. 16,
                         5016 Córdoba, Argentina
Peter Politzer   Department of Chemistry, University of New Orleans, New
                         Orleans, Louisiana 70148, USA
V. Prakash Reddy   Department of Chemistry, University of Miami, Coral Gables,
                         Florida 33124, USA
Roberto A. Rossi   Departamento de Química Organica, Facultad de Ciencias
                         Químicas, Universidad Nacional de Córdoba, CC 61, Suc. 16,
                         5016 Córdoba, Argentina
Shlomo Rozen   School of Chemistry, Raymond and Beverly Sackler Faculty of
                         Exact Sciences, Tel Aviv University, Ramat Aviv, Tel Aviv
                         69978, Israel
Yoel Sasson   Casali Institute of Applied Chemistry, The Hebrew University
                         of Jerusalem, Jerusalem 91904, Israel
Suzanne W. Slayden   Department of Chemistry, George Mason University, 4400
                         University Drive, Fairfax, Virginia 22030-4444, USA
Peter Taylor   Department of Chemistry, The Open University, Walton Hall,
                         Milton Keynes, MK7 6AA, UK
László Vasáros   KFKI Atomic Energy Research Institute, P.O. Box 49, H-1525
                         Budapest 114, Hungary

Umberto Vettori	Centro Studio Stabilità e Reattività dei Composti dei Coordinazione del CNR, Via Marzolo 1, 35131 Padova, Italy
Ian H. Williams	School of Chemistry, University of Bath, Bath, BA2 7AY, UK
Mieczysław Zieliński	Isotope Laboratory, Faculty of Chemistry, Jagiellonian University, ul. Ingardena 3, 30-060 Krakow, Poland
Marko Zupan	Laboratory for Organic and Bioorganic Chemistry, Department of Chemistry and 'J. Stefan' Institute, University of Ljubljana, Ljubljana, Slovenia

Contents

Contents

List of abbreviations used

Ac	acetyl (MeCO)
acac	acetylacetone
Ad	adamantyl
AIBN	azoisobutyronitrile
Alk	alkyl
All	allyl
An	anisyl
Ar	aryl
Bz	benzoyl (C_6H_5CO)
Bu	butyl (also t-Bu or Bu^t)
CD	circular dichroism
CI	chemical ionization
CIDNP	chemically induced dynamic nuclear polarization
CNDO	complete neglect of differential overlap
Cp	η^5-cyclopentadienyl
Cp*	η^5-pentamethylcyclopentadienyl
DABCO	1,4-diazabicyclo[2.2.2]octane
DBN	1,5-diazabicyclo[4.3.0]non-5-ene
DBU	1,8-diazabicyclo[5.4.0]undec-7-ene
DIBAH	diisobutylaluminium hydride
DME	1,2-dimethoxyethane
DMF	N,N-dimethylformamide
DMSO	dimethyl sulphoxide
ee	enantiomeric excess
EI	electron impact
ESCA	electron spectroscopy for chemical analysis
ESR	electron spin resonance
Et	ethyl
eV	electron volt
FC	ferrocenyl
FD	field desorption
FI	field ionization

FT	Fourier transform
Fu	furyl (OC_4H_3)
GLC	gas–liquid chromatography
Hex	hexyl (C_6H_{13})
c-Hex	cyclohexyl (C_6H_{11})
HMPA	hexamethylphosphortriamide
HOMO	highest occupied molecular orbital
HPLC	high performance liquid chromatography
i-	iso
Ip	ionization potential
IR	infrared
ICR	ion cyclotron resonance
LAH	lithium aluminium hydride
LCAO	linear combination of atomic orbitals
LDA	lithium diisopropylamide
LUMO	lowest unoccupied molecular orbital
M	metal
M	parent molecule
MCPBA	m-chloroperbenzoic acid
Me	methyl
MNDO	modified neglect of diatomic overlap
MS	mass spectrum
n	normal
Naph	naphthyl
NBS	N-bromosuccinimide
NCS	N-chlorosuccinimide
NMR	nuclear magnetic resonance
Pc	phthalocyanine
Pen	pentyl (C_5H_{11})
Pip	piperidyl ($C_5H_{10}N$)
Ph	phenyl
ppm	parts per million
Pr	propyl (also i-Pr or Pr^i)
PTC	phase transfer catalysis or phase transfer conditions
Pyr	pyridyl (C_5H_4N)
R	any radical
RT	room temperature
s-	secondary
SET	single electron transfer
SOMO	singly occupied molecular orbital

t-	tertiary
TCNE	tetracyanoethylene
TFA	trifluoroacetic acid
THF	tetrahydrofuran
Thi	thienyl (SC_4H_3)
TLC	thin layer chromatography
TMEDA	tetramethylethylene diamine
TMS	trimethylsilyl or tetramethylsilane
Tol	tolyl (McC_6H_4)
Tos *or* Ts	tosyl (*p*-toluenesulphonyl)
Trityl	triphenylmethyl (Ph_3C)
Xyl	xylyl ($Me_2C_6H_3$)

In addition, entries in the 'List of Radical Names' in *IUPAC Nomenclature of Organic Chemistry*, 1979 Edition. Pergamon Press, Oxford, 1979, p. 305–322, will also be used in their unabbreviated forms, both in the text and in formulae, instead of explicitly drawn structures.

CHAPTER **16**

Recent advances in the photochemistry of the carbon–halogen bond

GERRIT LODDER and JAN CORNELISSE

Leiden Institute of Chemistry, Gorlaeus Laboratories, Leiden University, P.O. Box 9502, 2300 RA Leiden, The Netherlands

I. INTRODUCTION

In this Series, two previous chapters about the photochemistry of the carbon–halogen bond appeared, which covered literature data up to 1971 and to 1981, respectively[1,2]. Before 1971, the photobehaviour of halocarbon compounds in solution was explained solely in terms of homolytic cleavage of the C—X bond, followed by reactions character-

Supplement D2: The chemistry of halides, pseudo-halides and azides
Edited by S. Patai and Z. Rappoport   © 1995 John Wiley & Sons Ltd

istic of the radical intermediates formed. Later, the importance of heterolytic and electron-transfer mediated cleavage processes became clear. Since then, many more examples of the latter types of photobehaviour of organic halides have been discovered and they are emphasized in the present update.

Apart from References 1–4, which cover, more or less, the whole field of the photochemistry of the C—X bond, a number of other reviews have appeared which deal with the photoreactions of subclasses of organic halides or with subclasses of photoreactions of organic halides[5–18]. Their subjects, in order of appearance in this chapter, are: photobehaviour of alkyl halides in solution: radical, carbocation and carbene intermediates[5]; radical anion reactions of nitro compounds[6]; the $S_{RN}1$ reaction of organic halides[7]; free radical chain processes in aliphatic systems involving an electron transfer[8]; reactivity of substituted aliphatic nitro compounds with nucleophiles[9]; photoinduced nucleophilic substitution at sp^3-carbon[10]; photosolvolyses and attendant photoreactions involving carbocations[11]; photoactivation of distal functional groups in polyfunctional molecules[12], chloroacetamide photocyclization and other aromatic alkylations initiated by photoinduced electron transfer[13]; the photochemistry of aryl halides and related compounds[14]; aromatic substitution by the $S_{RN}1$ mechanism[15]; recent advances in the photochemistry of quinones[16]; nucleophilic aromatic photosubstitution[17]; and the photochemistry of polyhalo compounds, dehalogenation by photoinduced electron transfer, new methods of toxic waste disposal[18].

II. ALIPHATIC HALIDES

A. Alkyl Halides

Irradiation of alkyl iodides and, to a lesser extent, of alkyl bromides in solution is by now a well-recognized, convenient and powerful method to generate carbocations. The reaction is reviewed in Reference 5. The range of bridgehead iodides which undergo photosolvolysis, next to reductive deiodination, has been expanded[19,20] beyond 1-iodonorbornane for which ionic photoreactions of *tert*-alkyl halides were first discovered. At present, structurally, one of the more extreme cases is the photomethanolysis of iodocubane and 1,4-diiodocubane (equation 1)[20] which indicates that the cubyl cation is also accessible via this photochemical method. For a homologous series of bridgehead bicyclic iodides, varying from the [2.2.1] to the [3.3.1] compound, the only ionic photoproduct is the nucleophilic trapping product[19]. No 1,2- or 1,3-deprotonation, which would produce a bridgehead alkene or propellane intermediate, is observed, even though other photolytically generated carbocations have a marked propensity for such behaviour[5]. Various further examples of such efficient loss of HX have been reported for secondary[21–25] (e.g. equation 2)[22] and primary[24,25] (e.g. equation 3)[25] alkyl halides, capable of elimination. *Inter alia*, the photochemical elimination of HBr from bromohydrins converts them efficiently in one step into ketones[24]. The predominance of elimination over nucleophilic substitution, even in nucleophilic media, has been thought to be due to the formation of a 'hot' carbocation, unencumbered by solvent molecules, which loses a β-proton rather than reacting with the solvent. The deuterium labeling experiments in equation 2, however, show that the predominance is also due to the formation of a carbene[22]. A substantial

$$ \text{(1)} $$

10–15%   50–55%

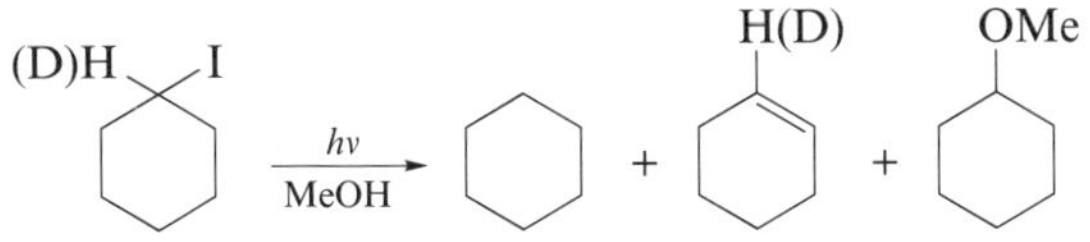

$$\text{(D)H} \quad \xrightarrow[\text{MeOH}]{hv} \quad + \quad + \quad \tag{2}$$

$$\Phi \text{dis} = 0.08 \qquad\qquad 2\% \qquad 92\% \qquad 6\%$$

$$100\% \, d_1: \qquad\qquad 29\% \, d_0, \, 71\% \, d_1$$

$$\Phi\alpha\text{-elimination} = 0.023$$
$$\Phi\beta\text{-elimination} = 0.057$$

percentage of the alkene product is formed via α-elimination. The photolysis of iodocyclohexane in argon or acetonitrile matrices at 15 K also yields cyclohexene[23]. Here, no cationic intermediate seems to be involved: the only reaction intermediate observed is the radical pair R'I'.

Both a better material balance and a larger proportion of ionic products are obtained for secondary and primary alkyl bromides if the irradiation is carried out in methanol with NH_4OH as HBr scavenger instead of in methanol[25]. Hydroxide ion is a poor electron donor, so photoinduced electron transfer to the halide, leading to reduction, is minimized compared to other scavengers. A striking example is the photoreaction of 1-bromooctane, which in MeOH yields exclusively the reduction product octane, but in the presence of OH⁻ gives mainly ionic products with the 1,2 elimination product 1-octene predominating (equation 3)[25]. Even under these optimized conditions, alkyl bromides yield a substantially smaller percentage of ion-derived products than the corresponding iodides (equation 3). As the ions are proposed to be formed via homolytic cleavage of the C—X bond followed by electron transfer in the initially formed radical pair in competition with its dissociation into free radicals (equation 4), such behaviour is contrary to the expected greater ease of electron transfer for Br' than for I'. The larger percentage of radical products from RBr may be due to the radical pair R'Br', remaining with greater excess energy after cleavage of the C—X bond than the R'I' pair, which results in more rapid escape from the cage.

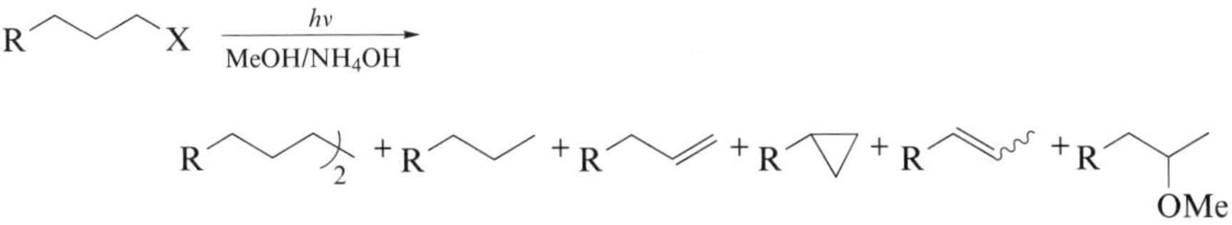

$$R \diagdown X \xrightarrow[\text{MeOH/NH}_4\text{OH}]{hv}$$

	R$\diagdown$$_2$	+ R	+ R	+ R	+ R	+ R$\diagdown$OMe
X = Br	16%	18%	30%	2%	4%	2%
X = I	—	3%	67%	2%	6%	1%
R = C$_5$H$_{11}$						

$$\tag{3}$$

$$RX \xrightarrow{hv} RX^* \longrightarrow [R^{\cdot}X^{\cdot}] \underset{R^{\cdot}+X^{\cdot}}{\overset{ET}{\longrightarrow}} [R^+ X^-] \longrightarrow R^+ + X^- \tag{4}$$

Photolysis of iodo sugars[26,27] and perfluorohaloadamantanes[28] only yields radical-derived and no ion-derived products. Apparently, the generation of electron-deficient intermediates in molecules containing many electronegative atoms is energetically too difficult.

The initial dynamics of the photodissociations of methyl iodide[29] in hexane and alkyl iodides[30] in cyclohexane have been studied by resonance Raman spectroscopic techniques. The dominant motion during the first 10 fs after excitation involves stretching of the C—I bond.

Bridgehead halides such as the 1-haloadamantanes **1**[31–34] also undergo facile nucleophilic photosubstitution upon their irradiation in liquid ammonia in the presence of, e.g., diphenylphosphide anion (equation 5)[31,32]. These reactions have been established to occur via the radical chain substitution mechanism of equations 6–9, the $S_{RN}1$ mechanism. The initiation step is photoinduced electron transfer from the nucleophile to the substrate forming a radical anion, which decomposes into an alkyl radical and a halide anion. The radical reacts with the nucleophile to give a new radical anion, which transfers its extra electron to the substrate. For reviews see References 6–10. In accordance with such initiation and propagation steps the photostimulated reaction of **1**, X = I with a series of carbon nucleophiles (in DMSO) is more catalysed when the base used is stronger (i.e. a better electron donor)[34] and the relative reactivities of **1** are **1**, X = I > **1**, X = Br > **1**, X = Cl, in parallel with their ease of reduction[32]. The formation of adamantane and 1,1′-biadamantyl are probably the result of termination steps. Other bridgehead halides which have been reported to undergo $S_{RN}1$ reactions under the conditions above are 1-halo- (**2**) and 1,4-dihalobicyclo[2.2.2]octanes[35] and 4-tricyclyl iodide (**3**)[36]. Their relative $S_{RN}1$ reactivity is **1 > 2 > 3**, i.e. the same order as the ease of formation of the intermediate radical and their ease of reduction, which both decrease as the bridgehead systems become more constrained. Interestingly, 1,4-diiodo- and 1,4-bromoiodobicyclo[2.2.2]octane only yield the disubstitution product[35]. The monosubstitution product is not an intermediate, which must mean that the radical anion thereof (XRNu⁻·) transfers an electron faster intramolecularly (to yield ⁻XRNu, followed by cleavage) than intermolecularly to another molecule of starting material, necessary for monosubstitution.

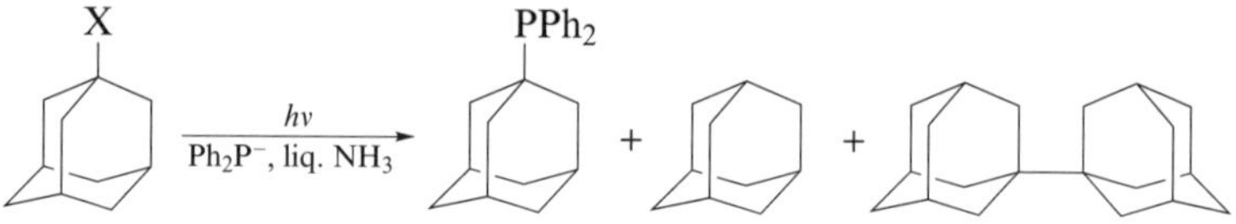

(5)

(**1**)   X = Cl, Br, I

$$RX + Nu^- \xrightarrow{h\nu} RX^{-\cdot} + Nu^\cdot \tag{6}$$

$$RX^{-\cdot} \longrightarrow R^\cdot + X^- \tag{7}$$

$$R^\cdot + Nu^- \longrightarrow RNu^{-\cdot} \tag{8}$$

$$RNu^{-\cdot} + RX \longrightarrow RNu + RX^{-\cdot} \tag{9}$$

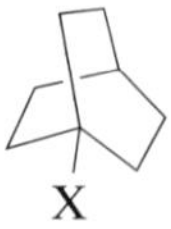

(**2**) X = halogen    (**3**)

t-Butyl chloride also undergoes $S_{RN}1$ substitution reactions[37], *tert*-butyl bromide only yields (thermal) elimination. Upon replacing one of the methyl groups in *tert*-butyl chloride by a 5-hexenyl group which serves as a radical probe, only the cyclized substitution product **4** is formed (equation 10), which shows that the tertiary radicals formed in the propagation step are trapped faster by the C=C bond than they react with the nucleophile. This is not the case for the secondary radical formed from **5** (equation 11) where the intermolecular reaction with the nucleophile can compete with the intramolecular reaction of the C=C bond, with only the former depending on the concentration of the nucleophile[38].

$$\text{(10)}$$

$$\text{(4)}$$

$$\text{(11)}$$

(5)			
0.6 mM	0.6 mM	5%	93%
19.9 mM	18.8 mM	61%	39%

$S_{RN}1$ reactions of the above substrates in liquid ammonia also occur with a fair variety of other nucleophiles: Ph_2As^-[31], $PhSe^-$ and $PhTe^-$[32] and PhS^-[33]. With carbanionic nucleophiles, however, only reductive dehalogenation products are found. Substitution with enolate anions is successful in DMSO as solvent[34].

Other examples of alkyl halides that do not readily undergo nucleophilic substitution via the S_N1 or S_N2 mechanism, but do so via the photostimulated $S_{RN}1$ mechanism (for a review, see Reference 39), are the secondary halide systems: chloro- and bromocyclohexane (with Ph_2P^- ions)[38] and 7-bromobicyclo[4.1.0]heptane (with Ph_2P^-[40] or PhS^-)[41] in liquid ammonia; (carbanionic nucleophiles give only[40] or mainly[41] reduction), and the primary halide systems: neopentyl halides (with Ph_2P^-, PhS^-, $PhSe^-$ and $PhTe^-$)[42,43] and perfluoroalkyl iodides or bromides[44–47]. Due to their good electron-accepting abilities, which favour electron transfer in the initiation and in the propagation, the perfluoroalkyl halides react with a larger range of nucleophiles than the alkyl halides. For instance, substitution occurs with nitronate ions (in DMF or DMSO)[44], the anion of ethyl acetoacetate in DMF[45], imidazole anion (in DMF)[46] and various thiolates[47].

Irradiation of alkyl iodides in aromatic solvents such as benzene, toluene or anisole affords intermolecular alkylation. Methyl iodide does so in a rather high yield (*ca* 50%)[48], the higher primary alkyl iodides in a fairly low yield (*ca* 5%)[49] and secondary alkyl iodides in low yield (*ca* 2%)[49,50]. In the latter cases elimination of HI is the major reaction. The isomeric distributions of the products (e.g. *o:m:p* = 39:40:21[48] and 35:28:37[49] for the photomethylation and *n*-propylation of toluene, respectively) differ from those expected for an attack of an alkyl radical or cation on the aromatic compounds. Conspicuous is also the lack of side-chain rearrangement, expected for a Friedel–Crafts-type alkylation reaction. Supposedly 'hot' cations are involved which are (unselectively) trapped before they can undergo the common hydride rearrangement. Alternatively, an exciplex of RI and the aromatics may be the product directing species. If the exciplex has charge-transfer character, the RX part will be $RX^{-\bullet}$ like and the R part $R^\bullet$ like, not prone to rearrangement.

Photochemical alkylation of pyrimidine nucleosides with 2-iodoethanol in aqueous acetonitrile has been used as a direct synthetic route to prepare 5-alkylpyrimidine nucleosides in fair yield[51,52]. Photochemical alkylation with perfluoroalkyl iodides and bromides occurs quite readily[53–62] and provides a facile entry to the trifluoromethyl derivatives of biologically significant imidazoles[53–57] (equation 12). There is preference for attack at C-4 over that at C-5 and the reactions are facilitated by electron-releasing substituents and retarded by electronegative groups. Also, there is high selectivity for attack on imidazole relative to other heteroaromatic or benzenoid rings. These results are all ascribed to reaction via attack of the highly electrophilic $CF_3^{\cdot}$ radical formed by photohomolysis of CF_3I. However, in view of the good electron-donating capacity of the more reactive substrates and the good electron-accepting capacity of perfluoroalkyl halides, a photo-induced electron-transfer mechanism is certainly also feasible. Mechanistic evidence for this latter pathway (equation 13) has been obtained for the photoreaction of aminopyridines with perfluoroalkyl iodides[61] and of naphthalene with trifluoromethyl bromide[62]. The perfluoroalkylations occur solely at the *ortho* and *para* positions relative to the amino group and mainly at the α-position ($\alpha:\beta = 78:22$) of naphthalene, respectively. Relative reactivities correlate with the electron-donor properties of the substrates. According to this mechanism, regioselectivity is determined in the combination step of the radical cation and $R_f^{\cdot}$. This will preferentially take place at the position of highest electron-spin density, which happens to be the same position where electrophiles would react with the substrate.

$$HN{\longrightarrow}N \xrightarrow[CF_3I, MeOH]{h\nu} HN{\longrightarrow}N \underset{CF_3}{} + N{\longrightarrow}NH \qquad (12)$$

$$32\% \qquad\qquad 47\%$$

$$ArH + R_fX \xrightarrow{h\nu} [ArH^{+\cdot} + R_f^{\cdot} + X^-] \longrightarrow [ArR_fH]^+ \longrightarrow ArR_f \quad (13)$$

Contrary to earlier reports, photolysis of appropriately substituted arylalkyl halides does give intramolecular alkylation, next to elimination of HX and reduction. 3-Phenyl-1-propyl bromide (in CCl_4) yields indane[63] and 4-phenyl-1-butyl bromide and iodide (in MeOH)[64], THF[65] and MeOH[25] with zinc wool, Et_3N and Et_4NOH, respectively, as HX scavenger yield tetralin as the result of photocyclization.

Intramolecular photoalkylation of a C=C bond occurs in citronellyl iodide (equation 14)[66]. The percentage of cyclization is enhanced at the expense of elimination by the presence of CuCl, possibly due to a template effect. CuCl also promotes the very inefficient photochemical addition of simple alkyl halides to to olefins, and does especially so as $CuCl \cdot n\text{-}Bu_3P$ complex[67].

THF/Et₃N	: product composition	9%	36%	2%	36%	7%	10%
THF/Et₃N/CuCl	: product composition	10%	17%	0%	51%	14%	8%

$$(14)$$

Introduction of a halogen substituent at the β-carbon atom of an alkyl halide opens a new photochemical reaction pathway. The unsaturated products formed upon direct irradiation of vicinal dihalides are not only produced by loss of HX but also by loss of X_2[68–70]. For example, the laser flash photolysis of 1,2-dibromoethane in acetonitrile reveals the rapid formation of two bromine atoms as the primary photochemical step[70]. Upon irradiation in the presence of Et_3N only the X_2-elimination products are formed[71], with efficiencies correlating with reduction potentials, implying initiation of the reaction by photoinduced electron transfer. With phenothiazines as sensitizers, the occurrence of electron transfer to a vicinal dibromide compound was demonstrated through emission and transient absorption spectroscopy[72].

For the largest part the photochemical behaviour of geminal dihaloalkanes resembles that of the corresponding monohalo compounds. Photolysis of geminal diiodides in nucleophilic or polar media affords mainly solvolysis or Wagner–Meerwein rearrangement (equations 15[73] and 16[73,74]). In methanol, the ratio of 1,2-elimination to substitution is much smaller than for corresponding monoiodo compounds and the reaction is more efficient (cf equation 3). This probably reflects the larger stabilization in the α-iodo cation and the weaker C—I bond in the diiodide. Of the series of diiodoalkanes studied, diiodomethane is the only one which, upon photolysis in the presence of olefins, gives cyclopropanation[75]. It does so in high yield with a variety of alkenes and the reaction is significantly less subject to steric effects than the Simmons–Smith method. The α-iodocation CH_2I^+ is thought to be the methylene transfer species rather than the previously suggested carbene (carbenoid).

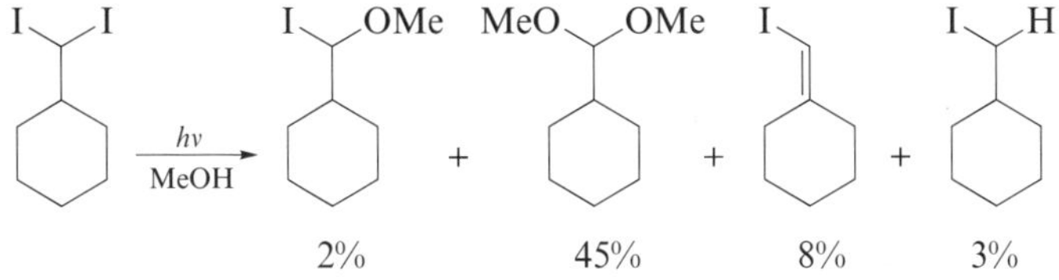

$$(15)$$

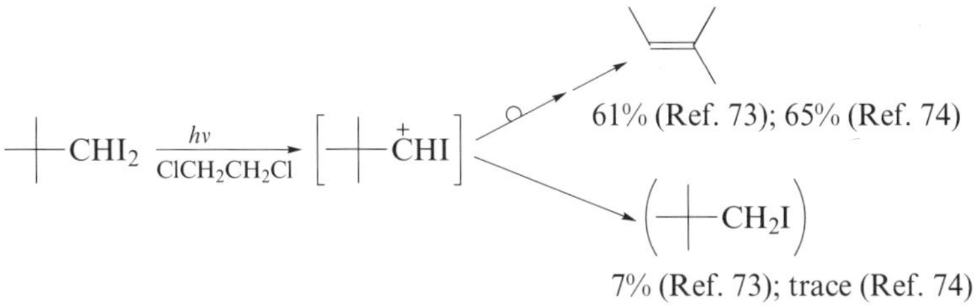

2%   45%   8%   3%

61% (Ref. 73); 65% (Ref. 74)

$$(16)$$

7% (Ref. 73); trace (Ref. 74)

Under appropriate conditions geminal dihaloalkanes give nucleophilic photosubstitution according to the $S_{RN}1$ mechanism[41,76,77]. They do so more efficiently than corresponding monohalo compounds because of their better electron-accepting properties. Noteworthy are the exclusive formation of the disubstituted product in the photostimulated reaction of 7,7-dibromobicyclo[4.1.0]heptane (6) with thiophenolate anion and the formation of the reduced monosubstituted product with n-butanethiolate ion (equation 17)[41a]. The primary product radical anion clearly loses the second Br faster than that it intermolecularly transfers an electron, and BuS⁻ apparently acts both as nucleophile and reducing agent. Alternatively, as established for PPh_2^- as anion[76], thermal attack of the nucleophile on bromine yields (after protonation) the monobromo compound, which subsequently undergoes $S_{RN}1$ substitution. α,α-Dibromo- and α,α-dichlorocyclopropyl derivatives such as 7 and 8 do not undergo C—X bond cleavage upon irradiation but C—C

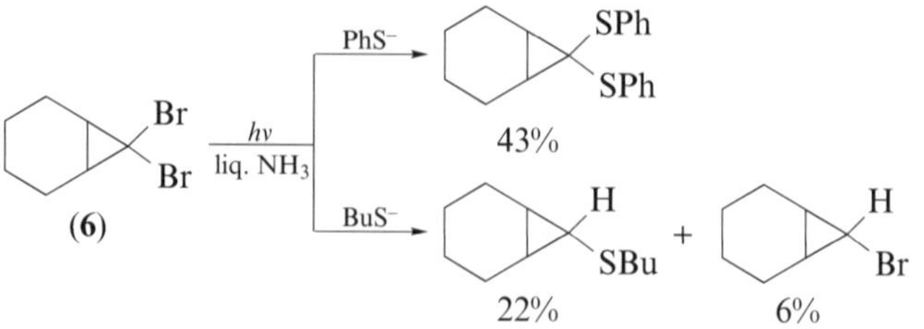

$$(17)$$

bond cleavage. These compounds constitute a convenient photochemical source of dibromo- and dichlorocarbene (equation 18)[78-80].

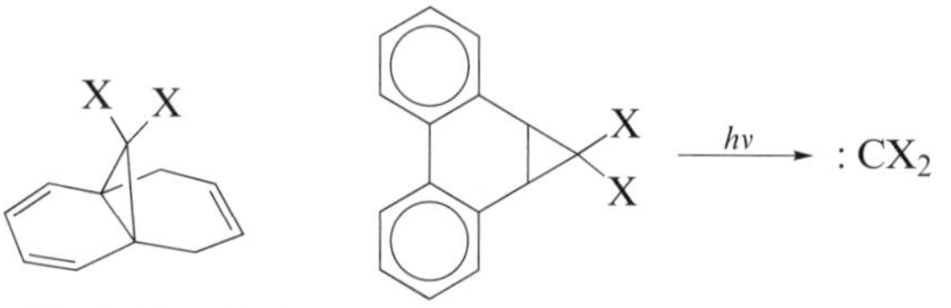

$$(18)$$

(7) X, X = Cl, Cl   (8) X = Cl
    X, X = Br, Br

Physical evidence for the existence of the species proposed as intermediate in the photocyclopropanation of alkenes by CH_2I_2 can be found in the results of the photolysis of various dihalomethanes in an argon matrix at 12 K[81,82]. Photoexcitation leads to the formation of isomers (equation 19) in which one halogen has been removed from the carbon atom and forms a weak bond with the other halogen. IR data and *ab initio* calculations indicate that the *iso*-species are best described as contact ion-pairs. Such photoisomerizations also occur with halodimethyl sulfides[83], trihalomethanes[82] and tetrahalomethanes[82], including CCl_4[84]. For liquid CCl_4 at room temperature, a transient was observed with picosecond time-resolved absorption spectroscopy, which is tentatively assigned as the solvated ion pair $CCl_3^+Cl^-$ [85]. By ESR, $CBr_4^{-\bullet}$ and CBr_3^- have been identified as paramagnetic species formed in the photolysis of neat CBr_4 at 77 K[86].

$$H_2C(I)X \xrightarrow[\text{Ar, 12 K}]{hv} H_2C-X\cdots I \left(\leftrightarrow H_2\overset{+}{C}-X\ I^- \leftrightarrow H_2C=\overset{+}{X}\ I^-\right) \quad (19)$$

X = Cl, Br, I

Due to their good electron-accepting properties the polyhalomethanes are especially susceptible to photoinduced electron transfer. Examples abound of C—X bond cleavage reactions of mainly $CHCl_3$, $CHBr_3$ and CCl_4 upon irradiation of a suitable donor compound in their presence. In many cases the reactions occur via electron transfer from the electronically excited donor to the alkyl halide with an exciplex as intermediate (equation 20a). As electron attachment to alkyl halides is dissociative, halide anion is released immediately or maybe even concertedly. Often, also the donor and acceptor pair forms a charge-transfer complex, which dissociates upon excitation (equation 20b). In a few cases, with alkyl amines as donors, the alkyl halide supposedly is the light-absorbing species and electron transfer occurs from the donor to the electronically excited alkyl halide (equation 20c). The alkyl radical formed may combine with the radical cation donor, leading to the

(a) $D^* + RX \longrightarrow (D\cdots RX)^* \leftrightarrow (D^{+\bullet}RX^{-\bullet})$

(b) $D + RX \rightleftharpoons (D\cdots RX) \xrightarrow{hv} D^{+\bullet} + R^\bullet + X^-$ (20)

(c) $D + RX^* \longrightarrow (D\cdots RX)^* \leftrightarrow (D^{+\bullet}RX^{-\bullet})$

formation of alkylation products, or it may abstract a hydrogen atom from a suitable source leading to reduction (see below).

In the period under review a variety of donor compounds has been used, ranging from dialkyl nitroxides[87,88], via tetrathiafulvalene[89], triethyl phosphite[90], aromatic compounds such as pyrene[91], uracil[92] and 2-acetylnaphthalene[93], tertiary aromatic amines[94–96], tertiary aliphatic amines[97,98], secondary aromatic amines[99–108] and primary aromatic amines[109,110], all the way to ynamines[111] and strained hydrocarbon compounds such as [1.1.1]propellane[112,113]. For a number of the systems the occurrence of a photoinduced electron-transfer step has been demonstrated by physical techniques. For instance, transient absorption spectroscopy shows that, upon laser flash photolysis at wavelengths within the charge-transfer band of dialkyl nitroxides and CCl_4, the oxoammonium ion and the $CCl_3^{\cdot}$ radical are formed essentially instantaneously (< 18 ps)[88]. For a series of halogenated solvents the efficiency of ion formation increases in the order $CH_2Cl_2 < CHCl_3 < CCl_4$ in parallel with the corresponding decrease in reduction potentials. Transient absorption spectroscopy also established the formation of the radical cation of di-p-tolylamine as primary intermediate in its photoreaction with $CHCl_3$ and $CHBr_3$ (cf equation 21)[101]. Time-resolved microwave conductivity[94] and time-resolved resonance Raman[95] spectroscopy showed the formation of the ion pair $TMPD^{+\cdot}Cl^-$ upon flash photolysis of N,N,N',N'-tetramethyl-p-phenylenediamine (TMPD) in CCl_4. Formation of $CCl_3^{\cdot}$ as the result of photoinduced electron attachment to CCl_4 was also inferred from its trapping with oxygen[88] and from its trapping with the anion of 2,6-di-t-butyl-4-cresol[90].

In exceptional cases, no electron transfer but energy transfer occurs from the electronically excited 'donor' to the alkyl halide. For example, the population of the T_2 state of 2-acetylnaphthalene by two-laser flash photolysis in CCl_4 yields $Cl^{\cdot}$ as transient (observed as $Cl^{\cdot}$ benzene π complex with added benzene)[93]. Possibly $Br^{\cdot}$, which is supposedly responsible for the rather efficient formation of 5-bromouracils upon the photolysis of 1-substituted and 1,3-disubstituted uracils in $CHBr_3$–CH_2Cl_2[92], is also formed via energy transfer.

The reactions in equations 21 and 22 are typical examples of the photoalkylation of secondary aromatic amines with polyhalomethanes[99–108]. UV irradiation of di-p-tolylamine with bromoform gives the photocyclized acridine product 9 via its flash photolytically detected radical cation (see above) (equation 21). The photoreactions of carbazole (10) with CCl_4/EtOH[102,103] and $CHCl_3$/H_2O[104] respectively afford mainly the 1- and 3-carboethoxy- and 1- and 3-formylcarbazoles (equation 22a). This regioselectivity is in accord with an electron transfer mechanism since the observed positions of alkylation, ortho and para to the nitrogen, are the positions of highest electron-spin density in the radical cation. Likewise, within this framework, the higher photoreactivity of methyl-carbazoles towards $CHCl_3$[105] as compared to carbazole results from the increase in electron-donor capacity upon introduction of methyl groups at an aromatic ring. Interestingly, β-carbolines such as 11, which are ring-nitrogen analogs of 10, do not yield photoalkylation products with CCl_4–EtOH[107] or $CHCl_3$–EtOH[109]. Photoinduced electron

(21)

(9)

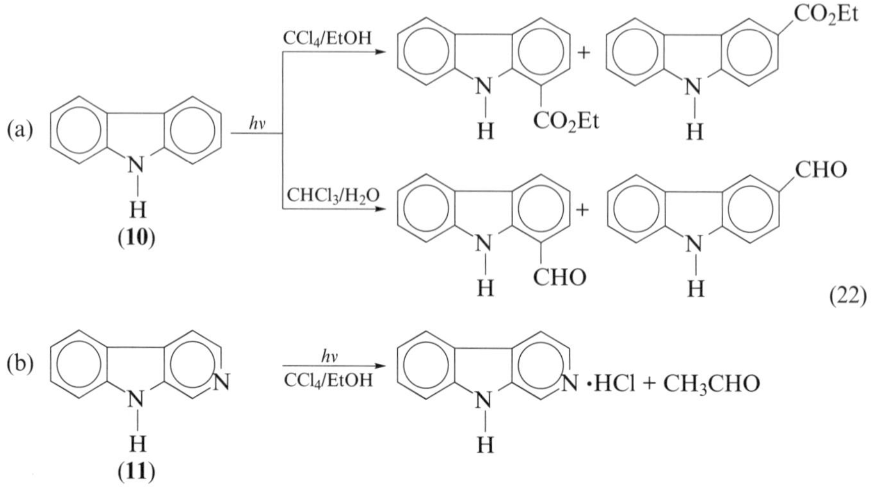

(22)

transfer does occur, as shown by the formation of **11** · HCl and acetaldehyde (produced via H-abstraction from EtOH by $CCl_3^{\cdot}$ (equation 22b), but apparently the coupling step between $CCl_3^{\cdot}$ and **11**$^{\cdot+}$ is suppressed by the lower electron density due to the presence of the extra ring-nitrogen.

By a deliberate choice of donor molecules and reaction conditions, alkyl radicals produced by photoinduced electron transfer can be selectively transformed into the corresponding alkanes. This method of photoreductive dehalogenation is much more efficient and more specific than the direct UV-irradiation of alkyl halides in hydrogen atom-donating solvents. For example, the photoreduction of methyl iodide to methane can be achieved in good yield by irradiating an NADH model compound in its presence[114]: As discussed in more detail in Section II.C (equations 39–42), the reaction occurs via an electron-transfer chain mechanism, initiated by electron transfer from the excited state of the NADH compound, which acts both as photocatalyst and reductant. The use of acridine derivatives as photocatalysts, with sodium borohydride as reductant, also reduces a variety of alkyl halides[115]. Other efficient photocatalysts are rhodium(I)[116] and iron(III) complexes[117].

With Et_3N as donor, the geminal dichloro moiety of aldrin and other cyclodiene insecticides is selectively photoreduced by UV-irradiation in high yield to the corresponding monochloro compound with a high degree of stereoselectivity[118–121]. Similar results have been obtained with 7,7-dibromobicyclo[4.1.0]heptane (**6**) in the presence of lithium aluminium hydride in diethyl ether[122] or tetrahydrofuran[123].

B. Allylic and Homoallylic Halides

Simple allylic chlorides and bromides undergo allylic isomerization and rearrangement to halocyclopropanes upon triplet-sensitized irradiation[11]. These characteristic reactions also occur on triplet-sensitization of **12**, X = Cl[124,125], but the bromine analogue **12**, X = Br[124] gives a [1,3]Br shift, Wagner–Meerwein rearrangement and solvolysis, and not the (slower) allyl to cyclopropyl isomerization (equation 23). The difference in photochemical behaviour is ascribed to the energetically more favourable intramolecular electron transfer from the electronically excited aromatic ring to the C—Br bond than to the less readily reduced C—Cl bond. As described in greater detail in Section II.D on homobenzylic

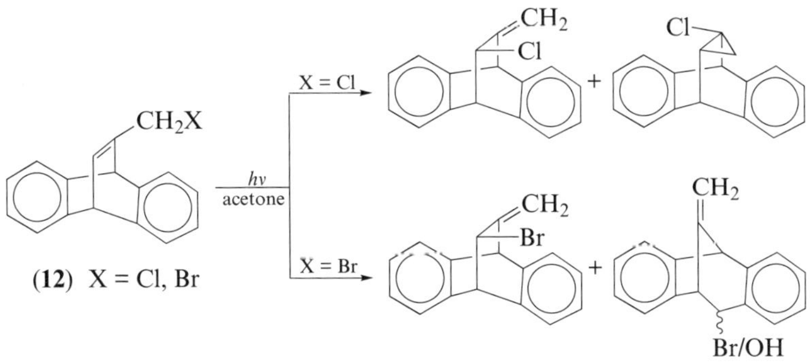

$$(23)$$

halides, such an electron transfer is supposedly a key requirement for the production of the allylic ion-derived products. In line with this explanation, the singlet excited states of **12**, X = Cl, Br, with their higher energies as compared to the triplets, both yield only ionic products.

Perfluoroallylic fluorides show 1,3-fluorine shifts upon direct irradiation[126].

Photoproducts consistent with cationic reactive intermediates are also formed in the singlet excited state reaction of the allylic iodides geranyl and neryl iodide in *n*-hexane (equation 24)[127]. The favourable 2-*Z* geometry in neryl iodide leads to a larger proportion of the intramolecular alkylation product compared to the 1,4-HI elimination. Use of tetrahydrofuran instead of *n*-hexane promotes the formation of the cyclization products[127], and so does the presence of Cu(I)[57], which probably acts as a template.

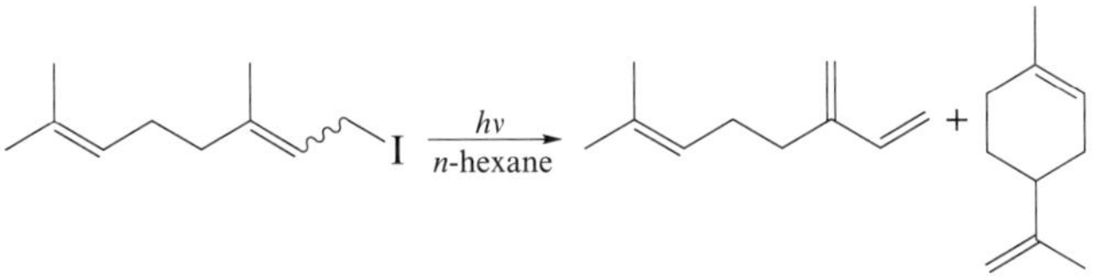

2-*E*:	71%	4%
2-*Z*:	5%	27%

$$(24)$$

The nucleophilic photosubstitution reactions of the nitro-substituted allylic bromide **13**[128] and allylic chloride **14**[129] with nitronate anion (equations 25 and 26) occur via a photo-induced radical-chain substitution process. Apparently, the allyl radical intermediate in

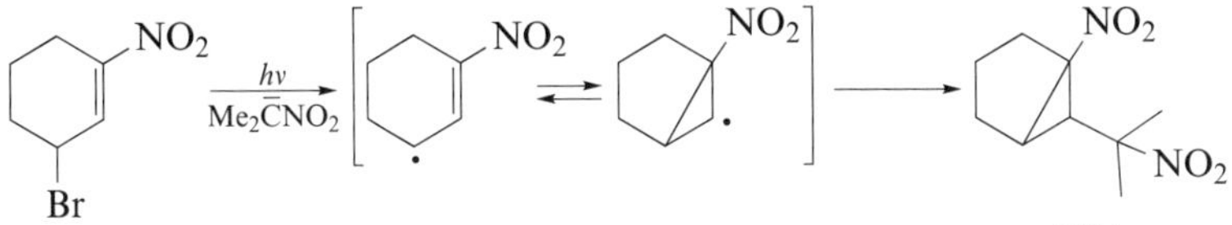

$$(25)$$

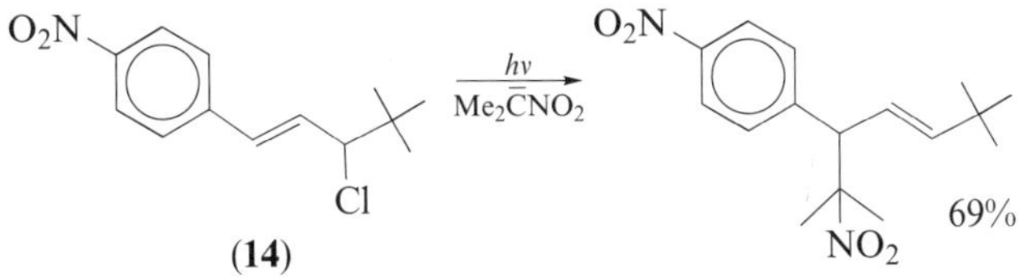

50%

$$(26)$$

69%

the $S_{RN}1$ reaction of **13** can undergo valence isomerization to a cyclopropyl radical which reacts faster than the stabilized allylic radical with the not very nucleophilic nitronate anion. The reaction in equation 26 constitutes a rare example of an $S_{RN}1$ reaction with allylic rearrangement: an $S_{RN}1'$ reaction.

Photolysis of hexachlorocyclopentadiene yields the pentachlorocyclopentadienyl radical, which reacts with a wide range of organic compounds to produce 5-substituted cyclopentadienes in fair yield[130]. The novel photodimerization of 2,3-dibromopropene involves (at least formally) its light-induced homolysis to an allyl radical and Br˙ and addition of these fragments to a second equivalent of starting material[131].

A number of allylic chlorides have been reported to yield photoproducts characteristic of the alkene moiety and not of the C—X bond[123]. Allyl and crotyl halides undergo photochemical reactions with chromium, molybdenum and tungsten complexes, which involve C—X bond cleavage[133].

Cyclopropanation of the C=C bond of an allylic halide as in the cyclopropane **15** greatly suppresses photoreactivity, which is restored by introduction of a vinyl group at the 2-position of the cyclopropyl group, as in **16** (equation 27)[134]. The shift in UV-absorption maxima of **15** and **16** indicates the presence of electronic interactions of the vinyl and bromomethyl moieties through the cyclopropane ring in **16**. A similar arrangement of chromophoric moieties is present in compound **17**, which on direct irradiation in acetic acid yields benzo[c]fluorene as sole product (equation 28)[135].

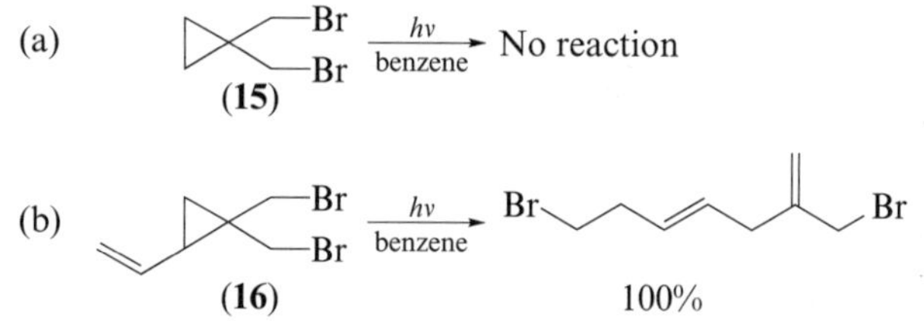

(27)

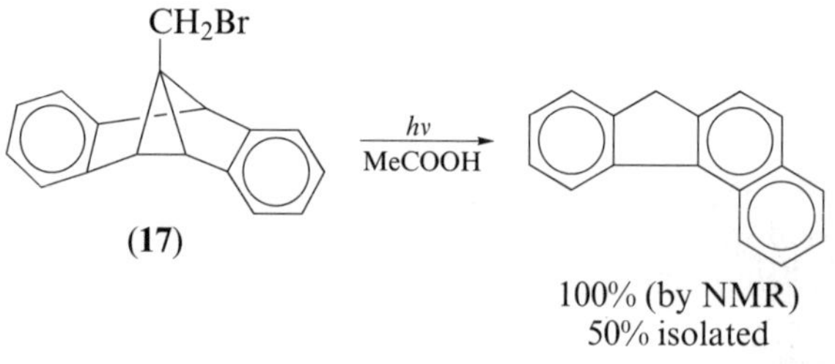

(28)

Homoallylic halides show enhanced photoreactivity of the C—X bond as compared to simple alkyl halides, dependent on the relative orientation of the C=C and C—X bonds[12]. For instance, the photolysis of both *exo*- and *endo*-**18** in hexane, containing an amine as acid scavenger to avoid the decomposition of the enol ethers, leads to homolysis of the C—Cl bond and clean reduction to 2-(trimethylsiloxy)norbornene, but the *exo*-isomer does so ninefold more efficiently[136]. This stereoelectronic effect is considered to be a manifestation of the better delocalization of excitation in the *exo*-isomer as the result of π^*/σ^* molecular orbital mixing[12,137]. The C—Cl bond order in *exo*-**18**, for example, is calculated to decrease from 0.93 in the S_0 state to 0.80 in the S_1 state, whereas the bond order in *endo*-**18** is virtually invariant (0.93) in both the S_0 and S_1 states. In support of the orbital mixing argument, *anti*-**19** and *syn*-**19** are comparably and minimally photoreactive[138]. Despite the similar relative orientation of the C—Cl bond and the C=C bond in *exo*-**18** and *anti*-**19**,

(18) *exo*: Φdis = 0.07
endo: Φdis = 0.008

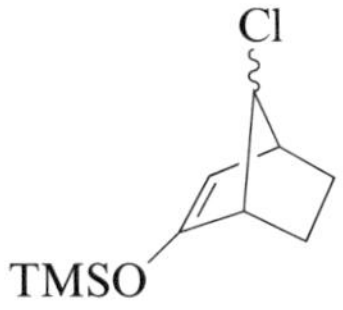

(19) *anti*: Φdis = 0.007
syn: Φdis = 0.003

calculations predict only a very small σ^* C—Cl contribution to the LUMO of *anti*-19 as well as to the LUMO of *syn*-19.

The irradiation of enone 20 in the presence of Et$_3$N or DABCO as electron donors yields 21 as major product and almost none of the valence tautomerized product 22 (equation 29)[139]. Photomediated electron transfer apparently differs here from alternative electron transfer processes, since in lithium/liquid ammonia and electrochemical reductions 22 is produced.

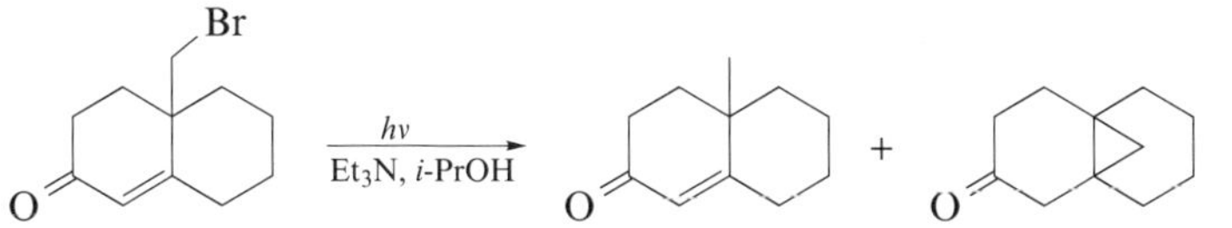

(20)

(21)
> 95%
ϕ = 0.22

(22)
< 3%

(29)

C. Benzylic Halides

Irradiation of benzylic halides in nucleophilic solvents yields both products character-istic of carbocation and radical intermediates (for a review see Reference 11). Quite interestingly, the ratio heterolytic:homolytic bond cleavage products depends on the way energy is supplied to the bond: The percentage of ionic product is larger upon photosensi-tization than upon direct irradiation. Further information about this phenomenon has been obtained for benzyl chlorides[140–142] in MeOH and *t*-BuOH and for 1-(chloro- and bromomethyl)naphthalene in MeOH[143,144] (equation 30)[144]. Quenching studies[140,143] and the use of *E/Z* isomerization of a *para* CH=CHMe group in benzyl chloride as internal probe for triplet reactivity[142] have shown that in the sensitized reactions the T_1 state is not respon-sible for the formation of cleavage products. Reaction is suggested to occur via T_2[140] or via an exciplex of the benzylic halide and the sensitizer[143] (equation 31). Unfortunately, a

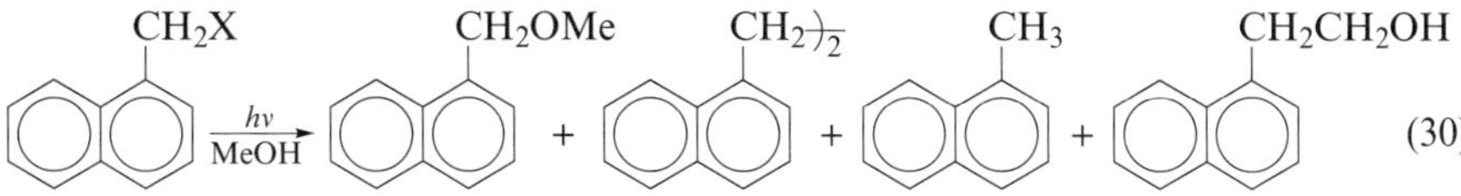

(30)

% ion product : % radical product

		% ion product	:	% radical product
X = Cl	*hv* direct:	31	:	56
	hv sensitized:	48	:	37
X = Br	*hv* direct:	74	:	16
	hv sensitized:	79	:	8

$$\text{Sens}^{*3} + \text{ArCH}_2\text{X} \longrightarrow (\text{Sens} \cdots \text{ArCH}_2\text{X})^{*3} \leftrightarrow (\overset{\delta-}{\text{Sens}} \overset{\bullet}{\cdots} \overset{\delta+}{\text{ArCH}_2} \overset{\delta}{\cdots} \overset{\bullet}{\text{X}})$$

$$\xrightarrow{\text{MeOH}} \text{Sens} + \text{ArCH}_2\text{OMe} + \text{HX} \tag{31}$$

CIDNP study of the photocleavage of benzyl chloride and bromide in methanol/acetone only gives signals for the radical-related products and not for the ion-related ones[145]. For concentrated solutions of 1-(iodomethyl)naphthalene in MeOH, yet another mechanistic pathway to photosolvolysis product occurs[146]. A complex formed between starting material and homolytically photogenerated I$^\bullet$ reacts with methanol to yield the ether product and I$_2^{-\bullet}$.

Nucleophilic photosubstitution reactions of benzylic chlorides have also been observed to occur with nucleophiles other than the alcohol solvents. n-Nucleophiles such as amine solvents[147] and halide ions and acetate ions[148], as well as π-nucleophiles such as toluene[149] have been used. The latter, a photoalkylation reaction, was achieved by irradiation of benzyl chloride absorbed within zeolite micropores in a slurry in cyclohexane. In cyclohexane itself only products of PhCH$_2^\bullet$ are formed. This large medium effect is due to the strong electrostatic fields experienced in the zeolite cavities[149].

Ion-derived products are also formed in the irradiation of bicyclic benzyl chlorides[150,151]. Compound **23**, which is both a benzylic and a homoallylic halide, shows a comparable direct versus sensitized irradiation dichotomy as simple, less activated benzylic halides do (equation 32)[150]. Also, in the direct isomerization, substantial epimerization occurs, which is not observed upon sensitization. This means that the R$^+$Cl$^-$ ion pairs produced in the singlet and triplet reaction are not identical and recombine or are captured before relaxing to identical species. Alternatively, the different behaviour may well be a manifestation of the intermediacy of an exciplex as product-directing intermediate in the sensitized reaction (cf equation 31).

$$\tag{32}$$

Introduction of an electron-withdrawing substituent at the α-carbon atom of benzylic halides as in PhCH(Br)CO$_2$Me[152] and PhCH(Br)CHBr(COPh)[153] does not impede photosolvolysis. In MeOH, nucleophilic photosubstitution occurs in good yield (80% and 50%, respectively), in the latter case accompanied by elimination of Br$_2$ to yield the corresponding alkene (37%).

Diphenylmethyl chloride in acetonitrile gives, as nucleophilic photosubstitution products, Ph$_2$CHNHC(O)Me and Ph$_2$CHOH in a combined quantum yield of 0.12 in good agreement with the quantum yield of 0.13 for Ph$_2$CH$^+$ ion formation determined by transient absorption spectroscopy (see below)[154]. Photosolvolysis also occurs here upon introduction of the electron-withdrawing carbomethoxy group[155], as depicted in equation 58 in Section II.F on α-haloesters, and upon transformation of the diphenyl moiety into a fluorenyl moiety as in 9-bromofluorene[156] and methyl 9-bromo-9-fluorenylcarboxylate[155], which makes the ion to be formed upon photoheterolysis a destabilized antiaromatic 9-fluorenyl cation.

Using time-resolved laser flash photolysis techniques, transient carbocations in the photolysis of benzyl halides have been widely observed. A variety of phenylmethyl halides[157–160], substituted diphenylmethyl halides[154,155,161–163] and substituted triphenyl-methyl halides[162,164] has been successfully used as precursor under various reaction conditions. The photogeneration of the transient cations is often accompanied by that of the corresponding transient radicals.

The unsubstituted benzyl cation is too reactive to be observed in water[157] or trifluoroethanol or even hexafluoroisopropanol (HFIP)[158] on the nanosecond time scale, but these weakly nucleophilic solvents permit the observation of the p-methoxybenzyl cation ($\lambda_{max} = 320$ nm, $k = 3 \times 10^2$ s^{-1} in HFIP) and the p-methylbenzyl cation ($\lambda_{max} = 310$ nm, $k = 2 \times 10^6$ s^{-1} in HFIP)[158]. Also, p-methoxybenzyl cations with an electron-withdrawing substituent, as the CF$_3$[159] or COOR[160] group, at the positive centre, have been observed in trifluoroethanol as transient intermediates by nanosecond transient spectroscopy. Change of the electron-withdrawing group from CF$_3$ to CF$_2$H to CH$_2$F has remarkably little effect on the amount of cation that forms as well as on the cation:radical ratio[159].

The quantum yields of cation formation of a large series of *para*-substituted diphenyl-methyl halides in acetonitrile increase with the electron-donating strength of the substituent, whereas the quantum yields of radical formation are rather independent of the nature of the substituent[154]. Also the ion:radical ratios correlate with the ionic leaving group power of the halide and not with the electron affinities of X·. All of this indicates that cleavage of the C—X bond to yield cation and halide proceeds, in the rate-determining step, by heterolysis (as depicted in equation 33) and not by homolysis followed by electron transfer in the radical pair, as proposed for alkyl halides (cf equation 4).

$$RX \xrightarrow{h\nu} RX^* \begin{array}{c} \longrightarrow [R^+ \, X^-] \longrightarrow R^+ + X^- \\ \\ \longrightarrow [R^· \, X^·] \longrightarrow R^· + X^· \end{array} \qquad (33)$$

Interestingly, the sequence of events in equation 34 was proposed as a model for the homolysis of triphenylmethyl chloride and bromide in non-polar solvents: Homolytic cleavage proceeding by initial bond heterolysis followed by electron transfer to generate the radical pair[164]. In the diphenylmethyl halide series, also, the antiaromatic 9-fluorenyl cation[161] and the 9-carbomethoxyfluoren-9-yl cation[155] were generated by laser flash excitation.

$$RX \xrightarrow{h\nu} [R^+ \, X^-] \longrightarrow [R^· \, X^·] \longrightarrow R^· + X^· \qquad (34)$$

The photochemical products of benzyl halides are solely radical-derived, if the irradiation is performed in a polar solvent but the substrate is not amenable to ion formation, as in the case of a 1,4-dichlorobenzonorbornene derivative[165], or if the irradiation is performed in an apolar solvent. The mechanism of the homolytic photodissociation has been investigated with a number of techniques: time-resolved laser flash photolysis[143,146,166–168], emission spectroscopy in rigid glasses[169] and fluid solutions[169,170], time-resolved ESR[171] and CIDNP[145]. The α-naphthylmethyl radical formed homolytically in the direct photolysis of the 1-(halomethyl)naphthalenes was detected by transient absorption[143,146,166,167] and transient emission[166,167] spectroscopy: I· (as I· mesitylene complex with added mesitylene) and Br· (as Br· benzene complex with added benzene or as Br$_2$·$^-$ with added Br$^-$) by transient absorption spectroscopy upon photolysis of 1-(iodomethyl)naphthalene[143,146] and *trans*-10,11-dibromodibenzosuberone[168], respectively. The benzyl radical has been observed in the photolysis of benzyl chloride in a rigid 2-methylpentane glass[169]. The measurement of the formation and decay kinetics of the α-naphthylmethyl radical by emission and absorption spectroscopy indicates that the major channel of dissociation is rapid intersystem crossing from the S_2 state to the dissociated T $(n\sigma^*)$ state[166,167]. The CIDEP patterns of the radicals produced in the direct photodissociation and triplet photosensitized reaction of 1-(chloromethyl)naphthalene, on the other hand, indicate that photodissociation mainly occurs from the triplet state following the excitation to the S_1 state[171]. CIDNP shows the involvement of the $PhCH_2 \cdot \cdot CH_2OH$ pair in the formation of $PhCH_2CH_2OH$ in the acetone-sensitized reaction of benzyl chloride or bromide in the presence of methanol[145].

The convenient entry to arylmethyl radicals via homolytic photodissociation of the corresponding halomethylaryl precursors has been used to study the ground and excited state reactivities of p-substituted benzyl radicals with O_2[172,173] and the excited state properties and reactivities of a series of arylmethyl radicals[174,175].

Bis(halomethyl)aryl compounds such as **24** yield biradical-derived products, i.e. acenaphthene (**25**), in a stepwise double homolytic bond cleavage (equation 35)[176–178]. Upon photolysis of suitable dihalides in a rigid organic glass at low temperature, biradicals are formed, which provides a simple and convenient method to prepare, for instance, the ground state triplet biradical m-xylylene from m-di(chloromethyl)benzene[179]. The double photodissociation of geminal dihalobenzyl compounds is a new entry to carbenes (equation 36)[180].

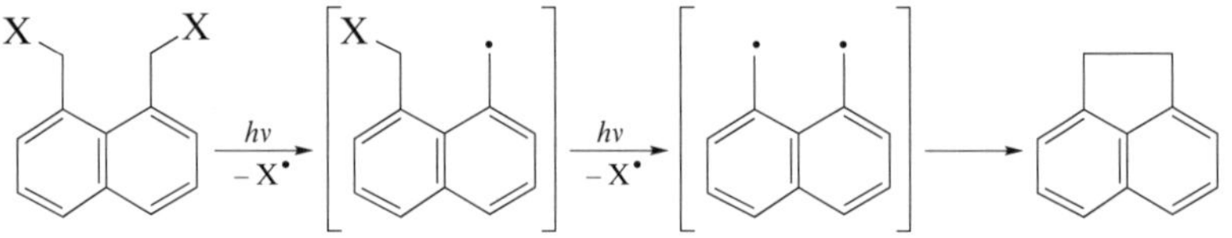

(24) X = Cl, Br   **(25)**   (35)

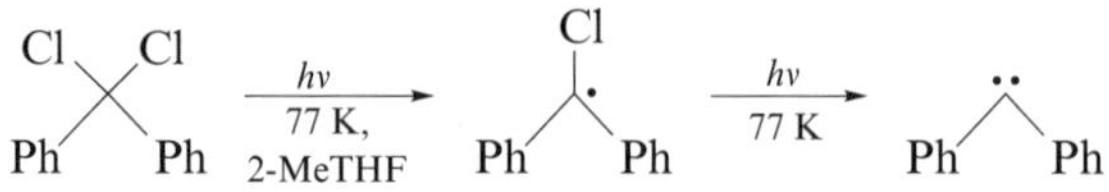

(36)

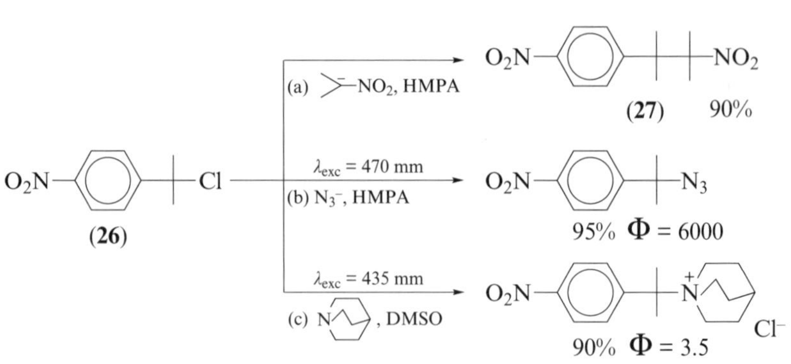

The nucleophilic photosubstitution reactions of benzylic halides with a nitro group at the *ortho*- or *para*-position of the aromatic ring such as p-nitrocumyl chloride (**26**) occur via the $S_{RN}1$ mechanism. The reactions have been reviewed in References 6–10. The initiation and propagation steps of this photoinduced radical-chain electron transfer process have already been depicted in this review in equations 6 to 9 in Section II.A. Equation 37 shows the excellent yields and the quantum yields for reaction of **26** with some of the wide range of organic and inorganic nucleophiles suitable for this substitution at a tertiary C-atom[181,182]. The exclusive formation of **27** with the ambient anion of 2-nitropropane as nucleophile shows that only C-alkylation occurs and that the reaction is rather insensitive to steric hindrance[181]. Light initiates the reactions by inducing an electron transfer from the nucleophile to the substrate (equation 6) and is more effective in doing so when its wavelength corresponds with the λ_{max} of the visible absorption of the charge transfer complex of the substrate with the nucleophile than when its wavelength corresponds to the absorption maximum of the substrate[182]. Clearly, the photochemical initiation proceeds preferentially via a charge-transfer intermediate. The quantum yield of the reaction of **26** with azide is *ca* 6000, with quinuclidine 3.5. Supposedly, the photochemical initiation involving a neutral nucleophile, which produces a radical anion and a radical cation, is less efficient, as the electrostatic interactions of this step are unfavourable compared to the corresponding step with a negatively charged nucleophile, where a radical anion and a radical are formed. In line with occasional reports that extra light accelerates the $S_{RN}1$ reactions of benzyl halides but decreases the yield of substitution product, the quantum yield of product formation in the reaction of α-*p*-dinitrocumene with azide anion decreases with increasing light intensity[182]. At high light intensity, the concentrations of the *p*-nitrocumyl radical and the radical of the nucleophile are relatively high and, as a result, chain-termination processes will occur more often.

In the case of primary and secondary *p*-nitrobenzyl halides, substitution with nucleophiles according to the S_N2 mechanism can compete with their photostimulated $S_{RN}1$ reactions. This alternative route will be least competitive when using substrates carrying the most difficulty displaced halide (i.e. Cl in this series of compounds) and the use of less nucleophilic anions. Since 1982, photostimulated $S_{RN}1$ reactions of *p*-nitrobenzyl chloride have been reported with nucleophiles such as functionalized[183] and heterocyclic[184,185] nitronate anions, nitroimidazole[186] and imidazole[187] anions, and dialkyl phosphite and thiophosphite ions[188]. Substitution proceeding by the photostimulated $S_{RN}1$ mechanism has also been described for analogous nitro-substituted heterocyclic benzylic chlorides such as **28** X = Cl (with nitronate)[189], **29** X = Cl (with nitronate[184,185,190] and nitroimidazole[186] anions) and **30** (with nitronate anion)[191]. As secondary nitro-substituted substrates, both the homocyclic **31**[192,193] and the heterocyclic **32**[194], **33**[195] and **34**[196] have been successfully used for $S_{RN}1$ reactions with a larger variety of nucleophiles than suitable for the primary systems.

O₂N—⟨furan⟩—CH₂X   O₂N—⟨imidazole⟩—CH₂X   ⟨imidazopyridine⟩—CH₂Cl
                                    | |
                                    Me   NO₂

(28) X = Cl, Br, I   **(29)** X = Cl, Br   **(30)**

O₂N—⟨benzene⟩—C(H)(R)—Cl   O₂N—⟨furan⟩—CH(Bu-*t*)(Cl)   O₂N—⟨thiophene⟩—CH(Bu-*t*)(Cl)   O₂N—⟨imidazole⟩—CH(Cl)CH₃ / Me

(31) R = CHMe₂, CMe₃   **(32)**   **(33)**   **(34)**

The nitro group does not necessarily have to be in the o- or p-position of the benzene ring, nor is the presence of a nitro group a prerequisite for nucleophilic photosubstitution via the $S_{RN}1$ mechanism to occur. The *meta*-nitro analogue 35 with a *tert*-butyl group at C_α displays $S_{RN}1$ behaviour (with a series of nucleophiles)[197] and so do the bridgehead halide 9-bromotriptycene (with phosphide and arsenide ions)[32], compound 36 (with nitronate and azide ions)[198], compound 37, X = Cl (with nitronate ions)[199] and compound 38 (with azide ion)[200]. For compound 37, X = Br, S_N2 substitution is competitive.

(35)　　　　　　(36)　　　　　　(37) X = Cl, Br　　　　　　(38)

As expected for $S_{RN}1$ reactions, almost all of the above compounds yield C-alkylates upon reaction with the ambient nitronate anions. Exclusive C-alkylation is no longer the case, though, for the reaction of 35 (with nitronate anions)[192,193] or for the reactions of 39 (with nitronate anion[192] or anions of β-keto or malonic esters[201]) (equation 38). Apparently, the limits of insensitivity to steric hindrance have been reached here. When two or more alkyl substituents are present at C_β in R or R^1, the product distribution changes from $S_{RN}1$ C-alkyl to $S_{RN}1$ O-alkyl derivatives, in addition to some reduction product (equation 38a). This change in regioselectivity is attributed to kinetic selection between C- and O-alkylation at the radical + anion association step[192], which has been shown to be irreversible[202]. Likewise, branching the alkyl group R and/or R^2 attached to the malonic ester anions changes the product distribution from C-alkyl to reductive dehalogenation products (equation 38b).

The radical anions of p-nitrobenzyl chloride, bromide and iodide[203], of p-nitrocumyl bromide[203], of 28, X = Br, I[204] and of 29, X = Cl, Br[205] postulated as reactive intermediates in the $S_{RN}1$ reactions of these compounds, have been observed by ESR spectroscopy, and so have the radicals formed by C—Br bond cleavage of the radical anions of p-nitrocumyl bromide and 29, X = Br. The radical anions of the primary homocyclic benzyl halides were stable under the reaction conditions used to generate them (γ-irradiation, 77 K, THF or MeOH matrix) and so were the radical anions of 28 and 29, X = Cl.

The dihalogen bridgehead compound 9,10-dibromotriptycene yields only the $S_{RN}1$ disubstituted product with diphenylphosphide anion in liquid ammonia[32]. The monosubstituted product is probably not an intermediate in this reaction, because its radical anion fragments faster than an electron is transferred to another molecule of starting material, necessary for monosubstitution to occur.

Irradiation of NADH model compounds in the presence of benzyl bromide or p-cyanobenzyl bromide in acetonitrile brings about reduction of the benzyl halides to the corresponding toluene compounds[114]. Like the $S_{RN}1$ substitution reaction, this photoreduction also occurs via an electron-transfer chain mechanism. Unlike in that case, though, here an electron transfer from the excited state of the NADH compound is solely responsible for the initiation step. In the propagation, the benzyl radical produced by C—Br bond cleavage in the radical anion abstracts hydrogen from the NADH compound. This yields a radical intermediate, from which electron transfer to benzyl bromide occurs readily (equations 39–42).

Similar electron transfer initiated chain reactions, with quantum yields up to 13, occur in the photosensitized debromination of vicinal, α,β-dibromobenzyl compounds such as 2,3-dibromo-3-phenylpropionic acid[206,207] and 1,2-dibromo-1-phenylethane[208] to the cor-

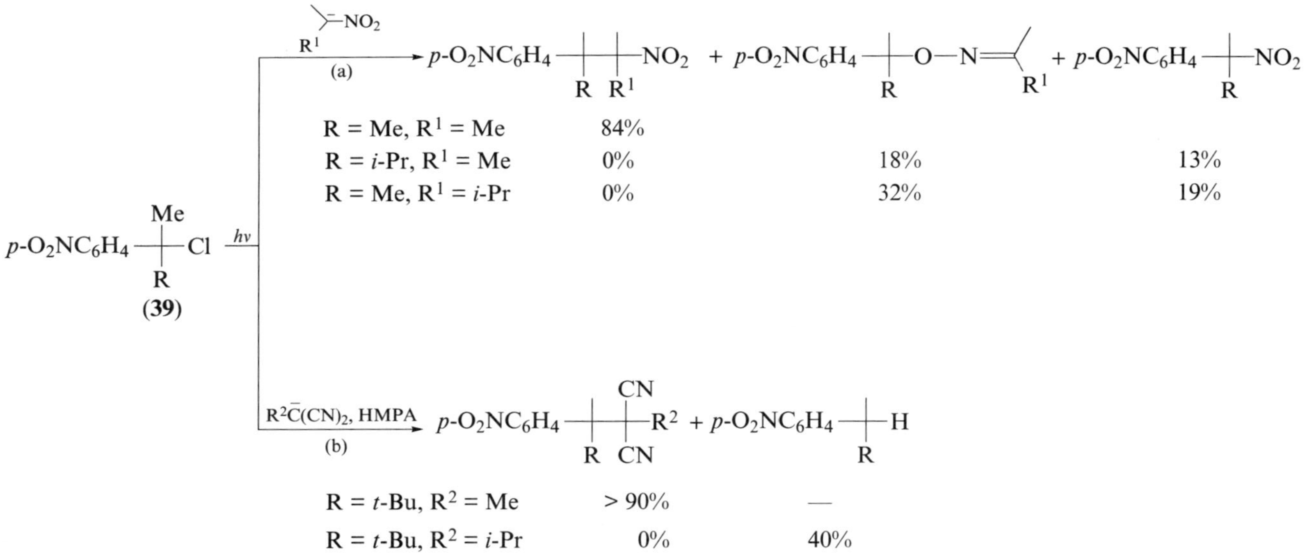

$$p\text{-}O_2NC_6H_4\overset{\displaystyle Me}{\underset{\displaystyle R}{\overset{\displaystyle |}{\underset{\displaystyle |}{C}}}}Cl \quad \xrightarrow{hv}$$

(39)

(a) $R^1\!\!=\!\!NO_2 \longrightarrow$

$$p\text{-}O_2NC_6H_4\overset{\displaystyle R}{\underset{\displaystyle R^1}{\overset{\displaystyle |}{\underset{\displaystyle |}{C}}}}NO_2 \;+\; p\text{-}O_2NC_6H_4\overset{\displaystyle R}{\underset{\displaystyle R^1}{\overset{\displaystyle |}{\underset{\displaystyle |}{C}}}}O\!-\!N\!\!=\!\!\overset{\displaystyle R}{\underset{\displaystyle R^1}{C}} \;+\; p\text{-}O_2NC_6H_4\overset{\displaystyle R}{\underset{\displaystyle |}{C}}NO_2$$

$R = Me, R^1 = Me$	84%	18%	13%
$R = i\text{-}Pr, R^1 = Me$	0%	32%	19%
$R = Me, R^1 = i\text{-}Pr$	0%		

(b) $\xrightarrow{R^2\bar{C}(CN)_2, \text{ HMPA}}$

$$p\text{-}O_2NC_6H_4\overset{\displaystyle R}{\underset{\displaystyle R^2}{\overset{\displaystyle CN}{\underset{\displaystyle |}{C}}}}CN \;+\; p\text{-}O_2NC_6H_4\overset{\displaystyle R}{\underset{\displaystyle |}{C}}H$$

$R = t\text{-}Bu, R^2 = Me$	> 90%	—
$R = t\text{-}Bu, R^2 = i\text{-}Pr$	0%	40%

(38)

$$\text{NADH}^* + \text{PhCH}_2\text{Br} \longrightarrow \text{NADH}^{+\bullet} + \text{PhCH}_2\text{Br}^{-\bullet} \qquad (39)$$

$$\text{PhCH}_2\text{Br}^{-\bullet} \longrightarrow \text{PhCH}_2^\bullet + \text{Br}^- \qquad (40)$$

$$\text{PhCH}_2^\bullet + \text{NADH} \longrightarrow \text{PhCH}_3 + \text{NAD}^\bullet \qquad (41)$$

$$\text{NAD}^\bullet + \text{PhCH}_2\text{Br} \longrightarrow \text{NAD}^+ + \text{PhCH}_2\text{Br}^{-\bullet} \qquad (42)$$

responding alkenes. Here Br$^\bullet$, produced by β-cleavage in the radical formed by C—Br bond cleavage of the radical anion of the dibromide, is the chain carrier in the propagation steps. The photosensitized photodebromination reaction is promoted by electron-withdrawing substituents at the aromatic ring with quantum efficiencies correlating with the reduction potentials of the dibromides, as expected for an electron transfer from the excited sensitizer to the substrate[206]. Ionic micelles accelerate the debromination reactions[207,208]. Interestingly, *erythro*-2,3-dibromo-3-phenylpropionic acid is debrominated 500 times faster than the *threo* compound[206]. This is not due to their difference in reduction potential, but to the difference in anchimeric assistance of the remaining bromine atom to the bromide elimination from the radical anion.

The debromination of 1,2-dibromodiarylethanes to the corresponding *trans*-diarylethenes, which occurs quantitatively upon sensitized irradiation of alkylviologens (AV^{2+}) in a phase transfer system, involves a different mechanistic pathway[209-211]. The active reductant here is AV, which is formed upon disproportionation of $\text{AV}^{-\bullet}$, produced upon photochemical sensitization of AV^{2+}.

The photochemical reduction of benzyl halides such as 1-bromo-1-phenylcyclopropanes[212], also occurs smoothly upon irradiation in the presence of an aliphatic tertiary amine (e.g. Et$_3$N). In this case the reaction proceeds via electron transfer to the electronically excited benzyl halide to yield the radical anion (equation 43) followed by loss of Br$^-$ and hydrogen abstraction. The occurrence of such an electron transfer process has been observed by transient absorption spectroscopy with diphenylmethyl chloride as acceptor and ferrocene as donor. Within 35 ps following photolysis, the diphenylmethyl radical and the ferricenium cation are detected[213]. If a halogen atom is present at the β-carbon atom as in PhCH(Br)CHBr(COPh), photoinduced electron transfer via irradiation in the presence of Et$_3$N leads to efficient loss of Br$_2$ to yield the corresponding alkene[153].

$$\text{PhCH}_2\text{X}^* + \text{Et}_3\text{N} \longrightarrow \text{PhCH}_2\text{X}^{-\bullet} + \text{Et}_3\text{N}^{+\bullet} \qquad (43)$$

D. Homobenzylic Halides

The presence of an aryl group at the β-carbon atom of an alkyl halide influences both its photochemical behaviour and its photoreactivity (for reviews, see Section VII of Reference 11 and Reference 12). The irradiation of 2-phenylethyl bromide and iodide (**40**) in nucleophilic solvents such as methanol (equation 44)[64] yields simpler and different product mixtures than those of 1-bromo- and 1-iodooctane, described in equation 3[25] in Section II.A. The only ion-derived product is **41**, while for the 1-halooctanes the main ionic

	X = Br	X = I		
	65%	80%	25%	—
	(40)		(41)	(42)

$$\qquad (44)$$

pathway is elimination of HX, and the radical-derived product is the carbinol **42**, and not the reductive dehalogenation product. As with alkyl halides, **40**, X = I gives more ion-derived product than the bromo compound. A large medium effect on the photochemical behaviour of **43** has been observed upon its adsorption within zeolites (equation 45)[214]. In a slurry of cyclohexane and zeolite the photoproducts are exclusively ion-derived, via a 1,2-phenyl migration and loss of a β–H$^+$ respectively, while in cyclohexane the photoproducts are mostly radical-derived. The promotion of the ionic pathway is ascribed to a combined contribution of local electrostatic fields inside the zeolites, and Lewis interaction with the zeolite metal cations enhancing the polarity of the C—Br bond. An increase in selectivity of product formation has also been observed for the irradiation of the homobenzylic trichloride DDT (2,2-bis(chlorophenyl)-1,1,1-trichloroethane) in micellar solutions as compared to various organic solvents[215]. In the former case the HCl-elimination product DDE (2,2-bis(4-chlorophenyl)-1,1-dichloroethene) is the sole photoproduct, whereas in the latter the reductive dechlorination product DDD (2,2-bis(4-chlorophenyl)-1,1-dichloroethane) and the oxidation product 4,4′-dichlorobenzophenone are also principal photoproducts. The homobenzylic *gem*-dichlorobenzobicyclo[5.1.0]octenone shows a cyclopropyl to allylic chloride photoconversion[216] not observed for simple *gem*-dihalocyclopropanes.

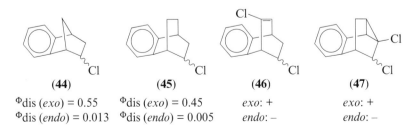

$$(45)$$

(**43**)   62%   38%

The activating effect of the β-aryl group is subject to stereoelectronic control[217–228]. For example, the *exo*-isomers of the benzobicyclo compounds **44**[217–219], **45**[219,220] and **46**[221], and of the benzotricyclo compound **47**[222] are all more photoreactive than the corresponding *endo*-isomers. Direct irradiation of both *exo*- and *endo*-**44** and *exo*- and *endo*-**45** (equation 46)[220] yields comparable reaction mixtures of radical- and ion-derived products, but the *exo*-isomer of **45** does so with a 900-fold higher quantum efficiency and 1400-fold higher rate. An unusual type of ionic product is **48**, which is the result of a *syn* migration and is not found in the thermal solvolysis. For **44**, it has been shown by deuterium labeling that a carbene intermediate is not involved in the formation of this unusual solvolysis product[218]. The *exo*-isomers of compounds **46** (equation 47)[221] and **47**, which are, next to homobenzylic, also homoallylic and cyclopropylcarbinylic respectively, only yield ionic-type products. *Endo*-**46** and *endo*-**47** dichlorides are virtually inert, and no products attributable to photosolvolysis or photo-Wagner–Meerwein rearrangements are produced. Also here, more of the type of products (**49**) which result from *syn* migration are

(**44**)	(**45**)	(**46**)	(**47**)
Φdis (*exo*) = 0.55	Φdis (*exo*) = 0.45	*exo*: +	*exo*: +
Φdis (*endo*) = 0.013	Φdis (*endo*) = 0.005	*endo*: −	*endo*: −

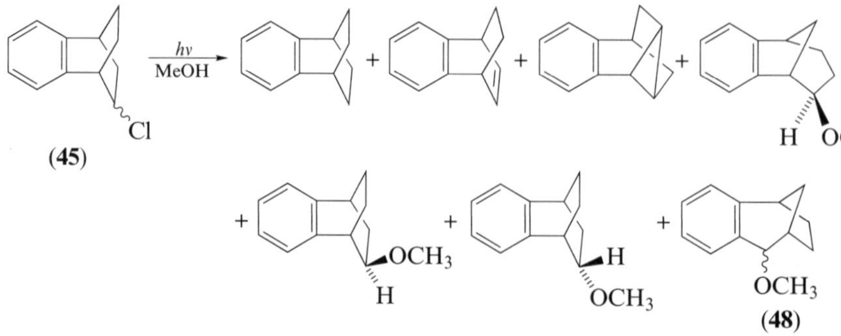

formed, than of the products (**50**) which result from *anti* migration. In ground-state heterolytic reactions of **46** only **50** is produced.

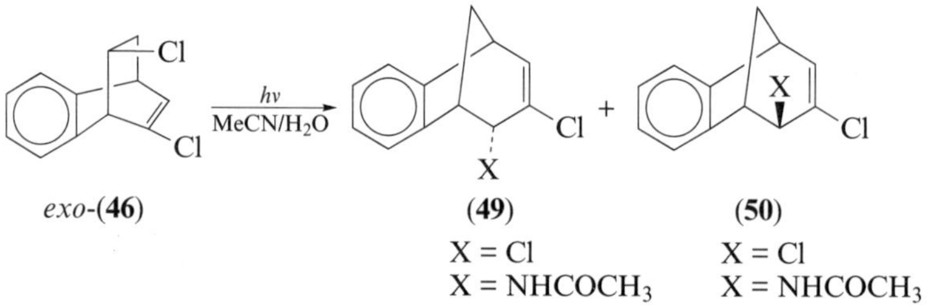

Two rationalizations have been offered for the high *exo/endo* photoreactivity ratios, and also two for the formation of the *syn* rearranged photoproducts (**48** and **49**). The relative reactivity is thought to be determined by either factor (a) or (b) below:

(a) The extent of orbital interaction in the transition state of the C—X bond cleavage reaction which, in terms of the natural correlation concept, results from a continuous transformation of the excited species from an initial $\pi\pi^*$ excited state into a dissociative, mainly $\sigma\sigma^*$, configuration. When the C—X unit is *anti* to the aryl group, this interaction is considerably larger and the barrier to reaction therefore considerably smaller[217].

(b) The difference in free energy of electron transfer of the π^* electron of the initial $\pi\pi^*$ excited state to the σ^* orbital of the C—X bond necessary for cleavage. This energy is lower for an *anti* C—X bond compared to that for a *syn* C—X bond because the σ^*-orbital of the *anti* bond has its major lobe close to the π orbital of the aromatic ring while the *syn* bond has not. Therefore, the zwitterionic biradical resulting from electron transfer has better coulombic stabilization[222,224].

The *syn* rearranged photoproducts are considered to be formed either via 'hot' carbocations generated photolytically[217] or to be formed in a frontside (*syn*) rearrangement which occurs concertedly with the loss of chloride from the intermediate zwitterionic biradical produced by electron transfer[224]. This process is in competition with C—X bond cleavage to a biradical cation which undergoes intramolecular back electron transfer to yield relaxed carbocations, which accounts for the *anti* migration products as formed in thermal solvolysis reactions.

Extensive information concerning the operation of stereoelectronic effects on the photoreactivity of homobenzylic halides has also been obtained for dibenzo[2.2.2]-systems[223–228] such as **51–54**. The photo-Wagner–Meerwein rearrangements and photo-solvolyses of these compounds also occur via exclusive[223] or preferential[224] reaction of the β-C—Cl bond *anti* to the excited aryl group. Electron transfer between these two moieties is considered to be the key step here too. Indeed, the thermodynamics of such a step seem

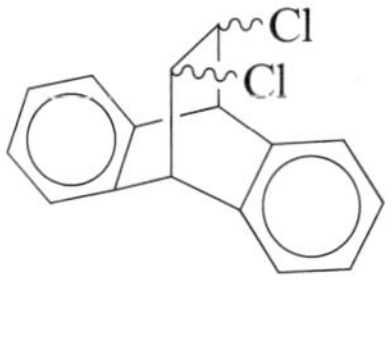

(51) 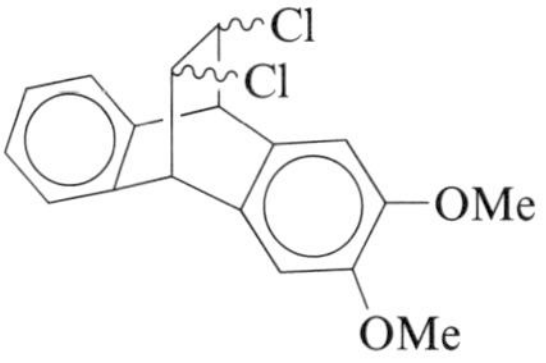 (52) 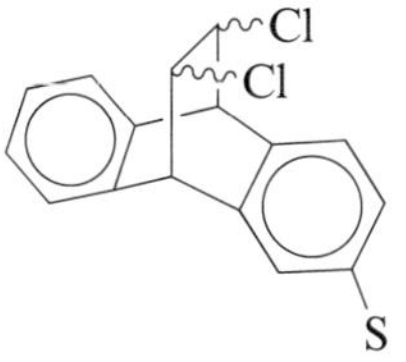 (53)

S = NO$_2$, COCH$_3$, CN

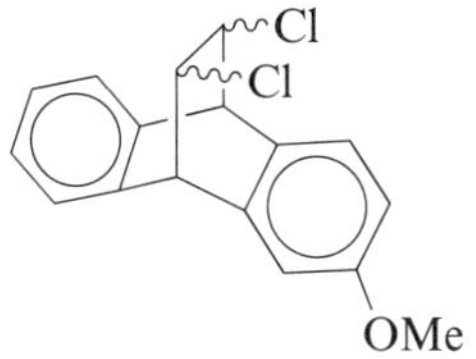

(54)

to play a crucial role in determining whether or not the reactions proceed. When the photoexcited (singlet) state is a relatively poor electron donor, as in **51**, only loss of the *anti* chlorine is observed; when it is a good donor, as in **52**, both *anti* and *syn* chlorine are lost[224], and when it is a poor donor system as in **53** there is no photoreaction at all[225]. Also, in most cases the triplet excited states with their lower energy are unreactive. Only the triplet states of **52**, for which the free energy of electron transfer can be calculated not to be prohibitive, are photoreactive[225].

Thermodynamics, however, is not the sole determining factor. Of the *m*-methoxy-substituted compounds **54** the most photoactive isomers are those in which the bridge Cl is *anti* to the methoxy-containing ring and attached to the side of the bridge which is *meta* rather than *para* to the methoxy-substituent[227]. This suggests that orbital overlap may also be an important factor.

The regioselectivity of the photo-Wagner–Meerwein rearrangements and photosolvolyses of **51–54** is similar to that observed for the monobenzo systems: preponderant or even stereospecific *syn* migration, in contrast to their ground-state reactions. The preference is ascribed similarly to a migration concerted with loss of halide in the zwitterionic diradical produced by electron transfer[224]. Interestingly, the stereochemistry of migration of the triplet-sensitized reactions of the *trans* dichloride **52** is quite different from that of its direct irradiation (equation 48)[225]. Virtually only the veratro-ring migration product is found, indicative of the intervention of a relaxed, ground-state solvolysis-type, carbocation as the migration product-determining intermediate in the triplet reaction.

A much more deep-seated photorearrangement than in the dibenzo[2.2.2] systems occurs in the dibenzo[3.2.1] compound **55** with chloride or bromine as nucleofugal group at C-8 (equation 49)[229]. 9-Functionalized phenanthrenes are formed, supposedly also by way of photoinduced intramolecular electron transfer in the (triplet) excited state.

Stereoelectronic effects also play a role in the photoreactivity of the chlorides **56**[230], in which the halide is γ to the aryl chromophore[230]. Their irradiation in methanol yields primarily products derived from carbocation intermediates, but here the *endo*-isomer is more photoreactive than the *exo*-isomer, in contrast with the situation observed for the β-aryl compounds. The inverted reactivity pattern is attributed to the favourable aryl/chlorine relationship in the *endo*-isomer.

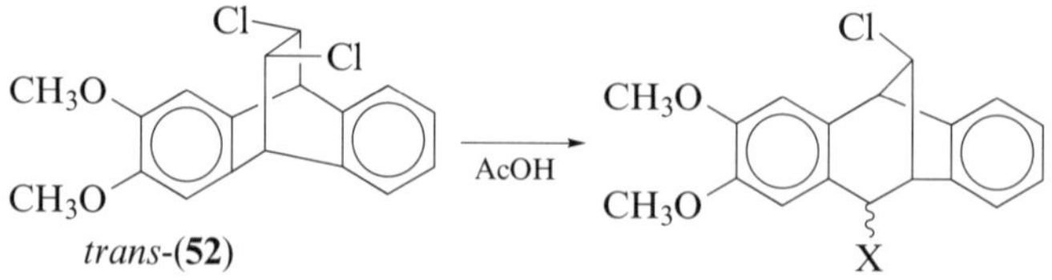

trans-(52)

AcOH: hv direct: X = Cl, OAc   66%

AcOH/acetone: hv sensitized: X = OAc   4%

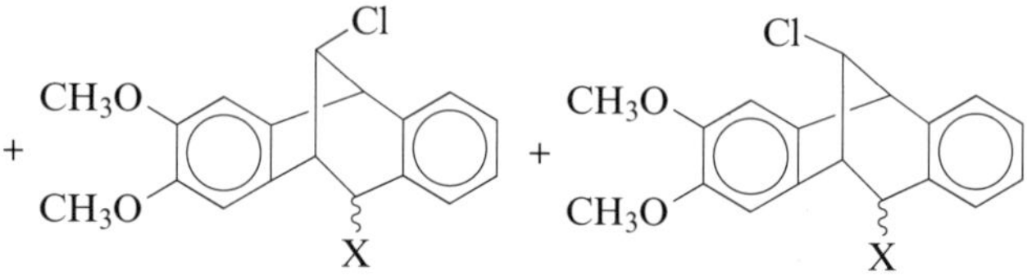

(48)

26%   6%

—   96%

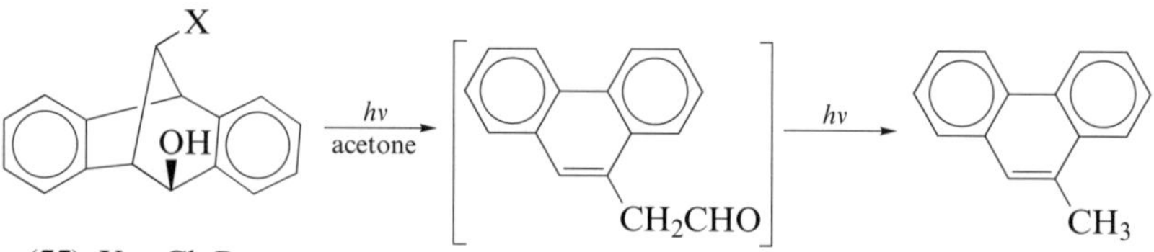

(49)

(55) X = Cl, Br

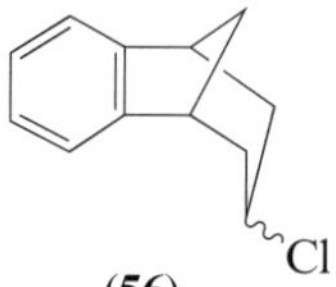

$\Phi\text{dis }(exo) = 7.6 \times 10^{-3}$

$\Phi\text{dis }(endo) = 8.1 \times 10^{-2}$

(56)

E. α-Halo Ketones

Elimination of HX and reductive dehalogenation are the major reactions observed upon irradiation of α-halo ketones such as the 2-X-cyclohexanones (X = Cl, Br, I) in cyclohexane[231]. The photoelimination presumably occurs via loss of a β-proton from an intermediate α-keto carbocation; the reductive dehalogenation occurs via homolytic cleavage of the C—X bond and hydrogen abstraction. As in the case of simple alkyl halides[25], the percentage of ionic product is largest for X = I, but even then the elimination is still only the minor process (30%). α-Fluorocyclohexanone in cyclohexane or tert-butanol yields no cyclohex-2-enone and only a minor amount of reductive defluorination product; its principal photoreactions are those characteristic of the keto group, i.e. ketone reduction and α-cleavage[232,233]. The photolytic cleavage of α-chlorocyclohexanones is subject to stereoelectronic control (equation 50)[234]. The axial-Cl isomer 57 **ax** is almost fourfold more photoreactive than the equatorial-Cl substrate 57 **eq**. The difference in photoreactivity is a manifestation of the stronger interaction of the axial relative to the

$$\text{t-Bu} \cdots \overset{O}{\underset{Cl}{\big|}} \quad \xrightarrow[\text{cyclopentane}]{hv} \quad \text{t-Bu} \overset{O}{\bigcirc} + \text{t-Bu} \overset{O}{\bigcirc} + \text{t-Bu} \overset{O}{\bigcirc} C_5H_9\text{-}c \tag{50}$$

	t-Bu	t-Bu	t-Bu
(57 eq) Φdis = 0.18	61%	7%	32%
(57 ax) Φdis = 0.69	67%	9%	24%

equatorial α-chlorine with the keto group. This results in a larger σ* component in the LUMO of the ground state and thus in the reactive electronically excited state.

The photochemistry of 2-halo-1-tetralones[235–237] and of the corresponding [b] benzo-fused five-[235,236] and seven-[238,239] membered alkyl ring compounds also involves competing ionic and radical processes. The percentages of ionic (= elimination) product formed increase in the order X = Cl < Br < I, and are larger than those found for the 2-X-cyclohexanones[236]. For example, 2-iodo-1-tetralone in cyclohexane yields 1-naphthol and α-tetralone in a ratio of 68:32. Clearly the introduction of a phenyl group in conjugation with the carbonyl group increases the ratio of ionic to radical product. For the α-fluoro-five-, six- and seven-membered alkyl ring systems in cyclohexane, no ketone photo-chemistry is observed. Instead, only reductive defluorination was found, with the percent-age of conversion increasing with ring size[238]. Replacement of the solvent cyclohexane by methanol does not increase the percentage of ionic product from 2-halo-1-tetralones, but the presence of metal(II) salts and of solid supports does[237]. For the five-[235] and seven-[239] membered alkyl ring compounds in methanol, substitution of X by methoxy instead of elimination is found to occur: nucleophilic trapping of the α-keto cation by the solvent competes here effectively with loss of a β-proton.

Acyclic alkyl α-chloroketones such as α-chloropropiophenone upon irradiation in methanol (equation 51)[239] or in aqueous acetone[240] yield 2-arylpropionic esters or acids. These products are the result of a Favorskii-type rearrangement, an apparent ionic process. Depending on the nature of the substituent in the benzene ring, competing photoreduction and nucleophilic substitution reactions also occur[240]. In compound 58a, the formation of products resulting from photoenolization (via γ-hydrogen abstraction by the nπ* excited carbonyl), namely 59a and 60a, competes with formation of the C—Cl bond cleavage products 61a and 62a (equation 52)[239]. On making the chlorine-bearing carbon atom primary, as in 58b, the C—Cl bond cleavage products are no longer observed. In laser nanosecond flash photolysis experiments with 58b in methanol, only one photoenol is detectable[241]. This species is believed to be the E enol responsible for indanone 59a formation; the Z enol, which has the preferred conformation for reketonization, will be very short-lived.

Irradiation of α-bromoacetophenones in methanol yields only ionic products (i.e. nucleophilic substitution and Favorskii-type rearrangement products) if there is an ortho-methoxy substituent in the aromatic ring[152]. This is due to stabilization of the intermediate cation as a result of interaction between the lone pair on the oxygen of the ortho-methoxy group and the vacant p-orbital of the cation. With an ortho-acetoxy substituent in the ring

$$\overset{O}{\bigcirc}\text{—CH}Cl\text{—CH}_3 \quad \xrightarrow[\text{MeOH}]{hv} \quad \bigcirc\text{—CH(CH}_3)\text{CO}_2Me + \overset{O}{\bigcirc}\text{—CH}_2\text{CH}_3 \tag{51}$$

| 41% | 27% |

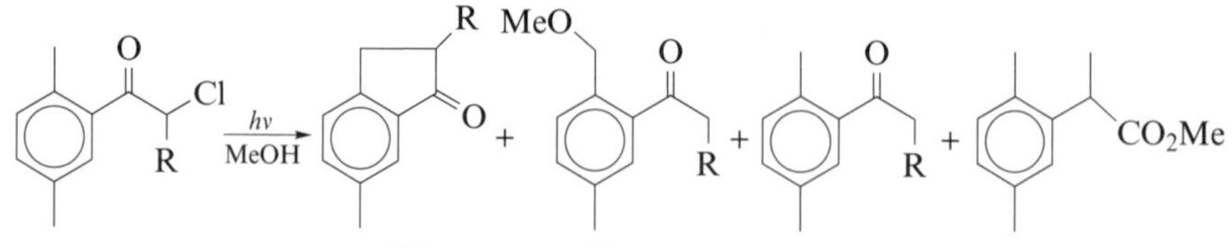

	(59)	(60)	(61)	(62)
(58a) R = CH$_3$, ϕ = 0.42	28%	28%	4%	24%
(58b) R = H, ϕ = 0.76	39%	57%	—	—

(52)

of α-bromoacetophenone, only ionic-type photobehaviour is observed in the highly polar solvent DMSO[242].

Laser flash photolysis of α-chloro-[243,244] and α-bromoacetophenone[244] in benzene is a convenient method to generate Cl and Br atoms (as their halogen atom–benzene π complexes). The C—X bond cleavage occurs with quantum yields of 0.88 and 0.41, respectively[244].

α-Bromoalkyl ketones easily undergo homolytic C—Br bond cleavage[245]. Interestingly, this is not the case for α-bromoalkyl 1,2-diketones with abstractable γ-hydrogen atoms. Such compounds only give photocyclization without loss of bromine[246].

Sterically hindered α-haloketones such as α-Cl- or α-Br-isobutyrophenone yield only a light-induced C-alkylation product with Me$_2$C=NO$_2^-$ in DMSO when the phenyl ring carries a p-nitro or a p-cyano substituent (equation 53)[247]. This photosubstitution meets the usual criteria for an S$_{RN}$1 reaction mechanism. So does the photoreaction of the bridgehead chloride 63 with diphenylphosphide ions in liquid ammonia (equation 54)[248]. The reactivity of 63 is larger than that of similar non-keto bridgehead chlorides in S$_{RN}$1 reactions.

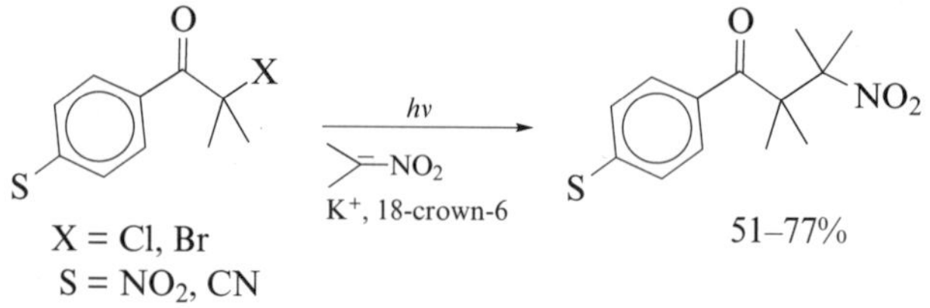

(53)

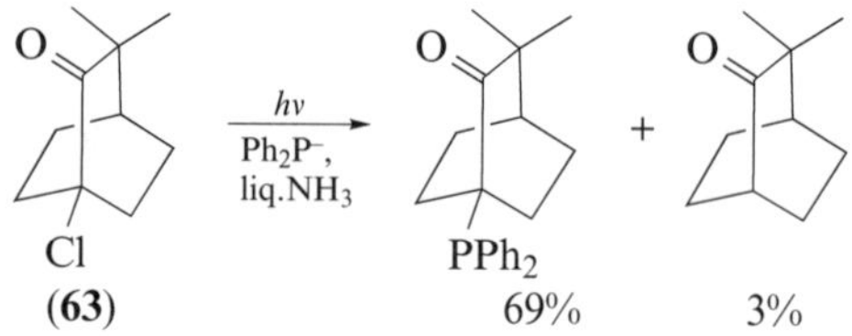

(54)

Photolysis of 2-fluorocyclohexanone in isopropanol selectively yields the reductive defluorination product[232]. In this solvent the cleavage of the C—F bond is assumed to occur heterolytically in the radical anion formed by electron transfer to the excited fluoroketone. Photoreduction of α-bromo- and α-chloroacetophenone by NADH model compounds also proceeds via photoinduced electron transfer[249-251]. Depending on the NADH analogue used, different mechanistic pathways are followed: a rate-determining photoinduced electron transfer from the sensitizer to the phenacyl halide or an electron transfer radical chain reaction. Electron transfer must also be important in the photoreaction of α-chloroacetophenone with alkenes in the presence of silver triflate (e.g. equation 55)[252]. In the absence of the silver salt, the photoreaction only proceeds with electron-rich alkenes. The silver ion may assist the interligand electron transfer.

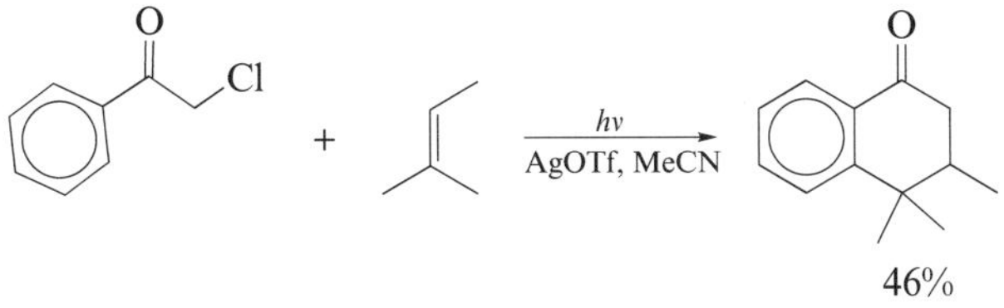

$$\text{(55)}$$

46%

Irradiation of α,β-dihaloketones such as 3,4-dibromobutyrophenone in methanol results in dehalogenation to the α,β-unsaturated ketone in addition to nucleophilic substitution and reduction of the Br at the carbon atom α to the carbonyl group[153]. Dehalogenation also occurs efficiently upon irradiation in benzene in the presence of triethylamine. Introduction of a second halogen atom at C-α in α-haloketones, as in e.g. 2,2-dibromo-1-tetralone, leads to an increased ratio of the ionic (= elimination of HBr): radical (= reductive debromination) pathways compared to the situation in the monobromo compound[236]. In 2-Br-2-F-1-tetralone, the C—Br bond is selectively cleaved; the percentage of ionic product is only a bit smaller than for the monobromo compound[238].

The photolysis of hexachloroacetone or tribromoacetaldehyde in methanol does not yield any alcoholysis product[253]. The major reaction is reduction. On the other hand, α,α,α-tribromoacetophenone yields the alcoholysis product methyl benzoylformate in good yield (equation 56)[254]. The initial photomethanolysis product is α,α-dibromo-α-methoxy-acetophenone, which in a dark reaction is converted into the benzoylformate. The methyl benzoate is formed by nucleophilic attack on the carbonyl carbon.

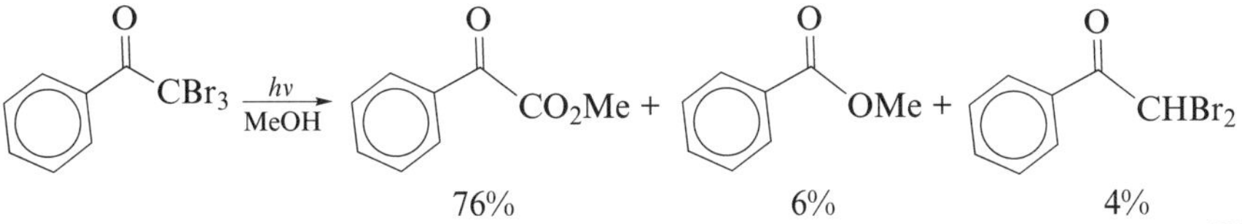

76%   6%   4%

$$\text{(56)}$$

In β-chloroketones the C—Cl and C=O groups show significant interaction, if the two entities are in the correct orientation with respect to each other[136]. This interaction results in an appreciable σ^* component in the LUMO of the β-chloroketone, and makes the C—Cl bond potentially photolabile in the excited state[12,136]. Indeed, irradiation of 4-chloro-2-butanone in methanol has been reported to afford the corresponding 4-methoxy ether in addition to the alcohol resulting from carbonyl reduction. The apparent nucleophilic photosubstitution product, however, turns out to be produced in a dark reaction, catalysed by acid which is formed when methanol solutions are irradiated[255].

The elimination of HX which occurs upon UV irradiation of δ-haloketones, such as the δ-X-valerophenones, is the result of a photoreaction characteristic of the C=O group: the reaction proceeds via loss of X$^•$ from the biradical formed by γ-hydrogen abstraction[256]. A comparable HX elimination does not occur with γ-haloketones, but does to some extent with an ε-iodoketone.

F. α-Halo Esters

Irradiation of alkyl α-chloroesters, such as **64**, in acetonitrile leads to photoelimination of HCl and formation of α,β-unsaturated esters in good yield (equation 57)[257]. Likewise, the corresponding α,α-dichloroester **65** also undergoes photoelimination. α-Bromoesters such as methyl α-bromophenyl acetate[152], **66** and **67**, which are also benzylic halides and

$$\text{(64) R = H} \quad \xrightarrow[\text{MeCN}]{hv} \quad \text{(CH}_2)_2\text{CH}{=}\text{CRCO}_2\text{Et} \tag{57}$$

(**64**) R = H

(**65**) R = Cl

63%

58%

are devoid of a β-hydrogen, smoothly yield nucleophilic substitution products in methanol in addition to reductive dehalogenation products (equation 58)[155]. Upon laser flash photolysis of **66** in acetonitrile and of **67** in hexafluoroisopropyl alcohol the corresponding carbocations, intermediates in the solvolysis process, were observed directly. The radicals which are intermediates in the reduction were also generated flash-photolytically. Comparison of the reactivities of the transient carbocations towards nucleophiles with those of the unsubstituted diphenylmethyl and 9-fluorenyl cations shows that replacement of an α-hydrogen by an electron-withdrawing carbomethoxy group brings about an increase in kinetic stability. The p-methoxyphenyl carbomethoxymethyl cation has also been observed by laser flash photolysis[160].

$$\begin{array}{c} \text{Ar} \quad \text{Br} \\ \diagdown\!\!\!\times\!\!\!\diagup \\ \text{Ar} \quad \text{CO}_2\text{Me} \end{array} \xrightarrow[\text{MeOH}]{hv} \begin{array}{c} \text{Ar} \quad \text{OMe} \\ \diagdown\!\!\!\times\!\!\!\diagup \\ \text{Ar} \quad \text{CO}_2\text{Me} \end{array} + \begin{array}{c} \text{Ar} \quad \text{H} \\ \diagdown\!\!\!\times\!\!\!\diagup \\ \text{Ar} \quad \text{CO}_2\text{Me} \end{array} \tag{58}$$

(**66**)   Ar = Ph   30%   20%

(**67**) Ar$_2$C = Fluorenyl   54%

If there is also a halogen substituent present at the β-position in α-bromoesters, as e.g. in compounds **68** (equation 59), photolysis in methanol gives efficient dehalogenation to the corresponding alkene[153]. No substitution product is found. Efficient debromination also occurs when the irradiation is carried out in benzene in the presence of Et$_3$N. Both dehalogenations are non-stereospecific. The reactions provide a useful method to regenerate, under mild conditions, C=C bonds which were protected as vicinal dihalides.

$$\text{PhCH(Br)CH(Br)CO}_2\text{Et} \xrightarrow{hv} \begin{array}{c} \text{Ph} \quad\quad \text{H} \\ \diagup\!=\!\diagdown \\ \text{H} \quad\quad \text{CO}_2\text{Et} \end{array}$$

(**68**)

			$E:Z$
threo	$\xrightarrow[\text{MeOH}]{hv}$	90%	3:2
erythro		46%	3:2
threo	$\xrightarrow[\text{Et}_3\text{N, benzene}]{hv}$	100%	3:2
erythro		75%	1:1

$$\tag{59}$$

UV irradiation of Cl$_3$CCO$_2$Me and Br$_3$CCO$_2$Me in methanol only results in the formation of reduction products[254]. This is also the case for the irradiation of perhalogenated chlorofluoropropionates in 2-propanol[258,259] and of alkyl perfluoroesters in hexamethylphosphortriamide (HMPA)[260,261]. The ester group exhibits a strong directing effect. In esters of the type CFXY—CClZ—COOR (X, Y, Z = Cl, F) the C—Cl bond α to the carboxyl group is selectively reduced in 2-propanol, and so are the α-C—F bonds of perfluoroesters in HMPA. Of esters with only an α-F and a β-Cl substituent, both bonds are reduced in 2-propanol at comparable rates. The reductions in both media do not occur via simple C—X bond homolyses. In 2-propanol, quantum yield values of over 200 have been measured, indicating a chain radical mechanism. In HMPA, the solvent probably functions as electron donor, yielding the radical anion of the perfluoroester which dissociates into fluoride anion and an α-C˙ radical. α-Chloroesters are also effectively reduced in 2-propanol by irradiation in the presence of ketones and benzoin derivatives[262]. In this case the reaction is proposed to occur via electron transfer from an easily oxidizable ketyl radical formed by α-cleavage.

Photolysis of diethyl bromomalonate in benzene in the presence of compounds with a benzylic hydrogen as possible H-atom source leads to reduction of the ester and bromination of the benzylic compound[263]. The malonyl radical and not the Br atom turns out to be the H-atom abstracting agent.

Irradiation of chloroacetonitrile in the presence of aromatic hydrocarbons leads to ring cyanomethylation[264]. The efficiency of this intermolecular alkylation reaction parallels the electron-donor capacity of the aromatic compounds. In contrast to the case of the photoalkylation of α-haloesters and α-haloamides, aromatic compounds without electron-donating substituents are also alkylated. The alkylation involves electron transfer, probably by way of an exciplex, from the electronically excited aromatic hydrocarbon to the good acceptor chloroacetonitrile, followed by cleavage of the C—Cl bond and combination of the radical with the aromatic radical cation (equation 60). The regioselectivity of the alkylation (e.g. for anisole, o:m:p = 67:20:12) is determined in the coupling step. Direct physical evidence for the resulting radical cation with p-dimethoxybenzene was obtained by laser flash photolysis.

$$ArH^* + ClCH_2CN \longrightarrow [exciplex] \longrightarrow [ArH^{+\bullet} + \;^{\bullet}CH_2CN + Cl^-] \longrightarrow products$$

$$(60)$$

Intramolecular alkylation reactions of α-haloesters appropriately substituted with an electron-rich arene, such as the 3-indolyl moiety, have been used as key steps in the syntheses of the 2,3-bridged indole alkaloids deethylcatharanthine[265] and catharanthine **70** ($Z = H_2$) from **69** (equation 61)[266] and in the syntheses of 3,4-bridged indoles such as the pyrrolobenzoxocine **72** (equation 62)[257]. Interestingly, the photochemical cyclization of **69** to the 2-position of the indole was only successful for **69**, Z = S and not for **69**, Z = O[266], despite the fact that the corresponding 20-deethyl compound with Z = O yields 5-oxo-20-deethylcatharanthine in moderate yield[265]. In the latter case, a comparable amount of the photocyclization product to the 4-position is also formed. The 3,4-bridged indole **72** is formed reasonably efficiently upon photolysis of the trichloroacetyl derivative **71**[257]. No cyclization to the 2-position is reported to occur. The photocyclization is prevented by placing an electron-withdrawing substituent at the indole nitrogen and consequently reducing the electron density in the ring. This observation is in line with a mechanism for the photocyclization as depicted in equation 60 for the intermolecular version of the reaction. So is the fact that the trichloroacetyl compound **71** gives a higher yield of

$$(61)$$

(**69**)   Z = O
         Z = S

(**70**)   Z = O   trace
         Z = S   41%
         Z = H₂

$$(62)$$

intramolecular alkylation product than the corresponding dichloro compound (42% and 19%, respectively). The additional Cl increases the yield by further stabilizing the intermediate radical. The yields with trichloroacetates are still poorer than with the corresponding trichloroacetamide derivatives, possibly due to the greater stability of ˙CH$_2$CONR vs ˙CH$_2$CO$_2$R radicals. Esters **64** and **65** (equation 57) do not photocyclize[257]. Only when the elimination of HCl is blocked by replacing the two β-hydrogens by methyl groups does photocyclization occur, albeit in poor yield.

G. α-Halo Amides

The N-chloroacetyl derivatives of amines appropriately substituted with electron-rich arenes often give intramolecular alkylation and not solvolysis upon irradiation in aqueous solution. These photocyclizations have proven to be quite useful and versatile methods to synthesize a variety of novel and natural aza-heterocyclic compounds. The reader is referred to Reference 13 for a comprehensive review of the scope and mechanism of these reactions.

Further information has been provided that for the N-chloroacetyl derivatives of anilines, the conformation of the amide bond determines their photochemical behaviour[267,268]. For instance, photolysis of compound **73** (R = H) in methanol yields (rearranged) solvolysis products, while **73** (R = CH$_2$Ph) gives photocyclization to a five-membered lactam (equation 63)[268]. Introduction of the alkyl group on the amide nitrogen changes the preferred conformation of the amide group from *trans* to *cis*, allowing close approach of the chloromethyl group to the aromatic moiety. An N-alkyl-9-anthrylchloroacetamide in acetonitrile/water also gives only photocyclization[269]. Trichloroacetanilide compounds, such as **74**, undergo substitution reactions at the side chain in methanol (equation 64)[254]; the photochemistry of the corresponding N-alkyl compounds has not (yet) been reported in the literature. A situation similar to the one for aniline compounds **73** also prevails for the N-chloroacetyl derivatives of benzylamines[267]. The N-nonalkylated amides yield side-chain photosubstitution products, the N-alkyl ones readily photocyclize. N-Alkylation apparently is not a necessary condition for the photochemical ring closure of N-chloroacetyl derivatives of 3-indolyl substituted alkylamines with one[270,271] or two[272,273] carbon atoms in the tether connecting the amino group and the indole ring such as **75** and **77**. The ring closure reactions of equations 65 and 66 show their synthetic utility to prepare 3,4-bridged indoles. Compound **75** only gives the 3,4-bridged azepino indole **76**, in fair yield[271]. This is not the case for the corresponding monochloroacetyl compound **77**, R = R′ = H with two carbons in the methylene chain, the archetype compound for which the photocyclization of α-haloacetamides was discovered. The 3,4-bridged indole product **78**, is formed in poor yield, which is partially due to competing formation of the 2,3-fused isomer **79**, formed by cyclization to the indole 2-position[272]. Selectivity for ring closure is much better with the dichloroacetyl and trichloroacetyl analogues **77**, R = Cl, R′ = H and

(63)

(64)

(65)

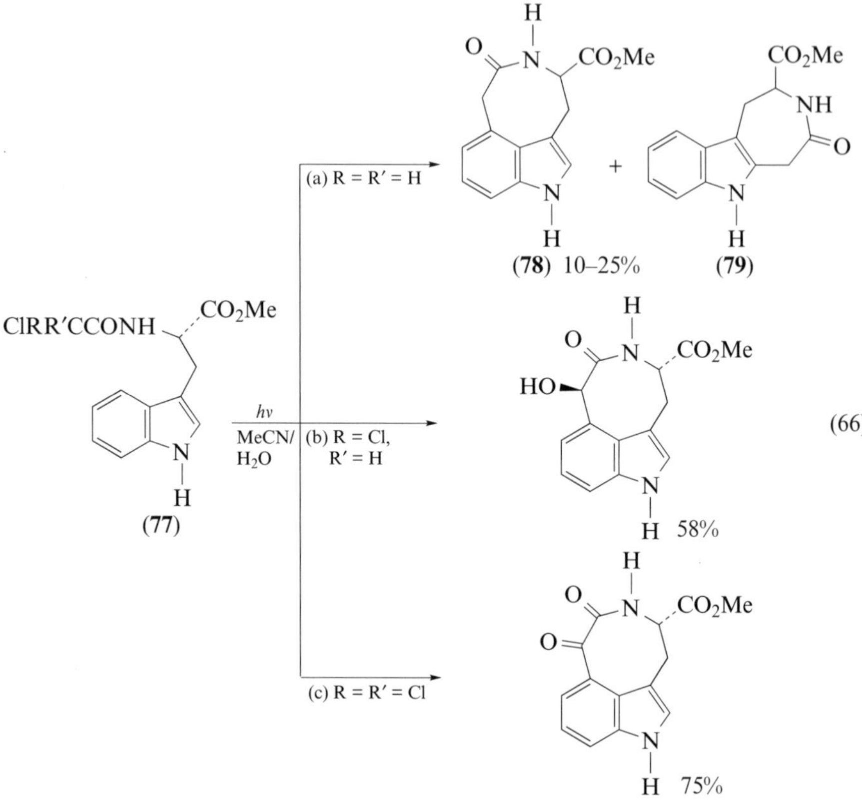

(66)

77, R = R' = Cl. Only photocyclization to the C-4 position occurs; also, the yields are higher.

H. α-Halo Nitroalkanes

Introduction of a nitro group into alkyl halides considerably enhances the facility of such compounds to act as electron acceptors. α-Halo nitroalkanes therefore readily participate in photostimulated $S_{RN}1$ reactions with a variety of nucleophiles. Various aspects of these reactions have been reviewed extensively and the reader is referred to References 6–10 for detailed accounts. In the period under review here $S_{RN}1$ reactions of α-halo nitroalkanes under irradiation have been reported with nucleophiles such as anions of β-diketones, β-keto esters and malonic esters[183,274–276], anions of monoketones[277–279], nitronate anions[280,281], thiolate anions[282–285], sulphinate anions[286], azide anions[287,288], (nitro) imidazole ions[186,187] and even alkoxide ions[289].

The representative examples of reaction with the first type of carbanions, depicted in equations 67[183] and 68[275], show that (functionalized) 2-chloro-2-nitropropane can be utilized as a ketone equivalent. Illumination, followed by elimination of the elements of HNO_2 (by using a second equivalent of anion as a base)[274] or of NO_2 and CO_2 (by heating with sodium bromide after the irradiation)[275] from the intermediate β-nitro products, constitutes a one-pot synthesis of alkenes. The method is especially useful to prepare highly substituted alkenes, because the radical reaction step in which the C—C bond is formed is

$$\underset{NO_2}{\overset{CH_2OTHP}{\underset{|}{\overset{|}{C}}}}Cl + H\bar{C}\underset{COMe}{\overset{CO_2Et}{<}} \xrightarrow[DMSO]{h\nu} \underset{NO_2}{\overset{THPOCH_2}{\underset{|}{\overset{H}{\underset{|}{C}}}}}\underset{COMe}{\overset{CO_2Et}{<}} \longrightarrow \underset{55\%}{\overset{THPOCH_2}{\underset{}{}}}\!\!=\!\!\underset{COMe}{\overset{CO_2Et}{<}} \quad (67)$$

$$\underset{NO_2}{\overset{|}{\underset{|}{C}}}Cl + Et\bar{C}(CO_2Et)_2 \xrightarrow{\underset{HMPA}{h\nu}} \underset{NO_2\ CO_2Et}{\overset{Et}{\underset{|}{\overset{|}{C}}}}CO_2Et \xrightarrow[NaBr/HMPA]{\Delta} \underset{63\%}{\overset{Et}{\underset{CO_2Et}{=}}} \quad (68)$$

less sensitive to steric hindrance than the usual ionic processes. A reaction as in equation 68 has been used for chain elongation of 1-C-nitroglycosyl halides[276]. In the photoinduced coupling of 2-chloro-2-nitropropane with enolates of monoketones, which are more easily oxidized than the above disubstituted carbanions, the nucleophilic substitution is accompanied by oxidative enolate dimerization (equation 69)[277–279]. The latter reaction is also a free radical chain process. As the product ratio depends on the nature of the nucleofugic group, in this case the reactions presumably involve bimolecular substitution ($S_{RN}2$) and electron transfer between the anion and the intermediate nitroalkane radical anion $ClCMe_2NO_2^{-\cdot}$.

$$O_2N\underset{}{\overset{|}{\underset{|}{C}}}Cl + Ph(OLi) = CHMe \xrightarrow[\substack{hexane}]{\underset{THF/}{h\nu}} \overset{O}{\overset{\|}{Ph\overset{}{C}}}-CH(Me)\underset{}{\overset{|}{\underset{|}{C}}}NO_2 + (\overset{O}{\overset{\|}{Ph\overset{}{C}}}-)_2 \quad (69)$$

$$ 48\% \qquad\qquad 37\%$$

The use of a 5-hexenyl or cyclopropyl group as radical probe to distinguish between $S_{RN}1$ and $S_{RN}2$ routes of α-halo nitroalkanes was not successful[280]. Unlike in the case of non-nitro alkyl halides (see equations 10 and 11 in Section II.A), even in well-established $S_{RN}1$ reactions of **80** and **81** no ring cyclization to the corresponding cyclopentylcarbinyl compound or ring opening to the 3-butenyl compound is observed. Apparently the nitro group stabilizes the intermediate radicals, which makes their rate of reaction with a nucleophile faster than their rates of ring closure or ring opening.

$$\underset{\textbf{(80)}}{\overset{NO_2}{\underset{Cl}{\diagdown\!\!\diagup\!\!\diagdown\!\!\diagup\!\!\diagdown}}} \qquad\qquad \underset{\textbf{(81)}}{\overset{}{\triangleright\!\!-\!\!\underset{NO_2}{\overset{|}{\underset{|}{C}}}Cl}}$$

The reactions of the stereo-pair of α-chloro nitrocyclohexanes **82** and **83** with a nitronate anion gave identical proportions of the epimeric substitution products (equation 70). This

$$\underset{\textbf{(82)}}{\overset{NO_2}{\underset{Cl}{\bigcirc}}} \ or\ \underset{\textbf{(83)}}{\overset{Cl}{\underset{NO_2}{\bigcirc}}} \xrightarrow[\substack{DMSO}]{\underset{>\!=\!NO_2Li^+}{h\nu}}$$

$$\underset{90}{\overset{NO_2}{\underset{CMe_2NO_2}{\bigcirc}}} + \underset{10}{\overset{CMe_2NO_2}{\underset{NO_2}{\bigcirc}}} \quad (70)$$
$$90 \qquad : \qquad 10$$

stereochemical result shows that both reactions involve the same, effectively planar, radical intermediate[281].

The photostimulated reactions of thiolate anions with 2-halo-2 -nitropropane derivatives yield both α-nitrosulphides via an $S_{RN}1$ pathway and disulphides (equation 71a)[282-284]. In contrast with the case of the oxidative dimerisation products of the mono-enolates, the disulphides are formed via an ionic mechanism: nucleophilic attack by the thiolate anion on the α-halogen and subsequent reaction of a second thiolate with the sulphenyl halide. As expected for such a process, disulphide formation is favoured (and thus α-nitrosulphide formation is disfavoured) the more nucleophilic the thiolate (i.e. derived from a less acidic thiol) and the easier the abstraction of the halo-substituent (i.e. I > Br > Cl). Use of the protic solvent methanol instead of the usual dipolar aprotic solvents for the reaction of equation 71a is detrimental to the yield of the $S_{RN}1$ substitution products; exclusively disulphides are formed[285] (equation 71b). Methanol solvation probably retards the dissociation of the radical anion intermediate in the $S_{RN}1$ reaction, into radical and anion, and hence retards the chain reaction relative to the ionic reaction. The non-nucleophilic methylsulphinate ion gives only an $S_{RN}1$ reaction product with 2-bromo-2-nitropropane[286].

$$Me_2C(X)NO_2 + RS^- \quad \begin{cases} \xrightarrow[DMF]{hv} Me_2C(SR)NO_2 + RSSR & (71a) \\ \\ \xrightarrow[MeOH]{hv} RSSR & (71b) \end{cases}$$
$$X = Cl, Br, I$$

Thiophosphite ion, even upon irradiation, only attacks the halogen atom of $Me_2C(Cl)NO_2$ in an S_N2 attack on the halogen[188]. With azide anion as nucleophile, only low yields of substitution products have been obtained[287,288]. Much more effective nitrogen-centred nucleophiles in photostimulated reactions with α-halo nitroalkanes are the anions of nitroimidazoles[186] and imidazoles[187]. With the ambident 4(5)-nitroimidazole anion, the 4-isomer is exclusively formed, with no indication of the 5-isomer (equation 72).

$$Me_2C(Br)NO_2 + \quad \xrightarrow[DMSO]{hv} \quad 92\% \quad (72)$$

Alkoxide anions are generally considered to be unreactive nucleophiles in $S_{RN}1$ reactions. However, photostimulated free radical chain reactions between alkoxide ions derived from primary alcohols and 2-chloro- or 2-bromo-2-nitropropane have been observed[289]. The products are acetals $Me_2C(OR)_2$ formed via trapping of $Me_2CNO_2^-$ by RO^-, followed by fragmentation of the resulting $ROCMe_2NO_2^{-\cdot}$ into NO_2^- and $ROCMe_2^{\cdot}$ which is oxidized to $ROCMe_2^+$ by the starting material.

III. VINYLIC HALIDES

The irradiation of vinylic halides in solution is by now as well-established as a convenient and powerful method to generate vinyl cations as the photolysis of alkyl halides is to create alkyl cations. In the period under review the scope of the method has been considerably expanded, and so has the knowledge of the relationship between the structure of vinylic halides and their photoreactivity. A review has appeared in which results up to 1990 have been tabulated[290]. The bulk of the studies concerns the photoreactions of vinylic halides with a hydrogen, an alkyl group or a (substituted) phenyl group as substituent at the α-position. Their photosolvolysis reactions are reported in References 291–312, which are ordered in increasing complexity of the α-substituent of the vinylic halide studied and sub-ordered in increasing complexity of the β-substituents, and therefore in increasing complexity of the chromophoric group. Photosolvolysis studies have been also reported of vinylic halides with an α-vinyl group[313], an α-halogen[291,294,314] and even with vinyl cation destabilizing α-carbonyl or α-cyano groups[315,316].

The photoreaction of compound **84** (equation 73)[291] exemplifies the behaviour of α-H-substituted vinyl halides upon electronic excitation. In such compounds solvolysis is often accompanied by Wagner–Meerwein rearrangement, which in the case of **84** leads to the ring-expanded ketal **85**. In addition, reduction occurs. The vinyl bromide **84**, X = Br yields a smaller percentage of ion-derived product than the corresponding iodide. A similar situation prevails for alkyl bromides versus alkyl iodides and, as for those compounds, seems not in line with the mechanism proposed for (vinyl) cation formation: Homolytic cleavage of the C—X bond followed by electron transfer in the initially formed radical pair, in competition with its dissociation into free radicals (see equation 4 in Section II.A). Electron transfer is expected to be easier for Br˙ than for I˙. Possibly the same cause as proposed for alkyl halides underlies this discrepancy. It is of interest to note that ω-iodo- and ω-bromocamphene, which are analogues of **84** with a bicyclo[2.2.1]heptane skeleton instead of a cyclohexane ring annulated to the β-side of the C=C bond, only give radical-derived and no ion-derived products upon photolysis in methanol[292,294].

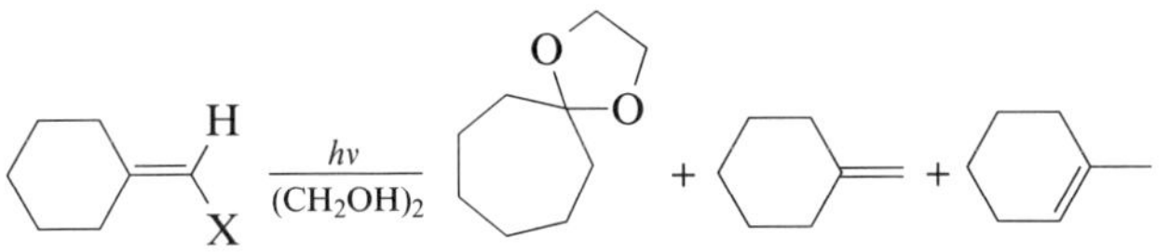

	(85)		
X = I	37%	24%	6%
X = Br	16%	54%	trace

The alicyclic vinyl halide 1-iodocyclooctene yields not only the direct nucleophilic trapping product **86** (and its acid-catalysed decomposition product **87**) but also **88** and **89** (equation 74)[291], which are produced via 1,2-cyclooctadiene as intermediate, as demon-strated by deuterium labelling for a related vinyl iodide. Deprotonation of the photo-generated vinyl cation is clearly more efficient than nucleophilic trapping in methanol. A

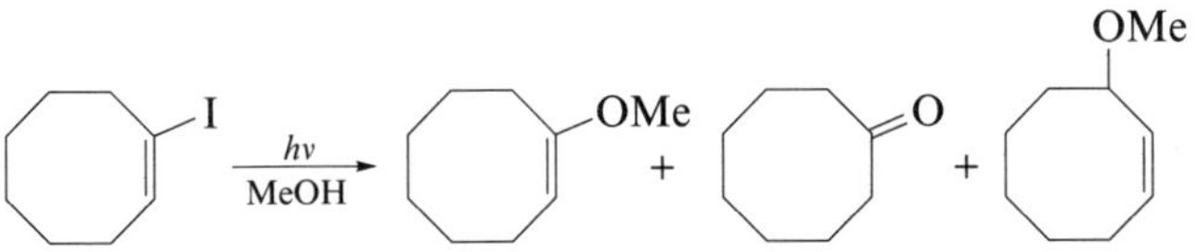

(86)	(87)	(88)
2%	16%	30%

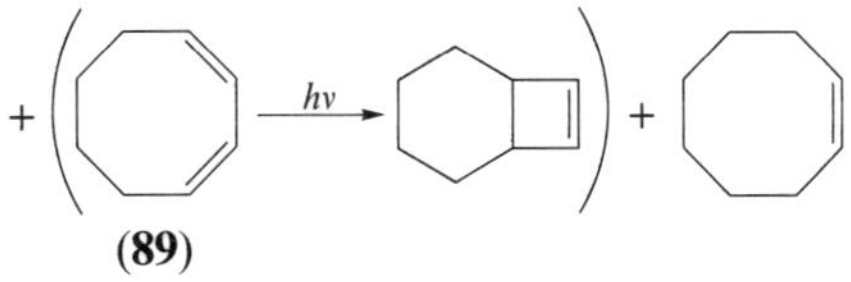

(89)		
4%	6%	32%

similar large propensity for loss of a β-proton has also been observed for photogenerated alkyl cations and ascribed to the 'hotness' of the ions formed. With 1-iodocyclohexene as substrate, the products derived from a 1,2-diene are no longer formed but the nucleophilic substitution products still are. Apparently a six-membered ring of cyclo-1,2-diene is prohibitively strained, but a vinyl cation can still be accommodated even though it prefers to be linear. This is no longer the case upon incorporation of the halocyclohexene ring in a rigid tricyclic (isolongifolene) compound: in this case radical-derived products are exclusively formed[293].

With α-H, β-aryl substituted vinyl halides such as **90** (equation 75) and the corresponding α-Me, β-aryl substituted compounds the efficiency of photochemical production of the vinyl cation-derived products **91** and **92** increases upon introduction of an o-substituent (methyl or methoxy) at the β-aryl ring[295]. Supposedly, the o-substituted β-aryl groups are twisted from the plane of the C=C bond, which favours the electron transfer in the radical pair between the aromatic ring and the halogen atom, to form the vinyl cation. In the photolysis of α-Me, β-aryl substituted vinyl halides such as **93**, deprotonation of the intermediate vinyl cation, yielding the allene **94**, is competitive with the 1,2-aryl shift across the C=C bond, yielding **95** (equation 76)[300]. No unrearranged vinyl ether product is formed, which shows that both intramolecular processes are faster than intermolecular trapping by the solvent. The percentages of ion-derived products increase in the order X = I < X = Br < X = Cl, in parallel with the electron affinities of the halogen atoms. This is the expected sequence and is thus supportive of the mechanism of vinyl cation formation via initial homolysis of the C—X bond followed by electron transfer in the radical pair (cf equation 4).

$$
\underset{\textbf{(90)}}{\overset{RC_6H_4 \quad H}{\underset{RC_6H_4 \quad Br}{>\!\!=\!\!<}}} \xrightarrow[\text{MeOH}]{h\nu} \underset{\textbf{(91)}}{\overset{RC_6H_4 \quad H}{\underset{MeO \quad C_6H_4R}{>\!\!=\!\!\cdot}}} + \underset{\textbf{(92)}}{RC_6H_4\!\!=\!\!\!=\!\!\!=\!\!C_6H_4R} + \underset{RC_6H_4 \quad H}{\overset{RC_6H_4 \quad H}{>\!\!=\!\!<}} \tag{75}
$$

	(91)	(92)	
R = p-Me	30%	39%	23%
R = o-Me	67%	8%	trace

$$
\underset{\textbf{(93)}}{\overset{Ph \quad Me}{\underset{Ph \quad X}{>\!\!=\!\!<}}} \xrightarrow[\text{MeOH}]{h\nu} \underset{\textbf{(94)}}{\overset{Ph}{\underset{Ph}{>\!\!=\!\!=}}} + \left(\underset{\textbf{(95)}}{\overset{Ph \quad Me}{\underset{MeO \quad Ph}{>\!\!=\!\!\cdot}}} \xrightarrow{h\nu} \text{(phenanthrene, MeO)} \right)
$$

	ionic : radical product
X = Cl	3.9
Br	1.1
I	0.6

$$
+ \quad \underset{Ph \quad H}{\overset{Ph \quad Me}{>\!\!=\!\!<}} \tag{76}
$$

The most simple α-aryl substituted vinyl halide studied, α-bromostyrene, flash photolytically yields the enol of acetophenone in aqueous solution, which has been used to measure rates of ketonization[302,303]. With α,β-diarylvinyl halides, loss of the β (vinylic) proton from the intermediate vinyl cation is competitive with nucleophilic trapping: Irradiation of compounds such as **96** in methanol gives about equal amounts of tolan and a vinyl ether product[305,306]. α-Bromo-β-fluorostilbene (97) affords significant formation of tolane, the FBr elimination product[307].

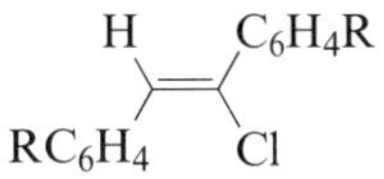

(96) R = H, o-vinyl 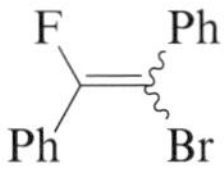   (97) 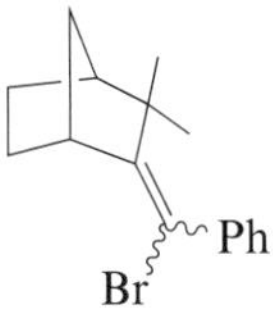   (98)

The efficiency of formation of vinyl cation-derived products is not only a function of the structure of the substrate, but also of the reaction conditions. For instance, in the irradiation of ω-bromo-ω-phenylcamphene (98) in a series of alcoholic solvents, the ratio of substitution product to reduction product increases in the order t-BuOH < EtOH < MeOH, in parallel with the polarity of the solvents[304]. For compounds 99, the product distribution is a function of the presence of oxygen (equation 77b vs 77a), and also of the wavelength of excitation[308,309]. The quantum yields of C—X bond cleavage are higher at $\lambda = 254$ nm than at $\lambda = 313$ nm, and there is relatively more heterolysis at the longer wavelength. The ratio of cation- to radical-derived product increases with the electron-donating strength of the substituent R: H < Me < MeO, and with the increased nucleofugality of the halides: F < Cl < Br < I. This indicates that the cleavage of the C—X bond to yield a vinyl cation and a halide anion proceeds here in the rate-determining step by heterolysis (cf equation 33 in Section II.C) and not by homolysis followed by electron transfer. Interestingly, the leaving group effect for the α-aryl substituted vinyl halides 99 is completely opposite to the one observed in the photolysis of the α-methyl substituted vinyl halides 93.

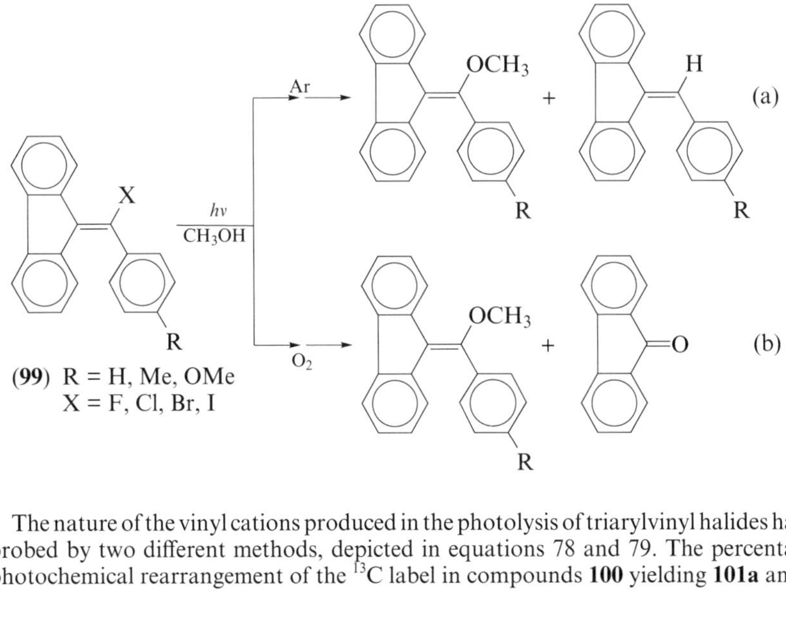

(99) R = H, Me, OMe
    X = F, Cl, Br, I

(77)

The nature of the vinyl cations produced in the photolysis of triarylvinyl halides has been probed by two different methods, depicted in equations 78 and 79. The percentages of photochemical rearrangement of the ^{13}C label in compounds 100 yielding 101a and 101b

$$
\underset{\textbf{(100)}}{\overset{Ar}{\underset{Ar}{\diagdown}}\overset{*}{\diagup}\overset{Ar}{\underset{Br}{\diagup}}} \xrightarrow[CF_3CH_2OH]{hv} \underset{\textbf{(101a)}}{\overset{Ar}{\underset{Ar}{\diagdown}}\overset{*}{\diagup}\overset{Ar}{\underset{OCH_2CF_3}{\diagup}}} + \underset{\textbf{(101b)}}{\overset{Ar}{\underset{Ar}{\diagdown}}\overset{*}{\diagup}\overset{Ar}{\underset{OCH_2CF_3}{\diagup}}} + \overset{Ar}{\underset{Ar}{\diagdown}}\diagup\overset{Ar}{\underset{H}{\diagup}} \quad (78)
$$

Ar = Ph, p-Tol (p-MeC$_6$H$_4$), p-An (p-MeOC$_6$H$_4$)

* = ^{13}C

$$
\underset{\substack{\textbf{(102)}\\ \alpha_{hv}^{23°}=23}}{\overset{Ph}{\underset{Ph}{\diagdown}}\diagup\overset{An}{\underset{Br}{\diagup}}} \xrightarrow[\substack{HOAc/\\NaOAc/\\Et_4NBr*}]{hv} \underset{\substack{\phi=0.28\\ \alpha_{hv}^{120°}=19}}{\overset{Ph}{\underset{Ph}{\diagdown}}\diagup\overset{An}{\underset{Br*}{\diagup}}} + \underset{\phi=0.015}{\overset{Ph}{\underset{Ph}{\diagdown}}\diagup\overset{An}{\underset{OAc}{\diagup}}} + \underset{\substack{\phi=0.001\\ \alpha_{\Delta}^{120°}=19}}{\overset{Ph}{\underset{Ph}{\diagdown}}\diagup\overset{An}{\underset{H}{\diagup}}} \quad (79)
$$

are 38, 72 and 92%, respectively, for Ar = Ph, Tol and An[310]. These extents of scrambling are in semi-quantitative agreement with the migratory aptitudes of the Ar groups observed in the corresponding thermal reactions. The selectivity α towards capture by bromide and acetate ion of the photogenerated vinyl cation from **102** is (after correction for the temperature difference) identical to the corresponding value in the thermal solvolysis under otherwise identical reaction conditions[311,312]. Both comparisons show that in the photochemical reactions, the same reactive intermediate is involved as in the thermal reaction, i.e. a 'cold' linear free vinyl cation. Such species are not the product-forming intermediates in the photosolvolysis of α-vinyl substituted vinyl halides as e.g. 2-bromo- and 2-chloro-1-phenyl-1,3-butadiene in methanol[313]. This is due to the relatively lower cation-stabilizing power of α-vinyl vs α-phenyl substituents. The quantum yields of formation of the Z and E vinyl ether products are not the same, as expected for reaction involving a free vinyl cation. The more efficient attack of the nucleophile at the side of the vinyl cation opposite the leaving halide anion suggests an ion pair as the product-forming intermediate.

Geminal vinyl dihalides such as **103** (equation 80)[291] undergo selective photocleavage of a single C—X bond[291,294,314]. As in the case of the corresponding monohalides (**84**) the percentage of ion-derived products increases in the order Cl < Br < I. Unlike for **84**, no products resulting from ring expansion are formed. This probably reflects the stabilizing effects of the α-iodo and α-bromo substituents.

$$
\underset{\textbf{(103)}}{\overset{X}{\underset{X}{\bigcirc\!\!=\!\!\diagup}}} \xrightarrow[MeOH]{hv} \underset{}{\overset{X}{\underset{OMe}{\bigcirc\!\!=\!\!\diagup}}} + \underset{}{\overset{X}{\underset{H}{\bigcirc\!\!=\!\!\diagup}}} \quad (80)
$$

(103) X =		
I	38%	38%
Br	3%	78%
Cl	—	67%

The range of vinyl halides from which vinyl cations can be generated seems even to include substrates with an electronegative α-substituent such as the α-formyl (**104**) (equation 81) or α-cyano group[315]. According to *ab initio* MO calculations, such α-substituted vinyl cations are among the least stable carbenium ions generated to date via solvolysis[316]. The formation of **105** and E and Z methyl cinnamates **106** supposedly occurs via expulsion of the β (vinylic) proton and the β (aldehyde) proton, respectively, from the α-formylvinyl cation. The latter process yields benzylidene ketene which reacts with the solvent, as demonstrated by the use of MeOD.

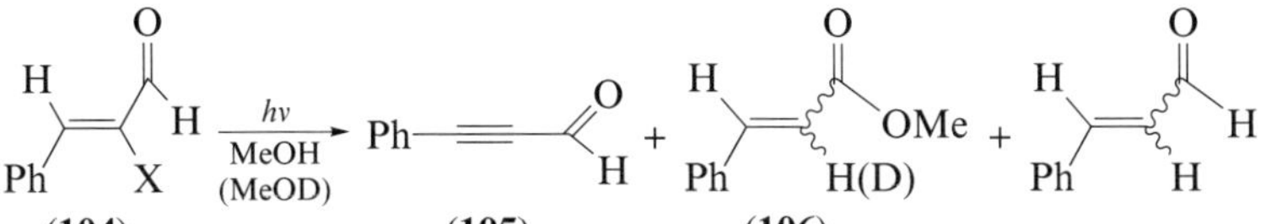

$$\text{(81)}$$

(104)
X = Br, Cl

(105)   (106)

In addition to the photosubstitution reactions of vinylic halides via intermediate vinyl cations with the nucleophiles already mentioned, similar reactions have also been reported to occur with acetonitrile[317], azide ion[318], cyanate and thiocyanate anions[319] and methyl phenyl sulphide[320]. The reactions of the first four nucleophiles are synthetically useful entries to heterocyclic compounds. For instance, photolysis of β-arylvinyl bromides such as **102** in acetonitrile does not yield the expected acetamide product RNHCOMe, as found for benzylic and homobenzylic halides (see equation 47 in Section II.D), but does yield an isoquinoline (**107**) (equation 82a) via cyclization of the intermediate nitrilium ion[317]. The vinyl azide product formed in the reaction of N_3^- with the photogenerated cation of **102** in a two-phase system is photochemically further converted into an azirine (**108**), which can be trapped by various dipolarophiles (equation 82b)[318]. The yields of the ion-derived products of **102** and analogues increase in the order α-Ph < α-Tol < α-An in parallel with the stability of the intermediate vinyl cations, and in the order β-CH$_3$ < β-Tol ~ β-An, as the result of the photodestruction of the β,β-dimethylvinyl cation-derived product[318]. With the ambient NCO$^-$ ion, only products resulting from N-site attack are formed, which are

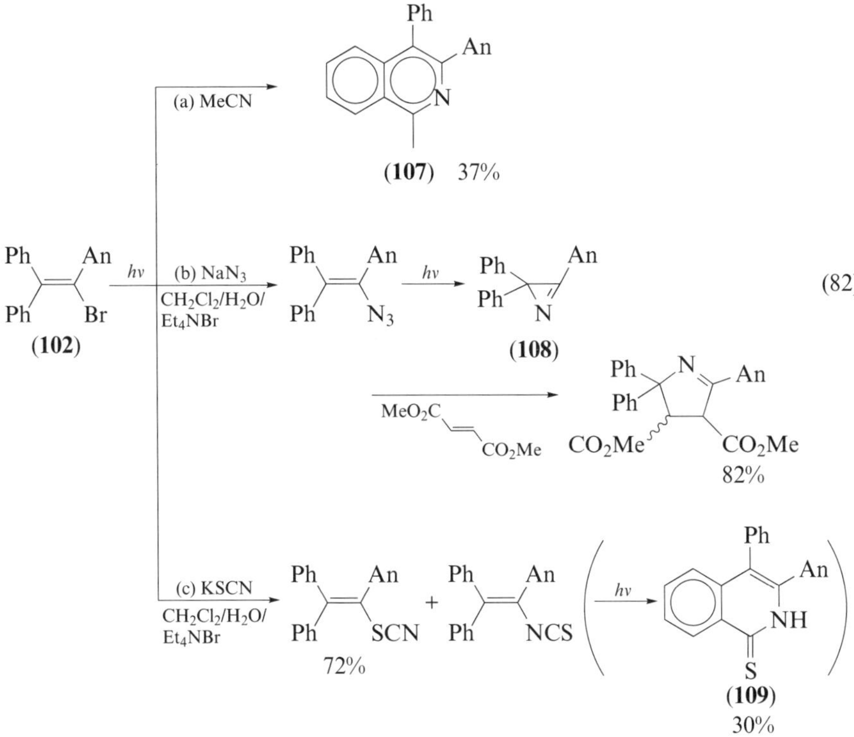

$$\text{(82)}$$

(107)   37%

(102)

(108)

82%

72%

(109)
30%

photoconverted into isoquinolinones. With the NCS⁻ ion both N- and S-site attack occur, with the regioselectivity of S vs N increasing in the order α-Ph ~ α-Tol < α-An. The N-site attack derived product is further converted into an isothioquinolinone (109) (equation 82c)[319]. The presence of a nucleophile does not always favour the formation of ion-derived products. Sodium methoxide affords enhanced photoreduction of 9-(α-bromobenzyli-dene) fluorene 99 (R = H, X = Br) in methanol, while it barely affects the quantum yield of enol ether formation[308]. Clearly, methoxide ion acts here as an electron donor to electron-ically excited 99 which, due to the presence of the fluorenyl moiety, is probably an extra good electron acceptor. The electron transfer yields a radical anion which easily loses Br⁻ to give a vinyl radical.

In all cases discussed thus far, the attack of the nucleophile occurs at the sp-carbon of the photogenerated vinyl cation. This is no longer exclusively so for 1-(p-alkoxyphenyl)vinyl bromides and cyanide anion[321] (e.g. equation 83a) or alkoxide anions[322–329] (e.g. equation 83b). The photogenerated vinyl cation derived, e.g., from 110 behaves towards these nucleophiles as an ambident cation, which results in the displace-ment of both the vinylic bromide and the alkoxy group in the α-phenyl ring. The ipso-adducts 111 and 112 have been observed by NMR[321,322]. Allenes, such as 112, have also been isolated from the reaction mixture[323] and are convenient precursors of vinyl cations under mild conditions by their acid-catalyzed decomposition[326]. The occurrence of 1,2-aryl migration, characteristic of vinyl cation behaviour, accompanying ipso-substitution in substrates such as An₂C=C(Br)Ph, serves to show that the ipso-substitution occurs via an intermediate vinyl cation[325] and not via direct aromatic photosubstitution (cf Section IV.C). Upon replacement of the β-phenyl groups in 110 by β-methyl groups, the ratio (ipso-adduct +ipso-substitution product) vs vinylic substitution product decreases drastically[327]. Supposedly, β-phenyl groups sterically hinder attack of the nucleophile at the vinylic posi-tion, so that ipso-attack becomes preferred. Ipso-adduct formation does not occur with alcohol instead of alkoxide (equation 83c vs 83b). A strong base, i.e. a hard nucleophile, is apparently needed because that favours attack at the position in the vinyl cation where the positive charge mostly resides[328,329]. According to MO calculations, this position is the ipso-position[329].

In suitable vinylic halides, nucleophilic photosubstitution occurs intramolecular-ly[295–297,324,330–332]. For instance, irradiation of 113 in methanol only yields benzofuran 114, and no vinyl ether product or 1,2-aryl rearrangement products are formed (equation 84a)[296]. Apparently, the reaction with an internal nucleophile is faster than the 1,2-aryl shift and also faster than the reaction with an external nucleophile. With the less stabilized vinyl cations derived from the α-methyl- and α-H-analogues of 113-OMe, 1,2-aryl shifts do

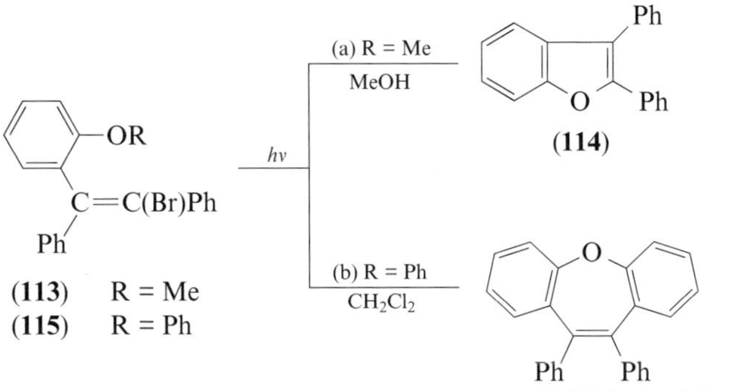

(84)

occur, but vinyl ethers are still not produced[296]. With an *o*-aryloxy instead of an *o*-methoxy group in the β-phenyl ring as in **115**, attack occurs not at the oxygen atom but at the *ortho*-position of the aryloxy group leading to oxepin (**116**) formation (equation 84b)[330]. On the other hand, both the SMe-analogue of **113**[297] and the SPh-analogue of **115**[331] only undergo cyclization at the sulphur atom, yielding benzothiophenes. An intramolecular *ipso*-substitution has also been reported[324]. Once again, *ipso*-attack only occurred if the side chain was an alkoxide as a functional group and not when it was an alcohol.

The photochemical formation of vinyl cations from vinylic halides has not only been investigated extensively by product studies, but also by time-resolved laser flash photolysis techniques[333-338]. Also, combined continuous irradiation and laser flash photolysis approaches have been reported[299,309]. Vinylic bromides with a *para*-substituted (mostly by methoxy) phenyl ring at the α-position in acetonitrile have been mainly used successfully as precursors to directly observable vinyl cations by transient absorption spectroscopy or electrical conductivity measurements. In the case of vinyl bromides with an α-H-[299] or α-Me-substituent[334] and β-anisyl groups, only products/transients in which a 1,2-anisyl shift across the C=C bond has occurred are observed on the nanosecond timescale. In the ns flash photolysis of **117**, both transient **118** and transient **119** are observed (equation 85) which enables the measurements of the rate of β-aryl migration across the C=C bond[334]. The relative rates of reaction of a series of differently β-substituted vinyl cations with nucleophiles varies in the order An$_2$C=C$^+$—An ~ Ph$_2$C=C$^+$—An < Me$_2$C=C$^+$-An[337]. As already mentioned regarding the influence of β-Me versus β-Ph on the percentage of *ipso*-substitution, the lesser shielding of the sp-carbocation by β-Me as compared to β-phenyl may be an important factor.

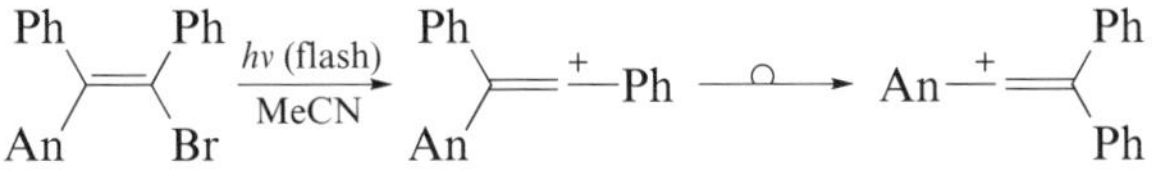

(85)

Although most laser flash photolysis studies have been performed in acetonitrile, transient vinyl cations have also been observed in acetonitrile/methanol mixtures and even in pure methanol[309]. In acetic acid, the solvent of choice for most thermal solvolysis studies of vinylic halides, the presence of polar additives such as sodium acetate is necessary for transient vinyl cations to be observed[335].

In addition to information about the identity and reactivity of transient vinyl cations, information is also available about their efficiency of formation[309]. The quantum yields of vinyl cation formation from compounds **99** as a function of R and X show that the cation formation from **99** increases with increasing electron-donor strength of the *p*-phenyl substituent R. The anionic leaving-group properties of the halides X are more related to the acidities of the conjugated acids HX than to the electron affinities of the X˙ radical. All of this is in good agreement with the results of the corresponding product analysis study[309] and leads to the same conclusion: Cleavage of the C—X bond to give C⁺ and X⁻ occurs here by heterolysis and not by homolysis followed by electron transfer in a radical pair. Moreover, the same wavelength dependence of the quantum yields and ratios of the vinyl cation and vinyl radical formation was observed as in the product studies.

Reports of nucleophilic vinylic photosubstitution reactions, which occur via the $S_{RN}1$ mechanism, are conspicuously scarce. One such example is the cobalt carbonyl catalysed photostimulated carbonylation of vinylic halides[339]. By this method 1-bromo and 1-chlorocyclohexene are converted into 1-cyclohexenecarboxylic acid in 98 and 97% yield, respectively. In a prototype vinylic $S_{RN}1$ reaction, of β-bromostyrene with the enolate anion $^-CH_2COCMe_3$, an ionic elimination–addition route seems to be followed along with the $S_{RN}1$ route[340].

Only a few additional examples of intermolecular photochemical vinylations of (hetero)aromatic compounds have been forthcoming. Coupling products are formed in the irradiation of dichloro- and dibromo-*N*-methylmaleimide in the presence of 1,3-dimethyluracils[341] and of 3-bromocoumarin in the presence of naphthalene, phenanthrene, 1-methylpyrrole and other aromatic compounds[342]. The former reaction is accompanied by cyclobutane adduct formation, which is the mode of reaction of *N*-methylmaleimide itself. The mechanism of these vinylation reactions is not clear, but most probably an exciplex (cf equation 20a) or a charge-transfer complex (cf equation 20b) is involved.

Photolysis of 2-bromo-4,4-dimethyl-2-cyclohexenone only affords reduction, even in a nucleophilic medium[343,344]. Apparently, this substrate is structurally not suitable to form a vinyl cation. Formation of vinyl radical-derived products is also the main process for all vinylic halides, if their irradiation is performed in an apolar medium. Such photochemical reductive dehalogenation and especially dechlorination reactions have been extensively studied in the past, not in the least because of their importance as abiotic transformation of persistent polychlorinated environmental pollutants. Examples are the cyclodiene insecticides aldrin and dieldrin, which contain a vicinal dichloroethene chromophore. In recent years only a few more, mainly supplementary, studies have been published in this area[120,345,346]. The photochemical reduction of vinyl chlorides and vinyl bromides is markedly accelerated by the presence of sodium borohydride[347] or of lithium aluminium hydride[122,123,347].

A number of compounds structurally related to vinylic halides also show vinyl halide to vinyl cation photoconversion-type photoreactivity. For example, the photochemical loss of halide ion from β-chloro- or β-bromo allyl anions[348,349] to yield a strained cyclic allene in the case of **120** (equation 86)[349] is, at least formally, a photoheterolysis of a vinylic halide with a strong carbocation-stabilizing substituent (i.e. a carbanion) at the α-position. Photolysis of triarylchloroallenes under conditions where thermal reaction does not occur, yields triarylallenyl cation derived products (e.g. equation 87)[350]. Methanol attacks at the same (γ)-position of the ambident allenyl ion as in the corresponding thermal solvolysis at higher temperatures. 1-Chloro- and 1-bromo-1-hexynes do not show ionic photobehaviour[351]. Even in polar solvents only radical-derived products are obtained. The instability of 1-alkynyl cations[290] must be prohibitively large.

(86)

(**120**)   50%

$$\text{(structure: } Ph_2C=C(Cl)Ph \text{)} \quad \xrightarrow[\substack{MeOH/CH_2Cl_2 \\ -30\,°C}]{h\nu} \quad Ph-\underset{\underset{76\%}{OMe}}{\overset{Ph}{\vert}}-\!\!\equiv\!\!-An \;+\; Ph_2C=C=CHAn \quad (87)$$

$$\text{trace}$$

IV. AROMATIC HALIDES

A. Reductive Dehalogenation Reactions

The reductive photodehalogenation of aryl halides has been actively investigated in recent years. Special attention has been given to (poly)halobenzenes and (poly)halo-biphenyls. The reactions are of interest in view of their mechanisms, and because of the importance of chlorinated aromatic hydrocarbons as environmental pollutants and the possibility of their photoinduced degradation. The photochemistry of aryl halides and related compounds in general[14] and the photochemistry of polyhaloarenes in particular[18] have been reviewed.

In non-polar solvents, photoreduction of chlorobenzene proceeds by simple homoly-sis[352]. Excimers are formed in concentrated solutions of chlorobenzene, but they are either completely unreactive or decompose with much lower efficiency than the monomer[353]. The preferred pathway for homolysis involves the π,π^* triplet state[353,354]. The mechanism can thus be described by equations 88–90.

$$ArCl \xrightarrow{h\nu} {}^1ArCl^* \longrightarrow {}^3ArCl^* \,(\pi,\pi^*) \qquad (88)$$

$$^3ArCl^* \longrightarrow Ar^{\bullet} + Cl^{\bullet} \qquad (89)$$

$$Ar^{\bullet} + RH \longrightarrow ArH + R^{\bullet} \qquad (90)$$

The photochemistry of chlorobenzene has also been investigated in ethanol–water solutions[355] and in methanol[356]. In addition to photosubstitution of chlorine by OH or OR, formation of benzene was observed. The latter process was shown to be initiated by homolytic rupture of the C—Cl bond in the excited singlet as well as in the excited triplet state.

The multiplicity of the photoexcited state responsible for the dechlorination of the monochlorobiphenyls in 2-propanol has also been determined[357]. The fluorescence as well as the dechlorination of 2- and 3-chlorobiphenyl are quenched by 1,3-cyclohexadiene and the quenching constants for both processes agree well. The reaction of 4-chlorobiphenyl is quenched by cis-1,3-pentadiene, but accelerated by 1,3-cyclohexadiene. Sensitization by acetone and acetophenone was inefficient with 2- and 3-chlorobiphenyl, but efficient with 4-chlorobiphenyl. It was concluded that dechlorination of 2- and 3-chlorobiphenyl occurs from the excited singlet state, while that of 4-chlorobiphenyl takes place predominantly from the excited triplet state. Quantum yields of dechlorination (in 2-propanol) are 0.3, 0.0047 and 0.0016 for the 2-chloro,3-chloro and 4-chloro isomer, respectively. The high reactivity of the 2-chloro isomer is ascribed to steric strain. It has, however, also been proposed that an electronic effect might be responsible[358]. In the planar excited state the ortho chlorine is in the strongest inductive electron-withdrawing position, while the adja-cent planar aromatic ring is in the strongest electron-resonance donating position. This might favour heterolytic cleavage of the C—Cl bond. The aryl cation, for which carbene resonance structures can be written, may take up an electron from the chloride ion and the ensuing radical leads to the reduced product. The preferential loss of *meta* chlorine versus *para* chlorine can also be explained by making use of carbene resonance structures.

Photodechlorination of 2,3-dichlorobiphenyl in 2-propanol gives 3-chlorobiphenyl, while 3,4-dichlorobiphenyl yields 4-chlorobiphenyl[359]. Both results are in agreement with the relative reactivities observed with the monochlorobiphenyls. The 2,3 isomer reacts exclusively from its singlet excited state, whereas 3,4-dichlorobiphenyl is dechlorinated in both excited singlet and triplet states in a ratio of 0.8:0.2.

4-Chloro-2′,5′-dimethylbiphenyl,   3-chloro-2′,5′-dimethylbiphenyl,   2-chloro-2′,5′-dimethylbiphenyl, 4-chloro-2-methylbiphenyl and 3-chloro-2-methylbiphenyl have been irradiated in cyclohexane and the quantum yields of their dechlorination determined[360]. The values are 0.0075, 0.010, 0.097, 0.005 and 0.010, respectively. As with the monochlorobiphenyls, the highest quantum yield was found when the chlorine atom occupies position 2. It was proposed that the photolability of *ortho*-chlorobiphenyls results from a combination of two factors, raising the energy of the excited state by steric compression, and relief of this steric strain when the *ortho*-substituent departs.

A study of the photodehalogenation of 1-halonaphthalenes[361], mainly by means of fluorescence measurements, has revealed that 1-iodonaphthalene undergoes photochemical homolytic dissociation with near-unit quantum yield from its triplet state, although the excited singlet state is not considered to be in principle unreactive. The efficiency of dissociation of 1-bromonaphthalene is much lower and the reaction cannot be sensitized by benzophenone. The triplet-state energy is too low to permit rupture of the C—Br bond and the singlet excited state undergoes efficient intersystem crossing competing with dissociation. 1-Chloronaphthalene does not undergo unassisted homolytic dissociation from either S_1 or T_1, but in the presence of triethylamine a fluorescent exciplex is observed from which naphthyl radicals and chloride ions are formed.

Lithium aluminium hydride markedly accelerates the photoreductive dehalogenation of chlorobenzene, bromobenzene and *para*-bromochlorobenzene[362]. Presumably, lithium aluminium hydride suppresses the recombination of alkyl and halogen radicals by means of efficient capture of the latter.

Photoinduced dehalogenation of dichlorobenzene isomers and their monosubstituted derivatives (30 compounds) is reported to give the corresponding monochlorobenzenes[363]. Upon irradiation in CD_3OD the compounds underwent exchange of one of the chlorine atoms with deuterium.

The photochemical dechlorinations and debrominations of a large number of polyhalobenzenes $C_6H_nX_{6-n}$ (X = Cl, Br) in hexane have been studied and the relative rates determined[364]. It was found that a halogen atom flanked by two other halogens (at positions 2 and 6) is the most reactive and a halogen atom having one *ortho* neighbour is the second most reactive. A halogen atom *meta* to the site of reaction accelerates the dehalogenation more than a *para* halogen, when other circumstances are similar. The effects are ascribed to steric and electronic effects and they have been related to Hammett σ-constants. In the series of polybromobenzenes, the reactivity increases in the following order: mono < *ortho*-di < *meta*-di < *para*-di < 1,3,5-tri < 1,2,4-tri < 1,2,3-tri < 1,2,4,5-tetra < 1,2,3,5-tetra < 1,2,3,4-tetra < penta < hexa. In the polychlorobenzenes, the reactivity is more sensitive to the number of neighbouring halogen atoms. The rates of dechlorination of isolated chlorine substituents (without *ortho* neighbours) are very low. The total number of halogen atoms in the polychlorobenzenes is less important than in the polybromobenzenes. A comparative study of the dechlorination of the three dichlorobenzenes, the three trichlorobenzenes, 1,2,4,5-tetrachlorobenzene, pentachlorobenzene and hexachlorobenzene in cyclohexane also revealed that quantum yields are higher when *ortho* chlorines are present, probably as a result of a steric effect[365]. In the same study, the quantum yields of dechlorination (φ_r) were compared with those of intersystem crossing (φ_{isc}). It was concluded that the results are incompatible with exclusively singlet reactivity. 1,2,3-

Trichlorobenzene has $\varphi_{isc} = 0.7 \pm 0.3$ and $\varphi_r = 1.0$; for 1,2,4,5-tetrachlorobenzene the values are $\varphi_{isc} = 1.0 \pm 0.1$ and $\varphi_r = 0.43$. Only in *ortho*-dichlorobenzene does singlet reactivity seem to be implicated definitively ($\varphi_{isc} = 0.4 \pm 0.1$; $\varphi_r = 0.72$).

The photolysis of the trichlorobenzenes in cyclohexane, 2-propanol and methanol was later studied by other authors[366] and their results suggest the excited singlet as the photochemically active excited state. The irradiations were performed in the presence of anthracene. On irradiation at 280 nm of a 0.05 M solution of trichlorobenzene containing 3.8×10^{-4} M of anthracene, the light is almost exclusively absorbed by the trichlorobenzenes. At this anthracene concentration more than 90% of the triplets are quenched, but the HCl quantum yields are unchanged within experimental error with respect to those in the absence of anthracene, indicating that the triplet state is not involved in the photolysis. Moreover, the sum of HCl and intersystem crossing quantum yields is near-unity in cyclohexane. It was confirmed in this report that the more reactive positions are those adjacent to other chlorine atoms.

In addition to product formation via direct homolysis of the excited singlet or triplet state of the aryl halide, a pathway via excimers leading to dechlorinated product is sometimes available in the absence of electron donors. In an investigation of the photodechlorination of pentachlorobenzene in acetonitrile[367], the intersystem crossing quantum yield was determined ($\varphi_{isc} = 0.8$), the quenching of the photodechlorination by fumaronitrile was measured and the dependence of fluorescence lifetime and quantum yield of photodechlorination upon the substrate concentration was studied. Quenching by fumaronitrile suggests the presence of a reactive excited singlet state which undergoes direct fission of the C—Cl bond. The possibility of reaction via a singlet excimer could be ruled out because the lifetime of the singlet (7.5 ns) remained constant over a concentration range of 0.005–0.04 M. Kinetic analysis has provided the relative contributions of three major mechanistic reaction pathways in the photodechlorination of pentachlorobenzene at various concentrations. At 0.05 M these contributions are 72.9% via excimer (formed from triplet excited aryl halide), 24.9% directly from the triplet and 2.1% from the singlet. The product-forming steps are given by equations 91–93.

$$\text{ArCl} \xrightarrow{h\nu} {}^1\text{ArCl}^* \longrightarrow \text{products} \tag{91}$$

$$^1\text{ArCl}^* \longrightarrow {}^3\text{ArCl}^* \longrightarrow \text{products} \tag{92}$$

$$^3\text{ArCl}^* + \text{ArCl} \longrightarrow (\text{ArCl}^{\cdot\delta-}\ \text{ArCl}^{\cdot\delta+})^* \longrightarrow \text{products} \tag{93}$$

The three products formed by dechlorination of pentachlorobenzene (**121**) in acetonitrile are 1,2,3,5-tetrachlorobenzene (**122**), 1,2,4,5-tetrachlorobenzene (**123**) and 1,2,3,4-tetrachlorobenzene (**124**) (equation 94).

$$\tag{94}$$

This reaction has also been studied in aqueous 0.100 M hexadecyltrimethylammonium bromide solution[368]. The total quantum yield depends on the concentration of pentachlorobenzene within the micelles and varies from 0.078 at 0.0173 M to 0.162 at 0.154 M. The ratio of the three products is only slightly dependent on the concentration; at the lowest concentration the ratio is 49.8:45.1:5.2 and this does not differ much from the ratio in acetonitrile (48.4:39.5:121 at 0.005 M pentachlorobenzene). Interestingly, three byproducts were found in the micellar solution, apparently formed by trapping of the excimer by bromide ion. They are 1-bromo-2,3,4,5-tetrachlorobenzene, 1-bromo-2,3,4,6-tetrachlorobenzene and 1-bromo-2,3,5,6-tetrachlorobenzene in a ratio of 10.0:66.0:22.8. The bromide ion attacks the $ArCl^{\delta+}$ moiety of the exciplex and a radical is formed which subsequently loses a chlorine atom. The preponderance of 1-bromo-2,3,4,6-tetra-chlorobenzene can be understood on the basis of stabilization of the pentadienyl radical by the chlorine atoms. It is estimated that excimer formation in the micellar solution is enhanced at least 100-fold relative to the acetonitrile solution.

Electron transfer between triplet excited and ground-state halide has also been proposed in the case of 1-chloro-2-naphthol[369]. The dehalogenation has been studied in methanol, ethanol, 1-propanol, 2-propanol and 2-butanol. The quantum yields at 313 nm in these solvents are 0.6 ± 0.05, 0.2 ± 0.05, 0.01 ± 0.005, 0.03 ± 0.005 and 0.007 ± 0.001, respectively. The quantum yields decrease with decreasing polarity of the medium. In the presence of air, the yields are greatly reduced. When the reaction is performed in methanol in the presence of potassium hydroxide, the quantum yield is found to be 1.5. The 2-hydroxy-1-naphthyl radical, formed by loss of chloride ion from the radical anion, abstracts a hydrogen atom from methanol. In the alkaline medium, the $\cdot CH_2OH$ radical is converted into $CH_2O^{-\cdot}$ and this species can transfer an electron to ground-state 1-chloro-2-naphthol, which leads to a chain reaction. In the case of of 1-bromo-2-naphthol, homolytic dissociation of the C—Br bond in the singlet excited state is proposed. This reaction is not very sensitive to changes in the polarity of the alcohol and the presence of oxygen has little effect on the quantum yield.

Polychlorobenzenes have been found to undergo photochemical rearrangement reactions in addition to dehalogenation in solvents that do not possess easily abstractable hydrogen atoms, such as acetonitrile and perfluorohexane[370-372]. Invariably, the rearrangement consisted of the migration of a chlorine atom to a *meta* position. The photolysis of 1,2,3,4-tetrachlorobenzene (**124**) in acetonitrile constitutes an example[372]. After 8 h of irradiation, two dehalogenation products were found: 1,2,4-trichlorobenzene (**125**) (14%) and 1,2,3-trichlorobenzene (**126**) (2%). In addition, 1,2,4,5-tetrachlorobenzene (**123**) (3%) and chloroacetonitrile (**127**) (2%) were detected (equation 95).

$$\text{(124)} \quad \xrightarrow[\text{CH}_3\text{CN}]{h\nu} \quad \text{(125)} \quad + \quad \text{(126)} \quad + \quad \text{(123)} \quad + \quad \text{ClCH}_2\text{CN} \quad (95)$$

The possibility of rearrangement via valence isomers of benzene was considered, but rejected on the grounds that isomerization via benzvalene would lead to *ortho* migration and isomerization via prismane to *ortho* and *para* migration. Tentative INDO-UHF molecular orbital calculations on *o*-halophenyl free radicals showed that a considerable spin density should be induced at the *meta*-position, in addition to the free radical centre (the *ipso*-position). The spin at the *meta*-position occupies the π-orbital, which favours interaction with the chlorine atom in the initial stage of 1,3-migration. The formation of chloroacetonitrile is considered to support this mechanism, because it suggests the coupling of long-lived chlorine atoms in the solvent cage with cyanomethyl radicals generated by hydrogen abstraction by the aryl radical.

Acetone-sensitized and unsensitized reductive dechlorination of tetra-, penta- and hexa-chlorobenzenes have been studied[373,374]. 1,2,4-Trichlorobenzene **125** is the major photo-product from 1,2,3,5-tetrachlorobenzene **122** in the unsensitized process, whereas the

1,3,5-isomer results in the presence of acetone[373]. Photoisomers of the tetrachlorobenzene are formed on direct irradiation.

4-Bromobiphenyl undergoes photoreduction from the triplet state[375]. The dependence of the quantum yield upon the concentration of the substrate does indicate the formation of an excimer. Since $\varphi_{isc} = 0.98$, it may be concluded that this excimer is formed via the triplet state. The linear solvation energy parameters indicate a weak polarization of the excimer, suggesting a weak radical anion and cation character in the two moieties. The charge separation is smaller than in the exciplex formed from 4-bromobiphenyl and triethylamine.

Direct photolysis of 2-halopyridines in methanol, ethanol or acetonitrile (with 1% water) affords pyridine, together with alkoxy-, or hydroxypyridines[376]. The formation of the substitution products will be discussed in Section IV.C. The reductive dehalogenation is considered to proceed via homolysis of the carbon–halogen bond. The efficiency of pyridine formation decreases in the order Br > Cl > I. The 3- and 4-halopyridines produce pyridine exclusively.

The relative reactivities of photodehalogenation of the three dibromobenzenes, the three bromotoluenes, the three bromoanisoles, the three bromophenols and the three iodotoluenes in hexane have been determined by irradiation under otherwise identical conditions at $\lambda > 290$ nm and at $\lambda > 230$ nm[377]. With the short-wavelength radiation, the dibromobenzenes and the bromoanisoles react fastest, having almost equal reactivity. The bromotoluenes react about three times as slow and the iodotoluenes are somewhat less reactive than their bromo analogues. The bromophenols are very unreactive. At $\lambda > 290$ nm, the highest yields are observed with the iodotoluenes, as expected on the basis of their absorption spectra.

The photoreduction of 2- and 3-chloroanisole in alcoholic solvents has been studied and is considered to be best accounted for by invoking methoxyphenyl radicals which abstract hydrogen atoms from the solvent[378]. 4-Chloroanisole probably reacts partly via a homolytic cleavage, but in view of the results of quenching and sensitization experiments, another pathway consists of electron transfer from the solvent (ROH) to excited aryl halide, followed by dissociation of the radical anion into chloride ion and aryl radical.

3- and 4-Chloroanisole undergo substitution of the chlorine atom, and reductive dechlorination, upon irradiation in methanol[378,379]. In the case of 3-chloroanisole the ratio of reduction versus substitution, R/S, has been determined under various conditions[379]. The acetone-sensitized irradiation gave R/S = 0.4; in the presence of high concentrations of cis-1,2-dichloroethene R/S = 5.7 and in the direct irradiation R/S = 3.12. The results permit the conclusion that the reduction proceeds mainly via the singlet state and that, in the direct irradiation, the majority of the molecules react via this state.

A wavelength effect and a solvent effect have been reported in the photochemistry of 2,4-dichloroanisole[380]. Photolysis in cyclohexane and in methanol produces a mixture of 4-chloroanisole (A) and 2-chloroanisole (B). In cyclohexane A/B = 30–35 at 254 nm and 70–80 at 300 nm. In methanol the ratio changes from 2–3 at 254 nm to 55 at 300 nm. The authors speculate that direct S–T absorption at 300 nm might be responsible for the wavelength effect.

Photoreactions of p-ClC$_6$H$_4$CH$_2$X (X = H, Cl, CN, COOH and OH) have been studied at wavelengths around 300 nm[381]. The acetone-sensitized photolyses of the compounds with X = H, CN, COOH and OH in deaerated methanol provided reductively dechlorinated photoproducts C$_6$H$_5$CH$_2$X with yields of 52% (X = H), 10% (X = CN), 40% (X = COOH) and 34% (X = OH). p-Chlorobenzyl chloride does not undergo dehalogenation but gives p-chlorobenzyl methyl ether as the main product. With direct irradiation, none of the five compounds studied undergoes reductive dehalogenation.

Quantum yields have been determined of photodehalogenation in aliphatic hydrocarbons of derivatives of chlorobenzene, ClC$_6$H$_4$R, with R = Cl, CH$_3$, F, CF$_3$, OCH$_3$, CN, CH$_2$CN and CH$_2$OH at ortho, meta and para positions[382]. The quantum yields of the aromatic photoproducts were found to agree very well with the quantum yields of HCl production. The highest values were found with the *ortho* derivatives; 1,2-dichloroben-zene: $\varphi = 0.68$ and 1-chloro-2-fluorobenzene: $\varphi = 0.70$. The presence of cyano and cyanomethyl groups drastically lowers the quantum yields. In view of the fact that quantum yields are very complex quantities, no attempts were made to correlate the numerical values with the nature of the substituents.

Irradiation of 2,4,5,6-tetrachloro-1,3-dicyanobenzene in ethanol leads to a complicated mixture of photoproducts[383]. Both Cl/H and CN/H exchange were found to take place.

The ultraviolet irradiation of halogenonitrobenzenes dissolved in ethyl ether or tetrahydrofuran leads to an increase in the electrical conductivity of the solution; relaxation of the conductivity is observed after the irradiation is stopped[384]. The kinetics appeared to be complicated: the structure of the compound, its concentration, the nature of the solvent, the temperature, the time of irradiation as well as the light intensity had an influence on the effects. The photodegradation of three nitrochlorobenzene isomers in pure water and river water under irradiation follows first-order reaction kinetics; the rate constants for the three isomers decrease in the order *p-* > *o-* > *m*-nitrochlorobenzene[385].

Irradiation of bromothiazoles in various organic solvents (methanol, ether, cyclohexane) produces thiazole (and the isomeric isothiazole) as the main reaction product[386]. The reactivity decreases in the order 2-bromothiazole > 5-bromothiazole >> 4-bromothiazole, and is probably related to the C—Br bond strengths. The mechanism is considered to involve homolytic cleavage of the C—Br bond, followed by abstraction of a hydrogen atom from the solvent. Oxygen does not exhibit any noticeable effect upon the reaction, but the debromination is accelerated by triethylamine.

Methyl 4-chloro-3-indoleacetate (**128**) undergoes a remarkably facile dehalogenation under the influence of light[387]. Methyl 3-indoleacetate is the major product (**130**; 70–80%). When the reaction was carried out in CD$_3$OD, the product was labelled at position 4. Homolytic cleavage of the C—Cl bond to **129** is proposed as the primary step following the excitation (equation 96).

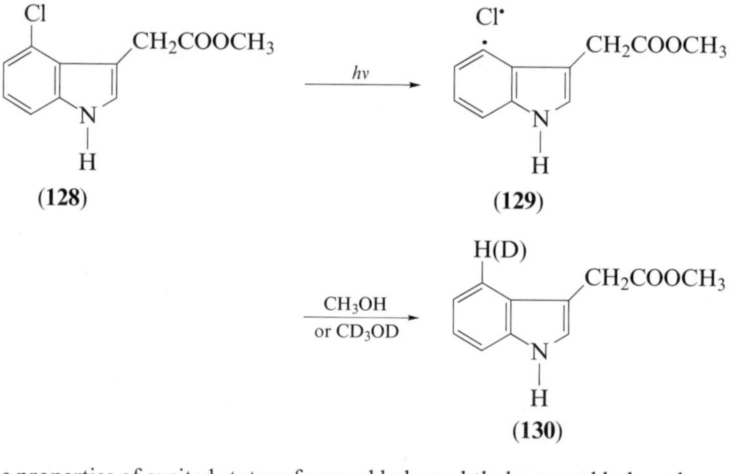

$$(96)$$

The properties of excited states of several halonaphthalenes and haloanthracenes have been investigated by means of picosecond absorption and emission spectroscopy[388,389]. Fluorescence measurements showed the expected heavy-atom effect due to spin–orbit coupling. Fluorescence lifetimes of the chloronaphthalenes in hexane were the longest, 2.4 and 3.3 ns for 1-chloro- and 2-chloronaphthalene, respectively. The 1-bromo and 2-bromo analogues had lifetimes of 75 ps and 150 ps, respectively. Iodo-substituted compounds had the shortest lifetimes, 14 ps and 35 ps for 2-iodoanthracene and 9-iodoanthracene, respectively. It can also be seen from these data that the intersystem crossing rate is heavily influenced by the position of the substituent on the ring. Photodissociation of these haloaromatics proceeds by fast relaxation from upper S_1 vibrational levels (6000 cm^{-1} excess vibrational energy) to S_1 ($v = 0$) and intersystem crossing to an upper T_1 ($v > 0$) state. Fast vibrational relaxation to T_1 ($v = 0$) is followed by predissociation of a $T(\sigma,\sigma^*)$ state.

In the presence of electron donors, photodechlorination of chlorobenzene[390], 4-chloroanisole[378] and other aryl halides is enhanced. The key step in this process is electron transfer from the donor to the excited halide. A kinetic study, based on determination of quantum yields and fluorescence quenching efficiencies, has revealed that in the triethylamine-assisted dehalogenation of chlorobenzene the product–forming interaction occurs with the triplet excited state (equations 97–99)[390].

$$^3ArCl^* + R_3N \longrightarrow (ArCl^{-\bullet} + R_3N^{+\bullet}) \tag{97}$$

$$(ArCl^{-\bullet} + R_3N^{+\bullet}) \longrightarrow R_3N^{+\bullet} + Ar^\bullet + Cl^- \tag{98}$$

$$Ar^\bullet + RH \longrightarrow ArH + R^\bullet \tag{99}$$

The same mechanism accounts for the results obtained with 4-chlorobiphenyl[390]. The rate constants for quenching of the triplet states of 4-chlorobiphenyl and 1-chloronaphthalene by various tertiary amines have been determined[391]. At an amine concentration of 2×10^{-4} M ca 50% of the triplets of 4-chlorobiphenyl are quenched and ca 90% are quenched at a 10^{-3} M amine concentration. At this latter concentration very little fluorescence quenching occurs. Since dechlorination occurs at amine concentrations in the range of 10^{-3} to 10^{-4} M, it appears reasonable to invoke a reaction which occurs via the triplet state.

When the amine-assisted photodechlorination of 4-chlorobiphenyl was performed in deuterium-containing solvents such as CH_3CN/D_2O or CD_3OD, incorporation of deuterium into biphenyl was observed[390]. This is considered to indicate the occurrence of another pathway for the transformation of $ArCl^-$ into ArH (equations 100–102).

$$ArCl^{-\bullet} + H^+ \longrightarrow ArClH^\bullet \tag{100}$$

$$ArClH^\bullet \longrightarrow ArH + Cl^\bullet \tag{101}$$

$$Cl^\bullet + RH \longrightarrow HCl + R^\bullet \tag{102}$$

Other authors have re-evaluated the fluorescence quenching data obtained with 4-chlorobiphenyl and triethylamine and their kinetic treatment indicates that the reaction from the singlet state, via a singlet exciplex, is more efficient than that from the triplet state, via a triplet exciplex[392]. The dehalogenation of 1-chloropyrene assisted by triethylamine in acetonitrile is also a singlet reaction proceeding via an exciplex[393]. The anion radical of 1-chloropyrene has been observed by means of transient absorption spectroscopy. They decay by dissociation into pyrene radicals and chloride ions as well as by reaction with the solvent, giving rise to hydrochloropyrenyl radicals. In the presence of cyanide ions, the hydrochloropyrenyl radicals are converted into hydrocyanopyrenyl radicals, which may lose a proton and yield cyanopyrenyl anion radicals. Transfer of an electron to 1-chloropyrene initiates a possible chain reaction.

The photodehalogenation of 2-(4-halophenyl)benzoxazoles has been investigated[394,395], particularly because derivatives of these compounds have therapeutic applications which are, however, hampered by undesirable phototoxic effects. 2-(4-X-Phenyl)benzoxazoles (X = Cl, Br, I) (**131**) are converted into 2-phenylbenzoxazole (**132**) upon irradiation in cyclohexane (equation 103). The fluoro derivative undergoes only dimerization and when X = Cl, dimerization accompanies the dehalogenation. Fluorescence studies and sensitization and quenching experiments have revealed that the triplet excited state is responsible for photodehalogenation of the bromo and iodo compounds, whereas the reaction of the chloro derivative occurs via the singlet state involving an excimer, as demonstrated by the influence of concentration of the substrate on the quantum yield. The photochemical reactivity of 2-(4-chlorophenyl)benzoxazole in the presence of various amines has been investigated[395]. Evidence was obtained for the formation of an exciplex between the singlet excited state of the benzoxazole derivative and the amine. In the case of tertiary amines the mechanism is described by equations 97–99, but for secondary amines an alternative reaction path consisting of hydrogen transfer between the amine and the excited triplet state of the (chlorophenyl)benzoxazole is also postulated in view of the discrepancy between the quenching efficiencies of fluorescence and dehalogenation. With n-butylmethylamine, the $k\tau$ value obtained by dehalogenation measurements is 35 times higher than that obtained from fluorescence.

$$\text{(benzoxazole)}-\text{X} \quad \xrightarrow[\text{RH}]{h\nu} \quad \text{(benzoxazole-phenyl)} \quad + \quad RX \qquad (103)$$

$$X = Cl, Br, I$$
$$(131) \qquad\qquad\qquad\qquad (132)$$

2-Iodo-4-triazotoluene has been irradiated in diethylamine and found to undergo dehalogenation to 4-triazotoluene[396].

In the dehalogenation of 4-chlorobiphenyl, 1-chloro- and 1-bromonaphthalene, 9-chloro- and 9-bromoanthracene and 4-chlorobenzonitrile, diethyl sulphide has been used as electron donor[397]. The involvement of radical anions in these reactions is evidenced by the incorporation of deuterium into the products when the reactions were performed in acetonitrile–deuterium oxide. Lithium diisopropylamide in hexane or tetrahydrofuran[398] and sodium methyl siliconate $(\text{MeSiO}_3\text{Na}_3)$[399] have also been used as electron transfer reagents in the photodehalogenation of 4-chlorobiphenyl.

The dechlorinations of 4-chlorobiphenyl[400,401] and of chlorobenzene[402,403], ortho-, meta- and para-dichlorobenzene and 3,4-dichlorotoluene[402] can be photosensitized by N,N-dimethylaniline. As with alkylamines, the reaction involves electron transfer but in this case it is not the halobenzene but the dimethylaniline (DMA) that is brought into an excited state. The reaction, supposed to proceed via a exciplex (Y), follows the steps indicated by equations 104–106.

$$^1\text{DMA}^* + \text{ArCl} \quad\longrightarrow\quad {}^1(\text{DMA}^{+\bullet} \cdots\cdots \text{ArCl}^{-\bullet}) \qquad (104)$$

$$^1(\text{DMA}^{+\bullet} \cdots\cdots \text{ArCl}^{-\bullet}) \quad\longrightarrow\quad \text{DMA}^{+\bullet} + \text{Cl}^- + \text{Ar}^\bullet \qquad (105)$$

$$\text{Ar}^\bullet + \text{RH} \quad\longrightarrow\quad \text{ArH} + \text{R}^\bullet \qquad (106)$$

By means of laser spectroscopic studies it has been established that chlorobenzene does not quench the triplet state of N,N-dimethylaniline and the reaction is exclusively singlet-mediated[402]. The radical anion of chlorobenzene could not be observed and its cleavage rate constant is considered to be very high[403]. Kinetic analysis has shown that the rate constants for the in-cage fragmentation of the radical anion and the back recombination of the ion are 1.8×10^9 s^{-1} and 5.1×10^9 s^{-1}, respectively. With the dichloro- and the trichlorobenzenes, however, quenching of the triplet state also takes place and the reaction involves both the singlet and the triplet states of the amine[402,404,405]. Rate constants of the quenching of the singlet excited states of N,N-dimethylaniline, 1-naphthylamine, 2-naphthylamine, triphenylamine, N,N-dimethyl-1-naphthylamine and 9-methylcarbazole by the three dichlorobenzenes and the three trichlorobenzenes as well as rate constants of triplet quenching of N,N-dimethylaniline by the dichloro- and the trichlorobenzenes have been determined[405]. The singlet and triplet reactivity do not conform to expectations based on the Rehm–Weller–Marcus theory and the authors speculate about extra inner-sphere contributions to the intrinsic barrier in the singlet state.

A detailed kinetic analysis involving quantum yield determinations and fluorescence quenching studies has been made of the reductive photodehalogenation of 4-chloro-biphenyl and a variety of other aryl halides, catalyzed by 10-methylacridine derivatives in acetonitrile-water in the presence of sodium borohydride[406,407]. The acridine derivative, 9,10-dihydro-10-methylacridine (AcrH$_2$) or acriflavine, transfers an electron from its excited singlet state to the aryl halide. Dissociation of RX$^{-\bullet}$ into R$^{\bullet}$ and X$^-$ may then be followed by fast hydrogen transfer from AcrH$_2^{+\bullet}$ to R$^{\bullet}$, yielding AcrH$^+$ and RH. In the presence of sodium borohydride, AcrH$_2$ is regenerated by hydride transfer from BH$_4^-$ to AcrH$^+$ (equations 107–110).

$$^1\text{AcrH}_2^* + \text{RX} \longrightarrow (\text{AcrH}_2^{+\bullet}\,\text{RX}^{-\bullet}) \tag{107}$$

$$(\text{AcrH}_2^{+\bullet}\,\text{RX}^{-\bullet}) \longrightarrow \text{AcrH}_2^{+\bullet} + \text{R}^{\bullet} + \text{X}^- \tag{108}$$

$$\text{AcrH}_2^{+\bullet} + \text{R}^{\bullet} \longrightarrow \text{AcrH}^+ + \text{RH} \tag{109}$$

$$\text{AcrH}^+ + \text{BH}_4^- \longrightarrow \text{AcrH}_2 + \text{BH}_3 \tag{110}$$

4-Chlorobiphenyl and 2-chloronaphthalene were irradiated in methanol containing β-naphthol and sodium methoxide and found to undergo photodechlorination[408]. The key step in the mechanism is electron transfer from photoexcited naphthoxide anion **133** to the aryl halide. The naphthoxyl radical is supposed to take up a hydrogen atom from the solvent and, by subsequent deprotonation, the naphthoxide anion is regenerated (equation 111).

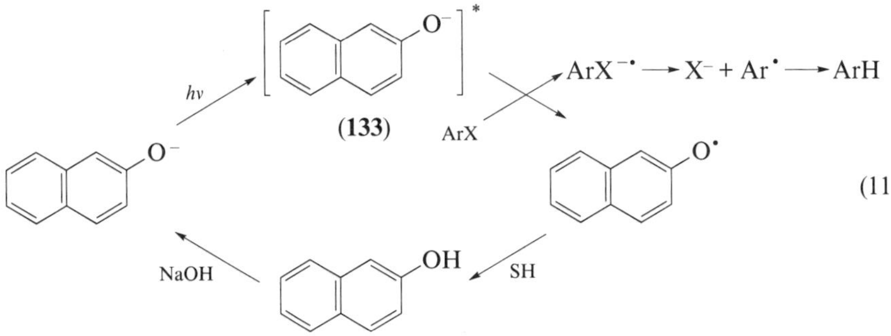

A photoexcited anion also plays a role in the photoelectrocatalytic reduction of 4-chlorobiphenyl using the anion radicals of anthracene and 9,10-diphenylanthracene[409]. The anion radicals were electrochemically generated and excited by means of visible light. Formation of the aryl anion radical then takes place either by direct electron transfer or by ejection of an electron from the excited anion radical, followed by reaction of the solvated electron with 4-chlorobiphenyl.

The photochemical dechlorination of 4-chlorobiphenyl in the presence of anthracene and triethylamine is supposed to proceed via two electron transfer steps, first from triethylamine to excited anthracene and then from the anthracene radical anion to the aryl halide[410]. This type of mechanism is also operative in the dechlorination of 4-chlorobiphenyl, 4,4′-dichlorobiphenyl, 1,3,5-trichlorobenzene, 1,2,3,5-tetrachlorobenzene and pentachlorobenzene photosensitized by visible dyes (protoporphyrin IX, acriflavin, rose bengal, zinc protoporphyrin, methylene green) in the presence of triethylamine[411]. The various steps of the mechanism are summarized in equations 112–115. A mechanistic alternative is that the excited photosensitizer transfers an electron directly to the aryl chloride and becomes regenerated by electron transfer from the amine.

$$Dye^* + R_3N \longrightarrow Dye^{-\bullet} + R_3N^{+\bullet} \tag{112}$$

$$Dye^{-\bullet} + ArCl \longrightarrow Dye + ArCl^{-\bullet} \tag{113}$$

$$ArCl^{-\bullet} \longrightarrow Ar^\bullet + Cl^- \tag{114}$$

$$Ar^\bullet + RH \longrightarrow ArH + R^\bullet \tag{115}$$

Haloanthracenes undergo reductive photodehalogenation upon irradiation in acetonitrile in the presence of triethylamine or N,N-dimethylaniline. 9,10-Dichloroanthracene[412–416], 9-chloroanthracene[416], 9,10-dibromoanthracene[417–419] and 9-bromoanthracene[418,419] in their lowest excited singlet states form exciplexes with ground-state amine. These exciplexes decompose into haloanthracene radical anions and amine radical cations. The absorption spectra of the radical anions of the chloroanthracenes[416] and the bromoanthracenes[419] have been observed following nanosecond pulse excitation. Quenching experiments using azulene and ferrocene show that the triplet state is not involved in these dehalogenations.

The photodechlorination of pentachlorobenzene (and of several tri- and tetrachlorobenzenes) has been studied in acetonitrile in the presence of triethylamine[420] and of sodium borohydride[421]. In the presence of triethylamine an electron is transferred from the amine to triplet excited (equation 97) or singlet excited pentachlorobenzene and a charge-transfer complex is produced. The regiochemistry of the dechlorination differs from that observed in the direct irradiation (vide supra); 1,2,4,5-tetrachlorobenzene is now the major product and the ratio of the three products (1,2,3,5-, 1,2,4,5- and 1,2,3,4-tetrachlorobenzene) is now 25.3:66.2:8.5. A rationale for the regiochemistry has been developed by considering the transition state (134) for dissociation of the radical anion (equation 116). The bent transition state resembles a Wheland intermediate and the relative stabilities can be estimated by calculating the sum of the charge densities on carbon atoms substituted with halogen in the HOMO. Good agreement has been obtained between calculated and observed relative rates. A similar approach was used in rationalizing the regiochemistry of the photochemical defluorination of pentafluorobenzene in the presence of triethylamine in acetonitrile and pentane[422]

$$\tag{116}$$

(134)

A non-planar transient complex was also proposed to account for the behaviour of the radical anions of halobenzoic acids[423]. The cleavage rates of these anions were measured using a photoelectrochemical method based on photoelectron injection from a metal into an electrolyte solution. The rates depend on the nature of the halogen and its position in the ring relative to the carboxylate group, increasing in the series F < Cl < Br and *meta* < *para* < *ortho* (Cl, Br) and *meta* < *ortho* < *para* (F).

The radical anion of pentachlorobenzene has been generated thermally by electron transfer from lithium *p,p'*-di-*tert*-butylbiphenyl radical anion (LiDBB)[424]. The relative ratios of the three dehalogenation products were in excellent agreement with the product ratios obtained photochemically in the presence of triethylamine.

A striking effect of triethylamine on the regiochemistry of photodehalogenation was also observed with three tetrachloronaphthalenes in acetonitrile[425]. In 1,2,3,4-tetra-chloronaphthalene the ratio of monodechlorination at positions 1 and 2 changes from 1:4 in the absence of amine to 1:10 in its presence. In the 1,4,6,7 isomer the change is more dramatic: for reaction at positions 1 and 6 the ratio goes from 4:1 to 1:50. 1,3,5,8-Tetrachloronaphthalene reacts at positions 1 and 8; addition of triethylamine causes the ratio to shift from 1:2.3 to 25:1. Kinetic studies show that, in the absence of amine, the dehalogenation proceeds through the triplet state, either directly or, dependent on the concentration, via an excimer. In the presence of amine, however, an exciplex with the amine is produced from the singlet excited state of the naphthalene derivative and radical anions are generated. Calculation of the charge densities in the bent transition state at carbon atoms bearing chlorine nicely rationalizes the observed regiochemistry.

The photodechlorination of 2,2',3,3',6,6'-hexachlorobiphenyl and of three commercial mixtures of polychlorinated biphenyls solubilized in an aqueous solution of poly(sodium styrenesulphonate-*co*-2-vinylnaphthalene) was studied with the use of solar-simulated radiation[426,427]. The reaction was found to be photosensitized by the naphthalene antenna units present in the copolymer. Exciplex formation and generation of radical anions lead to dechlorination.

In the presence of sodium borohydride, the quantum yield of photodechlorination of pentachlorobenzene increases with increasing concentration of the electron-transfer reagent and a mechanism is proposed in which BH_4^- transfers an electron to the triplet excited state of pentachlorobenzene[421]. The product distribution does not show the reversal in regioselectivity observed with triethylamine as donor, but is not very different from that obtained in the absence of $NaBH_4$. This is explained by assuming that the major reaction pathway of the charge-transfer complex is decomposition and formation of an aryl radical in the cage. The mechanism is summarized in equations 117–119.

$$ArCl \xrightarrow{\;h\nu\;} {}^1ArCl^* \longrightarrow {}^3ArCl^* \longrightarrow \text{products} \tag{117}$$

$$ {}^3ArCl^* + BH_4^- \longrightarrow (ArCl^{-\bullet} BH_4^\bullet) \tag{118}$$

$$ (ArCl^{-\bullet} BH_4^\bullet) \longrightarrow (Cl^- Ar^\bullet BH_4^\bullet) \longrightarrow ArH + BH_3 + Cl^- \tag{119}$$

A similar mechanism had been formulated earlier for the photoreductive dehalogenation of 1-bromo- and 1-chloro-4-cyanonaphthalene in acetonitrile/water containing $NaBH_4$[428]. In these cases, however, the excited singlet state of the arene was supposed to accept an electron from the borohydride.

The rate of photodehalogenation of 3-chlorobiphenyl, 4-chlorobiphenyl, 4,4'-dichloro-biphenyl and 2,2',5,5'-tetrachlorobiphenyl in acetonitrile/water is enhanced when the compounds are irradiated in the presence of sodium borohydride[429]. A mechanism is proposed involving direct attack of hydride on the photoexcited biphenyls. Photo-dehalogenation of *para*-chlorotoluene is also strongly stimulated by sodium borohydride, but a radical chain mechanism does not appear to be an important reaction pathway[430]. The photochemical dechlorination of Aroclor 1232, 1242, 1254 and 1260 was also examined in the presence and absence of sodium borohydride[431]. In both cases, the photoreduction proceeded to give dechlorinated biphenyls by sequential loss of chlorines, but the photoreactions in the presence of borohydride were much faster. A variety of aryl iodides and some bromides, but not chlorides or fluorides, undergo efficient dehalogenation by ultraviolet photostimulated reduction with sodium borohydride and a free radical initiator, di-*tert*-butyl hydroperoxide, in dimethylformamide[432].

Photochemical dehalogenation proceeding via electron transfer has been applied to halogenated purine and pyrimidine bases. 2-Amino-6-chloro-9-(2,3,5-tri-O-acetyl-β-D-ribofuranosyl)purine was irradiated in THF containing 10% triethylamine[433,434]. The dechlorinated nucleoside was obtained in 84% isolated yield; the reaction does not proceed in the absence of triethylamine. In competition experiments with triethylamine and benzene, only the dehalogenated product was isolated. Formation of an exciplex followed by electron transfer, C—Cl bond cleavage and hydrogen abstraction from the amine is considered a plausible mechanistic interpretation[433].

Upon irradiation of an aqueous solution of 5-bromouracil in the presence of N-acetylcysteine methyl ester, uracil was formed[435]. Excitation at 254 nm populates the π,π^* singlet state, whereupon C—Br homolysis creates a uracil-5-yl radical which abstracts a hydrogen atom from the SH group of cysteine. The triplet state of 5-bromouracil can be produced via excitation into the n,π^* singlet at 308 nm, using a XeCl excimer laser, followed by intersystem crossing. Electron transfer from N-Ac-Cys-OMe to the triplet is followed by proton transfer from the SH group of the radical cation to the oxygen atom of the radical anion. Loss of a bromine atom and tautomerization gives uracil. 5-Iodouracil was photochemically converted into uracil by irradiation in aqueous 2-propanol[436]. The quantum yield of uracil formation increases from 0.058 in neutral solution to 0.4 at pH 13 and depends linearly on the inverse square root of the light intensity. The conversion is significantly enhanced by acetone. A chain reaction is taking place, initiated by hydrogen atom abstraction by triplet acetone from 2-propanol. Electron transfer from $Me_2C^•OH$ radicals, or at high pH more efficiently from $Me_2C^•O^-$ radical anions, to 5-iodouracil generates acetone, iodide ion and the uracilyl radical, which regenerates the alcohol radical by hydrogen abstraction from 2-propanol. Similar results were obtained with 5-iodouridine.

Intramolecular electron transfer leading to a bromouracil radical anion is supposed to take place upon irradiation of the duplex $d(GCA^{Br}UGC)_2$ in sodium cacodylate buffer at pH 7.0[437]. The resulting uracilyl-5-yl radical can abstract the adjacent C—1' hydrogen of the adenosine radical cation. Adenine is then released by hydrolytic cleavage of the N-glycosidic bond.

Yet another way to achieve dehalogenation of aryl halides is electron transfer from an exciplex[438]. Upon excitation, naphthalene forms an exciplex with triethylamine and the fluorescence of this exciplex is quenched by 1,3-dichlorobenzene and the trichlorobenzenes. In ethyl acetate the quenching is accompanied by the production of hydrogen chloride. Hydrogen chloride was not formed during the irradiation of a naphthalene solution in the presence of trichlorobenzene in ethyl acetate. Irradiation of a solution of trichlorobenzene in the presence of triethylamine in ethyl acetate did not lead to production of HCl either. It was concluded that the simultaneous presence of naphthalene (N), triethylamine and trichlorobenzene is required. The key steps of the mechanism are summarized in equations 120–122. The net result is photoinduced transfer of an electron from triethylamine to the chlorobenzene without excitation of either partner. The quenching rate constants increase with an increase in chlorine substitution in the aromatic ring.

$$N \xrightarrow{\quad h\nu \quad} {}^1N \xrightarrow{\quad TEA \quad} EX \qquad\qquad (120)$$

$$EX + ArCl \longrightarrow (EX^{+\bullet}\cdots ArCl^{-\bullet})(Z) \tag{121}$$

$$Z \longrightarrow N + TEA^{+\bullet} + Ar^{\bullet} + Cl^{-} \tag{122}$$

Reductive photodehalogenation of halogenated quinones has been studied with benzoquinones, naphthoquinones and anthraquinones. N-Substituted 2-amino-3,5,6-tribromo-1,4-benzoquinone (**135**) with pyrrolidine, 2-, 3- and 4-methylpiperidine, morpholine and desipramine as substituents at position 2, are unstable in daylight[439,440]. Solutions of these compounds in dichloromethane were kept standing in daylight for periods varying from 4 h (pyrrolidine) to 290 h (morpholine). Debromination took place exclusively at position 3, yielding **136** (equation 123). N-Substituted 2-amino-3-bromo-1,4-naphthoquinones with desipramine, nor- and protriptyline, maprotiline and benzoctamine as 2-substituents likewise undergo debromination at position 3, neighbouring the amino group[441]. N-substituted 2-amino-3-chloro-1,4-naphthoquinones have also been studied, but these compounds undergo mainly dealkylation of the amino function[441,442].

$$(135) \qquad \xrightarrow[CH_2Cl_2]{h\nu} \qquad (136) \tag{123}$$

Detailed spectroscopic and kinetic investigations have been performed on the photochemical reactions in ethanol of haloanthraquinones, especially 1-chloro-, 2-chloro-, 1,5-dichloro- and 1,8-dichloroanthraquinone[443–445] and the corresponding bromo compounds[446]. Upon irradiation at 366 nm the following consecutive photochemical reactions take place: 1,5- or 1,8-dichloroanthraquinone → 1,5- or 1,8-dichloroanthrahydroquinone → 1-chloroanthraquinone → 1-chloroanthrahydroquinone → anthraquinone. A similar sequence of photoreactions occurs with the bromo derivatives. 2-Chloroanthraquinone is not dechlorinated, but reduced to 2-chloroanthrahydroquinone. The first step in the above sequences, photoreduction to haloanthrahydroquinones, is hydrogen abstraction from the solvent by the lowest triplet state of the haloanthraquinone, which has mixed $n,\pi^*-\pi,\pi^*$ or π,π^* character. No evidence of electron transfer from the solvent to triplet excited haloanthraquinone was obtained. In the second step, one halogen atom is expelled and the anthrahydroquinone is re-oxidized to an anthraquinone. This reaction, however, does not occur with 2-chloroanthrahydroquinone, but only with α-haloanthrahydroquinones. It is proposed that this reaction consists of intramolecular elimination of hydrogen halide for which the proximity of the hydroxyl group and the halide atom is essential.

Extensive investigations have been conducted in order to detect, identify and characterize the triplet states of haloanthraquinones[447–451]. The transient absorption spectra of 1,8-dichloroanthraquinone in various solvents (toluene, ethanol, carbon tetrachloride, EPA) at room temperature were measured using picosecond laser photolysis, and the build-up and decay of an absorption assigned to the second triplet state, with n,π^* character, were observed[449]. This triplet state has intramolecular charge-transfer character between the chlorine and oxygen atoms. A similar second excited n,π^* triplet state was detected in the case of 1,8-dibromoanthraquinone[451]. Intersystem crossing from $S_1(n,\pi^*)$ to $T_2(n,\pi^*)$ occurs in 15–20 ps and internal conversion from $T_2(n,\pi^*)$ to $T_1(\pi,\pi^*)$ takes 70–110 ps.

Picosecond laser spectroscopy has also been used to investigate the behaviour of 1,8-dichloroanthraquinone (DCAQ) upon irradiation in the presence of 2,5-dimethylhexa-

2,4-diene (DMHD)[452]. The two molecules form a ground-state complex and when this is excited by a picosecond 435 or 527 nm light pulse, an absorption is detected which is assigned to an excited singlet charge-transfer complex, $^1(DCAQ^{\delta-} \cdots DMHD^{\delta+})$, or a singlet ion pair, $^1(DCAQ^{\cdot-} \cdots DMHD^{\cdot+})$. This species decays by intramolecular proton transfer, yielding DCAQH$^\cdot$ and the DMHD radical, as well as by intramolecular charge recombination.

The reduction of 1-bromoanthraquinone[453] and 1-iodoanthraquinone[454] in acetonitrile in the presence of tetrabutylammonium perchlorate at gold electrodes irradiated with light (565 nm) produces the radical anion of anthraquinone. The first electron transfer yields the radical anion of the haloanthraquinone. Upon irradiation, this anion dissociates and the anthraquinone formed undergoes a further one-electron reduction at the electrode. The hydrogen atom abstracted in the formation of the anthraquinone arises from the tetrabutylammonium cations. The radical anions of 1-iodoanthraquinone, obtained through excitation at either 417 or 565 nm, show different dynamic behaviour. Expulsion is 7.4 times faster per photon absorbed with 565 nm radiation than with 417 nm light. Luminescence has been detected from two excited states of the radical anion. The slower iodide loss at 417 nm may result from the formation of a quartet state from which reaction is spin forbidden.

Many investigations on polyhalogenated, especially polychlorinated, aromatic molecules have been performed mainly because such compounds are ubiquitous environmental contaminants. The polychlorophenols occur in surface water and river sediments; they have also been found in mussels and in pike. Photoreactions of tetra- and pentachlorophenols have been investigated in water–acetonitrile mixtures[455,456]. 2,3,4,5-, 2,3,4,6-, 2,3,5,6-Tetrachlorophenol and pentachlorophenol undergo reductive dechlorination at $\lambda > 285$ nm. The 2,3,5,6-tetrachloro compound is unique among these compounds; in addition to dehalogenation, it yields hexa-, hepta- and octachlorodihydroxybiphenyls as well as heptachlorohydroxydiphenyl ether[455]. Perchloro-*o*-phenoxyphenol possesses the structural potential for conversion to polychlorodibenzo-*p*-dioxins. Irradiation of its sodium salt in methanol in the presence of the sensitizer *m*-methoxyacetophenone and excess triethylamine leads to octachlorodibenzo-*p*-dioxin as a major product, with ether cleavage and dechlorination as other important reaction pathways[457]. In the absence of triethylamine, ether cleavage and dechlorination are the major reaction pathways with no detectable cyclization to octachlorodibenzo-*p*-dioxin.

Several tetrachlorodibenzofuran isomers were photolysed by ultraviolet irradiation and the relative amounts of the various trichlorodibenzofurans were determined[458]. A simple set of guidelines for predicting the relative amounts of these photolysis products has been formulated. Two reports have appeared describing the results of the photolytic transformation of polychlorinated dibenzofurans and dioxins adsorbed onto fly ash[459,460]. The photodegradation of octachlorodibenzofuran in 1,4-dioxane was studied in various wavelength regions and found to be maximal at 279.1–332.0 nm[461]. The primary degradation products include hexa-, hepta- and pentachlorodibenzofurans with small amounts of tetrachlorodibenzofurans. The photolytic behaviour of brominated dibenzofurans and dibenzo-*p*-dioxins in methanol and *n*-hexane has been investigated[462]. Photolysis in methanol is nearly six times slower than in *n*-hexane. The rate of photolysis increases with increasing number of bromine atoms. The efficiency of photochemical dechlorination of tetrachlorodioxins[463,464], penta- and heptachlorodibenzo-*p*-dioxins[465] and octachlorodibenzo-*p*-dioxin[464] has been studied.

It has been argued[466] that the solution photolysis rates of polychlorinated dibenzo-*p*-dioxins may be explained by the preferential photodissociation of chlorine atoms from a lateral vs a non-lateral position to yield the corresponding aryl radical and/or aryl cation-aryl carbene intermediate.

The photodegradation mechanism of 2,3,7,8-tetrachlorodibenzo-p-dioxin has been studied with the PM3-MNDO method[467]. It was assumed that chlorine elimination followed by hydrogen addition is not a direct bond-cleavage process, but may rather proceed via the type of transition state that has been proposed in explaining substitution reactions of aromatic molecules. On this basis, the photochemical reactivity index for the reductive dechlorination can be represented in terms of the frontier electron density associated with the highest singly occupied orbital (SOMO*) of the first excited state, which would act as an electron-donor level to the proton-like reagent attacking the nucleophilic site. The photochemical reactivities predicted by the frontier electron index are quite in accord with the experimentally found photodegradation pathway. The method gives a correct prediction of the second photodegradation step in which 2,3,7-trichlorodibenzo-p-dioxin yields as main product 2,7-dichlorodibenzo-p-dioxin, as minor product 2,8-dichlorodibenzo-p-dioxin but no 2,3-dichlorodibenzo-p-dioxin.

Bromoxynil (a herbicide) undergoes photodebromination in water[468] and photodechlorination of vinclozolin in observed in methanol, but not in benzene[469].

Several reports have appeared on the photodegradation of adsorbed (poly)-halogenobenzenes and -biphenyls. The kinetics of the TiO_2-mediated photocatalyzed degradation of 4-chlorophenol[470,471], 2,4-dichlorophenol[471] and 2,4,5-trichlorophenol[471] have been investigated. Polychlorobiphenyls adsorbed on silica gel or montmorillonite were degraded mainly to less chlorinated biphenyls by ultraviolet irradiation at 254 nm[472]. The photolysis is enhanced by triethylamine. Ultraviolet irradiation-assisted catalytic dechlorination of o-, m- and p-dichlorobenzenes and 2′,3,4-trichlorobiphenyl in a prototype commercial reactor using fiberglass mesh-supported anatase has been reported[473]. Photodechlorination of chlorinated biphenyls has been found to occur in the presence of cyclodextrins[474].

Laser-induced decomposition of mixtures of polychlorinated biphenyls in the liquid phase has been investigated, employing radiation from three different excimer lasers (XeCl at 308 nm, KrF at 248 nm and ArF at 193 nm)[475]. The mixtures can be quantitatively and efficiently destroyed by means of UV radiation at 248 nm. A single-photon dissociation process, which leads to both HCl elimination and biphenyl bond rupture, is induced by the KrF laser radiation.

B. Arylation Reactions

Intermolecular photoreaction of an aryl halide with another aromatic compound may lead to the formation of biaryls. In this section several examples of such reactions will be discussed. In some cases, information concerning the reaction mechanism is available but the depth to which mechanisms have been investigated varies greatly. In many cases aryl radicals formed by homolysis of the carbon–halogen bond are the reactive species. Such radicals may also be produced via electron transfer, followed by departure of halide anion. In some cases aryl cations have been proposed as intermediates. Intermolecular bond formation may also be preceded by charge transfer within an exciplex or by formation of radical ion pairs.

Examples of reactions of halogenated quinones and of aryliodonium compounds, as well as arylation of alkenes and alkynes will also be mentioned in this section.

1. Intermolecular arylation

a. Homocyclic aromatic halides.   Photolysis of neat chlorobenzene at room temperature affords *ortho*-, *meta*- and *para*-chlorobiphenyls as main products in a ratio of 25:30:45, respectively, accompanied by a small amount of biphenyl[476]. This ratio differs from the ratio obtained from reactions of chlorobenzene with phenyl radicals generated from other sources (iodobenzene, dibenzoyl peroxide), which favour the formation of the *ortho* isomer. A similar high proportion of *para*-chlorobiphenyl was found upon irradiation of chlorobenzene in 1,1,2-trichlorotrifluoroethane at a concentration of 5×10^{-1} mol dm^{-3}. Decrease of the chlorobenzene concentration, however, changed the isomer ratio to give mainly the *ortho* isomer, approaching the normal value. It was proposed that in concentrated solutions, excited chlorobenzene interacts with ground-state chlorobenzene, forming a complex (presumably a triplet excimer) with a conformation (137) favouring the formation of the *para* isomer (equation 124).

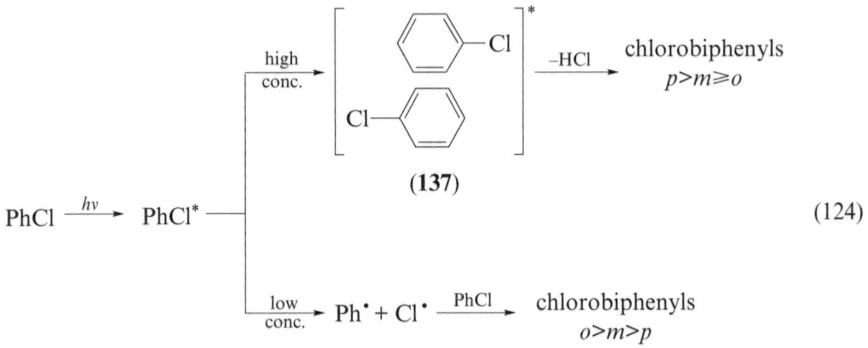

(124)

The photolysis of a large number of polychlorobenzenes, polybromobenzenes and polychloropolybromobenzenes $(C_6H_{6-n}X_n)$ in benzene solution has been studied[371]. In addition to dehalogenation, phenylation was observed, leading to (poly)halobiphenyls $(C_6H_{6-n}X_{n-1}$—$C_6H_5)$ and, by consecutive phenylation, to haloterphenyls.

The reaction of pentafluoroiodobenzene with aromatic compounds such as anilines, pyrroles, indoles, imidazoles, aromatic ethers and phenols leads to aryl—aryl coupling[477]. The reactions proceed via pentafluorophenyl radicals which are generated by photoinduced electron transfer (PET) and loss of iodide ion. Coupling between the pentafluorophenyl radical and the radical cation of the donor gives biaryl cations (138, 139) which lose a proton. The reaction is illustrated for N,N-dimethylaniline (equation 125).

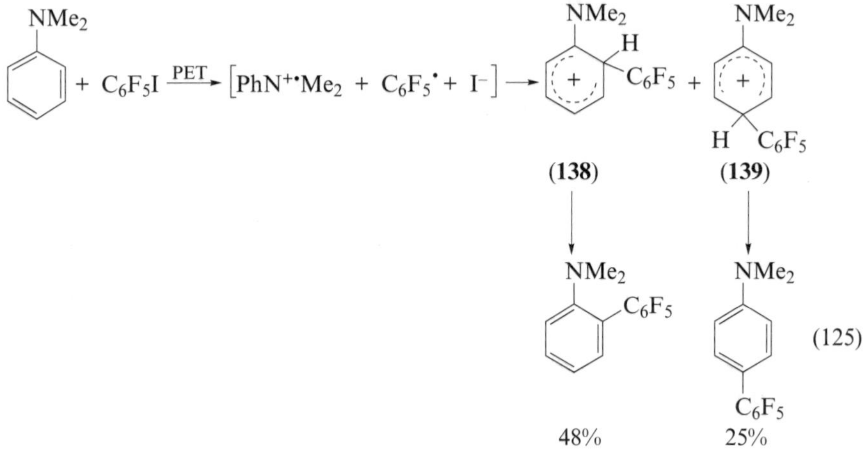

(125)

In this reaction the coupling takes place between the two partners in electron transfer. It is also possible to generate the aryl radical via electron transfer from a third component. Thus, the anion radical of anthraquinone was irradiated in the presence of 1, 2-dibromobenzene and anthracene[478]. Electron transfer followed by loss of bromide ion generated the 2-bromophenyl radical which reacted with anthracene, forming 9-(2-bromophenyl)anthracene. The same process occurs with 1,4-dibromobenzene, 4-bromobenzonitrile and 4-chlorobenzonitrile.

3-(4-Chlorophenyl)-1,1-dimethylurea was irradiated in a 50% solution of pyridine in water[479]. A regiospecific coupling at position 3 of pyridine was observed (92% yield). The product was further elaborated into 3-phenylpyridine.

Fluorobiphenyls are formed by irradiation of o-, m- and p-RC_6H_4I (R = OH, OCH_3, CH_3, H, F, Cl, $COOCH_3$, NO_2) in fluorobenzene and by irradiation of o-, m- and p-FC_6H_4I in C_6H_5R (R = OH, OCH_3, CH_3, F, Cl, $COOCH_3$, NO_2)[480]. Relative yields were determined and F-NMR spectroscopy was used for the identification of the substituted fluorobiphenyls. The results are discussed in terms of electronic and steric substituent effects.

The photochemistry of diphenyl- and bis(4-methylphenyl)iodonium salts has been investigated[481,482]. Diphenyliodonium halides (140, X = Cl, Br, I) exist as tight ion pairs in acetonitrile. Their photolysis gives almost exclusively iodobenzene by a homolytic cleavage from a charge-transfer excited state. In aqueous acetonitrile, however, the ion pairs are solvent-separated and substantial amounts of 2-, 3- and 4-iodobiphenyls (141) are formed in addition to iodobenzene (142), benzene (143), acetanilide (144) and biphenyl (145) (equation 126). In this medium the photodecomposition occurs via initial heterolysis of the molecule in its excited state, leading to iodobenzene and phenyl cation.

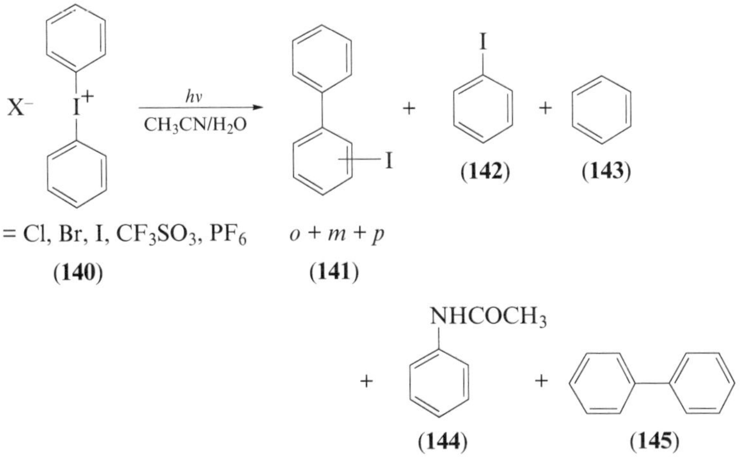

$X = Cl, Br, I, CF_3SO_3, PF_6$ o + m + p

(140)   (141)

(142)   (143)

NHCOCH$_3$

(126)

(144)   (145)

The photochemical behaviour of diphenyliodonium salts has been compared with that of the corresponding chloronium and bromonium compounds[483]. The diphenyliodonium photolysis produces relatively more 2-halobiphenyl and less 3-halobiphenyl than that of the chloronium and bromonium salts. The product distributions are related to stabilities of the proposed intermediates and to interaction of the halogen nuclear spin with the free-radical intermediates. The singlet excited state may undergo heterolytic cleavage to form the aryl cation–haloarene pair, or it may undergo intersystem crossing. The triplet may undergo homolysis to the triplet aryl radical–haloarene pair. The triplet excited state can also be formed by sensitization. The triplet aryl radical pair may undergo cage-escape reactions or intersystem crossing to the singlet radical pair, a process which is mediated by the spin–spin coupling between the haloarene odd electron and the nuclear spin of the halogen. From the singlet radical pair, the aryl cation-haloarene pair may be formed. The products observed are then derived from the intermediate pairs.

2,4-Dinitro-6-phenyliodonium phenolate (146) is a stable iodonium zwitterion[484]. It reacts under photolytic conditions with various alkenes, alkynes and aromatic compounds to afford 2,3-dihydrobenzo[b]furans, benzo[b]furans and 6-aryl-2,4-dinitrophenols. The mechanism involves hypervalent iodine compounds (iodinanes, 147) and is illustrated for the reaction with an aromatic compound (equation 127). Compounds 148 are the major products when ArH = PhH, PhOCH$_3$ or 1,4-dimethoxybenzene. With furan and thiophene, 149 is the principal product. The reaction does not proceed with chlorobenzene and nitrobenzene.

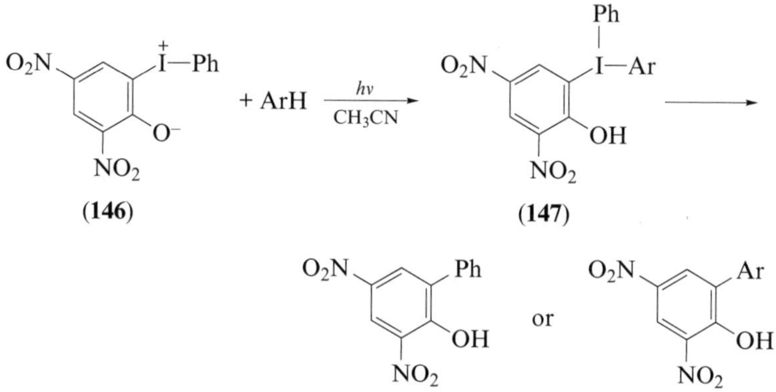

$$O_2N \underset{NO_2}{\overset{\overset{+}{I}-Ph}{\diagup\diagdown O^-}} + ArH \xrightarrow[CH_3CN]{h\nu} O_2N \underset{NO_2}{\overset{\overset{\underset{|}{Ph}}{I-Ar}}{\diagup\diagdown OH}} \longrightarrow$$

(146)   (147)

$$O_2N \underset{NO_2}{\overset{Ph}{\diagup\diagdown OH}} \quad \text{or} \quad O_2N \underset{NO_2}{\overset{Ar}{\diagup\diagdown OH}} \tag{127}$$

(148)   (149)

Phenyliodonium dimedonate (150) has been irradiated in the presence of PhCH=CHR in acetonitrile[485]. The reaction product is a 2-phenyl-3-R-4-oxo-2,3,4,5,6,7-hexahydrobenzo[b]furan (151) (R = H or PhCH=CH) (equation 128). The primary step is believed to be the decomposition of the iodonium ylide into a diketocarbene and PhI.

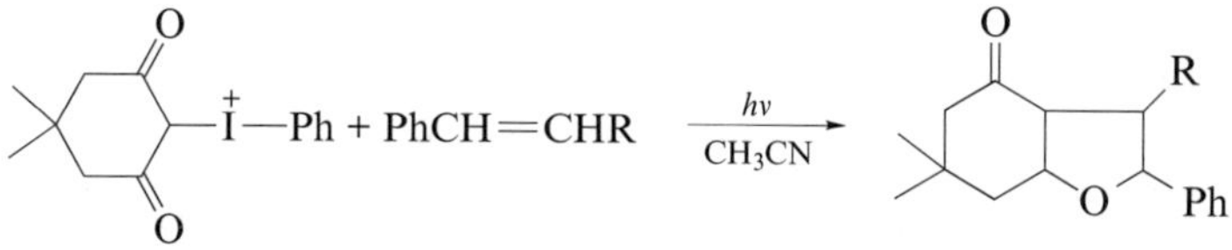

$$\underset{O}{\overset{O}{\diagdown}}\overset{+}{I}-Ph + PhCH=CHR \xrightarrow[CH_3CN]{h\nu} \text{(151)} \tag{128}$$

(150)   (151)

A recent volume in the *Advances in Photochemistry* series contains a chapter devoted to the photochemistry and photophysics of onium salts, including aryliodonium salts[486].

Many synthetic applications have been reported of the photochemical reactions of halogen-substituted 1,4-naphthoquinones with 1,1-diarylethenes[487–490], and with related electron-rich alkenes such as 1-aryl-1-trimethylsilyloxyethenes[491–493], 2-trimethylsilyloxy-1-alkenes[494], 2-methoxy-1-alkenes[495] and allyltributylstannane[496,497]. The process is exemplified by the reaction of 2-bromo-3-methoxy-1,4-naphthoquinone derivatives (152) with 1,1-diphenylethene[487] (equation 129).

The primary product (153) is photochemically converted into a derivative of benz[a]anthracene-7,12-dione (154). A variety of heteroatom-containing polycyclic aromatic compounds was synthesized by using a 1,1-disubstituted ethene in which one of the substituents is a phenyl group and the other a 2-furyl, 2-benzo[b]furyl, 2-thienyl, 3-

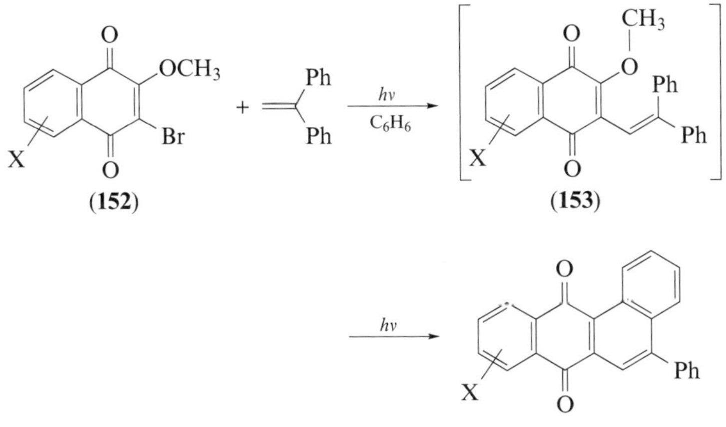

(152) + **(153)** → **(153)** (in brackets)

(129)

(154)

thienyl, 2-benzo[*b*]thienyl, 2-*N*-methylpyrrolyl or 2-*N*-methylindolyl group. In the cycliza-tion step, ring closure with the five-membered heterocyclic ring occurred preferentially[488]. By means of CIDNP it was demonstrated that electron transfer plays a role in the photoreaction of 2,3-dichloro-1,4-naphthoquinone with 1,1-diarylethenes[489] and this was later confirmed and more extensively investigated for various other substituted 1,4-naphthoquinones[490]. In the photoreaction of 2,3-dichloro-1,4-naphthoquinone with 1,1-bis(4-methoxyphenyl)ethene, a radical ion pair consisting of a quinone radical anion and an alkene radical cation has been observed upon flash photolysis in acetonitrile[498]. The decay constant of the solvent-separated ion pair is estimated to be $6 \times 10^{-5} \text{s}^{-1}$. Coupling of the ion radicals followed by elimination of hydrogen halide leads to formation of the adduct. In non-polar solvents such as benzene, the excited naphthoquinone forms an exci-plex (probably a triplet exciplex with charge transfer character) with the diarylethene and the adduct is formed from this exciplex. In polar solvents, solvent-separated ion radical pairs are formed only when the cation radicals are well stabilized, e.g. by *para*-methoxy groups. In other cases the adducts are formed via contact ion pairs.

In a variation on this process, 7-bromo-6-methoxyquinoline-5,8-dione[499] and 6-bromo-7-methoxyquinoline-5,8-dione[500] have been used in coupling reactions with 1,1-diarylethenes.

The photochemical reactions of naphthoquinones and quinolinediones with 1,1-diarylethenes have led to the synthesis of many new polycyclic aromatic and heteroaro-matic compounds. A review has appeared[16] in which a number of these results are tabulated.

A mechanism of electron transfer, radical ion coupling and loss of HX is also proposed to account for the photoinduced reactions of tetrafluoro- and tetrachlorophthalonitriles with anthracene and phenanthrene, leading to the 9-(3,5,6-trihalogeno-1,2-dicyanophenyl)arenes[501].

b. Heterocyclic aromatic halides. The photochemical reaction of 5-bromofuran-2-carboxaldehyde in various aromatic solvents produces 5-arylfuran-2-carboxaldehydes[502]. Quantum yields of product formation at 350 nm in acetonitrile have been determined for benzene ($\varphi = 0.25$), chlorobenzene ($\varphi = 0.36$), bromobenzene ($\varphi = 0.47$), toluene ($\varphi = 0.70$) and anisole ($\varphi = 0.40$). Spectroscopic studies lead to the conclusion that a triplet exciplex or a radical ion pair (in polar solvents) is involved, from which the product is formed via radical coupling with subsequent elimination of HBr. This type of reaction was used in the synthesis of various 3- and 5-aryl-2-furyl derivatives[503]. 5-Bromo-2-furyl dodecyl ketone was irradiated in benzene and in various other solvents and high yields of 5-aryl-2-furyl dodecyl ketones were obtained[504]. The comparison of reaction rates and quantum yields in various solvents gave results in agreement with the hypothesis of exciplex formation. On the basis of calculations using the Weller equation, it was expected that the coupling reaction would proceed with higher efficiency when using 5-iodofuran-2-carboxaldehyde instead of the 5-bromo derivative, and this indeed proved to be the case[505].

5-Bromo- and 4,5-dibromofurfural have been reported to give derivatives of 2,2'-bifuryl upon irradiation[506]. 5-Iodofuran-2-carboxaldehyde as well as 5-iodothiophene-2-carboxaldehyde and its methyl ketone react with indene upon irradiation in acetonitrile[507] to produce indene derivatives substituted at position 2. The photoreactions of 5-iodo- and 5-bromothiophene-2-carboxaldehyde (155) and of the corresponding methyl ketones resemble those of the furan analogues. Upon irradiation in benzene the corresponding 5-phenyl derivatives (156) are formed[508] (equation 130). The same reactivity for other halogenothiophenes is reported: 3,5-dibromothiophene-2-carboxaldehyde furnishes 3-bromo-5-phenylthiophene-2-carboxaldehyde, while the corresponding diiodo compound yields 3,5-diphenylthiophene-2-carboxaldehyde.

$$\text{Br} \diagup\!\!\!\diagdown_{\text{S}}\text{CHO} \quad\xrightarrow[\text{C}_6\text{H}_6]{h\nu}\quad \text{Ph}\diagup\!\!\!\diagdown_{\text{S}}\text{CHO} \qquad\qquad (130)$$

(155)   (156)

Bithienyl derivatives can be prepared by irradiating 2-acetyl-5-iodothiophene or 5-iodothiophene-2-carboxaldehyde in the presence of various thiophene derivatives[509,510]. Thiazolylthiophene derivatives can be formed by photochemical coupling between iodothiophene and thiazole derivatives[511].

5-Bromothiophene-2-carbonitrile turned out to be completely unreactive upon irradiation in benzene, while 5-iodothiophene-2-carbonitrile only underwent dehalogenation[512]. Methyl 5-iodothiophene-2-carboxylate, however, gives the corresponding aryl and heteroaryl derivatives in good yield when irradiated in the presence of various aromatic substrates (benzene, para-xylene, naphthalene, thiophene, 2-bromo- and 2-chlorothiophene). The reaction is proposed to proceed via an exciplex intermediate. The failure of 5-bromothiophene-2-carbonitrile to undergo the arylation reaction is ascribed to its low quantum yield of intersystem crossing. 5-Iodothiophene-2-carbonitrile is supposed to undergo C—I fission in the triplet state before it can form the exciplex.

Halofuran and halothiophene derivatives undergo photochemical reactions with arylalkenes and arylalkynes and with benzo[b]furan[513,514]. With the arylalkenes and arylalkynes, heteroarylation takes place at the terminal alkene or alkyne carbon atom, while benzo[b]furan is substituted at position 2. The experimental results are interpreted in terms of solvent-separated or contact radical ion pairs. Iodothiophene and iodofuran derivatives can also be used to synthesize derivatives of benzimidazole by means of photochemical coupling[515]. The reaction of 5-iodothiophene-2-carboxaldehyde (157) with benzimidazole (158) giving the coupling product 159 is illustrated in equation 131.

$$\text{I}\diagup\!\!\!\diagdown_{\text{S}}\text{CHO} \;+\; \text{(158)} \quad\xrightarrow[\text{CH}_3\text{CN}]{h\nu}\quad \text{(159)} \qquad\qquad (131)$$

(157)   (158)   (159)

Photochemical arylation reactions of furans and thiophenes have been reviewed[516].

Photoactivated chlorobithiazoles promote highly efficient DNA strand scission. In order to obtain insight into the possible role of photogenerated radicals, a chlorobithiazole derivative was irradiated in 1-octene. The nature of the adducts formed is consistent with a mechanism involving initial homolysis of the bithiazole C—Cl bond to produce Cl and bithiazole radicals[517].

Photochemical reaction in acetonitrile of 2′-deoxyuridine 5′-phosphate with the halo-heteroarenes 2-iodothiophene, 2-iodofuran, 1-methyl-2-iodopyrrole and 3-iodothiophene affords the C-5 heteroaryl substituted nucleotides[518]. 6-Aryluridines have been prepared by irradiation of 6-iodouridines in benzene, anisole, thiophene, N-methylpyrrole or 2-methyl-furan in the presence of triethylamine[519].

Irradiation of trans-stilbene in the presence of 2-chlorobenzothiazole leads to a variety of products among which are heteroarylated stilbenes and 9-heteroarylated phenanthrene[520].

Phenylpyridines are produced by irradiation of halopyridines in benzene. The effects of the nature and the position of the halogen atom as well as the effects of bases and solvents have been investigated. The reactivities of halopyridines in the reactions increased in the order $Cl < Br \cong I$. The reactivity also depends on the position of halogen on the ring: for bromo- and iodopyridines the observed order is $2 > 3 > 4$ and for chloropyridines $3 > 2 > 4$. A homolytic mechanism is proposed[521]. 2-Heteroarylpyridines can be synthesized via photoreaction of 2-iodo- or 2-bromopyridine with furan, pyrrole, N-methylpyrrole and thiophene[522] and similar photocouplings have been described for 4-iodopyridine[523]. With furan, pyrrole and N-methylpyrrole the reactions proceed regioselectively to afford the 4-(2-heteroaryl)pyridines, but in the reaction with thiophene this product was accompanied by a small amount of 4-(3-thienyl)pyridine. The behaviour of 2-iodopyridine in its reaction with benzene derivatives was investigated[524] and explained on the basis of the intermediacy of the 2-pyridyl cation[525]. The primary photochemical reaction is homolysis of the C—I bond, but this is followed by electron transfer from the 2-pyridyl radical to an electron acceptor (e.g. the iodine atom). The photochemical behaviour of 2-iodopyridine in CH_2Cl_2 (formation of nearly equal amounts of 2-chloropyridine and pyridine) and in CH_3OH (formation of 2-methoxypyridine as a major product) supports this hypothesis, as do the results obtained when the 2-pyridyl radical is generated by means of homolytic abstraction of the iodine atom from 2-iodopyridine. In the latter experiment, dibenzoyl peroxide was thermolysed in the presence of 2-iodopyridine in methyl benzoate. The 2-pyridyl radicals thus created reacted with methyl benzoate, yielding a mixture of 2-(x-carbomethoxyphenyl)pyridines ($x = 2, 3, 4$) in the same ratio as that of the homolytic phenylation of methyl benzoate.

Photoreaction of 2-, 3- and 4-iodoquinolines with five-membered heterocycles (pyrrole, N-methylpyrrole, furan and thiophene) affords the corresponding n-(2-heteroaryl)quinolines ($n = 2, 3, 4$) in appreciable yields[526]. 3-Halo-1-methylquinolin-2-ones can be converted into 3-aryl-1-methylquinolin-2-ones by photochemical coupling with various aromatic or heteroaromatic compounds[527,528].

4-Iodoindole-3-carboxaldehyde, when irradiated in benzene or furan in the presence of a trace amount of pyridine, gives the 4-phenyl or 4-(2-furyl) derivatives in 60% and 64% yields, respectively[529].

Several reports have appeared concerning photochemical coupling reactions of halouracils[530–538]. 6-Iodo- and 5,6-dihalouracil derivatives undergo efficient coupling reactions with N-methylpyrrole, benzene, N-phenylpyrrole and 1,1-diphenylethene[531]. The reaction of 6-iodo-1,3-dimethyluracil (160) with N-phenylpyrrole (161) starts with homolysis of the C—I bond and attack of the 6-uracilyl radical at the α-position of N-phenylpyrrole. The primary product (162) undergoes further photocyclization, giving the final product (163) (equation 132).

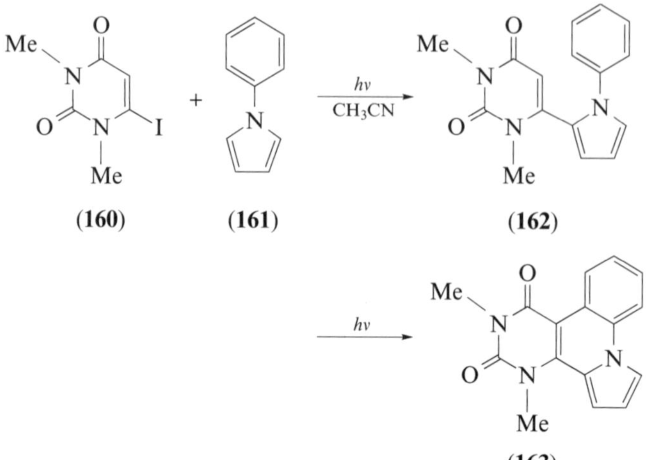

(160)   (161)   (162)

(132)

(163)

Photocoupling reactions of iodouracils with allylsilanes and alkenes have also been described[532,533]. The photoreaction of 5-bromo-1,3-dimethyluracil with some substituted benzenes (e.g. *p*-xylene) yields a 5-aryl substituted uracil derivative, but a 6-aryl substituted isomer was also unexpectedly formed[534,535]. The latter compound has been found to be the photoproduct of protonated starting material. Its formation increased upon addition of HBr and was completely blocked in the presence of base (pyridine or K_2CO_3). The photoinduced couplings of 5-chloro-1,3-dimethyluracil with benzene, toluene, *p*-xylene and *p*-chlorotoluene, however, affording the corresponding 5-aryl-1,3-dimethyluracils in appreciable yields, were significantly promoted by the addition of trifluoroacetic acid to the reaction mixture[538].

5-Aryl- and 5-heteroaryl-2′-deoxycytidines can be synthesized by photochemical reactions of 5-iodo-2′-deoxycytidine in the presence of arenes or heteroarenes[539]. The intermediacy of purin-6-yl radicals in the photoarylation of 6-iodo-9-ethylpurine in benzene was demonstrated by means of ESR[540]. 6-Iodo-9-benzylpurine has been irradiated in the presence of anisole[541]. The intermediate purin-6-yl radicals display selectivity in their attack on the arene: only the *para* (17.0% yield) and *meta* (6.4% yield) isomers were found, possibly as a result of steric control.

2. Intramolecular arylation

In intramolecular arylations, a new bond is created between two aromatic moieties of the same molecule or between an aromatic nucleus and an atom of a side-chain. Many intramolecular arylation reactions of homocyclic and heterocyclic aromatic halides have been studied mainly in view of their synthetic applications, and it is not always clear which mechanistic pathway is followed. The reaction may start with homolytic or heterolytic dissociation of the carbon–halogen bond and proceed by attack of the aryl radical or aryl cation on another part of the molecule. Electrocyclization followed by elimination of hydrogen halide is another possibility. Especially when heteroatoms such as nitrogen, sulphur or phosphorus are involved, the initial step may be a nucleophilic attack on the carbon atom bearing the halide atom.

a. Homocyclic aromatic halides. The photoreaction of 2-chlorophenyl phenyl ether (164) in degassed H_2O/10% CH_3CN leading to dibenzofuran (169) has been studied in detail[542]. The quantum yield of disappearance of starting material is 0.19 and, at 37% conversion, the product mixture contains 89% dibenzofuran; very small amounts of 2-hydroxyphenyl phenyl ether (170) were observed. In degassed hexane the photolysis is five times as efficient ($\varphi = 0.97$) as in water but the major product is now diphenyl ether (171) (84%) and dibenzofuran is formed in only 16% yield. The reaction is supposed to start in both solvents with homolytic dissociation of the C—Cl bond in the excited state (singlet or triplet) (equation 133). In the aqueous medium, electron transfer from the aryl radical (167) to the chlorine atom yields the aryl cation (166). Both the radical and the cation are supposed to be in equilibrium with the ring-closed forms. The dibenzofuran cation (165) can rapidly lose a proton in water, but the dibenzofuran radical (168) needs another radical (or oxygen) to lose a hydrogen atom and abstraction of a hydrogen atom from hexane by the diphenyl ether radical is considered to be faster. Cation 166 has also been represented as a resonance structure with a carbene at C-2 and the positive charge at oxygen[543]. A similar type of ring closure was observed with 2,2′,4,4′-tetrachlorodiphenyl ether which is converted with high efficiency into 2,4,8-trichlorodibenzofuran[544].

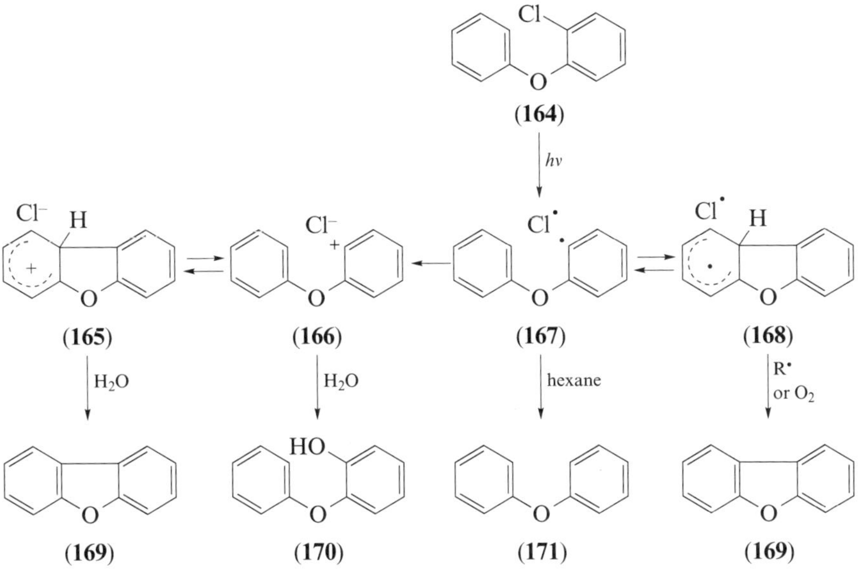

(133)

Photochemical cyclization reactions of iodinated diphenyl ethers to produce dibenzofuran derivatives have also been studied[545]. 1,3-Diiodo-2-(4′-methoxyphenoxy)-5-nitrobenzene affords 4-iodo-8-methoxy-2-nitrodibenzofuran upon irradiation in acetone.

Irradiation of N-(2-chlorobenzyl)pyridinium chloride (172) in aqueous solution yields pyrido[2,1-a]isoindolium chloride (173) (equation 134)[546–548].

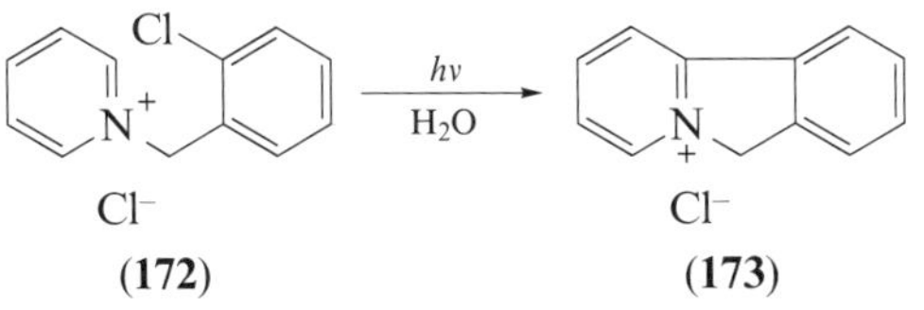

(134)

The reaction was also performed with the bromide salt which did not lead to a difference in the reaction rate. The 2-bromobenzyl analogue (174) (with Br⁻ as counterion), however, reacted four times slower. The reaction is proposed to proceed via a π-complex (176), formed from the triplet state (175), between the halogen atom and the pyridine ring (equation 135). The bulkiness of bromine is held responsible for less efficient complex formation than in the case of chlorine. This reaction type has also been studied with molecules having halogen atoms at position 2 of the pyridine ring; the results will be mentioned in the section on heterocyclic aromatic halides.

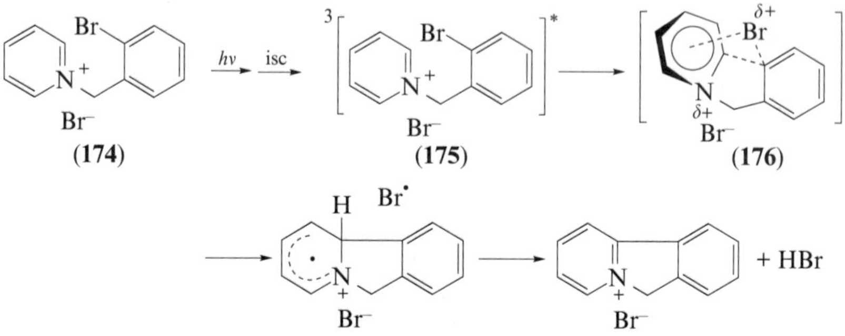

The diphenylamine derivative diclofenac, having two chlorine atoms at *ortho* positions in one of the phenyl rings, undergoes ring closure upon irradiation in water or methanol, yielding 8-chlorocarbazole-1-acetic acid[549]. Upon prolonged irradiation, the remaining chlorine atom is lost to yield the photoreduction product or the photosolvolysis product. Photocyclization of 1-(2-pyridylamino)-8-chloronaphthalene (177) proceeds non-selectively at the *peri* position of the naphthalene nucleus with formation of C—C and C—N bonds and leads to 7*H*-7,8-diazabenzl[*de*]anthracene (178) and pyrido[1,2-*a*]perimidine (179) (equation 136)[550,551].

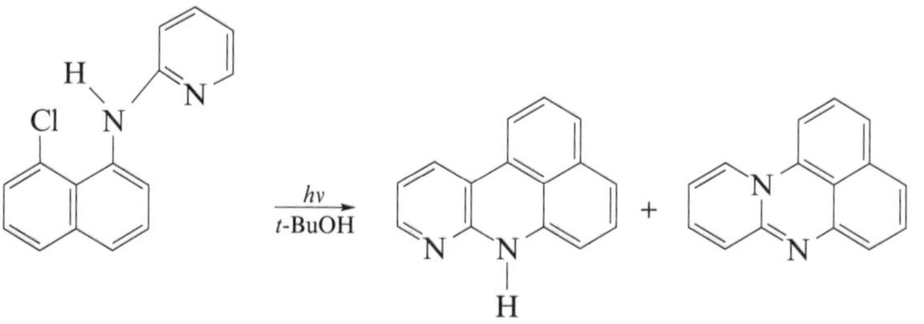

Several intramolecular arylations have been reported of molecules in which two aryl rings are connected by two saturated carbon atoms. In all these cases, one of these carbon atoms is part of a ring fused to one of the aryl rings. An example is the synthesis of (±)-oliveroline (180) (equation 137)[552,553].

The alkaloids (±)-ushinsinine[552], liriodenine[552,554] and (±)-oliveridine[553,555] were synthesized via similar procedures. Photochemical cyclization was also used as the key step in the synthesis of the aporphine alkaloids (±)-thaliporphine, (±)-*N*-methyllaurotetanine and (±)-isoboldine[556]. These compounds were prepared from *N*-ethoxycarbonyl-1-bromo-benzyltetrahydroisoquinolines. A so-called B-ring 'homoaporphine' was also photochemically synthesized from 1,2,4,5-tetrahydro-1-(*o*-iodobenzyl)-7,8-dimethoxy-3*H*-2-

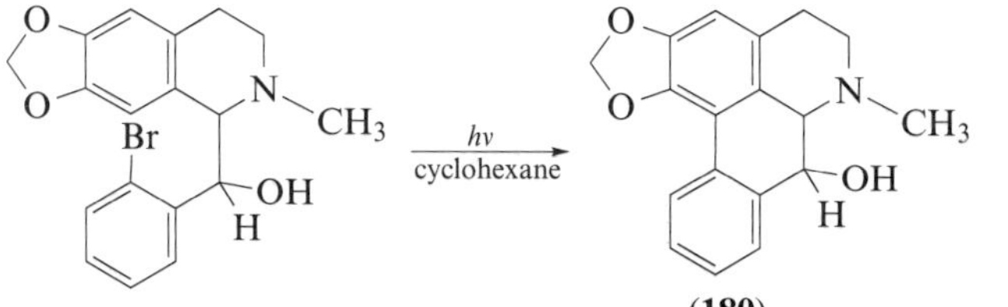

(137)

(180)

benzazepin-3-one (181) via an intramolecular aryl–aryl linkage (equation 138)[557]. The formation of lactam 182 was accompanied by the formation of a nearly equivalent amount of the unsaturated analogue 183, which is supposedly formed from 182 via an iodine-catalysed photodehydrogenation.

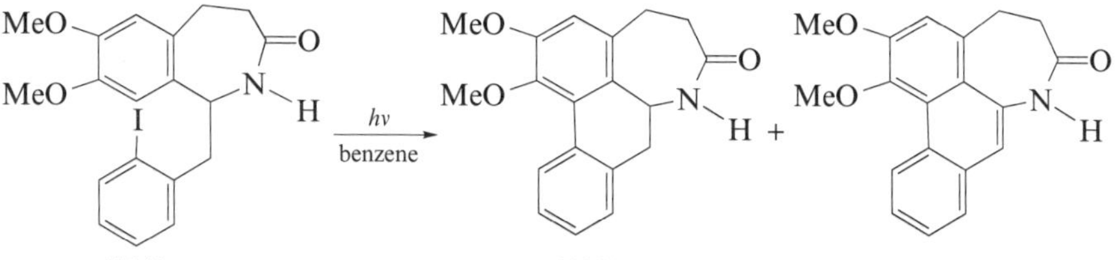

(138)

Photoreaction of 5-(2-iodophenyl)-1,3-diphenyl-Δ^2-pyrazoline (184) produces 1,3,5-triphenylpyrazoline (185) (24%) and 2-phenylpyrazolo[1,5-f]phenanthridine (186) (39%) (equation 139)[558,559]. The dehydrogenation to 186 takes place during the irradiation.

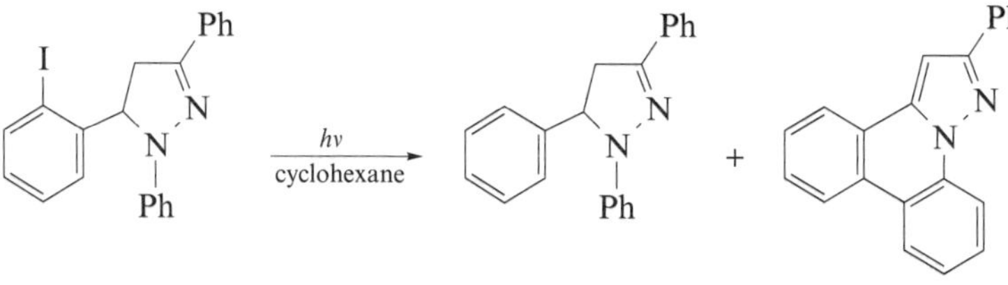

(139)

The corresponding chloro and bromo compounds as well as the 3-iodo and 4-iodophenyl compounds are photostable. A mechanism is proposed in which the primary excited S_1 state crosses into an $n\sigma^*$ state in the iodobenzene manifold. This is followed by homolysis of the C—I bond. The aryl radical may undergo ring closure or abstract a hydrogen atom from the solvent. The primary ring-closed product is photo-oxidized under the reaction conditions.

Irradiation of 5-(2-chlorophenyl)-1,3-diphenylpyrazole (187) also leads to ring closure and a mechanism has been considered[560] in which a π-complexed intermediate (188) and a cyclohexadienyl radical (189) play a role (equation 140).

A transient intermediate has been observed which decays with $k = 12.6 \, \text{s}^{-1}$ in a first-order process. In view of its long lifetime the transient is supposed to be the cyclohexadienyl radical. A similar reaction mechanism had earlier been proposed for the photochemical cyclization of 5-(2-chlorophenyl)-1,3-diphenyl-1,2,4[1H]-triazole[561]. Intramolecular aryl–aryl coupling has also been described for Δ^2-1,2,4-triazol-5(1H)-one derivatives containing an o-bromophenyl and a heteroaryl substituent in vicinal positions[562].

(187)   (188)   (189)

(140)

Other photocyclizations via which N-heterocyclic compounds are formed have been described for 2-chlorobenzanilides[563] and 3-(2-halogenoanilino)-1,3-diphenylprop-2-enones[564]. From 2-chlorobenzanilides with various substituents at positions 4′ the corresponding phenanthridones were obtained in 67–74% yield, whereas 2-bromo- and 2-iodobenzanilides undergo dehalogenative reduction. The reaction proceeds via the triplet state and a π-complex of the developing radicals is proposed as an intermediate or transition state. With carbon–halogen bonds weaker than the carbon–chlorine bond, direct bond fission from T_1 becomes possible and the ensuing radical does not have the correct conformation for cyclization[563]. The quantum yield of cyclization of the 3-(2-halogenoanilino)-1,3-diphenylprop-2-enones is strongly dependent on the halogen atom[564]. In hexane under air, quantum yields for the 2-iodo, 2-bromo and 2-chloro derivatives are 920×10^{-4}, 160×10^{-4} and 1.6×10^{-4}, respectively. The quantum yields decrease with increasing solvent viscosity which is in support of a homolytic pathway for cyclization. These photoreactions show a strong wavelength dependence. For the iodo and the bromo derivative, quantum yields of formation of the cyclized product increase by a factor of 100 on going from 370 nm to 254 nm, and this is taken as evidence for carbon–halogen bond fission from an upper excited state.

Photochemical cyclization of stilbene-like molecules containing an o-halogenoaryl group was used in the synthesis of the alkaloids cepharanone B[565], pontevedrine[566] and various other aporphines[567–571]. The reaction has also been applied in the synthesis of a phenanthrene derivative that was studied during attempts to synthesize steganone[572].

A study has been reported on the photocyclization of several ortho-halogenated stilbene derivatives (190) under both oxidative conditions (I_2, cyclohexane) and basic conditions (MeONa, MeOH)[573]. The major products were those anticipated from photodehydrogenation, i.e. 191, and photodehydrohalogenation (192), respectively (equation 141).

The pentacyclic marine alkaloid ascididemin (194) has been prepared by means of photocyclization of a quinone imine (193) (equation 142)[574,575].

The photocyclizations of halogenated N-benzyl-β-phenethylamines[576] are examples of reactions in which two aryl rings are connected by a chain of four atoms, one of which is a nitrogen atom. In the cases reported, one phenyl ring has a bromine atom and the other an iodine atom at the ortho position. As expected, products were formed via initial rupture of the carbon–iodine bond and these products still contained the bromine atom. In addition, however, some unexpected cyclization products were encountered, containing iodine instead of bromine. The formation of these products was ascribed to replacement of bromine by iodine in the intermediate cyclohexadienyl radicals.

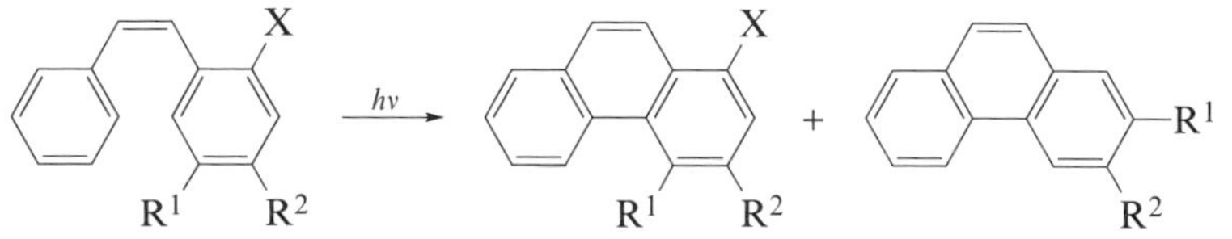

(141)

(190)   **(191)**   **(192)**

$X = Cl$, $R^1 = CH_3$, $R^2 = H$ oxidative   basic
$X = Br$, $R^1 = CH_3$, $R^2 = H$
$X = Br$, $R^1 = OCH_3$, $R^2 = H$
$X = Br$, R^1-$R^2 = OCH_2O$

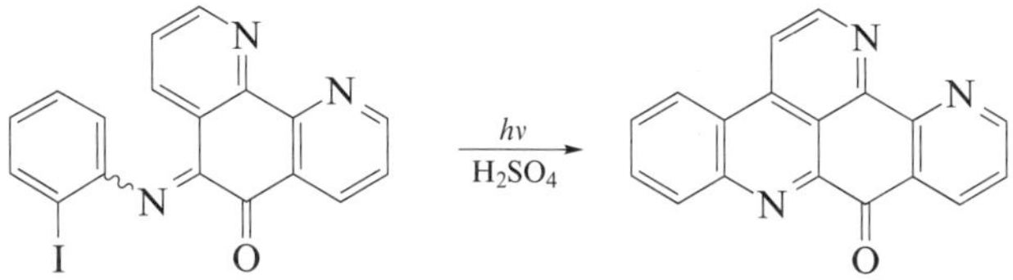

(142)

(193)   **(194)**

Crassifolazonine was prepared via intramolecular arylation in alkaline methanolic solution of a compound with a $CH_2C(O)NHCH_2CH_2$ chain between the aryl moieties[577]. The benzene ring connected to the ethylene group contained an *ortho* bromine atom. In addition to the desired product with a nine-membered ring, an indoline was found, formed by nucleophilic attack of the amide anion on the intermediate aryl radical[577]. Intramolecular arylation of a compound with a five-atom bridge containing an amide function was also used in the synthesis of 5,6-dihydro-4H,8H-pyrido[3,2,1-*de*]phenanthridin-8-ones[578]. A similar photocyclization was used as a key step in the synthesis of the dibenzazecine alkaloid dysazecine[579]. *Meta*-bridged aromatic lactams containing an eleven-membered ring were prepared by intramolecular aryl–aryl coupling of compounds having a $(CH_2)_2NHC(O)(CH_2)_2$ bridge between the aryl rings[580,581].

Intramolecular photochemical reactions in which a bond is created between an aryl group and the sulphur atom of a thioamide or thiourea group may formally be regarded as intramolecular arylations of the $C{=}S$ bond[582–585]. In the cases described, it is not always clear whether these processes occur via initial homolytic aryl–halogen bond cleavage or via nucleophilic attack of the $C{=}S$ group on the aryl halide. An example is the synthesis of benzothiazoles (**196**) from N-(2-chlorophenyl)-N'-phenylthiourea derivatives (**195**) (e.g. $R^1 = H$, $R^2 = Ph$ and $R^1 = R^2 = Ph$) (equation 143)[584]. Finally, examples have been described of intramolecular photoarylations of N-(haloarylethyl)-β- and -α-enaminoketones[586,587]. The reactions start with aryl radical generation and proceed through cyclization via bonding at the enamine terminal carbon or the nitrogen (equation 144).

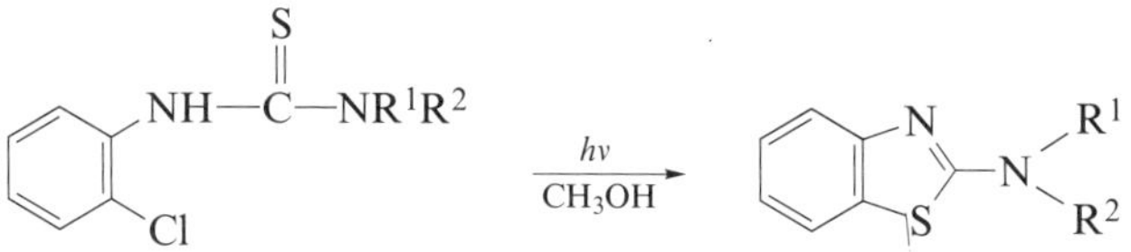

(143)

(195)   **(196)**

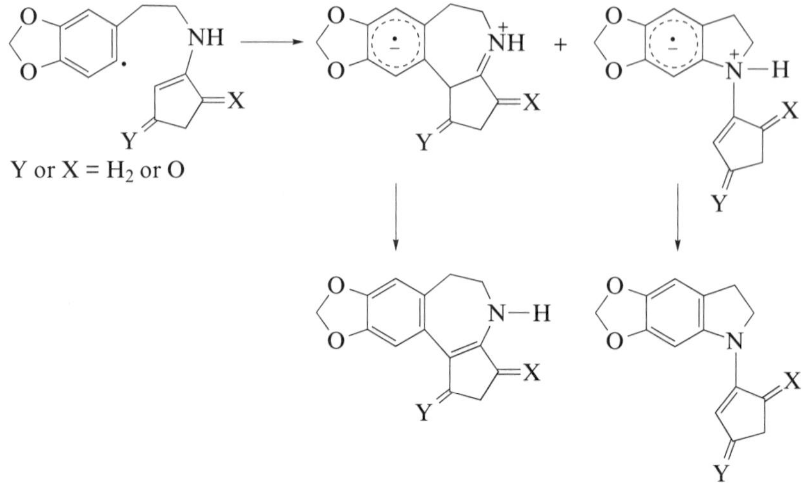

$$(144)$$

Y or X = H$_2$ or O

b. *Heterocyclic aromatic halides.* Photocyclizations of *N*-(2-halobenzyl)pyridinium salts have already been mentioned in the section on homocyclic aromatic halides. The reaction rates of *N*-benzyl-2-halopyridinium salts are significantly higher[546–548]. *N*-benzyl-2-chloropyridinium bromide and *N*-benzyl-2-bromopyridinium bromide react twice as fast as *N*-(2-chlorobenzyl)pyridinium bromide, yielding the same product, pyrido[2,1-*a*]isoindolium bromide. When the phenyl and the pyridinium ring both have an *ortho* halogen atom, the rate is further increased and the photocyclized product is formed by cleavage of the halogen atom from the pyridinium ring. This is ascribed to better π-complex formation of the chlorine atom of the pyridinium ring with the phenyl group than the other way around. The presence of a chlorine atom on the phenyl ring further assists the formation of the complex.

A five-membered ring in a heterocyclic system is also formed by photocyclization of 2-(1-naphthylamino)-3,5-dichloropyridine (**197**)[550]. Cyclization can occur at the *ortho* or *peri* position of the naphthalene ring, but irradiation in aqueous *tert*-butyl alcohol provided only 8-chlorobenz[*i*]-α-carboline (**198**), formed by ring formation at the *ortho* position (equation 145).

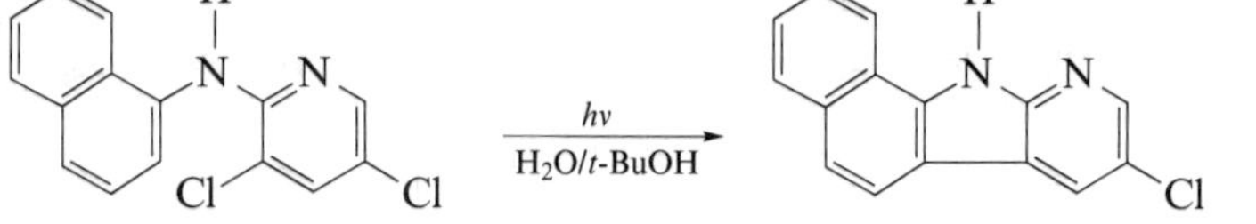

$$(145)$$

(**197**)   (**198**)

3,5-Dichloro-2,2'-dipyridylamine was irradiated in aqueous *tert*-butyl alcohol, yielding principally the cyclized product 8-chlorodipyrido[1,2-*a*;2',3'-*d*]imidazole (49%)[588]. In aqueous ethanol, however, formation of the reductive dehalogenation product predominates.

1,3-Disubstituted and 1,3,5-trisubstituted 1*H*-pyrazolo[3,4-*d*]pyridazin-4(5*H*)-ones (**200**) can be synthesized by irradiation of 5-(1-alkyl-2-alkylidenehydrazino)-4-chloro-3(2*H*)-pyridazinones (**199**) in benzene or acetone (equation 146)[589]. Eighteen derivatives of

200 were synthesized, e.g. with R = R^1 = Me, R^2 = Ph; R= R^2 = Ph, R^1 = Me and R = R^1 = PhCH$_2$, R^2 = Me.

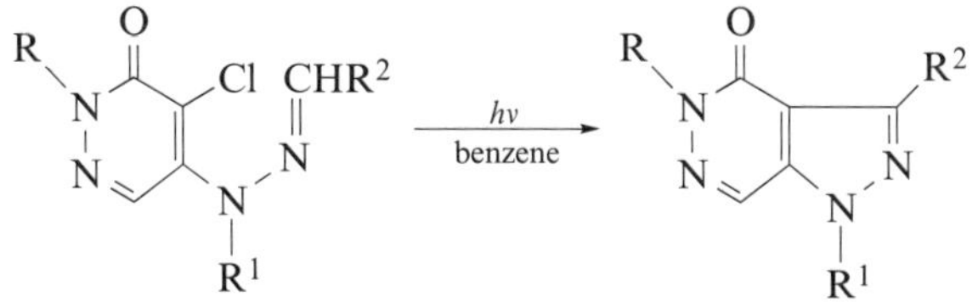

$$\text{(146)}$$

(199)   **(200)**

Intramolecular arylation of a C=S bond in a thioamide, also leading to formation of a five-membered ring, was performed with N-(3,5-dichloro-2-pyridyl)thiobenzamide (**201**), R = Ph and some analogues[590]. Excellent yields (*ca* 80%) of thiazolo[4,5-*b*]pyridines (**202**) were obtained (equation 147).

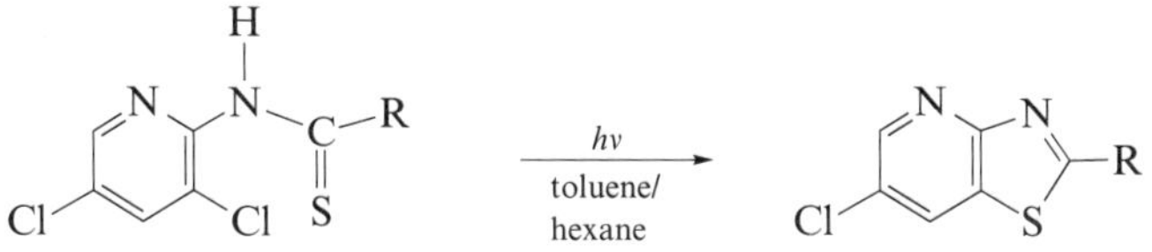

$$\text{(147)}$$

(201)   R = Ph, 2-thienyl, 2-furyl   **(202)**

Irradiation of 3-chloro-2-phenethyl-1-isoquinolinone (**203**) in methanol afforded 8-oxo-protoberberine (**205**) in 56% yield[591]. A solvent-caged biradical (**204**) formed by homolytic fission of the C—Cl bond is proposed as intermediate (equation 148). In support of this, irradiation of 3-chloro-2-methyl-1-isoquinolinone in benzene results in the formation of 2-methyl-3-phenyl-1-isoquinolinone in 79% yield, though a much longer reaction time was needed than for the intramolecular reaction. When **203** was irradiated in

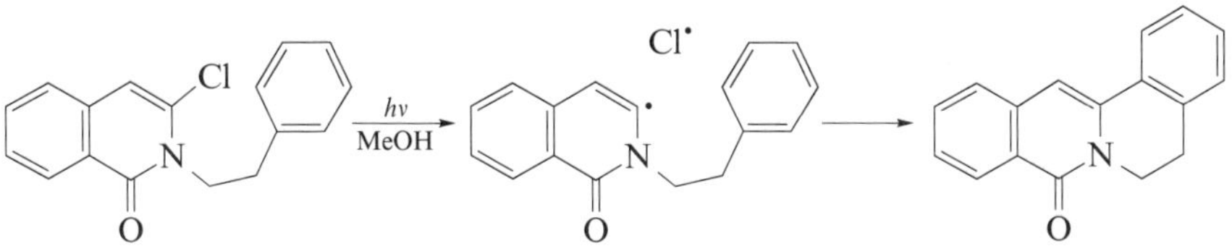

(203)   **(204)**   **(205)**

$$\text{(148)}$$

benzene, only **205** was formed, demonstrating the greater efficiency of the intramolecular arylation over the intermolecular one. 2-Benzyl-3-chloro-1-isoquinolinone did not afford the cyclized product with a five-membered ring, but instead yielded the 3-phenyl derivative upon irradiation in benzene.

In the synthesis of benzo[*c*]acridines, a photochemical ring closure of 2-chloro-3-(1-carboxyl-2-phenylethenyl)-4-quinolinone (**206**) was used (equation 149)[592]. Irradiation in methanol produced a lactone hydrate (**207**), believed to be formed by primary ring closure followed by elimination of HCl and concurrent formation of the lactone ring. Heating of the product in methanol containing a catalytic amount of concentrated sulphuric acid produces 6-methoxycarbonyl-7-oxo-7,12-dihydrobenzo[*c*]acridine, a compound which

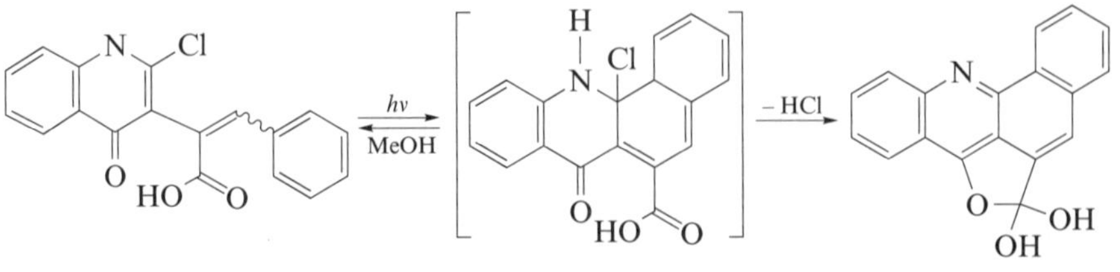

(206)   (207)

(149)

was also obtained directly upon irradiation of 2-chloro-4-hydroxy-3-(1-methoxycarbonyl-2-phenylethenyl)quinoline.

Many novel polycyclic heteroaromatic ring systems have been prepared via a photo-chemical ring closure of the type depicted in equation 150[593-604].

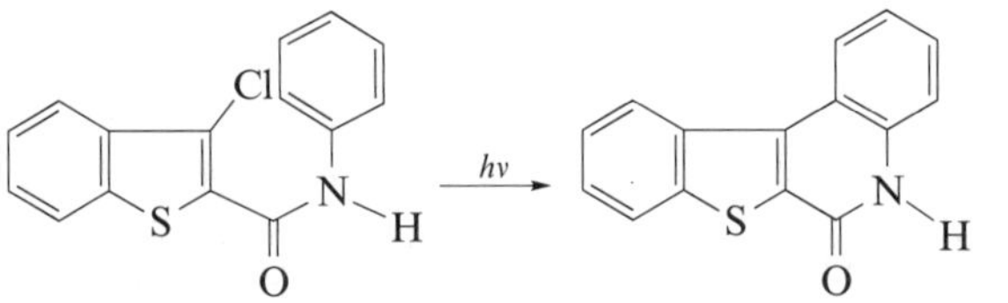

(150)

Irradiation of 3-chloro-N-phenylbenzo[b]thiophene-2-carboxamide and triethylamine in benzene/methanol (4/1) yielded [1]benzothieno[2,3-c]quinolin-6(5H)-one (92%)[597]. Other derivatives, analogues and related compounds were prepared by the use of N-tolyl[593,594], N-anisyl[595,596,602], N-(fluorophenyl)[599,602], N-pyridyl[604], N-naphthyl[598,601], N-quinolyl[600] and N-phenanthryl[603] groups at the nitrogen side of the amide, and chlorine substituted methoxybenzo[b]thiophene[596,602], fluorobenzo[b]thiophene[599], naphtho[2,1-b]thiophene[593,595,604], naphtho[1,2-b]thiophene[594] and thieno[3,2-b]thiophene[601] at the car-bonyl side.

A similar type of reaction, with bond formation between a chlorine substituted thiophene moiety and a phenyl ring connected to the thiophene via an amide bond, was used in the synthesis of 12-carbomethoxy-2-chloronaphtho[2″,1‴:2′,3′-b]thieno[4′,5′:2,3]-thieno[5,4-c]quinolin-6(5H)-one[605].

Irradiation of 1-(2′,5′-dibromo-3′-thenyl)-2-acetyl-6,7-dimethoxy-1,2,3,4-tetrahy-droisoquinoline (208) in methanol/water (9/1) and sodium hydroxide yielded 6-acetyl-1,2-dimethoxy-4H-5,6,6a,7-tetrahydrobenzo[de]thieno[2,3-g]quinoline (209) in 52% yield[606]. Both bromine atoms are eliminated during the photolysis (equation 151).

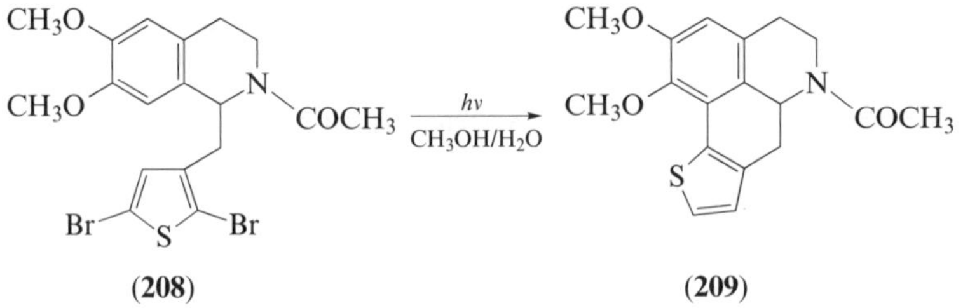

(208)   (209)   (151)

Photochemical arylation of the C=S bond of thioureas and monothiocarbamates by benzothiophenes has been used in the synthesis of benzothienothiazines[607–609]. Irradiation of N-(3-chloro-2-benzo[b]thienocarbonyl)-N',N'-disubstituted thioureas in acetone gives 2-substituted 4H-benzo[4,5]thieno[2,3-e]-1,3-thiazin-4-ones in yields varying from 55 to 80%[607]. The analogous N'-monosubstituted thioureas give the corresponding benzothienothiazine derivatives in lower yields and this is ascribed to intramolecular hydrogen bonding between the N—H and the C=O group in a conformation unfavourable for the cyclization[609]. O-Alkyl N-(3-chloro-2-benzo[b]thienocarbonyl)monothiocarbamates (**210**) are cyclized upon irradiation with light above 300 nm to give high yields (80–90%) of 2-alkoxy-4H-benzo[b]thieno[2,3-e]-1,3-thiazin-4-one derivatives (**211**) (equation 152)[608].

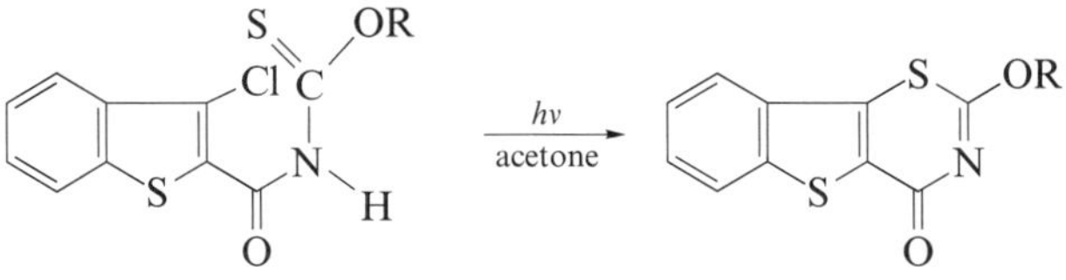

$$\text{(152)}$$

R = Me, Et, n-Pr, i-Pr
(**210**)   (**211**)

From a study of the absorption and emission spectra of the methoxy derivative, it was concluded that the S_1 state has an energy of 346 kJ mol^{-1} and the T_1 state 230 kJ mol^{-1}. The high value of the singlet–triplet splitting, the shift of the absorption maximum to shorter wavelengths on going to non-polar solvents and the intense absorption (log $\varepsilon \sim 3$) indicate the π,π^* character of the S_1 state. The energy of the S_1 state is lower than the C—Cl bond dissociation energy, which makes a mechanism proceeding via free radicals less likely. In agreement with this, no trace of the reduction product could be found, not even upon irradiation in cyclohexane or 2-propanol. The reduction product was synthesized via another route and irradiated. It proved to be photostable and this was taken as evidence that the reaction of the chlorine derivative does not involve an electrocyclic process. It was therefore presumed that the photocyclization proceeds via an intermediate in which the chlorine atom is partially bonded to sulphur and the sulphur atom to C-3. In the next step the chlorine atom is split off to give a radical, which is extensively stabilized by the free electron pairs of the adjacent heteroatoms. In the final step, a hydrogen atom is split off with formation of HCl and the photocyclization product.

A photochemical cyclization involving a phosphorus atom occurs in the reaction of diethyl 8-bromo-2',3',O-isopropylideneadenosine 5'-phosphite (**212**) upon irradiation in acetonitrile[610]. The primary step is homolysis of the C—Br bond and this is followed by intramolecular attack of the adeninyl 8-radical on the phosphite group with the formation

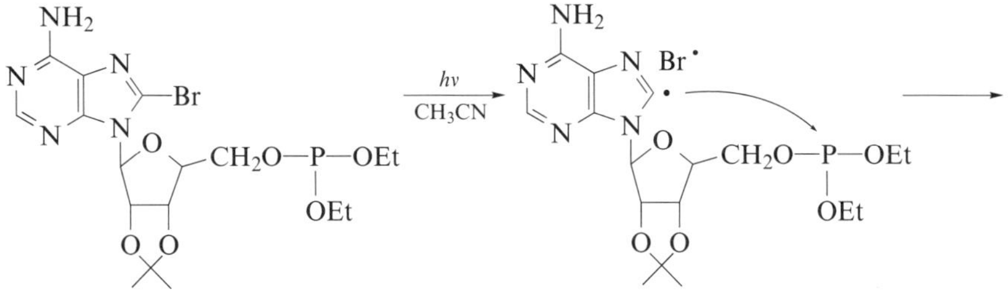

(**212**)   (*continued*)

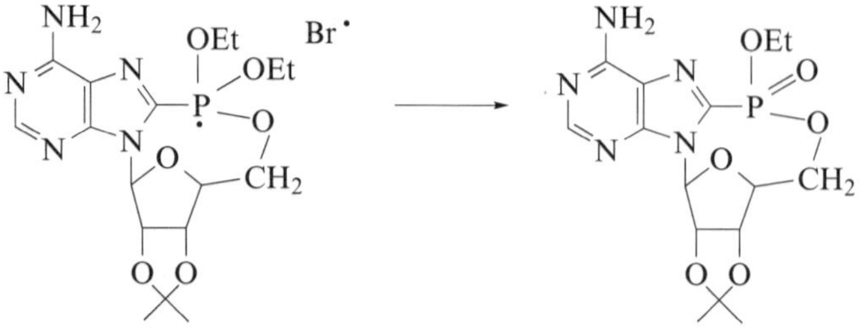

(153)

(213)   **(214)**

of an 8,5′-diethoxyphosphoranyl radical (**213**). The bromine atom abstracts an ethyl radical and the product (**214**) is formed as a mixture of diastereomers (equation 153).

C. Aromatic Substitution Reactions

In many photochemical substitution reactions of aryl halides the halide atom is replaced by a nucleophilic group. Nucleophilic aromatic photosubstitution reactions have recently been reviewed[17] and in all the cases described charged species play a role, either as reactants or as intermediates. Halide atoms may also be replaced in radical reactions, proceeding via electrically neutral species, initiated by primary homolytic photodissociation of the carbon–halogen bond or by attack of a radical on the halogen-bearing aromatic carbon. Many such reactions are treated in Section IV.B, but when the halogen atom is replaced by a relatively small group, such as OH, CN or OCH_3, they will be found in this section. Formally, replacement of halogen by hydrogen may also be regarded as aromatic substitution, but these processes are dissussed in Section IV.A.

Nucleophilic aromatic photosubstitution reactions have been divided into five mechanistic categories[17] and each of these mechanistic types has its representatives in the class of aryl halides. Which reaction pathway is followed in any particular case depends on a number of factors such as the nature of the leaving group, the presence of electron-donating or electron-withdrawing substituents on the aromatic ring, the solvent, the multiplicity and the lifetime of the reactive excited state and the presence or absence of electron donors or acceptors in the reaction medium. This renders it rather difficult to make predictions about the mechanistic course that will be followed under a given set of circumstances.

The S_N2 Ar* mechanism is characterized by direct interaction between the aromatic halide in its excited state (in many cases a triplet state) and the nucleophile, leading to a σ-complex. In its most simple form, the mechanism may be summarized as in equations 154 and 155. The photosubstitution of bromide by chloride in 3-bromonitrobenzene in

$$ArX^* + Nu^- \longrightarrow ArXNu^- \tag{154}$$

$$ArXNu^- \longrightarrow ArNu + X^- \tag{155}$$

acetonitrile-water or acetic acid-water containing LiCl is an example of this reaction type[611]. The photosubstitution is catalysed by hydronium ion; the limiting quantum yield of 0.021 at infinite [H^+] and 3.0 M [Cl^-] is, however, only twofold higher than the quantum yield of the uncatalysed photosubstitution at 3.0 M [Cl^-]. The reaction is proposed to proceed via the $^3\pi,\pi^*$ state, which is slightly higher in energy than the $^3n,\pi^*$ state. The mechanism is shown in equation 156. The n,π* and π,π* states are shown as modified valence bond representations.

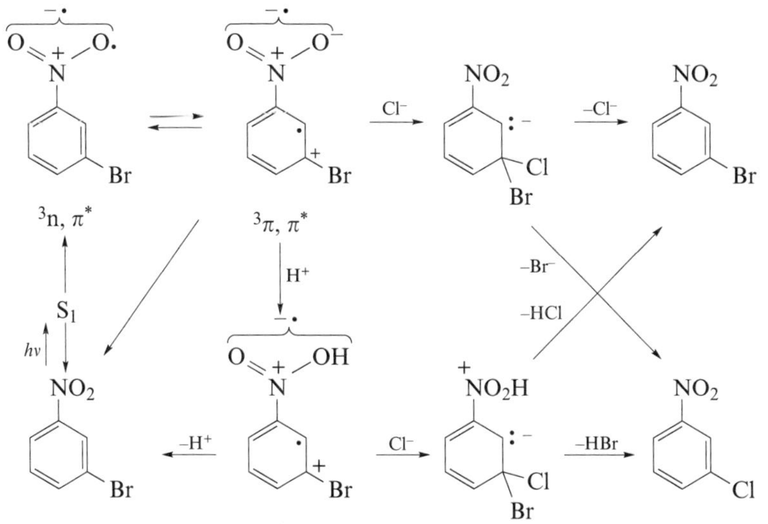

$$(156)$$

2-Bromonitrobenzene is unreactive towards photosubstitution by chloride ion, but irradiation of 4-bromonitrobenzene at 313 nm in aqueous solutions containing hydronium ion and relatively high concentrations of chloride ion cleanly but inefficiently forms 4-chloronitrobenzene[612]. This reaction is exclusively acid-catalysed and it is second order in chloride ion concentration. In this case it is considered likely that the reaction occurs from a triplet n,π^* state which forms an exciplex with chloride ion. The exciplex becomes protonated at the nitro group and undergoes intermolecular attack by Cl⁻ at position 4, followed by loss of HBr.

The photosubstitutions of 2-fluoro-4-nitroanisole with several amines have been studied in a search for new biochemical photoprobes[613–615]. The preparative yields of photosubstitution of fluoride by amines are relatively high, but the reactions display low selectivity. With methylamine and n-butylamine, for example, products are also found in which methoxy (*para* to the nitro group) instead of fluorine has been substituted. Furthermore, the reactions which are performed in H_2O/CH_3CN or $H_2O/MeOH$, always yield considerable amounts of products in which fluorine is displaced by OH or OCH_3 from the solvent. The photosubstitution of 2-fluoro-4-nitroanisole with n-hexylamine has been studied in order to elucidate its mechanism[616]. Detailed kinetic analysis has provided evidence for a dual mechanistic pathway. The substitution of fluoride by the amine occurs via the $S_N2\,^3Ar^*$ route, in which the π,π^* triplet state and the amine form a σ-complex which subsequently loses a proton from the nitrogen atom and a fluoride ion. In another route, the formation of a σ-complex is preceded by electron transfer from the amine to the n,π^* triplet state. The radical ion pair then collapses to a zwitterionic complex, but in this case the new bond is created at the carbon atom bearing the methoxy group and loss of a proton and a methoxide ion leads to a 2-fluoro-4-nitroaniline derivative. This mechanism is designated as $S_N(ET)$ Ar* and it is represented, for a negatively charged nucleophile, by equations 157–159. In the case of a neutral nucleophile, the complex formed by equation 158 may have either biradicaloid of zwitterionic character and it will lose the elements of HX in step 159.

$$ArX^* + Nu^- \longrightarrow ArX^{-\bullet} + Nu^\bullet \qquad (157)$$

$$ArX^{-\bullet} + Nu^\bullet \longrightarrow ArXNu^- \qquad (158)$$

$$ArXNu^- \longrightarrow ArNu + X^- \qquad (159)$$

Both routes, the S_N2 Ar* and the S_N(ET) Ar* pathway, thus proceed via intermediate σ-complexes of the nucleophile and the aromatic substrate. The fact that these complexes are formed in a different manner, in one step via S_N2 Ar* and in two steps via S_N(ET) Ar*, has important consequences. The nucleophilic photosubstitutions which involve one-step formation of the σ-complex are HOMO controlled. The dominant orbital interaction is between the HOMO of the substrates and the HOMO of the nucleophile[617]. This is in full agreement with the familiar *meta*-directing effect of the nitro group in nucleophilic aromatic photosubstitution[618]. In the S_N(ET) Ar* mechanism, the position of attack on the aromatic molecule may be different, because the dominant orbital interaction is now between the LUMO of the aromatic substrate (singly occupied in the radical anion) and the singly occupied HOMO of the nucleophile[617]. In many cases this interaction leads to replacement of substituents *para* to the nitro group.

The photoreactions of halopyridines with indole and with the 1-indolyl anion have been investigated[619–622]. The reactions are supposed to proceed via electron transfer from the indole or indolyl anion to the halopyridine. With 2-chloro-, 2-bromo- and 2-iodopyridine the reaction is not very selective and the pyridyl substituent is found at almost all the indole ring positions[620]. Photoreaction of 2-fluoropyridine gave 1-(2-pyridyl)indole regioselectively[619]. With the indolyl anion as nucleophile, the yields are significantly higher than with indole. Photoreaction of 3-fluoropyridine with the 1-indolyl anion in dimethylformamide afforded 1-(3-pyridyl)indole regioselectively, but in low yield (1.5%). Similar photolysis of 4-fluoropyridine and 1-indolyl anion yielded 1-(4-pyridyl)indole in 7.5% yield as the sole product, but in the dark at room temperature 45% was obtained after the same reaction time, implying that ultraviolet irradiation inhibits the thermal reaction. With neutral indole and 3- or 4-fluoropyridine, mixtures of 1- and 3-substituted indoles were formed[622].

A photoinduced substitution reaction initiated by electron transfer has also been proposed to account for the formation of 2-chloro-3-(2-thienyl)-1,4-naphthoquinones (**217**) from 2,3-dichloro-1,4-naphthoquinone (**215**) and thiophenes (**216**)[623] (equation 160).

$$\text{(215)} \qquad \text{(216)} \qquad \text{(217)}$$

$$R^1 = I, Br, Cl, H; R^2 = H$$
$$R^1 = H; R^2 = I, Br, Cl$$

The reaction starts with excitation of the quinone, followed by intersystem crossing and electron transfer from the thiophene to the triplet excited quinone. The ion radical pair collapses to a biradical which loses a chlorine and a hydrogen atom. Yields are high (65–78%) when R^1 = halogen and R^2 = H, fair (57%) when $R^1 = R^2$ = H and poor (2–17%) when R^1 = H and R^2 = halogen. The regioselectivity has been explained on the basis of calculated electron densities in the cation radicals of thiophenes.

Photochemical reactions of 2-bromo-3-X-1,4-naphthoquinones [X = H, Br, CH_3, OCH_3 or $N(CH_3)_2$] with *N*-methylpyrrole give regioselectively 2-(2-pyrrolyl)-1,4-naphthoquinone derivatives[624]. A ground-state charge-transfer complex seems to be involved and photoexcitation induces electron transfer from the pyrrole to the 1,4-naphthoquinone derivative. Chlorine atoms of chloranil can be photosubstituted by the aliphatic amines *n*-propylamine, *n*-butylamine, *n*-hexylamine and *n*-heptylamine[625,626]. A kinetic study has been performed[626] with chloranil and *n*-butylamine in cyclohexane and it was found that excited chloranil interacts with the amine to form a complex, probably of charge-transfer nature. With a second molecule of the amine, a second complex is formed and this slowly loses two molecules of HCl, thereby yielding 2,5-bis(n-butylamino)-3,6-dichlorocyclo-hexa-2,5-diene-1,4-dione.

Electron transfer from the nucleophile (e.g. an amine) to an excited arene may of course be preceded by the formation of an exciplex between the two species. An excited-state complex was mentioned as a possible intermediate in the photoreactions of a variety of substituted benzenes with aliphatic amines[627]. Among the aryl halides studied are chloro-benzene, *meta*- and *para*-fluorotoluene and *meta*- and *para*-fluoroanisole in their reactions with diethylamine and *tert*-butylamine. Chlorobenzene and the fluorotoluenes undergo halide substitution as well as 1,2 or 1,4 addition of the amine to the arene ring. No fluorine displacement was observed with the fluoroanisoles. Irradiation of a mixture of fluorobenzene and diethylamine produced N,N-diethylaniline and a mixture of two adducts[628]. From 1-fluoro-3,5-dimethylbenzene and diethylamine a complex assortment of products was obtained of which the substitution product was only a minor component. The reaction mechanism was investigated by performing fluorescence quenching studies, xenon quenching studies and irradiation in Et_2ND. It was concluded that the addition and the substitution reactions are related processes. An exciplex (**218**) is formed by bimolecular interaction between the amine and the S_1 state of the aromatic substrate ArF. Proton transfer leading to a radical pair (**219**) takes place within the exciplex, followed by formation of a 1,2-adduct (**220**). Elimination of HF gives the substitution product (**221**) and this occurs with a normal isotope effect, explaining the 83% incorporation of deuterium into the substitution product (equation 161).

In 1-fluoropyrene, the fluorine can be replaced by hydroxide ion (73% yield) or methox-ide ion (25% yield)[629] but the mechanisms of these reactions have not been further investigated.

2,5-, 2,6- and 2,3-Dichloro-1,4-benzoquinones undergo so-called self-substitution reactions when irradiated in the presence of certain hindered N,N-dimethylalkylamines[630,631]. On the basis of photo-CIDNP and electrochemical experiments a mechanism is proposed involving photoinduced electron transfer from the amine to the dichlorobenzoquinone as the initial step. The radical anion of the dichlorobenzoquinone attacks, with one of its oxygen atoms, a ground-state molecule and replaces a chlorine atom. In subsequent reactions the dimer yields trimers and pentamers.

In the third major type of aromatic photosubstitution, which is designated as $S_{R+N}1$ Ar*, a radical cation of the arene is the reactive species, interacting with the nucleophile. The radical cation may be solvated by the medium or it may be part of an excited donor–acceptor complex. The reactive excited state may be singlet or triplet. The electron expelled by excited ArX may be taken up by the solvent, by ground-state ArX, by oxygen or by an added electron acceptor. The radical formed by reaction of the radical cation with a nucleophilic anion (e.g. cyanide) may lose a halide ion and is thereby transformed into a radical cation of the substitution product. If this picks up an electron from the starting material, a chain reaction may result[618]. The mechanism is summarized in equations 162–166.

$$ArX^* + acceptor \longrightarrow ArX^{+\cdot} + acceptor^- \tag{162}$$

$$ArX^{+\cdot} + Nu^- \longrightarrow ArXNu^{\cdot} \tag{163}$$

$$ArXNu^{\cdot} \longrightarrow ArNu^{+\cdot} + X^- \tag{164}$$

$$ArNu^{+\cdot} + ArX \longrightarrow ArNu + ArX^{+\cdot} \tag{165}$$

$$or: ArXNu^{\cdot} \longrightarrow ArNu + X^{\cdot} \tag{166}$$

The photocyanation and the photohydrolysis of 4-chloroanisole and 4-fluoroanisole have been studied with time-resolved spectroscopy and by measurement of the photoconductivity in anhydrous and aqueous acetonitrile and *tert*-butyl alcohol solutions[632]. The mechanism is depicted in equation 167. The transient species which have been

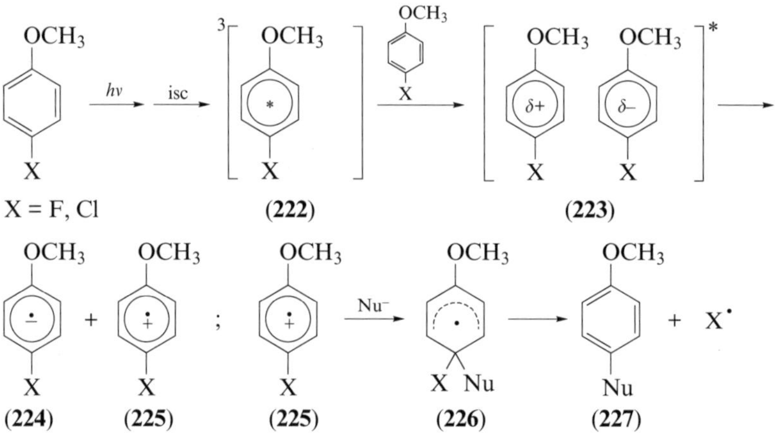

observed spectroscopically are $^3ArX^*$ (222), $[ArX^{\delta+} \cdots ArX^{\delta-}]^*$ (223), $ArX^{+\cdot}$ (225), and $ArXNu^{\cdot}$ (226) (Ar = p-CH$_3$OC$_6$H$_4$, X = F or Cl, Nu = OH or CN). The rate constants for the formation of the free ions in water–acetonitrile are $k_i = 4.1 \times 10^9$ M^{-1}s^{-1} and $k_i = 1.0 \times 10^{10}$ M^{-1}s^{-1} for 4-chloro- and 4-fluoroanisole, respectively. The radical anions (224) cannot be observed because of their rapid reaction with water.

At low water concentration, the lifetimes of the radical cations are estimated at 7 μs for 4-chloroanisole and 9 μs for 4-fluoroanisole. The rate constants k_N for reaction of the radical cation of 4-chloroanisole in 80% H$_2$O/CH$_3$CN are k_N(H$_2$O) = 5.8 $\times$ 10^3 M^{-1}s^{-1}, k_N(OH$^-$) = 5.2 $\times$ 10^7 M^{-1}s^{-1} and k_N(CN$^-$) = 0.21 $\times$ 10^9 M^{-1}s^{-1}. The radical cations ArCN$^{+\cdot}$ and ArOH$^{+\cdot}$ have not been observed. It is concluded that a halogen atom rather than a halide ion is released from the neutral radicals ArXCN$^{\cdot}$ and ArXOH$^{\cdot}$ to form the product (227) directly.

The photonucleophilic substitution of the monochloroanisoles in alcoholic solvents had been studied earlier[378] and it was then concluded that 4-chloroanisole reacts via a radical anion formed by electron transfer from the solvent to an excited molecule, whereas 3-chloroanisole undergoes substitution via an aryl cation. In another study[379], it was found, on the basis of quenching and sensitization experiments and on the basis of the ratios of dehalogenation and substitution, that 3-chloroanisole in methanol upon direct irradiation reacts mainly via its singlet excited state, giving mostly dehalogenation, whereas sensitization by acetone leads predominantly to substitution.

Aromatic nucleophilic photosubstitution reactions of 4-fluoroanisole with cyanide ion and water, leading to 4-cyanoanisole and 4-methoxyphenol, are suppressed by 80–100% if the fluoroanisole is complexed by α- or β-cyclodextrin (CD) in aqueous solutions[633]. In aqueous γ-cyclodextrin, however, the efficiency of photohydroxylation was increased slightly and the efficiency of photocyanation was unaffected. The larger γ-CD torus might allow water molecules and cyanide ions to enter and exit the torus. 2-Fluoroanisole behaves differently[634]. Complexation by α-CD strongly inhibits both photohydroxylation and photocyanation, but β-CD complexation inhibits photohydroxylation more than photocyanation. The differing propensities of the fluoroanisole isomers may be related to their hydrogen bonding to water, in which hydrogen bonding to the methoxy group would place the nucleophile near the fluorinated carbon of 2-fluoroanisole but far from that of 4-fluoroanisole. This was confirmed by a study of photohydroxylation and photocyanation of 2- and 4-fluoroanisoles in solvent mixtures of water and *tert*-butyl alcohol[635]. The ratio $k(CN)/k(H_2O)$ depends in a strange manner on $\chi(H_2O)$, the mole fraction of water. For 2-fluoroanisole the ratio increases from 90 ± 15 at $\chi(H_2O) = 1.0$ to 1020 ± 100 at $\chi(H_2O) = 0.92$ and then decreases; for 4-fluoroanisole the ratio decreases from 7500 ± 700 to 3700 ± 500 over the same $\chi(H_2O)$ range and then increases. Log $[k(CN)/k(H_2O)]$ in aqueous KCN is 1.9 for 2-fluoroanisole and 3.9 for 4-fluoroanisole. The value reported earlier for 4-fluoroanisole is 3.8[618] and Ritchie[636] finds N_+ values near 3.9 for cyanation of a number of cations in water. 2-Fluoroanisole appears to behave anomalously and this is ascribed to hydrogen bonding by water to the methoxy group which allows photohydroxylation to occur preferentially.

A radical cation mechanism is also supposed to be operative in the photosubstitution of chloride by cyanide in *N*-methyl-3-chlorocarbazole[637].

An investigation of the spectral and kinetic characteristics of radical cations of *para*-halogeno-*N*,*N*-dimethylanilines and *para*-halogenodiphenylamines in water/*tert*-butyl alcohol has been performed in view of the fact that such species are probable intermediates in photonucleophilic substitution of halogen[638]. Bathochromic shifts in the absorption maxima of the radical cations were observed in the order $H \cong F < Cl < Br$. The disappearance of the radical cations obeys second-order kinetics and the rate constants are close to the diffusion-controlled values. The radical cations arise from the singlet excited states of the halogenoarylamines and homolysis of the carbon–halogen bond competes with their formation.

An excited aryl halide may also transfer charge to oxygen and thereby become activated towards attack by nucleophiles. The photoamination of 1-amino-2,4-dibromoanthraquinone by *n*-butylamine, in which the 4-Br atom is replaced, is supposed to proceed via direct interaction between the triplet state and the amine when the reaction is performed under nitrogen. In air atmosphere, however, it proceeds via both the T_1 and S_1 states. From the singlet excited state and oxygen, an exciplex or collisional complex $[AQ^{\delta+} \cdots O_2^{\delta-}]$ is formed which undergoes the amination[639].

In the $S_{RN}1$ mechanism of aromatic substitution the initiating step is the formation of a radical anion. In order to distinguish the process from the route described above ($S_{R+N}1$) in which a radical cation plays a crucial role, the symbol $S_{R-N}1$ has been used[17]. Creation of the radical anion can occur by several procedures. The reaction can be electrochemically initiated, a solvated electron in a solution of alkali metal in liquid ammonia may be involved or a radical anion may be used as the source of electrons. The most common source of electrons is, however, the nucleophile itself involved in the substitution reaction. In many cases the electron transfer from nucleophile to substrate is light-catalysed and the process is then sometimes referred to as $S_{R-N}1$ Ar*. Although the nucleofugic group in $S_{R-N}1$

Ar* reactions is not limited to the halogens, halogen atoms are very frequently involved and the reaction is of special importance for aryl halides. A review on the $S_{RN}1$ reaction of organic halides[7] and an ACS Monograph on aromatic substitution by the $S_{RN}1$ mechanism[15] were both published in 1983. The mechanism of the $S_{R·N}1$ Ar* route can be represented by equations 168–171. An essential step in this mechanism is the dissociation of the radical anion into an aryl radical and a halide anion. Reaction between the aryl radical and the nucleophile creates the radical anion of the product. The chain reaction is propagated by transfer of an electron from this radical anion to the starting material. Reduction products that are sometimes observed may arise when the aryl radical abstracts a hydrogen atom from the medium (equation 172).

$$ArX + Nu^- \longrightarrow ArX^{-\bullet} + \text{residue} \tag{168}$$

$$ArX^{-\bullet} \longrightarrow Ar^\bullet + X^- \tag{169}$$

$$Ar^\bullet + Nu^- \longrightarrow ArNu^{-\bullet} \tag{170}$$

$$ArNu^{-\bullet} + ArX \longrightarrow ArNu + ArX^{-\bullet} \tag{171}$$

$$Ar^\bullet + RH \longrightarrow ArH + R^\bullet \tag{172}$$

This mechanism has recently been reconsidered[640,641]. It was argued that existing data as well as new experimental data support the alternative mechanism outlined in equations 173–175. The nucleophile is proposed to react directly with the radical anion, yielding the radical anion of the product which transfers an electron to the starting material, thereby continuing the chain. The alternative mechanism is designated as $S_{RN}2$. The arene is postulated to arise via addition of a second electron to the radical anion to give a dianion, ArX^{2-} or via the carbanion.

$$ArX^{-\bullet} + Nu^- \longrightarrow ArNu^{-\bullet} + X^- \tag{173}$$

$$ArNu^{-\bullet} + ArX \longrightarrow ArNu + ArX^{-\bullet} \tag{174}$$

$$ArX^{-\bullet} + e \longrightarrow Ar^- + X^- \rightleftharpoons ArX^{2-} \tag{175}$$

It was argued that the products and the isomer distributions observed in reactions proposed to proceed by the $S_{RN}1$ mechanism differ from those obtained by independent generation of aryl radicals and that the effects of leaving groups, counterions and substituents are not in accord with the intermediacy of an aryl radical.

This proposal has been critically discussed in two reports[624,643]. Important elements in the discussion are the stability and lifetime of the radical anions of aryl halides, the effects of leaving groups, the reactivities of nucleophiles and the nature of the transition state in the bimolecular step depicted in equation 173. It was concluded[643] that in most of the examples published in the literature with halobenzenes, and more in general with haloarenes featuring a rate of fragmentation of their radical anions in the order of $10^4 \, s^{-1}$ or higher, the reaction of the radical anions with nucleophiles postulated in the $S_{RN}2$ mechanism cannot compete with the $S_{RN}1$ mechanism. When radical anions have a relatively low rate of fragmentation, the $S_{RN}2$ mechanism might perhaps operate.

The photolysis of 1-iodoanthraquinone anion radical ($AQI^{-\bullet}$) in acetonitrile containing tetrabutylammonium iodide (TBA^+I^-) was found[644] to be unaffected by changes in $[TBA^+I^-]$ up to $0.2 \, \text{mol dm}^{-3}$. Two kinetic schemes were considered, one in which $AQI^{-\bullet}$ produces $AQ^{-\bullet}$ that can subsequently give AQH by reaction with TBA^+ or be quenched by I^- to give AQI⁻ and another one in which excited AQI˙ gives AQH by reaction with TBA⁺ and AQI˙ by reaction with I⁻. The results demonstrate that competition between reduction and quenching arises according to the first scheme. Thus, proof was obtained that aromatic anion radicals carrying a potential leaving group can dissociate unimolecularly to give the σ-aryl radical and the anionic leaving group.

The factors affecting the relative reactivity of aryl halides in $S_{RN}1$ reactions have been analysed and compared[645]. Competition experiments of pairs of substrates, in photo-stimulated reactions with pinacolone enolate ion in liquid ammonia, reveal a spread of reactivity exceeding three powers of ten. The ease of formation of the radical anion of the substrate appears to dominate the overall reactivity. The rate of dehalogenation of the radical anion may become important when its stability exceeds a certain threshold. When the fragmentation rate of the radical anion intermediate is fairly slow, the overall reactivity diminishes.

For three photostimulated $S_{RN}1$ reactions, the nature of the initiation step was investigated by means of wavelength-dependent quantum yield measurements[646]. Acetone enolate ion reacts with iodobenzene and with bromobenzene to give phenylacetone. The reagents form charge-transfer complexes and identical quantum efficiencies are observed upon excitation of the charge-transfer bands as with non-selective broad-band irradiation. This is consistent with electron transfer within an anion–aryl halide complex in the photoinitiation step. A charge-transfer complex is also formed from potassium diethyl phosphite and iodobenzene, but in this case the quantum yield of the photoreaction, which yields diethyl phenylphosphonate, is much lower upon excitation at $\lambda > 400$ than at wavelengths greater than 200 nm, implying a diminished importance of charge-transfer excitation. It is suggested that photoinduced homolysis of iodobenzene might be a competing pathway of formation of aryl radicals.

Fission of the carbon–halogen bond in the radical anion (equation 169) is a key step in the $S_{RN}1$ reaction. Fragmentation of that bond requires population of its σ* MO which may occur directly in the photostimulated initiation step. However, if initial population of the π* MO is energetically favoured, intramolecular electron transfer from π* to σ* must take place[15]. A nitro group on the aromatic ring so strongly stabilizes the π* MO that the electron transfer is prevented and $S_{RN}1$ reaction does not take place. It has been found, however, that *ortho*-iodonitrobenzene (228) gives an efficient nucleophilic substitution with pinacolone enolate ion (229) under photostimulation in liquid ammonia yielding 230 (equation 176)[647,648]. Steric inhibition of coplanarity of the nitro group by the bulky adjacent iodine atom reduces stabilization of the π* MO, so as to allow population of the σ* MO of the C—I bond and concomitant fragmentation. The effects of other *ortho* substituents (halogen, cyano, amino, hydroxyl, carboxyl, phenylsulphonyl, and trifluoromethyl) in $S_{RN}1$ reactions have also been studied[647].

$$
\underset{(228)}{\text{C}_6\text{H}_4(\text{I})(\text{NO}_2)} + \underset{(229)}{{}^-\text{CH}_2\text{CCMe}_3} \xrightarrow[\text{NH}_3]{h\nu} \underset{(230)}{\text{C}_6\text{H}_4(\text{CH}_2\text{CCMe}_3)(\text{NO}_2)} + \text{I}^- \quad (176)
$$

The trifluoromethyl substituent shows peculiar behaviour in $S_{RN}1$ processes[647,649]. Reaction of α,α,α-trifluoro-*o*-iodotoluene (231) with the enolate of 3,3-dimethylbutan-2-one (232) in liquid ammonia under standard photosensitization conditions yields only a minor amount (8%) of the expected substitution product (233), and mainly (43%) a more complex molecule (234) deriving from it (equation 177). Compound 233 is deprotonated

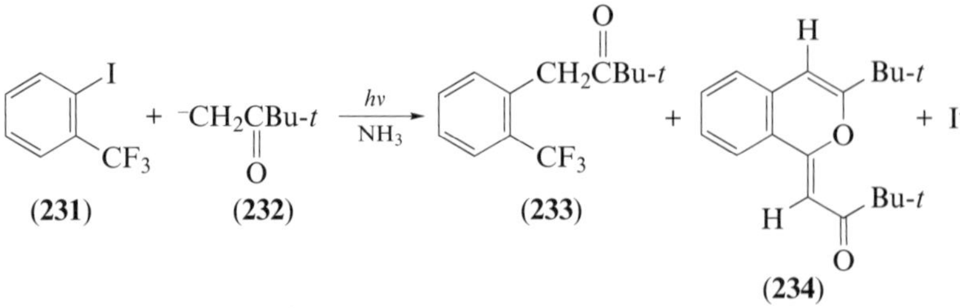

$$\begin{array}{ccc}
\text{(231)} & \text{(232)} & \text{(233)} \\
\end{array}$$

$$\text{(234)}$$

(177)

under the reaction conditions and, from its enolate anion, a fluoride ion is eliminated under formation of an *ortho*-quinonoid structure. This undergoes attack at the CF_2 group by **232** followed by intramolecular substitution of the second fluorine atom and elimination of HF^{649}, α,α,α-Trifluoro-*p*-iodotoluene also shows unusual behaviour and affords a rearranged product.

Fluorinated aromatic substrates have been used in the synthesis of fluorinated biaryl derivatives via $S_{RN}1$ reactions[650]. Substrates YC_6H_4Br ($Y = F$, CF_3 or OCF_3) were treated with the anions from 2,4-di-*tert*-butylphenol, 2,6-di-*tert*-butylphenol, *para*-methoxyphenol, *para*-(trifluoromethoxy)phenol and 2-naphthol leading to the biaryls YC_6H_4—ArOH by C-arylation at the carbon atom *ortho* to the deprotonated hydroxyl group (C-1 in 2-naphthol), but at the *para* carbon atom in 2,6-di-*tert*-butylphenol.

ortho-Functionalized aryl halides have been successfully used in the synthesis of a variety of compounds via the $S_{RN}1$ reaction. Treatment of primary or secondary *ortho*-halobenzamides (**235**) with various ketone enolates (**236**) affords the corresponding 3- and 4-substituted isocarbostyrils (**238**) directly (equation 178)[651]. The product of the $S_{RN}1$ reaction is **237**, which under the reaction conditions undergoes spontaneous cyclization to an σ-amino alcohol. This unstable product dehydrates readily at room temperature, yielding **238**.

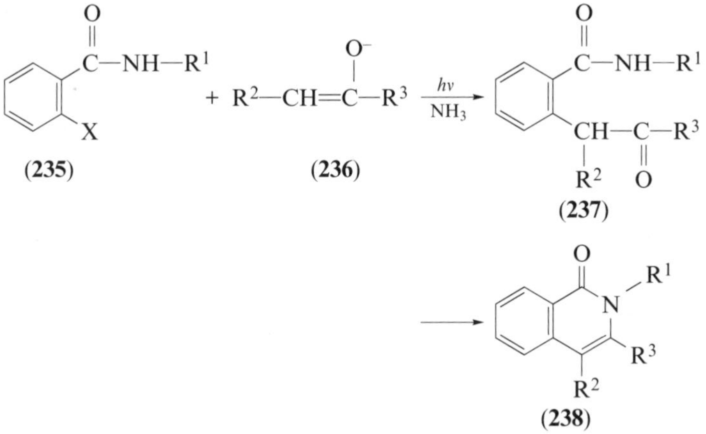

$$\begin{array}{ccc}
\text{(235)} & \text{(236)} & \text{(237)} \\
\end{array}$$

$$\text{(238)}$$

(178)

Similar methodology was used in the synthesis of 3-methyl derivatives of the alkaloids thalactamine, doryanine, and 6,7-dimethoxy-*N*-methyl-1(2*H*)-isoquinolone[652]. The $S_{RN}1$ reaction between *ortho*-halogenobenzylamines and enolates derived from a series of ketones and aldehydes affords 1,2-dihydroisoquinolines, from which the isoquinoline derivatives can be obtained by dehydrogenation and the 1,2,3,4-tetrahydroisoquinolines by reduction[653]. The products of the $S_{RN}1$ reactions of (2-halo-4,5-dimethoxyphenyl)acetic acid and various ketone enolates can be further elaborated into 3-benzazepines and 3-benzoxepines[654]. 2-Halobenzylamines and 2-halobenzoic acids react under $S_{RN}1$ conditions with enolates derived from substituted α-tetralone. The primary products undergo ring closure, either spontaneously or acid-catalysed, and the products are benzo[c]phenanthridine or benzo[c]phenanthridone derivatives[655]. A similar strategy was used in the synthesis of furo[3,2-h]quinolines and furo[3,2-b]pyridines[656], unsymmetrically substituted 2-phenyl-naphthalenes[657] and 2,2-binaphthyl derivatives[657,658]. The latter reaction is represented schematically in equation 179. Cyclization involving a substituent *ortho* to a ketone or aldehyde functionality introduced via an $S_{RN}1$ reaction has also been used in the synthesis of 4-azaindoles from 3-amino-2-chloropyridine[659].

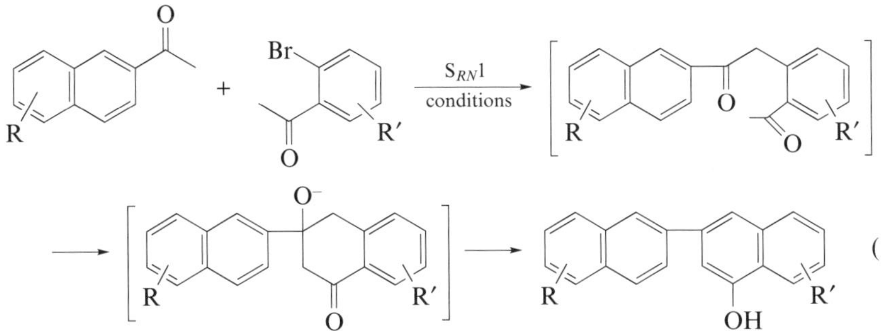

$$(179)$$

In the example of equation 179, one ring of one of the naphthyl moieties is created by spontaneous aldol condensation of the primary reaction product. A more direct and more versatile method for the synthesis of binaphthyls is the coupling of a naphthoxide anion with an iodonaphthalene under $S_{RN}1$ conditions. 1-Naphthoxide, 2-naphthoxide, 2-methyl-1-naphthoxide and 4-methoxy-1-naphthoxide anions were coupled with 1- and 2-iodonaphthalene, 2-iodo-6-isopropoxynaphthalene and 2-iodo-3,5-dimethoxynaphthalene, allowing new and efficient access to various unsymmetrically substituted 1,1 -, 1,2 - and 2,2 -binaphthyl derivatives[660]. The successful coupling of 2-naphthoxide ions with *para*-iodoanisole and 1-iodonaphthalene in liquid ammonia was reported by other authors almost simultaneously[661]. Naphthylquinolines and naphthylisoquinolines could be prepared by cross-coupling reactions of halobenzopyridine derivatives with anions from 2-naphthol, or conversely of iodonaphthalene with anions from hydroxyquinoline[662].

The photostimulated reaction of 2-naphthoxide ions (**239**) with *ortho*-diiodobenzene (**240**) in liquid ammonia in the presence of an excess of potassium *tert*-butoxide gives the substitution product **241** (20%), the cyclized product **242** (16%) and a small amount of PhI (equation 180)[663,664]. The substitution product **241** is formed via the usual $S_{RN}1$ reaction. In the solution, **241** is deprotonated and it is proposed that it receives another electron to give

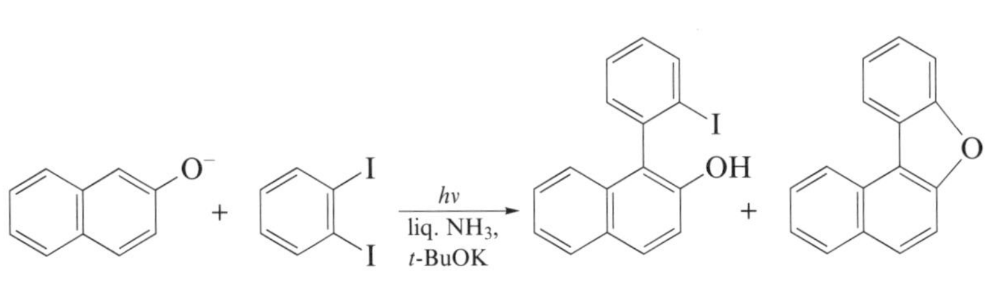

$$(180)$$

(**239**)   (**240**)   (**241**)   (**242**) a radical dianion which undergoes fragmentation of the C—I bond. The oxygen anion then reacts with the aryl radical, creating the radical anion of the cyclized product. A similar reaction was found to take place with **240** and 2-naphthaleneselenolate ions. Cyclization reactions occurring through the $S_{RN}1$ mechanism have also been reported to occur when *ortho*-dihaloaromatic compounds react with dithiolate ions[665]. *ortho*-Diiodo- and *ortho*-bromochlorobenzene react under photostimulation with 3,4-toluenedithiolate ion to give good yields (*ca* 60%) of the cyclic disubstituted compound 2-methylthianthrene. The monosubstituted compound is not an intermediate in the formation of the disubstituted one. After formation of the radical anion of the monosubstituted product, this species loses a halide anion and the resulting radical reacts with a second nucleophilic partner, in this case the second thiolate anion. Photostimulated reactions of 2,6-, 2,3-, 3,5- and 2,5-dihalopyridines with pinacolone potassium enolate in liquid ammonia lead to facile replacement of both halogens via an $S_{RN}1$ mechanism, which, as in the previous example, does not involve intermediate formation of monosubstitution products[666]. The reactions of several other dihalogenated nitrogen heterocycles were also investigated. The behaviour of bifunctional systems in the $S_{RN}1$ reaction has been studied in exploring the possibility of using this reaction to obtain polymers[667].

Most reactions discussed thus far involve enolate, phenolate or naphtholate anions as nucleophiles. Before turning to other nucleophiles, a few other examples belonging to this class will be mentioned. [*m.m*]Metacyclophadiones have been successfully synthesized from ω-(*m*-bromophenyl)alkan-2-ones by the photoinduced $S_{RN}1$ reaction[668]. Other photoinduced cyclizations involve amide enolate anions, such as the preparation of dihydroisoquinolinones from *ortho*-halogenated *N*-alkyl-*N*-acylbenzylamines[669,670] and the synthesis of oxindoles from *N*-alkyl-*N*-acyl-*ortho*-chloroanilines[670]. 1-Methyl-2-pyrrolidinone enolate ions were used in intermolecular arylations, e.g. with iodobenzene and 1-iodonaphthalene[671].

Examples in which heterocyclic haloarenes have been used include the photostimulated reactions of 2-chlorothiazole, 2-chloro-4-methylthiazole and 2-chloro-5-methylthiazole with pinacolone potassium enolate which lead to the formation of mono- and bis-2-thiazolyl ketones[672] and a study of the reactions of 3-halo-2-amino derivatives of benzo[*b*]thiophene with the potassium salts of acetophenone, pinacolone and cyclohexanone which indicated that, under $S_{RN}1$ conditions, mainly reduction products were formed[673].

The $S_{RN}1$ reaction has been used in the synthesis of purine derivatives[674,675]. 6-Iodo-9-ethylpurine reacts with a variety of enolate anions in liquid ammonia to give C-6-alkylated purines, which may be converted to other functionalized purine systems.

Biaryls were obtained by C-alkylation using 2,6-di-*tert*-butyl phenolate as nucleophile with *para*-chlorobenzonitrile and 2-chloro-5-cyanopyridine[676]. Bromobenzonitriles and bromocyanopyridines have been employed in a study in which it was demonstrated that these compounds can be successfully used to arylate monoanions of β-dicarbonyl compounds[677].

Photoreactions proceeding via the $S_{RN}1$ mechanism can also be carried out in DMSO solvent[678]. The enolate ions of acetone, acetophenone and anthrone gave good yields of substitution products with iodobenzene. The reaction fails when the enolates of acetylacetone, diethyl malonate and nitromethane are illuminated in the presence of iodobenzene. On the basis of experimental results and theoretical considerations, it was concluded that the reactivity of carbanion nucleophiles in photostimulated reactions with haloaromatic compounds depends on the nature of the reaction step. In the initiation step, the photostimulated electron transfer from the carbanion to the acceptor iodobenzene increases with the pK_a of the corresponding conjugated acids. In the propagation cycle, the reactivity of the carbanions would depend on their pK_a as well as on the SOMO of the radical anion intermediate, that is the HOMO–SOMO energy difference (loss in π energy).

The 2-naphthylamide anion (**243**) constitutes an example of an ambident nucleophile with negative charge on nitrogen and on carbon. It has been used in photostimulated reactions with iodobenzene, *para*-iodo- and *para*-bromoanisole, 1-iodo- and 1-bromonaphthalene[679]. The reaction with *para*-iodoanisole (**244**) is shown in equation 181. The C-arylated product (**245**) was formed in 63% yield and the *N*-arylated compound (**246**) in 6.3% yield, together with 1% of the reduction product anisole. The photoinduced reactions of the carbazole anion (in DMSO)[680] and of the anions from phenothiazine and benzimidazole[681] with aryl halides are reported to give *N*-arylated products.

$$ (181) $$

(**243**)   (**244**)   (**245**)   (**246**)

2-Bromopyridine and 2-chloroquinoline have been reported to react with nitrile-stabilized carbanions[682]. With potassiophenylacetonitrile, both compounds gave 88% yields of α-(2-pyridyl)phenylacetonitrile and α-(2-quinolyl)phenylacetonitrile, respectively. Photostimulated reaction of 2-bromopyridine with potassioacetonitrile produced the substitution product in 75% yield, along with 16% of 2-aminopyridine. The reaction of 2-chloroquinoline with potassioacetonitrile afforded 50% of the expected (2-quinolyl) acetonitrile and only traces of 2-aminoquinoline. The reaction with potassiophenylacetonitrile was later extended to a variety of halogenated pyridines, quinolines, pyrimidines, and pyrazines[683]. The secondary nitriles obtained could be converted in excellent yields to aryl hetaryl ketones by oxidative decyanation. The carbanion of cyclohexylideneacetonitrile may be compared with that of phenylacetonitrile; it has a cyclohexenyl group instead of a phenyl group. This species was studied in its reactions with *para*-bromoanisole, *para*-iodoanisole and 1-iodonaphthalene[684]. Good yields of the substitution products were obtained and it was found that the substitution occurs exclusively at the ring carbon atom Cγ.

The simplest carbanion containing nitrogen is cyanide ion[685]. Substitution of halogen by this anion has been investigated with 1-halogeno-2-naphthols and the chain reaction is proposed to take place via an S$_{RN}$1 mechanism.

Since carbon monoxide is isoelectronic with cyanide ion, it seems appropriate at this point to mention that irradiation under a slow stream of carbon monoxide of a stirred mixture of an aryl halide, benzene and 5 N aqueous sodium hydroxide in the presence of catalytic amounts of both Co$_2$(CO)$_8$ and Bu$_4$N$^+$Br$^-$ converts ArX into ArCOONa[686]. Very high yields were obtained from bromo-, chloro- and iodobenzene, 2- and 4-bromotoluene, 2- and 4-bromoanisole, 1- and 2-bromonaphthalene and various other aryl halides. A mechanism was proposed in which a charge transfer complex between ArX and [Co(CO)$_4^-$, Bu$_4$N$^+$] is excited and undergoes electron transfer. The aryl radical formed from ArX$^{-\cdot}$ may combine with $\cdot$Co(CO)$_4$, forming ArCo(CO)$_4$ which reacts with CO, or the aryl radical may react with Co(CO)$_4$Bu$_4$N$^+$ with formation of [ArCo(CO)]$^{-\cdot}$, Bu$_4$N$^+$ from which an electron can be transferred to ArX, in the typical S$_{RN}$1 fashion. For polycarbonylation of polyhalobenzenes an S$_{RN}$ mechanism has been considered to be unlikely (see below).

Irradiation of (triphenylmethyl)lithium in tetrahydrofuran containing bromobenzene or iodobenzene produced three major products: tetraphenylmethane (TPM), biphenylyldiphenylmethane (BDM) and 2-(triphenylmethyl)tetrahydrofuran (TTF)[687]. Product composition studies were performed and a mechanism was proposed in which excited triphenylmethyl anion transfers an electron to the aryl halide whereupon the radical anion loses halide ion. The aryl radical may attack the central carbon atom of Ph_3C^-, creating TPM$^{\cdot-}$, or a *para* carbon atom of one of the phenyl rings of Ph_3C^-, creating BDM$^{\cdot-}$. Both radical anions may transfer their electron to PhX, continuing the chain. Reaction between Ph$^{\cdot}$ and $Ph_3C^{\cdot}$ may also lead to TPM and BDM. Ph$^{\cdot}$ can abstract an α-hydrogen atom from THF and the THF radical may combine with $Ph_3C^{\cdot}$ or react with Ph_3C^-, leading to TTF. The combination of phenyl radical (radical–radical) with triphenylmethyl radical is a chain-termination step and diverts the process into an S_{ET} (radical-radical) pathway. The observed ratio of BDM to TPM is 2.9 and it provides a 'fingerprint' of the mechanism. Independent generation of Ph$^{\cdot}$ in the presence of $Ph_3C^{\cdot}$ leads to a BDM/TPM ratio of 1.0 and this leads to the conclusion that reaction via the S_{ET} mechanism is not a major pathway. Irradiation of indenyl anion in DMSO in the presence of bromobenzene yields 3-phenylindene, 1-phenylindenyl anion yields 1,3- and 1,1-diphenylindene and 2-phenylindenyl anion produces 2,3-diphenylindene[688]. 1,3- and 1,2-Diphenylindenyl anion were also investigated. The regiochemistry is attributed to radical attack at the most basic site. Related to these reactions is the observation[689] that irradiation of alkali metal salts of cyclooctadiene and bromo- or chlorobenzene leads to phenylcyclooctadienes.

Diethyl phosphite anion reacts with ethers derived from 5-chloro-7-iodo-8-isopropoxy-quinoline exclusively at position 7 by $S_{RN}1$ substitution of iodine[690]. Diphenyl- and dibenzylphosphinite ions, Ph_2PO^- and $(PhCH_2)_2PO^-$, can be prepared by reaction of triphenyl- and tribenzylphosphine oxides with alkali metals in liquid ammonia[691]. These ions react under photostimulation with aryl halides by the $S_{RN}1$ mechanism to give aryldiphenyl- and aryldibenzylphosphine oxides in good yields.

In the $S_{RN}1$ reactions with nucleophiles of the types RS$^-$, PhS$^-$, PhSe$^-$, PhTe$^-$, Ph_2As^- and Ph_2Sb^-, fragmentation reactions of the radical anions of the substitution products are often observed. A review on the phenomenon of radical anion fragmentation in the course of aromatic $S_{RN}1$ reactions has appeared in 1982[692]. The mechanism of fragmentation was also very clearly described in a report on photostimulated reactions of haloarenes with benzeneselenate ions[693]. In the photostimulated reaction of *para*-iodoanisole (247) with benzeneselenate ions (248) not only the straightforward substitution product 249 was obtained, but also the symmetrical diphenyl selenide (250) and the di-(*p*-anisyl) selenide (251) (equation 182).

$$p\text{-}IC_6H_4OMe + PhSe^- \xrightarrow{\;h\nu\;}$$
$$\quad\;(247)\qquad\quad(248)$$

$$p\text{-}PhSeC_6H_4OMe + Ph_2Se + p\text{-}MeOC_6H_4\!-\!Se\!-\!C_6H_4OMe\text{-}p \qquad (182)$$
$$\quad\;(249)\qquad\qquad(250)\qquad\qquad(251)$$

Scrambling of aromatic rings results from reversible fragmentation and coupling of radical anion intermediates during the $S_{RN}1$ process. Therefore, the following steps, illustrated for the case of PhSe$^-$, should be added to the mechanism described by equations 168–172. Equations 170a and 171a are similar to equations 170 and 171, but for PhSe$^-$ equation 170a is proposed to be an equilibrium. These reactions may then be followed by equations 183–185.

$$Ar^{\cdot} + PhSe^- \;\rightleftharpoons\; (ArSePh)^{\cdot-} \qquad (170a)$$

$$(ArSePh)^{\cdot-} + ArX \;\longrightarrow\; ArSePh + ArX^{\cdot-} \qquad (171a)$$

$$(ArSePh)^{-\bullet} \rightleftharpoons ArSe^- + Ph^\bullet \qquad (183)$$

$$Ar^{\bullet} + ArSe^- \rightleftharpoons (Ar_2Se)^{-\bullet} \xrightarrow{ArX} Ar_2Se + ArX^{-\bullet} \qquad (184)$$

$$Ph^\bullet + PhSe^- \rightleftharpoons (Ph_2Se)^{-\bullet} \xrightarrow{ArX} Ph_2Se + ArX^{-\bullet} \qquad (185)$$

The fragmentation of the radical anion $(ArSePh)^{-\bullet}$ occurs because the energy levels of the antibonding π^* MO of the aromatic system and the antibonding σ^* MO of the C—Se bonds are close enough to permit fragmentation to occur from the latter MO (equation 186).

$$Ar^* + {}^-SePh \rightleftharpoons [Ar\overset{\bullet}{-}SePh \longleftrightarrow ArSe\overset{\bullet}{-}Ph] \rightleftharpoons ArSe^- + Ph^\bullet \qquad (186)$$

$$\sigma^* \text{ radical anion}$$

An example of such scrambling is found in the photostimulated reactions of *para*-chloro-, *para*-bromo- and *para*-iodotoluenes with potassium diphenylarsine[694]. Four products were found: triphenylarsine, *p*-tolyldiphenylarsine, di-*p*-tolylphenylarsine and tri-*p*-tolylarsine. With 4-chlorobenzophenone only the straightforward substitution product is found, probably because the low-lying π^* MO of the benzophenone moiety prevents C—As bond breaking in the radical anion. Four products were likewise found from the reactions of 1-bromonaphthalene and 9-bromophenanthrene with diphenylarsenide ion[695]. With potassium diphenylstibide, 4-chlorobenzophenone also gave four products. Competition between electron transfer and fragmentation in the radical anion intermediate was also found in the reactions of RS⁻ ions (R = methyl, *n*-butyl, *tert*-butyl, and benzyl) with haloarenes[696].

Photoinduced nucleophilic thiylation of bromobenzene, chlorobenzene, *meta*- and *para*-bromotoluene using PhSNa and PhSH in liquid ammonia could be realized in the presence of tetrabutylammonium hydroxide or triethylamine as catalysts[697]. Tetrabutylammonium hydroxide was also found to be a catalyst in the photostimulated thiylation of aryl halides by 2- and 4-pyridinethiolates in liquid ammonia[698]. Several nitroaryl halides were among the compounds undergoing substitution. Various functionalized aromatic halides ArXY (Ar = C_6H_4; X = Br, I; Y = OCH_3, $CONH_2$, CN, $COCH_3$, CHO, COC_6H_5) undergo $S_{RN}1$ reactions with sulphur anions RS⁻, either simple (R = C_2H_5, $CH_2C_6H_5$) or functionalized [R = $(CH_2)_2OH$, $(CH_2)_2COOEt$, CH_2COOEt][699]. The formation of products ArYS⁻ by fragmentation of the radical anion ArYSR⁻$^\bullet$ is related to the redox potential of the aryl moiety ArY and the energy of the bond S—R. Competition experiments have been carried out in liquid ammonia to determine the relative rate constants of the coupling reactions of the nucleophiles PhO⁻, PhS⁻, PhSe⁻ and PhTe⁻[700]. An increasing reactivity is found in going down the group 6A of elements: PhO⁻ (0.0), PhS⁻ (1.00), PhSe⁻ (5.8), PhTe⁻ (28). Methyl and isopropyl ethers of 5-chloro-7-iodo-8-hydroxy-quinoline have been studied in their reactions with a variety of sulphanions and they react almost exclusively at position 7 by substitution of iodine[690]. The photoreactions of 4-bromobenzophenone and 4-haloacetophenones with phenylthiolate have been studied quantitatively in various solvents and the results have been compared with those from electrochemical experiments[701]. The similarities are the same approximate yield, the same primary termination step, the same initiation step and the comparable chain length. The main difference is the production of small amounts of diaryl disulphide in the photochemical experiment, together with about 3% of the reduction product. This gives a definite advantage to the electrode process. $S_{RN}1$ reactions between benzenethiolate and 2-chloro-, 2-bromo-, 2-iodo- and 3-bromothiophene in acetonitrile lead to complex mixtures of which phenyl 2- or 3-thienyl sulphide is the main component[702]. Fragmentation of ThSPh⁻˙ appeared to be a problem, but optimization of the yield of ThSPh could be achieved by the employment of suitable electron acceptors, although at the expense of the overall reaction rate.

From bromonaphthalenediazonium tetrafluoroborates, the corresponding diazosulphides are formed by reaction with sodium benzenethiolate in DMSO[703]. From these diazosulphides, dinitriles deriving from substitution of both the diazo group and the bromine atom can be formed by reaction with excess tetrabutylammonium cyanide in DMSO under photostimulation by a sunlamp. Derivatives of *para*-mercaptophenylalanine and iodotyrosine or diiodohydroxyphenylglycine are efficiently converted into thioethers by $S_{RN}1$ reaction in ammonia[704]. The problem of racemization of the amino acid was overcome by using the free base of the diiodohydroxyphenylglycine, because the rate of racemization is known to be highest when the nitrogen is least basic.

The mechanism of the reduction process (formation of benzene) in the photostimulated reaction of benzeneselenate ion with iodobenzene was found to be photostimulated electron transfer from the nucleophile to the substitution product (equations 187–189)[705].

$$PhSe^- + Ph_2Se \xrightarrow{h\nu} (Ph_2Se)^{-\bullet} + PhSe^\bullet \tag{187}$$

$$(Ph_2Se)^{-\bullet} \longrightarrow PhSe^- + Ph^\bullet \tag{188}$$

$$Ph^\bullet \longrightarrow PhH \tag{189}$$

This reaction is important when the aryl radical reacts under reversible conditions with the nucleophile. When it reacts irreversibly this mechanism of reduction does not operate significantly.

An S_N1 Ar* mechanism has been proposed to account for the behaviour of 2-nitrofuran in nucleophilic photosubstitution[706]. The mechanism is characterized by departure of the leaving group in the primary step following excitation; the photosubstitution of the nitro group of 2-nitrofuran by cyanide or water is often cited as the only example of this type of mechanism. In the photoreactions of chlorobenzene and 4-chlorobiphenyl with water and alcohols, however, a similar mechanism has been considered. The primary step after excitation may be either homolysis of the C—Cl bond, followed by electron transfer from the aryl radical to the chlorine atom or direct heterolytic fission leading to aryl cation and chloride ion. If this is followed by attack of the nucleophile on the cation, the reactions may be classified as S_N1 Ar*, although strictly speaking, if only the primary reaction step following excitation is considered, reactions that start with homolysis of the carbon–halogen bond should be classified as S_H1 Ar*, even if this is followed by electron transfer and attack of a nucleophile on the aryl cation. A sharp distinction between the two mechanistic types cannot always be made.

Irradiation of chlorobenzene at 254 nm in ethanol–water mixtures produces hydrogen chloride, phenol, benzene and phenetole with quantum yields (in 75% H_2O–25% EtOH) of 0.13, 0.051, 0.078 and 0.006, respectively[355]. The effects of added KOH, of oxygen and of the concentration of chlorobenzene were studied and a mechanism was proposed in which a reactive intermediate is formed from the singlet excited state. The intermediate could be either a phenyl cation or a chlorobenzene radical cation. With the help of transient absorption spectroscopy and fast kinetic measurements it was later demonstrated[356] that in methanol the singlet excited state undergoes ionization with a quantum yield of 0.03 ± 0.01. The characteristic absorption of the solvated electron could be observed. The radical cation reacts with methanol to produce anisole, a proton and a chlorine atom. The triplet state undergoes homolytic C—Cl bond fission and this reaction leads to benzene. In other reports[707–709], the triplet state is held responsible for the photochemical production of phenol ($\varphi = 0.10$) from chlorobenzene in aqueous solution and the mechanism in equation 190 was presented. The C—Cl bond is considered to be polarized in the triplet state (252) and it undergoes further polarization upon approach of a water molecule (254). Rupture of the C—Cl bond may precede bond formation between carbon and oxygen (253), or the reaction may be concerted. Electron-withdrawing substituents on the benzene ring counteract the polarization of the C—Cl bond and the quantum yields of phenol production from di- and trichlorobenzenes and from 4-nitrochlorobenzene are significantly lower than that from chlorobenzene.

The photoreaction of chlorobenzene in methanol has been studied by means of emission spectroscopy[710]. The quantum yield of formation of anisole is 0.049. Quenching experiments, study of the temperature dependence of the quantum yield and spectroscopic considerations lead to the conclusion that the nucleophilic photosubstitution occurs in the $^3(\pi,\sigma^*)$ state. In the photolysis of 2-, 3- and 4-chlorobiphenyl[711] and 2- and 4-chlorodiphenyl ethers[542] in water, the formation of the corresponding phenols is interpreted on the basis of an intermediate aryl cation. The triplet excited state of the chlorobiphenyls undergoes bond homolysis and the triplet radical pair is assumed to equilibrate with a singlet radical pair which then yields chloride anion and biphenyl cation. The photochemistry of 4-chlorobiphenyl in water is unusual because equimolar amounts of 3- and 4-hydroxybiphenyl are produced[712]. Apparently, 4-chlorobiphenyl photoisomerizes into 3-chlorobiphenyl, a reaction which may proceed via a benzvalene intermediate. Both isomers are supposed to undergo C—Cl bond homolysis in the excited state and the resulting biphenyl radicals cannot easily abstract a hydrogen atom from water. They may either revert to starting material by recombination or, facilitated by the high dielectric constant of water, transfer an electron to the chlorine atom. Reaction of the diphenyl cations with water leads to the hydroxybiphenyls.

The cation, formed by heterolysis of the C—Cl bond in 2-chlorobiphenyl, has been represented as a carbene-like resonance structure with the unshared electron pair at position 2 and the positive charge in the other ring[358].

The intermediacy of a cation, formed by electron transfer within a photochemically created radical pair, was also invoked to explain the results obtained upon photolysis of 2-bromo-, 2-chloro- and 2-iodopyridine in methanol, ethanol and acetonitrile–water[376]. The major products are 2-methoxypyridine, 2-ethoxypyridine and 2-acetamidopyridine + 2-hydroxypyridine. In all cases pyridine was the minor reaction product, in contrast with the 3- and 4-halopyridines which produce pyridine exclusively, via a radical process. It is proposed that the unshared electron pair on the nitrogen atom assists in the formation of the 2-pyridyl cation. The presence of cupric salts increases the relative amounts of products formed via ionic reactions because Cu^{2+} can accept an electron from the 2-pyridyl radical.

The formation of methylanisoles in the photolysis of chlorotoluenes in methanol has been studied[713,714]. The quantum yields are 0.033, 0.039 and 0.017 for the *ortho*, *meta* and *para* isomer, respectively, compared to 0.049 for methanolysis of chlorobenzene[710]. Quenching experiments with biacetyl indicate that the reaction takes place in the triplet state, but also in the initial excited levels. In the triplet route, the yields of the *ortho* and *para* isomers are smaller (0.009 and 0.010) than that of the *meta* isomer (0.027) due to the electron-donating effect of the methyl group. The rate of formation of methylanisole from the triplet state was derived to be in the order of $10^4 \, s^{-1}$.

2- and 3-Chlorophenol undergo photochemical replacement of the chlorine atom upon irradiation in aqueous solution[715-717]. The direct irradiation of 3-chlorophenol yields resorcinol and the reaction can be sensitized by phenol. A mechanism of direct photo-hydrolysis from the triplet state is proposed[717]. From 2-chlorophenol, pyrocatechol is formed, but in addition to this reaction, ring contraction is observed, which leads to cyclopentadienecarboxylic acid[715,716,718]. The total quantum yield is 0.032 ± 0.004. The reactions are presumed to proceed via a common intermediate, a cation (**255**) formed by heterolysis of the C—Cl bond (equation 191). Reaction with water leads to the substitution product (**256**) and loss of a proton generates a carbene (**257**) from which, via the Wolff rearrangement, the cyclopentadienecarboxylic acid (**258**) is formed. Ring contraction of

$$(191)$$

the 2-chlorophenolate anion proceeds with tenfold higher quantum yield and a similar reaction takes place with 2-bromophenolate with a quantum yield of 0.45. The formation of pyrocatechol and cyclopentadienecarboxylic acid from 2-halogenophenols can be sensitized in acidic solution by hydroquinone and proceeds via the triplet state[719].

The photoreactions of 4-chlorophenol in aqueous solutions[720,721] and of substituted 4-chlorophenols in aqueous alkaline solutions[722] have been investigated. The primary photoproduct from 4-chlorophenol is 1,4-benzoquinone, which is further photochemically converted into hydroquinone and 2-hydroxy-1,4-benzoquinone[720]. An EPR study has shown that in the pH range 6.5–10.5 two primary radicals are produced, one by homolytic fission of the O—H bond and the other by homolytic fission of the C—Cl bond[721]. The aryl radical formed in the latter process reacts with oxygen and, via this route, 1,4-benzoquinone is produced. Radical species were also observed in the photolysis of 2-methyl-, 3-methyl-, 2-cyclohexyl- and 2-benzyl-4-chlorophenol[722]. The C—Cl bond breaks in the primary step and, in the presence of oxygen, a semiquinone-type radical is formed.

4-Bromophenol and 3-bromophenol behave quite differently in aqueous undegassed solutions[723]. 1,4-Benzoquinone is the major product from 4-bromophenol, while 3-bromophenol yields resorcinol exclusively.

When (±)-*N*-formyl-4-fluorotryptophan methyl ester is irradiated in methanol, replacement of fluorine by methoxy takes place[387]. Because of the strength of the C—F bond, homolysis cannot occur under photolytic conditions. Heterolysis, however, may occur in a polar medium and the aryl cation is probably stabilized by resonance involving the lone electron pair on the indole nitrogen atom. The chlorine atom in the diuretic furosemide is replaced by a hydroxyl group upon irradiation in buffer solutions (pH between 1.2 and 12.0) using artificial light sources or direct or indirect sunlight[724].

Substitution of halogen by oxygen has been used in the synthesis of nucleosides. Photolysis of 2-iodoadenosine in water provides isoguanosine[725]. 2′-Deoxyisoguanosine and some base-modified analogues were prepared via photochemical replacement of chlorine or bromine by water[726].

Intramolecular photochemical replacement of bromine by the oxygen atom of a peptide linkage was observed and studied with (5-bromo-6-uracilyl)-N-ethylacetamide and N-[(5-bromo-6-uracilyl)acetyl]-D,L-threonine N-ethylamide[727]. At 254 nm, excitation leads to photodebromination, but when the $^1n,\pi^*$ state is selectively populated with a XeCl excimer laser, it does not undergo C—Br bond homolysis in competition with intersystem crossing. The initial step is electron transfer from the peptide linkage to the triplet excited bromouracil chromophore. Bond formation takes place prior to intersystem crossing and is followed by departure of a bromine and a hydrogen atom (equation 192).

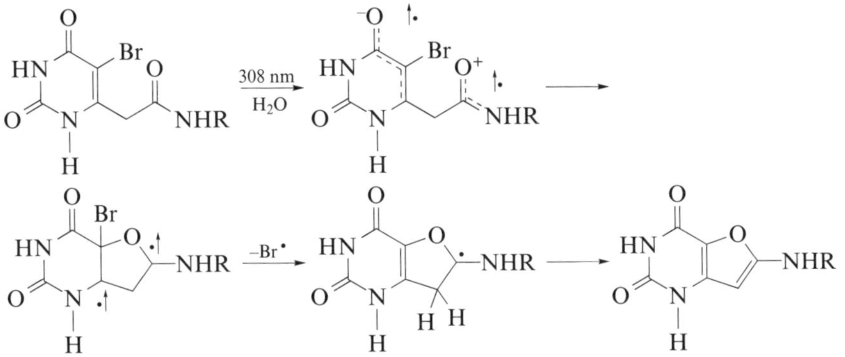

(192)

Chlorine atoms in aryl halides can also be replaced by hydroxyl groups in photoreactions catalysed by metal oxides such as ZnO[728–731] and TiO_2, ZrO_2 and MoO_3[732,733]. Such reactions which have been performed with dichlorobenzenes and (di)chlorophenols, start with complex formation between an ·OH radical and the aryl halide. The ·OH radicals are generated by interaction of photochemically produced positive holes on the metal oxide surface with water. The position of substitution is directed by any hydroxyl groups already present in the ring. If this leads to replacement of chlorine, the reaction may be regarded as photoinduced substitution of Cl by OH, but electronically excited states of the aryl halides are usually not involved. The case of 3-chlorophenol is illustrative[728]. Upon direct excitation in dilute aqueous solution, conversion into resorcinol is observed. The indirect phototransformation of 3-chlorophenol induced by the ·OH radicals formed at the surface of ZnO excited at 365 nm involves mainly primary hydroxylation to chlorohydroquinone (35%), 4-chloropyrocatechol (11%) and 3-chloropyrocatechol (8%).

In the $S_{R^+N}1$ Ar* and the S_N1 Ar* mechanisms, the aryl halide becomes reactive towards nucleophiles by loss of an electron or a halide anion. Another means of activation is protonation, followed by excitation. This mechanistic pathway has been proposed for the photohydroxydehalogenation of 1-chloroanthraquinones in 97% sulphuric acid[734], which leads to high yields (60–70%) of 1-hydroxyanthraquinones. The yields of the photochemical replacement of chlorine by pyridine in 1- and 2-chloroanthraquinone are much lower (3–6%)[735]. The reactions were performed with pyridine as solvent and the products were converted into aminoanthraquinones by treatment with alkali. Pyridine as nucleophilic reagent was also used in intramolecular photosubstitutions of halogen in the synthesis of benzo[c]quinolizinium salts from trans-2-[2-(2-chlorophenyl)vinyl]pyridines[736] and of pyrido[1,2-a]benzimidazoles from 2-[(2-halophenyl)amino]pyridine[737,738].

Nucleophiles containing elements of the second row have also been used in photosubstitution reactions of aryl halides. The selenocyanate ion[739], selenourea[740] and thiourea[741] are capable of replacing halogen atoms in various derivatives of benzene, naphthalene, pyridine and pyrimidine. Both chlorine atoms in ortho-dichlorobenzene are substituted upon irradiation in $(MeO)_3P$ at 60 °C for 5 days and a 50% yield of 1,2-bis(phosphino)-benzene is obtained[742].

Sulphite ion has been successfully used to replace halogen in 4-chloro-1-naphthol[743,744], 1-chloro- and 1-bromo-2-naphthol[744-747] and 1-iodo-2-naphthol[744], in direct and in dye-sensitized photoreactions. In the direct irradiation, quantum yields are greater than unity and depend on the concentration of sodium sulphite, the concentration of the substrate and the pH of the medium[745]. It has been established that 1-chloro- and 1-bromo-2-naphthol react via different mechanisms. In the case of the chloro compound, the reaction is initiated by electron transfer between triplet-excited and ground-state chloronaphthol, leading to a radical ion pair. This is the same primary step as in the $S_{R^+N}1$ Ar* reaction, but in that case it is the radical cation which reacts with the nucleophile. The radical cation of 1-chloro-2-naphthol, however, rapidly loses a proton and becomes a naphthoxyl radical. In this case, it is the radical anion that starts the propagation chain, in the same way as in the $S_{RN}1$ reaction (sometimes called $S_{R^-N}1$). This initiation of a photo-$S_{RN}1$-type reaction is rather unusual: in most cases such reactions start by photostimulated electron transfer from the nucleophile to the aromatic substrate. Chain initiation in 1-bromo-2-naphthol is supposed to be the homolytic photodissociation of the carbon–bromine bond. The dye-sensitized chain substitution of halogen by a sulpho group in halogenonaphthols is initiated by electron transfer from the sulphite dianion to triplet excited sensitizer (eosin, erythrosine or fluorescein)[744]. The sulphite radical anion then reacts with halogenonaphtholate anion, yielding $[\text{OAr}(SO_3^-)X]^{\cdot}$. This dianion radical may lose either X⁻ or X⁻. In the first case, X⁻ will react with SO_3^{2-} to give X⁻ and $SO_3^{\cdot}$. and in the second instance, $[\text{OAr}(SO_3^-)]$ will abstract an electron from SO_3^{2-}, and the chain will be propagated. The kinetics and mechanism of the photochemical chain substitution of chlorine by a sulpho group in 1-chloro-2-naphthol, photoinitiated by a ruthenium bipyridyl complex, have also been studied[748].

Halogen atoms in nucleosides can be replaced photochemically by sulphur as has been demonstrated in the photoinduced alkylthiolation of halogenated purine nucleosides[433,749], in the synthesis of 2-(methylthio)adenosine from 2-iodoadenosine[750] and in the photochemical formation of a cysteine-uracil adduct from 5-bromouracil and N-acetylcysteine methyl ester[435]. Photochemical synthesis of phosphonopyrimidine and phosphonopurine nucleosides by reaction of bromonucleosides with triethyl phosphite has also been reported[751].

Aryl halides can undergo photostimulated polycarbonylation catalysed by $Co_2(CO)_8$ in aqueous sodium hydroxide solution or in methanol containing sodium methoxide (equation 193)[752-754]. Such reactions have been performed with chloro-, bromo- and dichloro-

$$\text{ArX} + \text{CO} + \text{NaOMe} \xrightarrow[h\nu\ (350\ nm),\ 65\ °C]{\text{NaOMe/MeOH, Co}_2(CO)_8,\ CO} \text{ArCOOMe} + \text{NaX} \qquad (193)$$

benzoic acids, di- and trihalobenzenes, and dihalobenzoic esters in excellent yields. A mechanism has been proposed in which the halogen atom is replaced photochemically by $Co(CO)_4$. The next step is insertion of CO into the aryl–cobalt bond and this is followed by replacement of $Co(CO)_4$ by hydroxide or methoxide[753]. An S_{RN} mechanism, as proposed earlier[686], is not considered to be likely[754].

The chlorine atom in 4-chloropyridine can be replaced photochemically by the dimethyl ketyl radical[755]. Irradiation of 4-chloropyridine in a 4:1 mixture of 2-propanol and water gives a low yield (2%) of 2-(4′-pyridyl)-2-propanol. Sensitization by benzophenone increases the yield to 25%, but the product is now accompanied by 6% of diphenyl-(4-pyridyl)methanol. The major product is believed to be formed via hydrogen abstraction from 2-propanol by photoexcited pyridine. Protonated pyridines do not undergo this abstraction process, and accordingly the product yield decreases under acidic conditions. The radical (259) formed from the pyridine will combine (at position 4) with the dimethyl ketyl radical (260) and elimination of HCl from the adduct (261) completes the reaction (equation 194).

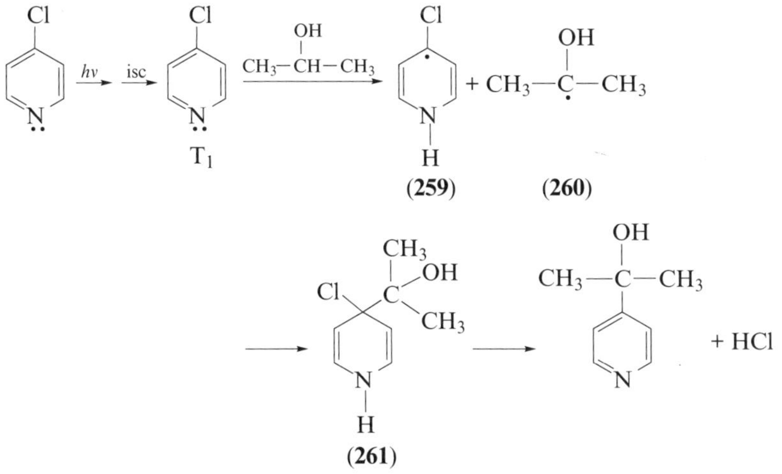

A mechanism in which ketyl radicals play a role was also postulated for the formation of cyclohexylpentafluorobenzene upon irradiation of a cyclohexane solution of hexafluorobenzene in the presence of benzophenone[756]. In this case, however, the excited benzophenone abstracts a hydrogen atom from the solvent and the cyclohexyl radical attacks hexafluorobenzene. The resulting radical is transformed further into the substitution product by loss of fluorine and into an addition product by abstraction of a hydrogen atom. Irradiation of a cyclohexane solution of pentafluoropyridine in the presence of benzophenone resulted in the formation of 4-cyclohexyltetrafluoropyridine, while no addition product was observed.

Another substitution reaction that is initiated by photochemical hydrogen abstraction is the replacement of the bromine atom in 2-bromo-8-methoxy-1,4-naphthoquinone by an acyl group[757]. Irradiation of a solution in benzene of the quinone, butyraldehyde or capraldehyde and pyridine yields mixtures of acylated quinone and acylated hydroquinone. In the first step, the excited quinone abstracts the aldehyde hydrogen atom and this is followed by bond formation between the acyl radical and C-2 of the quinone. The radical that is formed after departure of a bromine atom may either lose a hydrogen atom and yield acylated quinone or take up a hydrogen atom and become acylated hydroquinone.

Three special cases of aromatic substitution deserve to be mentioned. The first one is the substitution by alkoxide of the halogen atom in 3-chloro- and 3-bromo-1-isoquinolinone[758]. The unusual feature of this substitution is that, upon irradiation in methanol or ethanol containing triethylamine, the starting material (262) first undergoes ring opening (equation 195). The nucleophile then attacks the imine carbon atom (263), a proton is taken up by the neighbouring carbon atom and simultaneously the ring closes. Elimination of HX from the adduct (264) then gives rise to 3-alkoxy-1-isoquinolinone (265), which is the major product (*ca* 70%). A second product (*ca* 10%) is formed by elimination of HX from the ring-opened molecule. This is followed by reaction of the ketene with alcohol, which leads to a 2-(cyanomethyl)benzoate.

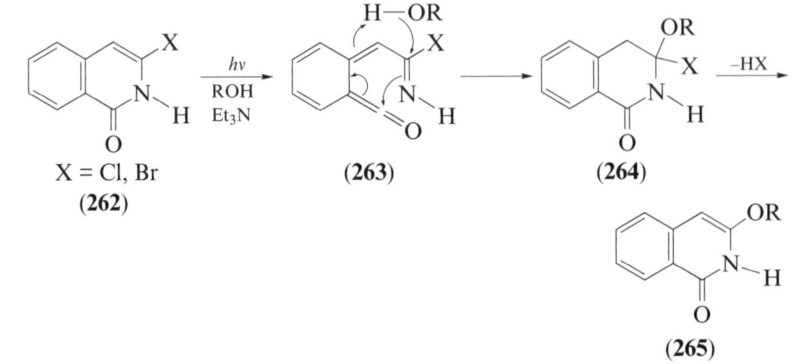

The second special case is formed by the photoreactions of 2,4-dinitro-6-(phenyliodonio)phenolate (266) with several nucleophiles[759]. Upon irradiation of this fairly stable zwitterion in methanol, 6-methoxy-2,4-dinitrophenol is formed in 65% yield. Photoreaction with pyridine affords 2,4-dinitro-6-pyridiniophenolate (85%) and irradiation in the presence of phenyl isothiocyanate in acetonitrile affords a mixture of two stereoisomeric 2-phenylimino-5,7-dinitro-1,3-benzoxathioles (269) (71%) which could not be separated. The reaction starts with attack of the nucleophile on the positively charged iodine atom, leading to an iodinane (267) and proceeds by expulsion of iodobenzene. The mechanism is illustrated for phenylisothiocyanate, a case in which the substitution product (268) can undergo further photocyclization to a benzoxathiole derivative (269) (equation 196).

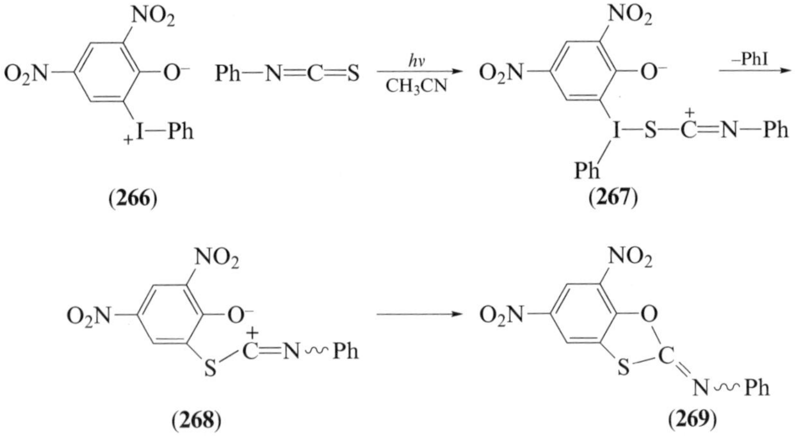

The third example concerns halide interchange via carbenes, formed from halogenated carbanions that undergo loss of halide ion upon irradiation (equation 110)[348]. In addition to halide interchange, these carbenes also undergo CH insertion and addition to electron-rich olefins. 2-Chloro- and 2-bromo-1,3-diphenylindene (**270**) were converted into their conjugate bases (**271**) by deprotonation, e.g. by potassium *tert*-butoxide in dimethyl sulphoxide (equation 197). Irradiation of both carbanions leads to loss of halide ion and formation of a hypovalent intermediate which can be described as 1,3-diphenylisoindenylidene (**272**) or alternatively as 1,3-diphenyl-1,2-dehydroindene (**273**). This intermediate can be captured by added halide ion or undergo return of halide ion within the cage. A greater efficiency for chloride as opposed to bromide loss is found upon photoexcitation, which is in contrast to the typical nucleofugic order. This probably reflects efficient quenching of the photochemistry by bromide ion return.

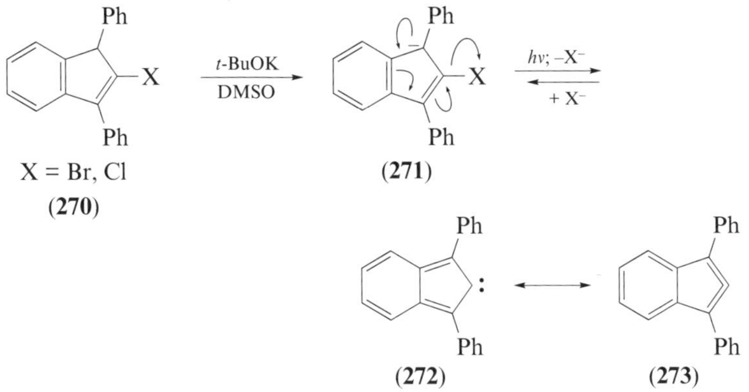

X = Br, Cl

(**270**)

(**271**)

(**272**)   (**273**)   (197)

V. REFERENCES

1. P. G. Sammes, in *The Chemistry of the Carbon–Halogen Bond* (Ed. S. Patai), Chap. 11, Wiley, Chichester, 1973.
2. G. Lodder, in *The Chemistry of Functional Groups: Supplement D* (Eds. S. Patai and Z. Rappoport), Chap. 29, Wiley, Chichester, 1983.
3. H. Dürr, in *Houben-Weyl's Methoden der Organischen Chemie*, Teil 4/5a, Thieme Verlag, Stuttgart, 1975, p. 627.
4. K. M. Saplay and N. P. Damodaran, *J. Sci. Ind. Res.*, **42**, 602 (1983).
5. P. J. Kropp, *Acc. Chem. Res.*, **17**, 131 (1984).
6. N. Kornblum, in *The Chemistry of Functional Groups: Supplement F* (Ed. S. Patai), Chap. 10, Wiley, Chichester, 1982.
7. R. K. Norris, in *The Chemistry of Functional Groups: Supplement D* (Eds. S. Patai and Z. Rappoport), Chap. 16, Wiley, Chichester, 1983.
8. G. A. Russell, *Adv. Phys. Org. Chem.*, **23**, 271 (1987).
9. W. R. Bowman, *Chem. Soc. Rev.*, **17**, 283 (1988).
10. W. R. Bowman, in *Photoinduced Electron Transfer, Part C* (Eds. M. A. Fox and M. Chanon), Elsevier, Amsterdam, 1988, p. 487.
11. S. J. Cristol and T. H. Bindel, in *Organic Photochemistry*, Vol. 6 (Ed. A. Padwa), Marcel Dekker, New York, 1983, p. 327.
12. H. Morrison, *Rev. Chem. Intermediates*, **8**, 125 (1987).
13. R. J. Sundberg, in *Organic Photochemistry*, Vol. 6 (Ed. A. Padwa), Marcel Dekker, New York, 1983, p. 121.
14. R. S. Davidson, J. W. Goodin and G. Kemp, *Adv. Phys. Org. Chem.*, **20**, 191 (1984).
15. R. A. Rossi and R. H. de Rossi, *Aromatic Substitution by the $S_{RN}1$ Mechanism*, ACS Monograph 178, Americal Chemical Society, Washington, DC, 1983.
16. K. Maruyama and A. Osuka, in *The Chemistry of Quinonoid Compounds*, Vol. II, Chap. 13 (Eds. S. Patai and Z. Rappoport), Wiley, Chichester, 1988.

17. F. Terrier, *Nucleophilic Aromatic Displacement: The Influence of the Nitro Group*, Chap.6, VCH Publishers, New York, 1991.
18. P. K. Freeman and S. A. Hatlevig, *Top. Curr. Chem.*, **168**, 47 (1993).
19. P. J. Kropp, P. R. Worsham, R. I. Davidson and T. H. Jones, *J. Am. Chem. Soc.*, **104**, 3972 (1982).
20. D. S. Reddy, G. P. Sollott and P. E. Eaton, *J. Org. Chem.*, **54**, 722 (1989).
21. P. H. McCabe, C. I. de Jenga and A. Stewart, *Tetrahedron Lett.*, **22**, 3681 (1981).
22. P. J. Kropp, J. A. Sawyer and J. J. Snyder, *J. Org. Chem.*, **49**, 1583 (1984).
23. H. Vančik, V. Gabelica, V. Rogan and D. E. Sunko, *J. Chem. Res., Synop.*, 92 (1990).
24. O. Piva, *Tetrahedron Lett.*, **33**, 2459 (1992).
25. P. J. Kropp and R. L. Adkins, *J. Am. Chem. Soc.*, **113**, 2709 (1991).
26. (a) R. W. Binkley, *Adv. Carbohydr. Chem. Biochem.*, **38**, 105 (1981).
    (b) R. W. Binkley and D. Bankaitis, *J. Carbohydr. Chem.*, **1**, 1 (1982).
27. R. C. Roth and R. W. Binkley, *J. Org. Chem.*, **50**, 690 (1985).
28. J. L. Adcock and H. Luo, *J. Org. Chem.*, **59**, 1115 (1994)
29. F. Markel and A. B. Myers, *Chem. Phys. Lett.*, **167**, 175 (1990).
30. D. L. Phillips and A. B. Myers, *J. Chem. Phys.*, **95**, 226 (1991).
31. R. A. Rossi, S. M. Palacios and A. N. Santiago, *J. Org. Chem.*, **47**, 4654 (1982).
32. S. M. Palacios, A. N. Santiago and R. A. Rossi, *J. Org. Chem.*, **49**, 4609 (1984).
33. S. M. Palacios, R. A. Alonso and R. A. Rossi, *Tetrahedron*, **41**, 4147 (1985).
34. G. L. Borosky, A. B. Pierini and R. A. Rossi, *J. Org. Chem.*, **55**, 3705 (1990).
35. A. N. Santiago, V. S. Iyer, W. Adcock and R. A. Rossi, *J. Org. Chem.*, **53**, 3016 (1988).
36. A. N. Santiago, D. G. Morris and R. A. Rossi, *J. Chem. Soc., Chem Commun.*, 220 (1988).
37. A. N. Santiago and R. A. Rossi, *J. Chem. Soc., Chem. Commun.*, 206 (1990).
38. S. M. Palacios and R. A. Rossi, *J. Phys. Org. Chem.*, **3**, 812 (1990).
39. R. A. Rossi, A. B. Pierini and S. M. Palacios, *J. Chem. Educ.*, **66**, 720 (1989).
40. R. A. Rossi, A. N. Santiago and S. M. Palacios, *J. Org. Chem.*, **49**, 3387 (1984).
41. (a) G. F. Meijs, *J. Org. Chem.*, **49**, 3863 (1984).
    (b) G. F. Meijs, *J. Org. Chem.*, **51**, 606 (1986).
42. A. B. Pierini, A. B. Peñéñory and R. A. Rossi, *J. Org. Chem.*, **50**, 2739 (1985).
43. E. R. N. Bornancini, S. M. Palacios, A. B. Penenory and R. A. Rossi, *J. Phys. Org. Chem.*, **2**, 255 (1989).
44. (a) A. E. Feiring, *J. Org. Chem.*, **48**, 347 (1983).
    (b) A. E. Feiring, *J. Org. Chem.*, **50**, 3269 (1985).
45. Q.-Y. Chen and Z.-M. Qiu, *J. Fluorine Chem.*, **35**, 343 (1987).
46. Q.-Y. Chen and Z.-M. Qiu, *J. Chem. Soc., Chem. Commun.*, 1240 (1987).
47. Q.-Y. Chen and M. Chen, *J. Fluorine Chem.*, **51**, 21 (1991).
48. Y. Ogata, K. Tomizawa and K. Furuta, *J. Org. Chem.*, **46**, 5276 (1981).
49. M. E. Kurz, T. Noreuil, J. Seebauer, S. Cook, D. Geier, A. Leeds, C. Stronach, B. Barnickel, M. Kerkemeyer, M. Yandrasits, J. Witherspoon and F. J. Frank, *J. Org. Chem.*, **53**, 172 (1988).
50. M. Kurz and M. Rodgers, *J. Chem. Soc., Chem. Commun.*, 1227 (1985).
51. M. E. Hassan, *Coll. Czech. Chem. Commun.*, **50**, 2319 (1985).
52. M. E. Hassan, *Recl. Trav. Chim. Pays-Bas*, **105**, 30 (1986).
53. H. Kimoto, S. Fujii and L. A. Cohen, *J. Org. Chem.*, **47**, 2867 (1982).
54. H. Kimoto, S. Fujii and L. A. Cohen, *J. Org. Chem.*, **49**, 1060 (1984).
55. S. Fujii, Y. Maki, H. Kimoto and L. A. Cohen, *J. Fluorine Chem.*, **35**, 437 (1987).
56. V. M. Labroo, R. B. Labroo and L. A. Cohen, *Tetrahedron Lett.*, **31**, 5705 (1990).
57. M. Nishida, H. Kimoto, S. Fijii, Y. Hayakawa and L. A. Cohen, *Bull. Chem. Soc. Jpn.*, **64**, 2255 (1991).
58. D. Naumann, B. Wilkes and J. Kischkewitz, *J. Fluorine Chem.*, **30**, 73 (1985).
59. D. Naumann and J. Kischkewitz, *J. Fluorine Chem.*, **46**, 265 (1990).
60. K. L. Kirk, M. Nishida, S. Fujii and H. Kimoto, *J. Fluorine Chem.*, **59**, 197 (1992).
61. Q.-Y. Chen and Z.-T. Li, *J. Chem. Soc., Perkin Trans. 1*, 1443 (1992).
62. T. Akiyama, K. Kato, M. Kajitani, Y. Sakaguchi, J. Nakamura, H. Hayashi and A. Sugimori, *Bull. Chem. Soc. Jpn.*, **61**, 3531 (1988).
63. A. A. Abdel-Wahab, M. T. Ismail, O. S. Mohamed and A. A. Khalaf, *Gazz. Chim. Ital.*, **115**, 591 (1985).
64. V. K. Bhalerao, B. S. Nanjundiah, H. R. Sonawane and P. M. Nair, *Tetrahedron*, **42**, 1487 (1986).

65. K. V. Subbarao, N. P. Damodaran and S. Dev, *Tetrahedron*, **43**, 2543 (1987).
66. K. V. Subbarao, N. P. Damodaran and S. Dev, *Indian J. Chem.*, **26B**, 1008 (1987).
67. (a) M. Mitani, M. Nakayama and K. Koyama, *Tetrahedron Lett.*, **21**, 4457 (1980).
    (b) M. Mitani, I. Kato and K. Koyama, *J. Am. Chem. Soc.*, **105**, 6719 (1983).
68. M. Hamada, E. Kawano, S. Kawamura and M. Shiro, *Agric. Biol. Chem.*, **45**, 659 (1981).
69. (a) H. Parlar and F. Korte, *Chemosphere*, **12**, 927 (1983).
    (b) H. Parlar, *Chemosphere*, **17**, 2141 (1988).
70. J. C. Scaiano, M. Barra, G. Calabrese and R. Sinta, *J. Chem. Soc., Chem. Commun.*, 1418 (1992).
71. K. Takagi and Y. Ogata, *J. Org. Chem.*, **48**, 1966 (1983).
72. M. Barra, R. W. Redmond, M. T. Allen, G. S. Calabrese, R. Sinta and J. C. Scaiano, *Macromolecules*, **24**, 4972 (1991).
73. P. J. Kropp and N. J. Pienta, *J. Org. Chem.*, **48**, 2084 (1983).
74. E. Moret, C. R. Jones and B. Grant, *J. Org. Chem.*, **48**, 2090 (1983).
75. P. J. Kropp, N. J. Pienta, J. A. Sawyer and R. P. Polniaszek, *Tetrahedron*, **37**, 3229 (1981).
76. G. F. Meijs, *Tetrahedron Lett.*, **26**, 105 (1985).
77. R. A. Rossi and A. N. Santiago, *J. Chem. Res., Synop.*, 172 (1988).
78. J. F. Hartwig, M. Jones, Jr. R. A. Moss and W. Lawrynowicz, *Tetrahedron Lett.*, **27**, 5907 (1986).
79. (a) P. M. Warner, S. L. Lu and R. Gurumurthy, *J. Phys. Org. Chem.*, **1**, 281 (1988).
    (b) N. A. Le, M. Jones, Jr. F. Bickelhaupt and W. H. de Wolf, *J. Am. Chem. Soc.*, **111**, 8491 (1989).
80. J. E. Chateauneuf, R. P. Johnson and M. M. Kirchhoff, *J. Am. Chem. Soc.*, **112**, 3217 (1990).
81. G. Maier and H. P. Reisenauer, *Angew. Chem., Int. Ed. Engl.*, **25**, 819 (1986).
82. G. Maier, H. P. Reisenauer, J. Hu, L. J. Schaad and B. A. Hess, Jr. *J. Am. Chem. Soc.*, **112**, 5117 (1990).
83. G. Maier, U. Flögel, H. P. Reisenauer, B. A. Hess Jr. and L. J. Schaad, *Chem. Ber.*, **124**, 2603 (1991).
84. G. Maier, H. P. Reisenauer, J. Hu, B. A. Hess Jr. and L. J. Schaad, *Tetrahedron Lett.*, **30**, 4105 (1989).
85. H. Miyasaka, H. Masuhara and N. Magata, *Chem. Phys. Lett.*, **118**, 459 (1985).
86. (a) R. Stösser, B. Pritze, W. Abraham, B. Dreher and D. Kreysig, *J. Prakt. Chem.*, **327**, 310 (1985).
    (b) R. Stösser, B. Pritze, W. Abraham, B. Dreher and D. Kreysig, *J. Prakt. Chem.*, **327**, 317 (1985).
87. J. S. Keute, D. R. Anderson and T. H. Koch, *J. Am. Chem. Soc.*, **103**, 5434 (1981).
88. J. Chateauneuf, J. Lusztyk and K. U. Ingold, *J. Org. Chem.*, **55**, 1061 (1990).
89. B. Vessal and J. G. Miller, *J. Phys. Chem.*, **86**, 2695 (1982).
90. S. Bakkas, M. Julliard and M. Chanon, *Tetrahedron*, **43**, 501 (1987).
91. W. M. Wiczk and T. Latowski, *Pol. J. Chem.*, **64**, 373 (1990).
92. C. Moltke-Leth and K. A. Jørgensen, *J. Chem. Soc., Perkin Trans. 2*, 1487 (1993).
93. T. Gannon and W. G. McGimpsey, *J. Org. Chem.*, **58**, 5639 (1993).
94. J. M. Warman and R. J. Visser, *Chem, Phys. Lett*, **98**, 49 (1983).
95. H. Isaka, S. Suzuki, T. Ohzeki, Y. Sakaino and H. Takahashi, *J. Photochem.*, **38**, 167 (1987).
96. G. Yang, W. Cao and X. Feng, *Sci. Sin. Ser. B. (Engl. Ed.)*, **30**, 816 (1987).
97. Sh. A. Markaryan, *Arm. Khim. Zh.*, **35**, 281 (1982); *Chem. Abstr.*, **97**, 126924a (1982).
98. S. C. Blackstock, J. P. Lorand and J. K. Kochi, *J. Org. Chem.*, **52**, 1451 (1987).
99. (a) V. A. Sazhnikov, A. G. Strukov, M. Stunzas, S. P. Efimov, O. M. Andreev and M. V. Alfimov, *Dokl. Akad. Nauk SSSR*, **288**, 172 (1986); *Chem. Abstr.*, **105**, 15124d (1986).
    (b) M. F. Budyka, O. D. Laukhina, V. A. Sazhnikov and M. G. Stunzhas and M. V. Alfimov, *Dokl. Akad. Nauk SSSR*, **309**, 1126 (1989); *Chem. Abstr.*, **112**, 178638h (1990).
100. M. F. Budyka, O. D. Laukhina, A. A. Korkin and M. V. Alfimov, *Izv. Akad. Nauk SSSR, Ser. Khim.*, 564 (1991); *Chem. Abstr.*, **116**, 127940a (1992).
101. M. F. Budyka, G. V. Zakharova, O. D. Laukhina and M. V. Alfimov, *J. Photochem. Photobiol., A: Chem.*, **66**, 205 (1992).
102. B. Zelent and G. Durocher, *J. Org. Chem.*, **46**, 1496 (1981).
103. (a) B. Zelent and G. Durocher, *Can. J. Chem.*, **60**, 945 (1982).
     (b) B. Zelent and G. Durocher, *Can. J. Chem.*, **60**, 2442 (1982).
     (c) P. D. Harvey and G. Durocher, *Can. J. Chem.*, **63**, 1723 (1985).
104. B. K. Chowdhury, C. Saha, G. Podder and P. Bhattacharyya, *Indian J. Chem., Sect. B.*, **29**, 405 (1990).

105. B. K. Chowdhury, C. Saha, G. Podder and P. Bhattacharyya, *Indian J. Chem., Sect. B.*, **30**, 63 (1991).
106. R. Erra-Balsells and A. R. Frasca., *Tetrahedron Lett.*, **25**, 5363 (1984).
107. M. C. Biondic and R. Erra-Balsells, *J. Photochem. Photobiol., A: Chem.*, **51**, 341 (1990).
108. M. C. Biondic and R. Erra-Balsells, *J. Photochem. Photobiol., A: Chem.*, **77**, 149 (1994).
109. W. Boszczyk and T. Latowski, *Z. Naturforsch.*, **44b**, 1585 (1989).
110. W. Boszczyk and T. Latowski, *Z. Naturforsch.*, **44b**, 1589 (1989).
111. I. Rico, D. Cantacuzene and C. Wakselman, *Tetrahedron Lett.*, **22**, 3405 (1981).
112. A. V. Maleev, K. A. Potekhin, A. I. Yanovsky, Yu. T. Struchkov, V. A. Vasin, I. Yu. Bolusheva, L. S. Surmina and N. S. Zefirov, *Dokl. Akad. Nauk* SSSR, **327**, 345 (1992); *Chem. Abstr.*, **118**, 147216p (1993).
113. M. A. Tyurekhodzhaeva, A. A. Bratkova, A. V. Blokhin, V. K. Brel, A. S. Kozmin and S. Zefirov, *J. Fluorine Chem.*, **55**, 237 (1991).
114. S. Fukuzumi, K. Hironaka and T. Tanaka, *Chem. Lett.*, 1583 (1982); S. Fukuzumi, K. Hironaka and T. Tanaka, *J. Am. Chem. Soc.*, **105**, 4722 (1983).
115. M. Ishikawa and S. Fukuzumi, *J. Am. Chem. Soc.*, **112**, 8864 (1990).
116. C.-M. Che and W.-M. Lee, *J. Chem. Soc., Chem. Commun.*, 616 (1986).
117. C. Bartocci, A. Maldotti, G. Varani, V. Carassiti, P. Battioni and D. Mansuy, *J. Chem. Soc., Chem. Commun.*, 964 (1989).
118. P. Dureja and S. K. Mukerjee, *Tetrahedron Lett.*, **26**, 5211 (1985).
119. P. Dureja, S. Walia and S. K. Mukerjee, *Indian J. Chem.*, **25B**, 741 (1986).
120. S. Walia, P. Dureja and S. K. Mukerjee, *Tetrahedron*, **43**, 2493 (1987).
121. W. V. Turner and S. Gäb, *Tetrahedron*, **45**, 3711 (1989).
122. A. L. J. Beckwith and S. H. Goh, *J. Chem. Soc., Chem. Commun.*, 907 (1983).
123. N. Shimizu, K. Watanabe and Y. Tsuno, *Bull. Chem. Soc. Jpn.*, **57**, 885 (1984).
124. S. J. Cristol, D. Braun, G. C. Schloemer and B. J. Vanden Plas, *Can. J. Chem.*, **64**, 1081 (1986).
125. S. J. Cristol and B. J. Vanden Plas, *J. Phys. Org. Chem.*, **4**, 541 (1991).
126. M. R. Bryce, R. D. Chambers and G. Taylor, *J. Chem. Soc., Chem. Commun.*, 1457 (1983).
127. K. M. Saplay, N. P. Damodaran and S. Dev, *Tetrahedron*, **39**, 2999 (1983).
128. W. R. Bowman, D. S. Brown, C. T. W. Leung and A. P. Stutchbury, *Tetrahedron Lett.*, **26**, 539 (1985).
129. S. D. Barker and R. K. Norris, *Aust. J. Chem.*, **36**, 527 (1983).
130. N. S. Zefirov, M. A. Kirpichenok and T. G. Shestakova, *Dokl. Akad. Nauk SSSR*, **262**, 890 (1982); *Chem. Abstr.*, **96**, 217304u (1982); N. S. Zefirov, M. A. Kirpichenok and T. G. Shestakova, *Zh. Org. Khim.*, **19**, 897 (1983) [Engl. Transl.: *J. Org. Chem. USSR*, **19**, 795 (1983)].
131. M. E. Krafft, *J. Photochem.*, **38**, 391 (1987).
132. (a) W. Oppolzer, F. Zutterman and K. Bättig, *Helv. Chim. Acta*, **66**, 522 (1983).
     (b) H. D. Roth, M. L. M. Schilling and C. C. Wamser, *J. Am. Chem. Soc.*, **106**, 5023 (1984).
133. (a) J. L. Davidson and G. Vasapollo, *J. Organomet. Chem.*, **291**, 43 (1983).
     (b) R. H. Hill, A. Becalska and N. Chiem, *Organometallics*, **10**, 2104 (1991).
134. N. S. Zefirov, S. I. Kozhushkov, T. S. Kuznetsova and I. M. Sosonkin, *Zh. Org. Khim.*, **22**, 666 (1986) [Engl. Transl.: *J. Org. Chem. USSR*, **22**, 596 (1986)]; N. S. Zefirov, S. I. Kozhushkov and T. S. Kuznetsova, *Zh. Org. Khim.*, **25**, 910 (1989). [Engl. Transl: *J. Org. Chem. USSR*, **25**, 816 (1989)].
135. S. J. Cristol and B. J. Vanden Plas, *J. Org. Chem.*, **54**, 1209 (1989).
136. H. Morrison, T. V. Singh, L. de Cardenas, D. Severance, K. Jordan and W.Schaefer, *J. Am. Chem. Soc.*, **108**, 3862 (1986).
137. B. D. Maxwell, J. J. Nash, H. Morrison, M. L. Falcetta and K. D. Jordan, *J. Am. Chem. Soc.*, **111**, 7914 (1989).
138. J. J. Nash and H. Morrison, *J. Org. Chem.*, **55**, 1141 (1990).
139. R. S. Givens and B. W. Atwater, *J. Am. Chem. Soc.*, **108**, 5028 (1986).
140. S. J. Cristol and T. H. Bindel, *J. Am. Chem. Soc.*, **103**, 7287 (1981).
141. G. G. Choudhry, A. A. M. Roof and O. Hutzinger, *J. Chem. Soc. Perkin Trans. 1*, 2957 (1982).
142. S. S. Hixson and V. R. Rao, *J. Chem. Soc., Chem. Commun.*, 65 (1984).
143. G.H. Slocum and G.B. Schuster, *J. Org. Chem.*, **49**, 2177 (1984).
144. B. Arnold, L. Donald, A. Jurgens and J. A. Pincock, *Can. J. Chem.*, **63**, 3140 (1985).
145. G. P. Gardini, J. L. Charlton and J. Bargon, *Tetrahedron Lett.*, **23**, 987 (1982).
146. G. H. Slocum, K. Kaufmann and G. B. Schuster, *J. Am. Chem. Soc.*, **103**, 4625 (1981).

147. S. C. Shim and S. J. Choi, *Bull. Korean Chem. Soc.*, **3**, 30 (1982).
148. S. Jaarinen, J. Niiranen and J. Koskikallio, *Int. J. Chem. Kinet.*, **17**, 925 (1985).
149. M. Alvaro, A. Corma, H. Garcia, M. A. Miranda and J. Primo, *J. Chem. Soc., Chem. Commun.*, 1041 (1993).
150. S. J. Cristol and W. A. Dickenson, *J. Org. Chem.*, **51**, 2973 (1986).
151. S. J. Cristol, R. J. Opitz and E. O. Aeling, *J. Org. Chem.*, **50**, 4834 (1985).
152. Y. Izawa, Y. Watoh and H. Tomioka, *Chem. Lett.*, 33 (1984).
153. Y. Izawa, M. Takeuchi and H. Tomioka, *Chem, Lett.*, 1297 (1983).
154. J. Bartl, S. Steenken, H. Mayr and R. A. McClelland, *J. Am. Chem. Soc.*, **112**, 6918 (1990).
155. L. J. Johnston, P. Kwong, A. Shelemay and E. Lee-Ruff, *J. Am. Chem. Soc.*, **115**, 1664 (1993).
156. P. Wan and E. Krogh, *J. Am. Chem. Soc.*, **111**, 4887 (1989).
157. J. Lilie and J. Koskikallio, *Acta Chem. Scand., Sect. A*, **38**, 41 (1984).
158. R. A. McClelland, C. Chan, F. Cozens, A. Modro and S. Steenken, *Angew. Chem., Int. Ed. Engl.*, **30**, 1337 (1991).
159. R. A. McClelland, F. L. Cozens, S. Steenken, T. L. Amyes and J. P. Richard, *J. Chem. Soc., Perkin Trans. 2*, 1717 (1993).
160. N. P. Schepp and J. Wirz, *J. Am. Chem. Soc.*, **116**, 11749 (1994).
161. S. L. Mecklenburg and E. F. Hilinski, *J. Am. Chem. Soc.*, **111**, 5471 (1989).
162. R. A. McClelland, V. M. Kanagasabapathy, N. S. Banait and S. Steenken, *J. Am. Chem. Soc.*, **113**, 1009 (1991).
163. J. Bartl, S. Steenken and H. Mayr, *J. Am. Chem. Soc.*, **113**, 7710 (1991).
164. L. E. Manring and K. S. Peters, *J. Phys. Chem.*, **88**, 3516 (1984).
165. H. Morrison, T. V. Singh and B. Maxwell, *J. Org. Chem.*, **51**, 3707 (1986).
166. D. F. Kelley, S. V. Milton, D. Huppert and P. M. Rentzepis, *J. Phys. Chem.*, **87**, 1842 (1983).
167. E. F. Hilinski, D. Huppert, D. F. Kelley, S. V. Milton and P. M. Rentzepis, *J. Am. Chem. Soc.*, **106**, 1951 (1984).
168. T. Gannon and W.G. McGimpsey, *J. Org. Chem.*, **58**, 913 (1993).
169. T. Takemura and M. Fujita, *Bull. Chem. Soc. Jpn.*, **60**, 399 (1987).
170. L. van Haelst, E. Haselbach and P. Suppan, *Chimia*, **42**, 231 (1988).
171. A. Kawai, T. Okutsu and K. Obi, *Chem. Phys. Lett.*, **174**, 213 (1990).
172. K. Tokumura, H. Nosaka and T. Ozaki, *Chem. Phys. Lett.*, **169**, 321 (1990).
173. K. Tokumura, T. Ozaki, H. Nosaka, Y. Saigusa and M. Itoh, *J. Am. Chem. Soc.*, **113**, 4974 (1991).
174. L. J. Johnston and J. C. Scaiano, *J. Am. Chem. Soc.*, **107**, 6368 (1985).
175. D. Weir, L. J. Johnston and J. C. Scaiano, *J. Phys. Chem.*, **92**, 1742 (1988).
176. A. Ouchi and A. Yabe, *Tetrahedron Lett.*, **31**, 1727 (1990).
177. W. Adam and A. Ouchi, *Tetrahedron Lett.*, **33**, 1875 (1992).
178. A. Ouchi and A. Yabe, *Tetrahedron Lett.*, **33**, 5359 (1992).
179. K. Haider, M. S. Platz, A. Despres, V. Lejeune, E. Migirdicyan, T. Bally and E. Haselbach, *J. Am. Chem. Soc.*, **110**, 2318 (1988).
180. K. W. Haider and M. S. Platz, *J. Phys. Org. Chem.*, **2**, 623 (1989).
181. N. Kornblum, L. Cheng, T. M. Davies, G. W. Earl, N. L. Holy, R. C. Kerber, M. M. Kestner, J. W. Manthey, M. T. Musser, H. W. Pinnick, D. H. Snow, F. W. Stuchal and R. T. Swiger, *J. Org. Chem.*, **52**, 196 (1987).
182. P. A. Wade, H. A. Morrison and N. Kornblum, *J. Org. Chem.*, **52**, 3102 (1987).
183. R. Beugelmans, A. Lechevallier and H. Rousseau, *Tetrahedron Lett.*, **24**, 1787 (1983).
184. M. P. Crozet and P. Vanelle, *Tetrahedron Lett.*, **26**, 323 (1985); M. P. Crozet and P. Vanelle, *Tetrahedron*, **45**, 5477 (1989).
185. M. P. Crozet, G. Archaimbault, P. Vanelle and R. Nouguier, *Tetrahedron Lett.*, **26**, 5133 (1985).
186. A. T. O. M. Adebayo, W. R. Bowman and W. G. Salt, *Tetrahedron Lett.*, **27**, 1943 (1986); *J. Chem. Soc., Perkin Trans. 1*, 2819 (1987).
187. R. Beugelmans, A. Lechevallier, D. Kiffer and P. Maillos, *Tetrahedron Lett.*, **27**, 6209 (1986).
188. G. A. Russell, F. Ros, J. Hershberger and H. Tashtoush, *J. Org. Chem.*, **47**, 1480 (1982).
189. C. D. Beadle and W. R. Bowman, *J. Chem. Res. (S)*, 150 (1985).
190. M. P. Crozet, J.-M. Surzur, P. Vanelle, G. Ghiglione and J. Maldonado, *Tetrahedron Lett.*, **26**, 1023 (1985).
191. P. Vanelle, J. Maldonado, N. Madadi, A. Gueiffier, J.-C. Teulade, J.-P. Chapat and M. P. Crozet, *Tetrahedron Lett.*, **31**, 3013 (1990).

192. R. K. Norris and D. Randles, *J. Org. Chem.*, **47**, 1047 (1982).
193. R. K. Norris and D. Randles, *Aust. J. Chem.*, **35**, 1621 (1982).
194. M. S. K. Lee, P. J. Newcombe, R. K. Norris and K. Wilson, *J. Org. Chem.*, **52**, 2796 (1987).
195. F. I. Flower, P. J. Newcombe and R. K. Norris, *J. Org. Chem.*, **48**, 4202 (1983).
196. M. P. Crozet, P. Vanelle, O. Jentzer and J. Maldonado, *Tetrahedron Lett.*, **31**, 1269 (1990).
197. S. D. Barker and R. K. Norris, *Aust. J. Chem.*, **36**, 81 (1983).
198. H. Feuer, J. K. Doty and N. Kornblum, *J. Heterocycl. Chem.*, **18**, 783 (1981).
199. M. P. Crozet, O. Jentzer and P. Vanelle, *Tetrahedron Lett.*, **28**, 5531 (1987); M. Crozet, P. Vanelle, O. Jentzer, S. Donini and J. Maldonado, *Tetrahedron*, **49**, 11253 (1993).
200. X. Creary, A. F. Sky and G. Phillips, *J. Org. Chem.*, **55**, 2005 (1990).
201. B. D. Jacobs, S.-J. Kwon, L. D. Field, R. K. Norris, D. Randles, K. Wilson and T. A. Wright, *Tetrahedron Lett.*, **26**, 3495 (1985).
202. R. K. Norris and T. A. Wright, *Aust. J. Chem.*, **38**, 1107 (1985).
203. M. C. R. Symons and W. R. Bowman, *J. Chem. Soc., Chem. Commun.*, 1445 (1984); M. C. R. Symons and W.R. Bowman, *J. Chem. Soc., Perkin Trans. 2*, 583 (1988).
204. M. C. R. Symons and W. R. Bowman, *J. Chem. Soc., Perkin Trans. 2*, 1133 (1987).
205. M. C. R. Symons and W. R. Bowman, *J. Chem. Soc., Perkin Trans. 2*, 1077 (1988).
206. K. Takagi, N. Miyake, E. Nakamura, Y. Sawaki, N. Koga and H. Iwamura, *J. Org. Chem.*, **53**, 1703 (1988).
207. K. Takagi, N. Miyake, E. Nakamura, H. Usami, Y. Sawaki and H. Iwamura, *J. Chem. Soc., Faraday Trans. 1*, **84**, 3475 (1988).
208. K. Yamashita, H. Ishida and K. Ohkubo, *Chem. Lett.*, 1637 (1991).
209. K. M. O'Connell and D. H. Evans, *J. Am. Chem. Soc.*, **105**, 1473 (1983).
210. (a) Z. Goren and I. Willner, *J. Am. Chem. Soc.*, **105**, 7764 (1983).
    (b) R. Maidan, Z. Goren, J. Y. Becker and I. Willner, *J. Am. Chem. Soc.*, **106**, 6217 (1984).
211. R. Maidan and I. Willner, *J. Am. Chem. Soc.*, **108**, 1080 (1986).
212. H. Tomioka and O. Inoue, *Bull. Chem. Soc. Jpn.*, **61**, 3725 (1988).
213. J. D. Simon and K. S. Peters, *Organometallics*, **2**, 1867 (1983).
214. A. Corma, H. García, M. A. Miranda, J. Primo and M. J. Sabater, *J. Org. Chem.*, **58**, 6892 (1993).
215. A. K. Singh and M. Singh, *Bull. Soc. Chem. Belg.*, **95**, 1131 (1986).
216. J. L. Wood, P. J. Carroll and A. B. Smith, *J. Chem. Soc., Chem. Commun.*, 1433 (1992).
217. H. Morrison and A. Miller, *J. Am. Chem. Soc.*, **102**, 372 (1980); H. Morrison, A. Miller and B. Bigot, *J. Am. Chem. Soc.*, **105**, 2398 (1983).
218. H. Morrison and L. M. de Cardenas, *Tetrahedron Lett.*, **25**, 2527 (1984).
219. H. Morrison, A. Miller, B. Pandey, G. Pandey, D. Severance, R. Strommen and B. Bigot, *Pure Appl. Chem.*, **54**, 1723 (1982).
220. H. Morrison, K. Muthuramu, G. Pandey, D. Severance and B. Bigot, *J. Org. Chem.*, **51**, 3358 (1986).
221. S. J. Cristol, W. A. Dickenson and M. K. Stanko, *J. Am. Chem. Soc.*, **105**, 1218 (1983).
222. S. J. Cristol and W. A. Dickenson, *J. Org. Chem.*, **51**, 3625 (1986).
223. S. J. Cristol, R. J. Opitz, T. H. Bindel and W. A. Dickenson, *J. Am. Chem. Soc.*, **102**, 7977 (1980); S. J. Cristol and R. J. Opitz, *J. Org. Chem.*, **50**, 4558 (1985).
224. S. J. Cristol, D. G. Seapy and E. O. Aeling, *J. Am. Chem. Soc.*, **105**, 7337 (1983).
225. S. J. Cristol, T. H. Bindel, D. Hoffmann and E. O. Aeling, *J. Org. Chem.*, **49**, 2368 (1984).
226. S. J. Cristol and E. O. Aeling, *J. Org. Chem.*, **50**, 2698 (1985).
227. S. J. Cristol, E. O. Aeling and R. Heng, *J. Am. Chem. Soc.*, **109**, 830 (1987).
228. S. J. Cristol, E. O. Aeling, S. J. Strickler and R. D. Ito, *J. Am. Chem. Soc.*, **109**, 7101 (1987).
229. S. J. Cristol and M. Z. Ali, *J. Org. Chem.*, **50**, 2502 (1985).
230. H. Morrison, K. Muthuramu and D. Severance, *J. Org. Chem.*, **51**, 4681 (1986).
231. P. C. Purohit and H. R. Sonawane, *Tetrahedron*, **37**, 873 (1981).
232. K. Reinholdt and P. Margaretha, *Helv. Chim. Acta*, **66**, 2534 (1983).
233. K. Reinholdt and P. Margaretha, *J. Fluorine Chem.*, **36**, 119 (1987).
234. H. Morrison and L. de Cardenas, *J. Org. Chem.*, **52**, 2590 (1987).
235. B. Šket and M. Zupan, *Bull. Chem. Soc. Jpn.*, **60**, 4489 (1987).
236. B. Šket and M. Zupan, *Coll. Czech. Chem. Commun.*, **53**, 1745 (1988).
237. N. Zupancic and B. Šket, *J. Photochem. Photobiol., A: Chem.*, **63**, 303 (1992).
238. B. Šket, N. Zupančič and M. Zupan, *J. Fluorine Chem.*, **45**, 313 (1989).

239. W. R. Bergmark, C. Barnes, J. Clark, S. Paparian and S. Marynowski, *J. Org. Chem.*, **50**, 5612 (1985).
240. H. R. Sonawane, D. G. Kulkarni and N. R. Ayyangar, *Tetrahedron Lett.*, **31**, 7495 (1990); H. R. Sonawane, N. S. Bellur, D. G. Kulkarni and N. R. Ayyangar, *Tetrahedron*, **50**, 1243 (1994).
241. J. C. Netto-Ferreira and J. C. Scaiano, *J. Am. Chem. Soc.*, **113**, 5800 (1991).
242. H. García, R. Martínez-Utrilla and M. A. Miranda, *J. Chem. Res. (S)*, 350 (1982); H. García, R. Martínez-Utrilla and M. A. Miranda, *Ann. Chem.*, 589 (1985).
243. N. J. Bunce, K. U. Ingold, J. P. Landers, J. Lusztyk and J. C. Scaiano, *J. Am. Chem. Soc.*, **107**, 5464 (1985).
244. W. G. McGimpsey and J. C. Scaiano, *Can. J. Chem.*, **66**, 1474 (1988).
245. H. E. Zimmerman and L. W. Linder, *J. Org. Chem.*, **50**, 1637 (1985).
246. N. K. Hamer, *Tetrahedron Lett.*, **23**, 473 (1982); N. K. Hamer, *J. Chem. Soc., Perkin Trans. 1*, 61 (1983).
247. G. A. Russell and F. Ros, *J. Am. Chem. Soc.*, **104**, 7349 (1982); G. A. Russell and F. Ros, *J. Am. Chem. Soc.*, **107**, 2506 (1985).
248. A. N. Santiago, K. Takeuchi, Y. Ohga, M. Nishida and R. A. Rossi, *J. Org. Chem.*, **56**, 1581 (1991).
249. S. Fukuzumi, S. Mochizuki and T. Tanaka, *Chem. Lett.*, 1983 (1988).
250. S. Fukuzumi, S. Mochizuki and T. Tanaka, *J. Chem. Soc., Perkin Trans. 2*, 1583 (1989).
251. S. Fukuzumi, S. Mochizuki and T. Tanaka, *J. Phys. Chem.*, **94**, 722 (1990).
252. T. Sato and K. Tamura, *Tetrahedron Lett.*, **25**, 1821 (1984).
253. Y. Izawa, K. Ishiguro and H. Tomioka, *Bull. Chem. Soc. Jpn.*, **56**, 951 (1983).
254. Y. Izawa, K. Ishiguro and H. Tomioka, *Bull. Chem. Soc. Jpn.*, **56**, 1490 (1983).
255. L. de Cardenas, B. D. Maxwell, T. V. Singh and H. Morrison, *J. Org. Chem.*, **53**, 219 (1988).
256. P. J. Wagner, M. J. Lindstrom, J. H. Sedon and D. R. Ward, *J. Am. Chem. Soc.*, **103**, 3842 (1981).
257. A. L. Beck, M. Mascal, C. J. Moody and W. J. Coates, *J. Chem. Soc., Perkin Trans. 1*, 813 (1992).
258. O. Paleta, R. Ježek and V. Dědek, *Coll. Czech. Chem. Commun.*, **48**, 766 (1983).
259. O. Paleta, V. Dadák, V. Dědek and H.-J. Timpe, *J. Fluorine Chem.*, **39**, 397 (1988).
260. C. Portella, *Ann. Chim. Fr.*, **9**, 689 (1984); C. Portella and J. P. Pete, *Tetrahedron Lett.*, **26**, 211 (1985).
261. C. Portella and M. Iznaden, *Tetrahedron*, **45**, 6467 (1989).
262. H.-J. Timpe, R. Wagner, R. Dusi, O. Paleta and V. Dadak, *Z. Chem.*, **26**, 256 (1986).
263. S. V. Truksa, A. Nibler, B. S. Schatz, K. W. Krosley and G. J. Gleicher, *J. Org. Chem.*, **57**, 2967 (1992).
264. (a) S. Lapin and M. E. Kurz, *J. Chem. Soc., Chem. Commun.*, 817 (1981).
    (b) M. E. Kurz, S. C. Lapin, K. Mariam, T. J. Hagen and X. Q. Qian, *J. Org. Chem.*, **49**, 2728 (1984).
265. C. Szántay, T. Keve, H. Bölcskei and T. Acs, *Tetrahedron Lett.*, **24**, 5539 (1983).
266. S. Raucher and B. L. Bray, *J. Org. Chem.*, **50**, 3236 (1985); S. Raucher, B. L. Bray and R. F. Lawrence, *J. Am. Chem. Soc.*, **109**, 442 (1987).
267. T. Hamada, Y. Okuno, M. Ohmori, T. Nishi and O. Yonemitsu, *Chem. Pharm. Bull.*, **29**, 128 (1981).
268. B. Kumar, R. M. Metha, S. C. Kalra and N. Kaur, *J. Chem. Soc., Perkin Trans. 1*, 1387 (1984).
269. J. B. Bremner, W. Jaturonrusmee, L. M. Engelhardt and A. H. White, *Tetrahedron Lett.*, **30**, 3213 (1989).
270. J. Bosch, M. Amat, E. Sanfeliu and M. A. Miranda, *Tetrahedron*, **41**, 2557 (1985).
271. S. E. Klohr and J. M. Cassady, *Synth. Commun.*, **18**, 671 (1988).
272. (a) M. Mascal and C. J. Moody, *J. Chem. Soc., Chem. Commun.*, 587 (1988).
    (b) A. L. Beck, M. Mascal, C. J. Moody, A. M. Z. Slawin, D. J. Williams and W. J. Coates, *J. Chem. Soc., Perkin Trans. 1*, 797 (1992).
273. (a) M. Mascal and C. J. Moody, *J. Chem. Soc., Chem. Commun.*, 589 (1988).
    (b) M. Mascal, C. J. Moody, A. M. Z. Slawin and D. J. Williams, *J. Chem. Soc., Perkin Trans. 1*, 823 (1992).
274. G. A. Russell, B. Mudryk and M. Jawdosiuk, *Synthesis*, 62 (1981).
275. N. Ono, R. Tamura, H. Eto, I. Hamamoto, T. Nakatsuka, J. Hayami and A. Kaji, *J. Org. Chem.*, **48**, 3678 (1983).
276. (a) B. Aebischer, R. Meuwly and A. Vasella, *Helv. Chim. Acta*, **67**, 2236 (1984).
    (b) R. Meuwly and A. Vasella, *Helv. Chim. Acta*, **68**, 997 (1985).

277. G. A. Russell, B. Mudryk and M. Jawdosiuk, *J. Am. Chem. Soc.*, **103**, 4610 (1981).
278. G. A. Russell, B. Mudryk, F. Ros and M. Jawdosiuk, *Tetrahedron*, **38**, 1059 (1982).
279. G. A. Russell, B. Mudryk, M. Jawdosiuk and Z. Wrobel, *J. Org. Chem.*, **47**, 1879 (1982).
280. G. A. Russell and D.F. Dedolph, *J. Org. Chem.*, **50**, 2498 (1985).
281. R. K. Norris and R. J. Smyth-King, *Tetrahedron*, **38**, 1051 (1982).
282. W. R. Bowman and G. D. Richardson, *Tetrahedron Lett.*, **22**, 1551 (1981).
283 W. R. Bowman, D. Rakshit and M. D. Valmas, *J. Chem. Soc., Perkin Trans. 1*, 2327 (1984).
284. A. Amrollah-Madjdabadi, R. Beugelmans and A. Lechevallier, *Tetrahedron Lett.*, **28**, 4525 (1987).
285. S. I. Al-Khalil and W. R. Bowman, *Tetrahedron Lett.*, **25**, 461 (1984).
286. W. R. Bowman and M. C. R. Symons, *J. Chem. Soc., Perkin Trans. 2*, 25 (1983).
287. S. I. Al-Khalil and W. R. Bowman, *Tetrahedron Lett.*, **23**, 4513 (1982).
288. S. I. Al-Khalil, W. R. Bowman and M. C. R. Symons, *J. Chem. Soc., Perkin Trans. 1*, 555 (1986).
289. G. A. Russell and W. Baik, *J. Chem. Soc., Chem. Commun.*, 196 (1988).
290. M. Hanack and L. R. Subramanian, in *Carbokationen, Carbokation-Radikale, Houben-Weyl's Methoden der organischen Chemie*, Band E19c (Ed. M. Hanack). Thieme Verlag, Stuttgart, 1990, pp. 97–119.
291. P. J. Kropp, S. A. McNeely and R. D. Davis, *J. Am. Chem. Soc.*, **105**, 6907 (1983).
292. H. R. Sonawane, B. S. Nanjundiah and S. I. Rajput, *Indian J. Chem.*, **23B**, 331 (1984).
293. H. R. Sonawane, B. S. Nanjundiah and S. I. Rajput, *Indian J. Chem.*, **23B**, 339 (1984).
294. H. R. Sonawane, B. S. Nanjundiah and M. D. Panse, *Tetrahedron Lett.*, **26**, 3507 (1985).
295. T. Kitamura, T. Muta, S. Kobayashi and H. Taniguchi, *Chem. Lett.*, 643 (1982).
296. T. Suzuki, T. Kitamura, T. Sonoda, S. Kobayashi and H. Taniguchi, *J. Org. Chem.*, **46**, 5324 (1981).
297. T. Kitamura, S. Kobayashi and H. Taniguchi, *Chem. Lett.*, 1637 (1988).
298. T. Kitamura, S. Kobayashi and H. Taniguchi, *J. Am. Chem. Soc.*, **108**, 2641 (1986).
299. N. Johnen, W. Schnabel, S. Kobayashi and J. P. Fouassier, *J. Chem. Soc., Faraday Trans.*, **88**, 1385 (1992).
300. T. Kitamura, S. Kobayashi and H. Taniguchi, *J. Org. Chem.*, **47**, 2323 (1982).
301. T. Kitamura, T. Muta, T. Tahara, S. Kobayashi and H. Taniguchi, *Chem. Lett.*, 759 (1986).
302. J. Andraos, A. J. Kresge and P. A. Obraztsov, *J. Phys. Org. Chem.*, **5**, 322 (1992).
303. A. J. Kresge and N. P. Schepp, *J. Chem. Soc., Chem. Commun.*, 1548 (1989).
304. H. R. Sonawane, B. S. Nanjundiah, M. Udaykumar and M. D. Panse, *Indian J. Chem.*, **24B**, 202 (1985).
305. P. M. op den Brouw and W. H. Laarhoven, *Recl. Trav. Chim. Pays-Bas*, **101**, 58 (1982).
306. (a) P. M. op den Brouw and W. H. Laarhoven, *J. Org. Chem.*, **47**, 1546 (1982).
    (b) P. M. op den Brouw and W. H. Laarhoven, *J. Chem. Soc., Perkin Trans. 2*, 1015 (1983).
    (c) P. M. op den Brouw, P. de Zeeuw and W. H. Laarhoven, *J. Photochem.*, **27**, 327 (1984).
307. A. Gregorcic and M. Zupan, *J. Fluorine Chem.*, **41**, 163 (1988).
308. J. M. Verbeek, J. Cornelisse and G. Lodder, *Tetrahedron*, **42**, 5679 (1986).
309. J. M. Verbeek, M. Stapper, E. S. Krijnen, J. D. van Loon, G. Lodder and S. Steenken, *J. Phys. Chem.*, **98**, 9526 (1994).
310. T. Kitamura, S. Kobayashi, H. Taniguchi, C. Y. Fiakpui, C. C. Lee and Z. Rappoport, *J. Org. Chem.*, **49**, 3167 (1984).
311. F. I. M. van Ginkel, J. Cornelisse and G. Lodder, *J. Am. Chem. Soc.*, **113**, 4261 (1991).
312. F. I .M. van Ginkel, J. Cornelisse and G. Lodder, *J. Photochem. Photobiol. A:Chem.*, **61**, 301 (1991).
313. E. S. Krijnen, H. Zuilhof and G. Lodder, *J. Org. Chem.*, **59**, 8139 (1994).
314. A. Gregorcic and M. Zupan, *J. Fluorine Chem.*, **34**, 313 (1987).
315. E. S. Krijnen and G. Lodder, *Tetrahedron Lett.*, **34**, 729 (1993).
316. Y. Apeloig, R. Biton, H. Zuilhof and G. Lodder, *Tetrahedron Lett.*, **35**, 265 (1994).
317. T. Kitamura, S. Kobayashi and H. Taniguchi, *Chem. Lett.*, 1351 (1984).
318. T. Kitamura, S. Kobayashi and H. Taniguchi, *J. Org. Chem.*, **49**, 4755 (1984).
319. T. Kitamura, S. Kobayashi and H. Taniguchi, *Chem. Lett.*, 1523 (1984); T. Kitamura, S. Kobayashi and H. Taniguchi, *J. Org. Chem.*, **55**, 1801 (1990).
320. T. Kitamura, T. Kabashima and H. Taniguchi, *J. Org. Chem.*, **56**, 3739 (1991).
321. T. Kitamura, M. Murakami, S. Kobayashi and H. Taniguchi, *Tetrahedron Lett.*, **27**, 3885 (1986).

322. T. Kitamura, T. Kabashima, S. Kobayashi and H. Taniguchi, *Chem. Lett.*, 1951 (1988).
323. T. Kitamura, T. Kabashima, S. Kobayashi and H. Taniguchi, *Tetrahedron Lett.*, **29**, 6141 (1988).
324. T. Kitamura, I. Nakamura, T. Kabashima, S. Kobayashi and H. Taniguchi, *J. Chem. Soc.,
     Chem. Commun.*, 1154 (1989).
325. T. Kitamura, I. Nakamura, T. Kabashima, S. Kobayashi and H. Taniguchi, *Chem. Lett.*, 9 (1990).
326. T. Kitamura, I. Nakamura, T. Kabashima, S. Kobayashi and H. Taniguchi, *J. Am. Chem. Soc.*,
     112, 6149 (1990).
327. T. Kitamura, S. Soda, I. Nakamura, T. Fukuda and H. Taniguchi, *Chem. Lett.*, 2195 (1991).
328. T. Kitamura, T. Kabashima, I. Nakamura, T. Fukuda and H. Taniguchi, *J. Am. Chem. Soc.*,
     113, 7255 (1991).
329. K. Hori, H. Kamada, T. Kitamura, S. Kobayashi and H. Taniguchi, *J. Chem. Soc., Perkin Trans.
     2*, 871 (1992).
330. T. Kitamura, S. Kobayashi and H. Taniguchi, *Chem. Lett.*, 547 (1984); T. Kitamura, S.
     Kobayashi, H. Taniguchi and K. Hori, *J. Am. Chem. Soc.*, **113**, 6240 (1991).
331. T. Kitamura, H. Kawasato, S. Kobayashi and H. Taniguchi, *Chem. Lett.*, 839 (1986).
332. T. Ikeda, S. Kobayashi and H. Taniguchi, *Synthesis*, 393 (1982).
333. S. Kobayashi, T. Kitamura, H. Taniguchi and W. Schnabel, *Chem. Lett.*, 1117 (1983).
334. S. Kobayashi, T. Kitamura, H. Taniguchi and W. Schnabel, *Chem. Lett.*, 2101 (1984).
335. F. I. M. van Ginkel, R. J. Visser, C. A. G. O. Varma and G. Lodder, *J. Photochem.*, **30**, 453
     (1985).
336. S. Kobayashi, Q. Q. Zhu and W. Schnabel, *Z. Naturforsch.*, **43b**, 825 (1988).
337. S. Kobayashi and W. Schnabel, *Z. Naturforsch.*, **47b**, 1319 (1992).
338. Y. Chiang, R. Eliason, J. Jones, A.J. Kresge, K.L. Evans and R.D. Gandour, *Can. J. Chem.*, **71**,
     1964 (1993).
339. J. J. Brunet, C. Sidot and P. Caubere, *J. Org. Chem.*, **48**, 1166 (1983).
340. C. Galli and P. Gentili, *J. Chem. Soc., Chem. Commun.*, 570 (1993); C. Galli, P. Gentili and Z.
     Rappoport, *J. Org. Chem.*, **59**, 6786 (1994).
341. G. Szilagyi and H. Wamhoff, *Angew. Chem., Int. Ed. Engl.*, **22**, 161 (1983).
342. J.-B. Meng, M.-G. Shen, D.-C. Fu, Z.-H. Gao, R.-J. Wang, H.-G. Wang and T. Matsuura,
     Synthesis, 719 (1990).
343. H. Hombrecher and P. Margaretha, *Helv. Chim. Acta*, **65**, 2313 (1982).
344. H. S. Lee, K. S. Namgung, J. K. Sung and W. K. Chae, *Bull. Korean Chem. Soc.*, **12**, 247 (1991).
345. H. Parlar, M. Mansour, D. Kotzias, M. Herrmann, P. N. Moza, G. Hartung and K. Hustert,
     Chem. Ztg., **108**, 335 (1984).
346. P. Dureja and S. K. Mukerjee, *Tetrahedron*, **41**, 4905 (1985).
347. N. Zupančič and B. Šket, *Tetrahedron*, **47**, 9071 (1991).
348. L. M. Tolbert and S. Siddiqui, *J. Am. Chem. Soc.*, **104**, 4273 (1982); L. M. Tolbert and S.
     Siddiqui, *J. Am. Chem. Soc.*, **106**, 5538 (1984).
349. L. M. Tolbert, Md. N. Islam, R.P. Johnson, P.M. Loiselle and W.C. Shakespeare, *J. Am. Chem.
     Soc.*, **112**, 6416 (1990).
350. T. Kitamura, S. Miyake, S. Kobayashi and H. Taniguchi, *Chem. Lett.*, 929 (1985).
351. Y. Inoue, T. Fukunaga and T. Hakushi, *J. Org. Chem.*, **48**, 1732 (1983).
352. D. R. Arnold and P. C. Wong, *J. Am. Chem. Soc.*, **99**, 3361 (1977).
353. N. J. Bunce, J. P. Bergsma, M. D. Bergsma, W. De Graaf, Y. Kumar and L. Ravanal, *J. Org.
     Chem.*, **45**, 3708 (1980).
354. T. Ichimura, Y. Mori, H. Shinohara and N. Nishi, *Chem. Phys. Lett.*, **122**, 51 (1985).
355. V. Avila, H. E. Gsponer and C. M. Previtali, *J. Photochem.*, **27**, 163 (1984).
356. C. M. Previtali and T. W. Ebbesen, *J. Photochem.*, **30**, 259 (1985).
357. T. Nishiwaki, T. Shinoda, K. Anda and M. Hida, *Bull. Chem. Soc. Jpn.*, **55**, 3569 (1982).
358. A. Mamantov, *Chemosphere*, **14**, 901 (1985).
359. T. Nishiwaki, T. Shinoda, K. Anda and M. Hida, *Bull. Chem. Soc. Jpn.*, **55**, 3565 (1982).
360. N. J. Bunce, C. T. DeSchutter and E. J. Toone, *J. Chem. Soc., Perkin Trans. 2*, 859 (1983).
361. E. Haselbach, Y. Rohner and P. Suppan, *Helv. Chim. Acta*, **73**, 1644 (1990).
362. N. Shimizu, K. Watanabe and Y. Tsuno, *Bull. Chem. Soc. Jpn.*, **57**, 885 (1984).
363. M. Mansour, H. Parlar and F. Korte, *Chemosphere*, **9**, 59 (1980).
364. M. Nakada, S. Fukushi, H. Nishiyama, K. Okubo, K. Kume, M. Hirota and T. Ishii, *Bull. Chem.
     Soc. Jpn.*, **56**, 2447 (1983).
365. N. J. Bunce, P. J. Hayes and M. E. Lemke, *Can. J. Chem.*, **61**, 1103 (1983).

366. V. Avila, C. A. Chesta, J. J. Cosa and C. M. Previtali, *J. Photochem. Photobiol., A: Chem.*, **47**, 337 (1989).
367. P. K. Freeman, N. Ramnath and A. D. Richardson, *J. Org. Chem.*, **56**, 3643 (1991).
368. P. K. Freeman and Y.-S. Lee, *J. Org. Chem.*, **57**, 2846 (1992).
369. V. L. Ivanov and L. Éggert, *Zh. Org. Khim.*, **22**, 1933 (1986) [Engl. Transl.: *J. Org. Chem. USSR*, **22**, 1736 (1986)].
370. K. Morisaki, Y. Miura, K. Abe, M. Hirota and M. Nakada, *Chem. Lett.*, 1589 (1987).
371. M. Nakada, C. Miura, H. Nishiyama, F. Higashi, T. Mori, M. Hirota and T. Ishii, *Bull. Chem. Soc. Jpn.*, **62**, 3122 (1989).
372. M. Hirota and M. Nakada, *Bull. Chem. Soc. Jpn.*, **65**, 2926 (1992).
373. G. G. Choudhry and O. Hutzinger, *Environ. Sci. Technol.*, **18**, 235 (1984).
374. G. G. Choudhry and G. R. B. Webster, *Toxicol. Environ. Chem.*, **9**, 291 (1985).
375. P. K. Freeman, J.-S. Jang and N. Ramnath, *J. Org. Chem.*, **56**, 6072 (1991).
376. K. Ohkura, K. Seki, M. Terashima and Y. Kanaoka, *Chem. Pharm. Bull.*, **39**, 3168 (1991).
377. H. Parlar, D. Kotzias, M. Mansour and F. Korte, *Chem.-Ztg.*, **105**, 303 (1981).
378. J. R. Siegman and J. J. Houser, *J. Org. Chem.*, **47**, 2773 (1982).
379. J. Vermeulen and J. Ph. Soumillion, *Bull. Soc. Chim. Belg.*, **94**, 1045 (1985).
380. J. Ph. Soumillion and J. Vermeulen, *Bull. Soc. Chim. Belg.*, **91**, 474 (1982).
381. G. G. Choudhry, A. A. M. Roof and O. Hutzinger, *J. Chem. Soc., Perkin Trans. 1*, 2957 (1982).
382. W. Augustyniak, J. Wojtczak and M. Sikorski, *J. Photochem. Photobiol., A: Chem.*, **43**, 21 (1988).
383. A. G. Giumanini, G. Verardo and P. Strazzolini, *J. Photochem. Photobiol., A: Chem.*, **48**, 129 (1989).
384. J. Szychliński, M. Jarosiewicz and M. Doroszkiewicz, *J. Photochem.*, **22**, 345 (1983).
385. X. Zhu and P. Lang, *Huanjing Huaxue*, **10**, 21 (1991); *Chem. Abstr.*, **115**, 262856 x (1991).
386. C. Párkányi, G. Vernin, R.-M. Zamkostian and J. Metzger, *Heterocycles*, **22**, 1077 (1984).
387. N. C. Yang, A. Huang and D. H. Yang, *J. Am. Chem. Soc.*, **111**, 8060 (1989).
388. D. Huppert, S. D. Rand, A. H. Reynolds and P. M. Rentzepis, *J. Chem. Phys.*, **77**, 1214 (1982).
389. E. F. Hilinski, S. V. Milton, C. Ramana and P. M. Rentzepis, *Can. J. Chem.*, **61**, 999 (1983).
390. N. J. Bunce, *J. Org. Chem.*, **47**, 1948 (1982).
391. R. A. Beecroft, R. S. Davidson and D. Goodwin, *Tetrahedron Lett.*, **24**, 5673 (1983).
392. M. Ohashi and K. Tsujimoto, *Chem. Lett.*, 423 (1983).
393. H. Lemmetyinen, R. Ovaskainen, K. Nieminen, K. Vaskonen and I. Sychtchikova, *J. Chem. Soc., Perkin Trans. 2*, 113 (1992).
394. S. Fery-Forgues and N. Paillous, *J. Org. Chem.*, **51**, 672 (1986).
395. S. Fery-Forgues, D. Lavabre and N. Paillous, *J. Org. Chem.*, **52**, 3381 (1987).
396. D. S. Watt, K. Kawada, E. Leyva and M. S. Platz, *Tetrahedron Lett*, **30**, 899 (1989).
397. R. S. Davidson, J. W. Goodin and J. E. Pratt, *Tetrahedron Lett.*, **23**, 2225 (1982).
398. Y. Tanaka, K. Tsujimoto and M. Ohashi, *Bull. Chem. Soc. Jpn.*, **60**, 788 (1987).
399. J. Hawari, J. Tronczynski, A. Demeter, R. Samson and D. Mourato, *Chemosphere*, **22**, 189 (1991).
400. N. J. Bunce and J. C. Gallacher, *J. Org. Chem.*, **47**, 1955 (1982).
401. C. A. Chesta, J. J. Cosa and C. M. Previtali, *J. Photochem.*, **39**, 251 (1987).
402. C. A. Chesta, J. J. Cosa and C. M. Previtali, *J. Photochem.*, **32**, 203 (1986).
403. C. A. Chesta, J. J. Cosa and C. M. Previtali, *J. Photochem Photobiol., A: Chem.*, **45**, 9 (1988).
404. A. T. Soltermann, C. A. Chesta, J. J. Cosa and C. M. Previtali, *An. Asoc. Quim. Argent.*, **77**, 487 (1989).
405. V. Avila, J. J. Cosa, C. A. Chesta and C. M. Previtali, *J. Photochem. Photobiol., A: Chem.*, **62**, 83 (1991).
406. M. Ishikawa and S. Fukuzumi, *Chem. Lett.*, 963 (1990).
407. M. Ishikawa and S. Fukuzumi, *J. Am. Chem. Soc.*, **112**, 8864 (1990).
408. J. Ph. Soumillion, P. Vandereecken and F. C. De Schryver, *Tetrahedron Lett.*, **30**, 697 (1989).
409. S. S. Shukla and J. F. Rusling, *J. Phys. Chem.*, **89**, 3353 (1985).
410. Y. Tanaka, T. Uryu, M. Ohashi and K. Tsujimoto, *J. Chem. Soc., Chem. Commun.*, 1703 (1987).
411. G. A. Epling, Q. Wang and Q. Qiu, *Chemosphere*, **22**, 959 (1991).
412. O. M. Soloveichik, A. Klimakova and V. L. Ivanov, *Zh. Org. Khim.*, **18**, 1033 (1982) [Engl. Transl.: *J. Org. Chem. USSR*, **18**, 894 (1982)].
413. M. G. Kuz'min, N. A. Sadovskii and I. V. Soboleva, *J. Photochem.*, **23**, 27 (1983).

414. O. M. Soloveichik, V. L. Ivanov and M. G. Kuz'min, *Khim. Vys. Energ.*, **18**, 131 (1984); *Chem. Abstr.*, **100**, 208953d (1984).
415. O. M. Soloveichik, V. L. Ivanov and M. G. Kuz'min, *Khim. Vys. Energ.*, **23**, 351 (1989); *Chem. Abstr.*, **111**, 123585 n (1989).
416. K. Hamanoue, T. Nakayama, K. Ikenaga and K. Ibuki, *J. Phys. Chem.*, **96**, 10297 (1992).
417. W. G. McGimpsey and J. C. Scaiano, *J. Am. Chem. Soc.*, **111**, 335 (1989).
418. T. Nagamura, T. Nakayama and K. Hamanoue, *Chem. Lett.*, 2051 (1991).
419. K. Hamanoue, T. Nakayama, K. Ikenaga, K. Ibuki and A. Otani, *J. Photochem. Photobiol., A: Chem.*, **69**, 305 (1993).
420. P. K. Freeman, R. Srinivasa, J.-A. Campbell and M. L. Deinzer, *J. Am. Chem. Soc.*, **108**, 5531 (1986).
421. P. K. Freeman and N. Ramnath, *J. Org. Chem.*, **53**, 148 (1988).
422. P. K. Freeman and R. Srinivasa, *J. Org. Chem.*, **52**, 252 (1987).
423. V. V. Konovalov, A. M. Raitsimring, Yu. D. Tsvetkov and I. I. Bilkis, *Chem. Phys. Lett.*, **157**, 257 (1989).
424. P. K. Freeman and N. Ramnath, *J. Org. Chem.*, **56**, 3646 (1991).
425. P. K. Freeman, G. E. Clapp and B. K. Stevenson, *Tetrahedron Lett.*, **32**, 5705 (1991).
426. M. Nowakowska, E. Sustar and J. E. Guillet, *J. Am. Chem. Soc.*, **113**, 253 (1991).
427. E. Sustar, M. Nowakowska and J. E. Guillet, *J. Photochem. Photobiol., A: Chem.*, **63**, 357 (1992).
428. M. Kropp and G. B. Schuster, *Tetrahedron Lett.*, **28**, 5295 (1987).
429. G. A. Epling and E. Florio, *Tetrahedron Lett.*, **27**, 675 (1986).
430. G. A. Epling and E. Florio, *J. Chem. Soc., Chem. Commun.*, 185 (1986).
431. G. A. Epling, W. M. McVicar and A. Kumar, *Chemosphere*, **17**, 1355 (1988).
432. A. N. Abeywickrema and A. L. J. Beckwith, *Tetrahedron Lett.*, **27**, 109 (1986).
433. V. Nair, D. A. Young and R. DeSilvia, Jr., *J. Org. Chem.*, **52**, 1344 (1987).
434. V. Nair, D. A. Young and R. DeSilvia, Jr., *Nucleic Acid Chem.*, **4**, 174 (1991).
435. T. M. Dietz and T. H. Koch, *Photochem. Photobiol.*, **49**, 121 (1989).
436. H. Görner, *J. Photochem. Photobiol., A: Chem.*, **72**, 197 (1993).
437. H. Sugiyama, Y. Tsutsumi and I. Saito, *J. Am. Chem. Soc.*, **112**, 6720 (1990).
438. A. T. Soltermann, J. J. Cosa and C. M. Previtali, *J. Photochem. Photobiol., A: Chem.*, **60**, 111 (1991).
439. H.-J. Kallmayer and W. Fritzen, *Arch. Pharm.*, **320**, 769 (1987).
440. H.-J. Kallmayer and C. Tappe, *Pharmazie*, **41**, 832 (1986).
441. H.-J. Kallmayer and C. Tappe, *Arch. Pharm.*, **319**, 791 (1986).
442. H.-J. Kallmayer and N. Petesch, *Pharm. Acta Helv.*, **66**, 130 (1991).
443. K. Hamanoue, K. Yokoyama, T. Miyake, T. Kasuya, T. Nakayama and H. Teranishi, *Chem. Lett.*, 1967 (1982).
444. K. Hamanoue, T. Nakayama, A. Tanaka, Y. Kajiwara and H. Teranishi, *J. Photochem.*, **34**, 73 (1986).
445. K. Hamanoue, K. Sawada, K. Yokoyama, T. Nakayama, S. Hirase and H. Teranishi, *J. Photochem.*, **33**, 99 (1986).
446. K. Hamanoue, T. Nakayama, K. Sawada, Y. Yamamoto, S. Hirase and H. Teranishi, *Bull. Chem. Soc. Jpn.*, **59**, 2735 (1986).
447. K. Hamanoue, Y. Kajiwara, T. Miyake, T. Nakayama, S. Hirase and H. Teranishi, *Chem. Phys. Lett.*, **94**, 276 (1983).
448. K. Hamanoue, K. Nakajima, Y. Kajiwara, T. Nakayama and H. Teranishi, *Chem. Phys. Lett.*, **110**, 178 (1984).
449. K. Hamanoue, T. Nakayama, M. Shiozaki, Y. Funasaki, K. Nakajima and H. Teranishi, *J. Chem. Phys.*, **85**, 5698 (1986).
450. K. Hamanoue, T. Nakayama, Y. Kajiwara, T. Yamaguchi and H. Teranishi, *J. Chem. Phys.*, **86**, 6654 (1987).
451. K. Hamanoue, T. Nakayama and M. Ito, *J. Chem. Soc., Faraday Trans.*, **87**, 3487 (1991).
452. K. Hamanoue, T. Nakayama, H. Sasaki and K. Ibuki, *J. Photochem. Photobiol., A: Chem.*, **76**, 7 (1993).
453. R. G. Compton, B. A. Coles, M. B. G. Pilkington and D. Bethell, *J. Chem. Soc., Faraday Trans.*, **86**, 663 (1990).
454. R. G. Compton, A. C. Fisher, R. G. Wellington, D. Bethell and P. Lederer, *J. Phys. Chem.*, **95**, 4749 (1991).

455. G. G. Choudhry, F. W. M. Van der Wielen, G. R. B. Webster and O. Hutzinger, *Can. J. Chem.*, **63**, 469 (1985).
456. G. G. Choudhry, N. J. Graham and G. R. B. Webster, *Can. J. Chem.*, **65**, 2223 (1987).
457. P. K. Freeman and R. Srinivasa, *J. Org. Chem.*, **51**, 3939 (1986).
458. T. Mazer and F. D. Hileman, *Chemosphere*, **11**, 651 (1982).
459. M. Tysklind and C. Rappe, *Chemosphere*, **23**, 1365 (1991).
460. C. J. Koester and R. A. Hites, *Environ. Sci. Technol.*, **26**, 502 (1992).
461. M. Koshioka, H. Iizuka, J. Kanazawa and T. Murai, *Agric. Biol. Chem.*, **51**, 949 (1987).
462. D. Lenoir, K.-W. Schramm, O. Hutzinger and G. Schedel, *Chemosphere*, **22**, 821 (1991).
463. G. G. Choudhry and G. R. B. Webster, *Chemosphere*, **15**, 1935 (1986).
464. S. Kieatiwong, L. V. Nguyen, V. R. Hebert, M. Hackett, G. C. Miller, M. J. Miller and R. Mitzel, *Environ. Sci. Technol.*, **24**, 1575 (1990).
465. K. J. Friesen, D. C. G. Muir and G. R. B. Webster, *Environ. Sci. Technol.*, **24**, 1739 (1990).
466. A. Mamantov, *Chemosphere*, **14**, 897 (1985).
467. M. Makino, M. Kamiya and H. Matsushita, *Chemosphere*, **24**, 291 (1992).
468. J. Kochany, G. G. Choudhry and G. R. B. Webster, *Int. J. Environ. Anal. Chem.*, **39**, 59 (1990).
469. T. Clark and D. A. M. Watkins, *Chemosphere*, **13**, 1391 (1984).
470. M. Barbeni, E. Pramauro, E. Pelizzetti, E. Borgarello, M. Graetzel and N. Serpone, *Nouv. J. Chem.*, **8**, 547 (1984).
471. H. Al-Ekabi, N. Serpone, E. Pelizzetti, C. Minero, M. A. Fox and R. B. Draper, *Langmuir*, **5**, 250 (1989).
472. G. Occhiucci and A. Patacchiola, *Chemosphere*, **11**, 255 (1982).
473. Y. Xu, P. E. Menassa and C. H. Langford, *Chemosphere*, **17**, 1971 (1988).
474. K. Fukunishi, W. Kawahara, H. Yamanaka, M. Kuwabara and M. Nomura, *Chem. Express*, **4**, 101 (1989).
475. R. Fantoni, R. Larciprete, S. Piccirillo, G. Bertoni, R. Fratarcangeli and M. Rotatori, *Chem. Phys. Lett.*, **143**, 245 (1988).
476. M. Kojima, H. Sakuragi and K. Tokumaru, *Chem. Lett.*, 1539 (1981).
477. Q.-Y. Chen and Z.-T. Li, *J. Chem. Soc., Perkin Trans. 1*, 1705 (1993).
478. P. Nelleborg, H. Lund and J. Eriksen, *Tetrahedron Lett.*, **26**, 1773 (1985).
479. F. S. Tanaka, R. G. Wien and B. L. Hoffer, *Synth. Commun.*, **13**, 951 (1983).
480. J. Kelm and K. Strauss, *Spectrochim. Acta*, **37A**, 689 (1981).
481. J. L. Dektar and N. P. Hacker, *J. Org. Chem.*, **55**, 639 (1990).
482. N. P. Hacker, D. V. Leff and J. L. Dektar, *J. Org. Chem.*, **56**, 2280 (1991).
483. J. L. Dektar and N. P. Hacker, *J. Org. Chem.*, **56**, 1838 (1991).
484. S. P. Spyroudis, *J. Org. Chem.*, **51**, 3453 (1986).
485. L. P. Hadjiarapoglou, *Tetrahedron Lett.*, **28**, 4449 (1987).
486. R .J. DeVoe, P. M. Olofson and M. R. V. Sahyun, *Adv. Photochem.*, **17**, 313 (1992).
487. K. Maruyama, M. Tojo, H. Iwamoto and T. Otsuki, *Chem. Lett.*, 827 (1980).
488. K. Maruyama, T. Otsuki, K. Mitsui and M. Tojo, *J. Heterocycl. Chem.*, **17**, 695 (1980).
489. K. Maruyama, S. Tai and T. Otsuki, *Chem. Lett.*, 843 (1983).
490. K. Maruyama, T. Otsuki and S. Tai, *J. Org. Chem.*, **50**, 52 (1985).
491. K. Maruyama, M. Tojo, K. Matsumoto and T. Otsuki, *Chem. Lett.*, 859 (1980).
492. K. Maruyama, M. Tojo, S. Tai and T. Otsuki, *Heterocycles*, **16**, 190 (1981).
493. K. Maruyama, S. Tai, M. Tojo and T. Otsuki, *Heterocycles*, **16**, 1963 (1981).
494. K. Maruyama, S. Tai and H. Imahori, *Bull. Chem. Soc. Jpn.*, **59**, 1777 (1986).
495. K. Maruyama, T. Otsuki and S. Tai, *Chem. Lett.*, 371 (1984).
496. K. Maruyama, H. Imahori, A. Osuka, A. Takuwa and H. Tagawa, *Chem. Lett.*, 1719 (1986).
497. K. Maruyama and H. Imahori, *Bull. Chem. Soc. Jpn.*, **62**, 816 (1989).
498. K. Maruyama and H. Imahori, *J. Chem. Soc., Perkin Trans. 2*, 257 (1990).
499. K. Maruyama, S. Tai and T. Otsuki, *Chem. Lett.*, 1565 (1981).
500. K. Maruyama, S. Tai and T. Otsuki, *Heterocycles*, **20**, 1031 (1983).
501. K. A. K. Al-Fakhri and A. C.Pratt, *Proc. R. Ir. Acad., Sect. B*, **83**, 5 (1983).
502. M. D'Auria and F. D'Onofrio, *Gazz. Chim. Ital.*, **115**, 595 (1985).
503. R. Antonioletti, M. D'Auria, A. De Mico, G. Piancatelli and A. Scettri, *J. Chem. Soc., Perkin Trans. 1*, 1285 (1985).
504. R. Antonioletti, M. D'Auria, A. De Mico, G. Piancatelli and A. Scettri, *Tetrahedron*, **41**, 3441 (1985).

505. M. D'Auria, R. Antonioletti, A. De Mico and G. Piancatelli, *Heterocycles*, **24**, 1575 (1986).
506. Ya. R. Tymyanskii, V. M. Feigel'man, O. A. Zubkov, V. N. Novikov, M. I. Knyazhanskii and V. S. Pustovarov, *Zh. Org. Khim.*, **24**, 459 (1988) [Engl. Transl.: *J. Org. Chem. USSR*, **24**, 407 (1988)].
507. M. D'Auria, A. De Mico and F. D'Onofrio, *Heterocycles*, **29**, 1331 (1989).
508. R. Antonioletti, M. D'Auria, F. D'Onofrio, G. Piancatelli and A. Scettri, *J. Chem. Soc., Perkin Trans. 1*, 1755 (1986).
509. M. D'Auria, A. De Mico, F. D'Onofrio and G. Piancatelli, *J. Chem. Soc., Perkin Trans. 1*, 1777 (1987).
510. M. D'Auria, A. De Mico, F. D'Onofrio and G. Piancatelli, *J. Org. Chem.*, **52**, 5243 (1987).
511. M. D'Auria, A. De Mico, F. D'Onofrio and G. Piancatelli, *Gazz. Chim. Ital.*, **119**, 381 (1989).
512. M. D'Auria, A. De Mico, F. D'Onofrio, D. Mendola and G. Piancatelli, *J. Photochem. Photobiol., A: Chem.*, **47**, 191 (1989).
513. M. D'Auria, A. De Mico, F. D'Onofrio and G. Piancatelli, *Gazz. Chim. Ital.*, **119**, 201 (1989).
514. M. D'Auria, G. Piancatelli and T. Ferri, *J. Org. Chem.*, **55**, 4019 (1990).
515. M. D'Auria, *Heterocycles*, **32**, 1059 (1991).
516. M. D'Auria, *Gazz. Chim. Ital.*, **119**, 419 (1989).
517. J. C. Quada, Jr., M. J. Levy and S. M. Hecht, *J. Am. Chem. Soc.*, **115**, 12171 (1993).
518. M. E. Hassan, *Coll. Czech. Chem. Commun.*, **50**, 2319 (1985).
519. K. Satoh, H. Tanaka, A. Andoh and T. Miyasaka, *Nucleosides Nucleotides*, **5**, 461 (1986).
520. G. Kaupp and E. Ringer, *Chem. Ber.*, **119**, 1525 (1986).
521. M. Terashima, K. Seki, C. Yoshida and Y. Kanaoka, *Heterocycles*, **15**, 1075 (1981).
522. K. Seki, K. Ohkura, M. Terashima and Y. Kanaoka, *Heterocycles*, **22**, 2347 (1984).
523. K. Seki, K. Ohkura, M. Terashima and Y. Kanaoka, *Heterocycles*, **24**, 799 (1986).
524. M. Terashima, K. Seki, C. Yoshida, K. Ohkura and Y. Kanaoka, *Chem. Pharm. Bull.*, **33**, 1009 (1985).
525. K. Ohkura, K. Seki, M. Terashima and Y. Kanaoka, *Tetrahedron Lett.*, **30**, 3433 (1989).
526. K. Ohkura, K. Seki, M. Terashima and Y. Kanaoka, *Heterocycles*, **30**, 957 (1990).
527. J.-B. Meng, M.-Q. Shen, X.-H. Wang, C.-H. Kao, R.-J. Wang, H.-G. Wang and T. Matsuura, *J. Heterocycl. Chem.*, **28**, 1481 (1991).
528. J.-B. Meng, M.-Q. Shen, X.-H. Wang, Z.-H. Gao, H.-G. Wang and T. Matsuura, *Chin. Sci. Bull.*, **36**, 2056 (1991).
529. M. Somei, H. Amari and Y. Makita, *Chem. Pharm. Bull.*, **34**, 3971 (1986).
530. S. Ito, I. Saito and T. Matsuura, *Tetrahedron*, **37**, 45 (1981).
531. H. Ikehira, T. Matsuura and I. Saito, *Tetrahedron Lett.*, **26**, 1743 (1985).
532. I. Saito, H. Ikehira and T. Matsuura, *Tetrahedron Lett.*, **26**, 1993 (1985).
533. I. Saito, H. Ikehira and T. Matsuura, *J. Org. Chem.*, **51**, 5148 (1986).
534. K. Seki, Y. Bando and K. Ohkura, *Chem. Lett.*, 195 (1986).
535. K. Seki, K. Matsuda and K. Ohkura, *Chem. Lett.*, 175 (1987).
536. K. Seki, K. Ohkura and N. Sasaki, *Nucleic Acids Symp. Ser.*, **22**, 7 (1990).
537. K. Seki, N. Kanazashi and K. Ohkura, *Heterocycles*, **32**, 229 (1991).
538. K. Ohkura, K. Matsuda and K. Seki, *Heterocycles*, **32**, 1371 (1991).
539. M.E. Hassan, *Nucleosides Nucleotides*, **10**, 1277 (1991).
540. V. Nair, S. G. Richardson and R. E. Coffman, *J. Org. Chem.*, **47**, 4520 (1982).
541. T. C. McKenzie and J. W. Epstein, *J. Org. Chem.*, **47**, 4881 (1982).
542. D. Dulin, H. Drossman and T. Mill, *Environ. Sci. Technol.*, **20**, 72 (1986).
543. A. Mamantov, *Chemosphere*, **14**, 905 (1985).
544. Y.-S. Chang, J.-S. Jang and M. L. Deinzer, *Tetrahedron*, **46**, 4161 (1990).
545. C. J. Chandler, D. J. Craik and K. J. Waterman, *Aust. J. Chem.*, **42**, 1407 (1989).
546. Y.-T. Park, C.-H. Joo and L.-H. Lee, *Bull. Korean Chem. Soc.*, **11**, 270 (1990).
547. Y.-T. Park, C.-H. Joo, C.-D. Choi and K.-S. Park, *J. Heterocycl. Chem.*, **28**, 1083 (1991).
548. Y.-T. Park, C.-H. Joo, C.-D. Choi and K.-S. Park, *Bull. Korean Chem. Soc.*, **12**, 163 (1991).
549. D. E. Moore, S. Roberts-Thomson, D. Zhen and C. C. Duke, *Photochem. Photobiol.*, **52**, 685 (1990).
550. A. N. Frolov and M. V. Baklanov, *Zh. Org. Khim.*, **27**, 2424 (1991). [Engl. Transl.: *J. Org. Chem. USSR*, **27**, 2153 (1991)].
551. A. N. Frolov and M. V. Baklanov, *Mendeleev Commun.*, 22 (1992).
552. S. V. Kessar, Y. P. Gupta, V. S. Yadav, M. Narula and T. Mohammad, *Tetrahedron Lett.*, **21**, 3307 (1980).

553. S. V. Kessar, Y. P. Gupta and T. Mohammad, *Indian J. Chem., Sect. B*, **20**, 984 (1981).
554. Y. P. Gupta, V. S. Yadav and T. Mohammad, *Indian J. Chem., Sect. B*, **22**, 429 (1983).
555. S. V. Kessar, T. Mohammad and Y. P. Gupta, *Indian J. Chem., Sect. B*, **22**, 321 (1983).
556. S. Gupta and D. S. Bhakuni, *Synth. Commun.*, **18**, 2251 (1988).
557. B. L. Jensen and K. Chockalingam, *J. Heterocycl. Chem.*, **23**, 343 (1986).
558. J. Grimshaw and A. P. De Silva, *J. Chem. Soc., Chem. Commun.*, 1236 (1980).
559. J. Grimshaw and A. P. De Silva, *J. Chem. Soc., Perkin Trans. 2*, 1679 (1983).
560. J. Grimshaw and J. Trocha-Grimshaw, *J. Chem. Soc., Chem. Commun.*, 744 (1984).
561. J. Grimshaw and S. A. Hewitt, *Proc. R. Ir. Acad., Sect. B*, **83**, 93 (1983).
562. M. Rama Devi, J. M. Rao and V. R. Srinivasan, *Synth. Commun.*, **19**, 2345 (1989).
563. J. Grimshaw and A. P. De Silva, *J. Chem. Soc., Perkin Trans. 2*, 857 (1982).
564. J. Grimshaw and A. P. De Silva, *J. Chem. Soc., Perkin Trans. 2*, 1010 (1981).
565. L. Castedo, E. Guitián, J. M. Saá and R. Suau, *Heterocycles*, **19**, 279 (1982).
566. L. Castedo, R. J. Estévez, J. M. Saá and R. Suau, *J. Heterocycl. Chem.*, **19**, 1319 (1982).
567. L. Castedo, C. Saá, J. M. Saá and R. Suau, *J. Org. Chem.*, **47**, 513 (1982).
568. S. Nimgirawath and W. C. Taylor, *Aust. J. Chem.*, **36**, 1061 (1983).
569. S. Nimgirawath and W. C. Taylor, *J. Sci. Soc. Thailand*, **9**, 73 (1983).
570. H. Soicke, G. Al-Hassan, U. Frenzel and K. Görler, *Arch. Pharm.*, **321**, 149 (1988).
571. S. Gupta and D. S. Bhakuni, *Synth. Commun.*, **19**, 393 (1989).
572. N. S. Narasimhan and I. S. Aidhen, *Tetrahedron Lett.*, **29**, 2987 (1988).
573. R. J. Olsen and S. R. Pruett, *J. Org. Chem.*, **50**, 5457 (1985).
574. C. J. Moody, C. W. Rees and R. Thomas, *Tetrahedron Lett.*, **31**, 4375 (1990).
575. C. J. Moody, C. W. Rees and R. Thomas, *Tetrahedron*, **48**, 3589 (1992).
576. S. Kobayashi, M. Kihara and Y. Miyake, *Heterocycles*, **23**, 159 (1985).
577. J. M. Boente, D. Domínguez and L. Castedo, *Heterocycles*, **23**, 1069 (1985).
578. O. Hoshino, H. Ogasawara, A. Hirokawa and B. Umezawa, *Chem. Lett.*, 1767 (1988).
579. H. Tanaka, Y. Takamura, K. Ito, K. Ohira and M. Shibata, *Chem. Pharm. Bull.*, **32**, 2063 (1984).
580. O. Hoshino, H. Ogasawara, A. Takahashi and B. Umezawa, *Heterocycles*, **23**, 1943 (1985).
581. O. Hoshino, H. Ogasawara, A. Takahashi and B. Umezawa, *Heterocycles*, **25**, 155 (1987).
582. T. Nishiwaki, E. Kawamura, N. Abe, Y. Sasaoka and H. Kochi, *Heterocycles*, **16**, 1203 (1981).
583. R. Paramasivam, S. Muthusamy and V. T. Ramakrishnan, *Indian J. Chem., Sect. B*, **28**, 597 (1989).
584. S. Muthusamy, R. Paramasivam and V.T. Ramakrishnan, *J. Heterocycl. Chem.*, **28**, 759 (1991).
585. S. Muthusamy and V. T. Ramakrishnan, *Synth. Commun.*, **22**, 519 (1992).
586. T. Tiner-Harding and P. S. Mariano, *J. Org. Chem.*, **47**, 482 (1982).
587. M. A. Brumfield, P. S. Mariano and U. C. Yoon, *Tetrahedron Lett.*, **24**, 5567 (1983).
588. M. V. Baklanov and A. N. Frolov, *Zh. Org. Khim.*, **26**, 1141 (1990) [Engl. Transl.: *J. Org. Chem. USSR*, **26**, 983 (1990)].
589. K. Kaji, H. Nagashima, Y. Ohta, K. Tabashi and H. Oda, *J. Heterocycl. Chem.*, **21**, 1249 (1984).
590. A. Couture, P. Grandclaudon and E. Huguerre, *J. Heterocycl. Chem.*, **24**, 1765 (1987).
591. C. Kaneko, T. Naito and C. Miwa, *Chem. Pharm. Bull.*, **30**, 752 (1982).
592. L. Jayabalan and P. Shanmugam, *Synthesis*, 789 (1990).
593. H. Kudo, R. N. Castle and M. L. Lee, *J. Heterocycl. Chem.*, **21**, 1761 (1984).
594. H. Kudo, R. N. Castle and M. L. Lee, *J. Heterocycl. Chem.*, **22**, 211 (1985).
595. S. Pakray and R. N. Castle, *J. Heterocycl. Chem.*, **24**, 231 (1987).
596. S. L. Castle, P. J. Buckhaults, L. J. Baldwin, J. D. McKenney, Jr. and R. N. Castle, *J. Heterocycl. Chem.*, **24**, 1103 (1987).
597. J. D. McKenney, Jr. and R. N. Castle, *J. Heterocycl. Chem.*, **24**, 1525 (1987).
598. J.-K. Luo and R. N. Castle, *J. Heterocycl. Chem.*, **27**, 1031 (1990).
599. J.-K. Luo, S. L. Castle and R. N. Castle, *J. Heterocycl. Chem.*, **27**, 2047 (1990).
600. M. J. Musmar and R. N. Castle, *J. Heterocycl. Chem.*, **28**, 203 (1991).
601. J.-K. Luo, A. S. Zektzer and R. N. Castle, *J. Heterocycl. Chem.*, **28**, 737 (1991).
602. J.-K. Luo. M. J. Musmar and R. N. Castle, *J. Heterocycl. Chem.*, **28**, 1309 (1991).
603. J.-K. Luo and R. N. Castle, *J. Heterocycl. Chem.*, **28**, 1825 (1991).
604. R. N. Castle, S. Pakray and G. E. Martin, *J. Heterocycl. Chem.*, **28**, 1997 (1991).
605. G. Karminski-Zamola, D. Pavličič, M. Bajič and N. Blaževič, *Heterocycles*, **32**, 2323 (1991).
606. S. Jeganathan and M. Srinivasan, *Indian J. Chem., Sect. B*, **19**, 1028 (1980).

607. P. Kutschy, J. Imrich and J. Bernát, *Synthesis*, 929 (1983).
608. P. Kutschy, J. Imrich, J. Bernát, P. Kristian and I. Fedoriková, *Coll. Czech. Chem. Commun.*, **51**, 2002 (1986).
609. P. Kutschy, J. Imrich, J. Bernát, P. Kristian, O. Hritzová and T. Schöffmann, *Coll. Czech. Chem. Commun.*, **51**, 2839 (1986).
610. T. Maruyama, Y. Adachi and M. Honjo, *J. Org. Chem.*, **53**, 4552 (1988).
611. G. G. Wubbels, D. P. Susens and E. B. Coughlin, *J. Am. Chem. Soc.*, **110**, 2538 (1988).
612. G. G. Wubbels, E. J. Snyder and E. B. Coughlin, *J. Am. Chem. Soc.*, **110**, 2543 (1988).
613. Y. Hatanaka, E. Yoshida, M. Taki, H. Nakayama and Y. Kanaoka, *Photomed. Photobiol.*, **10**, 215 (1988).
614. R. Pleixats, M. Figueredo, J. Marquet, M. Moreno-Mañas and A. Cantos, *Tetrahedron*, **45**, 7817 (1989).
615. M. Figueredo, J. Marquet, M. Moreno-Mañas and R. Pleixats, *Tetrahedron Lett.*, **30**, 2427 (1989).
616. R. Pleixats and J. Marquet, *Tetrahedron*, **46**, 1343 (1990).
617. K. Mutai, R. Nakagaki and H. Tukada, *Bull. Chem. Soc. Jpn.*, **58**, 2066 (1985).
618. J. Cornelisse, G. Lodder and E. Havinga, *Rev. Chem. Intermed.*, **2**, 231 (1979).
619. K. Seki, K. Ohkura, M. Terashima and Y. Kanaoka, *Heterocycles*, **26**, 3101 (1987).
620. K. Seki, K. Ohkura, M. Terashima and Y. Kanaoka, *Chem. Pharm. Bull.*, **36**, 940 (1988).
621. K. Seki, K. Ohkura, K. Matsuda, M. Terashima and Y. Kanaoka, *Chem. Pharm. Bull.*, **36**, 4693 (1988).
622. K. Ohkura, K. Seki, M. Terashima and Y. Kanaoka, *Heterocycles*, **31**, 1833 (1990).
623. K. Maruyama and H. Tamiaki, *Bull. Chem. Soc. Jpn.*, **60**, 1847 (1987).
624. K. Maruyama, T. Otsuki and H. Tamiaki, *Bull. Chem. Soc. Jpn.*, **58**, 3049 (1985).
625. P. C. Dwivedi and A. K. Banga, *J. Phys. Chem.*, **85**, 1768 (1981).
626. I. S. Monzó, J. Palou, J. Roca and R. Valero, *J. Chem. Soc., Perkin Trans. 2*, 1995 (1988).
627. A. Gilbert, S. Krestonosich and D. L. Westover, *J. Chem. Soc., Perkin Trans. 1*, 295 (1981).
628. N. J. Bunce and S. R. Cater, *J. Chem. Soc., Perkin Trans. 2*, 169 (1986).
629. C. Tintel, F. J. Rietmeyer and J. Cornelisse, *Recl. Trav. Chim. Pays-Bas*, **102**, 224 (1983).
630. M.-X. Zhang, Z.-L. Liu, L. Yang and Y.-C. Liu, *J. Chem. Soc., Chem. Commun.*, 1054 (1991).
631. Y.-C. Liu, M.-X. Zhang, L. Yang and Z.-L. Liu, *J. Chem. Soc., Perkin Trans. 2*, 1919 (1992).
632. H. Lemmetyinen, J. Konijnenberg, J. Cornelisse and C. A. G. O. Varma, *J. Photochem.*, **30**, 315 (1985).
633. J. H. Liu and R. G. Weiss, *Israel J. Chem.*, **25**, 228 (1985).
634. J. H. Liu and R. G. Weiss, *J. Photochem.*, **30**, 303 (1985).
635. J. H. Liu and R. G. Weiss, *J. Org. Chem.*, **50**, 3655 (1985).
636. C. D. Ritchie, *Acc. Chem. Res.*, **5**, 348 (1972).
637. A. N. Frolov, *Zh. Org. Khim.*, **20**, 924 (1984) [Engl. Transl.: *J. Org. Chem. USSR*, **20**, 838 (1984)].
638. A. N. Frolov, A. V. El'tsov, O. V. Kul'bitskaya and A. I. Ponyaev, *Zh. Org. Khim.*, **16**, 2467 (1980) [Engl. Transl.: *J. Org. Chem. USSR*, **16**, 2099 (1980)].
639. H. Inoue, T. Shinoda and M. Hida, *Bull. Chem. Soc. Jpn.*, **53**, 154 (1980).
640. D. B. Denney and D. Z. Denney, *Tetrahedron*, **47**, 6577 (1991).
641. D. B. Denney, D. Z. Denney and A. J. Perez, *Tetrahedron*, **49**, 4463 (1993).
642. J. F. Bunnett, *Tetrahedron*, **49**, 4477 (1993).
643. R. A. Rossi and S. M. Palacios, *Tetrahedron*, **49**, 4485 (1993).
644. D. Bethell, R. G. Compton and R. G. Wellington, *J. Chem. Soc., Perkin Trans. 2*, 147 (1992).
645. C. Galli, *Gazz. Chim. Ital.*, **118**, 365 (1988).
646. M. A. Fox, J. Younathan and G. E. Fryxell, *J. Org. Chem.*, **48**, 3109 (1983).
647. J. F. Bunnett, E. Mitchel and C. Galli, *Tetrahedron*, **41**, 4119 (1985).
648. C. Galli, *Tetrahedron*, **44**, 5205 (1988).
649. J. F. Bunnett and C. Galli, *J. Chem. Soc., Perkin Trans. 1*, 2515 (1985).
650. R. Beugelmans and J. Chastanet, *Tetrahedron Lett.*, **32**, 3487 (1991).
651. R. Beugelmans and M. Bois-Choussy, *Synthesis*, 729 (1981).
652. R. Beugelmans, H. Ginsburg and M. Bois-Choussy, *J. Chem. Soc., Perkin Trans. 1*, 1149 (1982).
653. R. Beugelmans, J. Chastanet and G. Roussi, *Tetrahedron*, **40**, 311 (1984).
654. R. Beugelmans and H. Ginsburg, *Heterocycles*, **23**, 1197 (1985).
655. R. Beugelmans, J. Chastanet, H. Ginsburg. L. Quintero-Cortes and G. Roussi, *J. Org. Chem.*, **50**, 4933 (1985).

970   G. Lodder and J. Cornelisse

656. R. Beugelmans and M. Bois-Choussy. *Heterocycles*, **26**, 1863 (1987).
657. R. Beugelmans, M. Bois-Choussy and Q. Tang, *Tetrahedron*, **45**, 4203 (1989).
658. R. Beugelmans, M. Bois-Choussy and Q. Tang, *J. Org. Chem.*, **52**, 3880 (1987).
659. R. Fontan, C. Galvez and P. Viladoms, *Heterocycles*, **16**, 1473 (1981).
660. R. Beugelmans, M. Bois-Choussy and Q. Tang, *Tetrahedron Lett.*, **29**, 1705 (1988).
661. A. B. Pierini, M. T. Baumgartner and R. A. Rossi, *Tetrahedron Lett.*, **29**, 3429 (1988).
662. R. Beugelmans and M. Bois-Choussy, *J. Org.Chem.*, **56**, 2518 (1991).
663. M. T. Baumgartner, A. B. Pierini and R. A. Rossi, *Tetrahedron Lett.*, **33**, 2323 (1992).
664. M. T. Baumgartner, A. B. Pierini and R. A. Rossi, *J. Org. Chem.*, **58**, 2593 (1993).
665. A. B. Pierini, M. T. Baumgartner and R. A. Rossi, *J. Org. Chem.*, **52**, 1089 (1987).
666. D. R. Carver, T. D. Greenwood, J. S. Hubbard, A. P. Komin, Y. P. Sachdeva and J. F. Wolfe, *J. Org. Chem.*, **48**, 1180 (1983).
667. R. A. Alonso and R. A. Rossi, *J. Org. Chem.*, **45**, 4760 (1980).
668. S. Usui and Y. Fukazawa, *Tetrahedron Lett.*, **28**, 91 (1987).
669. S. V. Kessar, P. Singh, R. Chawla and P. Kumar, *J. Chem. Soc., Chem. Commun.*, 1074 (1981).
670. R. R. Goehring. Y. P. Sachdeva, J. S. Pisipati, M. C. Sleevi and J. F. Wolfe, *J. Am. Chem. Soc.*, **107**, 435 (1985).
671. R. A. Alonso, C. H. Rodriguez and R. A. Rossi, *J. Org. Chem.*, **54**, 5983 (1989).
672. S. C. Dillender, Jr., T. D. Greenwood, M. S. Hendi and J. F. Wolfe, *J. Org. Chem.*, **51**, 1184 (1986).
673. L. Beltran, C. Gálvez, M. Prats and J. Salgado, *J. Heterocycl. Chem.*, **29**, 905 (1992).
674. V. Nair and S. D. Chamberlain, *J. Am. Chem. Soc.*, **107**, 2183 (1985).
675. V. Nair and S. D. Chamberlain, *J. Org. Chem.*, **50**, 5069 (1985).
676. C. Combellas, H. Gautier, J. Simon, A. Thiebault, F. Tournilhac, M. Barzoukas, D. Josse, I. Ledoux, C. Amatore and J.-N. Verpeaux, *J. Chem. Soc., Chem. Commun.*, 203 (1988).
677. R. Beugelmans, M. Bois-Choussy and B. Boudet, *Tetrahedron*, **38**, 3479 (1982).
678. G. L. Borosky, A. B. Pierini and R. A. Rossi, *J. Org. Chem.*, **57**, 247 (1992).
679. A. B. Pierini, M. T. Baumgartner and R. A. Rossi, *Tetrahedron Lett.*, **28**, 4653 (1987).
680. C. Xia, Z. Chen and Z. Zhang, *Chin. Chem. Lett.*, **2**, 131 (1991).
681. C. Xia, Z. Chen and Z. Zhang, *Chin. Chem. Lett.*, **2**, 429 (1991).
682. M. P. Moon, A. P. Komin, J. F. Wolfe and G. F. Morris, *J. Org. Chem.*, **48**, 2392 (1983).
683. C. F. K. Hermann, Y. P. Sachdeva and J. F. Wolfe, *J. Heterocycl. Chem.*, **24**, 1061 (1987).
684. R. A. Alonso, E. Austin and R. A. Rossi, *J. Org. Chem.*, **53**, 6065 (1988).
685. V. L. Ivanov and A. Kherbst, *Zh. Org. Khim.*, **24**, 1709 (1988) [Engl. Transl.: *J. Org. Chem. USSR*, **24**, 1542 (1988)].
686. J.-J. Brunet, C. Sidot and P. Caubere, *J. Org. Chem.*, **48**, 1166 (1983).
687. L. M. Tolbert and D. P. Martone, *J. Org. Chem.*, **48**, 1185 (1983).
688. L. M. Tolbert and S. Siddiqui, *J. Org. Chem.*, **49**, 1744 (1984).
689. M. A. Fox and N. J. Singletary, *J. Org. Chem.*, **47**, 3412 (1982).
690. R. Beugelmans and M. Bois-Choussy, *Tetrahedron*, **42**, 1381 (1986).
691. E. R. N. Bornancini and R. A. Rossi, *J. Org. Chem.*, **55**, 2332 (1990).
692. R. A. Rossi, *Acc. Chem. Res.*, **15**, 164 (1982).
693. A. B. Peñéñory and R. A. Rossi, *J. Phys. Org. Chem.*, **3**, 266 (1990).
694. R. A. Rossi, R. A. Alonso and S. M. Palacios, *J. Org. Chem.*, **46**, 2498 (1981).
695. R. A. Alonso and R. A. Rossi, *J. Org. Chem.*, **47**, 77 (1982).
696. R. A. Rossi and S. M. Palacios, *J. Org. Chem.*, **46**, 5300 (1981).
697. I. A. Rybakova, R. I. Shekhtman and E. N. Prilezhaeva, *Izv. Akad. Nauk SSSR, Ser. Khim.*, 2414 (1982); *Chem. Abstr.*, **98**, 34331 x (1983).
698. I. A. Rybakova, R. I. Shekhtman and E. N. Prilezhaeva, *Izv. Akad. Nauk SSSR, Ser. Khim.*, 833 (1987); *Chem. Abstr.*, **108**, 221560 h (1988).
699. R. Beugelmans, M. Bois-Choussy and B. Boudet, *Tetrahedron*, **39**, 4153 (1983).
700. A. B. Pierini, A. B. Peñéñory and R. A. Rossi, *J. Org. Chem.*, **49**, 486 (1984).
701. M. Julliard and M. Chanon, *J. Photochem.*, **34**, 231 (1986).
702. M. Novi, G. Garbarino, G. Petrillo and C. Dell'Erba, *J. Org. Chem.*, **52**, 5382 (1987).
703. M. Novi, G. Garbarino, G. Petrillo and C. Dell'Erba, *Tetrahedron*, **46**, 2205 (1990).
704. D. W. Hobbs and W. C. Still, *Tetrahedron Lett.*, **28**, 2805 (1987).
705. A. B. Peñéñory, A. B. Pierini and R. A. Rossi, *J. Org. Chem.*, **49**, 3834 (1984).
706. M. B. Groen and E. Havinga, *Mol. Photochem.*, **6**, 9 (1974).

707. A. Tissot, P. Boule and J. Lemaire, *Chemosphere*, **12**, 859 (1983).
708. A. Tissot, P. Boule and J. Lemaire, *Chemosphere*, **13**, 381 (1984).
709. P. Boule, A. Tissot and J. Lemaire, *Chemosphere*, **14**, 1789 (1985).
710. S. Nagaoka, T. Takemura and H. Baba, *Bull. Chem. Soc. Jpn.*, **58**, 2082 (1985).
711. J. Orvis, J. Weiss and R. M. Pagni, *J. Org. Chem.*, **56**, 1851 (1991).
712. T. Moore and R.M. Pagni, *J. Org. Chem.*, **52**, 770 (1987).
713. T. Ichimura, M. Iwai and Y. Mori, *J. Photochem.*, **39**, 129 (1987).
714. T. Ichimura, M. Iwai and Y. Mori, *J. Phys. Chem.*, **92**, 4047 (1988).
715. P. Boule, C. Guyon and J. Lemaire, *Chemosphere*, **11**, 1179 (1982).
716. C. Guyon, P. Boule and J. Lemaire, *Nouv. J. Chim.*, **8**, 685 (1984).
717. P. Boule, C. Guyon and J. Lemaire, *Toxicol. Environ. Chem.*, **7**, 97 (1984).
718. C. Guyon, P. Boule and J. Lemaire, *Tetrahedron Lett.*, **23**, 1581 (1982).
719. K. Oudjehani and P. Boule, *New J. Chem.*, **17**, 567 (1993).
720. E. Lipczynska-Kochany and J. R. Bolton, *J. Chem. Soc., Chem. Commun.*, 1596 (1990).
721. E. Lipczynska-Kochany, J. Kochany and J.R. Bolton, *J. Photochem. Photobiol., A: Chem.*, **62**, 229 (1991).
722. J. C. Evans, C. C. Rowlands, L. A. Turkson and M. D. Barratt, *J. Chem. Soc., Faraday Trans. 1*, **84**, 3249 (1988).
723. E. Lipczynska-Kochany, *Chemosphere*, **24**, 911 (1992).
724. N. Yagi, H. Kenmotsu, H. Sekikawa and M. Takada, *Chem. Pharm. Bull.*, **39**, 454 (1991).
725. V. Nair and D. A. Young, *J. Org. Chem.*, **50**, 406 (1985).
726. Z. Kazimierczuk, R. Mertens, W. Kawczynski and F. Seela, *Helv. Chim. Acta*, **74**, 1742 (1991).
727. T. M. Dietz, R. J. von Trebra, B. J. Swanson and T. H. Koch, *J. Am. Chem. Soc.*, **109**, 1793 (1987).
728. T. Sehili, G. Bonhomme and J. Lemaire, *Chemosphere*, **17**, 2207 (1988).
729. T. Sehili, P. Boule and J. Lemaire, *J. Photochem. Photobiol., A: Chem.*, **50**, 103 (1989).
730. T. Sehili, P. Boule and J. Lemaire, *J. Photochem. Photobiol., A: Chem.*, **50**, 117 (1989).
731. T. Sehili, P. Boule and J. Lemaire, *Chemosphere*, **22**, 1053 (1991).
732. J.-C. D'Oliveira, G. Al-Sayyed and P. Pichat, *Environ. Sci. Technol.*, **24**, 990 (1990).
733. G. Al-Sayyed, J.-C. D'Oliveira and P. Pichat, *J. Photochem. Photobiol., A: Chem.*, **58**, 99 (1991).
734. K. Seguchi and H. Ikeyama, *Chem. Lett.*, 1493 (1980).
735. V. A. Loskutov, S. M. Lukonina, A. V. Konstantinova and E. P. Fokin, *Zh. Org. Khim.*, **17**, 584 (1981) [Engl. Transl.: *J. Org. Chem. USSR*, **17**, 500 (1981)].
736. S. Arai, K. Tabuchi, H. Arai, T. Yamagishi and M. Hida, *Chem. Lett.*, 1355 (1991).
737. M. V. Baklanov and A. N. Frolov, *Zh. Org. Khim.*, **27**, 638 (1991) [Engl. Transl.: *J. Org. Chem. USSR*, **27**, 550 (1991)].
738. M. V. Baklanov and A. N. Frolov, *USSR SU* 1,641,821, (1991); *Chem. Abstr.*, **116**, 21051w (1991).
739. A. N. Frolov, E. V. Smirnov, O. V. Kul'bitskaya and A. V. El'tsov, *Zh. Org. Khim.*, **16**, 2302 (1980). [Engl. Transl.: *J. Org. Chem. USSR*, **16**, 1963 (1980)].
740. E. M. Klokova, A. N. Frolov and A. V. El'tsov, *Zh. Org. Khim.*, **17**, 2171 (1981). [Engl. Transl.: *J. Org. Chem. USSR*, **17**, 1935 (1981)].
741. A. N. Frolov, E. M. Klokova and A. V. El'tsov, *Zh. Org. Khim.*, **17**, 2161 (1981). [Engl. Trnsl.: *J. Org. Chem. USSR*, **17**, 1926 (1981)].
742. E. P. Kyba, S. T. Liu and R. L. Harris, *Organometallics*, **2**, 1877 (1983).
743. V. L. Ivanov and L. Éggert, *Zh. Org. Khim.*, **19**, 2373 (1983). [Engl. Transl.: *J. Org. Chem, USSR*, **19**, 2075 (1983).
744. V. L. Ivanov, J. Aurich, L. Eggert and M. G. Kuz'min, *J. Photochem. Photobiol., A: Chem.*, **50**, 275 (1989).
745. V. L. Ivanov, L. Éggert, A. Kherbst and M. G. Kuz'min, *Zh. Org. Khim.*, **20**, 1735 (1984). [Engl. Transl.: *J. Org. Chem. USSR*, **20**, 1581 (1984)].
746. V. L. Ivanov, L. Eggert and M. G. Kuz'min, *Zh. Org. Khim.*, **22**, 1493 (1986). [Engl. Transl: *J. Org. Chem. USSR*, **22**, 1348 (1986)].
747. V. L. Ivanov, I. Aurikh, Kh. Langbain and M. G. Kuz'min, *Zh. Org. Khim.*, **25**, 2142 (1989). [Engl. Transl.: *J. Org. Chem. USSR*, **25**, 1935 (1989)].
748. V. S. Savvina, V. L. Ivanov and M. G. Kuz'min, *Khim. Fiz.*, **7**, 1667 (1988); *Chem. Abstr.*, **111**, 22850 q (1989).
749. V. Nair and D.A. Young, *Synthesis*, 450 (1986).

750. V. Nair and G.S. Buenger, *J. Am. Chem. Soc.*, **111**, 8502 (1989).
751. T. Maruyama and M. Honjo, *Nucleosides Nucleotides*, **7**, 203 (1988).
752. T. Kashimura, K. Kudo, S. Mori and N. Sugita, *Chem. Lett.*, 299 (1986).
753. T. Kashimura, K. Kudo, S. Mori and N. Sugita, *Chem. Lett.*, 483 (1986).
754. T. Kashimura, K. Kudo, S. Mori and N. Sugita, *Chem. Lett.*, 851 (1986).
755. J. P. Cosgrove and B. M. Vittimberga, *J. Heterocycl. Chem.*, **21**, 1277 (1984).
756. B. Šket, M. Zupan, N. Zupančič and B. Pahor, *Tetrahedron*, **47**, 5029 (1991).
757. N. Tachikawa, *Chemistry Express*, **1**, 587 (1986).
758. C. Kaneko, T. Naito and C. Miwa, *Heterocycles*, **19**, 2275 (1982).
759. S. P. Spyroudis, *Justus Liebigs Ann. Chem.*, 947 (1986).

CHAPTER **17**

Radiation chemistry

ZEEV B. ALFASSI
Department of Nuclear Engineering, Ben Gurion University of the Negev, Beer Sheva, Israel

ABBREVIATIONS

D_g	gel dose
DCBP	dichlorobiphenyl
DCE	1,2-dichloroethane
DFBP	decafluorobiphenyl
FBTHF	perfluoro-2-butyltetrahydrofuran
HFP	hexafluoropropene
M_n	number average molecular weight
MBMTP	2,2-methylene-bis(4-methyl-6-t-butylphenol)
MCH	methylcyclohexane
5MD	5-methyldecane
MEK	methyl ethyl ketone
3MP	3-methylpentane
MTHF	2-methyltetrahydrofuran
PBN	phenyl-t-butylnitrone
PCB	polychlorinated biphenyl
PVDF	poly(vinylidene fluoride)
SOMO	singly occupied molecular orbital
TCBP	tetrachlorobiphenyl
TMS	tetramethylsilane
t-NB	t-nitrosobutane
VDF	vinylidene fluoride

Supplement D2: The chemistry of halides, pseudo-halides and azides
Edited by S. Patai and Z. Rappoport © 1995 John Wiley & Sons Ltd

I. INTRODUCTION

Radiation chemistry is the study of the chemical effects produced in a system by the absorption of ionizing radiation from radioactive sources, high-energy charged particles and short-wavelength (less than about 400 Å)[1] electromagnetic radiation from accelerators. The principal characteristic of high-energy radiation is that it causes ionization in all materials. This distinguishes between radiation chemistry and photochemistry[2,3], which deals with longer-wavelength radiations possessing energy lower than about 30 eV. In radiation chemistry the whole energy is not absorbed by a single molecule, as in photochemistry, but rather distributed over several molecules, along the track of the ionizing particle or photon. The high-energy photons and particles are not selective and may ionize, excite or dissociate any molecule lying in their path. This is in contrast to photochemistry where only some compounds may interact with the radiation, in accordance with the energy of the photons.

The high-energy photons or particles lose energy in successive events and produce ions and primary electrons, which in turn form several secondary electrons with lower energies[4]. The chemical effects of ionizing radiation occur almost exclusively through the secondary electrons, most of which have energies less than 100 eV. These electrons will ionize and excite the surrounding molecules and will lose energy until they reach thermal energies. In many solvents these thermal electrons polarize the solvent molecules and are bound in a stable quantum state to them, these electrons are called solvated electrons. On the average, half of the absorbed energy leads to ionization while the other half leads to excited molecules.

The study of radiation chemistry can be divided into two parts. The first is the study of unstable intermediates which have short lifetimes and thus cannot be studied by usual chemical methods. The second part is the study of the final products of the radiolysis which are measured by common chemical techniques.

One way to make the short-lived intermediates amenable to study is to increase their lifetime, usually by irradiation in the solid state and/or at very low temperatures. The intermediates can be then detected at the end of the irradiation by ESR or optical absorption spectroscopy. The ESR of radicals in the solid state is carried out on single crystals, polycrystalline samples or frozen aqueous solutions. In the two latter cases the identification of radicals from the ESR spectra is frequently difficult and, for better identification, the ESR should be done on irradiated single crystals.

Another method of increasing the lifetime of radicals is by adding compounds which react with the radicals to produce long-lived radicals; this method is called spin trapping[5]. A diamagnetic spin-trap is used to convert short-lived radicals into long-lived ones. For example, using nitroso compounds (such as t-nitrosobutane, t-NB) the short-lived radicals form long-lived nitroxide radicals (the spin-adduct) according to reaction 1[5].

$$R + R'N{=\!=}O \longrightarrow \underset{R}{\overset{R'}{\diagdown}}N{-}O^{\bullet} \longleftrightarrow \underset{R}{\overset{R'}{\diagdown}}{}^{+}N{-}O^{-} \qquad (1)$$

The method was developed for the liquid phase and was later extended to polycrystalline solids; the latter are γ-irradiated and subsequently dissolved in a solution containing the spin-trap. Most commonly it is aqueous solution and several spin-traps were used[6].

More common in the liquid phase is pulse radiolysis[7]. In this technique, electron accelerators which can deliver intense pulses of electrons lasting a very short time (ns up to μs) are used. Each single pulse can produce intermediates with sufficiently high concentration to be studied by various methods, such as light absorption sepectroscopy or electrical conductivity.

The radiolytic yields are always expressed by the G value, which is defined as the number of particles (molecules, radicals, ions) produced or consumed per 100 eV of energy absorbed in the system.

The units for the absorbed energy (dose) are the rad, defined by 1 rad = 100 erg g^{-1} = 6.243 × 10^{13} eV g^{-1}, and the Gray (Gy) defined by 1 Gy = 100 rad.

When radiolysing a solution, the radiation interacts mainly with solvent molecules which are in large excess, since the radiation interacts with the molecules unselectively. Consequently, the radiation chemistry of a solution is the combination of the production of initial intermediates from the solvent, which will be the same as in pure solvents, and the reactions of those intermediates with the solute. The most common solvent is water.

The radiolysis of water produces hydrated electrons (e$_{aq}^-$, $G = 2.9$), hydrogen atoms ($G = 0.55$) and hydroxyl radicals ($G = 2.8$) which react with the solute molecules. In addition, the radiolysis of aqueous solutions also yields H$_2$O$_2$ ($G = 0.75$), gaseous hydrogen ($G = 0.45$) and hydronium ions (H$_3$O$^+$, $G = 2.9$). In most cases the molecular products do not interfere with the reactions of the radicals. To study the reaction of one radical with the solute without interference from other radicals, scavengers for the other radicals should be added[7-10].

e$_{aq}^-$ can be eliminated and even converted to the other radicals OH and H, by the addition of either N$_2$O or H$^+$ to the system (equations 2 and 3, respectively).

$$e_{aq}^- + N_2O \xrightarrow{\text{H}_2\text{O}} N_2 + OH^- + OH \quad (k = 9 \times 10^9 \text{ M}^{-1}\text{s}^{-1}) \tag{2}$$

$$e_{aq}^- + H^+ \longrightarrow H \quad (k = 2.3 \times 10^9 \text{ M}^{-1}\text{s}^{-1}) \tag{3}$$

Therefore, in aqueous N$_2$O saturated solutions, OH radicals are the predominant species (90%), while in acidic aqueous solution H and OH radicals exist, but not hydrated electrons. The small fraction (ca 10%) of hydrogen atoms in N$_2$O saturated aqueous solution does not interfere significantly with the measurements of the reactions of OH radicals. Frequently, H atoms and OH radicals react with solutes in a similar manner, e.g. by addition to a double bond or by hydrogen atom abstraction, however with different rate constants.

The reaction of H atoms can be studied in acidic solution if the OH radicals are scavenged by t-butyl alcohol, in a very fast reaction, while hydrogen atoms react only very slowly with this alcohol (equations 4 and 5). The radical produced in reaction 4 is relatively unreactive and does not interfere with the study of the reaction of H atoms with the solute.

$$\dot{O}H + (CH_3)_3COH \longrightarrow H_2O + \dot{C}H_2C(CH_3)_2OH \quad (k = 5 \times 10^8 \text{ M}^{-1}\text{s}^{-1}) \tag{4}$$

$$H + (CH_3)_3COH \longrightarrow H_2 + \dot{C}H_2C(CH_3)_2OH \quad (k = 1.7 \times 10^5 \text{ M}^{-1}\text{s}^{-1}) \tag{5}$$

Hydrated electrons are obtained as predominant radicals by removing the OH radicals with t-butyl alcohol. The removal of both H and OH radicals is accomplished by isopropanol (equations 6 and 7).

$$H + (CH_3)_2CHOH \longrightarrow H_2 + (CH_3)_2\dot{C}OH \quad (k = 7.4 \times 10^7 \text{ M}^{-1}\text{s}^{-1}) \tag{6}$$

$$\dot{O}H + (CH_3)_2CHOH \longrightarrow H_2O + (CH_3)_2\dot{C}OH \text{ (86%)} + (CH_3)_2CH\dot{O} \text{ (1%)} +$$
$$+ \dot{C}H_2CH(OH)CH_3 \text{ (13%)} \quad (k = 1.9 \times 10^9 \text{ M}^{-1}\text{s}^{-1}) \tag{7}$$

Two previous reviews on radiolysis of haloorganic compounds were published in this series[11,12]. Consequently, this review is only for the period starting in 1983.

II. RADIATION CHEMISTRY OF POLYMERS WITH CARBON–HALOGEN BONDS

Ionizing radiations break different bonds in polymers, leading to the formation of both small radicals and polymer radicals. The polymer radicals can react with polymer molecules leading to cross-linking, transforming linear polymers to di- and tri-dimensional polymers with much higher molecular weight. On the other hand, the small radicals can recombine, leading to destruction of the polymer and to formation of low molecular weight products. The processes of chain scission and cross-linking take place concurrently and the fate of each polymer depends on the ratio of these two processes. Most of the earlier studies on the radiation chemistry of polyhalocarbons were carried out on polytetrafluoroethylene, and were reviewed by Florin[13] and by Dole[14]. Another fluoro-polymer studied was poly(vinylidene fluoride) (PVDF) described in patents for the production of heatshrinkable materials, but also studied by Timmerman and Greyson[15]. Further work on the cross-linking of PVDF and the dehydrofluorination of the product was done by Makuchi and coworkers[16], who found that the irradiation of PVDF leads, besides cross-linking, also to elimination of HF molecules. The presence of the formed HF deaccelerates the rate of further cross-linking, as was found by measuring the gel fraction for the polymer alone and for the polymer in the presence of KOH pellets. The gel fraction increased in the presence of KOH due to its reaction with the HF.

The measure of the cross-linking is determined by the gel fraction—the fraction of the polymer which, after irradiation, is not soluble any more in methyl ethyl ketone (MEK) due to the presence of two- and three-dimensional net. The gel fraction is denoted usually by g, and as a function of the absorbed dose is given usually by the Charlesby–Pinner equation[17], relating the sol fraction, s, with the dose:

$$S + \sqrt{s} = \frac{G(s)}{2G(c)} + \frac{\alpha}{G(c) \cdot M_n \cdot D}$$

The sol fraction, s, is the fraction which is dissolved in MEK and hence $s = 1-g$; $G(s)$ and $G(c)$ are the radiolytic yields (molecules per 100 eV energy absorbed) of chain scission and cross-linking, respectively, while M_n is the number-average molecular weight of the polymer before the irradiation, D is the absorbed irradiation dose and α is a constant having the value 4.8×10^5 for the dose in units of Mrad (10 kGy). The presence of KOH was found to increase both scission and cross-linking; however, the increase in the cross-linking is larger than the increase in bond scission [$G(c)/G(s)$ increases]. In the absence of KOH, SiF$_4$ was found in the irradiated ampules due to the reaction of HF with the SiO$_2$ of the pyrex ampule. In the presence of KOH no SiF$_4$ was detected. However, the reaction of HF with SiO$_2$ was not fast enough to stop its deactivating effect on the cross-linking. It was found that SiF$_4$ also had a deactivating effect on the cross-linking of PVDF, but considerably lower than that of HF. The effect of HF was found to be stronger on thinner films of the homopolymer. Some experiments were conducted with added HF to prove its effect. It was found that HF is present even if not added, probably due to HF elimination. The result of this process should be formation of conjugated double bonds (polyenes) along the PVDF chains. This can be proven by absorption bands at 227 and 272 mm in very thin (50 μm) foils.

The free radicals produced in PVDF by irradiation were studied by the ESR spectra of samples irradiated at 77 K. It was found that the total free radical yield is about 3 and the ESR spectrum is broad and diffuse with several shoulders. Warming the irradiated sample to 195 K (warming an irradiated solid is usually referred to as 'annealing') leads to a decrease in the radical concentration to two- thirds and a small portion of the spectrum at that temperature can be assigned to a terminal —CF$_2$ radical. Keeping the irradiated sample one hour at 195 K reduced further the radical concentrations (to about one-third) and the spectrum now contains a double sextet which can be expected from a radical with 4 β hydrogen atoms $-CH_2\dot{C}F-CH_2-$. On warming the sample to 70°C or keeping it for a long time at room temperature, the ESR spectrum is transformed into a singlet, due probably to polyenyl radicals. Makuchi and coworkers[16] could not find in the ESR spectrum evidence for a $-CF_2-\dot{C}H-CF_2-$ radical, although this is expected to be the main radical formed in the very first step, since the C—H bond is weaker than the C—F bond. However, this radical is very reactive and probably reacted during the annealing. Few radicals usually remain at 195 K and the reason that $-CH_2\dot{C}F-CH-$ does not react is due to stabilization by the strong electronegativity of the fluorine atom.

$$C—F \longleftrightarrow C^{+\delta}—F^{-\delta}$$

The authors suggest that the effect of HF is due to reaction with the radicals formed by the radiolysis ($-CH_2-\dot{C}F-CH_2-$ and $-CF_2-\dot{C}H-CF_2-$), converting the radicals into stable molecules. They explain the reaction of the radicals with HF as due to the high electronegativity of the covalently linked fluorine atom.

$$H^{\delta+}—F^{\delta-} + {}^{\delta+}\dot{C}—F^{\delta-} \longrightarrow \left[H\cdots F\cdots\overset{|}{\underset{|}{C}}—F \right] \longrightarrow \dot{H} + F—\overset{|}{\underset{|}{C}}—F$$

They suggested the following tentative scheme of reactions:

$$-H_2C-\dot{C}F-CH_2-+HF \longrightarrow -H_2C-CF_2-CH_2-+H$$

$$-F_2C-\dot{C}H-CF_2-+HF \longrightarrow -F_2C-CH_2-CF_2-\cdot+F$$

They ignored that the last reaction is very endothermic, due to H—F being a stronger bond than C—H, even in the presence of neighbouring CF_2 groups, and suggested that the atoms disappear by atom combination:

$$H+H \longrightarrow H_2 \quad \text{or} \quad H+F \longrightarrow HF$$

Carenza and coworkers[18] studied the effect of ionizing radiation on copolymers of vinylidene fluoride (VDF) with hexafluoropropene (HFP). These are non-crystalline fluoroelastomers with a wide temperature range (–40 to 300 °C) and resistance to chemicals, solvents and fuels which make them commercially useful for several purposes. They used a random polymer of VDF: HFP (81:19 molar ratio) which has a T_g of –20 °C ($M_w = 8 \times 10^4$ and $M_n = 3.4 \times 10^4$). The samples were irradiated by ^{60}Co γ-rays at room temperature in both the presence and absence of air. For measurements of radicals, samples were irradiated at 77 K in vacuo, and then the samples were heated inside the ESR spectrometer. It was found that both in vacuo and in air, the gel fraction increases with the absorbed dose until it reaches a plateau. However, the dependence on dose rate for vacuum irradiation is different than for irradiation in air. For irradiation in vacuo the gel fraction is independent of the dose rate and depends only on the total absorbed dose (before reaching the plateau). For irradiation in the presence of air the gel fraction increases both with dose rate and with total absorbed dose. The gel fraction even in the highest dose rate studied 0.62 Gy s^{-1}, is lower than for irradiation in vacuo. It was found that in the case of irradiation in the presence of air, the gel fraction is smaller for thin polymers than for thicker sheets. The least gel fraction was observed with polymers irradiated in a divided from. These findings agree with the observation that the gel fraction is considerably smaller in the presence of air. In the thickest (10 mm) sheet most of the radiolysis is done without O_2, although there is air surrounding the sample, due to the slow diffusion of oxygen. In anaerobic conditions there is much more cross-linking than in the presence of oxygen. For the highest dose studied (0.62 mrad), it was found that the gel fraction on irradiation in either vacuum or in air follows the Charlesby–Pinner equation for dose dependence. The value of D for which $s + \sqrt{s} = 2$ is called the 'gel dose'; D_g was found by extrapolation and found to be the same for the two systems. However, the plots for vacuum and air irradiation have different slopes and intercepts. Both $G(c)$ and $G(s)$ are larger in the presnce of air than in vacuo. The lower yield of gel fraction in the presence of air is due to the larger increase in the bond scission than in cross-linking; $G(s)/G(c)$ increased with the presence of air.

The results found by Carenza and coworkers[18] in vacuum, $G(c) = 1.37$ and $G(s) = 0.82$, are quite different than those found previously by Yoshida and workers[19], the latter being closer to those obtained by Carenza and coworkers in vacuo.

To find the nature of the radicals formed during the irradiation, Carenza and coworkers[18] irradiated the sample under vacuum at 77 K and studied the ESR spectra, which are dominated by a very broad peak with two minor additional peaks. A doublet peak of 516 G is attributed to trapped hydrogen atoms and two more peaks (separation of 420 G) are characteristic of RCF_2^- type radicals. H atoms were not detected in the irradiated poly-VDF, probably due to their tendency to undergo H-atom abstraction even at 77 K. In the VDF–HFP copolymer, suitable trapping sites must exist in proximity of units without C—H bonds available for H abstraction. A major question is the origin of the major structure-less peak. Assuming this to be a macroradical with the radical on a non-terminal carbon, there are several possibilities; either $-CF_2-\overset{\cdot}{C}H-CF_2-$ or $-H_2C-\overset{\cdot}{C}$ F$-CH_2-$ or $-CF_2CF-CF_2-$. Carenza and coworkers assumed that only the first two are possible and, by computer simulation of analogous radicals, showed that the $-CF_2\overset{\cdot}{C}H-CH_2-$ agrees with the observed spectra, whereas the calculated spectra of $-CH_2-\overset{\cdot}{C}F-CH_2-$ will be completely different, with broad doublets due to the splitting of the 4 β-hydrogen atoms. The $-CF_2-\overset{\cdot}{C}F-CF_2-$ radicals fit the observed spectra, but were not considered as a major contributors due to the lower concentration of HFP.

On warming the irradiated sample to $-40\,°C$, the main ESR signal is a singlet. This was found previously for the homopolymer poly-VDF, and was assigned to the polyenyl radicals.

According to these results the macroradical $-CF_2-\overset{\cdot}{C}H-CF_2-$ plays the major role in cross-linking of VDF–HFP copolymer. The dehydrofluorination is witnessed by the findings of polyenyl-type radicals. In the presence of oxygen, the alkyl radicals react with O_2 to form peroxy radicals, which intervene both in cross-linking and in chain-scission reactions.

The cross-linking is supposed to be increased in the presence of oxygen due to the much higher reactivities of the peroxyl radicals[20–24], and due to a possible increase in the number of radicals due to decomposition of labile oxidation products. Chain scission is expected also to be enhanced by the presence of oxygen due to bimolecular disproportionation and β-scission of the resulting alkoxy radical intermediates.

The radiolysis of polymers (both without and with halogen atoms) under oxidizing conditions leads to considerable degradation of the polymers, leading to drastic changes in their mechanical properties, such as elongation and tensile strength[25]. In order to stabilize polymeric materials in an environment of ionizing radiations various approaches were taken, ranging from the synthesis of new macromolecular structures having inherent radiation resistance, to the use of low molecular weight additives[26,27]. Clough and Gillen[28] studied enhancing the resistance of poly-chloropropene to degradation by γ-radiation by the addition of scavengers for excited molecules (pyrene) and for radicals (substituted hindered phenols). The stabilizing effect was studied in the presence of air, since the degradation is larger under oxidizing conditions. As a radical scavenger they use 2,2-methylene-bis(4-methyl-6-t-butylphenol) (MBMTP), which is used as antioxidant in commercial polymeric materials. Scavengers of excited molecules are usually aromatic molecules, which can rapidly absorb the excitation energy, and are subsequently deactivated with high efficiency to their ground state without any decomposition.

In the case of reaction in the presence of air, care should be taken to choose an aromatic compound which in its excited state, will not react with oxygen to produce singlet oxygen, since the latter can actually increase the oxidation of the polymer, especially in the case of polymers with high concentration of alkenic bonds (i.e. rubbers such as polychloropropene). It was found that addition of 1.7% (weight fraction) of pyrene enhances the resistance of polychloropropene, increasing the dose required for losing 50% of the elongation from 1.35×10^5 to 2.0×10^5 Gy. Addition of 1.7% MBMTP increases the radiation dose required for loss of 50% of elongation to 2.75×10^5 Gy. Addition of both 1.7% pyrene and 1.7% MBMTP leads to loss of 50% elongation only after 3.3×10^5 Gy, increasing the resistance to radiation by more than a factor of 2. Clough and Gillen[28] studied, besides the loss of elongation, also the loss of pyrene and MBMTP due to radiation. It was found that only less than 10% of the pyrene is consumed, but 97% of the MBVMTP was decomposed for 3×10^5 Gy. These observations fit the stabilization mechanisms expected for the two compounds. The hindered phenol scavenges radicals by reacting chemically with them, leading to products different than MBMTP. In contrast, the excitation energy transferred to pyrene does not lead to a chemical reaction but rather to an energy transfer. The excited pyrene molecules revert to their ground state either by fluorescence (and/or phosphorescence) or by radiationless decay to vib-rotational energy, and then to translational energy.

The authors suggested that the slow disappearance of pyrene is due to its reaction with radicals (although pyrene is less reactive toward the radicals than the hindered phenol). Another possibility suggested a low-efficiency reaction of excited pyrene competing with the radiation and the radiationless decay. When pyrene is added together with MBMTP, pyrene somewhat decreases the rate of disappearance of the MBMTP molecules, e.g. at 3×10^5 Gy from 97% to 94%.

γ-Radiation not only degrades polymer molecules but also cross-links them, and irradiation of two homopolymers together leads to the formation of graft polymers.

Radiation-induced grafting[29] was used for the preparation of low-cost polyelectrolyte membranes needed for fuel-cells[30]. Guzman-Garcia and coworkers[29] studied the grafting of poly(styrenesulphonic acid)on polystyrene on a copolymer of poly(tetrafluoroethylene) with polystyrene, and with a copolymer of poly(tetrafluoroethylene) with poly(perfluoropropylene).

III. RADIOLYSIS IN THE SOLID STATE AT LOW TEMPERATURES

Radiolysis at low temperature leads to ejection of an electron leaving a radical cation of the parent molecule. The electron usually reacts even at low temperatures due to the high electron affinity of the halogen atoms. The structure of the formed radical cation of the halogenated organic radical has been studied intensively by electron spin resonance and optical absorption studies. ESR studies show that in the case of alkyl halides and freons the radical usually forms a σ^* dimer by making a three- electron bond with a halogen atom in a neighbouring molecule[31–33], although the singly-occupied molecular orbital (SOMO) is a non-bonding orbital of the halogen atoms. In the case of $CBr_nH_{4-n}^{+}$ radical cation with $n \geq 2$, the SOMO has been reported recently to be an intramolecular Br $\cdots$ Br antibonding orbital composed of the non-bonding orbitals of two equivalent bromine atoms[34]. Symons[35] studied the ESR powder pattern of CCl_4^{+}, in low-temperature γ-irradiated CCl_4, and found that the radical cation has a structure of C_{3v} symmetry, composed of an intramolecular antibonding combination of three equivalent Cl atoms of in-plane p nonbonding orbitals. This assignment was questioned by Muto and coworkers[36] on the grounds that the total spin density estimated from the reported Cl hyperfine tensors is too small (ca 0.3). They also suggested that there is a misleading assignment of the isotopic combination (^{35}Cl and ^{37}Cl), which is due to the poorly resolved complex spectrum of the powder. In order to clarify the electronic and geometrical structures, they combined the analysis of the angular dependence of ESR spectra with spectra simulations of the powder pattern, including the isotopic combination of ^{35}Cl and ^{37}Cl. It was found that the cation gives a spectrum consisting of two chlorine atoms with different hyperfine coupling factors of $A(^{35}Cl_1) = 11.2, 0.5, 1.3$ mT and $A(^{35}Cl_2) = 6.2, 0.8, 1.8$ mT. The g tensor is approximately axially symmetric and shows a large positive $g_\perp$ shift, different from the rhombic g tensor of σ^* dimer cations of alkyl halides and freons. From a consideration of the anisotropy of g, combined with the optical atomic spectrum. $CCl_4^{+\bullet}$ was suggested to be a Cl atom-like species, in which one of the C—Cl bonds in the parent CCl_4^+ is released and the atom-like chlorine nuclei formed make a three-electron bond with a chlorine atom in the CCl_3^+ released species. This result is different from that found previously for other alkyl radical cations in which the cation is not dissociative.

Muto and collaborators[36] explained this difference as due to the weaker bond of C—Cl in CCl_4 (bond strength = 70.4 kcal mol^{-1}) as compared to the C—Cl bond in alkyl halides (81–84 kcal mol^{-1}). This agrees with findings that the only freon found to have a similar Cl atom-like cation is $CFCl_3$, which also has weak C—Cl bonds (bond strength Cl—$CFCl_2$ = 71 kcal mol^{-1}).

The electron ejected in the radiolysis reacts with the parent molecule. Optical absorption studies of the radiolysis of CCl_4 in the liquid phase (see detailed description later) show three absorption lines. These were assigned to the cation CCl_4^+ and to the ion pair $CCl_3^+ \cdots Cl^-$ or CCl_3^+/solvent/Cl$^-$. However, optical studies only show the existence of intermediates and not their structures. One way to study the structure of intermediates is by ESR of irradiated frozen samples at very low temperatures. Irradiation of neat CCl_4 enables the study of the cation formed in the radiolysis, whereas radiolysis of CCl_4 in tetramethylsilane allows the measurement of the ESR spectra of the anion radical of CCl_4 formed upon radiolysis. Radicals having many halogen atoms show complex and broad ESR spectra due to the nuclear spin multiplets, many kinds of magnetic interactions and overlapping of isotopic combinations. In order to help in elucidating the structure of the $CCl_4^{-\bullet}$ anion radical, Muto and Nunome[37] studied the ESR spectra of both $^{12}CCl_4$ and $^{13}CCl_4$ in tetramethylsilane (TMS) matrix irradiated at 4 or 77 K. Besides the ESR spectra they also measured the optical spectrum of $CCl_4(1\%)$/TMS polycrystalline samples irradiated at 4 and 77 K. Both temperatures gave essentially the same spectrum with two UV absorption bands at 265 and 370 nm, which are not observed in irradiated pure TMS. The band at 370 nm was observed with many organic solids and liquids. The 270 nm band was not reported for CCl_4/alkane mixtures but was observed in pure solid CCl_4. Warming the CCl_4/TMS to 150 K or illuminating by UV light leads to decay of these optical absorptions simultaneously with the disappearance of $CCl_4^{-\bullet}$ ESR spectra and the appearance of the ESR spectra of CCl_3 radicals. The ESR spectra of $CCl_4^{-\bullet}$ show that this radical has the following characteristics:

(a) The carbon atom has a large spin density ($\rho_{2p} = 0.62$, $\rho_{2s} = 0.18$) and an orbital with a large s character (s/p = 1/3.4).

(b) From the four chlorine atoms involved in the radical, three have equivalent spin densities ($\rho_{3s} = 0.008$, $\rho_{3p} = 0.108$) whereas the fourth has only about half of that spin density.

(c) All atoms have the orientations of the maximum hyperfine tensor element, nearly parallel to the direction of g_{min}.

(d) The g tensor has a large anisotropy ($g_{max} = 2.015$), and g_{min} is slightly larger than that of the usual π radicals ($g_{min} = 2.004$).

Characteristic (a) indicates that the unpaired electron is mainly localized in a carbon orbital with near-sp^3 hybridization. Characteristics (b) and (c) show that the radical has

C_{3v} symmetry. Some of the unpaired electron is delocalized on the p_x atomic orbitals of the three equivalent Cl atoms. These characteristics fit a predissociating $CCl_4^{-\cdot}$ radical anion in which one of the C—Cl bonds is slightly released by the capture of the additional electron $(CCl_3\cdots Cl)^-$. The small spin density on the fourth chlorine atom is probably due to the higher electron affinity of the Cl^- ion compared to the CCl_3^- ion (3.61 and 1.22 eV, respectively).

Shida and Takemura[38] have detected $CCl_4^{+\cdot}$ in both CCl_4 and CCl_3F matrices irradiated at low temperatures, measuring ESR and optical spectra. They found a value of 400 nm for λ_{max} with irradiation of neat CCl_4. This value lies between extreme values of 340 and 425 nm found in the liquid phase (see Section IV). Shida and Takemura pointed out that none of these values fits the photoelectron spectroscopic data[39], but they do not have an explanation for this. They found differences in the ESR spectra assigned to $CCl_4^{+\cdot}$ in CCl_4 and in CCl_3F, suggesting that the degree of loss of the T_d symmetry of CCl_4 depends on the surrounding matrix.

Gamma radiolysis of CCl_4/alkane glasses produces two absorption bands, one at 366 nm and one at 470 nm[40]. Results on the absorptions of these glasses containing cation scavengers showed that the band at 470 nm is due to a positive ion, which was suggested to be the CCl_3^+ cation[40,41]. The other band at 366 nm is present at 20 K for both 3-methylheptane/CCl_4 and 3-methylpentane/CCl_4 glasses. However, at 77 K it is present only in 3-methylheptane/CCl_4 but not in 3-methylpentane/CCl_4[42]. This was explained as due to the lower viscosity of the 3-methylpentane/CCl_4 glass. Willard and coworkers[42] attributed the 366 nm band to $CCl_4^{-\cdot}$ but, in a note added in proof, declared that this assignment might be wrong. Louwrier and Hamill[40] assumed the 366 nm band to be the precursor of the 470 nm cation band and assigned it to a positive ion of CCl_4. They suggested that this might be $CCl_4^{+\cdot}$ although they were aware of the fact that the ionization potential of CCl_4 is higher by about 1.5 eV than that of the alkanes, and that this difference in the ionization potentials ought to result in positive charge transfer from CCl_4^+ to the alkanes.

Suwalski[43] has attributed the 365 nm in pulse-irradiated methylcyclohexane/CCl_4 glass to $CCl_4^{-\cdot}$. However, his evidence was indirect by scavenging experiments as will be detailed later, and $CCl_4^{-\cdot}$ was a major participant in his overall mechanism. Brickenstein and Khairutdinov[44] used photo-sensitized solute in alcohol and ether glasses containing CCl_4. They found that after photoionization, charge recombination can be induced by illumination at 365 nm, indicating that $CCl_4^{-\cdot}$ absorbs at that wavelength. Klassen and Ross[45] studied pulse radiolysis and optical absorption in a 0.4 M CCl_4/3-methylpentane glass at 75–77 K. In order to subtract the contribution from 3-methylpentane (3MP), they also pulse-radiolysed a N_2O/3MP glass (with N_2O added to scavenge the electrons, as is done by CCl_4). They found an absorption band at 370 nm. No absorption was found at 470 nm, which however was found to be the only absorption for CCl_4/3MP glass irradiated continuously at 77 K[42].

The 370 nm absorption decayed by 45% over the first millisecond, but no other new absorption was observed to grow during this time. Since Louwrier and Hamill[40] found the fully formed 470 nm band after ^{60}Co radiolysis of 3MP/CCl_4, whereas Klassen and Ross found decay of the 370 nm band which was not followed by growth of the 470 nm absorption, it must be concluded that there is an intermediate between the 370 nm species and the 470 nm species. Klassen and Ross[45] measured the dependence of the initial 370 nm absorption as a function of the CCl_4 concentration in the glass. They observed that the dependence is sublinear and approaches saturation. This dependence suggests that the 370 nm species is a product of scavenging by CCl_4. It is known that CCl_4 in the concentrations used by Klassen and Ross scavenges efficiently electrons. It was common to assume that the resulting adduct, $CCl_4^{-\cdot}$ decomposes promptly to Cl^- and $CCl_3^\cdot$. However, it is possible that this decomposition is not instantaneous in the glassy alkane matrix. To test the hypothesis that the 370 nm is due to $CCl_4^{-\cdot}$. Klassen and Ross pulse-radiolysed CCl_4/3MP glasses which contain N_2O, an efficient electron scavenger, in a constant N_2O concentration with varying concentrations of CCl_4. They found that under these circumstances the absorption fits a simple competition between N_2O, CCl_4 and other species (e.g. positive ions and traps). The suggestion that the 366–370 nm absorbing species is the radical anion $CCl_4^{-\bullet}$ fits very well Gremlich and Bühler's suggestion[41] for the mechanism, which explains also the existence of an intermediate between the 370 nm absorbing species and that of 470 nm species:

$$3MP \xrightarrow{\quad\sim\sim\sim\quad} 3MP^{+\bullet} + e^-$$

$$CCl_4 \xrightarrow{\quad\sim\sim\sim\quad} CCl_4^{+\bullet} + e^-$$

$$CCl_4^{+\bullet} + 3MP \longrightarrow CCl_4 + 3MP^{+\bullet}$$

$$CCl_4 + e^- \longrightarrow CCl_4^{-\bullet} \; (370 \text{ nm})$$

$$CCl_4^{-\bullet} \longrightarrow CCl_3^{\bullet} + Cl^-$$

$$3MP^{+\bullet} + CCl_3^{\bullet} \longrightarrow 3MP + CCl_3^+ \; (470 \text{ nm})$$

Suwalski and Kroh[46] measured the formation and decay of the species produced by γ-radiolysis of methylcyclohexane (MCH)/CCl_4 glass at 4.2 K (liquid helium temperature). The glass, radiolysed and kept at 4.2 K, has a sharp absorbing band at 354 nm ($G\varepsilon = 12{,}500$) and a wide shoulder at 420–435 nm. Heating the irradiated glass leads to a decrease in the 354 nm band and a shift of the peak. At 77 K there is a peak at 370 nm instead of at 354 nm. From their figure it can be concluded that at 95 K the peak at 370 nm disappears completely and there is only a broad peak at 470 nm. However, later in the paper they stated that the 370 nm absorption remained constant till 173 K. It is possible that there is an error in the figure and the temperature should be 195 K. Although the absorption at 450–470 nm increases upon heating, this increase is not directly connected with the decrease in the 354 nm band. The decay of the absorption at 354 nm upon heating starts at about 20–25 K and is not accompanied by any absorption at 470 nm. Samples irradiated at 4.2 K and heated to 77 K show two absorption lines, one at 370 nm ($G\varepsilon = 3200$) and another one at 450–470 nm ($G\varepsilon = 2900$) [It should be noted that this is what Suwalski and Kroh[46] stated in their text, but the statement does not seem to fit their data.]. In a previous paper[47] Suwalski and Kroh found that the 370 nm band of the MCH/CCl_4 glass irradiated at 77 K is decreased by electron scavengers such as N_2O, SF_6 and CO_2. The addition of scavengers such as isopropanol leads to an increase in the 370 nm absorption[43,47]. From these results they concluded that the absorption at 370 nm is due to $CCl_3^{\bullet}$ radicals produced by the dissociation of Cl^- from the $CCl_4^{-\bullet}$ anion radical,

$$CCl_4^{-\bullet} \longrightarrow Cl^- + CCl_3^{\bullet}$$

Adding 1.28 M isopropanol to MCH/CCl_4 glass and irradiating at 4.2 K leads to an increase in the 354 nm peak by 1.49 (to $G\varepsilon = 18{,}600$). Heating the MCH/CCl_4/isopropanol irradiated glass from 4.2 to 77 K decreased the 354 nm band and shifted the band to 370 nm, so that the ratio of the absorption at 370 nm at 77 K of glas with isopropanol and glass without isopropanol is also 1.49, the same as at 4.2 K. Thus it can be concluded that the absorption at 354 nm observed at 4.2 K is the precursor of the absorption at 370 nm for 77 K. As the absorption is increased in the presence of isopropanol, the species responsible for it should be of anionic nature, most probably $CCl_4^{-\bullet}$. To confirm that the temperature increase 4.2 K → 77 K leads to chemical changes rather than to a change in the extinction coefficient of $CCl_3^{\bullet}$ radicals. Suwalski and Kroh irradiated MCH/CCl_4 glass at 77 K and cooled it after irradiation to 30 K and stored it at 30 K for several hours. No changes of absorption were observed, indicating that the changes due to heating of the 4.2 K irradiated sample are chemical ones. At 4.2 K, $CCl_4^{-\bullet}$ cannot dissociate to $CCl_3^{\bullet} + Cl^-$, probably due to matrix molecules acting as potential wells for Cl^- which prevents $CCl_4^{-\bullet}$ from dissociation.

TABLE 1. The characteristics of species observed in the irradiated MCH/CCl$_4$ system[47]

Wavelength (nm)	354	365–370	420–440	470
Species	CCl$_4^-$	CCl$_3$	CCl$_4^+$	CCl$_4$Cl
Condition of detection[a]	4.2K γR 87–107K PR	77K γR 87–173K PR	4.2K γR 87–95K PR	77K γR 87–173KPR
Intensity in the presence of isopropanol	increases	increases	decreases	decreases
Intensity in the presence of N$_2$O; SF$_6$; CO$_2$		decreases		decreases
Gε (MCH/0.2 M CCl$_4$)	12500	3100 ± 100	1280	7800 ± 300
G (MCH/0.2 M CCl$_4$)	2.57	2.57	—	1.105
ε	4900	1200 ± 100	—	6900 ± 300

[a] γR denotes gamma radiolysis and PR, pulse radiolysis.

In the presence of isopropanol the 4.2 K irradiated glass did not show any absorption at 450–470 nm neither at this temperature nor after heating to 77 K. This together with the constant ratio of absorption growth due to isopropanol (1.49 both at 354 nm and 4.2 K and at 370 nm and 77 K) leads to the conclusion that $CCl_3^{\cdot}$ radicals do not contribute to the 450–470 nm absorption, nor take part in the formation of the species responsible for it. Suwalski and Kroh stated that the constant absorption at 370 nm ($G\varepsilon = 3100$) irrespective of temperature (at the range 77 K–173 K) and independent of the experimental conditions (steady state or pulse irradiations, glassy or liquid phases) is further evidence that $CCl_3^{\cdot}$ radicals do not take part in the formation of the 450–470 nm band (up to 173 K). They stated that this does not agree with Bühler's postulate[48] that this absorption is due to a $CCl_3^+ \parallel Cl^-$ species, although $CCl_3^{\cdot}$ might the formed in other processes too.

Other evidence for the independence of the 450–470 nm absorption from the 370 nm band (due to $CCCl_3^{\cdot}$) comes from their ratios for MCH/CCl_4 irradiated at different temperatures (≤ 77 K) and later heated to 77 K. The absorption at 370 nm for 77 K is independent of the irradiating temperature while the absorption at 450–470 nm for 77 K decreases considerably with decreasing temperature ($G\varepsilon = 2860$ for 4.2 K increasing monotonously up to 7650 at 77 K, the largest increase being for 45–55 K). Suwalski and Kroh suggested that the 450–470 nm absorption is due to $CCCl_4^{\cdot}Cl$ formed by the reaction of $CCl_4^{+\cdot}$ with Cl^-. The formation of this species depends on the rate of decomposition of $CCl_4^{-\cdot}$ to supply the Cl^- anion. At 4.2 K, $CCl_4^{-\cdot}$ does not decompose (the 354 nm absorption is due to $CCl_4^{-\cdot}$) and hence the small absorption observed at 420–435 nm in MCH/CCl_3 glass irradiated at 4.2 K cannot be due to $CCl_4^{\cdot}Cl$. This weak absorption is not present in the presence of isopropanol, thus it is most likely due to $CCl_4^{+\cdot}$. The various species present in irradiated MCH/CCl_4 glass are summarized in Table 1.

Sumiyoshi and coworkers[49] studied the picosecond-pulse radiolysis of liquid CCl_4 at room tempeature and observed a narrow band at 325 nm immediately after the irradiation pulse. They assigned this absorption to $CCl_4^{+\cdot}$, since it is completely suppressed by 0.5 M cyclohexane. However, Suwalski and Kroh[47] suggested that, due to the similarity of the picosecond-pluse radiolysis of the liquid with that of CCl_4/MCH glass at 4.2 K, the 325 nm band is also due to the $CCl_4^{-\cdot}$ radical anion. They said that while the alkane molecules can react with $CCl_4^{+\cdot}$, the alkane positive ion can react with $CCl_4^{-\cdot}$. In the glassy state at 4.2 K, the growth of the 354 nm absorption in the presence of isopropanol indicates the presence of anionic species, and ther is no evidence for cationic species.

Klassen and Ross[50], in a wish to strengthen their claim that $CCl_4^{-\cdot}$ absorbs at 370 nm, studied the effect of several electron and positive ion scavengers on the yield and decay of the 370 nm band in 3MP/CCl_4 glasses and cold liquids. It should be pointed out that they did not consider at all Suwalski and Kroh's results at 4.2 K irradiated glass. They found, as in the previous study[45], that 100 ns spectra of 3MP/CCl_4 glass irradiated at 75 or 77 K contains in the range 300–800 nm only one band at 370 nm (when the absorption found for the blank CCl_4/N_2O is subtracted). At 77 K the 370 nm band decays by about 50% over the first millisecond, during which time there is no formation of absorption at 470 nm. At 95 K the initial absorption at 370 nm is almost the same as at 77 K, but the rate of decay is faster and the absorption decays completely within 0.1 ms. At the same time there appears a new absorption at 470 nm. There is some lag between the decay of the absorption at 370 nm and the formation at 470 nm, which is complete only after about 0.5 ms. At 123 K there is almost no absorption at 370 nm and most of the absorption is at 470 nm. The absorption at 470 nm increases with time till 1 μs and then decays. Thus a change of the band at 370 nm to the 470 nm band is most suitably studied at 95 K. The decay of the absorption at 370 nm at both 77 K and 95 K is linear with log of time ($OD/OD_0 = -\log kt$) in contrast to what is expected from a first-order reaction ($\log OD/OD_0 = -kt$). Thus, if the species responsible for the 370 nm band is $CCl_4^{-\cdot}$, it must be concluded that the decay is not of first order. If the decay goes by the dissociation $CCl_4^{-\cdot} \rightarrow Cl^- + CCl_3^{\cdot}$, which is expected to be first order, it must be concluded that the rate is limited by the diffusion of Cl^- and CCl_3^+ due to high viscosities at these low temperatures. It was found that the parameter k of the equation $OD/OD_0 = -\log kt$ is independent of the concentration of CCl_4 in both the absence and presence of N_2O, indicating that the rate of decay is limited by the molecular motion possible in the 3MP glass.

In order to prove the identity of the species responsible for the 370 nm and 470 nm bands, Klassen and Ross studied the effect of various scavengers on the absorption and its decay. *cis*-2-Pentene (0.6 M) at 75 K did not change $G\varepsilon_{370}$ at 100 ns, however it decreases the rate of decay of the 370 nm absorption. It is unclear whether this slower decay is due to a change in the viscosity of the glass, or whether another absorption is intervening. The addition of 3% cyclopropane to the glass before irradiating at 95 K leads to a 22% decrease in the 370 nm absorption and a 17% increase in the 470 nm band. No explanations were suggested for these findings. The presence of 2-methyltetrahydrofuran (MTHF) in a ^{60}Co -irradiated (continuous irradiation) 3MP/CCl_4 glass increased the absorption at about 350 nm but strongly suppressed the absorption at 470 nm, indicating that the 470 nm band is due to a radical cation from CCl_4, probably $CCl_3^{+\bullet}$, but it can also be $CCl_4^{+\bullet}$. The alkane radical cation produced on radiolysis of 3MP transfers a proton to MTHF, which traps the positive charge, preventing the formation of a cation from CCl_4. The 100 ns spectra of MTHF/0.2 M CCl_4 glass at 75 K is very similar to the spectrum in an alkane glass, except that the maximum of the absorption band is 10 nm lower, at 360 nm. The decay of the absorption at 370 nm for MTHF/CCl_4 glass is slightly faster than that for 3MP/CCl_4 glass, but no explanation is given to this faster decay.

Klassen and Ross argued with the previous suggestion of Suwalsky and Kroh[43,46,51], before their 4.2 K study, that the 365 nm absorption in MCH/CCl_4 pulse-irradiated at low temperatures is due to both CCl_3^+ and $CCl_4^{-\bullet}$, and claimed that it is only due to $CCl_4^{-\bullet}$. They based their arguments on the observation that this absorption was not observed in the liquid system, where CCl_3^+ is known to be formed. They also argued that Suwalski and Kroh did not subtract the contribution of a blank due to the hydrocarbon itself in the presence of the electron scavenger.

Truszkowski and Ichikawa[52] measured the ESR and optical spectra of irradiated glasses of 3MP containing a small amount of 5-methyldecane (5MD) with CCl_4, CCl_3Br and CCl_3F at 77 K, in order to clarify the nature of the species absorbing at 470 nm in γ-irradiated 3MP/CCl_4 glass[53]. 77 K γ-radiolysis of 5MD–3MP glasses with an electron scavenger produces the radical cation 5MD$^{+\bullet}$, which absorbs at 710 nm[54,55]. ESR studies[56] show that 77 K radiolysis of *n*-hexane with 1-chloropentane (an electron scavenger) and *n*-octane (a positive-charge scavenger) produced pentyl radicals ($RCH_2^\bullet$.) and *n*-octane radical cations ($nC_8^{+\bullet}$). Photobleaching by IR light (illuminating with $\lambda \geq 1200$ nm) of the $nC_8^{+\bullet}$ leads also to decrease in the pentyl radicals, suggesting that the positive charge liberated from $nC_8^{+\bullet}$ oxidizes $RCH_2^\bullet$, or, in other words, $RCH_2^\bullet$. captures a hole from the $nC_8^{+\bullet}$. In this work they looked for the same reaction of 5MD and CCl_4 in 5MD–CCl_4–3MP.

$$5MD^{+\bullet} + CCl_3^- \longrightarrow 5MD + CCl_3^+$$

Since there is in the proximity of the CCl_3^+ cation also a Cl^- ion which can form an ion pair $CCl_3^+ \cdots Cl^-$, there is a high probability for this reaction. They found that the absorption band at 710 nm due to the radical cation of 5-methyldecane (5MD$^{+\bullet}$) converts into the 470 nm in the presence of CCl_4. This conversion is increased by illumination with $\lambda = 900$ nm. In the absence of 5MD the band at 470 nm is formed directly by the γ-irradiation. Photobleaching with 900 nm increased this peak only very litttle. ESR studies of the matrix before and after the photobleaching with 900 nm shows the formation of CCl_3^+ radical spectra (as obtained from the difference between the spectra of neat 3MP and 3MP–CCl_4 after normalization). This result suggests the existence of the reaction 5MD$^{+\bullet}$ + CCl_3^+ → 5MD + CCl_3^+.

If CCl_4 is replaced by CCl_3Br or by CCl_3F in 3MP glasses, absorption bands are found at 480 and 438 nm, respectively, instead of the 470 nm band in the case of CCl_4. Scavenger studies and the study with 5MD showed similar patterns to what was found with CCl_4. The scavenger studies showed that the responsible species is a cation, possibly CCl_3^+. Since the maximum absorption is different for different X in CCl_3X (X = F, Cl, Br) it must be concluded that the species included also X. However, in CCl_3Br the weakest bond is C—Br and thus in the case of CCl_3Br it will also be CCl_3^+ as in CCl_4.

The 10 nm shift is explained by assuming the absorbing species to be $CCl_3^+ \cdots Cl^-$ and $CCl_3^+ \cdots Br^-$, as suggested by Reed and coworkers[57], who calculated theoretically a transition of about 500 nm for the charge-transfer stabilized contact ion pair $Cl^- \cdots CCl_3^+$. In this case it is clear that the replacement of Cl^- by Br^- will change the absorption band. It should be mentioned that Ha and Bühler[58] claimed that their theoretical calculation did not give a transition at about 500 nm for the contact-ion pair model. More details on these theoretical calculations will be given in Section IV. However, Truszkowski and Ichikawa argued that if the absorbing species is Cl^- (solvent) CCl_3^{+52}, it is not reasonable that the change of Cl^- with Br^- will lead to change in the absorption band.

Shoute and Mittal[60] suggested the use of perfluoro-2-butyltetrahydrofuran (FBTHF) for measurements of the optical spectra of perfluoro aromatic cations formed radiolytically. Many glassy matrices are suitable for measurement of radiolytic generation of anions, however only few matrices can be used to study cations, since in many matrices the holes produced by irradiation are rapidly immobilized by an ion–molecule reaction. Radiolytic generation and stabilization of the cation requires the matrix ionization potential to be sufficiently higher than that of the solute molecules. Shiotani and coworkers[61] demonstrated that perfluorocarbons are superior to freon matrices in resolving the ESR spectrum of solute cations. However, many perfluorocarbons matrices were found to be unsuitable for optical absorption studies since they formed cracked glasses. Shoute and Mittal[60] found that FBTHF forms a clear glass at 77 K which is transparent from 220 to 2600 nm. The gamma radiolysis of a pure FBTHF matrix at 77 K produces a species absorbing at 248 nm. Using this matrix, Shoute and Mittal measured the absorption spectra of radiolysis-generated cations from C_6F_6, $C_6F_5CF_3$, $C_6F_5CH{=}CH_2$, perfluorobiphenyl and perfluoronaphthalene. On either increasing the solute concentration or upon warming, they could not observe a band which can be attributed to the dimer or polymer cation, indicating that perfluoroaromatic cations do not associate, unlike their hydrocarbon counterparts.

IV. RADIOLYSIS IN THE LIQUID PHASE

A. Intermediates

Bühler and coworkers studied the radiolysis of CCl_4 in non-polar liquid systems, such as CCl_4, $CFCl_2CF_2Cl$ and in hydrocarbons. All this work was summarized in 1983[62] in a paper which showed evidence to support the existence of the cations as $(CCl_3^+ \| Cl^-)_{solvent}$ ion pairs. At least four different groups found that pulse radiolysis of pure CCl_4 reveals two typical absorption bands at 335 nm and 500 nm. Both decay initially by first-order kinetics followed by a slower decay. The half-lives of the decay of the bands are considerably different (100 ns and 14 ns, respectively), showing that they belonged to different species reacting with the solvent but not one with the other. However, the different groups of researchers have different opinions, concerning the species responsible for these bands. The band at 335 nm was assigned by Mehnert and coworkers[63] to CCl_4^{+*}, while Bühler is skeptical about this assignment and prefers to assign it to dichlorocarbene, although he admits that this assignment still needs further support. Bühler suggested that the 500 nm

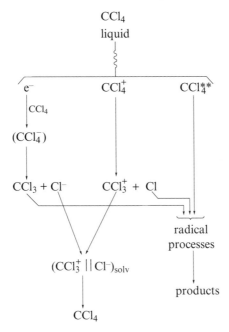

FIGURE 1. The mechanism of radiolysis of pure CCl_4. Reproduced with permission from Reference 62

band is due to CCl_3^+, which may exist as the separate cation or as a solvent-separated ion pair $(CCl_3^+ \, || \, Cl^-)_{solvent}$. This ion pair eventually disappears by unimolecular kinetics due to the ion-pair collapse, to form the neutral molecule. Bühler[62] suggested an ionic mechanism for the radiolysis of pure CCl_4 (Figure 1), and summarized it in the following words: "The parent cation CCl_4^+ is initially produced with $G_{tot}(ions) = 4.8 \pm 0.4$ and has a lifetime of the order of about 200 ps (up to 10 times less if a resonance charge-transfer process should be involved in liquid CCl_4). The fragment cation CCl_3^+ absorbs at 500 nm with an absorption coefficient of $\varepsilon_{500} = 6000 \pm 1200$ M^{-1} cm^{-1}. The initial yield is $G_{init}(CCl_3^+) = 2.5 \pm 0.5$. The difference to the total ion yield of 4.8 ± 0.4 is due to neutralization of CCl_4^+ before its decay into CCl_3^+ occurred.

The ion pair $(CCl_3^+ \, || \, Cl^-)_{solv}$ being formed on ion recombination most likely corresponds to a solvent separated ion pair. It therefore does not show an extra band (no charge-transfer band). This solvent separated ion pair probably has a low dipole moment due to dielectric shielding from the solvation shell and therefore may not be detected by the microwave absorption technique.

Initial ion separation. As the lifetime of CCl_4^+ is of the order of 200 ps (or smaller) the CCl_3^+-absorption is a very sensitive probe for the very early time profile of the ion kinetics and therefor allows to test various initial ion separations. It is concluded, that from the two possible distributions, compatible with conductivity data, only the exponential distribution with $G_0 = 5$ and width $b = 36$ Å is able to fit the experimental facts."

Brede and coworkers[64] commented on Bühler's reaction model and spectral assignment on the basis of their scavenger studies[65] and claimed that the 475 (500) nm band is due to a solvent-separated ion pair but with $CCl_4^{+\bullet}$ and not CCl_3^+ [Bühler $(CCl_3^+ \, || \, Cl^-)_{solvent}$, Brede $(CCl_4^+ \cdots Cl^-)$]. They found that some scavengers accelerated the decay of the visible band of irradiated liquid CCl_4 and claimed that this proved the involvement of charge transfer. In a reply to this comment[66] Bühler argued that Brede's studies were done mostly with high solute concentration ($\geq 10^{-1}$ M), which can lead to direct radiation effect, where the energy is absorbed directly by the solute, and even to changes in the solvation properties. In case of high solute concentrations, the solute may take part in the solvation shell of the ion pair, affecting the stability of the ion pair. Brede and coworkers claimed that the correlation of the acceleration of the decay rate constant with the ionization potential of the solute represents conclusive evidence for the charge-transfer reaction of the solutes. Bühler, however, argued that this correlation is not very good and cannot be used as evidence, especially since the solutes with the largest accelerating effects, methanol and cyclic ethers, deviate strongly from the correlation. He also claimed that the observation that molecules similar to CCl_4, such as $CHCl_3$, CH_2Cl_2 and CH_2ClCH_2Cl, have only a very small effect on the decay rate, favours the explanation that the solutes influence the solvation of the ion-pair, rather than being involved in a charge-transfer reaction.

Brede and coworkers[64] claimed that the fast part of the absorption decay is due to geminate ion recombination of the ion pair, whereas the slow part is due to free ion recombination. However, Bühler argued that once the ion pairs are formed, their structure is the same and should have the same decay behaviour. He also claimed that solvent-separated ion pairs are not displaying charge transfer bands and thus $CCl_4^+ \cdots Cl^-$ should not have a different band than CCl_4^+ (the UV band), and he saw convincing evidence for the suggestion that the 500 nm is not due to $CCl_4^+ \cdots Cl^-$ in the observation that there is no correlation of the decays of the two bands, the second of which (330 nm) is assigned by Brede to $CCl_4^{+\cdot}$. However, if the 330 nm band is really due to CCl_2: as Bühler suggested, then the 500 nm can be due to $CCl_4^+ \cdots Cl^-$.

Gremlich and Bühler[59] studied the pulse radiolysis of liquid alkane–CCl_4 systems, which leads to the formation of a visible band at 470 nm. From the comparison of the spectra and the kinetics of the decays of the visible absorption of pure CCl_4 and of MCH–CCl_4, they concluded that they are the same with a 30 nm shift in the absorption band, except for one significant difference, namely, that in the second the decay is entirely first order, without any contribution from a slow second-order process, whereas in the first there is a small second-order contribution. It should be noted that actually the widths of the bands differ by 23% which is not insignificant. Moreover, there is a difference in the decay kinetics. Bühler based the similarity in the kinetics not on rate constants but on Arrhenius parameters taken from data at different temperatures (153 K–213 K for MCH–CCl_4, but 251 K–293 K for CCl_4), and these parameters were calculated for MCH–CCl_4 from only three experimental points and for CCl_4 from only two points. Even so, calculating the rate constants from these Arrhenius parameters shows that they differ by more than one order of magnitude. This comparison of kinetics, by comparison of Arrhenius parameters in different temperature ranges, can be very misleading. The small second-order contribution in pure CCl_4 is explained as being due to free CCl_3^+ (as contrasted to the ion pair) which does not exist in MCH–CCl_4. The 470 nm band does not appear in MCH containing other electron scavengers or even containing $CF_2ClCFCl_2$, which proves that it is due to a species formed from CCl_4.

Cation scavengers such as aniline, N,N'-tetramethyl-p-phenylenediamine (TMPD) and cyclohexene reduced the absorption at 470 nm, indicating that the 470 nm is either due to a cation or has a cationic precursor. Cyclopropane does not exchange charge with MCH^+, but reacts with it by H or H_2 transfer. It was found to reduce the yield of the absorption at 470 nm without affecting the rate of the decay of this absorption, indicating that MCH^+ is the precursor for the 470 nm species. Ethanol, which is known to scavenge alkane cations by proton transfer, was found not to scavenge the precursor of the 470 nm band, which did not change with the addition of ethanol. However, the decay rate of this band increased in the presence of ethanol, by a factor of 5–8. This decay is of first order, but has a much higher A factor, indicating that it is due to a different process. SF_6, an electron scavenger, reduces the yield of the 470 nm absorption, but the rate of the decay of the absorption remains the same, indicating that SF_6 reacts with the the precursor of the 470 nm species, but not with the species itself. Similar results are obtained with $CF_2ClCFCl_2$, which is also an electron scavenger. These observations lead to the conclusion that the 470 nm band has both a cation and an anion as its precursors. A possible assignment can be the charge-transfer complex $Cl \cdot CCl_4$[40,43] produced from $Cl^- + CCl_4^+$ recombination. However, the ionization potential of CCl_4 is higher than that of MCH and it is difficult to assume that the endothermic reaction $MCH^+ + CCl_4 \rightarrow CCl_4^+ + MCH$ will occur to a large extent. Gremlich and Bühler[59] also argued that there is a correlation of charge-transfer band energy with the ionization energy of the electron donor using a fixed electron acceptor (Cl^-) and this correlation indicates that $CCl_4 \cdot Cl$ will have a CT band at 280 nm.

Gremlich and Bühler used the similarity they claimed for the 500 nm band in pure CCl_4 and the 470 nm band in $MCH–CCl_4$ to assign the latter to CCl_3^+ within a sovent-separated ion pair $(CCl_3^+ \| Cl^-)_{solv}$. The 30 nm shift is due to the effect of the solvent. CCl_3^+ is produced from charge transfer between $MCH^{+\bullet} + CCl_3^\bullet$, which can be looked upon as disproportionation of two radicals $MCH^{+\bullet} + CCl_3^\bullet \rightarrow MCH + CCl_3^+$, in which two radicals react to give two even–electron species. The model of this reaction is given in Figure 2.

In two reviews[67] Bühler generalized the aspects used in this model of complex formation, solvent and caging effects, geminate ion fragmentation, cross geminate charge transfer and delayed germinate ion neutralization through ion-pair formation for the radiolysis of all halocarbons. Bühler and coworkers[67b] studied the pulse radiolysis of solutions of $CFCl_3$ in liquid MCH at low temperatures.

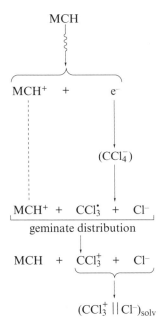

FIGURE 2. Reaction model for the charged species in liquid methylcyclohexane. Reproduced with permission from Reference 41

The samples exhibit an optical absorption with λ_{max} at 435 nm; the position of the maximum absorption is slightly dependent on the temperature and on the concentration of $CFCl_3$. It is 5 nm red-shifted from the value obtained for 1% $CFCl_3$/3MP glasses[51]. After 20 μs the absorption decays by a first-order process with low exponential parameters (log $A = 8.24 \pm 0.34$, $E = 12.6 \pm 1.0\,\text{kJ mol}^{-1}$). N_2O, an electron scavenger, decreases the absorption of the initial absorption at 435 nm by ca 35% and increases slightly the rate of decay of the absorption. Thus it must be concluded that the 435 nm species must have a negatively charged precursor. Cation scavengers such as cyclopropane, quadricyclane and MTHF both reduce the initial absorption at 435 nm and increase the rate of the decay of the remaining absorption, indicating that both the precursor and the 435 nm species carry a positive charge. These scavenger effects are similar to those observed for the 470 nm species in the case of CCl_4/MCH. Bühler and coworkers suggested that, similarly to the case of CCl_4, the 435 nm band is due to the $CFCl_2^+$ cation formed by a transfer of positive charge from the solvent radical cation $MCH^{+\bullet}$ to the $CFCl_2^\bullet$ radical, originated in a fast reaction of anion fragmentation:

$$e^- + CFCl_3 \longrightarrow CFCl_3^- \longrightarrow Cl^- + CFCl_2^\bullet$$

$$MCH^{+\bullet} + CFCl_2^\bullet \longrightarrow MCH + CFCl_2^+$$

$$\left.\vphantom{\begin{array}{c}a\\a\end{array}}\right\} \longrightarrow (CFCl_2^+ | Cl^-)$$

The cation formed is forming a solvent-separated ion pair. The $CFCl_2^\bullet$ radical absorbs in the range 365–415 nm with a maximum at about 385 nm and is not expected to interfere with the 435 nm band.

The decay of the 435 nm absorption is dose–independent up to ca 50 Gy. However, for higher doses the ion-pair decay rate constant changes almost linearly with the dose, indicating an additional reaction of the ion pair with some transient the concentration of which changes with the dose. Bühler and coworkers suggested[67] that this intermediate is the MCH radical produced from primary solvent excitation, but their explanation why this reaction starts only at doses ≥ 50 Gy is not convincing.

Bühler and coworkers[67b] argued that the relatively low A factor for the decay of the ion pair and the large rate constant for the reaction with quadricyclane favour a solvent-separated ion pair, and not a contact ion pair.

Tabata and coworkers[68] compared pulse radiolysis with laser flash photolysis and, from the similarities, concluded that the 330 nm is due to the CCl_2: carbene biradical. Their results indicate that the visible band (in their work this is at 480 nm and not at 500 nm as stated by Bühler) is best assigned to the charge transfer complex ($CCl_3^{+\delta} Cl^{-\delta}$). The 330 nm absorption was found to be formed also in flash photolysis with 266 nm photons. The energy of this photon (4.6 eV) is not sufficient to produce $CCl_4^{+\bullet}$, and thus the species should be a radical, and since CCl_3 has no absorption in this range, it was assigned to CCl_2:.

Washio and collaborators[69] studied the radiolysis of the CCl_4 by pico- and nanosecond pulse radiolysis measuring optical absorption in the UV using a system with time resolution up to 20 ps. They observed two absorbing species, the first one with a half-life of about 60 ps, whereas the other species has about 100 ns lifetime. The addition of 10 mM cyclohexane to CCl_4 decreased its absorption at 330 nm by about 30% as measured 2 ns after the irradiation pulse (pulse width 10 ps). The addition of 100 mM cyclohexane eliminated completely the 330 nm absorption immediately after the pulse of the electrons. It was found also that cyclohexane reacts with the long-life species with rate constant $7 \times 10^9\,\text{M}^{-1}\text{s}^{-1}$. On the other hand, while cyclohexane decreases the yield of the 480 nm band it does not affect the lifetime of this species. These observations show that cyclohexane reacts with the precursor of the absorbing species, but not with the species itself.

With pure CCl_4, at the end of the 10 ps electron pulse there is strong absorption at 330 nm and only weak absorption at 480 nm. 1 ns after the pulse, the absorption at 480 nm increased considerably and this rise is completed within 20 ps. At the same time, the absorption at 330 nm decreased with a lifetime of 60 ps, to about two-thirds of the initial absorption.

The kinetics of the formation of the absorption at 480 nm was analysed using the solution of the Smoluchowski diffusion equation. It was found that assuming an initial exponential distribution of the geminate pair formed in the radiolysis gives a better fit to the experimental results than assuming a gaussian initial distribution, although neither gives satisfactory agreement.

Tabata and his coworkers[69] also studied the transient absorption spectra in picosecond pulse radiolysis of $CHCl_3$. Two absorption bands were observed around 330 and 550 nm. The first similar to that in CCl_4, was assigned to CCl_2:, as it is formed also by photolysis, and CCl_3 and Cl do not have an absorption at 330 nm. The 330 nm band decays faster in $CHCl_3$ than in CCl_4, probably due to the fast reaction of CCl_2: with the $CHCl_3$ molecules. The broad absorption at 550 nm decays very rapidly, possibly owing to geminate decay of $CHCl_3^{+\cdot}$. However, the plot of the absorption vs $1/\sqrt{}$ time leads to a negative value for G_{fi}, (i.e., for the radiolytic yield of the free ions) showing that it is not a genuine geminate process. Thus, Tabata and his coworkers concluded that $CHCl_3^{+\cdot}$ decays by other reactions as well as geminate recombination.

Weinhold and coworkers[57] conducted a theoretical *ab initio* search for a singlet isomer of CCl_4 corresponding to the planar C_{2v} ion-pair structure $CCl_3^+ \cdots Cl^-$ where the Cl^- approached CCl_3^+ in the plane of the planar CCl_3^+. Previous calculation of Bühler and coworkers[70] considered only a non-planar C_{3v} geometry with Cl^- sitting above CCl_3^+ along its C_{3v} axis. Bühler's group, using SCF calculations with STO-3G basis set, could not find any barrier for the Cl^- attacking the carbon of CCl_3^+. The RHF calculations of Weinhold and coworkers[57] with small basis sets (STO-3G, 3-21G) also gave no minimum for the planar C_{2v}. However, using a large basis set (6-31G) the calculation leads to a true local C_{2v} minimum with charge distribution characteristic of an ion pair. Studies of spin coupling and correlation effects suggested that the ion-pair singlet is greatly stabilized (compared to dissociating products and the competing complex $CCl_2 \cdots Cl_2$) by electron correlation. Consequently, they suggested that the CCl_3^+, Cl^- ion pair can be the cause for the 500 nm absorption. Ha and Bühler[58] performed an *ab initio* Cl calculation on $CCl_3^+ \cdots Cl^-$ in the C_{2v} symmetry structure and calculated the theoretical transition energies for such a contact ion-pair. They found that the main transition should be of 1.3 eV and not of 2.50 eV (500 nm), and concluded that the $CCl_3^+ \cdots Cl^-$ contact ion-pair is not formed. However, their calculations did not show the reason for this, albeit there is a minimum energy state in Weinhold and coworkers' calculations.

Sumiyoshi and coworkers[71,72] studied also the pulse radiolysis of other halo aliphatic compounds. They found, by picosecond pulse radiolysis[71], that butyl chloride and butyl bromide formed a visible band which decayed with a half-life time of about 2 ns. For the butyl chloride the picosecond pulse radiolysis shows decay of the visible band with a half-life of 1.7 ± 0.3 ns, whereas the nanosecond pulse radiolysis shows a band at 500 nm with a half-life of 50 ns, as was found earlier by Brede and coworkers[73]. The absorption ($G\varepsilon$) at the picosecond pulse radiolysis is 14000 M^{-1} cm^{-1} at 520 nm, whereas the ns pulse radiolysis gives maximal absorption at 500 nm of 2000 M^{-1} cm^{-1}. This difference is attributed to a very short-lived vibrationally excited butyl chloride cation, as was suggested in a low-temperature glassy matrix[49]. In the case of butyl bromide, ps pulse radiolysis shows three bands (390 nm, 450 nm and 600 nm) whereas ns shows only the first two. The ns pulse radiolysis shows a half-life of 4 μs, whereas the ps pulse radiolysis shows a half-life of 2ns, with a higher $G\varepsilon$ value.

Sumiyoshi suggested that the rapid formation of the visible band (< 30 ps) excludes the contribution of an ion-pair intermediate ($BuCl^+ \cdots Cl^-$) and supported the idea of vibrationally excited cations.

In a later paper[72] they studied the ps and ns pulse radiolysis of pure liquid 1,2-dichloroethane (DCE), and observed a broad visible absorption with a maximum around 550 nm. This absorption decayed by first-order kinetics with $t_{1/2} = 0.2$ ns in ps pulse radiolysis and $t_{1/2} = 200$ ns in ns pulse radiolysis. In the latter there is an additional peak at 270 nm, which is formed during and after the pulse and decays by seocnd-order kinetics with $2k/\varepsilon = 2 \times 10^7$ cm s^{-1}.

An electron scavenger, such as N_2O, neither reduces the absorption of the bands nor increases their rate of decay. Solutes with an ionizaion potential lower than that of DCE, such as, for example, cyclohexane, increase the decay rate of the 550 nm band and decrease the absorbance of the 270 nm band. This indicates that the visible band is due to cationic species, and that the UV band is at least partly due to reaction with this cationic primary species. By measuring the decay rate as a function of solute concentration, the rate constant of the charge transfer reaction can be deduced. Sumiyoshi and coworkers showed that there is a quite good correlation between the ionization potential of the solutes and the logarithm of the charge transfer rate constants. However, methanol and ethanol are exceptions, having exceptionally high charge transfer rate constants in the studies of Brede's group in the case of CCl_4[65], as commented by Buhler[66]. Sumiyoshi and coworkers suggested, based on scavenger studies, that the visible band is due to a cation. The $G\varepsilon$ at 550 nm is much higher at the end of a 30 ps pulse than at that of a 10 ns pulse (2.4×10^4 vs 1.8×10^3 M^{-1} cm^{-1}), indicating that either there are two cationic forms as they suggested for the butyl halides with different ε values, or that more than 90% of the primary DCE radical cations disappears by geminate ion reactions, as they suggested in this paper. Dorfman and coworkers[74] reported that two cationic species are formed in the radiolysis of DCE, which are kinetically distinguishable. They suggested that one species is a DCE radical cation and the other is a carbocation formed after C—Cl bond rupture. Following this suggestion, Sumiyoshi and coworkers suggested that the radical cation is a fast decaying species, whereas the other cationic species decays more slowly, resulting in a shift of the maximum absorption during the decay.

de Haas and coworkers[75] measured the d.c. conductivity resulting from pulse radiolysis of liquid CCl_4 with a resolution time of 10 ns. They observed that at short times ($< 1 \mu s$) the transient conductivity is much larger than would be expected from free ions alone, and concluded that at least 83% of the d.c. conductivity is due to ion pairs which undergo eventually (in times longer than 1 μs) geminate recombinaion.

Halpern[76] studied the radicals formed in irradiated liquid CCl_4, using phenyl-t-butylnitrone (PBN) as spin trap followed by ESR studies. He clearly proved the existence of Cl atoms ($CCl_4 \rightarrow CCl_3 + Cl$), by observing the well resolved ^{35}Cl and ^{37}Cl couplings. The ratio of the splitting constants agrees with the ratio of the magnetic moments of these nuclei. In the radiolysis of CCl_4 containing CH_3OH, Cl atoms were not observed, probably due to their fast reaction with CH_3OH. This agrees with the observation[77] that the presence of a small concentration (2×10^{-3} M) of methanol in CCl_4 completely suppresses the formation of molecular Cl_2 in γ-radiolysis of CCl_4.

In the presence of CH_3OH, an intense triplet of doublets characteristic of the PBN–CCl_3 adduct dominates the spectrum. Using low-energy X-rays to radiolyse CCl_4 it was found that Sc–X-rays (4.12 keV) lead to 3.5 times more CCl_3 radicals than S–X-rays (2.32 keV) with the same dose. This is explained as due to these energies being on both sides of the K-absorption edge of the absorbing Cl atoms (2.82 keV). Thus, Sc–X-rays are absorbed mainly via the photoeffect with the K-shell electrons of chlorine, whereas S–X-rays can interact only with the L-shell electrons, which have considerably lower cross section.

Kovalev and coworkers[77] studied the effect of methanol on the radiolysis of liquid CCl_4. In pure CCl_4, γ-radiolysis leads to the formation of Cl_2 and C_2Cl_6 with a radiolytic yield of $G = 0.70 \pm 0.03$. When methanol is added, the yield of the first product falls to zero already at 2×10^{-3} M methanol, while $G(C_2Cl_6)$ increases to 4.8 ± 0.5. New products found are formaldehyde with equal yield to C_2Cl_6, and HCl with double that yield. They found constant yields of HCl. C_2Cl_6 and formaldehyde over a large range of concentrations of CH_3OH (2×10^{-3} M–0.1 M), which indicates complete scavenging of the Cl atoms formed by the radiolysis. They suggested the following mechanism:

$$CCl_4 \xrightarrow{\quad\wedge\wedge\wedge\quad} CCl_3 + Cl$$

$$Cl + CH_3OH \longrightarrow HCl + \dot{C}H_2OH$$

$$\dot{C}H_2OH + CCl_4 \longrightarrow \dot{C}Cl_3 + HCl + CH_2O$$

$$2\,\dot{C}Cl_3 \longrightarrow C_2Cl_6$$

The yields of the final products lead to radicals yields of

$$G(Cl) = G(CCl_3) = 4.8 \pm 0.5$$

However, these results do not agree with previous results of Alfassi and Feldman[78], who found for several scavengers a constant $G(C_2Cl_6)$ of 3.8 ± 0.3 over quite a large range of temperaturs, simliar to the results of Rajbenbach and coworkers[79], Kovalev and coworkers ignored all the previous studies with various additives. With ethanol at 27 °C Alfassi and Feldman[78] found $G(C_2Cl_6) = 3.8 \pm 0.2$ and with n-pentanol at 44 °C they obtained 3.6 $\pm$ 0.2. These studies were reviewed by Horowitz[12].

In pure CCl_4, the yield of 0.70 for Cl_2 and C_2Cl_6 is much less than if calculated from $G(CCl_3)$ and $G(Cl^\bullet)$, assuming that the rate constants for cross-recombination is the geometric average of the pure recombination rate constants[80]:

$$k_{Cl+CCl_3} = \sqrt{k_{Cl+Cl} \cdot k_{CCl_3+CCl_3}} \Rightarrow G\big(Cl+CCl_3\big) = 2\sqrt{G\big(Cl_2\big) \cdot G\big(C_2Cl_6\big)} = 1.4$$

Kovalev and coworkers[77] concluded that the rate of cross-combination is larger than that with identical particles. However, they ignore the possibility of the reaction of $CCl_3^\bullet$ radicals with Cl_2, which can lead to a reduction in the yield of the molecular products in pure CCl_4:

$$CCl_3^\bullet + Cl_2 \longrightarrow CCl_4 + Cl^\bullet$$

This reaction is strongly exothermic and should be very fast.

Asmus and coworkers[81] studied the pulse radiolysis of polychlorinated biphenyls (PCB) in N_2-saturated DCE solutions. The transient absorption spectrum at 5 μs after a 1 μs pulse showed either two distinct absorption maxima (as, e.g., for 3,3′, 4,4′-tetrachlorobiphenyl) or a broad absorption band (in the case of 2,2′, 4,4′,5,5′-hexachlorobiphenyl) or a broad maximum with an additional shallow shoulder (for 3,5-dichlorobiphenyl) in the range of 375–430 nm. It appears that the number and position of chlorine substituents determine the absorption category. Another absorption band was indicated in the near IR region, but could not be studied. The observed absorption in the 375–430 nm range is attributed to the radical cations from PCB, formed by charge transfer from the radical cation of the solvent:

$$DCE^{+\bullet} + PCB \longrightarrow DCE + PCB^{+\bullet}$$

This assignment is corroborated by the observation that the absorption is not affected by the presence of oxygen, an electron and radical anion scavenger, ruling out PCB radical anion as the species contributing to the absorption. This also rules out any alkyl or chloroalkyl radical from $C_2H_4Cl_2$, since they react with O_2. The yields measured in terms of $G\varepsilon$ (absorption units) depend strongly on the concentration of the PCB, the absorption being a linear function of the concentration, which agrees with a competition between the charge transfer and the decay of the solvent radical cation.

The kinetics of the decay of the radicals was measured by following the decay of their absorption at the peak in the spectrum. The measurement of the first half-lives were conducted at different applied radiation doses (i.e. different initial concentration of radicals) in order to distinguish between first-order and second-order decays. It was found that the number and position of the chlorine substituents influence not only the sepctra of the transient, but also its mode of decay. Radical cations which have two absorption maxima at 400 and 430 nm, such as 3,3',4,4'-TCBP (tetrachlorobiphenyl), decay bimolecularly. The absorption at 430 nm and in the IR decay with the same rate, however the decay of the absorption at 400 nm is slower, indicating that at least two different species are formed upon oxidation of 3,3',4,4'-TCBP. Radical cations which have absorption at 375 nm and a shoulder at 395 nm, such as 3,5-DCBP (dichlorobiphenyl), decay monomolecularly. Most PCBs tested decay bimolecularly, and only 3,5-DCBP$^{+\bullet}$ decayed purely monomolecularly. Some PCB cations decay by mixed orders.

The yields of the PCB radicals were measured using promethazine, assuming that the extinction coefficient in DCE is the same as that measured in aqueous solution (9500 M^{-1} cm^{-1}). Since the spectra of unoxidized promethazine is identical in both solvents (H$_2$O and DCE), it can be assumed that ε of the oxidized form is similar in the two. It was found that G values of the radical cations are 0.86 for 4-chlorobiphenyl, 0.65 for 3,3',4,4'-TCBP and 0.36 for 2,2', 4,4', 5,5'-HCBP (hexachlorobiphenyl). Asmus and coworkers[81] suggested that all optically observable species are PCB radical cations and have practically identical chemical properties. This suggestion is based on the observation that all absorption bands disappeared at the same rate in the presence of an oxidizable substrate such as promethazine. They suggested that when the PCB molecule transfers an electron to the solvent (DCE) cation radical, two different radical cations are formed, with one and two electronic transitions in the 250–800 nm range. The different radical cations are due to different conformations. The low-energy transition is probably due to a planar species, with maximum π-orbital overlap. Although PCBs with two *ortho*-chlorine atoms would not normally assume a planar conformation, their radical cations may do it, as is indicated by their near-IR absorbance. Apparently, the energy gain in establishing the expanded π system is larger than that due to repulsion of the *ortho* substituents. The high-energy absorption bands are attributed to PCB$^{+\bullet}$ radical cations with two twisted phenyl rings. The population of the two conformers depends on the substituents. PCBs without *ortho* substituents will have higher abundance of the planar conformer than those having *ortho* substitution. *Para* substitution, on the other hand, increases the population of the planar radical, which is usually shorter-lived than the twisted one, suggesting that the bimolecular decay is disproportionation between two radical cations PCB$^{+\bullet}$, rather than reaction of the PCB$^{+\bullet}$ with another species resulting from the radiolysis. The overlapping of two planar radicals is better than that of twisted ones. Monomolecular decay occurs by deprotonation, a process which may be accelerated by *meta*-chlorine substituents.

B. Final Products

Sagret and coworkers[82] studied the products formed in the γ-radiolysis of liquid C$_6$F$_6$ using GC/MS with high resolution. The GC shows two major peaks with an additional seven products of relatively large yields and several more minor ones. Only one product was definitely identified, namely the dimer, decafluorobiphenyl (DFBP). By analogy with benzene radiolysis, the expected major products should be DFBP accompanied by perfluorocyclohexadiene or some other fluorinated monomer products. However, perfluorocyclohexadiene was not observed at all, and although DFBP was found it was not one of the two major peaks. The three products with the largest yields all have mass spectra with the highest m/Z value of 558, assumed to be the molecular ion. This molecular weight fits $C_{18}F_{18}$. The largest peak in the mass spectra of all the three products is 353, corresponding to $C_{12}F_{11}^+$. The 18 carbon atoms in the molecular ions indicate that these are trimers. Trimers from perfluorobenzene should have the structure $C_6F_5-C_6F_4-C_6F_5$ $(C_{18}F_{14})$, and hence to account for the 18 fluorine atoms in $C_{18}F_{18}$ two of the double bonds have to be saturated with 4 fluorine atoms. There is no explanation why all the three different trimers have been saturated exactly with 4 fluorine atoms, except that this can be due to migration of F atoms rather than C—F bond breaking. An additional product is a dimer of the formula $C_{12}F_{14}$ (molecular ion $m/Z = 410$), i.e. a dimer with an extra four fluorine atoms. Three other peaks are of the tetramers $C_{24}F_{24}$. Some minor peaks show spectra that are probably of octamers. All these products are exact oligomers of C_6F_6 without C—F bond ruptures.

Assuming that the detector signals are proportional to the masses of the products, the yield of DFBP is proportional to the absorbed dose; however, the yields of the major products are less than proportional above 10–15 kGy. Some of the minor products show yields that are more than proportional at this range, and become more prominent at larger doses. If the detector sensitivity is assumed to be constant per monomer unit, i.e. the sensitivity of the tetramers is twice that of the dimer and 4/3 times that of the trimer, then the yield of all the products can be calculated. This total yield was found to be linear with the dose for the whole range studied (up to 25 kGy). Absolute yield can be measured only for DFBP, for which a standard exists, and it is 0.0465 molecule/100 eV. Measuring the formation of 'total polymer' by the total ion current in the GC/MS gave a total yield of 1.7 molecules/100 eV, and similar yields are obtained by the electron capture detector.

The observation that the yields of the major products are linear and less than linear shows that they are initial, not secondary, products, in the sense that they are not formed by radiolysis of products or by reaction with chemicals formed by the radiolysis. There is no proof of their direct formation in the initial event of radiation absorption, but they were the major products in the lower dose that made analysis possible. To prove further that the trimers and tetramers are initial products, the radiolysis of 10^{-4} M DFBP solution in C_6F_6 was studied. The yields of trimers and tetramers were the same without DFBP, indicating that they are not formed by the reaction with DFBP.

The large differences between the products of the radiolysis of C_6H_6 and of C_6F_6 can be explained due to the stronger C—F bond. The excited C_6F_6 molecule, formed either directly by the radiation or from ion–electron recombination, cannot in most cases lead to C—F bond rupture, so that there is very little formation of C_6F_5 radicals. In the case of C_6H_6 there is formation of C_6H_5 and C_6H_7 radicals, which by disproportionation and combination give $C_{12}H_{10}$ and C_6H_8. In the case of C_6F_6, the reactions are only of the ionized or excited molecules and hence most of the products are exact oligomers. C_6F_6 attached electrons in both the gas and the liquid phase[83–85], but it did not detach fluorine atoms as did other halo compounds. The entities $C_6F_6^+ + e^-$ formed directly by the γ-radiolysis are converted to $C_6F_6^+ + C_6F_6^-$. Since most products are trimers and tetramers, it seems that the reaction couple $C_6F_6^+ + C_6F_6^-$ needs at least one more molecule to absorb the excess energy of the neutralization, but there is no way yet to exclude the possibility that the products are formed by reaction of molecules in their excited states. The photolysis of perfluorobenzene leads to other products than the radiolysis, mainly to isomerization[86,87]. The final products of the γ-radiolysis of perfluorobenzene were also studied by X-ray photoelectron spectroscopy[88]. The results show not only cross-linkage but also formation of saturated carbon centres as indicated by the presence of CF_2 and CF_3 groups. The relative abundance of CF, CF_2 and CF_3 groups in the irradiated PFB was estimated to be about 86, 9 and 5%, respectively.

Szymanski and coworkers[89,90] studied the radiation yields of stable products from various chlorides (1-chlorobutane, 1-chlorobutene-2, 1-chloropropane, 2-chloropropane, 1,3-

dichloropropane, 1,4-dichlorobutane and 1,4-dichlorobutene-2). The main products in the gas phase are H_2, HCl and the alkane obtained after the detachment of the chlorine atoms. In the case of alkenes no H_2 was found, probably due to addition of H atoms to the olefinic double bond.

In the radiolysis of 1-chlorobutane the main products are HCl (G = 2.38 ± 0.15), H_2 (G = 1.9–4.9 decreasing with increasing dose), butane (G = 16–30 decreasing with increaseing dose), 2-chlorobutane (G = 74.9 ± 6.7) and 1,3-dichlorobutane (G = 1.8–2.9 increasing with increasing dose). The high values of G indicate a chain reaction, since the usual G values are not higher than 4–5. Szymanski and coworkers suggested the following scheme of reactions to explain the observed products:

$$CH_3CH_2CH_2Cl \xrightarrow{\gamma} CH_3CH_2CH_2\dot{C}H_2 + Cl^\bullet$$
$$\text{or } CH_3\dot{C}HCH_2CH_2Cl + H$$
$$\text{or } CH_3CH_2\dot{C}HCH_2Cl + H$$

$$Cl^\bullet + CH_3CH_2CH_2CH_2Cl \longrightarrow CH_3CHCH_2CH_2Cl + HCl$$

$$CH_3CH_2CH_2\dot{C}H_2 + HCl \longrightarrow Cl^\bullet + C_4H_{10}$$

$$CH_3CH_2CH_2\dot{C}H_2 + CH_3CH_2CH_2CH_2Cl \longrightarrow C_4H_{10} + CH_3CH_2\dot{C}HCH_2Cl$$

$$CH_3CH_2\dot{C}HCH_2Cl \longrightarrow CH_3CH_2CHCl\dot{C}H_2$$

$$CH_3CH_2CHCl\dot{C}H_2 + HCl \longrightarrow Cl^\bullet + \text{2-chlorobutane}$$

$$CH_3CH_2CHCl\dot{C}H_2 + \text{1-chlorobutane} \longrightarrow \text{2-chlorobutane} + CH_3CH_2\dot{C}HCH_2Cl$$
$$\text{or } CH_3\dot{C}HCH_2CH_2Cl$$

$$CH_3\dot{C}HCH_2CH_2Cl + HCl \longrightarrow CH_3CH_2CH_2CH_2Cl + Cl^\bullet$$
$$CH_3\dot{C}HCH_2CH_2Cl + \text{1-chlorobutane} \longrightarrow CH_3CHClCH_2CH_2Cl + C_3H_7\dot{C}H_2$$

Alfassi and Heusinger[91,92] studied the products of the radiolysis in liquid CF_2Cl_2 (under high pressure) and $CFCl_3$. The main products in the radiolysis of $CFCl_3$ are CF_2Cl_2, $C_2F_4Cl_2$, CCl_4 and C_2FCl_5. Alfassi[91] found that the yields of these four products are linear with the time of irradiation (absorbed dose), indicating that they are primary products and not formed by a reaction between primary products. These products are the same as those found by Yamamoto and Ootsuka[93], but the yields found in the two studies are very different. Yamamoto and Ootsuka[93] found that more than three times as much $C_2F_4Cl_2$ is formed than CF_2Cl_2 (G values of 0.44 and 0.12, respectively), compared to almost the same value found by Alfassi (0.24 and 0.26 G values). Alfassi suggested that Yamamoto and Ootsuka did not calibrate their gas chromatograms and used only a general rule for the relative sensitivity, and probably used the inverse proportion suggested by Alfassi and Heusinger[91,92].

The yields of the products in the radiolysis of highly scavenged $CFCl_3$ (with Br_2) were studied by Alfassi and Heusinger[91] and Tominaga and coworkers[94,95], although the latter measured only the yields of Br-containing products. Thus they measured only the yield of the radicals, but not that of the 'molecular products', formed by geminate combination or through monomolecular reactions of excited species. The radiolytic yields of the molecular products in the radiolysis of pure $CFCl_3$ and of highly scavenged $CFCl_3$ are given in the following table (in G units):

Product	CF_2Cl_2	CCl_4	$C_2F_2Cl_4$	C_2FCl_5	$G(-CFCl_3)$
Pure $CFCl_3$	0.26	0.18	0.24	0.036	1.00
Highly scavenged $CFCl_3$	0.12	0.035	>0.01	>0.01	6.50

The yields of the C_2 compounds are considerably lower than those found for CCl_4, where in a highly scavenged system it was found to be $G(C_2Cl_6) = 0.4^{96,97}$, indicating that the source for these compounds in pure $CFCl_3$ is by combination of radicals. Since the C_2 compounds are formed only by combination of $CFCl_2$ and CCl_3 radicals, the yields of the C_2 compounds allow the calculation of the ratio of the steady state (ss) concentration of the radicals: $([CFCl_2^{\cdot}]/[CCl_3^{\cdot}])_{ss} = 2G(C_2F_2Cl_4)/G(C_2FCl_5) = 13.3$.

CCl_4 is most probably expected to be formed via a $CCl_3^{\cdot}$ radical abstracting a chlorine atom from the parent molecule, $CFCl_3$. However, the yield of CCl_4 in pure $CFCl_3$ is much higher than the yield of CCl_3Br in $CFCl_3 + Br_2$. This discrepancy is in the formation of $CCl_3^{\cdot}$ radicals from the $CFCl_2^{\cdot}$ radicals in the pure system, whereas in the scavenged system all the $CFCl_2^{\cdot}$ radicals are captured by Br_2. Thus $CFCl_2^{\cdot}$ reacts with $CFCl_3$,

$$\dot{C}FCl_2 + CFCl_3 \longrightarrow CF_2Cl_2 + CCl_3^{\cdot}$$

followed by the fast reaction,

$$CCl_3^{\cdot} + CFCl_3 \longrightarrow CCl_4 + CFCl_2^{\cdot}$$

Assuming this to be the only source for CCl_4 and assuming $C_2F_2Cl_4$ to be formed by two $CFCl_2$ radicals reacting at 2×10^9 $M^{-1}s^{-1}$, the rate constant for fluorine atom transfer is 0.76 $M^{-1}s^{-1}$, about 3–4 orders of magnitude slower than transfer of a chlorine atom^{98,99}. This suggested mechanism of F atom transfer is further supported by the same decrease in the G values of CF_2Cl_2 and CCl_4 (0.14 G units) by the addition of Br_2 as a scavenger. This mechanism predicts that for every formed molecule of CCl_4 there is also formation of one molecule of CF_2Cl_2. The higher yield of CF_2Cl_2 indicates that CF_2Cl_2 is formed also by another route.

Alfassi and Heusinger^{92} studied the radiolysis of liquid CF_2Cl_2 (at 40 atmospheres in room temperature) in the pure form and in the presence of several scavengers. The main products are $CFCl_3$ ($G = 0.77$), CF_3Cl ($G = 0.59$), $(CF_2Cl)_2$ ($G = 0.28$), $CF_2ClCFCl_2$ ($G = 0.08$), CCl_4 ($G = 0.03$) and $(CFCl_2)_2$ ($G = 0.01$). The yields are proportional to the irradiation time (absorbed dose) and hence these products are primary products of the radiolysis. Another product, CF_3CF_2Cl, was found only at long irradiation times and its yield is not proportional to the irradiation time, indicating that it is formed by reaction with primary radiolysis products or by radiolysis of primary products. The total $G(-CF_2Cl_2)$ (2.1) is similar to that found by Yamamoto and Ootsuka (1.9)^{93} at lower temperatures, –100 °C to –30 °C. However, the product distribution is different. Alfassi and Heusinger explain this discrepancy not as due to the temperature effect, but rather to the wrong sensitivities used by Yamamoto and Ootsuka, who did not use known samples for calibration but used 'a rough proportionality of the detector to the square root of the molecular weight of the relevant compound'. Alfassi and Heusinger studied the yields also at –30 °C and found only negligible differences compared to ambient temperature. The effect of four scavengers (O_2, N_2O, I_2 and Br_2) were studied in order to clarify the mechanism of the formation of the various products. N_2O, an electron scavenger, did not affect the yield, most probably due to CF_2Cl_2 itself being a good electron scavenger. Oxygen reduces the yield, but not to a large extent. Since O_2 is not a better scavenger than N_2O, its effect is most probably due to its reaction with the radicals to form peroxyl radicals. Two new products were identified in the presence of oxygen, CO_2 and $COFCl$ (with G values of 1.09 and 0.04, respectively).

The ratio of CF_2ClI and $CFCl_2I$ and similarly of CF_2ClBr and $CFCl_2Br$ shows that the ratio of C—Cl to C—F bond rupture in radiolysis of CF_2Cl_2 is about 25 : 1. The formation of CF_2Br_2 and $CFClBr_2$ indicated a simultaneous rupture of two atoms from the same molecule. Also, in this case the ratio of C—Cl to C—F bond rupture is 20 : 1.

The G values for the main products in the radiolysis of CF_2Cl_2 + scavengers are given in Table 2.

TABLE 2. Values of G for the main products in the radiolysis of CF_2Cl_2 + scavengers

Product	Unscavenged		At ambient temperature			
	ambient	−30°C	+ 1 atm air	+ 1 atm N_2O	Saturated with I_2	5–7% Br_2
$CFCl_3$	0.77	0.73	0.69	0.85	0.65	0.58
CF_3Cl	0.59	0.55	0.55	0.59	0.39	0.28
$CF_2Cl—CF_2Cl$	0.28	0.24	0.18	0.22	0.10	0.03
$CF_2Cl—CFCl_2$	0.08	0.08	0.05	0.09	<0.01	<0.01
CF_2ClI					5.7	
$CFCl_2I$					0.2	
CF_2ClBr						6.5
CF_2Br_2						1.2
$CFCl_2Br$						0.3
$CFClBr_2$						0.06

The yield of $CFCl_2Br$ (0.3) is too low to explain the yield of $CFCl_3$ in the unscavenged system as due to the reaction

$$CFCl_2^{\cdot} + CF_2Cl_2 \longrightarrow CFCl_3 + CF_2Cl^{\cdot}$$

It must be concluded that $CFCl_3$ is formed also from excited dimer or, more probably, by F atom transfer from CF_2Cl_2 to CF_2Cl.

$$CF_2Cl^{\cdot} + CF_2Cl_2 \longrightarrow CF_3Cl + CFCl_2^{\cdot}$$

followed by fast Cl transfer to $CFCl_2^{\cdot}$,

$$CFCl_2^{\cdot} + CF_2Cl_2 \longrightarrow CFCl_3 + CF_2Cl^{\cdot}$$

This mechanism suggest that $G(CFCl_3) = G(CF_3Cl) + G(CFCl_2^{\cdot})$, which fifts the observed data with $G(CFCl_2^{\cdot}) = G(CF_2ClBr)$ or $G(CF_2ClI)$.

The studies of the radiolysis of freons arose mainly from the suggestion of using freons at low temperatures to extract radioactive krypton (together with xenon) in the reprocessing of burned nuclear elements. In these reprocessing processes, butanol is one of the main impurities. Alfassi and Heusinger[92] studied the effect of the presence of butanol on the decomposition of CF_2Cl_2 by γ-radiolysis. They found that butanol accelerated considerably the degradation of CF_2Cl_2, with production of CHF_2Cl ($G=38.5$) through the following reactions, which lead to considerable degradation through a chain process:

Initiation: $CF_2Cl_2 \longrightarrow^{\hspace{-0.3cm}\sim\sim\sim} CF_2Cl^{\cdot} + Cl$

               $Cl + RCH_2OH \longrightarrow HCl + R\dot{C}HOH$

           $R\dot{C}HOH + CF_2Cl_2 \longrightarrow HCl + RCHO + CF_2Cl^{\cdot}$

Propagation: $CF_2Cl^{\cdot} + RCH_2OH \longrightarrow CHF_2Cl + R\dot{C}HOH$

           $R\dot{C}HOH + CF_2Cl_2 \longrightarrow RCHO + HCl + CF_2Cl$

Termination: $2CF_2Cl^{\cdot} \longrightarrow (CF_2Cl)_2$

The effect of RH on the radiolysis of RH + CF_2Cl_2 mixtures was studied extensively by Mosseri and coworkers[100]. They studied the yield of the final products in radiolysis of CF_2Cl_2 + RH for various concentrations of ethanol, cyclohexane and cyclohexene. Similar studies of radiolysis of CCl_4 with various RH[101] show that the yield of CHX_3 and C_2X_6 fit the equations:

$$\frac{[CHX_3]}{[C_2X_6]} = \frac{k_1}{k_3^{1/2}}[RH] \cdot t^{1/2} \quad \text{or} \quad \frac{G(CHX_3)}{G^{1/2}(C_2X_6)} = \frac{k_1}{k_3^{1/2}}[RH] \cdot \alpha^{1/2}$$

where k_1 is the rate constant for $CX_3 + RH \rightarrow CHX_3 + R^\bullet$, k_3 is the rate constant for combination of two CX_3 radicals and α is the factor transforming from rate of formation to G values:

$$G = N/10 \cdot I \cdot \rho$$

N being the Avogadro number, I the dose rate in $eV\,g^{-1}\,s^{-1}$ and ρ the density of solvent g cm^{-3}.

While this behaviour was found for CCl_4 with various RH including alkanes, cycloalkanes, alkenes and alcohols, for CF_2Cl_2 it was found only for ethanol. The proportionality between $G(CHF_2Cl)/G^{1/2}(C_2F_4Cl_2)$ and the RH concentration was not observed for cyclohexane, or cyclohexene. In these systems the ratio $G(CHF_2Cl)/G^{1/2}(C_2F_4Cl_2)$ was found to be independent of the concentration of the hydrogen-containing substrate, RH. Such unexpected behaviour was also found previously in the reaction of the CF_2Cl radical with cyclopentane in the gas phase[102]. Majer and coworkers[102] observed a linear relation of the above-mentioned ratio with the square root of the cyclopentane concentration, and suggested that this strange behaviour is due to radical combination by triple collision, assuming only RH to be an efficient third body for energy removal:

$$CF_2Cl^\bullet + CF_2Cl^\bullet + RH \rightarrow (CF_2Cl)_2 + RH$$

However, it is difficult to understand why CF_2Cl is so different from CF_3 or CCl_3, and their suggestion in any case cannot be reasonable for liquid-phase reactions. Mosseri and coworkers[100] proposed the presence of charge transfer complexes of the $CF_2Cl^\bullet$ radical with hydrogen-containing substrates, $[CF_2ClRH]^\bullet$. The charge transfer complex can subsequently dissociate back to its components, decompose to $CF_2ClH + R^\bullet$ or dimerize with the exclusion of RH. It was shown that in the case of a strong complex (i.e. one with fast formation and slow decomposition), the ratio $G(CHF_2Cl)/G^{1/2}(C_2F_4Cl_2)$ is independent of [RH], whereas for a weak complex the ratio is proportional to the $RH^\bullet$ concentration.

In the case of cyclohexane and cyclohexene it was found that the ratio is independent of [RH], indicating a strong complex. This suggested different behaviour of $CF_2Cl^\bullet$ than for $CCl_3^\bullet$ is further supported by the comparison of cyclohexane and cyclohexene[73]. In the case of CCl_3 radicals, $G(CHCl_3)$ is larger for cyclohexene as can be expected from the weaker C—H bond, due to the allylic stabilization of the radical. For CF_2Cl radicals $G(CHF_2Cl)$ is larger in the case of cyclohexene (fivefold at room temperature), probably due to the stronger CT complex formed by the double bond. The observation that $G(CHF_2Cl)$ is larger for cyclohexane than for cyclohexene in the case of $CF_2Cl^\bullet$ indicates that the rate-determining step is not H atom abstraction, differing from what was found for CCl_3 and CF_3. This difference between CFCl on the one side and CF_3 or CCl_3 on the other is not yet understood.

V. REFERENCES

1. J. H. O'Donnell and D. F. Sangster, *Principles of Radiation Chemistry*, Edward Arnold, London, 1970.
2. J. W. T. Spinks and R. J. Wood, *An Introduction to Radiation Chemistry*, 2nd edn., Wiley, Chichester, 1976.
3. A. J. Swallow, *Radiation Chemistry—An Introduction*, Longman, London, 1973.
4. A. Mozumder, *Adv. Radiat. Chem.*, **1**, 1 (1969).

5. E. G. Janzen, *Acc. Chem. Res.*, **4**, 31 (1971).
6. C. Lagercrantz and S. Forshult, *Nature*, **218**, 1247 (1968).
7. P. K. Ludwig, *Adv. Radiat. Chem.*, **3**, 1 (1972).
8. J. H. Baxendale and F. Busi, *The Study of Fast Processes and Transient Species by Electron Pulse Radiolysis*, Reidel, Dordrecht, 1982.
9. V. Buxton, in *Radiation Chemistry, Principles and Applications* (Eds. Farhataziz and M. A. J. Roders), Verlag Chemie, Weinheim, 1987.
10. I. D. Draganic and Z. D. Draganic, *The Radiation Chemistry of Water*, Academic Press, New York, 1971.
11. R. E. Bühler, in *Chemistry of the Carbon–Halogen Bond* (Ed. S. Patai), Wiley, London, 1973, pp. 795–864.
12. A. Horowitz, in *Chemistry of the Carbon–Halogen Bond, Supplement D* (Eds. S. Patai and Z. Rappoport), Wiley, Chichester, 1983, pp. 369–403.
13. R. E. Florin, in *Fluoropolymers* (Ed. L. A. Wall,), Interscience, New York, 1972, p. 357.
14. M. Dole, in *The Radiation Chemistry of Macromolecules*, Vol. II (Ed. M. Dole), Academic Press, New York, 1973, P. 167.
15. R. Timmerman and W. Greyson, *J. Appl. Polym. Sci.*, **6**, 456 (1963),
16. K. Makuchi, M. Asano and T. Abe, *J. Polym. Sci., Polym. Chem. Ed.*, **14**, 617 (1976).
17. A. Charlesby, *Atomic Radiation and Polymers*, Pergamon Press, Oxford, 1960, pp. 146–148 and 167; A. Charlesby and S. H. Pinner, *Proc. R. Soc. London*, **A249**, 367 (1959).
18. M. Carenza, S. Lora, G. Pezzin, A. Faucitano and A. Buttafava, *Radiat, Phys. Chem.*, **35**, 172 (1990).
19. T. Yoshida, R. E. Florin and L. A. Wall, *J. Polym. Sci. A*, **3**, 1685 (1965).
20. Z. B Alfassi, A. Harriman, S. Mosseri and P. Neta, *Int. J. Chem. Kinet.*, **18**, 1315 (1986) ; S. Mosseri, Z. B Alfassi and P. Neta, *Int. J. Chem. Kinet.*, **19**, 309 (1987).
21. D. Brault and P. Neta, *J. Phys. Chem.*, **88**, 2857 (1984); J. Grodkowski and P. Neta, *J. Phys. Chem.*, **88**, 1205 (1984).
22. J. E. Packer, R. L. Wilson, D. Bahnemann and K. D. Asmus, *J. Chem Soc., Perkin Trans. 2*, 296 (1980); L. G. Forni, J. E. Packer, T. F. Slater and R. L. Wilson, *Chem. Biol. Interact.*, **45**, 171 (1983).
23. Z. B. Alfassi, S. Mosseri and P. Neta, *J. Phys. Chem.*, **91**, 3383 (1987); Z. B Alfassi, R. E Huie and P. Neta, *J. Phys. Chem.*, **97**, 7253 (1993).
24. Z. B Alfassi, S. Mosseri and P. Neta, *J. Phys. Chem.*, **93**, 1380 (1989); G. Nahor, P. Neta and Z. B. Alfassi, *J. Phys. Chem.*, **95**, 4419 (1991).
25. R. L. Clough, K. T. Gillen and C. A. Quintana, *J. Polym. Sci., Polym. Chem. Ed.*, **23**, 359 (1985).
26. R. L. Clough and K. T. Gillen, in *Oxidation Inhibition in Organic Materials* (Eds. P. Klemchuk and J. Pospisil), CRC Press, Boca Raton, 1990, p. 91.
27. R. L. Clough, in *Encyclopedia of Polymer Science and Engineering*, Vol. 13, Wiley, New York, 1988, p. 67.
28. R. L. Clough and K. T. Gillen, *Polym. Degrad. Stab.*, **30**, 309 (1990).
29. A. G. Guzman-Garcia, P. N. Pintauro, M. W. Verbrugge and E. W. Schneider, *J. Appl. Electrochem.*, **22**, 204 (1992).
30. E. A. Ticianelli, C. R. Derouin and S. Srinvisan, *J. Electroanal. Chem.*, **251**, 175 (1988).
31. G. W. Eastland, S. P. Maj, M. C. R. Symons, A. Hasegaw, C. Glidewell, M. Hayashi and T. Wakabayashi, *J. Chem. Soc., Perkin Trans. 2*, 1439 (1984).
32. M. C. R. Symons, B. W. Wren, H. Muto, K. Toriyama and M. Iwasaki, *Chem. Phys. Lett.*, **127**, 424 (1986).
33. M. C. R. Symons, H. Muto, K. Toryiama, K. Nunome and M. Iwasaki, *J. Chem. Soc., Perkin Trans. 2*, 2011 (1986).
34. M. C. R. Symons, *J. Chem. Res.*, 256 (1985).
35. M. C. R. Symons, *J. Chem. Soc., Faraday Trans. 1*, **78**, 2205 (1982).
36. H. Muto. K. Nunome and M. Iwasaki, *J. Chem. Phys.*, **90**, 6827 (1990).
37. H. Muto and K. Nunome, *J. Chem. Phys.*, **94**, 4741 (1991).
38. T. Shida and Y. Takemura, *Radiat. Phys.*, **21**, 157 (1983).
39. J. C. Green, M. L. H. Green, P. J. Joachim, A. F. Orchard and D. W. Turner, *Phil. Trans. R. Soc. London*, **A268**, 111 (1970).
40. P. W. F. Louwrier and W. H. Hamill, *J. Phys. Chem.*, **73**, 1702 (1969).
41. H. U. Gremlich and R. E. Bühler, *J. Phys. Chem.*, **87**, 3267 (1983).

42. R. F. C. Claridge, R. M. Iyer and J. E. Willard, *J. Phys. Chem.*, **71**, 3527 (1967).
43. J. P. Suwalski, *Radiat. Phys. Chem.*, **17**, 393 (1981).
44. E. Kh. Brickenstein and R. F. Khairutdinov. *Chem. Phys. Lett.*, **115**, 176 (1985).
45. N. V. Klassen and C. K. Ross, *Chem. Phys. Lett.*, **132**, 478 (1986).
46. J. P. Suwalski and J. Kroh, *Bull. Pol. Acad. Sci. Chem.*, **34**, 267 (1986).
47. J. P. Suwalski and J. Kroh, *Nukleonika*, **24**, 253 (1979).
48. B. Hurni and R. E. Bühler, *Radiat. Phys. Chem.*, **15**, 231 (1980).
49. T. Sumiyoshi, S. Sawamura, Y. Koshikawa and M. Katayama, *Bull. Chem. Soc. Jpn.*, **55**, 2341 (1982).
50. N. V. Klassen and C. K. Ross, *J. Phys. Chem.*, **91**, 3668 (1987).
51. J. P. Suwalski and J. Kroh, *Radiat. Phys. Chem.*, **20**, 365 (1982).
52. S. Truszkowski and T. Ichikawa, *J. Phys. Chem.*, **93**, 4522 (1989).
53. J. P. Guarino and W. H. Hamill, *J. Am. Chem. Soc.*, **86**, 777 (1964); J. B. Gallivan and W. H. Hamill, *J. Chem. Phys.*, **44**, 2378 (1964).
54. P. W. F. Louwrier and W. H. Hamill, *J. Phys. Chem.*, **72**, 3878 (1968).
55. T. Ichikawa and N. Ohta, *J. Phys. Chem.*, **91**, 3244 (1987).
56. T. Ichikawa, S. Shiotani, N. Ohta and S. Katsumata, *J. Phys. Chem.*, **93**, 3215 (1989).
57. A. E. Reed, F. Weinhold, R. Weiss and J. Macheleid, *J. Phys. Chem.*, **89**, 2688 (1985).
58. T. K. Ha and R. E. Bühler, *Radiat. Phys. Chem.*, **32**, 117 (1988).
59. H. U. Gremlich and R. E. Bühler, *J. Phys. chem.*, **87**, 3269 (1983).
60. L. T. Shoute and J. P. Mittal, *Spectrochim, Acta*, **45A**, 863 (1989).
61. M. Shiotani, Y. Nagata, M. Tasaki, J. Sohma and T. Shida, *J. Phys. Chem.*, **87**, 1170 (1983).
62. R. E. Bühler, *Radiat. Phys. Chem.*, **21**, 139 (1983).
63. R. Mehnert, O. Brede, J. Bös and W. Naumann, *Ber. Bunsenges. Phys. Chem.*, **83**, 992 (1979).
64. O. Brede, J. Bös and R. Mehnert, *Radiat. Phys. Chem.*, **23**. 739 (1984).
65. R. Mehnert, J. Bös and O. Brede, *Radiochem. Radioanal. Lett.*, **51**, 47 (1982).
66. R. E. Bühler, *Radiat. Phys. Chem.*, **23**, 741 (1984).
67. (a) R. E. Bühler, *J. Radioanal. Nucl. Chem.*, **101**, 329 (1986).
(b) A. S. Domazu, M. A. Quadir and R. E. Bühler, 'Ion pairs $(CFCl_2^+ \mid\mid Cl^-)_{solv}$ from geminate ion recombination in methylcyclohexane with $CFCl_3$: Formation, reactivity and stability', in press.
68. M. Washio, Y. Yoshida, N. Hayashi, H. Kobayashi, S. Tagawa and Y. Tabata, *Radiat. Phys. Chem.*, **34**, 115 (1989).
69. M. Washio, S. Tagawa and Y. Tabata. *Radiat. Phys. Chem.*, **21**, 239 (1983).
70. H. U. Gremlich, T. K. Ha, G. Zumofen and R. E. Bühler, *J. Phys. Chem.*, **85**, 1336 (1981).
71. T. Sumiyoshi, T. Yamada, A. Ohtaka, K. Tsugru and M. Katayama, *Chem. Lett.*, 307 (1986).
72. T. Sumiyoshi, N. Sugita, K. Watanabe and M. Katayama. *Bull. Chem. Soc. Jpn.*, **61**, 3055 (1988).
73. R. Mehnert, O. Brede and W. Naumann, *Ber. Bunsenges. Phys. Chem.*, **86**, 525 (1982).
74. Y. Wang, J. J. Tria and L. M. Dorfman, *J. Phys. Chem.*, **83**, 1946 (1979).
75. M. P. de Haas, J. M. Warman and B. Vojnovic, *Radiat. Phys. Chem.*, **23**, 61 (1984).
76. A. Halpern, *J. Chem. Soc., Faraday Trans. 1*, **83**, 289 (1987).
77. G. V. Kovalev, A. L. Kanasev, L. T. Bugaenko and E. P. Kalyazin, *High Energy Chem.*, **18**, 245 (1984).
78. Z. B. Alfassi and L. Feldman, *Int. J. Chem. Kinet.*, **13**, 517 (1981); Z. B. Alfassi and L. Feldman, *Int. J. Chem. Kinet.*, **13**, 771 (1981); L. Feldman and Z. B. Alfassi, *J. Phys. Chem.*, **85**, 3060 (1981); L. Feldman and Z. B. Alfassi, *Int. J. Chem. Kinet.*, **14**, 659 (1982); Z. B. Alfassi and L. Feldman, *Int. J. Chem. Kinet.*, **12**, 379 (1980).
79. M. G. Katz, G. Baruch and L. A. Rajbenbach, *Int. J. Chem. Kinet.*, **8**, 131 (1976); *Int. J. Chem. Kinet.*, **8**, 509 (1976). The data were reinterpreted in Z. B. Alfassi, *Int. J. Chem. Kinet.*, **12**, 217 (1980).
80. Z. B. Alfassi, 'Radical combination and disproportionation', in *Chemical Kinetics of Small Organic Radicals*, Vol. 1, (Ed. Z. B. Alfassi), CRC Press, Boca Raton, 1988, p. 129.
81. J. Monig, K. D. Asmus, L. W. Robertson and F. Oesch, *J. Chem. Soc., Perkin Trans. 2*, 891 (1986).
82. N. H. Sagret, J. C. Le Blanc, D. D. Wood, W. Kremers, J. B. Westmore and W. D. Buchannon, *Radiat. Phys. Chem.*, **38**, 399 (1981).
83. S. Chowdhury, E. P. Grimsurd, T. Heinis and P. Kebarle, *J. Am. Chem. Soc.*, **108**, 3630 (1986).

84. W. E. Wentworth, T. Limero and E. C. M. Chen, *J. Phys. Chem.*, **91**, 241 (1987).
85. L. C. T. Shoute and J. P. Mittal, *Radiat. Phys. Chem.*, **30**, 105 (1987).
86. I. Haller, *J. Chem. Phys.*, **47**, 1117 (1967).
87. J. L. Suijker, A. H. Huizer and C. A. G. O. Varma, *Laser Chem.*, **6**, 333 (1986).
88. S. Sunder, N. H. Sagret, D. D. Wood and N. H. Miller, *Can. J. Appl. Spectrosc.*, **35**, 137 (1990).
89. W. Szymanski, S. Truszkowski, G. Smietanska and M. Kowal, *J. Radioanal. Nucl. Chem.*, **172**, 57 (1993).
90. W. Szymanski, G. Smietanska and M. Kowal, *Radiat. Eff.*, **100**, 111 (1986).
91. Z. B. Alfassi, *Radiochem. Radioanal. Lett.*, **56**, 333 (1982).
92. Z. B. Alfassi and H. Heusinger, *Radiat. Phys. Chem.*, **22**, 995 (1983).
93. T. Yamamoto and N. Ootsuka, *J. Nucl. Sci. Technol (Jpn.*), **18**, 913 (1981).
94. T. Tominaga, Y. Makide, T. Fukumizu and T. Aoyama, *Radiochem. Radioanal. Lett.*, **25**, 137 (1976).
95. T. Tominaga, T. Fukumizu, T. Aoyama, A. Okuda and Y. Makide, *Bull. Chem. Soc. Jpn.*, **49**, 2185 (1976).
96. Z. B. Alfassi and L. Feldman. *Int. J. Chem. Kinet.*, **13**, 517 (1981); L. Feldman and Z. B. Alfassi, *Radiat. Phys. Chem.*, **15**, 687 (1980).
97. N. E. Bibler, *J. Phys. Chem.*, **75**, 24 (1971).
98. Z. B. Alfassi, *Int. J. Chem. Kinet.*, **12**, 217 (1980).
99. K. U. Ingold, in *Free Radicals*, Vol. I, (Ed. J. K. Kochi), Wiley, New York, 1972, p. 37.
100. S. Mosseri, Z. B. Alfassi, W. Fürst and H. Heusinger, *Radiat. Phys. Chem.*, **26**, 89 (1985).
101. Z. B. Alfassi and L. Feldman, *Int. J. chem. Kinet.*, **12**, 379 (1980); *Int. J. Chem. Kinet.*, **13**, 517 (1981).
102. J. R. Majer, D. C. Phillips and J. C. Robb, *Trans. Faraday Soc.*, **61**, 122 (1960).

CHAPTER **18**

Electrochemistry of the carbon–halogen bond

JOSEPH CASANOVA

Department of Chemistry, California State University, Los Angeles, California, 90032, USA

and

V. PRAKASH REDDY

Department of Chemistry, University of Miami, Coral Gables, Florida 33124, USA

Supplement D2: The chemistry of halides, pseudo-halides and azides
Edited by S. Patai and Z. Rappoport   © 1995 John Wiley & Sons Ltd

ABBREVIATIONS

bpy	2,2′-bipyridine
cv	cyclic voltammetry or cyclic voltammogram
DMA	N,N-dimethylacetamide
DMF	N,N-dimethylformamide
ECE	electrochemical–chemical–electrochemical reaction sequence
ET	electrons transfer
LFER	linear free-energy relationship
nce	normal calomel electrode
sce	standard calomel electrode
TBAB	tetra-n-butylammonium bromide
TBAF	tetra-n-butylammonium fluroborate
TBAP	tetra-n-butylammonium perchlorate
TEAB	tetraethylammonium bromide
TEAP	tetraethylammonium perchlorate
TMAP	tetramethylammonium perchlorate
TPP	tetraphenylporphinate

I. INTRODUCTION

Since the appearance of the original version of this chapter[1], the field of organic electrochemistry has experienced a slow growth while undergoing significant maturation. Mechanistic studies, novel syntheses and an improved understanding of electron transfer processes[2] are among the main areas of development during the two decades since this topic was reviewed by Casanova and Eberson in this series. A hope prevailed that judicious selection of reaction conditions (e.g. electrode material, electrode potential, solvent, supporting electrolyte) would permit highly selective oxidation or reduction of functional groupings in organic molecules. This has proven true in a few cases, but in general when a chemical reagent exists that can accomplish the same objective, chemists have elected to follow that route. Lack of laboratory experience by most organic chemists in electrochemical methods of synthesis has hampered the development of these techniques as standard tools for the organic chemist. This review will treat the electrochemistry of the carbon–halogen bond, important because of the ubiquitous nature of halogen compounds in organic chemistry.

Since our last effort a number of other reviews have appeared. Most notably among them is a thorough examination by Becker[3] of topics discussed in the first review of this series, expanded to include a broader range of direct and indirect anodic oxidations not covered in the original review.

It will be the objective of this chapter to provide a reasonably complete critical treatment of reductive reactions of the carbon–halogen bond that have been reported since the appearance of the Becker review, and a much less comprehensive examination of anodic oxidations inasmuch as Becker afforded this topic very thorough treatment earlier. Literature cited in the original version will not be referenced here except as it may be necessary to introduce or summarize the history of a topic. Section headings and sub-headings that were established in the original version have been preserved here where feasible in order to permit easier cross-referencing of topics. However, new topics have clearly emerged, and older ones have lost relevance in some cases. We will forego the important large and growing literature involving the electrochemistry of organometallic compounds, mostly transition metal complexes that contain halogen atoms in the coordination sphere of the metal, except as these materials are used as electron transfer agents to effect an electrochemical reaction on an organic halide. Nor shall we deal with the important area electroreductive coupling involving halogen-containing compounds, except as such coupling results directly from carbon–halogen bond reduction. Classification of references is based mostly on the structure of the starting halide for synthetic papers, and on topics for important general issues. Finally, since there are many good textbooks and reviews[4] which treat electrochemical techniques and their application in organic chemistry, we feel there is no need to include such background material here. A knowledge of the most elementary principles of electrochemistry, will be sufficient to follow the treatment in this chapter.

Modern electronic literature searching techniques have permitted the identification of some references for which the original literature was not available to us, but for which an abstract was available. We have, for the sake of completeness, included these references when appropriate.

II. GENERAL REACTION TYPES

Electrochemical reactions of the carbon–halogen bond will be divided according to whether the electrochemical step is reductive or oxidative in nature. Reduction of many carbon–halogen bonds is among the most easily achieved of any two electron bonds to carbon. The majority of work in the electrochemistry of the carbon–halogen bond deals with reduction of halides and this topic will be considered first. Oxidative reactions of organic halides remain limited in example and are largely confined to derivatives of iodo compounds. These center around detailed mechanisms of electroreduction, metal ion catalysis of electron transfer, electrocatalytic reactions and mechanisms of electron transfer. Important developments in the understanding of electron transfer processes, particularly the importance of the electroinduced $S_{RN}1$ reaction, have occurred in the last decade, largely due to the work of Eberson[5] and Savéant[6], and are discussed in some detail here. The general topic has been reviewed periodically and to varying degrees of completeness. Perhaps the most thorough review to appear recently on this subject is by Peters in the Baizer text[7]. Hills[8] has produced a brief review of reduction of carbon–halogen bonds as part of synthetic discussion. Two other brief reviews that focus on the application of controlled potential electroreduction for these compounds in synthesis have been reported[9]. A review by Fry describing electroorganic chemistry in general focuses in one limited section on the electroreduction of the carbon–halogen bond[10]. A brief popular review of important carbon–halogen electrochemistry has appeared[11], and a review has appeared in Croatian[12].

A. Direct Cathodic Reactions

A general formulation of the several possible steps in the direct electroreduction of a carbon-halogen bond is shown in Scheme 1. The Scheme differs little from that introduced

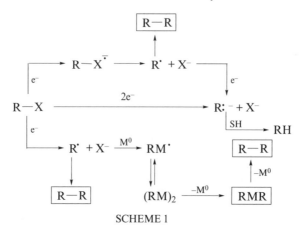

SCHEME 1

by Webb, Mann and Walborsky[13] more than twenty years ago. Which pathways to the main stable products (in boxes) are important depend on the potential, the identity of the electrode material, the solvent and the supporting electrolyte. The many questions that can arise regarding the mechanism of this reduction are apparent from an inspection of Scheme 1. The orientation of the carbon–halogen bond relative to the electrode surface, the concerted or stepwise nature of the first electron transfer and carbon–halogen bond rupture steps, the stereochemical fate of the carbanionic intermediate, the intervention of organometallic intermediates, and the origin of the coupling product are such questions. The role of the electrode material and its possible participation in the reaction (via adsorption and/or formation of organometallic species) is also an important and complicated problem. Electrolytic scission of the carbon–halogen bond is an irreversible process and requires a strongly negative cathodic potential if the carbon–halogen bond is not otherwise activated by substituents. Systematic examination of many of these questions has been hampered by the difficulty in applying conventional electroanalytical techniques.

B. Direct Anodic Reactions

Oxidation reactions of organic halides in which the halogen atom is involved are most easily accomplished with iodo compounds. Modern research on hypervalent iodine compounds has given some new insight into the behavior of these compounds[14]. More electronegative halogens render the isolated monosubstituted compounds so difficult to oxidize that they fall outside the accessible range of anodic potentials or cause oxidation to take place in other parts of the molecule. Geminal polyhalides may be an exception to this generalization and have exhibited some interesting oxidation chemistry. However, some interesting reactions of organofluorine derivatives have been reported in which the organic substrate is initially oxidized and causes fluorine incorporation. Anodic processes have been recently and thoroughly reviewed and will be addressed only passingly here[3].

C. Indirect Cathodic and Anodic Reactions

The indirect oxidation and reduction of the carbon–halogen bond has been an area of major activity. Mediated electrochemical reactions, in which the primary electron transfer

$$A \xrightarrow{\ e^-\ } \overset{\cdot}{A}^{-} \xrightarrow{\ S_1\ } A + \overset{\cdot}{S_1}^{-}$$

$$\overset{\cdot}{S_1}^{-} \xrightarrow{\hspace{3cm}} products$$

$$D \xrightarrow{\ -e^-\ } \overset{\cdot}{D}^{+} \xrightarrow{\ S_2\ } D + \overset{\cdot}{S_2}^{+}$$

$$\overset{\cdot}{S_2}^{+} \xrightarrow{\hspace{3cm}} products$$

SCHEME 2

step involves transfer of an electron to a carrier species, all illustrated in Scheme 2 and described in Section III.B.4. Such reactions show promise for rigorous study of reaction mechanism, since they are predominantly homogeneous electron transfer reactions and more amenable to conventional kinetic methods. But whether results from such studies can be applied to direct processes at the heterogeneous electrode/solution interface is unclear.

III. CATHODIC REACTIONS

A. Ease of Reduction by Halogen Type

With the exception of a few special cases of mixed halogen compounds, the ease of reduction of carbon–halogen bonds follows the expected order, I>Br>Cl>F, and becomes more facile as the number of halogen atoms on a given carbon atom increases. Accessible half-wave potentials range from -2.23 V (sce) (C—Cl) to -0.3 V (sce) (Br$_2$C—Br)[15]. The isolated C—F bond cannot be reduced electrochemically below the solvent cut-off limit. However, when the C—F functionality is adjacent to an aromatic or carbonyl moiety, or is geminal, reduction often does occur.

B. Ease of Reduction by Organic Moiety

It is helpful further to subdivide consideration of organic halides according to the nature of the organic fragment, inasmuch as this factor has a large effect on both the ease of reduction and the nature of the products. In Section III.B reactions will be classified predominantly according to the structural class of the starting halide.

1. Monohalides

Most monohalides (except simple fluorides and some alkyl chlorides) can be reduced at a mercury, platinum or carbon electrode in readily available protic or aprotic solvents such as DMF, DMSO or acetonitrile. Bromides are most often employed as substrates since they are usually chemically stable, easily prepared and reduced at potentials which are less cathodic than for the equivalent chloro compounds. The products are those of reduction of the carbon–halogen bond, either a 2e$^-$/molecule reduction in which the halide ion is replaced by a proton, or a 1e$^-$/molecule reduction in which coupling products are formed. The latter products tend to predominate in systems for which the stability of the corresponding radical/carbanion is relatively high due to conjugative stabilization. Occasionally, intermediate 1e$^-$ reduction species can react with the electrode material and form organometallic compounds. The role of organometallic species in these reactions is the subject of some controversy and is discussed in Section III.B.5. In general, the current efficiency of the reduction is moderate to low, particularly with isolated monohalides for which so negative a potential is required that reduction of the solvent and/or supporting electrolyte becomes a significant competing reaction. Whether the two electron reduction is stepwise or concerted has been the topic of some controversy and is reported in Section III.B.5. Stereochemistry of the reduction process which might reveal some detailed aspects of reaction/orientation at the electrode surface has not received much attention, with the exception of stereochemistry of some benzylic monohalides (Section III.B.1.b) and 1,3-dihaloketones (Section III.B.3.b).

a. *Compounds with isolated carbon–halogen bonds.*   Saturated aliphatic and alicyclic monobromides are reduced in the range of –2.1 V to –2.5 V versus sce in DMF/0.01 M tetraethylammonium bromide (TEAB) at a mercury cathode. This is a seemingly narrow potential range, but it should then be kept in mind that a difference of 0.31 V between ethyl bromide and l-bromobicyclo[2.2.2]octane, for example, corresponds to a maximum ratio between the two rate constants of roughly 10^{10} for a 2e⁻ process.

The halide radical anion stability has been calculated at the STO-3G level and found to have a low barrier to dissociation when solvated by water molecules[16]. In solution it cannot be observed (equation 1) because of the small barrier to its dissociation (*ca* 16 kcal

$$R\text{—}\underset{R}{\overset{R}{\diagdown}}\text{—}X \xrightarrow{\ e^-\ } \left[R\text{—}\underset{R}{\overset{R}{\diagdown}}\text{—}X \right]^{\overset{\cdot}{-}} \longrightarrow \underset{R\quad R}{\overset{R}{\diagdown}}\cdot + X^- \qquad (1)$$

mol⁻¹ at STO-3G), which corresponds to a lifetime of 4×10^{-4} to 10^{-8} s. The electroreduction of primary alkyl halides at mercury electrodes results in a one electron process the formation of RHg·, which is reduced to R⁻ after desorption[17].

Direct evidence for stepwise reduction was obtained in the electrolytic reduction of *t*-butyl bromide at a mercury cathode in DMF[18,19]. The polarogram or cyclic voltammogram shows two waves [$E_{1/2}$ = –1.23 V, –1.46 V (sce)] indicating stepwise reduction. Labelling experiments indicate that isobutylene and isobutane arise from disproportionation of *t*-butyl radicals, and the 2,2,3,3-tetramethylbutane arises from a coupling reaction (equation 2). Reduction at the second wave leads to a carbanion, evidence of which is provided by

$$Me_3CBr \xrightarrow{+e^-} Me_3C^\bullet \xrightarrow{+e^-} Me_3C:^- \xrightarrow{D_2O} Me_3CD \qquad (2)$$

$$Me_3C^\bullet \begin{cases} \longrightarrow Me_3CH + Me_2C=CH_2 \\ \qquad\quad 40\text{–}43\% \qquad 43\text{–}46\% \\ \\ \longrightarrow [Me_3C]_2 \\ \qquad\quad 9\text{–}10\% \end{cases}$$

trapping by D_2O. The coupling of anion radicals with sterically hindered alkyl halides has been reported by Lund[20].

Electrochemical reduction of diphenyliodonium salts and phenylmercuric halides to benzene in DMF typically shows three or four waves depending on the concentration of supporting electrolyte or substrate, showing that the process involves reduction to iodobenzene, followed by further reduction to benzene[21].

Polarography of alkyl halides at temperatures below –39 °C in DMF containing tetra-methylammonium perchlorate (TMAP) showed a prominent current minimum, which was attributed to the adsorption onto mercury of complex species composed of tetra-methylammonium and halide ions[22].

Cathodic reduction at zinc of chloroethyl β-naphthyl sulfide **1**, chloropropyl β-naphthyl sulfide **2** and chloroethyl β-naphthyl sulfone **3** in methanol (equation 3) gave a complex mixture of products including alkyl β-naphthyl sulfides **4** and alkyl β-naphthyl sulfones **5**[23].

Reduction of substituted 2-haloamides **6** appears to occur through an intermediate aziridinone **7** (equation 4). 2-Haloamide anions are formed by electroreduction of the corresponding NH-protic 2-haloamides and self-protonated. Such anions are labile species whose decay, in the case of 2-bromoamide anions, competes with their electroreduction in

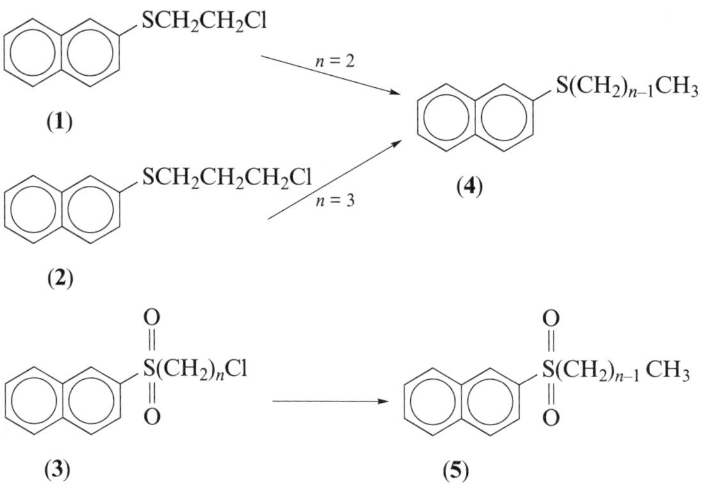

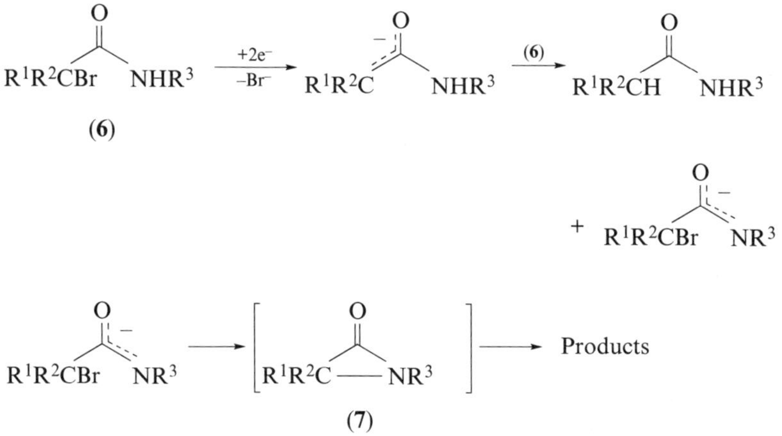

the voltammetric time scale. The first-order rate constant of the decay was determined for a series of 2-bromoamides. The lability orders point to an S_N2-type intramolecular substi-

TABLE 1. Product yields from 2-haloimide and 2-haloamide reduction (percent)

	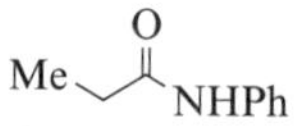 Me–CONHPh	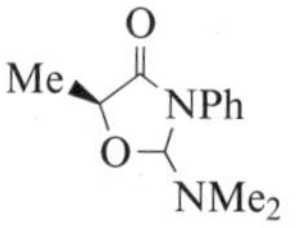 Me–(oxazolidinone)–NPh / NMe$_2$	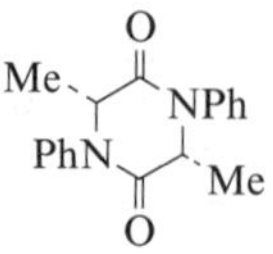 dioxopiperazine (Me, PhN, NPh)	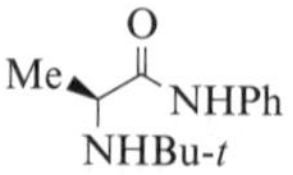 Me–CONHPh / NHBu-t
DMF, TBAP	50	28	22	
DMF, TBAP, t-BuNH$_2$	50	9	15	26

tution of bromine. Stereochemistry of the reaction using a chiral 2-bromopropanamide, an amine nucleophile, suggest a double inversion reaction. Yields are shown in Table 1. The decay of the 2-bromoamide anion lead to the formation of the retention product 2-aminoamide together with other optically active products, diastereomeric oxazolidin-4-ones, arising by cyclocondensation with DMF, and *cis*-2,5-dioxopiperazine[24]. Product mixtures (Table 1) were complex and could only be separated in small quantities by HPLC, suggesting that this does not represent a useful preparative method.

The electroreduction of *N*-(haloethyl)amides **8** at a mercury cathode in DMF or acetonitrile gives rise to the corresponding *N*-(2,2-dichlorovinyl)amides **9** in good to excellent yields (equation 5)[25]. The electrochemical reduction of fluorinated fullerene has been reported[26].

$$RCONHCHXCCl_3 \xrightarrow[\text{TEAP}]{\text{Hg, DMF}} RCONHCH=CCl_2$$

$$50\text{–}100\%$$

$$\text{(8)} \qquad\qquad \text{(9)}$$

$$\xrightarrow[\text{TEAP}]{\text{Hg, DMF}}$$

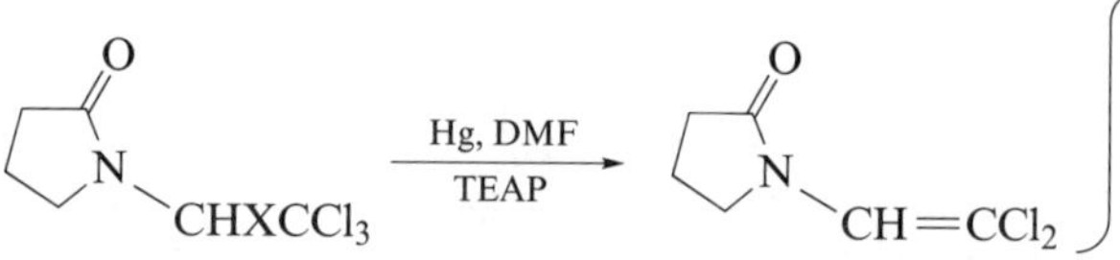

(5)

Electrocatalytic processes are discussed from a mechanistic point of view in Section III.B.5. However, some synthetic aspects as pertain to alkyl monohalides are relevant here. Electrocatalytic processes involve the formation of organometallic species characterized by the consumption of significantly less than a stoichiometric quantity of current, and can be generally illustrated by the process shown in Scheme 3.

The electrogenerated nickel(0), as 2,2′-bipyridinenickel in DMF has been used as catalyst for the electrochemical reduction and dimerization of aliphatic halides[27,28]. In the nickel-catalyzed dimerization of alkyl halides, it was shown that increasing the concentration of the supporting electrolyte TBAB increases the rate of reaction[29].

$$RX \xrightarrow{+e^-} R^{\bullet} + X^-$$

$$R^{\bullet} + M^0 \longrightarrow RM^{\bullet}$$

$$RM^{\bullet} + RX \longrightarrow RMX + R^{\bullet}$$

catalytic

SCHEME 3

Electrocatalytic haloarylation of olefins has been achieved by using zerovalent nickel complexes associated with triphenylphosphine and ethylene[30]. Some of the same materials may be used as electron transfer agents in homogeneous electron transfer reactions. Electrochemical radical cyclization of allyl β-haloalkyl ethers **10** (equation 6) or acetylenic-β-halo ethers **11** (equation 7) is achieved under mild conditions using nickel(II) complexes as electron transfer catalysts. The electrogenerated Ni(I) transfers an electron to the halogeno ethers forming the corresponding radicals with the loss of the halide ions. The radicals cyclize intramolecularly and stereospecifically to give the five-membered ring products. The reaction provides an attractive alternative to the radical cyclizations involving the use of tributyltin hydride, as typically high temperatures are required for the latter process[31]. Catalytic reduction of alkyl halides by carbon electrodes modified by iron porphyrins has also been described[32].

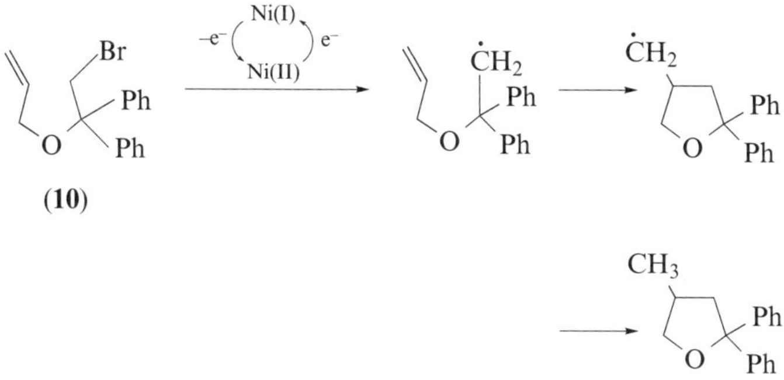

$$(6)$$

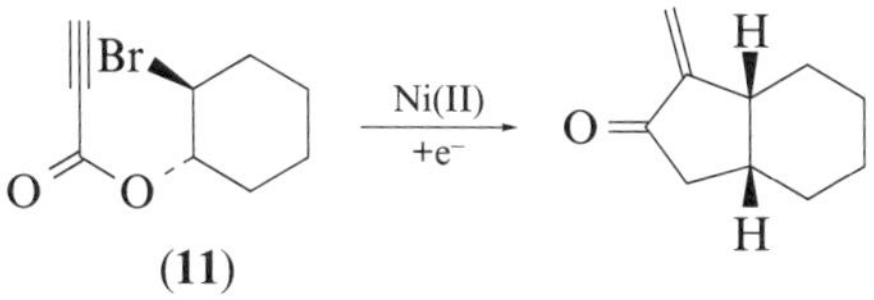

$$(7)$$

1-Bromooctane (**12**) is reported to give high yields of coupling products **13** from an electrochemically generated iron promoter (equation 8). Moderate yields of coupled prod-

$$\text{Me(CH}_2)_5\text{CH}_2\text{CH}_2\text{Br} \xrightarrow[\substack{\text{Al,TEAB} \\ -0.90 \text{ V [Cd(Hg)]}}]{\substack{\text{Fe(acac)}_3 \\ \text{DMF}}} [\text{Me(CH}_2)_5\text{CH}_2\text{CH}_2]_2 + \text{Me(CH}_2)_5\text{CH}_2\text{Me}$$

(**12**)   (**13**)
                                                                                   59%   24%

$$+ \ \text{Me(CH}_2)_5\text{CH}=\text{CH}_2$$

$$11\%$$

$$(8)$$

ucts were obtained with an aluminum electrode using Fe(acac)$_3$ as electron transfer agent in DMF even in the presence of β-protons[33]. The authors note that this is apparently not a homogeneous reaction, but one which occurs after deposition of the iron component on the cathode surface. Side products can be rationalized as resulting from disproportionation or from intermediate carbanion attack on unreacted bromide (equation 9).

$$RCH_2CH_2^+ \xrightarrow{+e^-} RCH_2CH_2^- \longrightarrow RCH_2CH_3 + RCH=CH_2 \quad (9)$$

$$\downarrow RCH_2CH_2^+$$

$$\longrightarrow [RCH_2CH_2]_2$$

Baizer and Chruma reported electrolytic reductive coupling, in a broad study in which reduction of organic halides was conducted in the presence of electrophiles as seen in equation 10[34]. Controlled potential electroreduction of reactive halides at a mercury cathode in the presence of olefinic substrates with electron-withdrawing groups (Michael receptors) gave moderate yields (50–75% in many cases) of carbanion addition products[35]. Current yields in excess of 100% in the case where chloroform was used as a cosolvent with carbon tetrachloride indicated the intervention of electrocatalytic reactions (equations 10 and 11).

$$CCl_4 + CH_2=CHE \xrightarrow{+2e^-} Cl_3CCH_2CH_2E \quad (10)$$
$$E = \text{electrophilic group}$$

$$RX + CH_2=CHCN \xrightarrow{+2e^-} RCH_2CH_2CN$$

$$RX + CH_2=CHCO_2Et \xrightarrow{+2e^-} RCH_2CH_2CO_2Et \quad (11)$$

$$RX= \begin{cases} Cl_3CCH_2CO_2Et \\ BrCH_2CO_2Et \\ BrCH_2CH=CH_2 \\ BrCH_2Ph \end{cases}$$

In the case of benzyl systems, added carbon dioxide resulted in the formation of benzyl benzoate, strongly implicating a carbanionic species (equations 12).

$$PhCH_2Cl + CO_2 \longrightarrow PhCH_2CO_2CH_2Ph \quad (12)$$
$$43\%$$

Electroformylation of organic halides in DMF solvent provides an efficient electrosynthesis of aldehydes **14**, which avoids the use of preformed Grignard or organolithium reagents (equations 13 and 14). A sacrificial magnesium or zinc electrode as the anode and stainless steel or nickel as the cathode gave optimal yields (30–85%).

$$PhBr \xrightarrow{DMF,e^-} PhCHO \quad (13)$$

$$p\text{-}CF_3C_6H_4Cl \xrightarrow{DMF,e^-} p\text{-}CF_3C_6H_4CHO \quad (14)$$
$$\textbf{(14)}$$

The anions resulting from electrolytic C—F bond cleavage of trifluoromethyl arenes **15** were also coupled with electrophiles such as DMF, CO_2, or acetone to give the corresponding aldehydes, esters (**16**), or alcohol (**17**) derivatives, as seen in equation 15[36].

$$PhCF_3 + CO_2 \quad \begin{array}{c} \overset{\text{Mg anode}}{\underset{\text{1. }+e^-; \text{ 2. MeOH,H}^+}{\diagup}} \quad PhCF_2CO_2Me \\ \qquad\qquad\qquad\qquad \textbf{(16)} \\ \overset{\text{Mg anode/acetone}}{\diagdown} \quad PhCF_2C(OH)Me_2 \\ \qquad\qquad\qquad\qquad \textbf{(17)} \end{array} \quad (15)$$

$$\textbf{(15)}$$

Controlled potential electroreduction of alkyl halides and $Fe(CO)_5$, followed by reaction with carbon monoxide and quenching with protic acids gave aldehydes in good yields (equation 16)[37]. This reaction was later extended to the preparation of ketones by reacting the iron acylate anion with alkyl halides. It is also found that the use of divided cells with stainless steel as the cathode and platinum anode significantly improves yields of the carbonyl compounds[38]. Electrosynthesis of aldehydes can also be achieved using DMF as the source of carbonyl group, and as the solvent. The alkyl halides were exhaustively electroreduced using a zinc, cadmium, or tin coated cathode and a sacrificial magnesium anode in DMF solvent at constant current to give the aldehydes[39]. The reaction conditions were optimized using 1-halogenopentanes. They can be electroreductively converted into hexanal using $[Fe(CO)_5]$. Good yields are obtained by using acetonitrile as solvent and tetraalkylammonium halides or p-toluenesulfonate as a supporting electrolyte in a divided cell. As compared with its chloro and iodo analogs, 1-bromopentane was found to be the best substrate for the carbonylation[40].

$$RX + Fe(CO)_5 \xrightarrow{+2e^-} RFe(CO)_4^- \xrightarrow{CO} R\overset{\overset{\displaystyle O}{\|}}{C}Fe(CO)_4^-$$

$$\xrightarrow{H^+} RCHO + Fe(CO)_4 \qquad (16)$$

Electrolytic reduction of benzylic (**18**) and allylic halides (equation 17) in the presence of anhydrides affords the corresponding ketones in good yields. The electrolysis was conducted in an undivided cell using aluminum or magnesium anode and under constant-current conditions. Similarly, benzylic halides were reported to react with acid chlorides under controlled potential conditions, in acetonitrile or DMF as solvent as shown in equation 18[41].

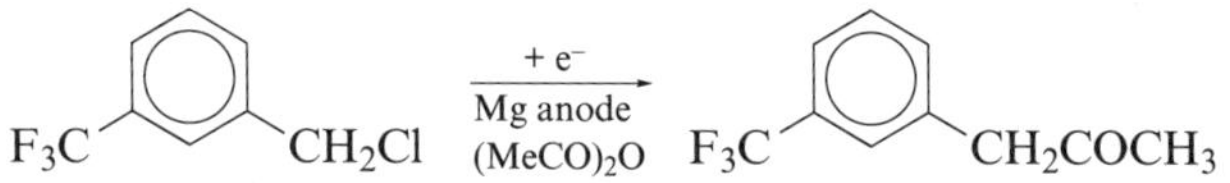

$$
\begin{aligned}
RX + 2e^- &\longrightarrow R^- + X^- \\
Mg &\longrightarrow Mg^{2+} + 2e^- \\
R^- + R'COCl &\longrightarrow R\overset{\overset{\displaystyle O}{\|}}{C}R'
\end{aligned}
\qquad (18)
$$

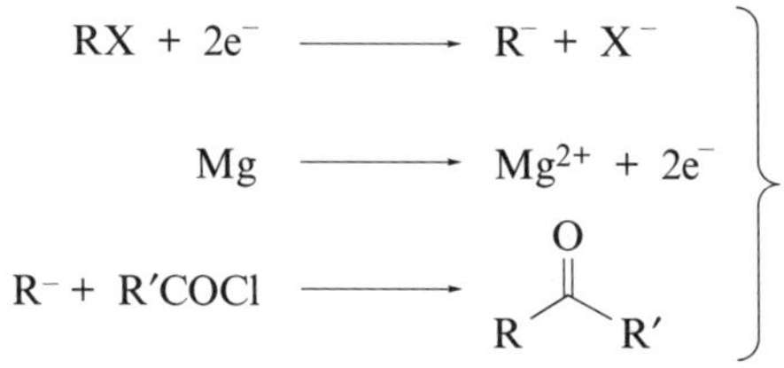

Symmetrical ketones are also prepared from organic halides and carbon dioxide catalyzed by 2,2′-bipyridine-nickel complexes[42]. Torii and coworkers found a synthetically useful electrochemical allylation, which involves the electroreductive generation of a diallyltin reagent in the presence of a catalytic amount of tin. A wide variety of aldehydes and ketones have been added in high yields under relatively mild conditions by this method (equation 19)[43]. The process was further improved by the addition of catalytic amounts of

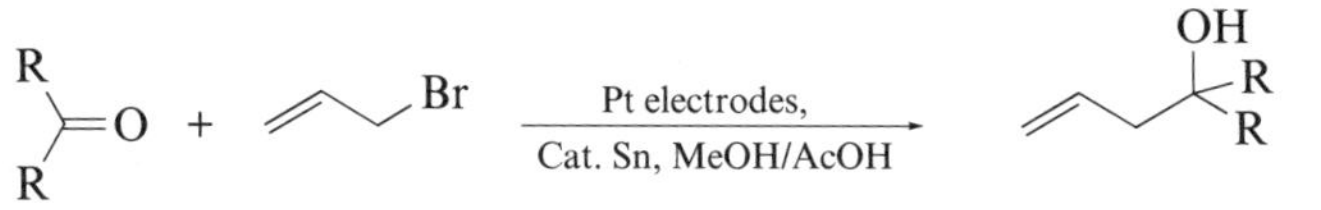

R=aliphatic/aromatic

NiBr$_2$(2,2'-bipyridine) complex[44]. Aryl halides, such as 4-bromobenzophenone, or 2-bromoquinoline could be transformed into diaryl diselenides **20** or ditellurides **21** electrochemically using sacrificial selenium or tellurium cathodes as can be seen in equation 20.

$$ ArBr + \xrightarrow{Se/Te} ArTeTeAr \ or \ ArSeSeAr \qquad (20) $$

$$ \textbf{(20)} \qquad \textbf{(21)} $$

A $S_{R'N}$ reaction mechanism involving the reaction of aryl radicals on the chalcogenide dianions was proposed for the NiBr$_2$(bpy) reaction. The reactions are particularly efficient when the aryl groups are substituted by electron-withdrawing unsaturated groups such as nitrile or carbonyl groups[45]. Symmetrical ketones can be synthesized electrochemically from various organic halides and carbon dioxide catalyzed by 2,2'-bipyridine–nickel complexes[46]. The reaction is suitable for alkyl, benzyl, and allyl halides (equation 21). The mechanism shown in equation 22 was postulated. The main advantage of this process is the use of CO$_2$ as an *in situ* source of CO, in controlled amounts to avoid the deactivation of the catalyst. The magnesium anode gives Mg^{2+}, which is used for trapping undesirable CO$_3^{2-}$ formed during reduction of CO$_2$. Large-scale production of carboxylic acids by electroreduction of organic halides in an atmosphere of carbon dioxide, and using a magnesium as the sacrificial anode was achieved[47].

$$ PhCH_2Br + CO_2 \xrightarrow[\text{Mg anode}]{\text{NiBr}_2\text{(bpy)}} PhCH_2COCH_2Ph $$

$$ C_6H_{13}Br + CO_2 \xrightarrow[\text{Mg anode}]{\text{NiBr}_2\text{(bpy)}} C_6H_{13}COC_6H_{13} $$

$$ (21) $$

$$ Ni(II) \xrightarrow{+e^-} Ni(0)(bpy) + CO_2 \longrightarrow Ni(bpy)CO \xrightarrow{RX} RNi(CO)(bpy)X $$
$$ \text{unstable} $$

$$ \xrightarrow{-Ni(II)} R_2Ni(CO)(bpy) \longrightarrow RCONi(bpy)R \xrightarrow{-Ni(0)(bpy)} RCOR \qquad (22) $$

b. Conjugated halides (allylic, benzylic, α-haloacids and esters and phenacyl halides).
(i) Benzylic halides. The effect of conjugation of the carbon–halogen bond with any other unsaturated functional group is to render the reduction potential more positive. An example of the conjugative effect can be seen in the reduction of α-bromoacetophenone, which undergoes 2e$^-$ reduction at –0.05 V (sce) to form acetophenone. Conjugation of the carbon-halogen bond with carbon–carbon double bonds or arene rings exerts a similar stabilizing effect. Benzyl bromide is reduced at –1.22 V (sce), about 1.0 V more positive than an isolated carbon-bromine bond. The presence of ring substituents in benzylic halides, regardless of their nature and orientation, renders reduction more facile.

Substituent effects have been reported previously and work on this topic continues. The halfwave potentials of a series of substituted benzyl chlorides and bromides gave excellent correlations with Hammett σ-substituent constants[48]. The positive ρ values from these Hammett LFERs ($\rho = 5.0$ and 2.8, respectively for chlorides and bromides) suggest that the potential-determining electrochemical process involves the formation of radical anion intermediates.

An interesting correlation has been found between the potential for the dissociative reduction of substituted 9-chlorofluorenes **22** and **23** and the redox potential of the ensuing radicals ($r = 0.88$) in DMSO at a glassy carbon electrode. The reduction potentials provide a link between the couple R·/R⁻ and the basicity of the electrogenerated carbanion

TABLE 2. Product yields (percent) from electroreduction of 1-phenylethyl bromide at different potentials

$-V$ (sce)	25	26 dl	26 meso	27	28
0.9	17	39.7	40.3	0.4	2.4
1.4	41	28.2	28.3	0.7	3.3
1.7	100				

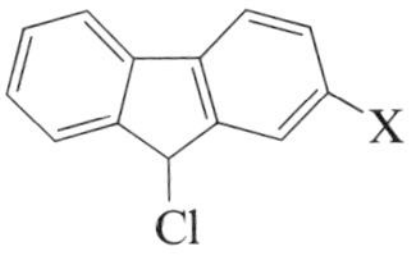

X=F,Br,CN

(22)

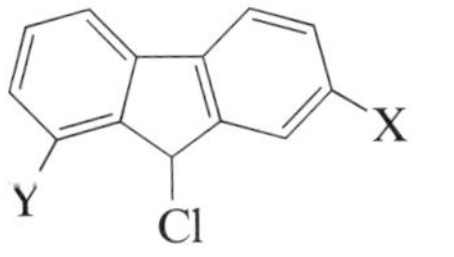

X=F,Br,CN   Y=Br,CN

(23)

within a family of related substrates[49]. The results are consistent with a concerted electron transfer bond-breaking mechanism for reduction. To the extent that this LFER is general, the observation provides a method for the evaluation of the relative basicity of a related set of benzylic halides.

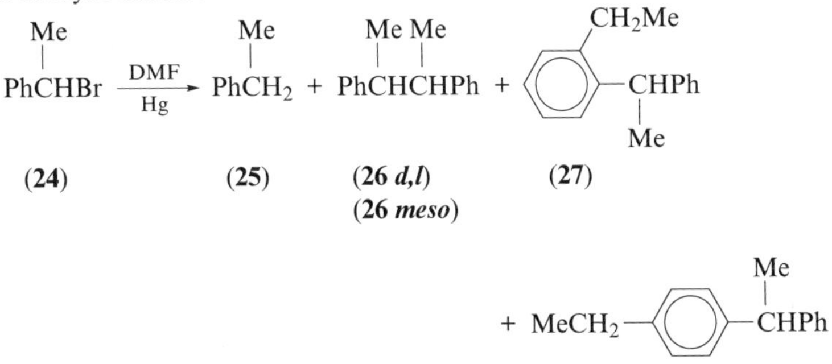

$$
\underset{(24)}{PhCHBr} \xrightarrow[Hg]{DMF} \underset{(25)}{PhCH_2} + \underset{\substack{(26\ d,l) \\ (26\ meso)}}{PhCHCHPh} + \underset{(27)}{}
$$

$$
+ \; MeCH_2 \text{—} \bigcirc \text{—} CHPh
$$

(28)   (23)

The products and stereochemistry of the electroreduction of 1-bromo-1-phenylethane (24) and ($\pm$)-1-deuterio-1-bromo-1-phenylethane (24-D) have been investigated using DMF at mercury and glassy carbon electrodes (equation 23)[50]. The products, obtained in nearly quantitative yield, were strongly potential dependent as shown in Table 2.

At low potential the products were those of radical coupling or decomposition of thermally unstable organomercurials to yield coupling products 26, 27, 28 in a non-stereospecific process. At more negative potential the products were those of two-electron reduction to the carbanion followed by protonation. When the reaction was conducted at higher potential using optically active 1-bromo-1-(deuterioethyl)benzene (24-D), the resulting hydrocarbon 25-D was racemic(equation 24). Using lithium perchlorate as sup-

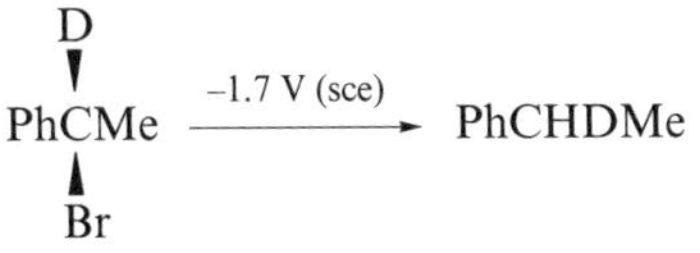

$$
\underset{(24\text{-}D)}{PhCMe} \xrightarrow{-1.7\ V\ (sce)} \underset{(25\text{-}D)}{PhCHDMe}
$$

(24)

porting electrolyte in aprotic solvents, a small amount of retention (11–15%) was observed in either acetonitrile or DMF. However, in protic solvents such as 2-methyl-2 propanol significant (13%) inversion was observed. These results suggest backside protonation of a carbanion at or near the electrode surface. The small net retention of optical activity, depending on the proton–donating ability of solvent, suggests that a carbanion is involved. When the reduction is conducted at low temperature and low potential [–1.4 V (sce)] the main product is a mixture of organomercurials that decomposes upon warming to give metallic mercury and the same product distribution as had been observed directly from the electroreduction.

Fry and Powers[51] examined the electroreduction of $PhCHBrCMe_3$ (29) on carbon electrodes and observed a similar product dependence on the electrolysis potential (equation 25). They observed an interesting increase in the **meso/dl** ratio of head-to-head coupled

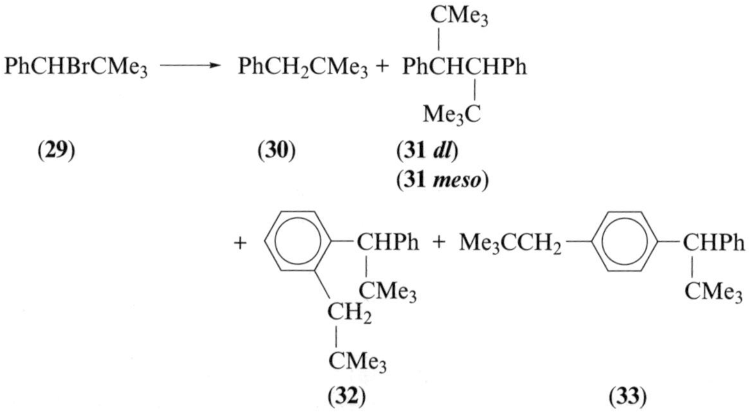

$$PhCHBrCMe_3 \longrightarrow PhCH_2CMe_3 + \underset{\underset{Me_3C}{|}}{\overset{\overset{CMe_3}{|}}{PhCHCHPh}}$$

(29)   (30)   (31 *dl*)
                                                   (31 *meso*)

(25)

(32)   (33)

products with increasing reduction potential. The temptation to speculate that **31dl** forms more easily from radical coupling and **31meso** forms best from an S_N2 process should be tempered by the fact that this is a highly hindered neopentyl system unlikely to exhibit significant S_N2 reactivity (equation 26).

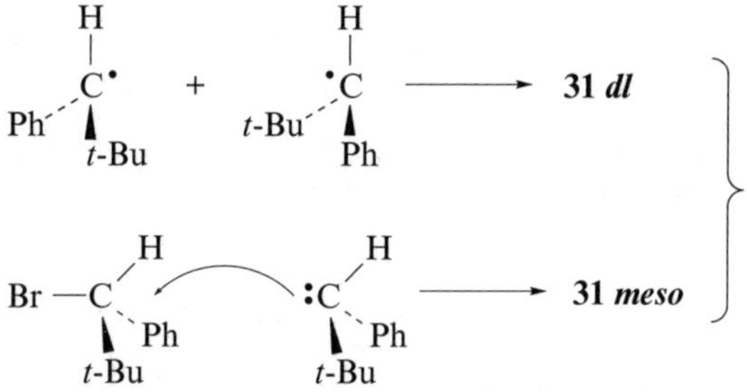

(26)

Benzyl chloride and its substituted derivatives are electrochemically reduced indirectly through the mediator, 1,4-dihydro-4-methoxycarbonyl-1-methylpyridine anion 35, (equation 27). The rates of the electron transfer between the mediators and benzyl halides, measured by cyclic voltammetry, were found to be about 961 ± 80 M^{-1} s^{-1}. The rate-determining step was proposed to involve a single electron transfer from the mediator to

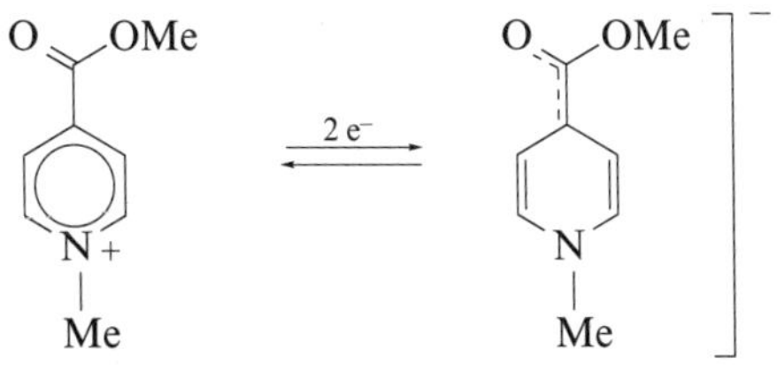

$$\text{(34)} \quad \xrightleftharpoons{2\,e^-} \quad \text{(35)} \tag{27}$$

(34)   (35)

$$\text{PhCH}_2\text{Cl} \xrightarrow{\text{(35)}} \text{PhCH}_2^- + \mathbf{34} \tag{28}$$

the aryl halides[57]. 2,2′-Dinitrodibenzyl (37) has been synthesized from the electrochemical reduction of 2-nitrobenzyl bromide (36), a reaction (equation 29) which involves one-electron reduction of 2-nitrobenzyl halide, expulsion of the halide anion and the dimerization of the resulting benzyl radicals[53]. Mechanistic and preparative efforts were part of this study.

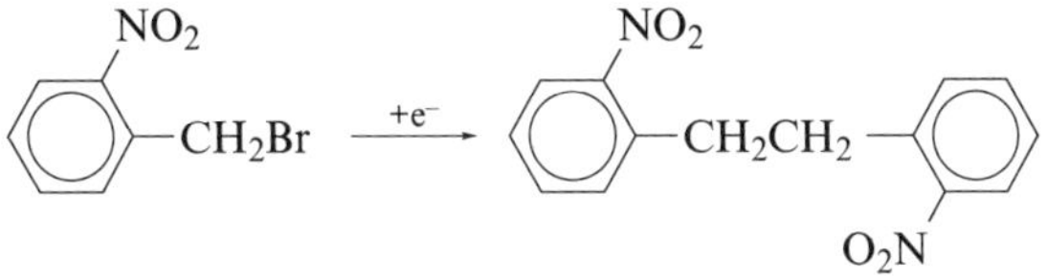

$$\tag{29}$$

(36)   (37)

(ii) Allylic halides. Allylic halides behave in a manner similar to benzyl halides toward electroreduction. Alkyl bromides, benzyl or allyl halides and α,β-unsaturated esters were transformed to their corresponding dimeric hydrocarbons in the presence of sacrificial Mg or Zn anodes using a divided cell[54]. Electrochemical reduction of allyl halides in the presence of α,β-unsaturated esters results in the conjugate addition of the allyl groups to the latter esters[35]. The electrolysis of allyl halides 38 in the presence of diethyl fumarate (39), for example, gives ethyl 3-(ethoxycarbonyl)-5-hexenoate (40), as shown by equation 30.

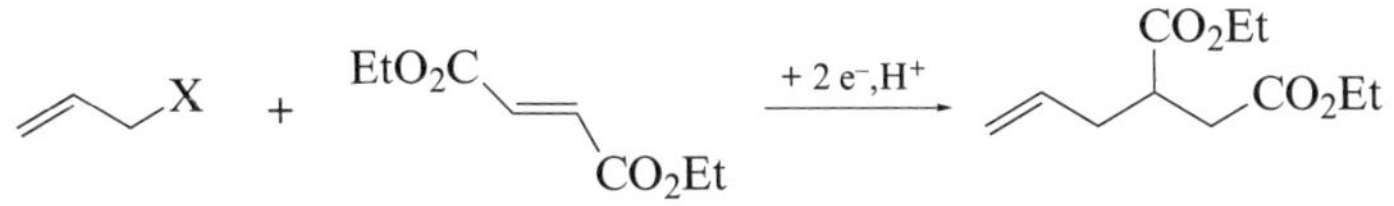

$$\xrightarrow{+\,2\,e^-,H^+} \tag{30}$$

X=Cl,Br,I

(38)   (39)   (40)

The C—X reduction potentials of allylic halides and unsaturated α-haloesters are similar, and vary depending on their substitution pattern. The regioselectivity of the addition, therefore, is also influenced by the relative reduction potentials. 1-Chloro-3-methyl-2-butene (41) reacts with diethyl fumarate (42) through attack from the primary carbon (equation 31), whereas with methyl crotonate (44) allylation takes place at the more highly substituted tertiary carbon (equation 32).

The reduction potential of 1-chloro-3-methyl-2-butene (41) is more positive than that of methyl crotonate which results in the initial 2-electron reduction of the allyl halide to give the allyl ion. This species subsequently undergoes Michael reaction through the tertiary carbon atom. On the other hand, the reduction potential of diethyl fumarate is less cathodic than that of **41**, and the anion radical formed from the diester reacts with **42** in a S_N2-like mechanism. These results are reminiscent of the Baizer study[34].

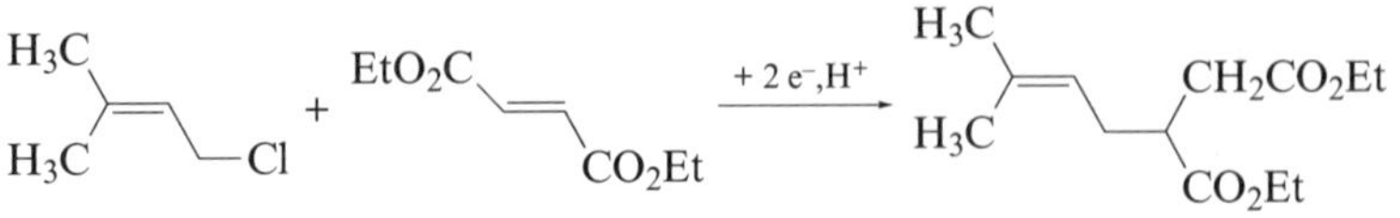

$$
\underset{(41)}{\overset{H_3C}{\underset{H_3C}{\bigvee}}\text{—Cl}} + \underset{(42)}{\overset{EtO_2C}{\bigvee}\text{CO}_2\text{Et}} \xrightarrow{+\,2\,e^-,H^+} \underset{(43)}{\overset{H_3C}{\underset{H_3C}{\bigvee}}\bigvee\overset{CH_2CO_2Et}{\underset{CO_2Et}{}}} \tag{31}
$$

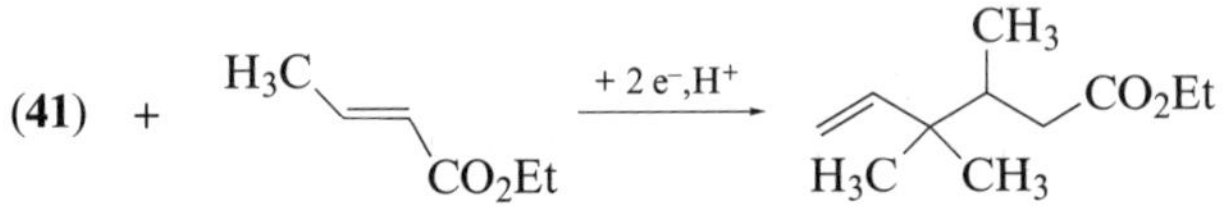

$$
(41) \; + \; \underset{(44)}{\overset{H_3C}{\bigvee}\text{CO}_2\text{Et}} \xrightarrow{+\,2\,e^-,H^+} \underset{(45)}{\overset{CH_3}{\underset{H_3C\;\;CH_3}{\bigvee\bigvee}\text{CO}_2\text{Et}}} \tag{32}
$$

Allyl halides (**46**) electrolyzed in the presence of carbonyl compounds such as acetone or benzaldehyde give the corresponding allylated products as seen in equation 33. The regiochemistry is dependent on the relative reduction potentials of the substrates. When the reduction potential of the carbonyl compound (e.g. acetone) is more negative than the allyl halides, allylation takes place at the more substituted terminus of (**46**). More easily reducible carbonyl substrates (eg. benzaldehyde) proceed through reaction at the less substituted center (**47**). The formation of **48** could be explained by assuming the formation of allylic carbanion involving a two electron reduction of the allyl halide. The resonance-stabilized allyl anion adds to the carbonyl group preferentially from the more substituted end. The formation of **47**, on the other hand, involves S_N2-type of displacement of the allyl halide by the initially formed carbonyl radical anion. (Eqn. 33).

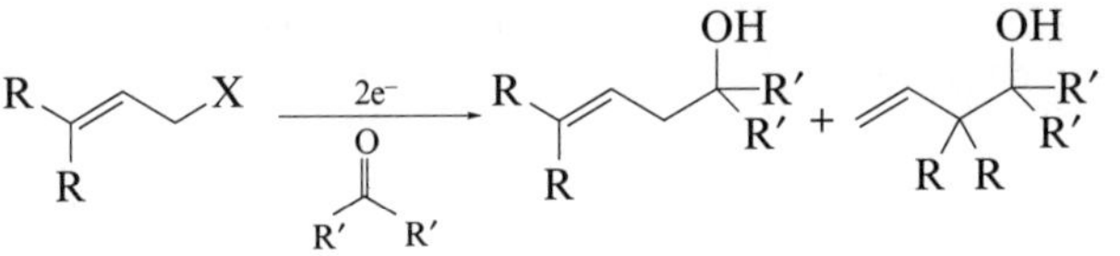

$$
\underset{(46)}{\overset{R}{\underset{R}{\bigvee}}\text{—X}} \xrightarrow[\underset{R'\;\;R'}{\overset{O}{\|}}]{2e^-} \underset{(47)}{\overset{R}{\underset{R}{\bigvee}}\bigvee\overset{OH}{\underset{R'}{\overset{|}{-}}}R'} + \underset{(48)}{\overset{OH}{\underset{R\;\;R}{\bigvee}\overset{|}{-}}R'} \tag{33}
$$

The regioselectivity was also dependent on the nature of the cathode material. Pt, Zn, Ni, or Al cathodes give allylation of acetone, in which the reaction occurs at the more highly substituted carbon of the allyl group whereas, using a mercury cathode, the reaction takes place at both the termini of the allylic groups[55]. Homoallylic alcohols could also be prepared in moderate to good yields from allyl chlorides and ketones, through a SmCl₃-catalyzed indirect electroreductive allylation of ketones (equation 34). This appears to be a useful reaction. The electrosyntheses are carried out in undivided cells using a sacrificial

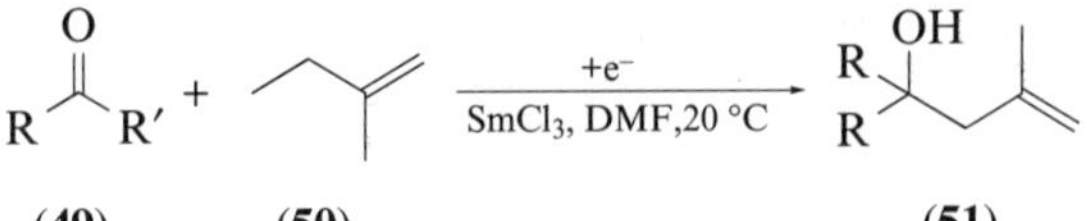

$$
\underset{(49)}{\overset{O}{\underset{R}{\overset{\|}{\bigwedge}}R'}} + \underset{(50)}{\bigvee} \xrightarrow[\text{SmCl}_3,\,\text{DMF},20\,°C]{+e^-} \underset{(51)}{\overset{OH}{\underset{R}{\overset{|}{\bigvee}}}\bigvee} \tag{34}
$$

magnesium anode[56]. The reaction involves electrochemical generation of Sm(II) species from SmCl$_3$, followed by its reaction with allyl chlorides and ketones to give homoallylic alcohols in good yields. The method is advantageous over chemical methods using stoichiometric amounts of SmI$_2$, which is air sensitive. A similar allylation of carbonyl compounds by electroreduction of allyl halides was also achieved using a sacrificial anode, such as Mg, Al, Zn or Fe and stainless steel cathodes (equation 35)[57].

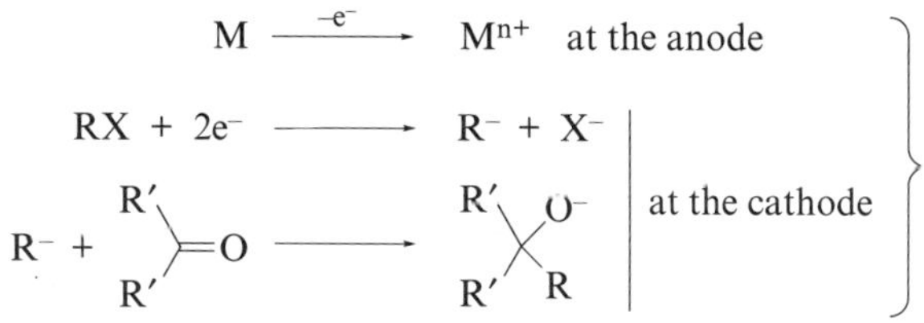

$$M \xrightarrow{-e^-} M^{n+} \quad \text{at the anode}$$

$$RX + 2e^- \longrightarrow R^- + X^-$$

$$R^- + \begin{matrix} R' \\ \diagdown \\ R' \end{matrix}{=}O \longrightarrow \begin{matrix} R' \\ \diagdown \\ R' \diagup \end{matrix}{\underset{R}{\overset{O^-}{\diagup}}} \quad \text{at the cathode}$$

$$(35)$$

Benzyl, allyl and vinyl halides were electrochemically transformed to their corresponding silanes (52–55) in the presence of trialkylsilyl chlorides as seen in equation 36.

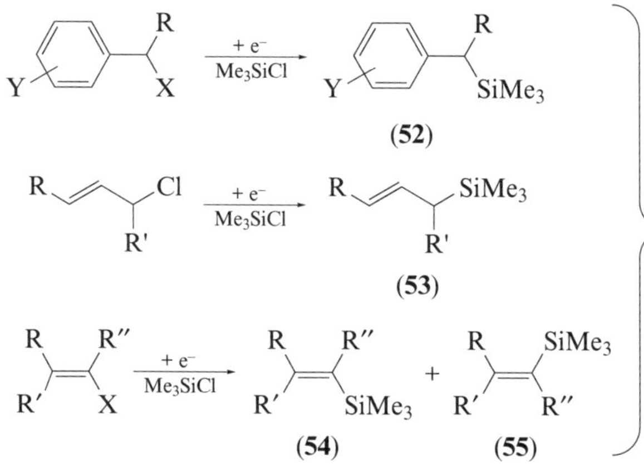

(52)

(53)

(54)   (55)

$$(36)$$

Constant-current electrolyses were done using platinum electrodes[58]. Stereochemical integrity was lost in the case of vinyl halides, presumably at the radical stage.

Electrochemical coupling of allyl halides through the agency of Ni[0] complexes has been achieved under very mild conditions. The Ni[0] species 57, generated electrochemically, undergoes oxidative addition by allyl halides giving π-bonded organometallic Ni[I] derivatives 58, the metal–carbon bond of which is cleaved at less negative potentials giving the coupled products, such as 1,5-hexadienes 59 (equation 37) or its derivatives[59]. The advan-

$$[Ni(II)(PPh_3)_2(CH_3CN)_4]^{2+} \xrightarrow[-4\,CH_3CN]{+\,2e^- + 2PPh_3} Ni(0)(PPh_3)_4$$

(56)   (57)

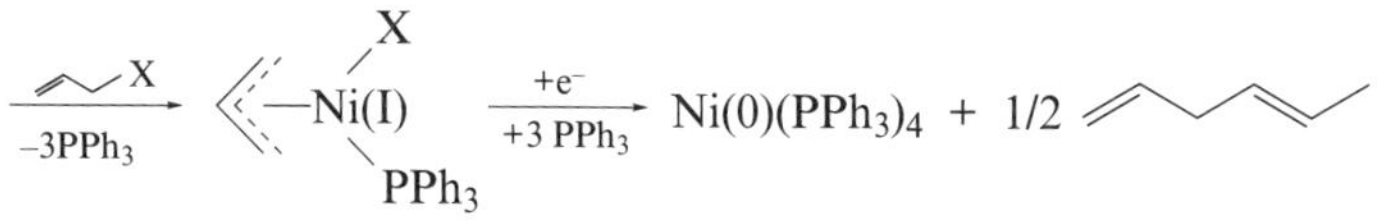

(58)   (59)

$$(37)$$

tage of the Ni[0] species lies mainly in lowering the E_a for the electroreduction of allyl halides. In the absence of the Ni[0] catalyst, the direct reduction can be achieved only at more cathodic potentials, where the primary one-electron reduction-derived radical is further reduced to the anion leading to the hydrocarbon rather than the coupled product. Reduction of allyl halides to 1,5-hexadiene has also been achieved with bipyridyl derivatives of cobalt in anionic and cationic micelles, which showed enhancement of reaction rates[60].

(iii) α-Halocarboxylic acids, esters, and phenacyl halides. The electroreduction of di- and monochloroacetic acid has been reported in another study[64]. The cathodic reduction of phenacyl halides **60** has been shown to depend on the substrate concentration under high dilution conditions. A carbene is postulated to be generated at a constant potential of –0.6 V (sce), which trimerizes to tribenzoylcyclopropane (**61**)[61,62] (equation 38).

$$PhCOCH_2Br \xrightarrow{e^-} PhCOCH: \longrightarrow$$

$$(60) \qquad\qquad\qquad\qquad (61) \qquad (38)$$

c. Acyl halides. The electrochemical reduction of phenylacetyl chloride (**62**) and hydrocinnamoyl chloride at a mercury cathode in acetonitrile with tetraalkylammonium perchlorates as supporting electrolyte is reported. Preparative cathodic reduction of the first wave led to a complex mixture of products. Phenylacetyl chloride yields 1,4-diphenyl-2-butene-2,3-diolbisdiphenylacetate (**63**), phenylacetaldehyde (**64**), 1,3-diphenylacetone (**65**) and toluene (**66**) (equation 39). Phenylacetic anhydride and phenylacetic acid are chemically formed byproducts. The direct products are best rationalized as resulting from radical combination (**63**), radical combination after decarbonylation (**65**), and hydrogen atom abstraction before (**64**) and after (**66**) decarbonylation as illustrated in equation 40.

$$PhCH_2CCl \longrightarrow PhCH_2C = CCH_2Ph + PhCH_2CHO$$

$$(62) \qquad\qquad (63) \qquad\qquad (64)$$

$$+ PhCH_2COCH_2Ph + PhCH_3 \quad (39)$$

$$(65) \qquad\qquad (66)$$

$$PhCH_2\overset{\bullet}{C}{=}O \longrightarrow CO + PhCH_2^{\bullet} \quad (40)$$

Electroreduction of hydrocinnamoyl chloride gave a similar array of products, but decarbonylation products were less prevalent[63]. Electrochemical reduction of heptanoyl chloride and benzoyl chloride at mercury and carbon electrodes in acetonitrile containing tetraalkylammonium salts gives the corresponding aldehydes. Using deuteriated acetonitrile and D_2O it could be shown that heptanal arises through heptanoyl radical, whereas both radical and anion species are involved in the reduction of benzoyl chloride (equation 41)[64]. The main product (17–59%) of hydrogen atom transfer from acetonitrile was the aldehyde (equation 42). A major side product, heptanoic anhydride, appears to form from partial hydrolysis of the starting material and subsequent reaction of the carboxylic acid with unreacted acid chloride. Under similar conditions benzoyl chloride shows quite different results, forming very low yields of benzaldehyde and a nearly quantitative yield

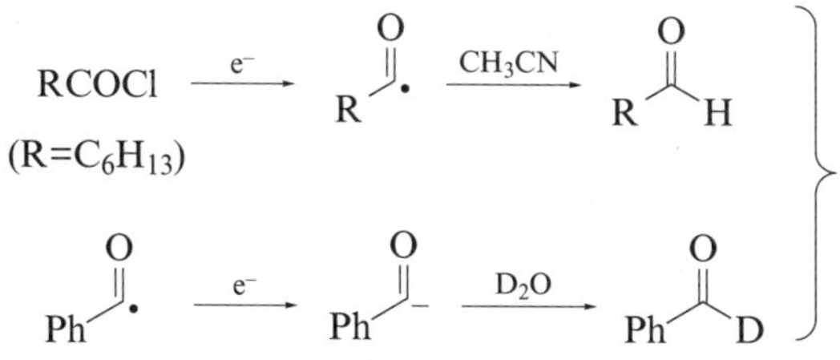

$$\text{RCOCl} \xrightarrow{e^-} \underset{R}{\overset{O}{\underset{\bullet}{\|}}} \xrightarrow{CH_3CN} \underset{R}{\overset{O}{\|}}\!\!H$$

(R=C$_6$H$_{13}$)

$$\underset{Ph}{\overset{O}{\underset{\bullet}{\|}}} \xrightarrow{e^-} \underset{Ph}{\overset{O}{\|}}^{-} \xrightarrow{D_2O} \underset{Ph}{\overset{O}{\|}}\!\!D$$

(41)

$$\text{Me(CH}_2)_5\text{COCl} \xrightarrow[\text{MeCN, Hg electrode}]{+\,e^-} \text{Me(CH}_2)_5\text{CHO}$$

(42)

of stilbenediol dibenzoate (**67**) (equation 43), presumably from the coupling of benzoyl radicals followed by reduction and nucleophilic reaction with benzoyl chloride. This is

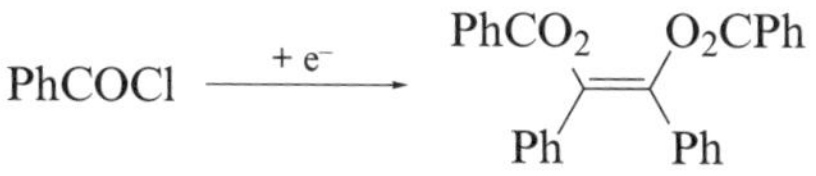

$$\text{PhCOCl} \xrightarrow{+\,e^-} \underset{\underset{Ph}{} \quad \underset{Ph}{}}{\overset{PhCO_2 \quad O_2CPh}{\diagup\!\!=\!\!\diagdown}}$$

(43)

(**67**)

shown in Scheme 4. This result may reflect the considerable difference in stability of the first-formed radicals. It is noteworthy that the formation of either aldehyde or diol diester requires one Faraday mole^{-1}.

Electroreduction of cyclohexanecarbonyl chloride (**68**) at a mercury cathode in acetonitrile leads to production of cyclohexanecarboxaldehyde (**69**) and 1,2-dicyclohexylethene-1,2-diol dicyclohexane carboxylate (**70**) as reduction products, reminiscent of phenylacetyl chloride (equation 39). The isolation of small amounts of dicyclohexyl ketone (**71**) and dicyclohexylethanedione (**72**) strongly suggests that cyclohexanecarboxaldehyde is formed exclusively via an acyl radical intermediate[65](equation 44).

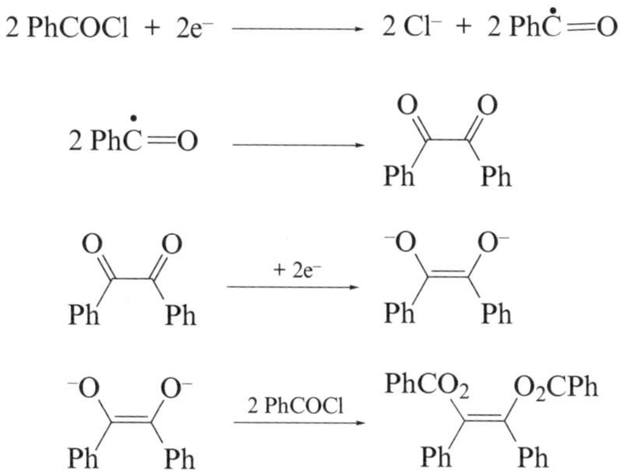

$$2\,\text{PhCOCl} + 2e^- \longrightarrow 2\,\text{Cl}^- + 2\,\text{Ph}\overset{\bullet}{\text{C}}\!\!=\!\!O$$

$$2\,\text{Ph}\overset{\bullet}{\text{C}}\!\!=\!\!O \longrightarrow \underset{Ph \quad Ph}{\overset{O \quad O}{\diagup\!\!\diagdown\!\!\diagup}}$$

$$\underset{Ph \quad Ph}{\overset{O \quad O}{\diagup\!\!\diagdown\!\!\diagup}} \xrightarrow{+\,2e^-} \underset{Ph \quad Ph}{\overset{^-O \quad O^-}{\diagup\!\!=\!\!\diagdown}}$$

$$\underset{Ph \quad Ph}{\overset{^-O \quad O^-}{\diagup\!\!=\!\!\diagdown}} \xrightarrow{2\,\text{PhCOCl}} \underset{Ph \quad Ph}{\overset{PhCO_2 \quad O_2CPh}{\diagup\!\!=\!\!\diagdown}}$$

(**67**)

SCHEME 4

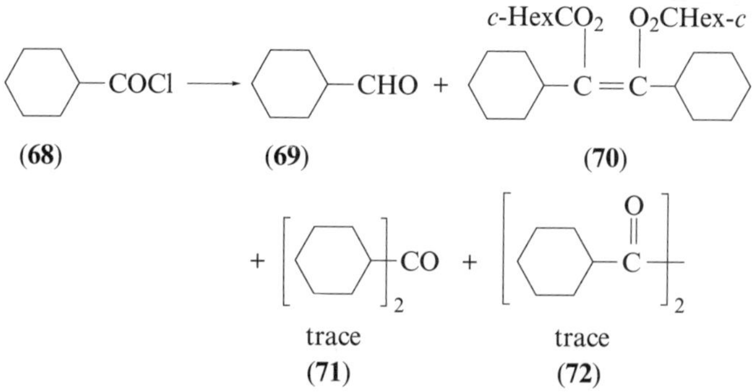

$$(44)$$

Electrochemical reduction of phthaloyl dichloride (73) at a carbon or mercury cathode in acetonitrile containing TEAP led to a complex array of products. Six cathodic waves observed in the CV for the reduction of phthaloyl dichloride arise from the reductions of different electrolysis products, as well as from hydrolytically formed phthalic anhydride (74),. caused by the presence of residual water in the solvent/supporting electrolyte (equation 45). From controlled potential electrolyses of phthaloyl dichloride, a variety of products including 3-chlorophthalide (75), phthalide (76), biphthalyl (77) and dihydrobiphthalide (78) can be obtained[69,70]. Reduction of glutaryl dichloride (79) at a mercury cathode in acetonitrile containing 0.1M TEAP results in the formation of 5-chlorovalerolactone (80) and valerolactone (81) as minor products, and a polymeric material (equation 46)[68].

The cathodic reduction of aroyl chlorides as seen earlier (equation 46) gives olefinic diethers involving reduction to aroyl radicals, combination of the radicals to give

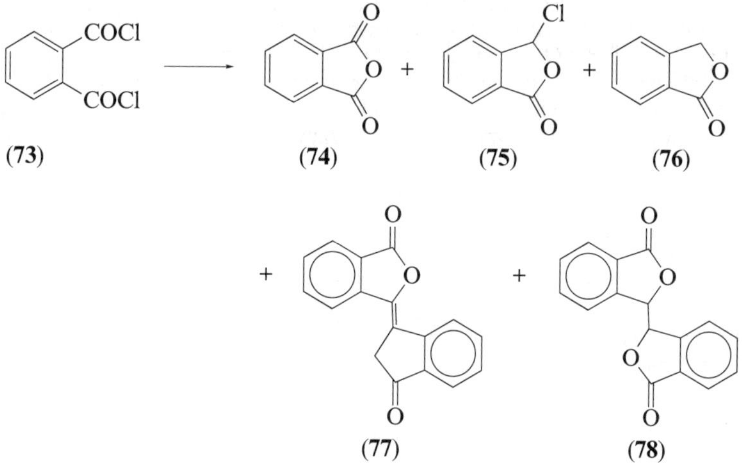

$$(45)$$

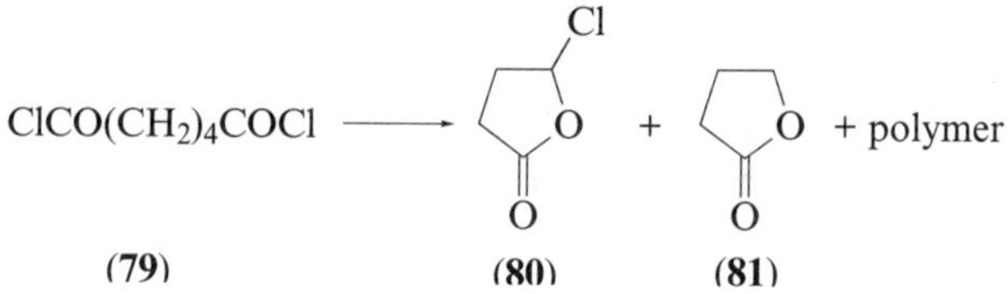

$$(46)$$

diketones and reduction of the diketones to form a dianion, followed by acylation as seen in equation 47[69]. When the aromatic ring is substituted by electron-withdrawing substituents, such as the nitro group at the *ortho* and *para* positions, the reduction potential becomes less negative than with the unsubstituted compound (equation 48). [–0.6 V (sce) for *p*-nitrobenzoyl chloride, –1.3 V (sce) for benzoyl chloride]. The initial reduction, however, involves the nitro group, and the reaction takes a different pathway. Thus, *p*-nitrobenzoyl chloride (**82**) gives *p*-nitrobenzoic 4-(*p*-nitrobenzoylamino)benzoic anhydride (**83**), formed through a series of anionic intermediates[70].

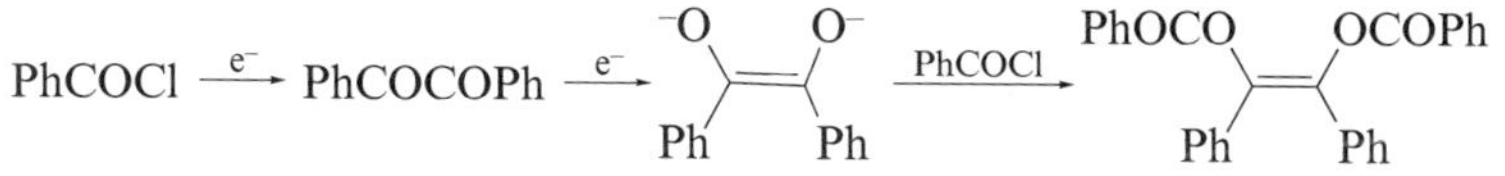

$$\text{PhCOCl} \xrightarrow{e^-} \text{PhCOCOPh} \xrightarrow{e^-} \begin{array}{c} {}^-O \quad O^- \\ \diagdown \quad / \\ Ph \quad Ph \end{array} \xrightarrow{\text{PhCOCl}} \begin{array}{c} \text{PhOCO} \quad \text{OCOPh} \\ \diagdown \quad / \\ Ph \quad Ph \end{array} \tag{47}$$

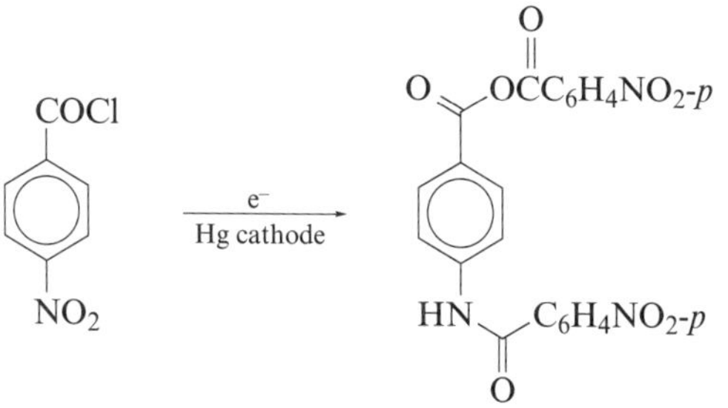

$$\tag{48}$$

(**82**)   (**83**)

An interesting and novel method for the preparation of isocyanides (**85**) has been reported and represents a new method for the preparation of these compounds in excellent yields. Thus with a variety of alkyl and aryl carbonimidoyl dichlorides (**84**) cathodic reduction in the range –1.25 V to –1.90 V (sce) provides a new, mild and reagent-free method for the preparation of isocyanides in 82–96% yields, as determined by gas chromatography (equation 49)[71].

$$\text{RN}{=}\text{CCl}_2 \xrightarrow{2\,e^-} \text{RN}{\equiv}\text{C} + 2\,\text{Cl}^- \tag{49}$$
$$\text{R=Ph, Bz, }c\text{-Hex}$$
$$(\textbf{84}) \qquad\qquad (\textbf{85})$$

Electrochemical synthesis of ketones by cross-coupling could be achieved with acid chlorides and alkyl or aryl halides using catalysis of 2,2'-bipyridinenickel complexes (equation 50) and a Zn rod as the sacrificial anode[72]. NiBr$_2$(bipy) was the best catalyst for these reactions. Symmetrical ketones were also prepared from organic halides and carbon dioxide by this method[42]. The electrogenerated Ni(0)(bpy)(CO) oxidatively adds to the alkyl halides, and a subsequent carbonyl insertion gives RCONi(bpy)R, which undergoes reductive elimination to give symmetrical ketones.

$$\text{PhCH}_2\text{X} + \text{PhCH}_2\text{COCl} \xrightarrow{+2e^-} (\text{PhCH}_2)_2\text{CO} + \text{Cl}^- + \text{X}^- \tag{50}$$

d. Aryl halides. Aryl halides have been reduced by a number of investigators, aryl iodides being the most studied ones. Mechanistic studies of aryl halides are discussed in Section III.B.5 as they have been central to a definitive mechanistic investigation. Electroreduction of aryl halides has proved to be a useful method for site-specific labelling of the aryl ring with deuterium. Using LiClO$_4$ at mercury cathode, iodoanisoles and iodotoluenes could be labelled in good yield at the site of the halogen atom in D$_2$O and unlabelled acetonitrile[73]. This result suggests proton rather that hydrogen atom transfer,

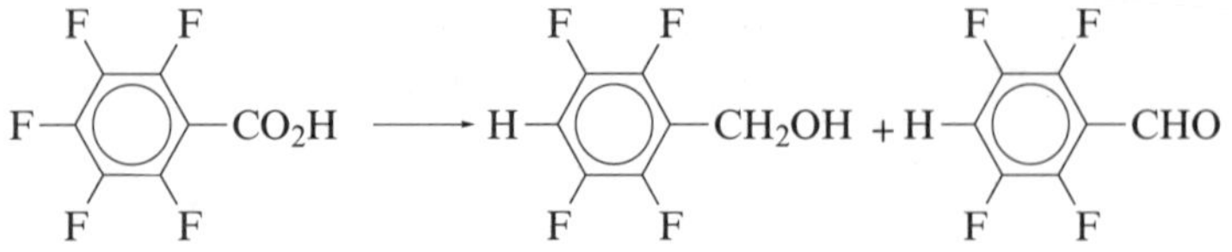

J. Casanova and V. P. Reddy

(51)

(86)   (87)   (88)

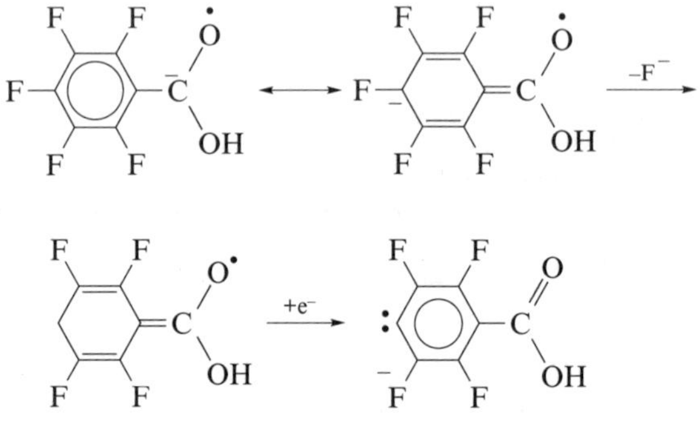

SCHEME 5

implicating an aryl carbanion. Reductive cleavage of aromatic carbon–halogen bonds in the presence of D_2O was also reported by Grimshaw[74].

2,3,4,5,6-Pentafluorobenzoic acid reduced selectively to either 2,3,5,6-tetrafluorobenzyl alcohol (87) or 2,3,5,6-tetrafluorobenzaldehyde (88) at a lead cathode in dilute sulfuric acid in the presence of quaternary ammonium salts[75] (equation 51). This is an interesting reaction that could involve loss of fluoride ion from an intermediate radical anion followed by further reduction (Scheme 5).

Electroreduction of 1-(4-fluorophenyl)-5-(2-halogenophenyl)tetrazoles (89), when the halogen substituent is Cl, Br or I, leads to cleavage of the carbon–halogen bond to give a phenyl radical (equation 52). Competition then follows between intramolecular radical substitution giving 7-fluorotetrazolo[1,5-f]phenanthridine (90). Further reduction of the

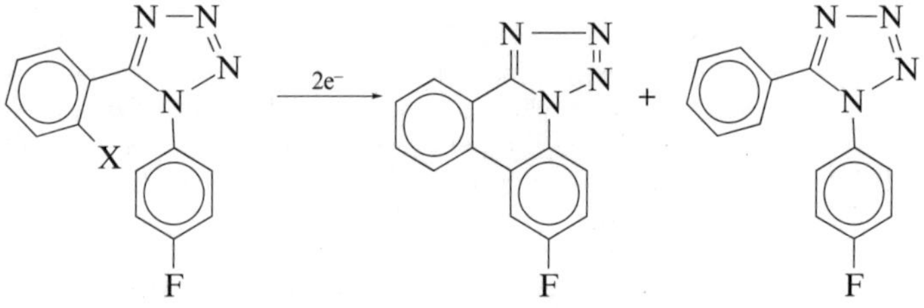

(52)

(89)   (90)   (91)

radical, then protonation, to give 1-(4-fluorophenyl)-5-phenyltetrazole (91). Substitution predominates, but reduction and protonation become more competing reactions when the halogen is Br or I. Carbon–halogen bond cleavage becomes the only reaction when the halogen is iodide[76]. Other aromatic radicals obtained by the electroreduction of aryl halides can also undergo intramolecular cyclization with a suitably oriented neighboring phenyl group[77], as can be seen in equation 53.

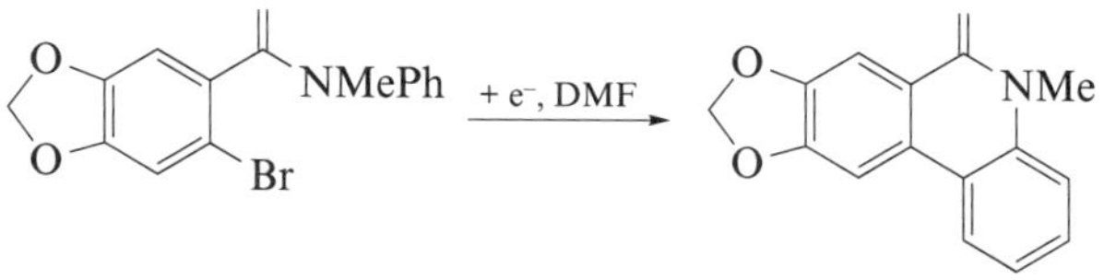

$$\text{(92)} \quad \xrightarrow{+\,e^-,\,\text{DMF}} \quad \text{(93)} \tag{53}$$

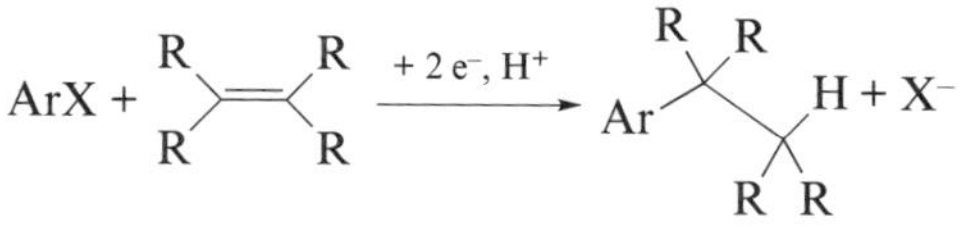

$$\text{ArX} + \underset{\text{(94)}}{\overset{R}{\underset{R}{\diagdown}}\overset{R}{\overset{}{\diagup}}\overset{R}{\underset{R}{}}} \xrightarrow{+\,2\,e^-,\,H^+} \underset{\text{(95)}}{Ar\overset{R\;\;R}{\underset{R\;\;R}{\diagdown}}H} + X^- \tag{54}$$

Electrochemical reduction of aryl halides in the presence of olefins (94), (equation 54) leads to the formation of arylated products (95)[78]. Electroreduction of several aralkyl halides at potentials ranging from -1.24 V to -1.54 V (sce) gives products which involve dimerization, cyclization, and reduction to the arylalkanes. Carbanions and/or free radicals were again postulated as intermediates[79]. Aryl radicals generated from the electrochemical reduction of aryl halides have been added to carbon–carbon double bonds[80,81]. Electrochemical reduction of aryl halides in the presence of olefins leads to the formation of arylated products[78]. Preparative scale electrolyses were carried out in solvents such as acetonitrile, DMF and DMSO at constant potential or in liquid ammonia at constant current. The reaction is proposed to involve an $S_{RN}1$ mechanism.

(i) Reductive carboxylations. Controlled potential electrolyses of p-iodoanisole (96) at a mercury pool cathode in the presence of carbon dioxide gave p-anisic acid (97), anisole (98) and bis(p-anisyl)mercury (99)[87]. Cyclic voltammograms for the reduction of p-iodoanisole at mercury exhibit two waves. p-Anisyl radicals appear to be formed in the first reduction step and are the source or the organomercuric product (Scheme 6). Further reduction to the anion provides the precursor to p-anisic acid. The anisole is a secondary product resulting from further reduction of the p-anisic acid. Reduction of aryl halides and electroformylation reactions have been studied in DMF on a cadmium modified gold microelectrode. Electrochemical studies and electrolyses of various aryl halides show that the formylation process implies reaction of the aryl anion with DMF[83].

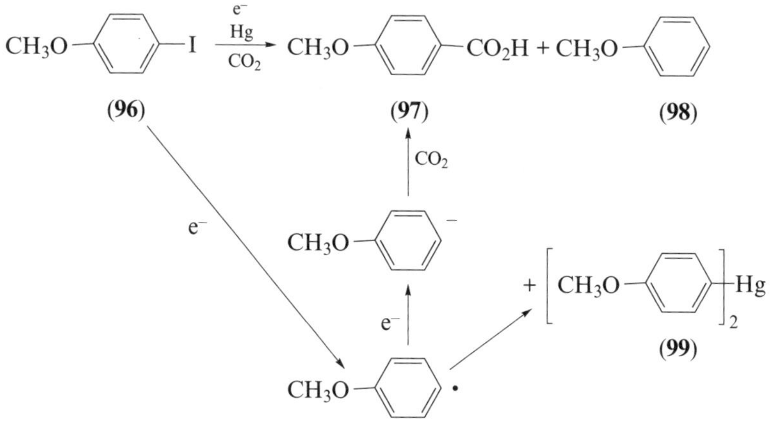

SCHEME 6

$$R-X + e^- \longrightarrow R-X^{\bar{\cdot}} \longrightarrow R^{\cdot} + X^-$$

$$R^{\cdot} + Nu^- \longrightarrow R-Nu^{\bar{\cdot}}$$

$$R-Nu^{\bar{\cdot}} + R-X \longrightarrow R-Nu + R-X^{\bar{\cdot}}$$

$$R-X^{\bar{\cdot}} \longrightarrow R^{\cdot} + X^-$$

SCHEME 7

(ii) Indirect reduction. 1,4-Diacetylbenzene dianion (**100**) generated cathodically acts as a reducing species toward aromatic halides to produce aryl radicals. It also reacts as an electrogenerated base with the starting point considered as a pronucleophile (a source of enolate). The concomitant presence in solution of aromatic radicals and electrogenerated nucleophiles imply $S_{RN}1$ type of reactions (Scheme 7)[84]. Redox catalysis can also be accomplished using thianthrene-9,9,10,10-tetraoxide dianion (**101**) for reaction with aromatic halides[85]. Whereas direct electrochemical reduction of 1,2-dibromobenzene (**102**) goes through transient bromobenzene radical **103** and benzyne species in very fast steps, the photochemical electron transfer involving anthraquinone anion radical (Aq⁻) anion results mainly in the formation of bromobenzene radical, which can be identified as an adduct with anthracene (Scheme 8)[86].

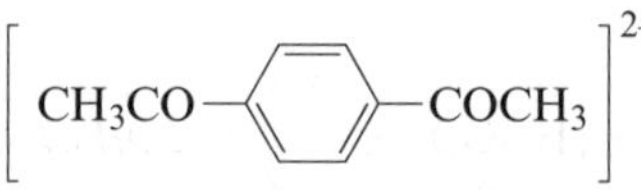

(100) 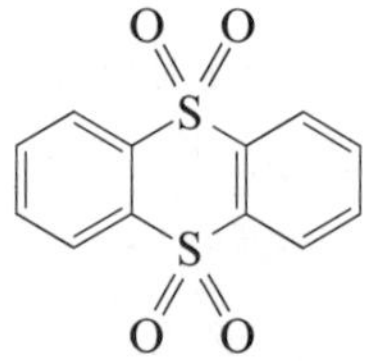   **(101)**

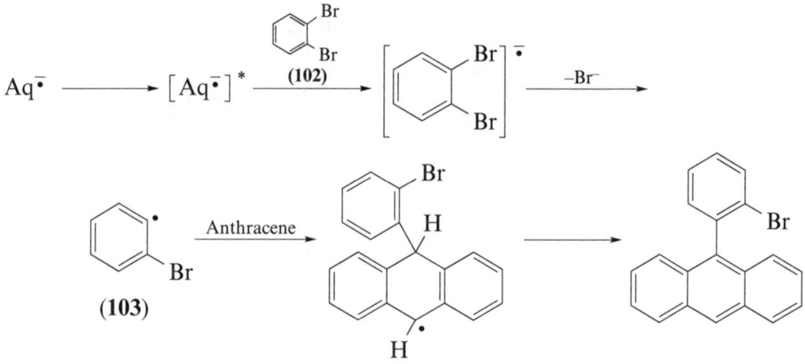

SCHEME 8

Electroreduction of triphenylphosphino- or 1,2-bisdiphenylphosphinoethanenickel(II) complexes in ethanol via zero valent complexes with halobenzenes has been described[87]. Arylation of olefins (equation 55) can be achieved by electrogenerated Ni(0) complex **106**, associated with triphenylphosphine and an alkene (**107**). The optimum conditions for the reaction include the use of one to three equivalents of triphenylphosphine and a base such as triethylamine[30].

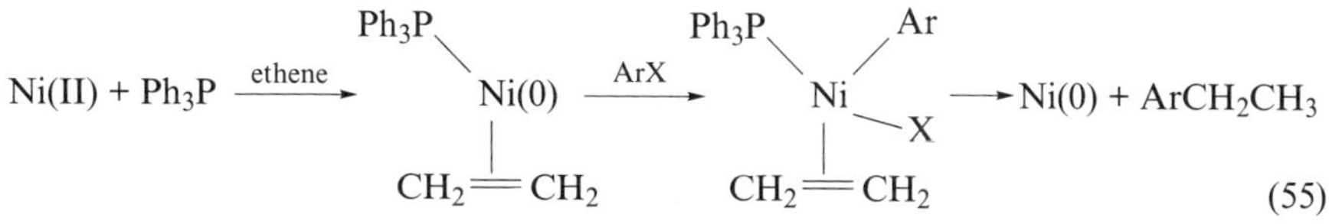

$$Ni(II) + Ph_3P \xrightarrow{\text{ethene}} \underset{\substack{| \\ CH_2 = CH_2 \\ \textbf{(106)}}}{\overset{Ph_3P}{Ni(0)}} \xrightarrow{ArX} \underset{\substack{| \\ CH_2 = CH_2 \\ \textbf{(107)}}}{\overset{Ph_3P\diagup \diagdown Ar}{Ni \diagdown X}} \longrightarrow Ni(0) + ArCH_2CH_3 \tag{55}$$

A number of biaryl and polynuclear aromatic halides have been electroreduced. Some of these reactions are discussed in a later section (III.B.10) in connection with the practical problem of detoxification of hazardous chemicals.

Fourier-transform faradaic admittance measurements (FT-FAM) on monohalobiphenyls indicate that the reduction to the biphenyls involves a competition between heterogeneous and homogeneous electron transfers, and hydrogen atom abstraction. When acetonitrile was used as the solvent, the competition favors the hydrogen atom abstraction; for example, the cleavage of the initially formed halobiphenyl radical anion proceeds with a rate constant of $1.5 \times 10^4 \, s^{-1}$[88]. Activation parameters for the decomposition of the anion radical of RX(R = 9-cyano-10-bromoanthracene, 9,10-dichloroanthracene and 9-chloroanthracene) were shown to be temperature dependent, indicating change of mechanism by increasing the temperature[84].

(iii) Reductive coupling. Electroreductive cross-coupling of haloaromatics or of vinyl halides (**110**) with α-chloroesters (**108**), using a nickel catalyst and a sacrificial aluminum anode in a one-compartment cell, gives the coupled products in moderate yield (equations 56 and 57)[90]. The sacrificial aluminum anode is essential to the success of this reaction. Aluminum(III) species produced in the anodic reaction may mediate the coupling process. The electrochemically generated {1,2-bis(di-2-propylphosphino benzene)nickel(0) (**113**) [from Ni(II) species **112**] was effective for the coupling of aryl or vinyl chlorides as seen in equation 58. Cyclization products **115–116** were obtained as the major products with

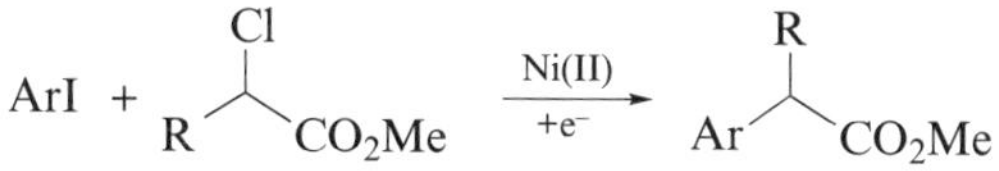

$$ArI + \underset{\substack{R \quad CO_2Me \\ \textbf{(108)}}}{\overset{Cl}{\bigwedge}} \xrightarrow[\substack{+e^-}]{Ni(II)} \underset{\substack{Ar \quad CO_2Me \\ \textbf{(109)}}}{\overset{R}{\bigwedge}} \tag{56}$$

(57)

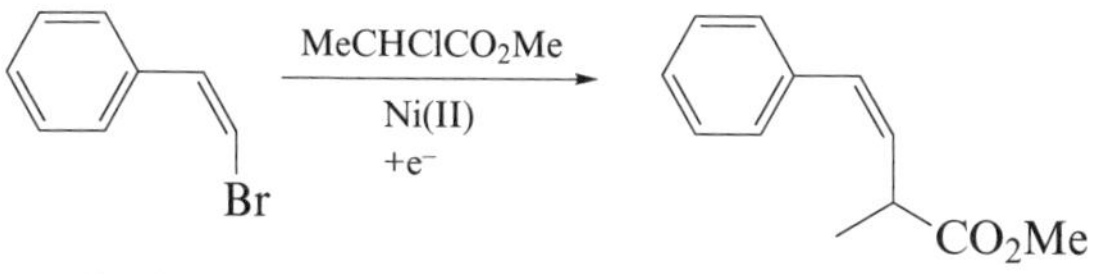

(110)   (111)

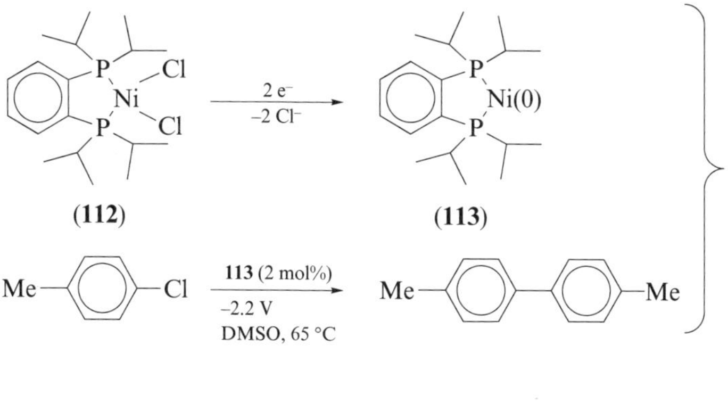

(58)

(112)   (113)

$$Me{-}\hexagon{-}Cl \xrightarrow[\substack{-2.2\,V \\ DMSO,\,65\,°C}]{113\,(2\,mol\%)} Me{-}\hexagon{-}\hexagon{-}Me$$

appropriately functionalized aryl (**114**) or vinyl halides using 1,2-bis(di-2-propylphosphino)benzene nickel(0) (equation 59)[91].

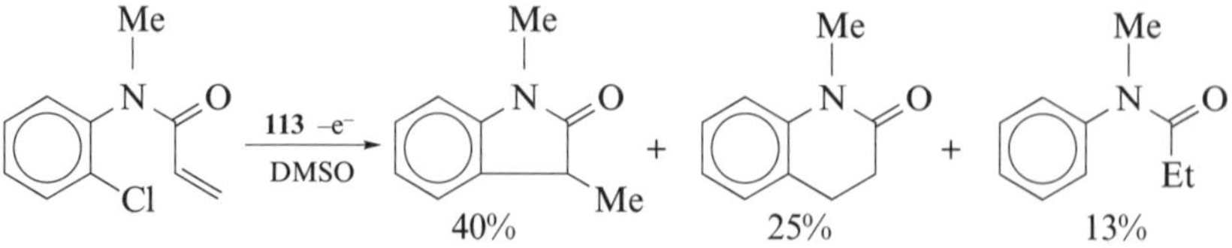

$$(59)$$

Whereas direct electroreduction of aromatic halides gives dehalogenated aromatics, in the presence of nickel catalysts biaryls are formed (equation 60)[92]. The formation of biaryls can be suppressed in the presence of CO_2 in favor of the formation of aromatic carboxylic acids (equation 61). 3',5'-Di-*tert*-butyl-4-chloro-4'-hydroxy-1,1'-biphenyl (**118**) is obtained selectively by an electroinduced $S_{RN}1$ reaction, starting from 1,4-dichlorobenzene and 2,6-di-tert-butylphenoxide as shown in equation 62. The reaction was induced by benzonitrile as a mediator, which is also a cosolvent[93]. In a mixed THF/HMPA solvent system, electrochemical reduction of NiX_2L_2 (X = Cl, L = PPh_3)gives Ni(0) complex, which oxidatively is added to ArX (X = I, Br, Cl), giving $ArNiXL_2$. Further electroreduction, oxidative addition of ArX and reductive elimination of Ar—Ar provides a catalytic method for coupling aryl halides[94].

$$2ArX + 2e^- \xrightarrow{\text{Ni(II)(dppe)Cl}_2} Ar—Ar \qquad (60)$$

$$\xrightarrow{CO_2} ArCO_2H \qquad (61)$$

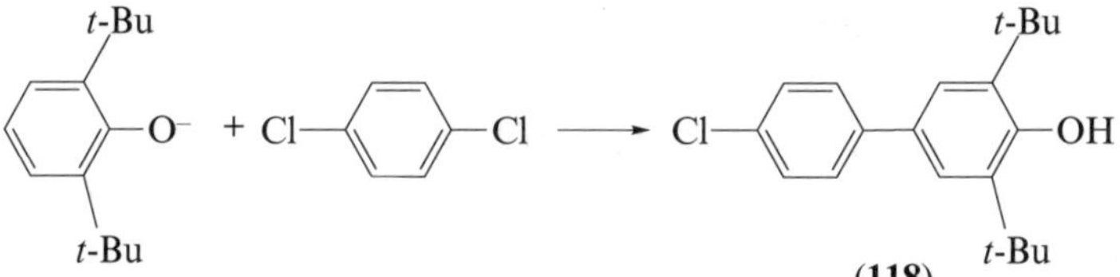

$$(62)$$

Trimethylsilyl chloride has been employed in some synthetically useful reductions of aryl halides. The electrochemical reductive trimethylsilylation of aryl chlorides is a good

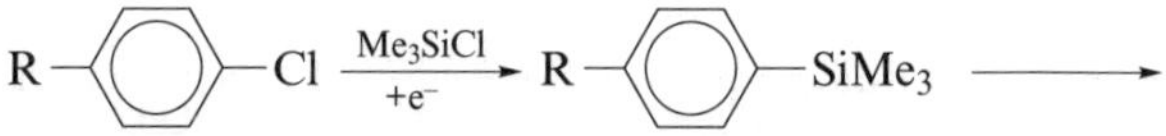

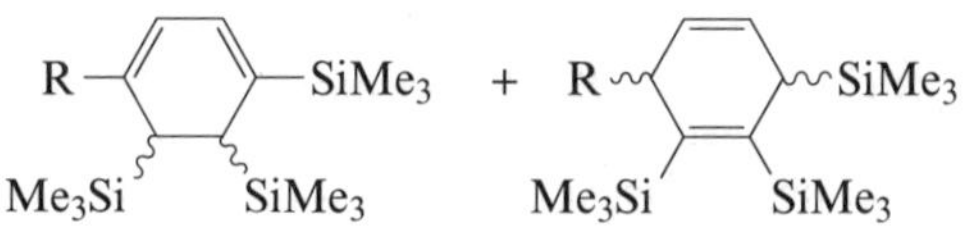

$$(63)$$

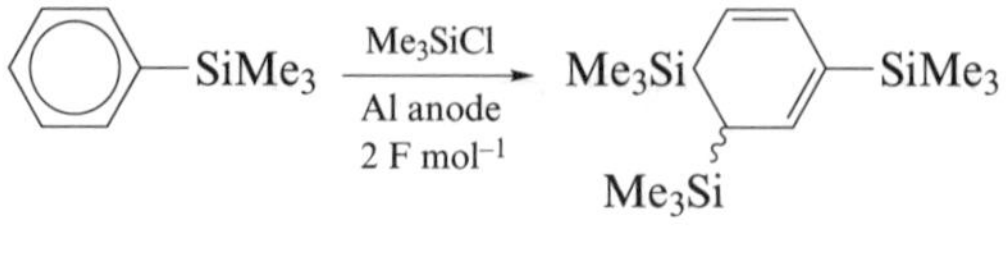

route to aryltrimethylsilanes and tris(trimethylsilyl)cyclohexadienes as seen in equation 63. The electroreductive trimethylsilylation of aryl chlorides RC_6H_4Cl **119** (R = H, o-Me, m-Me, p-Me) can be controlled to give either the corresponding aryltrimethylsilanes (**120**) or mixtures of *cis*- and *trans*-tris(trimethylsilyl)cyclohexa-1,3 (**121**) (or -1,4) (**122**) dienes, depending on how much current is passed during constant-current electrolysis. The reaction is carried out in an undivided cell equipped with a sacrificial aluminum anode in 80:20 THF/HMPA solution containing TEAB as the supporting electrolyte, and excess Me_3SiCl. The electroreductive trimethylsilylation of phenyltrimethylsilane gives a 62% yield of a stereoisomeric mixture of three 2,5,6-tris(trimethylsilyl)-cyclohexa-1,3-dienes, in which the *trans* isomer predominates. Such products cannot be obtained by the chemoreductive trimethylsilylation of phenyltrimethylsilane[95]. The electroreduction of aromatic esters and amides in dry acetonitrile in the presence of chlorotrimethylsilane affords aldehydes in moderate to good yields depending on the aromatic substituents[96].

2. Geminal polyhalides

The presence of more than one halogen atom lowers the potential at which reduction of the first carbon–halogen bond occurs. As early as 1939 it was reported that electroreduction of trichloromethyl derivatives led to a 50% yield of the dichloromethyl compounds.

The electroreduction of 1,1,1-trichloro-2-hydroxy-4-methyl-4-pentene (**123**) at a lead cathode led to two products, 1,1-dichloro-4-methyl-2,4-pentadiene (**124**) and 1,1-dichloro-2-hydroxy-4-methyl-4-pentene (**125**) (equation 64). Compound **124** was favored (yields > 75%) in acidic medium using methanol or 2-propanol as solvent[97].

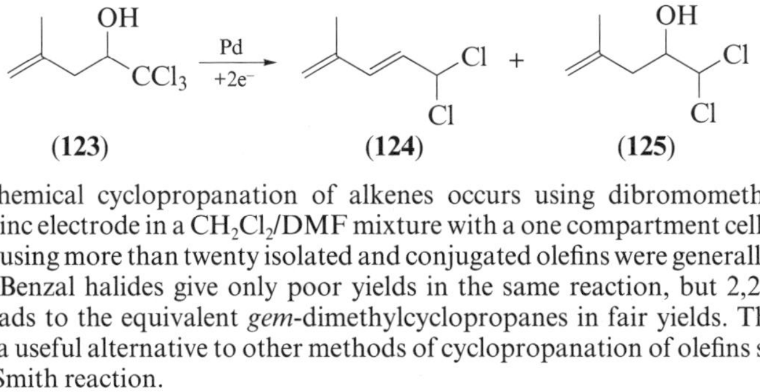

$$\text{(64)}$$

(**123**)   (**124**)   (**125**)

Electrochemical cyclopropanation of alkenes occurs using dibromomethanes at a sacrificial zinc electrode in a CH_2Cl_2/DMF mixture with a one compartment cell (equation 65). Yields using more than twenty isolated and conjugated olefins were generally good (30 to 70%)[98]. Benzal halides give only poor yields in the same reaction, but 2,2-dibromopropane leads to the equivalent *gem*-dimethylcyclopropanes in fair yields. The method represents a useful alternative to other methods of cyclopropanation of olefins such as the Simmons-Smith reaction.

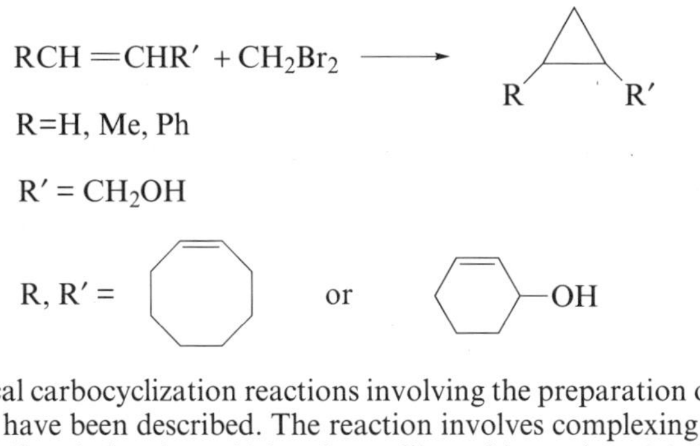

$$RCH = CHR' + CH_2Br_2 \longrightarrow$$

R=H, Me, Ph

$$R' = CH_2OH$$

$$\text{(65)}$$

$$R, R' =$$ or

Electrochemical carbocyclization reactions involving the preparation of 3-, 4-, 5-, or 6-membered rings have been described. The reaction involves complexing of olefinic compounds, such as dimethyl maleate **126** and α,ω-dibromide, such as 1,3-dibromopropane **127** in an undivided cell fitted with a sacrificial aluminum anode, in N-methylpyrrolidone at constant current (equation 66)[99]. The reaction is of special interest for the preparation of 5-membered ring compounds such as **128**, since these are less easily formed by conventional means than are 3- or 6-membered rings.

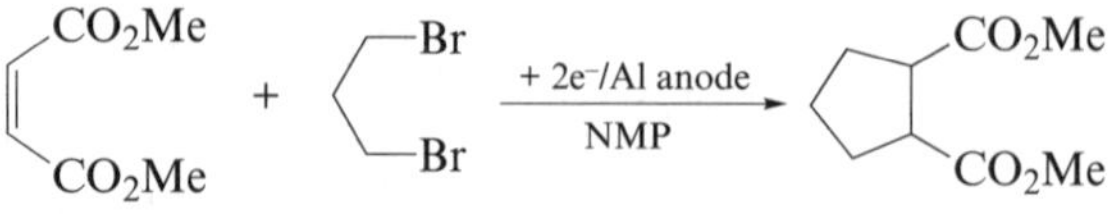

$$(66)$$

(126)　　　　**(127)**　　　　　　　**(128)**

The stereochemical parallel between electrochemical reductions at mercury and metal (zinc or lithium amalgam) reductions hold for a series of geminal dihalocyclopropanes (equation 67). For bromochloro substrates, reduction occurs with the exclusive removal of bromine and retention of configuration. Where the halogen atoms are the same, slight preference for *endo*-monohalide suggests preferential reduction from the *exo* side to produce a conformationally stable carbanion pictured in equation 68 and shown in Table 3^{100}. The electroreduction of 1,1-dihalo-2R-2-methylcyclopropanes (**129**) gave a mixture of stereoisomeric monobromo- and monochlorocyclopropanes (**130**) in yields of 60–70% for preparative electroreduction in methanol/LiClO$_4$(equation 69)[101].

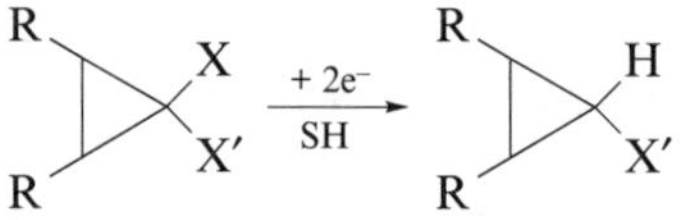

$$(67)$$

TABLE 3. Effect of structure, solvent and supporting electrolyte on *exo/endo* product ratio of [n.1.0]bicyclic geminal halides

Compound	Solvent	Electrolyte	*endo/exo* product ratio
	EtOH	TEAB	6:1
	DMF	TEAB	3.1:1
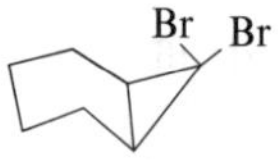	DMF	LiCl	1.2:1
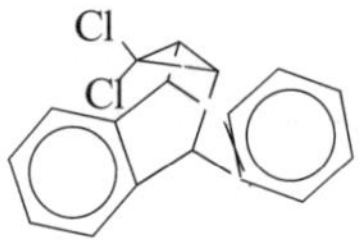	EtOH	TEAB	1:3.2
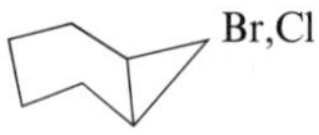	EtOH	LiCl	0.97:1
	EtOH	TEAB	3.5:1
	EtOH	LiCl	2.6:1
		TEAB	
	EtOH	LiCl	1:2.6

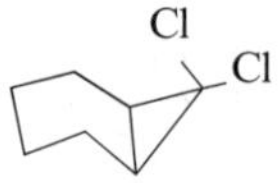

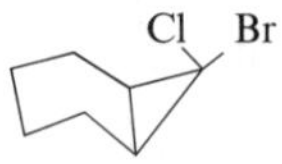

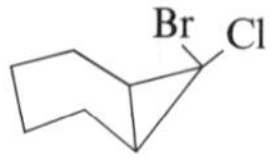

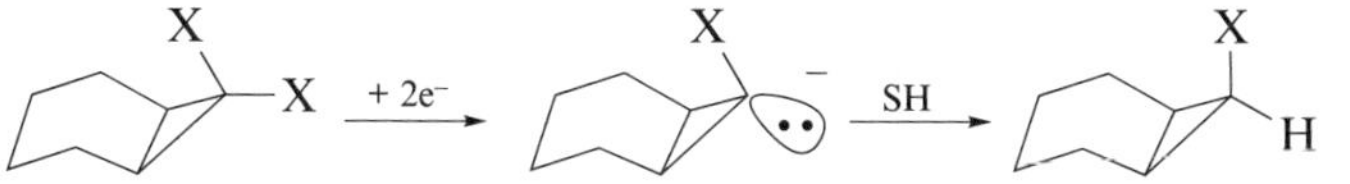

(68)

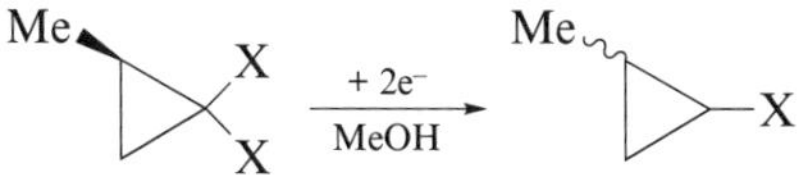

(69)

X = Cl, Br

(129)   (130)

Organic halides having large negative cathodic potentials such as 7,7-dichloronorcarane $[E_{1/2} = -2.9$ V (Ag/AgCl)] were reduced using soluble electrodes such as Al, Sn, and Fe, whose effectiveness decreases in that order[102].

(70)

(131)

Electroreduction of benzal chloride in the presence of carbon dioxide in DMF solvent gives very low yield of phenylacetic acid (131) (equation 70). The study suggests that the monoanion adds CO_2 to yield intermediate α-chlorophenylacetate which forms α-lactone. The acid is obtained following hydrolysis of the α-lactone by adventitious moisture in the DMF as shown in Scheme 9[103].

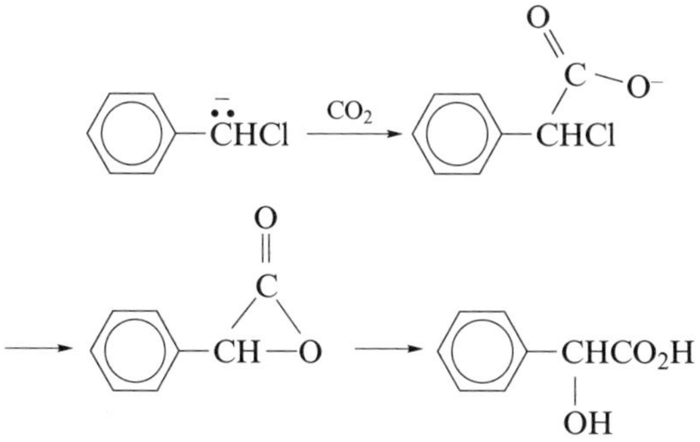

SCHEME 9

Recently, Fry and coworkers extended the above method to reduce selectively benzal chlorides in the presence of trimethylsilyl chloride to form either (α-chlorobenzyl)-trimethylsilanes (132) or benzal (geminal) disilanes (133) in good yields (equation 71). The reduction potentials of benzal chloride and the derived mono-α-chlorosilanes are sufficiently different (−2.23 V and −2.79 V, respectively vs Ag/0.1M $AgNO_3$) so that electrolysis can be interrupted to obtain α-chlorosilane before the second reduction takes place. Reactions were carried out in an undivided cell containing a stainless steel cathode and a sacrificial magnesium anode[104]. Other electrochemical and sonochemical reductive silylations of geminal dihalides have also been reported[105].

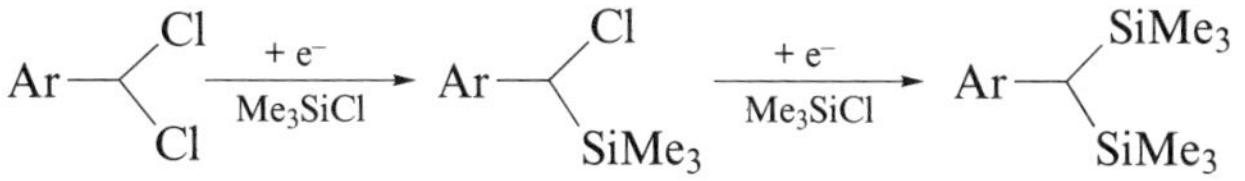

(71)

(132)   (133)

Benzal chloride is converted to a mixture consisting primarily of cis- and trans-stilbene (136) by the action of electrochemically generated cobalt(I)(salen)$^{-1}$(134). The process involves a sequence involving electrocatalytic conversion of benzal chloride to a mixture of the stereoisomeric 1,2-dichloro-1,2-diphenylethanes (135), followed by electrocatalyzed conversion of the latter to cis- and trans-stilbene. The mechanism shown in Scheme 10 was proposed. Voltammetric data and other chemical evidence have been interpreted in light of a previously suggested mechanism for stilbene formation[106].

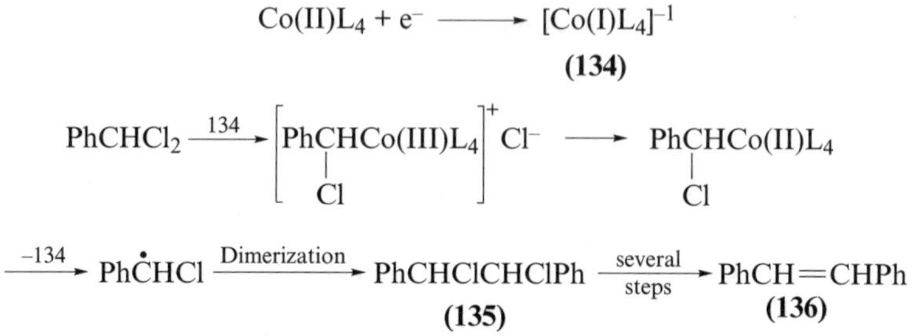

$$Co(II)L_4 + e^- \longrightarrow [Co(I)L_4]^{-1}$$
$$(134)$$

$$PhCHCl_2 \xrightarrow{134} \left[\underset{Cl}{PhCHCo(III)L_4} \right]^+ Cl^- \longrightarrow \underset{Cl}{PhCHCo(II)L_4}$$

$$\xrightarrow{-134} Ph\overset{\bullet}{C}HCl \xrightarrow{Dimerization} PhCHClCHClPh \xrightarrow[\text{steps}]{\text{several}} PhCH{=}CHPh$$
$$(135) \qquad\qquad\qquad (136)$$

SCHEME 10

Electrochemical reduction of diethyltrichloromethyl phosphonate gives the diethyl-dichloromethyl phosphonate anion (137), which could be protonated in protic medium[107]. Using aprotic media, and in the presence of carbonyl compounds, it undergoes Wittig–Horner reaction in moderate yields (50%) to give the corresponding 1,1-dichloroethylenes (138), resulting from the nucleophilic addition, followed by elimination of diethylphosphonate anion as seen in equation 72[108]. The anion could also be reacted with several alkyl halides in an S_N2 reaction (equation 73). It was found that the electrochemically generated carbanion shows more nucleophilic character than conventional lithium salts[109].

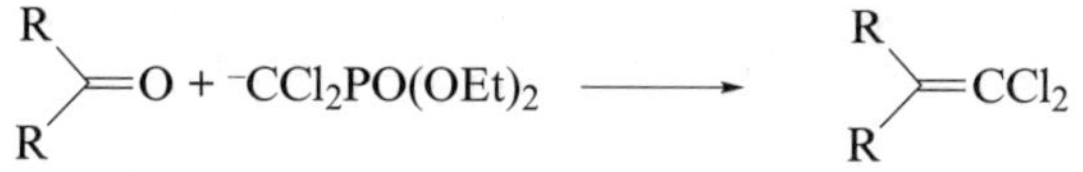

$$\underset{R}{\overset{R}{>}}{=}O + {}^-CCl_2PO(OEt)_2 \longrightarrow \underset{R}{\overset{R}{>}}{=}CCl_2 \qquad (72)$$
$$(137) \qquad\qquad\qquad (138)$$

$$^-CCl_2PO(OEt)_2 + RX \longrightarrow RCCl_2PO(OEt)_2 \qquad (73)$$
$$R = 1^0, 2^0 \qquad\qquad (139)$$
$$X = Cl, Br$$

The reduction of dichloroacetic acid in aqueous solutions in the presence of trace amounts of dissolved lead ions at a carbon electrode appears to inhibit the hydrogen evolution by decreasing the overpotential[110].

a. Polyfluoro substrates. In a polarographic study in which no products were isolated, Inesi and Rampazzo studied the electroreduction of mono-, di- and trifluoroacetic acid and their ethyl esters in DMF at a mercury cathode[111]. The monofluoroester did not reduce while the di- and trifluoro esters gave one-electron reduction at $E_{1/2} = -2.56$ and -2.36 V (sce), respectively. The acids behaved in a complex manner, suggesting involvement of the ionized carboxylate. Electroreduction of the α-fluoro bond gives protonated products or the esters. The relationship between $E_{1/2}$ and the number of α-fluorine atoms is linear. Ethyl perfluoropropionate is reduced at a mercury cathode. Usual reduction of esters (butyrate, caprylate and glutarate) involves loss of α-fluorine and replacement by hydrogen[112].

Controlled potential preparative electrolysis at glassy carbon cathodes results in the reduction of perfluoroalkyl iodides and bromides (equation 74). The reaction in the presence of phenolate ions in aprotic media gives products of perfluoroalkylation of phenol, and also the corresponding monohydroperfluoroalkanes[113].

$$R_fI \xrightarrow[\text{ArOH or ArSH}]{+ e^-} R_fH + I^- \tag{74}$$

3. Dihalides

Nongeminal dihalides behave differently depending on the number of saturated bonds separating the halogen atoms. They are discussed here separately according to that classification. When the dihalide is in a bicyclic system the classification can be ambiguous. Such compounds are classified according to the number of saturated bonds in the shortest bridge connecting the halides.

a. 1,2-Dihalides. Electrochemical reduction of vicinal dihalides is a facile and quantitative route to the parent olefin although of limited practical utility. Von Stackelberg and Stracke established early that electroreduction of 1,2-dihalides afforded olefinic hydrocarbons in a two-electron process[15]. Zavada and coworkers[114] found a marked dependence of the polarographic half-wave potential on the dihedral angle between the C—Br bonds. For a series of 21 1,2-bromides studied, $E_{1/2}$ varied between –0.82 V and –1.67 V (sce). Those dibromides for which the *anti*-periplanar arrangement of C—Br bonds were conformationally favored or fixed by a rigid structure were most easily reduced. Klein and Evans[115] presented an elegant example of the effect of conformation on reduction potential using *trans*-1,2-dibromocyclohexane, affording a method for the conformational analysis of this compound by low-temperature cyclic voltammetry (equation 75). The reaction is highly irreversible. Low-temperature fast-scan cyclic voltammetry of *trans*-1,2-dibromocyclohexane in butyronitrile with TBAP as supporting electrolyte reveals the onset of a rapid conformational equilibrium as the temperature is allowed to rise from –90 °C. The value of $\Delta G^{\circ \cdot}$ can be calculated from K and k using simulated cv scan and is shown in Table 4 for $k_{ee} \rightarrow k_{aa}$ and the reverse reaction. Savéant and coworkers examined the same system[116].

$$ \tag{75}$$

diequatorial (ee)   diaxial (aa)

The definitive study on this topic was reported by Savéant and coworkers[117]. They note a dichotomy between behavior of reagents believed to be involved in outer-sphere (redox) and inner-sphere (chemical) electron transfer processes. Outer-sphere reagents such as

TABLE 4. Kinetic and thermodynamic data for equilibration of diaxial and diequatorial 1,2-dibromocyclohexanes

Temperature (°C)	K	$k_{ee} \rightarrow k_{aa}$	ΔG° (kcal mol^{-1})
–90	0.55	0.53	10.8
–80	0.55	2.3	10.8
–70	0.55	5.2	11.1
–60	0.55	29.0	11.0
–55	0.52	—	—

carbon electrodes and aromatic radical anions react through a concerted electron transfer/bond-breaking process to produce a β-haloalkyl radical which is reduced rapidly in a second concerted electron transfer/bond-breaking process (equation 76). By contrast, inner-sphere reagents such as the mercury electrode, iron(I), iron (0) and cobalt(I) porphyrins react more rapidly and with complete stereospecificity, suggesting an E2 type concerted unimolecular elimination mechanism involving halonium ion transfer (equation 77). The behavior of more than 16 vicinal dibromides was investigated. Both polarogra-

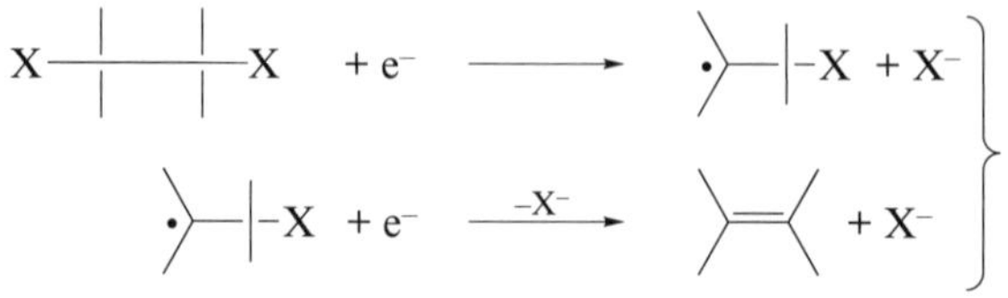

$$
\left.
\begin{array}{l}
X{-}{|}{-}{|}{-}X \ + \ e^- \longrightarrow \ \rangle{\cdot}{-}{|}{-}X \ + \ X^- \\[12pt]
\rangle{\cdot}{-}{|}{-}X \ + \ e^- \xrightarrow{-X^-} \ \rangle{=}\langle \ + \ X^-
\end{array}
\right\} \tag{76}
$$

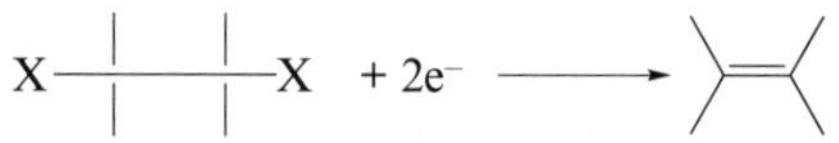

$$
X{-}{|}{-}{|}{-}X \ + \ 2e^- \longrightarrow \ \rangle{=}\langle \tag{77}
$$

phy and large-scale electroreduction at a stirred mercury pool in DMF show high yield of alkene. In a highly irreversible reaction no alkane was observed. Antiperiplanar cyclohexyl (**140**) and decalyl (**141**) dibromides were shown to reduce easily (equation 78). Similarity of the diffusion current constants (*ca* 3.4 for both compounds under several conditions) suggests a two-electron elimination reaction and a concerted elimination mechanism[118].

Preparative electroreduction of various vicinal dibromides at a stirred mercury cathode in the presence or absence of added proton source produced quantitative yields of olefins[119]. With *meso*- (**142 meso**) and *d,l*-2,3-dibromobutane (**142 dl**) *trans*- and *cis*-2-butene, respectively, were formed stereospecifically (equations 79 and 80). The absence of monobromide products suggests concerted antiperiplanar elimination or a very short-lived carbanion.

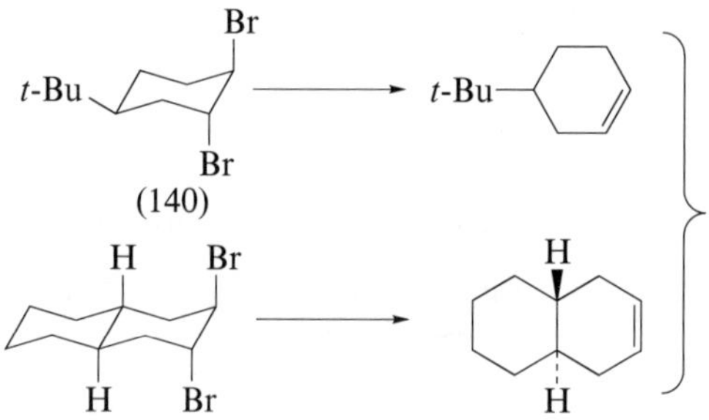

$$\tag{78}$$

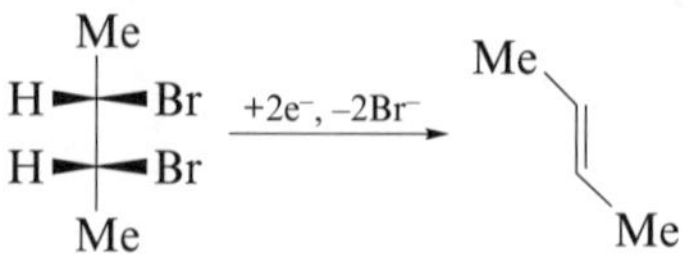

$$\tag{79}$$

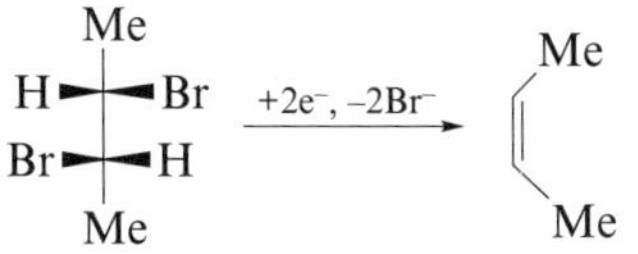

$$
\begin{array}{c}
\text{Me} \\
\text{H} \!-\!\!\!-\! \text{Br} \\
\text{Br} \!-\!\!\!-\! \text{H} \\
\text{Me} \\
\textbf{(142 dl)}
\end{array}
\quad \xrightarrow{+2e^-,\ -2Br^-} \quad
\begin{array}{c}
\text{Me} \\
\\
\\
\text{Me} \\
\sim\!100\%
\end{array}
\qquad (80)
$$

In a review of the older literature Fry[120] concluded that a concerted mechanism prevailed for the electroreduction of 1,2-dibromides but that a stepwise mechanism was indicated for reduction of 1,3- and 1,4-dihalides as well. Based on leaving-group effects, reduction potentials and product distribution studies of vicinal dihalides, such as 1,2-dihalo-1,2-diphenylethanes and trans-1,2-dihalocyclohexanes, a three-step mechanism involving successive addition of two electrons, followed by loss of second halide ion, was suggested. The intermediate anion species can undergo C—C bond rotation before the elimination of Br⁻, giving the observed mixture of cis and trans products, in case of the 1,2-diphenylethanes (equation 81). This does not establish the stepwise mechanism for the unstabilized 1,2-dihalides, as no stereochemical details are provided in these experiments[121].

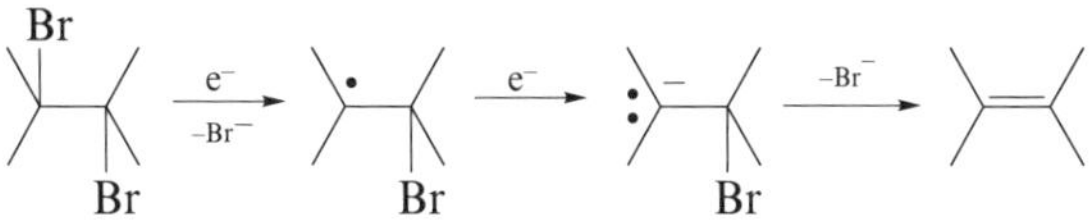

$$
\underset{\substack{\text{Br} \\ \text{Br}}}{\bigg\rangle\!\!-\!\!\bigg\langle} \ \xrightarrow[-Br^-]{e^-} \ \underset{\text{Br}}{\bigg\rangle\!\!\cdot\!\!-\!\!\bigg\langle} \ \xrightarrow{e^-} \ \underset{\text{Br}}{\bigg\rangle\!\!:^-\!\!-\!\!\bigg\langle} \ \xrightarrow{-Br^-} \ \bigg\rangle\!\!=\!\!\bigg\langle
\qquad (81)
$$

The electrochemical dehalogenation of a series of erythro- and threo-5,6-dihalodecanes at a mercury cathode showed an effect of leaving-group identity on $E_{1/2}$ values and on the steric course of reaction. Cis- and trans- decenes were formed in greater than 90% yield as shown in Table 5[122]. Stereospecificity of the product olefin formation depends on identity of substrate and halogen grouping, with less selectivity at higher potential. This suggests the possibility that the intermediate in the reduction is sufficiently long-lived to permit par-

TABLE 5. Half-wave potential for the reduction of 5, 6-dihalodecanes ($C_4H_9CHXCHYC_4H_9$) to olefins

X	Y	$-E_{1/2}$(Ag/AgCl)
Br	I	0.28
Cl	I	0.71
F	I	1.33
Br	Br	1.48
Cl	Br	1.95
F	Br	2.29
Cl	Cl	2.60
F	Cl	2.81

tial equilibration before product formation. Electroreduction of meso- and dl-3,4-dibromohexanes occurs at platinum, gold, copper amalgam or carbon electrodes in DMF or liquid ammonia to give products that result from antiperiplanar geometry, indicating again a concerted two-electron elimination process (Scheme 11)[123].

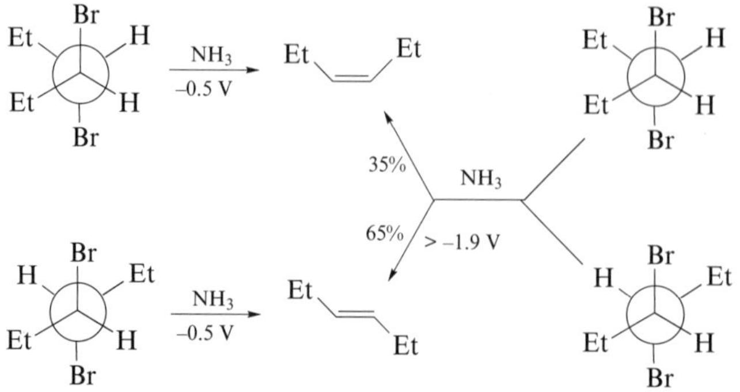

SCHEME 11

At very negative potentials (>−1.9 V) using liquid ammonia as solvent, the olefin product distribution becomes identical regardless of the starting diastereomer, suggesting that an equilibration process may be important. Stereoselective reduction of *meso-* and *dl*-1,2-dibromo-1,2-diphenylethane in DMF/TEAP at a mercury cathode gives 100% of *trans*- stilbene, even in the presence of a strong proton donor[124]. While the *meso*-isomer is expected to give *trans*- stilbene in a concerted process, the *d,l*-isomer should give *cis*-stilbene if the reaction were concerted as is shown in equation 82. No control experiment

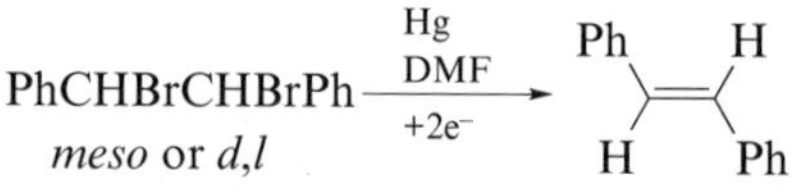

$$PhCHBrCHBrPh \xrightarrow[\substack{+2e^-}]{\substack{Hg \\ DMF}} \begin{array}{c} Ph \\ H \end{array}\diagdown=\diagup\begin{array}{c} H \\ Ph \end{array}$$

meso or *d,l*

(82)

to determine the stereochemical fate of *cis*-stilbene in the reaction medium was reported. It is possible that the produced stilbene serves as an electron transfer agent to equilibrate the initial product. These results differ significantly from those of Avraamides and coworkers[125], who reported that the electroreduction of *meso-* and *d,l*-1,2-dibromo-1,2-diphenylethane at a mercury cathode in acetonitrile/TBAP gave different products, depending on the identity of the starting stereoisomer. Whereas the *meso*-isomer always gave only *trans*- stilbene, the *d,l*-isomer gave olefin whose composition varied from pure *cis* to a nearly equimolar mixture of *cis*- and *trans*- isomers (Table 6). Although the authors invoke an equilibrating intermediate carbanion to explain the stereochemistry of the olefin, it is possible that the ground state population of conformers can explain these results (Scheme 12). The *meso-* isomer possesses a much more stable antiperiplanar conformer, whereas the *d,l-* isomer possesses three conformers of similar energy. The antiperiplanar conformer which leads to *cis*-stilbene is favored at low potential, whereas discrimination among conformers is lost at more negative potentials. A very interesting study of the effect of structure of the electron transfer agent on the stereochemistry of the product olefin in the reductive debromination of *meso-* and *d,l*-1,2-dibromo-1,2-diphenylethane by 9-substituted fluorenide ions (**143**) with 1,2-dibromo-1,2-

TABLE 6. *Z:E* ratio for the reductive debromination of 1,2-dibromo-1,2-diphenylethane as a function of potential

PhCHBrCHBrPh	$E(VAg^+/AgNO_3)$	PhCH=CHPh $Z:E$
meso-	−1.0	0 : 100
	−2.0	0 : 100
d,1-	−1.80	41 : 59

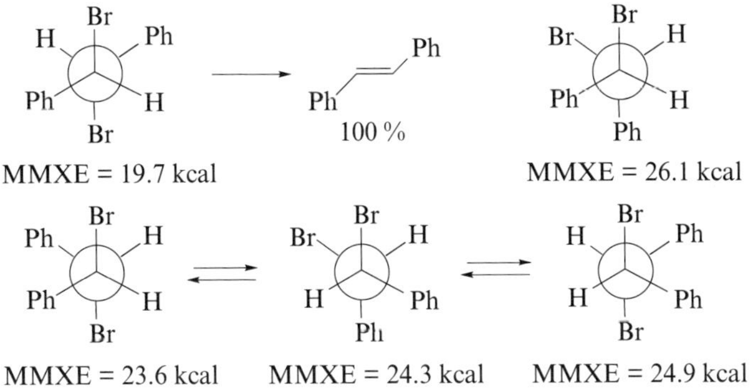
Br
H
Ph
Ph
H
Br
MMXE = 19.7 kcal
Ph
Ph
100 %
Br
Br
H
Ph
H
Ph
MMXE = 26.1 kcal
Br
Ph
H
Ph
H
Br
MMXE = 23.6 kcal
Br
Br
H
H
Ph
Ph
MMXE = 24.3 kcal
Br
H
Ph
H
Ph
Br
MMXE = 24.9 kcal

SCHEME 12

MMXE = heats of formation (kcal/mol) obtained from molecular mechanics force field calculations.

diphenylethane using DMSO has been reported by Lund and coworkers (equation 83)[126]. Reductive debromination showed that the ratio of (*E*:*Z*)-1,2-diphenylethene obtained for the reaction of ten 9-substituted fluorenide ions varied from 0.08 to 1.64 depending on the redox potential of the fluorenide anion and the bulkiness of the 9-substituent. Indirect electrochemical reduction of 1,2-dibromo-1,2-diphenylethane (*dl* and *meso*) can be achieved by using mediators such as chloranil or quinoxaline. Using the anion radicals of methylisonicotinoate (4-methoxycarbonylpyridine) as mediator, the transformation involves catalytic reduction[127].

PhCHBrCHBrPh (*meso* or *d,l*) —DMSO→ PhCH=CHPh (83)

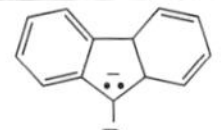
R

(143)

The high yield reduction of 1,2-dihalides to produce olefins has been employed to advantage to prepare reactive olefins. Electron transfer in electrochemistry is proportional to the diffusion coefficient, which is related in a much less sensitive way to temperature changes than is chemical reactivity. Thus it may become possible to synthesize and study electrochemically species whose chemical reactivity is high by working at low temperatures. Electroreduction of 1,2-dibromobenzocyclobutene (**144**) in acetonitrile or butyronitrile/TEAP or chemical reduction using the biphenyl radical anion resulted in the formation of benzocyclobutadiene (**145**)[128]. Efforts to observe the electrochemically generated anion radical or dianion of benzocyclobutadiene indicated that dimerization to **146** was faster than further reduction (equation 84).

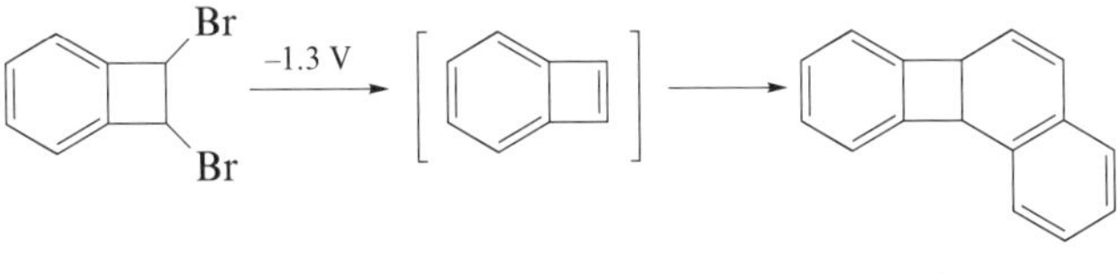
Br
Br
−1.3 V

(84)

(144) **(145)** **(146)**

Acenaphthylene dibromide (**147**) was also reduced to produce acenaphthylene (**148**) in quantitative yield (equation 85). A similar result was reported by Inesi and collaborators[129], and confirmed again by others[130]. Four-electron reduction of *ortho*-xylylene tetra-bromide (**149**) gives the same stable product, **146**, deduced to arise from benzocyclo-butadiene following a [4+2] cycloaddition and retro[2+2] rearrangement to rearomatize.

$$(85)$$

(**147**)　　　　　　　　　　(**148**)

Electroreduction of the bicyclic chlorobromide **150** in high vacuum at low temperature using DMF as solvent led to $\Delta^{1,4}$-bicyclo[2.2.0]hexene **151**[131] (equation 86). This exception-ally reactive olefin was isolated and characterized using vacuum line techniques[132].

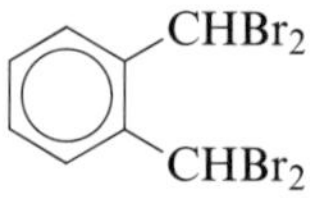

(**149**)

$$(86)$$

(**150**)　　　　　　　　　(**151**)

Electrochemical reduction of poly(trifluorochloroethylene) in the presence of LiBF$_4$ or NaBF$_4$ led to formation of unsaturated conjugated systems involving multiple elec-tron transfers and elimination of halide ions[133]. The reduction mechanism of 1, 2-dibromoethane was studied by coulometry and reverse pulse polarography. The tech-nique was also useful for the monitoring of residual trace amounts of hazardous alkyl halides[134].

Transformation of halogenated compounds into environmentally safe hydrocarbons can be achieved by electrochemical reduction. Vicinal dihalides can be electrochemically reduced to olefins by mediators such as cobalt corrin vitamin B$_{12}$[135], iron porphyrins[136], nickel coenzyme F$_{450}$[137] and cobalt or iron phthalocyanines in aqueous micellar CTAB solutions. These reactions proceed by inner-sphere electron transfer between the metallic species and alkyl halides, followed by dehalogenation of the resulting monohalo alkyl rad-icals to give olefins[138].

Electrochemically generated Co(I) species from vitamin B$_{12}$ has been used as a catalyst in the radical cyclization reactions. The stereochemistry of the product contrasts with that obtained by using Bu$_3$SnH/AIBN, and complements its synthetic usefulness[139]. The radi-cal type intermediates generated in these reactions can also be trapped by neighboring multiple bonds to form cyclic products[140]. These processes are illustrated in Scheme 13.

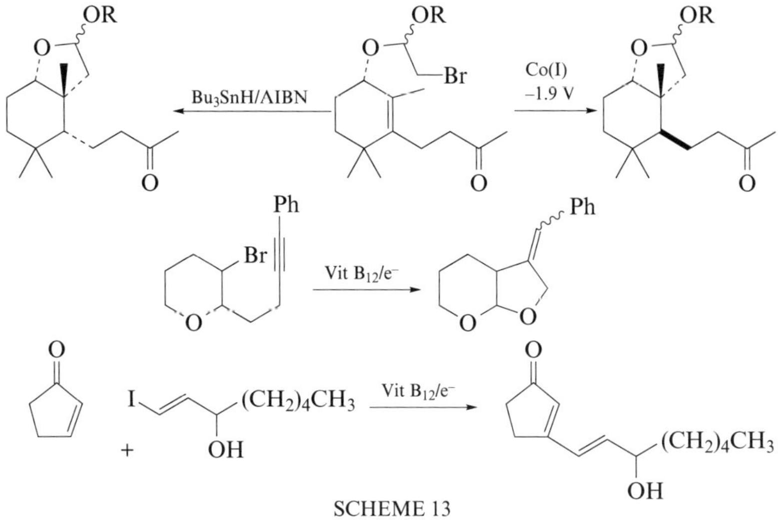

SCHEME 13

Dehalogenation of vicinal dibromides by electrogenerated polysulfide ions in DMA provides an indirect method to accomplish the same goal. A series of vicinal dibromides has been examined. These include methyl *erythro*-2,3-dibromo-3-phenylpropanoate (**152**),

PhCHBrCHBrCO$_2$Me	EtO$_2$CCHBrCHBrCO$_2$Et	PhCHBrCHBrPh
erythro	*meso*	*meso*
(**152**)	(**153**)	(**154**)
PhCHBrCHBrPh	MeCHBrCHBrMe	MeCHBrCHBrMe
dl	*meso*	*dl*
(**155**)	(**156**)	(**157**)
PhCHBrCH$_2$Br	C$_3$H$_5$CHBrCHBrC$_2$H$_5$	
(**158**)	(**159** *erythro* and *threo*)	

C$_4$H$_9$CHBrCHBrC$_4$H$_9$
meso
(**160**)

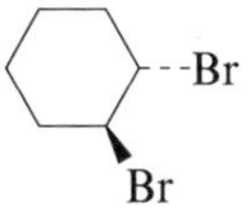

(**161**)

diethyl *meso*-2,3-dibromosuccinate (**153**), *meso*-1,2-dibromo-1,2-diphenylethane (**154**) *d,l*-1,2-dibromo-1,2-diphenylethane (**155**), *meso*-1,2-dibromobutane (**156**), *d,l*-2,3-dibromobutane (**157**), 1,2-dibromophenylethane (**158**), *erythro*-3,4-dibromoheptane (**159** *erthro*), *threo*-3,4-dibromoheptane (**159** *threo*), *meso*-5,6-dibromodecane (**160**) and *trans*-1,2-dibromocyclohexane (**161**)[141]. Quantitative formation of alkenes was observed with the all-substituted dibromides except with **156-158**, which showed some substitution (equation 87).

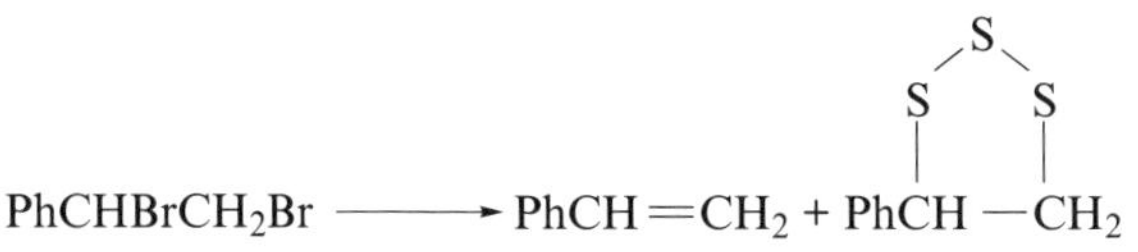

$$\text{PhCHBrCH}_2\text{Br} \longrightarrow \text{PhCH}{=}\text{CH}_2 + \text{PhCH}-\text{CH}_2 \qquad (87)$$

(**158**)

Electroreduction of *trans*- and *cis*-α,α'-difluorostilbene at a mercury cathode in DMF at −1.95 V (sce) consumes two Faradays of current and gives diphenylacetylene in high yield (measured spectrophotometrically) by what is proposed to be a concerted electrochemical process (equation 88)[142].

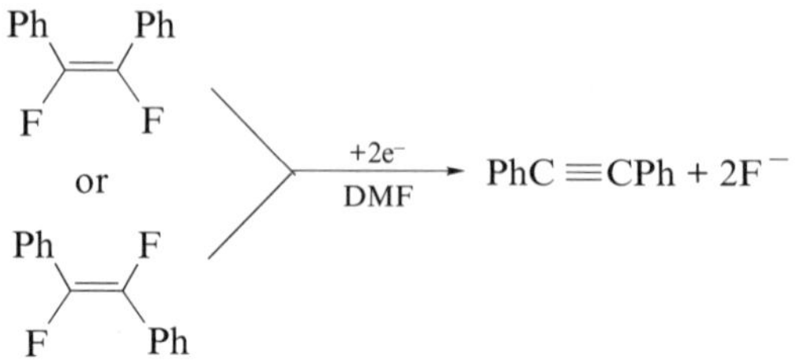

$$\text{PhC} \equiv \text{CPh} + 2\text{F}^- \qquad (88)$$

b. 1,3-Dihalides and 1,3-dihaloketones. The reduction of 1,3-dibromides occurs in good yields, to produce cyclopropane derivatives with little interference from a simple reduction, and appears to be a general synthetic method. Evidence is strong that the process is stepwise and involves radicals or organometallic species, depending on the electrode material, potential and solvent. Fry and Britton[143] provided evidence supporting a stepwise mechanism for the ring closure of 1,3-dibromopropanones. The mixture of *cis*- and *trans*-1, 2-dimethylcyclopropanones, formed during electroreduction of *meso*- and *dl*-dibromopentanones, was essentially the same regardless of the identity of the starting material (equation 89 and 90). Electroreduction of *meso*- and *dl*-2,4-dibromopentane in DMSO afforded roughly equal amounts of *cis*- and *trans*-1,2-dimethylcyclopropanones. Results of the electroreduction of (+)-(2S,4S)-2,4-dibromopentanone (**162**) suggested that the reaction could involve a stepwise reduction mechanism via the α-halo carbanion. Optically active dibromide gave dibromocyclopropane in high optical yield, suggesting cyclization via a semi-W transition state (equations 89 and 90)[144]. In a review Fry[120] concluded that a stepwise mechanism was indicated for reduction of 1,3- and 1,4-dihalides.

Wiberg and Epling[145] studied the electroreduction of 1,3-dibromopropane and 1,4-dibromobutane at platinum and mercury-coated platinum and found evidence for a two-step process. In DMF 1,3- and 1,4-dibromides at a mercury-coated platinum electrode give CV with two irreversible waves at slow scan. Reactions performed at a platinum electrode gave differing amounts of propane and cyclopropane and, depending on the potential employed, mostly cyclopropane (Table 7). The C_4 hydrocarbon system gave similar results. It is of interest that the preparative reactions were in the presence of benzyl methyl ether as a potential hydrogen atom source, and that when α-deuterioether was used, the product propane was found to contain 8.8% deuterium. These results all point to a stepwise mechanism involving the radical as illustrated in equation 91.

TABLE 7 Yields of cyclic and open-chain hydrocarbons as a function of potential and chain length

Br(CH$_2$)$_n$Br	−V(sce)	(CH$_2$)$_n$	H(CH$_2$)$_n$H
n= 3	2.15	91	9
3	1.45	86	14
4	2.30	90	10
4	1.75	26	74

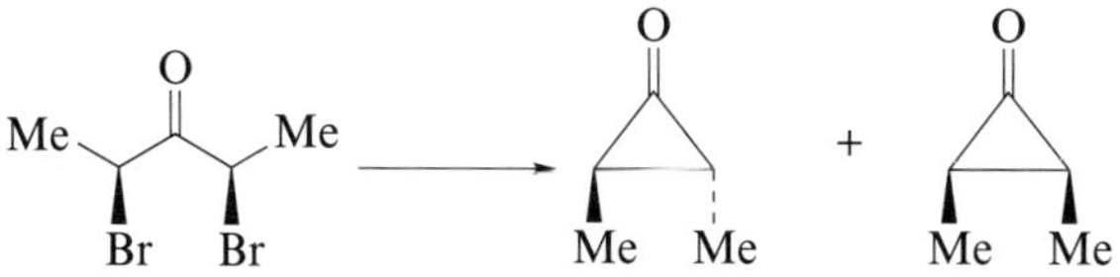

(162 *meso*) (163) (164) (89)

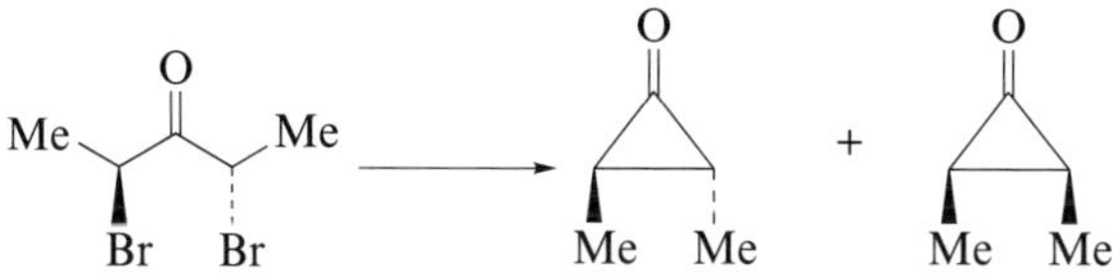

(162 *dl*) (90)

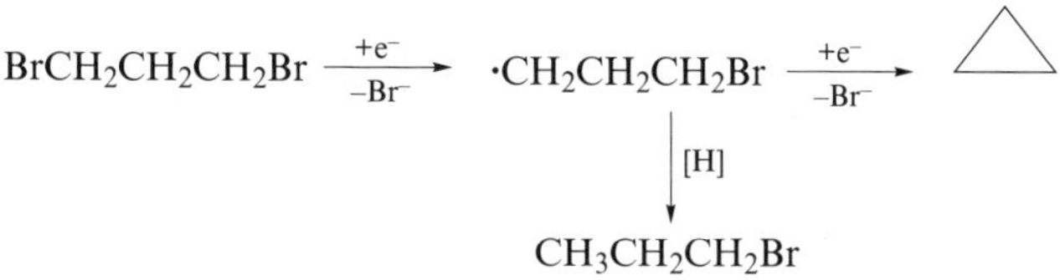
BrCH2CH2CH2Br
+e−
−Br−
·CH2CH2CH2Br
+e−
−Br−
[H]
CH3CH2CH2Br

(91)

Casanova and coworkers[130] studied the products of the electroreduction at mercury of 1,3-dibromopropane (**165**), 1,3-dibromo-1-phenylpropane (**166**), 1,3-dibromo-1, 3-diphenylpropane (**167**) and *endo*-4-*syn*-8-dibromodibenzobicyclo[3.2.1]octadiene (**168**); see equations 92–94. 1,3-Dibromopropane gave essentially quantitative yield of cyclopropane and no propane, even in the presence of proton sources. However, **166** gave both ring-opened **169** and ring-closed **170** products depending on the potential and presence or absence of proton donors. Dibromide **167** gave 1,2-diphenylcyclopropane **171** in nearly quantitative yield in an isomer ratio that did not depend on the *d,l* / *meso* ratio of **167**. Formation of modest yields of the tricyclic hydrocarbon **172** from **168** at a potential below the reduction potential of the bridgehead C—Br bond is reported (equation 94). These results also point toward a stepwise reduction mechanism for 1,3-ring closure.

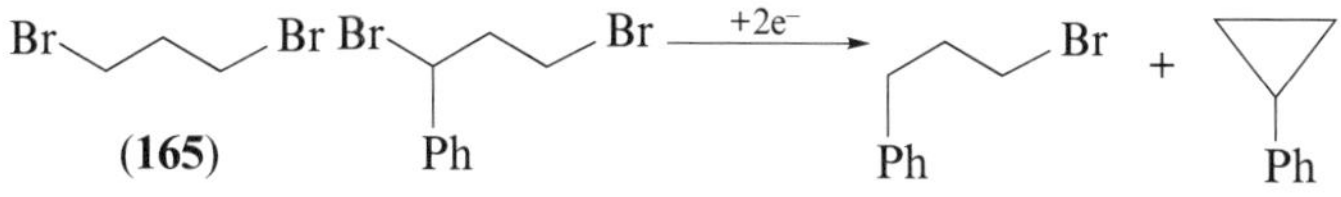

(165) (166) (169) (170) (92)

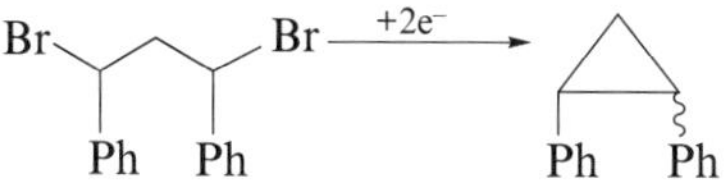

(167) (171) (93)

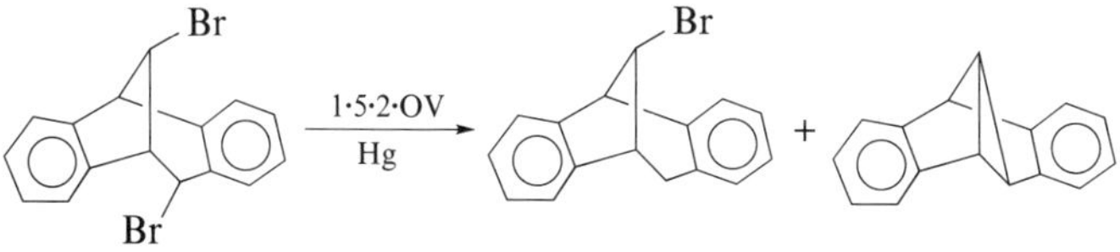
1·5·2·OV
Hg

(168) (173) (172) (94)

Electroreduction of the 1,3-dibromides *endo*-2, *endo*-6 (174) and *exo*-2,10-dibromobornane in aqueous ethanol and DMF was reported by Grimshaw and Grimshaw[146] to give high yields of tricyclene (175) and bornane (176) (80% yield) and *exo*-2,10-dibromobornane(177) produced bornane (176) and camphene (178) as products (75% yield); see equations 95 and 96.

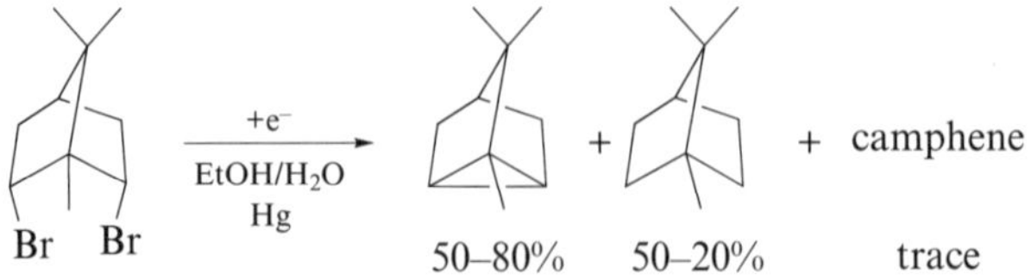

$$\underset{\textbf{(174)}}{}\xrightarrow[\substack{\text{EtOH/H}_2\text{O}\\ \text{Hg}}]{+e^-} \underset{\substack{\textbf{(175)}\\ 50\text{--}80\%}}{} + \underset{\substack{\textbf{(176)}\\ 50\text{--}20\%}}{} + \underset{\text{trace}}{\text{camphene}} \qquad (95)$$

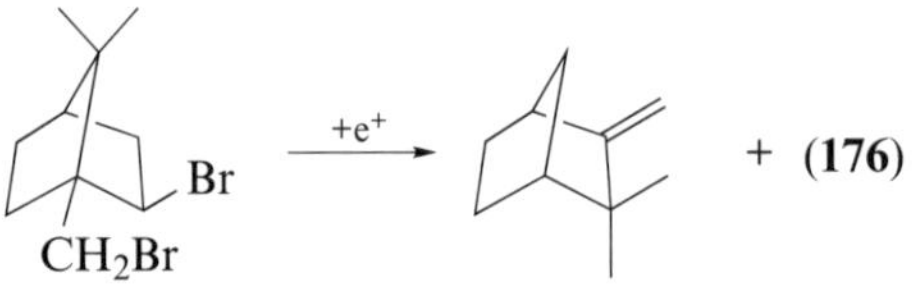

$$\underset{\textbf{(177)}}{} \xrightarrow{+e^+} \underset{\textbf{(178)}}{} + \textbf{(176)} \qquad (96)$$

The effectiveness of electroreduction as a method of ring closure has been applied to the synthesis of other strained ring compounds. Good yields of bicyclo[2.1.0]pentane (179) and bicyclo[3.1.0]hexane (180) have been obtained by the electroreduction of the corresponding 1,3-dibromocycloalkanes[147]. The results do not depend upon the stereochemistry of the starting material (equations 97 and 98). The technique has been applied to attempted synthesis of the highly strained [2.2.1]propellane 182 from 1,4-dibromonorbornane (181) at a mercury cathode in DMF[148] (equation 99). Evidence from the current consumption and polarograms excludes 1-bromonorbornane as an intermediate and points to an intermediate propellane. A high yield of a mixture of 183 and 184 was isolated as the stable products.

$$\xrightarrow{\hspace{2cm}} \underset{\textbf{(179)}}{} \qquad (97)$$

$$\xrightarrow{\hspace{2cm}} \underset{\textbf{(180)}}{} \qquad (98)$$

$$\underset{\textbf{(181)}}{} \xrightarrow[\substack{-34\,°\text{C}\\ \text{TBAP}}]{\text{Hg}} \underset{\textbf{(183)}}{} + \underset{\substack{\textbf{(184)}\\ \text{Hg}}}{\left[\quad\right]_2} \underset{\substack{98\%}}{} \quad \left[\underset{\textbf{(182)}}{}\right] \qquad (99)$$

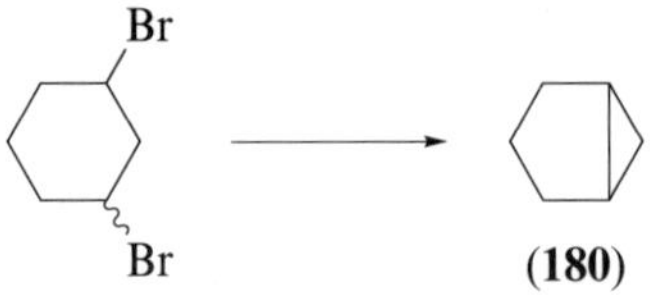

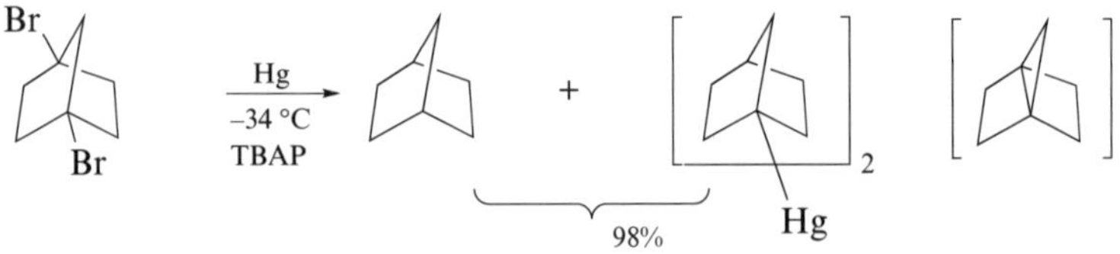

1,3-Dihaloketones. A number of studies of the electroreduction of α,α'-dibromoketones have been reportedly directed to obtaining evidence for the the intermediacy of cyclopropanone-derived intermediates (**186**). Reduction of 2,4-dibromo-2,4-dimethyl-3-pentanone (**185**) consumed two Faradays of current and gave a species which is briefly stable at –32 °C. The intermediate (**186**) could be intercepted by subsequent addition of a protonic trapping agent to the low-temperature catholyte (equation 100). Results have

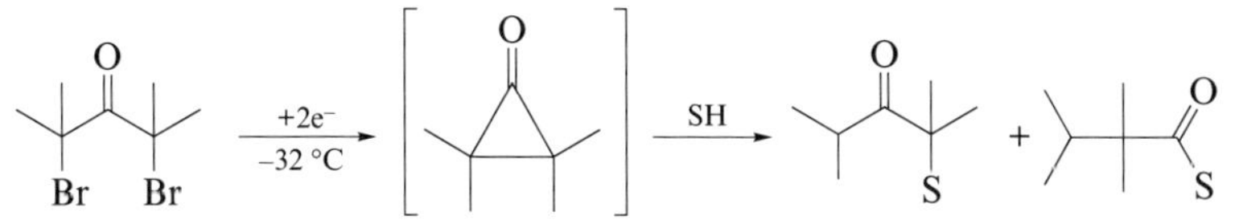

$$\underset{(\textbf{185})}{} \xrightarrow[-32\,°C]{+2e^-} \underset{(\textbf{186})}{[\quad]} \xrightarrow{SH} \underset{(\textbf{187})}{} + \underset{(\textbf{188})}{} \qquad (100)$$

been interpreted in terms of a zwitterionic intermediate or a cyclopropanone[149]. Fry and Scoggins [150] reduced 2,4-dibromo-2-methyl-3-pentanone (**185**) and 2,4-dibromo-2, 4-dimethyl-3-pentanone (**185**) in DMF using TEAB as the supporting electrolyte. Compound **185**, when reduced at low temperature, showed a characteristic cyclopropanone band at 1825 cm⁻¹. Compound **185** in methanol as a solvent gave an ester and ketoether, providing good evidence for the intermediacy of a cyclopropanone intermediate. Inesi and coworkers[151] reported that the electroreduction of 2, 4-dibromo-1,5-diphenyl-1,4-pentadiene-3-one (**189**) at a mercury cathode took a different course. Thus, when reduced in DMF it gives *cis*-1,5-diphenyl-1-penten-4-yn-3-one (**190**), proposed to result from an unstable dimethylenecyclopropanone (**191**). In methanol the reaction gave a 3,4-diphenylcyclopent-2-en-1-one (**192**), which was thought to arise from an intermediate bicyclic ketone followed by rearrangement. Regioselective reaction in the reductive acetoxylation of the α-position of α,α'-dibromoketones was reported by Fry and Lefor[152]. Fry and O'Dea[153] also carried out an exhaustive study of the electroreduction of 12 variously substituted α,α'-dibromoketones **193** in acetic acid using a mercury cathode. This synthesis represents a convenient route to highly branched α-acetoxyketones **195**. For the most highly branched system involving alkyl substituents, the major product was the α-acetoxyketone. With unsymmetrically substituted systems, prediction of the orientation of acetoxylation was not clear. The results could be rationalized as reduction to α-bromoenolate. Factors governing the competition between formation of α-acetoxyketone (in acetic acid) and double dehalogenation from substituted α,α'-dibromophenylacetones to yield (**194**) is reported[154]. The degree of substitution on the α-carbon atom is a key factor

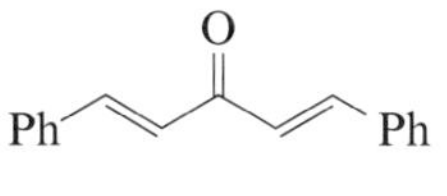

(**189**)

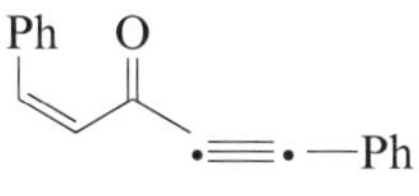

(**190**)

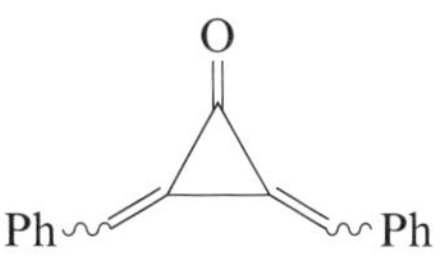

(**191**)

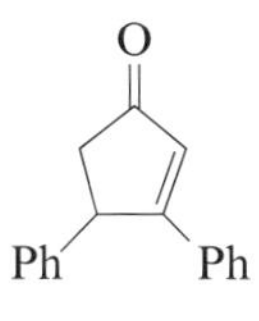

(**192**)

in the competition. Less highly substituted samples (R^1 = Ph, $R^2 = R^3 = R^4$ = H) gave mostly the fully reduced ketone (195), whereas more highly substituted samples (R^1=Ph, $R^2 = R^3$ = Me, R^4 = H) in (193) gave mostly monoacetoxy derivatives (194). Results are consistent with enol mechanism proposed by Fry earlier, but may be complicated by competing chemical solvolysis (equation 101).

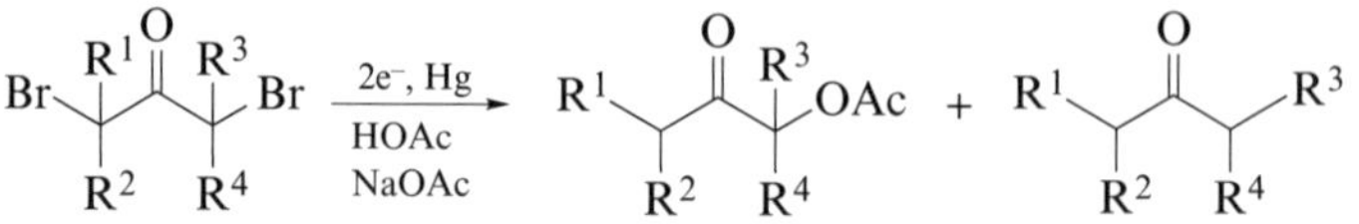

$$ (101) $$

(193)   (194)   (195)

c. *1,4-Dihalides.* Several interesting cases of the reduction of bridgehead 1,4-bicyclic dihalides have been reported. Such reactions represent a potential route to highly strained tricyclic compounds, although fragmentation sometime takes place. Cathodic reduction of bridgehead bicyclic dihalides was observed at a surprisingly unfavorable potential, suggesting that the process is more likely to be stepwise than concerted. Ginsburg and coworkers[155] attempted the synthesis of a substituted [2.2.2]propellane from triptycene derivatives (equation 102). Controlled potential electroreduction of 9,10-dibromotriptycene (196) in DMF at a mercury cathode gave no evidence for [2.2.2]propellane (197). Chemical reaction of 198 with strong bases also gave no evidence for transannular ring closure. No effort to trap propellane 197 was reported, and triptycene (199) was obtained after workup, leaving open the possibility that the propellane may have been present and undergone reductive ring opening. In nearly identical experiments the groups of Wiberg[156] and of Dannenberg[157] carried out the electroreduction of (200) to show the probable formation of [2.2.2]propellane 201. The electroreduction of 200 was conducted at –15 °C in DMF, then treated with chlorine *after* reduction and allowed to stand in the cold for a time. Workup gave 1,4-dimethylenecyclohexane (203) (elimination product) on gas chromatography (Wiberg), or dibromide 200 formed by addition of bromine (Dannenberg) (equation

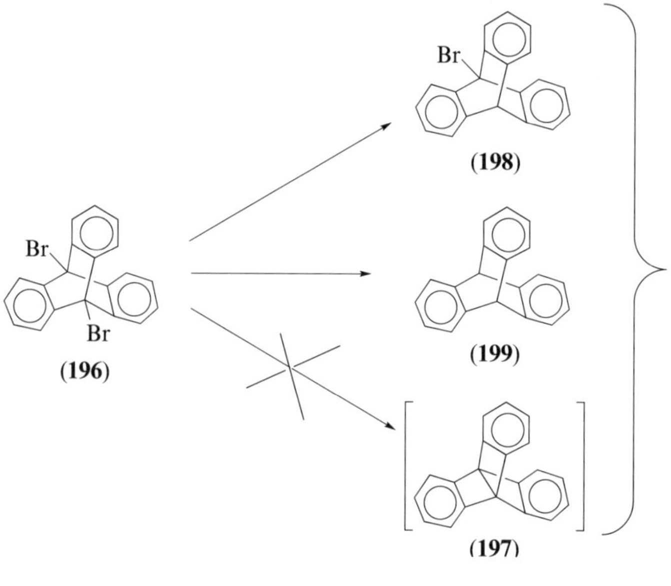

(198)

(196)

(199)

$$ (102) $$

(197)

103). As described earlier, Wiberg[145] reported the electroreduction of 1,4-dibromobutane at platinum and mercury-coated platinum giving evidence for a stepwise process, and a related study of the reduction of 1,4-dibromobutane at a reticulated vitreous carbon cathode in DMF containing TMAP was reported by Peters and collaborators[158].

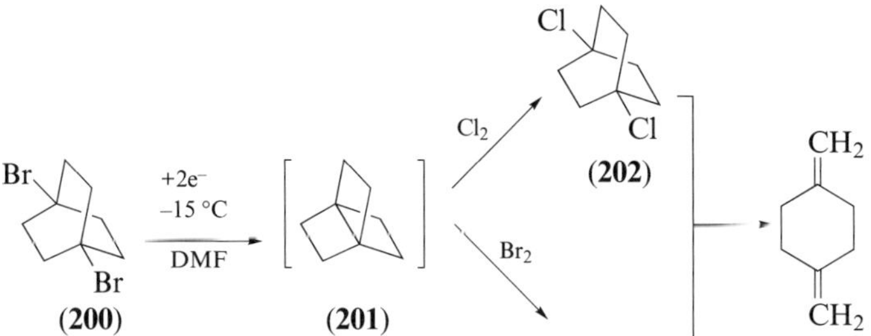

$$(103)$$

Dibromoesters. Electrochemical reduction of 1,2-, 1,3- and 1,4-dibromoesters has been examined (equations 104–106). Whereas reduction of 1,2-dibromoester **204** gave elimination cleanly, to yield **205** in high yield, and 1,3-dibromoester **206** gave a high yield of cyclopropanecarboxylic ester **207**, 1,4-dibromoester **209** gave similar amounts of cyclobutanecarboxylic ester **210** and ring-opened product **211** as shown by Inesi and collaborators[159]. Inesi also examined the influence of potential on the electrochemical reduction products of α,γ-dibromobutyrate in DMF at mercury. The yield of cyclo-

$$BrCH_2CHBrCO_2Et \xrightarrow[\text{-1.5 V sce}]{Hg} CH_2{=}CHCO_2Et \qquad (104)$$

$$\begin{array}{cc} & {\sim}98\% \\ \textbf{(204)} & \textbf{(205)} \end{array}$$

$$BrCH_2CH_2CHBrCO_2Et \xrightarrow[\text{-1.5 V sce}]{Hg} \triangleright{-}CO_2Et + Br(CH_2)_3CO_2Et \qquad (105)$$

$$\begin{array}{ccc} & {\sim}85{-}90\% & {\sim}10\% \\ \textbf{(206)} & \textbf{(207)} & \textbf{(208)} \end{array}$$

$$Br(CH_2)_3CHBrCO_2Me \xrightarrow[\text{-1.5 V sce}]{Hg} \square{-}CO_2Me + Br(CH_2)_4CO_2Me \quad (106)$$

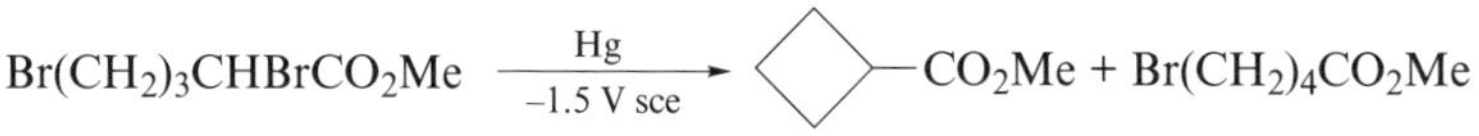

$$\begin{array}{ccc} & {\sim}45\% & {\sim}45\% \\ \textbf{(209)} & \textbf{(210)} & \textbf{(211)} \end{array}$$

propane derivatives was greater at higher potentials. Electrode-assisted cyclization at high negative potentials was suggested[160]. A similar result was found with long-chain diesters. Constant-current electroreduction of the dibromo diesters dimethyl 2,(ω-1)-dibromoalkandioates (**212**) in THF containing tetrabutylammonium salts at platinum cathode gave moderate yields of cyclic diesters by intramolecular reductive coupling. Cyclic diesters **213** containing three to seven atoms were formed in 50–60% yields generally, with fully reduced diesters **214** and monobromoesters **215** formed in smaller amounts (equation 107). In presence of a mercury cathode, yields were lower due to formation of organomercurials[161].

J. Casanova and V. P. Reddy

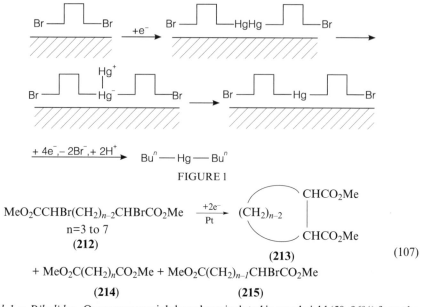

FIGURE 1

$$MeO_2CCHBr(CH_2)_{n-2}CHBrCO_2Me \xrightarrow[Pt]{+2e^-} (CH_2)_{n-2} \begin{array}{c} CHCO_2Me \\ | \\ CHCO_2Me \end{array}$$

n=3 to 7

(212)   **(213)**   (107)

$$+ MeO_2C(CH_2)_nCO_2Me + MeO_2C(CH_2)_{n-1}CHBrCO_2Me$$

(214)   **(215)**

d. 1,ω-Dihalides. Organomercurials have been isolated in good yield (50–96%) from the electroreduction of 1,ω-dibromides in DMF at a mercury cathode. Because symmetrical organomercurials are isolated in high yields from 1,ω-dibromides compared to organomercurials from the equivalent monobromides as shown in reaction 108, it has been sug-

$$Br(CH_2)_nBr \xrightarrow[\substack{-2.5V \\ +3e^-}]{+2e^-} 1/2Hg((CH_2)_{n-1}CH_3)_2 \qquad (108)$$

$4 \leq n \leq 12$ **(216)**

gested that adsorption of the dibromide on the mercury cathode surface is responsible for the synthesis (Figure 1)[162]. The adsorption postulate is further supported by the observation of organomercurials in the reduction of dibromodiesters.

4. Mediated reactions

A number of important synthetic processes involving mediated reactions of halides rather than direct electrochemical reduction have been reported. The reduction of aryl halides in the presence of alcohols with or without NH_3 as a cosolvent leads to the oxidation of the alcohols to the carbonyl compounds. The reaction involves an electrocatalytic process mediated by electron transfer from the initially reduced aryl halide. Alcohols such as benzhydrol and 2-propanol are converted to their respective ketones, on a preparative scale[163]. The proposed mechanism is shown in Schemes 14 and 15.

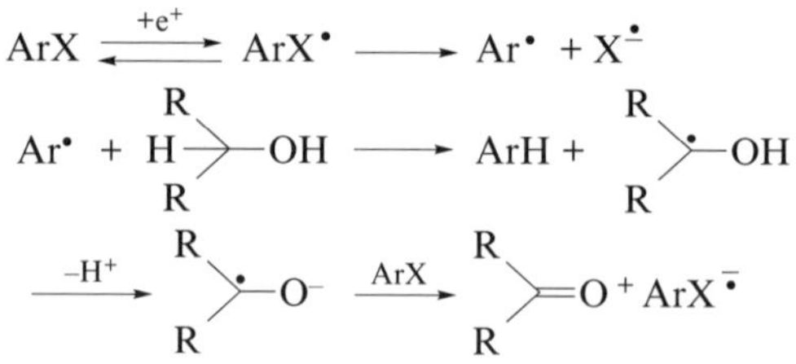

$$ArX \underset{}{\overset{+e^+}{\rightleftharpoons}} ArX^{\bullet} \longrightarrow Ar^{\bullet} + X^{\bullet-}$$

$$Ar^{\bullet} + H\overset{R}{\underset{R}{\rightarrow}}OH \longrightarrow ArH + \overset{R}{\underset{R}{\rightarrow}}OH$$

$$\xrightarrow{-H^+} \overset{R}{\underset{R}{\rightarrow}}O^- \xrightarrow{ArX} \overset{R}{\underset{R}{\rightarrow}}=O + ArX^{\bullet-}$$

SCHEME 14

$$p\text{-}BrC_6H_4COC_6H_5 \xrightarrow{\quad Ph_2CHOH,e^- \quad} Ph_2CO$$

Mechanism of oxidation:

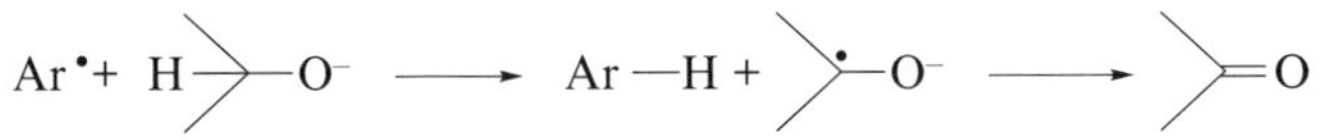

$$Ar^{\cdot} + H\text{—}\!\!\diagup\!\!\diagdown\text{—}O^- \longrightarrow Ar\text{—}H + \diagup\!\!\diagdown\text{—}O^- \longrightarrow \diagup\!\!\diagdown\!=\!O$$

SCHEME 15

Ni(0) obtained from electrochemical reduction of NiBr$_2$ in THF/HMPA acts as an efficient catalyst for the electroreductive coupling of ethylene with aryl halides to give 1,1-diarylethanes **217** (equation 109). By proper control of reaction conditions, such as reduction potential, solvent and supporting electrolyte, it can be shown that substituted olefins can be prepared from aromatic halides and alkenes[164].

$$PhBr + NiBr_2/ethylene \xrightarrow[\substack{HMPA/THF \\ LiClO_4,Bu_4N^+Br^-}]{+e^-} Ph_2CHMe \qquad (109)$$

$$(\mathbf{217})$$

The mechanism of the formation of biphenyls, catalyzed by electrogenerated coordinatively unsaturated nickel(0) complexes, has been studied kinetically and a chain mechanism is postulated. The reaction involves Ni(0), Ni(II) and Ni(III) complexes and the coupling reaction is initiated by Ni(0) species[165]. Electrocatalytic coupling of aryl halides could also be achieved with {1,2-bis(di-2-propylphosphino)benzene}nickel(0) as a catalyst[91]. Ni(I)octaethylisobacteriochlorin anion [Ni(I)(OE$_i$BC)$^-$]generated from Ni(II) species at –1.5 V (sce) reacts with alkyl halides or alkyl toluenesulfonates to give the corresponding alkanes. It was shown that [Ni(III)(OE$_i$BC)] was an intermediate. The reactivity order: I>Br>Cl and Me>n-Bu>sec-Bu>$tert$-Bu suggests an S$_N$2-like mechanism[166].

Nickel-2,2-bipyridine complexes are also used for the preparation of unsymmetrical biaryls such as 4-methoxy-4′-trifluoromethylbiphenyl by electroreduction of two aryl halides, one of which has electron-donating and the other electron-withdrawing groups in the aromatic ring as shown in equation 110. The reaction was carried out in N-methylpyrrolidinone at constant current in an undivided cell fitted with a sacrificial magnesium anode and excess of 2,2′-bipyridine[167].

$$L_2Ni(II) + 2e^- \longrightarrow L_2Ni(0) \xrightarrow{ArX} L_2Ni\diagup_{X}^{Ar} \xrightarrow{ArX} Ar\text{—}Ar + L_2Ni(II)X_2 \quad (110)$$

The mechanism of nickel-catalyzed electron transfer activation of aromatic halides in biphenyl electrosynthesis from bromobenzene has been reported[92]. In the presence of the electrogenerated nickel(I) salen [2,2′-ethylene-bis(nitrilomethylidyne)diphenolato]nickel(I) complex, generated at a carbon cathode in DMF containing TEAP, the acetylenic halides 6-iodo- and 6-bromo-1-phenyl-1-hexyne (**218**) are rapidly reduced catalytically to form a radical intermediate that cyclizes to produce benzylidenecyclopentane (**219**) in high yield. Direct electrolysis of 6-iodo- and 6-bromo-1-phenyl-1-hexyne at carbon cathodes, on the other hand, leads to undesired acyclic side-products and to benzylidenecyclopentane in only modest yields[168] (equation 111). In a somewhat similar process benzal chloride is converted to a mixture consisting primarily of *cis*- and *trans*-stilbene by the action of electrochemically generated cobalt(I)(salen)-9. The process involves the intermediate electrocatalytic conversion of benzal chloride to a mixture of the stereoisomeric 1,2-dichloro-1,2-diphenylethanes and subsequent further reduction[169].

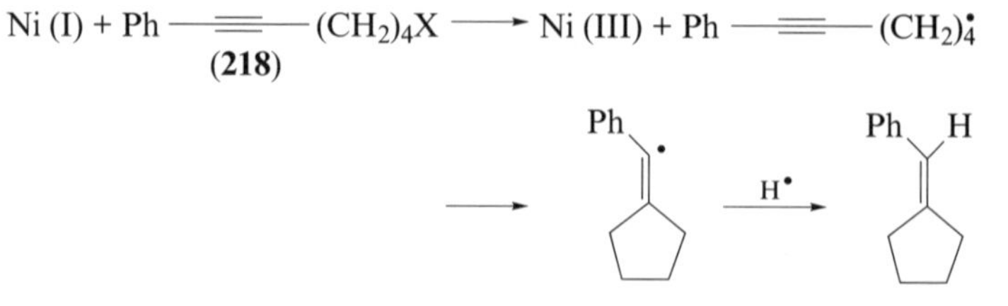

$$\text{Ni (I)} + \text{Ph}\!\!=\!\!(CH_2)_4X \longrightarrow \text{Ni (III)} + \text{Ph}\!\!=\!\!(CH_2)_4^{\bullet}$$
$$\textbf{(218)}$$

$$\hspace{6cm}(111)$$

$$\textbf{(219)}$$

The electrosynthesis of β-hydroxy esters (**220**), 2,3-epoxy esters (**221**) and β-hydroxy nitriles (**222**) was achieved under nickel-catalyzed conditions, obviating Reformatsky reaction[170](equation 112–114). The reaction proceeds in excellent yield when a sacrificial zinc rod is used as the anode. A mechanism has been proposed which involves reduction of a Ni(II) complex to a Ni(0) complex, oxidative addition of the α-chloroester to the Ni(0) complex, and a zinc (II)/Ni(II) exchange, leading to an organozinc reagent, in analogy to the Reformatsky reactions.

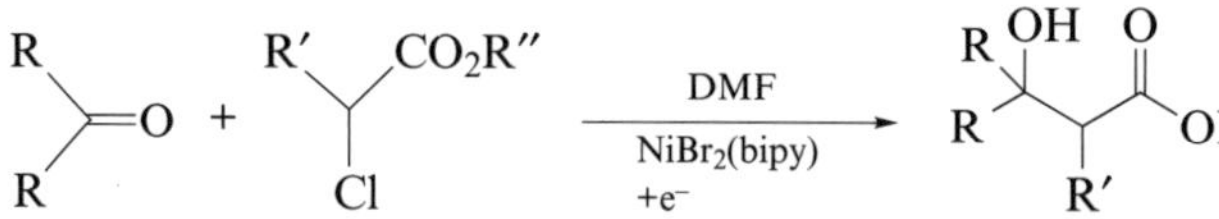

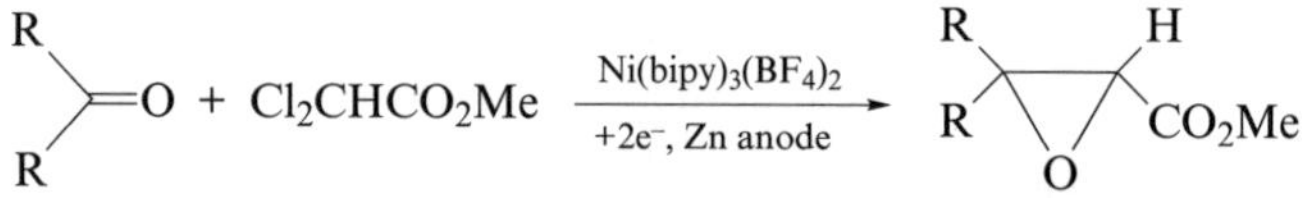

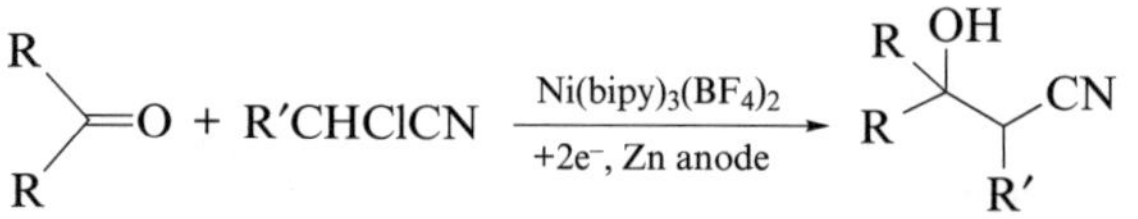

$$\hspace{6cm}(112)$$

$$\hspace{6cm}(113)$$

$$\hspace{6cm}(114)$$

Electrochemical reduction of nickel(I)(salen) in the presence of benzal chloride at a carbon electrode in DMF leads to a variety of monomeric and dimeric products[171]. Cyclic voltammetric studies using benzylic halides showed that carbanions can bring about autocatalytic processes by acting as electron-transfer catalysts (Scheme 16)[172].

The palladium(II) complex [PdCl₂(PPh₃)₂] has been used as a catalyst in the electrosynthesis of biaryls from aryl triflates. In the presence of carbon dioxide, aromatic carboxylic acids are obtained[173]. The effect of the σ-bonded ligand on the reactivity of porphyrins was probed by cyclic voltammetric studies with (tetraphenylporphinato)rhodium(III) complexes of the type (TPP)Rh(R) and (TPP)Rh(RX). The reduction potentials of the alkyl halide complexes are in general less cathodic than the uncomplexed alkyl halides. It was also found that the electrochemical behavior of these complexes is determined by the length of the R group. For longer chains, the initial step involves reduction of the C—X bond[174].

$$RX + 2e^- \longrightarrow R^- + X^-$$

$$R^- + RX \longrightarrow 2R^{\bullet} + X^-$$

$$R^{\bullet} + 1\,e^- \longrightarrow R^-$$

SCHEME 16

Electrochemically generated iron promoter has been used for coupling of alkyl halides. Low to moderate yields of coupled product are obtained from aluminum electrode using $Fe(acac)_3$ as the electron transfer agent in DMF^{33}. The electrogenerated alkoxide ions react with alkyl bromides (benzylic, allylic and methyl) in the presence of iron(0)pentacarbonyl to give the corresponding esters **224**, (equation 115). The iron(0)pentacarbonyl is required in only catalytic quantities (e.g. 0.1 eq.). Yields are superior to those involving equivalent chemical methods (e.g. 96% vs 66%)[175].

$$PhCH_2Br + Fe(CO)_5 + MeOH \xrightarrow[Me_4N^+ \; Cl^-]{\text{Pt electrode}} PhCH_2CO_2Me \qquad (115)$$

$$\textbf{(223)} \qquad\qquad\qquad\qquad\qquad \textbf{(224)}$$

5. Mechanisms

When considering mechanisms, several questions arise concerning the timing of electron transfer and the sequence of chemical and electrochemical events. The nature of the electron transfer event, whether it is delivered to the substrate directly from the electrode or indirectly through another chemical species (equation 116), and the involvement of the electrode material itself, especially if the electrode is a reactive metal, are important questions (equation 117). The lifetime of unstable ion radicals has been measured by homogeneous redox catalysis of electrochemical reactions. The electrochemical reduction of benzyl halides gives a single wave at about -1.4 to -2.2 V on Pt or carbon electrodes corresponding to an uptake of one electron per molecule as shown by chronoamperometry and coulometric measurements. It has been proposed that the one-electron stoichiometry is the apparent result of either direct or indirect reaction of the benzyl anion intermediate with unreacted benzyl halide (equation 118). 3-Phenylpropionitrile was also observed as one of the products when the reaction was carried out in acetonitrile, which presumably arises through electron transfer from the benzyl anion as illustrated in equation 119[176].

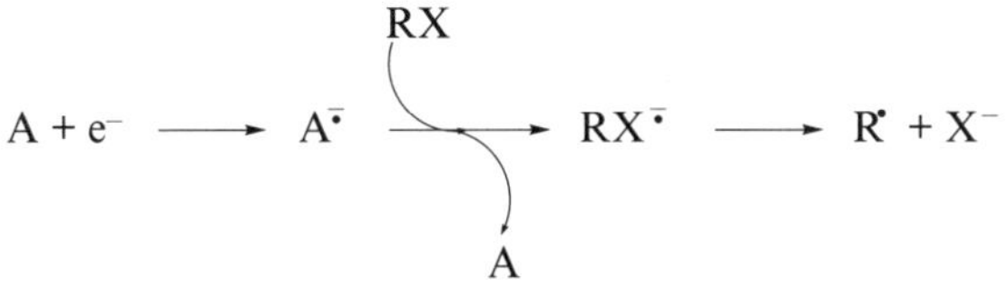

$$A + e^- \longrightarrow A^{\cdot} \xrightarrow{RX} \;\; RX^{\overline{\cdot}} \longrightarrow R^{\cdot} + X^- \qquad (116)$$
$$A$$

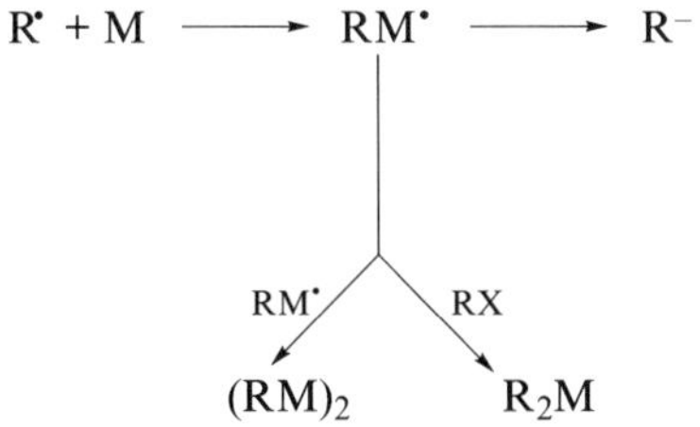

$$R^{\cdot} + M \longrightarrow RM^{\cdot} \longrightarrow R^- \qquad (117)$$

$$RM^{\cdot} \diagup \quad \diagdown RX$$
$$(RM)_2 \qquad R_2M$$

$$PhCH_2X \xrightarrow{2e^-} PhCH_2^- \left.\begin{array}{}\\[2em]\end{array}\right\}$$
$$PhCH_2^- \xrightarrow{PhCH_2X} PhCH_2^{\cdot} + (PhCH_2X)^{\cdot -} \qquad (118)$$

$$PhCH_2^- + CH_3CN \longrightarrow PhCH_3 + {}^-CH_2CN \left.\begin{array}{}\\[2em]\end{array}\right\}$$
$$^-CH_2CN + PhCH_2I \longrightarrow PhCH_2CH_2CN \qquad (119)$$

Aryl radicals obtained directly or through mediation of aromatic anion radicals react with olefins in solvents such as liquid NH_3, DMSO, CH_3CN or DMF to give the arylated products. The electrochemically generated aryl radicals react with styrene or its derivatives in aprotic media to give arylated addition products (**225**). The reaction, shown in equation 120, has been optimized to be synthetically useful. The rate constants of the key step, i.e. addition of the radical to the olefin, were determined by cyclic voltammetry[177].

$$ArX + e^- \longrightarrow ArX^{\cdot} \longrightarrow Ar^{\cdot} + X^-$$

$$
\left.
\begin{array}{c}
\begin{matrix} Me \\ \diagdown \\ Me \diagup \end{matrix} = \begin{matrix} Ar' \\ \diagup \\ \diagdown Me \end{matrix}
\qquad \xrightarrow{Ar^{\cdot}} \qquad
Ar \overset{Me}{\underset{Me}{\rule{0pt}{1.5em}}} \overset{H}{\underset{Me}{\rule{0pt}{1.5em}}} Ar'
\end{array}
\right\}
\tag{120}
$$

$$(\mathbf{225})$$

It has become clear in the past decade that electron transfers between electrodes and organic molecules are primarily characterized by single electron transfers[178], although in some cases the initial transfer may be followed by a rapid, second electron transfer to produce a carbanion. In a theoretical study of the electroreduction of halogenated aromatic compounds, the electrochemical dehalogenation of a variety of halogenated aromatic compounds was studied using semiempirical (AM1) and *ab initio* (RHF 3-21G*) calculations. Comparison of the electrochemical half-wave potential values ($E_{1/2}$) and calculated theoretical values indicates the formation of a π radical following the electron uptake[179]. Both theoretical and experimental studies have demonstrated important parameters in the electron transfer process. Seminal contributions to theory were advanced by Eberson[5, 180] in which processes involving outer-sphere electron transfer between organic species are discussed in the framework previously applied to electron transfer between inorganic complexes. The process is discussed both qualitatively and quantitatively using Marcus–Hush theory as a model for irreversible electron transfer processes. Marcus–Hush theory appears to be applicable to these processes, providing a circumstance in which experimentally accessible parameters can be evaluated to predict reaction rates. Savéant and coworkers[181] have invested a prodigious effort to develop this area in both a theoretical and experimental way for the electroreduction mechanism of the carbon–halogen bond. One special interest in their work has been progress in the understanding of the mechanism of dissociative electron transfer in electrocatalytic processes, which are characterized by the sequence shown in Scheme 17.

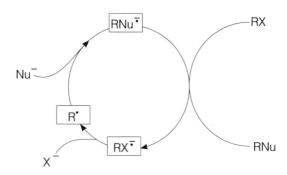

SCHEME 17

There is general agreement that bond cleavage of the saturated carbon–halogen bond is a primary electrochemical event. Platinum electrodes coated with poly-p-nitrostyrene catalyze the reduction of C—X bonds, involving electron transfer[182]. Conversion of 1,2-dibromo-1,2-diphenylethane to stilbene was studied by cyclic voltammetry and by preparative electrolysis using the polymer-coated electrodes. It was found that high efficiencies for the reduction of the dihalide in acetonitrile could be obtained with layers corresponding to a surface charge of 4 mM cm^{-2}. The catalyst has a high turnover number, of the order of 10^4.

Bunnett[183] has examined and rejected several conceivable mechanisms for the suggested direct displacement of halide ion from the radical anion by a nucleophile because of violation of quantum-mechanical principles or incompatibility with experimental observations. Energy relationships are nonlinear and obey a Marcus–Hush quadratic equation. Standard activation energy for electrochemical rates appear to be governed by ease of carbon–halogen bond stretching[184]. Savéant reported that judicious selection of operating potential permits electrocatalyzed nucleophilic substitution in aromatic systems that are not otherwise reactive by ionic pathways. An electrocatalytic pathway involving electroreduction in the substrate, decomposition, recombination and electron transfer gave good yields with poor turnover constants. [$k(t)$=1 to 8)[185]. Substituted fluorenyl halides which give stable radicals and carbanions were studied by cyclic voltammetry. Autocatalysis was demonstrated and an activation–driving force relationship has been studied[172]. The heterogeneous and homogeneous reductive cleavage of organic halides were kinetically studied by direct electrochemical methods and redox-catalysis methods. Aromatic halides differ from the aliphatic halides in their reactivity in the sense that the reduction of aromatic halides involves the intermediacy of radical anions, whereas aliphatic halides undergo the cleavage of the C—X bond in concert with the electron transfer process. There was also an approximate correlation between the rate constant for the cleavage of ArX$^{\cdot}$ and the RX/RX$^{\cdot}$ standard potentials[186]. Cyclic voltammetric studies at vitreous carbon electrodes suggested that primary alkyl halides such as 1-iododecane, 1-bromodecane and 2-bromodecane undergo simultaneous two-electron reduction to give carbanions, which are protonated by the solvent molecules. Secondary and tertiary alkyl halides, on the other hand, show stepwise generation of alkyl radicals and carbanions[187]. Similarly, reduction of alkyl radicals has been studied[188].

Fast-scan cyclic voltammetry (10 mVs^{-1} to 106 Vs^{-1}) was used to measure the rate constants of C—X cleavage, which are extremely fast. The technique was applied to measure rate constants of the order of submicrosecond half-lives. For example, the radical anions generated at the electrode surface were determined to have a half-life ranging from less than 100 ns in the case of p-bromoacetophenone to 70 ms for m-nitrobenzyl chloride[189]. The method complements the redox catalytic method developed by Savéant and coworkers[190].

Uncertainties arising from the estimation of the effect of solvent dynamics and reorganization, as well as of the resonance energy in the transition state have been minimized by utilizing information derived from conventional outer-sphere electron transfer data gathered in the same solvent. This strategy leads to good agreement between theory and experiment in the electrochemical case. In the homogeneous case, the agreement is equally satisfactory in the case of tertiary halides. With the sterically less hindered secondary and primary halides, the reaction possesses an S$_N$2 character involving small bonded interactions in the transition state in accord with earlier stereochemical results and reaction entropy determinations[191].

Daasbjerg and Lund provided an interesting probe of the borderline between direct and mediated reactions[192]. Rate constants (k_r) for the S$_N$2 reaction of superoxide (O$_2^{\cdot-}$) with alkyl halides (e.g. chlorobutane) were compared with the expected rate constants (k_{ET}) for electron transfer reaction between the same alkyl halide and an aromatic anion radical of the same standard oxidation potential as superoxide. The k_r / k_{ET} ratios show that the mechanism of the substitution reaction may shift from S_N2-like to ET-like upon changes in the steric hindrance and the acceptor ability of the alkyl halide[193]. A mechanism for the electrochemical cleavage of the C—X bond was analyzed by correlating the standard potential of the redox system with the rate constant for the formation of the product[194].

Direct and indirect electrochemistry of perfluoroalkyl bromides and iodides has also been reported using glassy-carbon electrodes, and aromatic anion radicals, respectively[195]. The reduction of these substrates does not proceed through the intermediacy of the anion radicals, but involves concerted electron transfer and C–X bond breaking to form the perfluoroalkyl radicals. The perfluoroalkyl radicals are good H-atom scavengers, unlike the alkyl radicals. Reactions of nucleophiles, radical anions or metals with perfluoroalkyl bromides or iodides, which follow an ET mechanism, were reviewed by Wakselman[196].

Autocatalysis of the electrochemical reductive cleavage of the carbon–halogen bond in alkyl halides to give stable radicals and carbanions has been studied kinetically[197]. The dependence of the potential on the carbanion and on the radical has been demonstrated [198]. Reduction of benzyl iodide, for example, at potentials insufficient for the formation of benzyl anion gives bibenzyl as the major product, arising from radical coupling. At more negative potentials (–1.8 V sce) products are derived from the benzyl anion.

Organometallic compounds. Involvement of the electrode material in many electroreductive processes has been well established, particularly in the case of mercury as the cathode and most certainly in the action of sacrificial electrodes. The best evidence for the manner in which metallic electrodes can be involved in the electrochemical reaction comes from studies of mercury as the cathode. A relation between reactivity and electrochemical redox potentials of organomercury compounds was put forward by Butin and coworkers[199], who developed an electrochemical classification of organomercury compounds. Many researchers have reported the electrocatalytic formation of aryl or alkyl mercuric halides at potentials below those necessary to bring about rapid reduction of the carbon–halogen bond. Kretchmer and Glowinski[200] determined that aryl mercuric halides in the presence of copper metal with palladium chloride at 115 °C in pyridine gave biaryl coupling (Ullmann) products in excellent yields. Polarograms of decylmercuric halides in DMF show two waves. Reduction at the first wave gives didecylmercury, whereas reduction at the second potential gives only decane[9]. The role of mercuric salts in the electroreduction of benzylic bromides has been studied in detail[201]. Electroreduction of benzylmercuric halides at a stirred mercury pool gave products which were weakly potential-dependent in contrast with the products of the reduction of benzyl bromide. When benzyl bromide was the starting material, dibenzyl products were formed at low potentials, but at high potentials the products were those of a carbanion intermediate (Figure 2). This result suggests that the mercuric halide intermediate could be involved in the formation of organomercurials.

1,3-Dibromo-1-phenylpropane (226) is the substrate in which the reduction potentials of the two C—Br bonds are very different. In this case the mercuric halide (227) as well as the organomercurial 228 and phenylcyclopropane 229 can be isolated at low potential [–0.85 V (sce)]. However, when the reaction is conducted in the presence of an electroinactive, strongly adsorbed species, formation of the 'dimeric' product (228) is strongly suppressed in favor of 'monomeric' (cyclopropane) products (229) (equation 121). This further suggests that one-electron intermediates dimerize at the electrode surface, particularly when their local concentrations are high. Involvement of the electrode in the electrochemical reduction process influences the coulometric n value. Primary alkyl halides at carbon electrodes give carbanions in a two-electron reduction whereas, at mercury electrodes, two waves are observed. For example, decylmercuric halides in DMF containing tetraalkyl ammonium perchlorates give didecylmercury at the first wave and the carbanion at the second one. The nature of the intermediates was confirmed by bulk electrolysis experiments[9].

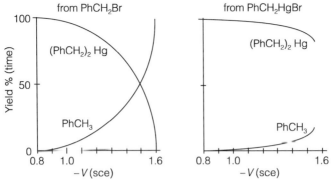

FIGURE 2. Product distribution as a function of reduction potential for benzyl bromide at mercury

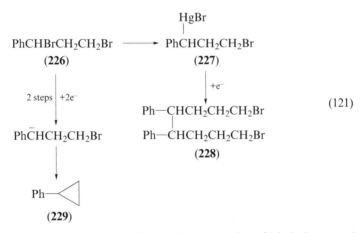

(121)

Little or no organomercurial was reported in the electroreduction of 1-iododecane and 1-bromodecane at a mercury cathode in DMF. This result is consistent with much earlier reports by Sease[262] of high potential electroreduction of monohaloalkanes. Decanol was also formed, and was claimed to arise from an S_N reaction assisted by the electrode[202]. Baizer[203] discussed the mechanism of electroreductive coupling, but did not invoke the possibility of organomercurous/ic halide, citing no evidence for RHgBr. He concluded that it is not possible to reduce the benzyl radical to benzyl carbanion at mercury in aqueous or alcohol solution.

It was shown that controlled potential electrolyses of 1,ω-dibromides (ω equal to or greater than 4) using a stirred mercury cathode yield symmetrical organomercurials in an overall three-electron reduction in high yields. In addition to dialkylmercury compounds, significant amounts of di-n-hexyldimercury were obtained in the case of 1,6-dibromo-hexane. A suggestion is made here again that adsorption of the dibromide on the mercury cathode surface is responsible for the good yields[162,204]. Indeed, the method of preparation is sufficiently general to have permitted the preparation of a variety of symmetrical organomercurials for ^{13}C NMR chemical shift and $^{13}C^{199}Hg$ coupling constant studies[205].

Cathodic reduction of o-dichlorobenzenes in the presence of furan gives α-naphthol as the major product as shown in equation 122. A benzyne intermediate mechanism was

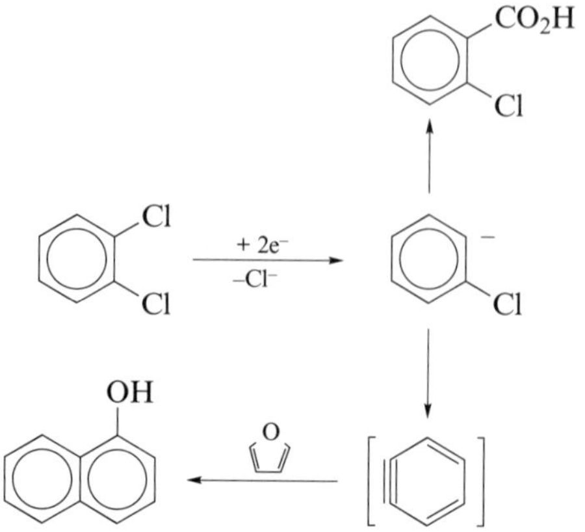

$$\text{o-C}_6\text{H}_4\text{Cl}_2 \xrightarrow[-\text{Cl}^-]{+2e^-} \text{(intermediate)} \qquad (122)$$

proposed for the reaction[206]. Whereas earlier workers favored concerted mechanism for the formation of benzyne, Barba and coworkers showed that the benzyne is formed in an ionic stepwise process, by isolation of o-dichlorobenzoic acid from o-dichlorobenzene in the electroreduction in the presence of carbon dioxide[207].

ESR study of the electrochemical dehalogenation of 2-halo-5-nitrofurans showed that the first step involves the formation of radical anions the dehalogenation of which proceeds via a dianion intermediate for chloro compounds, for the iodo compounds. Both routes were involved for the bromo compounds. The stability of the intermediate anion radicals of 2-halo-5-nitrofurans increased in the order $I < Br < Cl$[208].

6. Influence of solvents and electrolytes

The electrochemical behavior of dansyl chloride depends on the electron donor and the nature of the solvents. In water, DMF or DMSO, dansyl chloride is hydrolyzed to dansyl sulfonic acid in the bulk of the solution and the rate constants have been determined by electrochemical and spectrophotometric measurements. In acetonitrile, the hydrolysis process is so slow that it has been possible to follow the electrochemical reduction of dansyl chloride and to show that the overall process depends on the acidic properties of the solution. Moreover, the half-wave potential value of dansyl chloride in CH_3CN fits the structural correlations determined for a number of benzenesulfonamide derivatives in the same experimental conditions. This fact suggests that the half-wave potential is determined by the first electron transfer, which involves the same C—S—R moiety of the benzenesulfonamide derivatives, since the positive charge is localized on the sulfur atom[209]. Polarographic half-wave potentials are sensitive to supporting electrolyte. The half-wave potentials of alkyl halides were shifted to more negative values as the cation of supporting electrolyte was progressively increased in size. This is due in part to a decrease in the rate constant for electron transfer[210].

Polarograms of 1-iodonorbornane and 1-iododecane showed single irreversible waves corresponding to the electrolytic cleavage of the carbon–halogen bond. The halfwave potentials for these compounds vary depending on the supporting electrolyte. For example, switching from tetramethylammonium perchlorate to tetrabutylammonium perchlorate causes shifts of their potential to more cathodic ($\Delta E_{1/2} = 0.49$ V and 0.25 V respectively for 1-iododecane and 1-iodonobornane)[212]. The 1-norbornyl radical intermediate is either further reduced electrolytically or accepts a hydrogen atom from the solvent giving norbornane, whereas the 1-decyl radical undergoes adsorption onto mercury to give didecylmercury.

Using spin markers it could be shown that redox catalysis occurs in which the solvent itself plays the role of an electron carrier. Thus indirect reduction of aromatic halides having more negative potentials than benzonitrile has been achieved at the reduction potential of benzonitrile when it was used as a solvent[211].

7. Role of the electrode

The nature of the electrode plays a significant role in the direction and often the products of electrochemical processes, particularly reduction. Metals that can form relatively stable organometallics with the substrate under study often intervene directly to produce a product like that of direct organometallic reaction. The electroreduction of alkylmercury halides was studied on Pt, Hg and carbon electrodes. Whereas at Pt and carbon electrodes two-electron reduction was observed, at mercury-coated electrodes multistep reduction occurred and RHgHgX was observed[213].

The cyclic voltammograms for 1-iododecane, 1-bromodecane and 2-bromooctane at vitreous carbon electrodes in DMF containing TMAP show single irreversible reduction waves at –1.41 V, –1.85 V, and –1.6 V (sce), respectively, signaling a two-electron reduction to the carbanion. Large-scale electrolysis of 1-iododecane at vitreous carbon gives decane, eicosane, 1-decene and 1-decanol. The decane arises from protonation of the carbanion, whereas eicosane is obtained from S_N2-like reaction of decyl anion and bromodecane. The last two products, 1-decene and 1-decanol, are formed by reaction (E2 and S_N2) of hydroxide ion (generated from the reaction of water and the carbanion) with bromodecane.

The cyclic voltammograms at vitreous carbon electrodes for 2-iodooctane, t-butyl bromide and t-butyl iodide show two waves [e.g., –1.6 V and –1.8 V (sce) for t-butyl bromide] indicating stepwise generation of alkyl radicals and carbanions. The products of large-scale electrolyses of t-butyl bromide (isobutane, isobutylene, 2,2,3,3-tetramethylbutane) are indicative of the involvement of both radical and carbanion species[214].

Current methods for the preparation of glassy carbon have been described, together with a discussion of recent progress in the modification of the surface of glassy carbon[215]. Reduction of aryl halides and electroformylation reactions have been studied in DMF on a cadmium-modified gold microelectrode[83]. Voltammetric reductions of several organohalides in films of dodecyldimethylammonium bromide (DDAB) and clay-DDAB cast onto pyrolytic graphite electrodes have been examined[216]. In the low temperature polarography of alkyl halides in DMF, adsorption onto mercury of complex species composed of tetramethylammonium and halide ions has been reported[217].

8. Synthesis of the halocarbon bond

CCl_3^- obtained by electroreduction of CCl_4 at mercury or lead cathode at –0.9 to –1.3 V (sce) in DMF were reacted with CH_3I, acetone and cyclohexanone to give the corresponding trichloromethylated products (**230**)[218](equation 123).

Chlorofluoro compounds were electroreductively added to aldehydes to give the corresponding fluorinated alcohols (**231**) (equation 124)[214]. In this process platinum or lead are employed as the cathode, under constant-current conditions. The starting dihalides or trihalides (halogen = F, Cl or Br) can be prepared on an industrial scale by the reaction of the corresponding tri- or tetrahalide at a sacrificial anode (Zn, Al or Mg)[220].

$$CCl_4 \xrightarrow{+e^-} CCl_3^- \xrightarrow{MeCOMe} \begin{array}{c} Me \quad CCl_3 \\ \diagdown / \\ / \diagup \\ Me \quad OH \end{array} \qquad (123)$$

(**230**)

$$CCl_3CF_3 + RCHO \xrightarrow[DMF]{+e^-, \text{ Pt cathode}} \underset{\underset{(231)}{Cl \quad Cl}}{\overset{OH}{R \diagdown \diagup CF_3}} \qquad (124)$$

Electrolytic partial fluorination of organic compounds has been reported. Anodic partial fluorination of β-lactams has been performed for the first time. β-Lactams are monofluorinated in excellent yield with high current efficiency and in a highly regioselective manner. This is claimed to be superior to chemical methods[221]. Perfluoroalkyl halides, upon electroreduction give the perfluoroalkyl radicals, which are not further easily reduced at the potentials where they are generated. The perfluoroalkyl radicals react with imidazole anions (232) or their derivatives by the $S_{RN}1$ mechanism giving the perfluoroalkylated imidazoles (233) (equation 125). The electroreduction was carried out using an electron transfer agent, 4-nitropyridine-N-oxide, which causes the alkyl halide to be reduced at significantly less negative potentials[222].

$$\underset{(232)}{\underset{N}{\overset{N}{\diagup}}\diagdown_{Y}} + R_fX \xrightarrow[S_{RN}1]{e^-} \underset{(233)}{\overset{R_f}{\underset{N}{\overset{N}{\diagup}}}\diagdown_{\underset{H}{Y}}} + X^- \qquad (125)$$

The characteristics of the electrooxidation of fluorosulfate anions in the electrolysis of a potassium fluorosulfate solution in fluorosulfonic acid have been investigated. The formation of oxide layers on platinum and the modification of glassy carbon with fluorosulfate groups during anodic polarization in fluorosulfonic acid are studied. The reactions of fluoroolefin fluorosulfation are considered and a mechanism is suggested[223]. Trifluoromethylation of carbonyl compounds can be achieved using bromotrifluoromethane and a sacrificial electrode in solvents such as DMF/pyridine, and DMF/TMEDA, as seen in equation 126[224].

$$CF_3Br + \underset{R}{\overset{R}{\diagup}}\!\!=\!\!O \xrightarrow{\text{Zn electrodes}} \underset{R}{\overset{R}{CF_3\diagup}}\!\!-\!\!OH \qquad (126)$$

Electrolysis of fluorosulfonic acid in the presence of perfluoro-2-methyl-3-isopropyl-2-pentene yields the products of addition of HSO_3F to the double bond, that is, 2-hydro-3-(fluorosulfato)perfluoro-2-methyl-3-isopropylpentane and 2-fluorosulfato-3-hydroperfluoro-2-methyl-3-isopropylpentane. The formation of monofluorosulfates probably includes the stage of one-electron reduction of the initial olefin[225].

9. Fluoro derivatives: Applications in batteries and fuel cells

The chemical behavior of graphite fluoride samples in concentrated alkaline aqueous solutions (5M) was investigated. The concentration of liberated F^- ions was determined using a selective fluoride electrode. The discharge performances of two CF(x) cathodes were compared, for application in Zn/KOH/CF(x) batteries. In both cases, flat discharge curves were observed, even for high current intensities. Values higher than those obtained with conventional 1M $LiClO_4$ electrolytes may be used, because of the higher conductivity of 5 M KOH aqueous solutions[226]. Polymeric perfluoro bis-sulfonimides have been studied as possible fuel cell electrolytes. Linear polymeric chain perfluorinated bis-sulfonimides have been synthesized. They are stable strong acids with interesting properties as electrolytes. Like sulfonic acids, they are hydronium ion conductors which require the presence of water to obtain significant ionic conductivity. They are therefore not suitable for use as fuel cell electrolytes at temperatures much beyond $100\,^{\circ}C^{227}$. Perfluoroalkane-α,ω-disulfonic acids $[HSO_3(CF_2)_nSO_3H]$ are used as electrolytes in the fuel cells, because of their favorable conductivities and solubilities in water[228]. The $n = 2$ acid was found to be the best fuel cell electrolyte. The electrochemical properties of a homologous series of perfluoroalkane-α,γ-disulfonic acids were measured. Properties measured included the solubility and conductivity of aqueous solutions, the extent of anion-specific adsorption on platinum and oxygen reduction kinetics on platinum, for acids having $n = 2, 4$ and 6.

10. Electrochemical destruction of toxic compounds

Waste liquors from the chlorination of acetic acid, containing predominantly di- and trichloroacetic acid, can be reduced to chloroacetic acid on a magnetite electrode. Reviews of methods for electrochemical degradation of polychlorinated biphenyls were presented. The procedures use both cathodic and anodic processes[229, 230]. Degradation and dehalogenation of polychlorobiphenyls and halogenated aromatic molecules by superoxide ion and by electrolytic reduction has been studied[231].

Electron transfer properties of polyhalogenated biphenyls were investigated by cyclic voltammetry. The primary reduction peak of 4,4′-dichlorobiphenyl, involving replacement of halide with hydrogen in an irreversible ECE- type reaction, are under kinetic control of the initial ET step. Electrochemical transfer coefficients, standard potentials and standard heterogeneous rate constants were also estimated from the voltammetric data[230].

The environmentally hazardous polychlorobiphenyls can be dehalogenated electrochemically under anaerobic conditions to give biphenyls (equation 127). This method is especially suitable for the biphenyls having fewer than three halogen atoms, which are very difficult to reduce chemically, e.g. by superoxides[231]. The reduction potentials of polyhalogenated benzenes were dependent on the electrode material. Thus the reduction potentials of hexafluorobenzene at Pt, vitreous carbon, gold, and Hg electrodes are: –2.18, –2.05, –2.15 and –1.97, respectively. The ease of reduction at the cathodes is therefore of the order Hg>Au>C ~ Pt[232].

$$C_{12}Cl_{10} + 10\ H_2O + 20\ e^- \xrightarrow{-2.5\ V\ vs\ sce} C_{12}H_{10} + 10\ Cl^- + 10\ OH^- \qquad (127)$$

Direct electrolytic dechlorination of 9-chloroanthracene at a mercury electrode occurs at about –1.65 V (sce) in a layer of adsorbed cetyltrimethylammonium bromide on the electrode surface[233]. Similarly, electrochemical degradation of trichloroethylene in acetonitrile resulted in quantitative conversion to chloroacetylene, which was reduced further to acetylene at a more negative reduction potential (–2.8 V) in 96% yield[234]. Reductive destruction of 1,3,5-trichlorobenzene in the cathode compartment could be observed[235]. Electrochemical methods presumably can be used for decontamination of chemical warfare agents such as mustard derivatives as an alternative to the chemical methods such as base-catalyzed dehydrohalogenation[236].

11. Miscellaneous

The variation of the cathodic peak potential with the scan rate (0.3–0.4 mV precision on each determination, 1 mV reproducibility over the whole set of experiments) allows the determination of the rate constant with a relative error of 3–11%. The results are consistent with those derived from anodic-to-cathodic peak current ratios. Simulation of the whole voltammogram confirms the absence of significant systematic errors that could arise from the assumptions underlying the analysis of kinetic data. Activation parameters derived from weighted regression Arrhenius plots of the data points taken at 5 or 6 temperatures are determined with an uncertainty of 4 to 8 kJ mol^{-1} for the activation energy and 12–20 J K^{-1} mol^{-1} for the activation entropy[237].

Improvement and estimation of precision in the cyclic voltammetric determination of rate constants and activation parameters of coupled homogeneous second-order reactions has been claimed[238]. Variations of the single-scan peak current with scan rate and concentration offer a reliable source of data for the determination of the rate constants of second-order reactions involving the substrate and an electrogenerated reagent. The rate constants were determined for the reaction of n-butyl bromide with anthracene radical anion with a relative precision ranging between 3 and 10%. Activation parameters were estimated with a precision of ±2.4 kJ mol^{-1} for the activation energy and of 8 J K^{-1} mol^{-1} for the activation entropy. Repetitive cyclic voltammograms of irreversible systems tend towards a nonzero steady-state limit and can be explained in terms of interference of natural convection. Simulations of the current-potential responses obtained for the reduction of 1-bromonaphthalene in acetonitrile under these conditions, and also when the electrode is rotated, show good agreement between theory and experiment. The parameter controlling the shape of the cyclic voltammetric responses and the rate of their decay towards the steady-state limit is the ratio of the thickness of the diffusion-convection layer to that of a diffusion layer[239].

The cyclic voltammetry of the reductive cleavage of N-fluoro-, N-chloro- and N-bromosaccharin sultams and of their 4-nitro-substituted analogues was investigated in acetonitrile at an inert electrode. The two main factors governing the occurrence of a stepwise or a concerted mechanism in reductive cleavages are the energy of the π^* orbital and the strength of the breaking bond. The standard potential for the reduction of the nitrogen-centered sultam radicals could also be determined from the oxidation of the amide ion obtained from the reductive cleavage of the halosultams or from the deprotonation of the sultams[240].

The photoemission current was investigated on a mercury electrode in 0.5 M KCl containing an acceptor RX (X = Cl, Br, R = isopropyl). In the potential region where RX is not adsorbed by the electrode, the stoichiometric coefficient of photoemission increases with increasing negative potential, which can be related to the decrease in electrooxidation of the anion radical of RX[241, 242].

Aromatic bromides were electrochemically reduced in the presence of a spin marker, such as t-butylphenylnitrone, and studied by ESR. In weakly hydrogen donor solvent (DMSO), the radicals are cleanly trapped by nitrone **234** forming the nitroxide species **235**[243] (equation 128).

$$\text{PhCH}\!=\!\overset{\text{O}}{\overset{\uparrow}{\text{N}}}\,t\text{-Bu} \quad \xrightarrow{\text{R}^\bullet} \quad \text{Ph}\diagdown\underset{\overset{|}{\text{R}}}{\text{N}}\overset{\text{O}^\bullet}{\diagdown}\!t\text{-Bu} \tag{128}$$

$$\textbf{(234)} \qquad\qquad\qquad \textbf{(235)}$$

Polarographic study of diphenyliodonium halides in DMF containing TMAP showed three or four waves depending on the concentrations of substrate or supporting electrolyte. The products were characterized at each wave by bulk electrolysis[21].

Electrochemical reduction of tetra-O-acetyl-α-D-glucopyranosyl bromide (**236**) in dipolar aprotic solvents at a mercury electrode affords glycal **237** in quantitative yield (equation 129). The reaction presumably involves the dehalogenative carbanionic species followed by the loss of acetate group[244].

Organic halides have been converted electrochemically into tertiary phosphines using chlorophosphines. Electrochemical synthesis of a variety of tertiary mono- and diphos-

$$CHClBrCF_3 \xrightarrow{e^-} [CHClBrCF_3]^{\bullet} \xrightarrow{fast} [CHClCF_3]^{\cdot} + Br^-$$

$$\xrightarrow[fast]{+e^-} [CHClCF_3]^- \xrightarrow[fast]{H_2O} CH_2ClCF_3 + OH^-$$

SCHEME 18

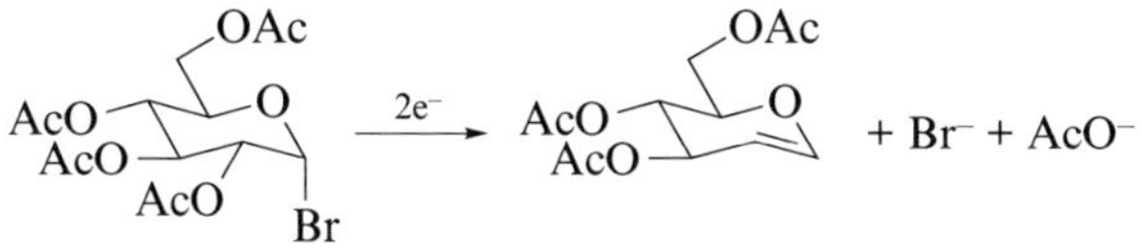

$$\text{(236)} \xrightarrow{2e^-} \text{(237)} + Br^- + AcO^- \tag{129}$$

(236)   **(237)**

phines has been achieved from organic halides and chlorophosphines using an undivided electrolytic cell with a sacrificial anode of magnesium as can be seen in equation 130[245].

$$Ph_2PCl + 2e^- \longrightarrow Ph_2P^- + Cl^-$$

$$Ph_2P^- + RX \longrightarrow Ph_2PR \tag{130}$$

$$RX = PhCH_2Cl, \ Br(CH_2)_3Br, \ etc.$$

The reduction of the anaesthetic 'halothane' at pH 13 was studied on a variety of electrode materials. The overall reaction has been shown to involve two electrons with the production of a bromide ion. The first electron transfer is the slowest electrochemical step. An overall reaction mechanism consistent with these observations is shown in Scheme 18. The first electron transfer involves two electrons and results in the production of a bromide ion. Of these, the first electron transfer is irreversible and the slowest[246, 247].

IV. ANODIC REACTIONS

Electrochemical oxidation, reduction and formation of the C—X bond by direct and in direct processes have been reviewed[3, 248].

Anodic oxidation of homo allyltrimethylsilylmethyl ethers **238** or homo allyl trimethyl-stannyl methyl ethers in the presence of tetrabutylammonium tetrafluoroborate results in the formation of fluorine- containing tetrahydropyrans **239**[249](equation 131). The process involves formation of a resonance stabilized carbocation and its intramolecular cycliza-tion by the participation of a neighboring vinyl group, followed by attack of fluoride ion. This process is a convenient way to form the C—F bond involving electrochemical steps.

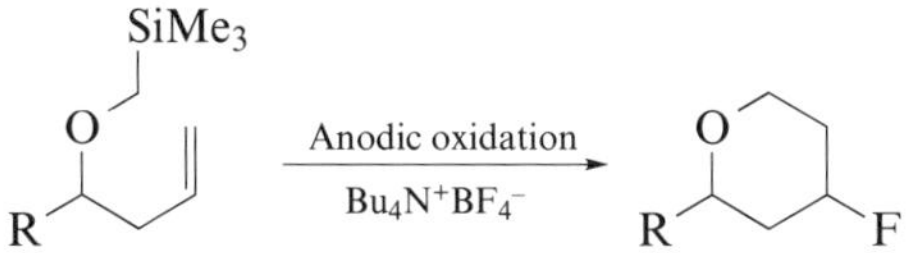

$$\text{(238)} \xrightarrow[Bu_4N^+BF_4^-]{\text{Anodic oxidation}} \text{(239)} \tag{131}$$

(238)   **(239)**

Electrooxidation of a mixture of iodine and aromatics in acetonitrile gave monoiodides in low yields from many aromatic substrates. However, good yields could be obtained by preoxidizing the I_2, followed by the addition of aromatics. Reactive iodinated species such as N-iodoacetonitrilium ion are implicated in this reaction[250]. Preparation and high-pressure liquid chromatography of iodinated diethylstilbestrols and some related steroids for medical application and their subsequent labeling with radioiodine by a cuprous chloride catalyzed exchange reaction has been reported[251, 253].

Poly(oxy-1,4-phenylene) is obtained by electrooxidative polymerization of *p*-bromo-phenol in aqueous NaOH solution. The yield increases when aqueous NaOH is replaced by aqueous KOH or when the reaction is conducted at higher temperature. In contrast, *p*-chlorophenol electrooxidatively dimerizes to give the biologically and pharma-cologically important dioxin, 2,7-dichlorodibenzo[*b*,*e*][1,4]dioxine[254]. In an effort to find protective chemical coatings, electrooxidative polymerization of *m*-chlorophenol and *m*-bromophenol was observed[255].

The characteristics of adsorption and electrooxidation of perfluorocarboxylic acids and organic compounds, which are radical acceptors, on a platinum electrode have been inves-tigated over a wide range of potentials (0 to 3 V). The binding of radicals produced during the electrolysis of perfluorocarboxylates by perfluoroolefins was studied using the electron paramagnetic resonance (EPR) technique. It was shown that, depending on the radical structure and the nature of the solvent, perfluoroalkyl radicals participate in the perfluoroalkylation of organic compounds adsorbed on the electrode surface or dissolved in the solution[256].

Regioselective carbon–carbon bond formation of 2,2,2-trifluoroethyl substituted ter-tiary amines initiated by electron transfer to photochemically excited enone has been studied. Photoaddition of *N*-(2,2,2-trifluoroethyl)amines to 2-phenylcyclohex-2-en-1-one occurred at the α-position to the trifluoromethyl group predominantly. The regioselectiv-ity of this photoaddition reaction was comparable to that of anodic methoxylation of *N*-(2,2,2-trifluoroethyl)anilines. This reaction is the first example of direct carbon substi-tution at α to the trifluoromethyl group and this promotion effect was general for other electron-withdrawing groups[257]. Highly regioselective electrochemical monofluorination of α-phenylsulfenyl lactams was carried out in good chemical and current yields. This is the first successful example of anodic monofluorination of aliphatic nitrogen-containing heterocycles.

Electrochemical oxidation of sulfides at Pt electrodes in acetonitrile in the presence of $Et_3N \cdot 3HF$ gives α-fluorination (equation 132)[258].

$$\underset{Ph}{\overset{O}{\diagdown}}\diagup SPh \xrightarrow[\substack{-2e^-, -H^+ \\ 87\%}]{Et_3N \cdot 3HF; \ E = 1.4 \ V} \underset{Ph}{\overset{O}{\diagdown}}\diagup \underset{F}{\overset{|}{SPh}} \tag{132}$$

The oxidation potentials of several haloaromatics were determined at a Pt electrode in acetonitrile containing TBAF. The halogens destabilize the aromatic cation-radicals by inductive effect. Thus the oxidation potentials increase with increase in the electronegativ-ity of the halogens, being most favorable with aryl iodides (equation 133).

$$PhI \xrightarrow{-e^-} PhI^{+\cdot} \tag{133}$$

It has been proposed that mediated electrochemical oxidation may be used for the ambient temperature destruction of hazardous waste. Using $Co(III)^-$-mediated electro-chemical oxidation in sulfuric acid, 1,3-dichloro-2-propanol and 2-chloro-1-propanol were oxidized to carbon dioxide[259]. A series of studies is being conducted on the use of the anodic oxidation of barium peroxide to produce an intermediate that leads to destruction of halogenated organic compounds.

The cyclic voltammetry of 1,2,3,4,5-pentachloro- and of decachloroferrocene have been studied in acetonitrile. A plot of redox potential $E(f)$ vs. σ_p (σ_p is the Hammett *para* coefficient for the chloro substituent) shows that the neutral molecules are stabilized with respect to the corresponding ferrocenium cations by 0.16 to 0.12 V per Cl^{260}.

From kinetic studies it was shown that nucleophilic substitutions involving sterically hindered substrates, such as bromoadamantane proceed through single electron transfer[261].

V. ACKNOWLEDGMENTS

We dedicate this chapter to the memory of Manuel M. Baizer, an outstanding human being and a scientist who has done more to advance the cause of organic electrochemistry than any other person we know. We acknowledge in general the encouragement of Professor George A. Olah and of the Loker Hydrocarbon Research Institute of the University of Southern California and in particular one of us (V. P. R.) for financial support during part of the time this chapter was written.

VI. REFERENCES

1. J. Casanova and L. Eberson, 'Electrochemistry of the carbon–halogen bond', in *The Chemistry of the Carbon–Halogen Bond* (Ed. S. Patai), Wiley, London, 1974, pp. 979–1047.
2. J.-M. Savéant, *Acc. Chem. Res.*, **26**, 455 (1993).
3. J. Y. Becker, in *The Chemistry of Halides, Pseudohalides and Azides* (Eds. S. Patai and Z. Rappoport), Wiley, Chichester, 1983; pp 203–286.
4. A. J. Fry, *Synthetic Organic Electrochemistry*, Wiley, New York, 1989.
5. L. Eberson, *Acta Chem. Scand.*, **B36**, 533 (1982).
6. J.-M. Savéant, *Bull. Soc. Chim. Fr.*, 225 (1988).
7. D. G. Peters, in *Organic Electrochemistry*, 3rd ed. (Eds. H. Lund and M. M. Baizer), Marcel Dekker, New York, 1991, 1576 pp.
8. G. J. Hills, in *Special Periodical Reports*, Vol. I, The Chemical Society, London, 1970, pp. 118–121.
9. T. Shono and H. Hamaguchi, *Kagaku (Kyoto)*, **36**, 560 (1981); *Chem. Abstr.*, **95**, 114668r (1981).
10. A. J. Fry, *Aldrichimica Acta*, **26**, 3 (1993).
11. N. Rabjohn, *Chem. Eng. News*, **43**, 57 (1984).
12. M. Tarle, *Kemija U. Industiji*, **11**, 647 (1974); *Chem. Abstr.*, **83**, 154439a (1974).
13. J. L. Webb, C. K. Mann and H. M. Walborsky, *J. Am. Chem. Soc.*, **92**, 2042 (1970).
14. P. J. Stang, *Angew Chem., Int. Ed. Engl.*, **31**, 274 (1992); R. M. Moriarty and R. K. Vaid, *Synthesis*, 431 (1990).
15. M. von Stackelberg and W. Stracke, *Z. Electrochem.*, **53**, 118 (1949).
16. E. Canadell, P. Karafiloglou and L. Salem *J. Am. Chem. Soc.*, **102**, 855 (1980).
17. R. Bilewicz and J. Osteryoung, *J. Electroanal. Chem. Interfacial Electrochem.*, **226**, 27 (1987).
18. K. L. Vieira and D. G. Peters, *J. Electroanal. Chem. Interfacial Electrochem.*, **196**, 93 (1984); K. Vieira, *Diss. Abstr. Int. B*, **45**, (1984); *Chem. Abstr.*, **102**, 35177 (1984).
19. K. L. Vieira and D. G. Peters, *J. Org. Chem.*, **51**, 1231 (1986).
20. K. Daasbjerg, J. N. Hansen and H. Lund, *Acta Chem. Scand.*, **44**, 711 (1990).
21. M. S. Mubarak and D. G. Peters, *J. Electroanal. Chem. Interfacial Electrochem.*, **152**, 183 (1983).
22. W. F. J. Carroll and D. G. Peters, *J. Electrochem. Soc.*, **127**, 2594 (1980).
23. M. T. Ismail and M. F. Elzohry, *J. Chem. Technol. Biotechnol.*, **56**, 135 (1993).
24. F. Maran, *J. Am. Chem. Soc.*, **115**, 6557 (1993).
25. M. A. Casadei, S. Cesa, F. M. Moracci and A. Inesi, *Gazz. Chim. Ital.*, **123**, 457 (1993).
26. T. Nakajima and Y. Matsuo, *Carbon*, **30**, 1119 (1992).
27. T. Komura, T. Terashima and K. C. Takahashi, *Denki Kagaku oyobi Kogyo Butsuri Kagaku*, **59**, 780 (1991); *Chem. Abstr.*, **116**, 79326y (1991).
28. S. Mabrouk, S. Pellegrini, J. C. Folest, Y. Rollin and J. Perichon, *J. Organomet. Chem.*, **301**, 391 (1986).
29. M. Troupel, Y. Rollin, O. Sock, G. Meyer and J. Perichon, *Nouv. J. Chim.*, **10**, 593 (1986).
30. M. Troupel, Y. Rollin, G. Meyer and J. Perichon, *Nouv. J. Chim.*, **9**, 487 (1985).
31. S. Ozaki, H. Matsushita and H. Ohmori, *J. Chem. Soc., Chem. Commun.*, 1120 (1992).

32. C. M. Elliott and C. A. Marrese, *J. Electroanal. Chem. Interfacial Electrochem.*, **119**, 395 (1980).
33. J. L. Hall, R. D. Geer and P. W. Jennings, *J. Org. Chem.*, **43**, 4364 (1978).
34. M. M. Baizer and J. L. Chruma, *J. Org. Chem.*, **37**, 1951 (1972).
35. S. Satoa, H. Suginome and M. Tokda, *Bull. Chem. Soc. Jpn.*, **54**, 3456 (1981).
36. C. Saboureau, M. Troupel, S. Sibille, E. D'Incan and J. Perichon, *J. Chem. Soc. Chem. Commun.*, 1138 (1989).
37. D. Vanhoye, F. Bedioui, A. Mortreux and F. Petit, *Tetrahedron Lett.*, **29**, 6441 (1988).
38. K. Yoshida, E. Kunugita, M. Kobayashi and S. Amano, *Tetrahedron Lett.*, **39**, 6371 (1989).
39. C. Saboureau, M. Troupel, S. Sibille, E. D'Incan and J. Perichon, *J. Chem. Soc., Chem. Commun.*, 895 (1989); M. Heintz, O. Sock, C. Saboureau, J. Perichon and M. Troupel, *Tetrahedron Lett.*, **44**, 1631 (1988); G. Silvestri, S. Gambino, G. Filardo and A. Gulotta, *Angew. Chem., Int. Ed. Engl.*, **23**, 979 (1984).
40. K. Yoshida, M. Kobayashi and S. Amano, *J. Chem. Soc., Perkin Trans. 1*, 1127 (1992).
41. T. Shono, I. Nishiguchi and H. Ohmizu, *Chem. Lett.*, 1021 (1977).
42. L. Garnier, Y. Rollin and J. Perichon, *J. Organomet. Chem.*, **367**, 347 (1989).
43. K. Uneyama, H. Matsuda and S. Torii, *Tetrahedron Lett.*, **25**, 6017 (1984).
44. S. Sibille, E. D'Incan and L. Leport, *Tetrahedron Lett.*, **28**, 55 (1987).
45. C. Degrand and R. Prest, *J. Electroanal. Chem.*, **282**, 281 (1990); C. Thobie-Gautier and C. Degrand, *J. Org. Chem.*, **56**, 5703 (1991).
46. L. Garnier, Y. Rollin and J. Perichon, *J. Organomet. Chem.*, **367**, 347 (1989); C. Thobie-Gautier and C. Degrand, *J. Org. Chem.*, **56**, 5703 (1991).
47. J. Chaussard, M. Troupel, Y. Robin, G. Jacb and J. P. Juhauz, *J. Appl. Electrochem.*, **19**, 345 (1989).
48. D. D. Tanner, J. A. Plambeck, D. W. Reed and T. W. Mojelsky, *J. Org. Chem.*, **45**, 5177 (1980).
49. F. Maran and E. Vianello, *Tetrahedron Lett.*, **31**, 5803 (1990).
50. R. B. Yamasaki, M. Tarle and J. Casanova, *J. Org. Chem.*, **44**, 4519 (1979).
51. A. J. Fry and T. A. Powers, *J. Org. Chem.*, **52**, 2498 (1987).
52. T. Lund and H. Lund, *Acta Chem. Scand.*, *Ser. B*, **B41**, 93 (1987).
53. B. Vieth, W. Jugelt and F. Pragst, *Z. Phys. Chem. (Leipzig)*, **270**, 338 (1989).
54. J. C. Folest, J. Y. Nedelec and J. Perichon, *J. Chem. Res. (S)*, 3394 (1989).
55. S. Satoh, H. Suginome and M. Tokda, *Bull. Chem. Soc. Jpn.*, **56**, 1791 (1983); M. Tokuda, S. Satoh and H. Suginome, *J. Org. Chem.*, **54**, 5608 (1989).
56. H. Hebri, E. Dunach and J. Perichon, *Tetrahedron Lett.*, **34**, 1475 (1993).
57. S. Sibille, E. D'Incan, L. Lepart and J. Perichon, *Tetrahedron Lett.*, **27**, 3129 (1986).
58. J. Yoshida, K. Funahashi and N. Kawabata, *J. Org. Chem.*, **51**, 3996 (1986); T. Shono, Y. Matsumura, S. Katoh, and N. Kise, *Chem. Lett.*, 463 (1985).
59. G. Bontempelli, S. Daniele and M. Fiorani, *J. Electroanal. Chem. Interfacial Electrochem.*, **160**, 249 (1984).
60. G. N. Kamau and J. F. Rusling, *J. Electroanal. Chem. Interfacial Electrochem.*, **240**, 217 (1988).
61. G. Montero, G. Quintanilla and F. Barba, *J. Electroanal. Chem.*, **345**, 457 (1993).
62. A. N. Zhuravlev, S. D. Shamshinov and I. A. Avrutskaya, *Sov. Electrochem. (Engl. Transl.)*, **28**, 709 (1992). *Chem. Abstr.*, **117**, 200409m (1992).
63. M. S. Mubarak, G. A. Urove and D. G. Peters, *J. Electroanal. Chem.*, **350**, 205 (1993).
64. G. A. Urove, D. G. Peters and M. S. Mubarak, *J. Org. Chem.*, **57**, 786 (1992).
65. G. A. Urove and D. G. Peters, *J. Electroanal. Chem.*, **140**, 932 (1993).
66. G. A. Urove and D. G. Peters, *J. Electroanal. Chem.*, **352**, 229 (1993).
67. G. A. Urove and D. G. Peters, *J. Electroanal. Chem., Electroorg. Synth.*, *Manuel M. Baizer Meml. Symp.*, U.C., Santa Barbara, 1991.
68. G. A. Urove and D. G. Peters, *Tetrahedron Lett.*, **34**, 1271 (1993).
69. A. Guirado, F. Barba, C. Manzanera and M. D. Velasco, *J. Org. Chem.*, **47**, 142 (1982).
70. F. Barba, A. Guirado, J. I. Lozano, A. Zapata and J. Escudero, *J. Chem. Res. (S)*, **10**, 290 (1991); *Chem. Abstr.*, **115**, 255738o (1991).
71. A. Guirado, A. Zapata and M. Fenor, *Tetrahedron Lett.*, **33**, 4779 (1992).
72. H. Marzouk, Y. Rollin, J. C. Folest, J. Y. Nedelec and J. Perichon, *J. Organomet. Chem.*, **369**, C47 (1989).

73. J. R. Cockrell and R. W. Murray, *J. Electrochem. Soc.*, **119**, 849 (1972).
74. J. Grimshaw and G. J. Trocha-Grimshaw, *J. Chem. Soc., Perkins Trans. 1*, 215 (1975).
75. N. Sato, A. Yoshiyama, P. C. Cheng, T. Nonaka and M. Sasaki, *J. Appl. Electrochem.*, **22**, 1082 (1992).
76. S. Donnelly, J. Grimshaw and G. J. Trocha-Grimshaw, *J. Chem. Soc., Perkin Trans.*, *1*, 1557 (1993).
77. J. Grimshaw, R. Hamilton and G. J. Trocha-Grimshaw, *J. Chem. Soc., Perkin Trans*, *1*, 229 (1982); R. Gottlieb and J. L. Neumeyer, *J. Am. Chem. Soc.*, **98**, 7108 (1976).
78. Z. Chami, M. Gareil, J. Pinson, J.-M. Savéant and A. Thiebault, *J. Org Chem.*, **56**, 586 (1991).
79. M. T. Ismail, W. A. A. Abdel, O. S. Mohamed and A. A. Khalaf, *Bull. Soc. Chim. Fr.*, 1174 (1985).
80. Z. Chami, M. Gareil, J. Pinson, J.-M. Savéant and A. Thiebault, *Tetrahedron Lett.*, **29**, 639 (1988).
81. Z. Chami, M. Gareil, J. Pinson, J.-M. Savéant and A. Thiebault, *Croat. Chem. Acta*, **63**, 239 (1990).
82. N. S. Murcia and D. G. Peters, *J. Electroanal. Chem.*, **326**, 69 (1992).
83. E. Lojou, M. Devaud, M. Heintz, M. Troupel and J. Perichon, *Electrochim. Acta*, **38**, 613 (1993).
84. J. Simonet and H. Dupuy, *J. Electroanal. Chem.*, **327**, 201 (1992).
85. J. Hoffmann, A. Belkasmioui and J. Simonet, *J. Electroanal. Chem. Interfacial Electrochem.*, **307**, 155 (1991)
86. P. Nelleborg., H. Lund and J. Eriksen, *Tetrahedron Lett.*, **26**, 1773 (1985).
87. S. Sibille, M. Troupel, J. F. Fauvarque and J. Perichon, *J. Chem. Res. (S)*, 147 (1980).
88. M. Grzeszczuk and D. E. Smith, *J. Electroanal. Chem. Interfacial Electrochem.*, **201**, 315 (1986).
89. J. Heinze and J. Schwart, *J. Electroanal. Chem. Interfacial Electrochem.*, **126**, 283 (1981); see also C. Amatore, J. Chaussard, J. M. Pinson, J.-M. Savéant and A. Thiebault, *J. Am. Chem. Soc.*, **101**, 6012 (1979).
90. A. Conan, S. Sibille, E. D'Incan and J. Perichon, *J. Chem. Soc. Chem. Commun.*, 48 (1990).
91. M. A. Fox, D. A. Chandler and C. Lee, *J. Org. Chem.*, **56**, 3246 (1991).
92. C. Amatore, A. Jutand and L. Mottier, *J. Electroanal. Chem. Interfacial Electrochem.*, **306**, 125 (1991); C. Amatore and A. Jutand, *J. Electroanal. Chem. Interfacial Electrochem.*, **306**, 141 (1991).
93. C. Combellas, H. Marzouk, C. Suba and A. Thiebault, *Synthesis*, 788 (1993).
94. M. Troupel, Y. Rollin, S. Sibille, J. Perichon and J. F. Fauvarque, *J. Organomet. Chem.*, **202**, 435 (1980).
95. M. Bordeau, C. Biran, P. Pons, M. P. Legerlambert and J. Dunogues, *J. Org. Chem.*, **57**, 4705 (1992).
96. P. R. Goetzschatowitz, G. Struth, J. Voss and G. Wiegand, *J. Prakt. Chem.*, *Chem. Ztg.*, **335**, 230 (1993).
97. L. F. Filimonova, S. M. Makarochkina, I. N. Chernykh, V. G. Soldatov, G. V. Motsak and A. P. Tomilov, *Sov. Electrochem. (Engl. Transl.)*, **28**, 725 (1992).
98. S. Durandetti, S. Sibille and J. Perichon, *J. Org. Chem.*, **56**, 3255 (1991).
99. Y. W. Lu, J. Y. Nedelec, J. C. Folest and J. Perichon, *J. Org. Chem.*, **55**, 2503 (1990).
100. R. E. Erickson, R. Annino, M. D. Scanlon and G. Zon, *J. Am. Chem. Soc.*, **91**, 1767 (1969).
101. Y. M. Kargin, E. I. Gritenko, V. V. Yanilkin, V. V. Plemenkov, L. K. Dubovik, N. I. Maksimyuk, B. M. Garifullin, S. K. Letypov and A. V. Ilyasov, *Bull. Russ. Acad. Sci. Ch. (Engl. Transl.)*, (1992); *Chem. Abstr.*, **118**, 80346k (1992).
102. V. A. Afanas'ev, T. G. Vaistub, L. S. Medoks, G. N. Nesterenko, T. G. Panina and B. G., Rogachev, *Izv. Akad. Nauk SSSR.*, *Ser. Khim.*, 955 (1989); *Chem. Abstr.*, **110**, 239086j.
103. S. Wawzonek and J. M. Shradel, *Electrochem. Sci. Technol.*, **126**, 401 (1979).
104. A. J. Fry and J. Touster, *J. Org. Chem.*, **54**, 4829 (1989); A. J. Fry, U. N. Sirisoma and A. S. Lee, *Tetrahedron Lett.*, **34**, 809 (1993).
105. J. Touster, *Denki Kagaku oyobi Kogyo Butsuri Kagaku*, **59**, 780 (1991); *Chem. Abstr.*, **116**, 59476b.
106. A. J. Fry and U. N. Sirisoma, *J. Org. Chem.*, **58**, 4919 (1993).

107. T. B. Brand and M. DeVaud, *Tetrahedron Lett.*, **28**, 3799 (1987).
108. F. Karrenbuck, H. J. Schafer and I. Langer, *Tetrahedron Lett.*, 2915 (1979); W. J. M. vanTilborg and C. J. Smith, *J. R. Neth. Chem. Soc.*, **99**, 202 (1980).
109. M. J. C. Le and J. Sarrazin, *J. Chem. Res. (S)*, 26 (1989).
110. R. Holze and U. Fette, *J. Electroanal. Chem.*, **339**, 247 (1992).
111. A. Inesi and L. Rampazzo, *J. Electroanal. Chem. Interfacial Electrochem.*, **49**, 85 (1974).
112. A. Inesi, L. Rampazzo and A. Zeppa, *J. Electroanal. Chem.*, **69**, 203 (1976).
113. S. D. Datsenko, N. V. Ignat'ev and L. M. Yagupolskii, *Elektrokhimiya*, **27**, 1674 (1991); *Chem. Abstr.*, **115**, 265300n (1991).
114. J. Zavada, T. Krupicka and J. Sicher, *Collect. Czech. Chem. Comman.*, **28**, 1644 (1963).
115. A. J. Klein and D. H. Evans, *J. Am. Chem. Soc.*, **101**, 757 (1979).
116. D. Lexa, J.-M. Savéant, K. B. Su and D. L. Wang, *J. Am. Chem. Soc.*, **109**, 6464 (1987).
117. D. Lexa, J.-M. Savéant, H. J. Schaefer, K. B. Su, B. Vering and D. L. Wang, *J. Am. Chem. Soc.*, **112**, 6162 (1990).
118. H. R. Koch and M. G. McKeon, *J. Electroanal. Chem. Interfacial Electrochem.*, **30**, 331 (1971).
119. J. Casanova and H. R. Rogers, *J. Org. Chem.*, **39**, 2408 (1974).
120. A. J. Fry, *Fortschr. Chem. Forschung*, **34**, 1 (1972).
121. P. Fawell, J. Avraamides and G. Heftner, *Aust. J. Chem.*, **44**, 791 (1991).
122. J. Zavada, J. Krupicka, O. Kocian and M. Pankova, *Collect. Czech. Chem. Commun.*, **48**, 3552 (1983).
123. O. R. Brown, P. H. Middleton and T. L. Threlfall, *J. Chem. Soc., Perkin Trans*, **2**, 955 (1984).
124. A. Inesi and L. Rampazzo, *J. Electroanal. Chem. Interfacial Electrochem.*, **54**, 289 (1974).
125. P. Fawell, J. Avraamides and G. Hefter, *Aust. J. Chem.*, **43**, 1421 (1990).
126. T. Lund, C. Bjorn, H. S. Hansen, A. K. Jensen and T. K. Thorsen, *Acta Chem. Scand.*, **47**, 877 (1993).
127. H. Lund and E. Hobolth, *Acta Chem. Scand., Ser. B*, **B30**, 895 (1987); T. Lund, S. U. Pedersen, H. Lund, K. M. Cheung and J. H. P. Utley, *Acta Chem. Scand., Ser. B*, **B41**, 285 (1987).
128. R. D. Rieke and P. M. Hudnall, *J. Am. Chem. Soc.*, **95**, 2646 (1973).
129. L. Rampazzo, A. Inesi and R. M. Bettolo, *J. Electronal. Chem. Interfacial Electrohem.*, **83**, 341 (1977).
130. J. Casanova, H. R. Rogers, J. Murray, R. Ahmed, O. Rasmy and M. Tarle, *Croat. Chim. Acta*, **63**, 225 (1990); *Chem. Abstr.*, **114**, 90537e (1990).
131. J. Casanova and H. R. Rogers, *J. Org. Chem.*, **39**, 3803 (1974).
132. J. Casanova, J. Bragin and F. D. Cottrell, *J. Am. Chem. Soc.*, **100**, 2264 (1978).
133. V. S. Kublanovskii, K. I. Litovchenko, E. A. Stezeryanskii, I. V. Plyuto and A. P. Shpak, *Dokl. Akad. Nauk Ukr. SSR, Ser. B: Geol., Khim. Biol. Nauki*, 36 (1989); *Chem. Abstr.*, **112**, 78100h.
134. R. Tokoro and J. Osteryoung, *An. Simp. Bras. Eletroquim. Eletroanal.*, *3rd*, **1**, 307 (1982); *Chem. Abstr.*, **98**, 61911t (1982).
135. T. F. Connors, J. V. Arena and J. F. Rushing, *J. Phys. Chem.*, **92**, 2810 (1988); A. Owlia, Z. Wang and J. F. Rushing, *J. Am. Chem. Soc.*, **111**, 5091 (1989).
136. R. S. Wade and C. E. Castro, *J. Am. Chem. Soc.*, **95**, 226 (1973).
137. C. A. Schanke and L. P. Wackett, *Environ. Sci. Technol.*, **26**, 830 (1991).
138. M. O. Iwunze, N. F. Hu and J. F. Rusling, *J. Electroanal. Chem.*, **333**, 353 (1992).
139. J. H. Hutchinson, G. Pattenden and P. L. Myers, *Tetrahedron Lett.*, **28**, 1313 (1987); M. J. Begley, H. Bhandal, J. H. Hutchinson and G. Pattenden, *Tetrahedron Lett.*, **28**, 1317 (1987).
140. S. Torii, T. Inokuchi and T. Yukawa, *J. Org. Chem.*, **50**, 5875 (1985).
141. G. Bosser and J. Paris, *J. Chem. Soc., Perkin Trans. 2*, 2057 (1992).
142. A. Inesi and L. Rampazzo, *Ann. Chim.*, **65**, 279 (1975).
143. A. J. Fry and W. E. Britton, *Tetrahedron Lett.*, 4363 (1971).
144. A. J. Fry and W. E. Britton, *J. Org. Chem.*, **38**, 4016 (1973).
145. K. B. Wiberg and G. A. Epling, *Tetrahedron Lett.*, 1119 (1974).
146. A. Grimshaw and J. Grimshaw, *J. Chem. Soc.*, 425 (1973).
147. J. Hoffmann and J. Voss, *Chem. Ber.*, **125**, 1415 (1992).
148. W. F. Carroll and D. G. Peters, *J. Am. Chem. Soc.*, **102**, 4127 (1980).
149. J. P. Dirlam, L. Eberson and J. Casanova, *J. Am. Chem. Soc.*, **94**, 240 (1972).

150. A. J. Fry and R. Scoggins, *Tetrahedron Lett.*, 4079 (1972).
151. L. Rampazzo, A. Inesi and Z. Alessandra, *J. Electroanal. Chem.*, **76**, 175 (1977).
152. A. J. Fry and A. T. Lefor, *J. Org. Chem.*, **44**, 1270 (1979).
153. A. J. Fry and J. J. O'Dea, *J. Org. Chem.*, **40**, 3625 (1975).
154. A. J. Fry and J. P. Bujanauskas, *J. Org. Chem.*, **43**, 3157 (1978).
155. H. Bohm, H. Kalo, H. C. Yarnitsky and D. Ginsburg, *Tetrahedron*, **30**, 217 (1974).
156. K. B. Wiberg, G. A. Epling and M. Jason, *J. Am. Chem. Soc.*, **96**, 912 (1974).
157. J. J. Dannenberg, T. M. Prociv and C. Hunt, *J. Am. Chem. Soc.*, **96**, 913 (1974).
158. W. A. Pritts, K. L. Vieira and D. G. Peters, *Anal. Chem.*, **65**, 2145 (1993).
159. C. Giomini, A. Inesi and E. Zeuli, *J. Chem. Res. (S)*, 280 (1983).
160. C. Giomini and A. Inesi, *Electrochim. Acta*, **29**, 1107 (1984).
161. S. Satoh, M. Itoh and M. Tokuda, *J. Chem. Soc., Chem. Commun.*, 481 (1978).
162. J. Casanova and H. R. Rogers, *J. Am. Chem. Soc.*, **96**, 1942 (1974).
163. C. P. Andrieux, J. Badoz-Lambling, C. Combellas, D. Lacombe, J.-M. Savéant, A. Thiebault and D. Zann, *J. Am. Chem. Soc.*, **109**, 1518 (1987).
164. Y. Rollin, G. Meyer, M. Troupel and J. F. Fauvarque, *Tetrahedron Lett.*, **23**, 3573 (1982); Y. Rollin, G. Meyer, M. Troupel, J. F. Fauvarque and J. Perichon, *J. Chem. Soc., Chem. Commun.*, 793 (1983); Y. Rollin, M. Troupel, J. Perichon and J. F. Fauvarque, *J. Chem. Res. (S)*, 322 (1981).
165. C. Amatore and A. Jutand, *Organometallics*, **7**, 2203 (1988).
166. A. M. Stolzenberg and M. T. Stershic, *J. Am. Chem. Soc.*, **110**, 5397 (1988).
167. G. Meyer, M. Troupel and J. Perichon, *J. Organomet. Chem.*, **393**, 137 (1990).
168. M. S. Mubarak and D. G. Peters, *J. Electroanal. Chem.*, **332**, 127 (1992).
169. A. J. Fry and U. N. Sirisoma, *J. Org. Chem.*, **58**, 4919 (1993).
170. A. Conan, S. Sibille and J. Perichon, *J. Org. Chem.*, **56**, 2018 (1991).
171. A. J. Fry and P. F. Fry, *J. Org. Chem.*, **58**, 3496 (1993).
172. C. P. Andrieux, A. Merz and J.-M. Savéant, *J. Am. Chem. Soc.*, **107**, 6097 (1985).
173. A. Jutand, S. Negri and A. Mosleh, *J. Chem. Soc., Chem. Commun. 1*, 1729 (1992).
174. J. E. Anderson, Y. H. Liu and K. M. Kadish, *Inorg. Chem.*, **26**, 4174 (1987).
175. S. Hashiba, T. Fuchigami and T. Nonaka, *J. Org. Chem.*, **54**, 2475 (1989).
176. D. A. Koch, B. J. Henne and D. E. Bartak, *J. Electrochem. Soc.*, **134**, 3062 (1988).
177. Z. Chami, M. Gareil, J. Pinson, J.-M. Savéant and A. Thiebault, *Tetrahedron Lett.*, **29**, 639 (1988).
178. C. P. Andrieux and J.-M. Savéant, 'Electrochemical reactions', in *Investigation of Rates and Mechanisms of Reactions, Techniques of Chemistry* (Ed. C. F. Bernasconi), Vol. VI/4E, Part 2, Wiley, New York, 1986, pp. 305–390.
179. G. B. Gavioli, M. Borsari and C. Fontanesi, *J. Chem. Soc., Faraday Trans.*, **89**, 3931 (1993).
180. L. Eberson, 'Electron transfer reactions in organic chemistry', in *Adv. Phys. Org. Chem.*, Vol. 18 (Eds. V. Gold and D. Bethel), Academic Press, London, 1982, pp. 79–185.
181. J.-M. Savéant, *Acc. Chem. Res.*, **26**, 456 (1993).
182. J. B. Kerr, L. L. Miller and M. R. Van De Mark, *J. Am. Chem. Soc.*, **102**, 3383 (1980).
183. J. F. Bunnett, *Tetrahedron*, **49**, 4477 (1993).
184. C. P. Andrieux, I. Gallardo, J.-M. Savéant and K. B. Su, *J. Am. Chem. Soc.*, **108**, 638 (1986).
185. J.-M. Savéant, *Acc. Chem. Res.*, **13**, 323 (1980).
186. C. P. Andrieux, J.-M. Savéant and K. B. Su, *J. Phys. Chem.*, **90**, 3815 (1985).
187. C. P. Andrieux, I. Gallardo, J.-M. Savéant and K. B. Su, *J. Electroanal. Chem. Interfacial Electrochem.*, **198**, 107 (1986).
188. J. A. Cleary, M. S. Mubarak, K. L. Viera, M. R. Anderson and D. G. Peters, *J. Am. Chem. Soc.*, **111**, 1620 (1989).
189. D. O. Wipf and R. M. Wightman, *J. Phys. Chem.*, **93**, 4286 (1989).
190. J.-M. Savéant and K. B. Su, *J. Electroanal. Chem.*, **196**, 1 (1985); C. P. Andrieux and J.-M. Savéant, *J. Electroanal. Chem.*, **205**, 43 (1986).
191. J.-M. Savéant, *J. Am. Chem. Soc.*, **114**, 10595 (1992).
192. C. P. Andrieux, P. Hapiot and J.-M. Savéant, *J. Phys. Chem.*, **92**, 5987 (1987).

193. K. Daasbjerg and H. Lund, *Acta Chem. Scand.*, **47**, 597 (1993).
194. V. G. Mairanovskii, *Elektrokhimiya*, **26**, 443 (1990); *Chem. Abstr.*, **112**, 242121 p (1990).
195. C. P. Andrieux, L. Gelis, M. Medebielle, J. Pinson and J.-M. Savéant, *J. Am. Chem. Soc.*, **112**, 3509 (1990).
196. C. Wakselman, *J. Fluorine Chem.*, **59**, 367 (1992).
197. C. P. Andrieux, A. Merz and J.-M. Savéant, *J. Am. Chem. Soc.*, **107**, 6097 (1985).
198. D. A. Koch, B. J. Henne and D. E. Bartak, *J. Electrochem. Soc.*, **134**, 3062 (1988).
199. K. P. Butin, T. V. Magdesieva, R. D. Rakhimov and O. A. Reutov, *Zh. Org. Khim.*, **23**, 2481 (1989).
200. R. A. Kretchmer and R. Glowinski, *J. Org. Chem.*, **41**, 2661 (1976).
201. M. Tarle, O. Rasmy, H. Rogers and J. Casanova, *Croat. Chim. Acta*, **63**, 239 (1990).
202. G. M. McNamee, B. C. Willett, D. M. La Perriere and D. G. Peters, *J. Am. Chem. Soc.*, **99**, 1831 (1977).
203. M. M. Baizer, and J. P. Petrovich, in *Prog. Phys. Org. Chem.* (Eds. A. Streitwieser and R. W. Taft), Vol. 7, 1970, pp. 221–223.
204. J. Casanova and H. R. Rogers, *Inorg. Synth.*, **17**, 752 (1975).
205. J. Casanova, H. R. Rogers and K. Servis, *Org. Magn. Res.*, **7**, 57 (1975).
206. S. Wawzonek and J. H. Wagenknecht, *J. Electrochem. Soc.*, **110**, 420 (1963); M. R. Rifi, in *Organic Electrochemistry* (Ed. M. M. Baizer), Marcel Dekker, New York, 1963.
207. F. Barba, A. Guirado and A. Zapata, *Electrochim. Acta*, **27**, 1335 (1982).
208. I. M. Sosonkin, G. N. Strogov, V. N. Novikov and T. K. Ponomareva, *Khim. Geterotsikl. Soedin*, **23** (1977); *Chem. Abstr.*, **86**, 154848v; *Nauchn. Tr. Kuban. Gos. Univ.*, **256**, 65 (1977); *Chem., Abstr.*, **89**, 41874 u (1997).
209. R. Andreoli, L. Benedetti, M. Borsari, C. Fontanesi, G. Gavioli and G. Grandi, *Ann. Chim. Rome*, **82**, 293 (1992).
210. A. J. Fry and R. L. Krieger, *J. Org. Chem.*, **41**, 54 (1976).
211. J. Simonet, B. M. C. El and G. Mabon, *J. Electroanal. Chem. Interfacial Electrochem.*, **281**, 289 (1990).
212 W. F. J. Carroll and D. G. Peters, *J. Org. Chem.*, **43**, 4633 (1978).
213. K. P. Butin, P. D. Rakhimov and I. V. Novikova, *Metalloorg. Khim.*, **2**, 849 (1989); *Chem. Abstr.*, **111**, 220936 k (1989).
214. J. A. Cleary, *Diss. Abstr. Int. B*, **43**, (1983); *Chem. Abstr.*, **58**,134217h; J. A. Cleary, M. S. Mubarak, K. L. Vieira, M. R. Anderson and D. G. Peters, *J. Electroanal. Chem. Interfacial Electrochem.*, **198**, 107 (1986).
215. N. L. Pocard, D. C. Alsmeyer, R. L. McCreery, T. X. Neenan and M. R. Callstrom, *J. Mater. Chem.*, **2**, 771 (1992).
216. H. P. Zhang and J. F. Rusling, *Talanta*, **40**, 741 (1993).
217. W. F. J. Carroll and D. G. Peters, *J. Electrochem. Soc.*, **127**, 2594 (1980).
218. B. Struebing, P. Jeroschewski and A. Deutsch, *Z. Chem.*, **20**, 442 (1980); *Chem. Abstr.*, **94**, 73569 t (1980).
219. T. Shono, N. Kise and H. Oka, *Tetrahedron Lett.*, **32**, 6567 (1991).
220. J. Y. Nedelec, J. C. Folest, J. Perichon, and M. H. A. Haddou, *Fr. Demande*, FR2, **623**, 525 (1989); *Chem. Abstr.*, **112**, p107378n.
221. S. Narizuka and T. Fuchigami, *J. Org. Chem.*, **58**, 4200 (1993).
222. M. Medebielle, J. Pinson and J.-M. Savéant, *Tetrahedron Lett.*, **31**, 1279 (1990).
223. V. A. Grinberg and Y. B. Vassiliev, *J. Electroanal. Chem.*, **325**, 185 (1992).
224. S. Sibille, E. D'Incan, L. Leport and J. Perichon, *Tetrahedron Lett.*, **27**, 3129 (1986); M. A. McClinton and D. A. McClinton, *Tetrahedron*, **48**, 6555 (1992).
225. E. A. Avetisyan, A. F. Aerov, V. F. Cherstkov, B. L. Tumanskii, S. R. Sterlin and L. S. German, *Izv. Akad. Nauk, Ser. Khim.*, (5), 1208 (1992); *Chem. Abstr.*, **118**, 124035q.
226. D. Devilliers, H. Groult and M. Chemla, *J. Fluorine Chem.*, **57**, 73 (1992).
227. A. J. Appleby, O. A. Velev, J. G. Lehelloco, A. Parthasarthy, S. Srinivasan, D. D. Desmarteau, M. S. Gillette and J. K. Ghosh, *J. Electrochem. Soc.*, **140**, 109 (1993).
228. H. Saffarian, P. Ross, F. G. Behr and H. Gard, *J. Electrochem. Soc.*, **139**, 2391 (1992).
229. P. Janderka and P. Broz, *Chem. Listy*, **87**, 594 (1993).

230. J. F. Rusling and J. V. Arena, *J. Electroanal. Chem. Interfacial Electrochem.*, **186**, 225 (1985).
231. H. Sugimoto, S. Matsumoto and D. T. Sawyer, *Environ. Sci. Technol.*, **22**, 1182 (1988); *Chem. Abstr.*, **109**, 134468 m (1988).
232. T. V. Magdesieva, V. N. Shishkin and K. P. Butin, *Zh. Obshch. Khim. SSSR*, **61**, 2403 (1991); T. V. Magdesieva, V. N. Shishkin and K. P. Butin, *Zh. Obshch. Khim. SSSR*, **61**, 2400 (1991).
233. A. Sucheta, I. U. Haque and J. F. Rusling, *Langmuir*, **8**, 1633 (1992).
234. T. Nagaoka, J. Yamashita, M. Kaneda and K. Ogura, *J. Electroanal. Chem.*, **335**, 187 (1992).
235. T. C. Franklin, G. Oliver, R. Nnodimele and K. Couch, *J. Electrochem. Soc.*, **139**, 2192 (1992).
236. Y. C. Yang, J. A. Baker and J. R. Ward, *Chem. Rev.*, **92**, 1729 (1992).
237. C. P. Andrieux, G. Delgado and J.-M. Savéant, *J. Electroanal. Chem.*, **348**, 123 (1993).
238. C. P. Andrieux, G. Delgado, J.-M. Savéant and K. B. Su, *J. Electroanal. Chem.*, **348**, 141 (1993).
239. C. P. Andrieux, P. Hapiot and J.-M. Savéant, *J. Electroanal. Chem.*, **349**, 299 (1993).
240. C. P. Andrieux, E. Differding, M. Robert and J.-M. Savéant, *J. Am. Chem. Soc.*, **115**, 6592 (1993).
241. N. M. Rufman and Z. A. Rotenberg, *Elektrokhimija*, **15**, 163 (1979); *Chem. Abstr.*, **94**, 22077 y (1979).
242. N. M. Rufman and Z. A. Rotenberg, *Elektrokhimija*, **16**, 1548 (1980).
243. J. Simonet, B. M. Chaquiq, and G. Mousset, *J. Electroanal. Chem. Interfacial Electrochem.*, **286**, 163 (1990).
244. F. Maran, E. Vianello, G. Catelani and F. D'Angeli, *Electrochim. Acta*, **34**, 587 (34).
245. J. C. Folest, J. Y. Nedelec and J. Perichon, *Tetrahedron Lett.*, **28**, 1885 (1988).
246. A. R. Mount, M. S. Appleton, W. J. Albery, D. Clark and C. Hahn, *J. Electroanal. Chem. Interfacial Electrochem.*, **334**, 155 (1992).
247. A. R. Mount, M. S. Appleton, W. J. Albery, D. Clark and C. Hahn, *J. Electroanal. Chem. Interfacial Electrochem.*, **340**, 287 (1992).
248. J. Y. Becker, *J. Electroanal. Chem. Interfacial Electrochem.*, **160**, 249 (1984).
249. J. Yoshida, Y. Ishichi and S. Isoe, *J. Am. Chem. Soc.*, **114**, 7594 (1992).
250. L. L. Miller, E. P. Kujawa and C. B. Campbell, *J. Am. Chem. Soc.*, **92**, 2821 (1970).
251. D. Maysinger, C. Marcus, W. Wolf, M. Tarle and J. Casanova, *J. Chromatogr.*, 2105 (1976).
252. D. Maysinger, W. Wolf, M. Tarle and J. Casanova, *Croat. Chim. Acta*, **49**, 123 (1977).
253. D. Maysinger, C. Marcus, W. Wolf, M. Tarle and J. Casanova, *J. Pharmacol*, 2315 (1977).
254. S. Taj, M. F. Ahmed and S. Sankarapapavinasam, *J. Chem. Res. (S)*, (1993).
255. S. Taj, M. F. Ahmed and S. Sankarapapavinasam, *J. Electroanal. Chem.*, **356**, 269 (1993).
256. V. A. Grinberg and Y. B. Vassiliev, *J. Electroanal. Chem.*, **325**, 167 (1992).
257. A. Konno and T. Fuchigami, *Chem. Lett.*, (1992).
258. A. Konno, W. Naito and T. Fuchigami, *Tetrahedron. Lett.*, **33**, 7017 (1992); J. Brigaud and E. Laurent, *Tetrahedron Lett.*, **31**, 2287 (1990); T. Fuchigami, M. Shimojo, A. Konno and L. Nakagawa, *J. Org. Chem.*, **55**, 6074 (1990).
259. J. C. Farmer, F. T. Wang, P. R. Lewis and L. J. Summers, *J. Electrochem. Soc.*, **139**, 3025 (1992).
260. A. F. Brown, K. N. Gulyas, P. T. Lay, P. A. M'calpine, N. S. Masters and L. Phillips, *J. Chem. Soc., Dalton Trans.*, **21**, 835 (1993).
261. T. Lund and H. Lund, *Tetrahedron. Lett.*, **27**, 95 (1986).
262. J. W. Sease, P. Chang and J. L. Groth, *J. Am. Chem. Soc.*, **36**, 3154 (1964).

CHAPTER **19**

Pyrolysis of organic halides

GABRIEL CHUCHANI

Department of Chemistry, Instituto Venezolano de Investigaciones Científicas (IVIC), Apartado 21827, Caracas 1020-A, Venezuela

Supplement D2: The chemistry of halides, pseudo-halides and azides
Edited by S. Patai and Z. Rappoport   © 1995 John Wiley & Sons Ltd.

I. INTRODUCTION

Several excellent reviews[1-4] give an account of most of the works on pyrolysis of organic halides described in the chemical literature. For details of experimental pyrolysis methods used for kinetic determinations, the above-mentioned reviews and the original papers may conveniently be consulted.

To believe that a consideration of more complex organic halides may not contribute basic and novel information as well as additional data to that already known from the study of simple halohydrocarbons is far from being correct. In this respect, many contributions in recent years have advanced our knowledge regarding the interpretation of substituent effects, linear free-energy relationships, *trans*-eliminations, elimination-cyclization, the occurrence of neighboring group participation and rearrangement reactions.

Heterogeneous or surface effects have been found to complicate the interpretation of kinetic experiments, which lead to erroneous Arrhenius parameters. However, with special precautions involving the use of seasoned vessels and the presence of a free-radical suppressor, the errors are minimized. Consequently, the present chapter will cover mostly homogeneous gas-phase processes. Studies on chemical activation, the use of catalysts, the bimolecular gas phase and heterogeneous reactions are not included. As an attempt to describe important pyrolyses data from 1972 to 1992, this review does not pretend to offer a complete coverage of the literature.

It is expected that the many interesting reactions cited in this review will provide ample insight of these gas-phase processes and stimulate further challenging work on pyrolyses of halogen compounds and of other organic molecules in general.

A. Historical

In order to offer continuity with important considerations established up to 1971 and connect them with what follows in this chapter, it seems necessary to provide a brief historical background on this subject. In this sense, an ample view of the evolution and new tendencies of these processes may be projected and will hopefully invite unfamiliarized investigators to participate in the resolution of some interesting problems.

Prior to 1953, few kinetic works on the homogeneous, unimolecular gas-phase pyrolysis or elimination of simple alkyl halides were reported. According to these experimental data the commonly accepted mechanism consisted of a concerted four-membered cyclic transition state yielding the corresponding olefin and hydrogen halide as shown in equation 1. For molecular dehydrohalogenation, the presence of a β-hydrogen adjacent to the C—X bond is necessary.

$$R^1R^2C-CR^3R^4 \longrightarrow R^1R^2C\cdots CR^3R^4 \longrightarrow R^1R^2C=CR^3R^4 + HX \quad (1)$$
$$\qquad | \quad | \qquad\qquad\qquad | \cdots |$$
$$\qquad H \quad X \qquad\qquad\qquad H\cdots X$$

Maccoll and coworkers[5] were able to demonstrate for a series of alkyl bromides a clear correlation between the activation energies for elimination and the C—Br bond strengths. At the same time, a change from primary to tertiary bromides resulted in a significant increase in rates. By 1955, Maccoll and Thomas[6] showed a clear correlation between the activation energy for elimination and the heterolytic bond dissociation energies of the carbon–halogen bond, whereas little or no correlation was found with the homolytic bond dissociation energies. In view of these observations, they have suggested that the activation consisted mainly of an elongation with subsequent polarization of the carbon–halogen bond, in the sense of $C^{\delta+}\cdots X^{\delta-}$, with some assistance of the adjacent polarized C—H bond. In later works, Maccoll and Thomas[7] have considered the very polar transition state in terms of an intimate ion-pair intermediate and have offered some evidence in support of this assumption: (1) The energies of activation decreased while the rates increased by α-substitution with electron-releasing substituents. (2) Only small increase in rates is obtained when electron-releasing groups substitute the β-position. (3) There is a strong correlation between the energies of activation and the heterolytic C—X bond dissociation energies. (4) A distinctive feature of carbenium ion behavior in solution, such as the Wagner–Meerwein rearrangement, has been observed in the gas-phase pyrolysis of neopentyl chloride[8,9] and of bornyl and isobornyl chlorides[10,11]. (5) A similar charged transition state or an intermediate appears to exist in the gas-phase elimination and in nucleophilic substitution (S_N1) and elimination reactions (E1) in polar solvents. This is deduced by the linear correlation of substituent effects on unimolecular pyrolysis and on solvolysis of alkyl halides of known mechanism[11].

Further studies on organic halide pyrolyses appeared to question the theory of an ion-pair intermediate. Thus, Maccoll and coworkers[12] have found that racemization of optically active D-(+)-2-chlorooctane does not take place in gas-phase pyrolysis, suggesting that the transition state is best regarded as a polarized C—Cl bond rather than as a carbenium–chloride ion pair.

The lactonization of γ-bromoesters, which involves neighboring group participation and an ionic intermediate in solution, was not observed in the gas phase. Kwart and Waroblak[13] reported that only HBr gas was produced from ethyl γ-bromobutyrate pyrolysis at 450 °C. Consequently, they doubted the heterolytic nature via an ion-pair intermediate of alkyl halides pyrolyses.

Herndon and Manion[14] have pointed out that the pyrolysis results of bornyl and isobornyl chlorides are subject to serious errors, since the product bornylene decomposes by a retro-Diels-Alder reaction at a rate comparable to HCl elimination, thus making quantitative estimations questionable. A new work many years later appears to support Herndon and Manion's observation since bornylene, the main primary product of isobutyl acetate elimination, was found to partially isomerize to camphene and tricyclene, together with a retro-Diels-Alder decomposition to yield trimethylcyclopentadienes and ethylene[15]. Because pure bornylene isomerizes to camphene and tricyclene in the gas phase[15], the occurrence of rearrangement via neighboring group participation from the C(1)—C(6) bond was therefore considered to be doubtful.

Benson and Bose[16] have proposed a semi-ion-pair transition state mechanism for organic halides pyrolysis in the gas phase. On the basis of this model involving a four-membered transition state, O'Neal and Benson[17,18] have presented calculations of energies and entropies of activation which were in good agreement with the experimental values.

According to the data reported in the literature up to 1972, the gas-phase pyrolysis of alkyl halides was to be described in terms of a discrete polar transition state which yields products directly without the formation of an ion-pair intermediate. The only apparent exception to this consideration is the Wagner–Meerwein rearrangement of neopentyl chloride[8,9].

II. ORGANIC FLUORO COMPOUNDS

A. Alkyl Fluorides

In the last twenty years, very few works on the homogeneous gas-phase elimination of alkyl fluorides having C_β—H with respect of C—F bond have been reported. This type of compounds is rarely examined in a static system, because one of the products, i.e. hydrogen fluoride gas, will attack the seasoned coating and subsequently the glass wall of the reaction vessel. Otherwise, the flow method and the shock tube technique have been used successfully in the study of fluorohydrocarbon pyrolyses. Several additional data are listed in Table 1, where large differences in the k values from various laboratories are presented. In spite of this fact, when comparing the results under a similar experimental method, the general trend of rate increase from primary to tertiary alkyl fluorides suggests a mechanism involving a polar four-membered cyclic transition as proposed by Maccoll[1] (equation 2).

$$RCH_2CH_2F \longrightarrow \underset{\underset{H\text{-----}F}{\overset{|}{\vert}}{\overset{\delta^+}{RCH}=\overset{}{CH_2}}} \longrightarrow RCH=CH_2+ HF \qquad (2)$$

TABLE 1. Comparative rates and Arrhenius parameters for pyrolysis of alkyl fluorides at 450 °C

Compound	Method[a]	k_1 (s^{-1})	E_a (kJ mol^{-1})	log A (s^{-1})	Ref.
CH$_3$CH$_2$F	SS	5.22×10^{-5}	243.5	13.31	19
	FS	3.45×10^{-5}	261.9	14.46	20
	FS	1.86×10^{-6}	247.6	12.16	21
	ST	2.06×10^{-5}	250.6	13.42	22
	SPST	4.65×10^{-5}	248.9	13.65	23
	CST, T	4.10×10^{-5}	252.2	13.84	23
CH$_3$CH$_2$CH$_2$F	ST	4.35×10^{-5}	243.9	12.36	22
(CH$_3$)$_2$CHF	FS	3.34×10^{-5}	225.7	11.83	21
	ST	1.13×10^{-3}	225.7	13.36	24
CH$_3$CH$_2$CH$_2$CH$_2$F	ST	1.30×10^{-4}	237.6	13.28	24
(CH$_3$)$_2$CHCH$_2$F	ST	4.00×10^{-5}	245.1	13.31	24
(CH$_3$)$_3$CF	FS	5.37×10^{-2}	210.8	13.96	21
	ST	7.20×10^{-3}	215.4	13.42	24
CH$_2$=CHF	FS	5.50×10^{-8}	279.4	12.93	25
	SPST	3.98×10^{-8}	296.2	14.00	26

[a]SS = static system. FS = flow method. ST = shock tube technique. SPST = single-pulse shock tube. CST, T = comparative shock tube study, temperature eliminated.

The polar nature of these reactions finds additional support in the pyrolysis kinetics of vinyl fluoride (Table 1) where the C—F bond polarization, in the sense $C^{\delta+}\cdots X^{\delta-}$, is the limiting factor. Due to the delocalization of the p electrons on the fluorine of vinyl fluoride (1a↔1b), the C_α—F bond polarization in the transition state is reduced. Cleavage of the C—F bond becomes more difficult, making the dehydrofluorination of vinyl fluoride slower than that of ethyl fluoride:

$$\overset{\frown}{CH_2}=CH\overset{\frown}{\underset{}{\ddot{F}}} \longleftrightarrow {}^-CH_2-CH=F^+$$
$$\qquad \text{(1a)} \qquad\qquad\qquad \text{(1b)}$$

Ethyl fluoride was investigated in a single-pulse shock tube using the comparative method (with *n*-propyl chloride as an internal standard) and the absolute rate (isolation-section) method[23]. The Arrhenius parameters are given in Table 1. This result is in closer agreement with the work of Sianesi and coworkers[20] and the static pyrolysis of Day and Trotman-Dickenson[19] than with Dastoor and Emovon's data[21].

B. Alkyl Polyfluorides

The shock tube technique employed in the pyrolytic decomposition of polyfluorohydrocarbons[22,27,28] showed that the elimination of molecular hydrogen fluoride is the predominant reaction. Yet, a side process of C—C bond breaking becomes important as the temperature is increased beyond 1300 K. Several fluoroethanes have been found to react by molecular dehydrofluorination in chemical activation process[29] and Table 2 summarizes the kinetic parameters for the gas-phase pyrolysis of this type of compound. In the case of 1,1,2-trifluoroethane, three olefin products were obtained (equations 3–5).

TABLE 2. Comparative rates and Arrhenius parameters for HF elimination of fluoroethanes at 1100 K

Compound	Method[a]	k_1 (s^{-1})	E_a (kJ mol^{-1})	log A (s^{-1})	Ref.
CH_3CH_2F	SS	55.5	243.5	13.31	19
	FS	104.8	261.9	14.46	20
	FS	2.5	247.6	12.16	21
	ST	32.9	250.6	13.42	22
	SPST	67.3	248.9	13.65	23
	CST,T	72.7	252.2	13.84	23
CH_3CHF_2	FS	10.2	258.9	13.31	20
	ST	6.8	271.9	13.74	22
	SPST	39.8	258.9	13.90	30
CH_3CF_3	FS	0.98	256.8	12.14	20
	ST	0.22	297.4	13.47	22
	SPST	2.2	287.4	14.00	31
CH_2FCH_2F	SS	7.8	263.1	13.39	32
CH_2FCHF_2	SPST	8.1	275.7	14.00	29
CHF_2CF_3	SPST	0.24	295.7	13.42	28
CHF_2CHF_2	SPST	0.32	290.3	13.30	33
CHF_2CF_3	SPST	0.23	299.5	13.60	34
$CH_2{=}CF_2$	SPST	0.05	335.1	13.10	35

[a]SS = static system. FS = flow method. ST = shock tube technique. SPST = single-pulse shock tube. CST, T = comparative shock tube.

$$CH_2FCHF_2 \begin{cases} \longrightarrow CH_2{=}CF_2 + HF & (3) \\ \longrightarrow \textit{cis-}CHF{=}CHF + HF & (4) \\ \longrightarrow \textit{trans-}CHF{=}CHF + HF & (5) \end{cases}$$

The comparative effect of the F atom in HF elimination under the same experimental conditions reveals that on increasing the number of halogens at the α-carbon, the energy of activation tends to increase, while the rate coefficient decreases in the following sequence:

$$CH_3CH_2F > CH_3CHF_2 > CH_3CF_3$$
$$CH_2FCH_2F > CH_2FCHF_2 > CH_2FCF_3$$
$$CHF_2CH_2F > CHF_2CHF_2 > CHF_2CF_3$$

G. Chuchani

This sequence suggests that fluoride incorporation at either one of the carbon atoms tends to reduce the C—F bond polarization in the transition state, thus making it more difficult to eliminate the leaving F group.

The presence of a F atom in vinyl fluoride, i.e. $CH_2=CF_2$, gave in a single-pulse shock tube[35], as expected, a marked decrease in the rate of molecular HF elimination. This result appears to support not only the polar nature of these reactions, but also the mechanistic argument for the $CH_2=CHF$ decomposition.

III. ORGANIC CHLORO COMPOUNDS

A. Alkyl and Aryl Chlorides

Methyl chloride was studied behind shock waves over the range 1680–2430 K. The mechanism for this unimolecular decomposition was described as a radical reaction[36]. This process was found to proceed in the fall-off region. Consequently, the low- and high- pressure rate coefficients (k_1 and k_∞) were estimated by the application of a refined RRKM theory involving a weak collision effect. The following Arrhenius parameters were obtained: $\log k_1$ (s^{-1}) = 15.56–247 kJ mol^{-1} $(2.303RT)^{-1}$ and $\log k_\infty$ (s^{-1}) = 13.86–383 kJ mol^{-1} $(2.303RT)^{-1}$. The collision efficiency factor β_c, was 0.02 at 2000 K. In this work it was also found that estimation of high- and low-pressure limit rate coefficients from fall-off data leads to erroneous results when, especially at high temperatures, the fall-off curve is used without correction for the weak collision effect.

When 3-chloro-1-butene was pyrolyzed in a static system at temperatures 776–835°K, the primary product was methane and some higher molecular weight hydrocarbons[37]. This result was ascribed to the initial dissociation of the substrate into chloroalkyl and methyl radicals followed by H abstraction (equations 6–9).

$$CH_3CH(Cl)CH=CH_2 \longrightarrow CH_3 + C_3H_4Cl \tag{6}$$

$$CH_3 + CH_3CH(Cl)CH=CH_2 \longrightarrow CH_4 + C_4H_6Cl \tag{7}$$

$$CH_3 + CH_3CH(Cl)CH=CH_2 \longrightarrow C_5H_{10}Cl \tag{8}$$

$$CH_3 + CH_3CH(Cl)CH=CH_2 \longrightarrow CH_3Cl + C_4H_7 \tag{9}$$

The methyl radical reacts either by addition to or by chlorine abstraction from the substrate as represented in steps 8 and 9. The free-radical process may not be a surprise, since the reaction was carried out in a silica vessel and in the absence of a chain suppressor.

The kinetic isotope effect in the unimolecular decomposition of ethyl chloride was studied in an attempt to characterize the nature of the transition state[38]. The value obtained, $^{35}k/^{37}k$ = 1.0015 ± 0.0002 at 723 K (450), suggested a moderate lengthening of the C—Cl bond, and only a little motion of the Cl atom in the transition state. Yet, in the same year, Christie estimated a ^{37}Cl isotope effect for the pyrolyses of $(CH_3)_2CHCl$, $(CH_3)_3CCl$, CH_3CHCl_2 and $CH_2ClCH_2CH_2Cl$[39]. The author came to the conclusion that these reactions proceeded by a rate-determining migration of H from C to Cl, followed by a facile detachment of HCl.

Hiberty[40] used the single determinant *ab initio* molecular orbital theory to study the unimolecular HCl elimination from ethyl chloride. The calculations of three potential energy surfaces corresponding to *syn*-elimination, planar *anti*-elimination, and nonplanar *anti*-elimination were performed and the dehydrochlorination process was predicted to be *syn*, and to proceed via a planar four-membered transition state. According to these estimations, the nonbonding pair of the chlorine atom played an important part in the reaction mechanism and enabled the formation of such a transition state.

The thermal decomposition of 1-(1-chloroethyl)-2-methylbenzene in the gas phase was examined at 285–325 °C[41]. Attempts to prove experimentally that the reaction proceeds via a six-membered transition state with the o-Me group as in 1-(1-chloromethyl)-2-methyl-benzene[42] were negative. In fact, the reaction proceeds through a four-membered transition state mechanism with moderate charge separation.

The effect of methyl groups in the aromatic ring of 1-phenylethyl chloride is shown in Table 3. The small but significant rate increase of the o-CH$_3$ group was reasonably ascribed to steric acceleration.

TABLE 3. Comparative rates of $RC_6H_4CHClCH_3$ at 335 °C

R	$10^4 k_1$ (s^{-1})	Relative rate
H	3.19	1.0
o-CH$_3$	25.70	8.1
m-CH$_3$	4.67	1.5
p-CH$_3$	9.12	2.9

The unimolecular decompositions of ethyl chloride, isopropyl chloride and tertiary butyl chloride were carried out in a static system, and the working temperature and pressure ranges were 576–767 K and 0.1–300 torr[43]. The Arrhenius parameters for the high-pressure limit were reported as follows: for CH_3CH_2Cl, log A (s^{-1})= 13.33. E_a = 235.5 kJ mol^{-1}; for $(CH_3)_2CHCl$, log A (s^{-1}) = 13.55, E_a = 212.9 kJ mol^{-1} and for $(CH_3)_3CCl$, log A (s^{-1}) = 13.93, E_a = 191.2 kJ mol^{-1}. The pressure dependence of the unimolecular first-order law (fall off) of these eliminations, starting with approximately equal k_∞ values and by proper choice of temperature, was found to shift to lower pressures with increasing molecular size.

Vinyl chloride pyrolysis in a shock tube proceeded via HCl elimination[44]. However, the C—Cl bond fission was expected to become important at higher temperature, being competitive with HCl formation at temperature ≥ 2400 K. The measured activation energy of the process $CH_2{=}CHCl \rightarrow CH{\equiv}CH + HCl$ was higher than the value estimated from the electrostatic model proposed by Benson and Haugen[45] by 18 kJ mol^{-1}. Therefore, the electrostatic model was believed to give slightly low values for molecular HCl elimination from C—C double bonds.

The pyrolysis of 3-chloropropionitrile in a conventional static system was very complex and a detailed kinetic and mechanistic study proved to be difficult[46]. The decomposition process yielded mainly vinyl chloride and hydrogen chloride. Yet, the reaction appeared to undergo a radical chain mechanism with some contributions from heterogeneous processes. Two years later, 3-chloropropionitrile was pyrolyzed under very low pressure pyrolysis (VLPP) giving exclusively a unimolecular elimination of HCl[47]. The application of RRKM theory gave the equation: log k_∞ (s^{-1}) = (31.2 ± 0.3) − (241.0 ± 4.0) kJ mol^{-1} (2.303RT)$^{-1}$. When the k value for this nitrile is compared to that of ethyl chloride, it was found that the β-CN substituent has a retarding effect on the rate of HCl elimination. This fact was therefore believed to be consistent with a polar transition state.

A comparison of two single-pulse shock-tube-technique experiments applied to the HCl elimination of ethyl chloride and n-propyl chloride in the temperature range of 960–1100 K was made by Evans and coworkers[48]. The observed rate coefficients were compared with those of previous works. In this investigation it was believed that the activation energy E_a of 242.6 kJ mol^{-1} for $CH_3CH_2Cl \rightarrow CH_2=CH_2 + HCl$ is more appropriate than the most commonly reported E_a values of 234.2–236.8 kJ mol^{-1}.

Deuterated ethyl chlorides CH_3CD_2Cl and CD_3CD_2Cl showed on pyrolysis a good quantitative agreement of both Forst calculations and RRKM calculations with the experimental results[49]. The reported data were found to be more accurate than previous experimental works and covered a larger range of pressures. An important aspect of this work is the possibility of describing the transition states of all the experimentally studied ethyl chlorides with the same planar cyclic four-membered activated complex model:

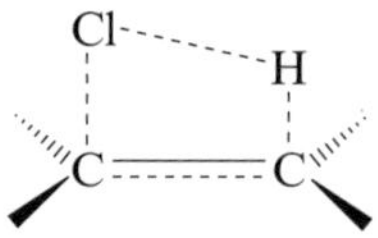

This work did not take into consideration the heavy isotope effects, and the conclusion reached by Christie and coworkers[38] from $^{35}Cl/^{37}Cl$ isotope effects in the ethyl chloride decomposition disagreed with the activated complex model depicted above.

The chlorine isotope effect was examined for the pyrolysis of CH_3CH_2Cl in a static system at 395–482 °C. The variation of the kinetic isotope effects with temperature was found to be slight, though significant. The isotope effect studied on the chlorine leaving group appear to be primary, with normal dependence on temperature: their magnitudes increase as the temperature decreases. The best agreement with the experimental $^{35}k/^{37}k$ values was given by a four-membered activated complex model which involves chlorine participation in the reaction coordinate with three degrees of freedom[50]. The reaction was represented schematically as shown in equation 10.

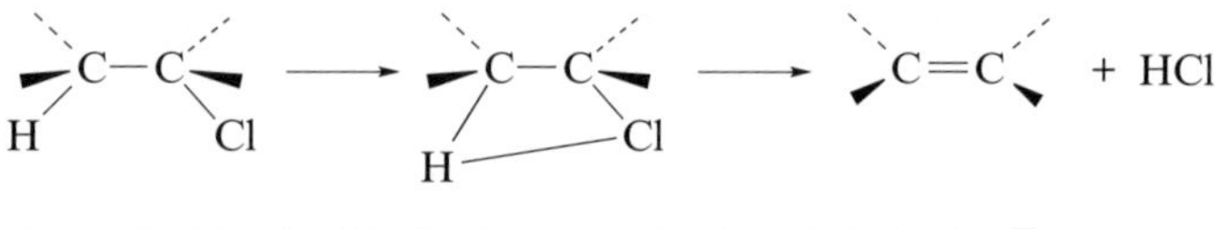

This study on the kinetic chlorine isotope effect in ethyl chloride[50] was extended to secondary and tertiary alkyl halides pyrolyses[51]. The isotope effects on isopropyl chloride and *tert*-butyl chloride pyrolysis were found to be primary and exhibited a definite dependence on temperature. They increased with increasing methyl substitution on the central carbon atom. The pyrolysis results and model calculations implied that all alkyl chlorides involve the same type of activated complex. The C—Cl bond is not completely broken in the activated complex, yet the chlorine participation involves a combination of bending and stretching modes.

The elimination of DCl from CD_3CH_2Cl decomposition was investigated by means of a very low pressure pyrolysis (VLPP) technique over the temperature range 975–1213 K (gas–wall collision only) and at 1173 K with CO_2 as a collision bath gas. The data implied a reaction proceeding via the four-membered transition state. The lack of HCl formation obviously excluded a three-membered transition state for elimination[52]. The application of the RRKM theory gave the extrapolated high-pressure parameters log $A(s^{-1}) = 13.6$ and $E_a = 243$ kJ mol^{-1}.

The Arrhenius expression for the overall decomposition of CH_2DCH_2Cl at high-pressure limits and in the temperature range 670–1100 K[53] was found to be log k_∞ (s^{-1}) =

$(13.33 \pm 0.10) - (239.3 \pm 2.1)\,kJ\,mol^{-1}(2.303RT)^{-1}$, in good agreement with the expression reported by Volker and Heydtmann[54]. The Arrhenius parameters were also shown to be consistent with those reported by Jonas and Heydtmann for pyrolysis of CH_3CH_2Cl, CH_3CD_2Cl and CD_3CD_2Cl[49]. The RRKM theory calculations for the pressure dependence were in good agreement with the experimental data and described quite successfully the pressure dependence of the intramolecular isotope effect in the pyrolysis of CH_2DCH_2Cl. The pyrolysis of this monodeuterated ethyl chloride was believed to proceed through the formation of two independent four-membered cyclic transition states. In this respect, as the decomposition reaction progresses there is a gradual loosening of the C—Cl and C—H bonds, the C—C bond is shortened and the H—Cl bond is formed. Therefore, the transition state was considered to have an intermediate configuration in relation to all four processes. The activated complexes for the HCl and DCl loss from CH_2DCH_2Cl pyrolysis were represented as:

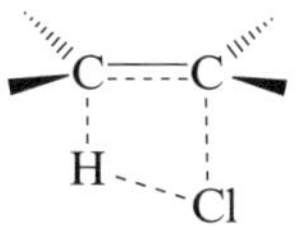

HCl elimination 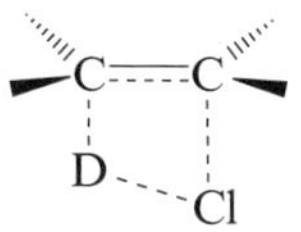   DCl elimination

Preparative flash vacuum thermolysis (FVT) of organic ester derivatives of 6-chloro-1-hexanol yielded as final product 1,5-hexadiene[55]. Thus, when R = H the process proceeds as described in equation 11.

$$ROCH_2CH_2CH_2CH_2CH_2CH_2Cl \longrightarrow HOCH_2CH_2CH_2CH_2CH=CH_2 + HCl \quad (11)$$

R = H

However, when RO is an acetate, tosylate and xanthate, which are better leaving groups than chloride in gas-phase pyrolysis[2,3] the process was found to occur as shown in equation 12.

$$RO(CH_2)_5CH_2Cl \longrightarrow CH_2=CH(CH_2)_4Cl + ROH$$

$$(12)$$

$$CH_2=CH(CH_2)_2CH=CH_2 + HCl$$

The authors advocate the use of the FVT method for obtaining good results in the isolation and characterization of highly reactive intermediates.

Recently, Huybrechts and coworkers[56] carried out a numerical integration for a hypothetical molecular and radical model for the HCl elimination from ethyl chloride in the gas phase. The simulation data indicated that a radical chain process does not contribute to the decomposition rate. Only ethylene and hydrogen chloride are formed in the molecular decomposition.

B. Cycloalkyl Chlorides

Dehydrochlorination of 3-chloro- and 4-chloro-cyclohexenes in a static seasoned vessel was found to be homogeneous and unimolecular[57]. The presence of the double bond at the 3,4-positions with respect to the C—Cl bond showed no effect on the rate of elimination, yet, for a C=C bond at the 2,3-positions the rate increased by a factor of 88.0 (Table 4). This large difference was accounted for by the allylic weakening of the C_β—H bond and increase of the incipient conjugation in the 1,3-diene.

TABLE 4. Arrhenius parameters and comparative rates for decomposition of chlorocyclohexenes at 360 °C

Substrate	$\log A$ (s^{-1})	E_a (kJ mol^{-1})	Relative rate
Chlorocyclohexane	13.77	209.2	1.0
3-Chlorocyclohexene	11.15	154.4	79.8
4-Chlorocyclohexene	13.17	202.5	0.9

In connection with the observation of neighboring olefinic double-bond participation (see Section IX. D), it may not be unreasonable to conceive the participation of the double bond of 3-chlorocyclohexene in the dehydrochlorination as described in equation 13.

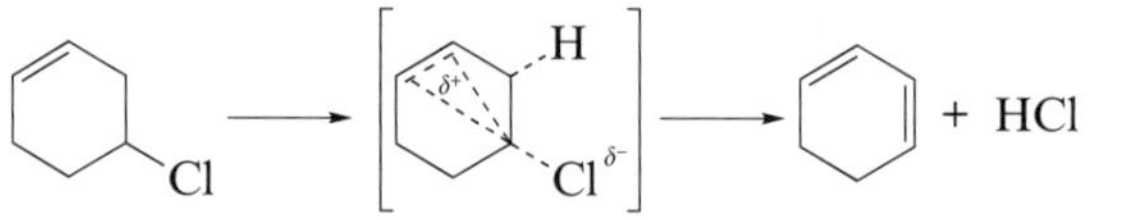

$$\text{(13)}$$

trans-1,2-Dichloro-3,3-difluorocyclopropane, at 192.8–242.7 °C, pyrolyzed by two pathways to produce the isomers *trans*-1,3-dichloro-3,3-difluoropropene and *cis*-1,2-dichloro-3,3-difluorocyclopropane[58] (equation 14).

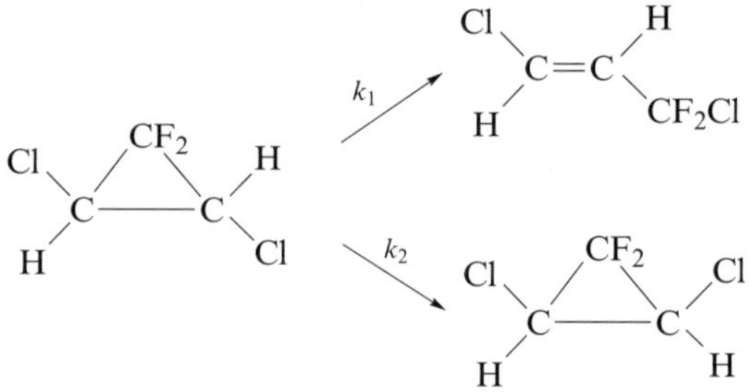

$$\text{(14)}$$

k_t = Overall rate constant
$\log k_t$ (s^{-1}) = (14.34 ± 0.23) − (178.3 ± 1.3) kJ mol^{-1} ((2.303RT)$^{-1}$
$\log k_1$ (s^{-1}) = (14.06 ± 0.25) − (176.7 ± 1.0) kJ mol^{-1} (2.303RT)$^{-1}$
$\log k_2$ (s^{-1}) = (14.20 ± 0.74) − (183.3 ± 3.3) kJ mol^{-1} (2.303RT)$^{-1}$

A diradical mechanism was suggested for the geometrical isomerization, while a concerted process with chlorine migration was suggested for the structural isomerization.

The first-order homogeneous nonradical chain decomposition of '*all cis*' 1α-chloro-2α, 3α-dimethylcyclopropane at 550–607 °C gave 1,3-pentadiene and HCl as products[59]. The Arrhenius parameters are $\log A$ (s^{-1}) = 13.92 and E_a = 199.6 kJ mol^{-1}. The isomer with *trans*-methyl groups, i.e. 1α-chloro-2α, 3β-dimethylcyclopropane, undergoes parallel two first-order decompositions to give *trans*-4-chloro-2-pentene [$\log A$ (s^{-1}) = 14.6, E_a = 199.5 kJ mol^{-1}] and 1,3 pentadiene [$\log A$ (s^{-1}) = 13.8, E_a = 190.2 kJ mol^{-1}]. The results suggest ring opening as the rate-determining step in formation of the olefins.

The thermal decomposition of 1,1-dichlorocyclopropane, in a static system and at 610–725 K, gave through parallel reactions the isomer products[60] shown in equation 15.

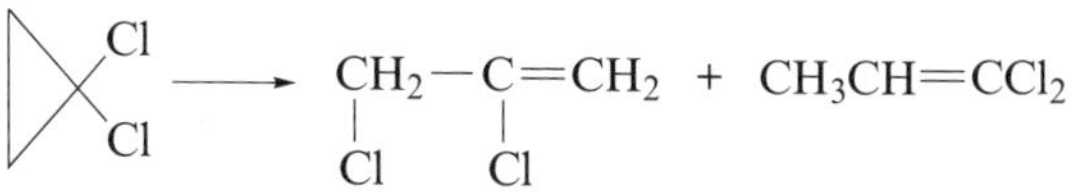

$$\text{[structure]} \longrightarrow \underset{\underset{Cl}{|}}{CH_2}-\underset{\underset{Cl}{|}}{C}=CH_2 \ + \ CH_3CH=CCl_2 \tag{15}$$

Chlorine and hydrogen migration were considered to take place during these parallel reactions. A further process of dehydrochlorination of the products was not described.

The pyrolyses of several β-chlorine-containing acetates and vinyl ethers proceeded by a *syn*-elimination mechanism, well established for the normal esters, and the chlorine atom took the role common for the β-hydrogen[61]. However, the stereochemical studies of cyclohexene ring systems such as the reactions described in equations 16 and 17, proved that the β-chlorine may compete with the β-hydrogen in a *syn*-elimination (equation 18).

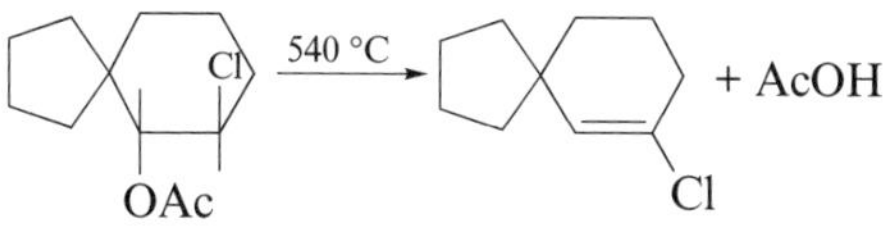

$$\text{[structure] OAc} \xrightarrow{540\,°C} \text{[structure] Cl} \ + \ AcOH \tag{16}$$

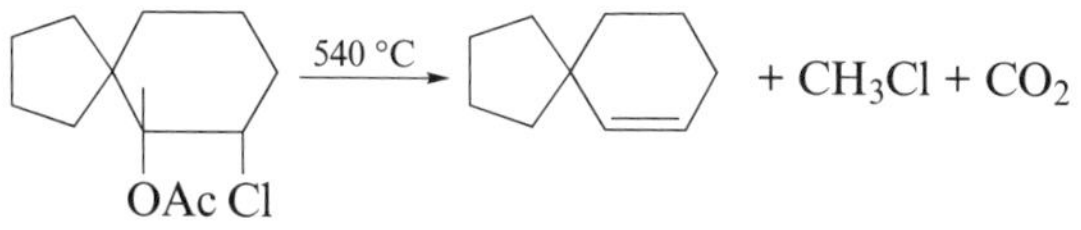

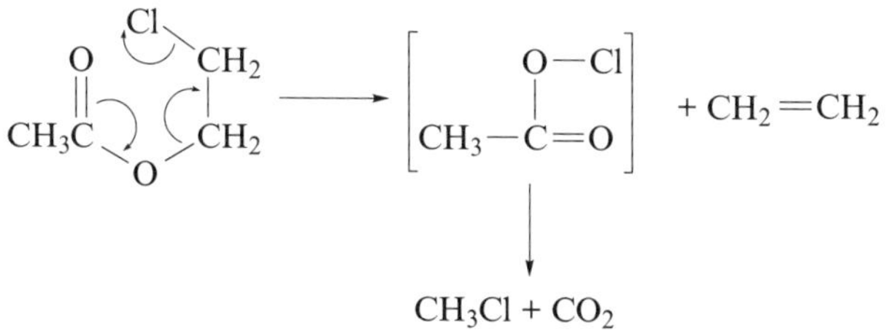

$$\text{[structure] OAc Cl} \xrightarrow{540\,°C} \text{[structure]} \ + \ CH_3Cl \ + \ CO_2 \tag{17}$$

$$\text{[structure]} \longrightarrow \left[\underset{CH_3-C=O}{\overset{O-Cl}{|}} \right] \ + \ CH_2=CH_2 \tag{18}$$

$$\downarrow$$

$$CH_3Cl + CO_2$$

C. Effect of Substituents

1. β-Substituents at the primary C—Cl bond

Chuchani and coworkers have examined the substituents effect in the gas-phase elimination kinetics of β-substituted ethyl chlorides[62]. The comparative effect of substituent Z in compounds ZCH_2CH_2Cl listed in Table 5 indicated that the inductive electron release of alkyl groups stabilize the moderate positive charge on the carbon atom in the transition state, and therefore enhanced the rate of the dehydrochlorination process. A Taft correlation of $\log k/k_s$ vs σ^* values was reasonably linear with a reaction constant ρ^* of -1.17 and a correlation coefficient $r = 0.971$ at $440\,°C$. However, electron-withdrawing β-substituents decreased the rates giving rise to an inflection point of the line at σ^* $(CH_3) = 0.00$ into another good straight line with $\rho^* = -0.30$ and $r = 0.992$ at $440\,°C$. The inflection and the observation of two slopes were attributed to a combination of two effects operating at the transition state, particularly with the electron-withdrawing substituents. As these groups destabilize the positive charge developed on carbon in the transition state, the hydrogen adjacent to Z in ZCH_2CH_2Cl may become more acidic and thus assists in the expulsion of the leaving chloride ion. This fact may sometimes result in a similar or slightly higher pyrolysis rate with respect to the parent unsubstituted ethyl chloride. Therefore, the appearance of two slopes was explained in terms of a slight alteration in the polarity of the transition state due to changes in the transmission of electronic effects to the carbon reaction center.

TABLE 5. Kinetic parameters for unimolecular elimination of HCl in pyrolysis of ZCH_2CH_2Cl at 440 °C

Z	$\log A$ (s^{-1})	E_a (kJ mol^{-1})	$10^4 k_1$ (s^{-1})	Ref.
H	13.83	241.8	1.34	48
CH_3	13.44	229.2	4.47	23
CH_3CH_2	13.63	230.7	5.37	63
$CH_3CH_2CH_2$	14.61	244.0	5.44	64
$CH_3CH_2CH_2CH_2$	14.09	236.2	6.13	62
$(CH_3)_2CHCH_2$	13.91	232.7	7.31	62
$(CH_3)_2CH$	14.12	235.3	7.64	65
$c\text{-}C_6H_{11}$	12.83	216.9	8.71	62
$c\text{-}C_5H_9$	13.87	231.8	7.76	62
$CH_3CH_2CH(CH_3)$	13.40	223.7	10.30	62
$(CH_3)_3C$	13.08	218.8	11.34	65
$CH_2{=}CHCH_2$	14.25	238.4	6.11	66
CH_3O	14.06	244.7	1.36	67
HO	12.80	229.6	0.96	68
Cl	9.80	192.4	0.91	69
NC	13.20	241.0	0.35	47

The negative reaction constant ρ^* for electron-donating substituents seemed to support Maccoll's view[1] on the heterolytic nature of the transition state for alkyl halide pyrolysis in the gas phase.

2. α-Substituents at the secondary C—Cl bond

Evidently, the approximately linear Taft correlation obtained for the elimination of β-substituted primary chlorides[62] suggested an examination of a possible linear free-energy relationship for aliphatic secondary alkyl chlorides pyrolyses, that is, for α-substituted ethyl chlorides $ZCH(Cl)CH_3$[70] (Table 6).

Plotting the $\log k_{rel}$ for alkyl substituents [Table 6, Z = H to $(CH_3)_3C$] against Taft σ^* values gave a very good linear correlation with $\rho^* = -3.58$ and $r = 0.996$ at 360 °C. Here also, the branching of alkyl groups at the α-carbon of ethyl chloride enhanced the dehydrochlorination rates due to their electron-releasing effect. The negative ρ^* suggests that formation of a partially positive charged carbon atom in the transition state is rate-determining in these processes. However, the Taft plot of the electron-withdrawing substituents listed in Table 6 (Z = F to CH_3OCOCH_2) again gave rise to an inflection point of the line at σ^* (CH_3) = 0.00 and to another approximate straight line with a reaction constant $\rho^* = -0.46$ and $r = 0.972$ at 360 °C. The small negative ρ^* value implied relatively small polarization of the C—Cl bond in the transition state. The Taft plot for these pyrolytic eliminations resembles the one discussed above for β-substituted ethyl chlorides[62] and was interpreted, as before, as due to a slight alteration in the polarity of the transition state due to the simultaneous operation of opposing effects during the process of elimination.

With respect to α-substituents bearing p- or π-electrons which are directly attached to the C—Cl bond (Table 6, Z = $CH_2{=}CH$ to CH_3CH_2O), these may delocalize their electrons through resonance or mesomeric effects with the positively charged carbon atom in the transition state. Because of this, they were not plotted in the Taft figure for α-substituted ethyl chlorides. Furthermore, the rates for these substituents also could not be correlated with the electrophilic substituent constants σ^+. The σ^+ parameters have been defined for substituents on the benzene ring which are far from the reaction site. Even though steric effects may interfere with the coplanarity and hence with delocalization, the effect of these substituents was believed to be polar in nature.

TABLE 6. Kinetic parameters for the unimolecular elimination of HCl from pyrolysis of ZCH(Cl)CH$_3$ at 360 °C

Z	log A (s^{-1})	E_a (kJ mol^{-1})	$10^4 k_1$(s^{-1})^a	Ref.
H	13.83	241.8	0.008	48
CH$_3$	13.47	211.6	0.51	71
CH$_3$CH$_2$	13.99	211.7	1.42	72
CH$_3$CH$_2$CH$_2$	14.10	212.1	1.71	73
(CH$_3$)$_2$CH	13.80	207.8	2.07	74
(CH$_3$)$_3$C	12.99	197.3	4.62	74
F	14.02	239.7	0.018	75
Cl	13.28	224.0	0.032	49
Cl$_3$C	13.50	225.1	0.057	76
CH$_3$CO	11.77	205.3	0.068	77
CH$_3$OCO	12.22	217.0	0.21	70
NC	13.45	236.1	0.009	70
CH$_3$OCOCH$_2$	13.65	214.9	0.20	70
CH$_2$=CH	13.42	203.8	4.02	78
CH$_2$=C(CH$_3$)	13.39	197.4	12.65	78
cis-CH$_3$CH=CH	9.85	131.3	1040.00	79
trans-CH$_3$CH=CH	13.86	196.8	41.85	79
CH$_3$CH=CHCH=CH	10.43	144.5	320.00	80
C$_6$H$_5$	12.63	187.8	13.62	81
CH$_3$O	11.44	139.0	9400.00	82
CH$_3$CH$_2$O	10.68	128.2	12600.00	83

^{a}Rate coefficient for formation of the terminal olefin.

3. α-Substituents at the tertiary C—Cl bond

The results of the Taft plot for the alkyl substituents of tertiary alkyl chlorides[84] [Table 7, Z = H to (CH$_3$)$_3$C] yielded a very good straight line with a slope $\rho^* = -4.75$ and $r = 0.993$ at 300 °C. In the case of the few electron-withdrawing substituents listed in Table 7 (Z =

TABLE 7. Kinetic parameters for the unimolecular elimination of HCl from pyrolysis of Z(CH$_3$)$_2$CCl at 300 °C

Z	log A (s^{-1})	E_a (kJ mol^{-1})	$10^4 k_1$ (s^{-1})^a	Ref.
H	13.64	213.8	0.017	85
CH$_3$	13.77	188.2	2.75	86
CH$_3$CH$_2$	13.77	184.1	6.54	87
(CH$_3$)$_2$CH	13.33	175.3	17.77	87
(CH$_3$)$_3$C	13.80	171.5	147.19	86
CH$_3$CO	12.56	190.8	0.15	88
CH$_3$OCO	13.81	215.2	0.016	84
ClCH$_2$	14.29	207.6	0.23	84
Cl	12.88	199.0	0.028	84
CH$_2$=CH	13.30	178.2	11.48	89

^{a}Rate constant for formation of the terminal olefin.

CH$_3$CO to Cl), a plot of their log k_{rel} versus σ^* values also gave rise to an inflection point of the line at σ^* (CH$_3$) = 0.00 and to another approximate straight line with $\rho^* = 0.73$ and $r = 0.912$ at 300 °C. Again, this figure resembles the Taft slope found in the gas-phase elimination of HCl of α- and β-substituted ethyl chlorides[62,70], and the small alteration in

TABLE 8. Extrapolated ρ^* values using alkyl and polar substituents

R = Alkyl	RCH_2CH_2Cl	$RCHClCH_3$	$R(CH_3)_2CCl$
Temp (°C)		ρ^*	
300	−1.46	−3.95	−4.75
360	−1.32	−3.58	−4.30
440	−1.17	−3.18	−3.82
Z = polar group	ZCH_2CH_2Cl	$ZCHClCH_3$	$Z(CH_3)_2CCl$
Temp (°C)		ρ^*	
300	−0.37	−0.51	−0.73
360	−0.32	−0.46	−0.63
440	−0.30	−0.41	−0.59

the polarity of the transition state was explained as mentioned above. The vinyl group was not included in the Taft correlation, because the double-bond electrons may be delocalized by resonance in addition to the polar effect.

The slopes of the lines obtained from the Taft correlations of aliphatic primary, secondary and tertiary chlorides obtained at different temperatures by the extrapolation $\rho^*_{T_2}/\rho^*_{T_1} = T_1/T_2$ indicate that the positive nature at the carbon reaction center of the C—Cl bond in the transition state increases from a primary to a tertiary carbon atom (Table 8)[70]. An additional fact is that for each type of alkyl halide, the degree of positive charge at the carbon reaction center tends to decrease as the temperature increases. This means that the pyrolytic eliminations tend to be more concerted and less polar at very high temperatures. These data support Maccoll's theory on the heterolytic character of the alkyl halides pyrolyses in the gas phase[1].

Several investigations[77,88,90] indicated that steric factors do not determine the velocity of dehydrochlorination in branched alkyl chlorides, and the paramount factor which determines the rate of HCl elimination is electronic in nature.

	CH_3CH_2Cl	$CH_3CH_2CH_2Cl$	$NCCH_2CH_2Cl$
Rel. rate 400 °C	1.0	8.1	0.75

	H | CH_3CHCH_2Cl	CH_3 | CH_3CHCH_2Cl	CN | CH_3CHCH_2Cl
Rel. rate 400 °C	1.0	1.99	0.28

	CH_3 | CH_3CHCl	CH_2CH_3 | CH_3CHCl	$COCH_3$ | CH_3CHCl
Rel. rate 400 °C	1.0	1.3	0.06

	CH_3 | CH_3CCl | CH_3	CH_2CH_3 | CH_3CCl | CH_3	$COCH_3$ | CH_3CCl | CH_3
Rel. rate 400 °C	1.0	1.4	0.04

This conclusion is demonstrated by the fact that for primary alkyl chlorides, an added CH_3 group at the β-carbon of ethyl chloride, i.e. propyl chloride, increases the rate 8.1-fold faster than ethyl chloride, while 3-chloropropionitrile of nearly equal chain length is significantly slower in rate. Moreover, on replacement of the β-hydrogen in n-propyl chloride by CH_3 and CN as shown above, the β-CN exerts a deactivating effect. Similar observations apply for β-substituted propyl chloride, and for the secondary and tertiary alkyl chlorides. These values demonstrate again the polar nature of the transition state where the C—Cl bond is polarized as in $C^{\delta+}\cdots C^{\delta-}$.

D. Alkyl Polychlorides

Methylene chloride, CH_2Cl_2, showed a complex reaction at 450–500 °C[91]. Its pyrolytic elimination involved initiation, propagation branching and termination steps. The initiation step was considered the limiting factor and has an activation energy of 261.4 kJ mol⁻¹. The radical propagation step has an activation energy of 5.5 kJ mol⁻¹. An explanation of the surface effect of the quartz reaction vessel was presented.

The kinetics of the pyrolysis of $CHCl_3$ was determined behind incident and reflected shock waves at 1050–1600 K[92]. This decomposition was reported to be unimolecular and the process is an α,α-elimination $CHCl_3 \rightarrow :CCl_2 + HCl$. The Arrhenius parameters for this process are $\log A = 14.3$, $E_a = 228$ kJ mol⁻¹.

Trichloroethylene and tetrachloroethylene highly diluted in argon were also investigated behind reflected shock waves[93]. Both halides were believed to dissociate by Cl abstraction and the pyrolysis rates are pressure-dependent. According to the products formed and the Arrhenius parameters, the mechanism was presented in terms of consecutive reactions.

Verbovskaya and Lebedev reported the thermal decomposition of several polychloropentanes[94]. The rate coefficients for the dehydrochlorination processes were k_1 (s⁻¹) = $(5.0 \pm 0.5) \times 10^{-4}$ for $Cl_3CCCl_2CH_2CCl_2CCl_3$ and k_1 (s⁻¹) = $(5.4 \pm 0.3) \times 10^{-5}$ for $Cl_3CCH(Cl)CH(Cl)CCl_2CCl_3$. The k value for HCl elimination and cyclization of $Cl_3CC(Cl)=CHCCl_2CCl_3$ was $(1.1 \pm 0.3) \times 10^{-5}$. The presence of the free radical initiator atomic Cl or O increased the rate coefficients by approximately an order of magnitude.

Using the toluene carrier technique in a stirred-flow system, the pyrolysis of pentachloroethane[95] was examined at 820–865 K. The reaction was first order in substrate with $\log A = 11.6 \pm 0.7$ and $E_a = 201.6 \pm 10.9$ kJ mol⁻¹. The surface effect of the reaction vessel was negligible. The initiation propagation and termination reactions were suggested to be homogeneous, and the termination involved both Cl and C_2Cl_5 radicals. However, autocatalysis was explained by interaction between chlorine and C_2Cl_5H rather than by dissociation of Cl_2. A radical mechanism was proposed for the formation of C_2HCl_3, CH_2Cl_2 and $HCCl_3$.

The thermal, unsensitized infrared laser, and SiF_4 laser sensitized decomposition of 1,2-dichloropropane were described by Tsang and coworkers[96]. The dichloro substrate yielded via four reaction channels 3-chloropropene, cis-1-chloropropene, $trans$-1-chloropropene and 2-chloropropene. These products have also been observed in thermal and laser-induced processes. Comparative data of the dichloropropane decomposition between the laser-induced experiment and the other pyrolytic methods indicated the former to be associated with complex phenomena with infrared multiphoton-induced decomposition. Consequently, a wide range of processes appeared to occur and the unimolecular rate parameters can be unambiguously interpreted only at definite high pressure.

An IR laser pyrolysis method was developed for the determination of Arrhenius parameters for unimolecular elimination of chloroalkanes[97]. This technique was reported to be a clean, efficient method for the measurement of high-temperature homogeneous gas-phase reaction rates. A pulse of CO_2 laser radiation was used to excite multiphoton SiF_4

sensitizer molecules producing temperatures in the range of 1100–1400 K. For molecular elimination, the following decompositions were reported:

$$Cl_3CCH_3 \longrightarrow Cl_2C{=}CH_2 + HCl$$

$$\log k_1(\text{s}^{-1}) = (13.1 \pm 0.3) - (207.1 \pm 5.4)\text{kJ mol}^{-1}(2.303RT)^{-1}$$

$$Cl_2CHCH_2Cl \longrightarrow ClCH{=}CHCl + HCl$$
$$\text{cis, trans}$$

$$cis\text{-}\log k_1(\text{s}^{-1}) = (14.1 \pm 0.4) - (244.7 \pm 8.4)\text{kJ mol}^{-1}(2.303RT)^{-1}$$
$$trans\text{-}\log k_1(\text{s}^{-1}) = (13.9 \pm 0.4) - (248.9 \pm 8.4)\text{kJ mol}^{-1}(2.303RT)^{-1}$$

IV. ORGANIC BROMO COMPOUNDS

A. Alkyl Bromides

It is a known fact that the gas-phase pyrolysis kinetics of alkyl bromides have not been extensively investigated due to the experimental difficulties as well as to the complexity of concurrent unimolecular and radical chain mechanisms. However, when these organic bromides are pyrolyzed under maximum inhibition, the reaction in the presence of a free radical suppressor is a molecular elimination. Sometimes, these organic bromides are pyrolyzed under maximum catalysis with HBr gas, and the process may proceed by an autocatalytic molecular mechanism.

Using the shock tube method and at a temperature above 1200 K, ethyl bromide and vinyl bromide were found to dehydrobrominate through a four-membered cyclic transition state[98]. The activation energies for these elimination reactions were 130 and 160 kJ mol^{-1}, respectively. The presence of a β-bromine in ethyl bromide, i.e. BrCH$_2$CH$_2$Br, decreased the activation energy for the molecular elimination of HBr gas to about 84 kJ mol^{-1}, and at the same time permitted a sequence of radical reactions initiated by a C—Br bond fission in the dibromoethane molecule. The introduction of a β-bromine into vinyl bromide to give BrCH=CHBr has a similar effect as in dibromoethane and E_a is reduced to 124 kJ mol^{-1} and a free radical process was suggested as the possible reaction mechanism.

The homogeneous decomposition of methyl and methylene bromide in the gas phase was investigated in a quartz tube using the flow method[99]. The first step of the reaction was thought to be the break of the C—Br bond. In the case of CH$_2$Br$_2$, the Br initiates a chain reaction yielding dibromoethylene, HBr and acetylene. In a similar type of mechanism CH$_3$Br was shown to give methane, HBr gas, acetylene and carbon. The strength of the C—Br bond in CH$_2$Br$_2$ and CH$_3$Br was estimated to be 267.7 and 309.9 kJ mol^{-1}, respectively. A previous work was revisited and extended to the CHBr$_3$ pyrolysis[100]. The most probable decomposition mechanism was similar to that suggested above, i.e., a chain reaction, based on thermodynamic considerations and calculations. In CHBr$_3$ the strength of the C—Br bond was estimated to be 213.3 kJ mol^{-1}.

The pyrolysis kinetics of alkyl bromides [RBr; R = CH$_3$CH$_2$CH$_2$, (CH$_3$)$_2$CH, CH$_3$CH$_2$CH$_2$CH$_2$, (CH$_3$)$_2$CHCH$_2$, (CH$_3$)$_3$C, CH$_3$CH$_2$(CH$_3$)CH] were determined over the temperature range of 600–720 K at 0.15 torr[101]. The rate coefficient of n-propyl bromide was greater than that of isopropyl bromide by a factor of 15. Because of the low activation energies obtained for isobutyl bromide and t-butyl bromide, the authors believed that the surface effect of the reaction walls was responsible for the heterogeneous processes. In contrast with this work, these alkyl bromides have already been pyrolyzed when highly pure, in seasoned reaction vessels and in the presence of a free-radical inhibitor. The dehydrobrominations are homogeneous and molecular in nature. Moreover, secondary alkyl bromides showed much higher rates than a primary alkyl bromide (see Section IV.D1–2).

Vinyl bromide in argon was examined by the reflected shock waves technique[102]. The working temperature range was between 1300 and 2000 K and the density range 5.2×10^{-6} -3.8×10^{-5} mol cm^{-3}. The bromide undergoes decomposition through molecular dehydrobromination, and the molecular process was under the experimental conditions in the low-pressure region. The low-pressure limit rate coefficient was obtained as $\log k_1$ (s^{-1}) = $(13.98 \pm 0.21) - (173.6 \pm 8.3)$ kJ mol^{-1} $(2.303\ RT)^{-1}$. The collision efficiency β_c (1.6×10^{-4}) was surprisingly found to be very low compared with that of vinyl chloride ($\beta_c = 0.04$).

Ethyl bromide, in a static system, was studied at 724.5–755.1 K[103]. The pressure dependence for the HBr elimination was observed in its fall-off region. Evaluation of the rate coefficients was performed by using the RRKM theory and the values were compared with the experimental observation. The work reported an activation energy of 216.3 kJ mol^{-1} and an Arrhenius A factor of $10^{12.5}$. The low-frequency factor was rationalized in terms of the formation of a tight activated complex and a molecular elimination as a prevalent reaction mode.

The pressure and temperature dependence of the isotope effect in a competitive parallel elimination of HBr and DBr from CH$_2$DCH$_2$Br was studied by Kang and coworkers[104]. The steady-state distribution function for DBr elimination shifted to higher energy levels due to the higher critical energy required for the process than for the HBr elimination. Because of this, the steady-state populations of the higher energy levels are depleted more rapidly. At low pressures, the overall k_D value for DBr elimination decreased faster than the k_H value for HBr elimination. This meant that although the isotope effect increased with the decrease of pressure, the overall and individual rate coefficients decreased on decreasing the pressure.

An interesting study of the stereochemistry of the gas-phase dehydrobromination of *threo*- and *erythro*-2-bromo-3-deuteriobutane has been described[105]. Deuterium analysis of the 2-butene isomers showed that thermal HBr elimination occurs mainly via a *syn* transition state, that is, the β-hydrogen and the leaving bromide are removed from the same side of the C—C bond. The *anti* contribution was small, perhaps zero. Primary and secondary deuterium isotope effects for *threo*-2-bromo-3-deuteriobutane were 2.07 ± 0.24 and 1.02 ± 0.13, respectively; and for *erythro*-2-bromo-3-deuteriobutane, 2.11 ± 0.3 and 0.98 ± 0.12, respectively, at 590 K.

The gas-phase dehydrobromination of 1,2-dibromopropane, in a static system and seasoned vessels, gave the bromopropenes[106] described in equation 19.

$$CH_3CHCH_2Br \longrightarrow CH_2{=}CHCH_2Br + CH_3CH{=}CHBr + CH_3C{=}CH_2 + HBr$$

$$\underset{Br}{|} \qquad\qquad\qquad\qquad\qquad \textit{cis, trans} \qquad\qquad \underset{Br}{|} \qquad (19)$$

The presence of the chain inhibitors cyclohexene and propene, and HBr catalyst showed that the dibromopropane decomposed by concurrent mechanisms of unimolecular elimination and a free-radical chain process. The rate coefficient at maximal inhibition was

TABLE 9. Arrhenius parameters for formation of the elimination products of CH$_3$CHBrCH$_2$Br

Product	E_a (kJ mol^{-1})	$\log A$ (s^{-1})
CH$_2$=CHCH$_2$Br	214.6	13.9
cis-CH$_3$CH=CHBr	216.3	13.5
trans-CH$_3$CH=CHBr	220.9	13.6
CH$_3$CBr=CH$_2$	228.8	13.5

given by the following Arrhenius equation: $\log k_1$ (s^{-1}) = (13.80 ± 2.5) – (210.8 ± 12.5) kJ mol^{-1} $(2.303RT)^{-1}$. The kinetic parameters for the formation of the four elimination products of 1,2-dibromopropane are given in Table 9.

The result of a fall-off curve under maximum inhibition was compared with the theoretical curve estimated by the RRKM formulation. From the foregoing observations, the authors offered a reaction mechanism which may explain satisfactorily the process of the 1,2-dibromopropane pyrolysis.

The two parallel thermal decompositions of bromopropane-1,1,2,2-d$_4$, together with the decomposition of ethyl bromide, have been investigated by using the very low-pressure pyrolysis (VLPP) method[107] (equations 20 and 21).

$$CD_2{=}CD_2 + HBr \qquad (20)$$

$$CHD_2CD_2Br$$

$$CHD{=}CD_2 + DBr \qquad (21)$$

The rate coefficients were determined both at pressures so low that only gas/wall collisions occur and, in dilute gases, with pressures up to P_a. The extrapolated high-pressure rate coefficients were given by the following Arrhenius equations: for HBr elimination, $\log k_1$ (s^{-1}) = 13.20 – 227.7 kJ mol^{-1} $(2.303RT)^{-1}$; for DBr elimination, $\log k_1$ (s^{-1}) = 13.15 – 230.2 kJ mol^{-1} $(2.303RT)^{-1}$; and for CH$_3$CH$_2$Br pyrolysis, $\log k_1$ (s^{-1}) = 13.6 – 221.0 kJ mol^{-1} $(2.303RT)^{-1}$, in good agreement with the result obtained by other techniques.

A flash vacuum thermolysis (FVT) study of 2-bromoethanol[108] reported the formation of 80% 1-bromo-1-(2-bromoethoxy)ethane and 10% 1-bromo-1-(1-bromoethoxy)ethane. However, at 900–950 °C, the pyrolysis product was exclusively the latter compound. The formation of several intermediates was considered responsible for these results.

The pyrolysis of 2-bromo-2-butene in a static system, with seasoned vessels, and even in the presence of a free radical inhibitor, was autocatalyzed by the HBr product[109]. However, under maximum catalysis with HBr gas, the reaction is molecular in nature and follows first-order kinetics. The overall rate coefficient was given by the following Arrhenius equation: $\log k_1$ (s^{-1}) = (13.57 ± 0.56) – (200.4 ± 6.8) kJ mol^{-1} $(2.303RT)^{-1}$. The mechanism was suggested to involve a six-membered cyclic transition state as described in equation 22.

$$CH_3CH{=}CCH_3 \xrightarrow{\ HBr\ } CH_3CH{\overset{\cdots}{=}}C{\overset{\cdots}{=}}CH_2 \longrightarrow CH_3CH{=}C{=}CH_2 + 2HBr \quad (22)$$

with Br and H----Br below.

The incident and reflected shock-wave technique was employed for a kinetic study of the thermal decomposition of t-butyl bromide[110]. The substrate dehydrobrominated even at the highest temperature of 1050 K via a unimolecular four-membered cyclic transition state. The A factor and the activation energy obtained in different investigations were compared and, because of the small temperature range in each individual study, these data were combined in order to estimate more reliable Arrhenius parameters between 500 K and 1050 K. Thus:

$$\log k_1 \text{ (s}^{-1}) = (14.4 \pm 0.2) - (180 \pm 3)\text{kJ mol}^{-1} (2.303RT)^{-1}$$

A previous investigation indicated that the pyrolysis of methyl 3-bromopropionate alone was autocatalyzed by the HBr gas product. However, under maximum free-radical inhibition the reaction was homogeneous, unimolecular and of first order[111]. Moreover, the acidity of the hydrogen adjacent to the COOCH$_3$ was held responsible for the faster rate of HBr elimination than that of CH_3CH_2Br. In an extension of this work, the HBr gas assisted as a catalyst the elimination process of methyl 3-bromopropionate[112]. This result was confirmed by the large difference (a factor of 38) between the maximally catalyzed and maximally inhibited dehydrobromination rates. The rate coefficient, under maximum catalysis was given by: $\log k_1 \,(s^{-1}) = (11.19 \pm 0.64) - (171.6 \pm 7.7)\text{kJ mol}^{-1}(2.303RT)^{-1}$. The proposed mechanism involved a six-membered cyclic transition state as described in equation 23.

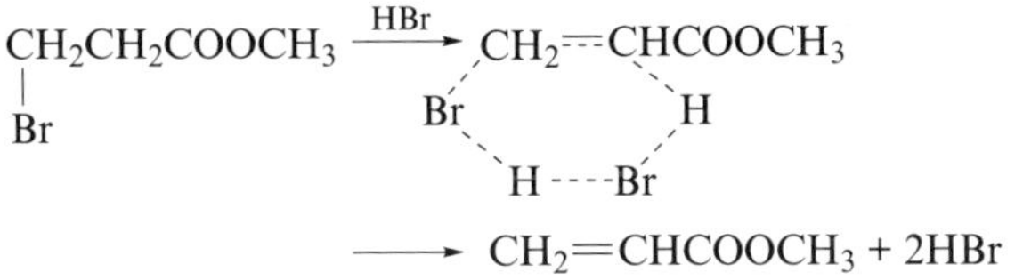

$$CH_2CH_2COOCH_3 \xrightarrow{\;HBr\;} CH_2\!=\!\!=\!CHCOOCH_3$$

$$\text{(structure with Br, H, H----Br)}$$

$$\longrightarrow CH_2\!=\!CHCOOCH_3 + 2HBr \qquad (23)$$

The pyrolysis of 3-bromobutyronitrile, at 370–420 °C yielded mostly cis- and trans-crotononitriles and small amounts of allyl cyanide[113]. The cyano substituent was shown to retard the HBr elimination, and this was attributed to its electron-withdrawing resonance effect. The reported Arrhenius equation was $\log k_1\,(s^{-1}) = (13.74 \pm 0.25) - (213.7 \pm 3.2)$ kJ mol^{-1} $(2.303RT)^{-1}$ The greater acidity of the C—H bond adjacent to the CN in 3-bromobutyronitrile plus the fact that the transition state is more stabilized by conjugation of the double bond with the CN group explain why the direction of elimination was in favor of formation of the crotononitriles rather than of allyl cyanide.

B. Cycloalkyl Bromides

The gas-phase elimination of bromocyclobutane was studied at 791–1224 K using the very low-pressure pyrolysis (VLPP) technique[114]. Hydrogen bromide and 1,3-butadiene formed from isomerization of the product cyclobutene (equation 24) were the products obtained under these experimental conditions.

$$\square\text{—Br} \xrightarrow{\;-HBr\;} \square \longrightarrow CH_2\!=\!CHCH\!=\!CH_2 \qquad (24)$$

A possible alternative decomposition as described in equation 25 was not observed. A four-membered cyclic transition state and an Arrhenius factor similar to that of the HCl elimination from chlorocyclobutane was assumed for the RRKM calculations. The experimental unimolecular rate coefficients are consistent with the Arrhenius equation: $\log k_1$ $(s^{-1}) = (13.6 \pm 0.3) - (217.5 \pm 4.2)$ kJ mol^{-1} $(2.303RT)^{-1}$. The E_a value was found to be higher than that of the open-chain analog 2-bromobutane. The result of this work seems to be consistent with that obtained from the corresponding cyclic chloro and iodo compounds.

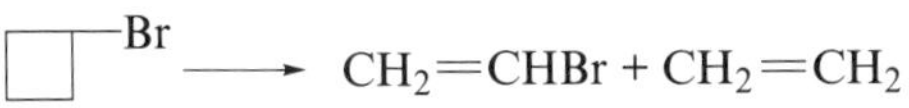

$$\square\text{—Br} \longrightarrow CH_2\!=\!CHBr + CH_2\!=\!CH_2 \qquad (25)$$

C. Aryl Bromides

Kominar and coworkers[115] used the toluene carrier technique for the pyrolysis of bromobenzene. The elimination process was found to be homogeneous and of first order. The Arrhenius expression for equation 26 is: $\log k_1\,(s^{-1}) = (14.6 \pm 0.3) - [(317.000 \pm 5.4550)/19.15T]$. The result of this pyrolysis was discussed in terms of the following mechanism (equations 26–31):

$$C_6H_5Br \longrightarrow C_6H_5^* + Br^* \tag{26}$$

$$Br^* + C_6H_5CH_3 \longrightarrow C_6H_5CH_2^* + HBr \tag{27}$$

$$C_6H_5^* + C_6H_5CH_3 \longrightarrow C_6H_6 + C_6H_5CH_2^* \tag{28}$$

$$C_6H_5^* + C_6H_5CH_2^* \longrightarrow C_6H_5CH_2C_6H_5 \tag{29}$$

$$2\,C_6H_5CH_2^* \longrightarrow C_6H_5CH_2CH_2C_6H_5 \tag{30}$$

$$2\,C_6H_5^* \longrightarrow (C_6H_5)_2 \tag{31}$$

The products described in equations 29–31 were identified qualitatively by gas chromatography.

Years later, the same authors examined the decomposition of (bromomethyl)penta fluorobenzene, $C_6F_5CH_2Br$, using the toluene carrier technique at 727–800 °C[116]. The reaction was shown to be first order and homogeneous. According to the products formed, the following mechanism of equations 32–37 was proposed:

$$C_6F_5CH_2Br \longrightarrow C_6F_5CH_2^* + Br^* \tag{32}$$

$$Br^* + C_6H_5CH_3 \longrightarrow C_6H_5CH_2^* + HBr \tag{33}$$

$$C_6F_5CH_2^* + C_6H_5CH_3 \longrightarrow C_6F_5CH_3 + C_6H_5CH_2^* \tag{34}$$

$$C_6F_5CH_2^* + C_6H_5CH_2^* \longrightarrow C_6F_5CH_2CH_2C_6H_5 \tag{35}$$

$$2\,C_6H_5CH_2^* \longrightarrow C_6H_5CH_2CH_2C_6H_5 \tag{36}$$

$$2\,C_6F_5CH_2^* \longrightarrow C_6F_5CH_2CH_2C_6F_5 \tag{37}$$

The Arrhenius equation was reported as $\log k_1\ (s^{-1}) = (14.54 \pm 0.42) - (225.000 \pm 6.000)/19.15T$. The A factor was found to be very close to the preferred value of 14.6 suggested by Benson and O'Neal for the pyrolysis of benzyl bromide[18]. Yet, the energy of activation of the process above was 12 kJ mol^{-1} less than the estimated value for $C_6H_5CH_2Br$, suggesting that the presence of fluorine in the aromatic ring weakens the C—Br bond.

D. Effect of Substituents

1. β-Substituents at the primary C—Br bond

The maximally inhibited HBr elimination kinetics of β-substituted ethyl bromides[117,118] are listed in Table 10. The Taft plot for the electron-releasing alkyl substituents gave a good straight line with slope $\rho^* = -1.87$ and correlation coefficient $r = 0.991$ at 400 °C. However, plotting the $\log k_{rel}$ values for the three groups with the electron-withdrawing CN substituent (Table 10) against σ^* values produced an inflection point at σ^* (CH$_3$) = 0.00 and gave for them another good straight line ($\rho^* = -0.24$, $r = 0.988$ at 400 °C). This result of one slope for electron-releasing groups and another slope initiating at σ^* (CH$_3$) = 0.00 for electron-withdrawing CN–substituted groups resembles the Taft correlation obtained in the pyrolytic elimination of alkyl and polar substituted primary[62], secondary[70], and tertiary[84] alkyl chlorides. The explanation is similar to that previously suggested, involving a slight alteration in the polarity of the transition state due to changes in the balance of two effects affecting the positive carbon reaction center.

Even though the energy of activation may be theoretically related to bond dissociation energy, the comparative rates and correlation, adequate parameters were found for the interpretation of the substituent effect. An isomerization of equation 38 and an equilibrium process of the primary products formed from 4-bromobutyronitrile are shown in a

TABLE 10. Kinetic parameters for the unimolecular elimination of HBr from pyrolysis of ZCH_2CH_2Br at 400 °C

Z	$\log A\,(s^{-1})$	$E_a\,(kJ\,mol^{-1})$	$10^4 k_1\,(s^{-1})$	Ref.
H	13.19	224.6	0.58	85
CH_3	12.90	212.1	2.75	119
CH_3CH_2	13.18	212.9	4.57	119
$CH_3CH_2CH_2$	13.09	211.2	5.01	120
$CH_3CH_2CH_2CH_2$	13.14	211.2	5.62	120
$(CH_3)_2CHCH_2$	13.04	209.3	6.31	117
$(CH_3)_2CH$	13.57	214.0	9.12	117
$CH_3CH_2CH(CH_3)$	13.94	220.1	7.24	117
$(CH_3)_3C$	13.66	214.8	9.77	117
NC	13.56	231.9	0.36	118
$NCCH_2$	14.12	233.5	1.00	118
$NCCH_2CH_2$	13.79	225.7	1.86	118

mechanism proceeding through a six-membered HBr catalysed transition state, as described in equation 39.

$$BrCH_2CH_2CH_2CN \longrightarrow CH_2{=}CHCH_2CN + CH_3CH{=}CHCN + HBr \quad (38)$$
$$ 57\% \qquad\qquad cis\text{-}trans\ 43\%$$

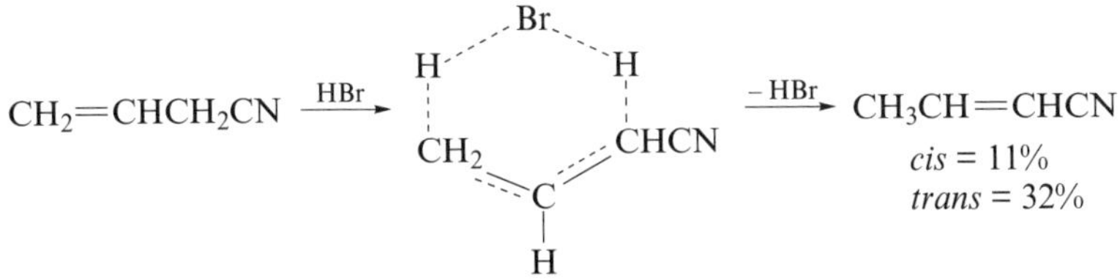

$$cis = 11\%$$
$$trans = 32\%$$

2. α–Substituents at the secondary C—Br bond

Kinetic data for the HBr elimination of secondary alkyl bromide, i.e., α-substituted ethyl bromide, in the gas phase are shown in Table 11[121]. Contrary to primary β-substituted ethyl bromides, the rate constants for these secondary halides could not be correlated by the use of the Taft equation. This limitation arised because the corresponding olefin products underwent rapid isomerization with HBr gas acting as a catalyst. The lack of a kinetic control prevented evaluation of the factor by which the Z substituent in $ZCH(Br)CH_3$ affected the direction of elimination. However, as the electron-releasing effect of Z increases (Table 11), a small but significant increase in the overall rate constant was obtained. In view of the catalysis by HBr in the isomerization process of the olefin products, a general mechanism for this process was suggested (equation 40).

TABLE 11. Kinetic parameters for the unimolecular elimination of HBr from pyrolysis of $ZCH(Br)CH_3$ at 340 °C

Z	$\log A\,(s^{-1})$	$E_a\,(kJ\,mol^{-1})$	$10^4 k_1\,(s^{-1})$	Ref.
H	13.19	224.6	0.012	85
CH_3	13.62	199.9	3.89	122
CH_3CH_2	13.04	188.3	9.88	123
$CH_3CH_2CH_2$	13.08	187.0	14.13	123
$CH_3CH_2CH_2CH_2$	13.08	185.7	18.20	121
$(CH_3)_2CHCH_2$	13.08	183.4	28.38	121

$$
\underset{\text{H}}{\overset{\text{H}}{\underset{|}{\text{Z}\overset{|}{\text{C}}}}}-\overset{|}{\underset{|}{\text{C}}}=\text{C}- \quad \underset{\text{HBr}}{\rightleftharpoons} \quad \left[\begin{array}{c}\text{Br} \\ \text{H} \quad \text{H} \\ \text{ZC} \qquad \text{C} \\ \text{C}\end{array}\right] \quad \rightleftharpoons \quad \underset{\text{H}}{\overset{\text{H}}{\underset{|}{\text{Z}\overset{|}{\text{C}}}}}=\text{C}-\overset{|}{\underset{|}{\text{C}}}- \; + \; \text{HBr} \qquad (40)
$$

V. ORGANIC IODO COMPOUNDS

A. Alkyl Iodides

A comprehensive picture of alkyl iodides pyrolyses has been presented by Benson[124]. These substrates are sensitive and difficult to handle in homogeneous molecular HI elimination studies and this is the reason for the comparatively few gas-phase investigations. Concurrent radical and unimolecular mechanisms are frequently observed in organic iodides decomposition.

The gas-phase thermal decomposition of methyl iodide was examined behind shock waves between 1050–1500 K[125]. The reaction was believed to initiate by the C—I bond fission. In this sense, the sequence of the reaction in the CH_3I decomposition was believed to proceed by a radical mechanism. The low-pressure limit rate coefficient gave the following Arrhenius equation: $\log k_1 \, (s^{-1}) = (15.40 \pm 0.26) - (178.0 \pm 6.3) kJ \, mol^{-1} (2.303RT)^{-1}$. With the application of the RRKM strong collision theory, the collision efficiency factor, β_c, was found to be 0.018 at 1300 K

The rate coefficients for the pressure-dependent unimolecular two-channel decomposition of 1-iodopropane (equations 41 and 42) were determined by using the very low-pressure pyrolysis (VLPP) technique[126].

$$CH_3CH_2CH_2CI \longrightarrow CH_3CH=CH_2 + HI \qquad (41)$$

$$CH_3CH_2CH_2CI \longrightarrow CH_3CH_2CH_2^{\bullet} + I^{\bullet} \longrightarrow CH_3^{\bullet} + CH_2=CH_2 + I^{\bullet} \quad (42)$$

Analysis of the data in this work and the use of finite diffusion rates suggested that weak gas–gas collision and strong gas–wall collision take place.

t-Butyl iodide decomposition was carried out by the shock-wave technique at temperatures between 465–1020 K[127]. The reaction was performed close to the high-pressure limit of the unimolecular elimination process. The mechanism for dehydroiodination did not change at high temperatures and, according to the experimental data, the reaction via a four-membered cyclic transition state led to the expression: $\log k_1 \, (s^{-1}) = (13.9 \pm 0.2) - (164 \pm 3) kJ \, mol^{-1} (2.303RT)^{-1}$.

B. Aryl Iodides

The pyrolysis of C_6F_5I was investigated by the toluene carrier technique[128]. When the molar ratio of $C_6H_5CH_3/C_6F_5I$ was greater than 150, the pentafluorobenzene radical abstracted the H atom of toluene to produce C_6F_5H. Molecular iodine, I_2, was formed to an insignificant amount and the major gaseous product was HI. The experiments showed lack of surface effect, which meant that the pyrolysis of C_6F_5I was homogeneous and followed a first-order rate law. The results were discussed in terms of a radical mechanism. Analysis of the data for the process $C_6F_5I \rightarrow C_6F_5^{\bullet} + I^{\bullet}$ led to the equation: $\log k_1 \, (s^{-1}) = (15.9 \pm 0.03) - (289.5 \pm 0.5) kJ \, mol^{-1} (2.303RT)^{-1}$. An approximate correction of the activation energy to 298 K estimated $D[C_6F_5—I]$ as 277.0 kJ mol^{-1}. Within the experimental error, $D[C_6F_5—I]$ was found to be somewhat greater than $D[C_6H_5—I]$. It was suggested by the authors that this reflects the tendency of F to backdonate electrons to the aromatic nuclei. Moreover, according to their preliminary results, they predicted that the difference $D[C_6F_5—X] – D[C_6H_5—X]$ is expected to increase rapidly as X changes in the order I, Br, Cl, F.

A previous work[128] gave $D[C_6F_5—I] = 277$ kJ mol^{-1} a significantly larger value than $D[C_6F_5—I]$ of 239 kJ mol^{-1}[129]. Due to the discrepancy, the pyrolysis of C_6H_5I was examined by the toluene carrier technique[130]. The reaction was found to be homogeneous and to obey a first-order rate law. Hydrogen iodide was the main gaseous product, while traces of I_2 and small amounts of H_2 and CH_4 were also detected. The Arrhenius equation for $C_6H_5I \rightarrow C_6H_5^· + I^·$ was expressed as: $\log k_1 (s^{-1}) = (15.0 \pm 0.28) - (269.6 \pm 4.8)$ kJ mol^{-1} $(2.303RT)^{-1}$. An estimated correction of the energy of activation to 298 K gave $D[C_6H_5—I] = 266.3 \pm 4.8$ kJ mol^{-1}. The mechanism was explained as a radical process. This investigation still showed that $D[C_6F_5—I]$ is to some extent larger than $D[C_6H_5—I]$, although the difference is lower than for other types of C_6F_5X/C_6H_5X pairs.

An interesting decomposition reaction is the pyrolysis of a series of arylphenyliodonium chlorides, bromides and iodides[131], for 5 min at 235 °C in Pyrex tubes sealed at one end and immersed in an oil bath[131]. These type of compounds gave upon heating two parallel reactions a and b as described in equations 43 and 44, where Z may be more than one methyl

$$ZC_6H_4I^+PhX^- \xrightarrow{\quad a \quad} PhI + ZC_6H_4X \tag{43}$$

$$\xrightarrow{\quad b \quad} PhX + ZC_6H_4I \tag{44}$$

substituent in the aromatic nuclei, and X^- are Cl^-, Br^- or $^{131}I^-$ anions. These and other reported data[132–134] were thought to be consistent with a nucleophilic aromatic substitution where X^- ion preferentially attacks the aryl ring best able to accommodate a developing negative charge. In equations 43 and 44, path b predominates over path a. Additional examples are described.

VI. COMPETITIVE DEHYDROHALOGENATION

The single-pulse shock tube method in the thermal decomposition of 1,1,1-trifluoro-2-chloroethane gave parallel eliminations[135]. The major reaction involved HCl elimination which was believed to be formed by an α,α-elimination (equation 45) with a concerted transfer of a fluorine atom, while to a lesser extent a slower α,β-elimination of HF also took place (equation 46).

$$F_2C=CHF + HCl \tag{45}$$

$$F_2C=CHCl + HF \tag{46}$$

Step 1: $\log k_a(s^{-1}) = (13.3 \pm 0.4) - (274.0 \pm 9.2)$ kJ mol^{-1} $(2.303RT)^{-1}$
Step 2: $\log k_b(s^{-1}) = (12.7 \pm 0.5) - (282.8 \pm 11.3)$ kJ mol^{-1} $(2.303RT)^{-1}$

The mechanism of the α,α-elimination (path a) implies substantial freezing of rotational modes of the transition state by the concerted transfer of the F atom and a three-membered cyclic transition state for elimination of HCl. These geometric factors should be reflected in high negative entropies of activation. However, the actual experimental $\Delta S^{\ddagger}$ value of -0.61 e.u. and the rearrangement of a F atom implies a more polar character of the transition state. In this respect, it seems reasonable that the fluorine 1,2-shift proceeds by an intimate ion-pair mechanism as described in equation 47. 2,2-Dichloro-1,1,1-trifluo-

$$F_3CCH_2Cl \longrightarrow F_3C\overset{\delta^+}{C}H_2 \cdots \overset{\delta^-}{Cl} \longrightarrow \left[F_2\overset{F}{\underset{}{C}}\text{---}CH_2 \, Cl^{\delta^-} \right]^{\delta^+} \longrightarrow F_2C{=}CHF + HCl \quad (47)$$

roethane, F_3CCHCl_2, was decomposed in a single-pulse tube and, below $1260\,K^{136}$, yielded $CF_2{=}CFCl$, $CF_2{=}CHF$ and CF_3CCl_3. The results were attributed to a C—Cl bond breaking in competition with a molecular HCl elimination. The rate coefficient for the α,α-elimination (equation 48) was deduced to be given by $\log k_1$ $(s^{-1}) = (13.4 \pm 0.7) - (263.9 \pm 15.9) kJ\,mol^{-1}\,(2.303RT)^{-1}$.

$$F_3CCHCl_2 \longrightarrow F_2C{=}CFCl + HCl \quad (48)$$

The overall rate coefficient for the formation of $F_2C{=}CHF$ was also estimated as $\log k_1$ $(s^{-1}) = (11.5 \pm 0.8) - (226.3 \pm 18.8)\,kJ\,mol^{-1}\,(2.303RT)^{-1}$. At temperatures above 1260 K the HF gas was detected, and under this condition a C—C bond breaking took place and the kinetics of the reaction became untractable.

The pyrolyses of 1-chloro-1-fluoroethane (equations 49 to 50) and 1-chloro-1,1-difluoroethane (equations 51 and 52) in a flow system and seasoned vessels[75], gave two concurrent eliminations where the HF elimination is less than 0.5%.

$$CH_3CHClF \overset{\nearrow\ CH_2{=}CHF + HCl \quad (49)}{\searrow\ CH_2{=}CHCl + HF \quad (50)}$$

$$CH_3CClF_2 \overset{\nearrow\ CH_2{=}CF_2 + HCl \quad (51)}{\searrow\ CH_2{=}CClF + HF \quad (52)}$$

A decrease in rate upon increase in the number of α-fluorines was expected according to the relative inductive and mesomeric effects of fluorine. However, the HCl elimination from F-substituted ethyl chloride gave a rate sequence of $CH_3CHClF > CH_3CClF_2 > CH_3CH_2Cl$ instead of the predicted order of $CH_3CH_2Cl > CH_3CHClF > CH_3CClF_2$.

Thermolysis of CF_3I was found to be unimolecular[137] and to obey the Arrhenius law: $\log k_1\,(s^{-1}) = 15.40 - (197.4 \pm 8.4)\,kJ\,mol^{-1}\,(2.303RT)^{-1}$. The E_a value is lower than the reported bond energy of $225.9\,kJ\,mol^{-1\,138}$, while $\log A = 15.4$ is significantly higher than that previously suggested[139], namely, $\log A = 13.0\,s^{-1}$. The mechanism was described by equations 53–56.

$$CF_3I \longrightarrow CF_3^{\bullet} + I^{\bullet} \quad (53)$$

$$2\,CF_3^{\bullet} \longrightarrow F_3CCF_3 \quad (54)$$

$$CF_3I + I^{\bullet} \longrightarrow CF_3^{\bullet} + I_2 \quad (55)$$

$$CF_3^{\bullet} + I_2 \longrightarrow CF_3I + I^{\bullet} \quad (56)$$

The total yield of the other carbon-containing products, i.e., CF_2I_2, C_2F_4 and CF_4, was less than 0.1%.

Qualitative work on the pyrolysis of bromochlorodifluoromethane, either pure or mixed with 1:1 air, at 400–800 °C showed it to proceed, as expected, through a complex free radical type mechanism[140]. The overall reaction yielded the products described in equation 57:

$$CBrClF_2 \longrightarrow CCl_2F_2 + CClF_2CClF_2 + CBrF_2CClF_2 + CBr_2F_2 + CBr_2F_2CBrF_2 \quad (57)$$

Homogeneous laser heating decomposed several ω-chloro-ω'-fluoroalkanes, $Cl(CH_2)_nF$, in two pathways[141] (equations 58 and 59).

$$Cl(CH_2)_nF \bigg\langle {\begin{array}{l} CH_2{=}CH(CH_2)_{n-2}Cl + HF \quad (58) \\[2mm] CH_2{=}CH(CH_2)_{n-2}F + HCl \quad (59) \end{array}}$$

The Arrhenius parameters for each decomposition pathway of these substrates are listed in Table 12. According to the kinetic data, with the exception of $ClCH_2CH_2F$, as expected, Cl was found to be a better leaving group than F at any of the temperature studies of dehydrohalogenation. In the case of 1-chloro-4-fluorobutane, butadiene was also formed from the further decomposition of chlorobutene and/or fluorobutene. The overall percent of butadiene in the product mixture was less than 1.5% in the temperature range studied and in the decomposition of $Cl(CH_2)_4F$ itself it did not exceed 1%. This result is not surprising, since Chuchani and coworkers[142] have already found that $CH_2{=}CHCH_2CH_2Cl$ loses HCl to give butadiene through a reaction involving double bond participation at a rate 25 times faster than that of CH_3CH_2Cl at 713 K.

TABLE 12. Kinetic parameters for the dehydrohalogenation of $Cl(CH_2)_nF$

Substrate	Temp (K)	Product	$\log A$ (s^{-1})	E_a (kJ mol^{-1})
$CH_3CH_2F^a$	683–738	$CH_2{=}CH_2$	13.31	243.5
$ClCH_2CH_2F$	785–1020	$CH_2{=}CHF$	13.77	264.1
		$CH_2{=}CHCl$	13.40	255.8
$ClCH_2CH_2CH_2F$	700–800	$CH_2{=}CHCH_2F$	13.84	245.6
		$CH_2{=}CHCH_2Cl$	13.45	246.0
$ClCH_2CH_2CH_2CH_2F$	670–800	$CH_2{=}CHCH_2CH_2F$	13.34	229.3
		$CH_2{=}CHCH_2CH_2Cl$	12.45	226.8

^{a}Reference 19.

VII. FIVE-MEMBERED TRANSITION STATE

Gas-phase decomposition of chlorocyclobutane, using the very low-pressure pyrolysis technique[143], proceeds via two competing unimolecular eliminations, one to yield 1,3-butadiene and HCl in larger amounts (equation 60) and the other to give vinyl chloride and ethylene (equation 61). Under the assumption that glass–wall collisions are strong, the RRKM calculations showed that the experimental unimolecular rate coefficients are consistent with the high-pressure Arrhenius parameters: for $CH_2{=}CHCH{=}CH_2$ formation, $\log k_1$ (s^{-1}) = (13.6 ± 0.3) – (233.0 ± 4.2) kJ mol^{-1} (2.303RT)$^{-1}$ and for $CH_2{=}CHCl$ forma-

$$\square{-}Cl \bigg\langle {\begin{array}{l} \xrightarrow{k_1} CH_2{=}CHCH{=}CH_2 + HCl \quad (60) \\[2mm] \xrightarrow{k_2} CH_2{=}CHCl + CH_2{=}CH_2 \quad (61) \end{array}}$$

tion, $\log k_1 \ (\mathrm{s}^{-1}) = (14.8 \pm 0.3) - (255.6 \pm 4.2) \ \mathrm{kJ \ mol}^{-1} \ (2.303RT)^{-1}$. Two transition states for chlorocyclobutane pyrolysis were previously proposed[144] depending on whether a biradical is formed via cleavage at C_1—C_2 or C_2—C_3. However, the work of King and coworkers[143] assumed for their calculations the five-membered transition state structure shown below:

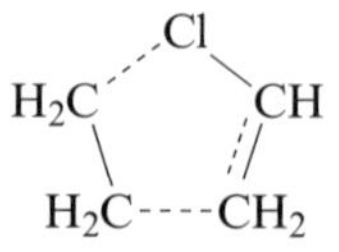

After careful consideration of the many previous experimental investigations of chloro-cyclobutane pyrolyses, Frey and Watts[145] reported in a detailed study the ratio k_2/k_1 as a function of pressure well into the fall-off region at 720.5 K. The ratio was found to fall from a high-pressure value of about 2.08 to 4.64 at 0.014 torr. The RRKM calculation assuming the strong collision hypothesis showed agreement with experiment at high pressure. However, in the fall-off region serious discrepancies between experimental and theoretical results were obtained. In using the step-ladder model with a step size (the step size is equal to the magnitude of the energy transferred in a collision between an energized molecule and its collision partner), the calculated values fitted the experimental results rather well and much better than the strong collision results.

The elimination kinetics of 2-chloropropionic acid[146] and 2-bromopropionic acid[147] have been studied in seasoned vessels and in the presence of a free radical inhibitor. The reactions proved to be homogeneous, unimolecular and to obey a first-order rate law. According to the dehydrohalogenation products, i.e., acetaldehyde and carbon monoxide in both cases, the hydrogen atom of the carboxylic COOH group was considered to assist the expulsion of the leaving halogen through a polar five-membered cyclic transition state (equation 62).

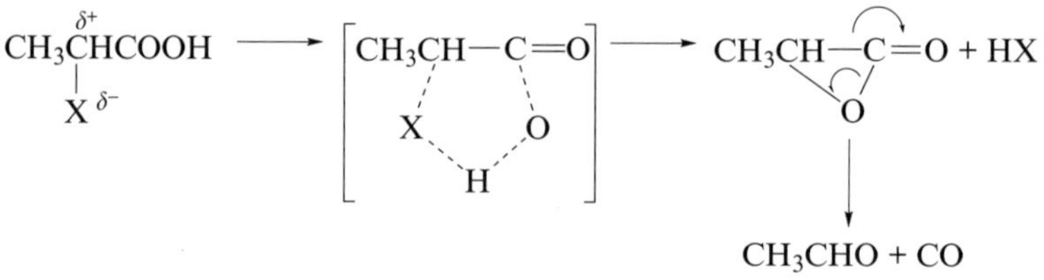

Apparently, the carboxylic acid proton of $\mathrm{p}K_a \approx 4.8$ assists readily the halogen expulsion as compared with the H atom of the CH_3 group ($\mathrm{p}K_a \approx 48$). In order to support this conclusion, the H of the COOH was replaced by a methyl group to give methyl 2-halopropionates. The comparative kinetic data are shown in Table 13. The methyl haloesters pyrolyze much slower than the free acids and they probably eliminate HX through a normal four-membered transition state.

TABLE 13. Arrhenius parameters and comparative rates at 360 °C

Substrate	$\log A \ (\mathrm{s}^{-1})$	$E_a \ (\mathrm{kJ \ mol}^{-1})$	Relative rate
$CH_3CH(Cl)COOCH_3$	12.22	217.0	1.0
$CH_3CH(Cl)COOH$	12.53	186.9	609.5
$CH_3CH(Br)COOCH_3$	13.01	211.4	1.0
$CH_3CH(Br)COOH$	12.41	180.3	91.6

VIII. SIX-MEMBERED TRANSITION STATE

In relation to the six-membered transition state proposed for the pyrolysis of α-chloro-o-xylene which gives benzocyclobutene and HCl[148], the gas-phase elimination of α-phenylethyl chloride was carried out in the presence of $CH_2=CH_2$ in order to see whether any 1,2,3,4-tetrahydronaphthalene could be obtained through the Diels–Alder addition[149]. This experiment to trap an o-quinodimethane intermediate gave negative results and the compound dehydrochlorinated through a four-membered transition state. Moreover, the effect of methyl substitution in the aromatic ring at 335 °C is a mild acceleration for m-Me, a larger one for p-Me and an appreciable one for o-Me (equation 63).

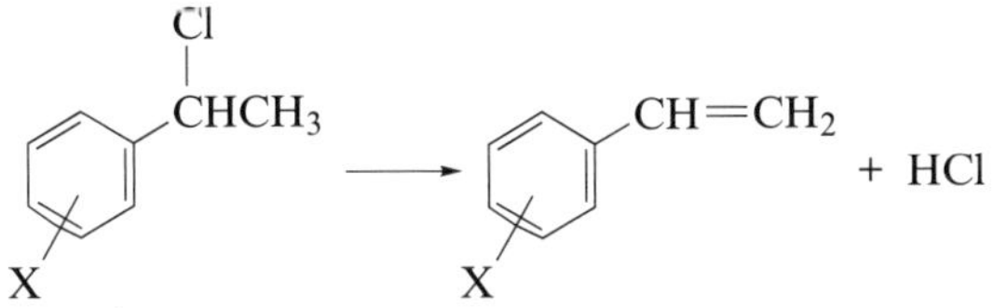

X	H	m-CH$_3$	p-CH$_3$	o-CH$_3$
Rel. rate	1.0	1.5	2.9	8.1

1,3-Pentadiene and HCl were produced on pyrolysis of a mixture of *cis*- and *trans*-4-chloro-2-pentene in the gas phase[150]. The homogeneous decomposition of the *cis*-isomer at 520–602 K and of the *trans*-isomer at 564–640 K gave the following Arrhenius equation: for *cis*, $\log k_1$ (s^{-1}) = (9.85 ± 0.46) – (131.3 ± 5.1) kJ mol^{-1} (2.303RT)$^{-1}$; for *trans*, $\log k_1$ (s^{-1})= (13.86 ± 0.26) – (196.8 ± 3.5) kJ mol^{-1} (2.303 RT)$^{-1}$. The A factor and the activation energy for the *cis*-chloropentene are much lower than for the *trans*-isomer, suggesting a six-membered cyclic transition state (equation 64). The *trans*-chloropentene was assumed to decompose by a normal four-centered cyclic transition state (equation 65). An alternate mechanism involving a rate-determining isomerization of the *trans*-chloropentene to the *cis*-isomer followed by a rapid HCl elimination was not ruled out.

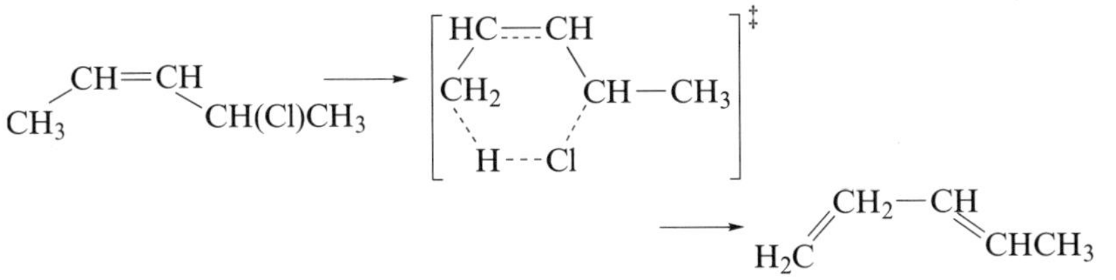

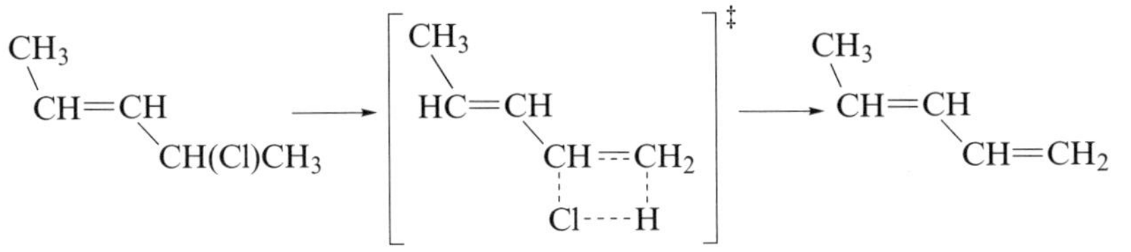

Further work to firmly establish which position loses a hydrogen atom in the gas-phase pyrolysis of α-chloro-o-xylene or o-methylbenzyl chloride[151] was undertaken by means of flash pyrolysis of o-methylbenzyl chloride-α,α-d$_2$ at 750 °C[152]. The results as described in equation 66 indicated that the hydrogen atom is abstracted from the o-CH$_3$ group. Consequently, the mechanism was rationalized in terms of a direct δ-elimination followed by cyclization (equation 67). An alternate mechanism was also proposed (equation 68), where an initial 1,3-sigmatropic shift of the chlorine atom is followed by the β-elimination of HCl.

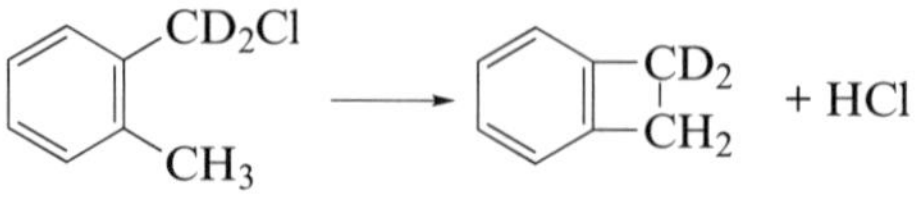

$$(66)$$

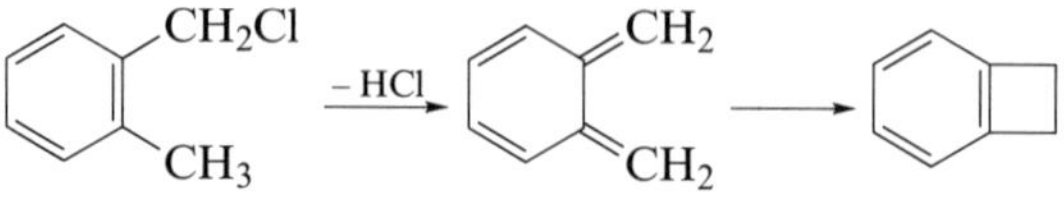

$$(67)$$

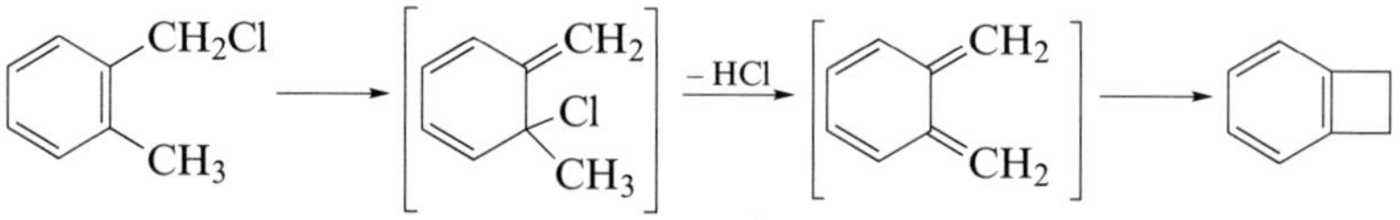

$$(68)$$

An interesting pyrolysis product of 3-chloro-2,2-bis (chloromethyl) propionic acid at 200–220 °C is 1,1-bis (chloromethyl)ethene[153]. This reaction may well be associated with the five-membered dehydrochlorination process of α-halopropionic acids[146,147]. In this respect, the HCl elimination may be considered to proceed through a mechanism involving a six-membered transition state as described in equation 69.

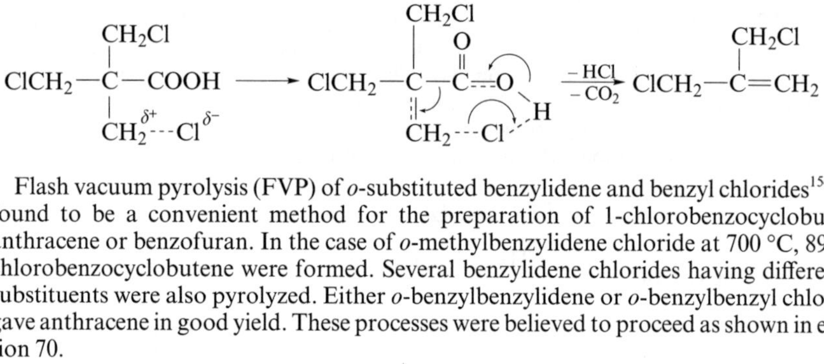

$$(69)$$

Flash vacuum pyrolysis (FVP) of o-substituted benzylidene and benzyl chlorides[154] was found to be a convenient method for the preparation of 1-chlorobenzocyclobutene, anthracene or benzofuran. In the case of o-methylbenzylidene chloride at 700 °C, 89% 1-chlorobenzocyclobutene were formed. Several benzylidene chlorides having different o-substituents were also pyrolyzed. Either o-benzylbenzylidene or o-benzylbenzyl chlorides gave anthracene in good yield. These processes were believed to proceed as shown in equation 70.

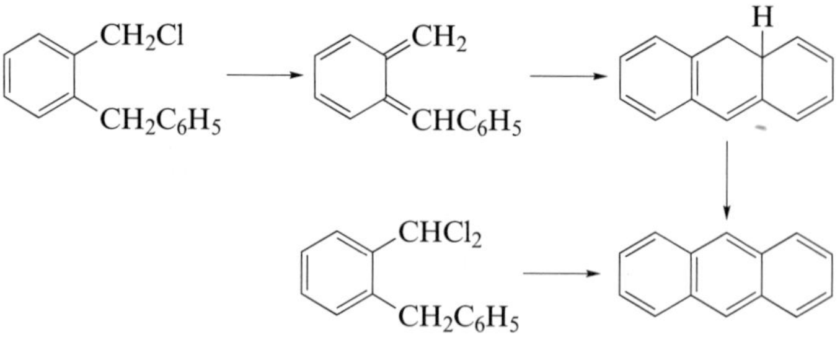

$$(70)$$

The FVP of o-methoxybenzylidene chloride gave mainly benzofuran, which suggested the formation of a diradical or a carbene (see opposite) as an intermediate. Evidence in support of the carbene intermediate was presented when obtaining unlabelled benzofuran from o-CH$_3$OC$_6$H$_4$CDCl$_2$ pyrolysis.

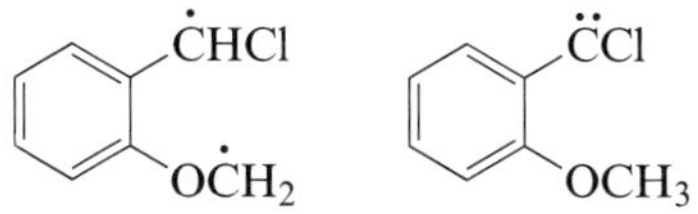

$$\text{(o-}\overset{\bullet}{\text{C}}\text{HCl, OCH}_2\text{)} \qquad \text{(o-}\overset{\bullet\bullet}{\text{C}}\text{Cl, OCH}_3\text{)}$$

Again, in association with the six-membered cyclic transition state proposed for the pyrolysis of α-chloro-o-xylene into benzocyclobutene or the dimer and HCl[151], the substituent effect on α-chloro-o-xylenes pyrolyses was examined under reduced pressure in a flow system[155]. The relative reactivities of the substituted α-chloro-o-xylenes with respect to α-chloro-o-xylene are given below and the Hammett plot against σ_p values gave a reaction constant $\rho = -0.46$ and $r = 0.99$. The small negative ρ value suggested development of a relatively small positive charge in the transition state.

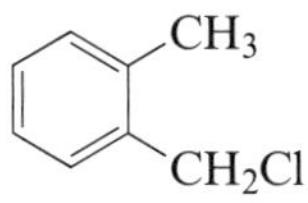

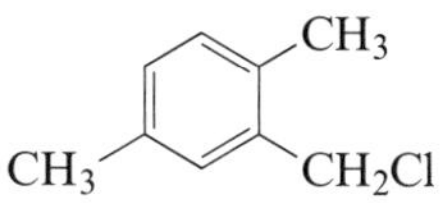

Relative reactivity   1.0   1.2

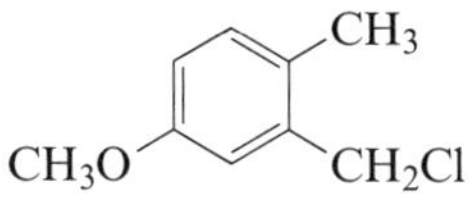

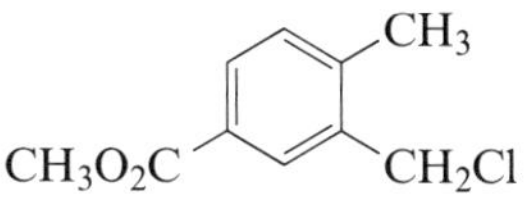

1.4   0.64

The reaction rates of 2,4-dimethylbenzyl chloride and 2,5-dimethylbenzyl chloride were also determined by pyrolyzing a mixture of both isomers. The results indicated that an electron-donating group *para* to the CH$_2$Cl enhanced the reactivity more than the corresponding *meta* group. In a competitive reaction of a mixture of 2-methyl-4-carbomethoxybenzyl chloride and 2-methyl-5-carbomethoxybenzyl chloride, the electron-withdrawing carbomethoxy substituent retarded the reaction more when it is *para* than at *meta*-position to the CH$_2$Cl group. According to the relative reactivities of 2-methylbenzyl chlorides, the chloride ion was believed to leave slightly ahead of the H of the 2 – CH$_3$ group in a 1,4-elimination of HCl (equation 71).

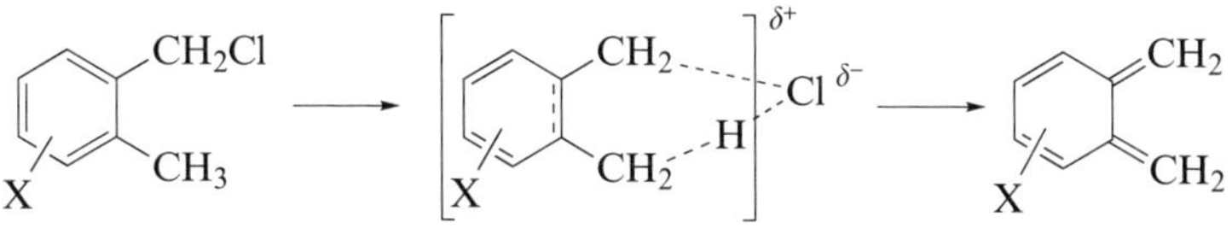

$$(71)$$

The substituent effects discussed above provide additional support to the C—Cl bond polarization, in the sense of $C^{\delta+}\cdots Cl^{\delta-}$, as a main contributing factor to the rate-determining step.

IX. NEIGHBORING GROUP PARTICIPATION

A. General Considerations

The first case of neighboring group participation in homogeneous unimolecular gas-phase pyrolysis was deduced from the significant rate increase in the thermal decomposition of 2-(methylthio)ethyl chloride[156]. Table 14 indicates that 2-chloro, 2-hydroxy and 2-methoxy ethyl chlorides pyrolyzed at a rate slower or slightly higher than that of ethyl chloride due to the electron-withdrawing effect of these substituents (see also Table 5). The greatly enhanced rate of $CH_3SCH_2CH_2Cl$ was believed to be due to participation by the sulfur atom in the transition state which was suggested to involve a *trans*-elimination by way of an intimate ion-pair as shown below.

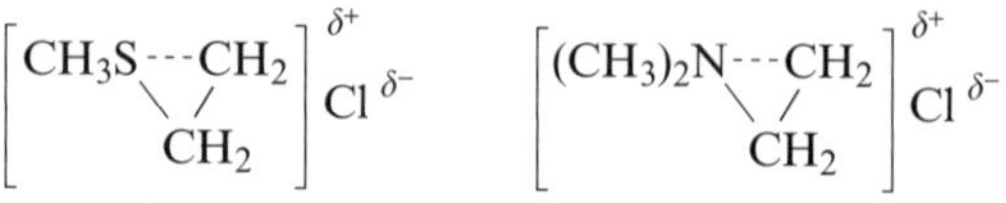

$$\left[\begin{array}{c} CH_3S\text{---}CH_2 \\ \diagdown \diagup \\ CH_2 \end{array} \right]^{\delta+} Cl^{\delta-} \qquad \left[\begin{array}{c} (CH_3)_2N\text{---}CH_2 \\ \diagdown \diagup \\ CH_2 \end{array} \right]^{\delta+} Cl^{\delta-}$$

TABLE 14. Arrhenius parameters of ZCH_2CH_2Cl pyrolysis at 400 °C[a]

Z	$\log A$ (s^{-1})	E_a (kJ mol^{-1})	Relative rate per H
H	13.16	236.2	1.0
Cl	11.90	225.9	0.4
HO	12.80	229.6	2.2
CH_3O	14.06	244.7	2.4
C_6H_5	13.07	220.9	18.5
$p\text{-}CH_3OC_6H_4$	13.81	228.4	26.1
CH_3S	13.90	212.5	558.7
$(CH_3)_2N$	13.22	203.7	560.0

[a]See Table 5.

An additional example of neighboring group participation in the gas-phase pyrolysis of 2-substituted ethyl chlorides was the elimination kinetics of 2-dimethylaminoethyl chloride[67]. The magnitude of the effect of Me_2N on the dehydrochlorination rate (Table 14) led to a similar consideration to that for $CH_3SCH_2CH_2Cl$ by assuming the transition state for the elimination as an intimate ion-pair as represented above.

According to these previous results, a generalization was proposed[67] for an effective gas-phase neighboring group participation in reactions of organic molecules as follows:

(a) The transition state must be highly polar.
(b) The participating atom has to be large and capable of overlap with the reaction center.
(c) The participating atom must be highly polarizable.

B. Hydrogen Participation

The theory of Maccoll and Thomas on ion-pair intermediates suggested the possibility of both *cis*- and *trans*-elimination[6,7], while Benson and Bose's semi-ion pair transition state[16] is only limited to a *cis*-elimination. This mechanistic problem was examined during

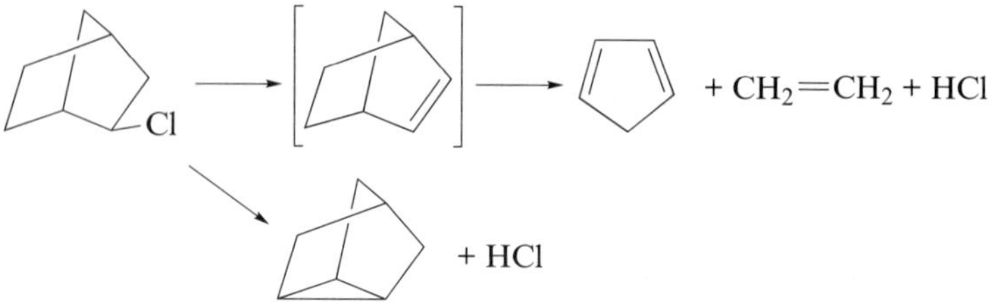

$$+ CH_2{=}CH_2 + HCl \qquad (72)$$

$$+ HCl \qquad (73)$$

the gas-phase pyrolysis of *exo*-2-norbornyl chloride labelled with deuterium[157]. This work showed norbornene to be produced by a *cis*-1,2-elimination (equation 72), whereas nortricyclene was formed via a *trans*-1,3-elimination through participation of the *endo*-6 H atom (equation 73). Gas-phase Wagner–Meerwein rearrangement was not involved in the formation of these products.

C. Aromatic Participation

When the substituent Z in ZCH_2CH_2Cl is an aromatic ring, a moderate but significant increase in rate relative to ethyl chloride was obtained[158] (Table 14). This result suggests that the neighboring phenyl group provides anchimeric assistance in the HCl elimination. The alternative explanation that the acidic C_β—H adjacent to the benzene ring assists the leaving Cl atom expulsion and thus causes a faster rate was refuted by introducing the *p*-CH_3O group into 2-chloroethylbenzene. Since the *p*-CH_3O group decreases the acidity of the C_β—H, the rate of dehydrochlorination of *p*-methoxyethylbenzene should in this case be much slower than that of 2-chloroethylbenzene. Since the rate was enhanced, a partial positive charge delocalization by the *p*-CH_3O group via participation of the aromatic nuclei in the elimination augments the rate (Table 14). The transition state was suggested to be similar to that proposed with the 2-(methylthio)ethyl chloride, as shown below for the unsubstituted and the MeO-substituted systems.

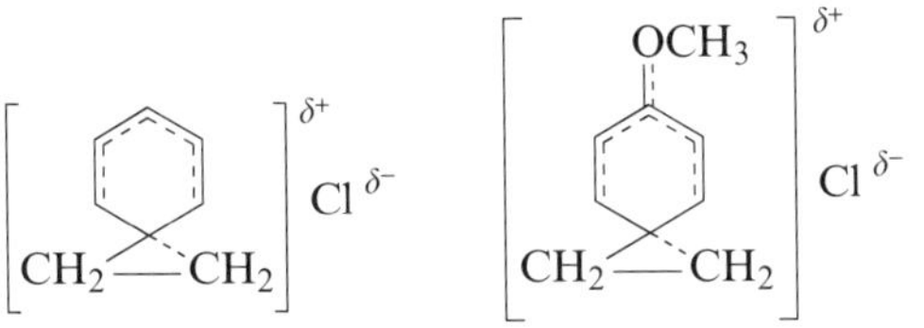

The rates of pyrolysis of *o*-, *m*- and *p*-methylphenylethyl chloride have been compared with that of ethyl chloride (Table 15)[159]. Even though the phenyl group participated in the HCl elimination, electron release of the methyl group at the three isomeric position of the aromatic ring was found to be, within the experimental errors, ineffective on the rates when compared to the unsubstituted phenylethyl chloride. Consequently, the effect of the CH_3 substituent was too small for the reinforcement of the phenyl assistance as reflected by the pyrolysis rate.

TABLE 15. Arrhenius parameters of ZCH_2CH_2Cl pyrolysis at 440 °C

Z	$\log A$ (s^{-1})	E_a (kJ mol^{-1})	Relative rate per H
H	13.84	241.8	1.0
C_6H_5	13.07	220.9	8.6
o-$CH_3C_6H_4$	13.76	231.4	7.2
m-$CH_3C_6H_4$	12.84	219.2	6.7
p-$CH_3C_6H_4$	14.03	234.1	8.4

Chuchani and coworkers[160] examined the participation by a more remote benzene ring in ω-phenylalkyl chlorides. Analysis of the data of Table 16 indicated that participation of C_6H_5 at the 2-, 4- and 5-positions is more favored than in the 3-position which seems unimportant. The relative sequence found in this work is somewhat similar to that found in the formolysis and acetolysis of ω-phenylalkyl *p*-bromobenzenesulfonates by Heck and Winstein[161].

TABLE 16. Relative rates and Arrhenius parameters for ω-phenylalkyl chloride pyrolysis at 440 °C

Compound	$\log A$ (s^{-1})	E_a (kJ mol^{-1})	Relative rate per H
CH_3CH_2Cl	13.84	241.8	1.0
$C_6H_5CH_2CH_2Cl$	13.07	220.9	8.9
$C_6H_5(CH_2)_2CH_2Cl$	13.99	238.4	3.7
$C_6H_5(CH_2)_3CH_2Cl$	13.07	220.5	9.2
$C_6H_5(CH_2)_4CH_2Cl$	13.75	231.2	7.2

The pyrolysis rate of 2-(bromoethyl)benzene was measured only at 385.5 K[81] and the rate coefficient was found to be 0.85 ($\pm$ 0.07) $\times$ 10^{-4}. A later work on the decomposition of this substrate[162] in a flow system gave a reaction rate higher by a factor of 120. Because of these two contradictory results, the elimination kinetics of 2-(bromoethyl)benzene and of 1-bromo-3-phenylpropane were carried out in a static system, with seasoned vessels and maximally inhibited with a free radical suppressor[163]. The rate of HBr elimination from 2-(bromoethyl)benzene is significantly higher than that of the corresponding unsubstituted ethyl bromide (Table 17). This observation confirmed the anchimeric assistance of the phenyl ring in a similar way to that in 2-(chloroethyl)benzene[158] and ω-phenylalkyl chlorides[160]. When the C_6H_5 group is at the 3-position in 3-bromo-1-phenylpropane for a C-4 conformation, it was thought that neighboring aromatic participation was absent. The mechanism of pyrolysis of these organic bromides was explained as before[159,164].

TABLE 17. Arrhenius parameters for pyrolysis of 2-phenyl and 3-phenylalkyl bromides at 420 °C

Substrate	$\log A$ (s^{-1})	E_a (kJ mol^{-1})	Relative rate per H
CH_3CH_2Br	13.19	224.6	1.0
$C_6H_5CH_2CH_2Br$	13.04	210.8	11.6
$C_6H_5CH_2CH_2CH_2Br$	14.09	227.7	6.9

D. Olefinic Double-bond Participation

The influence of an olefinic double bond in alkenyl chlorides pyrolyses[66] has been assessed and described in Table 18. The vinyl substituent showed a significant rate-enhanc-

TABLE 18. Kinetic parameters and comparative rates for pyrolyses of saturated and doubly bonded chlorides at 440 °C

Substrate	$\log A$ (s^{-1})	E_a (kJ mol^{-1})	Relative rate per H
CH_3CH_2Cl	13.84	241.8	1.0
$CH_3CH_2CH_2CH_2Cl$	13.63	230.7	6.1
$CH_2{=}CHCH_2CH_2Cl$	13.79	223.8	28.5
$CH_3CH_2CH_2Cl$	13.48	227.6	1.0
$CH_3CH_2CH_2CH_2Cl$	13.81	231.4	1.1
$CH_2{=}CHCH_2CH_2CH_2Cl$	14.25	238.4	1.0
$CH_3CH_2CH_2CH_2Cl$	13.63	230.7	1.0
$CH_3CH_2CH_2CH_2CH_2Cl^a$	13.81	231.4	1.4
$CH_2{=}CHCH_2CH_2CH_2CH_2Cl$	12.38	209.6	2.0

^{a}n-Hexyl chloride pyrolysis has not yet been reported; n-pentyl chloride pyrolysis was therefore used for approximate comparison.

TABLE 19. Arrhenius parameters and comparative rates of alkenyl chlorides pyrolysis at 440 °C

Substrate	$\log A$ (s^{-1})	E_a (kJ mol^{-1})	Relative rate per H
CH_3CH_2Cl	13.84	241.8	1.0
$CH_2=CHCH_2CH_2Cl$	13.79	223.8	28.5
$CH_2=C(CH_3)CH_2CH_2Cl^a$	12.20	201.6	30.6
$(CH_3)_2C=CHCH_2CH_2Cl$	13.43	215.0	53.2

^{a}Data from J. Mauger and J. Maurin, *Bull. Soc. Chim, Fr.*, 2332 (1970).

ing effect in 4-chloro-1-butene. The authors have considered, with the aid of adequate evidence, that the acidity of the allylic C_β—H bond and steric factors are not important in affecting the rate of dehydrochlorination of $CH_2=CHCH_2CH_2Cl$. However, they assumed that the neighboring olefinic double bond provided anchimeric assistance to the HCl elimination of this compound.

In the case that the $CH_2=CH$ substituent of 4-chloro-1-butene and the chlorine were separated by one or two methylene groups, i.e. in 5-chloro-1-pentene and 6-chloro-1-hexene (Table 18), the olefinic double bond did not provide assistance to the pyrolysis. In this respect, the formation of a three-membered conformation appeared to favor participation of the double bond in the transition state.

A classical study giving evidence of neighboring double-bond participation was described in a work on acetolysis of 4-methyl-3-pentenyltosylate[165]. In association with this convincing result, the pyrolysis of the tosylate analog, the 5-chloro-2-methyl-2-pentene was undertaken[142]. Even though this substrate does not yield the cyclic product 2-cyclopropylpropene presumably obtained by the anchimerically assisted process, the k value for its reaction is considerably greater than that of ethyl chloride at 440 °C (Table 19). 4-Chloro-2-methyl-1-butene shows a small rate increase relative to 4-chloro-1-butene, which suggested a slight effect of the electron-releasing CH_3 at the terminal carbon of the $CH_2=CH$ substituent. This work supported a previous investigation[66] on the anchimeric assistance of the olefinic double bond in the gas-phase pyrolysis of alkenyl chlorides. A three-membered ring was therefore regarded as the most favored structure for the transition state of the participation process as shown below:

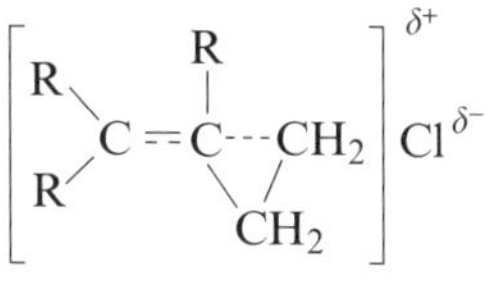

$$\left[\begin{array}{c} R \\ R \diagdown \\ \diagdown C = C \overset{R}{} \text{---} CH_2 \\ R \diagup \diagdown \diagup \\ CH_2 \end{array} \right]^{\delta+} Cl^{\delta-}$$

E. Carbonyl Ester Participation

On examination of the reported lactonization of ethyl γ-bromobutyrate at 200 °C in various media[13], there was no reaction in the gas phase and only HBr elimination in a flow

TABLE 20. Comparative rates and Arrhenius parameters for pyrolysis of ZCH_2CH_2Cl at 440 °C

Z	$\log A$ (s^{-1})	E_a (kJ mol^{-1})	Relative rate per H
H	13.84	241.8	1.0
CH_3OCO	13.22	231.5	2.0
CH_3OCOCH_2	13.41	226.8	7.0
$CH_3OCOCH_2CH_2$	13.20	227.6	3.7

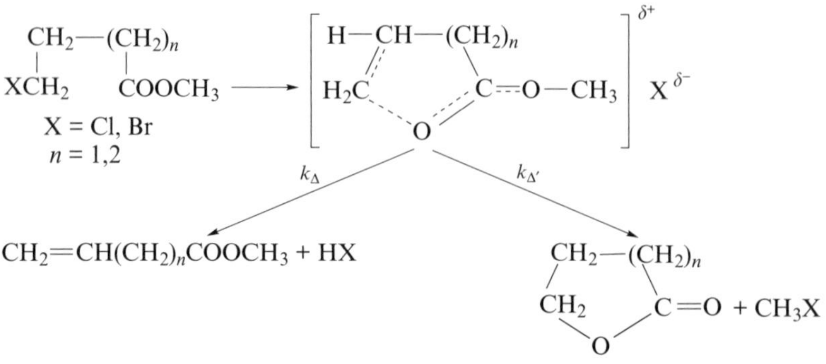

$$CH_2\!=\!CH(CH_2)_nCOOCH_3 + HX \qquad\qquad CH_2\!-\!(CH_2)_n \atop CH_2 \quad C\!=\!O + CH_3X$$

(74)

system at 450 °C was described. The fact that only HBr was formed appeared to be surprising, since at 450 °C the ethyl group of the ester should have been eliminated more easily than the dehydrobromination process. In relation to this work, the gas-phase elimination of methyl ω-chloroesters was undertaken[166]. The results given in Table 20 suggested anchimeric assistance of the COOCH$_3$ group in the elimination of the methyl esters of 4-chlorobutyric and 5-chlorovaleric acids. This suggestion was confirmed by the fact that besides the normal formation of a dehydrochlorination product, ring-closed products were also obtained. The small but significant rate enhancements suggested that a neighboring COOCH$_3$ group assisted these reactions in which a five- or six-membered cyclic ring are being formed.

These elimination processes were described in terms of an intimate ion-pair mechanism involving a neighboring COOCH$_3$ group participation and an intramolecular solvation of the chloride ion (X = Cl), which may proceed via two directions giving noncyclic and cyclic products (equation 74).

The rates and the Arrhenius parameters for each of the pyrolysis products of the chloroesters were reported. The moderately higher rate of methyl 3-chloropropionate relative to CH$_3$CH$_2$Cl (Table 20) was rationalized by the acidity of the hydrogen atom adjacent to the COOCH$_3$ group. Consequently, the acidic hydrogen assists the expulsion of the chloride leaving group. The sequence 5-COOCH$_3$ > 6-COOCH$_3$ > 4-COOCH$_3$ for the extent of participation is in agreement with the order of formation of the most favorable conformation in neighboring group participation in solvolysis reactions[167].

According to previous work[166] and the gas-phase data for ethyl 4-bromobutyrate[13], a reexamination of the gas-phase elimination kinetics of the latter was undertaken[168,169].

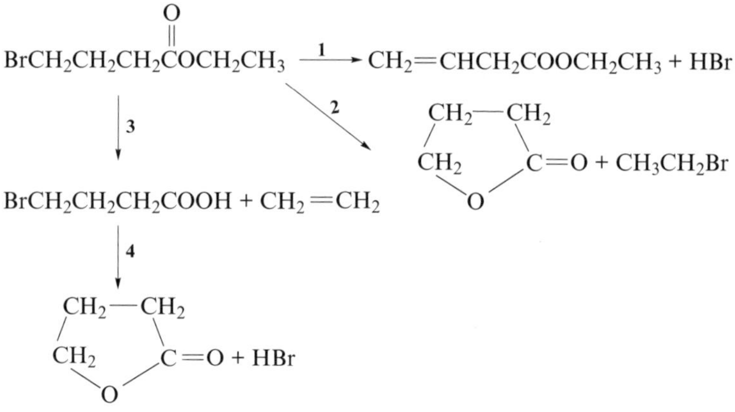

(75)

Parallel eliminations were found to occur during the homogeneous, unimolecular decomposition of the bromoester at 354–375 °C (equation 75). The overall rate coefficient was given by the following equation: $\log k_1$ (s^{-1}) = $(13.71 \pm 0.60) - (209.9 \pm 7.3)$ kJ mol^{-1} $(2.303RT)^{-1}$

The reported analyses of the products suggested that the COOCH$_2$CH$_3$ substituent provided anchimeric assistance for the reaction paths **2** and **3**. The mechanism was explained in terms of an intramolecular solvation of the bromide ion through an intimate ion-pair intermediate which decomposes in two different directions (equation 74, where X = Br and CH$_3$CH$_2$ replaces CH$_3$). The formation of bromobutyric acid and ethylene (path **3**) indicated a normal six-membered transition state for their formation as found in primary ethyl esters pyrolyses in the gas phase. Bromobutyric acid, which is known to be unstable at room temperature, rapidly produced butyrolactone. The consecutive reaction of path **4**, under the experimental conditions, was ascribed to a similar mechanism where an intimate ion-pair intermediate is formed through COOH participation (equation 76).

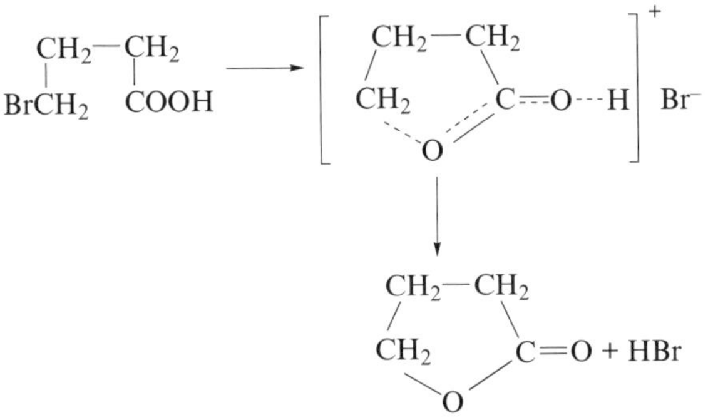

$$ (76) $$

TABLE 21. Arrhenius parameters and relative rates of ZCH$_2$CH$_2$Br pyrolysis at 400 °C

Z	long A (s^{-1})	E_a (kJ mol^{-1})	Relative rate per H
H	13.19	224.6	1.0
CH$_3$OCO	12.94	214.5	5.1
CH$_3$OCOCH$_2$	13.38	216.7	9.5
CH$_3$OCOCH$_2$CH$_2$	13.84	228.9	3.1

TABLE 22. Arrhenius paramers for γ-bromoester pyrolysis at 360 °C

Substrate	$\log A$ (s^{-1})	E_a (kJ mol^{-1})	Relative rate per H
CH$_3$OCOCH$_2$CH$_2$CH$_2$Br	13.38	216.7	1.0
CH$_3$OCOCH=CHCH$_2$Br	13.30	185.2	327.2

Evidently, on changing the ethyl ester to methyl ester in bromoacids, the pyrolytic eliminations will be limited only to a debromination process. In this respect, Table 21 describes the results on the pyrolysis of methyl bromoesters under maximum inhibition[170] and the data provide further evidence for COOCH$_3$ participation in the elimination from methyl ω-bromoesters. By analogy with previous papers[166,168,169] the mechanisms of these reactions were also explained in terms of an intimate ion-pair intermediate as for equation

74 and X = Br. Since neighboring group participation is less favorable when a four-membered ring is formed, the higher rate of methyl 3-bromopropionate compared with ethyl bromide was thought to be due to a greater acidity of the α-hydrogen adjacent to the $COOCH_3$ group[111].

The overall rate of elimination of methyl 4-bromocrotonate was found to be significantly higher than that of methyl 4-bromobutyrate (Table 22)[171]. The participation by the $COOCH_3$ group is more effective as predicted[67] because of the high polarity of the allylic C—Br bond in the transition state. The pyrolysis, under maximum inhibition, gave parallel reactions as described in equation 77.

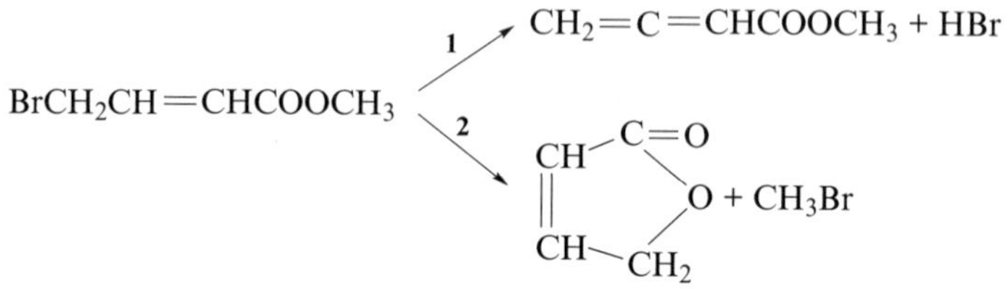

$$BrCH_2CH{=}CHCOOCH_3 \begin{array}{c} \nearrow^{1} CH_2{=}C{=}CHCOOCH_3 + HBr \\ \\ \searrow_{2} \quad \begin{array}{c} CH{-}C{=}O \\ \| \qquad O + CH_3Br \\ CH{-}CH_2 \end{array} \end{array} \tag{77}$$

Participation of the $COOCH_3$ in the pyrolysis of methyl o-(bromomethyl)benzoate was believed to be the reason why it decomposes much faster than methyl 4-bromocrotonate to yield phthalide and CH_3Br. Unfortunately, attempts to prepare the benzoate yielded phthalide. An intimate ion-pair mechanism as in methyl ω-chloro and ω-bromo esters (equation 74) was suggested.

By analogy with ethyl 4-bromobutyrate thermolysis, the kinetics of the pyrolysis and the product formed from ethyl 4-chlorobutyrate were examined[172]. At the same time, the pyrolysis of the expected stable product 4-chlorobutyric acid product (equation 78), which results in butyrolactone formation, was also carried out.

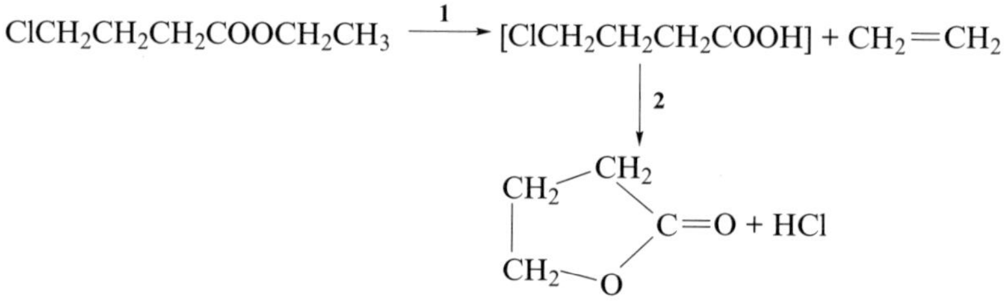

$$ClCH_2CH_2CH_2COOCH_2CH_3 \xrightarrow{1} [ClCH_2CH_2CH_2COOH] + CH_2{=}CH_2$$

$$\downarrow 2$$

$$\begin{array}{c} CH_2{-}CH_2 \\ CH_2 \qquad \quad C{=}O + HCl \\ CH_2{-}O \end{array} \tag{78}$$

The reaction proceeds via the normal six-membered transition state mechanism of ester elimination. The intermediate chlorobutyric acid yielded butyrolactone through participation of the COOH group under the reaction conditions. The pure chlorobutyric acid was pyrolyzed and gave butyrolactone in quantitative yield. The mechanism is similar to that of bromobutyric acid pyrolysis (equation 76). The data are shown in Table 23.

TABLE 23. Kinetic parameters and comparative rates for chloroester and chloroacid pyrolysis at 400 °C

Substrate	$\log A$ (s^{-1})	E_a (kJ mol^{-1})	Relative rate per H
$CH_3COOCH_2CH_3$	12.55	200.4	1.0
$ClCH_2CH_2CH_2COOCH_2CH_3$	12.21	197.6	0.7
CH_3CH_2Cl	13.83	241.8	1.0
$ClCH_2CH_2CH_2COOH$	12.28	172.0	7738.0

F. Carbonyl Ketone Participation

Since the C=O of esters has been found to assist the elimination reactions of substituted alkyl halides, the kinetic data on the unimolecular pyrolysis of chlorobutanones[88,173,174] were determined. Table 24 gives the data and comparison between primary, secondary and tertiary chlorobutanones with respect to their corresponding saturated hydrocarbons and reference compounds. The significantly higher rate of $CH_3COCH_2CH_2Cl$ relative to CH_3CH_2Cl was believed to be due to increased acidity of the β-hydrogen of the acetyl group. The acidic hydrogen may assist the Cl expulsion in the elimination process of 4-chloro-2-butanone, as shown below.

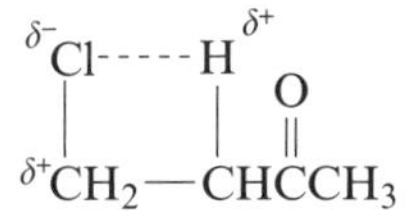

$$\overset{\delta^-}{Cl}-----\overset{\delta^+}{H} \quad O$$
$$| \qquad\qquad | \quad \|$$
$$\overset{\delta^+}{CH_2}-CHCCH_3$$

TABLE 24. Kinetic and Arrhenius parameters of haloketone pyrolyses at 360 °C

Compound	$\log A$ (s^{-1})	E_a (kJ mol^{-1})	Relative rate per H^a
CH_3CH_2Cl	13.84	241.8	1.0
$CH_3CH_2CH_2CH_2Cl$	13.63	230.7	7.6
$CH_3COCH_2CH_2Cl$	13.67	225.2	23.5
$CH_3CHClCH_3$	13.64	213.8	1.0
$CH_3CH_2CHClCH_3$	13.71	208.3	1.4
$CH_3COCHClCH_3$	11.80	205.8	0.07
$CH_3CCl(CH_3)_2$	13.77	188.0	1.0
$CH_3CH_2CCl(CH_3)_2$	13.77	184.1	1.4
$CH_3COCCl(CH_3)_2$	12.56	190.8	0.04
CH_3CH_2Cl	13.84	241.8	1.0
$CH_3CH_2CH_2CH_2CH_2Cl$	13.81	231.4	10.0
$CH_3COCH_2CH_2CH_2Cl$	13.12	207.8	181.2
CH_3CH_2Cl	13.84	241.8	1.0
$C_6H_5CH_2CH_2CH_2CH_2Cl$	13.07	220.5	14.4
$C_6H_5COCH_2CH_2CH_2Cl$	12.02	182.4	1794.4

^{a}For formation of the 1-olefin.

3-Chloro-2-butanone and 3-chloro-3-methyl-2-butanone gave a significant decrease in the rates. These results were attributed to the electronic deactivation of the acetyl group with respect to CH_3 and CH_3CH_2 groups. Since a five-membered ring is a very favorable structure for neighboring group participation, the elimination kinetics of 5-chloro-2-pentanone and 4-chloro-1-phenyl-1-butanone were also determined. These compounds were found to be much faster than the corresponding saturated haloalkane and the reference

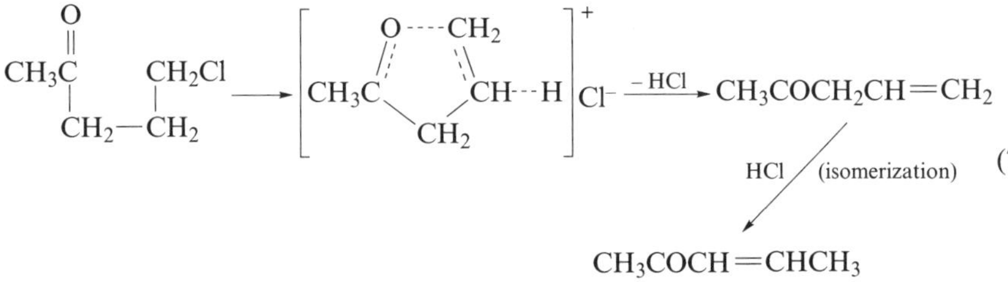

$$CH_3\overset{O}{\overset{\|}{C}} \quad CH_2Cl \longrightarrow \left[CH_3\overset{O----CH_2}{C}\overset{\diagdown}{\underset{CH_2}{CH---H}} \right]^+ Cl^- \xrightarrow{-HCl} CH_3COCH_2CH=CH_2$$
$$| \qquad |$$
$$CH_2-CH_2$$

$$HCl \diagup (isomerization) \qquad (79)$$

$$CH_3COCH=CHCH_3$$

compound. Consequently, these reactions were explained in terms of an intimate ion-pair mechanism through the assistance of the C=O group (equation 79).

The higher rate of HCl elimination of 4-chloro-1-phenyl-1-butanone with respect to 5-chloro-2-pentanone by a factor of 9.9 confirmed the participation of the C=O group. The π-electron delocalization of the benzene ring increases the basicity and nucleophilicity of the carbonylic oxygen and led to greater stabilization of the transition state through assistance to the C—Cl bond cleavage in the transition state (equation 80).

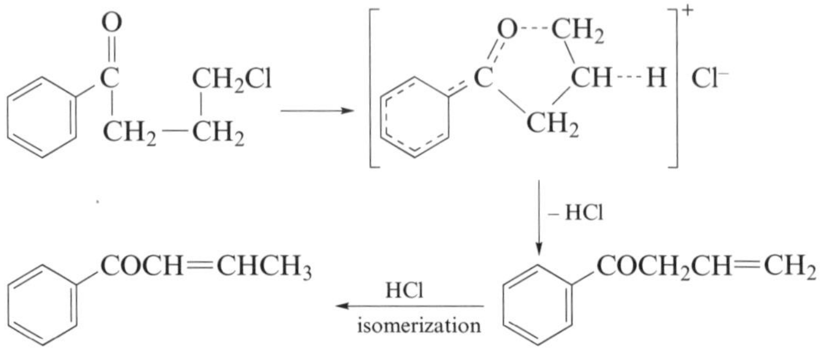

$$(80)$$

G. Alkoxy Oxygen Participation

The homogeneous unimolecular gas-phase pyrolysis of 3-methoxy-1-chloropropane and 4-methoxy-1-chlorobutane[175] was described as in equations 81 and 82.

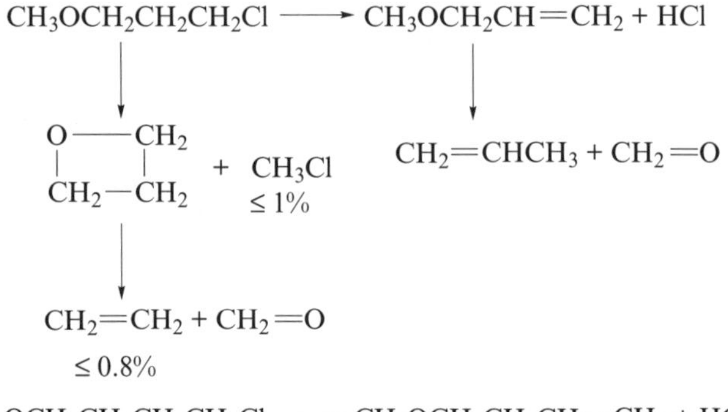

$$CH_3OCH_2CH_2CH_2Cl \longrightarrow CH_3OCH_2CH=CH_2 + HCl$$

$$(81)$$

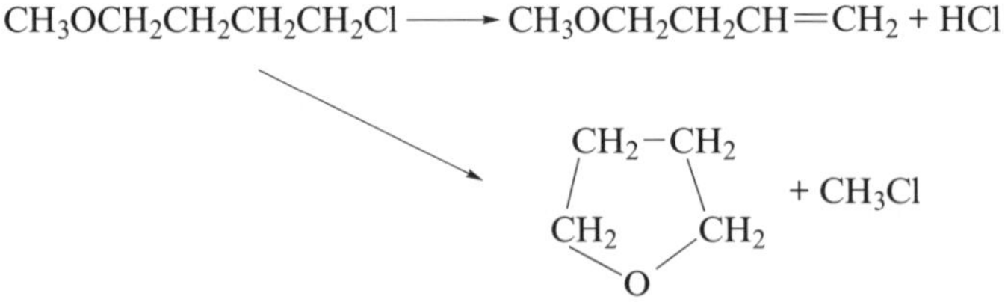

$$CH_2=CHCH_3 + CH_2=O$$

$$+ CH_3Cl \quad \leq 1\%$$

$$CH_2=CH_2 + CH_2=O$$
$$\leq 0.8\%$$

$$CH_3OCH_2CH_2CH_2CH_2Cl \longrightarrow CH_3OCH_2CH_2CH=CH_2 + HCl$$

$$(82)$$

$$+ CH_3Cl$$

The effect of the CH_3O substituent on the elimination rates of chloroethers relative to ethyl chloride listed in Table 25 indicate that the CH_3O group is providing anchimeric assistance in the elimination of 4-methoxy-1-chlorobutane. This conclusion is further supported by the formation of the ring-closed product, tetrahydrofuran, in addition to the formation of the normal corresponding methoxyalkene. According to these findings, an intimate ion-pair mechanism with intramolecular solvation of the chloride ion which leads to products by two pathways was suggested (equation 83).

TABLE 25. Relative rates and Arrhenius parameters of chloroether pyrolysis at 440 °C

Substrate	$\log A$ (s^{-1})	E_a (kJ mol^{-1})	Relative rate per H
CH_3CH_2Cl	13.83	241.08	1.0
$CH_3OCH_2CH_2Cl$	14.06	244.7	1.5
$CH_3OCH_2CH_2CH_2Cl$	12.92	226.0	2.6
$CH_3OCH_2CH_2CH_2CH_2Cl$	12.89	218.1	9.0

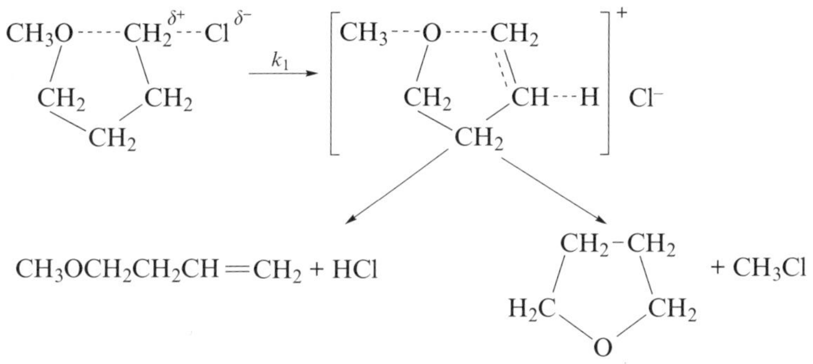

$$(83)$$

The very small yield of CH_3Cl and $CH_2{=}CH_2$ from 3-methoxy-1-chloropropane (equation 81) was rationalized as due to a weak participation of the CH_3O substituent, since a strained four-membered cyclic transition state is not a favorable conformation in neighboring group participation.

Due to the fact that the p-CH_3O group in 2-phenylethyl chloride[158] caused a significant increase in the dehydrochlorination rate, o-methoxy- and m-methoxy-2-phenylethyl chlorides were investigated in order to find out whether the resonance effect of this substituent strengthens the C_6H_5 participation. The pyrolysis of m-methoxy-2-phenylethyl chloride showed the normal process of formation of HCl and m-methoxystyrene presumably via a four-membered transition state[164]. However, o-methoxy-2-phenylethyl chloride yielded, besides o-methoxystyrene and HCl gas, also the cyclic dihydrobenzofuran and CH_3Cl.

The data presented in Table 26 indicate that the resonance effect of the o-CH_3O and p-CH_3O increase the rate augmentation by the phenyl ring. However, the m-CH_3O does not interact resonatively with the reaction center, as shown by the elimination rate of HCl which is similar to that of the unsubstituted phenylethyl chloride.

TABLE 26. Comparative rates and Arrhenius parameters for pyrolysis of ω-methoxyphenylalkyl chlorides at 440 °C

Compound	$\log A$ (s^{-1})	E_a (kJ mol^{-1})	Relative rate per H
CH_3CH_2Cl	13.84	241.8	1.0
$C_6H_5CH_2CH_2Cl$	13.07	220.9	8.6
o-$CH_3OC_6H_4CH_2CH_2Cl$	13.81	225.6	21.3
o-$CH_3OC_6H_4CH_2CH_2CH_2Cl$	13.97	231.4	11.6
m-$CH_3OC_6H_4CH_2CH_2Cl$	12.87	218.7	7.9
p-$CH_3OC_6H_4CH_2CH_2Cl$	13.81	228.4	13.3
p-$CH_3OC_6H_4CH_2CH_2CH_2Cl$	13.52	229.5	5.7
p-$CH_3OC_6H_4CH_2CH_2CH_2CH_2Cl$	14.14	233.3	12.5

In spite of the reduced nucleophilicity of o-CH$_3$O by resonance interaction with the aromatic nuclei, the p electrons of the oxygen assist the participation of the aryl ring in the partial C—Cl bond cleavage in the transition state, which eventually leads to formation of the corresponding cyclic product, dihydrobenzofuran. The mechanism for the two parallel processes is given in equations 84 and 85.

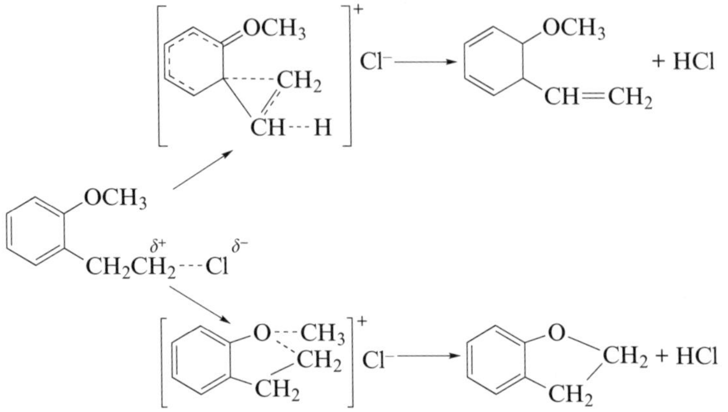

Further examination of the anchimeric assistance by the remote anisyl group in ω-alkyl chlorides[176] showed that 3-(o-methoxyphenyl)-1-chloropropane gave a slower rate than the analogous chloroethane. (Table 26) presumably due to the unfavorable four-membered ring required for neighboring group participation. Yet, since the k value is a little higher than for phenylethyl chloride, the oxygen atom of o-CH$_3$OC$_6$H$_4$(CH$_2$)$_3$Cl was believed to assist directly in the C—Cl bond cleavage through a six-membered transition state.

The presence of CH$_3$O at the p-position in ω-phenylchloroalkanes gave enhanced rates which suggested participation of the p-anisyl ring through a transition state with three- and five-membered rings. The exception was 3-(p-methoxyphenyl)-1-chloropropane, which gave the lowest rate of all the p-methoxyphenyl derivatives studied.

H. Hydroxy Oxygen Participation

In connection with the methoxy participation, the gas-phase pyrolytic elimination of 4-chloro-1-butanol was investigated[177]. The products are tetrahydrofuran, propene, formaldehyde and HCl. It is implied that the OH group provides anchimeric assistance from the fact that, besides formation of the normal unstable dehydrochlorinated intermediate 3-buten-1-ol, a ring-closed product, tetrahydrofuran, was also obtained. The higher rate of chlorobutanol pyrolysis with respect to chlorethanol and ethyl chloride (Table 27) confirmed the participation of the OH group through a five-membered ring in the transition state.

TABLE 27. Arrhenius parameters for pyrolysis of ω-chloro alcohols at 440 °C

Compound	log A (s^{-1})	E_a (kJ mol^{-1})	Relative rate per H
CH$_3$CH$_2$Cl	13.83	241.8	1.0
HOCH$_2$CH$_2$Cl	12.80	229.6	1.1
HOCH$_2$CH$_2$CH$_2$CH$_2$Cl	13.34	221.1	15.3

The suggested mechanism of this process involves an ion-pair intermediate which partitions to two pathways (equation 86).

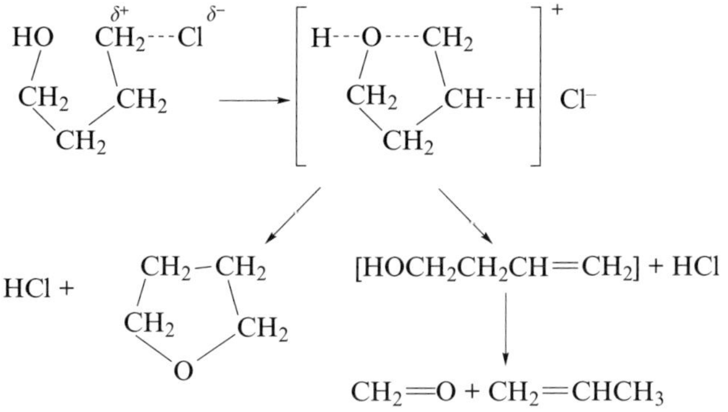

$$(86)$$

The OH was found to be a better participating group than the CH_3O group, as revealed by the comparative pyrolysis rates in Tables 25 and 27.

The homogeneous unimolecular decomposition of *o*-hydroxy-2-phenylethyl chloride yielded mainly benzodihydrofuran, HCl and much less *o*-hydroxystyrene[178]. The rate was significantly higher than that of phenylethyl chloride and ethyl chloride (Table 28). According to the nature of the product formation and the kinetic data, the OH provided anchimeric assistance in the elimination process. The mechanism proposed is described in equations 87 and 88.

TABLE 28. Kinetic parameters for pyrolysis of ZCH_2CH_2Cl at 440 °C

Z	$\log A$ (s^{-1})	E_a (kJ mol^{-1})	Relative rate per H
H	13.84	241.8	1.0
C_6H_5	13.07	220.9	8.6
o-$CH_3OC_6H_4$	13.81	225.6	21.2
o-HOC_6H_4	13.85	225.7	22.7

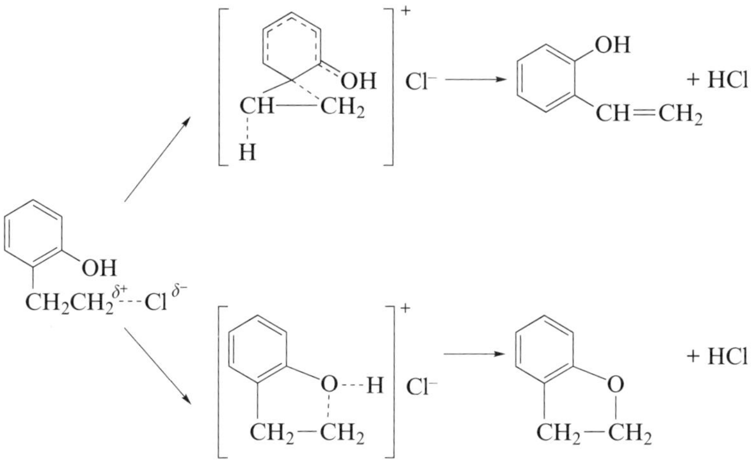

$$(87)$$

$$(88)$$

It is interesting to point out that the o-OH group in 2-phenylethyl chloride produced the cyclic benzodihydrofuran in 87% yield, while the o-CH$_3$O derivative gave the latter in about 25% yield[164]. This difference was reasonably ascribed to the more favored participation ability of the elimination transition state, thus producing more of the corresponding cyclic product.

I. Sulfur Participation

An interesting work on neighboring group participation in the gas phase is the pyrolysis kinetics of 1-chloro-4-(methylthio)butane which forms quantitatively tetrahydrothiophene and methyl chloride[179]. In addition to the formation of the cyclic product, the rate of decomposition was also rather high as shown in Table 29.

TABLE 29. Comparative Arrhenius parameters for CH$_3$S participation in pyrolyses at 400 °C

Compound	$\log A$ (s^{-1})	E_a (kJ mol^{-1})	Relative rate per H
CH$_3$CH$_2$Cl	13.83	241.8	1.0
CH$_3$SCH$_2$CH$_2$Cl	13.90	212.5	338.0
CH$_3$SCH$_2$CH$_2$CH$_2$CH$_2$Cl	12.23	175.7	5105.0

There is no doubt that the CH$_3$S group is providing anchimeric assistance through the favorable cyclic five-membered transition state. In this respect, the reaction mechanism via an ion-pair intermediate suggested intramolecular solvation of the leaving chloride ion to form only tetrahydrothiophene and CH$_3$Cl (equation 89).

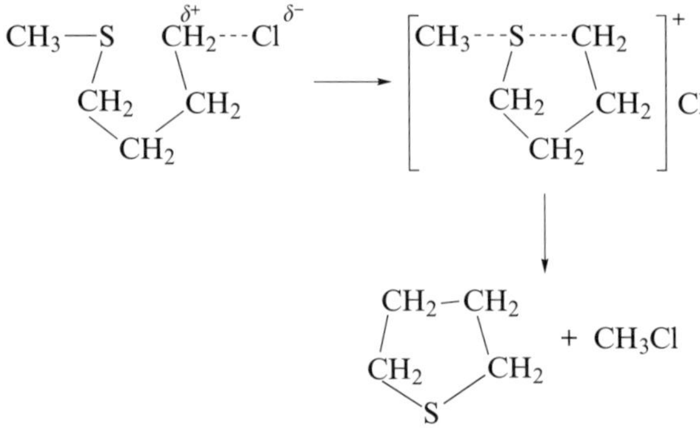

$$(89)$$

X. REARRANGEMENTS

In a short communication, without a detailed description of the experimental conditions, 1,3-dichloropropyl-2-acetate was reported to rearrange when heated at 180 °C to an equilibrium mixture of 37% of 1,3-dichloropropyl-2-acetate and 63% of 1,2-dichloropropyl-3-acetate. The corresponding dibromo and bromochloro esters gave also rearrangement products as a mixture of isomeric acetates. A halide migration via intermediate formation of an acetoxonium ion was proposed[180]. This mechanism (equation 90) may be related to the neighboring group participation of the oxygen carbonyl (Section IX. E) of the acetate, followed by halide migration.

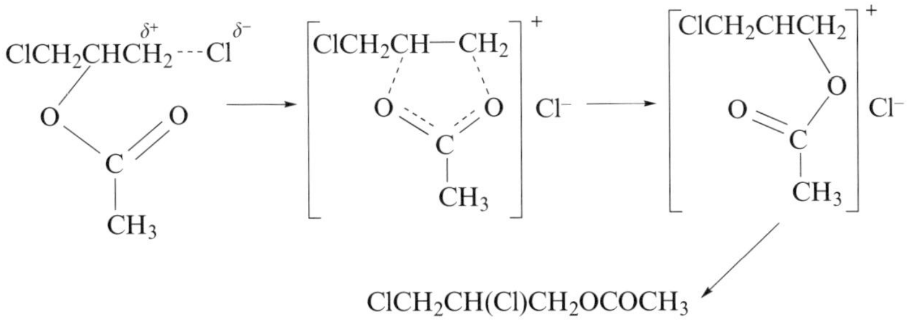

$$\overset{\delta+}{ClCH_2}\overset{}{CHCH_2}\cdots\overset{\delta-}{Cl} \longrightarrow \left[\begin{matrix}ClCH_2CH-CH_2\\O\cdots\quad\cdots O\\C\\CH_3\end{matrix}\right]^+ Cl^- \longrightarrow \left[\begin{matrix}ClCH_2CHCH_2\\O\\O=C\\CH_3\end{matrix}\right]^+ Cl^- \quad (90)$$

$$ClCH_2CH(Cl)CH_2OCOCH_3$$

The gas-phase pyrolysis of neopentyl bromide was studied over the temperature range 389–444 °C. The reaction in a static seasoned vessel, and under maximal inhibition of cyclohexene, was a homogeneous molecular process of first-order rate law[181]. The following Arrhenius parameters of $E_a = 247 \pm 5\,kJ\,mol^{-1}$ and $\log A(s^{-1}) = 14.2 \pm 0.3$ were obtained. The activation energy of this bromide is much higher than for other primary alkyl bromides (see Table 10), suggesting a more strained transition state due to the lack of a β-hydrogen. The proposed mechanism is described in equation 91.

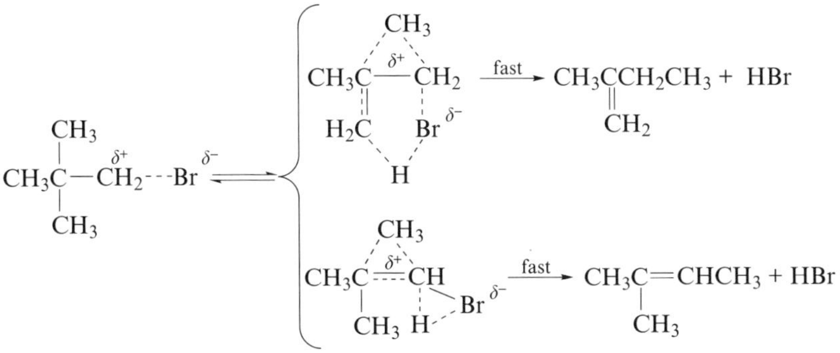

$$\overset{CH_3}{\underset{CH_3}{\overset{|}{\underset{|}{CH_3C}}}}\overset{\delta+}{-CH_2}\cdots\overset{\delta-}{Br} \rightleftharpoons \left\{\begin{matrix}CH_3\\CH_3\overset{\delta+}{C}-CH_2\\H_2C\quad \overset{\delta-}{Br}\\H\end{matrix} \xrightarrow{fast} \begin{matrix}CH_3CCH_2CH_3 + HBr\\ \| \\ CH_2\end{matrix}\right.$$

$$\left.\begin{matrix}CH_3\\CH_3\overset{\delta+}{C}\cdots CH\\CH_3\ H\ \overset{\delta-}{Br}\end{matrix} \xrightarrow{fast} \begin{matrix}CH_3C=CHCH_3 + HBr\\ | \\ CH_3\end{matrix}\right\} \quad (91)$$

The nature of the products evidently suggests a Wagner–Merwein rearrangement. Moreover, preliminary examination in a single-pulse shock tube was reported to confirm the molecular nature of the reaction and the rearrangement of neopentyl bromide to 2-methyl-1-butene and 2-methyl-2-butene. When this bromide was thermally decomposed in the absence of the chain inhibitor cyclochexene, the reaction prcceeded in a radical process.

The decomposition of 2-chloroethanol at 430–496 °C yielded acetaldehyde and HCl[182]. The reaction obeys a first-order rate law: $\log k_1(s^{-1}) = (12.8 \pm 1.0) - (229.7 \pm 4.0)\,kJ\,mol^{-1} (2.303RT)^{-1}$. The acetaldehyde formed in the presence of HCl decomposed further, as shown in equation 92.

$$HOCH_2CH_2Cl \longrightarrow CH_3CHO + HCl$$
$$\downarrow \qquad (92)$$
$$CH_4 + CO + HCl$$

A 1,2-hydrogen shift was part of the mechanism for the formation of acetaldehyde (equation 93).

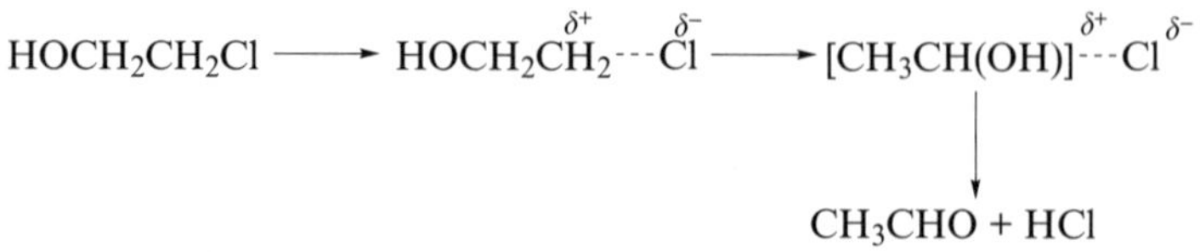

$$HOCH_2CH_2Cl \longrightarrow HOCH_2\overset{\delta+}{C}H_2\overset{\delta-}{\cdots}\overset{}{C}l \longrightarrow [CH_3CH(OH)]\overset{\delta+}{\cdots}\overset{\delta-}{C}l$$

$$\downarrow$$

$$CH_3CHO + HCl$$

$$(93)$$

Deuteriated neopentyl chloride -α,α-d$_2$was pyrolyzed under maximum inhibition of cyclohexene at 445 °C. 2-Methyl-1-butene-3,3-d$_2$ and 2-methyl-2-butene-3-d$_1$ were found to be unequivocally the products of a rearrangement process[183]. Consequently, the reaction mechanism is consistent with involvement of an intimate ion-pair intermediate (equation 94).

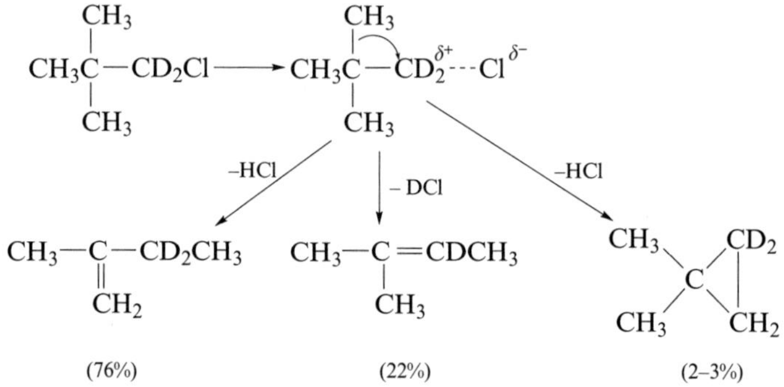

$$(94)$$

Neopentyl chloride was also decomposed under maximum inhibition of cyclohexene at 424–478 °C with the Arrhenius results of $E_a = 258.7$ kJ mol^{-1} and log $A(s^{-1}) = 13.78$. A preliminary study of neopentyl chloride in the presence of cyclohexene and under a single-pulse shock tube showed, by qualitative analysis, the formation of only the rearrangement products 2-methyl-1-butene and 2-methyl-2-butene, thus providing further evidence of the unimolecular nature of this elimination.

In a study of the effect of alkyl and polar groups on the gas-phase pyrolyses of α-substituted ethyl chlorides[70], pinacolyl chloride was found to produce dimethylbutenes by kinetic control. The main alkene product, i.e. 3,3-dimethyl-1-butene, was formed through the normal four membered transition state for HCl elimination. However, the formation of

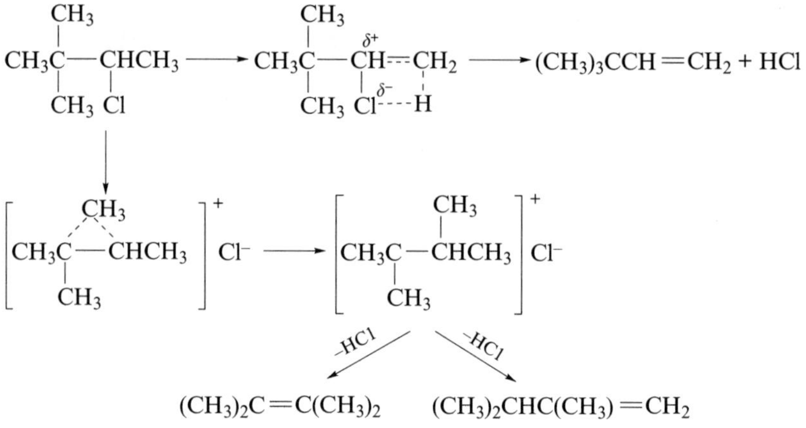

$$(95)$$

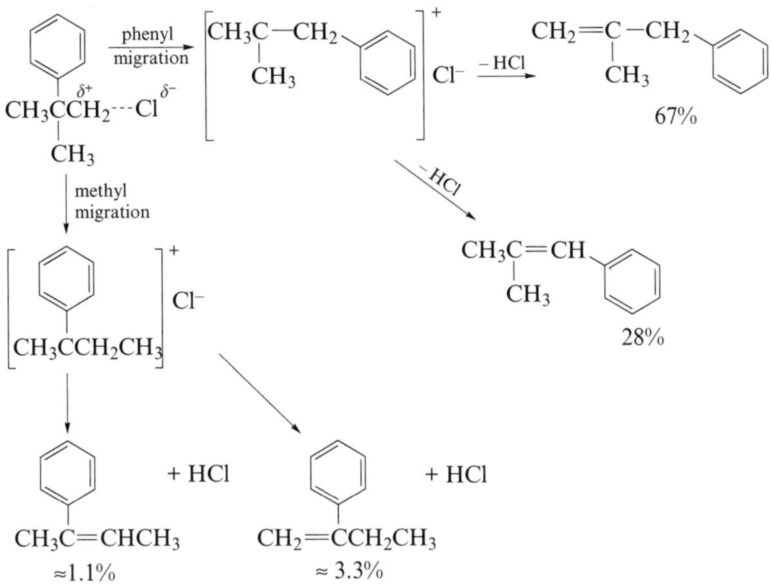

$$\text{(96)}$$

about 12% of 2,3-dimethyl-1- and -2-butenes suggested a Wagner–Meerwein rearrangement and the process was described as proceeding by two pathways (equation 95).

In connection with the rearrangement of neopentyl chloride and bromide, since a phenyl ring is a better migrating group than methyl, the pyrolysis of 2-methyl-2-phenyl-1-chloropropane(neophyl chloride) was examined[184]. This substrate, at 420–475 °C, yielded mostly phenyl migration products and very little methyl migration products (equation 96). The Arrhenius parameters derived from the overall rates are $E_a = 227.8$ kJ mol^{-1} and log $A(s^{-1}) = 13.47$. This work corroborates a mechanism involving ion-pair intermediate in homogeneous pyrolyses of some types of alkyl halides in the gas phase.

XI. ACID HALIDES

Very little work has been reported on the gas-phase pyrolysis of acid halides and two excellent reviews are available[4,185]. Acetyl chloride was decomposed in a static system at 242–491 °C[186]. The reversible reaction (equation 97) occurs at 242–350 °C, where the equilibrium lies to the left. The equilibrium constant, K_p, was found to be invariable with initial pressure, and temperature-dependent according to the van't Hoff equation: 8.314 ln K_p (−100.3 ± 2.0) 10^3/T + (132.9 ± 3.2). Addition of HCl reduced the extent of the reaction but did not alter the value of K_p. However, at 270–329 °C the reaction is found to be homogeneous, molecular, and to obey a first-order rate law. The rate coefficients were given by

$$CH_3COCl \underset{}{\overset{K_p}{\rightleftharpoons}} CH_2=C=O + HCl \tag{97}$$

$$\begin{array}{c} R^1 \\ \diagdown \\ R^2 \diagup \end{array} CH-COCl \longrightarrow \begin{array}{c} R^1 \\ \diagdown \\ R^2 \diagup \end{array} C=C=O + HCl \tag{98}$$

the following Arrhenius equation: $\log k_1(s^{-1}) = (12.42 \pm 0.23) - (168.9 \pm 2.6)$ kJ mol^{-1} $(2.303RT)^{-1}$. At the temperature range 299–491 °C, and especially above 400 °C, the following decarbonylation process became significant and addition of HCl enhanced its rate.

$$CH_2=C=O + HCl \rightleftharpoons CH_3COCl \rightarrow CH_3Cl + CO$$

Several acyl chlorides which were thermolyzed at 0.1 torr, gave elimination of HCl and formed the corresponding ketene[187] (equation 98).

A radical process involving initiation of Cl atom as chain carrier and a ketene intermediate formed via R_2CCOCl was suggested. The elimination of HCl together with CO was also observed (equations 99 and 100).

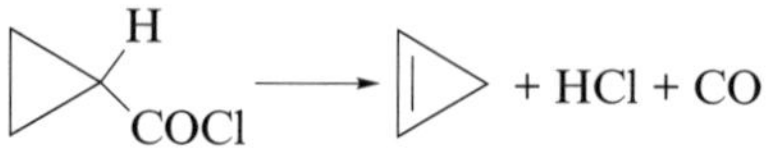

$$ (99) $$

$$CH_2=CHCOCl \longrightarrow HC\equiv CH + HCl + CO \qquad (100)$$

β,γ-Unsaturated acyl chlorides, when pyrolyzed in the gas phase at 400–600 °C, yielded the corresponding alkenyl ketenes and HCl, respectively[188]. These alkenyl ketenes are then transformed into $\alpha,\beta,\gamma,\delta$-unsaturated aldehydes by intramolecular 1,5-hydrogen shift.

Benzocyclobutenone was obtained from decomposition of *o*-alkyl substituted benzoyl chlorides[189]. Thus, with 2-propylbenzoyl chloride at 600 °C, the *o*-quinonoid ketene intermediate is stabilized by 1,5-hydrogen shift to give the aldehyde (equation 101). In this case, the *o*-alkyl contains a β-hydrogen. However, if the *o*-substituent lacks a β-hydrogen, i.e. it is a methyl group, then the cyclobutenone is obtained (equation 102).

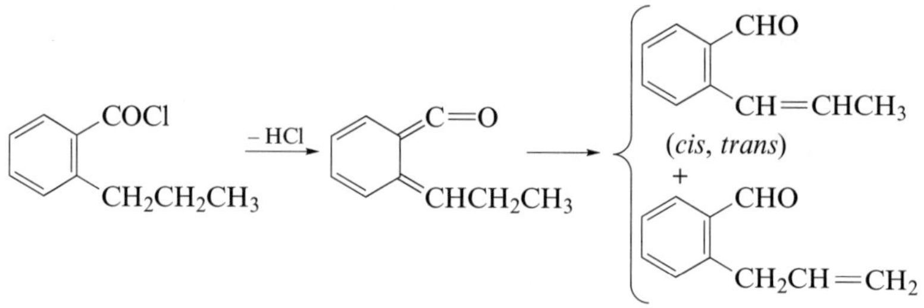

$$ (101) $$

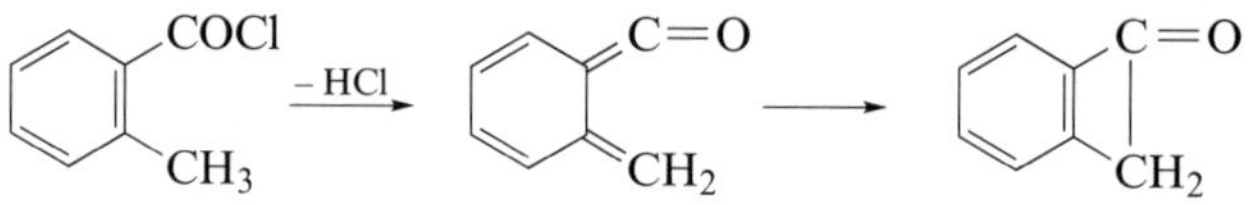

$$ (102) $$

A list of substituted benzoyl chlorides which yield the corresponding benzocyclobutenone was given[189].

XII. CHLOROFORMATES

The molecular decomposition of chloroformates in the gas phase has been very well reviewed by Smith and Kelly[2], and by Egger and Cocks[4].

Ethyl chloroformate was pyrolyzed in the gas phase at 286–353 °C[190]. The reaction, in a seasoned vessel and in the presence of an inhibitor, was found to be homogeneous and

$$CH_2{=}CH_2 + HCl + CO_2 \qquad (103)$$

$$ClCOOCH_2CH_3$$

$$CH_3CH_2Cl + CO_2 \qquad (104)$$

molecular, and followed the rate expression: $k = 10^{12.64}\exp(-183{,}600/8.314T)$ s^{-1}. The decomposition proceeds via two pathways (equations 103 and 104).

The formation of $CH_2{=}CH_2$, as the main product, gave a first-order rate coefficient which was invariable to initial pressure and surface conditions. Yet, the formation of CH_3CH_2Cl was variable and increased in a packed vessel. Consequently, the production of ethyl chloride is heterogeneous. A polar six-membered cyclic transition state was suggested for the elimination of HCl or ClCOOH, a mechanism already advanced by Clinch and Hudson[191].

$$ClCOOCH_3 \longrightarrow CH_3Cl + CO_2 \qquad (105)$$

When methyl chloroformates, which lack a β-hydrogen at the alkyl side of the ester, were pyrolyzed in a seasoned vessel and in the presence of a radical suppressor at 425–480 °C, they gave consistently the elimination products shown in equation 105[192].

The reaction was found to be homogeneous and unimolecular, and it followed a first-order rate law. The rate coefficient was given by the equation: $k_1 = 10^{(14.26 \pm 0.20)}\exp[(-251{,}000 \pm 2000)/8.314T]$ s^{-1}. Olefin formation is impossible and only CH_3Cl was obtained. The reaction rate is 4000 times slower than the overall rate of ethyl chloroformate decomposition[190], and the mechanism proposed involves a cyclic four-membered transition state.

The homogeneous and unimolecular decomposition of isopropyl chloroformate at 229–277 °C gave two parallel reactions[193] (equations 106 and 107).

$$CH_3CH{=}CH_2 + HCl + CO_2 \qquad (106)$$

$$ClCOOCH(CH_3)_2$$

$$(CH_3)_2CHCl + CO_2 \qquad (107)$$

The temperature dependence of the rate coefficient was given by the following Arrhenius equation: $k_1 = 10^{(13.94 \pm 0.2)}\exp[(-174{,}000 \pm 2000)/8.314T]$ s^{-1}. The relative rate ratio isopropyl chloroformate: ethyl chloroformate was 160 at 280 °C. The enhanced rate due to an α-substitution was associated with the degree of polarity of the alkyl halides[7]. Therefore, the transition state for chloroformate ester decomposition must also be polar. The mechanism was believed to involve a cyclic six-membered transition state in which formation of HCl is assisted, as shown below.

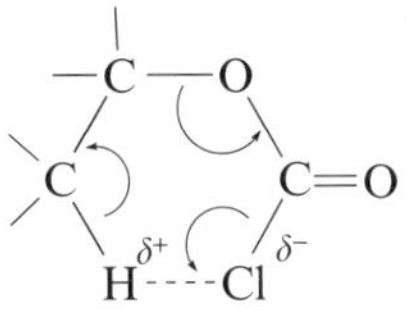

This view appeared to support the carbenium ion mechanism suggested for the olefin formation in liquid phase[191].

XIII. ACKNOWLEDGMENT

The author wishes to express his appreciation to Mrs Dilcia Sanchez and M.Sc. Alexandra Rotinov for their help in preparation of this manuscript.

XIV. REFERENCES

1. A. Maccoll, *Chem. Rev.*, **69**, 33 (1969).
2. G. G. Smith and F. W. Kelly, *Prog. React. Kinet.*, **8**, 75 (1971).
3. W. H. Saunders, Jr. and A. F. Cockerill, *Mechanisms of Elimination Reactions,* Chapter VIII, *Wiley-Interscience,* New York, 1973, p 378 ff.
4. K. W. Egger and A. T. Cocks, in *The Chemistry of the Carbon–Halogen Bond* (Ed. S. Patai) Chapter 10, Wiley, Chichester, 1973, p. 703 ff.
5. J. R. Green, G. D. Harden, A. Maccoll and P. J. Thomas, *J. Chem. Phys.*, **21**, 178 (1953).
6. A. Maccoll and P. J. Thomas, *Nature*, **176**, 392 (1955).
7. A. Maccoll and P. J. Thomas, *Progr. React. Kinet.*, **4**, 119 (1967).
8. A. Maccoll and E. S. Swinbourne, *J. Chem. Soc.*, 149 (1964).
9. J. S. Shapiro and E. S. Swinbourne, *Can. J. Chem.*, **46**, 1341 (1968).
10. R. C. L. Bricknell and A. Maccoll, *Chem. Ind. (London)*, 1912 (1961).
11. H. M. R. Hoffman and A. Maccoll, *J. Am. Chem. Soc.*, **87**, 3774 (1965).
12. C. J. Harding, A. Maccoll and R. A. Ross, *Chem. Commun.*, 289 (1967).
13. H. Kwart and M. T. Waroblak, *J. Am. Chem. Soc.*, **89**, 7145 (1967).
14. W. C. Herndon and J. M. Manion, *Tetrahedron Lett.*, 6327 (1968).
15. G. Chuchani, J. A. Hernandez, G. D. Morris and A. G. Shepherd, *J. Chem. Soc., Perkin Trans. 2*, 917 (1982).
16. S. W. Benson and A. H. Bose, *J. Chem. Phys.*, **39**, 3463 (1963).
17. H. E. O'Neal and S. W. Benson, *J. Phys. Chem.*, **71**, 2903 (1967).
18. S. W. Benson and H. E. O'Neal, *National Bureau of Standards, National Data Reference System,* U.S. Printing Office, Washington, D.C., (1970).
19. M. Day and A. F. Trotman-Dickenson, *J. Chem. Soc. (A)*, 233 (1969).
20. D. Sianesi, G. Nelli and R. Fontanelli, *Chim. Ind. (Milan)*, **50**, 619 (1968).
21. P. N. Dastoor and E. U. Emovon, *J. Chem. Soc., Faraday Trans. 1*, 2098 (1972).
22. P. Cadman, M. Day, A. W. Kirk and A. F. Trotman-Dickenson, *Chem. Commun.*, 203 (1970).
23. K. Okada, E. Tschiukow-Roux and P. J. Evans, *J. Phys. Chem.*, **84**, 467 (1980).
24. P. Cadman, M. Day and A. F. Trotman-Dickenson, *J. Chem. Soc. (A)*, 248 (1971).
25. V. F. Kochubei, A. P. Gavrili, F. B. Moin and Y. A. Pazderskii, *Kinet. Katal.*, **20**, 1041 (1979); *Chem. Abstr.*, 91, 192390 a (1979).
26. J. M. Simmie, W. J. Quiring and E. Tschiukow-Roux, *J. Phys. Chem.*, **74**, 992 (1970).
27. D. C. Montagne and R. Walsh, *Annu Rep. A, Chem. Soc. London*, 175 (1971).
28. G. E. Millward and E. Tschiukow-Roux, *J. Phys. Chem.*, **76**, 292 (1972).
29. M. V. C. Sekhar and E. Tschiukow-Roux, *J. Phys. Chem.*, **78**, 472 (1974).
30. E. Tschiukow-Roux, W. J. Quiring and J. M. Simmie, *J. Phys. Chem.*, **75**, 3195 (1971).
31. E. Tschiukow-Roux and W. J. Quiring, *J. Phys. Chem.*, **75**, 295 (1971).
32. J. A. Kerr and D. M. Timilin, *Int. J. Chem. Kinet.*, **3**, 427 (1971).
33. G. E. Millward, R. Harting and E. A. Tschiukow-Roux, *J. Phys. Chem.*, **75**, 3195 (1975).
34. E. A. Tschiukow-Roux, G. E. Milward and W. J. Quiring, *J. Phys. Chem.*, **75**, 3493 (1975).
35. J. M. Simmie and E. Tschiukow-Roux, *J. Chem. Soc. (D)*, 773 (1970).
36. O. Kondo, K. Saito and I. Murakami, *Bull. Chem. Soc. Jpn.*, **53**, 2133 (1980).
37. A. B. Trenwith, *Int. J. Chem. Kinet.*, **5**, 67 (1973).
38. J. R. Christie, W. D. Johnson, A. G. London, A. Maccoll and M. N. Mruzek, *J. Chem. Soc., Faraday Trans. 1*, **71**, 1937 (1975).
39. J. R. Christie, *Mol. Rate Process Pap. Symp.*, B5, 1975, 5 pp.
40. P. C. Hiberty, *J. Am. Chem. Soc.*, **97**, 5975 (1975).
41. A. Maccoll and M. Umaña, *J. Chem. Soc., Perkin Trans. 2*, 1194 (1975).
42. A. G. Loudon, A. Maccoll and S. K. Wong, *J. Am. Chem. Soc.*, **91**, 7577 (1969).
43. H. Heydtmann, B. Dill and R. Jonas, *Int. J. Chem. Kinet.*, **7**, 973 (1975).
44. F. Zabel, *Int. J. Chem. Kinet.*, **9**, 651 (1977).

45. S. W. Benson and G. R. Haugen, *J. Phys. Chem.*, **70**, 3336 (1966); *J. Am. Chem. Soc.*, **87**, 4036 (1965).
46. N. Barroeta and M. Rincon, *Acta Cient. Venez.*, **27**, 247 (1976); *Chem. Abstr.*, **87**, 22045p (1977).
47. K. D. King, *J. Chem. Soc., Faraday Trans. 1*, **74**, 912 (1978).
48. P. J. Evans, T. Ichimura and E. Tschiukow-Roux, *Int. J. Chem. Kinet.*, **10**, 855 (1978).
49. R. Jonas and H. Heydtmann, *Ber. Bunsenges Phys. Chem.*, **82**, 823 (1978).
50. A. Maccoll and M. N. Mruzek, *J. Chem. Soc., Faraday Trans. 1*, **74**, 2714 (1978).
51. A. Maccoll, M. N. Mruzek and M. A. Baldwin, *J. Chem. Soc., Faraday Trans. 1*, **76**, 838 (1980).
52. J. S. Francisco, J. I. Steinfeld, K. D. King and R. G. Gilbert, *J. Phys. Chem.*, **85**, 4106 (1981).
53. P. J. Papagiannakopoulos and S. W. Benson, *Int. J. Chem. Kinet.*, **14**, 63 (1982).
54. G. W. Volker and H. Heydtmann, *Z. Naturforsch.*, **23B**, 1407 (1981).
55. L. W. Jeuneskens, C. A. M. Hoefs and U. E. Wiersum, *J. Org. Chem.*, **54**, 5811 (1989).
56. G. Huybrechts, Y. Hubin and B. Van Mele, *Int. J. Chem. Kinet.*, **24**, 671 (1992).
57. J. L. Holmes and M. Dakubu, *J. Chem. Soc., Perkin Trans. 2*, 2110 (1972).
58. J. C. Ferrero, E. A. R. de Starico and E. H. Starico, *J. Phys. Chem.*, **79**, 1242 (1975).
59. P. J. Robinson and M. J. Waller, *Int. J. Chem. Kinet.*, **11**, 937 (1979).
60. H. Heydtmann and B. Korbiitzer, *Z. Phys. Chem., N. F.*, **125**, 255 (1981).
61. W. J. Bailey and A. Onopchenko, *J. Org. Chem.*, **56**, 846 (1991).
62. G. Chuchani, I. Martin, A. Rotinov, J. A. Hernandez and N. Reikonnen, *J. Phys. Chem.*, **88**, 1563 (1984).
63. H. Hartman, H. G. Bosch and H. Heydtmann, *Z. Phys. Chem. (Frankfurt am Main)*, **42**, 329 (1964).
64. R. C. S. Grant and E. S. Swinbourne, *J. Chem. Soc.*, 4423 (1965).
65. G. Chuchani, J. A. Hernandez and I. Avila, *J. Phys. Chem.*, **82**, 2767 (1978).
66. G. Chuchani, J. A. Hernandez and I. Martin, *Int. J. Chem. Kinet.*, **11**, 1279 (1979).
67. G. Chuchani, I. Martin, G. Martin and D. B. Bigley, *Int. J. Chem. Kinet.*, **11**, 109 (1979).
68. D. C. Skingle and V. R. Stimson, *Aust. J. Chem.*, **29**, 609 (1976).
69. D. H. R. Barton and K. E. Howlett, *J. Chem. Soc.*, 155 (1949).
70. R. M. Dominguez, A. Rotinov, and G. Chuchani, *J. Phys. Chem.*, **90**, 6277 (1986).
71. D. H. R. Barton and A. J. Head, *Trans. Faraday Soc.*, **46**, 114 (1950).
72. H. Heydtmann and G. Rinck, *Z. Phys. Chem., N. F.*, **30**, 250 (1961).
73. V. Chytry, B. Obereigner and D. Lim, *Chem. Ind.*, 470 (1970).
74. G. Chuchani, I. Martin and M. E. Alonso, *Int. J. Chem. Kinet.*, **9**, 819 (1977).
75. G. Martens, M. Godfroid, R. Decelle and J. Verbeyst, *Int. J. Chem. Kinet.*, **4**, 645 (1972).
76. R. M. Dominguez and G. Chuchani, *Int. J. Chem. Kinet.*, **17**, 217 (1985).
77. M. Dakubu and A. Maccoll, *J. Chem. Soc. (B)*, 1248 (1969).
78. P. J. Thomas, *J. Chem. Soc. (B)*, 1238 (1967).
79. P. J. Robinson, G. C. Skelhorne and M. J. Waller, *J. Chem. Soc., Perkin Trans. 2*, 349 (1978).
80. V. Chytry, B. Obereigner and D. Lim. *Eur. J. Polym.*, 379 (1969).
81. M. R. Bridge, D. H. Davies, A. Maccoll, R. A. Ross, B. Stephenson and O. Banjoko, *J. Chem. Soc. (B)*, 805 (1968).
82. P. J. Thomas, *J. Chem. Soc. (B)*, 136 (1961).
83. R. L. Frailes and V. R. Stimson, *Aust. J. Chem.*, **20**, 1553 (1967).
84. G. Chuchani, A. Rotinov, I. Martin, I. Avila and R. M. Dominguez, *J. Phys. Chem.*, **89**, 4134 (1985).
85. W. Tsang, *J. Chem. Phys.*, **41**, 2487 (1964).
86. A. Maccoll and S. W. Wong, *J. Chem. Soc. (B)*, 1492 (1968).
87. G. Chuchani and I. Martin, *J. Phys. Chem.*, **84**, 3188 (1980).
88. G. Chuchani and J. A. Hernandez, *J. Phys. Chem.*, **85**, 4139 (1981).
89. C. D. Harding and A. Maccoll, *J. Chem. Soc. (B)*, 634 (1967).
90. G. Chuchani and A. Rotinov, *React. Kinet. Catal. Lett.*, **12**, 333 (1979).
91. E. Santacesaria, A. Morini and S. Carra, *Chim. Ind. (Milan)*, **56**, 747 (1974).
92. K. P. Schug, H. G. Wagner and F. Zabel, *Ber. Bunseges Phys. Chem.*, **83**, 167 (1979).
93. F. Zabel, *Ber. Bunseges Phys. Chem.*, **78**, 232 (1974).
94. L. M. Verbovskaya and D. D. Lebedev, *Zh. Prikl. Khim. (Leningrad)*, **48**, 382 (1975); *Chem. Abstr.*, **82**, 124373 m (1975).
95. T. J. Houser and T. Cuzcano, *Int. J. Chem. Kinet.*, **7**, 331 (1975).

96. W. Tsang, J. A. Walker and W. Braun, *J. Phys. Chem.*, **86**, 719 (1982).
97. H. L. Dai, E. Specht, M. R. Berman and C. D. Moore, *J. Chem. Phys.*, **77**, 4494 (1982).
98. J. H. Lippiatt and D. E. Wells, *Dyn. Mass Spectrom.*, **4**, 273 (1976).
99. O. Kaposi, A. B. Kiss, M. M. Riedel and T. Deutsch, *Thermal Analysis*, Vol. 2, *Proceedings 4th ICTA* (Ed. I. Buzós), Budapest, 1974, 297 pp.
100. A. B. Kiss, T. Deutsch, O. Kaposi and L. Lelik, *Acta Chim. Acad. Sci. Hungar.*, **93**, 221 (1977).
101. H. G. Garzayan, T. A. Garibyan and A. B. Nalbandyan, *Arm. Khim. Zh.*, **31**, 27 (1978); *Chem. Abstr.*, **89**, 5726p (1978).
102. K. Saito, T. Yokubo, T. Fuse, H. Tahara, O. Kondo, T. Higashihara and I. Murakami, *Bull. Chem. Soc. Jpn.*, **52**, 3507 (1979).
103. T. J. Park and K. H. Jung, *Bull. Korean Chem. Soc.*, **1**, 30 (1980).
104. S. H. Kang, H. S. Yoo, and K. H. Jung, *Bull. Korean Chem. Soc.*, **2**, 35 (1981).
105. S. D. Palsley and B. E. Holmes, *J. Phys. Chem.*, **87**, 3042 (1982).
106. K. H. Jung, S. J. Yun and D. S. Huh, *J. Chem. Soc., Faraday Trans. 2.*, **83**, 971 (1987).
107. T. T. Nguyen, K. D. King and R. G. Gilbert, *J. Phys. Chem.*, **87**, 494 (1983).
108. L. W. Jeuneskens, U. E. Viersum and J. L. Ripall, *Tetrahedron Lett.*, **29**, 6489 (1988).
109. G. Chuchani and R. M. Dominguez, *Int. J. Chem. Kinet.*, **22**, 211 (1990).
110. H. Hippler, A. Riedl, J. Troe, and J. Willner, *Z. Phys. Chem.*, **171**, 161 (1991).
111. G. Chuchani and R. M. Dominguez, *React. Kinet. Catal. Lett.*, **27**, 185 (1985).
112. G. Chuchani and R. M. Dominguez, *React. Kinet. Catal. Lett.*, **43**, 217 (1991).
113. R. M. Dominguez and G. Chuchani, *React. Kinet. Catal. Lett.*, **47**, 247 (1992).
114. K. D. King and R. G. Gilbert, *Int. J. Chem. Kinet.*, **12**, 339 (1980).
115. R. J. Kominar, M. J. Krech and S. J. W. Price, *Can. J. Chem.*, **56**, 1589 (1978).
116. R. J. Kominar, M. J. Krech and S. J. W. Price, *Can. J. Chem.*, **58**, 1906 (1980).
117. G. Chuchani, A. Rotinov, R. M. Dominguez and I. Martin, *Int. J. Chem. Kinet.*, **19**, 781 (1987).
118. G. Chuchani and R. M. Dominguez, *J. Phys. Chem.*, **93**, 203 (1989).
119. A. Maccoll and P. J. Thomas, *J. Chem. Soc.*, 5033 (1957).
120. J. H. S. Green, A. Maccoll and P. J. Thomas, *J. Chem. Soc.*, 184 (1960).
121. G. Chuchani, I. Martin, A. Rotinov and R. M. Dominguez, *Int. J. Chem. Kinet.*, **22**, 1249 (1990).
122. A. Maccoll and P. J. Thomas, *J. Chem. Soc.*, 969 (1955).
123. G. Chuchani, R. M. Dominguez, A. Rotinov and I. Martin, *J. Phys. Chem.*, **93**, 206 (1989).
124. S. W. Benson, *J. Phys. Chem.*, **38**, 1945 (1963).
125. K. Saito, H. Tahara, O. Kondo, T. Yokudo, T. Higashihara and I. Murakami, *Bull. Chem. Soc. Jpn.*, **53**, 1335 (1980).
126. B. J. Gaynor, R. G. Gilbert and K. D. King, *Chem. Phys. Lett.*, **58**, 591 (1978).
127. H. Hippler, A. Riedl, J. Troe and J. Willner, *Z. Phys. Chem.*, **171**, 161 (1991).
128. M. J. Krech, S. J. W. Price and W. F. Yared, *Int. J. Chem. Kinet.*, **6**, 257 (1974).
129. M. Szwarc, *Chem. Rev.*, **47**, 75 (1950).
130. R. J. Kominar, M. J. Krech and S. J. W. Price, *Can. J. Chem.*, **54**, 2981 (1976).
131. K. L. Lancer and G. H. Wiegand, *J. Org. Chem.*, **41**, 3360 (1976).
132. Y. Yamada and M. Okawara, *Bull. Chem. Soc. Jpn.*, **45**, 1680 (1972).
133. Y. Yamada and M. Okawara, *Bull. Chem. Soc. Jpn.*, **45**, 2515 (1972).
134. Y. Yamada, K. Kashuma and M. Okawara, *Bull. Chem. Soc. Jpn.*, **47**, 3179 (1974).
135. G. E. Millward and E. Tschiukow-Roux, *Int. J. Chem. Kinet.*, **4**, 559 (1971).
136. M. V. C. Sekhar, G. E. Millward and E. Tschiukow-Roux, *Int. J. Chem. Kinet.*, **5**, 363 (1973).
137. O. B. Danilov, V. V. Elagin, V. Y. Zalesskii and I. L. Yachnev, *Kinet. Katal.*, **16**, 302 (1975); *Chem. Abstr.*, **83**, 78241 n (1975).
138. R. K. Boyd, G. W. Downs, J. S. Gow and C. Horrex, *J. Chem. Phys.*, **67**, 719 (1963).
139. V. Y. Zalesskii, *Zh. Eksp. Teor. Fiz.*, **61**, 892 (1971); *Chem. Abstr.*, **76**, 8745 m (1972).
140. Z. Bruzenski and R. A. Kolinki, *Rocz. Chem.*, **51**, 1259 (1977).
141. A. V. Baklanov, V. A. Gostygin and A. K. Petrov, *Kinet. Katal*, **23**, 886 (1982).
142. G. Chuchani, I. Martin, M. E. Alonso and P. Jano, *Int. J. Chem. Kinet.*, **13**, 1 (1981).
143. K. D. King, B. J. Gaynor and R. G. Gilbert, *Int. J. Chem. Kinet.*, **11**, 11 (1979).
144. A. T. Cocks and H. M. Frey, *J. Am. Chem. Soc.*, **91**, 7583 (1969).
145. H. M. Frey and H. P. Watts, *Curr. Top. Mass Spect., Chem. Kinet. Proc. Symp.*, 101 (1981).
146. G. Chuchani and A. Rotinov, *Int. J. Chem. Kinet.*, **21**, 367 (1990).
147. G. Chuchani, R. M. Dominguez and A. Rotinov, *Int. J. Chem. Kinet.*, **23**, 779 (1991).
148. C. J. Harding, A. Maccoll and R. A. Ross, *J. Chem. Soc. (B)*, 634 (1969).

149. A. Maccoll and M. Umaña, *J. Chem. Soc., Perkin Trans. 2*, 1194 (1975).
150. P. J. Robinson, G. G. Skelthorne and M. J. Wallar, *J. Chem. Soc., Perkin Trans. 2*, 349 (1978).
151. A. G. Loudon, A. Maccoll and S. Wong, *J. Am. Chem. Soc.*, **91**, 7577 (1969).
152. M. J. Morello and W. S. Trahanovsky, *Tetrahedron Lett.*, 4435 (1979).
153. A. N. Gafarov, T. S. Kharasova and V. F. Nikolskaya, *Zh. Org. Khim.*, **16**, 2543 (1980).
154. A. Hussain and J. Parrick, *Tetrahedron Lett.*, **24**, 609 (1983).
155. Y. H. So, *J. Org. Chem.*, **52**, 1615 (1987).
156. G. Chuchani, I. Martin and D. B. Bigley, *Int. J. Chem. Kinet.*, **10**, 649 (1978).
157. J. L. Holmes, D. L. McGillivray and D. Yuan, *Can. J. Chem.*, **57**, 2621 (1979).
158. J. A. Hernandez and G. Chuchani, *Int. J. Chem. Kinet.*, **10**, 923 (1978).
159. G. Chuchani, J. D. Medina, I. Martin and J. A. Hernandez, *J. Phys. Chem.*, **85**, 3900 (1981).
160. G. Chuchani, A. Rotinov and I. Martin, *J. Phys. Chem.*, **89**, 551 (1985).
161. R. Heck and S. Winstein, *J. Am. Chem. Soc.*, **79**, 3144 (1957).
162. S. Daren, M. Levy and V. Vopsi, *Br. Polym. J.*, **7**, 247 (1975).
163. G. Chuchani and I. Martin, *J. Phys. Org. Chem.*, **3**, 77 (1990).
164. R. M. Dominguez, A. Rotinov, J. A. Hernandez, I. Martin, J. D. Medina and G. Chuchani, *React. Kinet. Catal. Lett.*, **33**, 285 (1987).
165. J. B. Rogan, *J. Org. Chem.*, **27**, 3910 (1962).
166. G. Chuchani, R. M. Dominguez and A. Rotinov, *Int. J. Chem. Kinet.*, **18**, 203 (1986).
167. S. H. Unger and C. Hansch, *Prog. Phys. Org. Chem.*, **12**, 91 (1976).
168. G. Chuchani and R. M. Dominguez, *Int. J. Chem. Kinet.*, **15**, 795 (1983).
169. G. Chuchani, R. M. Dominguez and I. Martin, *React. Kinet. Catal. Lett.*, **30**, 77 (1986).
170. G. Chuchani and R. M. Dominguez, *J. Phys. Chem.*, **91**, 1883 (1987).
171. G. Chuchani and I. Martin, *Int. J. Chem. Kinet.*, **20**, 1 (1988).
172. G. Chuchani and A. Rotinov, *Int. J. Chem. Kinet.*, **19**, 789 (1987).
173. R. M. Dominguez and G. Chuchani, *Int. J. Chem. Kinet.*, **13**, 403 (1981).
174. G. Chuchani and R. M. Dominguez, *Int. J. Chem. Kinet.*, **15**, 1275 (1983).
175. G. Chuchani and I. Martin, *J. Phys. Chem.*, **90**, 431 (1986).
176. G. Chuchani and A. Rotinov, *React. Kinet. Catal. Lett.*, **38**, 1 (1989).
177. G. Chuchani and I. Martin, *Int. J. Chem. Kinet.*, **19**, 183 (1987).
178. G. Chuchani and A. Rotinov, *Bull. Soc. Chim. Fr.*, 322 (1988).
179. G. Chuchani, I. Martin and R. M. Dominguez, *Int. J. Chem. Kinet.*, **19**, 683 (1987).
180. R. Garth Pews and R. A. Davis, *J. Chem. Soc., Chem. Commun.*, 269 (1973).
181. R. L. Failes, Y. M. A. Mollah and J. S. Shapiro, *Int. J. Chem. Kinet.*, **13**, 7 (1981).
182. D. C. Skingle and V. R. Stimson, *Aust. J. Chem.*, **29**, 609 (1976).
183. R. L. Failes, Y. M. A. Mollah and J. S. Shapiro, *Int. J. Chem. Kinet.*, **11**, 1271 (1979).
184. G. Chuchani and R. M. Dominguez, *J. Chem. Soc. Perkin Tras. 2*, in press.
185. R. Taylor, in *Supplement B: The Chemistry of Acid Derivatives* (Ed. S. Patai), Chapter 15, Wiley, Chichester, 1979.
186. V. R. Stimson and J. W. Tilley, *Aust. J. Chem.*, **30**, 81 (1977).
187. H. Bock, T. Hirabayashi, S. Mohmand and B. Solouki, *Angew. Chem., Int. Ed. Engl.*, **16**, 105 (1977).
188. P. Schiess and P. Radimerski, *Helv. Chim. Acta*, **57**, 2853 (1974); *Chimia*, **26**, 652 (1972).
189. P. Schiess and M. Heitzmann, *Angew. Chem., Int. Ed. Engl.*, **16**, 469 (1977).
190. L. Johnson and V. R. Stimson, *Aust. J. Chem.*, **29**, 1389 (1976).
191. P. W. Clinch and H. R. Hudson, *J. Chem. Soc. (B)*, 747 (1971).
192. R. L. Johnson and V. R. Stimson, *Aust. J. Chem.*, **30**, 1917 (1977).
193. R. L. Johnson and V. R. Stimson, *Aust. J. Chem.*, **35**, 849 (1982).

CHAPTER **20**

Positive halogen compounds

HISHAM G. LUTFI and CAL Y. MEYERS

Department of Chemistry and Biochemistry, Southern Illinois University, Carbondale, Illinois 62901–4409, USA

Supplement D2: The chemistry of halides, pseudo-halides and azides
Edited by S. Patai and Z. Rappoport © 1995 by John Wiley & Sons Ltd

I. INTRODUCTION

'Positive halogen compounds', historically and in the context of this chapter, refers to halogen-containing organic compounds which can transfer their halogen to a variety of nucleophilic species; namely, they are halogenating agents. The halogens in such compounds have included fluorine, chlorine, bromine and iodine as substituents on carbon as well as on heteroatoms. Although numerous efforts have been made to describe the mechanism of specific halogenation reactions with specific nucleophiles, several general mechanisms mentioned below have been alluded to most frequently. The terminology used in this chapter is: X will denote a halogen atom, Nu a nucleophile, Y an unspecified group or moiety, R an alkyl group and Ar an aryl group, except when a different notation is used in material taken directly from cited references.

1. The simplest and, unfortunately, perhaps most generally perceived mechanism suggests that the halonium ion, X^+, itself is the halogenating agent which is either innately present in the positive halogen compound or formed *in situ* from such compounds (equation 1). While substantial data now support the existence of X^+ in the gas phase[1], there is

$$Y—X \text{ is } Y^- |X^+ \quad \text{or} \quad Y—X \longrightarrow Y^- + X^+ \tag{1}$$

$$Nu^- \searrow \qquad \swarrow Nu^-$$

$$Nu —X + Y^-$$

little support of their existence in solution under conditions generally used for organic halogenations. The term 'halogen cations', introduced by Arotsky and Symons[2], has been used interchangeably with the term 'halogen cation compounds' and applies to complexes composed of several halogen atoms. Many of these complexes have been studied and, indeed, exist in solution and solid phase. The structure and reactivity of such complexes have been reviewed by Stepin[3], who has pointed out that: 'The compounds of halogen cations are mainly ionic species with a certain amount of covalent character in the chemical bonds, which is manifested by the formation of halide bridge bonds between the cation and the anion'.

2. Another frequently presumed mechanism appropriates the classical S_N2 pathway and interprets the halogenation as a simple nucleophilic substitution reaction, the displacement of Y in a direct attack by Nu on X of Y—X (equation 2). The precedence for this mecha-

$$Nu^- \quad \overset{\frown}{X—Y} \longrightarrow \left[\overset{\delta+}{Nu}-----X----\overset{\delta-}{Y} \right]^{\ddagger} \longrightarrow Nu—X + Y^- \tag{2}$$

nism, of course, is the S_N2 displacement reaction on an sp^3-hybridized carbon in compounds possessing good leaving groups, reactions which generally proceed with inversion at the carbon attacked. Since there is no stereochemical equivalent in the halogenation reactions, it is easier to invoke this mechanism than to support it in these cases.

3. An increasing number of reports over the past 10 years or so have denoted mechanisms involving single-electron transfer (ET) (equation 3) to describe halogenations of nucleophiles with so-called positive halogen compounds. In such cases a postulated transition state is the tightly paired caged radical (from the nucleophile)/anion radical (from

$$Nu^- \quad \overset{\frown}{X:Y} \longrightarrow [Nu^. \quad (X:Y)^{.-}]^{\ddagger} \longrightarrow Nu:X + Y^- \tag{3}$$

the organohalogen compound) in which a halogen atom is transferred to the nucleophile moiety and the residual organic moiety leaves as an anion[4-10]. Nucleophiles with high oxidation potentials (e.g. RS⁻, carbanions, etc., sometimes called soft bases/nucleophiles) can be halogenated this way with organic halogen compounds having high reduction potentials (e.g. CX_4, perhaloethanes, sulfonyl halides, some acyl halides, etc.). In some instances Y instead of X is transferred to the nucleophile. This pathway has been coined the 'RARP' mechanism from 'Radical Anion-Radical Pair'[7-9]. It has been suggested that in these cases Nu˙ is sufficiently stable to leave the ET transition cage, the anion radical dissociates into X⁻ and Y˙, and the product, Nu—Y, is formed either by direct combination of Nu˙ and Y˙ or through a chain reaction involving the anion radical Nu—Y⁻˙ [7-9,11-13].

A. Organization of Chapter

The broad scope of coverage under the umbrella 'positive halogen compounds' allowed any number of arrangements to be used in organizing this chapter. Since readers will peruse the chapter with different objectives, no one arrangement would be completely satisfactory to all. It therefore was deemed most reasonable to implement the arrangement already familiar, that used in the previous coverage of this subject in this series. The positive halogen compounds are classified as halogenating (or halogen-transfer) agents and categorized as organo-C—X, -N—X, -O—X and -S—X compounds, wherein X in each category is F, Cl, Br and I.

B. Literature-search Methodology

The basic *ILLINET Online Network* (Illinois Online Network) was used as the major computer-assisted information retrieval program. Within this network *IBIS* (Illinois Bibliographic Information Service), which provides abstracts taken from journal articles published from 1991 through December 1993, and *CARL UnCover* (Colorado Alliance of Research Libraries), which provides titles to journal articles published in 1988 through December 1993, afforded much of the current and recent information reported on here. Abstracts of the pertinent literature published from 1982 through mid-1993 were also obtained directly from *CA* (Chemical Abstracts). It is hoped that the overlap provided this way assures the broadest retrieval of published results. All of these sources were searched by means of the following *key words*:

1. Key words by compound type

positive halogen	halonium	alkyl hypohalites	hypohalites	haloamines	halo amides
positive fluorine	fluoronium	alkyl hypofluorites	hypofluorites	fluoroamines	fluoro amides
positive chlorine	chloronium	alkyl hypochlorites	hypochlorites	chloroamines	chloro amides
positive bromine	bromonium	alkyl hypobromites	hypobromites	bromoamines	bromo amides
positive iodine	iodonium	alkyl hypoiodites	hypoiodites	iodoamines	iodo amides

halo imides	halo imidines	halo sulfonamides	perhalo carbons	polyhalo alkanes
fluoro imides	fluoro imidines	fluoro sulfonamides	perfluoro carbons	polyfluoro alkanes
chloro imides	chloro imidines	chloro sulfonamides	perchloro carbons	polychloro alkanes
bromo imides	bromo imidines	bromo sulfonamides	perbromo carbons	polybromo alkanes
iodo imides	iodo imidines	iodo sulfonamides	periodo carbons	polyiodo alkanes

electrophilic halogenating reagents	halogen compounds	halogens	sulfonium halides
electrophilic fluorinating reagents			
electrophilic chlorinating reagents	chlorine azide	sulfenyl chloride	N-chlorobenzotriazole
electrophilic brominating reagents	bromine azide		
elecrophilic iodinating reagents	iodine azide	iodine isocyanate	
electrophiles	haloalkylation		halogen sulfur compounds

2. Key words by reaction type

halogenation	electrophilic halogenation	electron-transfer halogenation reactions
fluorination	electrophilic fluorination	electron-transfer fluorination
chlorination	electrophilic chlorination	electron-transfer chlorination
bromination	electrophilic bromination	electron-transfer bromination
iodination	electrophilic iodination	electron-transfer iodination
		electron-transfer organic reactions
		electron-transfer deiodination

II. OXYGEN-HALOGEN COMPOUNDS

A. Alkyl Hypohalites

1. Reactions with amines

Zwanenburg and coworkers[14] reported that aziridine-2-carboxylic esters (**1**) when treated with *t*-butyl hypochlorite give *N*-chloroaziridines (**2**) (equation 4). Referring to

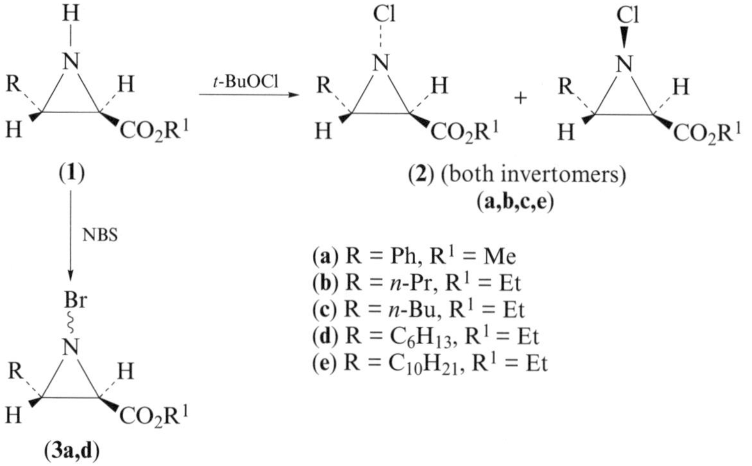

(1)

NBS

(3a,d)

(2) (both invertomers)
(a,b,c,e)

(**a**) R = Ph, R^1 = Me
(**b**) R = *n*-Pr, R^1 = Et
(**c**) R = *n*-Bu, R^1 = Et
(**d**) R = C_6H_{13}, R^1 = Et
(**e**) R = $C_{10}H_{21}$, R^1 = Et

(4)

previous work by Schreiber and colleagues[15], Zwanenburg and coworkers said that when NCS was employed for chlorination of **1a** an equilibrium mixture of starting material and *N*-chloroaziridine compound **2a** was obtained. They also said that, apparently, the Cl^+ donating ability of *N*-chloroaziridine **2a** is of the same order as that of NCS, and suggested that 'a stronger Cl^+ donor is required' to accomplish complete chlorination of **1** and such a reagent was *t*-BuOCl. The products consisted of mixtures of both invertomers of **2** in variable ratios. N-bromination leading to **3** was readily accomplished with NBS. This work overlaps that of Kostyanovsky and coworkers[16] who prepared (1*R*, 2*S*)-1-bromoaziridine-2-carboxylic acid methyl ester (**4**) (and some N-brominated aziridine esters) from NBS and (2*S*)-aziridine-2-carboxylic acid methyl ester.

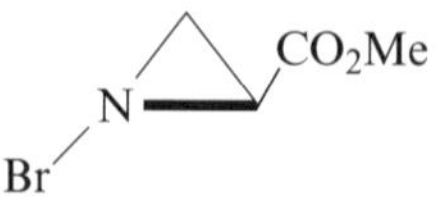

(4)

2. Reactions with alkenes and dienes

Iodine electrophiles including t-BuOI/BF$_3$, AcOI, ICl and IBr undergo 1,2- and 1,4-addition reactions with 1,3-butadiene (equation 5). Reactions of t-BuOBr and t-BuOCl

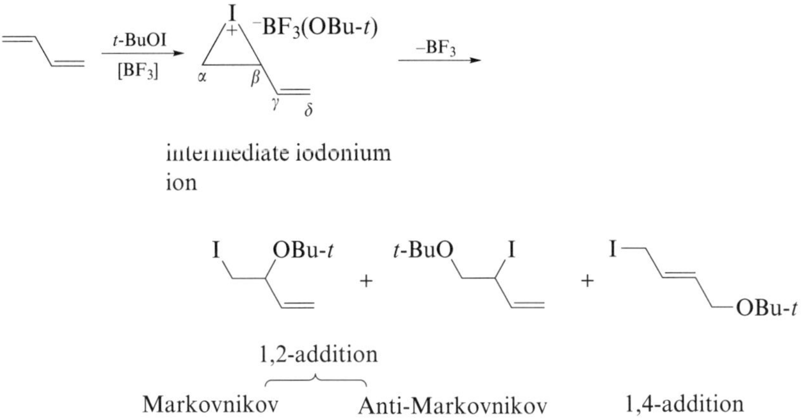

intermediate iodonium
ion

$$\text{1,2-addition} \quad\quad \text{1,4-addition} \tag{5}$$

Markovnikov   Anti-Markovnikov   1,4-addition

with 1,3-butadiene in the presence of BF$_3$ are reported for comparison. Under ionic conditions all the electrophiles give mixtures of 1,4-additions and Markovnikov 1,2-additions. Radical reactions were also reported for t-BuOI, t-BuOBr and t-BuOCl, in which cases the major products are the 1,4-adducts along with some anti-Markovnikov 1,2-adducts. These authors report that 'greater 1,4-addition occurs for the electrophiles with anions of lower basicity (IBr and ICl), and the halonium ion with increased dispersal of charge (t-BuOCl). The results are rationalized from the viewpoints of charge density and ion-pair stability'[17]. Catalysis by trimethyl borate promotes rapid ionic addition reactions of methyl hypobromite with alkenes and dienes in CH$_2$Cl$_2$ (equation 6)[18]. Boron trifluoride likewise cat-

$$\begin{array}{c}\diagdown \\ \diagup\end{array}\!C\!=\!C\!\begin{array}{c}\diagup \\ \diagdown\end{array} + \text{ MeOX} \xrightarrow{\text{[B(OMe)}_3]} X\!-\!\overset{|}{\underset{|}{C}}\!-\!\overset{|}{\underset{|}{C}}\!-\!\text{OMe} \tag{6}$$

alyzes ionic addition reactions of methyl hypochlorite with α,β-unsaturated esters in CH$_2$Cl$_2$[19]. 'Fluoro chloride' incorporation occurs in competition with methoxide; it is obvious that the fluorine incorporated is derived from the catalyst itself as outlined below. It has been shown that the BF$_3$-hypohalite reaction is a general reaction of olefinic compounds with both alkyl hypochlorites and hypobromites[20]. Olefinic compounds such as cyclohexene, 1-hexene, trans-1,2-dichloroethene, methyl acrylate, methyl crotonate, methyl isocrotonate, 1,3-butadiene, methyl vinyl ketone and styrene were studied. These reactions and their regiochemistry are illustrated by the examples in equation 7.

It was suggested that the reaction proceeds via a halonium trifluoroalkoxyborate ion pair intermediate (equation 8). The cation could capture either fluorine or alkoxy from the trifluoroalkoxyborate anion through a variety of mechanisms, although the investigators specified that fluoride and alkoxide per se were the species transferred. The mechanism shown in equation 9 was also considered by these investigators as well as by others. However, this pathway was discarded by the investigators on several grounds.

The addition of t-butyl hypoiodite to selected olefins has been investigated[21]. In the absence of BF$_3$ and in the dark, t-BuOI reacts with styrene to give the anti-Markovnikov regioisomer (i.e. by a free radical mechanism). In the presence of BF$_3$, β-t-butoxy-β-

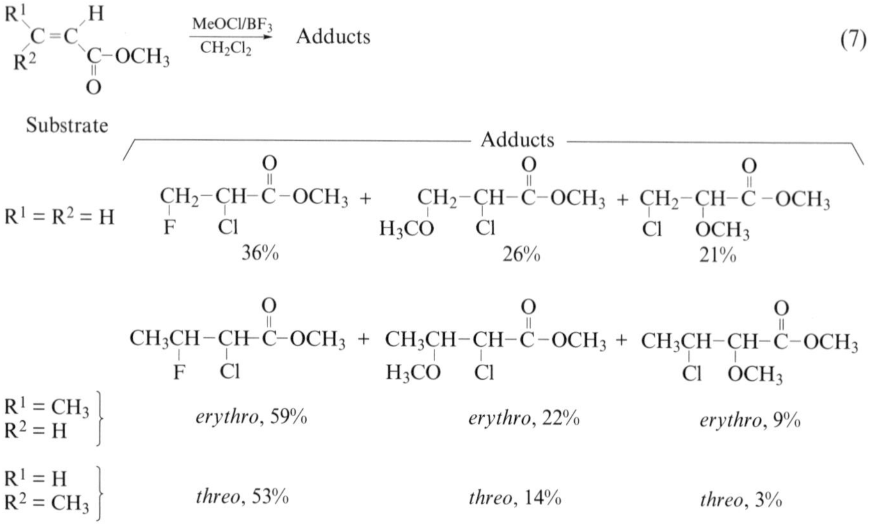

$$\underset{R^2}{\overset{R^1}{\diagdown}}C=C\underset{\underset{O}{\overset{\|}{C}-OCH_3}}{\overset{H}{\diagup}} \xrightarrow[CH_2Cl_2]{MeOCl/BF_3} \text{Adducts} \qquad (7)$$

Substrate

Adducts

$R^1 = R^2 = H$

$$\underset{\underset{36\%}{F \quad Cl}}{CH_2-CH-\overset{\overset{O}{\|}}{C}-OCH_3} + \underset{\underset{26\%}{H_3CO \quad Cl}}{CH_2-CH-\overset{\overset{O}{\|}}{C}-OCH_3} + \underset{\underset{21\%}{Cl \quad OCH_3}}{CH_2-CH-\overset{\overset{O}{\|}}{C}-OCH_3}$$

$$\underset{F \quad Cl}{CH_3CH-CH-\overset{\overset{O}{\|}}{C}-OCH_3} + \underset{H_3CO \quad Cl}{CH_3CH-CH-\overset{\overset{O}{\|}}{C}-OCH_3} + \underset{Cl \quad OCH_3}{CH_3CH-CH-\overset{\overset{O}{\|}}{C}-OCH_3}$$

$\left.\begin{array}{l}R^1 = CH_3\\R^2 = H\end{array}\right\}$ *erythro*, 59%   *erythro*, 22%   *erythro*, 9%

$\left.\begin{array}{l}R^1 = H\\R^2 = CH_3\end{array}\right\}$ *threo*, 53%   *threo*, 14%   *threo*, 3%

$$\underset{\diagup}{\overset{\diagdown}{}}C=C\underset{\diagdown}{\overset{\diagup}{}} \xrightarrow[BF_3]{ROX} \underset{X}{\overset{\diagdown}{}}C\overset{+}{\diagdown}C\overset{\diagup}{} \left[F-\underset{\underset{F}{|}}{\overset{\overset{F}{|}}{B}}-OR \right]^- \underset{\diagdown}{\diagup} \begin{array}{c} \underset{X \quad OR}{-\overset{|}{C}-\overset{|}{C}-} \\[2ex] \underset{X \quad F}{-\overset{|}{C}-\overset{|}{C}-} \end{array} \qquad (8)$$

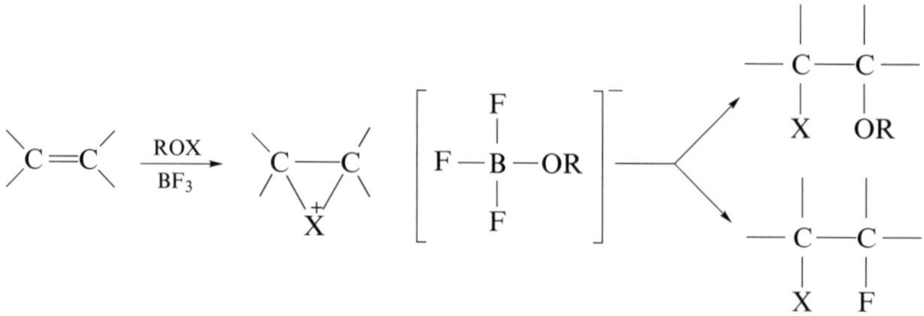

$$ROX + BF_3 \rightleftharpoons ROBF_2 + XF \qquad (9)$$

phenylethyl iodide (Markovnikov product) was produced. It was noted that the function of BF_3 in the ionic reaction is the same as that reported earlier for the reactions of t-BuOCl[19,20] and for t-BuOBr[20] where it was postulated that BF_3 attacks the oxygen, forming an ion pair **5**. According to these investigators fluoride apparently is not nucle-

$$C_6H_5CH\overset{\overset{\displaystyle I}{\overset{+}{\diagup\diagdown}}}{}CH_2[F_3BOBu\text{-}t]^-$$

(5)

ophilic in this ion pair since no fluoroiodides were observed. However, the reaction of t-BuOCl[19,20] and t-BuOBr[20] with styrene and BF_3 produced the fluorochloride and fluoro-bromide compounds, respectively. The iodonium ion may be less reactive and less selective than other halonium ions. Another interesting reaction is that of t-BuOI with 1-hexene in the presence of BF_3 in which Markovnikov and anti-Markovnikov products are obtained in a ratio independent of the lighting conditions[21]. This observation eliminates a free-radical mechanism. Since the reaction must therefore be ionic, these workers add that the formation of more anti-Markovnikov regioisomer 'is probably due to greater steric hindrance as the bulky t-butoxide ion attacks the intermediate iodonium ion'. However, the difference in stability of a primary vs secondary carbocation is very small, and neither is substantial. This fact rules out an S_N1 pathway and this reaction undoubtedly involves an S_N2 mechanism, which makes it obvious that the preferred attack is on CH_2[22].

Chloroxy compounds such as $ClOCF_2CFXSO_2F$ (X = F; CF_3) undergo reaction with simple olefins to give ethers (equation 10)[23]. With unsymmetrical olefins the reaction

$$ClOCF_2CFXSO_2F + RR^1C=CR^2R^3 \longrightarrow ClRR^1C-CR^2R^3(OCF_2CFXSO_2F)$$

$$X = F \text{ or } CF_3$$
$$R, R^1, R^2, R^3 = F \text{ or } Cl \text{ or } H \tag{10}$$

mainly follows an 'electrophilic cis addition with the positive polarized chlorine in a Markovnikov manner', according to this report.

Aryl and alkyl hydrazones of propanal and of alkanones are transformed into 1-chloroalkylazo compounds by reaction with t-butyl hypochlorite (equation 11)[24]. This

$$\begin{array}{c} R^2 \\ \diagdown \\ \diagup C=N-NHR^3 \\ R^1 \end{array} \xrightarrow[\text{CHCl}_3 \text{ or } \text{CH}_2\text{Cl}_2]{t\text{-BuOCl}} \begin{array}{cc} R^2 & Cl \\ & \diagup \\ \diagup & \diagdown \\ R^1 & N=NR^3 \end{array} \tag{11}$$

method, which was introduced by Moon[25], seems to be generally applicable to hydrazones of ketones. In contrast, chlorination with Cl_2 sometimes leads to mixtures of compounds.

B. Acetyl Hypofluorite

Fluorinating reagents, including acetyl hypofluorite, have been reviewed by Rozen[26]. Rozen and coworkers discuss quite effectively 'electrophilic fluorine' and 'electrophilic fluorination'. They prepared acetyl hypofluorite from the reaction of sodium acetate with 10% fluorine in nitrogen in 10:1 $CFCl_3/AcOH$ (equation 12), and it has been used $in\ situ$

$$NaOAc + F_2 \xrightarrow{CFCl_3/AcOH} AcOF \tag{12}$$

without any isolation or purification[27]. According to these investigators acetyl hypofluorite reacts efficiently and selectively with nucleophilic aromatic systems, particularly phenol and aniline derivatives whose functional groups have been protected. The net result of the reaction is partly according to classical aromatic electrophilic substitution. Unlike such substitution, however, the electrophilic fluorine atom generally replaces an $ortho$ hydrogen; only small amounts of $para$-fluoro derivatives are occasionally found. Rozen and coworkers believe this reaction follows an addition–elimination pathway in which the adduct containing the fluorine and acetate moieties is a real intermediate from which the elements of HOAc are subsequently eliminated ($\mathbf{A}$, equation 13). However, we suggest that the formation of such an intermediate probably does not occur when the 'substituent' to be displaced by F is H. That is , we believe the pathway involves transfer of F from FOAc to form a tight π-complex cation with the aromatic compound leading to $para$- and $ortho$-σ-complex cations and then to product by proton abstraction ($\mathbf{B}$, equation 13). The only base present is AcO^- which, in both cases, is localized near the MeO^+ group. It is reasonable to believe, therefore, that proton abstraction from the $para$-σ-complex cation is slow relative to the reversible reaction (loss of F^-), while for the same reason proton abstraction from the *ortho-σ*-complex cation is fast. Rearomatization, therefore, leads predominantly to the *ortho* product. The mechanism presented by Rozen and colleagues (**A**, equation 13) does not account for the largely *ortho* product. Substantial evidence has now evolved

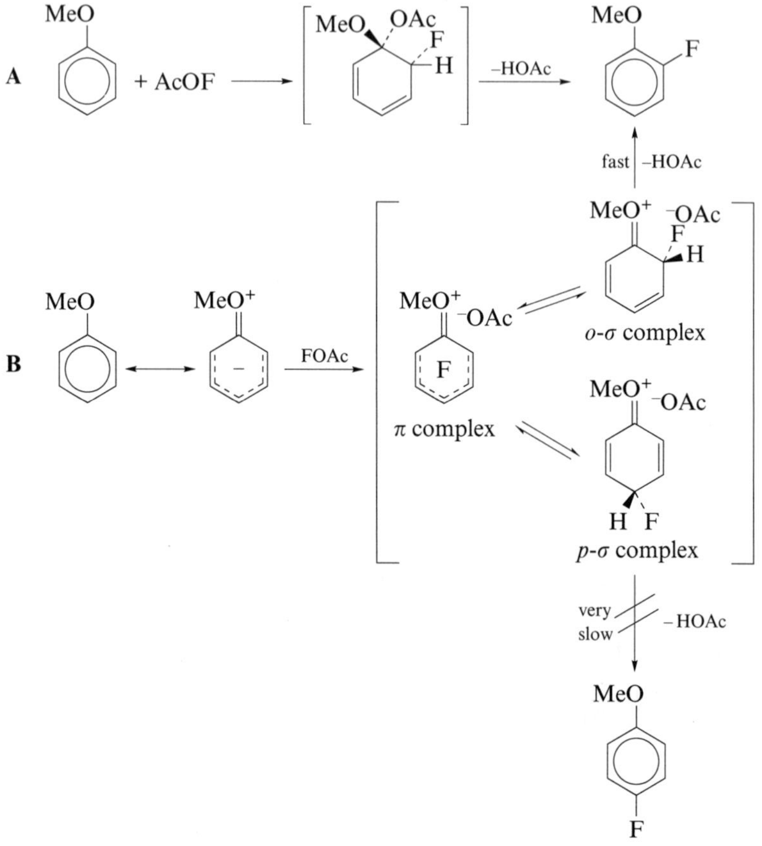

(13)

showing that stable aromatic-ring π-cationic complexes are of even greater mechanistic significance than previously believed; e.g. Auerbach and colleagues[28] have reported convincing calculations that Me_4N^+ binds to benzene more favorably than to chloride ion.

In their report Rozen and coworkers attempt to show that a true intermediary adduct containing both F and OAc moieties is formed prior to the production of the final product (i.e. mechanism **A**). While they demonstrated the formation of such an adduct when direct elimination of an H^+ from the σ-cationic complex is not possible, they have not demonstrated that such an adduct is formed when direct elimination of an H^+ from the σ-cationic complex *is* possible (i.e. mechanism **B**). We believe that this direct aromatization (mechanism **B**) is by far the most energetically favourable pathway.

The precise mechanism of fluorine transfer from AcOF to the aromatic compound in these reactions is still a controvertial issue, although in all cases the initially formed moiety is probably the same π-cation complex. Three 'reasonable' routes leading to this complex should be considered (equation 14): (1) The aromatic π electrons simply act like a base to 'abstract F^+' which leaves its binding electron pair with OAc (as ^-OAc); (2) the

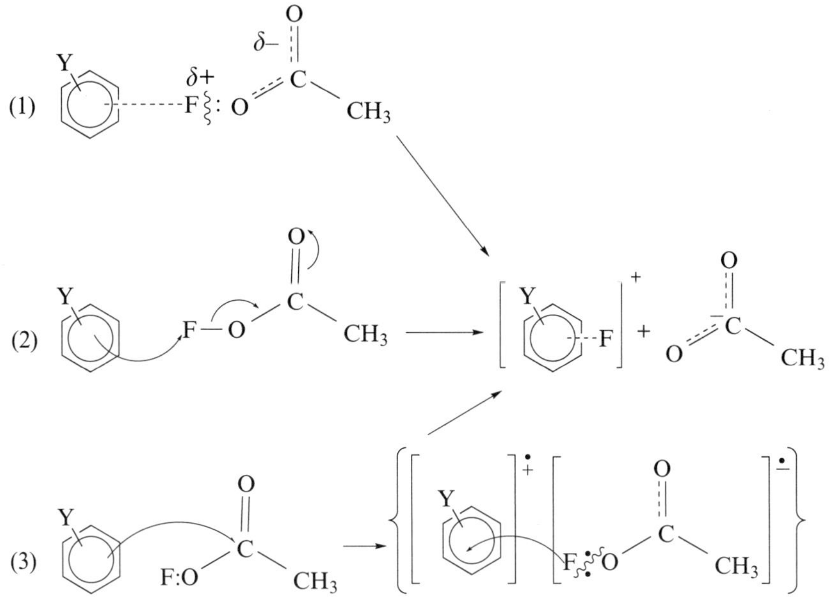

$$(14)$$

aromatic π electrons act like a nucleophile to attack F in an 'S_N2 manner', with ^-OAc as the leaving group; and (3) the aromatic π electrons donate a single electron to AcOF which forms an intimate cation-radical/anion-radical pair from which a fluorine atom (F·) is transferred to the Ar cation radical (an electron-transfer mechanism incorporating a RARP-like transition state originally proposed by Meyers and coworkers[7-9]). Chanon and coworkers[10] have discussed the RARP transition state in comparison with other mechanistic possibilities in the halogenation of carbanions with CCl_4 and related perhaloalkanes. Differding and coworkers[4-6] likewise have discussed similar possible mechanisms in related halogen-transfer reactions. Visser and coworkers[29] have also suggested a single-electron-transfer mechanism for aromatic fluorination with AcOF (*vide infra*).

Rozen and coworkers[30] have also reported that acetonitrile can replace trichlorofluoromethane in the preparation and *in situ* reactions of acetyl hypofluorite without sacrificing the efficiency and the regio- and stereoselectivity of the addition of AcOF to olefins. Most other solvents investigated proved to be unsatisfactory for the synthesis of AcOF, despite the fact that fluorine does react with the dispersed sodium acetate present in the solvent.

Treatment of pyridine with acetyl hypofluorite results in the formation of N-fluoropyridinium acetate whose α-positions, according to Rozen and coworkers[31], are activated towards nucleophilic attack by AcO$^-$, or by '$Cl^{\delta-}$, $Br^{\delta-}$ and $RO^{\delta-}$ originating from solvents such as CH_2Cl_2, and various primary alcohols'. Subsequently, according to their report, dehydrofluorination rapidly affords these respective 2-substituted pyridines. Thus, in CH_2Cl_2, 2-chloropyridine was formed in 70% yield and 2-acetoxypyridine was formed only in 15% yield. However, when the solvent was $CFCl_3$, 2-acetoxypyridine was formed in 'excellent' yield, and likewise when $CHCl_3$ or CCl_4 was used as a solvent only 2-acetoxypyridine (> 80% yields) was formed. As indicated in the following scheme, Rozen and coworkers depicted the origin of the chlorine in the chlorinations carried out in CH_2Cl_2 as arising from complexation of CH_2Cl_2 with pyridine and, ultimately, abstraction of Cl^-

by the N-fluoropyridinium cation. They noted that precedence for the CH_2Cl_2–pyridine complex was derived from the work of Nevstad and Songstad[32]. Rozen and coworkers' mechanistic scheme for the reactions in CH_2Cl_2 is shown in equation 15. At least in part

$$(15)$$

this pathway, suggested by Rozen and coworkers to account for the high yields of 2-chloropyridine when CH_2Cl_2 was the solvent and of 2-bromopyridine when CH_2Br_2 or CH_3Br was the solvent, seems quite improbable in light of the facts. Their high yields were obtained in a matter of a few minutes from reactions carried out at room temperature. These results are hardly supported by those of Nevstad and Songstad[32] who reported that pyridine '...was found to react exceedingly slowly with CH_2Cl_2...the half-lives of these reactions were several months at room temperature.' Moreover, no substantiated precedent could be found of a chlorination or bromination in which a cation simply abstracted Br^- or Cl^- from a haloalkane like CH_3Br or a 1,1-dihaloalkane like CH_2Cl_2, or CH_2Br_2. The real question is then: what mechanism accounts for these combined results? In their patents for the preparation of CD_2Cl_2 (> 99% D) Meyers and Chan-Yu-King[33] reported that CH_2Cl_2 treated with Na_2O in D_2O (i.e $NaOD/D_2O$) in the presence of <1% $R_4N^+Cl^-$ undergoes deprotonation leading to equilibrium D/H exchange, with only trace amounts of Cl^- being formed during the overall exchange process. This result was 'novel and unexpected' in light of studies by Hine and coworkers[34,35] which showed that CH_2Cl_2 readily undergoes S_N2 displacement of Cl^- when treated with $NaOCH_3$ in CH_3OH to form CH_3OCH_2Cl and $CH_2(OCH_3)_2$ and, more apropos, as shown by Atkinson and coworkers[36,37], when it is treated with AcO^- in Ac_2O to form $AcOCH_2Cl$ and $CH_2(OAc)_2$. The Atkinson group also reported that CH_2Br_2 as well as CH_2I_2 similarly undergo S_N2 attack by AcO^- with displacement of the respective halide anion. They took advantage of this reaction with these three dideuteriated dihalomethanes to produce CD_2O from the subsequent hydrolysis of the formed diacetates[36,37]. In light of these results it is reasonable to believe that the halogen source in the formation of the 2-halopyridines is the halide ion derived from the respective solvent, i.e. CH_2Cl_2, CH_2Br_2 and CH_3Br, via S_N2 displacement by AcO^-, as illustrated in equation 16.

A regiospecific electrophilic fluorination of the tyrosine ring of the N-terminal tetrapeptide amide (Tyr-D-Ala-Phe-Gly-NH$_2$) sequence of the opiate peptide dermorphin has been achieved in good yields by Labroo and coworkers using acetyl hypofluorite (equation 17)[38]. They also investigated the effect of fluorination of the phenolic ring of the Tyr residue in opiate peptides on the affinity and selectivity of these peptides for various subtypes of opiate receptors (e.g. μ, δ and k) and the consequent effects on biological activities. Another aim was to develop receptor-specific and metabolically stable fluorinated analogues for possible use as positron emission tomography (PET) scanning agents. A regioselective aromatic-ring fluorination of substituted veratroles was achieved by Luxen and

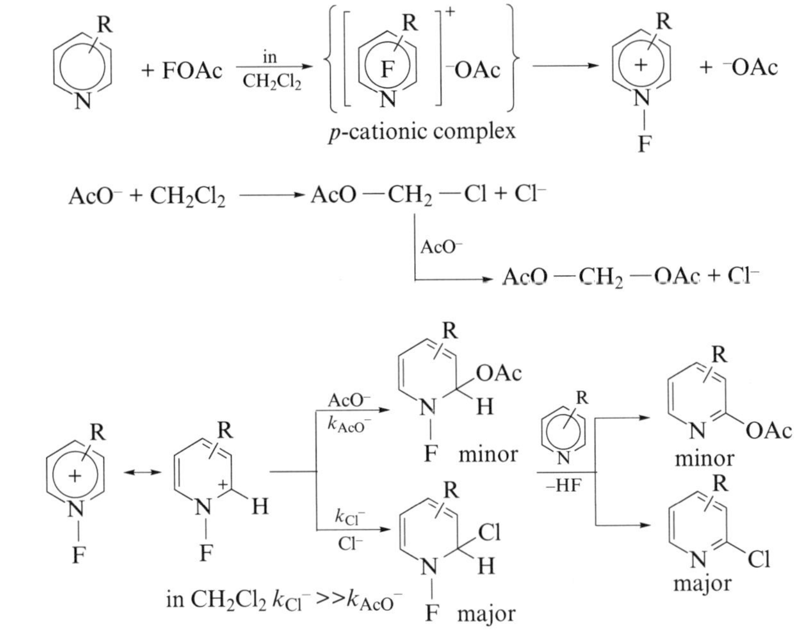

$$R\text{-pyridine} + \text{FOAc} \xrightarrow[\text{CH}_2\text{Cl}_2]{\text{in}} \left\{ \left[\begin{array}{c} R \\ F\text{-N} \end{array} \right]^{+} \text{OAc} \right\} \longrightarrow \begin{array}{c} R \\ +\text{N} \\ F \end{array} + {}^{-}\text{OAc}$$

p-cationic complex

$$\text{AcO}^- + \text{CH}_2\text{Cl}_2 \longrightarrow \text{AcO}-\text{CH}_2-\text{Cl} + \text{Cl}^-$$

$$\downarrow \text{AcO}^-$$

$$\text{AcO}-\text{CH}_2-\text{OAc} + \text{Cl}^-$$

(16)

in CH$_2$Cl$_2$ $k_{\text{Cl}^-} \gg k_{\text{AcO}^-}$

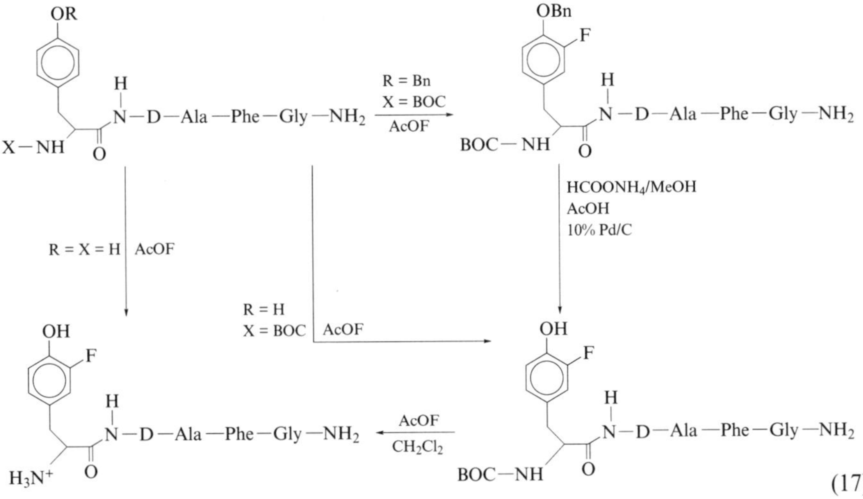

(17)

Barrio[39] via a mercuration–fluorodemercuration reaction. Direct fluorination using dilute molecular fluorine or acetyl hypofluorite was not satisfactory because mixtures of isomers were always obtained (equation 18). The hydroxy derivatives of compounds **6a–e** are of pharmacological and medicinal interest[40].

Investigation of the use of acetyl hypofluorite in acetic acid for the regiocontrolled monofluorination of aromatic compounds starting from the corresponding mercurated derivatives has been carried out by Visser and coworkers[29] (equation 19). On the basis of the observed fluorinated (**7**), acetoxylated (**8**) and methylated (**9**) products, a one-electron-transfer mechanism leading to an intermediate radical cation was proposed which might contribute to a better understanding of electrophilic fluorination (equation 20). This type of one-electron-transfer mechanism is tantamount to the RARP mechanism proposed by Meyers and coworkers[7-9] to explain a variety of similar reactions and also used in this context (*vide supra*).

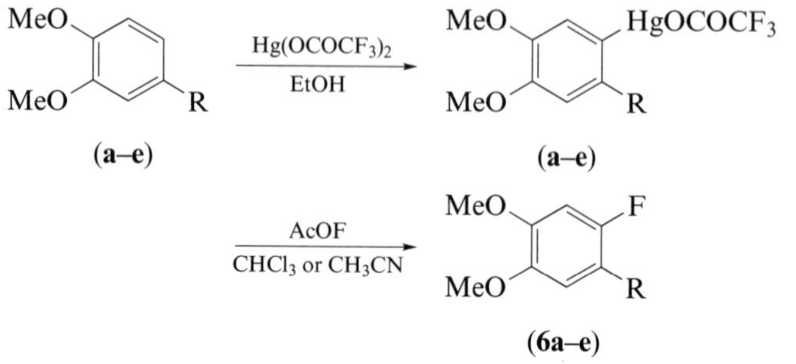

$$(18)$$

(a) R = L-CH$_2$CHCOOEt   (b) R = D-CH$_2$CHCOOEt
        | |
       NHCOCF$_3$ NHCOCF$_3$

(c) R = CH$_2$COOEt   (d) R = CH$_2$CH$_2$NHCOCF$_3$
(e) R = CHO

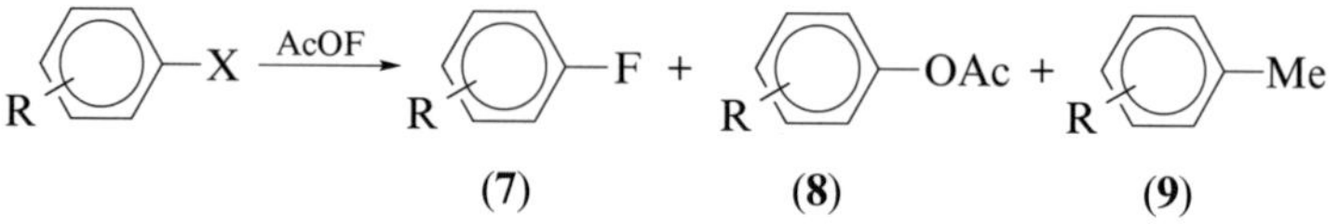

$$(19)$$

Examples: X = H; R = OMe, OH, NH$_2$, NHCOCH$_3$, Cl, H
            X = HgOAc; R = *p*-MeO, *p*-H$_2$N, *p*-NHCOCH$_3$, H
            X = HgCl; R = *o*-OH, *p*-OH, *m*-CH$_3$, H

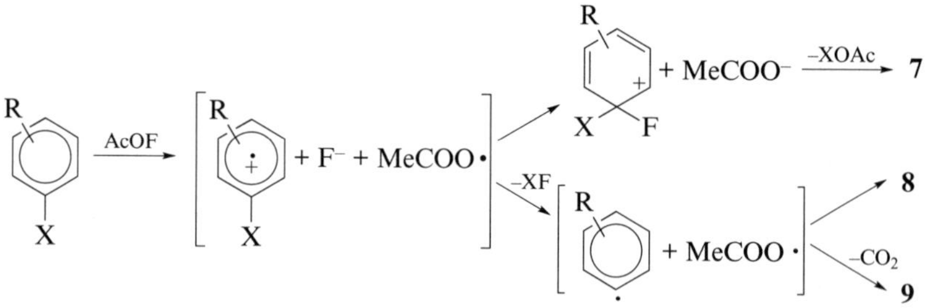

$$(20)$$

The fluorination and stereochemistry of the reaction of pyranoid and furanoid glycals with acetyl hypofluorite are described[41]. In related work[42] the synthesis of 2-deoxy-2-fluoro-D-galactapyranose by treatment of 2-deoxy-D-galactapyranose with acetyl hypofluorite is described and the ^{18}F-labelled product was similarly prepared through the use of AcO[18]F.

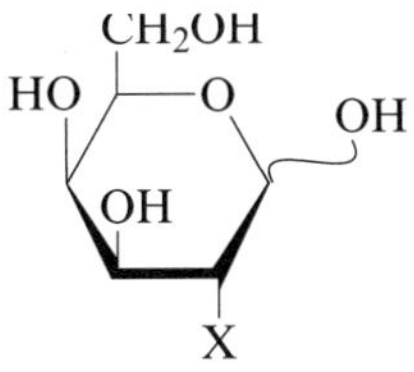

$$X = F \text{ or } {}^{18}F$$

An interesting paper[43a] discusses the use of ${}^{18}F$-DOPA as a tracing agent to follow the pathway of dopamine (a neurotransmitter) in the human brain by using PET. One route to ${}^{18}F$-DOPA utilized AcO${}^{18}F$. Millicurie quantities of ${}^{18}F$-dopa[43b,c] and ${}^{18}F$-m-tyrosine[43d] were prepared this way (equation 21). These reactions produced mixtures of mono-

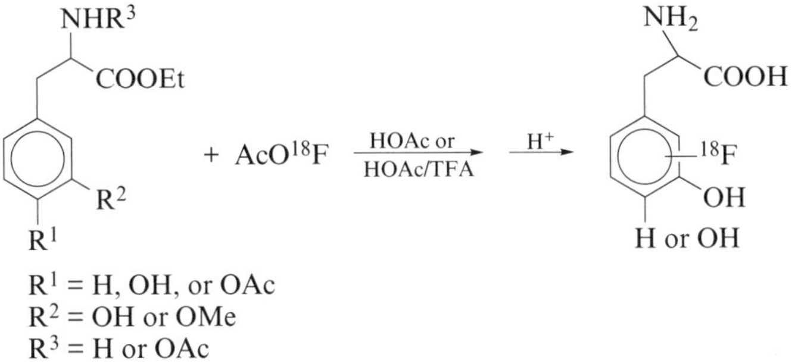

$$R^1 = H, OH, \text{ or } OAc$$
$$R^2 = OH \text{ or } OMe$$
$$R^3 = H \text{ or } OAc$$

fluorinated products, the desired isomers being separated by HPLC. Acetyl hypofluorite has also been used for the regioselective synthesis of ${}^{18}F$-DOPA via an aryl mercury compound (equation 22)[43e].

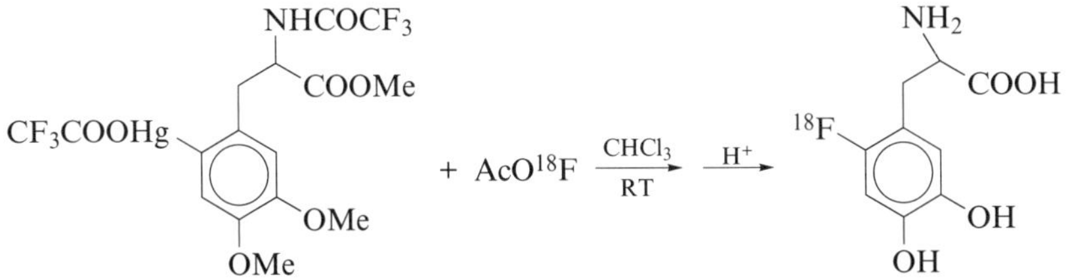

$$6\text{-}{}^{18}F\text{-DOPA}$$

In a similar context, 6-${}^{18}F$-3,4-dihydroxyphenylalanine (6-${}^{18}F$-DOPA) was synthesized by direct fluorination of L-3-(3-hydroxy-4-pivaloxyphenyl)alanine (m-P-DOPA) with AcO${}^{18}F$ in acetic acid resulting in the 2- and 5-${}^{18}F$ isomers. Hydrolysis of the reaction mixture in hydrochloric acid followed by HPLC separation gave 6-${}^{18}F$-DOPA (equation 23)[44]. Another application of AcO${}^{18}F$ was reported in the synthesis of a trimethyl tin precursor of 2-oxoquazepam, 7-chloro-1-(2,2,2-trifluoroethyl)-1,3-dihydro-5-(2-fluorophenyl)-2H-1,4-benzodiazepin-2-one, a benzodiazepine agonist, and its conversion to [${}^{18}F$]-2-oxoquazepam by reaction with AcO${}^{18}F$ (equation 24)[45].

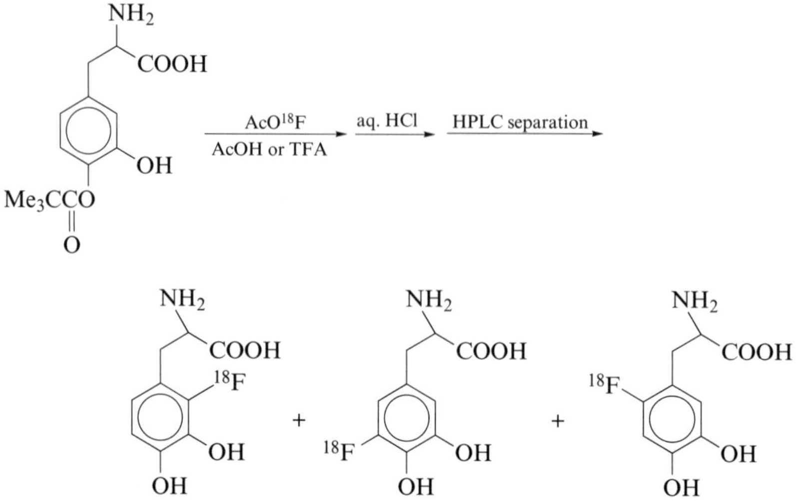

$$(23)$$

2-^{18}F-DOPA   5-^{18}F-DOPA   6-^{18}F-DOPA

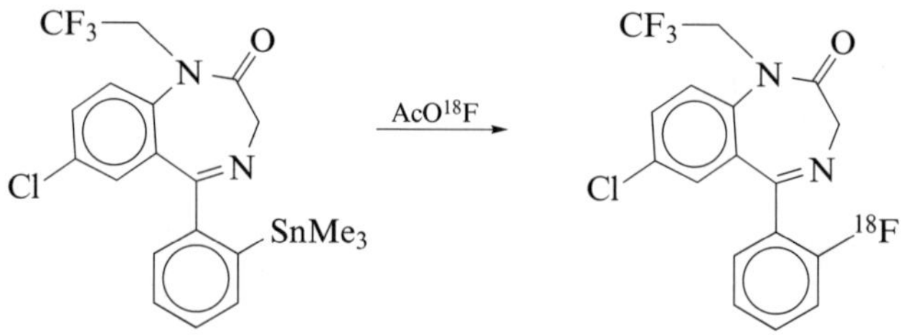

$$(24)$$

C. Miscellaneous Reactions

Alkyl hypochlorites add to isocyanides; hydrolysis of the adducts leads to carbamates (equation 25)[46].

$$R—N^+\equiv C^- + t\text{-BuOCl} \xrightarrow{\text{ZnCl}_2} R—N=C—OBu\text{-}t$$
$$\overset{|}{Cl}$$

$$\xrightarrow{H_2O} R—NH—C—OBu\text{-}t \qquad (25)$$
$$\overset{\|}{O}$$

Yoshida and coworkers[47] reported that they readily oxidized 'bromide and chloride anions into positive halogens' with p-nitrobenzenesulfonyl peroxide (equation 26). They found that oxyhalogenated products were formed from the reaction of these positive halogens with olefins, and proposed epihalonium ion intermediates which, in turn, were trapped by oxygen nucleophiles inter- or intramolecularly to afford the oxyhalogenated products (equation 27).

$$(p\text{-}O_2NC_6H_4SO_3)_2 + \bigcirc \xrightarrow[\text{MeOH--MeCN}]{\text{KBr}} \text{[structure with OMe and Br]} \qquad (26)$$

$$(ArSO_3)_2 + KBr \longrightarrow ArSO_3Br + ArSO_3K$$

$$ArSO_3Br + \bigcirc \longrightarrow \left[\bigcirc \overset{+}{\rhd} Br \quad ArSO_3^- \right]$$

$$Ar = p\text{-}O_2NC_6H_4$$

$$\xrightarrow{\text{MeOH}} \text{[structure with OMe and Br]} + ArSO_3H \qquad (27)$$

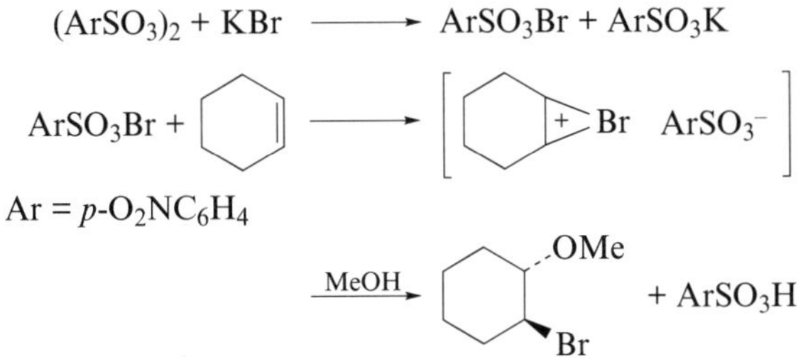

Musin and coworkers[48] reported that the reaction of methyl (diethylamino)sulfenate (10) with *t*-butyl hypochlorite provides a mixture of diethylaminosulfinate esters 11 and 12 in a ratio of 1:4 (equation 28); these esters could not be separated by distillation. It was

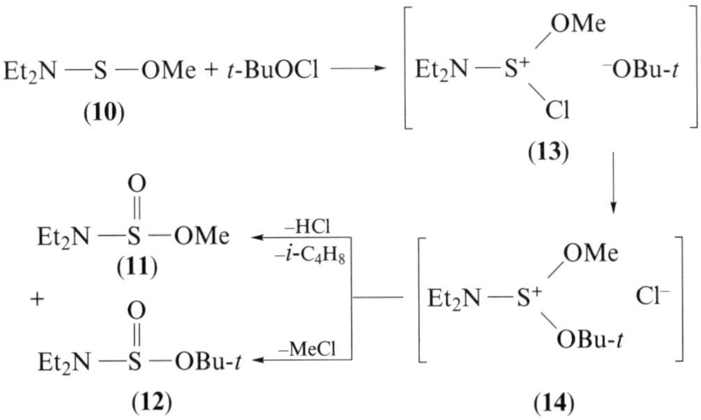

$$Et_2N-S-OMe + t\text{-BuOCl} \longrightarrow \left[Et_2N-\overset{+}{S}\begin{smallmatrix} OMe \\ \\ Cl \end{smallmatrix} \quad {}^-OBu\text{-}t \right]$$

$$(10) \qquad\qquad (13)$$

$$
\begin{array}{c}
\overset{\displaystyle O}{\underset{\displaystyle \|}{}} \\
Et_2N-S-OMe \\
(11) \\
+ \\
\overset{\displaystyle O}{\underset{\displaystyle \|}{}} \\
Et_2N-S-OBu\text{-}t \\
(12)
\end{array}
\quad
\begin{array}{c}
\xleftarrow{\substack{-HCl \\ -i\text{-}C_4H_8}} \\[2em]
\xleftarrow{-MeCl}
\end{array}
\quad
\left[Et_2N-\overset{+}{S}\begin{smallmatrix} OMe \\ \\ OBu\text{-}t \end{smallmatrix} \quad Cl^- \right]
\qquad (28)
$$

$$(14)$$

suggested by these workers that the reaction takes place via initial formation of sulfonium salt 13, which is probably converted into more stable salt 14 by anion exchange. The chloride anion can then abstract a proton from the *t*-Bu group effecting an E2 elimination to yield 11, as well as nucleophilically attack the Me group to produce 12.

D. Special Cases

Ionica, Georgescu and Dinulescu[49] have surveyed the literature and report that C-bonded acidic hydrogens can be substituted by halogens through reaction with hypohalites. They derive this mechanism from experimental results obtained in the haloform reaction, e.g. equation 29. They explained the formation of Cl^+ and HO^- through the sequence of reactions in equation 30. They noted that two equivalents of OH^- are generated from each ClO^- used up, and they said that the haloform reaction supported their mechanism: 'Experimentally, by treating one mole of *p*-bromoacetophenone with three

$$R-H \xrightarrow{HO^-} R^- \xrightarrow{Cl^+} R-Cl \qquad (29)$$

$$Cl-O^- \rightleftharpoons Cl^+ + O^{-2}$$

$$O^{-2} + H_2O \longrightarrow 2\,HO^- \tag{30}$$

moles of NaOBr (free of NaOH) p-bromobenzoic acid was obtained in 80% yield... (and) in the reaction medium together with sodium p-bromobenzoate, approximately 0.8–1.0 moles NaOH was evidenced'. We find no precedent for invoking such a mechanism or an intrinsic halonium ion (X^+) (see Section I). What we do find reasonable is a simplified pathway following documented reactions[7-9], for example that illustrated in equation 31. This pathway likewise leads to the production of the same amount of hydroxide as noted in the report cited.

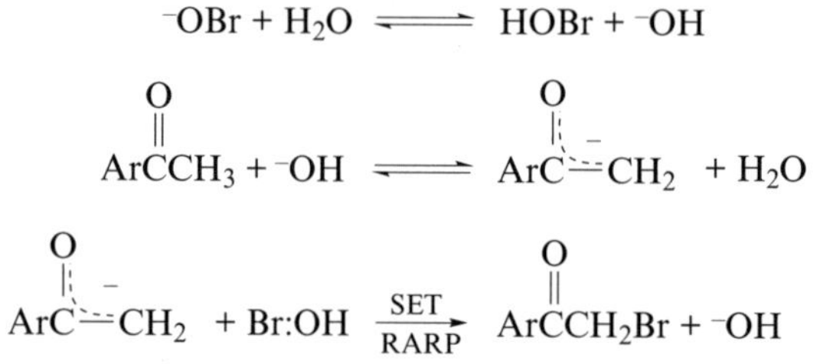

$$^-OBr + H_2O \rightleftharpoons HOBr + {}^-OH$$

$$\underset{\overset{\|}{O}}{ArCCH_3} + {}^-OH \rightleftharpoons \underset{\overset{|}{O}}{ArC{\cdots}CH_2^-} + H_2O$$

$$\underset{\overset{|}{O}}{ArC{\cdots}CH_2^-} + Br{:}OH \xrightarrow[\text{RARP}]{\text{SET}} \underset{\overset{\|}{O}}{ArCCH_2Br} + {}^-OH \tag{31}$$

Ionica, Filip and Dinulescu[50] studied hypohalite halogenation of N-bonded acidic hydrogens, which proceeds considerably faster than corresponding carbon–H halogenations. They used Hoffmann degradation of amides as a proof of their mechanism (equation 32). Based on the overall stoichiometry of the reaction they suggested that if their

$$Br-O^- \rightleftharpoons Br^+ + O^{-2}$$

$$O^{-2} + H_2O \longrightarrow 2\,HO^-$$

$$R-CONH_2 + Br^+ + 2\,HO^- \longrightarrow [R-CO\ddot{N}-Br]^- + 2\,H_2O$$

$$[R-CO\ddot{N}-Br]^- \xrightarrow{-Br^-} R-CO-\ddot{N} \longrightarrow O{=}C{=}N-R$$

$$\xrightarrow{H_2O} HO_2CNHR \xrightarrow{-CO_2} RNH_2$$

$$CO_2 + 2\,NaOH \longrightarrow Na_2CO_3 + H_2O \tag{32}$$

halogenation mechanism with hypohalites were correct, one equivalent of hypohalite and two equivalents of hydroxide would be necessary for the reaction of an amide (equation

$$R-CONH_2 + BrO^- + 2\,HO^- \longrightarrow R-NH_2 + Br^- + CO_3^{2-} + H_2O \tag{33}$$

33). By treating benzamide with one equivalent of NaOBr and two of NaOH, 0.85 equivalent of aniline and 0.85 equivalent of sodium carbonate were formed. In spite of the good intentions of these investigators, these results would be the same regardless of the 'mechanism' of the bromination... which was the goal of their study. We object mainly to the initial reaction they suggested, i.e. the formation of $Br^+ + O^{-2}$. The straightforward mechanism we have noted above for their haloform reaction of p-bromoacetophenone with BrO^- would apply as well here.

III. NITROGEN–HALOGEN COMPOUNDS

N-halo electrophiles (N-halo amines, N-halo amides and N-halo imides) react with alkenes such as 1-hexene, cyclohexene and styrene in the presence of boron trifluoride to give halofluorides and N-halo adducts (equation 34)[51]. These types of reactions have been dis-

$$\ce{>=<} \quad \xrightarrow[\text{BF}_3]{\ce{\backslash N-X/}} \quad \underset{\underset{X}{|} \quad \underset{F}{|}}{-C-C-} \quad + \quad \underset{\underset{X}{|} \quad \underset{\overset{N}{\diagup\diagdown}}{|}}{-C-C-} \tag{34}$$

cussed in a review by Labeish and Petrov which emphasizes the work of Soviet chemists[52]: reactions involving the addition of N-halo- and N,N-dihalo amides of carboxylic, sulfonic, sulfuric, and phosphoric acids, as well as the esters of N-halo- and N,N-dihalocarbamic acids to various types of unsaturated compounds. In many cases the addition takes place with a regiospecificity different from that for other compounds with N-halogen bonds, which has been ascribed to different reaction mechanisms. The adducts of halo amides with unsaturated compounds can be used in the synthesis of 'specialty' amides and aziridines as well as other heterocyclic compounds. Classes of unsaturated compounds that have been incorporated into these addition reactions and discussed in this review are: aliphatic alkenes, alicyclic alkenes, aryl alkenes, substituted alkenes, alkadienes, acetylenes and alkenynes.

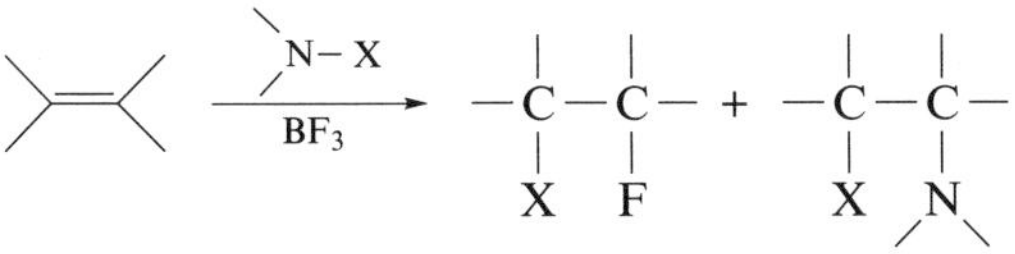

A. *N*-Halo Amides and *N*-Halo Imides

A useful preparation of N-acylureas initiated by the addition of N-halo amides to isocyanides was described (equation 35)[53]. β-Keto-α-(phenylthio)alkyl p-tolyl sulfones in

$$\ce{R-N^+#C^- + MeCONHBr} \longrightarrow \ce{R-N=C-NHCOMe} \\ \underset{\text{Br}}{|}$$

$$\xrightarrow{\text{H}_2\text{O}} \quad \ce{R-NH-C-NHCOMe} \tag{35} \\ \underset{\text{O}}{\|}$$

approximately 60% yields were obtained from β-ketoalkyl phenyl sulfides by halogenation with NCS followed by treatment with sodium p-toluenesulfinate; the corresponding β-ketoalkyl p-tolyl sulfones were always found as side products (equation 36)[54a]. The reaction with p-Tol SO$_2$Na was carried out in refluxing (60 °C) acetone-benzene in the presence of about 8% tetrabutylammonium hydrogen sulfate, a phase-transfer catalyst. Since no aqueous phase was present it is difficult to understand the function of this catalyst under the conditions reported. Yields were not reported for the β-ketoalkyl p-tolyl sulfone side

$$\underset{\underset{\|}{O}}{\ce{R-C-CH_2-SPh}} \quad \xrightarrow{\text{NCS/CCl}_4} \quad \underset{\underset{\|}{O}}{\ce{R-C-CHCl-SPh}} \quad \xrightarrow[\substack{\text{acetone–benzene} \\ \text{Bu}_4\text{N}^+\text{SO}_4\text{Hcat.}}]{p\text{-TolSO}_2\text{Na}}$$

$$\underset{\underset{\|}{O} \quad \underset{\text{SPh}}{|}}{\ce{R-C-CH-S-Tol-}p} \quad + \quad \underset{\underset{\|}{O}}{\ce{R-C-CH_2-S-Tol-}p} \tag{36} \\ \text{side product}$$

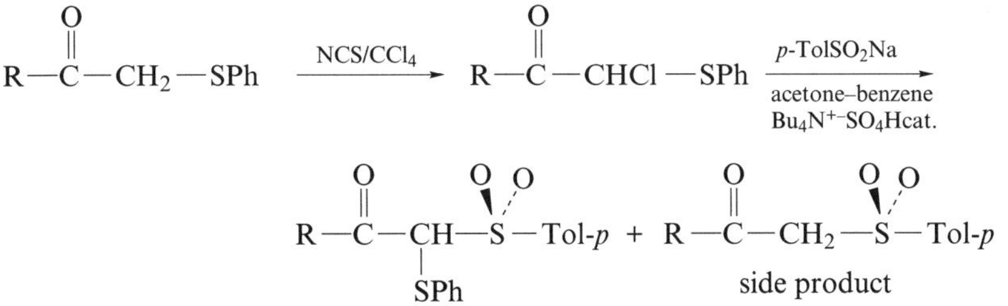

products, nor was it suggested how they were formed. It was noted that these side products could not be eliminated but their quantity was reduced "after careful studies of the reaction time, temperature and solvents" of the second step. According to the report, the chlorinated keto sulfides formed in the first step were "practically pure and used in the next step without any purification", viz. the reaction mixtures were filtered and the evaporated filtrates were treated directly with sodium p-toluenesulfinate.

Based on the conditions used and the relatively small yields of desired products, small amounts of residual keto sulfide substrate were most likely present in the second step so that their reaction with sodium p-toluenesulfinate could be responsible for the formation of the side products. Simple S_N2 displacements of thiolate anions from thioethers are at best very uncommon. It is well known that nucleophilic displacement of α substituents in ketones is greatly facilitated via attack at carbonyl[7,54b]. We are suggesting that formation of the side products occurs via attack of sulfinate anions at the carbonyl of residual starting sulfide. As illustrated in equation 37a, under the conditions noted, it is quite possible that a displacement of benzenethiolate anion occurred and that this anion subsequently abstracted the very acidic proton of the product so formed, inhibiting the reverse reaction and allowing the byproduct to be isolated upon addition of water. Good evidence that displacement of thiolate anions from α-arylthio ketones occurs via attack at carbonyl can be interpreted by a comparison of the reactions in equations 37b and 37c[54c].

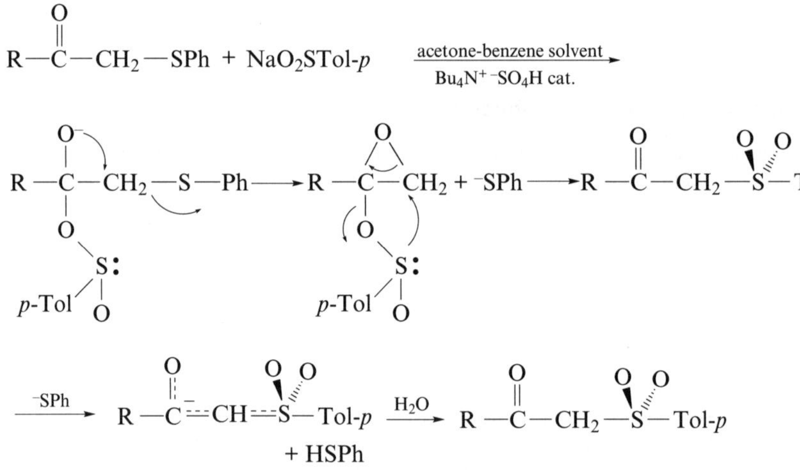

$$(37a)$$

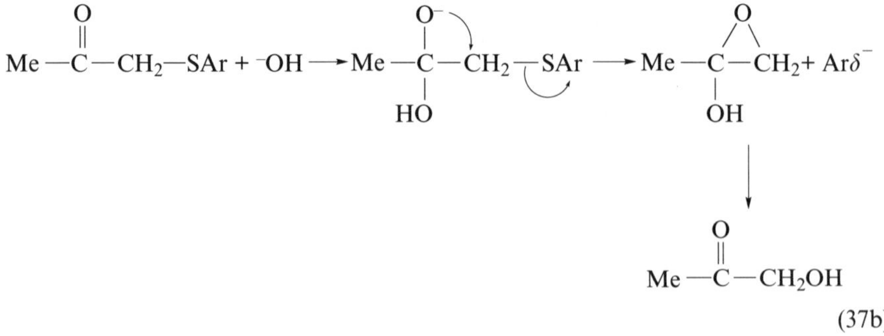

$$(37b)$$

$$\text{Ph}-\overset{\overset{\displaystyle O}{\|}}{\underset{\underset{\displaystyle Ph}{|}}{C}}-\text{CH}-\text{SEt} + {}^-\text{OH} \longrightarrow \text{Ph}-\overset{\overset{\displaystyle O^-}{|}}{\underset{\underset{\displaystyle OH\ \ Ph}{|\ \ |}}{C}}\overset{\frown}{\underset{}{}}\text{CH}-\text{SEt} \longrightarrow \text{Ph}-\overset{\overset{\displaystyle O}{\|}}{\underset{\underset{\displaystyle OH\ \ Ph}{|\ \ |}}{C}} + {}^-\text{CH}-\text{SEt}$$

$$\downarrow$$

$$\text{PhCO}_2{}^- + \text{PhCH}_2\text{SEt}$$

(37c)

According to a report by Goldberg and Alper[55], protonation of NXS (X = Cl or Br) with a strong, highly lipophilic acid such as perchloric acid can generate positive halonium species capable of halogenating activated aromatics (e.g. mesitylene) and heteroaromatics (e.g. thiophene) in a catalytic manner in low-polarity media. They provided the scheme shown in equation 38. Since this is a biphasic system, perchloric acid should act as phase-

$$
\begin{array}{ccc}
\text{NXS} & \text{H}^+\text{ClO}_4{}^- & \text{Ar}-\text{X} \\
& & \\
\text{NHS} & \text{X}^+\text{ClO}_4{}^- & \text{Ar}-\text{H}
\end{array}
$$

(38)

transfer agent transferring *in situ*-generated halonium species into the bulk of the organic phase. Goldberg and Alper derived their idea of generating a halonium species from a report by Fraser-Reid and coworkers[56] on the application of the NXS (X = I)/CF$_3$SO$_3$H pair for the generation of iodonium ion.

Along similar lines, Mozek and Sket[57] reported that aromatic compounds could be chlorinated or brominated using an *N*-chlorosaccharin/pyridinium poly(hydrogen fluoride) or *N*-bromosaccharin/pyridinium poly(hydrogen fluoride) system (NXSac = *N*-halosaccharin). From the results obtained they concluded that 'pyridinium poly (hydrogen fluoride) probably served not only as a solvent but together with NXSac also as a source of XF species which acted as the reagent for electrophilic aromatic chlorination and bromination'. Referring to the work of Rozen and coworkers[58], Mozek and Sket continued: 'It is also possible that the HF presence in NBSac/pyridinium poly (hydrogen fluoride) system formed a potential highly ionic species Br$^+$HF$_2{}^-$'.

Olefins undergo 'halofluorination' in a highly regio- and stereoselective manner when treated with hexafluoropropene diethylamine (HFP-DA) and *N*-halo imides as described by Fujisawa and coworkers (equation 39)[59]. The *N*-halo imides used were *N*-iodosuccin-

$$\underset{R^2}{\overset{R^1}{\diagup}}\!\!=\!\!\underset{R^4}{\overset{R^3}{\diagdown}} \quad \xrightarrow[\text{HMPA, PhCH}_3]{\text{HFP-DA, H}_2\text{O, X}^+} \quad X\!\!\rightarrow\!\!\underset{R^2}{\overset{R^1}{\diagup}}\!\!\underset{R^4}{\overset{R^3}{\diagdown}}\!\!-\text{-F}$$

(39)

$$X = I, Br, Cl$$

imide, 1,3-dibromo-5,5-dimethylhydantoin (DBH) and 1,3-dichloro-5,5-dimethylhydan-toin (DCH). These workers write the halogenating species (other than fluorine) as X$^+$ and that it is derived from the halo imides. The fluorine atom was always introduced into the more substituted carbons and a *trans* orientation of fluorine with respect to the other halogen was always observed. The reaction proceeds most probably via generation of a limited amount of hydrogen fluoride from the HFP-DA complex and H_2O followed by formation of the XF species with NIS or DBH. Fujisawa and coworkers, implicating the presence of X^+, illustrate this reaction as noted in equation 40.

$$Et_2NCF_2CHFCF_3 + H_2O \longrightarrow Et_2NCOCHFCF_3 + 2HF$$

$$HF + X^+ \longrightarrow XF + H^+ \tag{40}$$

Likewise, Kuroboshi and Hiyama[60] reported that regio-, stereo- and chemoselective halofluorination of alkenes is achieved using N-halo imides and tetrabutylammonium dihydrogen trifluoride (equation 41). They also noted that the fluorine is always introduced

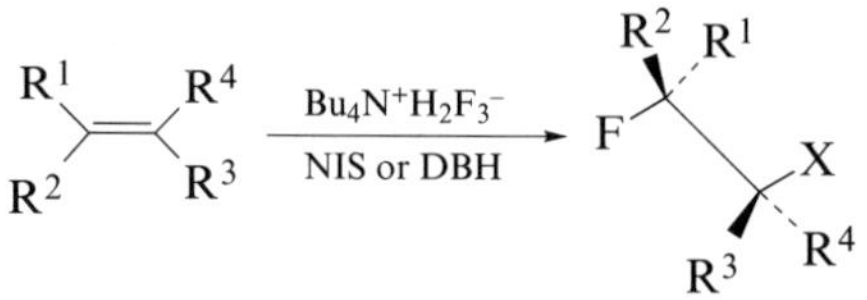

$$\underset{R^2}{\overset{R^1}{\diagdown}} \underset{R^3}{\overset{R^4}{\diagup}} \quad \xrightarrow[\text{NIS or DBH}]{Bu_4N^+H_2F_3^-} \quad F \underset{R^3}{\overset{R^2 \quad R^1}{\diagdown}} \underset{R^4}{\overset{X}{\diagup}} \tag{41}$$

$$X = I \text{ or } Br$$

at the olefinic carbon which contains more electron-donating substituents and can stabilize the transient positive charge more efficiently. They reported that the stereochemistry of the addition of fluorine and iodine is *anti* for the olefins studied.

Under free-radical conditions (e.g. in CCl_4) the cyclic tryptophan tautomer **15** undergoes bromination with NBS at the 3a-position, but under electrophilic conditions, aromatic bromination occurs at C-5 as reported by Crich and coworkers (equation 42)[61]. Their

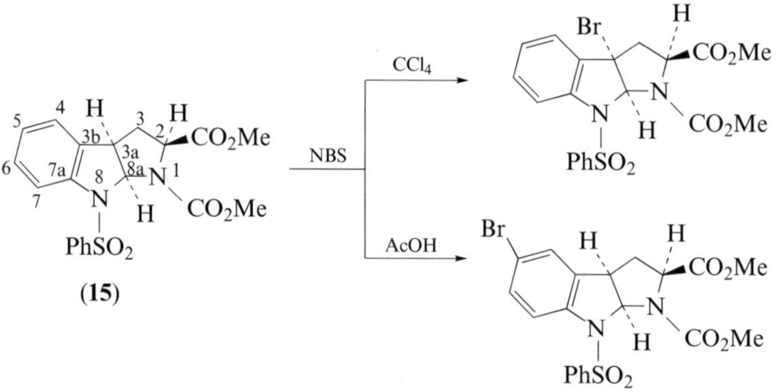

$$\tag{42}$$

results are in agreement with those of Hino and coworkers[62] who found that the N-8 acetamido analogue of **15** undergoes clean bromination with NBS/acetic acid at C-5.

B. *N*-Fluoro Sulfonamides

Selective fluorination of a wide variety of carbanions has been carried out very effectively with *N*-fluoro-*N*-alkyl sulfonamides by Barnette[63]. Treatment of carbanions of malonates, enolates of ketones, acids and amides, and alkyl and aryl organometallics results in the transfer of fluorine from nitrogen to carbon (equation 43). Yields are fair to

$$RSO_2NFR^1 + [R^2]^- \longrightarrow R^2F + [RSO_2NR^1]^- \tag{43}$$

good. These reagents are easily prepared by treatment of N-alkyl sulfonamides with elemental fluorine diluted in nitrogen (equation 44)[64]. Sulfonamides of amines possessing a

$$RSO_2NHR^1 \xrightarrow{F_2/N_2} RSO_2NFR^1 \tag{44}$$

R = p-Tol; R^1 = Me, t-Bu, exo-2-norbornyl, $endo$-2-norbornyl,
cyclohexyl, neopentyl
R = n-Bu; R^1 = neopentyl

primary alkyl substituent afford higher yields of the fluorosulfonamides than do amines with secondary- or tertiary-alkyl groups. This difference arises from the competing fluorination of the initially formed fluoro sulfonamide (equation 45). The synthesis and appli-

$$RSO_2NFR^1 \xrightarrow{F_2/N_2} RSO_2F + F_2NR^1 \tag{45}$$

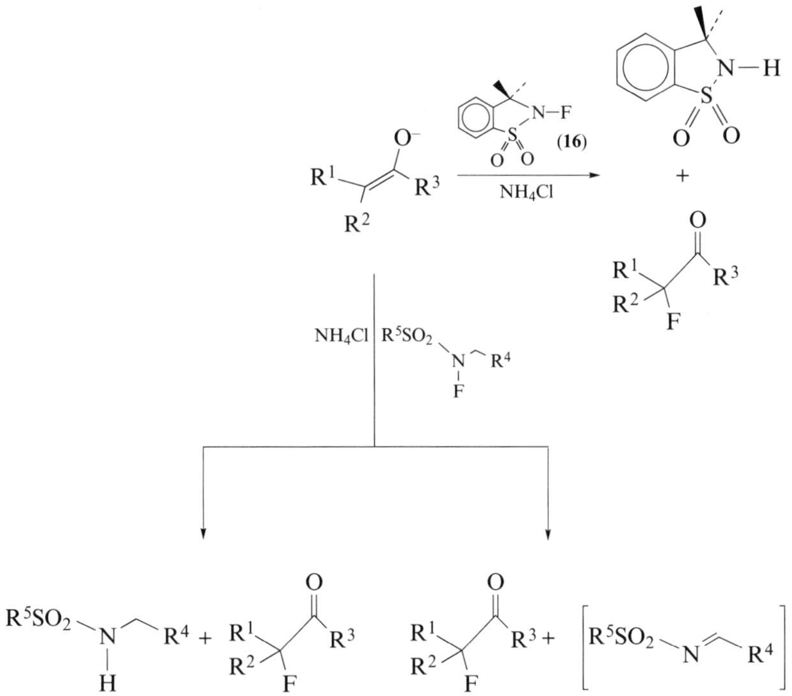

cation of a saccharin-derived N-fluoro sultam (16) has been described by Differding and Lang[65]. In a comparative study with commercially available N-fluoro sulfonamides, 16 was found to be a superior fluorinating reagent for the preparation of α-fluorocarbonyl compounds. Its superiority is ascribed to the absence of H atoms α to N, which prevents base-induced HF elimination. Differding and Lang provided the scheme in equation 46 for the use of **16**. Differding and coworkers subsequently reported[66] that *selective* mono- and difluorination of enolates can be performed in a one-pot procedure using N—F sultam **16** (equation 47). The yields and selectivity strongly depend on the counterion of the enolate

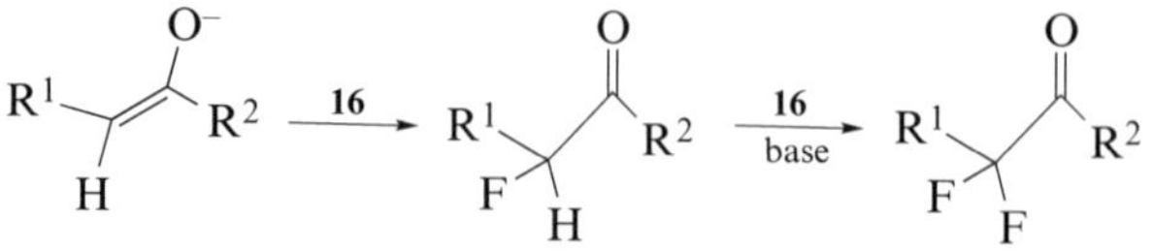

$$(47)$$

and the acidity of the α proton to the carbonyl group. Differding and Lang[67] also described the crystalline, optically pure *N*-fluoro sultams [(−)-**17** and (+)-**18**] derived from camphor as the first examples of enantioselective fluorinating reagents. Unprecedented enantiomeric excesses up to 70% were observed when various prochiral metal enolates were fluorinated with these new reagents, two of which are illustrated in equation 48.

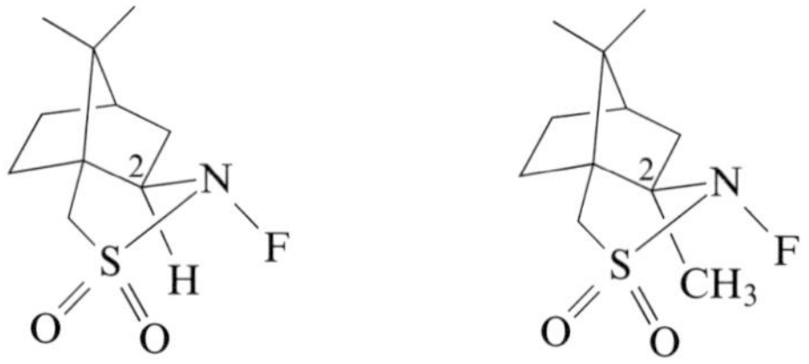

[(−)-**17**]   [(+)-**18**]

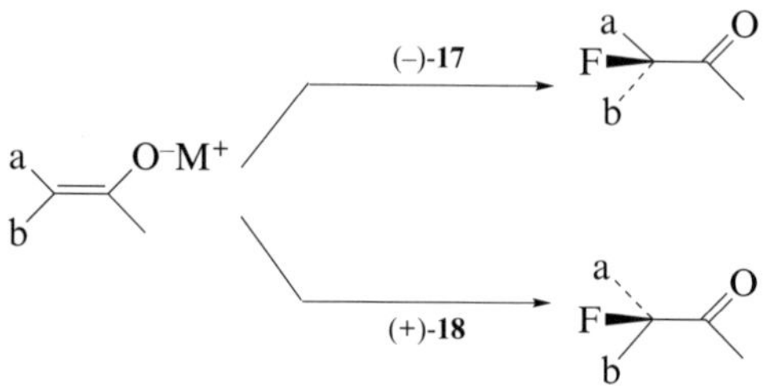

$$(48)$$

C. *N*-Fluoro Sulfonimides

N-Fluoro(perfluoroalkane)sulfonimides constitute a new class of electrophilic fluorinating agents[68,69]. They are effective in the fluorination of aromatics and carbanions and in the α fluorination of functionalized carbonyl compounds. In particular, *N*-fluoro(trifluoromethane)sulfonimide (**19**)[70] is an attractive fluorinating reagent. It has excellent long-term stability, favorable physical properties and high reactivity. It reacts with many types of alkenes in solvents of varying nucleophilicity to yield products whose formation can be rationalized on the basis of fluoro carbocation intermediates, according to DesMarteau and coworkers[70]. They postulate in a preferred mechanism an electron-transfer step leading to fluoro carbocation intermediates, illustrated in equation 49. The

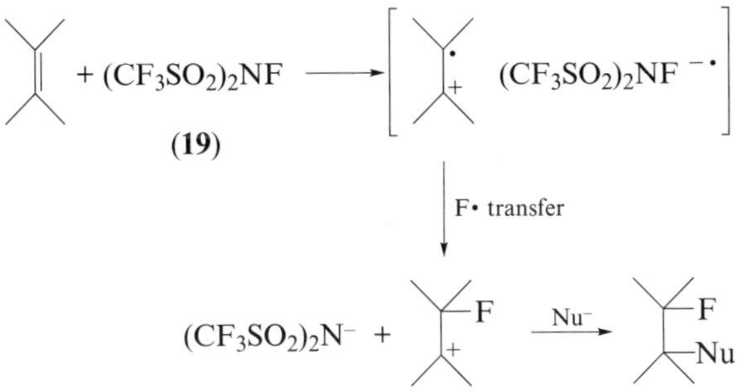

$$\text{(19)} \quad + (CF_3SO_2)_2NF \longrightarrow \left[\overset{+}{\diagup} \quad (CF_3SO_2)_2NF^{-\bullet} \right]$$

$$\downarrow \text{F}\bullet \text{ transfer}$$

$$(CF_3SO_2)_2N^- + \overset{+}{\diagup}\!\!-F \xrightarrow{Nu^-} \diagup\!\!\begin{array}{l}-F\\-Nu\end{array} \qquad (49)$$

electron-transfer step to provide a transient cation radical/anion radical parallels the RARP mechanism of electron-transfer halogenation of carbanions with perhaloalkanes described by Meyers and coworkers[7-10]. DesMarteau and coworkers noted that ultimate proof of their proposed mechanism must await detection and identification of the intermediates. Compound 19 has been used for the fluorination of 1,3,5(10)-estratrien-3-ols[71], as a very selective fluorinating agent for different aromatic compounds such as toluene, phenol, m-cresol, p-cresol, m-xylene and naphthalene[68] and in the preparation in good yields of α-fluoro derivatives of carbonyl and β-dicarbonyl compounds from lithium enolate esters, amides and ketones. The β-diesters, β-diamides, β-ketoesters and β-diketones have been fluorinated directly or as the metal enolates[72].

N-Fluoro-o-benzenesulfonimide (21) is readily prepared in greater than 90% yield by treatment of o-benzenesulfonimide (20) with 10% molecular fluorine (equation 50)[73].

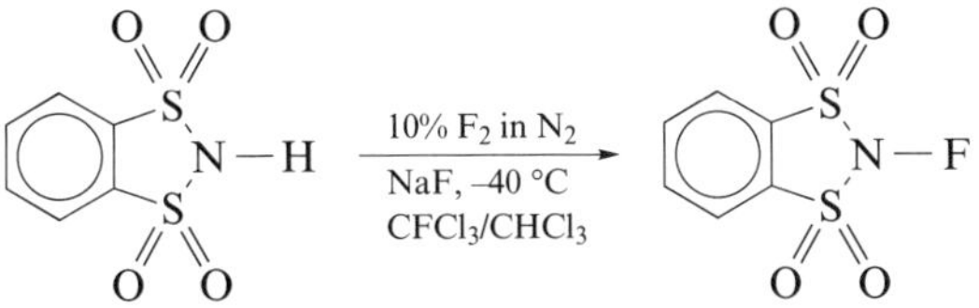

$$\text{(20)} \qquad \xrightarrow[\substack{\text{NaF, }-40\,°C\\ \text{CFCl}_3/\text{CHCl}_3}]{10\% \text{ F}_2 \text{ in N}_2} \qquad \text{(21)} \qquad (50)$$

Compound 21 is an efficient source of electrophilic fluorine which fluorinates enolates, azaenolates and carbanions in good to excellent yields.

N-Fluorobenzenesulfonimide [(PhSO₂)₂NF], an electrophilic fluorinating agent prepared from benzenesulfonimide in one step, allows the fluorination of neutral and carbanionic nucleophiles ranging from slightly activated aromatic compounds to highly reactive aryl and vinyl lithium derivatives in good yield[74]. This reagent has the advantage over the fluoro sultam prepared by the same authors[65,66] in being more reactive towards less reactive nucleophiles such as enol ethers and aromatics. α-Fluoro- and α,α-difluorophosphonates also can be prepared in good yield from the corresponding phosphonates through the use of this reagent (equation 51)[75].

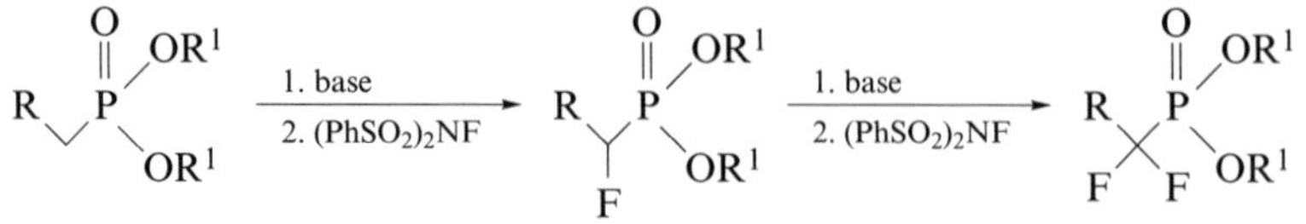

$$\text{(51)}$$

D. Halogen Azides and Iodine Isocyanate

1. Addition of XN₃

March[76] has provided a valuable summary of the work on additions of halogen azides to olefinic double bonds to provide β-halo azides (e.g. equation 52); the summary is based

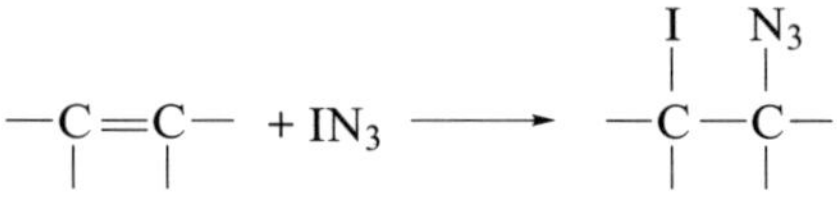

$$\text{(52)}$$

on references generally between 1968 and 1990. The addition is stereospecific and *anti*, suggesting that the mechanism involves a cyclic iodonium ion intermediate. Similar reactions can be performed with BrN_3 and ClN_3. The reaction has been carried out with many categories of compounds containing double bonds, including allenes and α,β-unsaturated ketones. 1,4-Addition has been found to occur with acyclic conjugated dienes. With ClN_3 the additions are mainly free radical, while with BrN_3 both electrophilic and free-radical mechanisms are important. IN_3 adds to triple bonds to give β-iodo-α,β-unsaturated azides.

The functionalization of ring A in diazatetracyclic analogs of anthracyclinones, potential anticancer reagents, via electrophilic reactions with positive-halogen sources has been investigated by Gómez-Contreras and coworkers[77]. Bromine azide, iodine azide, NBS/H₂O/DMSO and NBS/EtOH were the reactants of choice. These reactions are illustrated in equation 53. Napolitano and Fiaschi[78] have utilized the reaction of iodine azide

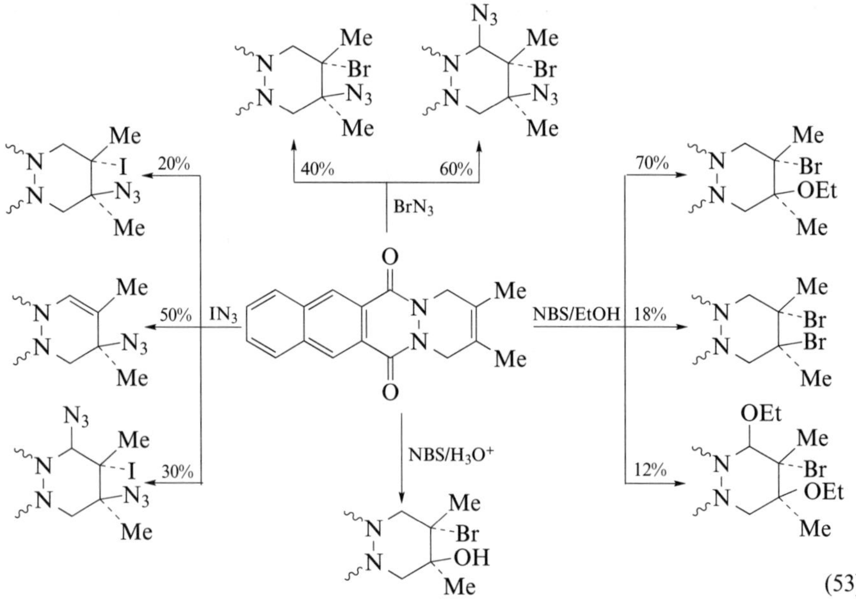

$$\text{(53)}$$

with 2-alken-1-ols, followed by mild treatment with base to effect dehydroiodination, in the stereospecific synthesis of 3-azido-1,2-epoxides. These results are consistent with a stereoselective *anti* addition of iodine azide to the double bond. It should be noted that Hassner and coworkers[79] previously showed that the addition products of iodine azide to several 1-arylcyclohexenes are 1-azido-2-iodo-1-arylcyclohexanes, *not* 2-azido-1-iodo-1-arylcyclohexanes. Further investigations into the regiochemistry and stereochemistry of the addition reactions of iodine azide were carried out by Sivasubramanian and coworkers using 2-aryl-2-cyclohexen-1-one oximes and were based on ^{1}H and ^{13}C NMR studies (equation 54)[80,81]. Attempted dehydroiodination of the adducts by treatment with base was not realized; instead, the starting oximes were regenerated.

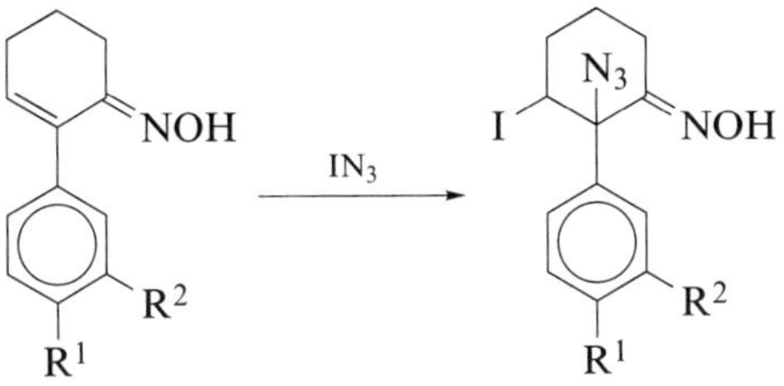

$$R^1 = R^2 = H, Me$$
$$R^1 = Me, t\text{-}Bu, OMe; R^2 = H$$

2. Addition of iodine isocyanate (IN=C=O)

March[82] has summerized the addition reactions of iodine isocyanate (IN=C=O) to C=C functions which were reported in the period 1963–1983. These reactions are similar to those with iodine azide described above. With IN=C=O, β-iodo isocyanates are formed (equation 55). As is the case with IN$_3$, the addition is stereospecific and *anti*. The

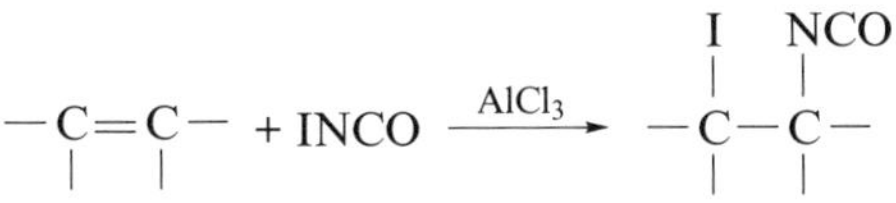

reaction has been applied successfully to mono-, di-, and some trisubstituted olefins. The orientation generally follows Markovnikov's rule: the positive iodine adds to the less-substituted carbon. Unlike the halo azides (*vide supra*), iodine isocyanate does *not* react with α,β-unsaturated carbonyl compounds. Allenes react to give β-iodo-β,γ-unsaturated isocyanates, and compounds containing the C≡C function provide low yields of β-iodo-α,β-unsaturated isocyanates.

E. *N*-Chlorobenzotriazole (BT-Cl)

As initially reported by Greci and coworkers[83] and later by Greci, Eberson and coworkers[84], 1-methyl-2-phenylindole (**22**) treated with *N*-chlorobenzotriazole (BT-Cl, **23**) is monochlorinated to **24** and also yields dimeric structures, some of which themselves are chlorinated and dichlorinated (equation 56). Product **27** was mentioned only in the more recent paper. The mechanism in equation 57 was suggested in the initial paper. The formation of products **25–28** was ascribed to the interaction between various substituted indole-radical cations and adventitious water in the medium. The later study confirmed that the substituted indole-radical cation **22a** is an intermediate and that the formation of

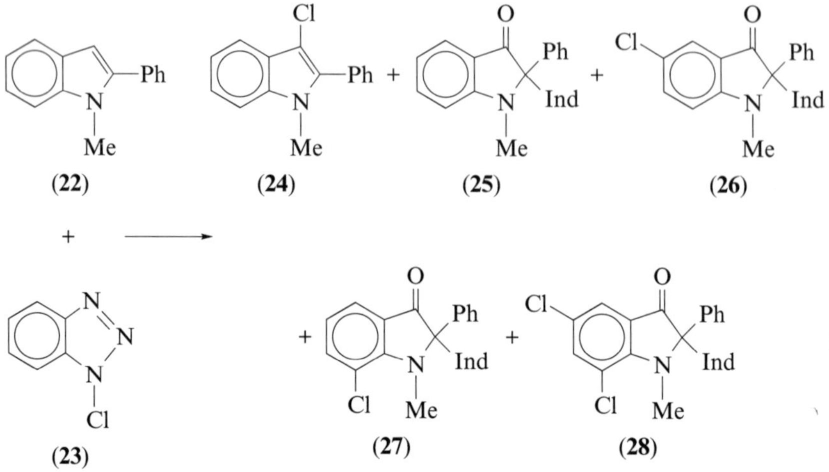

Ind = 1-methyl-2-phenylindol-3-yl

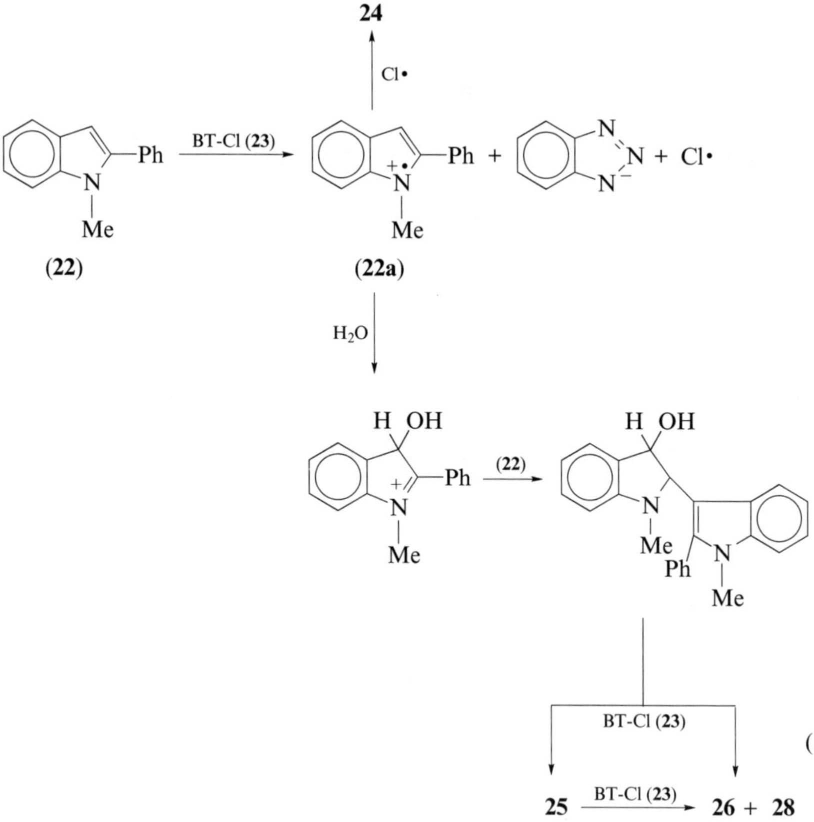

(56)

(57)

the indoxyls is due to the presence of dioxygen and oxygenated nucleophiles in the medium. However, while the report by Greci and coworkers[83] noted that benzotriazole *anion* and a chlorine *atom* (Cl·) are formed along with the indole-radical cation **22a** and are responsible for the formation of **24** as illustrated in equation 57, the subsequent report by Greci, Eberson and coworkers[84] reversed the order and stated that benzotriazole *radical* and chloride *anion* were formed along with the indole-radical cation **22a** (equation 58).

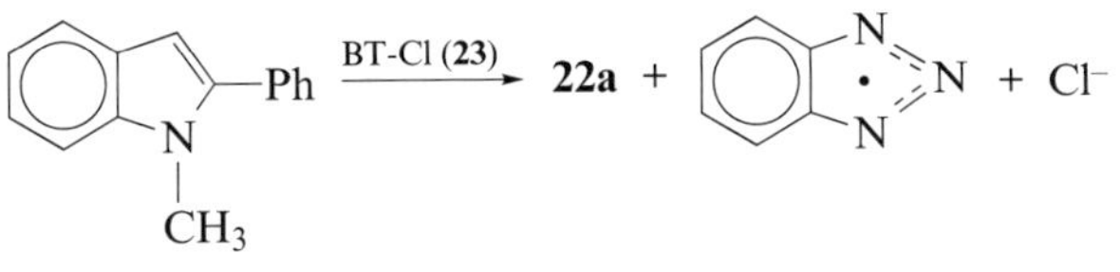

$$\text{(22)} \qquad \xrightarrow{\text{BT-Cl (23)}} \quad \textbf{22a} \; + \quad \underset{}{\text{benzotriazolyl radical}} \; + \; \text{Cl}^- \qquad (58)$$

(22)

According to Ryan and Harpp[85], the reaction of diallyl selenide (**29**) with *N*-chlorobenzotriazole (**23**) affords the first selenium-transfer reagent **30** as an analog of SeX_2 (equation 59). It was suggested that the mechanism of the conversion of **31** to **33** may proceed directly via a [2,3] sigmatropic rearrangement, or from **32** with displacement of the allyl group by chloride.

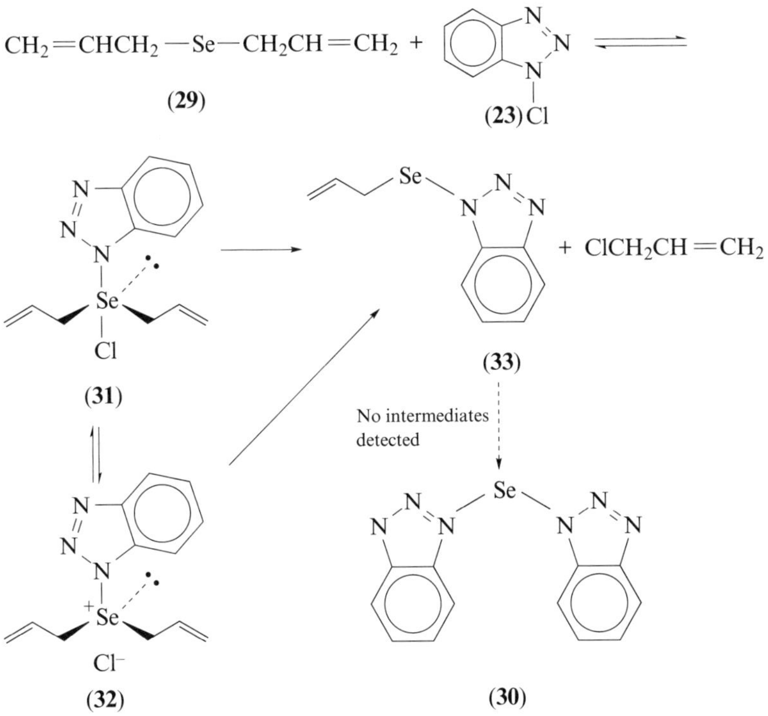

$$CH_2=CHCH_2-Se-CH_2CH=CH_2 \; + \qquad \text{(23)Cl} \qquad \rightleftharpoons$$

(29)

(31)

No intermediates detected

(32)

$$+ \; ClCH_2CH=CH_2$$

(33)

(30)   (59)

F. N-Halo Amines

Perfluoro-*N*-fluoropiperidine (**34**) acts as a site-selective electrophilic fluorinating agent towards carbanionic substrates, according to Banks and coworkers (equation 60)[86]. In

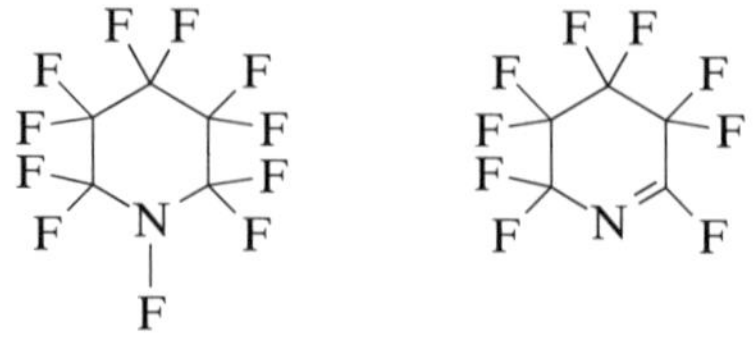

(34)   (35)

$$Me_2CNaNO_2 \xrightarrow{34} Me_2CFNO_2 + NaF + 35$$

$$PhMgBr \xrightarrow{34} PhF + MgBrF + 35$$

$$PhCNa(CO_2Et)_2 \xrightarrow{34} PhCF(CO_2Et)_2 + NaF + 35$$

$$EtCNa(CO_2Et)_2 \xrightarrow{34} EtCF(CO_2Et)_2 + NaF + 35$$

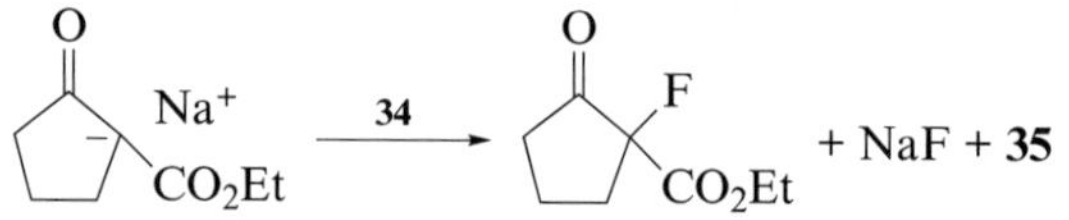

(60)

doing so **34** is converted into perfluoro-1-azacyclohex-1-ene (**35**), which itself is highly electrophilic. Thus, **35** competes with **34** for the carbanionic substrate as exemplified by the formation of 2-substituted octafluoro-1-azacyclohex-1-enes (not shown here) through an addition–elimination reaction.

Banks and coworkers refer to possible S_N2 or ET mechanisms. They emphasized that a feature common to both mechanisms is the generation of nitranion (**36**) and hence the release of the highly electrophilic perfluoro-1-azacyclohex-1-ene (**35**). Their schemes are illustrated in equation 61. However, we believe another mode of electron transfer should

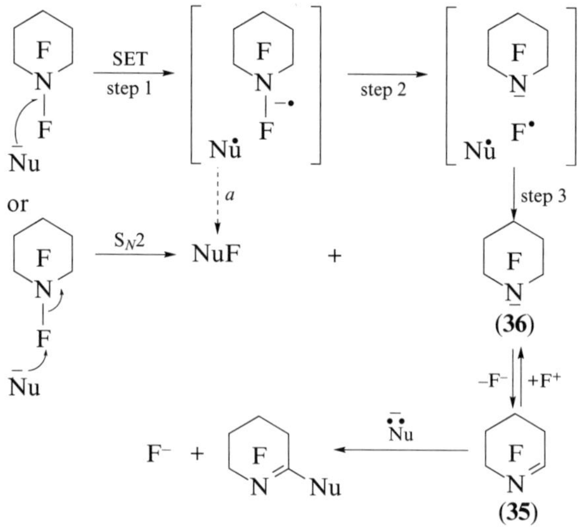

(61)

be considered in describing this type of fluorination reaction. That is, it is difficult to believe that an *intact intermediate*, as illustrated in the stepwise scheme of the Banks group, is generated in this case by electron transfer: (1) the fluorine atoms, unlike other halogens, cannot promote their electrons readily to vacant higher orbitals, and (2) there is no possibility of electron-resonance delocalization in this system. It is altogether possible, however, that a *transition state* is involved which closely resembles the RARP[7–9] inasmuch as in both cases an electron is transferred from the anion to the halogenating agent and a halogen atom is transferred, all in a single step. With all of the hydrogen atoms replaced by fluorine, the F—N bond of **34** must be very highly polarized, the electron density shifting *away* from this fluorine and towards the highly electronegative perfluorinated ring. We are suggesting the mechanism shown in equation 62 to explain this type of fluorination.

transition state

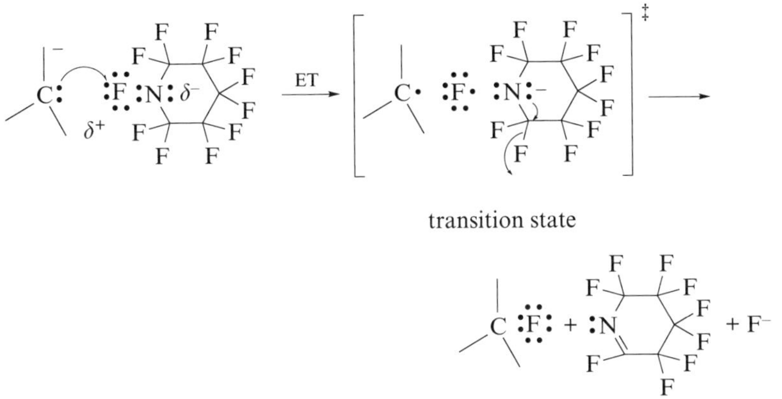

$$\tag{62}$$

Differding and coworkers[4–6] carried out a variety of investigations, including electrochemical studies, of these kinds of fluorination reactions and drew similar mechanistic conclusions.

Gowda and Sherigara[87] carried out a kinetic investigation of positive-halogen oxidation reactions of semicarbazide by chloramine-T (TsNClNa) and dichloramine-T (TsNCl$_2$), both of which are 'sources of positive-halogen species and hypohalites'. These reagents furnish different reactive species depending upon the pH of the reaction medium (e.g. RNHCl, RNCl$_2$, HOCl at low [H$^+$] and possibly RNH$_2$Cl$^+$ and H$_2$OCl$^+$ under highly acidic conditions). Their suggested mechanism of the oxidation of semicarbazide by the reactive species HOCl is shown in equation 63. Minisci and Platone and their coworkers[88] studied

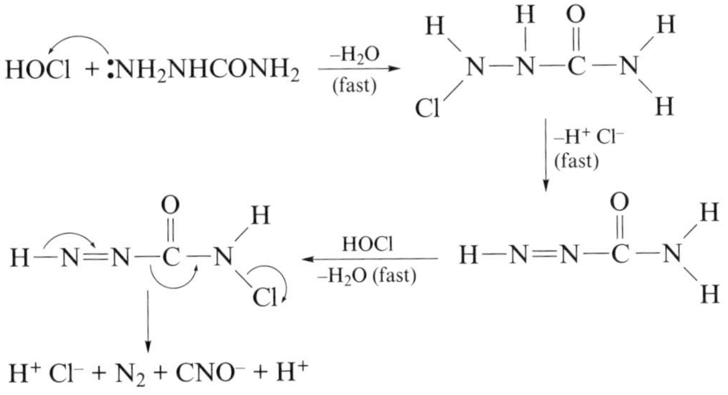

$$\tag{63}$$

the regioselectivity of chlorination and bromination of aromatic compounds with variously substituted *N*-halo amines. They found that with toluene the *para/ortho* halogenation ratio is enhanced considerably in going from simple *N*-halo amine reagents (e.g. dicyclohexyl, di-isobutyl, piperidino) to those with electron-withdrawing groups in their β positions [e.g. morpholino, piperazino, bis-(2-cyanoethyl)amino]. In contrast, steric effects emanating from the reagent seem to have little influence, namely *N*-chloromorpholine and 2,6-dimethyl-*N*-chloromorpholine provided almost the same *para/ortho* chlorination ratios. *m*-Xylene is converted to >97% 4-chloro-*m*-xylene by treatment with *N*-chloromorpholine, which is to be expected in view of the highly sterically hindered 2-position, low '*meta*' reactivity of the 5-position, and high '*ortho/para*' activity of the 4-(6-) position. According to the report, the low and high *para* selectivity might be explained by suggesting a different mechanism operating in each case. With the monoalkylbenzenes, the formation of molecular chlorine could be responsible for the low *para* selectivity (equation 64). These workers reported that '...the importance of

$$R_2\overset{+}{N}HCl + HCl \longrightarrow R_2\overset{+}{N}H_2 + Cl_2$$

$$ArH + Cl_2 \longrightarrow ArCl + HCl \qquad (64)$$

the polar characteristics of the aromatic substrate and of the *N*-chloro amine in determining the high *para* selectivity suggests an alternative interpretation: a charge-transfer complex could result from the initial interaction between the *N*-chlorodialkylammonium ion and the more reactive aromatic substrates. The high selectivity should be related to the collapse of the charge-transfer complex' (equation 65). The latter pathway, of course, is an

$$ArH + R_2\overset{+}{N}HCl \rightleftharpoons \left[\begin{array}{c} Ar\text{---}\overset{+}{H}\cdot R_2\overset{..}{N}H \ Cl\cdot \\ \text{or} \quad R_2\overset{+}{N}H \ Cl^- \end{array} \right] \longrightarrow ArCl + R_2\overset{+}{N}H_2 \qquad (65)$$

charge-transfer complex

electron-transfer mechanism of the RARP type, which we have suggested to explain a number of halogenations discussed here. While these authors suggested a charge-transfer complex apparently as a true intermediate, we believe a transition state is closer to reality in this case, i.e. that there is only one step (*vide supra*).

G. N-Halo Ammonium Salts

Lal[89a] has discussed the application of a new family of 'N—F'-type electrophilic-fluorination reagents consisting of 1-alkyl-4-fluoro-1,4-diazabicyclo[2.2.2]octane salts (**37**) whose preparation, patented by Banks[89b], utilizes the reaction of elemental fluorine

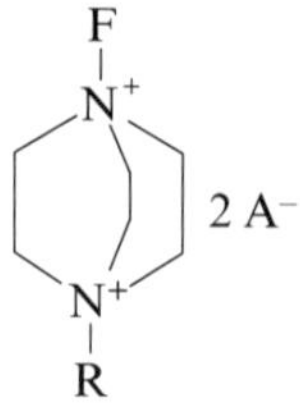

R = alkyl, CH$_2$Cl
A = BF$_4$, OTf

(37)

with 1-alkyl-1,4-diazabicyclo[2.2.2]octane salts. These salts have been found very effective for the fluorination of a wide variety of organic substrates, such as steroidal enol acetates and steroidal silyl enol ethers, phenyl-substituted olefins, sulfides bearing α–H atoms, certain carbanions and mildly activated aromatic compounds. The products were obtained in good yield and regioselectivity under very mild reaction conditions. The chemical reactivity can be modified by changing the alkyl substituent (R) in the reagent. It has been reported that an increase in the electronegativity of the quaternary group increases the ease of 'F$^+$' transfer[90]. However, Lal reported that no significant reactivity differences have been observed by variation of the counteranion (A$^-$), and he also reported that the compounds are soluble in fairly polar solvents such as CH_3CN or DMF because of their ionic character[89a]. Vinyl stannanes are electrophilically fluorinated with **37** (equation 66)[91].

$$R^1 \diagup SnBu_3 \diagdown F \quad + \quad \mathbf{37} \quad \xrightarrow{\quad} \quad R^1 \diagup F \diagdown F \qquad (66)$$
$$R^2 \qquad\qquad R = CH_2Cl \qquad R^2$$
$$A = BF_4$$

Similar reagents, N-fluoropyridinium triflate and its derivatives (**38**), which are stable, nonhygroscopic crystals, are widely applicable in mild and selective fluorinations of a

$$\begin{array}{c} R^3 \\ R^2 \diagup\diagdown R^4 \\ R^1 \diagdown N^+ \diagup R^5 \\ | \\ F \end{array} \quad {}^-OTf$$

(**a**) R^{1-5} = H
(**b**) $R^{1,3,5}$ = Me, $R^{2,4}$ = H
(**c**) $R^{1,3,5}$ = H, $R^{2,4}$ = Cl
(**d**) $R^{1,5}$ = COOMe, $R^{2,3,4}$ = H

(**38**)

variety of organic substrates including aromatics, enol ethers and acetates, highly enolized active methylene compounds, anions of active methylene compounds, and alkyl and aryl organometallics[92].

Concurrent iodination and 'functionalization' (with Nu) of alkynyl compounds has been carried out by Barluenga and coworkers with somewhat related reagents. They found that alkynes[93] and alkynyl sulfides[94] react with reagents consisting of mixtures of bis(pyridine)iodonium(I) tetrafluoroborate (**39**) and nucleophiles (Nu) such as $MeCO_2H$, Cl$^-$, Br$^-$, I$^-$ or pyridine to give 1-iodo-2-Nu-alkenes and 1-Nu-2-iodovinyl sulfides, respectively. The stereochemistry of these reactions is discussed in both reports. An ionic mechanism through a vinyliodonium ion is proposed to explain the regio- and stereochemical results. This same group reported that **39** and the associated nucleophiles react with 1-bromo-1-alkynes in a stereoselective manner to give 1-bromo-1-iodo-2-Nu-1-alkenes (equation 67)[95a].

$$R \text{—}\equiv\text{—} Br + I(Py)_2 \cdot BF_4 + NuH\,[Nu^-] \xrightarrow[\substack{20\,°C,\ 14–80\ h \\ 65–80\%}]{2\ HBF_4,\ solvent} \begin{array}{c} R \diagup I \\ Nu \diagdown Br \end{array} \qquad (67)$$

$$R = Ph;\ C_6H_{13} \qquad (\mathbf{39})$$

1-Iodo-1-alkynes undergo an unprecedented head-to-tail dimerization in the presence of **39**/HBF$_4$ in CH_2Cl_2 (equation 68)[95b]. In discussing the mechanism, the authors state that the formation of the dimer should require regeneration of an equivalent amount of 'I$^+$', and they propose the mechanism in equation 69. Aromatic compounds treated with **39** in the

$$2\ \text{Ar}-\text{C}\equiv\text{C}-\text{I} \quad \xrightarrow[\text{CH}_2\text{Cl}_2]{\substack{\text{cat. amt. of} \\ \textbf{39}/\text{HBF}_4}} \quad \begin{array}{c} \text{Ar}\diagdown \quad \diagup\text{I} \\ \text{C}=\text{C} \\ \diagup \quad \diagdown\text{I} \\ \text{C} \\ \!\!\!\!\!\!\!\!\parallel\!\!\!\!\!\!\!\! \text{C} \\ \diagup \\ \text{Ar} \end{array} \tag{68}$$

$$\text{Ar} = \text{C}_6\text{H}_5;\ 4\text{-Y-C}_6\text{H}_4\ \ (\text{Y} = \text{Me, Cl, OMe})$$

$$\text{Ar}-\text{C}\equiv\text{C}-\text{I} \xrightarrow[\text{catalyst}]{\textbf{39}/\text{HBF}_4} \begin{array}{c}\text{Ar}\diagdown \ \diagup\text{I}\\ \text{C}=\text{C}\\ \diagup \ \overset{+}{\diagdown}\\ \text{I}\end{array} \xrightarrow{\text{Ar}-\text{C}\equiv\text{C}-\text{I}}$$

$$\begin{array}{c}\text{Ar}\diagdown\quad\overset{\displaystyle\text{I}}{\underset{\displaystyle \,}{\text{C}}}\\ \text{C}=\text{C}\diagdown\text{I}\\ \diagup\\ \text{I}-\text{C}\diagdown\\ \quad\ \ \text{C}^+\\ \quad\quad\diagdown\text{Ar}\end{array} \xrightarrow{-\text{`I}^+\text{'}} \begin{array}{c}\text{Ar}\diagdown\ \diagup\text{I}\\ \text{C}=\text{C}\\ \diagup\ \ \diagdown\text{I}\\ \text{C}\\ \!\!\!\parallel\!\!\!\\ \text{C}\\ \diagup\\ \text{Ar}\end{array} \tag{69}$$

presence of HBF_4 or $\text{CF}_3\text{SO}_3\text{H}$ in CH_2Cl_2 at room temperature undergo monoiodination with excellent regioselectivity and yields (equation 70)[96]. The scope of the synthetic appli-

$$\text{ArH} + \textbf{39} \quad \xrightarrow[\text{CH}_2\text{Cl}_2]{\text{H}^+,\ \text{rt}} \quad \text{ArI} \tag{70}$$

$\text{ArH} = $ e.g. benzene, m-xylene, phenol, anisole H^+ is from HBF_4 or $\text{CF}_3\text{SO}_3\text{H}$

cation of **39** is broadened via a new methodology to selectively attach iodine and/or fluorine to an alkene (equation 71)[97].

$$\diagup\!\!=\!\!\diagdown\ +\ \textbf{39} \quad \xrightarrow[\substack{\text{CH}_2\text{Cl}_2\\ -60\,^\circ\text{C}}]{\text{HBF}_4} \quad \begin{array}{c}\qquad\ \text{F}\\ \,\diagup\!\!\overset{|}{\underset{|}{+\!\!+}}\!\!\diagdown\\ \text{I}\qquad\ \end{array} \tag{71}$$

H. Mechanistic Studies of Fluorination with Electrophilic Fluorinating Reagents

Differding and coworkers[4–6] have investigated the mechanism of electrophilic fluorination. His report with Rüegg[4] discusses what they term the 'considerable dispute' about the possible mechanism of electrophilc-fluorine transfer[98,99a]. They pointed out particularly the high enthalpy of F^+ in the gas phase (420 kcal mol^{-1}) and the identification of radical intermediates or radical derived products, which have been taken as arguments for a two-step pathway where electron transfer (ET) precedes fluorine transfer. Possible pathways in the reaction $\text{Nu}^- + \text{Y}-\text{F}$ (e.g. $\text{Y}-\text{F} = N$-fluoro sultams suggested by Differding and coworkers are illustrated in equation 72. They studied this problem by means of a 'radical clock' employing citronellic ester enolate[4]. When the latter was allowed to react with different electrophilic fluorinating reagents (with the exception of XeF_2) *only the open-chain* prod-

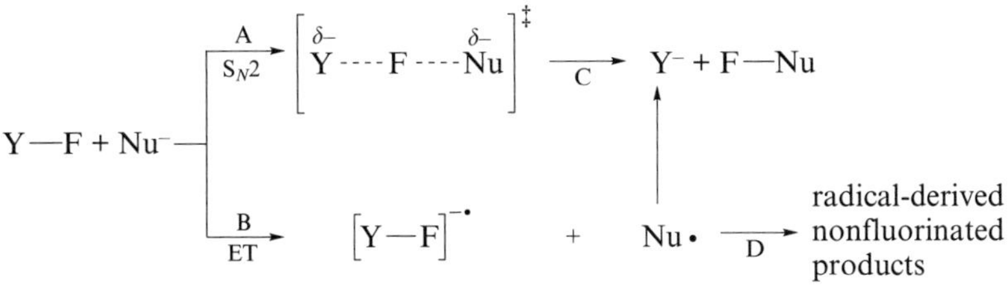

$$Y-F + Nu^- \begin{cases} \xrightarrow[S_N2]{A} \left[\overset{\delta-}{Y} \cdots F \cdots \overset{\delta-}{Nu} \right]^{\ddagger} \xrightarrow{C} Y^- + F-Nu \\ \\ \xrightarrow[ET]{B} \left[Y-F \right]^{-\bullet} + Nu\bullet \xrightarrow{D} \text{radical-derived nonfluorinated products} \end{cases} \quad (72)$$

$$\begin{array}{c} \xrightarrow{\substack{1.\ KHMDS,\ THF \\ -78\ °C}} \\ \xrightarrow{2.\ Y-F,\ -78\ °C-RT} \end{array}$$

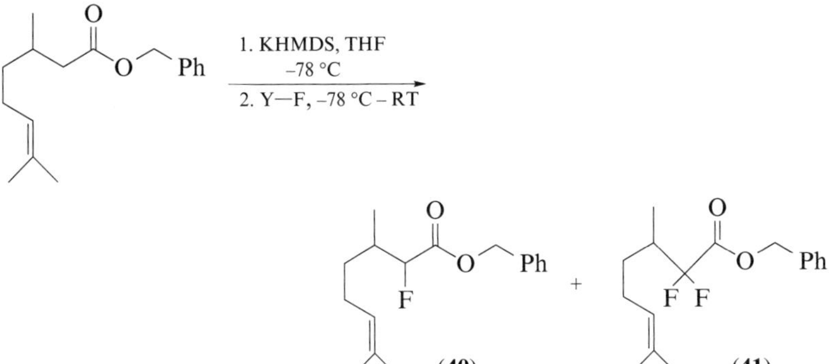

(40) + (41) (73a)

Y—F = fluorinating agent, e.g. N-fluoro sultams
KHMDS = potassium hexamethyldisilazide

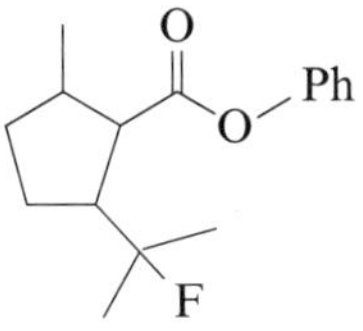

(42)

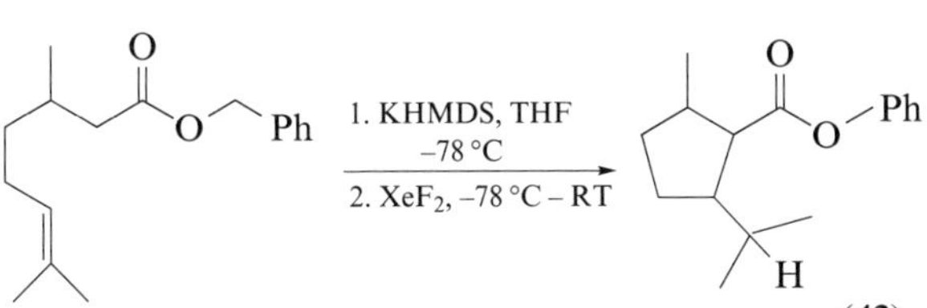

$$\xrightarrow{\substack{1.\ KHMDS,\ THF \\ -78\ °C}}_{2.\ XeF_2,\ -78\ °C-RT}$$

(43) + 40 + 41 (73b)

ucts (**40** and **41**) were found (equation 73a). Differding and Rüegg concluded that 'the absence of cyclic fluorinated products (e.g. **42**) is a proof that fluorination does not occur via free-radical intermediates'. Their reactions with XeF_2 (equation 73b) likewise produced *the cyclic nonfluorinated* compound **43** in addition to the open-chain fluorinated compounds **40** and **41**. The formation of **43** was explained by an ET pathway. Based on these results they suggested that the ET route does *not* lead to *cyclic fluorinated* products but is competitive with the ionic S_N2 route which *does* produce the open-chain fluorinated products: 'The absence of rearranged fluorinated products in this system, a potential precursor to a 5-hexenyl-type radical clock, indicates that free radicals are not intermediates on the path to fluorinated products'. They continue: 'Although these results exclude free radical intermediates, it is important to point out that they can not exclude ET followed by fast in cage recombination: they merely fix a lower limit to the rate constant with which a radical/radical-anion pair—if ever formed—would have to react inside the solvent cage by a fluorine radical transfer in order to escape diffusion (i.e. 10^{10} s^{-1}) and subsequent isomerization'.

In a joint publication Differding and Wehril conclude: 'A comparison between the observed rate constants of electrophilic fluorinations and the calculated rates for electron transfer gives for the first time kinetic proof that nucleophilic attack at fluorine has to occur in order to explain the high reaction rates. The low yields of fluorinated products under conditions where electron transfer becomes important are an indication that S$_N$2 and ET are competing in different pathways'[5]. They based these conclusions on their fluorination results which they say showed that 'direct nucleophilic attack at fluorine is the only pathway which can account for the observed rate constants and that under certain conditions ET is a competitive and different pathway in the reaction of a variety of nucleophiles with an N-fluoro sultam'. However, Differding does not wholly rule out an ET route in electrophilic fluorination. He and Bersier[6] investigated the cyclovoltammetric and polarographic behavior of compounds containing nitrogen–fluorine bonds, some which were electrophilic fluorinating agents. They found that the reduction potentials were strongly affected by the substituents on the nitrogen atom and by the electrode material, and reported that all the reductions were irreversible on the investigated time-scale. Their results demonstrated that compounds containing a nitrogen–fluorine bond activated by electron-withdrawing, highly electronegative substituents, including those in which the nitrogen atom itself carries a positive charge, possess a fairly high oxidizing power. They concluded that it should not be surprising, then, to see ET (fluorination) reactions with a variety of nucleophiles.

Earlier investigations pointing to ET mechanisms in reactions of nucleophiles should be consulted in this context. The studies of Meyers and coworkers[7–9] on radical anion–radical pair (RARP) transitions in halogenations with perhaloalkanes and nucleophilic (ionic) vs electron-transfer substitution reactions of trityl chloride with thiolates and other anions, and the studies of Chanon and coworkers[10,99b] which utilized radical-clock traps in efforts to monitor the intermediacy of free radicals in the ET pathways suggested by Meyers and colleagues.

IV. CARBON–HALOGEN COMPOUNDS

A. Polyhalo Alkanes

1. General

The chemistry of fluorinated perhalo alkanes containing chlorine and/or bromine atoms has been reviewed by Wakselman and Kaziz[100]. From the point of view of positive-halogen compounds and chemistry, they discuss the reactions of these perhalogenated alkanes with nucleophilic phosphorus, sulfur, selenium, oxygen, carbon and halogen compounds. The controversy among possible mechanisms, e.g. carbene-mediated ionic chain, radical-anion chain (S$_{RN}$1), is addressed but not solved: '...the orientation between the ionic-chain mechanism and the radical-chain process is not well understood, hence the need for further studies'. Iodofluorocarbons have been reviewed along similar lines[101,102].

Li and coworkers have published a series of reports describing their investigations of the reactions of perhalofluoro alkanes, e.g., XYZC—CFYZ (X = Cl, Br, I; Y, Z = F, Cl, Br,

CF_3), with different kinds of nucleophiles such as RO^-, RS^-, R_2N^- and R_3N, enamines, enolates and phosphorous ylides[103–108]. They reported that these compounds 'have been found to react spontaneously' affording decent preparations of perhalofluoroalkyl compounds like R_fOR, R_fSR, R_fNR_2. Although these workers reported that 'different and competitive pathways have been found in some other cases' they *say* that they have shown that most of their reactions proceed via an ionic process. For example, they note that, in contrast to earlier reports, they found that dibromoperfluoro alkanes actually reacted spontaneously with PhOK to yield fluoro alkyl phenyl ethers, and they conclude: 'Evidence for an anionic chain mechanism involving a bromophilic attack by the phenoxide is presented'.

While some of their mechanistic conclusions may seem to be reasonable, their concepts are narrow and not always realistic. Moreover, the absence of sufficient data leaves unequivocal product-structure characterization open to question in a number of instances, especially in the reactions with ArO^- and RO^- nucleophiles for which only ^{19}F NMR and mass spectral data were actually reported. More data undoubtedly were in their 'unpublished results' noted in at least eight references. Giving them the beneifit of the doubt as to product structure, we can suggest an electron-transfer mechanism as a more plausible initiation step, in contrast to their ionic-halophilic route.

In the example below, our suggested ET initiation is illustrated. Example 1 incorporates a RARP, step 1, in which a radical is sufficiently stable to separate from the RARP cage i.e. 'pathway b'[109]. This type of mechanism is well established in the reaction of BHT (2,6-

Example 1

$$RO^- + ClF_2C-CCl_3 \longrightarrow RO-CF_2CCl_3$$

R = *p*-tolyl, *t*-amyl   R = *p*-tolyl, 91%
                                 RO = *t*-amyl, 41%

1

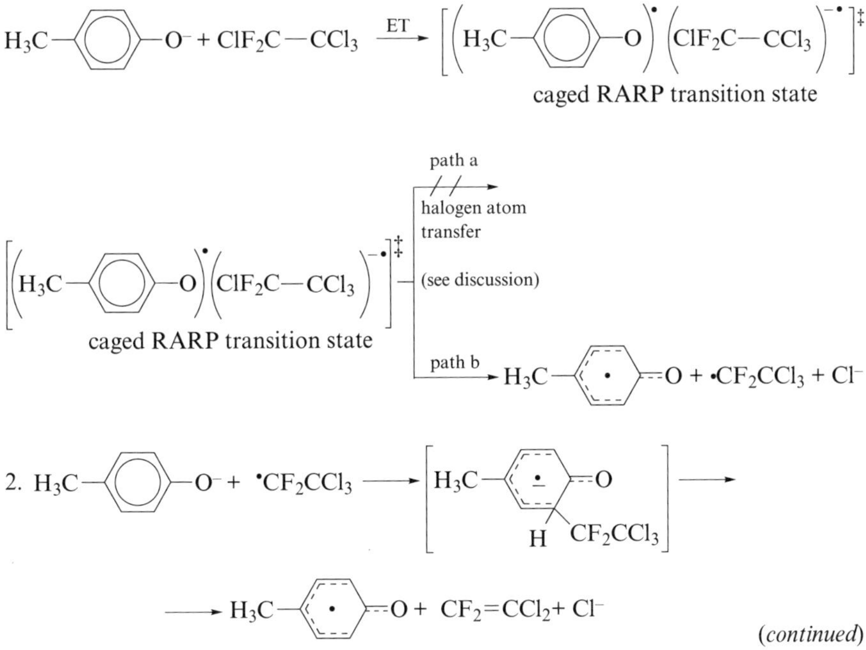

caged RARP transition state

path a
halogen atom transfer
(see discussion)

path b

2.

(continued)

Example 1 (*continued*)

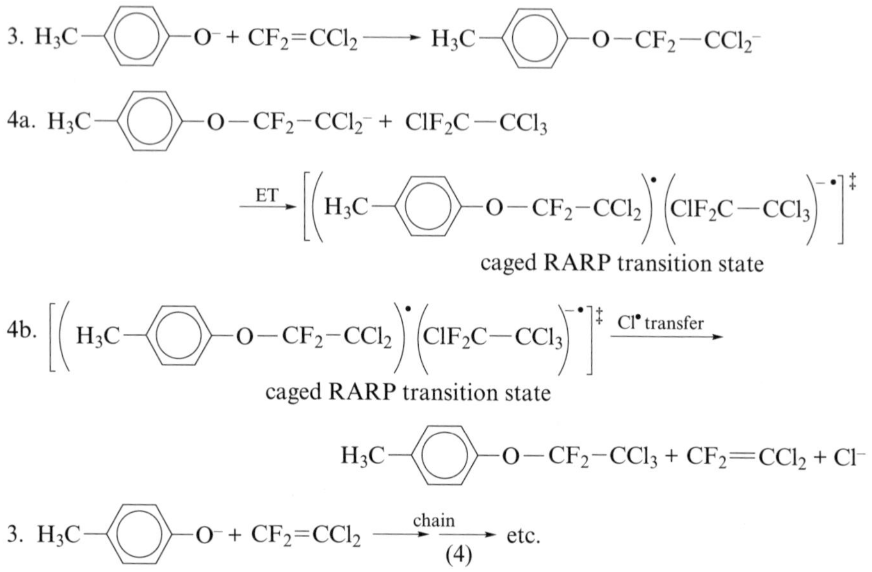

3. $H_3C-\langle\bigcirc\rangle-O^- + CF_2{=}CCl_2 \longrightarrow H_3C-\langle\bigcirc\rangle-O-CF_2-CCl_2^-$

4a. $H_3C-\langle\bigcirc\rangle-O-CF_2-CCl_2^- + ClF_2C-CCl_3$

$$\xrightarrow{ET} \left[\left(H_3C-\langle\bigcirc\rangle-O-CF_2-\overset{\bullet}{C}Cl_2\right)\left(ClF_2C-CCl_3\right)^{-\bullet}\right]^{\ddagger}$$

caged RARP transition state

4b. $\left[\left(H_3C-\langle\bigcirc\rangle-O-CF_2-\overset{\bullet}{C}Cl_2\right)\left(ClF_2C-CCl_3\right)^{-\bullet}\right]^{\ddagger} \xrightarrow{Cl^\bullet\ transfer}$

caged RARP transition state

$$H_3C-\langle\bigcirc\rangle-O-CF_2-CCl_3 + CF_2{=}CCl_2 + Cl^-$$

3. $H_3C-\langle\bigcirc\rangle-O^- + CF_2{=}CCl_2 \xrightarrow[(4)]{chain} etc.$

di-*t*-butyl-*p*-cresol) with $CBrCl_3$ in KOH/*t*-BuOH[110] and the Kornblum[12] and Bunnett[13] and Russell[11] series of electron-transfer-chain reactions. Chanon and coworkers recently have also investigated the possibility of a 'pathway b' mechanism in related reactions[10]. Step 2 illustrates the reactions of the free radical ($\cdot CF_2CCl_3$), formed in step 1, at an *ortho* position of substrate ArO^- to form a resonance-stabilized radical anion[12] which decomposes into $ArO\cdot$ and, via a concerted pathway, to $CF_2{=}CCl_2 + Cl^-$. Step 3 shows the simple nucleophilic addition of ArO^- to $CF_2{=}CCl_2$ to form the anionic intermediate whose negative charge is stabilized by the geminal chlorines (α-fluorines destabilize carbanions[111a]). This intermediate is then chlorinated (step 4) via electron-transfer RARP reaction by substrate ClF_2C-CCl_3 which provides the isolated product plus $CF_2{=}CCl_2$, the latter entering into step 3 again to continue the chain reaction.

This pathway is supported by related results. For example, clean halogenations of carbanions with perhaloethanes, reported by De Waard and coworkers[111b] for the chlorination of the α-carbanion of sulfolane with hexachloroethane, were demonstrated and worked out by Meyers' coworkers[112,113]. They have the advantage over perhalomethanes in that they do not form dihalocarbenes as byproducts which lead to contaminated mixtures. Li and coworkers themselves proved unequivocally that the addition reaction in step 3 readily occurs; when more $CF_2{=}CCl_2$ was added, their reaction was greatly accelerated.

Example 2

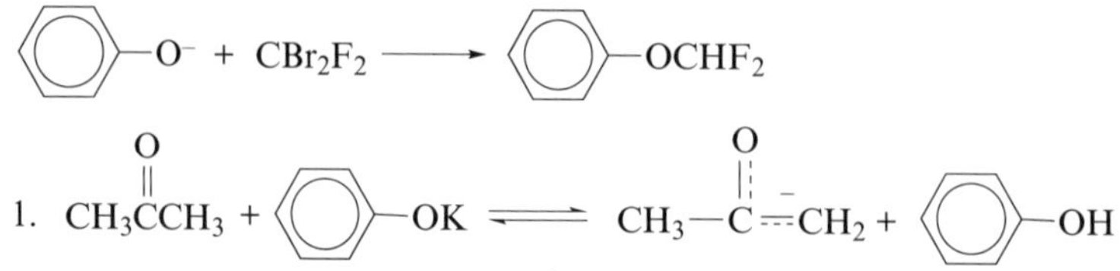

$\langle\bigcirc\rangle-O^- + CBr_2F_2 \longrightarrow \langle\bigcirc\rangle-OCHF_2$

1. $CH_3\overset{O}{\overset{\|}{C}}CH_3 + \langle\bigcirc\rangle-OK \rightleftharpoons CH_3-\overset{O}{\overset{\|}{C}}{=}CH_2^- + \langle\bigcirc\rangle-OH$

(*continued*)

Example 2 (*continued*)

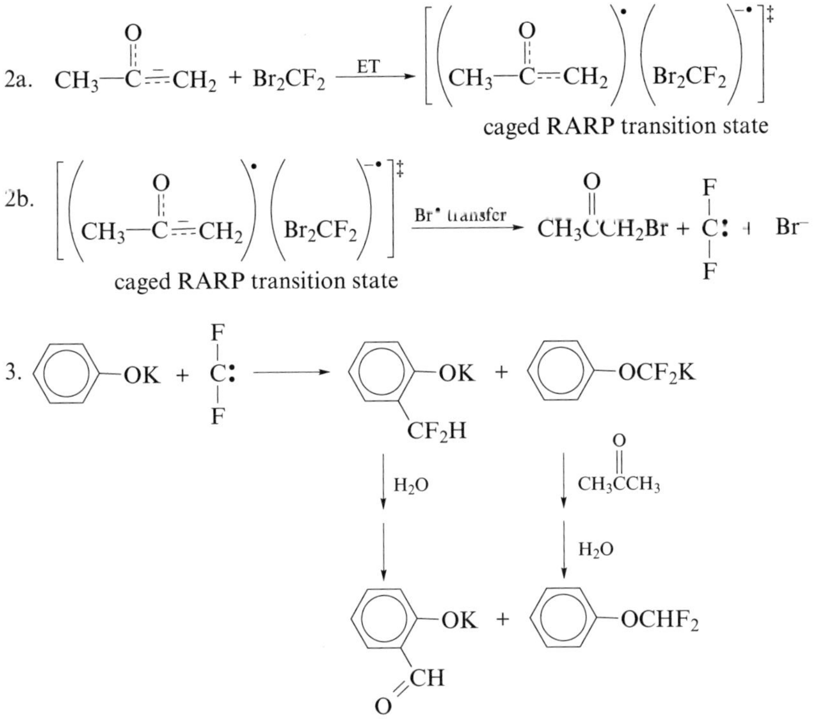

2a. $CH_3-\overset{\overset{\displaystyle O}{\|}}{C}=CH_2 + Br_2CF_2 \xrightarrow{ET} \left[\left(CH_3-\overset{\overset{\displaystyle O}{\|}}{C}=CH_2\right)^{\cdot}\left(Br_2CF_2\right)^{-\cdot}\right]^{\ddagger}$

caged RARP transition state

2b. $\left[\left(CH_3-\overset{\overset{\displaystyle O}{\|}}{C}=CH_2\right)^{\cdot}\left(Br_2CF_2\right)^{-\cdot}\right]^{\ddagger} \xrightarrow{Br^{\cdot} \text{ transfer}} CH_3\overset{\overset{\displaystyle O}{\|}}{C}CH_2Br + \overset{\overset{\displaystyle F}{|}}{\underset{\displaystyle F}{C}}\text{:} + Br^-$

caged RARP transition state

3.

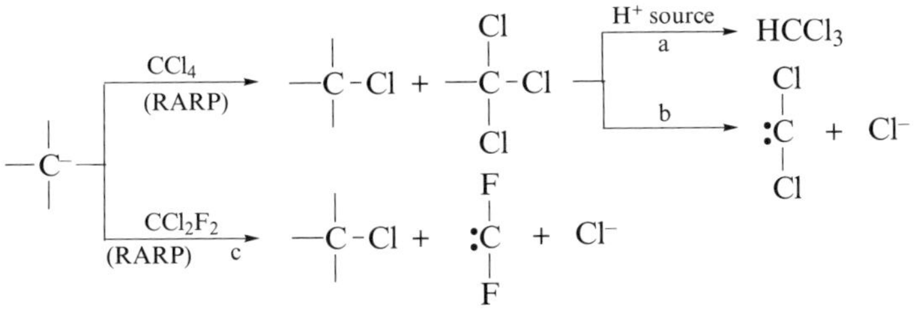

We also believe (Example 2) that the formation of phenyl difluoromethyl ether from the reaction of potassium phenoxide with dibromodifluoromethane as noted by Li and coworkers[104b] is a result of an electron-transfer RARP reaction. This reaction, reported by Clark and Simons[114], was carried out in acetone. It is well known that enolizable ketones are readily halogenated with perhalomethanes in the presence of base (steps 1 and 2)[7-9,115]. Furthermore, these reactions specifically with CBr_2F_2 have been studied thoroughly[116]. With perhalomethanes containing one or more fluorines, halogenation involves transfer of a halogen *other than* fluorine. Generally, residual trihalomethyl anion can (a) abstract a proton to form the haloform or (b) lose a halide to form the dihalocarbene. When the perhalomethane has two fluorines, no trihalomethyl anion is formed (the α-fluorine atoms destabilize carbanions[111a]), but difluorocarbene is formed directly in a concerted process (c)[116] (equation 74). This likewise is the case here (steps 2a,b). Step 3, the addition of phenoxide anion to difluorocarbene, is reminiscent of the reaction of chloroform with phenols in base to form 2- and 4-hydroxybenzaldehydes via the intermediacy of dichloro-carbene (i.e. Reimer–Tiemann reaction). While phenolic-O attack probably competes with C attack in this case too, the formation of dichloromethyl phenyl ether from O attack is reversible because of the high acidity promoted by the geminal chlorines. In contrast, gem-inal fluorines greatly reduce acidity, so the reverse reaction is unlikely, and the ether can be isolated.

During their preparation of fluorinated ethers $ArOCF_2CHFROR_f$ from the reactions of sodium phenoxides with perfluoroalkyl vinyl ethers, namely $CF_2=CFOR_f$ [$R_f = C_3F_7$, $C_3F_7OCF(CF_3)=CF_2$, and $MeO_2CCF_2CF_2OCF(CF_3)=CF_2$], Feiring and Wonchoba[117] found that brominated and chlorinated products were also formed when hexachloroethane and 1,2-dibromotetrafluoroethane, respectively, were present (equation 75). While they

$$\underset{F}{\overset{F}{>}}C=C\underset{OR_f}{\overset{F}{<}} + CCl_3CCl_3 + ArONa \xrightarrow{DMF} ArOCF_2CFClOR_f + Cl_2C=CCl_2 + NaCl$$

$$\underset{F}{\overset{F}{>}}C=C\underset{OC_3F_7}{\overset{F}{<}} + BrCF_2CF_2Br + ArONa \xrightarrow{DMF}$$

$$ArOCF_2CFBrOC_3F_7 + (F_2C=CF_2) + NaBr \quad (75)$$

observed the coformation of tetrachloroethylene in the first reaction, they presumed that tetrafluoroethylene was likewise formed in the second reaction although they neither detected it nor were able to detect the formation of an adduct of it. They concluded that the highly volatile gaseous tetrafluoroethylene (bp−78.5 °C) was simply lost. Feiring and Wonchoba referred to the formation of the chlorinated and brominated products as involving 'electrophilic attack by carbanion on halogen in highly halogenated molecules'. They continued: 'Hexachloroethane is an efficient reagent for transfer of Cl^+ since the tetra-chloroethylene byproduct is relatively inert'. They must have erred by referring to the attack by carbanion on halogen as 'electrophilic', regardless of mechanism. They found that 1,2-dibromotetrafluoroethane works efficiently for bromination of carbanions, although, noting earlier work[118], they said that the presumed byproduct in these reactions, tetrafluoroethylene, 'could also react with phenoxide'. It should be noted that halogena-tions of carbanions with perhaloethanes like those carried out by Feiring and Wonchoba were reported much earlier by De Waard and coworkers (1973)[111] and the mechanisms were likewise investigated some time ago by Meyers' students (1979[112] and 1986[113]) (*vide supra*).

Sasson and Webster[119] reported some examples of monoprotic carbon acids, for exam-ple alkylmalonic esters and phenylacetylene, that are directly halogenated in high yield by polyhalo alkanes under mild conditions when catalyzed by tetrabutylammonium fluoride (TBAF) (equation 76). They do not discuss the halogenation mechanism other than to say that the reaction is 'apparently taking place via nucleophilic attack on the tetra-halomethane' (equation 77). It is reasonable to believe that it is the carbanion of the sub-

$$RH + ZCX_3 \xrightarrow[\text{cat.}]{Bu_4N^+F^-} RX + ZCHX_2 \quad (76)$$

RH = carbon acid with one acidic hydrogen
X = halogen (chlorine, bromine or iodine)
Z = electron-withdrawing substituent, e.g. halogen, carboxylic ester or perhaloalkyl

$$CCl_4 + MeCH(CO_2Me)_2 \underset{cat.}{\overset{Bu_4N^+F^-}{\rightleftarrows}} CHCl_3 + MeCCl(CO_2Me)_2 \qquad (77)$$

strate (*vide infra*) that is involved in the halogenation step *per se*, and that this step proceeds via the RARP mechanism as described for related halogenation of carbanions with perhalomethanes[7-9]. However, Sasson and Webster postulated that the attacking nucleophile is the hydrogen-bonded complex of TBAF with the acidic substrate as shown:

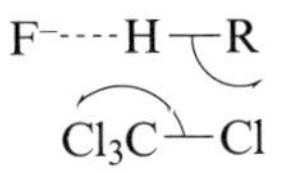

This apparently means that under their conditions (absence of solvent) the 'naked' F⁻ of the TBAF catalyst is a strong base, strong enough to abstract the acidic proton from the substrate to form its anion. Support for this mechanism is provided by several experiments. Chloroform was isolated as a product, and they showed that proton equilibration between Cl₃C⁻ and substrate esters occurs. Thus, it is almost certain that their mechanism is a chain reaction, initiated by proton abstraction by the catalyst, as they suggested, and propagated by the chlorination with CCl₄ and generation of Cl₃C⁻ which, directly or indirectly, continues the proton-abstraction step of the chain process. The propagation, like other RARP halogenation processes, is reversible (equation 78)[7-9]. Sasson and Webster reported that

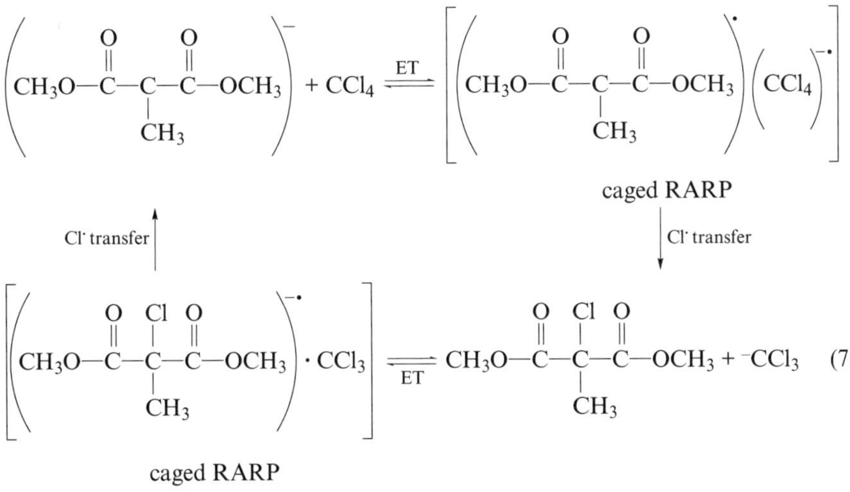

caged RARP

$$\qquad (78)$$

caged RARP

while they successfully chlorinated dimethyl methylmalonate [$pK_a(H_2O)$ *ca* 13][120] and phenylacetylene [$pK_a(H_2O)25$][121] in very high yields, they failed in their attempts to halogenate 2-nitropropane [$pK_a(H_2O)$ 7.67][120] and methyl isobutyrate. We can suggest an explanation for the failure in the case of 2-nitropropane: The expected product, 2-chloro-2-nitropropane, would be a better chlorinating agent than CCl₄ inasmuch as the aci-nitronate anion is a far more stable and better leaving group than is trichloromethyl anion. The thermodynamic consequence would favor the 'reisolation' of 2-nitropropane and CCl₄ in such a reversible process (equation 79). The failure of Sasson and Webster to chlorinate methyl isobutyrate by their method is more difficult to understand. We could not find a reported pK_a value for this ester. If it resembles methyl acetate, its $pK_a(H_2O)$

$$\left(\begin{array}{c} H_3C \\ \diagdown \\ H_3C \end{array}C{-}NO_2\right)^{-} + CCl_4 \xrightleftharpoons{ET} \left[\left(\begin{array}{c} H_3C \\ \diagdown \\ H_3C \end{array}C{-}NO_2\right)^{\cdot} (CCl_4)^{-\cdot}\right]^{\ddagger}$$

caged RARP

Cl⁻ transfer ↑   ↓ Cl⁻ transfer

$$\left[\left(\begin{array}{c} H_3C \quad Cl \\ \diagdown\diagup \\ C{-}NO_2 \\ H_3C \end{array}\right)^{-\cdot} \cdot CCl_3\right]^{\ddagger} \xrightleftharpoons{ET} \begin{array}{c} H_3C \quad Cl \\ \diagdown\diagup \\ C{-}NO_2 \quad + {}^-CCl_3 \\ H_3C \end{array} \qquad (79)$$

caged RARP

would be close to 24^{120} and thus would be similar to that of phenylacetylene which *was* chlorinated. The failure might be associated with steric hindrance. However, isobutyrophenone, whose enolate anion is sterically hindered like that of methyl isobutyrate, is easily chlorinated with CCl_4 in powdered KOH/t-BuOH via the RARP mechanism[7–9,115]. It is likely that in the process of Sasson and Webster, which we view as a chain process (*vide supra*), Cl_3C^- is generated and, as n-Bu$_4$N$^+$ $^-CCl_3$, it is the proton acceptor that propagates the chain. Abstraction of the sterically hindered α-H of methyl isobutyrate by sterically encumbered n-Bu$_4$N$^+$ $^-CCl_3$ undoubtedly is a very unfavorable and slow process. Instead, the $^-CCl_3$ decomposes into Cl$^-$ and dichlorocarbene, thereby precluding the required chain-propagating proton-abstraction step. If this is the cause of the failure, chlorination of this ester with CCl_4 in KOH/t-BuOH should *not* be inhibited, since the base is always in excess, and the reaction should proceed readily under these conditions as it does with isobutyrophenone. It should be added that although proton abstraction of the α-H of dimethyl methylmalonate by n-Bu$_4$N$^+$ $^-CCl_3$ should also be sterically encumbered, Sasson and Webster found that this compound was easily chlorinated under their conditions. We suggest that it is the *enolic* proton that is abstracted from this β-diester since a fair quantity of this enol is present at equilibrium and the enolic proton is unhindered and is the more acidic[120,121]. The chain reaction is, therefore, not precluded here.

2. Reaction with phosphorous compounds

The reaction of triethyl phosphite with CCl_4 has been studied from a mechanistic point of view by Chanon and coworkers using a variety of techniques from radical clocks to electrochemical methods[99b]. They isolated diethyl trichloromethanephosphonate and chloroethane as major products and diethyl chlorophosphate and 1,1,1-trichloropropane as minor products (equation 80). Under certain circumstances they also observed radical

$$(EtO)_3P + CCl_4 \longrightarrow (EtO)_2\overset{\overset{\displaystyle O}{\|}}{P}CCl_3 + EtCl + (EtO)_2\overset{\overset{\displaystyle O}{\|}}{P}Cl + EtCCl_3 \qquad (80)$$

intermediates. They proposed a variety of different mechanisms to explain the formation of the observed products, namely nucleophilic attacks on Cl, free-radical reactions, RARP pathways, $S_{RN}1$ reactions, etc. Chanon and coworkers concluded that: '...the thermal reaction of triethyl phosphite with CCl_4 which displays some experimental earmarks of an $S_{RN}1$ reaction is in fact an S_NCl^+ substitution'. They continue: 'The radical intermediates observed in the medium would mainly result from an E.T. between Cl_3C^- and CCl_4. Depending on the relative concentration of other reagents this $\cdot CCl_3$ (radical) may or may not initiate a radical chain reaction whose final products are the same as those obtained in the ionic mechanism. Under usual conditions, the ionic channel is the main one but when special conditions are applied (hv, AIBN) the radical pathway participation increases'.

Phosphorous ylides **44** react with carbon tetrahalides (CCl_4, CBr_4) by the replacement of the hydrogen atoms on the α-carbon with halogen atoms to form halo ylides **45** and **46**, according to Kolodyazhnyi and Golokhov[122] (equation 81). They stated that 'halogena-

$$R_2^1(R^2)P{=}CHR^3 \xrightarrow[-CHX_3]{CX_4} R_2^1(R^2)P{=}C(R^3)X$$

$$\textbf{(44)} \qquad\qquad\qquad\qquad \textbf{(45)}$$

$R^1 = Et_2N$ (**45a, 46a, b**), $C_5H_{10}N$ (**46c**); $R^2 = F$ (**45a, 46c**),
  Bu (**46b**), $R^3 =$ Me (**45a**);
$X = Cl$ (**45a, 46b, c**), Br (**46a**).

$$R_2^1(R^2)P{=}C(R^3)X \xrightarrow[-CHX_3]{CX_4} R_2^1(R^2)P{=}CX_2 \qquad (81)$$

$$\textbf{(45)} \qquad\qquad\qquad\qquad \textbf{(46a–c)}$$

$$R^3 = H$$

tion probably proceeds through halogenophilic attack by the negatively charged ylide carbon atom on CX_4 with the formation of intermediate **47** from which haloform is readily eliminated' (equation 82). While Kolodyazhnyi and Golokhov apparently use the term

$$R_2^1(R^2)\overset{+}{P}{-}\overset{-}{C}HR^3 + X{-}CX_3 \longrightarrow [R_2^1(R^2)\overset{+}{P}CH(R^3)X]\,^-CX_3 \xrightarrow{-CHX_3} \textbf{45} \quad (82)$$

$$\textbf{(47)}$$

'halogenophilic attack' to explain the halogenation mechanism, it simply says 'what' and not 'how'. We believe the halogenation pathway of phosphorous ylides with CX_4 parallels that of carbanions in that a caged RARP mechanism is responsible[7–9]. In the case of the phosphorous ylides, the halogen transfer *and* the proton transfer probably take place in the caged RARP transition state, as illustrated in equation 83.

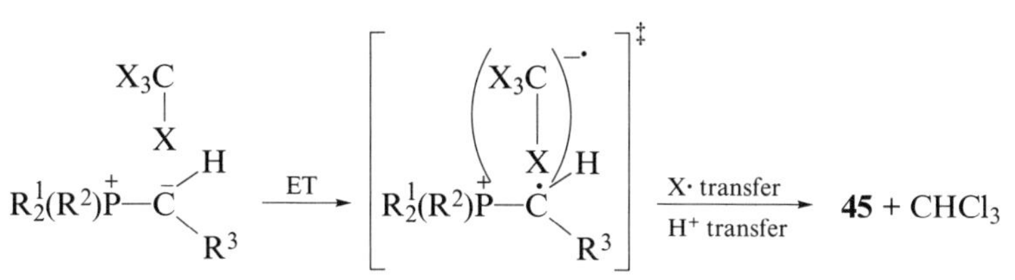

caged RARP transition state

Along similar lines, Pinchuk and collaborators[123] reported that the system formed by the reaction of a phosphorous triamide or trialkylphosphine with CCl_4, which is known to provide a source of Cl_3C^{-}[124], can be successfully used for the 'nucleophilic trichloromethylation' of silicon chlorides[125], mono- and disubstituted derivatives of phosphorous trichloride[126,127], carboxylic acid chlorides[128a] and for the formation of N'''-(trihaloacetyl)-hexaalkyl phosphorimidic triamides[128b]. Some examples are shown in equations 84–87.

$$(R_2N)_2PCl + (R_2N)_3P + CCl_4 \longrightarrow (R_2N)_2PCCl_3 + (R_2N)_3\overset{+}{P}\!-\!Cl \; Cl^- \quad (84)^{126}$$
$$R = Et, \textit{i-}Pr$$

$$RPCl_2 + (Et_2N)_3P + CCl_4 \longrightarrow RP(Cl)CCl_3 + (Et_2N)_3\overset{+}{P}\!-\!Cl \; Cl^- \quad (85)^{127}$$

$$R = Me_2N, \; Et_2N, \; \textit{i-}Pr_2N, \; \textit{t-}Bu_2N, \quad \text{(ring)} \quad N, \; \textit{t-}Bu, \; MeO, \; \textit{t-}BuO$$

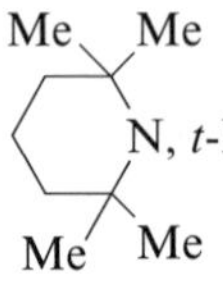

$$(\textit{t-}Bu_3)P + CCl_4 \longrightarrow (\textit{t-}Bu)_3\overset{+}{P}\!-\!Cl \; CCl_3^- \xrightarrow[-(\textit{t-}Bu)_3PCl_2]{RC(O)Cl} R\!-\!\overset{\displaystyle O}{\overset{\|}{C}}\!-\!CCl_3 \xrightarrow{(\textit{t-}Bu)_3P}$$
$$R = Me, \; Et, \; Ph$$

$$(\textit{t-}Bu)_3\overset{+}{P}\!-\!Cl \; Cl_2\overset{-}{C}\!-\!\overset{\displaystyle O}{\overset{\|}{C}}\!-\!R \xrightarrow[-(\textit{t-}Bu)_3PCl_2]{RC(O)Cl} RC(O)OC(R)\!=\!CCl_2 \quad (86)^{128a}$$

$$2(R_2N)_3P + 2CCl_4 \xrightarrow[-(R_2N)_3PCl_2]{Me_3SiNCO} (R_2N)_3P\!=\!NC(O)CCl_3 + Me_3SiCCl_3 \quad (87)^{128b}$$
$$R = Me, \; Et$$

These researchers also stated that 'Other polychloroalkanes containing a 'positivized' chlorine atom should react in the same way'. They therefore investigated the reactions of 1,1,1-trichloro-2,2,2-trifluoroethane (equation 88). They concluded that the reaction

$$(R_2N)_3P + Cl_3C\!-\!CF_3 + (R_2N)_2PCl \longrightarrow (R_2N)_2PCCl_2CF_3 + (R_2N)_3PCl_2 \quad (88)^{123}$$

$$R = Me, \; Et, \; Pr$$

mechanisms are similar in all of their cases (with CCl_4 and CCl_3CF_3) and that the first step is a 'halogenophilic attack' which they illustrated in equation 89. However, this terminol-

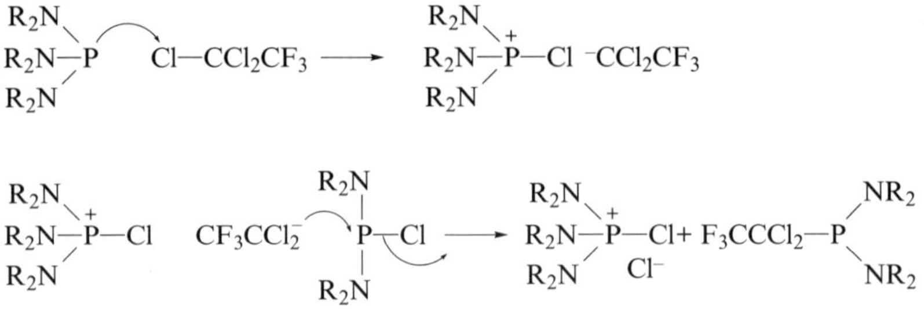

$$\begin{array}{ccc} R_2N & & R_2N \\ R_2N\!-\!P & Cl\!-\!CCl_2CF_3 \longrightarrow & R_2N\!-\!\overset{+}{P}\!-\!Cl \; ^-CCl_2CF_3 \\ R_2N & & R_2N \end{array}$$

$$R_2N\!-\!\overset{+}{P}\!-\!Cl \quad CF_3CCl_2^- \quad P\!-\!Cl \longrightarrow R_2N\!-\!\overset{+}{P}\!-\!Cl + F_3CCCl_2\!-\!P \quad (89)$$

ogy does not describe a mechanism but only the fact that a nucleophilic substrate is chlorinated by electrophilic chlorine. As we have suggested many times, the mechanism most likely involves electron transfer via a caged RARP transition state (*vide supra*).

Pinchuk and coworkers also investigated reactions of diamido halophosphites (R_2PX; $R = R_2N$, alkyl, aryl; $X = Cl$ or F) with CCl_4 which they referred to as 'reaction of nucleophilic polyhalogenalkylation in two- and three-component systems'[129]. They stated that when their studies were begun it was believed that 'interaction of $(R_2N)_2PX$ (X = halogen) with CCl_4 can not be possible because of the low nucleophilicity of (the) phosphorus'. But they 'determined for the first time that diamido chlorophosphites react with CCl_4 at 20 °C'. It is difficult to ascertain from this report the scope of compounds actually used in their investigation. An example of their reactions and products is illustrated in equation 90. Reactions of other less nucleophilic phosphine derivatives with CCl_4 were also investigated in this study.

$$3(R_2N)_2PCl + CCl_4 \longrightarrow$$

$$[(R_2N)_2\overset{|}{\underset{|}{P}}\overset{\overset{Cl}{|}}{=\!\!=}C\overset{|}{\underset{|}{=\!\!=}}\overset{}{\underset{Cl}{P}}(NR_2)_2]^+Cl^- + (R_2N)_2\overset{+}{P}Cl_2 \ \ Cl^- \quad (90)$$

B. Carbon–Halogen Compounds with Electronegative Substituents (EWG-C-X)

It was reported by Rozhkov and Chaplina[130] that under mild conditions perfluorinated *t*-alkyl bromides (*t*-R_fBr) in nonpolar solvents can be added across the π bond of terminal alkenes, alkynes and butadiene. Slow addition to alkenes at 20 °C is accelerated in proton-donating solvents and is catalyzed by readily oxidizable nucleophiles. Bromination of the π bond and formation of reduction products of *t*-R_fBr, according to Rozhkov and Chaplina, suggest a radical-chain mechanism initiated by electron transfer to the *t*-R_fBr molecule. Based on their results they proposed a scheme invoking nucleophilic catalysis for the addition of *t*-R_fBr across the π bond. The first step of the reaction consists of electron transfer from the nucleophilic anion of the catalyst ($Bu_4N^+Br^-$, $Na^+NO_2^-$, K^+SCN^-, $Na^+N_3^-$) to *t*-R_fBr with formation of an anion-radical (*t*-R_fBr)$^{-\cdot}$. Dissociation of this anion radical produces a perfluorocarbanion and Br·, and the latter adds to the π bond thereby initiating a radical-chain process (equation 91).

$$R_fBr + Nu^- \longrightarrow R_fBr^{-\cdot} + Nu^\cdot$$
$$R_fBr^{-\cdot} \longrightarrow R_f^- + Br^\cdot$$
$$R_f^- + SH \longrightarrow R_fH + S^-$$
$$Br^\cdot + H_2C=\!\!=CHR \longrightarrow BrCH_2\overset{\cdot}{C}HR$$
$$BrCH_2\overset{\cdot}{C}HR + R_fBr \longrightarrow BrCH_2CH(Br)R + R_f^\cdot$$
$$R_f^\cdot + CH_2=\!\!=CHR \longrightarrow R_fCH_2\overset{\cdot}{C}HR$$
$$R_fCH_2\overset{\cdot}{C}HR + R_fBr \longrightarrow R_fCH_2CH(Br)R + R_f^\cdot, \text{ etc.} \quad (91)$$

$$R_f = C_3F_7(CF_3)_2C; \ R = Bu; \ Nu^- = Br^- \text{ of } Bu_4N^+Br^-; \ SH = \text{solvent}$$

According to a report by Allen[131], reaction of 2-bromothiazole as well as 2-iodothiazole with a tertiary phosphine or phosphorus(III) ester in alcohol solvents involves nucleophilic attack by phosphorus on halogen, with subsequent formation of thiazole and the phosphine oxide or phosphorus(V) ester, respectively. An example as illustrated in Allen's report is shown in equation 92. In a thiol solvent, the related phosphine sulfide is formed.

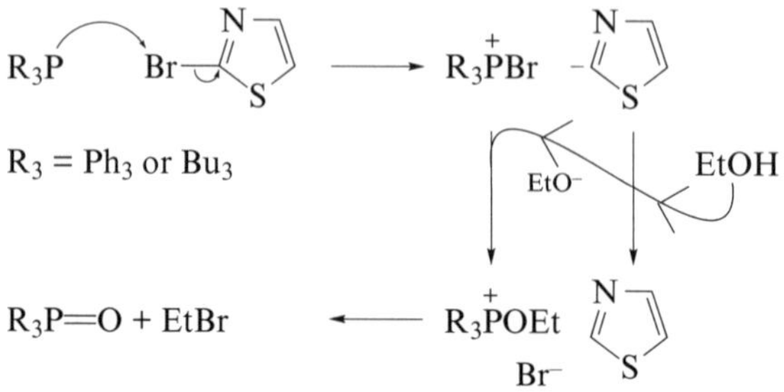

$$R_3 = Ph_3 \text{ or } Bu_3$$

(92)

The latter reaction fails with trialkyl phosphites. Allen justifies his mechanism of a 'nucleophilic attack by phosphorus at halogen' of the bromothiazole by stating that 'nucleophilic attack by trivalent phosphorus at 'positive' halogen is well established in phosphorous chemistry particularly in situations in which the halogen is bound to carbon-bearing electron-withdrawing substituents'. There is now convincing evidence which discounts this concept of simple S_N2-type attack on a 'positive halogen', but invokes an electron-transfer pathway to explain such displacements: when the nucleophile is anionic the RARP (radical anion–radical pair) transition state is called upon[7-9], and when the nucleophile is neutral, the CRARP (cation radical-anion radical pair) transition state is invoked (vide supra). Adaptation of this mechanism to subject reactions is illustrated(equation 93).

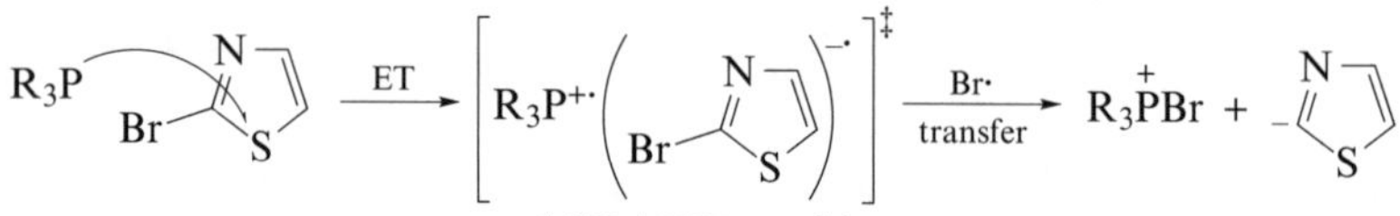

caged CRARP transition state

(93)

The mechanisms of the reduction of α-chloro- and α-bromoacetophenone with triphenyltin hydride (Ph₃SnH) has been investigated and described by Tanner and Singh[132]. They concluded that these reductions follow the same pathways as previously reported for the reduction of α-fluoroacetophenone[133]. Homolytic as well as heterolytic reductions could be recognized: the homolytic reactions yielded acetophenone and the heterolytic reactions yielded α-(halomethyl)benzyl alcohol. According to Tanner and Singh the homolytic reductions proceed by a free-radical-chain process where the initiation step and one of the propagation steps involve electron-transfer reactions. Small amounts of a radical initiator (AIBN) enhance the rate and small amounts of a radical-anion trap (m-dinitrobenzene) greatly reduce the rate and yields. It was also reported that the reduction reaction apparently does not proceed by a direct halogen transfer since no secondary deuterium isotope effect was observed on reduction of α,α-dideuterio-α-haloacetophenone to α,α-dideuterioacetophenone: the rates of dehalogenation of the α,α-diprotio- and α,α-dideuteriohaloacetophenones were virtually identical, whereas a secondary isotope effect would have reduced the rate of the dideuteriated substrate by about 26–44%, according to Tanner and Singh. Their homolytic reduction mechanism is illustrated in equation 94.

Their electron-transfer homolytic mechanism is quite reasonable and is supported by other recent related investigations (vide supra). However, a modification in their propagation mode would bring this mechanism even more in line with that of similar ones, namely

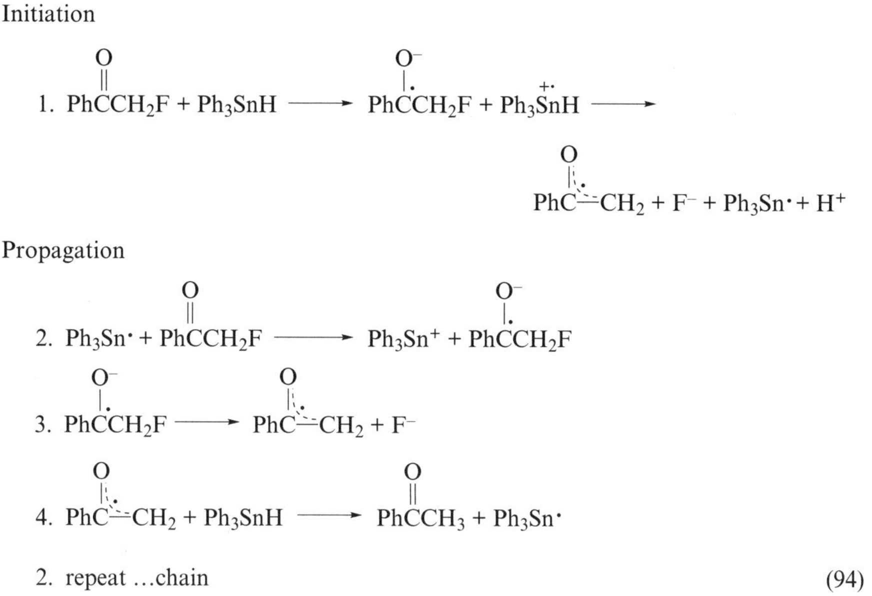

Initiation

$$1. \ PhCCH_2F + Ph_3SnH \longrightarrow Ph\overset{O^-}{\underset{|\cdot}{C}}CH_2F + Ph_3Sn\overset{+\cdot}{H} \longrightarrow$$

$$Ph\overset{O}{C}\!\!=\!\!CH_2 + F^- + Ph_3Sn\cdot + H^+$$

Propagation

$$2. \ Ph_3Sn\cdot + Ph\overset{O}{\underset{\|}{C}}CH_2F \longrightarrow Ph_3Sn^+ + Ph\overset{O^-}{\underset{|\cdot}{C}}CH_2F$$

$$3. \ Ph\overset{O^-}{\underset{|\cdot}{C}}CH_2F \longrightarrow Ph\overset{O}{C}\!\!=\!\!CH_2 + F^-$$

$$4. \ Ph\overset{O}{C}\!\!=\!\!CH_2 + Ph_3SnH \longrightarrow Ph\overset{O}{\underset{\|}{C}}CH_3 + Ph_3Sn\cdot$$

2. repeat ...chain (94)

a caged transition state bridging the electron transfer and halogen transfer, which can be termed a 'CARP'(cation anion-radical pair), as shown in equation 95. It should be point-

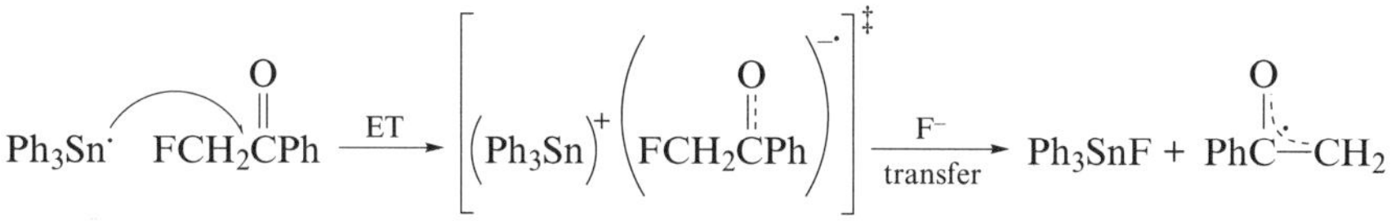

$$Ph_3Sn\cdot \ \ F\overset{O}{\underset{\|}{C}H_2CPh} \overset{ET}{\longrightarrow} \left[(Ph_3Sn)^+ \left(F\overset{O}{\underset{\|}{C}H_2CPh} \right)^{-} \right]^{\ddagger} \xrightarrow[\text{transfer}]{F^-} Ph_3SnF + Ph\overset{O}{C}\!\!=\!\!CH_2$$

caged CARP transition state (95)

ed out that in this CARP mechanism the halogen transfer is realized as a rapid fluoride (F^-) transfer from the *high-energy anion radical* to Ph_3Sn^+, not a simple 'direct' fluorine-atom ($F\cdot$) abstraction by $Ph_3Sn\cdot$, and therefore would not impart a secondary isotope effect (*vide supra*). Tanner and Singh also point out that while a competitive hydride-transfer mechanism operates for the reaction of the fluoro and chloro ketones in polar solvents, in the reduction of bromoacetophenone, the strongest electron acceptor, only the homolytic pathway is followed, irrespective of the solvent polarity.

Tin hydride reagents have been commonly used in radical-chain reductions in synthetic organic chemistry. It was reported by Vedejs and coworkers[134] that the amino tin hydride reagent [*o*-(dimethylaminomethyl)phenyl]dimethyltin hydride (**48**) did *not* reduce esters,

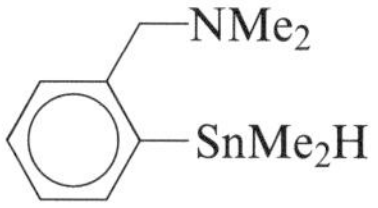

—NMe₂

—SnMe₂H

(**48**)

simple enolates, nitriles, epoxides, sulfones, saturated nitro alkanes or simple halides, in methanol solution. They attributed the failure, at least in part, to the destruction of the reagent itself within a few hours by the methanol (protic solvent). However, in THF(aprotic polar solvent) halides and other typical radical-chain reduction substrates did react with this reagent: phenacyl bromide, 1-bromoethylbenzene, γ-iodobutyrophenone,3-bromo-1-phenylpropene, p-bromoanisole and p-iodoanisole were dehalogenated, respectively, to acetophenone, ethylbenzene, butyrophenone, β-methylstyrene and the last two entries to anisole. In accounting for these dehalogenation reactions an ET-initiated radical-chain mechanism is consistent with available evidence for the dehalogenations. It was pointed out many years ago that an electron-transfer mechanism is also responsible for the reductive dehalogenation of α-halo sulfones and ketones with $CHCl_3$ in basic media (i.e. Cl_3C^-), although these were not chain reactions (*vide supra*)[7–9,113]. The amino tin hydride reagent **48** can serve as a selective *nucleophilic hydride source* in protic solvents or as a *radical-chain reducing agent* under aprotic conditions, according to Vedejs and coworkers. They also point out that the room temperature reactivity of **48** should be an advantage with thermally sensitive substrates, that no initiator is necessary for the radical reduction of halides and that the tin byproduct in dehalogenations is the crystalline tin halide, which is easily recycled.

V. SULFUR–HALOGEN COMPOUNDS

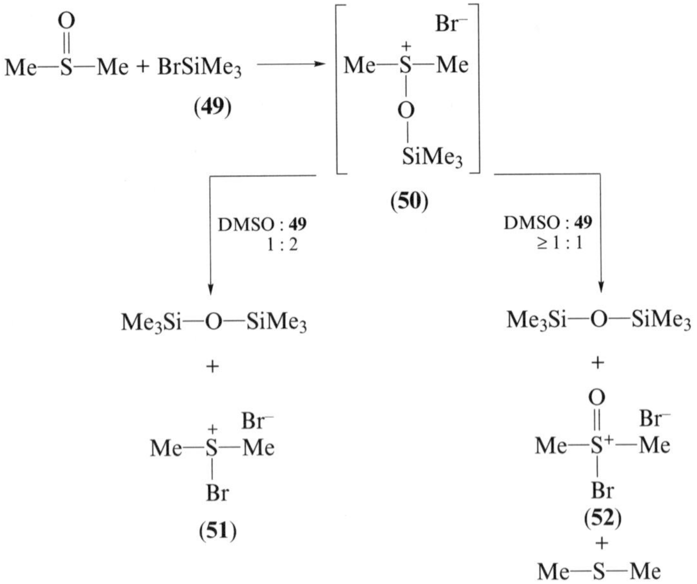

Megyeri and Keve[135] observed the formation of bromodimethylsulfoxonium bromide (**52**) along with bromodimethylsulfonium bromide (**51**), from the reaction of DMSO with bromotrimethylsilane (**49**). They unequivocally characterized **52** and noted that it had not been previously reported in the literature. It should be mentioned, however, that the corresponding chlorodimethylsufoxonium chloride was prepared and reported some time ago by Corey and Kim[136]. Megyeri and Keve also described a system prepared by the reaction of DMSO with HBr. The mechanism they presented for the formation of **51** and **52** is illustrated in equation 96. As noted throughout this chapter, halogens attached to heteroatoms are generally 'positive halogens' and the compounds are halogenating agents. When those heteroatoms carry a positive charge, the compounds are *very reactive* halogenating reagents. Megyeri and Keve used **51**, **52** and the DMSO/HBr system as halogenation reagents for indole alkaloids. They prepared and used the reagents seperately, or in one-pot halogenations. Chlorodimethylsulfonium chloride and the corresponding bromo bromide (**51**) have been used for α-halogenation of carbonyl compounds[137-140] and for halogen addition to alkenes[141]. A mechanism was proposed for the chlorination of ketones by Fraser and Kong[140], who used a reagent prepared from the bromide-ion-catalyzed reaction of trimethylchlorosilane with dimethyl sulfoxide under PTC conditions. They illustrtated the one-pot preparation of reagent and chlorination of a ketone shown in equation 97. Note that the final step requires the displacement of α-Br by Cl⁻ according to Fraser and Kong.

$$Me_3Si\!-\!Cl + Me_2S\!=\!O \longrightarrow Me_3Si\!-\!O\!-\!\overset{+}{S}Me_2 \; Cl^-$$

$$Me_3Si\!-\!O\!-\!\overset{+}{S}Me_2 \; Cl^- \xrightarrow[\text{cat.}]{n\text{-Bu}_4N^+Br^-} Me_2\overset{+}{S}\!-\!Br + Me_3Si\!-\!O^-$$
$$+ \, n\text{-Bu}_4N^+Cl^-$$

$$Me_2\overset{+}{S}Br + RCOCH_2R^1 \longrightarrow Me_2S + RCOCH(Br)R^1 + H^+$$

$$RCOCH(Br)R^1 + Cl^- \longrightarrow RCOCH(Cl)R^1 + Br^- \tag{97}$$

A mechanism is reported[141] for the transformation of alkenes and cycloalkanes by trimethylchlorosilane and dimethyl sulfoxide into 2-chloroalkyl methyl sulfide and 2-chlorocycloalkyl methyl sulfide (equation 98). The regioselectivity and stereochemistry of

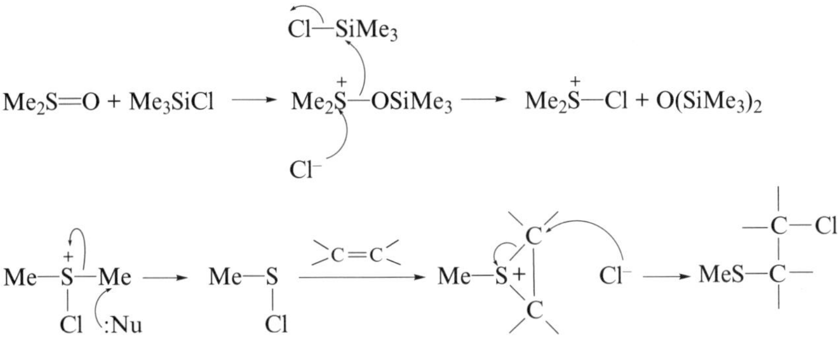

this conversion are debatable, although under nonpolar conditions the predominant formation of anti-Markovnikov adducts with exclusive *trans* geometry is generally observed, while in ionizing solvents Markovnikov adducts predominate as mixtures of *cis* and *trans* isomers.

VI. REFERENCES

1. P. Spanel, M. Tichy and D. Smith, *J. Chem. Phys.*, **98**, 8660 (1993) and references cited therein.
2. J. Arotsky and M. C. R. Symons, *Quart. Rev. Chem. Soc.*, **16**, 282 (1962).
3. B. D. Stepin, *Usp. Khim.*, **56**, 1273 (1987); *Russ. Chem. Rev. (Engl. Transl.)*, **56**, 726 (1987).
4. E. Differding and G. Rüegg, *Tetrahedron Lett.*, **32**, 3815 (1991).

5. E. Differding and M. Wehrli, *Tetrahedron Lett.*, **32**, 3189 (1991).
6. E. Differding and P. M. Bersier, *Tetrahedron*, **48**, 1595 (1992).
7. C. Y. Meyers, W. S. Matthews, L. L. Ho, V. M. Kolb and T. E. Parady, in *Catalysis in Organic Syntheses* (Ed. G. V. Smith), Academic Press, New York, 1977, pp.197–278.
8. C. Y. Meyers and V. M. Kolb, *J. Org. Chem.*, **43**, 1985 (1978); M.-L. Hsu, Ph.D. Dissertation, Southern Illinois University, Carbondale, IL, 1975; C. Y. Meyers and M.-L. Hsu, American Chemical Society, 170th National Meeting, Chicago, 1975, Abstract ORGN-45; C. Y. Meyers, R. T. Arnold and A. P. Wahner, American Chemical Society, Midwest Regional Meeting, Carbondale, IL, 1975, Abstract ORG 510; C. Y. Meyers, Symposium on Electron-transfer Chemistry—An Interdisciplinary Review and Survey of Current Trends, American Chemical Society, Midwest Regional Meeting, Carbondale, IL, 1985, Abstract 660.
9. C.Y. Meyers, in *Topics in Organic Sulfur Chemistry* (Ed. M. Tišler), University Press, Ljubljana, Yugoslavia, 1978, pp. 207–260.
10. J.-M. Mattalia, B. Vacher, A. Samat and M. Chanon, *J. Am. Chem. Soc.*, **114**, 4111 (1992).
11. G. A. Russell and R. K. Khanna, in *Nucleophilicity*, Advances in Chemistry Series, **215** (Eds. J.M. Harris and S. P. Manus), American Chemical Society, Washington, DC, 1987, pp. 355–368.
12. N. Kornblum, *Angew. Chem., Int. Ed. Engl.*, **14**, 734 (1975); N. Kornblum, XXIIIrd International Congress of Pure and Applied Chemistry, **4**, 81 (1971).
13. J. F. Bunnett, *Acc. Chem. Res.*, **11**, 413 (1978).
14. J. Legters, L. Thijs and B. Zwanenburg, *Recl. Trav. Chim. Pays-Bas*, **111**, 75 (1992).
15. K. Schreiber, A. Walther and H. Ronsch, *Tetrahedron*, **20**,1939 (1964).
16. G. V. Shustov, G. K. Kadorkina, R. G. Kostyanovsky and A. Rauk, *J. Am. Chem.Soc.*, **110**, 1719 (1988).
17. V. L. Heasley, L. S. Holstein III, R. J. Moreland, J. W. Rosbrugh, Jr. and D. F. Shellhamer, *J. Chem. Soc., Perkin Trans. 2*, 1271 (1991).
18. G. E. Heasley, M. Duke, D. Hoyer, J. Hunnicutt, M. Lawrence, M. J. Smolik, V. L. Heasley and D. F. Shellhamer, *Tetrahedron Lett.*, **23**, 1459 (1982).
19. V. L. Heasley, D. F. Shellhamer, R. K. Gipe, H. C. Weise, M. L. Oakes and G. E. Heasley, *Tetrahedron Lett.*, **21**, 4133 (1980).
20. V. L. Heasley, R. K. Gipe, J. L. Martin, H. C. Weise, M. L. Oakes and D. F. Shellhamer, *J. Org. Chem.*, **48**, 3195 (1983).
21. V. L. Heasley, B. R. Berry, S. L. Holmes, L. S. Holstein III, K. A. Milhoan, A. M. Sauerbery, B. R. Teegarden, D. F. Shellhamer and G. E. Heasley, *J. Org. Chem.*, **53**,198 (1988).
22. G. A. Olah and J. A. Olah, *Carbonium Ions*, Vol. 2 (Eds. G. A. Olah and P. v. R. Schleyer), John Wiley & Sons, New York, 1970, pp. 715–782.
23. W. Storzer and D. D. Des Marteau, *J. Fluorine Chem.*, **58**, 56 (1992).
24. Q. Wang, J. C. Jochims, S. Köhlbrandt, L. Dahlenburg, M. Al-Talib, A. Hamed and A. Ismail, *Synthesis*, 710 (1992).
25. M. W. Moon, *J. Org. Chem.*, **37**, 386 (1972).
26. S. Rozen, *Acc. Chem. Res.*, **21**, 307 (1988).
27. O. Lerman, Y. Tor, D. Hebel and S. Rozen, *J. Org. Chem.*, **49**, 806 (1984).
28. J. Gao, L. W. Chou and A. Auerbach, *Biophys. J.*, **65**, 43 (1993) and references cited therein.
29. G. W. M. Visser, C. N. M. Bakker, B. W. v. Halteren, J. D. M. Herscheid, G. A. Brinkman and A. Hoekstra, *J. Org. Chem.*, **51**, 1886 (1986).
30. S. Rozen, Y. Bareket and M. Kol, *J. Fluorine Chem.*, **61**, 141 (1993).
31. D. Hebel and S. Rozen, *J. Org. Chem.*, **56**, 6298 (1991).
32. G. O. Nevstad and J. Songstad, *Acta Chem. Scand.*, **B 38**, 469 (1984).
33. C. Y. Meyers and R. Chan-Yu-King, U.S. Patent 4,967,021, October 30, 1990; European Patent E. P. O. Patent No. 0 246 805, May 9, 1990; Canadian Patent 1,283,131, April 16, 1991.
34. J. Hine, R. B. Duke and E. F. Glod, *J. Am. Chem. Soc.*, **91**, 2316 (1969).
35. J. Hine, *Physical Organic Chemistry*, 2nd ed., McGraw-Hill, New York,1962, p. 487, Table 24-1.
36. J. G. Atkinson and D. W. Cillis, U.S. Patent 3,737,464, June 5, 1973.
37. J. G. Atkinson, D. W. Cillis and R. S. Stuart, *Can. J. Chem.*, **47**, 477 (1969).
38. D. Hebel, K. L. Kirk, L. A. Cohen and V. M. Labroo, *Tetrahedron Lett.*, **31**, 619 (1990).
39. A. Luxen and J. R. Barrio, *Tetrahedron Lett.*, **29**, 1501 (1988).
40. K. L. Kirk and C. R. Creveling, *Med. Res. Rev.*, **4**, 189 (1984).
41. K. Dax. B. I. Glänzer, G. Schulz and H. Vyplel, *Carbohydr. Res.*, **162**, 13 (1987).

42. M. Tada, T. Matsuzawa, K. Yamaguchi, Y. Abe, H. Fukuda, M. Itoh, H. Sugiyama, T. Ido and T. Takahashi, *Carbohydr. Res.*, **161**, 314 (1987).
43. (a) R. Chirakal, E. S. Garnett,G. J. Schrobilgen, C. Nahmias and G. Firnau, *Chem. Br.*, 47(1991).
    (b) R. Chirakal, G. Firnau, J. Couse and E. S. Garnett, *Int. J. Appl. Radiat. Isot.*, **35**, 651 (1984).
    (c) M. J. Adam, B. Abeysekera, T. J. Ruth, J. R. Grierson and B. D. Pate, *J. Nucl. Med.*, **26**, P125 (1985).
    (d) O. T. DeJesus, J. Mukherjee and E. H. Appleman, *J. Labelled Compd. Radiopharm.*, **26**, 133 (1989).
    (e) A. Luxen et al., *J. Labelled Compd. Radiopharm.*, **23**, 34 (1986).
44. K. Ishiwata, S.-I. Ishi, M. Senda, Y. Tsuchiya and K. Tomimoto, *Appl. Radiat. Isot.*, **44**, 755 (1993).
45. T. Duelfer, P. Johnström, S. Stone-Elander, A. Holland,C. Halldin, M. Haaparanta, O. Solin, J. Bergman, M. Steinma and G. Sedvall, *J. Labelled Compd. Radiopharm.*, **29**, 1223 (1991).
46. M. Okano, Y. Ito, T. Shono and R. Oda., *Bull. Chem. Soc. Jpn.*, **36**, 1314 (1963).
47. M. Yoshida, H. Mohizuki, T. Suzuki and N. Kamigata, *Bull. Chem. Soc. Jpn.*, **63**, 3704 (1990).
48. B. M. Musin, P. V. Yudina, V. B. Ivanov, V. M. Nekhoroshkov and Y. Y. Efermov, *Izv. Akad. Nauk SSSR, Ser. Khim.*, **41**, 1229 (1992); *Bull. Acad. Sci. USSR, Div. Chem. Sci. (Engl.Transl.)*, **41**, 971 (1992).
49. I. Ionica, E. G. Georgescu and I. G. Dinulescu, *Rev. Roum. Chim.*, **37**, 729 (1992).
50. I. Ionica, P. Filip and I. G. Dinulescu, *Rev. Roum. Chim.*, **37**, 595 (1992).
51. G. E. Heasley, J. M. Jones, S. R. Stark, B. L. Robinson, V. L. Heasley and D. F. Shellhamer, *Tetrahedron Lett.*, **26**, 1811 (1985).
52. N. N. Labeish and A. A. Petrov, *Usp. Khim.*, **58**, 1844 (1989); *Russ. Chem. Rev. (Engl. Transl.)*, **58**, 1048 (1989).
53. H. Yamada, Y. Wada, S. Tanimoto, and M. Okano, *Bull. Chem. Soc. Jpn.*, **55**, 2480 (1982).
54. (a) C. C. Fortes and T. A. Coimbra, *Synth. Commun.*, **21**, 2039 (1991);
    (b) P. D. Robinson, C. Y. Meyers, V. M. Kolb, J. L. Tunnell and J. D. Ferrara, *Acta Crystallogr.*, **C48**, 1033 (1992); J. Tyrrell, V. M. Kolb and C. Y. Meyers, *J. Am. Chem. Soc.*, **101**, 3497 (1979);
    (c) E. E. Reid, *Organic Chemistry of Bivalent Sulfur*, Vol. II, Chemical Publishing Co., Inc., New York, 1960, p. 302.
55. Y. Goldberg and H. Alper, *J. Org. Chem.*, **58**, 3072 (1993).
56. P. Konradsson, D. R. Mootoo, R. E. McDevitt and B. Fraser-Reid, *J. Chem. Soc., Chem. Commun.*, 270 (1990).
57. I. Mozek and B. Sket, *Synth. Commun.*, **22**, 2513 (1992).
58. S. Rozen, M. Brand and R. Lidor, *J. Org. Chem.*, **53**, 5545 (1988).
59. M. Shimizu, M. Okamura and T. Fujisawa, *Bull. Chem. Soc. Jpn.*, **64**, 2596 (1991).
60. M. Kuroboshi and T. Hiyama, *Tetrahedron Lett.*, **32**, 1215 (1991).
61. M. Bruncko, D. Crich and R. Samy, *Heterocycles*, **36**, 1735 (1993).
62. M. Taniguchi, A. Gonsho, M. Nakagawa and T. Hino, *Chem. Pharm. Bull.*, **31**, 1856 (1983). See also: T. Hino, U. Hideya, M. Takashima, T. Kawata, H. Seki, R. Hara, T. Kuramochi and M. Nakagawa, *Chem. Pharm. Bull.*, **38**, 2632 (1990); R. K. Dua and R. S. Philips, *Tetrahedron Lett.*, **33**, 29 (1992).
63. W. E. Barnette, *J. Am. Chem. Soc.*, **106**, 452 (1984).
64. D. H. R. Barton, R. H. Hesse, M. M. Pechet and H. T. Toh, *J. Chem. Soc., Perkin Tran.1*, 732 (1974); D. H. R. Barton and R. H. Hesse, U.S. Patent 3, 917,688, Nov. 4, 1975; M. Seguin, J. C. Adenis, C. Michand and J. J. Basselier, *J. Fluorine Chem.*, **15**, 201 (1980).
65. E. Differding and R. W. Lang, *Helv. Chim. Acta*, **72**, 1248 (1989).
66. E. Differding, G. M. Rüegg and R. W. Lang, *Tetrahedron Lett.*, **32**, 1779 (1991).
67. E. Differding and R.W. Lang, *Tetrahedron Lett.*, **29**, 6087 (1988).
68. S. Singh, D. D. DesMarteau, S. S. Zuberi, M. Witz and H.-N. Huang, *J. Am. Chem. Soc.*, **109**, 7194 (1987).
69. M. Witz and D. D. DesMarteau, *J. Fluorine Chem.*, **52**, 7 (1991).
70. D. D. DesMarteau, Z.-Q. Xu and M. Witz, *J. Org. Chem.*, **57**, 629 (1992).
71. W. T. Pennington, G. Rensati and D. D. DesMarteau , *J. Org. Chem.*, **57**, 1536 (1992).
72. G. Resnati and D. D. DesMarteau, *J. Org. Chem.*, **56**, 4925 (1991).

73. F. A. Davis and W. Han, *Tetrahedron Lett.*, **32**, 1631 (1991).
74. E. Differding and H. Ofner, *Synlett*, 187 (1991).
75. E. Differding, R. O. Duthaler, A. Krieger, G. M. Rüegg and C. Schmidt, *Synlett.*, 395 (1991).
76. J. March, *Advanced Organic Chemistry*, 4th ed., Wiley-Interscience, New York, 1992, p. 818 and references cited therein.
77. M. C. Cano, F. Gómez-Contreras, A. M. Sanz and M. J. R. Yunta, *Tetrahedron.*, **49**, 243 (1993).
78. E. Napolitano and R. Fiaschi, *Gazz. Chim. Ital.*, **122**, 233 (1992).
79. A. Hassner, G. J. Mathews and F. W. Fowler, *J. Am. Chem. Soc.*, **91**, 5045 (1969).
80. S. Sivasubramanian, S. Aravind, L. T. Kumarsingh and N. Arumugam, *J. Org. Chem.*, **51**, 1985 (1986).
81. S. Sivasubramanian, S. Aravind, S. Muthusubramanian and N. Arumugam, *Indian J. Chem., Sect. B.*, **31**,125 (1992).
82. Reference, 76, p. 819 and references cited therein.
83. C. Berti, L. Greci, R. Andruzzi and A. Trazza, *J. Org. Chem.*, **47**, 4895 (1982).
84. P. Carloni, L. Eberson, L. Greci, P. Stipa and G. Tosi, *J. Chem. Soc., Perkin Trans. 2*, 1779 (1991).
85. M. D. Ryan and D. N. Harpp, *Tetrahedron Lett.*, **33**, 2129 (1992).
86. R. E. Banks, V. Murtagh and E. Tsiliopoulos, *J. Fluorine Chem.*, **52**, 389 (1991).
87. B. T. Gowda and B. S. Sherigara, *J. Indian Chem. Soc.*, **66**, 91 (1989).
88. F. Minisci, E. Vismara, F. Fontana, E. Platone and G. Faraci, *J. Chem. Soc., Perkin Trans. 2*,123 (1989).
89. (a) G. S. Lal, *J. Org. Chem.*, **58**, 2791 (1993).
    (b) R. E. Banks, U.S. Patent 5,086,178 (1992).
90. R. E. Banks, S. N. Mohialdin-Khaffaf, G. S. Lal, I. Sharif and R. G. Syvret, *J. Chem. Soc., Chem. Commun.*, 595 (1992).
91. D. P. Matthews, S. C. Miller, E. T. Jarvi, J. S. Sabol and J. R. McCarthy, *Tetrahedron Lett.*, **34**, 3057 (1993).
92. T. Umemoto, K. Kawada and K. Tomita, *Tetrahedron Lett.*, **27**, 4465 (1986).
93. J. Barluenga, M. A. Rodríguez and P. Campos, *J. Org. Chem.*, **55**, 3104 (1990).
94. J. Barluenga, P. Campos, F. López, I. Llorente and M. A. Rodríguez, *Tetrahedron Lett.*, **31**, 7375 (1990).
95. (a) J. Barluenga, M. A. Rodríguez and P. Campos, *Synthesis*, 270 (1992).
    (b) J. Barluenga, J. M. González, I. Llorento and P. J. Campos, *Angew. Chem., Int. Ed. Engl.*, **32**, 893 (1993).
96. J. Barluenga, J. M. González, M. A. García-Martín, P. J. Campos and G. Asensio, *J. Org. Chem.*, **58**, 2058 (1993).
97. J. Barluenga, P. J. Campos, J. M. González, J. L. Suárez and G. Asensio, *J. Org. Chem.*, **56**, 2234 (1991).
98. M. M. Cartwright and A. A. Woolf, *J. Fluorine Chem.*, **25**, 263 (1984).
99. (a) K. O. Christe, *J. Fluorine Chem.*, **25**, 269 (1984).
    (b) S. Bakkas, M. Julliard and M. Chanon, *Tetrahedron*, **43**, 501 (1987).
100. C. Wakselman and C. Kaziz, *J. Fluorine Chem.*, **33**, 347 (1986).
101. P. Tarrant, *J. Fluorine Chem.*, **25**, 69 (1984).
102. A. Commeyras, *Ann. Chim.(Paris)*, **9**, 673 (1984).
103. X.-Y. Li, X.-K. Jiang, H.-Q. Pan, J.-S. Hu and W.-M. Fu, *Pure Appl. Chem.*, **59**, 1015 (1987).
104. (a) X.-Y. Li, H.-Q. Pan and X.-K. Jiang, *Tetrahedron Lett.*, **25**, 4937 (1984).
     (b) Reference 104 (a), note 6.
105. X.-Y. Li, H.-Q. Pan and X.-K. Jiang, *Acta Chimica Sinica (Engl. Ed.)*,180 (1984).
106. X.-Y. Li, X.-K. Jiang, Y.-F. Gong and H.-Q. Pan, *Acta Chimica Sinica (Engl. Ed.)*, 228 (1985).
107. X.-Y. Li, H.-Q. Pan, X.-K. Jiang and Z.-Y. Zhan, *Angew. Chem., Int. Ed. Engl.*, **24**, 871 (1985).
108. X.-Y. Li, H.-Q. Pan, W.-M. Fu and X.-K. Jiang, *J. Fluorine Chem.*, **31**, 213 (1986).
109. Reference 7, pp. 270 ff.
110. Reference 7, pp. 263 and 265.
111. (a) Reference 35, pp. 486–488.
     (b) E. R. De Waard, J. Kattenberg and H. O. Huisman, *Tetrahedron*, **29**, 4149 (1973).
112. D. H. Hua, Ph.D. Dissertation, Southern Illinois University, Carbondale, IL, 1979.
113. R. Chan-Yu-King, Ph.D. Dissertation, Southern Illinois University, Carbondale, IL, 1986.

114. R. F. Clark and J. H. Simons, *J. Am. Chem. Soc.*, **77**, 6618 (1955).
115. C. Y. Meyers, A. M. Malte and W. S. Matthews, *J. Am. Chem. Soc.*, **91**, 7510 (1969); V. M. Kolb, Ph.D. Dissertation, Southern Illinois University, Carbondale, IL, 1976.
116. T. E. Parady, M. S. Thesis, Southern Illinois University, Carbondale, IL, 1977.
117. A. E. Feiring and E. R. Wonchoba, *J. Org. Chem.*, **57**, 7014 (1992).
118. W. J. Pummer and L. A. Wall, *Soc. Plast. Eng., Trans.*, 220 (1963); E. L. Zaitseva, E. A. Nikanorova, A. T. Monakhova and A. Ya. Yakubovich, *Zh. Org. Khim.*, **8**, 974 (1972); *J. Org. Chem. USSR (Engl. Transl.)*, **8**, 981 (1972).
119. Y. Sasson and O. W. Webster, *J. Chem. Soc., Chem. Commun.*, 1200 (1992).
120. O. A. Reutov, I. P. Beletskaya and K. P. Butin, *CH-Acids*, Pergamon Press, New York, 1978.
121. D. J. Cram, *Fundamentals of Carbanion Chemistry*, Academic Press, New York, 1965.
122. O. I. Kolodyazhnyi and D. B. Golokhov, *Zh. Obshch. Khim.*, **58**, 491 (1988); *J. Gen. Chem. USSR (Engl. Transl.)*, **58**, 426 (1988).
123. G. N. Koidan, V. A. Oleinik, A. A. Kudryavtsev, A. P. Marchenko and A. M. Pinchuk, *Zh. Obshch. Khim.*, **59**, 2675 (1989); *J. Gen. Chem. USSR (Engl. Transl.)*, **59**, 2391 (1989).
124. W. Reid and H. Appel, *Justus Liebigs Ann. Chem.*, **679**, 51 (1964); B. Castro, Y. Chapleur, B. Gross and C. Sleeve, *Tetrahedron Lett.*, 5001 (1972).
125. A. P. Marchenko, V. V. Miroshnichenko, G. N. Koidan and A. M. Pinchuk, *Zh. Obshch. Khim.*, **50**, 1897 (1980); *Chem. Abstr.*, **93**, 239512a (1980).
126. A. P. Marchenko, I. S. Zal'tsman and A. M. Pinchuk, *Zh. Obshch. Khim.*, **56**, 1910 (1986); *J. Gen. Chem. USSR (Eng. Transl.)*, **56**, 1687 (1986).
127. V. A. Oleinik, G. N. Koidan, A. P. Marchenko and A. M. Pinchuk, *Zh. Obshch. Khim.*, **58**, 482 (1988); *J. Gen. Chem. USSR (Engl. Transl.)*, **58**, 418 (1988).
128. (a) G. N. Koidan, S. I. Shaposhnikov, A. P. Marchenko and A. M. Pinchuk, *Zh. Obshch. Khim.*, **58**, 1185 (1988); *J. Gen. Chem. USSR (Engl. Transl.)*, **58**, 1054 (1988).
    (b) A. P. Marchenko, S. I. Shaposhnikov, G. N. Koidan, A. V. Kharchenko and A. M. Pinchuk, *Zh. Obshch. Khim.*, **58**, 2230 (1988); *J. Gen. Chem. USSR (Engl. Transl.)*, **58**, 1985 (1988).
129. A. M. Pinchuk, A. P. Marchenko and G. N. Koidan, *Phosphorus, Sulfur, Silicon Relat. Elem.*, **49/50**, 195 (1990).
130. I. N. Rozhkov and I. V. Chaplina, *Izv. Akad. Nauk SSSR, Ser. Khim.*, **40**, 2810 (1991); *Bull. Acad. Sci. USSR, Div. Chem. Sci (Engl. Transl.)*, **40**, 2451 (1991).
131. D. W. Allen, *Phosphorus, Sulfur, Silicon Relat. Elem.*, **66**, 73 (1992).
132. D. D. Tanner and H. K. Singh, *J. Org. Chem.*, **51**, 5182 (1986).
133. D. D. Tanner, G. E. Diaz and A. Potter, *J. Org. Chem.*, **50**, 2149 (1985).
134. E. Vedejs, S. M. Duncan and A. R. Haight, *J. Org. Chem.*, **58**, 3046 (1993).
135. G. Megyeri and T. Keve, *Synth. Commun.*, **19**, 3415 (1989).
136. E. J. Corey and C. U. Kim, *J. Am. Chem. Soc.*, **94**, 7586 (1972); *Tetrahedron Lett.*, 919 (1973).
137. E. Armani, A. Dossena, R. Marchelli and G. Casnati, *Tetrahedron*, **40**, 2035 (1984).
138. F. Bellesia, F. Ghelfi, R. Grandi and U. M. Pagnoni, *J. Chem. Res. (S)*, 428 (1986).
139. F. Bellesia, F. Ghelfi, R. Grandi and U. M. Pagnoni, *J. Chem. Res. (S)*, 426 (1986).
140. R. R. Fraser and F. Kong, *Synth. Commun.*, **18**, 1071 (1988).
141. F. Bellesia, F. Ghelfi, U. M. Pagnoni and A. Pinetti, *J. Chem. Res. (S)*, 238 (1987).

CHAPTER **21**

Halonium ions

GERALD F. KOSER

Department of Chemistry, The University of Akron, Akron, Ohio 44325–3601, USA

Supplement D2: The chemistry of halides, pseudo-halides and azides
Edited by S. Patai and Z. Rappoport   © 1995 John Wiley & Sons Ltd

I. INTRODUCTION

Since the 1983 publication of the chapter on halonium ions in *The Chemistry of Functional Groups, Supplement D*[1], some of the most significant developments in halonium ion chemistry have been associated with the alkynyliodonium and vinyliodonium salts. Much new information has been generated, and because of the considerable promise of alkynyl- and vinyliodonium compounds as synthetic reagents, this chapter is devoted to a thorough review of their synthesis and chemical reactivity. Earlier reviews of this subject include articles by M. Ochiai (1989)[2] and P. J. Stang (1992)[3] and a chapter section in A. Varvoglis' 1992 monograph, *The Organic Chemistry of Polycoordinated Iodine*[4].

II. ALKYNYLIODONIUM SALTS

A. Introduction

Alkynyliodonium ions, **1** and **2**, are hypervalent iodine species in which one or two alkynyl ligands are bound to a positively charged iodine(III) atom. They are sensitive to nucleophiles, especially at the β-carbon atom(s) of the alkynyl ligand(s), and for that reason, the isolation of *stable* alkynyliodonium salts generally requires the incorporation of nucleofugic anions. A list of known alkynyliodonium compounds (i.e. as of 4/1/94), containing 134 iodonium salts derived from 103 iodonium ions, and references (5–45) to their preparation and characterization are presented in Table 1. Among these compounds, alkynyl(phenyl)iodonium sulfonates and tetrafluoroborates are the most common, while alkynyl(alkyl)iodonium salts of any kind are unknown.

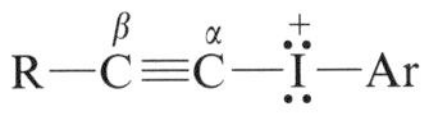

$$R-\overset{\beta}{C}\equiv\overset{\alpha}{C}-\overset{+}{\overset{..}{\underset{..}{I}}}-Ar \qquad R-\overset{\beta}{C}\equiv\overset{\alpha}{C}-\overset{+}{\overset{..}{\underset{..}{I}}}-\overset{\alpha}{C}\equiv\overset{\beta}{C}-R$$

$$(1) \qquad\qquad\qquad (2)$$

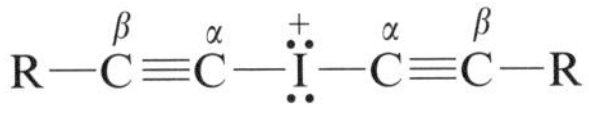

TABLE 1. Structures of known alkynyliodonium salts

$$R-C\equiv C-\overset{+}{I}-Ar, A^-$$

R	Ar	A⁻	Characterization data	References
Me	Ph	TsO⁻	mp, IR, ¹H, MS, CH	5,6,7
Et	Ph	TsO⁻	mp, IR, ¹H, MS, CH	7
n-Pr	Ph	TsO⁻	mp, IR, ¹H, MS, CH	7
i-Pr	Ph	TsO⁻	mp, IR, ¹H, CHI	8
n-Bu	Ph	TsO⁻	mp, IR, ¹H, MS	5,6,7
i-Bu	Ph	TsO⁻	mp, IR, ¹H, CHI	8
s-Bu	Ph	TsO⁻	mp, IR, ¹H, ¹³C, CHI	6,8
t-Bu	Ph	TsO⁻	mp, IR, ¹H, CHI	6,8,9
Cyclopentyl	Ph	TsO⁻	mp, ¹H, ¹³C, CH	10
Cyclohexyl	Ph	TsO⁻	mp, IR, ¹H, CHI	8,11
Ph	Ph	TsO⁻	mp, IR, ¹H, CHI, X-ray	6,8,9,11
4-Tol	Ph	TsO⁻	mp, ¹H, ¹³C, CH	10
4-ClC₆H₄	Ph	TsO⁻	mp, IR ¹H, MS, CH	12
4-An	Ph	TsO⁻	mp, IR ¹H, MS, CH	12
Me₃Si	Ph	TsO⁻	mp, IR ¹H, MS, CH	7
t-Bu	2-Tol	TsO⁻	mp, ¹H, CHI	9
t-Bu	3-Tol	TsO⁻	mp, ¹H, CHI	9
t-Bu	4-Tol	TsO⁻	mp, ¹H, CHI	9
t-Bu	2-FC₆H₄	TsO⁻	mp, ¹H, CHI	9
t-Bu	3-FC₆H₄	TsO⁻	mp, ¹H, CHI	9
t-Bu	4-FC₆H₄	TsO⁻	mp, ¹H, CHI	9
t-Bu	3-ClC₆H₄	TsO⁻	mp, ¹H, CHI	9
t-Bu	4-ClC₆H₄	TsO⁻	mp, ¹H, CHI	9
Me	Ph	NsO⁻	mp, IR, ¹H, MS,	7
Me	Ph	MsO⁻	IR, ¹H	6

(continued)

TABLE 1. (*continued*)

R	Ar	A⁻	Characterization data	References
n-Bu	Ph	MsO⁻	mp, IR, ^{1}H	6
t-Bu	Ph	MsO⁻	mp, IR, ^{1}H	6
H	Ph	TfO⁻	IR, ^{1}H, ^{13}C, ^{19}F, X-ray	13
Me	Ph	TfO⁻		14,15
Me	4-FC$_6$H$_4$	TfO⁻	mp, IR, ^{1}H, ^{13}C, ^{19}F, MS	15
Me	4-CF$_3$C$_6$H$_4$	TfO⁻	mp, IR, ^{1}H, ^{13}C, ^{19}F, MS	15
Me	3,5-(CF$_3$)$_2$C$_6$H$_3$	TfO⁻	mp, IR, ^{1}H, ^{13}C, ^{19}F, MS	15
n-Bu	Ph	TfO⁻	mp, IR, ^{1}H, ^{13}C, ^{19}F, CHS	6,16
t-Bu	Ph	TfO⁻	mp, IR, ^{1}H, ^{13}C, ^{19}F, CHS	6,16
ClCH$_2$	Ph	TfO⁻	mp, IR, ^{1}H, ^{13}C, ^{19}F, MS, CHS	17
BrCH$_2$	Ph	TfO⁻	mp, IR, ^{1}H, ^{13}C, ^{19}F, CHS	17
MeOCH$_2$	Ph	TfO⁻	mp, IR, ^{1}H, ^{13}C, ^{19}F, CHS	17,18
Me(Ph)C(OH)	Ph	TfO⁻	oil, IR, ^{1}H, ^{13}C	19
Ph$_2$C(OH)	Ph	TfO⁻		20
4-An$_2$C(OH)	Ph	TfO⁻		20
(3-CF$_3$C$_6$H$_4$)$_2$C(OH)	Ph	TfO⁻		20
Me$_3$Si	Ph	TfO⁻	mp, IR, ^{1}H, ^{13}C, ^{19}F, CHS	16
(*i*-Pr)$_3$Si	Ph	TfO⁻		19
1-Cyclohexenyl	Ph	TfO⁻	mp, IR, ^{1}H, ^{13}C, ^{19}F	17
Ph	Ph	TfO⁻	mp, IR, ^{1}H, ^{13}C, ^{19}F, CHS	16,21
Cl	Ph	TfO⁻	mp, IR, ^{1}H, ^{13}C, ^{19}F, MS, CHS	17
NC	Ph	TfO⁻	mp, IR, ^{1}H, ^{13}C, ^{19}F, MS, X-ray	17,18
Ts	Ph	TfO⁻	mp, IR, ^{1}H, ^{13}C, ^{19}F, CHS	17,18
n-Pr	Ph	(PhO)$_2$PO$_2^-$	mp, CH	22
s-Bu	Ph	(PhO)$_2$PO$_2^-$		23
t-Bu	Ph	(PhO)$_2$PO$_2^-$	mp, IR, ^{1}H, ^{13}C	23
t-Bu	Ph	(Et$_2$O)$_2$PO$_2^-$		23
n-C$_8$H$_{17}$	Ph	IO$_4^-$	IR, ^{1}H	24,25
t-Bu	Ph	IO$_4^-$	mp, IR, ^{1}H, ^{13}C, MS, CHI, X-ray	24,25
Cyclopentyl–CH$_2$	Ph	IO$_4^-$	IR, ^{1}H	24,25
Cyclohexyl	Ph	IO$_4^-$	IR, ^{1}H	24,25
Ph	Ph	Cl⁻	mp, CHICl	26
H	Ph	BF$_4^-$	mp, IR, ^{1}H, ^{13}C, MS	27
Me	Ph	BF$_4^-$	IR, ^{1}H, ^{13}C, MS	25,28
n-Pr	Ph	BF$_4^-$	oil, ^{1}H	29
t-Bu	Ph	BF$_4^-$	mp, IR, ^{1}H, ^{13}C, MS, CHI	25,30
Et(Me)CHCH$_2$	Ph	BF$_4^-$	mp, IR, ^{1}H, ^{13}C, MS, CHI	25,28
n-C$_8$H$_{17}$	Ph	BF$_4^-$	mp, IR, ^{1}H, ^{13}C, MS, CHI	25,31
Cyclohexyl	Ph	BF$_4^-$	mp, IR, ^{1}H, ^{13}C, MS, CH	25,31
Cyclohexyl–CH$_2$	Ph	BF$_4^-$	mp, IR, ^{1}H, ^{13}C, MS, CH (hemihydrate)	25
Cyclohexyl–CH$_2$CH$_2$	Ph	BF$_4^-$	mp, IR, ^{1}H, ^{13}C, MS, CHI	25,28
Cyclopentyl–CH$_2$	Ph	BF$_4^-$	mp, IR, ^{1}H, ^{13}C, MS, CHI	25,28
Ph	Ph	BF$_4^-$	oil, IR, ^{1}H, ^{13}C, MS	25,31
PhCH$_2$	Ph	BF$_4^-$	oil, IR, ^{1}H, ^{13}C, MS	25,31
PhCH$_2$CH$_2$	Ph	BF$_4^-$	oil, IR, ^{1}H, ^{13}C, MS	25,31
PhCH$_2$CH$_2$CH$_2$	Ph	BF$_4^-$	IR, ^{1}H, ^{13}C, MS	25,32
HOCH$_2$CH$_2$	Ph	BF$_4^-$	oil, IR, ^{1}H, ^{13}C, MS	25,30
BrCH$_2$	Ph	BF$_4^-$	IR, ^{1}H, ^{13}C, MS	25
BrCH$_2$CH$_2$	Ph	BF$_4^-$	IR, ^{1}H, ^{13}C, MS	25
Ph$_3\overset{+}{P}$CH$_2$CH$_2$	Ph	BF$_4^-$	IR, ^{1}H, ^{13}C, MS, CH	25
Me$_3$Si	Ph	BF$_4^-$	mp, IR, ^{1}H, ^{13}C, MS, CHI	25,27

$$\underset{\substack{\|\\O}}{R-\overset{\displaystyle O}{\overset{\|}{C}}-C\equiv C-\overset{+}{I}Ph, \ OTf^-}$$

R	Characterization data	References
t-Bu	mp, IR, ^{1}H, ^{13}C, ^{19}F, CHS	17,18
Et(Me)$_2$C	mp, IR, ^{1}H, ^{13}C, ^{19}F, MS, CHS	33
1-Adamantyl	mp, IR, ^{1}H, ^{13}C, ^{19}F, MS, CHS	17
Cyclopropyl	mp, IR, ^{1}H, ^{13}C, ^{19}F, MS	17
Cyclobutyl	mp, IR, ^{1}H, ^{13}C, ^{19}F, MS, CHS	33
Cyclohexyl	mp, IR, ^{1}H, ^{13}C, ^{19}F, MS, CHS	33
Ph	mp, IR, ^{1}H, ^{13}C, ^{19}F, CHS	17,18
2-Furyl	mp, IR, ^{1}H, ^{13}C, ^{19}F, MS	17
2-Thienyl	mp, IR, ^{1}H, ^{13}C, ^{19}F, MS	17
MeO	mp, IR, ^{1}H, ^{19}F	17,18
Me$_2$N	mp, IR, ^{1}H, ^{13}C, ^{19}F, MS	17
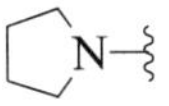 $\underset{}{\boxed{}N{-}}$	mp, IR, ^{1}H, ^{13}C, ^{19}F, MS	33
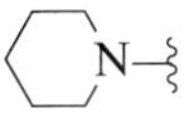 $\underset{}{\hexagon N{-}}$	mp, IR, ^{1}H, ^{13}C, ^{19}F, MS, CHNS	33
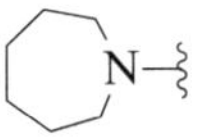 $\underset{}{\heptagon N{-}}$	mp, IR, ^{1}H, ^{13}C, ^{19}F, MS, CHNS	33
 $O\!\!\boxed{}\!N{-}$	mp, IR, ^{1}H, ^{13}C, ^{19}F, MS, CHNS	33

$$R-C\equiv C-C\equiv C-\overset{+}{I}Ph, \ OTf^-$$

R	Characterization data	References
Me	IR, ^{1}H, ^{13}C	34
n-Bu	IR, ^{1}H, ^{13}C	34
t-Bu	IR, ^{1}H, ^{13}C	34
Me$_3$Si	IR, ^{1}H, ^{13}C	34
Ph	IR, ^{1}H, ^{13}C	34

$$(R-C\equiv C)_2\overset{+}{I}, \ OTf^-$$

R	Characterization data	References
t-Bu	mp, IR, ^{1}H, ^{13}C, ^{19}F, MS	35
Me$_3$Si	mp, IR, ^{1}H, ^{13}C, ^{19}F, MS	35
(*i*-Pr)$_3$Si	mp, IR, ^{1}H, ^{13}C, ^{19}F, MS, X-ray	35

(*continued*)

TABLE 1. (*continued*)

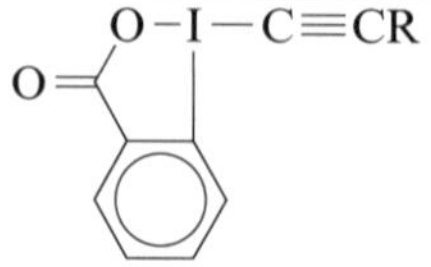

$$O=\overset{\displaystyle O-I-C\equiv CR}{\underset{\displaystyle \text{(benzene ring)}}{\big|}}$$

R	Characterization data	References
t-Bu	mp, IR, ^{1}H, MS	36
n-C$_8$H$_{17}$	oil, ^{1}H, ^{13}C, MS	36
Cyclohexyl	mp, IR, ^{1}H, ^{13}C, MS, CH, X-ray	36

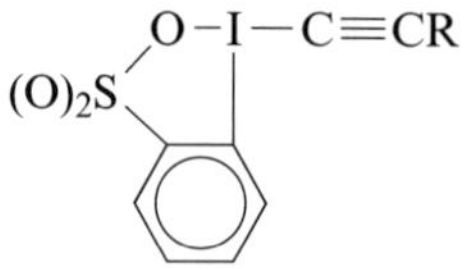

$$(O)_2S\overset{\displaystyle O-I-C\equiv CR}{\underset{\displaystyle \text{(benzene ring)}}{\big|}}$$

R	Characterization data	References
n-Pr	mp, IR, ^{1}H, ^{13}C, CH	37
n-Bu	mp, IR, ^{1}H, ^{13}C, CH	37
i-Bu	mp, IR, ^{1}H, ^{13}C, CH	37
s-Bu	mp, IR, ^{1}H, ^{13}C, CH	37
t-Bu	mp, IR, ^{1}H, ^{13}C, CH	37
n-C$_5$H$_{11}$	mp, IR, ^{1}H, ^{13}C, CH	37
n-C$_6$H$_{13}$	mp, IR, ^{1}H, ^{13}C, CH	37
Cyclohexyl	mp, IR, ^{1}H, ^{13}C, CH	37
Ph	mp, IR, ^{1}H, ^{13}C, CH	37

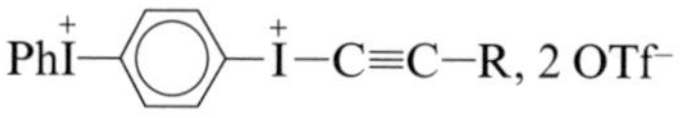

$$Ph\overset{+}{I}-\langle\bigcirc\rangle-\overset{+}{I}-C\equiv C-R,\ 2\ OTf^-$$

R	Characterization data	References
n-Bu	mp, IR, ^{1}H, ^{13}C, CH	38
t-Bu	mp, IR, ^{1}H, ^{13}C, CH	38
n-C$_6$H$_{13}$	mp, IR, ^{1}H, ^{13}C, CH	38
n-C$_{10}$H$_{21}$		39
Me$_3$Si	mp, ^{1}H, ^{13}C, CH	38
Ph	mp, IR, ^{1}H, ^{13}C, CH	38

$$Ph\overset{+}{I}-C\equiv C-\overset{+}{I}Ph,\ 2A^-$$

A$^-$	Characterization data	References
TfO$^-$	mp, IR, Raman, ^{1}H, ^{13}C, ^{19}F, CHS	40,41
n-C$_4$F$_9$SO$_3^-$	mp, IR, ^{1}H, ^{13}C, ^{19}F	40

$$\overset{+}{PhI}-C\equiv C(CH_2)_nC\equiv C-\overset{+}{IPh}, 2A^-$$

n	A^-	Characterization data	References
0	TfO$^-$	IR, ^{1}H, ^{13}C, ^{19}F, $-15\,°C/-10\,°C$ (dec)	40
2	TfO$^-$	mp, IR, ^{1}H, ^{13}C, ^{19}F, MS, CH	42
4	TfO$^-$	mp, IR, ^{1}H, ^{13}C, ^{19}F, MS, CH	42
5	TfO$^-$	mp, IR, ^{1}H, ^{13}C, ^{19}F, MS	42
6	TfO$^-$	mp, IR, ^{1}H, ^{13}C, ^{19}F, CHS	43
8	TfO$^-$	mp, IR, ^{1}H, ^{13}C, ^{19}F, CHS	43
4	TsO$^-$	mp, IR, ^{1}H, ^{13}C, MS	43
5	TsO$^-$	mp, IR, ^{1}H, ^{14}C, MS	43
6	TsO$^-$	mp, IR, ^{1}H, ^{13}C, MS	43
8	TsO$^-$	mp, IR, ^{1}H, ^{13}C, MS	43

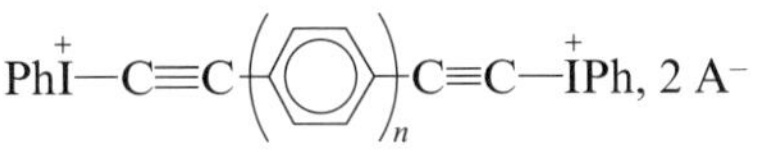

$$\overset{+}{PhI}-C\equiv C\!\!\left(\!\!\left\langle\bigcirc\right\rangle\!\!\right)_{\!n}\!\!C\equiv C-\overset{+}{IPh}, 2\,A^-$$

n	A^-	Characterization data	References
1	CF$_3$COO$^-$	mp, IR	44
1	TfO$^-$	mp, IR, ^{1}H, ^{13}C, ^{19}F, MS, CHS	42
2	TfO$^-$	mp, IR, ^{1}H, ^{13}C, ^{19}F, CHS	42

Structure	Characterization data	References
$\overset{+}{PhI}-C\equiv CCH_2OCH_2C\equiv C-\overset{+}{IPh}, 2\,OTf^-$	oil, IR, ^{1}H, ^{13}C, ^{19}F, MS	42
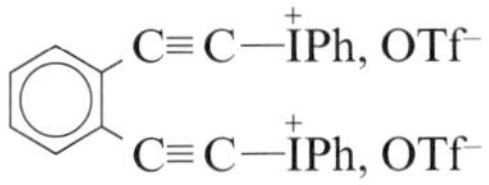 ⬡ with $C\equiv C-\overset{+}{IPh}, OTf^-$ and $C\equiv C-\overset{+}{IPh}, OTf^-$		45

The first alkynyliodonium salt, (phenylethynyl)phenyliodonium chloride, synthesized in low yields from (dichloroiodo)benzene (**3**) and lithium phenylacetylide (equation 1), was reported in 1965[26]. This chloride salt is unstable and readily decomposes to a 1:1 mixture of chloro(phenyl)acetylene and iodobenzene. It was not until the 1980s, however, that alkynyliodonium salts became generally available. This was made possible by the introduction of sulfonyloxy-λ^3-iodanes as synthetic reagents[46] and by the recognition that iodosylbenzene (**4**) can be activated either with boron trifluoride etherate or with triethyloxonium tetrafluoroborate[31]. These reagents are now widely employed for the conversion of terminal alkynes and their 1-silyl and 1-stannyl derivatives to alkynyliodonium salts (equations 2 and 3). A more exhaustive survey of iodine(III) reagents that have been

$$PhC\equiv CLi + PhICl_2 \xrightarrow[0-5\,°C]{Et_2O, hexane} PhC\equiv C\overset{+}{IPh}, Cl^- \xrightarrow{rt} PhC\equiv CCl + PhI \quad (1)$$
$$\textbf{(3)}$$

$$R^1C\equiv CG + PhI(L)OSO_2R^2 \xrightarrow{solvent} R^1C\equiv C-\overset{+}{IPh}, {}^-OSO_2R^2 + GL \quad (2)$$
$$G = H, Me_3Si, R_3Sn$$

$$RC\equiv CSiMe_3 + Ph-I=O \xrightarrow[CH_2Cl_2]{Et_2O\cdot BF_3} \xrightarrow[H_2O]{NaBF_4} RC\equiv C-\overset{+}{IPh}, BF_4^- \quad (3)$$
$$\textbf{(4)}$$

TABLE 2. Selected preparative approaches to iodine(III) reagents employed for the synthesis of alkynyliodonium salts

Iodine(III) reagent	Starting material, Reagents, Solvent	References
$PhICl_2$	PhI, $Cl_2(g)$, $CHCl_3$	47
$PhI{=}O$	(A) $PhICl_2$, (1) anhyd. $NaHCO_3$, ice (2) 5 N NaOH	48
	(B) $PhI(OAc)_2$, 3 N NaOH	49
$PhI(OAc)_2$	(A) $PhI{=}O$, CH_3CO_2H	50
	(B) PhI, 30% H_2O_2, Ac_2O	51
	(C) PhI, 40% CH_3CO_3H	52
$PhI(O_2CCF_3)_2$	(A) $PhI(OAc)_2$, CF_3CO_2H	53,54
	(B) PhI, 80% H_2O_2, $(CF_3CO)_2O$, CH_2Cl_2	15
$PhI(OH)OTs$	$PhI(OAc)_2$, $TsOH \cdot H_2O$, MeCN or $ClCH_2CH_2Cl$	55,56
$PhI(OH)OMs$	$PhI(OAc)_2$, CH_3SO_3H, H_2O, MeCN or $CHCl_3$	6,57
$PhI{-}O{-}I{-}Ph$ with OTf OTf	(A) $PhI{=}O$, Tf_2O, CH_2Cl_2 (25 min, rt)	58
	(B) $PhI(OAc)_2$, TfOH, $CHCl_3$	59
$PhI(CN)OTf$	(A) $PhI(O_2CCF_3)_2$, (1) Me_3SiOTf (2) Me_3SiCN, CH_2Cl_2	15
	(B) $PhI{=}O$, (1) Me_3SiOTf (2) Me_3SiCN, CH_2Cl_2	60
$PhI(CN)OTs$	$PhI{=}O$, (1) Me_3SiOTs (2) Me_3SiCN, CH_2Cl_2	60
$TfOI{=}O$	HI_3O_8 (or HIO_3 or I_2O_5), I_2, TfOH	61
$TfOI(CN)_2$	$TfOI{=}O$, Me_3SiCN, CH_2Cl_2	62
$PhI{-}O\bar{B}F_3$ (with + on I)	$PhI{=}O$, $Et_2O \cdot BF_3$, CH_2Cl_2	31
$PhI{-}OEt$, BF_4^- (with + on I)	$PhI{=}O$, Et_3O^+ BF_4^-, CH_2Cl_2	31
$(PhI)_2O$, 2 BF_4^- (with +)	$PhI(OAc)_2$, 40% $HBF_4(aq)$, $CHCl_3$	29
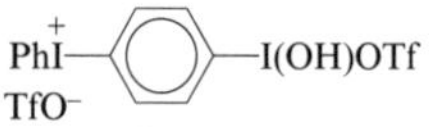 $PhI{-}\bigcirc{-}I(OH)OTf$, TfO^- (with + on PhI)	(A) $PhI{=}O$, Tf_2O, CH_2Cl_2 (12h, rt)	63
	(B) $PhI{=}O$, 2 TfOH, CH_2Cl_2 (4h, rt)	63
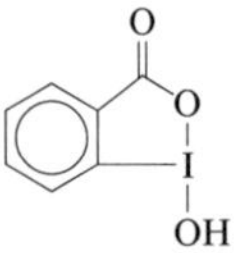 [benziodoxolone structure with O, I, OH]	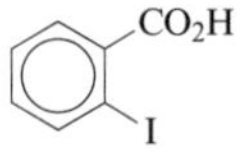 [benzoic acid structure with CO_2H, I] (1) $Cl_2(g)$, $CHCl_3$ (2) NaOH(aq)	64
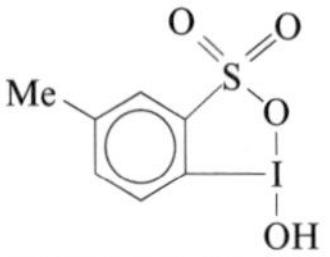 [benzothiazole-S-dioxide structure with Me, O, S, O, I, OH]	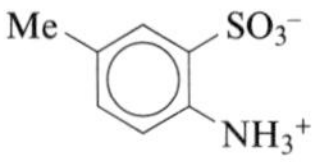 [structure with Me, SO_3^-, NH_3^+] (1) $NaNO_2$, HCl, H_2O (2) NaI (3) HCl (4) 35% CH_3CO_3H (5) H_2O	37

employed for the synthesis of alkynyliodonium salts and references (47–64) for their preparation are given in Table 2.

1. Iodanes

Since λ^3-iodanes figure prominently in the synthesis of alkynyliodonium salts, a brief description of this class of iodine(III) compounds is presented here.

λ^3-Iodanes, **5**, are hypervalent 10-I-3 compounds[65] which, except for the iodosyl congeners, contain one equatorial ligand, two apical ligands and two nonbonded electron pairs superimposed on a pseudotrigonal bipyramidal (ψ-TBP) framework[66]. λ^3-Aryliodanes, **6**, the most common members of the iodane family, can be symmetrical ($L^1 = L^2$) or unsymmetrical ($L^1 \neq L^2$), but in both types the electronegative heteroligands *invariably* occupy apical sites in the ψ-TBP. In symmetrical species such as (diacetoxyiodo)benzene (**7**), the apical ligands are held in place by 'hypercovalent' bonds and are roughly equidistant from the iodine(III) center[67,68]. On the other hand, mixed aryliodanes such as [hydroxy(tosyloxy)iodo]benzene (**8**, HTIB)[69] with one basic and one nucleofugic heteroligand are highly polarized and reside at the structural borderline between symmetrical iodanes and iodonium salts. Indeed, from the standpoint of its chemical reactivity, HTIB may be regarded as 'hydroxy(phenyl)iodonium tosylate', [PhIOH, $^-$OTs][46]. Both symmetrical and unsymmetrical λ^3-iodanes exhibit electrophilic behavior but the latter to a greater degree.

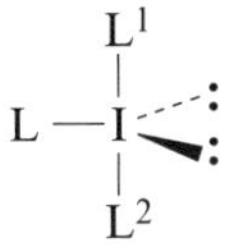

(5) 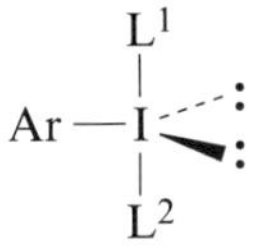   **(6)** 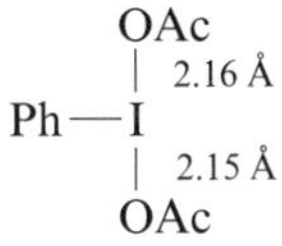   **(7)**[68] 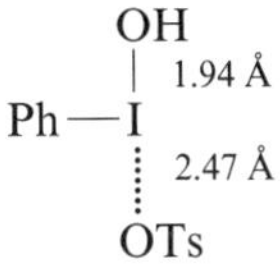   **(8)**

2. Iodosylbenzene

Iodosylbenzene (**4**), traditionally referred to as iodosobenzene, is a pale yellow amorphous solid characterized by its low solubility in common organic solvents. It does 'dissolve' in methanol, but this is due to the formation of (dimethoxyiodo)benzene, $PhI(OMe)_2$[70]. Iodosylbenzene is sometimes represented as $(PhIO)_n$ since it is thought to be polymeric in the solid state[71], exhibiting an oxygen-bridged structure, **9**, reminiscent of the structures of μ-oxo-λ^3-iodanes, **10**[72-74]. Upon heating, iodosylbenzene disproportionates to iodobenzene and iodylbenzene (i.e. iodoxybenzene, $PhIO_2$)[75], a transformation consistent with its strongly associated solid state structure.

(4) 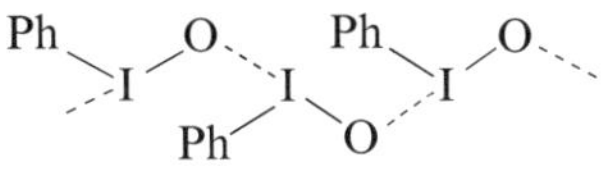   **(9)** 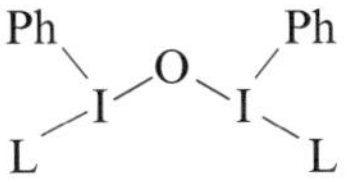   **(10)**

When iodosylbenzene is treated with boron trifluoride etherate, it is both 'depolymerized' and rendered more electrophilic, presumably because of the formation of a Lewis acid-base adduct (equation 4).

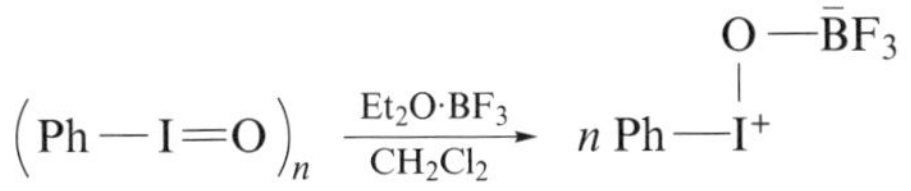

$$\left(Ph-I{=}O\right)_n \xrightarrow[\text{CH}_2\text{Cl}_2]{\text{Et}_2\text{O·BF}_3} n \; Ph-\overset{\overset{\displaystyle O-\bar{B}F_3}{\displaystyle |}}{I^+} \tag{4}$$

3. Mechanistic considerations

The conversion of terminal alkynes to alkynyl (phenyl) iodonium tosylates with HTIB, considered here as an instructive example, is best understood as a polar process initiated by electrophilic attack of HTIB on the triple bond of the alkyne (equation 5)[46]. Loss of H_2O

from the resulting iodanylvinyl cation, either directly (path a) or indirectly (path b), would give an alkynyliodonium tosylate.

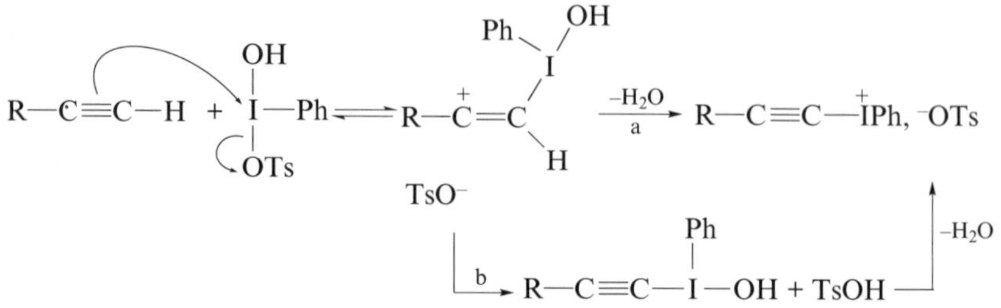

$$R—C\equiv C—H \; + \; \overset{+}{I}—Ph \rightleftharpoons R—\overset{+}{C}=C \quad \xrightarrow[a]{-H_2O} \quad R—C\equiv C—\overset{+}{I}Ph, \; {}^-OTs \quad (5)$$

$$\xrightarrow{b} \; R—C\equiv C—\overset{Ph}{\underset{}{I}}—OH + TsOH$$

As might be expected from such a mechanism, the capacity of trialkysilyl groups[76] and, to an even greater degree, trialkylstannyl groups[77] to stabilize β-carbocations has been exploited to advantage for alkynyliodonium salt synthesis. For example, alkynylstannanes condense readily with [cyano(trifluoromethyl(sulfonyl)oxy)iodo]benzene in dichloromethane at $-42\ °C$, thus providing access to alkynyliodonium triflates with electron-withdrawing groups in the acetylenic moiety (e.g. equation 6)[17].

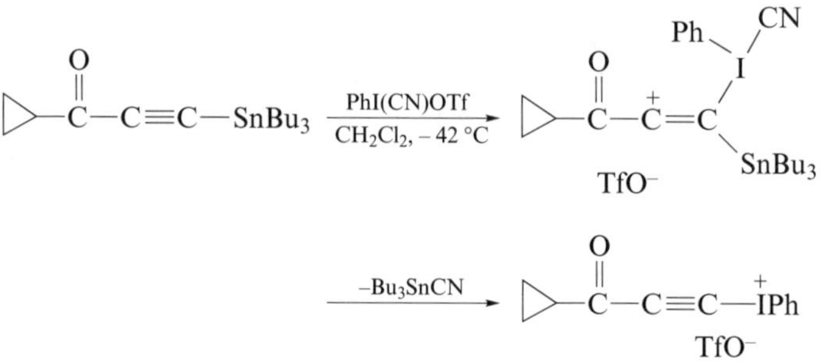

$$\xrightarrow[CH_2Cl_2, \, -42\ °C]{PhI(CN)OTf}$$

$$\xrightarrow{-Bu_3SnCN} \quad (6)$$

The production of alkynyliodonium tetrafluoroborates from alkynylsilanes with $PhIO·BF_3$ may be rationalized in a similar way (equation 7)[31].

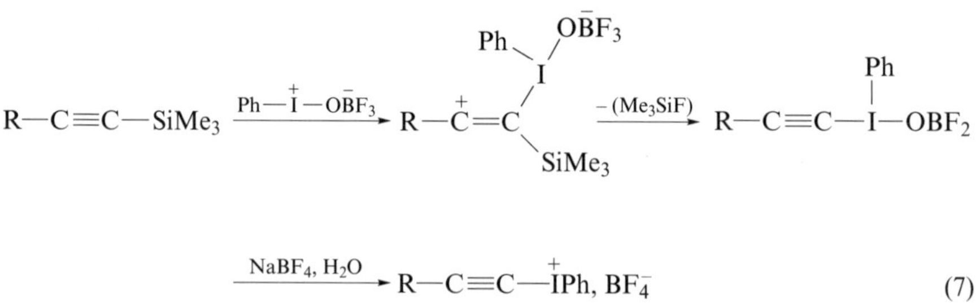

$$R—C\equiv C—SiMe_3 \xrightarrow{Ph—\overset{+}{I}—O\bar{B}F_3} R—\overset{+}{C}=C \xrightarrow{-(Me_3SiF)} R—C\equiv C—I—OBF_2$$

$$\xrightarrow{NaBF_4, \, H_2O} R—C\equiv C—\overset{+}{I}Ph, \; BF_4^- \qquad (7)$$

Finally, it is important to note that in each of the foregoing examples, the iodine(III) reagent bears a ligand which exhibits a high affinity for the atom or group attached to C–1 of the alkyne.

B. Synthesis

1. Alkynyliodonium tosylates and mesylates

The synthesis of the first alkynyliodonium tosylates was achieved by the treatment of terminal alkynes with [hydroxy(tosyloxy)iodo]benzene (HTIB) (equation 8)[8,10,11]. Such reactions are generally conducted with an excess of alkyne in chloroform at reflux, although they can be carried out at room temperature, and dichloromethane can be employed as solvent. This procedure is, however, restricted to terminal alkynes in which R is either an aryl group or a bulky alkyl group. With linear alkyl groups (i.e. R = n-Pr, n-Bu, n-C$_5$H$_{11}$), phenyl(β-tosyloxyvinyl)iodonium tosylates are obtained instead (equation 9)[8]. In some cases (R = i-Pr, i-Bu), mixtures of alkynyl- and (β-tosyloxyvinyl)iodonium tosylates are produced[8]. tert-Butylacetylene appears to be the optimum substrate for this approach and has been employed with a series of [hydroxy(tosyloxy)iodo]arenes for the synthesis of various aryl(tert-butylethynyl)iodonium tosylates (equation 10)[9].

$$RC\equiv CH + PhI(OH)OTs \xrightarrow{CHCl_3} RC\equiv C-\overset{+}{I}Ph, \ ^-OTs + H_2O \qquad (8)$$

R (yield): i-Pr (15%), i-Bu (33%), s-Bu (50.5%), t-Bu (74%),
cyclopentyl (27%), cyclohexyl (47%), Ph (61%), 4-MeC$_6$H$_4$ (44%)

$$RC\equiv CH + PhI(OH)OTs \xrightarrow[\Delta]{CHCl_3} \underset{TsO}{\overset{R}{\diagdown}}C=CH-\overset{+}{I}Ph, \ ^-OTs \qquad (9)$$

R (yield): n-Pr (58%), n-Bu (52%), n-C$_5$H$_{11}$ (26%),
i-Pr (11%), i-Bu (29%)

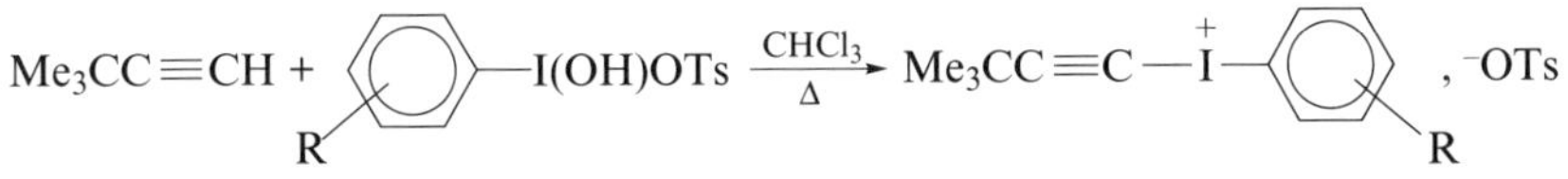

$$Me_3CC\equiv CH + \underset{R}{\diagup}\!\!\langle\bigcirc\rangle\!\!-I(OH)OTs \xrightarrow[\Delta]{CHCl_3} Me_3CC\equiv C-\overset{+}{I}-\langle\bigcirc\rangle\underset{R}{\diagdown}, \ ^-OTs$$

R (yield): 2-Me (80%), 3-Me (79%), 4-Me (56%), 2-F (64%), 3-F (65.5%), $\qquad$ (10)
4-F (76%), 3-Cl (69%), 4-Cl (63%)

The incorporation of propyne and 1-hexyne into alkynyliodonium salts with HTIB and its mesyloxy analog has been accomplished with the aid of a silica bead desiccant (equation 11)[5,6], but the yields of the products are low. A better method for the synthesis of alkynyliodonium tosylates in which R is a linear alkyl group entails the treatment of (trimethylsilyl)alkyne/iodosylbenzene mixtures in chloroform with boron trifluoride etherate. When aqueous sodium tosylate is added to the resulting *solutions*, alkynyliodonium tosylates are produced and can be isolated from the organic phase (equation 12)[7]. The

$$RC\equiv CH + PhI(OH)OSO_2R' \xrightarrow[rt]{CH_2Cl_2, \ desiccant} RC\equiv C-\overset{+}{I}Ph, \ ^-OSO_2R' \qquad (11)$$

R' = Me; R (yield): Me (35%), n-Bu (27%), t-Bu (72%)
R' = p-tolyl; R (yield): Me (19%), n-Bu (12%), Ph (61%)

$$RC\equiv CSiMe_3 + PhI=O \xrightarrow[0\,°C \ to \ rt]{Et_2O \cdot BF_3, \ CHCl_3} \xrightarrow[2. \ NaOTs, \ H_2O]{1.\ 0\,°C} RC\equiv C-\overset{+}{I}Ph, \ ^-OTs$$

R (yield): Me (62%), Et (81%), n-Pr (89%), n-Bu (76%), Me$_3$Si (70%) $\qquad$ (12)

initial stage of this procedure is borrowed from one of the standard approaches to alkynyliodonium tetrafluoroborates[31]. It is noteworthy that [(trimethylsilyl)ethynyl]-(phenyl)iodonium tosylate can be made from bis(trimethylsilyl)acetylene by this method.

2. Alkynyliodonium triflates

The first alkynyliodonium triflates were obtained by the passage of alkynyl(phenyl)-iodonium tosylates through a triflate-loaded anion exchange resin (equation 13)[6]. However, they are more generally prepared by the treatment of alkynylsilanes or alkynyl-stannanes with iodine(III) triflate reagents. In one approach, 1-trimethylsilyl- or 1-(trialkylstannyl)alkynes are mixed with diphenyl-μ-oxodiiodine(III) bistrifluoromethane-sulfonate, generated *in situ* from iodosylbenzene and triflic anhydride, in dichloromethane at 0 °C (equations 14 and 15)[13,16]. Once the solvent is removed, the alkynyl(phenyl)-iodonium triflates can be precipitated from the residual material with pentane or ether and purified if necessary by recrystallization[16]. This method has been employed for the synthe-sis of the parent compound[13].

$$RC\equiv C\overset{+}{-}IPh,\ ^-OTs \xrightarrow{\text{TfO}^-\text{--resin, MeOH (aq)}} RC\equiv C\overset{+}{-}IPh,\ ^-OTf \qquad (13)$$

R (yield): *n*-Bu (85%), *t*-Bu (88%)

$$RC\equiv CSiMe_3 + \quad \begin{array}{c} Ph\ \diagdown\ O\ \diagup\ Ph \\ \diagup I\ \diagdown\ I\ \diagdown \\ TfO \qquad OTf \end{array} \quad \xrightarrow{CH_2Cl_2,\ 0\,°C} RC\equiv C\overset{+}{-}IPh,\ ^-OTf + (Me_3Si)_2O$$

R (yield): *t*-Bu (76%), Ph (67%), Me$_3$Si (88%) $\qquad (14)$

$$RC\equiv CSnBu_3 + \quad \begin{array}{c} Ph\ \diagdown\ O\ \diagup\ Ph \\ \diagup I\ \diagdown\ I\ \diagdown \\ TfO \qquad OTf \end{array} \quad \xrightarrow{CH_2Cl_2,\ 0\,°C} RC\equiv C\overset{+}{-}IPh,\ ^-OTf + (Bu_3Sn)_2O$$

$\qquad (15)$

R (yield): H (56%), *n*-Bu (45%)

A related but more versatile method for the synthesis of alkynyl(phenyl)iodonium triflates entails the treatment of (trialkylstannyl)alkynes with [cyano(trifluoromethyl-(sulfonyl)oxy)iodo]benzene, typically in dichloromethane at low temperatures. This approach is applicable to an exceptionally wide range of alkynylstannane structures including those with electron-withdrawing groups (equations 16 and 17)[17–19,33]. The preparation of alkynyliodonium triflates from alkynyl ketones and alkynamides by this method is particularly noteworthy[17,33]. [Cyano(trifluoromethyl(sulfonyl)oxy)iodo]-benzene has been similarly employed for the synthesis of 1,3-diynyl(phenyl)iodonium triflates (equation 18)[34], aromatic and aliphatic bis(phenyliodonium)diyne ditriflates (equations 19 and 20)[40,42,43] and the bisphenyliodonium ditriflate of acetylene, a novel 'C$_2$-transfer agent' (equation 21)[40,41]. The related λ^3-iodanes, PhI(CN)O$_3$SC$_4$F$_9$ and ArI(CN)OTf, have been used in a few instances for the preparation of alkynyliodonium nonaflates[40] and aryl(1-propynyl)iodonium triflates (equation 22)[15]. The synthesis of bis(phenyliodonium)-1,7-octadiyne and -1,8-nonadiyne tosylates from distannanes with [cyano(tosyloxy)iodo]benzene, PhI (CN) OTs, has also been reported[43].

The treatment of alkynylsilanes in acetonitrile with an iodine(III) reagent prepared from iodosylbenzene and two molar equivalents of triflic acid, believed to exhibit structure

$$RC\equiv CSnR'_3 + PhI(CN)OTf \xrightarrow[-40\,°C\ or\ -42\,°C]{CH_2Cl_2} RC\equiv C\overset{+}{-}IPh,\ ^-OTf + R'_3SnCN \quad (16)$$

R (yield): 1-cyclohexenyl (73%), MeC(OH)Ph (80%), MeOCH$_2$ (77%), ClCH$_2$ (59%), BrCH$_2$ (76%), Cl (72%), $-$CN (72%), Ts (85%)

$$\overset{\displaystyle O}{\overset{\|}{RC}}-C\equiv CSnR'_3 + PhI(CN)OTf \xrightarrow[-42\,°C]{CH_2Cl_2} \overset{\displaystyle O}{\overset{\|}{RC}}-C\equiv C\overset{+}{-}IPh,\ ^-OTf + R'_3SnCN \quad (17)$$

R (yield): t-Bu (82%), Me$_2$(Et)C (75%), cyclopropyl (59%), cyclobutyl (58%), cyclohexyl (47%), 1-adamantyl (52%), Ph (77%), 2-furyl (75%), 2-thienyl (88%),

MeO (42%), Me$_2$N (89%), ☐N— (45%), ⬠N— (79%), ⬡N— (55%),

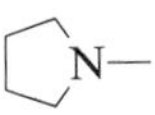

O☐N— (82%)

$$RC\equiv C-C\equiv CSnBu_3\text{-}n + PhI(CN)OTf$$

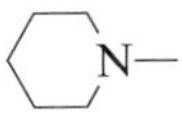

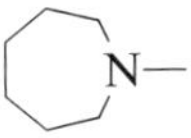

$$\xrightarrow[-40\,°C]{CH_2Cl_2} RC\equiv C-C\equiv C\overset{+}{-}IPh,\ ^-OTf + (n\text{-}Bu)_3SnCN \quad (18)$$

R (yield): Me (72%), n-Bu (80%), t-Bu (84%), Me$_3$Si (96%), Ph (80%)

$$R_3SnC\equiv C(CH_2)_nC\equiv CSnR_3 + 2PhI(CN)OTf \xrightarrow[-78\,°C\ to\ 25\,°C]{CH_2Cl_2}$$

$$Ph\overset{+}{I}-C\equiv C(CH_2)_nC\equiv C\overset{+}{-}I\ Ph,\ 2\ ^-OTf \quad (19)$$

n (yield): 0 (85%, T = –30 °C), 2 (90%), 4 (93%), 5 (90%), 6 (89%), 8 (90%)

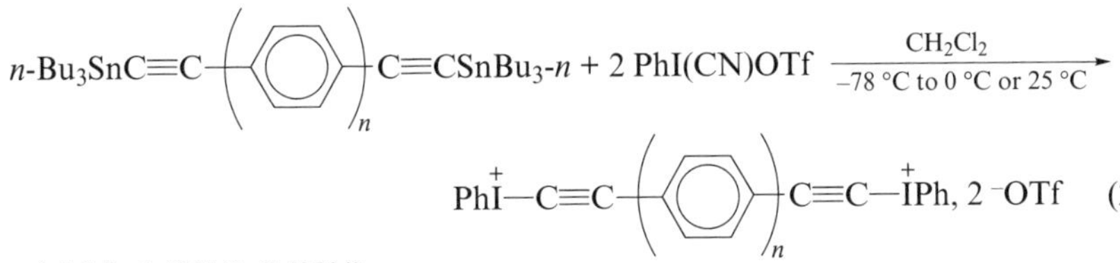

$$n\text{-}Bu_3SnC\equiv C\!-\!\!\left(\!\!\left\langle\!\bigcirc\!\right\rangle\!\!\right)_{\!\!n}\!\!-\!C\equiv CSnBu_3\text{-}n + 2\ PhI(CN)OTf \xrightarrow[-78\,°C\ to\ 0\,°C\ or\ 25\,°C]{CH_2Cl_2}$$

$$Ph\overset{+}{I}-C\equiv C\!-\!\!\left(\!\!\left\langle\!\bigcirc\!\right\rangle\!\!\right)_{\!\!n}\!\!-\!C\equiv C\overset{+}{-}IPh,\ 2\ ^-OTf \quad (20)$$

n (yield): 1 (82%), 2 (92%)

$$n\text{-}Bu_3SnC\equiv CSnBu_3\text{-}n + 2\ PhI(CN)OTf \xrightarrow[-78\,°C\ to\ 0\,°C]{CH_2Cl_2} Ph\overset{+}{I}-C\equiv C\overset{+}{-}IPh,\ 2\ ^-OTf \quad (21)$$

$$(81\%)$$

$$MeC\equiv CSnBu_3\text{-}n + ArI(CN)OTf \xrightarrow[-40\,°C\ to\ rt]{CH_2Cl_2} MeC\equiv C\overset{+}{-}IAr,\ ^-OTf \quad (22)$$

Ar (yield): 4-FC$_6$H$_4$ (73%), 4-CF$_3$C$_6$H$_4$ (70%), 3,5-(CF$_3$)$_2$C$_6$H$_3$ (40%), Ph (85%)

11[39,63], affords monoalkynyl monophenyl bisiodonium(p-phenylene) ditriflates (equation 23)[38]. The iodine(III) reagent must be isolated before it can be used for this purpose; i.e. when it was generated in situ (in CH$_2$Cl$_2$) and treated with 1-trimethylsilyl-1-hexyne, a vinyliodonium salt was also obtained[38].

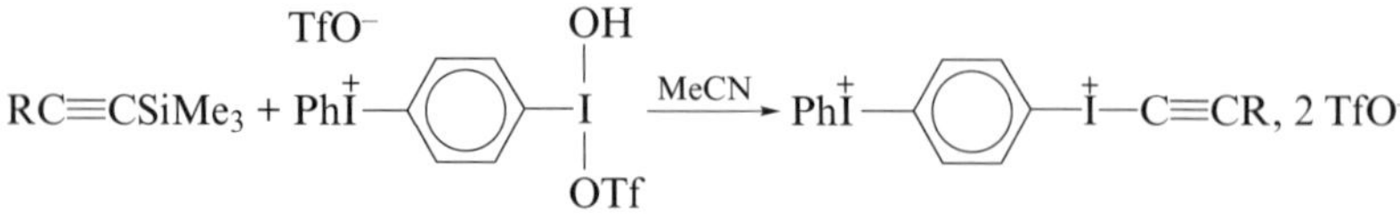

$$RC{\equiv}CSiMe_3 + PhI{\overset{\overset{\displaystyle TfO^-}{+}}{-}}{\underset{\underset{\displaystyle OTf}{|}}{\overset{\overset{\displaystyle OH}{|}}{\underset{}{I}}}} \xrightarrow{MeCN} Ph{\overset{+}{I}}{-}{\underset{}{\underset{}{}}}{-}{\overset{+}{I}}{-}C{\equiv}CR,\ 2\ TfO^- \quad (23)$$

(11)

R (yield): n-Bu (76%), t-Bu (83%), n-C$_6$H$_{13}$ (49%), Me$_3$Si (76%), Ph (82%)

General methods for the *direct* conversion of terminal alkynes (i.e. without silyl or stannyl activation) to alkynyliodonium triflates have not been described. The preparation of (*tert*-butylethynyl)phenyliodonium triflate from *tert*-butylacetylene with a 1:1 molar mixture of iodosylbenzene and triflic acid [PhIO–TfOH] has been reported[78], but with other terminal alkynes, this procedure affords β-(trifluoromethanesulfonyloxyvinyl)-iodonium triflates[78].

3. Alkynyliodonium tetrafluoroborates

In 1985, two general methods for the synthesis of alkynyl(phenyl)iodonium tetrafluoroborates, based on coupling reactions between alkynylsilanes and iodosylbenzene, were communicated[31]. Iodosylbenzene by itself is unreactive with 1-trimethylsilyl-1-decyne, and presumably with other silylalkynes, in dichloromethane[31]. However, when such mixtures are exposed to triethyloxonium tetrafluoroborate, alkynyliodonium tetrafluoroborates are produced (equation 24)[31]. In a related approach requiring shorter reaction times, boron trifluoride etherate is employed as the coupling reagent, and aqueous sodium tetrafluoroborate is subsequently introduced (equation 25)[27,30,31]. Among the known alkynyliodonium tetrafluoroborates, five were reported in the 1985 communication. Later publications, in which newer members of this family are introduced, are largely focused on their chemical reactivity, and yields and characterization data for the iodonium compounds are not given. However, for most of the tetrafluoroborate salts listed in Table 1, such information is presented in Reference 25.

$$RC{\equiv}CSiMe_3 + PhI{=}O \xrightarrow[CH_2Cl_2,\ rt]{Et_3O^+,\ BF_4^-} RC{\equiv}C{-}{\overset{+}{I}}Ph,\ BF_4^- \quad (24)$$

R (yield): n-C$_8$H$_{17}$ (70%), PhCH$_2$ (56%), PhCH$_2$CH$_2$ (75%), cyclohexyl (64%), Ph (65%)

$$RC{\equiv}CSiMe_3 + PhI{=}O \xrightarrow[CH_2Cl_2,\ rt]{Et_2O{\cdot}BF_3} \xrightarrow{NaBF_4,\ H_2O} RC{\equiv}C{-}{\overset{+}{I}}Ph,\ BF_4^- \quad (25)$$

R (yield): n-C$_8$H$_{17}$ (85%), PhCH$_2$ (53%), Ph (79%), Me$_3$Si (83%)[27], HOCH$_2$CH$_2$ (68 %, 0 °C)[30]

The treatment of (trimethylsilyl)acetylene with iodosylbenzene and boron trifluoride etherate does *not* give ethynyl(phenyl)iodonium tetrafluoroborate, but leads instead to (*E*)-1-trimethylsilyl-2-ethoxyethynyl(phenyl)iodonium tetrafluoroborate (equation 26)[79].

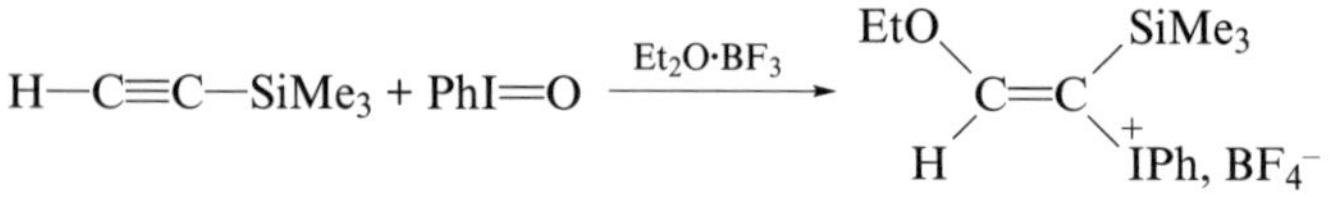

$$H{-}C{\equiv}C{-}SiMe_3 + PhI{=}O \xrightarrow{Et_2O{\cdot}BF_3} \underset{H}{\overset{EtO}{\diagdown}}C{=}C\underset{\overset{+}{I}Ph,\ BF_4^-}{\overset{SiMe_3}{\diagup}} \quad (26)$$

42%

(68% with Et$_3$O$^+$, BF$_4^-$)

However, when [(trimethylsilyl)ethynyl]phenyliodonium tetrafluoroborate is treated with 46% aqueous hydrofluoric acid, protodesilylation occurs, and ethynyl(phenyl)iodonium tetrafluoroborate is obtained in high yield (equation 27)[27].

$$Me_3Si—C≡C—\overset{+}{I}Ph, BF_4^- \quad \xrightarrow[\text{rt, 30 min}]{\text{46% HF (2 equiv), CH}_2\text{Cl}_2} \quad H—C≡C—\overset{+}{I}Ph, BF_4^- \quad (27)$$
$$94\%$$

The BF$_3$-induced reactions of iodosylbenzene with 1-trimethylsilyl-3-aryl-1-propynes in dioxane follow a divergent course and afford aryl silylethynyl ketones (equation 28)[31]. The ketone derived from 1-trimethylsilyl-3-phenyl-1-propyne is also generated in dichloromethane, but as a minor product. These oxidations are apparently facilitated by the presence of benzylic hydrogen atoms, since a similar alkynyl ketone is not obtained from 1-trimethylsilyl-1-decyne with PhIO/BF$_3$ in dioxane[31].

$$ArCH_2C≡CSiMe_3 + PhI=O \quad \xrightarrow[\text{dioxane, rt}]{Et_2O·BF_3} \quad Ar—\overset{\overset{\displaystyle O}{\|}}{C}—C≡CSiMe_3 \quad (28)$$
$$\text{Ar (yield): Ph (69\%), 4-An (52\%)}$$

Two methods for the 'direct' conversion of terminal alkynes to alkynyl(phenyl)iodonium tetrafluoroborates have been reported, but their generality has not been documented in the literature. The *tert*-butylethynyl salt has been prepared by the generation of lithium *tert*-butylethynyl(trifluoro)borate and its coupling with iodosylbenzene by the standard method (equation 29)[80]. The treatment of 1-pentyne and phenylacetylene with diphenyl-μ-oxodiiodine(III) bistetrafluoroborate in dichloromethane likewise affords the corresponding alkynyliodonium tetrafluoroborates (equations 30)[29].

$$Me_3CC≡CH \quad \xrightarrow[Et_2O·BF_3]{n\text{-BuLi}} \quad Me_3CC≡C—\overset{-}{B}F_3, Li^+$$

$$\xrightarrow[\text{2. NaBF}_4, \text{H}_2\text{O}]{\text{1. PhIO, Et}_2\text{O·BF}_3, \text{CH}_2\text{Cl}_2} \quad Me_3CC≡C—\overset{+}{I}Ph, BF_4^- \quad (29)$$
$$85\%$$

$$RC≡CH + Ph\overset{+}{I}—O—\overset{+}{I}Ph, 2\ BF_4^- \quad \xrightarrow{\text{CH}_2\text{Cl}_2, \text{rt}} \quad RC≡C—\overset{+}{I}Ph, BF_4^- \quad (30)$$

$$\text{R (yield): } n\text{-Pr (42\%), Ph (—)}$$

4. Heterocyclic alkynyliodonium salts

Carboxylate ions (i.e. ArCO$_2^-$, RCO$_2^-$) are generally too nucleophilic to serve as counterions in alkynyliodonium salts[81]. 1,4-Bis[(phenyliodonium)ethynyl]benzene ditrifluoroacetate, prepared from 1,4-diethynylbenzene and [bis(trifluoroacetoxy)iodo]-benzene in *dry* chloroform (equation 31)[44], is the only reported example of an isolable acyclic alkynyliodonium carboxylate. The only other known carboxylate 'salts' are those in which the carboxylate group is incorporated into a heterocyclic iodine(III) substructure[36]. When mixtures of 1-hydroxy-1,2-benziodoxol-3(1H)-one ('o-iodosylbenzoic acid') and 1-(trimethylsilyl)alkynes in dichloromethane are exposed to boron trifluoride etherate, 1-(1-alkynyl)-1,2-benziodoxol-3 (1H)-ones are obtained (equation 32)[36]. The stability of these compounds is due at least in part to their heterocyclic nature, a known stabilizing phenomenon for λ^3-iodanes[66,82,83]. Furthermore, Michael addition of the

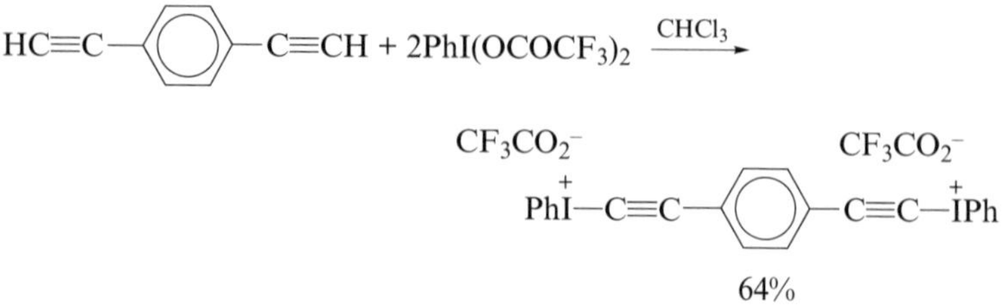

$$HC\equiv C-\text{C}_6H_4-C\equiv CH + 2PhI(OCOCF_3)_2 \xrightarrow{CHCl_3}$$

$$\underset{PhI}{\overset{CF_3CO_2^-}{\underset{+}{PhI}}}-C\equiv C-\text{C}_6H_4-C\equiv C-\underset{+}{\overset{CF_3CO_2^-}{IPh}} \quad (31)$$

64%

$$\text{(benziodoxolone, } O-I-OH) + Me_3SiC\equiv CR \xrightarrow[\text{CH}_2\text{Cl}_2, \text{ rt}]{Et_2O\cdot BF_3} \text{(benziodoxolone, } O-I-C\equiv CR) \quad (32)$$

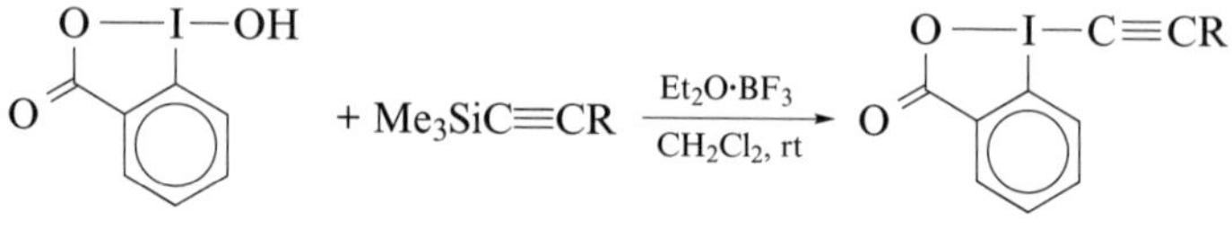

R (yield): cyclohexyl (34%), n-C$_8$H$_{17}$ (22%), t-Bu (35%)

carboxylate function to the triple bond, a common mode of reactivity between alkynyl-iodonium ions and nucleophiles, would be constrained in the benziodoxolones to a cyclic seven-membered intermediate.

The treatment of $1H$-1-hydroxy-1,2,3-benziodoxathiole-3,3-dioxide, a heterocyclic analog of HTIB, with various terminal alkynes in the presence of p-toluenesulfonic acid (MeCN, reflux) affords $1H$-1-alkynyl-1,2,3-benziodoxathiole-3,3-dioxides (equation 33)[37]. It is noteworthy that these reactions permit the introduction of linear alkyl groups into alkynyl(aryl)iodonium arenesulfonates without the aid of a desiccant or the use of silyl-substituted acetylenes. The hydroxybenziodoxathiole is less reactive than HTIB with terminal alkynes, presumably because of its low solubility in hot acetonitrile and its more stable heterocyclic structure[37]. It seems likely that the role of tosylic acid is to provide the ring-opened species **12**, which is both more electrophilic at iodine and more soluble in the hot solvent[37]. Finally, the yields of the alkynylbenziodoxathioles appear to depend on the alkyne/hydroxyiodane ratios; i.e. the greater the excess of alkyne, the higher the yield.

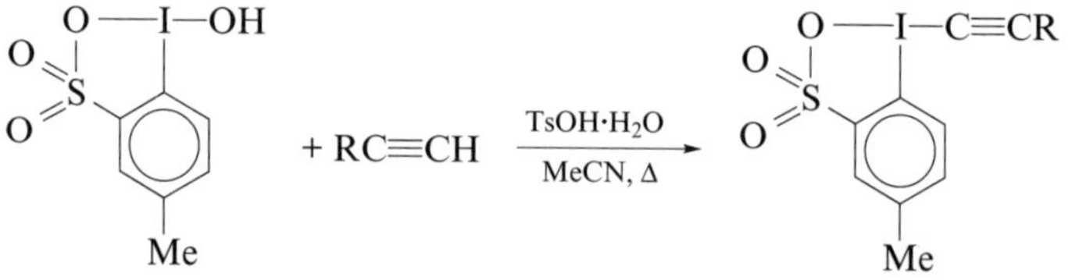

$$\text{(benziodoxathiole-dioxide, } O-I-OH) + RC\equiv CH \xrightarrow[\text{MeCN, }\Delta]{TsOH\cdot H_2O} \text{(benziodoxathiole-dioxide, } O-I-C\equiv CR) \quad (33)$$

R (yield): n-Pr (61%), n-Bu (54%), n-C$_5$H$_{11}$ (41%), n-C$_6$H$_{13}$ (51%), i-Bu (66%), s-Bu (32%), t-Bu (70%), cyclohexyl (26%), Ph (51%)

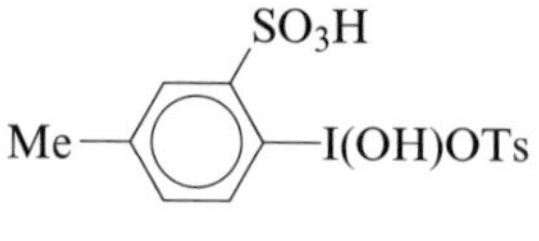

$$Me-\text{C}_6H_3(SO_3H)-I(OH)OTs$$

(12)

5. Dialkynyliodonium salts

The coupling of aryl λ^3-iodanes, most notably (diacetoxyiodo)benzene and its substituted analogs, with aromatic substrates is an important method for the synthesis of diaryliodonium salts (equation 34)[1]. Although a similar procedure for the preparation of symmetrical and unsymmetrical dialkynyliodonium salts might be envisioned (equation 35), alkynyl λ^3-iodanes are *unknown*, thereby precluding such an approach. The synthesis of symmetrical dialkynyliodonium triflates can, however, be accomplished by the treatment of alkynylsilanes with iodosyl triflate (equation 36)[35] or alkynylstannanes with dicyanoiodonium triflate (equation 37)[62].

$$ArI(OAc)_2 + Ar'H \xrightarrow{\text{H}^+} \xrightarrow{\text{MX}} Ar\overset{+}{I}Ar', X^- \qquad (34)$$

$$RC\equiv CIL^1L^2 + R'C=CH \xrightarrow[\text{(unknown)}]{} RC\equiv C-\overset{+}{I}-C\equiv CR, X^- \qquad (35)$$

$$2\,RC\equiv CSiMe_3 + TfO-I=O \xrightarrow[-78\ °C\ to\ rt]{CH_2Cl_2} \left(RC\equiv C\right)_2\overset{+}{I}, TfO^- \qquad (36)$$

R (yield): *t*-Bu (46%), Me$_3$Si (43%), (*i*-Pr)$_3$Si (83%)

$$2\,RC\equiv CSnBu_3\text{-}n + TfO-I(CN)_2 \xrightarrow[-40\ °C\ to\ 20\ °C]{CH_2Cl_2} \left(RC\equiv C\right)_2\overset{+}{I}, TfO^- \qquad (37)$$

R (yield): *t*-Bu (55%), (*i*-Pr)$_3$Si (63%)

C. Structure

Single-crystal X-ray structures of six alkynyliodonium compounds, all containing oxyanions and including four alkynyl(phenyl)iodonium salts, one alkynylbenzodoxolone and one dialkynyliodonium salt, have been reported. Selected structural data for these compounds are given in Table 3.

The alkynyl(phenyl)iodonium salts and the alkynylbenzodoxolone are roughly T-shaped in the solid state. Like the λ^3-iodanes, they are generally regarded as ψ-TBP species in which the iodine(III) atom is bonded to one equatorial and two apical ligands. Given the preference of the anion in such compounds for an apical site, two geometric isomers of a given alkynyl(phenyl)iodonium salt might be expected, one in which the alkynyl ligand is apical (structure 13) and one in which the aryl ligand is apical (structure 14). However, alkynyl groups are more electronegative than aryl groups, and in all cases studied so far, only one geometric variant (i.e. 13) has been observed. (Phenylethynyl)phenyliodonium tosylate (15) is typical, exhibiting $C_{sp}-I-C_{sp^2}$, $C_{sp}-I-O$ and $C_{sp^2}-I-O$ angles of 95.0(10)°, 170.9(8)° and 76.8(8)°, respectively[6]. As expected for an iodonium salt, the I—OTs bond in 15 (2.56 Å) is 0.57 Å longer than the computed covalent bond distance, thereby reflecting its ionic character[6].

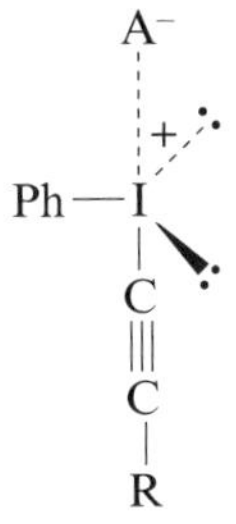

(13) 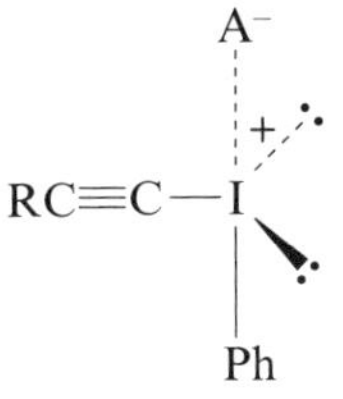   (14) 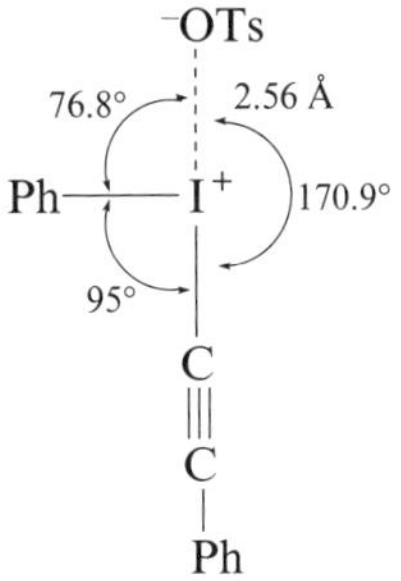   (15)

TABLE 3. Selected X-ray structural data for alkynyliodonium salts

Structure	Bond angles (deg.)				Bond distances (Å)				Reference
	$C_{sp}IC_{sp^2}$	$C_{sp}IO$	$C_{sp^2}IO$		$C_{sp}I$	$C_{sp^2}I$	IO		
PhC≡C—I—OTs Ph	95.0(10)	170.9(8)	76.8(8)		1.969(30)	2.120(21)	2.556(19)		6
HC≡C—I—OTf Ph	93.2(2)	170.9(1)	—		2.017(4)	2.108(4)	2.620(3)		13
NCC≡C—I—OTf Ph	92.1(4) 93.6(5)	— —	— —		2.00(1) 2.01(1)	2.11(1) 2.10(1)	2.56 2.62		17
			(two crystallographically distinct species)						
t-BuC≡C—I—OIO₃ Ph	93.1(4)	172.4(4)	81.1(3)		2.006(13)	2.124(9)	2.618(9)		24
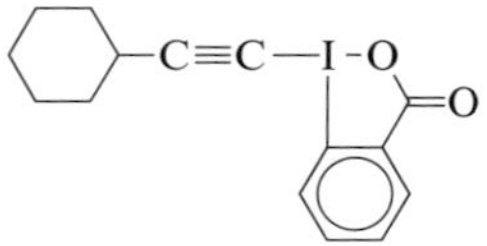 [cyclohexyl C≡C—I—O benzofuranone]	90.9(6)	166.7(5)	75.8(4)		2.03(2)	2.14(1)	2.34(1)		36
i-Pr₃SiC≡C—I—OTf i-Pr₃Si—C≡C	92.6(4)	~90° ~180°	— —		2.02(1) 2.01(1)	—	2.7		35

D. Reactions with Nucleophilic Species

1. Historical perspective

In their 1965 paper[26], Beringer and Galton described two reactions of (phenylethynyl)-phenyliodonium chloride which, taken together, anticipated discoveries to come twenty years later. When the iodonium salt was mixed with sodium 2-phenyl-1,3-indandionate in tert-butyl alcohol, alkynylation of the enolate ion occurred, and 2-phenyl-2-phenylethynyl-1,3-indandione was obtained in 73% yield (equation 38)[26]. This transformation is not unexpected, since iodobenzene is an excellent leaving group, and nucleophilic substitution at the *alpha*–carbon atom of the acetylenic group by an addition–elimination (Ad-E) sequence (equation 39) can be envisioned. However, when the alkynyliodonium chloride was treated with aqueous NaBF$_4$ in an attempt to prepare the corresponding tetrafluoroborate salt, a(β-chlorovinyl)iodonium compound was obtained instead (equation 40)[26]. This was the first indication that alkynyliodonium ions might serve as viable partners in Michael reactions.

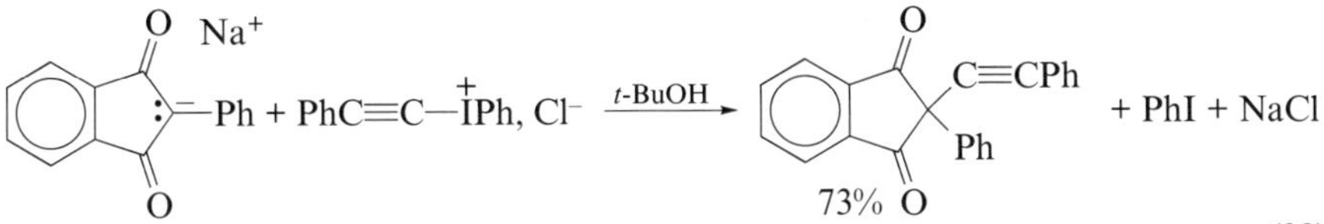

$$\text{(indandione)}\!=\!\text{Ph} + \text{PhC}\!\equiv\!\text{C}-\overset{+}{\text{I}}\text{Ph, Cl}^- \xrightarrow{t\text{-BuOH}} \text{(indandione)}\begin{smallmatrix}\text{C}\equiv\text{CPh}\\\text{Ph}\end{smallmatrix} + \text{PhI} + \text{NaCl}$$

73% (38)

$$\text{RC}\!\equiv\!\text{C}-\overset{+}{\text{I}}\text{Ph} + :\text{Nu}^- \;\rightleftharpoons\; \text{R}-\overset{..}{\text{C}}\!=\!\text{C}\begin{smallmatrix}\overset{+}{\text{I}}\text{Ph}\\\text{Nu}\end{smallmatrix} \longrightarrow \text{RC}\!\equiv\!\text{C}-\text{Nu} + \text{PhI} \quad (39)$$

$$\text{Ph}-\text{C}\!\equiv\!\text{C}-\overset{+}{\text{I}}\text{Ph, Cl}^- \xrightarrow{\text{NaBF}_4,\ \text{H}_2\text{O}} \begin{smallmatrix}\text{Ph}\\\text{Cl}\end{smallmatrix}\!\text{C}\!=\!\text{CH}-\overset{+}{\text{I}}\text{Ph, BF}_4^- \quad (40)$$

50%

Despite the synthetic possibilities suggested by this early study, the chemistry of the alkynyliodonium salts lay dormant until the mid-1980s. In 1986, Ochiai and his coworkers published an important communication which shaped much of the later thinking on the reactions of alkynyliodonium ions with nucleophiles[28]. When β-dicarbonyl enolates are treated with alkynyliodonium tetrafluoroborates containing a long (≥ three carbons) alkyl chain, derivatives of cyclopentene are produced. This is illustrated in equation 41 for the

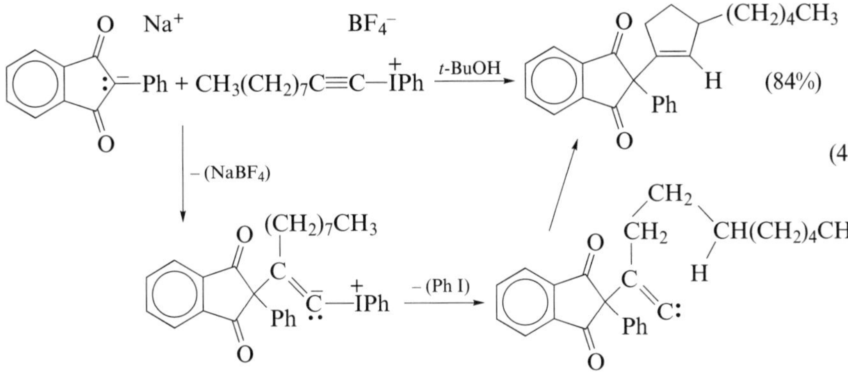

$$\text{(indandione)}\!=\!\text{Ph} + \text{CH}_3(\text{CH}_2)_7\text{C}\!\equiv\!\text{C}-\overset{+}{\text{I}}\text{Ph}\ \ \text{BF}_4^- \xrightarrow{t\text{-BuOH}} \text{(product with (CH}_2)_4\text{CH}_3)\quad (84\%)$$

(41)

$$\Big\downarrow -(\text{NaBF}_4)$$

$$\text{(indandione intermediate, (CH}_2)_7\text{CH}_3,\ \overset{..}{\text{C}}-\overset{+}{\text{I}}\text{Ph}) \xrightarrow{-(\text{Ph I})} \text{(carbene intermediate)}$$

reaction of sodium 2-phenyl-1,3-indandionate with 1-decynyl(phenyl)iodonium tetrafluoroborate in *tert*-butyl alcohol[28]. A mechanism for such reactions (i.e. the 'MCI' mechanism), involving Michael addition of the carbanion to the β-carbon atom of the alkynyliodonium ion to give an intermediate iodonium ylide, was proposed. Elimination of iodobenzene from the ylide to give an alkylidenecarbene and insertion of the carbenic center into a γ-CH bond of the *n*-octyl side chain would give the cyclopentene derivative.

The alkyl chain in such MCI reactions need not be restricted to the alkynyliodonium component. For example, the treatment of sodium 2-*n*-hexyl-1,3-indandionate with 1-propynyl(phenyl)iodonium tetrafluoroborate in THF affords the spiro-nonene system shown in equation 42[28].

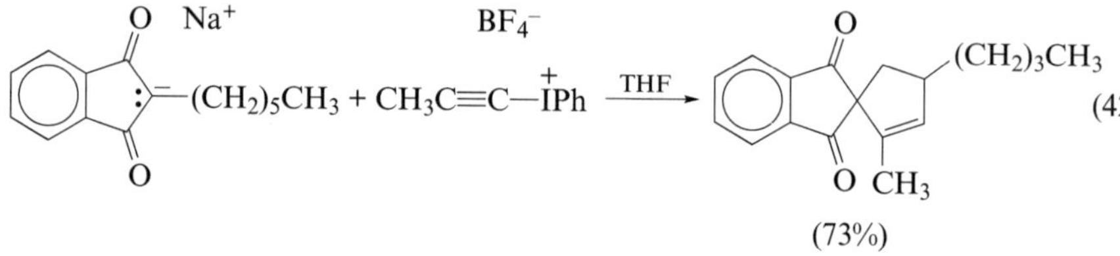

$$(42)$$

$$(73\%)$$

One of the most notable features of Ochiai's 1986 paper is the demonstration that alkynylation reactions, such as that shown in equation 38, might also proceed via alkylidenecarbenes. Thus, when the 2-*n*-hexylindandionate ion was allowed to react with (β-^{13}C-phenylethynyl)phenyliodonium tetrafluoroborate, 2-*n*-hexyl-2-phenylethynyl-1,3-indandione, largely enriched with ^{13}C at the α-carbon atom of the ethynyl group, was obtained (equation 43). At most, only 6% of the alkynylation product could have arisen by the Ad-E pathway[28]. Apparently, the migratory aptitude (i.e. 'tropophilicity') of the phenyl group is sufficiently high that migration in the alkylidenecarbene pre-empts insertion (equation 44).

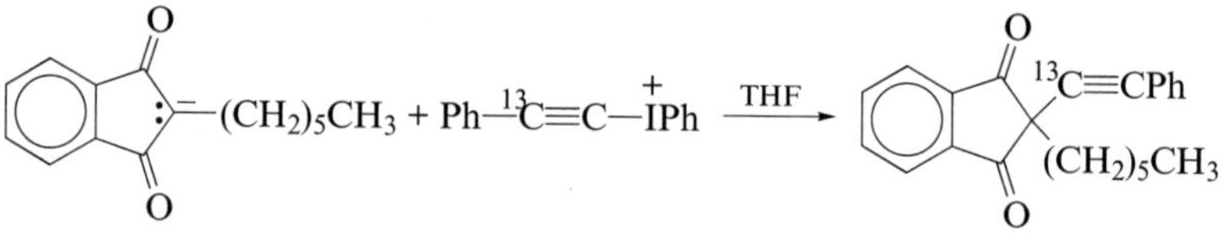

$$(43)$$

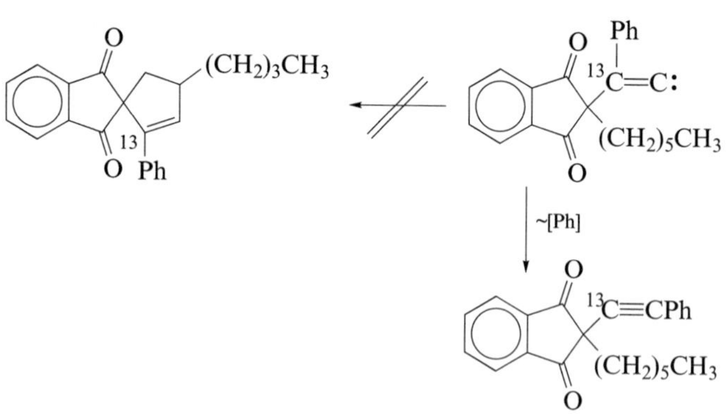

$$(44)$$

A variety of soft nucleophiles have since been shown to react with alkynyliodonium salts via the Michael addition–carbene (MC) pathway (equation 45). If the tropophilicities of R

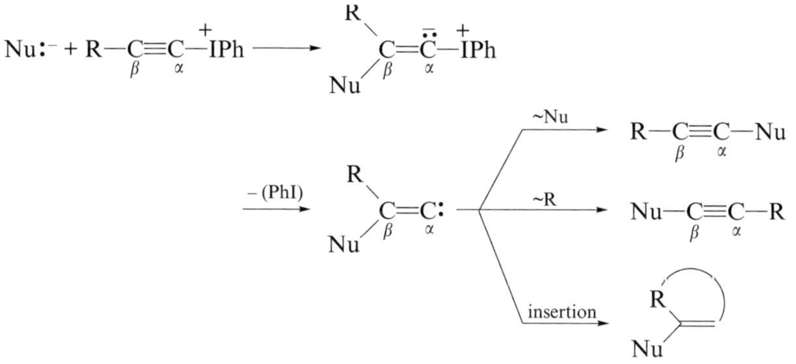

$$\text{Nu:}^- + \underset{\beta}{R}-\underset{}{C}\equiv\underset{\alpha}{C}-\overset{+}{I}\text{Ph} \longrightarrow \underset{Nu}{\overset{R}{\diagdown}}\underset{\beta}{C}=\underset{\alpha}{\overset{+}{C}}-\overset{+}{I}\text{Ph}$$

$$\xrightarrow{-\,(\text{PhI})} \underset{Nu}{\overset{R}{\diagdown}}\underset{\beta}{C}=\underset{\alpha}{C:}$$

$$\xrightarrow{\sim Nu} \underset{\beta}{R}-C\equiv\underset{\alpha}{C}-Nu$$

$$\xrightarrow{\sim R} \underset{\beta}{Nu}-C\equiv\underset{\alpha}{C}-R \quad (45)$$

insertion

and/or Nu are high, substituted alkynes are obtained, but if the tropophilicities are low, cyclopentenes are produced.

2. Sulfur nucleophiles

a. Sulfinic acids and sulfinate salts. Sulfinic acids and sulfinate salts behave as sulfur nucleophiles with alkynyliodonium ions. Such reactions give sulfones and are best understood in the context of the Michael addition–alkylidenecarbene (MC) pathway. For example, various alkynyl(phenyl)iodonium tetrafluoroborates, when mixed with sulfinic acids in methanol, afford high yields of (Z)-(β-sulfonylvinyl)iodonium tetrafluoroborates (equation 46)[32,84]. The reaction between 1-decynyl(phenyl)iodonium tetrafluoroborate and benzenesulfinic acid has been studied in detail; when water or aprotic solvents (i.e. PhH, Et$_2$O, CH$_2$Cl$_2$, dioxane) are employed instead of methanol, cyclopentenyl- and alkynyl sulfones are also generated (equation 47)[32]. It seems likely that all three types of products originate from an intermediate vinylidene-iodonium ylide (equation 48). In methanol, protonation of the ylide, presumably with HBF$_4$, is faster than the elimination of iodobenzene, but in solvents such as benzene, (β-sulfonyl)alkylidenecarbene formation

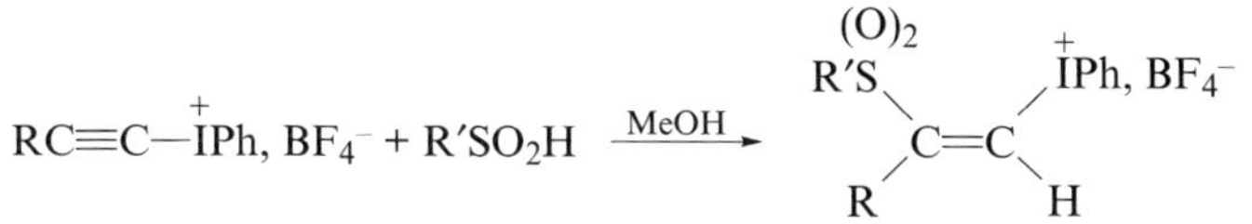

$$R\overset{+}{C}\equiv C-\overset{+}{I}\text{Ph, BF}_4^- + R'SO_2H \xrightarrow{\text{MeOH}} \underset{R}{\overset{\overset{(O)_2}{R'S}}{\diagdown}}C=C\underset{H}{\overset{\overset{+}{I}\text{Ph, BF}_4^-}{\diagup}} \quad (46)$$

R′, R(yield): Ph, Me (77%); Ph, n-C$_8$H$_{17}$ (100%);
Ph, cyclohexyl-CH$_2$CH$_2$ (88%); Ph, cyclopentyl-CH$_2$ (99%);
Ph, PhCH$_2$CH$_2$CH$_2$ (93%); Ph, , HOCH$_2$CH$_2$ (64%); 4-O$_2$NC$_6$H$_4$,
n-C$_8$H$_{17}$ (90%); 4-An, n-C$_8$H$_{17}$ (91%); n-Bu, Me (89%)

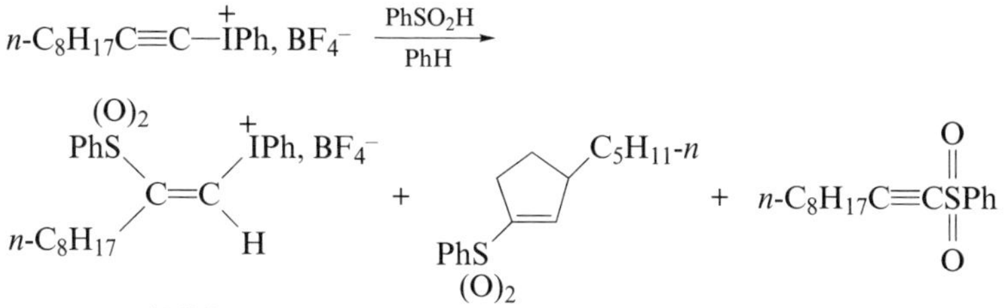

$$n\text{-C}_8\text{H}_{17}C\equiv C-\overset{+}{I}\text{Ph, BF}_4^- \xrightarrow[\text{PhH}]{\text{PhSO}_2\text{H}}$$

$$\underset{n\text{-C}_8\text{H}_{17}}{\overset{\overset{(O)_2}{\text{PhS}}}{\diagdown}}C=C\underset{H}{\overset{\overset{+}{I}\text{Ph, BF}_4^-}{\diagup}} \quad + \quad \underset{\overset{PhS}{(O)_2}}{\diagdown}\diagup C_5H_{11}\text{-}n \quad + \quad n\text{-C}_8\text{H}_{17}C\equiv C\overset{\overset{O}{\|}}{\underset{\underset{O}{\|}}{S}}\text{Ph} \quad (47)$$

(56%) (27%) (11%)

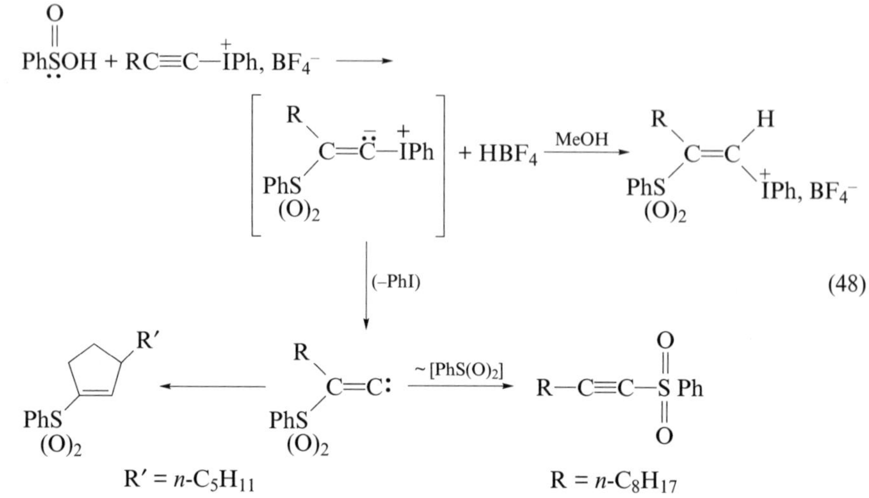

$$\text{Ph}\overset{\overset{\displaystyle O}{\|}}{\underset{\cdot\cdot}{S}}\text{OH} + RC{\equiv}C{-}\overset{+}{I}\text{Ph, BF}_4^- \longrightarrow$$

(48)

is competitive. The migrating species in such carbenes is thought to be the sulfonyl group[32]. The intrusion of carbenic products in water, although a bit surprising, is probably due to the low solubility of benzenesulfinic acid in that solvent[32]. Finally, the factors that govern the stereoselective production of (Z)-$(\beta$-sulfonylvinyl)iodonium ions in these reactions are not clear. Perhaps, dipolar interactions between the sulfonyl oxygen atoms and the iodonium center in the intermediate ylides stabilize the (Z)-configuration **16** (equation 49).

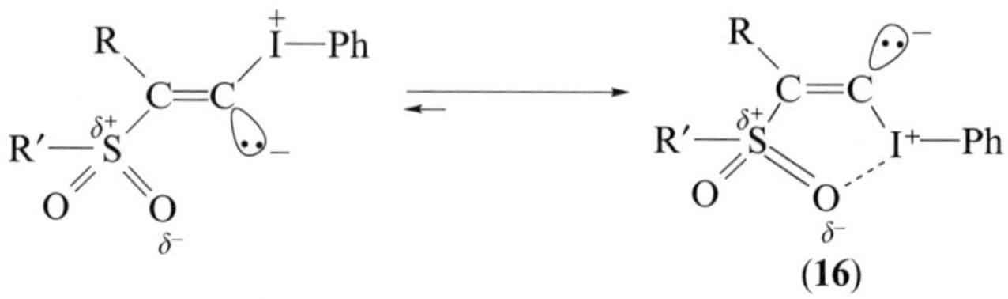

(49)

(**16**)

When alkynyliodonium compounds are mixed with sulfinate salts, the vinyliodonium pathway is supressed, and alkynyl- and/or cyclopentenyl sulfones are obtained. If the R group of the alkynyl moiety has no γ-CH bonds, alkynyl sulfones are formed exclusively[19,85]. For example, the (3,3-dimethyl-1-butynyl)iodonium ion reacts with sodium p-toluenesulfinate under various conditions to give good yields of 3,3-dimethyl-1-butynyl p-tolyl sulfone (equation 50)[19,85], This is not surprising, since the carbene insertion pathway would be restricted to the production of a cyclobutenyl sulfone. Alkynyl sulfones that have been prepared in this way are given in Table 4.

When γ-CH bonds are present in the R group of the alkynyliodonium ion, cyclopentenyl sulfones predominate. For example, the treatment of 5-phenyl-1-pentynyl(phenyl)-iodonium tetrafluoroborate with tetra-n-butylammonium benzenesulfinate in THF (i.e. homogeneous conditions) affords a moderate yield of 1-phenylsulfonyl-3-phenylcyclopentene and a low yield of the corresponding alkynyl sulfone (equation 51)[32]. With appropriately constructed alkynyliodonium ions, annulated cyclopentenyl sulfones are obtained (equations 52 and 53)[32].

The insertion manifold of MCI reactions is most generally observed with aliphatic CH bonds. However, several diaryl alkynyl carbinols have been reported to react with sodium

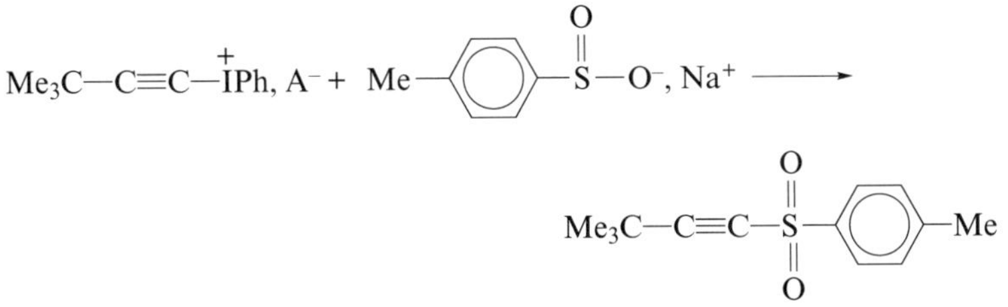

$$Me_3C-C\equiv C-\overset{+}{I}Ph, A^- + Me-\underset{\underset{O^-}{\overset{O}{\parallel}}}{\overset{O}{\underset{\parallel}{S}}}-O^-, Na^+ \longrightarrow$$

$$Me_3C-C\equiv C-\underset{\overset{\parallel}{O}}{\overset{\overset{O}{\parallel}}{S}}-\underset{}{\bigcirc}-Me \quad (50)$$

A^-	Conditions	isolated yield
TsO⁻	CHCl₃, H₂O, TEBA, rt	72%[85]
TfO⁻	CH₂Cl₂, H₂O, 0 °C	89%[19]
TfO⁻	CH₂Cl₂, 25 °C	95%[19]

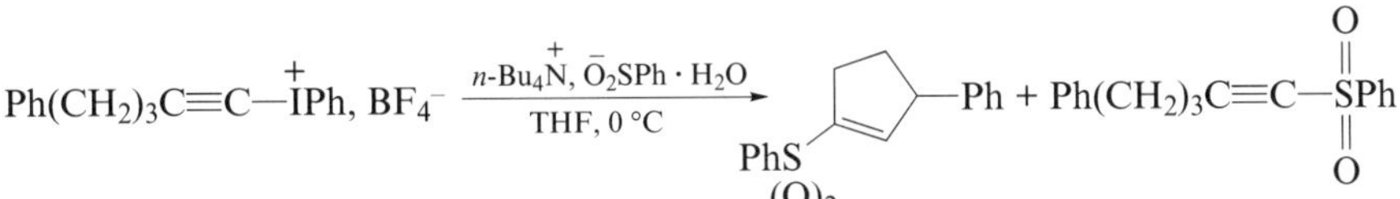

$$Ph(CH_2)_3C\equiv C-\overset{+}{I}Ph, BF_4^- \xrightarrow[\text{THF, 0 °C}]{n\text{-Bu}_4N, \bar{O}_2SPh \cdot H_2O} \underset{\underset{(O)_2}{PhS}}{\bigcirc}-Ph + Ph(CH_2)_3C\equiv C-\underset{\overset{\parallel}{O}}{\overset{\overset{O}{\parallel}}{S}}Ph$$

$$96:4 \ (43\%) \qquad\qquad (51)$$

TABLE 4. Synthesis of alkynyl sulfones from alkynyl(phenyl)iodonium salts

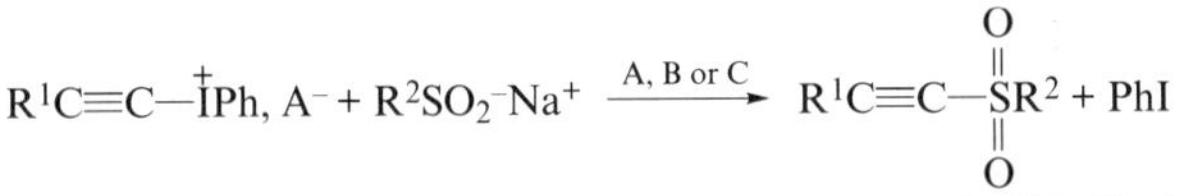

$$R^1C\equiv C-\overset{+}{I}Ph, A^- + R^2SO_2^-Na^+ \xrightarrow{A, B \text{ or } C} R^1C\equiv C-\underset{\overset{\parallel}{O}}{\overset{\overset{O}{\parallel}}{S}}R^2 + PhI$$

(A) CHCl₃, H₂O, TEBA, rt; (B) CH₂Cl₂, H₂O, 0 °C; (C) CH₂Cl₂, 25 °C

R^1	A^-	R^2	conditions, Isolated yield (Reference)
H	TfO⁻	4-Tol	B, 46% (19) C, 87% (19)
Me	TfO⁻	4-Tol	B, 71% (19) C, 87% (19)
t-Bu	TfO⁻	4-Tol	B, 89% (19) C, 95% (19)
t-Bu	TsO⁻	Ph	A, 81% (85)
t-Bu	TsO⁻	4-Tol	A, 72% (85)
t-Bu	TsO⁻	4-ClC₆H₄	A, 80% (85)
t-Bu	TsO⁻	3-NO₂C₆H₄	A, 84% (85)
Ph	TfO⁻	4-Tol	B, 42% (19) C, 78% (19)
Ph	TsO⁻	Ph	A, 75% (85)
Ph	TsO⁻	4-Tol	A, 68% (85)
Ph	TsO⁻	4-ClC₆H₄	A, 74% (85)
Ph	TsO⁻	3-NO₂C₆H₄	A, 77% (85)
(i-Pr)₃Si	TfO⁻	4-Tol	B, 80% (19) C, 79% (19)
ClCH₂	TfO⁻	4-Tol	B, 43% (19) C, 86% (19)
BrCH₂	TfO⁻	4-Tol	B, 31% (19) C, 60% (19)
Ph(Me)COH	TfO⁻	4-Tol	C, 81% (19)

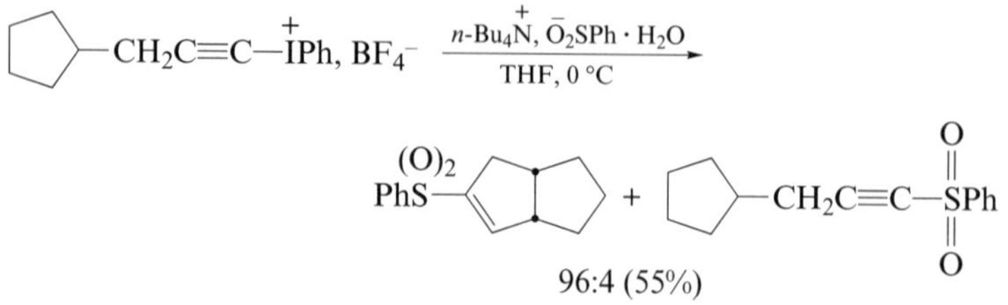

$$\text{cyclopentyl}-CH_2C\!\!\equiv\!\!C\overset{+}{-}IPh, \, BF_4^- \quad \xrightarrow[\text{THF, 0 °C}]{n\text{-}Bu_4\overset{+}{N}, \, \bar{O}_2SPh \cdot H_2O}$$

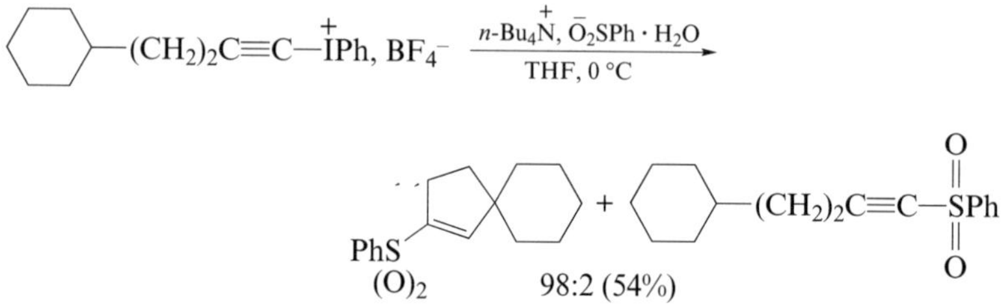

$$\underset{PhS-}{(O)_2}\text{(bicyclic)} \quad + \quad \text{cyclopentyl}-CH_2C\!\!\equiv\!\!C\overset{O}{\underset{O}{\overset{\|}{-}}}SPh \qquad (52)$$

$$96:4 \, (55\%)$$

$$\text{cyclohexyl}-(CH_2)_2C\!\!\equiv\!\!C\overset{+}{-}IPh, \, BF_4^- \quad \xrightarrow[\text{THF, 0 °C}]{n\text{-}Bu_4\overset{+}{N}, \, \bar{O}_2SPh \cdot H_2O}$$

$$\underset{PhS}{\underset{(O)_2}{}}\text{(spiro)} \quad + \quad \text{cyclohexyl}-(CH_2)_2C\!\!\equiv\!\!C\overset{O}{\underset{O}{\overset{\|}{-}}}SPh \qquad (53)$$

$$98:2 \, (54\%)$$

p-toluenesulfinate in dichloromethane to give cyclopentenyl sulfones derived from carbenic insertion into the *ortho*-CH bonds of the aromatic rings (equation 54)[20]. It is noteworthy that insertion into the CH bonds of the (*m*-trifluoromethyl)phenyl ring occurs regioselectively at the *ortho*-carbon closer to the trifluoromethyl group.

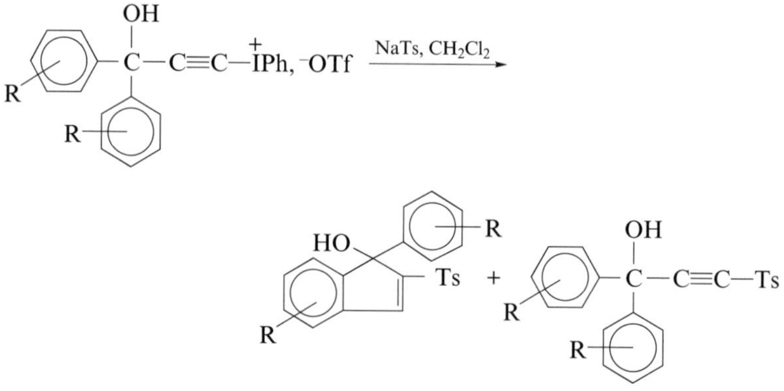

$$\underset{R}{\text{(aryl)}}\overset{OH}{\underset{\underset{R-\text{(aryl)}}{|}}{\overset{|}{C}}}-C\!\!\equiv\!\!C\overset{+}{-}IPh, \, ^-OTf \quad \xrightarrow{NaTs, \, CH_2Cl_2}$$

$$\underset{R}{\overset{HO}{\text{(bicyclic)}}}-R \quad - \, Ts \quad + \quad \underset{R-\text{(aryl)}}{\overset{OH}{\underset{\underset{R}{|}}{\overset{|}{C}}}}-C\!\!\equiv\!\!C-Ts \qquad (54)$$

R (yield): H (50%, 18%); *p*-MeO (48%, 36%); *m*-CF$_3$ (18%, 34%)

The reactions of β-ketoethynyl- and β-amidoethynyl(phenyl)iodonium triflates, **17** and **18**, with sodium *p*-toluenesulfinate illustrate the synthetic potential of alkynyliodonium salts[33]. Although the direct attachment of a carbonyl group to the β-carbon atom of the triple bond in alkynyliodonium ions might be expected to facilitate alkynyl sulfone formation via the Ad-E mechanism, this mode of reactivity has not been observed. Instead, the MC pathway with carbenic insertion dominates and affords sulfones containing the

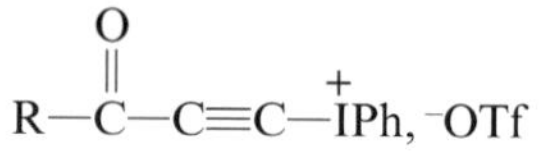

$$R-\overset{O}{\overset{\|}{C}}-C\!\!\equiv\!\!C\overset{+}{-}IPh, \, ^-OTf \qquad\qquad R^1R^2N-\overset{O}{\overset{\|}{C}}-C\!\!\equiv\!\!C\overset{+}{-}IPh, \, ^-OTf$$

$$\textbf{(17)} \qquad\qquad\qquad\qquad\qquad\qquad \textbf{(18)}$$

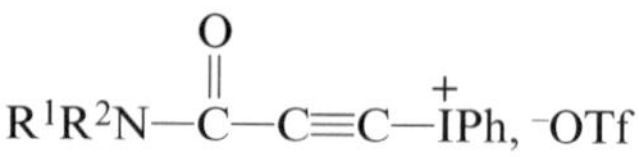

cyclopentenone and γ-lactam ring systems. When the R and R^1R^2N groups in **17** and **18** are cyclic, fused bicyclic structures can be readily assembled (equations 55 and 56)[33]. This transformation appears to be general and has been applied to the alkynyliodonium triflates shown in Table 5.

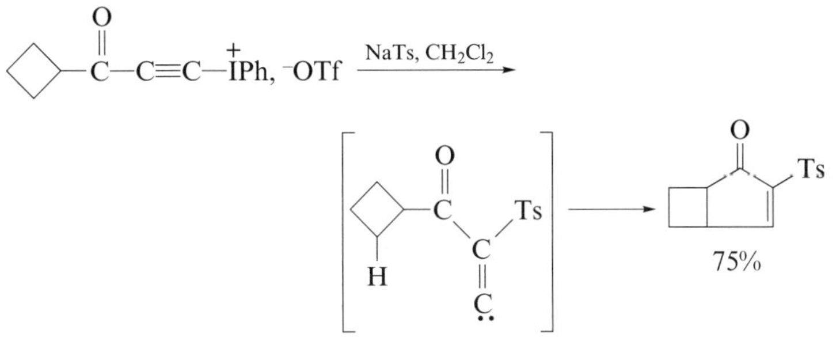

(55)

TABLE 5. Cyclopentenyl sulfones and sulfonyl lactams from (ketoethynyl)iodonium and (amido-ethynyl)iodonium triflates **17** and **18** with sodium p-toluenesulfinate in dichloromethane[a]

R in **17**	Product (yield)	R^1R^2N in **18**	Product (yield)
t-Bu	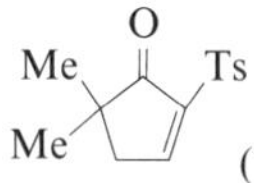 (72%)	Me_2N-	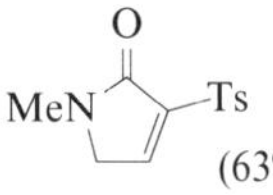 (63%)
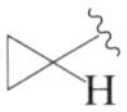	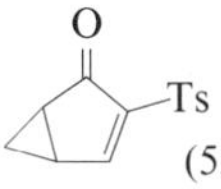 (53%)	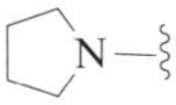	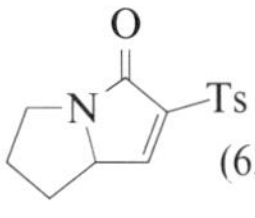 (63%)
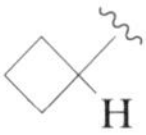	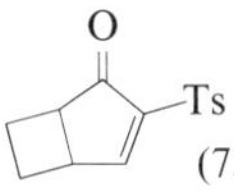 (75%)	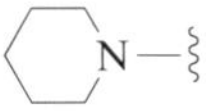	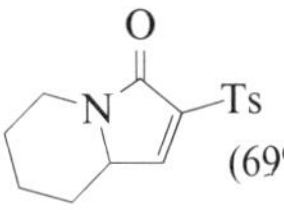 (69%)
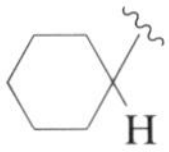	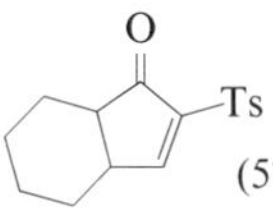 (57%)	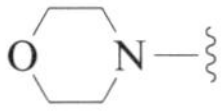	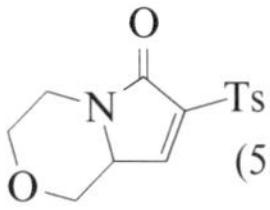 (53%)
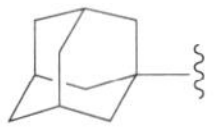	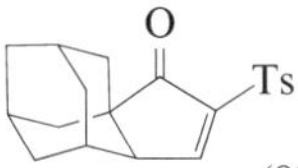 (82%)	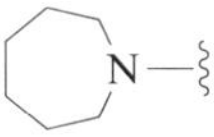	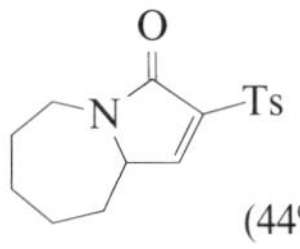 (44%)

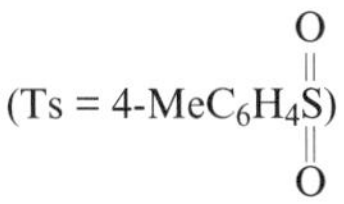

$(Ts = 4\text{-}MeC_6H_4\overset{O}{\underset{O}{S}})$

[a] Modified with permission from Reference 33. Copyright (1994) American Chemical Society

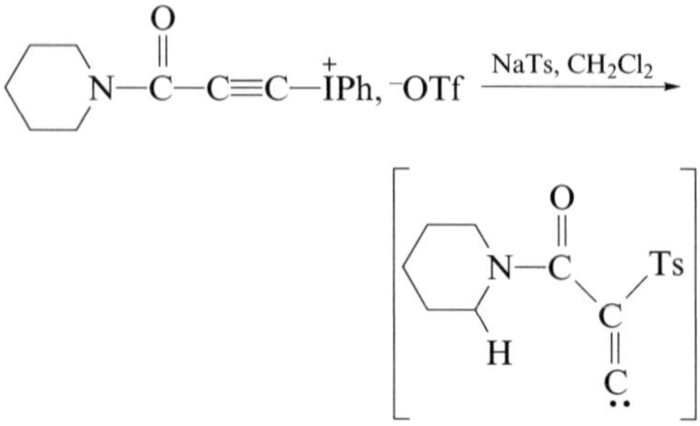

$$(56)$$

69%

The treatment of bis(phenyliodonium)diyne triflates, **19** (n = 5,6,8), with sodium p-toluenesulfinate provides access to disulfones possessing spiroannulated or tethered cyclopentene rings (equation 57)[86]. However, because migration of the p-toluenesulfonyl group in the intermediate carbenes is competitive with insertion, the yields of the bis-cyclopentene compounds are relatively low, and disulfones of general structures **20** and **21** are also obtained.

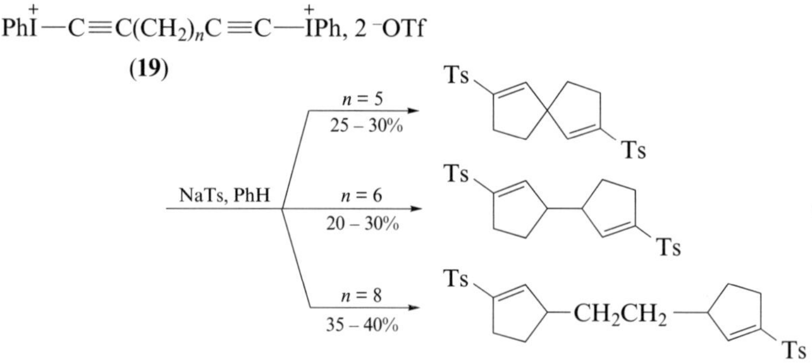

$$(57)$$

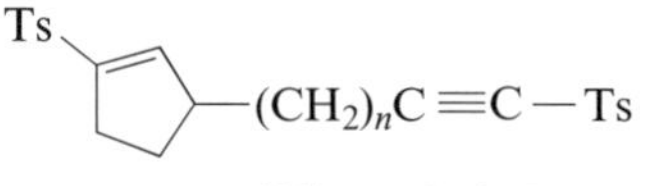

(**20**) n = 2, 3, 5   (**21**) n = 5, 6, 8

b. Thiocyanate ion, thiotosylate ion and dialkyl phosphorodithioate ions. Various alkynyl(phenyl)iodonium triflates react with sodium thiocyanate in aqueous dichloromethane to give good yields of alkynyl thiocyanates (equation 58)[14]. It is noteworthy that the 1-hexynyl salt (R = n-Bu) does not afford a cyclopentenyl thiocyanate despite the presence of a γ-CH bond in the n-butyl group. Thus, if the alkynyl thiocyanates arise via the MC pathway, the thiocyanate group must be sufficiently tropophilic that migration pre-empts insertion in the intermediate carbenes. Evidence for the MC pathway is provided by the production of cyclopentenyl and alkynyl thiocyanates when 4,4-dimethyl-3-oxo-1-pentynyl(phenyl)iodonium triflate and sodium thiocyanate are mixed in dichloromethane (equation 59)[14]. Alkynyl thiocyanates have also been prepared from alkynyl phenyl bisiodonium(p-phenylene) triflates (equation 60)[38]. Whether phenyl(p-iodophenyl)iodonium triflate is a by-product (i.e. as opposed to iodobenzene; equation 58) in such reactions was not reported. However, the authors did state that 'the pure product (> 95%) is obtained only by extraction with ether'[38].

$$RC\equiv C\overset{+}{-}IPh,\ OTf^- + NaSCN \xrightarrow{CH_2Cl_2,\ H_2O,\ 20\ °C} RC\equiv C-SCN + PhI \quad (58)$$

R (yield): Me (70%), *n*-Bu (81%), *t*-Bu (72%), Me$_3$Si (76%), MeOCH$_2$ (74%), Ph (94%)

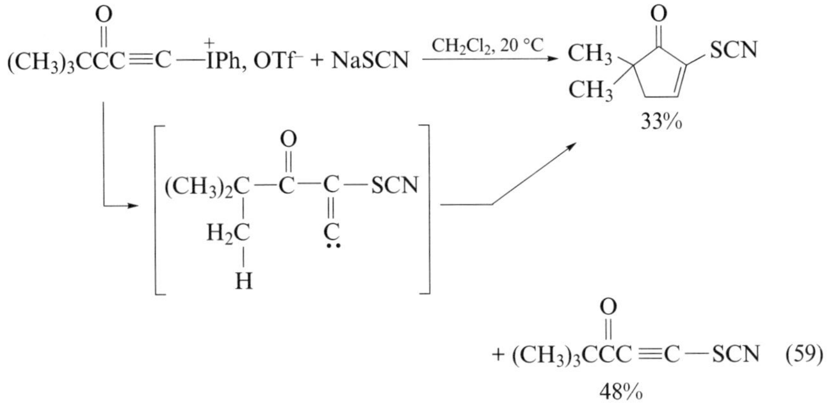

$$(CH_3)_3CCC\equiv C\overset{+}{-}IPh,\ OTf^- + NaSCN \xrightarrow{CH_2Cl_2,\ 20\ °C}$$

33%

$$\left[(CH_3)_2\underset{\underset{H}{\overset{|}{\underset{|}{H_2C}}}}{C}-\overset{\overset{O}{\parallel}}{C}-\underset{\overset{|}{\underset{\cdot\cdot}{C}}}{C}-SCN \right]$$

$$+ (CH_3)_3CCC\equiv C-SCN \quad (59)$$

48%

$$PhI^+\!-\!\!\!\bigcirc\!\!\!-\overset{+}{I}-C\equiv CR,\ 2\ ^-OTf \xrightarrow{KSCN,\ DMF} RC\equiv C-SCN \quad (60)$$

R (yield): *n*-Bu (94%), *t*-Bu (90%), *n*-C$_6$H$_{13}$ (97%), Ph (87%)

The propensity of the thiocyanate ion for alkynylation with alkynyliodonium ions has also been demonstrated with a series of bis(phenyliodonium)diyne triflates (equations 61 and 62)[43]. The efficient production of diynediyl dithiocyanates in these reactions may be contrasted with the favored formation of mono- and bis-cyclopentenyl sulfones from bisiodonium diyne salts and sodium *p*-toluenesulfinate (see equation 57)[86].

$$\overset{+}{PhI}-C\equiv C(CH_2)_nC\equiv C\overset{+}{-}IPh,\ 2\ ^-OTf \xrightarrow[CH_2Cl_2,\ MeCN,\ H_2O]{NaSCN}$$

$$NCS-C\equiv C(CH_2)_nC\equiv C-SCN \quad (61)$$

n (yield): 2 (72%), 4 (69%), 6 (80%), 8 (68%)

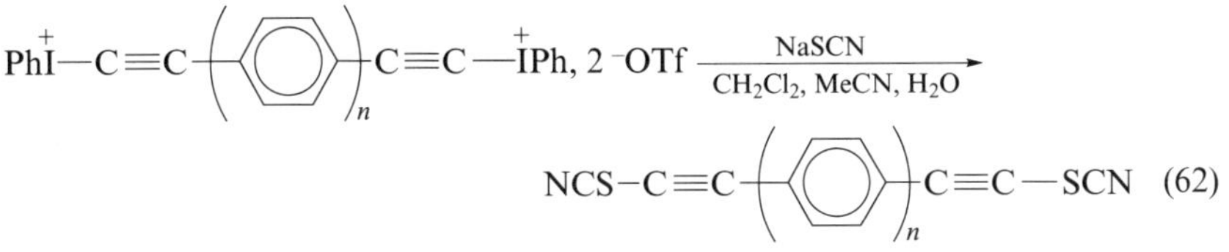

$$\overset{+}{PhI}-C\equiv C\!\!\left(\!\!\bigcirc\!\!\right)_{\!\!n}\!\!C\equiv C\overset{+}{-}IPh,\ 2\ ^-OTf \xrightarrow[CH_2Cl_2,\ MeCN,\ H_2O]{NaSCN}$$

$$NCS-C\equiv C\!\!\left(\!\!\bigcirc\!\!\right)_{\!\!n}\!\!C\equiv C-SCN \quad (62)$$

n (yield): 1 (66%), 2 (75%)

The reactions of potassium thiotosylate and potassium *O*,*O*-dialkyl phosphorodithioates with alkynyl(phenyl)iodonium salts afford good yields of alkynyl thiotosylates and alkynyl phosphorodithioates (equations 63 and 64)[87,88]. Cyclopentenyl products have not been reported in either case.

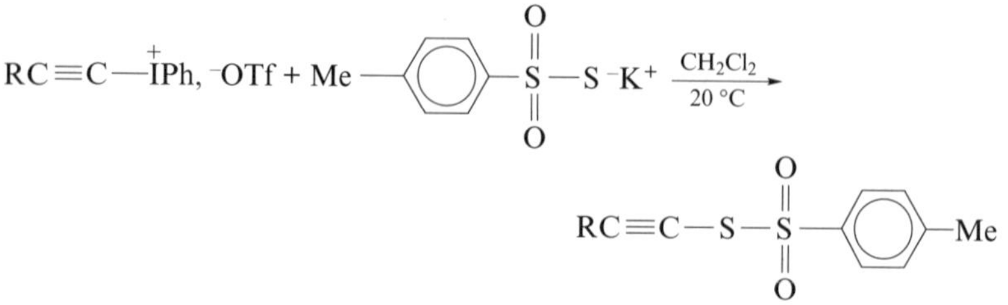

$$RC \equiv C - \overset{+}{I}Ph, \ ^-OTf + Me - \underset{\underset{O}{\overset{O}{\|}}}{\overset{\overset{O}{\|}}{\underset{\|}{S}}} - S \ ^-K^+ \xrightarrow[20\ ^\circ C]{CH_2Cl_2}$$

$$RC \equiv C - S - \underset{\underset{O}{\overset{O}{\|}}}{\overset{\overset{O}{\|}}{S}} - \underset{}{\bigcirc} - Me \quad (63)$$

R (yield): Me (56%), n-Bu (67%), t-Bu (67%), ClCH$_2$ (50%), MeOCH$_2$ (51%), Me$_3$Si (66%), Ph (58%)

$$R^1C \equiv C - \overset{+}{I}Ph, \ ^-OTs + (R^2O)_2PS_2^-K^+ \xrightarrow{CHCl_3, H_2O, TEBA, rt} R^1C \equiv C - \overset{\overset{S}{\|}}{S}P(OR^2)_2$$

R^1 = Ph; R^2 (yield): Me (85%), Et (83%), n-Pr (96%), n-Bu (87%), PhCH$_2$ (91%), (64)
Ph (93%): R^1 = t-Bu; R^2 (yield): Et (94%), PhCH$_2$ (89%), Ph (95%)

c. Thiophenoxide ion. Among the sulfur nucleophiles, alkyl and aryl thiolates are the most obvious candidates for reactivity assessments of alkynyliodonium ions. However, reported studies of this type are limited to the treatment of bis(phenyliodonium)ethyne and bis(phenyliodonium)butadiyne triflates with sodium thiophenoxide[40,41]. These iodonium compounds serve as 'C$_2$ and C$_4$ transfer agents' and, with two equivalents of sodium thiophenoxide, afford the corresponding bis(phenylthio)alkynes (equation 65).

$$Ph\overset{+}{I} + (C \equiv C)_n \overset{+}{I}Ph, \ 2\ ^-OTf + 2\ NaSPh \xrightarrow{MeCN} PhS + (C \equiv C)_n SPh + 2\ PhI \quad (65)$$

n (yield): 1 (66%), 2 (67%)

3. Phosphorus nucleophiles

a. Phosphines. When alkynyl(phenyl)iodonium tetrafluoroborates are treated with triphenylphosphine in THF at −78 °C and the mixtures are exposed to sunlight, (alkynyl)triphenylphosphonium tetrafluoroborates are generated in 'quantitative' yields (equation 66)[80]. In the absence of sunlight, the 1-decynyliodonium salt did not give any of the 1-decynylphosphonium salt[80]. Although the absorbing species has not been identified and quantum yields have not been measured, a radical chain mechanism initiated by photoinduced electron transfer from triphenylphosphine to the alkynyliodonium ions has been proposed (equations 67–69)[80].

$$RC \equiv C - \overset{+}{I}Ph, BF_4^- + Ph_3P \xrightarrow{sunlight, THF, -78\ ^\circ C} RC \equiv C - \overset{+}{P}Ph_3, BF_4^- + PhI \quad (66)$$

R = Me, n-C$_8$H$_{17}$, t-Bu, cyclopentyl - CH$_2$, cyclohexyl

$$RC \equiv C - \overset{+}{I}Ph + :PPh_3 \xrightarrow{h\nu} RC \equiv C - \overset{\cdot}{I}Ph + \overset{+}{\cdot}PPh_3 \quad (67)$$

$$RC \equiv C - \overset{\cdot}{I}Ph + :PPh_3 \longrightarrow \underset{\overset{|}{\cdot PPh_3}}{RC \equiv C - IPh} \longrightarrow RC \equiv C - \overset{\cdot}{P}Ph_3 + PhI \quad (68)$$

$$RC \equiv C - \overset{\cdot}{P}Ph_3 + RC \equiv C - \overset{+}{I}Ph \longrightarrow RC \equiv C - \overset{+}{P}Ph_3 + RC \equiv C - \overset{\cdot}{I}Ph \quad (69)$$

The regiospecific coupling of triphenylphosphine with the alkynyl ligands of alkynyl(phenyl)iodonium ions in such photosubstitution reactions is remarkable. This may be contrasted with the less selective and somewhat unpredictable photolytic decomposition (i.e. high pressure mercury lamp, pyrex) of alkynyl(phenyl)iodonium salts in the absence of nucleophiles (e.g. see equation 70)[89].

$$t\text{-BuC}\equiv\text{C}\overset{+}{-}\text{IPh, A}^- \xrightarrow{hv} t\text{-BuC}\equiv\text{C}-\text{I} + \text{PhI} + t\text{-BuC}\equiv\text{C}-\text{Ph} \quad (70)$$

A$^-$, solvent (% yield-GC of iodoalkyne, PhI, phenylalkyne): TsO$^-$, THF (62, 9, trace); TsO$^-$, EtOH (52, 10, trace); TsO$^-$, CH$_2$Cl$_2$, (12, 51, 8); TfO$^-$, EtOH (59, 4, 1); TfO$^-$, CH$_2$Cl$_2$ (10, 60, 8)

Photochemical activation is not universally required for the conversion of alkynyliodonium salts into alkynylphosphonium compounds. For example, triphenylphosphine reacts with various alkynyl(phenyl)iodonium triflates in dichloromethane to give high yields of alkynylphosphonium triflates (equation 71)[90]. Since 'these reactions occur readily in the dark' and are not inhibited by molecular oxygen or 2,6-di-*tert*-butyl-4-methylphenol, they are thought to proceed via the MC mechanism (equation 45)[90].

$$\text{RC}\equiv\text{C}\overset{+}{-}\text{IPh, }^-\text{OTf} + \text{Ph}_3\text{P} \xrightarrow{\text{CH}_2\text{Cl}_2, -78\,°\text{C to rt}} \text{RC}\equiv\text{C}\overset{+}{-}\text{PPh}_3, {}^-\text{OTf} + \text{PhI} \quad (71)$$

R (yield): Me (97%), *n*-Bu (88%), *t*-Bu (98%), Me$_3$Si (98%), Ph (85%)

The contrasting behavior of alkynyliodonium tetrafluoroborates and alkynyliodonium triflates with triphenylphosphine (i.e. photochemical vs thermal activation) documented in the foregoing studies seems unlikely to be due to the different anions. Perhaps the thermal alkynylations are simply too slow to be observed at −78 °C in coordinating solvents such as THF, the conditions employed in the photochemical study.

The treatment of [bis(phenyliodonium)]ethyne ditriflate with triphenylphosphine has also been investigated[40,41]. The products depend on the stoichiometric ratios of the reactants. Thus, with one equivalent of triphenylphosphine, a monoiodonium-monophosphonium derivative of acetylene is obtained, but with two equivalents of the phosphine, [bis(triphenylphosphonium)]ethyne ditriflate is produced (equation 72). When three equivalents of triphenylphosphine are employed in acetonitrile spiked with H$_2$O or D$_2$O, a *trans*-alkenediyl bisphosphonium salt is generated (equation 73)[41]. The reduction of the triple bond and formation of triphenylphosphine oxide in this reaction is thought to proceed at the bisphosphonium alkyne stage[41].

$$\text{Ph}\overset{+}{\text{I}}-\text{C}\equiv\text{C}-\overset{+}{\text{I}}\text{Ph, 2}\,^-\text{OTf}$$

$$\xrightarrow{\text{1 PPh}_3,\text{ CCl}_4,\text{ rt}} \text{Ph}\overset{+}{\text{I}}-\text{C}\equiv\text{C}-\overset{+}{\text{P}}\text{Ph}_3, \text{2}\,^-\text{OTf}$$
76%, crude $\quad (72)$

$$\xrightarrow[\substack{\text{MeCN}, -30\,°\text{C to }25\,°\text{C}\\ \text{or CCl}_4,\text{ rt}}]{\text{2 PPh}_3} \text{Ph}_3\overset{+}{\text{P}}-\text{C}\equiv\text{C}-\overset{+}{\text{P}}\text{Ph}_3, \text{2}\,^-\text{OTf}$$
99% (MeCN)
59%, crude (CCl$_4$)

$$\text{Ph}\overset{+}{\text{I}}-\text{C}\equiv\text{C}-\overset{+}{\text{I}}\text{Ph, 2}\,^-\text{OTf} + \text{3 Ph}_3\text{P} \xrightarrow[-30\,°\text{C to rt}]{\text{MeCN, H}_2\text{O or D}_2\text{O}}$$

$$\underset{\text{(D)H}}{\overset{\text{Ph}_3\overset{+}{\text{P}}}{\diagdown}}\text{C}=\text{C}\underset{\overset{+}{\text{PPh}_3}}{\overset{\text{H(D)}}{\diagup}} , \text{2}\,^-\text{OTf} + \text{Ph}_3\text{P}=\text{O} \quad (73)$$

88% (H$_2$O)
80% (D$_2$O)

The reactions of various bisiodonium diyne triflates with two equivalents of triphenylphosphine in dichloromethane similarly result in alkynylation of the nucleophile and afford good yields of bisphosphonium diyne triflates (equations 74 and 75)[42].

$$\overset{+}{PhI}—C{\equiv}C(CH_2)_nC{\equiv}C—\overset{+}{IPh},\ 2\ ^-OTf + 2Ph_3P \xrightarrow[-78\ °C\ to\ rt]{CH_2Cl_2}$$

$$\overset{+}{Ph_3P}—C{\equiv}C(CH_2)_nC{\equiv}C—\overset{+}{PPh_3},\ 2\ ^-OTf \quad (74)$$

n (yield): 2 (70%), 4 (67%), 5 (78%)

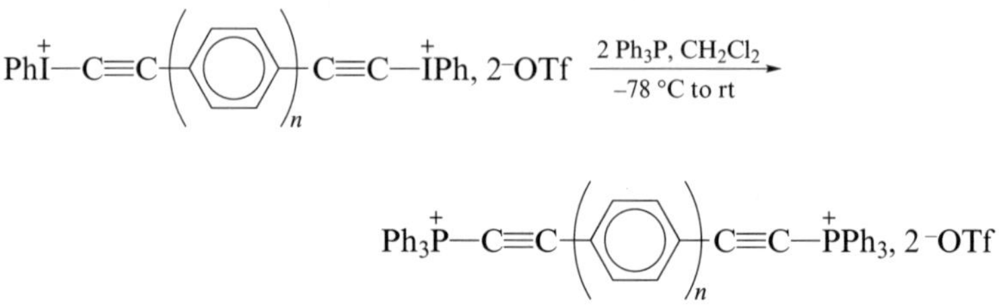

$$\overset{+}{PhI}—C{\equiv}C\!\!\left(\!\!\bigcirc\!\!\right)_{\!\!n}\!\!C{\equiv}C—\overset{+}{IPh},\ 2\ ^-OTf \xrightarrow[-78\ °C\ to\ rt]{2\ Ph_3P,\ CH_2Cl_2}$$

$$\overset{+}{Ph_3P}—C{\equiv}C\!\!\left(\!\!\bigcirc\!\!\right)_{\!\!n}\!\!C{\equiv}C—\overset{+}{PPh_3},\ 2\ ^-OTf \quad (75)$$

n (yield); 1 (94%), 2 (76%)

When 5,5-dimethyl-1,3-hexadiynyl(phenyl)iodonium triflate is mixed with two equivalents of triphenylphosphine in toluene, a mixture of bisphosphonium triflates is produced (equation 76)[34]. This transformation apparently involves the initial formation of (5,5-dimethyl-1,3-hexadiynyl)triphenylphosphonium triflate followed by competing Michael and conjugate Michael reactions of this compound with a second mole of triphenylphosphine. The (E)-geometry about the carbon–carbon double bonds in both products has been established by NMR analysis[34].

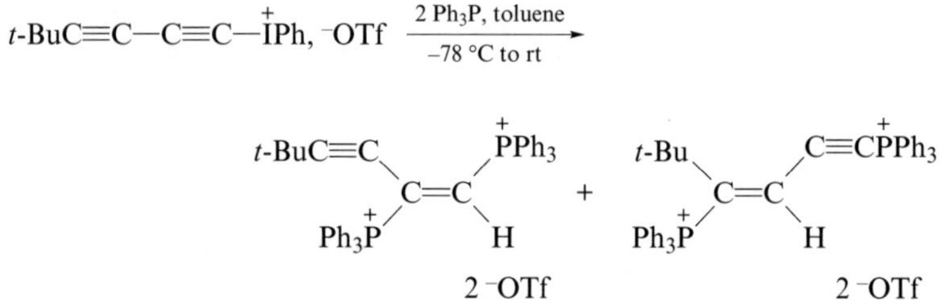

$$t\text{-}BuC{\equiv}C—C{\equiv}C—\overset{+}{IPh},\ ^-OTf \xrightarrow[-78\ °C\ to\ rt]{2\ Ph_3P,\ toluene}$$

$$\begin{array}{cc}
\overset{t\text{-}BuC{\equiv}C}{\underset{\overset{+}{Ph_3P}}{\Large\diagdown}}C{=}C\overset{\overset{+}{PPh_3}}{\underset{H}{\Large\diagup}} & +\quad \overset{t\text{-}Bu}{\underset{\overset{+}{Ph_3P}}{\Large\diagdown}}C{=}C\overset{C{\equiv}C\overset{+}{PPh_3}}{\underset{H}{\Large\diagup}}
\end{array} \quad (76)$$

$$2\ ^-OTf \qquad\qquad\qquad 2\ ^-OTf$$

3:2 (31%)

Certainly one of the most unusual nucleophiles to be treated with alkynyliodonium salts is the tetra-tert-butyltetraphosphacubane shown in equation 77[91]. However, despite its unique structure, this compound reacts with ethynyl(phenyl)- and 1-propynyl(phenyl)-iodonium triflates in the now expected way to give good yields of alkynylphosphonium salts[91].

3,3-Dimethyl-1-butynyl(phenyl)iodonium tosylate has been employed with bis-(diphenylphosphino)methane to give the (tert-butyl)phospholium tosylate shown in equation 78[92]. The initial formation of an alkynylphosphonium ion with a free phosphino group has been proposed. Intramolecular cyclization of this intermediate followed by a 1,3-prototropic shift would lead to the observed product. Evidence for the probability of such an intermediate is provided by the fact that the alkynyliodonium tosylate, like the

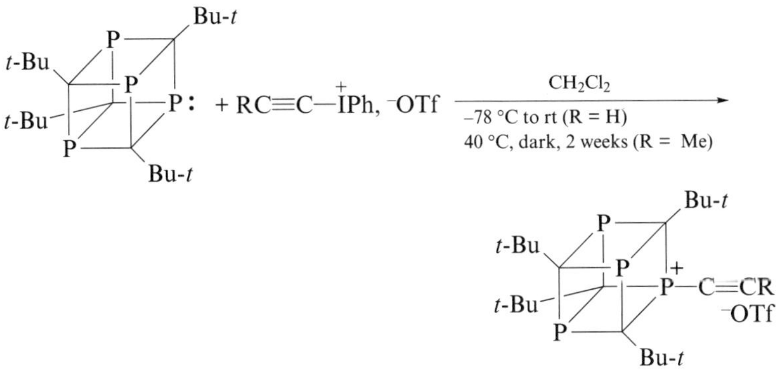

$$R \text{ (yield): H (67\%), Me (65\%)} \qquad (77)$$

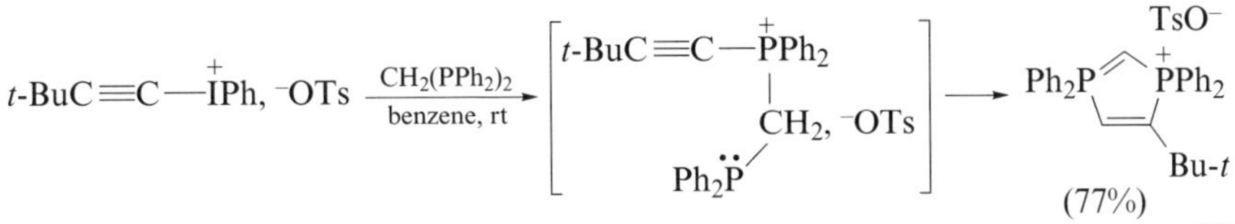

$$(78)$$

corresponding triflate (equation 71), reacts with triphenylphosphine to give a 97% yield of (*tert*-butylethynyl)triphenylphosphonium tosylate[92].

b. *Phosphites.* The treatment of various alkynyl(phenyl)iodonium tosylates with excess trimethyl phosphite in the absence of solvent gives dimethyl alkynylphosphonates, methyl tosylate and iodobenzene (equation 79)[10]. The high yields of iodobenzene, determined to be 92% and 101% for the *p*-tolylethynyl and *sec*-butylethynyl substrates, clearly manifest the regiopreference of trimethyl phosphite for the alkynyl ligand of alkynyl(phenyl)-iodonium ions. When the *sec*-butylethynyl and phenylethynyl salts were allowed to react with trimethyl phosphite in dichloromethane, the yields of the alkynylphosphonates (i.e. 55% and 46%) were about the same as those of the neat reactions[10]. However, the yield of methyl tosylate (51%) was much lower in the phenylethynyl case, and phenyl (*β*-phenyl-*β*-tosyloxyvinyl)iodonium tosylate was obtained in 20% yield. Admixture of 3,3-dimethyl-1-butynyl(phenyl)iodonium tosylate with neat triethyl and triisopropyl phosphites similarly affords the corresponding diethyl and diisopropyl alkynylphosphonates (equation 80)[10].

$$RC\equiv C-\overset{+}{I}Ph, \; {}^-OTs + (MeO)_3P: \xrightarrow{\text{neat, rt}} RC\equiv C-\overset{O}{\underset{\|}{P}}(OMe)_2 + MeOTs + PhI \quad (79)$$

$$82\text{-}100\%$$

R (yield): *i*-Pr (50%), *s*-Bu (63%), *t*-Bu (90%), cyclopentyl (44%), Ph (42%),
p-tolyl (34%)

$$t\text{-}BuC\equiv C-\overset{+}{I}Ph, \; {}^-OTs + (RO)_3P \xrightarrow{\text{neat, } \Delta} t\text{-}BuC\equiv C-\overset{O}{\underset{\|}{P}}(OR)_2 + ROTs + PhI \quad (80)$$

R = Et	81%	93%
R = *i*-Pr	58%, 45%	86%

These Arbusov-like reactions are thought to involve the intermediate formation and S_n collapse of (alkynyl)trialkoxyphosphonium salts, **22**. The production of **22** by the MC pathway (equation 45) seems plausible[10] and is at least consistent with the high regioselectivity of alkynyl(phenyl)iodonium ion cleavage with trimethyl phosphite. It is noteworthy in this regard that the (*tert*-butylethynyl)iodonium salt, when treated with trimethyl phosphite in methanol, still gave an 88% yield of dimethyl (*tert*-butylethynyl)phosphonate. Thus, if alkylidenecarbenes, **23**, are generated, the trimethoxyphosphonio group must be tropophilic enough to preclude external trapping of the carbene.

$$RC\equiv C-\overset{+}{P}(OMe)_3, \ ^-OTs \qquad\qquad \begin{array}{c} R \\ \diagdown \\ C=C\mathbf{:} \\ \overset{+}{\diagup} \\ (MeO)_3P \end{array}$$

$$(22) \qquad\qquad\qquad (23)$$

4. Oxygen nucleophiles

a. Sulfonate ions. In view of the nucleofugic character of sulfonate ions, the decomposition of alkynyl(phenyl)iodonium tosylates and mesylates into alkynyl sulfonates and iodobenzene might be expected to require thermal activation or catalysis. Actually, two modes of ligand coupling are available, one resulting in alkynylation and the other resulting in phenylation of the sulfonate ion (equation 81). No quantitative data on the relative efficacy of these pathways have been published, although it has been noted that alkynyl(phenyl)iodonium sulfonates decompose in benzene at reflux to iodoalkynes and phenyl sulfonates and give 'only minor amounts (if any) of the alkynyl esters'[6]. At first glance, this may seem surprising since nucleophiles such as $^-$SCN and :P(OMe)$_3$ typically react with high regioselectivity at the alkynyl ligands of alkynyl (phenyl)iodonium ions. However, if the solid state configurations of the alkynyl(phenyl)iodonium sulfonates are retained in the benzene medium, coupling of the sulfonate ion with the equatorial phenyl ligand should be favored over similar coupling with the apical alkynyl ligand (i.e. the so-called '*ortho* effect')[93].

$$\begin{array}{c} Ph \\ | \\ RC\equiv C-I^+ \cdots\cdots \ ^-OSO_2R' \end{array} \xrightarrow{\Delta} \Big\langle \begin{array}{l} PhOSO_2R' + RC\equiv C-I \\[2mm] RC\equiv COSO_2R' + PhI \end{array} \tag{81}$$

The presence of catalytic amounts of cuprous triflate or silver(I) sulfonates exerts a remarkable influence on the activation energy and regiochemistry of alkynyl(phenyl)-iodonium tosylate and mesylate decompositions[5,6]. Such reactions proceed in acetonitrile at room temperature and afford moderate yields of alkynyl tosylates and mesylates (equations 82 and 83)[5,6]. It is noteworthy, however, that the treatment of alkynyliodonium triflates (R = *n*-Bu, *t*-Bu) with cuprous triflate in acetonitrile does *not* afford alkynyl triflates[6]. Silver(I) catalysis has similarly been applied to the conversion of bis(alkynyliodonium) tosylates to bisalkynyl tosylates (equation 84)[43]. As might be expected, monotosylate esters are also produced in these reactions.

$$RC\equiv C-\overset{+}{I}Ph, \ ^-OTs \xrightarrow{\text{Cu(I) or Ag(I), MeCN, 25 °C}} RC\equiv C-OTs + PhI \tag{82}$$

R (catalyst, yield): Me (CuOTf, 38%), *n*-Bu (CuOTf, 88%),
s-Bu (AgOTs, 72%), *t*-Bu (AgOTf, 37%), Ph (CuOTf, —)

$$RC{\equiv}C{\overset{+}{-}}IPh,\ {}^-OMs \xrightarrow{\text{CuOTf, MeCN, 25 °C}} RC{\equiv}C{-}OMs + PhI \quad (83)$$

R (yield): Me (60%), n-Bu (54%), t-Bu (50%, reflux)

$$Ph\overset{+}{I}{-}C{\equiv}C(CH_2)_nC{\equiv}C{\overset{+}{-}}IPh,\ 2\ {}^-OTs \xrightarrow[\text{CH}_2\text{Cl}_2,\ 25\ °C]{\text{AgOTf}}$$

$$TsO{-}C{\equiv}C(CH_2)_nC{\equiv}C{-}OTs + TsO{-}C{\equiv}C(CH_2)_nC{\equiv}CH \quad (84)$$

n (yield of ditosylate); 4 (21%), 5(18%), 6(26%), 8(16%)

Several mechanisms for the catalytic action of Cu(I) and Ag(I) have been considered[6]. Among these, the metal-assisted addition–elimination sequence shown in equation 85 and illustrated with cuprous triflate was deemed most consistent with various control studies. A mechanism not discussed but equally plausible is the metal-assisted MC sequence depicted in equation 86. The greater separation of iodonium-sulfonate ion pairs in acetonitrile versus benzene should provide the tosylate (or mesylate) ions with sufficient mobility to add to the β-carbon atom of the alkynyliodonium ion.

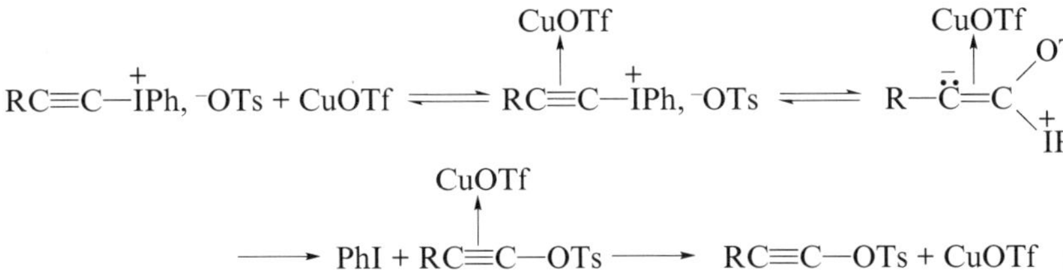

$$RC{\equiv}C{\overset{+}{-}}IPh,\ {}^-OTs + CuOTf \rightleftharpoons RC{\equiv}C{\overset{+}{-}}IPh,\ {}^-OTs \rightleftharpoons R{-}\overset{..}{C}{=}C\overset{OTs}{\underset{\overset{+}{I}Ph}{\diagdown}}$$

$$\longrightarrow PhI + RC{\equiv}C{-}OTs \longrightarrow RC{\equiv}C{-}OTs + CuOTf \quad (85)$$

$$RC{\equiv}C{\overset{+}{-}}IPh,\ {}^-OTs + CuOTf \rightleftharpoons RC{\equiv}C{\overset{+}{-}}IPh,\ {}^-OTs \rightleftharpoons \overset{R}{\underset{TsO}{\diagup}}C{=}\overset{..}{\underset{}{C}}{\overset{+}{-}}IPh$$

$$\longrightarrow PhI + \overset{R}{\underset{TsO}{\diagup}}C{=}C: \longrightarrow CuOTf + \overset{R}{\underset{TsO}{\diagup}}C{=}C: \longrightarrow RC{\equiv}C{-}OTs \quad (86)$$

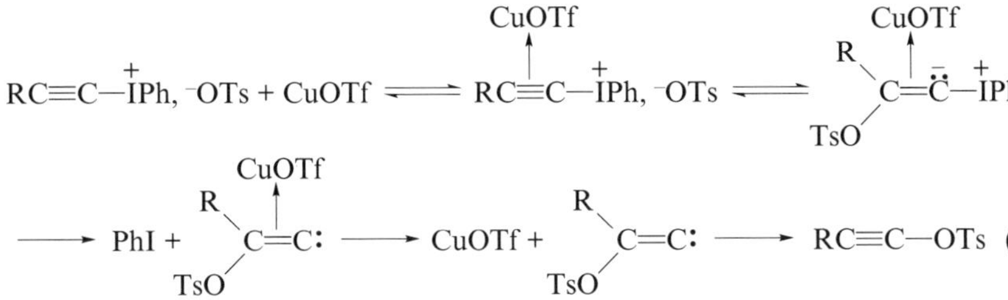

b. *Carboxylate ions.* When alkynyl(phenyl)iodonium tosylates are passed with dichloromethane through a benzoate-loaded ion exchange resin, alkynyl benzoates and iodobenzene are produced (equation 87)[81,94]. These reactions are believed to proceed through alkynyl(phenyl)iodonium benzoates, although no such intermediates have yet been isolated.

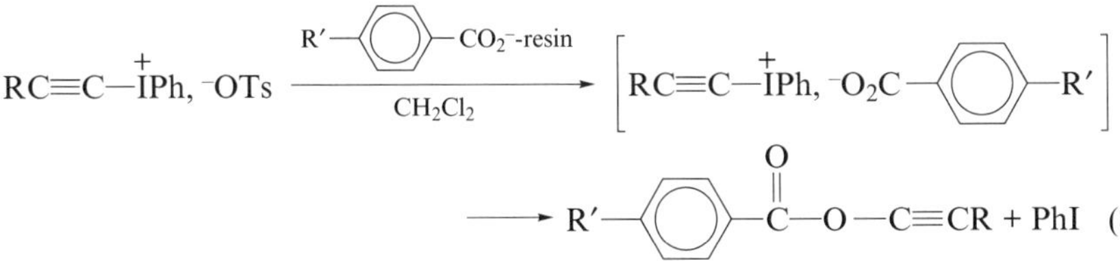

$$\longrightarrow R'{-}\langle O \rangle{-}\overset{\overset{O}{\|}}{C}{-}O{-}C{\equiv}CR + PhI \quad (87)$$

R, R' (yield): s-Bu, H(16%); t-Bu, H(40%); s-Bu, MeO (16%); t-Bu, MeO (10%)

A more general and efficient approach to alkynyl carboxylates, also thought to involve alkynyliodonium carboxylate intermediates, entails the treatment of bis(acyloxyiodo)-arenes with alkynyllithium reagents (equation 88)[81]. These reactions are best conducted in the presence of 2-nitroso-2-methylpropane in order to suppress oxidative coupling of the lithium acetylides by the acyloxyiodanes.

$$R^1C{\equiv}CLi + PhI(OCR^2)_2 \xrightarrow[-78\ °C\ to\ rt]{t\text{-}BuNO,\ THF} \left[R^1C{\equiv}C{-}\overset{+}{I}\text{-Ph},\ {}^-O_2CR^2 \right]$$

$$\xrightarrow{(-PhI)} R^1C{\equiv}C{-}\overset{\overset{\displaystyle O}{\|}}{O}CR^2 \quad (88)$$

R^2 = Ph, R^1 (yield): Me (57%), i-Pr (32%), n-Bu (44%), t-Bu (50%), MeO(Me)$_2$C (48%)
R^2 = 4-An; R^1 (yield): Me (44%)
R^2 = 3,5-(MeO)$_2$C$_6$H$_3$; R^1 (yield) n-Bu (33%), t-Bu (35%)
R^2 = 4-O$_2$NC$_6$H$_4$; R^1 (yield): Me (45%), n-Bu (56%), t-Bu (35%)
R^2 = Me; R^1 (yield): n-Bu (15%), s-Bu (9%)
R^2 = t-Bu; R^1 (yield): t-Bu (5.5%)
R^2 = PhCH$_2$CH$_2$; R^1 (yield): Me (2%), t-Bu (6%)
R^2 = Ph$_2$CH; R^1 (yield): Me (24%), t-Bu (25%)

The synthesis of bisalkynyl benzoates from bisalkynyliodonium triflates with sodium benzoate or p-nitrobenzoate in dichloromethane has also been reported (equation 89)[43]. The bisalkynyl esters are sensitive to hydration of the triple bonds, especially in solution, and have been isolated only in low yields. When R = H, monobenzoate derivatives are also obtained.

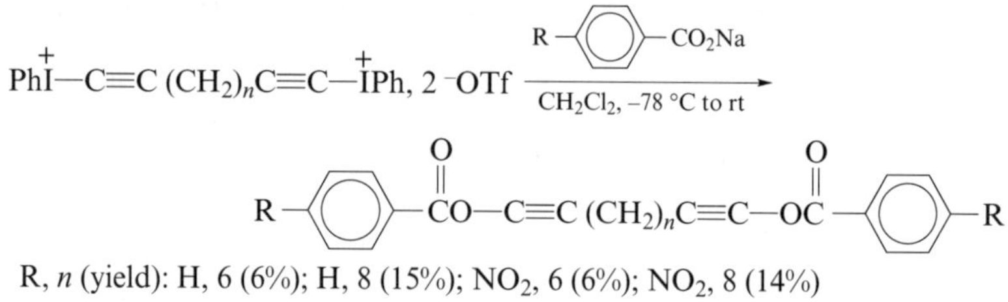

$$Ph\overset{+}{I}{-}C{\equiv}C\,(CH_2)_n C{\equiv}C{-}\overset{+}{I}Ph,\ 2\ {}^-OTf \xrightarrow[CH_2Cl_2,\ -78\ °C\ to\ rt]{R-\!\!\bigcirc\!\!-CO_2Na}$$

$$R{-}\!\!\bigcirc\!\!{-}\overset{\overset{\displaystyle O}{\|}}{C}O{-}C{\equiv}C\,(CH_2)_n C{\equiv}C{-}O\overset{\overset{\displaystyle O}{\|}}{C}\!\!{-}\!\!\bigcirc\!\!{-}R \quad (89)$$

R, n (yield): H, 6 (6%); H, 8 (15%); NO$_2$, 6 (6%); NO$_2$, 8 (14%)

The conversion of alkynyl(phenyl)iodonium ions to alkynyl esters with carboxylate ions via the MC mechanism (equation 45) has been proposed[81]. Evidence for the viability of this process is provided by the generation of 3,3-dimethyl-1-cyclopentenyl benzoate in addition to the expected alkynyl benzoate when the iodonium triflate shown in equation 90 is mixed with sodium benzoate in dichloromethane (yields not reported)[3].

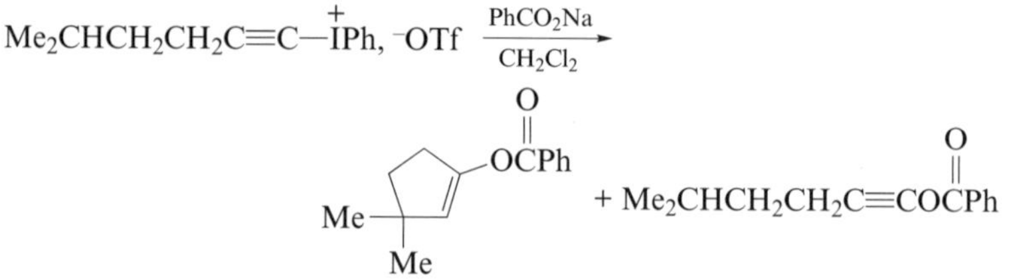

$$Me_2CHCH_2CH_2C{\equiv}C{-}\overset{+}{I}Ph,\ {}^-OTf \xrightarrow[CH_2Cl_2]{PhCO_2Na}$$

$$+ Me_2CHCH_2CH_2C{\equiv}C\overset{\overset{\displaystyle O}{\|}}{O}CPh \quad (90)$$

Various alkynyl(phenyl)iodonium tetrafluoroborates react with aliphatic sodium carboxylates in aqueous THF to give α-(acyloxy)ketones (equation 91)[30]. The presence of

$$R^1C{\equiv}C{-}\overset{+}{I}Ph,\ BF_4^- + R^2CO_2Na \xrightarrow[\text{0 °C to rt}]{\text{THF, H}_2\text{O (2 : 1)}} R^1\overset{\overset{\displaystyle O}{\|}}{C}CH_2O\overset{\overset{\displaystyle O}{\|}}{C}R^2 \qquad (91)$$

R^1,R^2 (yield): n-C$_8$H$_{17}$, Me (72%, 78%); n-C$_8$H$_{17}$, t-Bu (75%); t-Bu, Me (30%, 47%); cyclohexyl, Me (48%, 58%); Ph, Me (24%, 29%)

water is essential for the success of these transformations. Thus, when the 1-decynyliodonium salt was treated with sodium acetate in dry THF (heterogeneous mixture), 1-decynyl iodide (n-C$_8$H$_{17}$C≡CI) was obtained in 58% yield. α-(Acyloxy)ketones are also generated when solutions of the alkynyliodonium tetrafluoroborates in acetic acid are heated at 80 °C (equation 92)[30].

$$RC{\equiv}C{-}\overset{+}{I}Ph,\ BF_4^- \xrightarrow[\text{80 °C}]{\text{CH}_3\text{COOH}} R\overset{\overset{\displaystyle O}{\|}}{C}CH_2OAc \qquad (92)$$

R (yield): t-Bu (36%), n-C$_8$H$_{17}$ (86%), cyclohexyl (55%), Ph (11%)

   c. Phosphate ions. Standard methods for the synthesis of alkynyliodonium sulfonates, when applied with phosphate regents [i.e. (RO)$_2$PO$_2^-$], generally lead directly to alkynyl phosphates[23,94]. These include (1) the elution of alkynyl(phenyl)iodonium tosylates through a phosphate-loaded ion exchange resin (equation 93)[23,94], (2) admixture of terminal alkynes with [hydroxy(phosphoryloxy)iodo]arenes (equation 94)[23] and (3) the treatment of alkynyliodonium tetrafluoroborates (generated *in situ* from alkynylsilanes, iodosylbenzene and Et$_2$O·BF$_3$) with aqueous sodium dialkyl phosphates (equation 95)[23]. Alkynyl(phenyl)iodonium phosphates are believed to be intermediates in all of these reactions and a few have been isolated, but they decompose readily, presumably via the MC mechanism, to iodobenzene and alkynyl phosphates (equation 96)[23].

$$RC{\equiv}C{-}\overset{+}{I}Ph,\ ^-OTs \xrightarrow{\text{(EtO)}_2\text{PO}_2^-\text{-resin, CH}_2\text{Cl}_2} RC{\equiv}C{-}\overset{\overset{\displaystyle O}{\|}}{O}P(OEt)_2 + PhI \qquad (93)$$

R (yield): n-Bu (17%), s-Bu (25%), t-Bu (18%, iodonium phosphate isolated first)

$$R^1C{\equiv}CH + PhI(OH)\ \overset{\overset{\displaystyle O}{\|}}{O}P(OR^2)_2 \xrightarrow[\text{desiccant}]{\text{CH}_2\text{Cl}_2} R^1C{\equiv}C{-}\overset{\overset{\displaystyle O}{\|}}{O}P(OR^2)_2 + PhI \quad (94)$$

R^1, R^2 (yield): n-Bu, Et (19%); s-Bu, Et (46%); t-Bu, Et (33%); s-Bu, Me (23%); t-Bu, Me (37%)

$$R^1C{\equiv}CSiMe_3 \xrightarrow[\text{CHCl}_3,\ \text{rt}]{\text{PhIO, Et}_2\text{O·BF}_3} R^1C{\equiv}C{-}\overset{+}{I}Ph,\ BF_4^-$$

$$\xrightarrow[\text{H}_2\text{O}]{\text{(R}^2\text{O)}_2\text{PO}_2\text{Na}} R^1C{\equiv}C{-}\overset{\overset{\displaystyle O}{\|}}{O}P(OR^2)_2 + PhI \quad (95)$$

R^1R^2 (yield): n-Pr, Me (41%); n-Bu, Me (36%); t-Bu, Me (58%); n-C$_6$H$_{13}$, Me (38%); Me, Et (40%); n-Pr, Et (44%); n-Bu, Et (31%); t-Bu, Et (55%); n-C$_6$H$_{13}$, Et (42%); n-Pr, PhCH$_2$ (45%); t-Bu, PhCH$_2$ (57%)

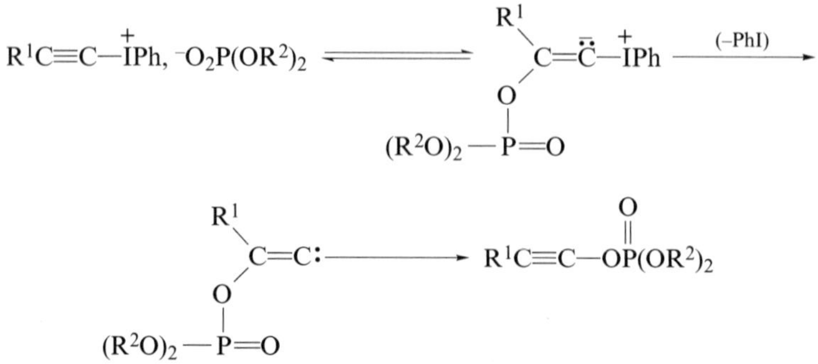

$$R^1C{\equiv}C{-}\overset{+}{I}Ph, \ ^-O_2P(OR^2)_2 \ \rightleftharpoons \ \underset{\underset{\displaystyle (R^2O)_2-P{=}O}{|}}{\overset{\displaystyle \underset{R^1}{\diagdown}}{\underset{\displaystyle O}{C{=}\overset{..}{C}{-}\overset{+}{I}Ph}}} \ \xrightarrow{\ (-PhI)\ }$$

$$\underset{\underset{\displaystyle (R^2O)_2-P{=}O}{|}}{\overset{\displaystyle \underset{R^1}{\diagdown}}{\underset{\displaystyle O}{C{=}C{:}}}} \ \longrightarrow \ R^1C{\equiv}C{-}O\overset{\displaystyle O}{\overset{\|}{P}}(OR^2)_2 \qquad (96)$$

It is interesting to note that the yields of alkynyl phosphates in these reactions tend to be highest when the R group in the alkyne moiety is *tert*-butyl. This might be regarded as evidence against the MC mechanism, since the addition of phosphate ions to the β-carbon atom of the (*t*-butylethynyl)iodonium ion should be sterically impeded. However, when (*t*-butylethynyl)phenyliodonium tosylate is allowed to react with ammonium phenyl phosphate in ethanol, a stable internal (β-phosphoryloxyvinyl)iodonium salt is obtained (equation 97)[95], thus providing direct evidence for the viability of such Michael additions.

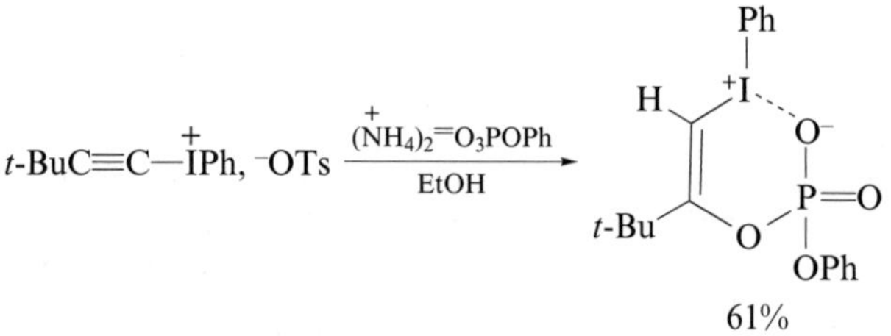

$$t\text{-BuC}{\equiv}C{-}\overset{+}{I}Ph, \ ^-OTs \ \xrightarrow[\text{EtOH}]{\overset{+}{(NH_4)_2}{=}O_3POPh} \quad (97)$$

61%

d. *Phenoxide ion.* Literature reports on the reactions of alkynyliodonium salts with phenoxide ion are restricted to the conversion of [bis(phenyliodonium)]ethyne ditriflate to diphenoxyacetylene with lithium phenoxide (equation 98)[41] and the production of benzo-furans from alkynyl(*p*-phenylene)bisiodonium ditriflates with sodium phenoxide in methanol (equation 99)[39]. The formation of the benzofurans is consistent with the MC

$$Ph\overset{+}{I}{-}C{\equiv}C{-}\overset{+}{I}Ph, \ 2 \ ^-OTf \ \xrightarrow[-78\ ^\circ C \text{ to rt}]{PhOLi, CH_2Cl_2} \ PhO{-}C{\equiv}C{-}OPh + 2PhI \qquad (98)$$

57%

$$Ph\overset{+}{I}{-}\!\!\left\langle\bigcirc\right\rangle\!\!{-}\overset{+}{I}{-}C{\equiv}CR, \ 2 \ ^-OTf \ \xrightarrow{\ PhONa\ }_{MeOH}$$

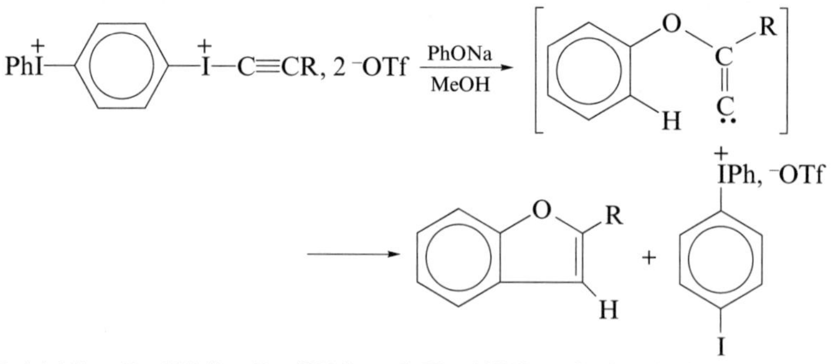

$$\longrightarrow \qquad + \qquad \qquad (99)$$

R (yield): *n*-Bu (62%), *t*-Bu (59%), *n*-C$_6$H$_{13}$ (62%), *n*-C$_{10}$H$_{21}$ (49%)

pathway and the intermediate existence of (β-phenoxy)alkylidenecarbenes. Perhaps the most unexpected property of such carbenes is their preference for intramolecular insertion into the *ortho*-CH bonds of the phenoxy ring, despite the presence of γ-CH bonds in the *n*-butyl, *n*-hexyl and *n*-decyl side chains (i.e. 1-phenoxycyclopentenes were not reported[39]). In the carbene generated from the (phenylethynyl)iodonium substrate (i.e. R = Ph), migration of the phenyl group is competitive with intramolecular insertion (equation 100)[39].

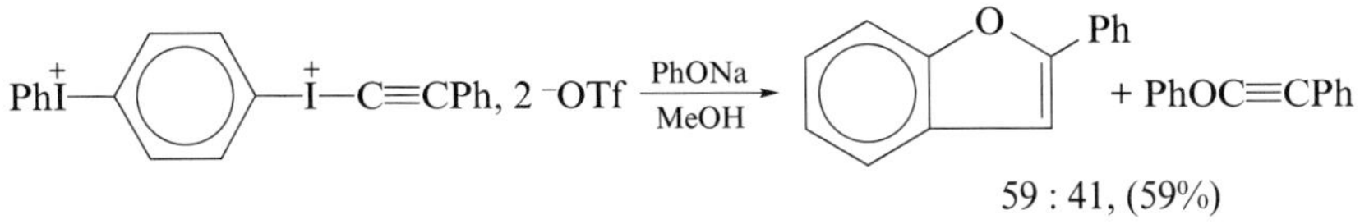

$$\text{PhI}^+\!\!-\!\!\langle\bigcirc\rangle\!-\!\overset{+}{\text{I}}\!-\!\text{C}\!\equiv\!\text{CPh}, 2\ ^-\text{OTf} \xrightarrow[\text{MeOH}]{\text{PhONa}} \quad + \text{PhOC}\!\equiv\!\text{CPh}$$

$$59 : 41, (59\%)$$

$$(100)$$

e. Water. In a 1979 publication, it was noted that [bis(phenyliodonium)] *p*-phenylene diethynyl bistrifluoroacetate, upon treatment with water, undergoes hydrolysis in two stages to 1,4-[bis(hydroxyacetyl)]benzene (equation 101)[44]. The synthesis of mono and bis α-hydroxyketones from 1,4-diethynylbenzene with one and two molar equivalents of [bis-(trifluoroacetoxy) iodo]benzene in wet chloroform was also reported (equation 102)[44].

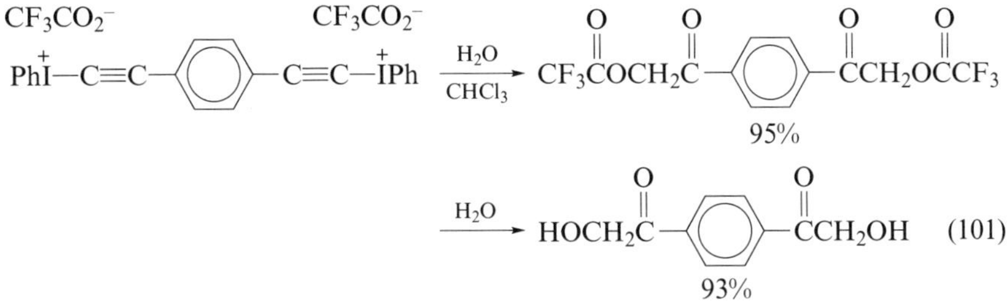

$$\text{PhI}^+\!\!-\!\!\text{C}\!\equiv\!\text{C}\!-\!\!\langle\bigcirc\rangle\!-\!\text{C}\!\equiv\!\text{C}\!-\!\overset{+}{\text{I}}\text{Ph} \xrightarrow[\text{CHCl}_3]{\text{H}_2\text{O}} \text{CF}_3\text{COCH}_2\text{C}\!-\!\!\langle\bigcirc\rangle\!-\!\text{CCH}_2\text{OCCF}_3$$

(CF$_3$CO$_2^-$ ... CF$_3$CO$_2^-$)

$$95\%$$

$$\xrightarrow{\text{H}_2\text{O}} \text{HOCH}_2\text{C}\!-\!\!\langle\bigcirc\rangle\!-\!\text{CCH}_2\text{OH} \quad (101)$$

$$93\%$$

$$\text{HC}\!\equiv\!\text{C}\!-\!\!\langle\bigcirc\rangle\!-\!\text{C}\!\equiv\!\text{CH}$$

$$\xrightarrow[\text{CHCl}_3, \text{H}_2\text{O}]{\text{PhI (OCOCF}_3)_2} \text{HC}\!\equiv\!\text{C}\!-\!\!\langle\bigcirc\rangle\!-\!\text{CCH}_2\text{OH}$$

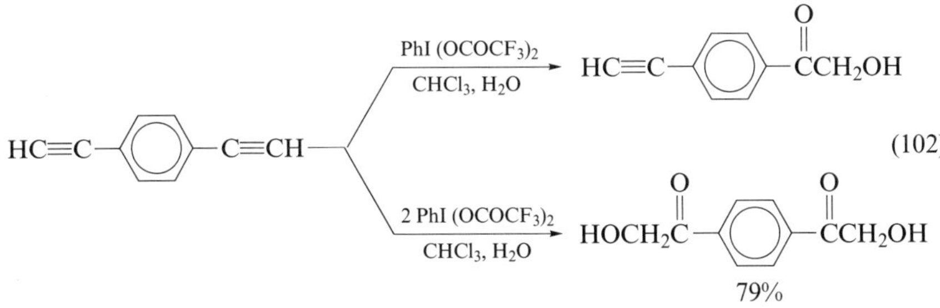

$$(102)$$

$$\xrightarrow[\text{CHCl}_3, \text{H}_2\text{O}]{2\ \text{PhI (OCOCF}_3)_2} \text{HOCH}_2\text{C}\!-\!\!\langle\bigcirc\rangle\!-\!\text{CCH}_2\text{OH}$$

$$79\%$$

This overall pattern of reactivity has since been developed into a general method for the synthesis of dihydroxyketones from ethynylcarbinols, including steroid and tetrahydronaphthacene-5,12-dione representatives (equations 103–105)[96–98]. The ethynylcarbinol is

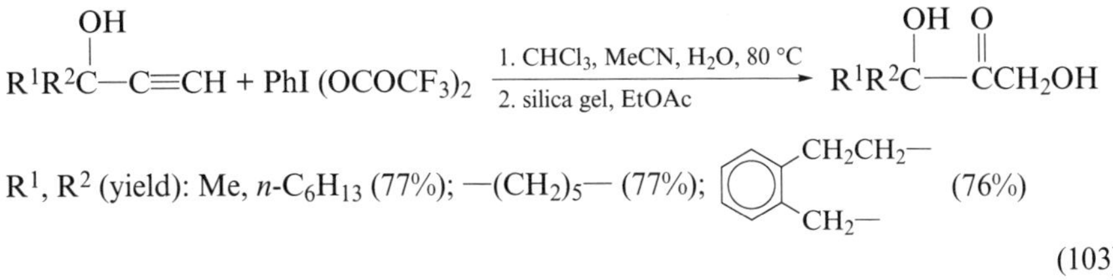

$$\text{R}^1\text{R}^2\overset{\text{OH}}{\underset{|}{\text{C}}}\!-\!\text{C}\!\equiv\!\text{CH} + \text{PhI (OCOCF}_3)_2 \xrightarrow[\text{2. silica gel, EtOAc}]{\text{1. CHCl}_3, \text{MeCN}, \text{H}_2\text{O}, 80\,°\text{C}} \text{R}^1\text{R}^2\overset{\text{OH}}{\underset{|}{\text{C}}}\!-\!\overset{\text{O}}{\underset{\|}{\text{C}}}\text{CH}_2\text{OH}$$

$$\text{R}^1, \text{R}^2 \text{ (yield): Me, } n\text{-C}_6\text{H}_{13}\ (77\%);\ -(\text{CH}_2)_5-\ (77\%);\ \langle\bigcirc\rangle\!\!\overset{\text{CH}_2\text{CH}_2-}{\underset{\text{CH}_2-}{}}\ (76\%)$$

$$(103)$$

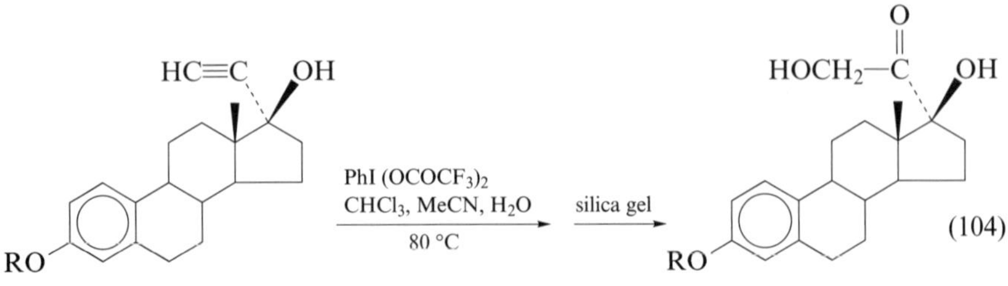

(104)

R (yield) : Ac (53%), Ts (60%)

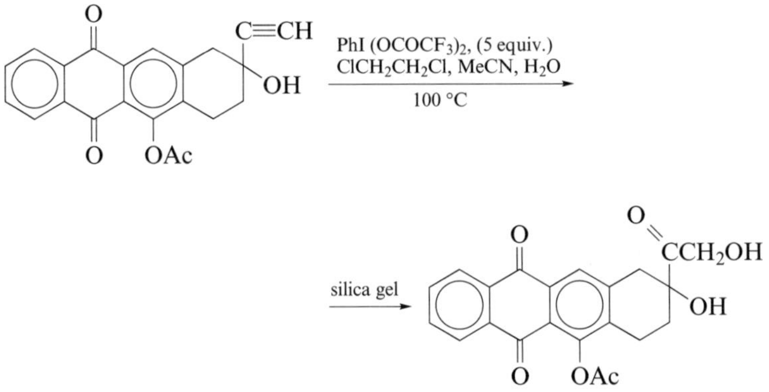

(105)

(85%)

first treated with [bis(trifluoroacetoxy)iodo]benzene in an aqueous organic medium. The organic layer is then concentrated, and the crude product, a mixture of the dihydroxyketone and its trifluoroacetate, is eluted through dry silica gel with ethyl acetate, thus completing the hydrolysis[97]. The direct conversion of three terminal alkynes to α-hydroxyketones by this method has also been demonstrated[96,97].

A mechanism for these reactions involving the initial formation of alkynyliodonium trifluoroacetates (i.e. $R^1R^2C(OH)C\equiv C-\overset{+}{I}Ph, CF_3CO_2^-$) has been proposed[97].

5. Nitrogen nucleophiles

a. Azide ion. The azide ion is a useful probe for reactivity assessments of intermediates associated with the MCI pathway (equation 45). For example, when alkynyl(phenyl)iodonium tosylates are mixed with sodium azide and 18-crown-6 in dichloromethane, both intra- and intermolecular capture of $RC (N_3)=C$: can be demonstrated (equation 106)[99]. With the phenylethynyl substrate, protonation (with moisture ?) occurs at the ylide stage, even when Et_3SiH is present, and a vinyliodonium salt is obtained.

The treatment of alkynyliodonium salts not amenable to cyclopentene formation with sodium azide in methanol affords vinyliodonium salts and/or enol ethers (equation 107)[99]. Enol ether formation also occurs when glyme is employed as the solvent (equation 108)[99]. Finally, regeneration of the vinylidene-iodonium ylide, $Ph\ddot{C} (N_3)=\overset{+}{C}-\overset{+}{I}Ph$, from (Z)-(β-azido-β-phenylvinyl)phenyliodonium tosylate with potassium *t*-butoxide in glyme likewise affords an enol ether (equation 109).

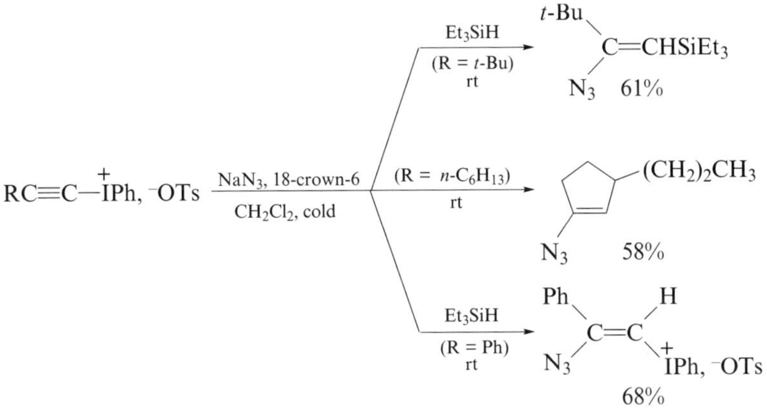

$$RC{\equiv}C{-}\overset{+}{I}Ph,\ ^-OTs \xrightarrow[\text{CH}_2\text{Cl}_2,\ \text{cold}]{\text{NaN}_3,\ 18\text{-crown-6}}$$

Et₃SiH
(R = t-Bu)
rt

$$\underset{N_3}{\overset{t\text{-Bu}}{\diagdown}}C{=}CHSiEt_3 \quad 61\%$$

(R = n-C₆H₁₃)
rt

cyclopentene with (CH₂)₂CH₃ and N₃, 58%

Et₃SiH
(R = Ph)
rt

$$\underset{N_3}{\overset{Ph}{\diagdown}}C{=}C\underset{\overset{+}{I}Ph,\ ^-OTs}{\overset{H}{\diagup}} \quad 68\%$$

(106)

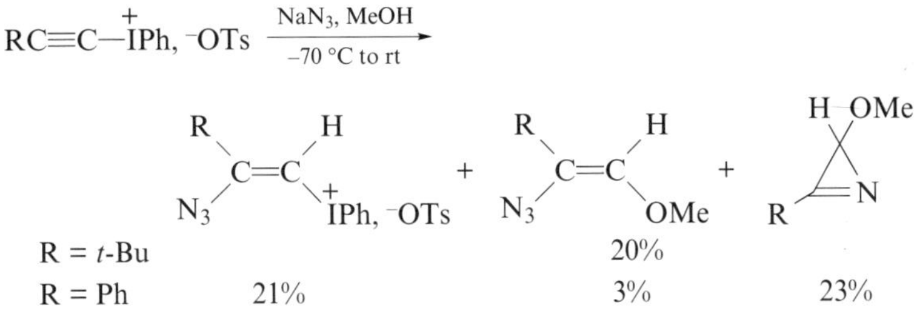

$$RC{\equiv}C{-}\overset{+}{I}Ph,\ ^-OTs \xrightarrow[-70\ °C\ \text{to rt}]{\text{NaN}_3,\ \text{MeOH}}$$

$$\underset{N_3}{\overset{R}{\diagdown}}C{=}C\underset{\overset{+}{I}Ph,\ ^-OTs}{\overset{H}{\diagup}} \quad + \quad \underset{N_3}{\overset{R}{\diagdown}}C{=}C\underset{OMe}{\overset{H}{\diagup}} \quad + \quad \underset{R}{\overset{H\ \ OMe}{\diagup}}{=}N$$

(107)

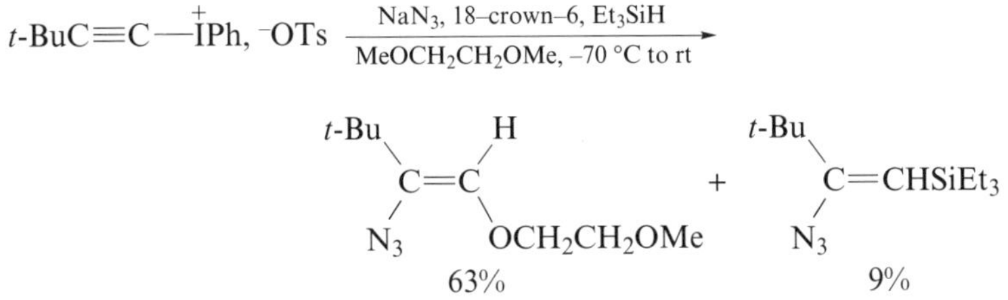

R = t-Bu

R = Ph   21%   20%   3%   23%

$$t\text{-BuC}{\equiv}C{-}\overset{+}{I}Ph,\ ^-OTs \xrightarrow[\text{MeOCH}_2\text{CH}_2\text{OMe},\ -70\ °C\ \text{to rt}]{\text{NaN}_3,\ 18\text{-crown-6},\ \text{Et}_3\text{SiH}}$$

$$\underset{N_3}{\overset{t\text{-Bu}}{\diagdown}}C{=}C\underset{OCH_2CH_2OMe}{\overset{H}{\diagup}} \quad + \quad \underset{N_3}{\overset{t\text{-Bu}}{\diagdown}}C{=}CHSiEt_3 \quad (108)$$

63%   9%

$$\underset{N_3}{\overset{Ph}{\diagdown}}C{=}C\underset{\overset{+}{I}Ph,\ ^-OTs}{\overset{H}{\diagup}} \xrightarrow[\text{cold}]{t\text{-BuOK}} \underset{N_3}{\overset{Ph}{\diagdown}}C{=}C\underset{OCH_2CH_2OMe}{\overset{H}{\diagup}}$$

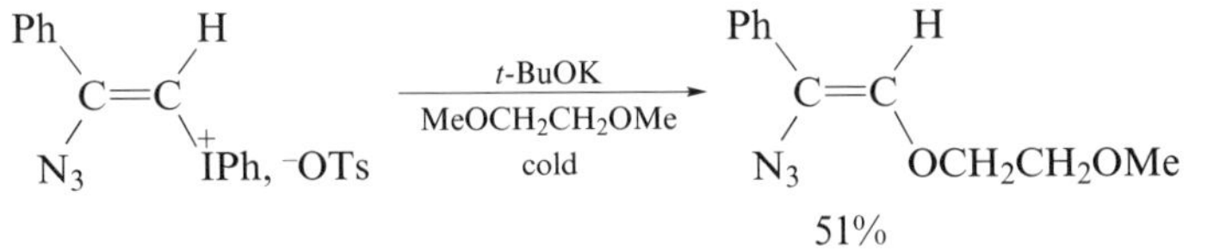

(109)

51%

Since glyme is unreactive with free alkylidenecarbenes[100] and since (β-aryl)alkyli-denecarbenes rearrange to arylalkynes too rapidly for intermolecular capture[100,101], the production of enol ethers is thought to proceed directly from vinylidene-iodonium ylides[99]. Although the nature of the 'ylide-transfer process' has not been established, transylidation to oxonium ylides **24** seems plausible. Similar transylidations of cyclopentadienyl and β-dicarbonyl iodonium ylides with nitrogen, sulfur, selenium and arsenic nucleophiles are well documented[66]. In the case of methanol, the production of enol ethers via the reductive elimination of iodobenzene from intermediate λ^3-methoxyiodanes, **25**, can also be envisioned.

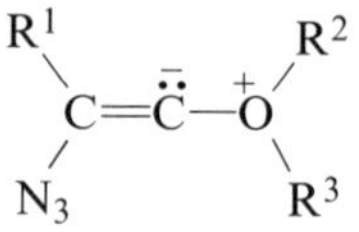

$$
\begin{array}{cc}
\underset{N_3}{\overset{R^1}{\diagup}}C=\overset{\cdot\cdot}{\underset{R^3}{\overset{+}{C}}}-\overset{R^2}{\underset{R^3}{\diagup}}O & R-\underset{H}{\overset{N_3}{\diagup}}C\overset{}{\diagup}C-\underset{Ph}{\overset{|}{I}}-OMe \\
(24) & (25)
\end{array}
$$

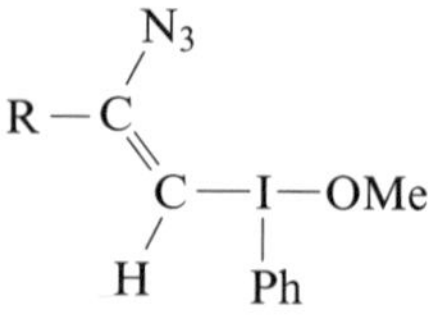

b. *Hydrazoic acid.* The treatment of alkynyl(phenyl)iodonium tetrafluoroborates with trimethylsilyl azide and H_2O in dichloromethane constitutes a stereoselective synthesis of (Z)-(β-azidovinyl)phenyliodonium tetrafluoroborates (equation 110)[102]. These reactions are believed to involve the *in situ* production of hydrazoic acid (HN_3) which adds in Michael fashion to the alkyliodonium ions.

$$
RC\equiv\overset{+}{C}-IPh,\ BF_4^- + Me_3SiN_3 \xrightarrow[-78\ °C\ to\ rt]{CH_2Cl_2,\ H_2O} \underset{N_3}{\overset{R}{\diagup}}C=\underset{\overset{+}{I}Ph,\ BF_4^-}{\overset{H}{\diagup}}C \tag{110}
$$

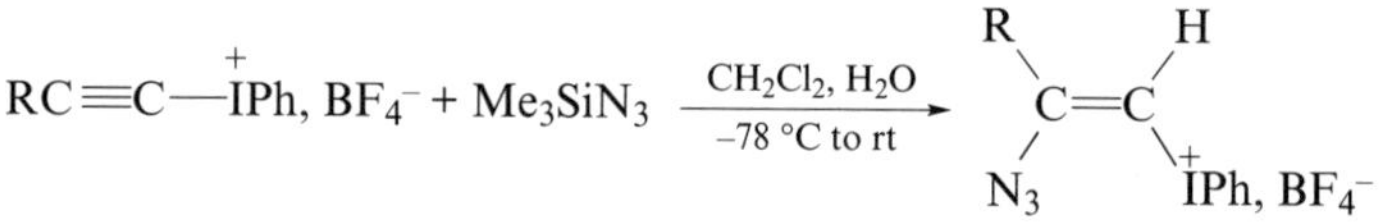

R (yield): Me (79%), t-Bu (50%), n-C_8H_{17} (91%), cyclopentyl –CH_2 (87%)

6. Halogen nucleophiles

a. *Halide ions.* When alkynyl(phenyl)iodonium tetrafluoroborates are mixed with a tenfold excess of lithium halides ($X^- = Cl^-$, Br^-) in acetic acid or with an excess of HX (X = Cl, Br) in methanol (followed by the introduction of aqueous KX), (Z)-(β-halovinyl)phenyliodonium halides are produced stereoselectively (equations 111 and 112)[103]. Efforts to prepare a (β-fluorovinyl)iodonium fluoride from the 1-decynyliodonium salt by both methods were unsuccessful.

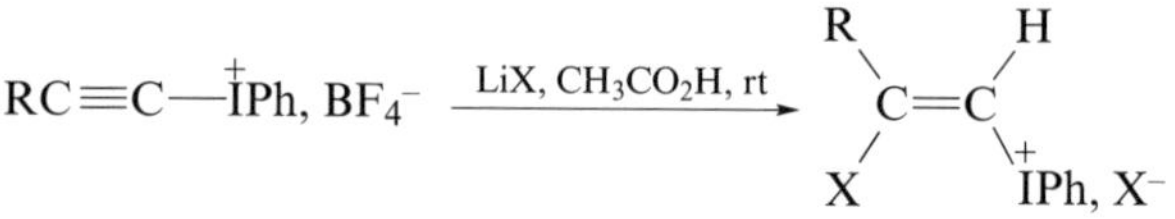

$$
RC\equiv\overset{+}{C}-IPh,\ BF_4^- \xrightarrow{LiX,\ CH_3CO_2H,\ rt} \underset{X}{\overset{R}{\diagup}}C=\underset{\overset{+}{I}Ph,\ X^-}{\overset{H}{\diagup}}C \tag{111}
$$

R, X (yield): t-Bu, Br (66%); n-C_8H_{17}, Br (86%); cyclopentyl–CH_2, Br (81%); Ph$(CH_2)_3$, Br (74%); n-C_8H_{17}, Cl (100%); cyclopentyl–CH_2, Cl (95%)

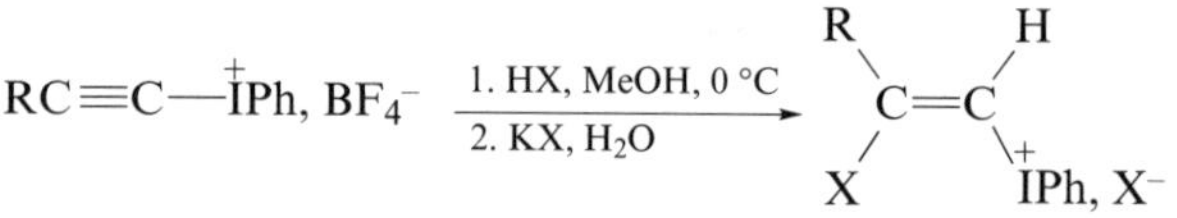

$$
RC\equiv\overset{+}{C}-IPh,\ BF_4^- \xrightarrow[2.\ KX,\ H_2O]{1.\ HX,\ MeOH,\ 0\ °C} \underset{X}{\overset{R}{\diagup}}C=\underset{\overset{+}{I}Ph,\ X^-}{\overset{H}{\diagup}}C \tag{112}
$$

R, X (yield): n-C_8H_{17}, Cl (65%); Ph$(CH_2)_2$, Cl (95%); n-C_8H_{17}, Br (85%)

Although the treatment of the foregoing alkynyliodonium tetrafluoroborates 'with halide ions under neutral conditions' has been reported to give 'a complex mixture of products'[103], there is at least one exception to this rule. Thus, when (2-^{13}C-phenylethynyl)phenyliodonium tetrafluoroborate is exposed to lithium halides ($X^- = Cl^-$, Br^-, I^-) in CH_2Cl_2–MeOH at –78 °C, 2-^{13}C-1-halo-2-phenylethynes are obtained in 'good yields' (equation 113)[104]. Since the formation of the halo(phenyl)acetylenes is thought to proceed through (β-halo)alkylidenecarbenes, these labeling studies imply that the halogen atoms are even more tropophilic than the phenyl group[104] (equation 114).

$$Ph — ^{13}C \equiv C — \overset{+}{I}Ph, \; BF_4^- + LiX \xrightarrow[-78\,°C]{CH_2Cl_2,\; MeOH} Ph — ^{13}C \equiv C — X + PhI \quad (113)$$

(99% ^{13}C)   (X = Cl, Br, I)   (98% ^{13}C)   > 98%

$$Ph — ^{13}C \equiv C — \overset{+}{I}Ph + X:^- \; \rightleftharpoons \; \overset{Ph}{\underset{X}{\diagdown}} {}^{13}C = \overset{..}{\underset{}{\overset{-}{C}}} — \overset{+}{I}Ph \xrightarrow{-PhI} \overset{Ph}{\underset{X}{\diagdown}} {}^{13}C = C:$$

$$\xrightarrow{\sim [X:]} \; Ph — ^{13}C \equiv C — X \qquad\qquad (114)$$

7. Carbon nucleophiles

a. Enolate ions. Among the reactions of alkynyliodonium salts with carbon nucleophiles, those with β-dicarbonyl enolates have been explored most extensively. However, even in this subclass, enolates derived from cyclic precursors have been favored, and most β-dicarbonyl compounds selected for deprotonation possess only one CH bond at the central atom. Reactions of simple enolates [i.e. $R^1C(O^-)=CR^2R^3$; R^2, R^3 = H, alkyl, aryl] with alkynyliodonium salts have not been described.

When β-dicarbonyl enolates are allowed to react with alkynyliodonium salts, typically in *tert*-butyl alcohol or THF, alkynyl- and/or cyclopentenyl-β-dicarbonyl compounds are obtained. The product compositions are largely regulated by the migratory aptitude of R in the alkynyl moiety and the availability of alkyl side chains for the MC-insertion (MCI) pathway (equation 45). These divergent modes of reactivity are nicely illustrated by the reactions of the 2-phenyl-1,3-indandionate ion with ethynyl(phenyl)- and 4-methyl-1-hexynyl(phenyl)iodonium tetrafluoroborates (equation 115)[27,28].

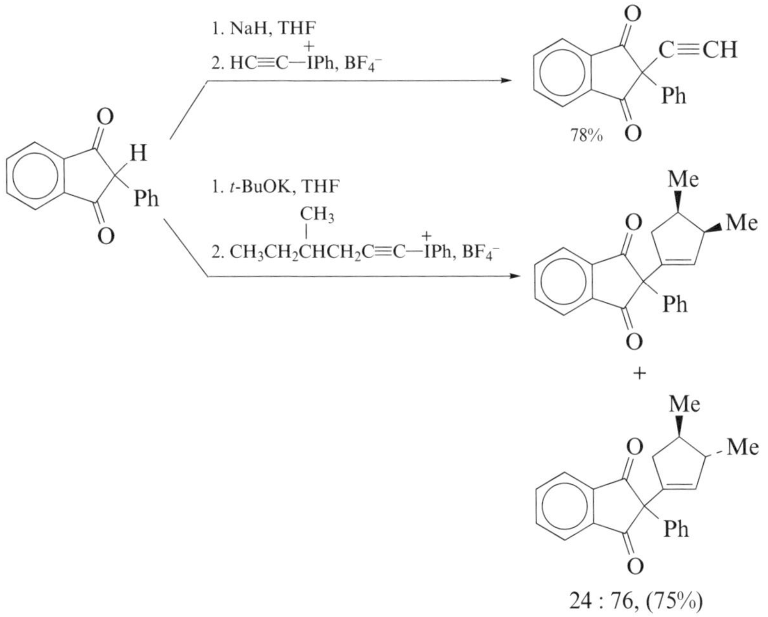

24 : 76, (75%)

Two MCI patterns have been recognized[28], one in which the alkyl chain for carbenic insertion is provided by the alkynyliodonium ion (i.e. '[5 + 0] cyclopentene annulation') and another in which the alkyl chain is provided by the enolate ion (i.e. '[2 + 3] cyclopentene annulation'). Generic transformations of each type are illustrated in equations 116 and 117[28].

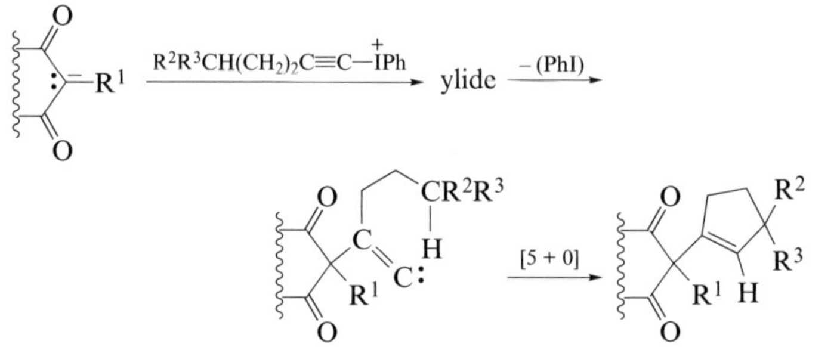

$$(116)$$

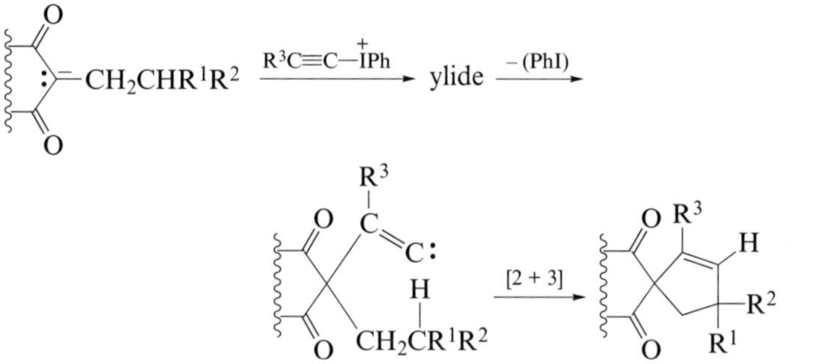

$$(117)$$

The MCI cyclopentene annulations proceed with moderate efficiency and have been utilized for the construction of polycyclic molecules with fused, spiro and tethered ring systems (e.g. equations 118–120)[28]. With bis(phenyliodonium) diyne salts, biscyclopentene annulations are observed (e.g. equation 121)[86].

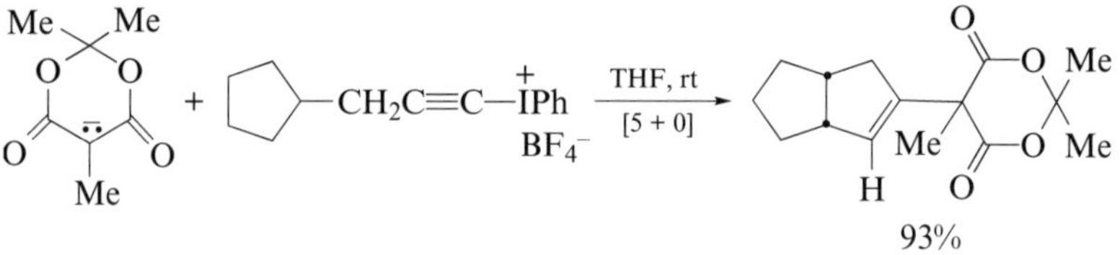

93%

$$(118)$$

52%

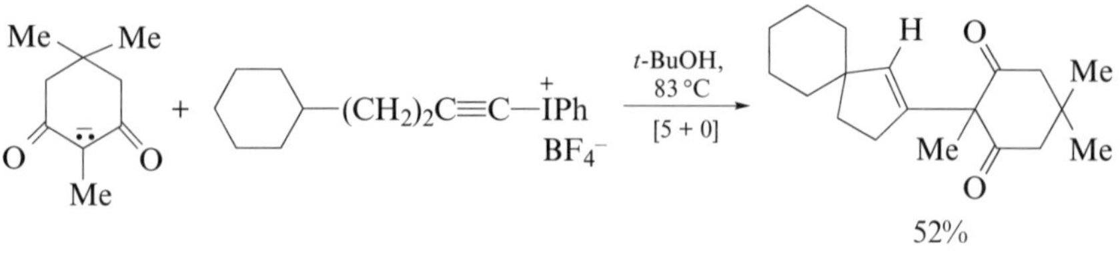

$$(119)$$

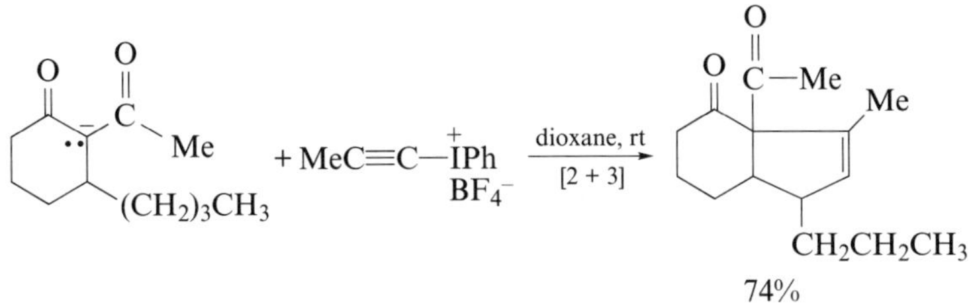

(120)

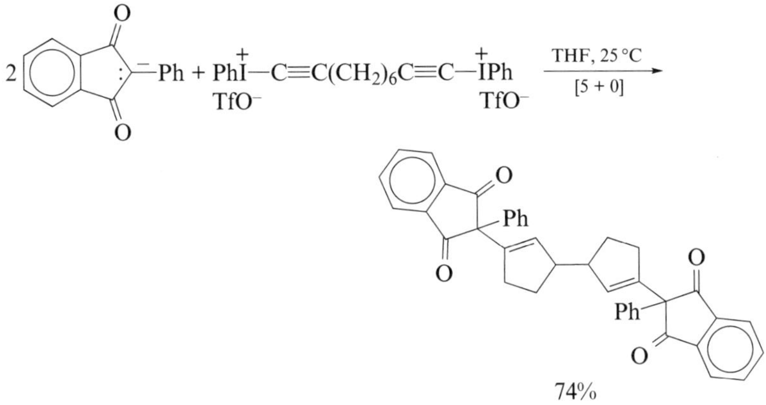

74%

(121)

74%

MCI reactions of alkynyliodonium salts with enolates derived from active methylene compounds containing two acidic CH bonds follow a divergent course that leads to furans, presumably via carbenic insertion into enolic OH bonds (equation 122)[28]. In the reaction of acetylacetonate ion with the 1-decynyl(phenyl)iodonium ion, CH insertion is competitive with OH insertion (equation 123)[28].

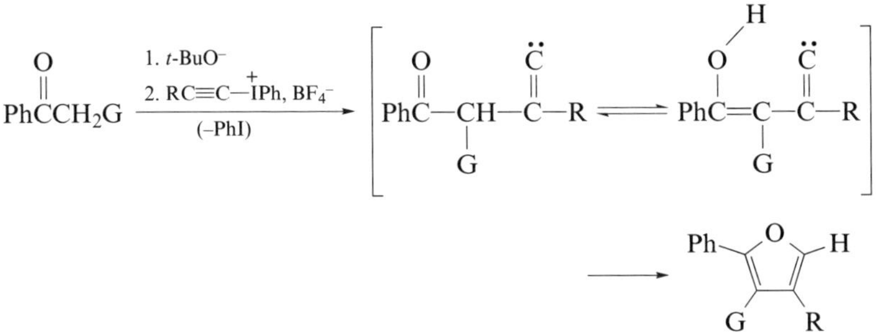

(122)

G, R (yield): $-S(O)_2Ph$, $n-C_8H_{17}$ (67%); $-CN$, Me (46%)

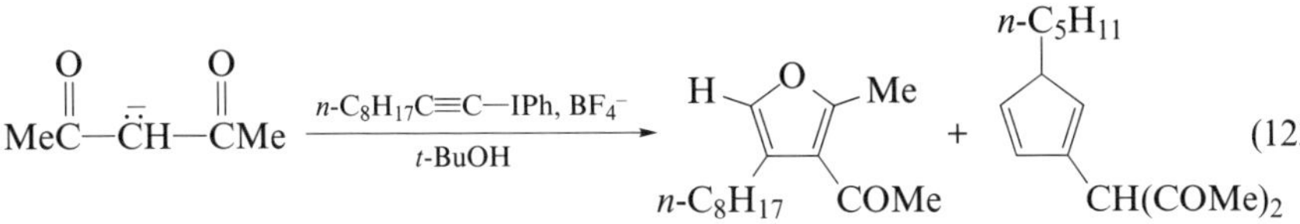

(123)

64 : 36 (61%)

Because the hydrogen atom and phenyl group migrate so readily, the reactions of β-dicarbonyl enolates with ethynyl- and (phenylethynyl)iodonium salts can be expected to result in alkynylation. It has already been noted that the 2-n-hexyl-1,3-indandionate ion undergoes alkynylation with (phenylethynyl)phenyliodonium tetrafluoroborate (equation 43), despite the availability of the n-hexyl group for [2 + 3] annulation. Ethynylations of six β-dicarbonyl enolates and the anion of 2-nitrocyclohexane with ethynyl(phenyl)iodonium tetrafluoroborate in THF have also been reported[27]. For example, admixture of the ethynyliodonium salt and the anion of ethyl 2-cyclopentanone-1-carboxylate in THF affords the 1-ethynyl derivative in 71% isolated yield (equation 124)[27].

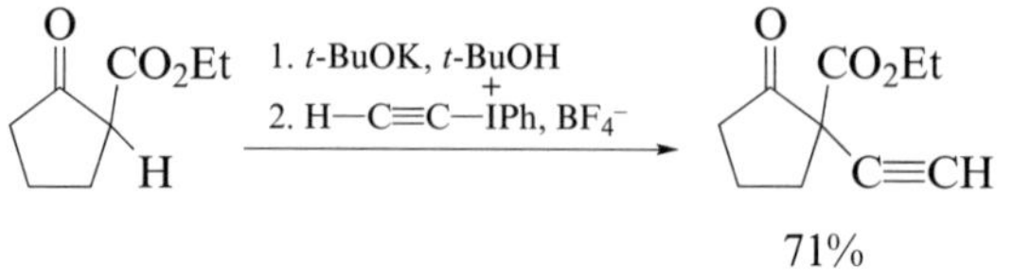

$$(124)$$

The reactions of the lithium enolate of diethyl 2-[(diphenylmethylene)amino]malonate with several alkynyliodonium triflates are rare examples of enolate alkynylations with iodonium species other than the ethynyl(phenyl)- and (phenylethynyl)phenyliodonium ions (equation 125)[16]. Two experimental protocols were followed, i.e. addition of the enolates to the iodonium salts and *vice versa*, the former procedure giving higher yields of alkynylmalonates. As with other enolate alkynylations, these reactions are thought to involve alkylidenecarbene intermediates. It has been proposed, however, that the carbenes rearrange with migration of the 'diethyl 2-[(diphenyl) amino] malonate anion'[16].

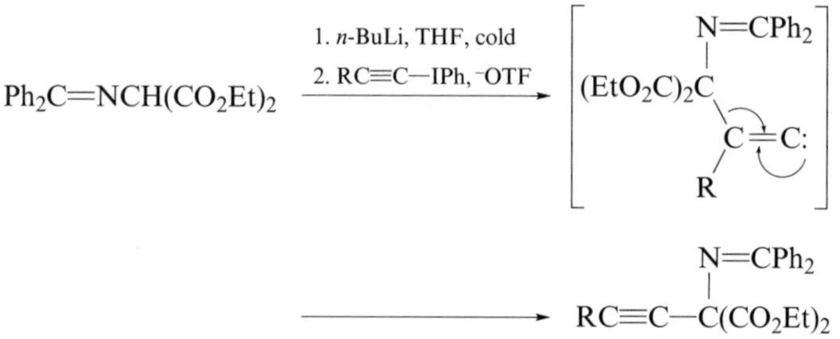

$$(125)$$

R (yield): n-Bu (40%, 30%); t-Bu (95%, 95%); Me$_3$Si (95%, 71%);
Ph (33%, 33%)

Phenyl[(trimethylsilyl)ethynyl]iodonium triflate has also been employed for alkynylations of diethyl 2-phthalimidomalonate and the (2-oxoazetidinyl)malonates shown in equation 126[105]. However, unlike the result with the [(diphenyl)amino]malonate system (equation 125), the trimethylsilyl group is lost, and the ethynyl group is ultimately introduced.

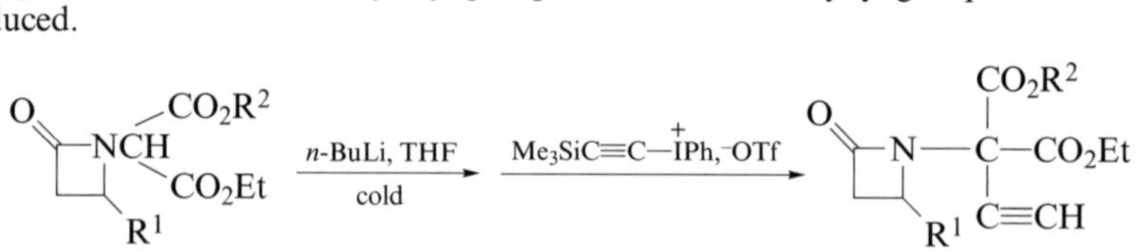

$$(126)$$

R^1, R^2 (yield): PhSeCH$_2$, Et (93%); t-BuS, t-Bu (92%)

A summary of reported reactions of enolate ions with alkynyliodonium salts is presented in Table 6. Those that result in alkynylation are denoted with an (A), while those that afford cyclopentenes are marked with a (C). Reactions that lead to furans are designated with an (F).

Some general comments. Those factors that regulate partitioning between the migration and insertion pathways in MC reactions between enolates and alkynyliodonium ions have not yet been thoroughly tested. Reaction partners of the [5 + 0] β-dicarbonyl enolates (i.e. no alkyl chain) have been largely restricted either to [5 + 0] alkynyliodonium ions (i.e. alkyl chains $\geq$ three carbon atoms) or to alkynyliodonium ions in which R is highly tropophilic (i.e. H, Ph). Reactions of [5 + 0] enolates with such iodonium species as RC≡C—İPh (R = Me, Et, *i*-Pr, *t*-Bu) have not been reported. What, for example, might be the fate of alkylidenecarbenes **26** and **27**, generated from the 2-phenyl-1,3-indandionate and iso-propylidene 2-methylmalonate ions with 1-propynyliodonium salts?

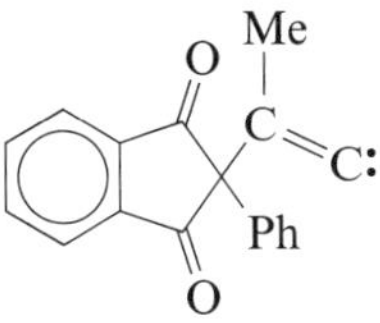

(**26**) 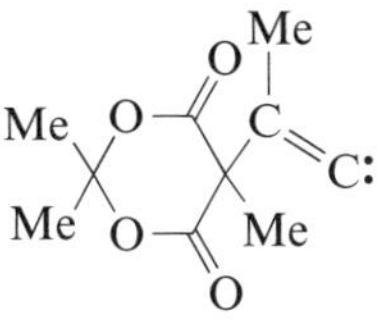   (**27**)

The enolates of unactivated monocarbonyl compounds are enigmatic. Although the products of their reactions with alkynyliodonium salts have not been described, it has been reported that 'Alkynylation of simple enolates does not occur'[3]. It appears that alkynylations via the MC pathway are promoted by the action of soft nucleophiles[3].

b. Lithiofurans and lithiothiophenes. The addition of aryl(*tert*-butylethynyl)iodonium tosylates as the solids to an excess of 2-lithiofuran in Et$_2$O/hexane and subsequent treatment of the mixtures with tosylic acid leads to aryl(2-furyl)iodonium tosylates (equation 127)[9]. Similar results are obtained when 2-lithiothiophenes are used instead of 2-lithiofuran (equation 128)[9]. Even (phenylethynyl)phenyliodonium tosylate with its highly tropophilic phenyl group affords a furyliodonium salt with 2-lithiofuran (equation 129).

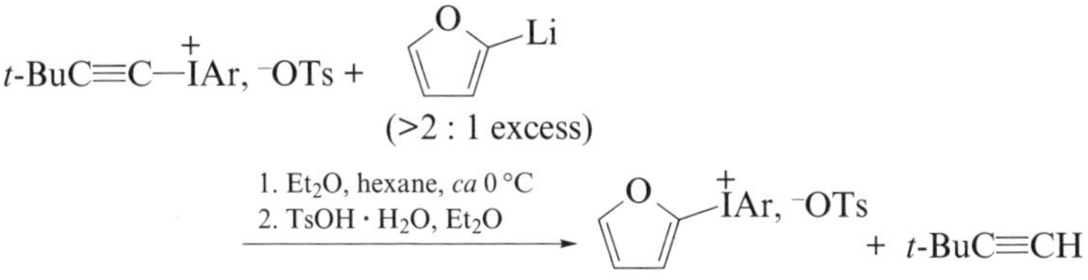

Ar (yield): 2-Tol (74%), 3-Tol (72%), 4-Tol (65%), 2-FC$_6$H$_4$ (45%),
3-FC$_6$H$_4$ (21%), 4-FC$_6$H$_4$ (32.5%); 3-ClC$_6$H$_4$ (21%)

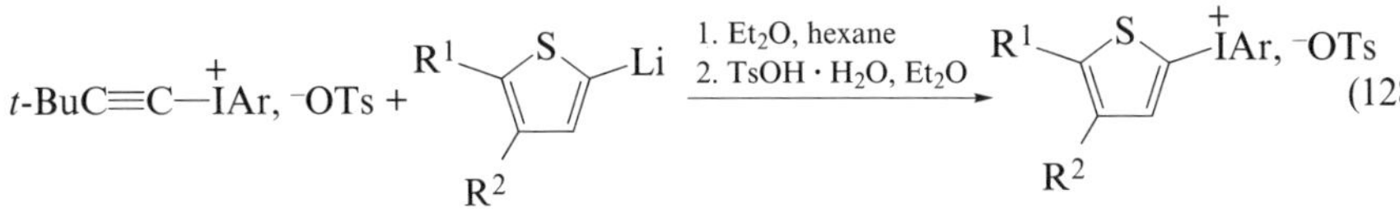

Ar, R^1, R^2 (yield): 2-Tol, H, H, (62.5%); 2-Tol, Me, H (64%); Ph, benzo (70.5%)

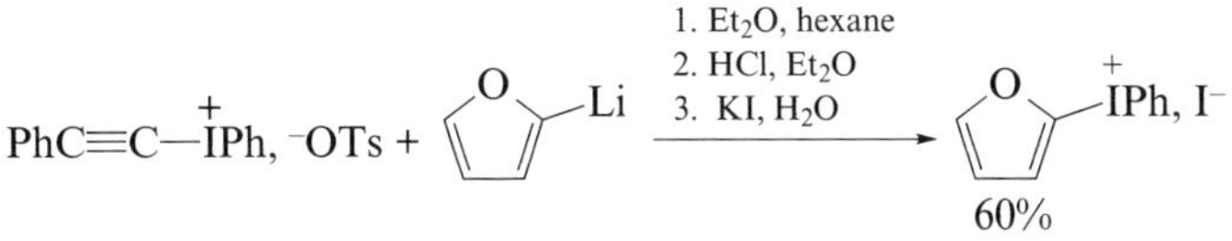

TABLE 6. Literature survey of reactions of alkynyliodonium salts with enolates

Enolate	Alkynyliodonium salt (Mode of reaction)	Reference
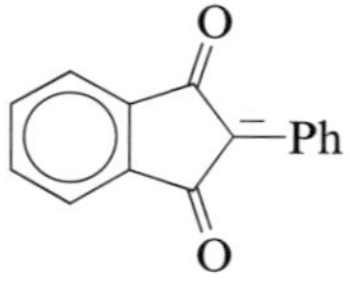 2-phenyl-1,3-indandione (Ph)	$PhC\equiv C{-}\overset{+}{I}Ph$, Cl^- (A)	26
	$HC\equiv C{-}\overset{+}{I}Ph$, BF_4^- (A)	27
	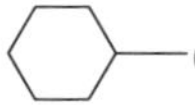 $RC\equiv C{-}\overset{+}{I}Ph$, BF_4^-; R = EtCH(Me)CH$_2$, n-C$_8H_{17}$, cyclohexyl–(CH$_2$)$_2$ (C)	28
	$PhI{-}\overset{+}{C}\equiv C(CH_2)_n C\equiv C{-}\overset{+}{I}Ph$, 2 TfO$^-$; $n = 5,6,8$ (C)	86
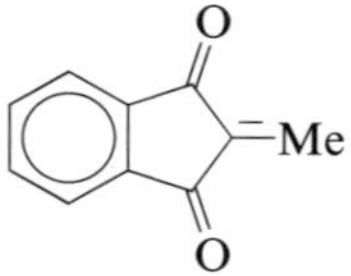 2-methyl-1,3-indandione (Me)	$RC\equiv C{-}\overset{+}{I}Ph$, BF_4^-; R = H, Me$_3$Si (A)	27
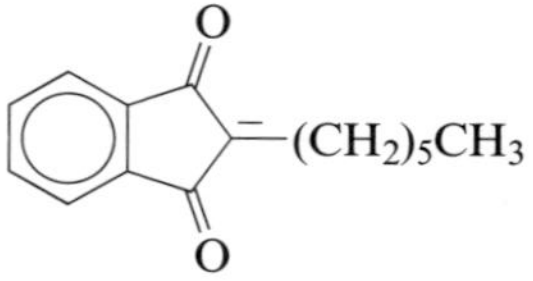 2-(CH$_2$)$_5$CH$_3$-1,3-indandione	$RC\equiv C{-}\overset{+}{I}Ph$, BF_4^-; R = Me, n-C$_8H_{17}$ (C)	28
	$^{13}Ph{-}C\equiv C{-}\overset{+}{I}Ph$, BF_4^-, $PhC\equiv C{-}\overset{+}{I}Ph$, BF_4^- (A)	28
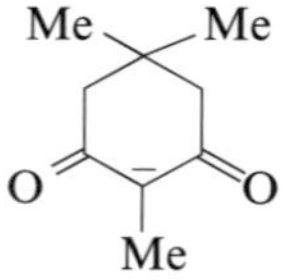 2,5,5-trimethylcyclohexane-1,3-dione (Me, Me, Me)	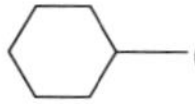 $RC\equiv C{-}\overset{+}{I}Ph$, BF_4^-; R = n-C$_8H_{17}$, cyclohexyl–(CH$_2$)$_2$ (C)	28

Substrate	Reagent	Ref.
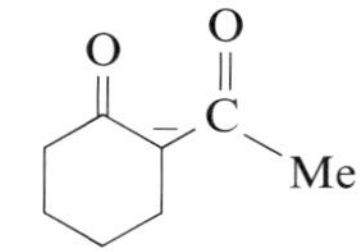 cyclohexanone enolate, $=O$, $-C(=O)Me$	$n\text{-}C_8H_{17}C\equiv C-\overset{+}{I}Ph,\ BF_4^-$ (C)	28
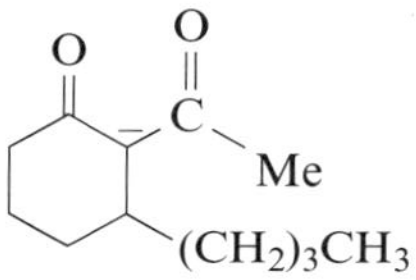 cyclohexanone enolate, $=O$, $-C(=O)Me$, $(CH_2)_3CH_3$	$MeC\equiv C-\overset{+}{I}Ph,\ BF_4^-$ (C)	28
$H-\overset{-}{C}(CMe)_2$, $=O$	$n\text{-}C_8H_{17}C\equiv C-\overset{+}{I}Ph,\ BF_4^-$ (F, C)	28
$PhC-\overset{-}{C}HSO_2Ph$, $=O$	$n\text{-}C_8H_{17}C\equiv C-\overset{+}{I}Ph,\ BF_4^-$ (F)	28
$PhC-\overset{-}{C}HCN$, $=O$	$MeC\equiv C-\overset{+}{I}Ph,\ BF_4^-$ (F)	28
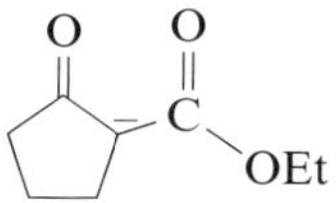 cyclopentanone enolate, $=O$, $-C(=O)OEt$	$HC\equiv C-\overset{+}{I}Ph,\ BF_4^-$ (A)	27
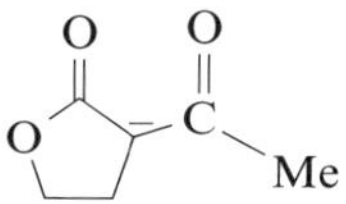 lactone enolate, $=O$, $-C(=O)Me$	$HC\equiv C-\overset{+}{I}Ph,\ BF_4^-$ (A)	27

(continued)

TABLE 6. (*continued*)

Enolate	Alkynyliodonium salt (Mode of reaction)	Reference
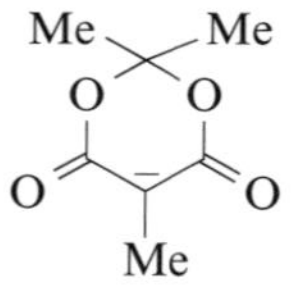 Me⟍Me (dioxanedione with Me)	HC≡C—$\overset{+}{\text{I}}$Ph, BF$_4^-$ (A)	27
	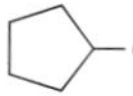 RC≡C—$\overset{+}{\text{I}}$Ph, BF$_4^-$; R = n-C$_8$H$_{17}$, cyclopentyl-CH$_2$ (C)	28
	PhI—C≡C(CH$_2$)$_n$C≡C—$\overset{+}{\text{I}}$Ph, 2TfO$^-$; n-5,6,8 (C)	86
Ph—$\overline{\text{C}}$(COOEt)$_2$	HC≡C—$\overset{+}{\text{I}}$Ph, BF$_4^-$ (A)	27
Ph$_2$C=N—$\overline{\text{C}}$(COOEt)$_2$	RC≡C—$\overset{+}{\text{I}}$Ph, $^-$OTf; R = n-Bu, t-Bu, Me$_3$Si, Ph (A)	16
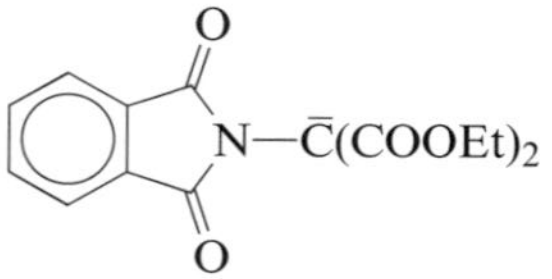 N—$\overline{\text{C}}$(COOEt)$_2$ (phthalimido)	Me$_3$SiC≡C—$\overset{+}{\text{I}}$Ph, $^-$OTf (A)	105
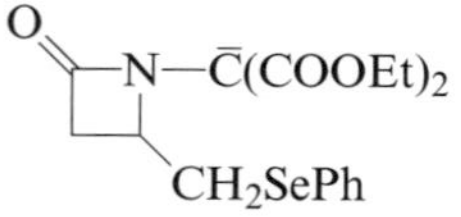 N—$\overline{\text{C}}$(COOEt)$_2$ CH$_2$SePh (azetidinone)	Me$_3$SiC≡C—$\overset{+}{\text{I}}$Ph, $^-$OTf(A)	105
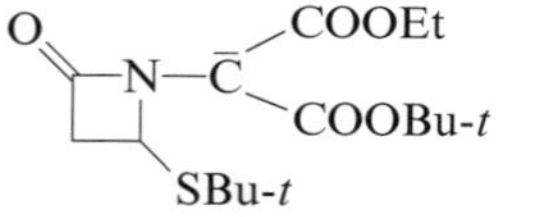 N—$\overline{\text{C}}$ COOEt / COOBu-t SBu-t (azetidinone)	Me$_3$SiC≡C—$\overset{+}{\text{I}}$Ph, $^-$OTf(A)	105

In view of the propensity of alkynyliodonium ions for Michael reactions with a wide range of nucleophiles, such displacements at iodine represent an unusual mode of reactivity. However, carbon ligand exchanges of this type at polyvalent iodine do find precedent in the literature[106] and probably proceed via the tetrasubstituted iodate ions shown in equation 130. A similar mechanism was first proposed by Beringer and Chang to account for interconversions of diaryliodonium salts with aryllithium reagents (equation 131)[106].

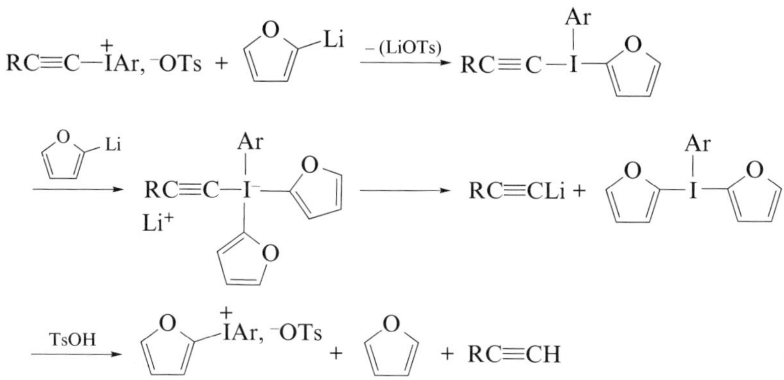

(130)

$$\overset{+}{Ar_2I}, Cl^- + Ar'Li \ (4 \ equiv.) \xrightarrow[\quad Et_2O \quad]{} \xrightarrow[\quad 2. \ MI \quad]{1. \ H^+} \overset{+}{Ar'_2I}, I^-$$

(131)

Why the thienyl- and furyllithium compounds attack the iodine atom instead of the β-carbon atom of alkynyliodonium ions is not entirely clear. However, they are much harder nucleophiles than the enolate salts of β-dicarbonyl compounds and might prefer the harder iodonium center[3]. The much higher basicity of the heteroaryllithium reagents might also facilitate the formation of iodate ions and the displacement of alkynyllithiums from such intermediates.

c. *Vinyl-, alkynyl- and alkylcopper reagents.* In contrast to 2-lithiofuran, vinylcopper reagents undergo alkynylation with alkynyliodonium tosylates, thus providing access to conjugated enynes (equation 132)[107]. These reactions are highly stereoselective (*ca.* 100%) and proceed with retention of configuration in the olefinic component. A nice demonstration of the stereochemical control afforded by this method is provided by the independent syntheses of the E- and Z-isomers of 1-phenyl-4-n-propyl-3-octen-1-yne[107]. Preparations of (E)-5-phenyldodec-5-en-7-yne and its alkynyliodonium precursor are presented in *Organic Syntheses*[108]. Although the mechanism of condensation is not known, the MC sequence seems unlikely, and these reactions may not be regulated by the nucleophilic behavior of the vinylcopper reagents. A process involving 'oxidative addition of the alkynyl species' to give a Cu(III) intermediate, followed by 'reductive elimination and coupling', has been proposed[107].

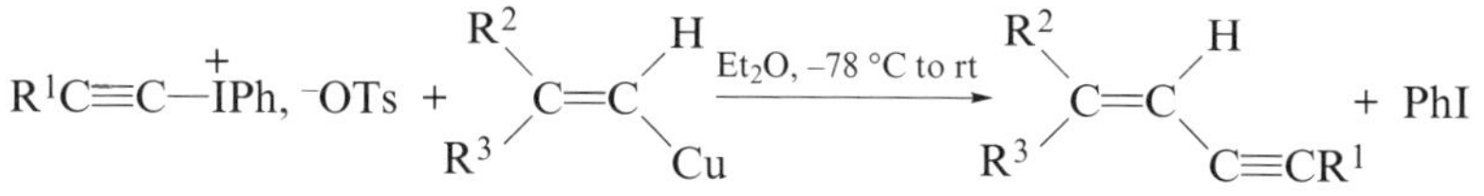

(132)

R^1, R^2, R^3 (yield): *n*-Bu, Ph, *n*-Bu (78%); *t*-Bu, Ph, Et (79%); *t*-Bu, Ph, *n*-Bu (94%); Ph, Me, Et (47%); Ph, *n*-Bu, Et (52%); Ph, *n*-Pr, *n*-Bu (46%); Ph, *n*-Bu, *n*-Pr (48%); Ph, Ph, Et (54%)

Lithium dialkynylcuprates behave similarly with alkynyliodonium tosylates and lead to conjugated diynes (equation 133)[109]. Unsymmetrical diynes can be prepared with moderate selectivity by this method, although they are accompanied by symmetrical diynes derived from the alkynyliodonium component. The treatment of lithium diphenyl- and dialkylcuprates with alkynyliodonium tosylates has also been investigated and affords alkyl- and phenyl-substituted alkynes (equation 134)[109].

$$R^1C\equiv C-\overset{+}{I}Ph, \ ^-OTs + (R^2C\equiv C)_2Cu(CN)Li_2$$

$$\xrightarrow{\text{THF, } -70\,°C \text{ to rt}} R^1C\equiv C-C\equiv CR^2 + R^1C\equiv C-C\equiv CR^1 \quad (133)$$

$$6-23\%$$

R^1, R^2 (yield): n-Bu, Ph (71%); n-Bu, 4-An (68%); t-Bu, n-Bu

(65%); n-C$_6$H$_{13}$, n-Pr (60%); Ph, 4-An (65%); 4-An, Ph (75%)

$$R^1C\equiv C-\overset{+}{I}Ph, \ ^-OTs + (R^2)_2CuLi \xrightarrow{\text{THF, } -70\,°C \text{ to rt}} R^1C\equiv CR^2 \quad (134)$$

R^1, R^2 (yield): t-Bu, n-Bu (62%); n-C$_6$H$_{13}$, n-Bu (77%); Ph, Me (52%); Ph, n-Bu (83%); Ph, Ph (90%)

d. Carbon monoxide. When alcoholic solutions of alkynyl(phenyl)iodonium tosylates are treated with carbon monoxide in the presence of palladium(II) acetate (0.2 molar equivalents) and an excess of triethyl or tri-n-butylamine, esters of alkynoic acids are obtained (equation 135)[110]. These 'alkoxycarbonylation' reactions occur readily at room temperature under one atmosphere of carbon monoxide and were terminated after one hour. In one experiment, the ethoxycarbonylation of the (phenylethynyl)iodonium salt was conducted at 0 °C and still gave a 60% yield of ethyl 3-phenyl-2-propynoate[110].

$$R^1C\equiv C-\overset{+}{I}Ph, \ ^-OTs + CO \ (1 \ atm) \xrightarrow[\text{Et}_3\text{N or } (n\text{-Bu})_3\text{N}]{R^2OH, \ Pd \ (OAc)_2, \ rt} R^1C\equiv C-\overset{O}{\overset{\|}{C}}OR^2 + PhI$$

R^1, R^2 (GC yield): n-Bu, Me (64%); Ph, Me (66%, 70%); 4-An, Me (70%); n-Bu, Et (59%); Ph, Et (80%); 4-An, Et (64%, 69%)

$$(135)$$

A list of nucleophilic reagents that have been employed with alkynyliodonium salts and appropriate references to the literature are given in Table 7.

E. Cycloaddition Reactions

1. Dipolar additions

In view of their capacity for Michael reactions with nucleophiles to give intermediate vinylidene-iodonium ylides, alkynyliodonium ions might be expected to behave as 1,3-dipolarophiles. Cycloadducts in which the nucleophilic end of the dipole is bound to the β-carbon atom of the starting alkynyliodonium ion (i.e. the β-adduct) might also be anticipated (equation 136).

Such studies have, thus far, been restricted to the reactions of selected alkynyliodonium salts with limited sets of nitrile oxides, nitrones, diazocarbonyl compounds and organo-

TABLE 7.   Directory of literature references for reactions of alkynyliodonium salts with nucleophilic reagents

---

1. *Sulfur nucleophiles*

RSO$_2$H (32, 84) p-MeC$_6$H$_4$S(O)$_2$S$^-$ (87)
RSO$_2^-$ (19, 20, 32, 33, 85,86)   (RO)$_2$PS$_2^-$ (88)
NCS$^-$ (14, 38, 43)   PhS$^-$ (40, 41)

2. *Phosphorus nucleophiles*

Ph$_3$P: (34, 40, 41, 42, 80, 90)   (Ph$_2$P)$_2$CH$_2$ (92)
tetra-t-butyltetraphosphacubane (91)   (RO)$_3$P: (10)

3. *Oxygen nucleophiles*

TsO$^-$ (5, 6, 43)   PhOPO$_3^{2-}$ (95)
MsO$^-$ (6)   PhO$^-$ (39, 41)
RCO$_2^-$ (30, 43, 81, 94)   H$_2$O (44, 96, 97, 98)
(RO)$_2$PO$_2^-$ (23, 94)

4. *Nitrogen nucleophiles*

N$_3^-$ (99)   Me$_3$SiN$_3$, H$_2$O [HN$_3$] (102)

5. *Halogen nucleophiles*

Cl$^-$ (103, 104)   I$^-$ (104)
Br$^-$ (103, 104)

6. *Carbon nucleophiles*

R$\bar{\text{C}}$(COR)$_2$ (26, 27, 28, 86)   Ph$_2$C=N—$\bar{\text{C}}$(CO$_2$Et)$_2$ (16)
R$\bar{\text{C}}$(CO$_2$R)$_2$ (27, 28, 86)   phthalimido—$\bar{\text{C}}$(CO$_2$Et)$_2$ (105)
RCO$\bar{\text{C}}$(R)CO$_2$R (27)   R^{1}R^{2}C=CHCu (107, 108)
RCO$\bar{\text{C}}$HSO$_2$Ph, PhCO$\bar{\text{C}}$HCN (28)   (RC≡C)$_2$Cu(CN)Li$_2$ (109)
(cyclo—C$_6$H$_{10}$NO$_2$)$^-$ (27)   R$_2$CuLi (109)

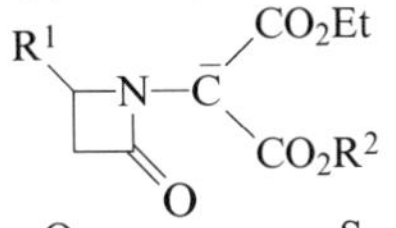

CO (110)

$$(136)$$

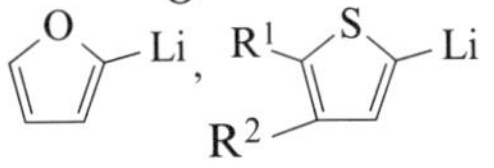

azides. Cycloadducts are indeed obtained, but the regiochemistry is not always in accord with that depicted in equation 136.

  a. Nitrile oxides, nitrones. Several cycloadditions of (arylethynyl)phenyliodonium tosylates with arenenitrile oxides to give isoxazolyl(phenyl)iodonium tosylates have been reported (equation 137)[12]. These reactions were conducted in dichloromethane at room temperature and in three cases gave only the 4-(phenyliodonium)isoxazolyl regioisomers (i.e. the β-adducts). However, the regiochemistry of dipolar addition does respond to substituent perturbations, and with the (p-anisylethynyl)phenyliodonium ion, a mixture of the 4- and 5-(phenyliodonium)isoxazolyl compounds is obtained.

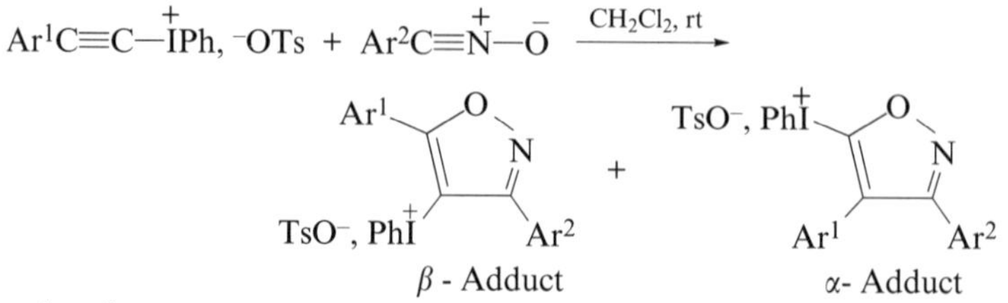

$$Ar^1C{\equiv}C-\overset{+}{I}Ph, \ ^-OTs \ + \ Ar^2C{\equiv}\overset{+}{N}-\overset{-}{O} \quad \xrightarrow{CH_2Cl_2, \ rt}$$

$$\beta \text{ - Adduct} \qquad\qquad \alpha\text{- Adduct}$$

(137)

Ar^1, Ar^2 (yield; β, α): Ph, mesityl (76%, —); 4-ClC_6H_4, mesityl (82%, —); Ph, 2,6-$Cl_2C_6H_3$ (71%, —); 4-An, mesityl (β, α mixture, 60%)

The treatment of (phenylethynyl)phenyliodonium tosylate with several nitrones in the benzaldehyde series [p-$RC_6H_4C{=}\overset{+}{N}(Me)O^-$, R = H, OMe, Cl, CN] has also been reported[12]. However, a cycloaddition product (β-adduct) was obtained only with the *para*-cyano derivative (equation 138).

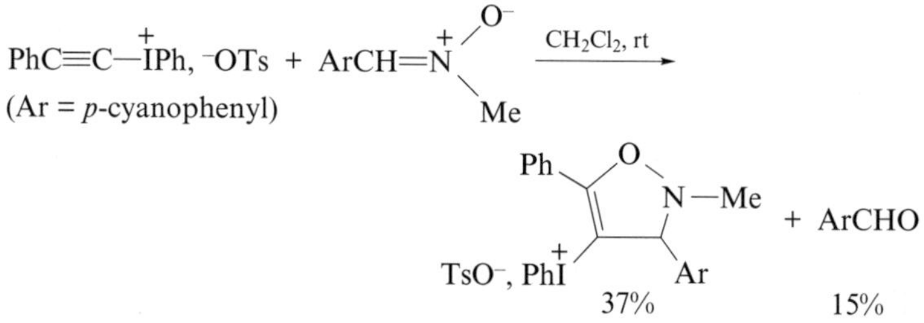

$$PhC{\equiv}C-\overset{+}{I}Ph, \ ^-OTs \ + \ ArCH{=}\overset{+}{N}\diagup^{O^-}_{\diagdown Me} \quad \xrightarrow{CH_2Cl_2, \ rt}$$

(Ar = p-cyanophenyl)

$$+ \ ArCHO \quad (138)$$

$$37\% \qquad\qquad 15\%$$

   b. Diazocarbonyl compounds. [(Trimethylsilyl) ethynyl]phenyliodonium triflate reacts with α-diazoesters and α-diazoketones in dichloromethane to give phenyl(4-pyrazolyl)-iodonium triflates (α-adducts, equation 139)[111]. This complete reversal of regiochemistry has been attributed to the steric bulk of the trimethylsilyl group and is consistent with the generation of an α-cycloadduct when phenyl(pivaloyl)iodonium triflate is mixed with methyl diazoacetate (equation 140)[111]. The treatment of ethynyl(phenyl)iodonium triflate with methyl diazoacetate, on the other hand, affords a (bisiodonium)pyrazolyl ditriflate, presumably via the initial formation of an α-cycloadduct (equation 141)[111]. The second step of this sequence in noteworthy, since it constitutes the only known example of a Michael reaction between an amine and an alkynyliodonium ion.

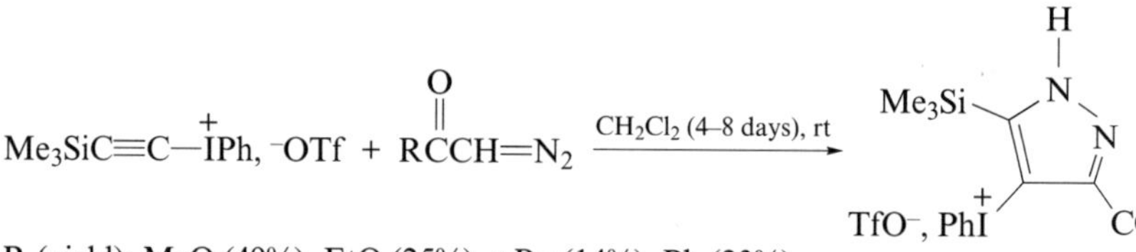

$$Me_3SiC{\equiv}C-\overset{+}{I}Ph, \ ^-OTf \ + \ RC\overset{\overset{O}{\|}}{C}CH{=}N_2 \quad \xrightarrow{CH_2Cl_2 \ (4\text{–}8 \ days), \ rt}$$

(139)

R (yield): MeO (49%), EtO (25%), t-Bu (14%), Ph (23%)

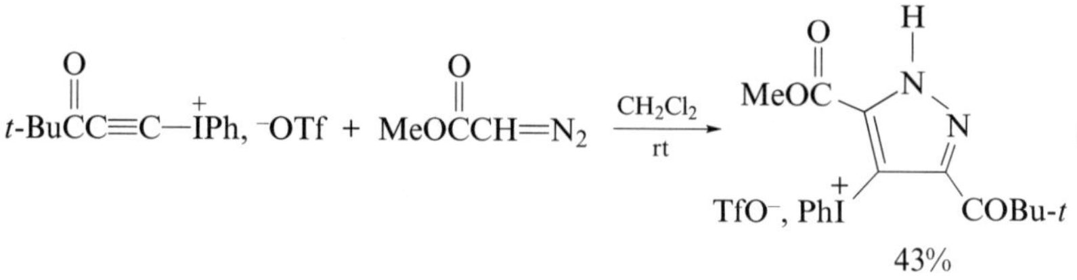

$$t\text{-}BuC\overset{\overset{O}{\|}}{C}{\equiv}C-\overset{+}{I}Ph, \ ^-OTf \ + \ MeOC\overset{\overset{O}{\|}}{C}CH{=}N_2 \quad \xrightarrow[rt]{CH_2Cl_2}$$

(140)

$$43\%$$

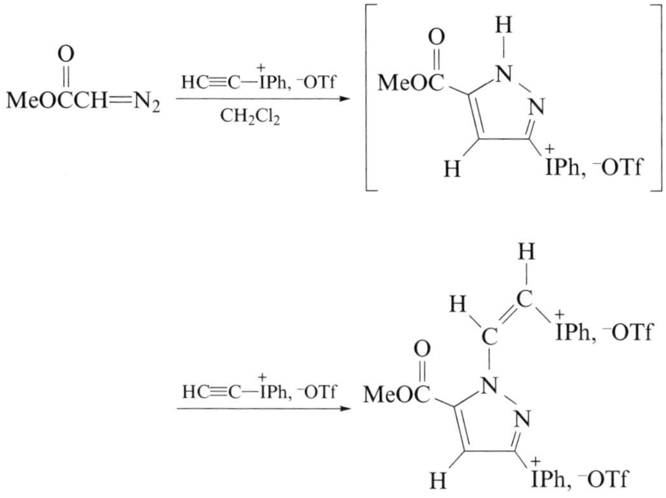

$$(141)$$

41%

c. Organoazides. Methyl- and phenyl azides are relatively unreactive with alkynyliodonium triflates. However, when such mixtures are heated in acetonitrile (75 °C) or THF (85 °C), dipolar additions occur, and 4-(1,2,3-triazolyl)phenyliodonium triflates (i.e. α-adducts) are obtained in low yields (equation 142)[111]. The β-regioisomers have not been detected.

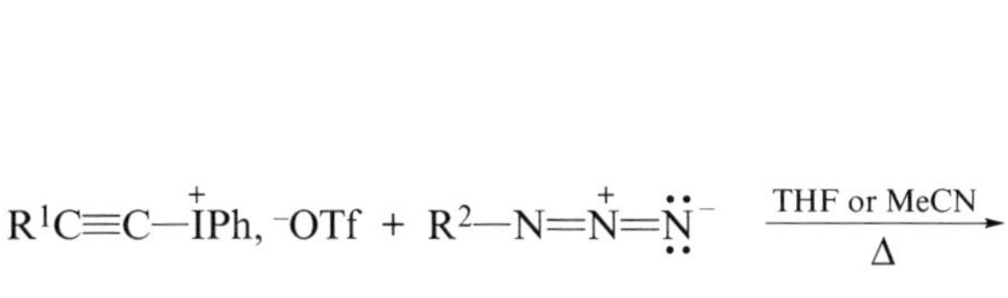

$$(142)$$

R^1, R^2 (yield): H, Me (25%); H, Ph (17%); t-Bu, Me (26%)

2. Diels–Alder reactions

Diels–Alder reactions of alkynyl(aryl) iodonium salts [RC≡C—$\overset{+}{I}$Ar, A⁻] in which R is hydrogen, alkyl or aryl have not been reported. However, Stang and his coworkers have found that alkynyl(phenyl)iodonium triflates bearing electrophilic groups in the alkynyl ligand (i.e. R= —CN, Ts, R′C=O) undergo facile cycloadditions with the dienes **28–31** in acetonitrile at 20 °C, thus providing access to mono-, bi- and tricyclic vinyliodonium salts[17]. Twenty such reactions have been demonstrated, an example set being shown in equation 143[17].

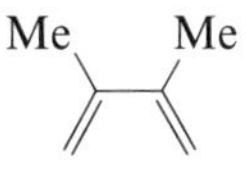

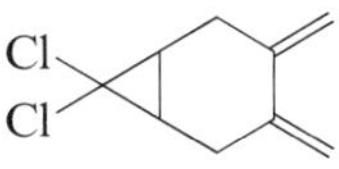

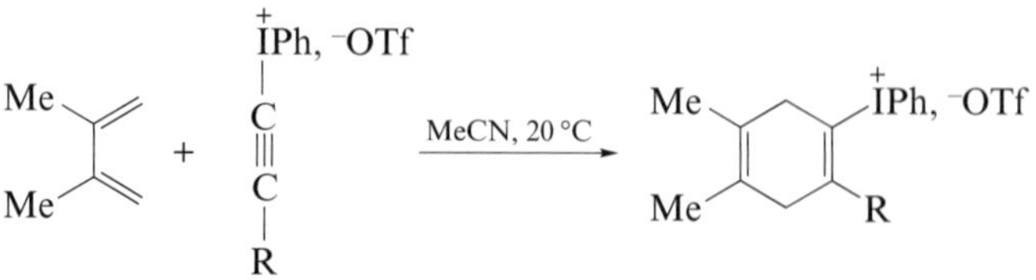

$$\text{(143)}$$

R (yield): 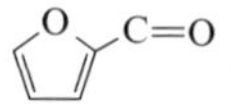 (78%), 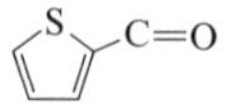 (88%), $t\text{-BuC}=O$ (73%),

—CN (79%), Ts (75%)

[Bis(phenyliodonium)]ethyne ditriflate reacts similarly with cyclopentadiene, furan and 2,5-diphenyl-3,4-benzofuran to give the corresponding bisiodonium salts (equation 144)[41].

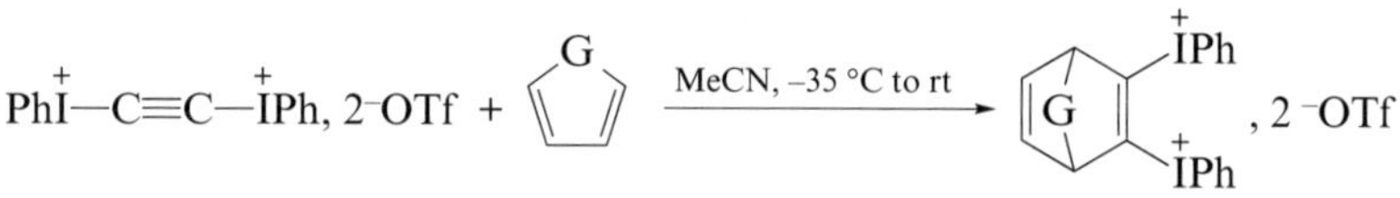

$$\text{PhI}-\text{C}\equiv\text{C}-\overset{+}{\text{IPh}}, 2^-\text{OTf} + \underset{G}{\diagdown} \xrightarrow{\text{MeCN}, -35\ ^\circ\text{C to rt}} \quad , 2^-\text{OTf} \quad \text{(144)}$$

G (yield): CH_2 (69%), O (73%)

F. Reactions with Transition Metal Complexes

1. Iridium and rhodium

Vaska's complex, the square planar iridium(I) compound shown in equation 145, readily interacts with alkynyl(phenyl)iodonium triflates in toluene at room temperature[21]. Such reactions proceed with loss of iodobenzene and deliver octahedral Ir(III) complexes possessing σ-alkynyl and trifluoromethanesulfonato ligands in a *trans*-relationship. The rhodium(I) analog of Vaska's complex behaves in a similar way[21].

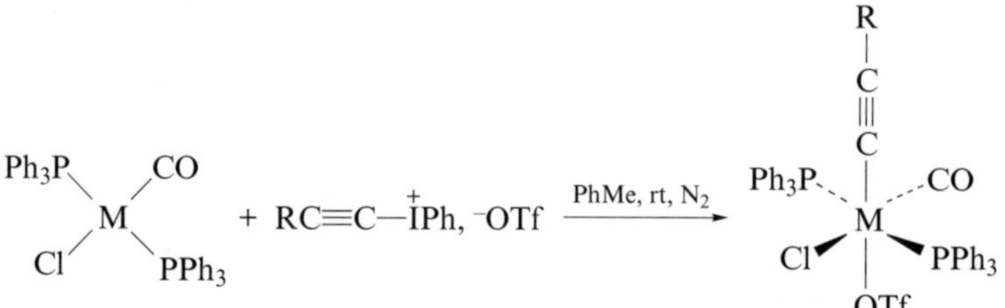

$$\text{(145)}$$

(M = Ir, Rh)

M, R (yield): Ir, H (90%); Ir, *t*-Bu (93%); Ir, Ph (90%); Rh, H (91%); Rh, *t*-Bu (96%); Rh, Ph (89%)

Since it is known that halo(phenyl)acetylenes add oxidatively to Vaska's complex to give σ-phenylethynyl iridium(III) halides, **32**[112], the intervention of phenyliodonium iridium(III) and rhodium(III) intermediates, **33**, in the alkynyliodonium reactions seems plausible. In any case, the production of σ-alkynyl complexes with alkynyl(phenyl)-iodonium triflates appears to be both more general and efficient[21].

'Rigid-rod' bisiridium(III) σ-diyne complexes have also been prepared from Vaska's complex and the bisphenyliodonium triflates of butadiyne and 1,2- and 1,4-diethynylbenzenes[45]. The reactions were conducted in acetonitrile and, under such conditions, the products are doubly charged bimetallic salts in which acetonitrile occupies one coordination

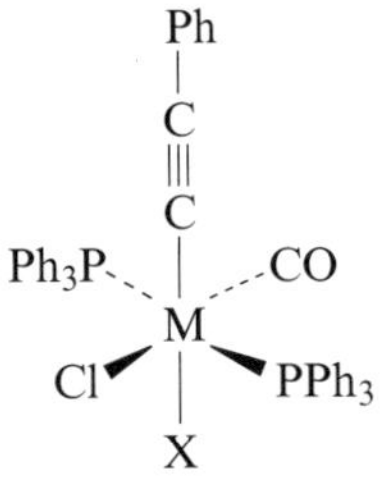

$$
\begin{array}{cc}
\text{(32) X = Cl, Br, I} & \text{(33)}
\end{array}
$$

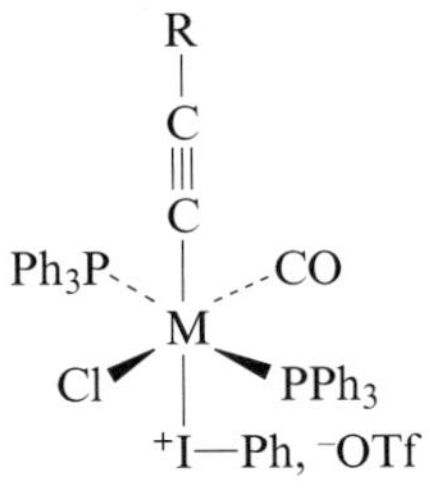

Structure (32): Ph–C≡C–, Ph₃P, CO, M, Cl, PPh₃, X (X = Cl, Br, I)

Structure (33): R–C≡C–, Ph₃P, CO, M, Cl, PPh₃, $^+$I—Ph, $^-$OTf

site at each metallic center (e.g. equation 146). The (p-phenylene)diethynyl ligand has been similarly incorporated into an analogous bisrhodium(III) complex[45].

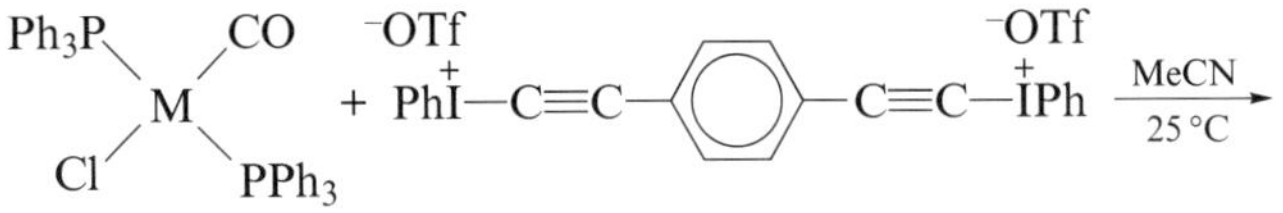

$$
\begin{array}{c}
\text{Ph}_3\text{P} \quad \text{CO} \\
\text{M} \\
\text{Cl} \quad \text{PPh}_3
\end{array}
\; + \; \text{PhI}\!-\!\text{C}\!\equiv\!\text{C}\!-\!\bigcirc\!-\!\text{C}\!\equiv\!\text{C}\!-\!\text{IPh} \;\;
\xrightarrow[\text{25 °C}]{\text{MeCN}}
$$

M (yield): Ir (85%), Rh (73%)

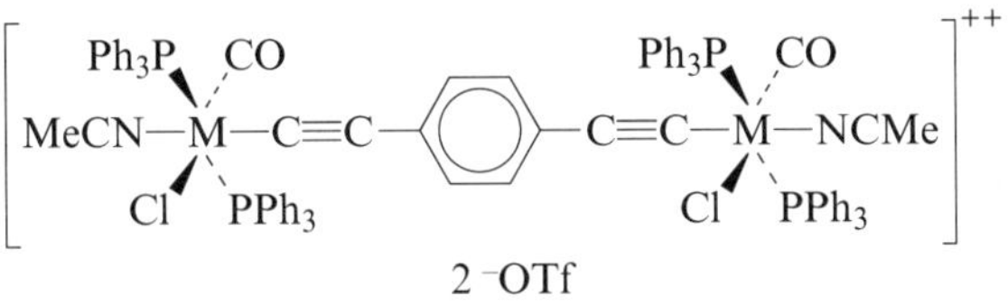

$$
\left[
\begin{array}{c}
\text{Ph}_3\text{P} \quad \text{CO} \qquad\qquad\qquad\qquad \text{Ph}_3\text{P} \quad \text{CO} \\
\text{MeCN}\!-\!\text{M}\!-\!\text{C}\!\equiv\!\text{C}\!-\!\bigcirc\!-\!\text{C}\!\equiv\!\text{C}\!-\!\text{M}\!-\!\text{NCMe} \\
\text{Cl} \quad \text{PPh}_3 \qquad\qquad\qquad\qquad \text{Cl} \quad \text{PPh}_3
\end{array}
\right]^{++}
\quad (146)
$$

2 $^-$OTf

2. Platinum

Alkynyliodonium triflates exhibit two distinct modes of reactivity with bis(triphenylphosphine)ethyleneplatinum(0)[113]. When (tert-butylethynyl)phenyliodonium triflate is mixed with the Pt(0) complex in degassed dichloromethane, ethylene insertion occurs, and a cationic η^3-propargylplatinum(II) triflate is obtained (equation 147). Employment of 1-propynyl(phenyl)iodonium triflate in degassed toluene, on the other hand, leads to a square planar σ-propynylplatinum(II) complex.

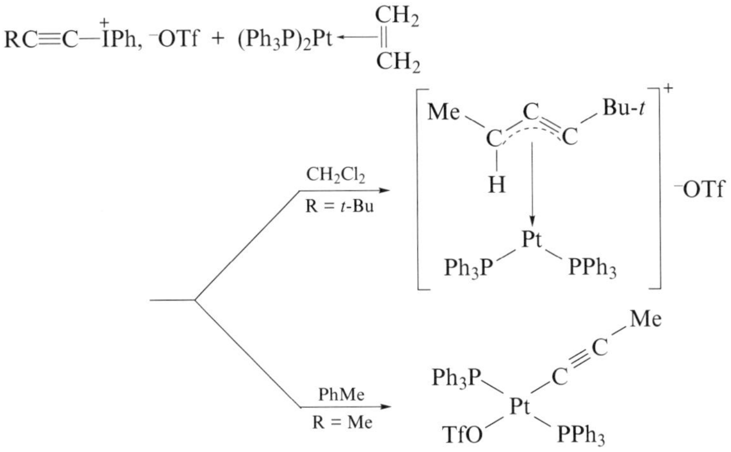

$$
\text{RC}\!\equiv\!\text{C}\!-\!\overset{+}{\text{I}}\text{Ph}, \; ^-\text{OTf} \; + \; (\text{Ph}_3\text{P})_2\text{Pt}\!\longleftarrow\!\overset{\displaystyle \text{CH}_2}{\underset{\displaystyle \text{CH}_2}{\|}}
$$

$$
\xrightarrow[\text{R = }t\text{-Bu}]{\text{CH}_2\text{Cl}_2}
\left[
\begin{array}{c}
\text{Me} \qquad\qquad \text{Bu-}t \\
\text{C}\!=\!\!=\!\text{C} \\
\text{H} \\
\text{Pt} \\
\text{Ph}_3\text{P} \qquad \text{PPh}_3
\end{array}
\right]^{+}
\; ^-\text{OTf}
$$

$$
\xrightarrow[\text{R = Me}]{\text{PhMe}}
\begin{array}{c}
\text{Ph}_3\text{P} \qquad \text{C}\!\equiv\!\text{C}\!-\!\text{Me} \\
\text{Pt} \\
\text{TfO} \qquad \text{PPh}_3
\end{array}
$$

$$(147)$$

Some clarification of the factors governing the σ- and η^3-pathways is provided by an NMR study of the reactions of the four alkynyliodonium triflates shown in equation 148 with $(Ph_3P)_2PtC_2H_4$ in degassed CD_2Cl_2[113]. The reactions were conducted under three conditions [(1) degassed solvent, (2) argon bubbling, (3) ethylene bubbling], product compositions being determined by integration of appropriate ^{31}P resonances. Although bulky groups in the alkynyl ligands clearly favor the production of η^3-complexes, all of the alkynyliodonium ions are capable of giving both types of products. In general, the argon sweep promotes σ-complex formation, while the ethylene sweep favors η^3-complex formation (i.e. except when R = n-Bu).

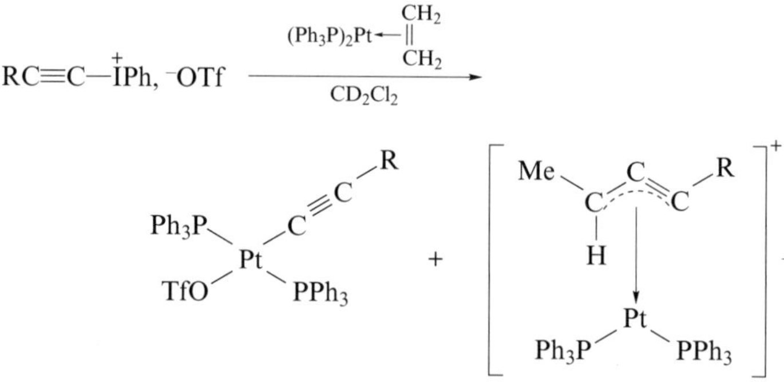

$$RC\equiv C\overset{+}{-}IPh,\ ^-OTf \quad \xrightarrow[\quad CD_2Cl_2 \quad]{\begin{array}{c} CH_2 \\ (Ph_3P)_2Pt\leftarrow\| \\ CH_2 \end{array}}$$

(148)

CD_2Cl_2; R (σ/η^3): Me (100:0), n-Bu (31 : 69), t-Bu (0 :100), Me_3Si (17 : 83)
CD_2Cl_2–argon; R (σ/η^3): n-Bu (~100 : trace), t-Bu (25 : 75), Me_3Si (41 : 59)
CD_2Cl_2–ethylene; R (σ/η^3): Me (78 : 22), n-Bu (50 : 50), Me_3Si (0 : 100)

The influence of argon and ethylene bubbling on the product ratios is consistent with expected shifts of the dissociative equilibrium shown in equation 149[113]. Thus, the σ-alkynyl complexes are thought to arise via reactions of the alkynyliodonium triflates with the coordinatively unsaturated bis(triphenylphosphine)platinum(0) species **34**[113], a pathway that should be optimized as ethylene is removed (i.e. by argon). The η^3-complexes, on the other hand, apparently originate from the undissociated platinum(0)–ethylene complex[113], their yields being maximized as ethylene is introduced.

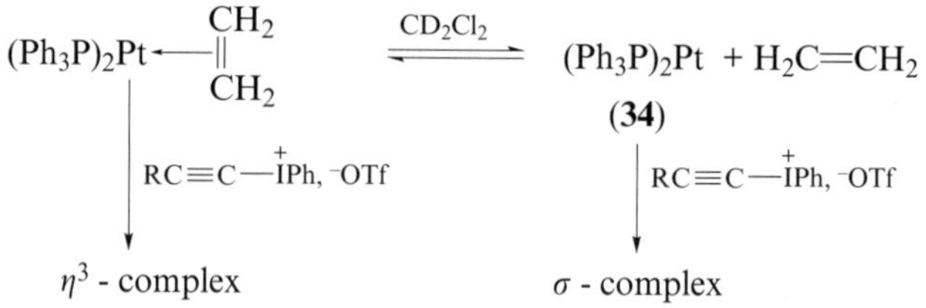

(149)

G. Miscellaneous Observations

The treatment of alkynyl(phenyl)iodonium tetrafluoroborates in dichloromethane with aqueous sodium periodate affords the corresponding alkynyliodonium periodates[24]. However, except for the *tert*-butyl analog, which has been characterized by X-ray analysis (Table 3), these compounds are unstable to autooxidation and readily 'decompose' to carboxylic acids (equation 150).

$$RC\equiv C-\overset{+}{I}Ph, \ IO_4^- \ \xrightarrow{-PhI} \ RCO_2H \ \xrightarrow{CH_2N_2} \ RCO_2Me \tag{150}$$

R (yield, ester): n-C_8H_{17} (35%), cyclohexyl (37%), cyclopentyl–CH_2 (40%)

It has already been noted that the enolates of unactivated monocarbonyl compounds do not undergo alkynylation with alkynyliodonium salts[3]. It is therefore particularly intriguing that [bis(phenyliodonium)]ethyne ditriflate reacts with the silyl enol ether (SEE) of acetophenone to give an allenic diketone (equation 151)[41]. Except for the SEE of cyclohexanone, which gives a 'black tar' with the bisiodonium compound[41], similar studies of other SEEs have not been reported.

$$PhI\!-\!C\equiv C\!-\!\overset{+}{I}Ph, \ 2 \ ^-OTf \ + \ 2PhC\!\!\overset{\overset{\displaystyle OSiMe_3}{|}}{=}\!\!CH_2 \ \xrightarrow[-35\,°C \ to \ rt]{MeCN}$$

$$\overset{O}{\overset{||}{PhC}}\!-\!CH\!=\!C\!=\!CH\!-\!CH_2\!-\!\overset{O}{\overset{||}{CPh}} \tag{151}$$

84%

Finally, various iodonium compounds, mono- and bisalkynyliodonium salts among them, have recently been found to be inhibitors of the redox cofactor methoxatin (PQQ)[114].

III. VINYLIODONIUM SALTS

A. Introduction

Vinyliodonium ions, **35** and **36**, are hypervalent iodine species in which one or two alkenyl ligands are bound to a positively charged iodine(III) atom. Although they are reactive with nucleophilic reagents, they are less labile than alkynyliodonium ions, and stable halide salts of vinyliodonium ions can be prepared. The first vinyliodonium compounds [i.e. (α, β-dichlorovinyl)iodonium salts] were synthesized by the treatment of silver acetylide–silver chloride complexes with (dichloroiodo)arenes or 1-(dichloroiodo)-2-chloroethene in the presence of water (equation 152). The early work was summarized by Willgerodt in 1914[115]. This is, of course, a limited and rather impractical synthetic method, and some time elapsed before the chemistry of vinyliodonium salts was developed. Contemporary synthetic approaches to vinyliodonium compounds include the treatment of (1) vinylsilanes and vinylstannanes with λ^3-iodanes, (2) terminal alkynes with λ^3-iodanes, (3) alkynyliodonium salts with nucleophilic reagents and (4) alkynyliodonium salts with dienes.

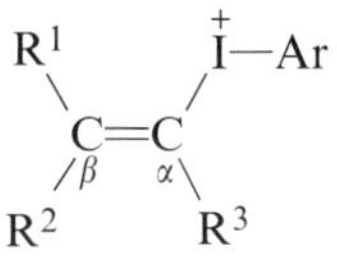

$$\begin{matrix} R^1 & \overset{+}{I}\!-\!Ar \\ \diagdown & \diagup \\ & C\!=\!C \\ \diagup_\beta & {}_\alpha\diagdown \\ R^2 & R^3 \end{matrix} \qquad \begin{matrix} R^1 & & R^3 & & R^3 & & R^1 \\ \diagdown & & \diagup & & \diagdown & & \diagup \\ & C\!=\!C & & & & C\!=\!C \\ \diagup_\beta & {}_\alpha\diagdown & & & \diagup_\alpha & & {}_\beta\diagdown \\ R^2 & & & \overset{+}{I} & & & R^2 \end{matrix}$$

$$(35) \qquad\qquad\qquad (36)$$

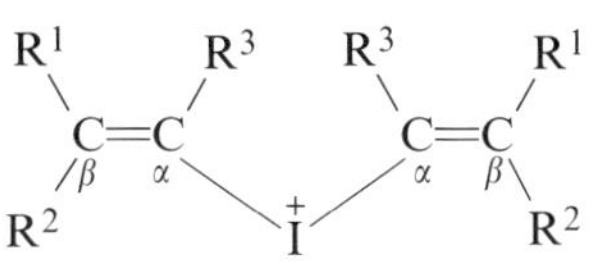

$$RC\equiv CAg\cdot AgCl \ + \ ArICl_2 \ \xrightarrow{H_2O} \ R\overset{\overset{\displaystyle Cl}{|}}{C}\!=\!\overset{\overset{\displaystyle Cl}{|}}{C}\!-\!\overset{+}{I}Ar, \ Cl^- \tag{152}$$

(R = H, Me)

B. Synthesis

1. Vinylorganometallic compounds and vinylsilanes with λ^3-iodanes

a. The early period (1945–1981). The utility of coupling reactions between vinylorganometallic compounds and λ^3-iodanes for the synthesis of vinyliodonium salts, although recognized by 1945, was exploited only sporadically prior to 1985. One of the earliest procedures involved the treatment of *trans*-chlorovinylmercuric chloride with iodine trichloride (equation 153)[116] or with various (dichloroiodo)arenes (equation 154)[117,118] in dilute hydrochloric acid. The (β-chlorovinyl)iodonium compounds prepared in this way are obtained in low to moderate yields as the trichloromercurate salts but can be converted to the chloride salts by the action of hydrogen sulfide. This approach to vinyl(aryl)-iodonium salts is most efficient when electron-donating substituents are present in the (dichloroiodo)arene nucleus. The use of vinylmercuric compounds for vinyliodonium salt synthesis has not been generalized, the only other example being the condensation of 2-(phenylethenyl)mercuric bromide with (dichloroiodo)benzene (equation 155)[119].

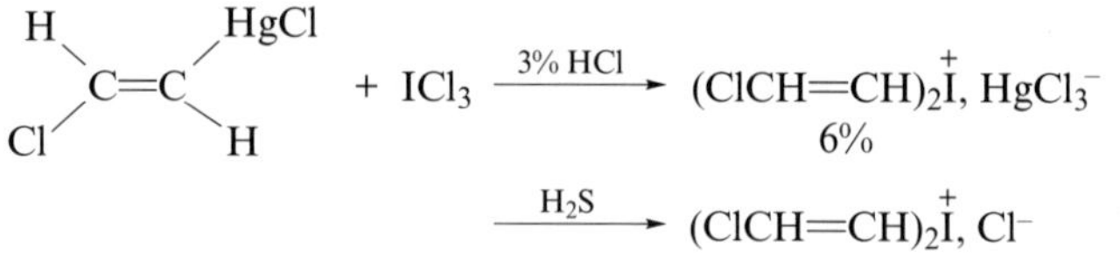

$$\underset{Cl}{\overset{H}{>}}C=C\underset{H}{\overset{HgCl}{<}} + ICl_3 \xrightarrow{3\% \text{ HCl}} \underset{6\%}{(ClCH=CH)_2\overset{+}{I}, HgCl_3^-}$$

$$\xrightarrow{H_2S} (ClCH=CH)_2\overset{+}{I}, Cl^- \qquad (153)$$

$$\underset{Cl}{\overset{H}{>}}C=C\underset{H}{\overset{HgCl}{<}} + \text{R}{-}\bigcirc{-}ICl_2 \xrightarrow{3\% \text{ HCl}} ClCH=CH-\overset{+}{I}-\bigcirc-\text{R} , HgCl_3^-$$

$$\xrightarrow{H_2S} ClCH=CH-\overset{+}{I}-\bigcirc-\text{R} , Cl^- \quad (154)$$

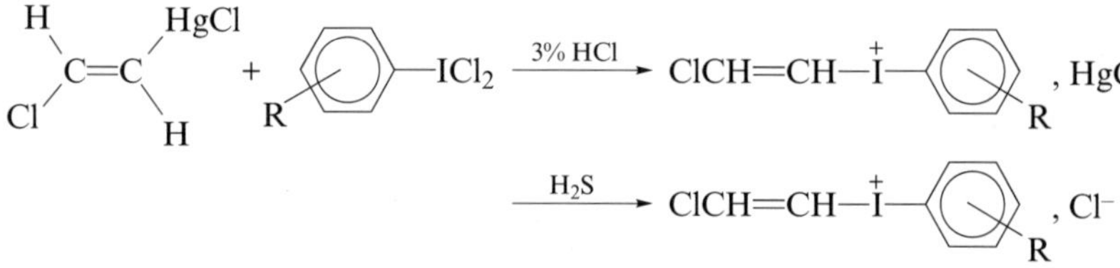

R (yield, HgCl$_3^-$ salt)[117]: H (31%), *o*-Me (40%), *p*-Me (—), *p*-Cl (6%), *m*-NO$_2$ (5%)
R (yield, Cl$^-$ salt)[118]: *o*-Me (31%), *m*-Me (45%), *p*-Me (43%), *o*-Cl (8%, 10%), *m*-Cl (20%), *p*-Cl (18%), *o*-MeO (43%, 50%), *p*-MeO (53%)

$$PhCH=CH-HgBr + PhICl_2 \xrightarrow[\text{2. NaI, acetone}]{\text{1. 3\% HCl}} \underset{4\%}{PhCH=CH-\overset{+}{I}Ph, I^-}$$

$$\xrightarrow[\text{MeNO}_2]{Et_3\overset{+}{O}, BF_4^-} PhCH=CH-\overset{+}{I}Ph, BF_4^- \quad (155)$$

The preparation of vinyliodonium salts from vinyl(trichloro)stannanes with (dichloroiodo)arenes was also explored during this period (equations 156 and 157)[119–122]. By this approach, the vinyliodonium compounds are available in unexceptional yields as the hexachlorostannate salts; other counterions (X$^-$, BF$_4^-$) can be introduced by anion metathesis.

$$CH_2=CHSnCl_3 + 4\text{-RC}_6\text{H}_4ICl_2 \xrightarrow{\text{THF}}$$

$$(4\text{-RC}_6\text{H}_4-\overset{+}{I}-CH=CH_2)_2, SnCl_6^= \quad (156)^{120,121}$$

R (yield): H (5%), Me (7%), MeO (12%)

$$RCH{=}CHSnCl_3 + PhICl_2 \xrightarrow{\text{THF}}$$

$$(RCH{=}CH{-}\overset{+}{I}{-}Ph)_2, SnCl_6^{=} \quad (157)^{121,122}$$

R (yield): Ph (62%)[a], (E)-Me (24%)[b], (Z)-Me (5%)[b]
[a] 5% yield of I⁻ salt from preparation in concentrated HCl[119].
[b] prepared from stannanes of *ca* 85% isomeric purity.

The geometry (i.e. *E* or *Z*) about the carbon–carbon double bonds in the β-chlorovinyl-
and (β-phenylvinyl)iodonium salts was not established in these early studies.

The preparation of acyclic vinyliodonium salts by the treatment of λ^3-iodanes with vinyl-
lithium reagents has not been reported. However, low yield syntheses of the cyclic
vinyliodonium compounds, 3-butyl-2-phenylbenziodolium chloride and 2,3,4,5-tetra-
phenyliodolium chloride, from *trans*-1-(dichloroiodo)-2-chloroethylene and the appropriate vinyl dilithium reagents have been described (equations 158 and 159)[123]. These
reactions are thought to proceed through trivalent iodine intermediates which decompose
with loss of acetylene. Various salts of the tetraphenyliodolium species have also been
made from 1,1-dimethyl-2,3,4,5-tetraphenylstannole[124]. For example, treatment of the
stannole in acetone first with iodine trichloride and then with sodium iodide affords the
tetraphenyliodolium iodide monohydrate in yields of 12–18% (equation 160).

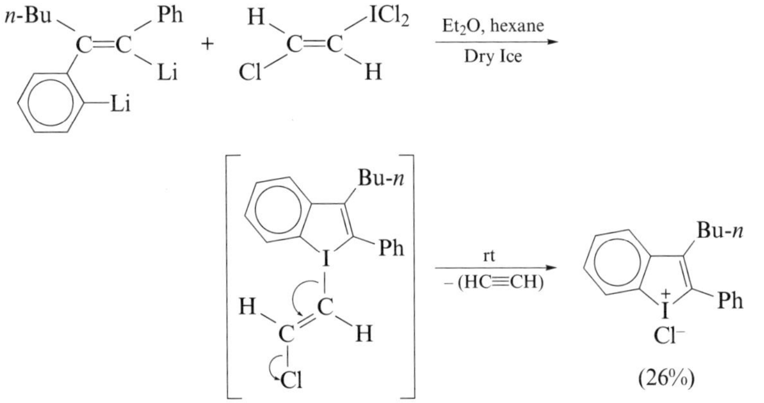

(158)

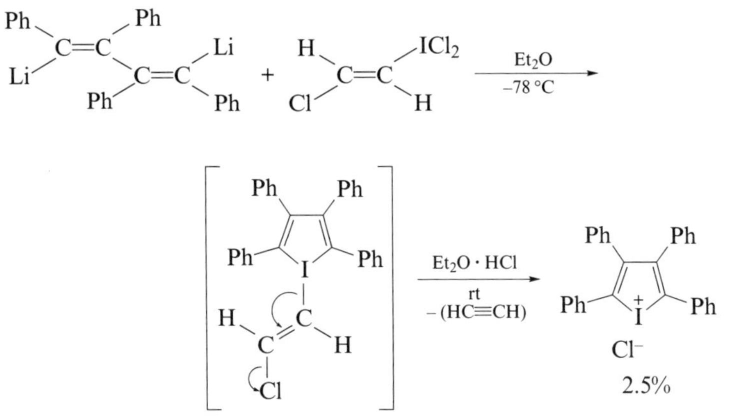

(159)

$$\text{(structure: tetraphenyl stannole with Me}_2\text{)} + ICl_3 \xrightarrow[\text{2. NaI}]{\text{1. acetone, 0 °C}} \text{(tetraphenyl iodonium structure)} \cdot H_2O \quad (160)$$

12–18% I^-

b. *Recent developments (1985–1994).* In recent years, methods for the synthesis of alkynyliodonium and vinyliodonium salts have run on a parallel track; i.e. methods that give the former from alkynylsilanes and -stannanes give the latter when applied to vinylsilanes and -stannanes. Furthermore, the early use of dichloro-λ^3-iodanes and halogenated metallovinyl compounds as coupling components has been supplanted by the use of trialkylsilyl- and (trialkylstannyl)vinyl compounds with oxyiodanes.

Vinyl(phenyl)iodonium tetrafluoroborates can be prepared by the treatment of vinylsilane/iodosylbenzene mixtures (in CH_2Cl_2) with triethyloxonium tetrafluoroborate (equations 161 and 162)[125,126] or with boron trifluoride etherate followed by aqueous sodium tetrafluoroborate (equations 162 and 163)[126–128]. The coupling reagents and iodosylbenzene are typically employed in excess. These transformations occur with retention of configuration (i.e. *E* or *Z*) in the alkenyl moiety and appear to be general, although applications of this methodology to the parent vinylsilane (e.g. $H_2C\text{=CHSiMe}_3$) and α,β,β-trisubstituted vinylsilanes have not, to our knowledge, been reported. The 1-decenyl system provides an instructive example of the impact of stereochemistry on the stability of (β-alkylvinyl)iodonium tetrafluoroborates. Thus, while the *trans*-vinylsilane gives a stable *trans*-vinyliodonium salt with $PhIO/Et_3O^+$, BF_4^-, the *cis*-vinylsilane is converted to 1-decyne (equation 164)[126]. The production of the alkyne has been shown (^{1}H NMR) to proceed through an unstable *cis*-vinyliodonium intermediate.

$$\begin{array}{c} R^1 \\ R^2 \end{array}\text{C=C}\begin{array}{c} SiMe_3 \\ H \end{array} + PhI\text{=O} \xrightarrow{Et_3\overset{+}{O}, BF_4^-; CH_2Cl_2} \begin{array}{c} R^1 \\ R^2 \end{array}\text{C=C}\begin{array}{c} \overset{+}{I}Ph, BF_4^- \\ H \end{array} \quad (161)$$

R^1, R^2 (yield): H, n-C_8H_{17} (72%); H, $PhCH_2$ (69%); H, Ph (61%); Me, $PhCH_2CH_2$ (71%)*; $PhCH_2CH_2$, Me (89%)* [*Z : E mixtures of silanes and iodonium salts]

$$R-\langle\rangle-SiMe_3 + PhI\text{=O} \begin{cases} \xrightarrow{Et_3\overset{+}{O}, BF_4^-; CH_2Cl_2} R-\langle\rangle-\overset{+}{I}Ph, BF_4^- \\ \qquad\qquad R = t\text{-Bu (74%)} \\ \\ \xrightarrow[\text{2. NaBF}_4, H_2O]{\text{1. Et}_2O \cdot BF_3, CH_2Cl_2} R-\langle\rangle-\overset{+}{I}Ph, BF_4^- \\ \qquad\qquad R = H \text{ (94%)}, t\text{-Bu (80%)} \end{cases} \quad (162)$$

$$\begin{array}{c} R^1 \\ R^2 \end{array}\text{C=C}\begin{array}{c} SiMe_3 \\ H \end{array} + PhI\text{=O} \xrightarrow[\text{2. NaBF}_4, H_2O]{\text{1. Et}_2O \cdot BF_3, CH_2Cl_2} \begin{array}{c} R^1 \\ R^2 \end{array}\text{C=C}\begin{array}{c} \overset{+}{I}Ph, BF_4^- \\ H \end{array} \quad (163)$$

R^1, R^2 (yield): H, $PhCH_2CH_2$ (77%); H, $4\text{-ClC}_6H_4CH_2CH_2$ (75%); H, $4\text{-BrC}_6H_4OCH_2$ (72%); Me, $PhCH_2CH_2$ (92%); Me, cyclopentyl–CH_2 (72%); Me, cyclohexyl–CH_2 (85%)

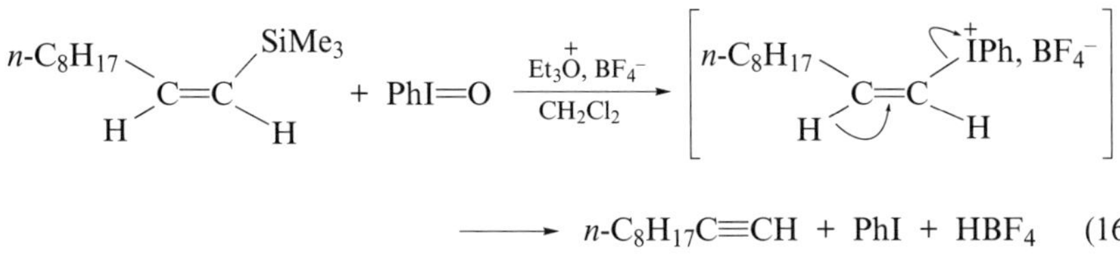

$$n\text{-}C_8H_{17}\diagdown_{H}C{=}C\diagup^{SiMe_3}_{H} \quad + \text{ PhI}{=}O \xrightarrow[\text{CH}_2\text{Cl}_2]{\overset{+}{\text{Et}_3}\text{O, BF}_4^-} \left[n\text{-}C_8H_{17}\diagdown_{H}C{=}C\diagup^{\overset{+}{\text{I}}\text{Ph, BF}_4^-}_{H} \right]$$

$$\longrightarrow \quad n\text{-}C_8H_{17}C{\equiv}CH \ + \ \text{PhI} \ + \ \text{HBF}_4 \qquad (164)$$

$$90\%, \text{ GC}$$

The species generated from iodosylbenzene and BF$_3$–etherate, presumably PhI$^+$–OBF$_3$, is sufficiently electrophilic to react with activated and silicon-bound aromatic rings. Thus, the vinylsilane **37** undergoes phenyliodination at the *ortho*- and *para*- carbons of the trialkoxybenzene ring, while (trimethylsilyl)benzene [PhSiMe$_3$] is converted to diphenyliodonium tetrafluoroborate [Ph$_2\overset{+}{\text{I}}$, BF$_4^-$][126]. The alkenyl linkage of phenyl(β-alkylvinyl)silanes is more labile than the phenyl linkage unless an electron-withdrawing group is present (equation 165)[126], and such silanes have been employed for stereoselective syntheses of (*E*)- and (*Z*)-2-*n*-butyl-1-decenyl(phenyl)iodonium tetrafluoroborates (equation 166)[128].

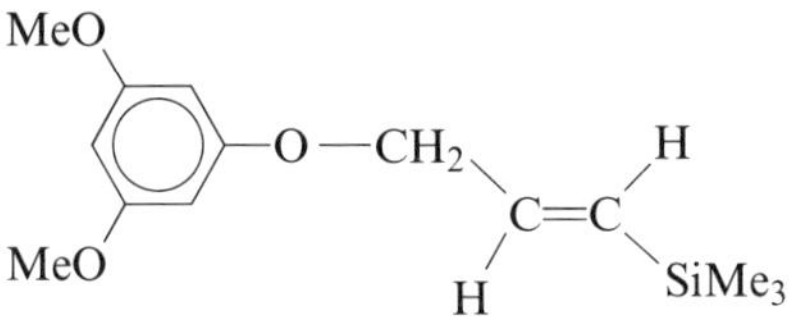

(**37**)

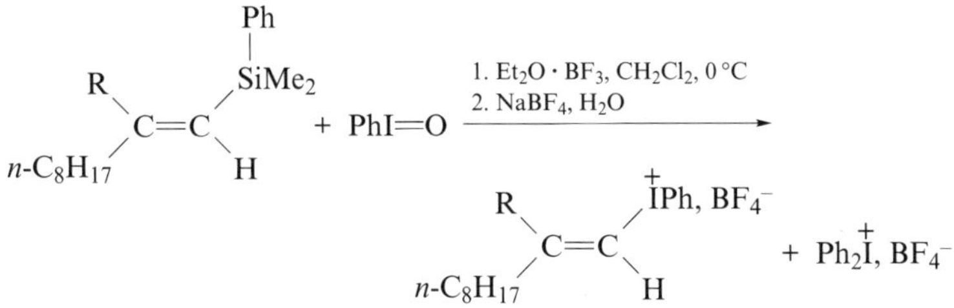

$$R\diagdown_{n\text{-}C_8H_{17}}C{=}C\diagup^{\overset{\displaystyle Ph}{\underset{\displaystyle |}{\text{SiMe}_2}}}_{H} \quad + \text{ PhI}{=}O \xrightarrow{\begin{array}{l}1.\ \text{Et}_2\text{O}\cdot\text{BF}_3,\ \text{CH}_2\text{Cl}_2,\ 0\,°\text{C}\\2.\ \text{NaBF}_4,\ \text{H}_2\text{O}\end{array}}$$

$$R\diagdown_{n\text{-}C_8H_{17}}C{=}C\diagup^{\overset{+}{\text{I}}\text{Ph, BF}_4^-}_{H} \quad + \ \text{Ph}_2\overset{+}{\text{I}}, \text{BF}_4^- \qquad (165)$$

R = H   93 : 7 (82%)
R = COOMe --   68%

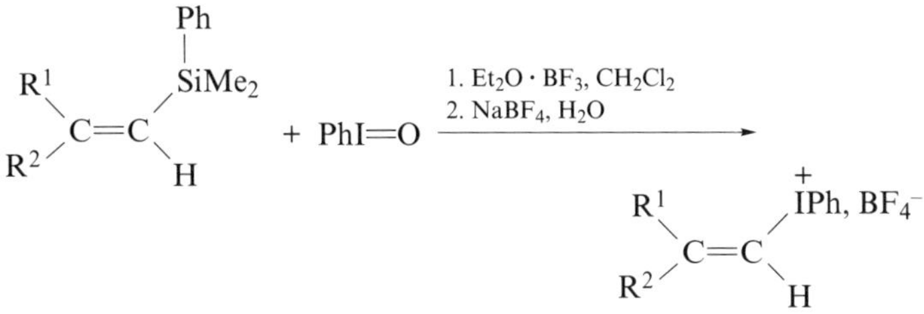

$$R^1\diagdown_{R^2}C{=}C\diagup^{\overset{\displaystyle Ph}{\underset{\displaystyle |}{\text{SiMe}_2}}}_{H} \quad + \text{ PhI}{=}O \xrightarrow{\begin{array}{l}1.\ \text{Et}_2\text{O}\cdot\text{BF}_3,\ \text{CH}_2\text{Cl}_2\\2.\ \text{NaBF}_4,\ \text{H}_2\text{O}\end{array}}$$

$$R^1\diagdown_{R^2}C{=}C\diagup^{\overset{+}{\text{I}}\text{Ph, BF}_4^-}_{H} \qquad (166)$$

R^1, R^2 (yield): *n*-Bu, *n*-C$_8$H$_{17}$ (84%); *n*-C$_8$H$_{17}$, *n*-Bu (78%)

The synthesis of (*Z*)-alkenyliodonium perchlorates with a single alkyl or phenyl group at β-carbon can be achieved by the sequential treatment of vinylsilane/(diacetoxyiodo)benzene mixtures with BF$_3$–etherate and aqueous sodium perchlorate (equation 167)[129].

$$\underset{H}{\overset{R}{}}C=C\underset{H}{\overset{SiMe_3}{}} + PhI(OAc)_2 \xrightarrow[\text{2. NaBF}_4, \text{H}_2\text{O}]{\text{1. Et}_2\text{O} \cdot \text{BF}_3, \text{CH}_2\text{Cl}_2, 0\,^\circ\text{C}} \underset{H}{\overset{R}{}}C=C\underset{H}{\overset{\overset{+}{I}Ph, ClO_4^-}{}} \tag{167}$$

R (yield): n-C$_8$H$_{17}$ (42%), Me$_2$CH(CH$_2$)$_4$ (22%), cyclopentyl–CH$_2$ (39%),
Ph(CH$_2$)$_3$ (53%), Ph (30%)

Given the propensity of (Z)-1-decenyl(phenyl)iodonium tetrafluoroborate for β-elimination (cf. equation 164), it is interesting that the Z-perchlorate salts can be isolated. They do, however, readily decompose to alkynes in CDCl$_3$[129].

The reactions of vinyl(tri-n-butyl)stannanes with the appropriate [cyano(sulfonyloxy)-iodo]benzenes in dichloromethane constitute a facile stereoselective approach to β,β-disubstituted vinyl(phenyl)iodonium tosylates and triflates (equation 168)[130,131]. Although this method has not yet been used for the preparation of α,β-disubstituted and α,β,β-trisubstituted analogs, it has been applied to the synthesis of ethenyl(phenyl)iodonium triflate[131]. The efficiency of this approach may be contrasted with the early work on chlorostannanes and chloroiodanes.

$$\underset{R^2}{\overset{R^1}{}}C=C\underset{H}{\overset{Sn(n\text{-}Bu)_3}{}} + PhI(CN)L \xrightarrow[-23\,^\circ\text{C}]{\text{CH}_2\text{Cl}_2} \underset{R^2}{\overset{R^1}{}}C=C\underset{H}{\overset{\overset{+}{I}Ph, L^-}{}} + (n\text{-Bu})_3\text{SnCN}$$

$$(\text{L}^- = \text{TsO}^-, \text{TfO}^-) \tag{168}$$

L$^-$ = TsO$^-$; R^1,R^2 (yield): Me, Me (86%); Et, n-Bu (74%); n-Bu, Et (55%); n-Bu, n-Bu (62%); Et, Ph (63%)
L$^-$ = TfO$^-$; R^1,R^2 (yield): H, H (75%, $-40\,^\circ$C to rt); Me, Me (89%); Me, Et (90%); Et, Me (81%); Et, n-Bu (86%); n-Bu, Et (60%); n-Bu, n-Bu (86%); Et, Ph (60%)

Finally, vinyl(tri-n-butyl)stannanes have also been utilized with iodosylbenzene and Meerwein's reagent[126] for the synthesis of the ($tert$-butylcyclohexenyl)iodonium tetrafluoroborate shown in equation 162 and with iodosylbenzene/BF$_3$–etherate to give the (ketovinyl)iodonium salt shown in equation 169[128].

$$\xrightarrow[\text{2. NaBF}_4, \text{H}_2\text{O}]{\text{1. PhIO, Et}_2\text{O} \cdot \text{BF}_3, \text{CH}_2\text{Cl}_2,} \tag{169}$$

80%

2. Alkynes with iodanes

While vinylsilanes and -stannanes have been used primarily for the synthesis of vinyliodonium salts with one or two β-alkyl substituents in the vinyl moiety, the treatment of alkynes with oxyiodanes permits the introduction of oxygen functionality at β-carbon. The conversion of terminal alkynes with [hydroxy(tosyloxy) iodo]benzene (HTIB) to alkynyliodonium tosylates (equation 8) and/or (β-tosyloxyvinyl)iodonium tosylates [TsOC(R)=CHPh, $^-$OTs: R = n-Pr, n-Bu, n-C$_5$H$_{11}$, i-Pr, i-Bu] (equation 9), depending on the size of R, has already been discussed[8,11]. In at least three cases, E:Z mixtures were obtained[8,132]. The reactions of internal alkynes with HTIB proceed in a similar way and afford (β-tosyloxyvinyl)iodonium tosylates fully substituted (i.e. α,β,β) in the vinyl ligands (equation 170)[11]. However, the stereochemistry of these compounds has not been reported, and the regiochemistry of the 2-heptyne and 1-phenylpropyne adducts is uncertain. When (trimethylsilyl) acetylene is employed as the substrate ($CHCl_3$, reflux), desilylation occurs, and phenyl(β-tosyloxyethenyl)iodonium tosylate is obtained in low yield (equation 171)[8]. The geometry of this compound, not initially assigned, was later shown to be the E-configuration[132].

$$R^1C\equiv CR^2 + PhI(OH)OTs \xrightarrow[\text{or } CH_2Cl_2, \Delta]{\text{neat, rt } (R^1, R^2 = Me)} \begin{array}{c} R^1 \\ \diagdown \\ TsO \end{array} C=C \begin{array}{c} \overset{+}{IPh,} {}^-OTs \\ \diagup \\ R^2 \end{array} \qquad (170)$$

R^1, R^2 (yield): Me, Me (62%); n-Bu, Me (42% crude, 32% pure); Ph, Me (56.5%)

$$Me_3SiC\equiv CH + PhI(OH)OTs \xrightarrow{CHCl_3, \Delta} \begin{array}{c} H \\ \diagdown \\ TsO \end{array} C=C \begin{array}{c} \overset{+}{IPh,} {}^-OTs \\ \diagup \\ H \end{array} \qquad (171)$$

22%

The introduction of acetylene, terminal alkynes or internal alkynes into methylene chloride solutions of iodosylbenzene and triflic acid [1:1, *in situ* generation of PhI(OH)OTf] results in the production of [β-(trifluoromethanesulfonyloxy)vinyl]iodonium triflates (equation 172)[78,133]. Since the E-configuration has been determined for selected members of this series (R^1, $R^2 = n$-Pr, H; n-Bu, H), these reactions appear to proceed via stereoselective anti-additions of PhI(OH)OTf to the alkynes.

$$R^1C\equiv CR^2 \xrightarrow{PhI=O, TfOH (1:1); CH_2Cl_2} \begin{array}{c} R^1 \\ \diagdown \\ TfO \end{array} C=C \begin{array}{c} \overset{+}{IPh,} {}^-OTf \\ \diagup \\ R^2 \end{array} \qquad (172)$$

22%

R^1, R^2 (yield): H, H (72%); n-Pr, H (87%); n-Bu, H(100%); n-C_6H_{13}, H (100%); Ph, H (53%); Me, Me (80%); Et, Et (67%)

When iodosylbenzene and one equivalent of triflic anhydride are mixed in dichloromethane, a (p-phenylene)bisiodine(III) species, similar to that obtained from iodosylbenzene with two equivalents of triflic acid (structure 11, equation 23)[38,63], is generated[134]. The addition of terminal alkynes to such mixtures affords (p-phenylene)bisiodonium ditriflates with one phenyl and one TfOC(R)=CH— ligand (stereochemistry unspecified)[134]; see equation 173.

$$RC\equiv CH \xrightarrow{PhI=O, Tf_2O (1:1); CH_2Cl_2} \begin{array}{c} R \\ \diagdown \\ TfO \end{array} C=CH-\overset{+}{I}-\langle\bigcirc\rangle-\overset{+}{IPh}, 2\ {}^-OTf \qquad (173)$$

R (yield): n-Pr (70%); n-Bu (68%); n-C_6H_{13} (61%)

Although mechanisms for the production of (β-sulfonyloxyvinyl)iodonium salts from terminal alkynes via alkynyliodonium salts can be envisioned (e.g. equation 174), they are not consistent with similar transformations of internal alkynes. The generation of vinyl cations, or iodine-bridged counterparts[78], and their capture with sulfonate ions to give

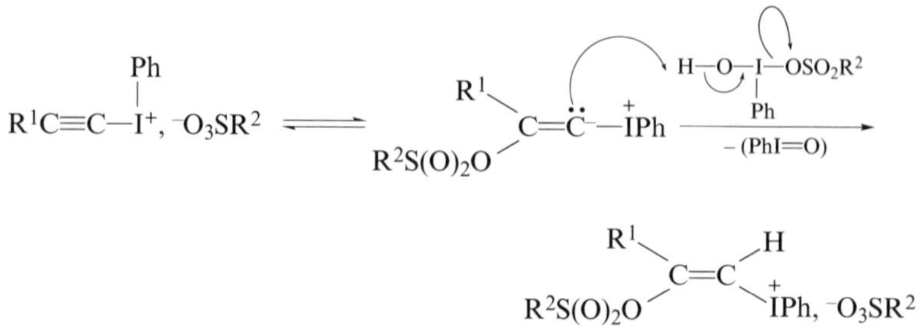

$$R^1C{\equiv}C{-}I^+, \ {}^-O_3SR^2 \ \rightleftharpoons \ \underset{R^2S(O)_2O}{\overset{R^1}{\diagdown}}C{=}\ddot{C}{-}\overset{+}{I}Ph \ \xrightarrow[-(PhI{=}O)]{} $$

$$\underset{R^2S(O)_2O}{\overset{R^1}{\diagdown}}C{=}C\underset{\overset{+}{I}Ph, \ {}^-O_3SR^2}{\overset{H}{\diagup}} \qquad (174)$$

intermediate (sulfonyloxyvinyl)iodonium 'hydroxides' seems more likely (equations 175). When R^2 = H, proton loss from the vinyl cations to give alkynyliodonium salts is sometimes competitive (equations 5 and 8), although this pathway is much more favorable when R^2 = Me$_3$Si or R$_3$Sn. Regardless of the mechanism, it is clear that two moles of the sulfonyloxyiodanes are required to deliver one mole of product. This presents an interesting puzzle since iodosylbenzene, the expected byproduct in such reactions, has not been reported.

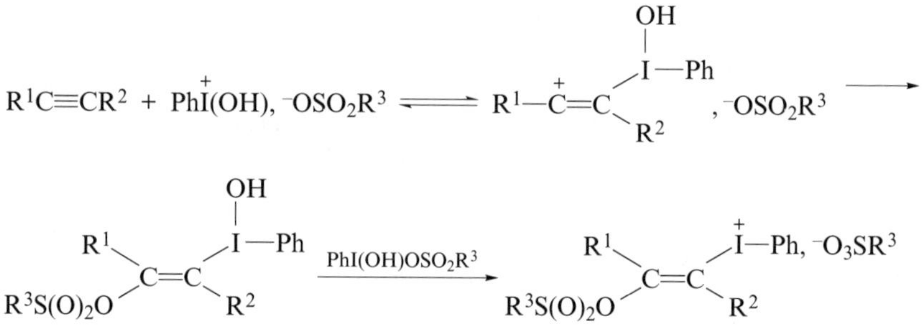

$$R^1C{\equiv}CR^2 + Ph\overset{+}{I}(OH), \ {}^-OSO_2R^3 \ \rightleftharpoons \ R^1{-}\overset{+}{C}{=}C\underset{R^2}{\overset{\overset{\displaystyle OH}{\displaystyle |}}{\diagup}}\!\!\!\!\!\!\!\!\!\!\!\!\!\!\overset{I{-}Ph}{}, \ {}^-OSO_2R^3 \ \longrightarrow $$

$$\underset{R^3S(O)_2O}{\overset{R^1}{\diagdown}}C{=}C\underset{R^2}{\overset{\overset{\displaystyle OH}{\displaystyle |}}{\diagup}}\!\!\!\!\!\!I{-}Ph \ \xrightarrow{PhI(OH)OSO_2R^3} \ \underset{R^3S(O)_2O}{\overset{R^1}{\diagdown}}C{=}C\underset{R^2}{\overset{\overset{+}{I}{-}Ph, \ {}^-O_3SR^3}{\diagup}}$$

$$+ \ H_2O + [PhI{=}O] \qquad (175)$$

With the exception of (trimethylsilyl)acetylene (equation 26)[79], BF$_3$–etherate methodology has not been applied to typical terminal alkynes [RC≡CH]. However, when various alkynoic acids are subjected to the action of iodosylbenzene and BF$_3$–etherate, intramolecular participation of the carboxyl group occurs, and lactonic vinyliodonium fluoroborates are obtained[135]. Some examples are presented in equation 176. ^{1}H NMR studies of these compounds 'showed no appreciable NOE enhancement between the vinylic and allylic protons'[135], consistent with the E-configuration about the carbon–carbon double bonds.

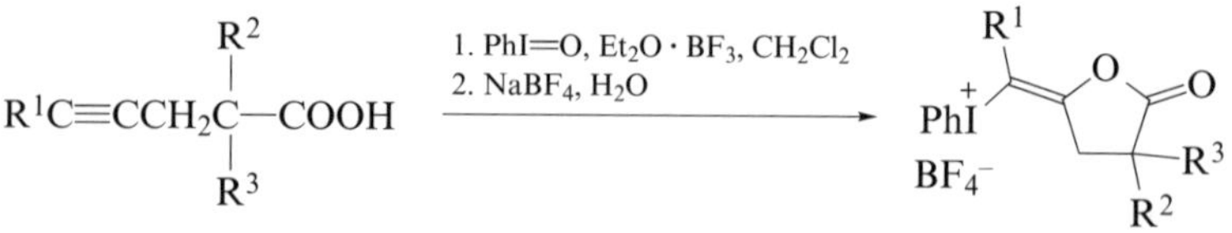

$$R^1C{\equiv}CCH_2\underset{R^3}{\overset{\overset{\displaystyle R^2}{\displaystyle |}}{C}}{-}COOH \quad \xrightarrow[\text{2. NaBF}_4, \ H_2O]{\text{1. PhI=O, Et}_2O \cdot BF_3, \ CH_2Cl_2} \quad \underset{BF_4^-}{\overset{+}{Ph}I}\!\!\diagup\!\!\overset{R^1}{\diagdown}\!\!\overset{O}{\diagdown}\underset{\underset{R^2}{\overset{|}{}}}{\overset{O}{\diagup}}{R^3} \qquad (176)$$

R^1, R^2, R^3 (yield): H, H, H (75%); H, H, Ph (78%); H, Me, CO$_2$H (75%); Me, H, H (96%); Me$_3$Si, H, H (77%)

3. Addition reactions of alkynyliodonium salts

As discussed in Section II.D, the ability of alkynyliodonium salts to undergo Michael additions with nucleophilic reagents provides access to β-functionalized vinyliodonium salts (equation 177). However, this approach will not succeed unless the intermediate vinylidene-iodonium ylides can be captured by protonation. Thus, the best results are obtained when the nucleophile bears an acidic hydrogen or when the reactions are conducted in an acidic medium.

$$
\begin{array}{c}
\text{Nu:}^- \longrightarrow \quad
\overset{R}{\underset{Nu}{}} \!\!\! C \!=\! \overset{\cdot\cdot}{C} \!-\! \overset{+}{I}Ph \qquad \xrightarrow{\;H^+\;} \\[3mm]
RC\!\equiv\!C\!-\!\overset{+}{I}Ph \\[3mm]
\text{HNu:} \longrightarrow \quad
\overset{R}{\underset{H-\overset{+}{N}u}{}} \!\!\! C \!=\! \overset{\cdot\cdot}{C} \!-\! \overset{+}{I}Ph
\end{array}
\qquad
\overset{R}{\underset{Nu}{}} \!\!\! C \!=\! C \!\!\! \overset{H}{\underset{\overset{+}{I}Ph}{}}
\qquad (177)
$$

$$
Nu^- = {}^-SO_2R,\ N_3{}^-,\ Cl^-,\ Br^-,\ {}^-O_2\overset{\overset{\displaystyle O}{\|}}{P}(OPh)
$$

So far, this method has been employed for the synthesis of (1) (Z)-(β-sulfonylvinyl)-iodonium tetrafluoroborates from alkynyliodonium tetrafluoroborates with sulfinic acids in methanol (equation 46)[32,84], (2) (Z)-(β-azidovinyl)iodonium tetrafluoroborates from alkynyliodonium tetrafluoroborates with Me_3SiN_3, H_2O (i.e. HN_3) in dichloromethane (equation 110)[102] and (3) (Z)-β-bromo- and (β-chlorovinyl)iodonium halides from alkynyliodonium tetrafluoroborates with lithium halides in acetic acid or with HX in methanol (equation 111 and 112)[103].

Syntheses of an internal (β-phosphoryloxyvinyl)iodonium salt from (t-butylethynyl)phenyliodonium tosylate with ammonium phenyl phosphate (equation 97)[95] and a (β-azidovinyl)iodonium salt from (phenylethynyl)phenyliodonium tosylate with sodium azide (equation 106)[99] have also been reported.

These reactions invariably afford β-substituted vinyliodonium salts in the Z-configuration (i.e. anti-introduction of the nucleophile and proton) and are generally regarded as nucleophilic additions. However, it seems plausible in the case of HX in methanol that addition is initiated by protonation of the triple bond.

Diels–Alder reactions of alkynyl(phenyl)iodonium triflates (i.e. containing electron-withdrawing groups in the alkynyl moiety) and [bis(phenyliodonium)] ethyne ditriflate have been employed for the synthesis of cyclic vinyliodonium salts (equations 143 and 144)[17,41]. The availability of such compounds offers considerable potential for the elaboration of densely functionalized cyclic molecules.

4. Miscellaneous methods

Attempts to prepare a (β-fluorovinyl)iodonium salt from 1-decynyl(phenyl)iodonium tetrafluoroborate by the procedures employed for β-chloro- and (β-bromovinyl)iodonium halides have been unsuccessful[103]. Admixture of the decynyliodonium salt and lithium fluoride in acetic acid ultimately (2 days, rt) leads to 1-acetoxy-2-decanone[103]. (Z)-(β-Fluoro-β-perfluoroalkylvinyl)iodonium triflates, on the other hand, can be made by the treatment of ($1H$, $1H$-perfluoroalkyl)phenyliodonium triflates with sodium hydride (equation 178)[136]. Apart from 2-fluoro-1-hexadecenyl(phenyl)iodonium chloride (synthesis not described)[104], these are the only reported examples of (β-fluorovinyl)iodonium salts.

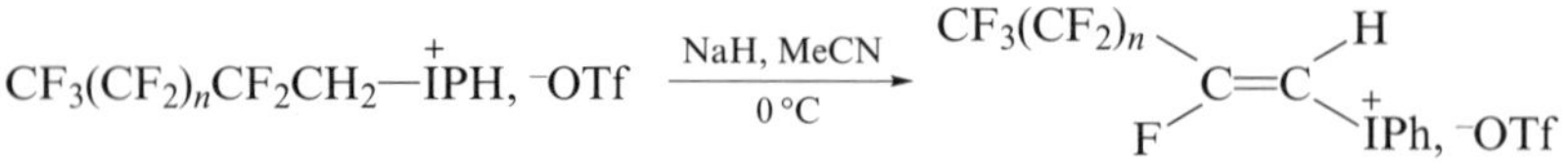

$$CF_3(CF_2)_nCF_2CH_2-\overset{+}{I}PH,\ ^-OTf \xrightarrow[0\ ^\circ C]{NaH,\ MeCN} \begin{array}{c} CF_3(CF_2)_n \\ \\ F \end{array}\!\!C=C\!\!\begin{array}{c} H \\ \\ \overset{+}{I}Ph,\ ^-OTf \end{array} \quad (178)$$

n (yield): 0 (80%), 1 (86%), 5 (87%)

Stable allenyliodonium salts have not been reported. However, allenyliodine(III) species have been posited as intermediates in reactions of (diacetoxyiodo)arenes with propargyl-silanes[137]. For example, when (diacetoxyiodo)benzene–propargylsilane mixtures (CH_2Cl_2) are treated with BF_3–etherate, 1-iodo-2-propargylbenzenes are produced (equation 179). The parent propargyl system has also been installed with stannane and germane compounds. These transformations are thought to involve the initial formation of allenyl-(phenyl)iodonium acetates, **38**, and their collapse to the observed products by a 'reductive iodonio-Claisen rearrangement' (equation 180)[137]. The function of boron trifluoride is to render (diacetoxyiodo)benzene more electrophilic, thereby facilitating S_E2' desilylation of the propargyl component[137].

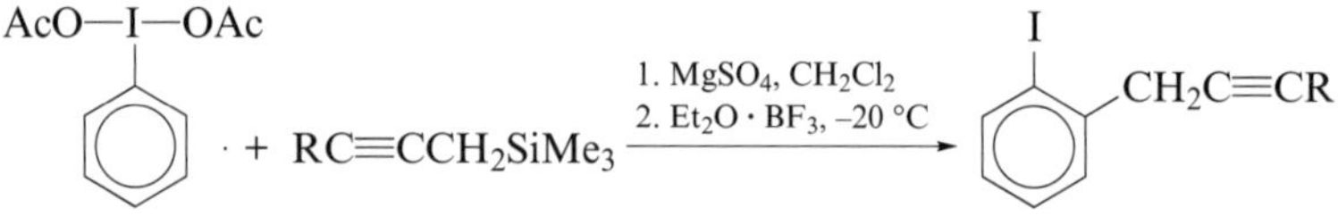

$$(179)$$

R (yield): H (66%), t-Bu (51%), n-C_5H_{11} (80%), n-C_8H_{17} (88%), Me_3Si (82%), cyclohexyl (90%)

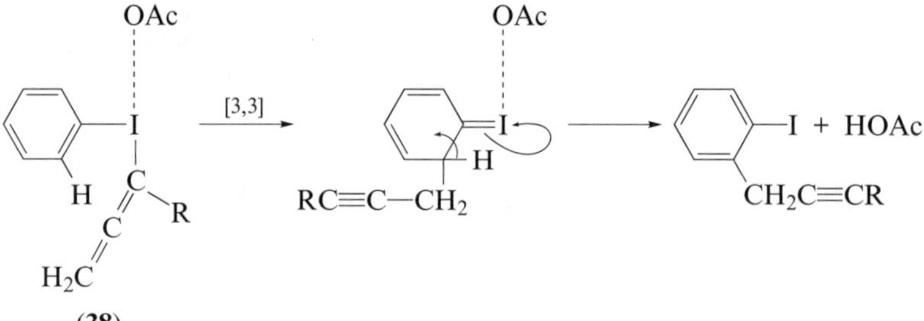

$$(180)$$

(38)

The same results can be achieved with iodosylbenzene, and similar reactions of propargylsilane (R = H) with HTIB and 1-hydroxy-1,2-benziodoxol-3(1H)-one have been demonstrated[137]. p-Iodosyltoluene and various methyl derivatives of (diacetoxyiodo)-benzene have also been studied[137].

The treatment of alkenes with iodine(III) reagents usually results in functionalization of the carbon–carbon double bond. However, 1,1-diphenylethylene affords a low yield of (1,1-diphenylethenyl)phenyliodonium tosylate with HTIB (equation 181)[11,138]. The cyclic dithiolylidene derivative of malonic acid, shown in equation 182, undergoes decarboxylation with [bis(trifluoroacetoxy)iodo]benzene in methanol and gives an unusual vinyliodonium trifluoroacetate[139]. Finally, when the allenylphosphonate shown in equation 183 is added to a mixture of (difluoroiodo)benzene and BF_3–etherate in dichloromethane, a

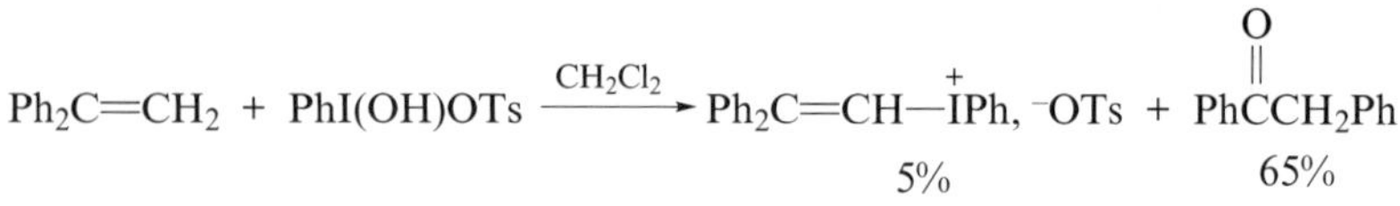

$$Ph_2C=CH_2 + PhI(OH)OTs \xrightarrow{CH_2Cl_2} Ph_2C=CH-\overset{+}{I}Ph,\ ^-OTs + \overset{O}{\underset{||}{Ph\overset{}{C}CH_2Ph}} \quad (181)$$

5%   65%

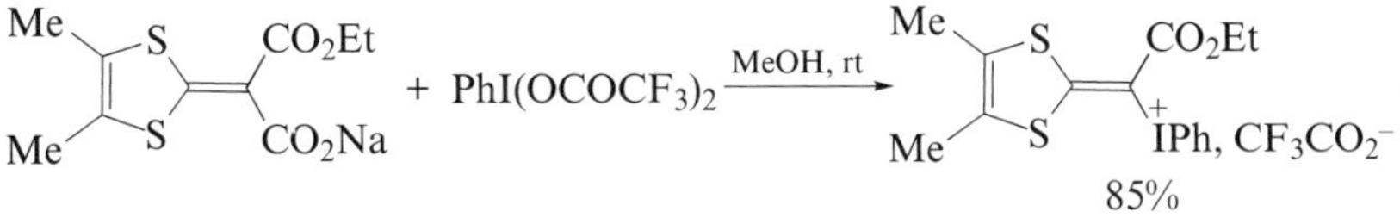

$$(182)$$

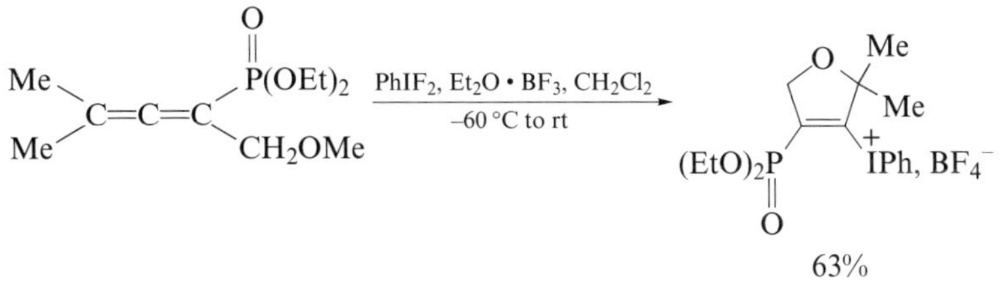

85%

$$(183)$$

63%

dihydrofuryl(phenyl)iodonium tetrafluoroborate is obtained[140]. This transformation can also be effected with iodosylbenzene/BF_3–etherate, the subsequent addition of aqueous $NaBF_4$ or $LiClO_4 \cdot 3H_2O$ giving the tetrafluoroborate and perchlorate salts[140].

C. Configurations and Structure

Structures of vinyliodonium salts that have appeared in the literature since 1945 are shown in Table 8. For compounds that were reported earlier, the reader is directed to Willgerodt's monograph[115] and to the compilation of polyvalent iodine compounds by Beringer and Gindler[141]. Table 8 includes a number of vinyliodonium ions whose stereoisomeric configurations (i.e. E or Z) have been reported. Most of these assignments are based on NMR information and require some discussion.

TABLE 8. Structures of known vinyliodonium salts

1. Vinyl(aryl)iodonium salts of assigned stereochemistry

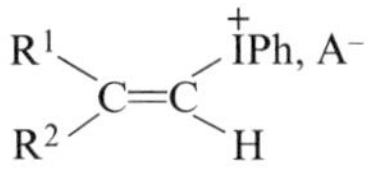

R^1	R^2	A^-	References
H	H	$SnCl_6^-$	120,122
		Br^-, I^-, BF_4^-	120
		TfO^-	131
Me	H	$SnCl_6^-, Br^-, BF_4^-$	122
H	Me	$SnCl_6^-, Br^-, BF_4^-$	122
H	n-C_8H_{17}	BF_4^-	125,126
		ClO_4^-	129
$Me_2CH(CH_2)_4$	H	ClO_4^-	129
Cyclopentyl–CH_2	H	ClO_4^-	129
H	$PhCH_2$	BF_4^-	125,126
H	$PhCH_2CH_2$	BF_4^-	126,127
H	4-$ClC_6H_4CH_2CH_2$	BF_4^-	126,127
H	4-$BrC_6H_4OCH_2$	BF_4^-	126,127
$PhCH_2CH_2CH_2$	H	ClO_4^-	129
H	Ph	BF_4^-	125,126
Ph	H	ClO_4^-	129

(continued)

TABLE 8. (*continued*)

R^1	R^2	A$^-$	References
H	EtO	BF$_4^-$	79
H	TsO	TsO$^-$	8,132
H	TfO	TfO$^-$	78
Me	Me	TsO$^-$, TfO$^-$	130
n-Bu	*n*-Bu	TsO$^-$, TfO$^-$	130
Et	*n*-Bu	TsO$^-$, TfO$^-$	130
n-Bu	Et	TsO$^-$, TfO$^-$	130
Et	Me	TfO$^-$	130
Me	Et	TfO$^-$	130
Et	Ph	TsO$^-$, TfO$^-$	130
n-Bu	*n*-C$_8$H$_{17}$	BF$_4^-$	128
n-C$_8$H$_{17}$	*n*-Bu	BF$_4^-$	128
PhCH$_2$CH$_2$	Me	BF$_4^-$	125,126
Me	PhCH$_2$CH$_2$	BF$_4^-$	125,126
Me	Cyclopentyl–CH$_2$	BF$_4^-$	128
Me	Cyclohexyl–CH$_2$	BF$_4^-$	126
Me	Me(CH$_2$)$_5$C(O)CH$_2$CH$_2$CH$_2$	BF$_4^-$	128
Me	Cyclopentyl–C=O	BF$_4^-$	128
Ph	Ph	TsO$^-$	11,138
F	*n*-C$_{14}$H$_{29}$	Cl$^-$	104
F	CF$_3$	TfO$^-$	136
F	CF$_3$CF$_2$	TfO$^-$	136
F	CF$_3$(CF$_2$)$_4$CF$_2$	TfO$^-$	136
Cl	*n*-C$_8$H$_{17}$	Cl$^-$	103
Cl	Cyclopentyl–CH$_2$	Cl$^-$	103
Cl	PhCH$_2$CH$_2$	Cl$^-$	103
Br	*n*-C$_8$H$_{17}$	Br$^-$	103
Br	*n*-Bu	Br$^-$	103
Br	Cyclopentyl–CH$_2$	Br$^-$	103
Br	PhCH$_2$CH$_2$CH$_2$	Br$^-$	103
N$_3$	Me	BF$_4^-$	102
N$_3$	*t*-Bu	BF$_4^-$	102
N$_3$	*n*-C$_8$H$_{17}$	BF$_4^-$	102
N$_3$	Cyclopentyl–CH$_2$	BF$_4^-$	102
N$_3$	Ph	TsO$^-$	99
TsO	*n*-Bu	TsO$^-$	8,132
n-Bu	TsO	TsO$^-$	8,132
n-Pr	TfO	TfO$^-$	78,133
n-Bu	TfO	TfO$^-$	78,133
n-C$_6$H$_{13}$	TfO	TfO$^-$	78,133
Ph	TfO	TfO$^-$	78
PhS(O)$_2$	Me	BF$_4^-$	32
PhS(O)$_2$	*n*-C$_8$H$_{17}$	BF$_4^-$	32
PhS(O)$_2$	*n*-C$_{16}$H$_{33}$	BF$_4^-$	83
PhS(O)$_2$	cyclopentyl–CH$_2$	BF$_4^-$	32
PhS(O)$_2$	cyclohexyl–CH$_2$CH$_2$	BF$_4^-$	32
PhS(O)$_2$	PhCH$_2$CH$_2$CH$_2$	BF$_4^-$	32
PhS(O)$_2$	HOCH$_2$CH$_2$	BF$_4^-$	32
4-NO$_2$C$_6$H$_4$S(O)$_2$	*n*-C$_8$H$_{17}$	BF$_4^-$	32
4-MeOC$_6$H$_4$S(O)$_2$	*n*-C$_8$H$_{17}$	BF$_4^-$	32
n-BuS(O)$_2$	Me	BF$_4^-$	32
PhS	cyclopentyl–CH$_2$	BF$_4^-$	128
PhS(O)	cyclopentyl–CH$_2$	BF$_4^-$	128

TABLE 8. (*continued*)

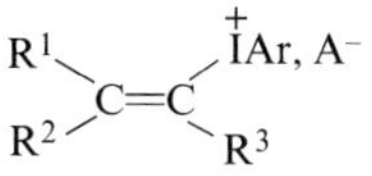

$$R^1R^2C=C(R^3)\overset{+}{I}Ar,\ A^-$$

R^1	R^2	R^3	Ar	A^-	References
H	H	H	4-Tol	$SnCl_6^-$, Br^-, BF_4^-	121
H	H	H	4-An	$SnCl_6^-$, Br^-, BF_4^-	121
D	n-C_8H_{17}	H	Ph	BF_4^-	128,146
H	n-C_8H_{17}	D	Ph	BF_4^-	128,146
H	EtO	D	Ph	BF_4^-	79
$ClCH_2$	FSO_3	H	C_6F_5	FSO_3^-	142
H	EtO	$SiMe_3$	Ph	BF_4^-	79
Me	TfO	Me	Ph	TfO^-	78
Et	TfO	Et	Ph	TfO^-	78

2. Vinyl(aryl)iodonium salts of unconfirmed stereochemistry

$$R^1R^2C=C(R^3)\overset{+}{I}\!-\!Ar,\ A^-$$

R^1	R^2	R^3	Ar	A^-	References
Cl	H	H	Ph	Cl^-, $HgCl_3^-$	116,117
Cl	H	H	2-Tol	Cl^-, $HgCl_3^-$	117,118
Cl	H	H	3-Tol	Cl^-, $HgCl_3^-$	118
Cl	H	H	4-Tol	Cl^-, $HgCl_3^-$	117,118
Cl	H	H	2-ClC_6H_4	Cl^-, $HgCl_3^-$	118
Cl	H	H	3-ClC_6H_4	Cl^-, $HgCl_3^-$	118
Cl	H	H	4-ClC_6H_4	Cl^-, $HgCl_3^-$	117,118
Cl	H	H	2-An	Cl^-, $HgCl_3^-$	118
Cl	H	H	4-An	Cl^-, $HgCl_3^-$	118
Cl	H	H	3-$NO_2C_6H_4$	Cl^-, $HgCl_3^-$	117
Ph	H	H	Ph	$SnCl_6^=$, Br^-, I, BF_4^-	119,121
Ph	Cl	H	Ph	Cl^-, I^-, BF_4^-	26
Me	TsO	Me	Ph	TsO^-	11
n-Bu	TsO	Me	Ph	TsO^-	11
Ph	TsO	Me	Ph	TsO^-	11
n-Pr	TsO	H	Ph	TsO^- (E and Z)	8
n-C_5H_{11}	TsO	H	Ph	TsO^-	8,11
i-Pr	TsO	H	Ph	TsO^-	8
i-Bu	TsO	H	Ph	TsO	8
n-Pr	$(PhO)_2PO_2$	H	Ph	$(PhO)_2PO_2^-$	22

(*continued*)

TABLE. 8 (*continued*)

3. Vinyl(aryl)iodonium salts with cyclic vinyl ligands

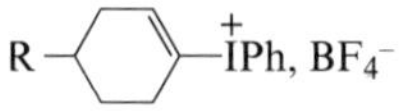

R = H (Reference 126)
R = *t*-Bu (Reference 125 and 126)

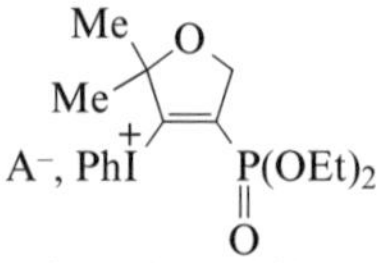

A⁻ = BF₄⁻, ClO₄⁻
(Reference 140)

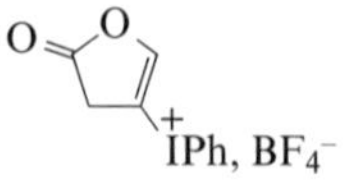

(Reference 135)

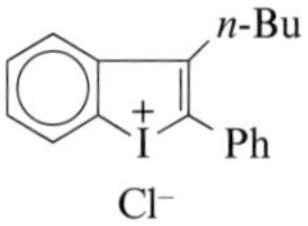

(Reference 123)

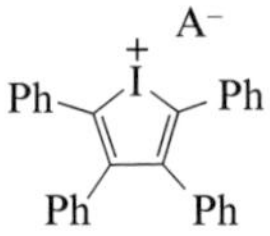

A⁻ = Cl⁻, I⁻, I⁻·H₂O, ⁻BPh₄, SbCl₆⁻

(Reference 123 and 124)

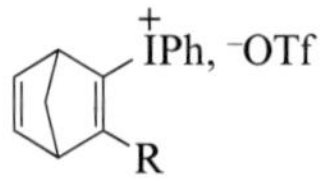

R = –CN, 4-Tol S(O)₂, PhC=O, 2-furyl-C=O, 2-thienyl-C=O
(Reference 17)

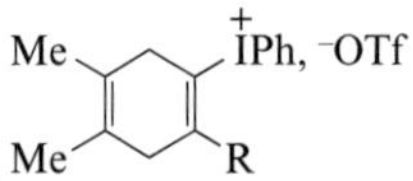

R = –CN, 4-Tol S(O)₂, *t*-BuC=O, 2-furyl-C=O, 2-thienyl-C=O
(Reference 17)

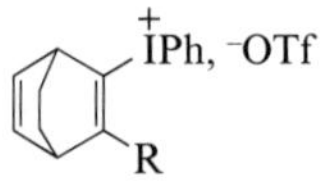

R = –CN, 4-Tol S(O)₂, *t*-BuC=O, 2-furyl-C=O, 2-thienyl-C=O
(Reference 17)

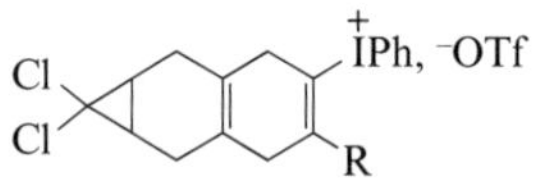

R = –CN, 4-Tol S(O)₂, *t*-BuC=O, 2-furyl-C=O, 2-thienyl-C=O
(Reference 17)

(Reference 41)

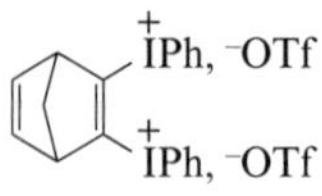

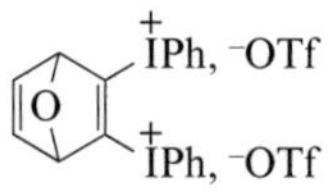

(Reference 41)

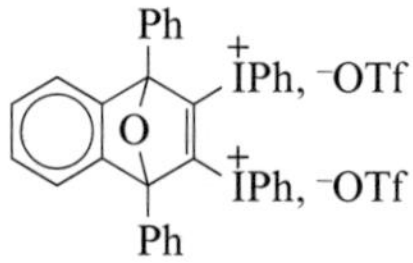

(Reference 41)

TABLE 8. (*continued*)

4. Vinyl(aryl)iodonium salts with exocyclic vinyl ligands

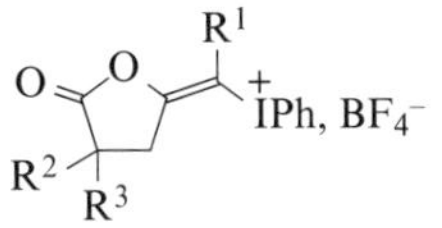

R^1	R^2	R^3	Reference
H	H	H	135
Me	H	H	135
Me$_3$Si	H	H	135
H	H	Ph	135
H	Me	CO$_2$H	135

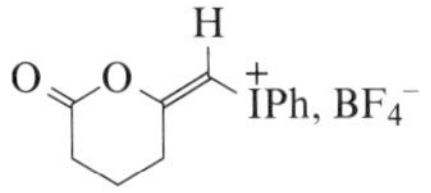

(Reference 135)

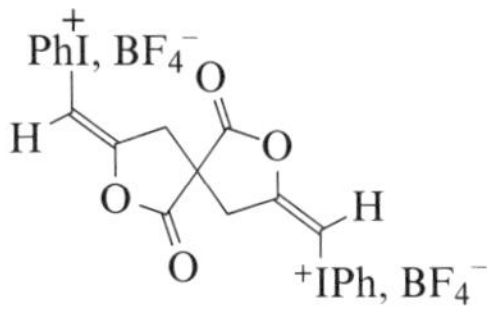

(Reference 135)

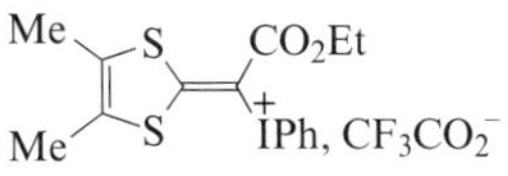

(Reference 139)

5. Miscellaneous vinyliodonium salts

R = n-Pr, n-Bu, n-C$_6$H$_{13}$ (Reference 134)

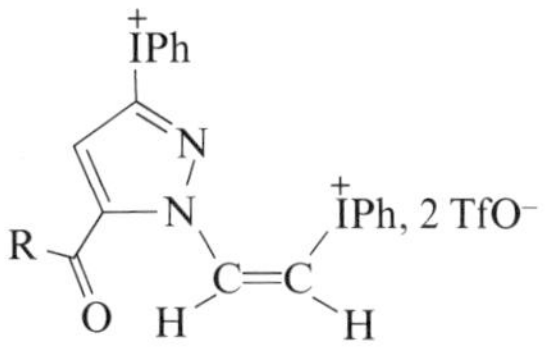

R = MeO, t-BuO
(Reference 111)

(ClCH=CH)$_2$Ï, A$^-$
A$^-$ = Cl$^-$, HgCl$_3^-$
(Reference 116)

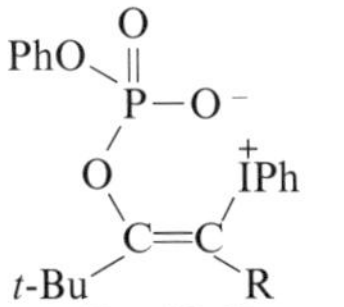

R = H, D
(Reference 95)

The geometric isomers, **39** and **40**, of vinyliodonium ions containing one β-substituent can be distinguished by the coupling constants for their vinyl proton interactions. It is generally assumed that J-values of 12 to 14 hertz manifest the E-configuration, while J-values of 6 to 7 hertz are evidence for the Z-configuration. Coupling constants for the vinyl hydrogen interactions in ethenyl(phenyl)iodonium triflate (**41**) are consistent with this view[131]. It is noted, however, that preparative methods for these compounds typically give a single stereoisomer and that J-values for $E:Z$ pairs are not usually available for comparison. The configurations of the (β-fluorovinyl)iodonium triflates, **42**, are similarly based on their vinyl H/F coupling constants (i.e. $J_{HF} = 33$ Hz in all cases)[136].

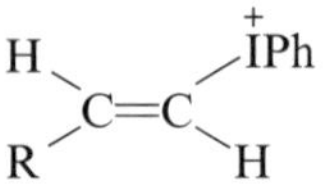

(39) 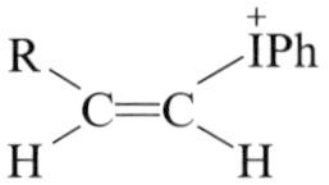   (40) 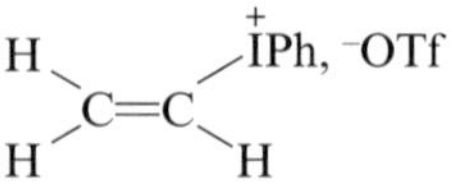   (41)   (42) $R_f = CF_3, C_2F_5, n\text{-}C_6F_{13}$

$$J_{gem} = 6.8 \text{ Hz}$$
$$J_{cis} = 5.2 \text{ Hz}$$
$$J_{trans} = 15.0 \text{ Hz}$$

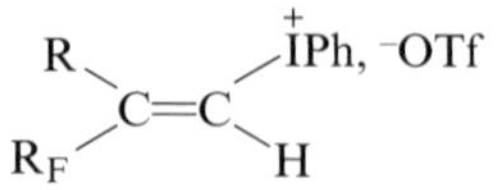

When allylic hydrogen atoms are present in the vinyl ligands, the geometric configurations of β,β-disubstituted vinyliodonium ions (e.g. **43** and **44**) can be probed by NOE experiments. Such studies are centered on interactions between the vinyl and allylic protons and the occurrence of larger NOE enhancements when these 'groups' are *cis* than when they are *trans*. Indeed, it appears to have been tacitly assumed that NOE enhancements will be nonexistent, or nearly so, for the latter arrangement. In most studies of vinyliodonium ions, E:Z pairs have not been available, and conclusions have been drawn from single stereoisomers. For example, the [(β-trifluoromethanesulfonyloxy)vinyl]-iodonium triflates, **45**, exhibit NOEs of less than 0.5% and have been assigned the *E*-configuration[78,133], while the (β-azidovinyl)iodonium tetrafluoroborates, **46**, give NOEs in the range 7.5–10.3% attributed to the *Z*-configuration[102]. NOEs of 2.7 to 7.3% for the (β-sulfonylvinyl)iodonium tetrafluoroborates, **47**, are also thought to be indicative of the *Z*-configuration[32].

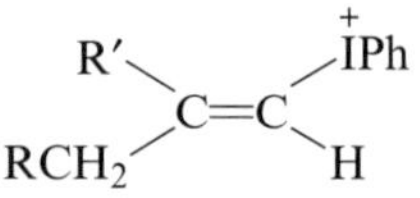

(43) 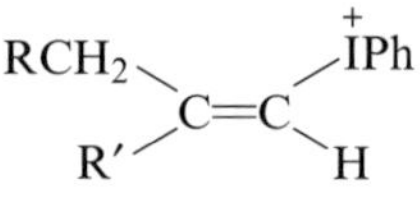   (44) 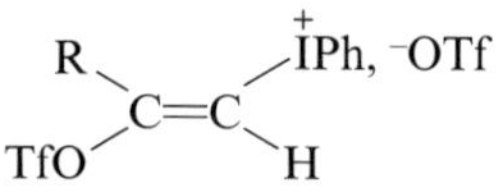   (45) R = *n*-Pr, *n*-Bu

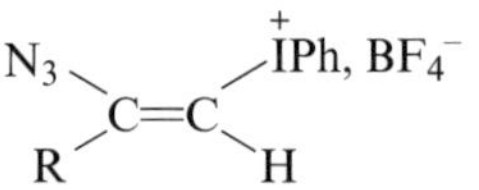

(46) R = Me, *n*-C$_8$H$_{17}$,  ⬡-CH$_2$ (47)

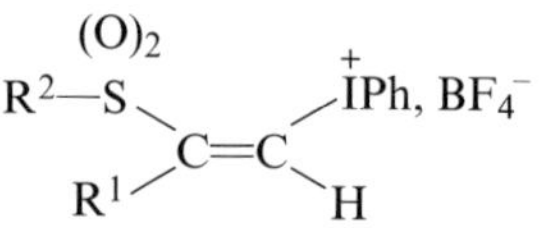

$$\left(\begin{array}{l} R^2 = Ph; R^1 = Me, n\text{-}C_8H_{17}, Ph(CH_2)_3, \square\text{-}CH_2, \hexagon\text{-}CH_2CH_2 \\ R^2 = 4\text{-}An, R^1 = n\text{-}C_8H_{17} \end{array} \right)$$

Although the foregoing assignments are probably valid, experimental details are lacking, and one is forced to attribute only qualitative significance to the reported NOEs. For example, sample preparations (solvents, concentrations, degassing) are not described, and the field strengths of the instruments used for the NOE measurements are not specified. Except for the study of **46** (enhancement of the vinyl proton resonance), it is not indicated whether the allylic and/or vinyl peaks were saturated. This is important for the interpretation of NOE magnitudes since allylic enhancements should be smaller than vinyl enhancements. For other vinyliodonium systems[103,135], it is simply reported that NOE enhancements were or were not observed, no other information being given. Finally, it is not at all clear for vinyliodonium ions such as G—C(R)=CH—İPh (where G is a heteroatom group) how low an NOE enhancement might be to *confidently* assign the Z-configuration without also analyzing the other stereoisomer.

The most complete NOE analysis of a vinyliodonium system is that reported for the geometric isomers of 2-methyl-1-butenyl(phenyl)iodonium triflate[130]. When the methyl doublet of the Z-isomer, **48**, was irradiated, the integrated intensity of the vinyl proton resonance increased by 21%. Saturation of the ethyl quartet, on the other hand, gave a vinyl proton NOE of 3%. Similar experiments with the E-isomer, **49**, led to vinyl NOEs of 4% (methyl saturation) and 15% (ethyl saturation). Aside from the fact that both isomers were examined, it is noteworthy that irradiation of the *trans*-related allylic multiplets in **48** and **49** resulted in significant vinyl enhancements.

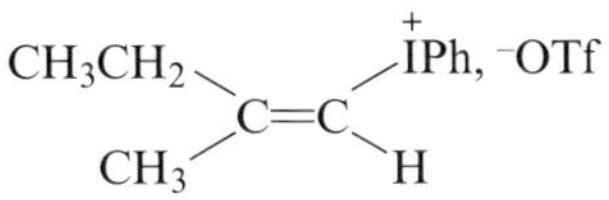

CH₃CH₂ and CH₃ structures

(**48**) 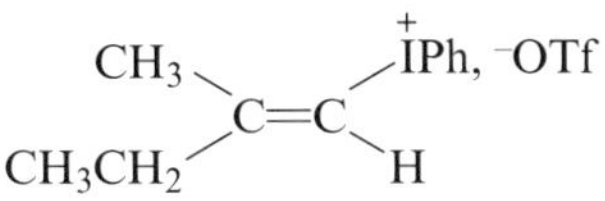   (**49**)

Other isomeric pairs that have been distinguished by NOE measurements include (Z)- and (E)-2-methyl-4-phenyl-1-butenyl(phenyl)iodonium tetrafluoroborates (13% vinyl proton NOE vs. no NOE)[126] and (Z)- and (E)-2-tosyloxy-1-hexenyl(phenyl)iodonium tosylates (NOEs not reported)[132].

As a final comment, it is noted that some of the configurations shown in Table 8 are not based on direct determinations, but were inferred from the configurations of selected vinyliodonium compounds in the same series. In other cases, structures have been presented without any supporting data.

Single-crystal X-ray structures for the nine vinyliodonium salts, **50–58** shown in Table 9, have been determined. Except for the benziodolium species, **54**, the C—İ—C angles in these compounds range from 91.6 to 99.2° and are consistent with the ψ-TBP model. However, whether the vinyl ligands in vinyl(phenyl)iodonium salts prefer to be apical or equatorial (i.e. **59** or **60**) is difficult to ascertain. For compounds **54, 57** and **58**, the anion locations are not reported, although there is some indication that **54** exhibits a dimeric structure with I⋯Cl contacts of 2.95 Å and 3.22 Å[123]. The vinyliodonium tetrafluoroborates, **50** and **53**, crystallize as centrosymmetric dimers[79,126], crudely illustrated by structure **61**, which may be regarded as a pair of strongly associated T-shaped molecules. In such dimeric units, the iodonium centers appear to be coordinated with fluorine atoms in both tetrafluoroborate ions. Thus, each carbon ligand is roughly collinear with one anion and

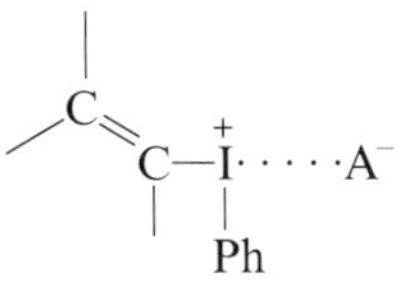

(**59**) 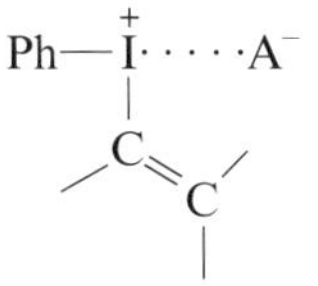   (**60**) 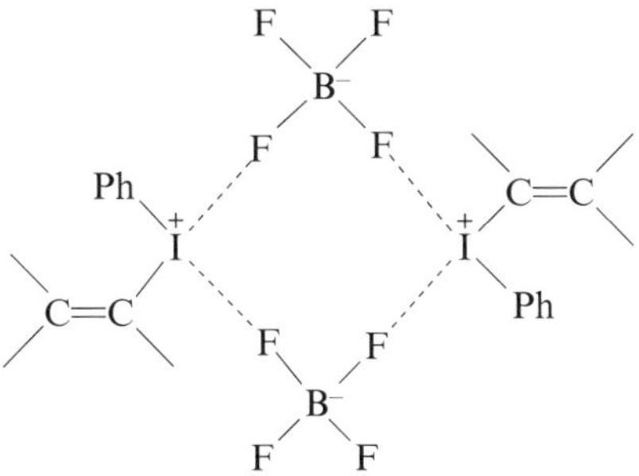   (**61**)

TABLE. 9 Single-crystal X-ray studies of vinyliodonium salts

Structure	Reference	Structure	Reference
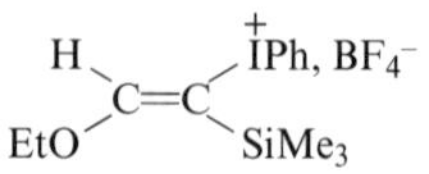 (50)	79	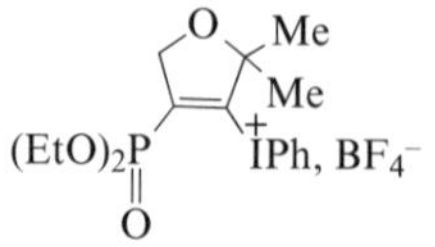 (55)	140
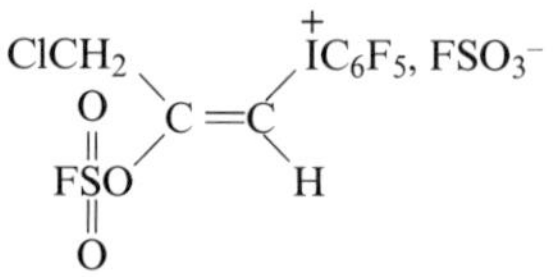 (51)	142	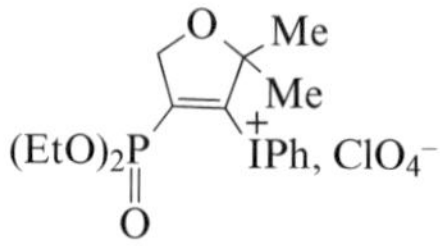 (56)	140
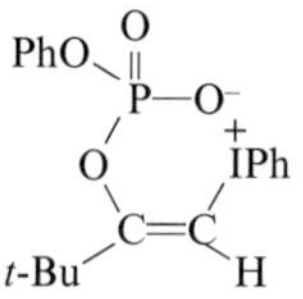 (52)	95	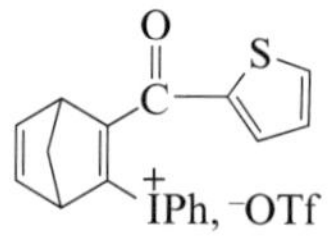 (57)	17
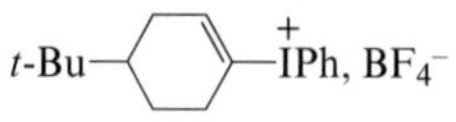 (53)	126	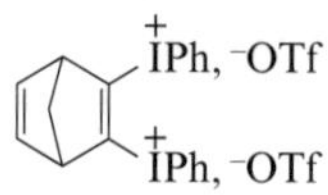 (58)	41
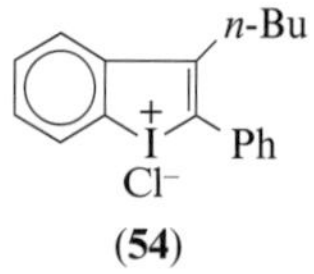 (54)	123		

perpendicular to the other. In compound **50**, the I$\cdots$F distances are 3.03(1) Å and 3.15(1) Å, the shorter (i.e. primary) contact being 'collinear' (172.5°) with the phenyl ligand. The I$\cdots$F distances in **53** are similar [3.03(7) Å and 3.189(9) Å], but in this case the shorter contact is 'collinear' with the vinyl ligand. A similar dimeric structure has been reported for the vinyl(pentafluorophenyl)iodonium fluorosulfate, **51**[142]. The fluorosulfate ion plays the same role as the fluoroborate ion in **50** and **53**, the I$\cdots$O 'bonds' being 2.654(6) Å and 2.640(5) Å.

The vinyliodonium salts **52, 55** and **56** exist as head-to-tail dimers in the solid state. The dimeric unit of **52** exhibits nearly equidistant intra- and intermolecular P$-$O$\cdots$I contacts and has been described as a pseudo-octahedral 12-I-4 system[95]. The dimeric structures of **55** and **56** are held together by P$=$O$\cdots$I contacts collinear with the vinyl ligands, while the fluoroborate and perchlorate ions are nearly collinear with the phenyl ligands[140].

D. Reactions with Nucleophilic Species

1. Nitrogen, oxygen, sulfur and phosphorus nucleophiles

Reactions of vinyliodonium salts with nitrogen, oxygen, sulfur and phosphorus nucleophiles have been reported only sporadically and have not been examined in a systematic way. The earliest studies are qualitative, while later investigations are based on rather atypical vinyliodonium structures. Such reactions usually result in the replacement of iodobenzene by the nucleophile (equation 184). The MC pathway, a common mode of reactivity for alkynyliodonium salts, has not been documented for any vinyliodonium compound.

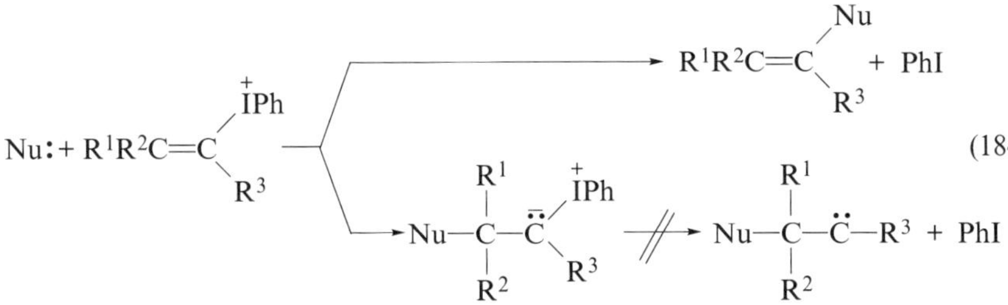

$$Nu\colon + R^1R^2C=C\overset{\overset{+}{I}Ph}{\underset{R^3}{\big\langle}}$$

$$\longrightarrow R^1R^2C=C\overset{Nu}{\underset{R^3}{\big\langle}} + PhI$$

$$\longrightarrow Nu-\underset{R^2}{\overset{R^1}{\underset{|}{C}}}-\overset{\overset{+}{I}Ph}{\overset{\cdot\cdot}{\underset{R^3}{C}}} \quad\diagup\!\!\!\!\diagup\quad Nu-\underset{R^2}{\overset{R^1}{\underset{|}{C}}}-\overset{\cdot\cdot}{C}-R^3 + PhI$$

(184)

Ethenyl(phenyl)iodonium tetrafluoroborate[120] and 2-phenyl-3-n-butylbenziodolium chloride[123] have been reported to give enol ethers upon treatment with alkoxide salts (equations 185 and 186). Ring opening of the iodolium ion proceeds nonstereospecifically and gives a mixture of geometric isomers[123]. When ethenyl(phenyl)iodonium iodide is mixed with aqueous silver(I) oxide, acetaldehyde is produced (equation 187)[120]. It seems likely that this reaction involves the initial formation and nucleophilic collapse of a (vinyl)hydroxyiodane, although metallic silver is also generated and cannot be accounted for by this pathway.

$$H_2C=CH-\overset{+}{I}Ph, BF_4^- \xrightarrow{\textit{n-BuONa, n-BuOH}} \textit{n-BuO}-CH=CH_2 + PhI \quad (185)$$
$$(1.1\!:\!1, \text{no yield given})$$

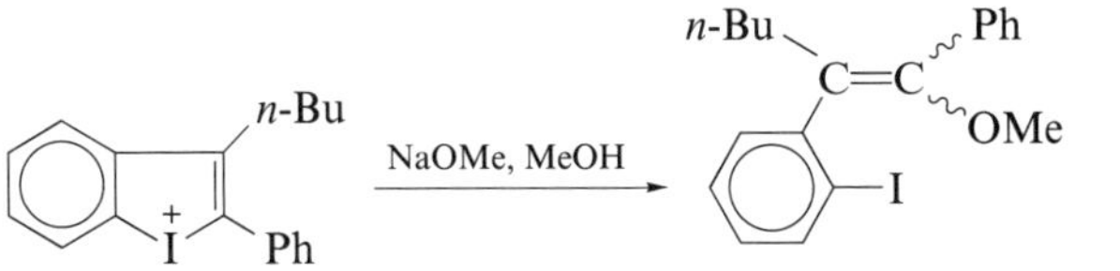

(186)

$$92\% \text{ (GC)}, ca\ 1\!:\!1\ E\!:\!Z$$

$$CH_2=CH-\overset{+}{I}Ph, I^- + Ag_2O \xrightarrow{H_2O} CH_3CHO + Ag^0 + AgI + PhI \quad (187)$$
$$78\%$$

$$\xrightarrow{?} H_2C=CH-\underset{Ph}{\overset{OH}{\underset{|}{I}}} \xrightarrow{(-PhI)} H_2C=CHOH$$

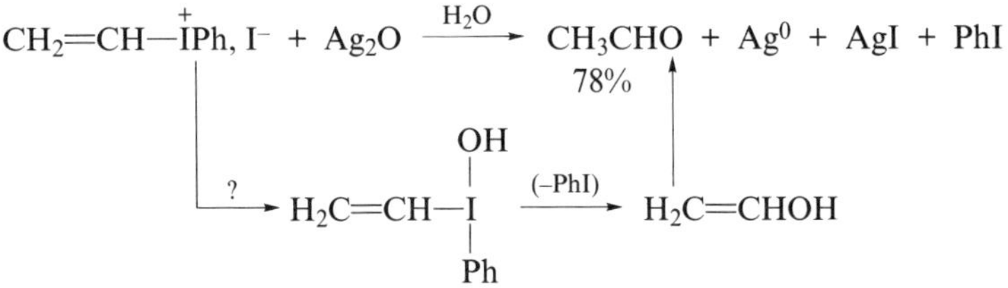

The nitrite ion behaves as a nitrogen nucleophile with vinyliodonium ions. Thus, admixture of the vinyliodonium tetrafluoroborates shown in equations 188 and 189 with sodium nitrite results in the production of nitroalkenes[119,125,126]. Cupric sulfate is apparently necessary for the success of the latter reaction, although its role has not been clarified. The (cyclohexenyl)iodonium salt also reacts with sodium thiophenoxide to give the corresponding vinyl phenyl sulfide[125,126].

$$\text{PhCH}=\text{CH}-\overset{+}{\text{I}}\text{Ph, BF}_4^- \xrightarrow[\Delta]{\text{NaNO}_2,\ \text{H}_2\text{O}} \text{PhI} + \text{PhCH}=\text{CHI} + \text{PhCH}=\text{CHNO}_2 \quad (188)$$

identified by TLC

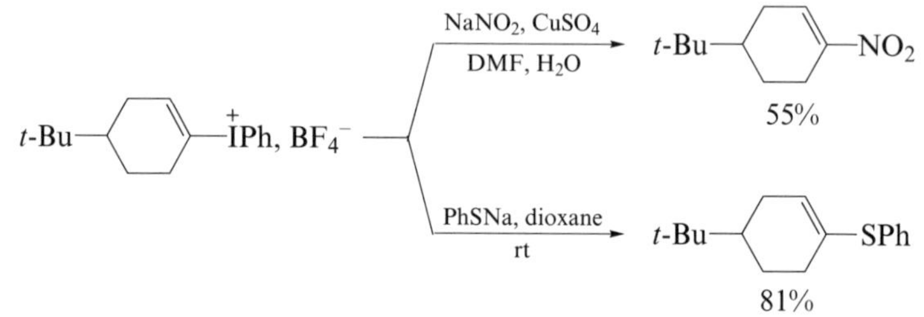

(189)

The p-toluenesulfinate ion affords sulfones with vinyliodonium salts. For example, (β-phenylethenyl)phenyliodonium fluoroborate (configuration not specified) is converted by sodium p-toluenesulfinate in water to (E)-β-phenylethenyl p-tolyl sulfone (equation 190)[121]. Under similar conditions, (E)-1-propenyl(phenyl)iodonium tetrafluoroborate undergoes competing elimination and substitution reactions leading to propyne, isopropenyl p-tolyl sulfone and 1,2-bis (p-toluenesulfonyl)propane (equation 191)[122]. The Z-isomer, on the other hand, gives a low yield of the disulfone and a much higher yield of propyne (equation 192)[122]. It has been suggested that the sulfones derived from the (E)-propenyliodonium salt originate from an intermediate (β-sulfonylpropyl)iodonium species generated via Michael addition of the sulfinate ion to the vinyliodonium ion

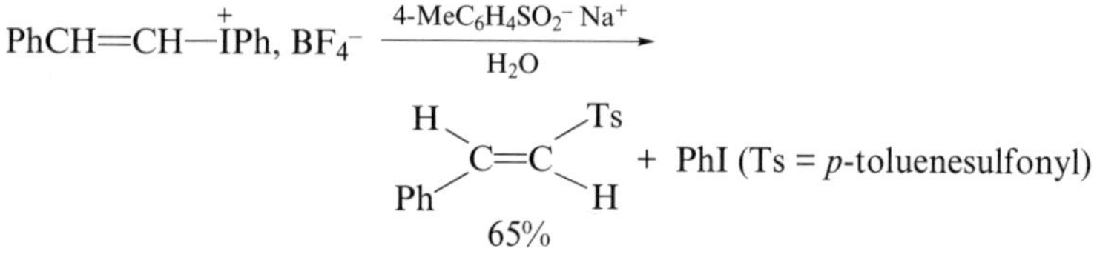

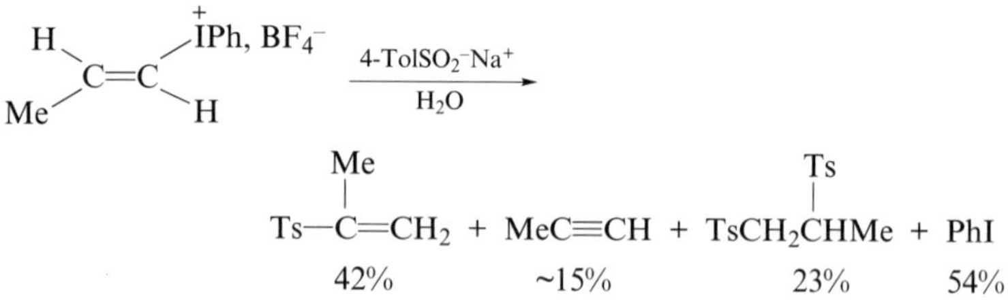

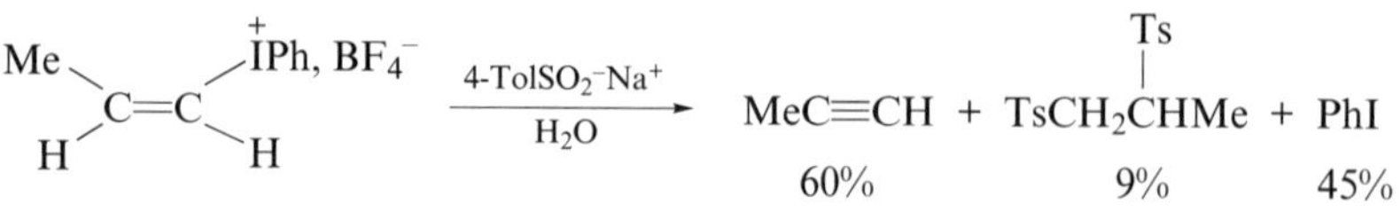

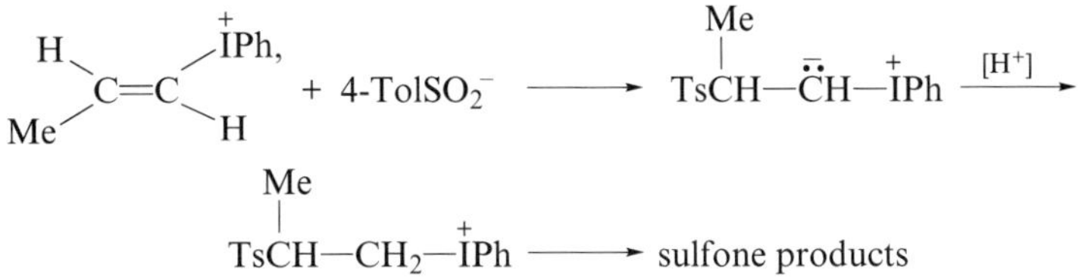

$$\underset{Me}{\overset{H}{}}C=C\underset{H}{\overset{\overset{+}{IPh,}}{}} + 4\text{-TolSO}_2^- \longrightarrow \text{TsCH}\overset{\overset{Me}{|}}{\overset{\cdot\cdot}{C}H}\overset{+}{I}Ph \xrightarrow{[H^+]}$$

$$\text{TsCH}\overset{\overset{Me}{|}}{}\text{CH}_2\overset{+}{I}Ph \longrightarrow \text{sulfone products} \qquad (193)$$

(equation 193)[122]. However, it was recognized that the addition of *p*-toluenesulfinic acid to isopropenyl *p*-tolyl sulfone might also give the disulfone.

The lactonic vinyliodonium tetrafluoroborate shown in equation 194 is biphilic and undergoes ring opening *and* displacement of iodobenzene when it is treated with water, methanol or diethylamine[135]. Because these reactions are initiated by relatively weak nucleophiles (i.e. H_2O, MeOH) and proceed at room temperature, it seems doubtful that they involve displacements at vinylic carbon. A sequence initiated by nucleophilic opening of the lactone ring to give intermediate α-phenyliodonio ketones, species known to undergo S_N displacements with weak nucleophiles[46], has been proposed[135] and is illustrated with water in equation 195.

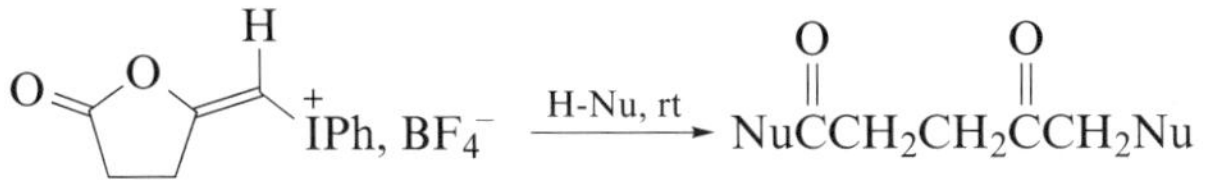

$$O\overset{O}{\underset{}{\diagdown}}\overset{\overset{H}{}}{=}\overset{+}{I}Ph,\ BF_4^- \xrightarrow{\text{H-Nu, rt}} \underset{O}{\overset{O}{\underset{\|}{\text{NuC}}}}\text{CH}_2\text{CH}_2\underset{\|}{\overset{O}{\text{C}}}\text{CH}_2\text{Nu} \qquad (194)$$

NuH (yield): H_2O (99%), MeOH (72%), Et_2NH (67%)

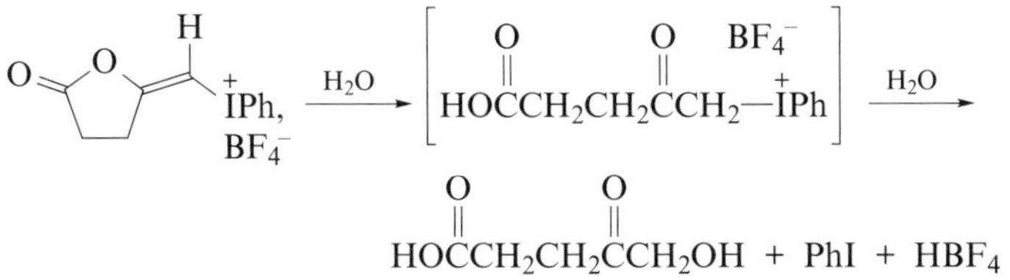

$$O\overset{O}{\underset{}{\diagdown}}\overset{\overset{H}{}}{=}\overset{\overset{+}{IPh,}}{\underset{BF_4^-}{}} \xrightarrow{H_2O} \left[\underset{O}{\overset{O}{\underset{\|}{\text{HOC}}}}\text{CH}_2\text{CH}_2\underset{\|}{\overset{O}{\text{C}}}\text{CH}_2\overset{+}{I}Ph\ \overset{BF_4^-}{}\right] \xrightarrow{H_2O}$$

$$\underset{O}{\overset{O}{\underset{\|}{\text{HOC}}}}\text{CH}_2\text{CH}_2\underset{\|}{\overset{O}{\text{C}}}\text{CH}_2\text{OH} + \text{PhI} + \text{HBF}_4 \qquad (195)$$

The displacement of iodobenzene from the phosphono-2, 5-dihydrofuryliodonium perchlorate shown in equation 196 with alkoxide and azide salts occurs readily at room

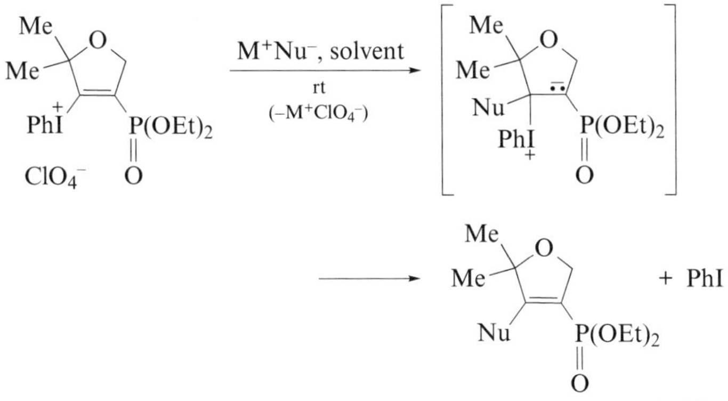

$$\begin{array}{c}\underset{Me}{\overset{Me}{\diagup}}\overset{O}{\underset{\overset{+}{PhI}\quad P(OEt)_2}{\diagdown}}\\ \underset{ClO_4^-\quad O}{}\end{array} \xrightarrow[\substack{\text{rt}\\(-M^+ClO_4^-)}]{M^+Nu^-,\ \text{solvent}} \left[\begin{array}{c}\underset{Me}{\overset{Me}{\diagup}}\overset{O}{\underset{\underset{PhI}{Nu}\quad P(OEt)_2}{\diagdown}}\\ \underset{+\quad\quad O}{}\end{array}\right] \xrightarrow{}$$

$$\longrightarrow \underset{Nu}{\overset{Me}{\diagup}}\overset{Me}{\underset{}{\diagdown}}\overset{O}{\underset{P(OEt)_2}{\diagdown}} + \text{PhI} \qquad (196)$$

MNu, solvent (yield): NaN_3, MeCN (98%); NaOMe, MeOH (83%);
NaOEt, EtOH (91%)

temperature[140]. These reactions are thought to proceed via an addition–elimination sequence at the α-carbon atom of the vinyl moiety facilitated by the diethylphosphono group. Cleavage of the iodonium ion with triphenylphosphine, on the other hand, is much less selective for vinyl carbon and affords a low yield of the expected product (equation 197)[140]. The production of tetraphenylphosphonium perchlorate and the 1-iodo-2,5-dihydrofuran in this reaction has been attributed to the attack of triphenylphosphine at the iodonium center of the starting compound. Decomposition of the resulting λ^3-iodane by competing reductive elimination–ligand coupling pathways would account for all of the observed products (equation 198 and 199)[140].

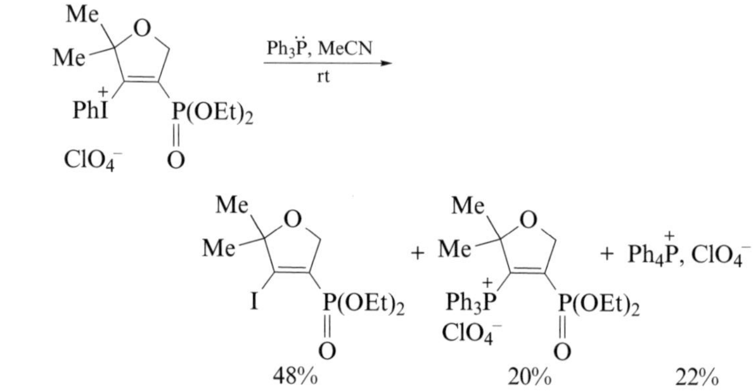

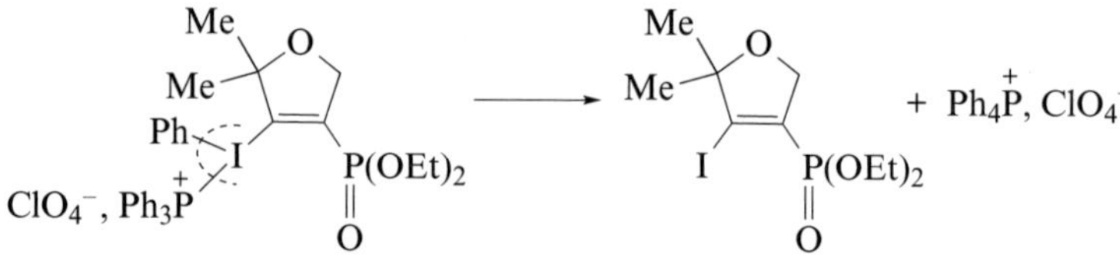

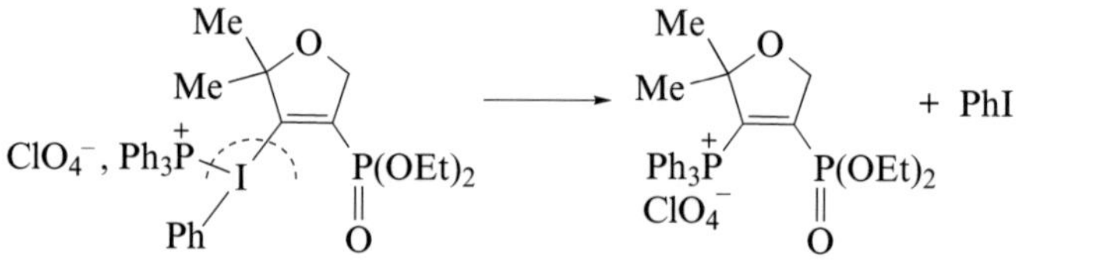

When (Z)-(β-fluoro-β-trifluoromethylvinyl)phenyliodonium triflate is mixed with sodium phenoxide or sodium 2-phenylethoxide in dichloromethane, substitution is directed to the β-carbon atom of the vinyl ligand, while the iodonium function remains intact (equation 200)[136]. With the 4-tert-butylbenzenethiolate ion, however, substitution occurs at both vinyl carbon atoms (equation 201)[136].

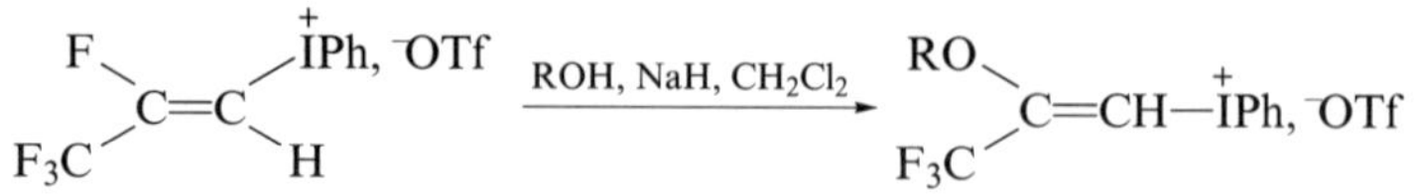

R (yield): Ph (42%, 1 isomer), PhCH$_2$CH$_2$ (32%; 2 isomers, 88:12)

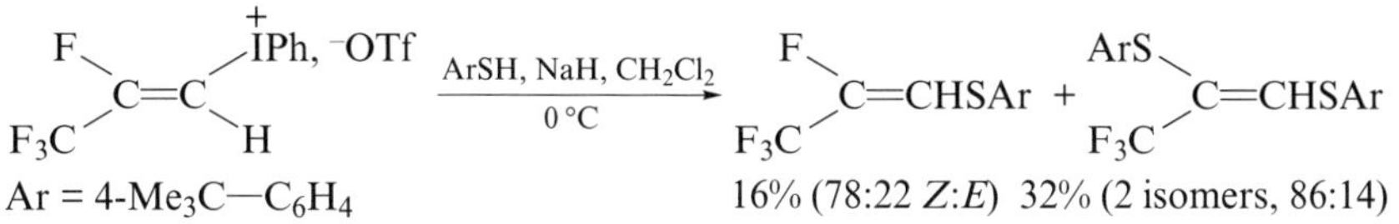

$$\underset{F_3C}{\overset{F}{>}}C=C\underset{H}{\overset{\overset{+}{I}Ph,\ ^-OTf}{<}} \xrightarrow[0\ ^\circ C]{ArSH,\ NaH,\ CH_2Cl_2} \underset{F_3C}{\overset{F}{>}}C=CHSAr\ +\ \underset{F_3C}{\overset{ArS}{>}}C=CHSAr$$

Ar = 4-Me₃C—C₆H₄   16% (78:22 Z:E)   32% (2 isomers, 86:14)

$$\text{Ar} = 4\text{-Me}_3\text{C—C}_6\text{H}_4 \qquad 16\%\ (78{:}22\ Z{:}E) \quad 32\%\ (2\ \text{isomers},\ 86{:}14)$$

(201)

2. Halogen nucleophiles

a. 2-Phenylsulfonyl-1-alkenyliodonium tetrafluoroborates with tetrabutylammonium halides. The treatment of various (Z)-[(β-phenylsulfonyl)vinyl]phenyliodonium tetrafluoroborates with tetrabutylammonium halides in dichloromethane results in the displacement of iodobenzene and affords 1-halo-2(phenylsulfonyl)alkenes (equation 202)[84]. These reactions have been reported to be 'completely stereoselective' (i.e. by 270 MHz ^{1}H NMR analysis) and to occur with retention of configuration in the vinyl moiety. The latter conclusion is based on NOE studies of the 1-halo-2-phenylsulfonyl-1-decenes ($R = n$-C_8H_{17}).

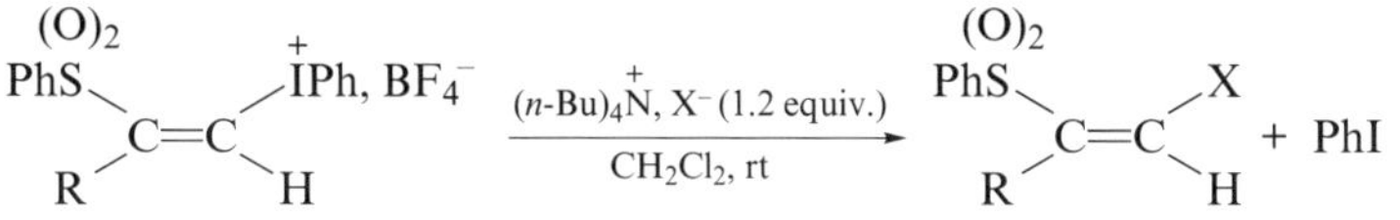

$$\underset{R}{\overset{(O)_2}{\underset{}{PhS}}}\overset{}{>}C=C\underset{H}{\overset{\overset{+}{I}Ph,\ BF_4^-}{<}} \xrightarrow[CH_2Cl_2,\ rt]{(n\text{-Bu})_4\overset{+}{N},\ X^-\ (1.2\ equiv.)} \underset{R}{\overset{(O)_2}{\underset{}{PhS}}}\overset{}{>}C=C\underset{H}{\overset{X}{<}}\ +\ PhI \quad (202)$$

	X, Isolated Yield (%)		
R	Cl	Br	I
t-Bu	100	97	83
n-C₈H₁₇	95	85	79
n-C₁₆H₃₃	90	—	—
Cyclopentyl–CH₂	75	91	83
Ph	100	96	93

The production of the (Z)-haloalkenes is thought to proceed via initial exchange of the tetrafluoroborate and halide ions and collapse of the resulting vinyliodonium halides by the addition–elimination (Ad–E) mechanism (equations 203 and 204)[84]. As with Ad–E reactions of moderately activated vinyl halides (X = Cl, Br), which typically occur with configurational retention (≥ 95%)[143–145], the intermediate carbanions apparently prefer a least motion rotation of 60° prior to the expulsion of iodobenzene. It has been demonstrated by an NMR study that anion exchange between (Z)-(2)-phenylsulfonyl-1-decenyl)-phenyliodonium tetrafluoroborate and tetrabutylammonium chloride occurs instantaneously in deuteriochloroform[84]. Furthermore, when authentic halide salts of the

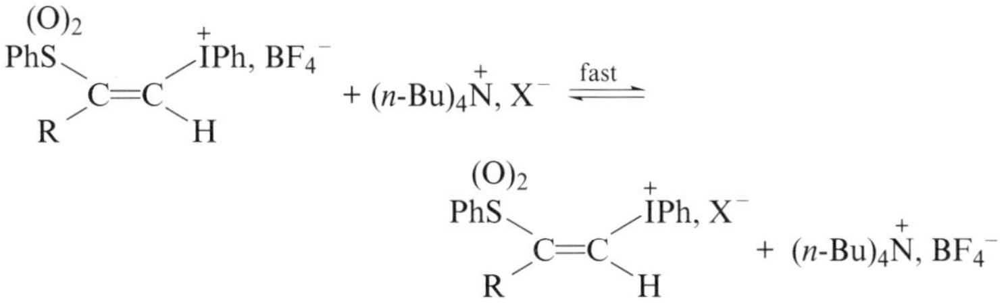

$$\underset{R}{\overset{(O)_2}{\underset{}{PhS}}}\overset{}{>}C=C\underset{H}{\overset{\overset{+}{I}Ph,\ BF_4^-}{<}}\ +\ (n\text{-Bu})_4\overset{+}{N},\ X^- \underset{}{\overset{fast}{\rightleftharpoons}}$$

$$\underset{R}{\overset{(O)_2}{\underset{}{PhS}}}\overset{}{>}C=C\underset{H}{\overset{\overset{+}{I}Ph,\ X^-}{<}}\ +\ (n\text{-Bu})_4\overset{+}{N},\ BF_4^- \quad (203)$$

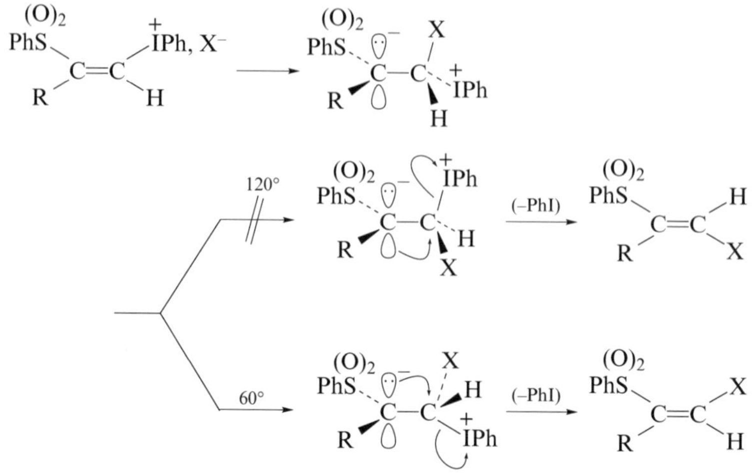

(204)

(sulfonyldecenyl)iodonium ion, prepared from the fluoroborate salt and MX in aqueous dichloromethane, are added to dichloromethane, (Z)-1-halo-2-phenylsulfonyl-1-decenes are obtained in yields of 56% to 88%[84]. Finally, the relative rates of the reactions shown in equation 202 follow the trend I$^-$ > Br$^-$ > Cl$^-$, at least in a qualitative sense.

The reaction of the decenyliodonium tetrafluoroborate with tetrabutylammonium fluoride has also been studied but follows a different course[84]. Thus, the relatively basic fluoride ion induces the α-elimination of iodobenzene and affords products derived from an intermediate alkylidenecarbene (equation 205).

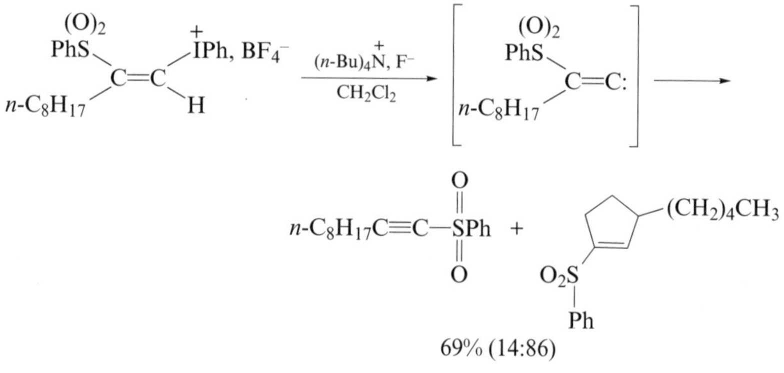

$$n\text{-}C_8H_{17}C\equiv C\text{-}SPh +$$

(205)

69% (14:86)

b. 1-Alkenyliodonium tetrafluoroborates with tetrabutylammonium halides. Admixture of (E)-1-decenyl(phenyl)- or (E)-5-phenyl-1-pentenyl(phenyl)iodonium tetrafluoroborate with ten molar equivalents of (n-Bu)$_4$NX in acetonitrile affords high yields of (Z)-1-haloalkenes (equation 206)[146]. Thus, unlike similar displacements of iodobenzene from (β-sulfonylvinyl)iodonium tetrafluoroborates (equation 202), these reactions occur with complete inversion of configuration. Alkynes are also generated in low yields, the product compositions depending to some extent on the nature of the halide ion [i.e. subst—elim (I$^-$ > Br$^-$ > Cl$^-$)].

A detailed study of the decenyliodonium tetrafluoroborate—tetrabutylammonium chloride (TBACl) system (CH$_2$Cl$_2$, rt) has shown that partitioning between the substitu-

$$\underset{R}{\overset{H}{\diagdown}}C=C\underset{H}{\overset{\overset{+}{IPh},BF_4^-}{\diagup}} + (n\text{-}Bu)_4\overset{+}{N},X^- \xrightarrow[\text{rt}]{\text{MeCN}} \underset{R}{\overset{H}{\diagdown}}C=C\underset{X}{\overset{H}{\diagup}} + RC\equiv CH + PhI \quad (206)$$
$$\underset{(10\ \text{equiv})}{}$$

R, X$^-$ (% yield-GC of haloalkene, alkyne, PhI): n-C$_8$H$_{17}$, Cl$^-$ (91, 9, 99); n-C$_8$H$_{17}$, Br$^-$ (95, 5, 100); n-C$_8$H$_{17}$, I$^-$ (88, 2, 91); Ph(CH$_2$)$_3$, Cl$^-$ (86, 14, 100); Ph(CH$_2$)$_3$, Br$^-$ (96, 3, 100); Ph(CH$_2$)$_3$, I$^-$ (99, 1, 100)

tion and elimination pathways is sensitive to the starting molar ratios of the reactants[146]. For example, when the iodonium salt was treated with one equivalent of TBACl, the haloalkene/alkyne ratio was 45:55, but when a fivefold excess of TBACl was employed, a 75:25 mixture was obtained. With ten or more equivalents of TBACl, haloalkene production was maximized, and the product ratios leveled off.

The reactions of (E)-1-deuterio- and (E)-2-deuterio-1-decenyl(phenyl)iodonium tetrafluoroborates with TBACl in dichloromethane have also been investigated (equations 207 and 208)[146]. The deuterium content of the alkyne, or lack thereof, and the effect of the β-deuterium atom on the chlorodecene/decyne ratio are consistent with alkyne production primarily via a *cis* β-elimination process.

$$\underset{n\text{-}C_8H_{17}}{\overset{H}{\diagdown}}C=C\underset{D}{\overset{\overset{+}{IPh},BF_4^-}{\diagup}} + (n\text{-}Bu)_4\overset{+}{N},Cl^- \xrightarrow[25\,°C]{CH_2Cl_2} n\text{-}C_8H_{17}C\equiv CD + \underset{n\text{-}C_8H_{17}}{\overset{H}{\diagdown}}C=C\underset{Cl}{\overset{D}{\diagup}}$$
$$\underset{(97\%\text{-}D)}{} \qquad\qquad\qquad \underset{(2\ \text{equiv})}{} \qquad\qquad\qquad 37\% \qquad\qquad 62\%$$
$$\text{(no deuterium loss)}$$
$$(207)$$

$$\underset{n\text{-}C_8H_{17}}{\overset{D}{\diagdown}}C=C\underset{H}{\overset{\overset{+}{IPh},BF_4^-}{\diagup}} + (n\text{-}Bu)_4\overset{+}{N},Cl^- \xrightarrow[25\,°C]{CH_2Cl_2} n\text{-}C_8H_{17}C\equiv CH + \underset{n\text{-}C_8H_{17}}{\overset{D}{\diagdown}}C=C\underset{Cl}{\overset{H}{\diagup}}$$
$$\underset{(93\%\text{-}D)}{} \qquad\qquad\qquad \underset{(2\ \text{equiv})}{} \qquad\qquad 18\%\ (7\%\text{-}D) \qquad 80\%\ (95\%\text{-}D)$$
$$(208)$$

Anion exchange between (E)-1-decenyl(phenyl)iodonium tetrafluoroborate and TBACl occurs 'instantaneously' in CDCl$_3$ at room temperature, while addition of the authentic chloride salt to dichloromethane (rt) affords a 45:55 mixture of (Z)-1-chlorodecene and 1-decyne[146]. Thus, the reactions shown in equations 206–208 are thought to involve the initial formation of vinyliodonium halides which then give rise to the observed products.

Elimination. Intramolecular collapse of the vinyliodonium halides to HX, alkyne and iodobenzene has been proposed, the iodonium center serving to deliver the halide 'base' to the β-CH bonds (equation 209)[146]. Although *cis* β-eliminations are unfavorable from a

$$\underset{R-C}{\overset{H}{\diagdown}}\overset{Ph}{\underset{\diagup}{\overset{|}{C}}}-I^+ \!\!-\!\! X^- \rightleftharpoons \underset{R-C}{\overset{H}{\diagdown}}\overset{Ph}{\underset{H\quad X^-}{\overset{|}{C}}}-I^+ \longrightarrow HX + RC\equiv CH + PhI \quad (209)$$

stereoelectronic vantage point, such unimolecular decompositions should enjoy a considerable entropic boost. Finally, if these reductive eliminations occur in concerted fashion from undissociated vinyliodonium halides or tight ion pairs, the chloride and vinyl ligands must presumably reside in an apical–equatorial relationship prior to decomposition.

Substitution. Because haloalkene formation (equation 206) proceeds with inversion instead of retention, these reactions are deemed inconsistent with the Ad–E manifold[146]. A bimolecular S_N2-like process between the vinyliodonium halides and TBAX, facilitated by the exceptional leaving ability of iodobenzene, has been proposed (equation 210)[84,146].

$$(n\text{-Bu})_4\overset{+}{\text{N}},\text{X}^- \quad \begin{array}{c} \text{H} \quad \text{Ph} \\ \diagdown \quad | \\ \text{C}-\text{I}^+\cdots\text{X}^- \\ \diagup \\ \text{R}-\text{C} \\ \diagdown \\ \text{H} \end{array} \longrightarrow (n\text{-Bu})_4\overset{+}{\text{N}},\text{X}^- + \begin{array}{c} \text{R} \quad \text{X} \\ \diagdown \diagup \\ \text{C}=\text{C} \\ \diagup \diagdown \\ \text{H} \quad \text{H} \end{array} + \text{PhI} \quad (210)$$

(E)-$(\beta$-phenylvinyl)phenyliodonium tetrafluoroborate exhibits a similar reactivity pattern with TBAX in acetonitrile (equation 211)[146]. However, elimination is more competitive, and although inversion predominates, mixtures of (Z)- and (E)-haloalkenes are obtained. The production of (E)-haloalkenes (i.e. retention) has been attributed to the possible 'intervention of a vinylidenephenonium ion' during substitution. It is noted that the reported yields of chloroalkene and iodobenzene are relatively low for the TBACl system. Perhaps this reaction was incomplete during the allotted time (10 hours).

$$\begin{array}{c} \text{H} \quad \overset{+}{\text{IPh}},\text{BF}_4^- \\ \diagdown \diagup \\ \text{C}=\text{C} \\ \diagup \diagdown \\ \text{Ph} \quad \text{H} \end{array} + (n\text{-Bu})_4\overset{+}{\text{N}},\text{X}^- \xrightarrow[\text{rt}]{\text{MeCN}} \begin{array}{c} \text{H} \\ \diagdown \\ \text{C}=\text{CHX} + \text{PhC}\equiv\text{CH} + \text{PhI} \quad (211) \\ \diagup \\ \text{Ph} \end{array}$$
$$\text{(10 equiv.)}$$

X$^-$ [% yield-GC of haloalkene (Z:E), alkyne, PhI]: Cl$^-$ [8 (66:34), 36, 64]; Br$^-$ [46 (88:12), 29, 89]; I$^-$ [73 (95:5), 17, 97]

Once again (cf equation 205), the behavior of tetrabutylammonium fluoride (TBAF) is exceptional. Thus, the treatment of (E)-1-decenyl(phenyl)iodonium tetrafluoroborate with ten equivalents of TBAF in acetonitrile leads exclusively to 1-decyne (equation 212)[146]. Furthermore, the reactions of the α- and β-deuterio isotopomers of the decenyliodonium salt with one equivalent of TBAF in dichloromethane are clearly consistent with alkyne formation via the α-elimination-alkylidenecarbene pathway (equations 213 and 214)[146].

$$\begin{array}{c} \text{H} \quad \overset{+}{\text{IPh}},\text{BF}_4^- \\ \diagdown \diagup \\ \text{C}=\text{C} \\ \diagup \diagdown \\ n\text{-C}_8\text{H}_{17} \quad \text{H} \end{array} + (n\text{-Bu})_4\overset{+}{\text{N}},\text{F}^- \xrightarrow[\text{rt}]{\text{MeCN}} n\text{-C}_8\text{H}_{17}\text{C}\equiv\text{CH} + \text{PhI} \quad (212)$$
$$\text{(10 equiv.)} \qquad\qquad 47\% \qquad\qquad 100\%$$

$$\begin{array}{c} \text{H} \quad \overset{+}{\text{IPh}},\text{BF}_4^- \\ \diagdown \diagup \\ \text{C}=\text{C} \\ \diagup \diagdown \\ n\text{-C}_8\text{H}_{17} \quad \text{D} \end{array} + (n\text{-Bu})_4\overset{+}{\text{N}},\text{F}^- \xrightarrow[-\text{(PhI)}]{\text{CH}_2\text{Cl}_2} \left[\begin{array}{c} \text{H} \\ \diagdown \\ \text{C}=\text{C:} \\ \diagup \\ n\text{-C}_8\text{H}_{17} \end{array} \right]$$
$$\text{(97\%- D)} \qquad \text{(1 equiv.)}$$

$$\xrightarrow{\sim\text{H}} n\text{-C}_8\text{H}_{17}\text{C}\equiv\text{CH} \quad (213)$$
$$86\% \ (0\%\text{–D})$$

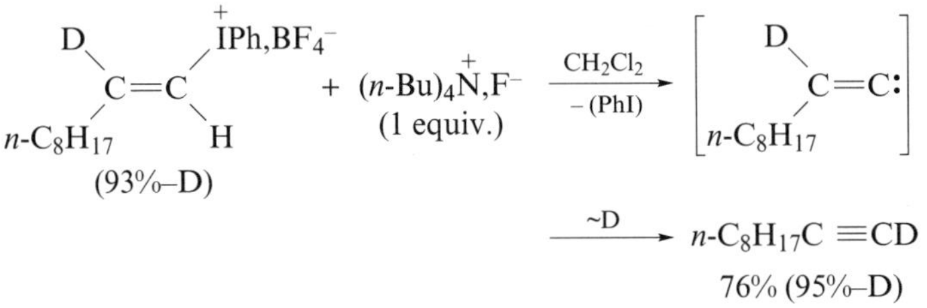

$$\underset{\substack{(93\%-D)}}{\underset{\substack{n\text{-}C_8H_{17}}}{\overset{D}{\diagdown}}C=C\overset{\overset{+}{I}Ph,BF_4^-}{\diagup}_H} + (n\text{-}Bu)_4\overset{+}{N},F^- \xrightarrow[-(PhI)]{CH_2Cl_2} \left[\underset{n\text{-}C_8H_{17}}{\overset{D}{\diagdown}}C=C\colon\right]$$

$$\xrightarrow{\sim D} n\text{-}C_8H_{17}C \equiv CD \qquad (214)$$
$$76\% \ (95\%-D)$$

3. Copper(I) reagents, carbon nucleophiles

In an effort to demonstrate the synthetic utility of vinyliodonium salts, small-scale reactions of (4-*tert*-butyl-1-cyclohexenyl)phenyliodonium tetrafluoroborate (**62**) with various nucleophilic species, especially copper(I) reagents, have been conducted[125,126]. The copper(I)-assisted reactions include the conversions of **62** to 1-cyano-, 1-halo-, 1-alkyl- and 1-phenyl-4-*tert*-butylcyclohexenes (equation 215). The alkylation and phenylation of the cyclohexenyl ligand in **62** with lithium diorganocuprates is noteworthy, since the treatment of **62** with *n*-butyllithium leads to fragmentation of the iodonium ion and affords only a 0.2% yield of 1-*n*-butyl-4-*tert*-butylcyclohexene (equation 216)[126].

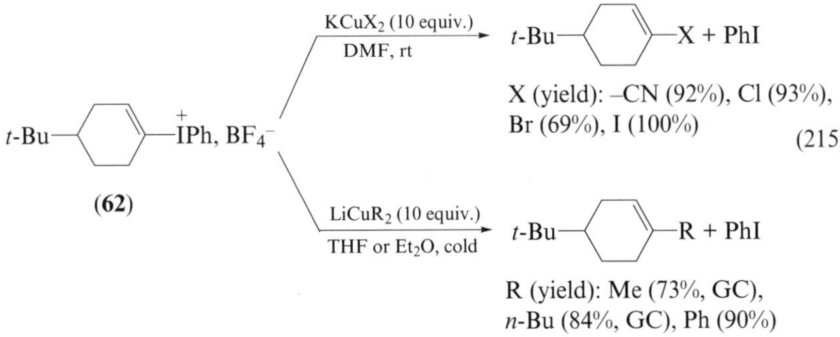

$$t\text{-}Bu-\!\!\!\!\bigcirc\!\!\!\!-\overset{+}{I}Ph, BF_4^-$$

(**62**)

$$\xrightarrow[\text{DMF, rt}]{KCuX_2 \ (10 \ equiv.)} t\text{-}Bu-\!\!\!\!\bigcirc\!\!\!\!-X + PhI$$

X (yield): −CN (92%), Cl (93%), Br (69%), I (100%)   (215)

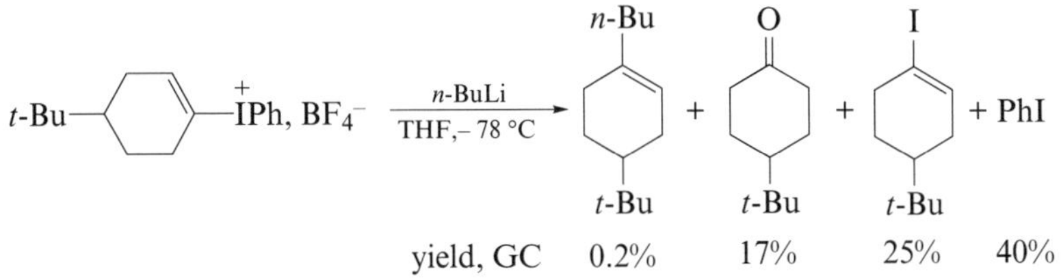

$$\xrightarrow[\text{THF or Et}_2O, \text{ cold}]{LiCuR_2 \ (10 \ equiv.)} t\text{-}Bu-\!\!\!\!\bigcirc\!\!\!\!-R + PhI$$

R (yield): Me (73%, GC), *n*-Bu (84%, GC), Ph (90%)

$$t\text{-}Bu-\!\!\!\!\bigcirc\!\!\!\!-\overset{+}{I}Ph, BF_4^- \xrightarrow[\text{THF},-78\,°C]{n\text{-}BuLi} \underset{t\text{-}Bu}{\overset{n\text{-}Bu}{\bigcirc}} + \underset{t\text{-}Bu}{\overset{O}{\bigcirc}} + \underset{t\text{-}Bu}{\overset{I}{\bigcirc}} + PhI \quad (216)$$

yield, GC   0.2%   17%   25%   40%

It has been demonstrated with three geometrically defined vinyliodonium salts that copper(I)-mediated halogenations[146] and phenylations[126] occur with nearly complete retention of configuration (equations 217–219). Since substitutive halogenations of unactivated vinyliodonium tetrafluoroborates with TBAX proceed with complete (R = *n*-C$_8$H$_{17}$, PhCH$_2$CH$_2$CH$_2$) or predominant (R = Ph) inversion, the copper(I)-promoted reactions appear to follow a fundamentally different pathway. Oxidative coupling of the cuprate and vinyliodonium ions to give λ^3-iodanes, **63**, with I(III)–Cu(III) bonds has been proposed (equation 220)[126]. Ligand coupling at iodine in **63** with retention of con-

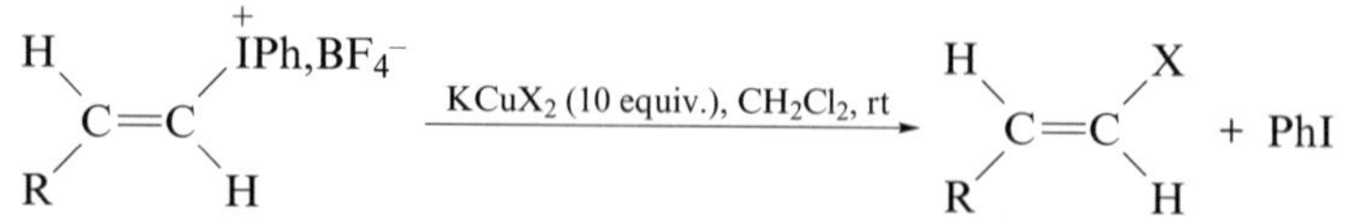

$$\underset{R}{\overset{H}{>}}C=C\underset{H}{\overset{\overset{+}{I}Ph,BF_4^-}{<}} \xrightarrow[\text{KCuX}_2 \ (10 \ \text{equiv.}), \ \text{CH}_2\text{Cl}_2, \ \text{rt}]{} \quad \underset{R}{\overset{H}{>}}C=C\underset{H}{\overset{X}{<}} \ + \ \text{PhI} \qquad (217)$$

R, X (yield): $n\text{-}C_8H_{17}$, Cl (94%); $n\text{-}C_8H_{17}$, Br (91%); Ph(CH$_2$)$_3$, Cl (70%);
Ph(CH$_2$)$_3$, Br (92%); Ph, Cl (77%); Ph, Br (80%)

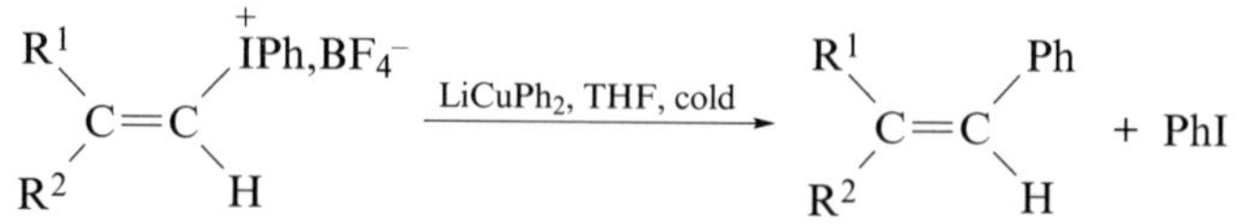

$$\underset{R^2}{\overset{R^1}{>}}C=C\underset{H}{\overset{\overset{+}{I}Ph,BF_4^-}{<}} \xrightarrow[\text{LiCuPh}_2, \ \text{THF, cold}]{} \quad \underset{R^2}{\overset{R^1}{>}}C=C\underset{H}{\overset{Ph}{<}} \ + \ \text{PhI} \qquad (218)$$

R^1, R^2 (yield): H, $n\text{-}C_8H_{17}$ (72%)[a]; Me, Ph(CH$_2$)$_2$, (82%)[b]
[a]7% of 1-decyne, [b]iodonium salt (E:Z = 97:3), product (E:Z = 96:4)

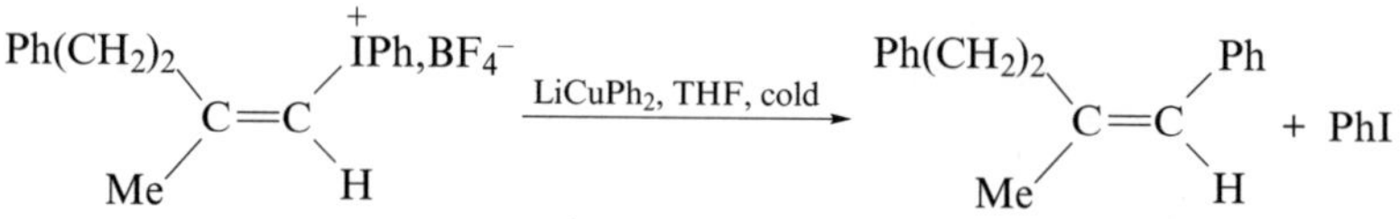

$$\underset{Me}{\overset{Ph(CH_2)_2}{>}}C=C\underset{H}{\overset{\overset{+}{I}Ph,BF_4^-}{<}} \xrightarrow[\text{LiCuPh}_2, \ \text{THF, cold}]{} \quad \underset{Me}{\overset{Ph(CH_2)_2}{>}}C=C\underset{H}{\overset{Ph}{<}} \ + \ \text{PhI} \qquad (219)$$

(Z:E = 88:12)   82% (Z:E = 88:12)

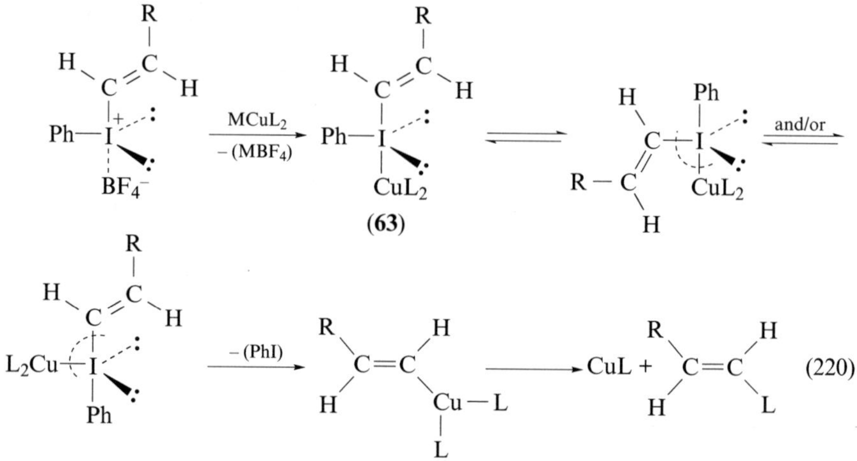

$$\qquad (220)$$

(64)

figuration at vinyl carbon and reductive elimination of iodobenzene would give vinylcopper(III) intermediates of general structure **64**. Ligand coupling at copper in **64**, again with retention in the vinyl moiety, and the reductive expulsion of CuL would lead to the observed products.

The stereoelectronic requirements for ligand coupling in species such as **63** are not clear. However, if the conversion of **63** to **64** is viewed as a concerted process, it seems likely that the vinyl and copper ligands should first assume an apical–equatorial relationship in the ψ-TBP. Thus, if both ligands originally occupy apical sites, pseudorotation about the iodine atom should precede reductive coupling. That pseudorotation can at least occur in λ^3-iodanes has been demonstrated with a chiral (diacetoxyiodo)binaphthyl system[147].

The synthetic potential of coupling reactions between vinyliodonium salts and organocuprates has not been exploited. However, some indication of their promise is provided by reported syntheses of bicyclic enediynes in the norbornadiene and 7-oxanor-bornadiene series from the appropriate bisiodonium triflates and lithium dialkynyl-cuprates (equation 221)[148].

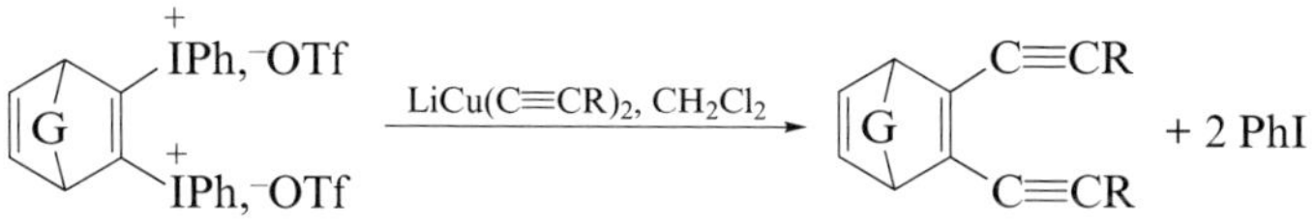

$$G = CH_2; R \text{ (yield)}: n\text{-Bu (69\%)}, t\text{-Bu (66\%)}, Ph (64\%), Me_3Si (41\%)$$
$$G = O; R \text{ (yield)}: n\text{-Bu (49\%)}, t\text{-Bu (52\%)}, Me_3Si (36\%)$$

Apart from copper(I)-mediated reactions, few studies of the treatment of vinyliodonium salts with 'carbanions' have appeared. The vinylations of the 2-phenyl- and 2-n-hexyl-1,3-indandionate ions shown in equations 222 and 223 are the only reported examples of vinyliodonium-enolate reactions known to this author[26,126]. (E)-1-Dichloroiodo-2-chloroethene has been employed with aryl- and heteroaryllithium reagents for the synthesis of symmetrical diaryliodonium salts (equation 224)[149,150]. These transformations are thought to occur via the sequential displacement of both chloride ions with ArLi to give diaryl (β-chlorovinyl)iodanes which then decompose with loss of acetylene (equation 225). That aryl(β-chlorovinyl)iodonium chlorides are viable intermediates in such reactions has been shown by the conversion of (E)-(β-chlorovinyl)phenyliodonium chloride to diaryliodonium salts with 2-naphthyl- and 2-thienyllithium (equation 226)[149,150].

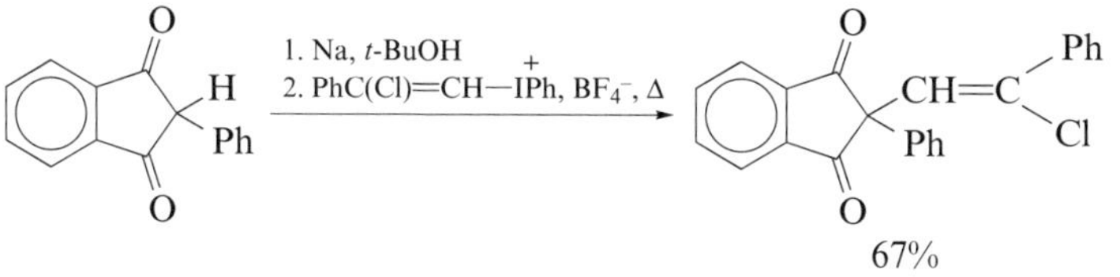

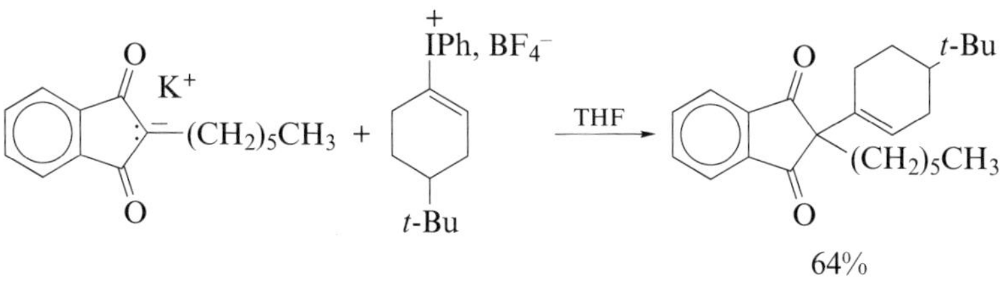

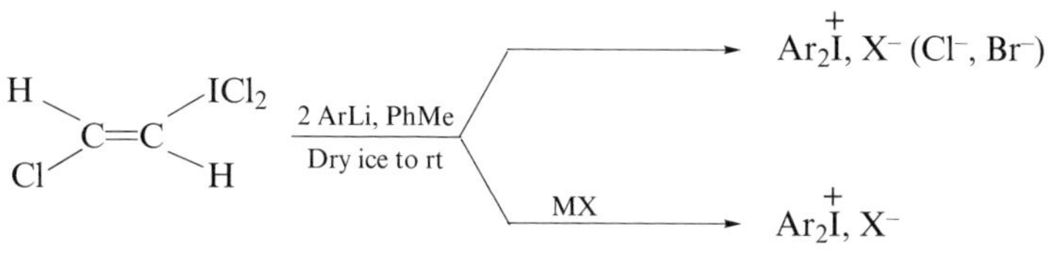

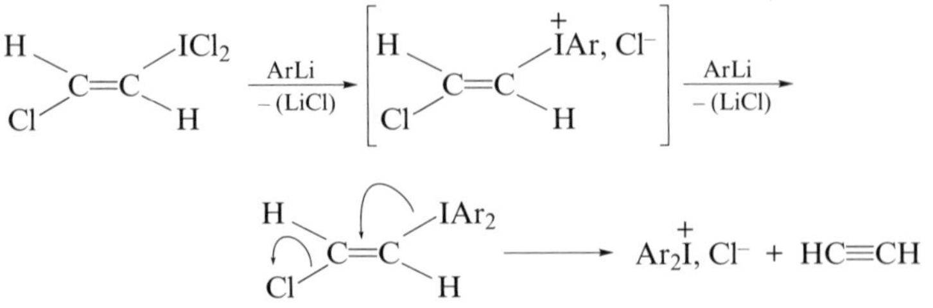

$$Ar_2\overset{+}{I}, Cl^- + HC\equiv CH \quad (225)$$

$$Ar-\overset{+}{I}-Ph, Cl^- \quad (226)$$

Ar (yield): 2-thienyl (38%), 1-naphthyl (27.5%)

E. Alpha Elimination Reactions

1. Introduction

The α-carbon–hydrogen bonds of vinyl(phenyl)iodonium ions are relatively acidic, two examples of base catalyzed hydrogen-deuterium exchange in β-oxavinyliodonium salts having been reported (equations 227 and 228)[79,95]. Although vinylidene-iodonium ylides $[\overset{\diagup}{C}=\overset{..}{C}-\overset{+}{I}Ph]$ have not been isolated, they are presumably generated in such reactions.

Thus, it is not surprising that vinyl(phenyl)iodonium salts of general structure **65** are prone to base induced α-elimination reactions thereby serving as progenitors of alkylidenecarbenes (equation 229). Once generated, the carbenes give alkynes and/or cyclopentenes depending on the migratory aptitudes of the β-substituents and the availability of side chains for intramolecular insertion (e.g. equations 230–232)[32,128]. Styrene capture of the alkylidenecarbene generated from 2-methyl-1-propenyl(phenyl)iodonium fluoroborate and potassium *tert*-butoxide has also been reported[2]. In its general aspects, the α-elimination chemistry of vinyliodonium salts converges with the Michael addition–alkylidenecarbene (MC) chemistry of alkynyliodonium salts.

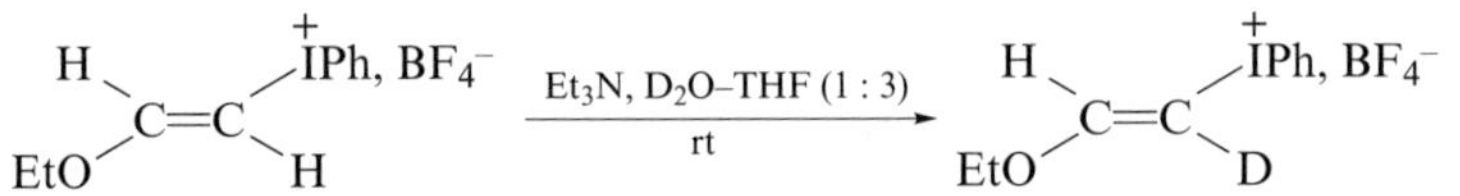

$$(227)^{79}$$

81% (90%-D)

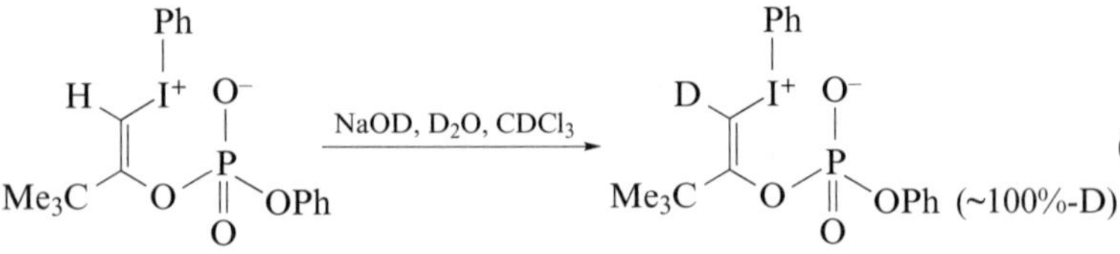

$$(228)^{95}$$

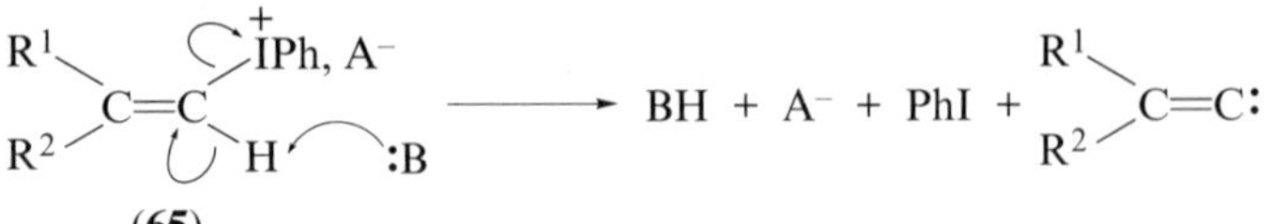

$$BH + A^- + PhI + \underset{R^2}{\overset{R^1}{\diagdown}}C=C: \quad (229)$$

(65)

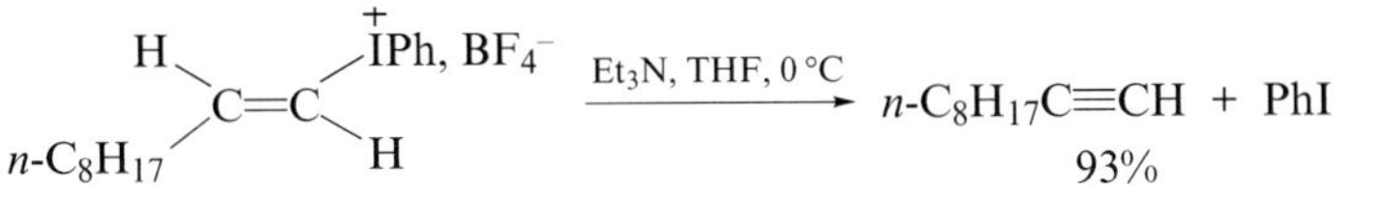

$$\underset{n\text{-}C_8H_{17}}{\overset{H}{\diagdown}}C=C\underset{H}{\overset{\overset{+}{I}Ph,\ BF_4^-}{\diagup}} \quad \xrightarrow{Et_3N,\ THF,\ 0\ °C} \quad n\text{-}C_8H_{17}C\equiv CH \ + \ PhI \qquad (230)^{128}$$
$$93\%$$

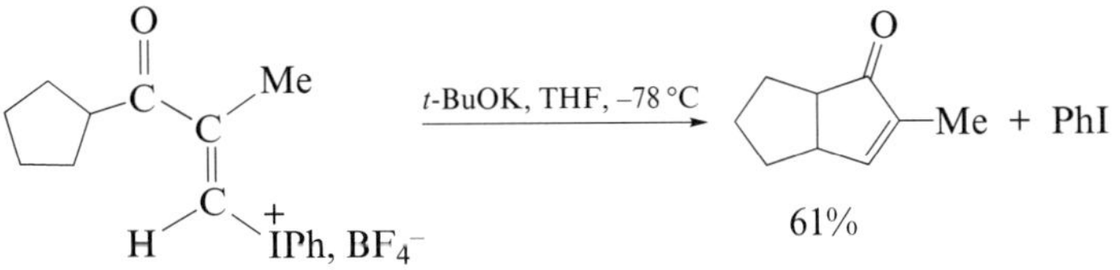

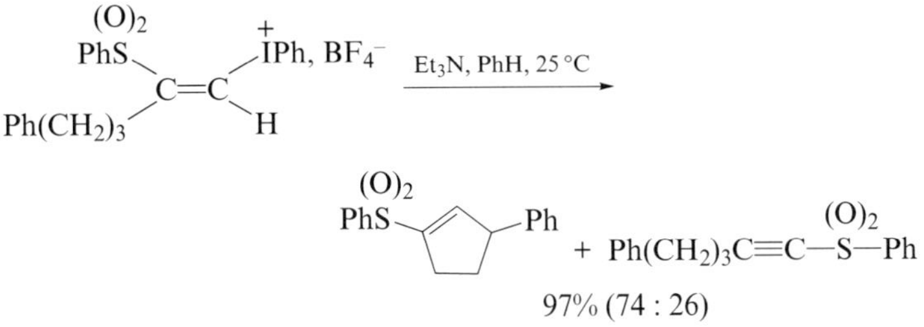

$$\xrightarrow{t\text{-BuOK, THF, }-78\ °C} \qquad \qquad \text{—Me} \ + \ PhI \qquad (231)^{128}$$
$$61\%$$

$$\underset{Ph(CH_2)_3}{\overset{\overset{(O)_2}{PhS}}{\diagdown}}C=C\underset{H}{\overset{\overset{+}{I}Ph,\ BF_4^-}{\diagup}} \quad \xrightarrow{Et_3N,\ PhH,\ 25\ °C}$$

$$\underset{PhS}{\overset{(O)_2}{\diagdown}}\text{—Ph} \quad + \ Ph(CH_2)_3C\equiv C\overset{(O)_2}{\overset{|}{—S—Ph}} \quad (232)^{32}$$
$$97\%\ (74:26)$$

2. Alkyne formation

Although alkyne formation from β,β-disubstituted vinyliodonium salts via the α-elimination-alkylidenecarbene pathway is expected, the generation of 1-decyne from (E)-1-decenyl(phenyl)iodonium fluoroborate (equation 230) might also occur via *syn* β-elimination. That α-elimination is dominant, even in this case, has been demonstrated by treatment of the α- and β-deuterio-1-decenyl isotopomers with triethylamine (equations 233 and 234)[128].

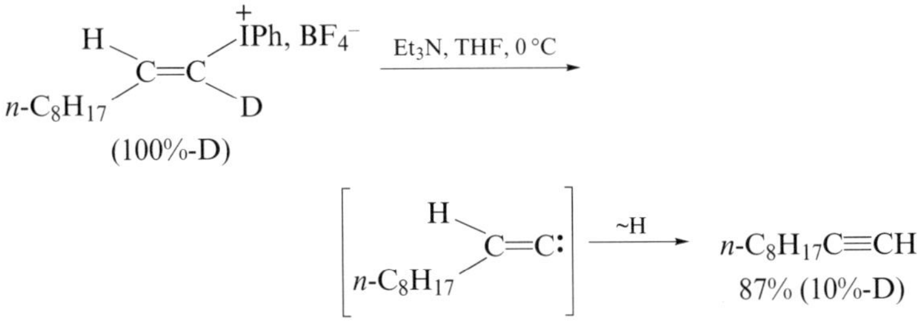

$$\underset{n\text{-}C_8H_{17}}{\overset{H}{\diagdown}}C=C\underset{D}{\overset{\overset{+}{I}Ph,\ BF_4^-}{\diagup}} \quad \xrightarrow{Et_3N,\ THF,\ 0\ °C}$$
$$(100\%\text{-D})$$

$$\left[\underset{n\text{-}C_8H_{17}}{\overset{H}{\diagdown}}C=C:\right] \xrightarrow{\sim H} \quad n\text{-}C_8H_{17}C\equiv CH \quad (233)$$
$$87\%\ (10\%\text{-D})$$

$$\underset{n\text{-}C_8H_{17}}{\overset{D}{\diagdown}}C=C\underset{H}{\overset{\overset{+}{I}Ph,\ BF_4^-}{\diagup}} \quad \xrightarrow{Et_3N,\ THF,\ 0\ °C}$$
$$(92\%\text{-D})$$

$$\left[\underset{n\text{-}C_8H_{17}}{\overset{D}{\diagdown}}C=C:\right] \xrightarrow{\sim D} \quad n\text{-}C_8H_{17}C\equiv CD \quad (234)$$
$$99\%\ (89\%\text{-D})$$

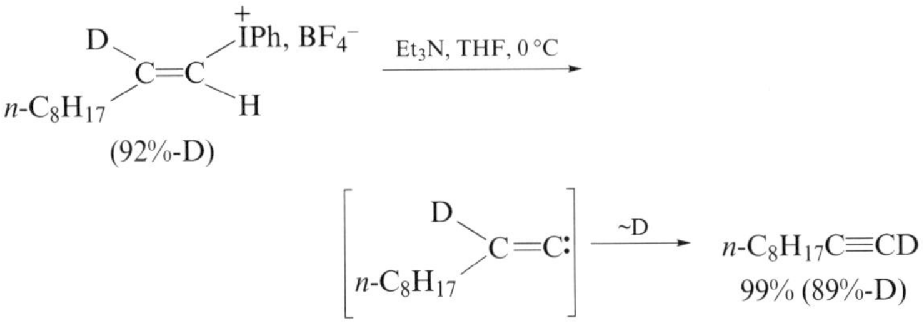

3. Cyclopentene formation

Intramolecular insertions of alkylidenecarbenes derived from vinyl(phenyl)iodonium salts are highly selective for the formation of five-membered rings. The production of cyclopropenes, cyclobutenes and larger cycloalkenes (> five carbons) has not been reported for any of these reactions. When nonequivalent β-alkyl groups of sufficient length are present in the vinyliodonium ions, isomeric cyclopentenes might be expected. This has been demonstrated with (Z)- and (E)-2-n-butyl-1-decenyl (phenyl)iodonium tetrafluoroborates (equation 235)[128]. The fact that the product compositions are nearly the same for both geometric isomers and under various conditions is consistent with the intermediacy of a 'free' alkylidenecarbene in these reactions[128].

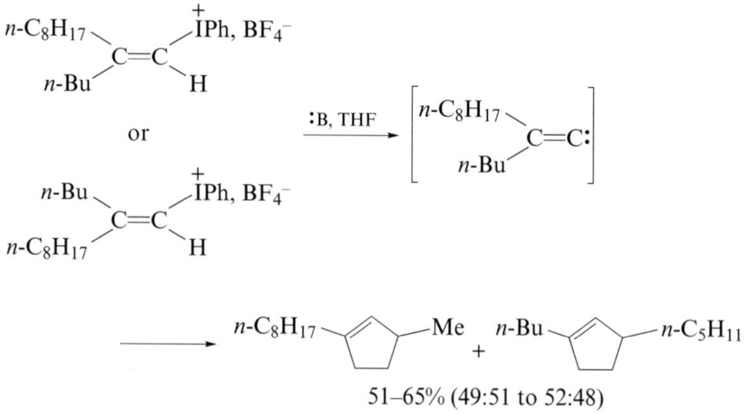

$$51–65\% \ (49{:}51 \ to \ 52{:}48) \tag{235}$$

Finally, the participating alkyl chain for such insertion reactions need not be attached directly to vinyl carbon. Thus, exposure of the [β-(n-butylsulfonyl)vinyl]iodonium salt shown in equation 236 to triethylamine gives a cyclic sulfone generated via carbenic insertion into the n-butyl group[32].

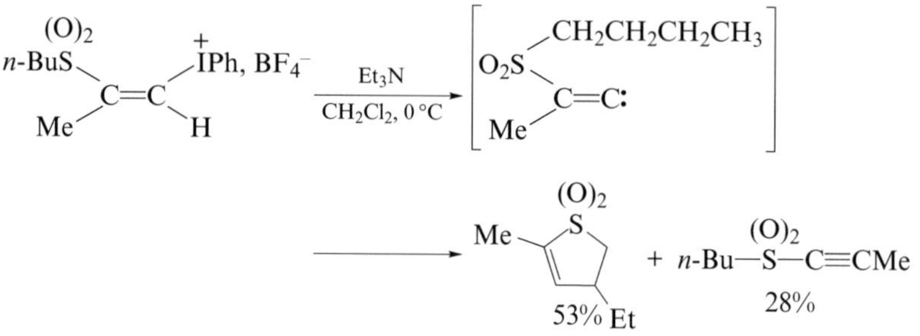

$$\tag{236}$$

4. Migration versus insertion

Reported α-elimination studies of vinyliodonium salts are limited to the substrates shown in Table 10. While aryl migrations might be expected, α-elimination reactions of (β-arylvinyl)iodonium salts have not been described. Thus far, migrations of β-hydrogen atoms (equations 230, 233 and 234)[128], β-halogen atoms (Cl, Br)[104], β-ArS(O)$_n$ groups ($n = 0,1,2$)[32] and the methyl group[128] have been reported.

TABLE 10. α-Elimination reactions of vinyliodonium salts

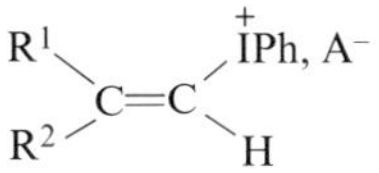

$$\begin{array}{c} R^1 \\ \diagdown \\ R^2 \diagup C=C \diagup \overset{+}{IPh},\ A^- \\ \diagdown H \end{array}$$

R^1	R^2	A^-	References
H	$n\text{-}C_8H_{17}$	BF_4^-	128,146
$n\text{-}Bu$	$n\text{-}C_8H_{17}$	BF_4^-	128
$n\text{-}C_8H_{17}$	$n\text{-}Bu$	BF_4^-	128
Me	Cyclopentyl–CH_2	BF_4^-	128
Me	Cyclopentyl–$C{=}O$	BF_4^-	128
Me	$CH_3(CH_2)_5C(O)(CH_2)_3$	BF_4^-	128
PhS	Cyclopentyl–CH_2	BF_4^-	128
PhS(O)	Cyclopentyl–CH_2	BF_4^-	128
PhS(O)$_2$	$n\text{-}C_8H_{17}$	BF_4^-	32,84
PhS(O)$_2$	$Ph(CH_2)_3$	BF_4^-	32
PhS(O)$_2$	Cyclopentyl–CH_2	BF_4^-	32
PhS(O)$_2$	Cyclohexyl–CH_2CH_2	BF_4^-	32
PhS(O)$_2$	$HOCH_2CH_2$	BF_4^-	32
4-AnS(O)$_2$	$n\text{-}C_8H_{17}$	BF_4^-	32
4-O$_2$NC$_6$H$_4$S(O)$_2$	$n\text{-}C_8H_{17}$	BF_4^-	32
$n\text{-}BuS(O)_2$	Me	BF_4^-	32
Br	$n\text{-}C_8H_{17}$	Br^-	104
Cl	$n\text{-}C_8H_{17}$	Cl^-	104
Br	$n\text{-}C_8H_{17}$	Cl^-	104
Cl	$n\text{-}C_8H_{17}$	Br^-	104
Br	$t\text{-}Bu$	Br^-	104
Cl	$t\text{-}Bu$	Cl^-	104
Br	Cyclopentyl–CH_2	Br^-	104
Cl	Cyclopentyl–CH_2	Cl^-	104
F	$n\text{-}C_{14}H_{29}$	Cl^-	104

a. *Vinyl(aryl)iodonium salts with sulfur groups at β-carbon.* The tendency of sulfur groups to migrate during base-promoted α-elimination reactions of vinyl(phenyl)-iodonium salts depends to some extent on the availability of lone pairs at sulfur. This has been demonstrated with a series of (Z)-vinyliodonium tetrafluoroborates, **66**, containing sulfur groups representing three different oxidation states of sulfur (equation 237)[32,128]. Thus, while the β-phenylsulfenyl and β-phenylsulfinyl compounds lead exclusively to alkynes, the β-phenylsulfonyl analog affords a mixture of alkynyl and cyclopentenyl

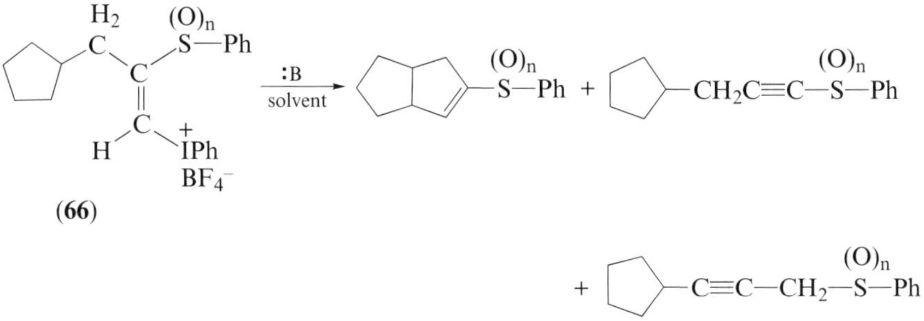

(237)

n	Conditions		Yield (%)	
0	t-BuOK, THF, −78 °C	0	72	0
1	t-BuOK, THF, −78 °C	0		84 (79:21)
2	Et₃N, PhH, rt		89 (80:20)	0

$$
\begin{array}{c}
\underset{\substack{\text{H} \\ \text{(99\%-}^{13}\text{C)}}}{\overset{\text{H}_2}{\underset{\text{BF}_4^-}{\text{C}}}} \quad \overset{(O)_n}{\underset{\text{IPh}}{\text{S—Ph}}}
\end{array}
\xrightarrow[\text{PhH, 25 °C}]{\text{Et}_3\ddot{\text{N}}}
\quad \overset{(O)_n}{\text{S—Ph}} \;+\; \overset{(O)_2}{\text{CH}_2\text{C}{\equiv}\text{C—S—Ph}}
$$

$$65\% \qquad\qquad 18\% \quad (99\% - {}^{13}\text{C}) \qquad (238)$$

sulfones. That the phenylsulfonyl group and not the β-alkyl group is the migrating species in the latter reaction has been verified by a study of the β-^{13}C labeled vinyliodonium isotopomer (equation 238)[32].

As evidenced by the treatment of three (Z)-(β-arylsulfonyl-1-decenyl)phenyliodonium fluoroborates with triethylamine, the presence of electronically diverse substituents (H, p-NO$_2$, p-MeO) in the arylsulfonyl groups appears to have little impact on their migratory aptitudes (equation 239)[32]. Indeed, among six [(β-alkyl-β-arylsulfonyl)vinyl]iodonium salts that have been investigated, the cyclopentene/alkyne ratios show little variability (i.e. 70:30 to 80:20)[32]. Finally, the presence of an appropriately placed alcohol function in the side chain of a (β-sulfonylvinyl)iodonium salt has been shown to completely suppress migration in favor of insertion (equation 240)[32].

$$
\underset{n\text{-C}_8\text{H}_{17}}{\overset{(O)_2}{\text{Ar—S}}}\text{C}{=}\text{C}\overset{\overset{+}{\text{IPh, BF}_4^-}}{\underset{\text{H}}{}}
\xrightarrow[\text{PhH, rt}]{\text{Et}_3\text{N}}
$$

$$
\underset{\text{Ar—S}}{\overset{(O)_2}{}}\!\overset{n\text{-C}_5\text{H}_{11}}{\bigtriangleup} \;+\; n\text{-C}_8\text{H}_{17}\text{C}{\equiv}\text{C—S—Ar} \overset{(O)_2}{} \qquad (239)
$$

Ar, yield (cyclopentene/alkyne): Ph, 97% (77:23); 4-An, 98% (79:21); 4-NO$_2$C$_6$H$_4$, 100% (72:28)

$$
\underset{\text{HOCH}_2\text{CH}_2}{\overset{(O)_2}{\text{Ph—S}}}\text{C}{=}\text{C}\overset{\overset{+}{\text{IPh, BF}_4^-}}{\underset{\text{H}}{}}
\xrightarrow{\text{Et}_3\text{N, PhH, rt}}
\quad \underset{\underset{61\%}{\text{S(O)}_2\text{Ph}}}{\text{(furan ring)}} \qquad (240)
$$

b. *(β-Halovinyl)iodonium salts.* The base-induced transformations of (β-alkyl-β-halovinyl) phenyliodonium halides provide a striking illustration of the influence of β-substituents on the vinyliodonium reaction manifold[104]. As might be expected, the treatment of (Z)-2-chloro-1-decenyl(phenyl)- and (Z)-2-chloro-3-cyclopentyl-1-propenyl-(phenyl)iodonium chlorides either with sodium bicarbonate or tetrabutylammonium fluoride affords mixtures of chloroalkynes and chlorocyclopentenes (equations 241 and 242). Each base gives identical product ratios with each iodonium salt indicative of the intermediacy of free (chloroalkylidene)carbenes, R(Cl)C=C:, in these reactions[104]. Because of its short alkyl chain, (Z)-2-chloro-3,3-dimethyl-1-butenyl(phenyl)iodonium chloride leads exclusively to the chloroalkyne (equation 243).

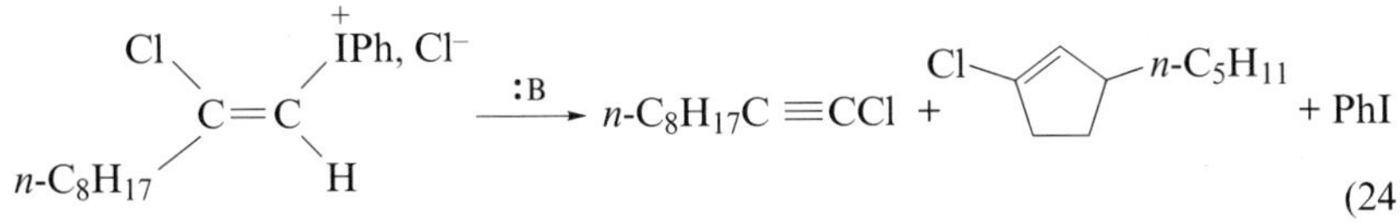

$$
\begin{array}{c}
\text{Cl}\diagdown\quad\diagup\overset{+}{\text{I}}\text{Ph, Cl}^-\\
\text{C}=\text{C}\\
n\text{-C}_8\text{H}_{17}\diagup\quad\diagdown\text{H}
\end{array}
\xrightarrow{\text{:B}}
n\text{-C}_8\text{H}_{17}\text{C}\equiv\text{CCl} +
\begin{array}{c}\text{Cl}\diagdown\!\!\!\!\!-\!\!\text{n-C}_5\text{H}_{11}\end{array}
+ \text{PhI}
$$

(241)

	Yields, GC		
Conditions	alkyne	cyclopentene	
NaHCO$_3$, CH$_2$Cl$_2$, MeOH, H$_2$O, 0 °C	52%	36%	(59:41)
(n-Bu)$_4$NF, CH$_2$Cl$_2$, rt	59%	41%	(59:41)

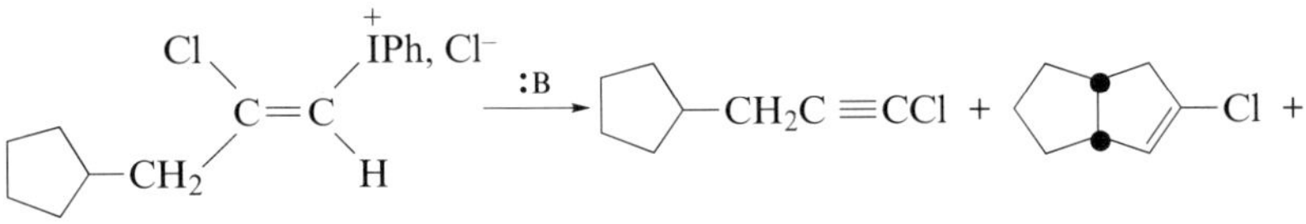

$$
\begin{array}{c}
\text{Cl}\diagdown\quad\diagup\overset{+}{\text{I}}\text{Ph, Cl}^-\\
\text{C}=\text{C}\\
\diagup\text{CH}_2\diagup\quad\diagdown\text{H}
\end{array}
\xrightarrow{\text{:B}}
\quad\text{CH}_2\text{C}\equiv\text{CCl} +
\quad\text{Cl} + \text{PhI}
$$

(242)

	Yields, GC		
Conditions	alkyne	cyclopentene	
NaHCO$_3$, CH$_2$Cl$_2$, MeOH, H$_2$O, 0 °C	54%	33%	(62:38)
(n-Bu)$_4$NF, CH$_2$Cl$_2$, rt	41%	25%	(62:38)

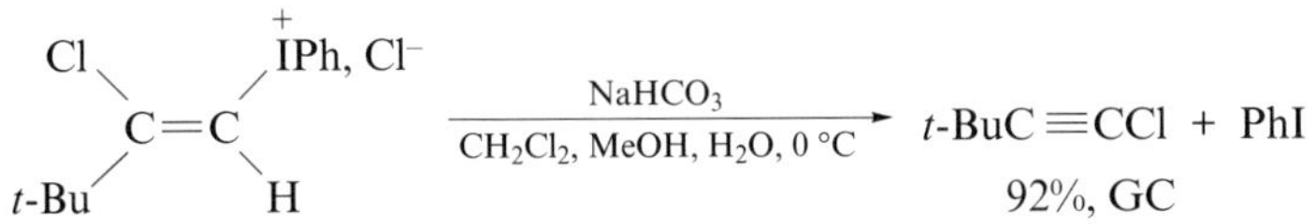

$$
\begin{array}{c}
\text{Cl}\diagdown\quad\diagup\overset{+}{\text{I}}\text{Ph, Cl}^-\\
\text{C}=\text{C}\\
t\text{-Bu}\diagup\quad\diagdown\text{H}
\end{array}
\xrightarrow[\text{CH}_2\text{Cl}_2,\ \text{MeOH},\ \text{H}_2\text{O},\ 0\ °\text{C}]{\text{NaHCO}_3}
\quad t\text{-BuC}\equiv\text{CCl} + \text{PhI}
$$

(243)

92%, GC

Treatment of the corresponding (β-bromovinyl)iodonium bromides with sodium bicarbonate leads only to bromoalkynes in all three cases (equation 244)[104]. The absence of bromocyclopentenes in these reactions has been attributed to the higher migratory aptitude of bromine, relative to chlorine, in the intermediate (bromoalkylidene)carbenes, R(Br)C=C:. However, while there can be little doubt that such carbenes are generated, they do not arise exclusively via the α-elimination pathway. Crossover experiments between alkynyl- and (β-bromovinyl)iodonium *tetrafluoroborates* indicate that bromoalkyne formation is initiated to an appreciable extent by the *anti* β-elimination of 'HBr' from the (bromovinyl)iodonium ions[104]. For example, sodium bicarbonate treatment of a 1:1 mixture of the (Z)-2-bromo-1-decenyl- and 3-cyclopentyl-1-propynyliodonium fluo-

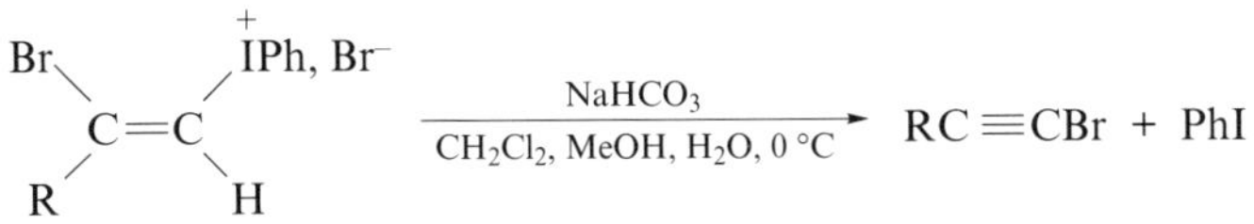

$$
\begin{array}{c}
\text{Br}\diagdown\quad\diagup\overset{+}{\text{I}}\text{Ph, Br}^-\\
\text{C}=\text{C}\\
\text{R}\diagup\quad\diagdown\text{H}
\end{array}
\xrightarrow[\text{CH}_2\text{Cl}_2,\ \text{MeOH},\ \text{H}_2\text{O},\ 0\ °\text{C}]{\text{NaHCO}_3}
\quad \text{RC}\equiv\text{CBr} + \text{PhI}
$$

(244)

R (GC yield): n-C$_8$H$_{17}$ (98%), cyclopentyl-CH$_2$ (95%), t-Bu (77%)

roborates shown in equation 245 affords two bromoalkynes. If bromoalkyne formation occurred solely via the α-elimination pathway, only 1-bromo-1-decyne would be obtained; i.e. the production of 3-cyclopentyl-1-bromopropyne is restricted to the MC manifold of the alkynyliodonium ion and requires the availability of bromide ion in the reaction medium. However, beginning with the tetrafluoroborate salts, bromide ion can only arise via its *anti* β-elimination from 2-bromo-1-decenyl(phenyl)iodonium tetrafluoroborate. These arguments are summarized in equations 246 and 247. Finally, it is noted that a 70:30 mixture of normal and crossover products (equation 245) is consistent with a dominant (≥ 60%) *anti* β-elimination process.

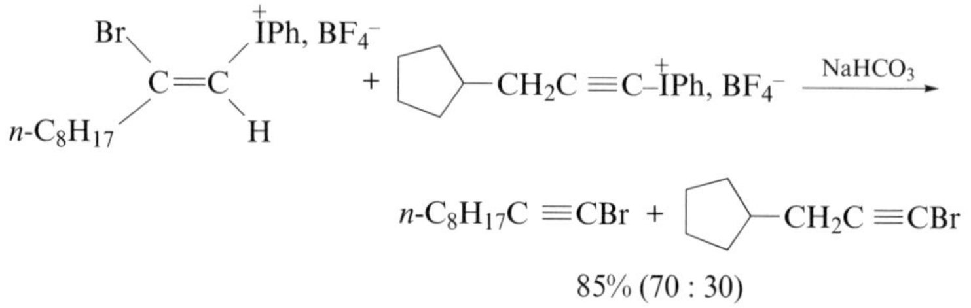

$$\underset{n\text{-}C_8H_{17}}{\overset{Br}{\diagdown}}C=C\underset{H}{\overset{\overset{+}{I}Ph,\,BF_4^-}{\diagup}} + \bigcirc\hspace{-0.3em}-CH_2C\equiv\overset{+}{C}\text{--}IPh,\,BF_4^- \xrightarrow{NaHCO_3}$$

$$n\text{-}C_8H_{17}C\equiv CBr + \bigcirc\hspace{-0.3em}-CH_2C\equiv CBr \quad (245)$$

$$85\%\ (70:30)$$

$$\underset{n\text{-}C_8H_{17}}{\overset{Br}{\diagdown}}C=C\underset{H}{\overset{\overset{+}{I}Ph,\,BF_4^-}{\diagup}} \xrightarrow{:B} n\text{-}C_8H_{17}C\equiv\overset{+}{C}\text{--}IPh,\,BF_4^- + BH + Br^- \quad (246)$$

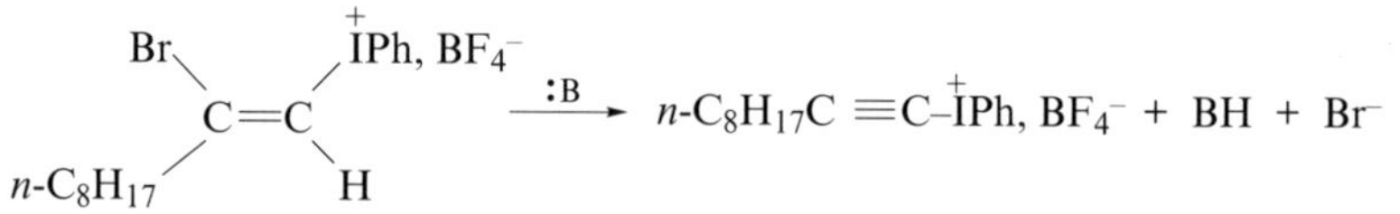

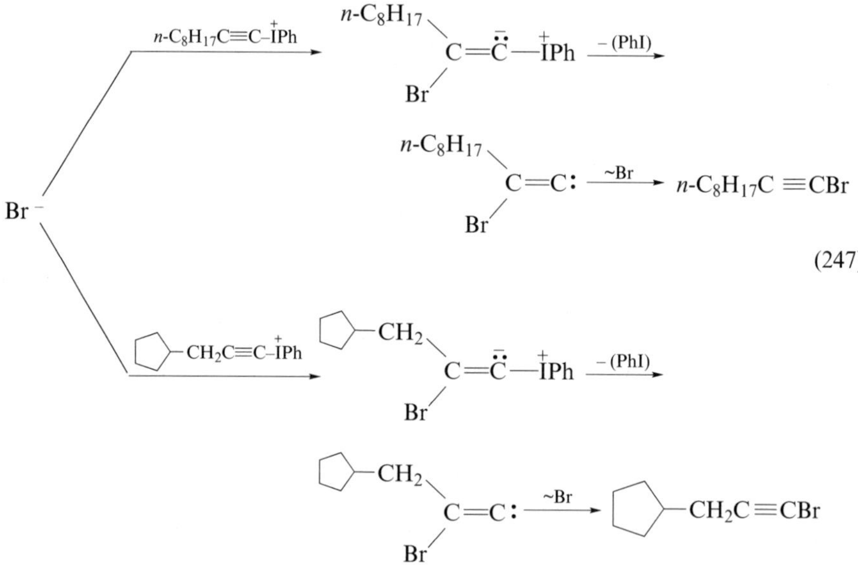

(247)

Crossover also occurs between alkynyl- and (β-chlorovinyl)iodonium tetrafluoroborates, but to a much lesser extent (equation 248)[104].

Further evidence for *anti* β-elimination is provided by the observation of halogen crossover when either (Z)-2-bromo-1-decenyl(phenyl)iodonium chloride or (Z)-2-chloro-1-decenyl(phenyl)iodonium bromide is mixed with sodium bicarbonate (equation 249)[104].

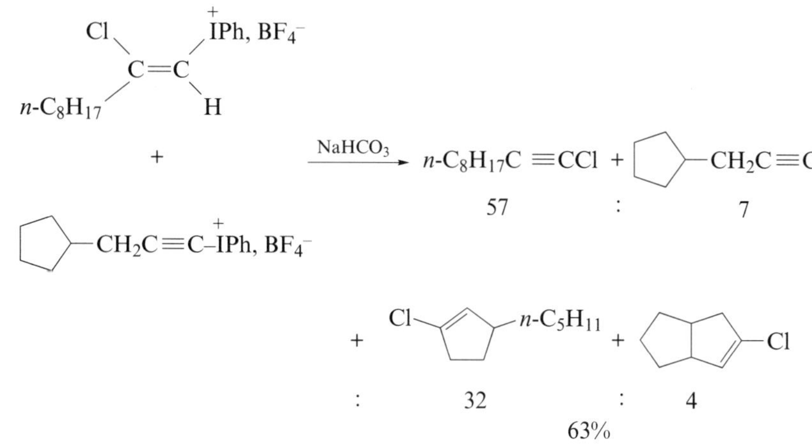

$$n\text{-C}_8\text{H}_{17}\diagdown \overset{+}{\underset{\diagup}{\text{C}}}=\overset{\text{Cl}}{\underset{\diagdown}{\text{C}}} \overset{\overset{+}{\text{IPh, BF}_4^-}}{\diagup}$$

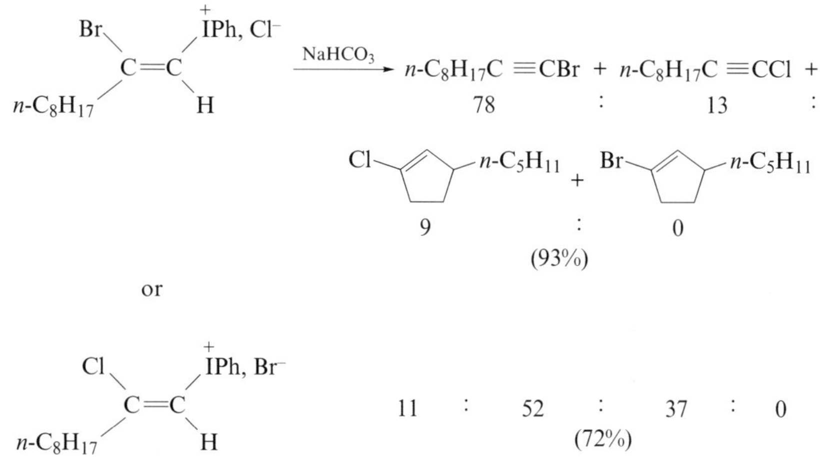

+

$$\bigbackslash\hspace{-1em}\bigcirc\hspace{-0.5em}-\text{CH}_2\text{C}\equiv\overset{+}{\text{C}}-\text{IPh, BF}_4^-$$

$$\xrightarrow{\text{NaHCO}_3} n\text{-C}_8\text{H}_{17}\text{C}\equiv\text{CCl} + \bigcirc\hspace{-0.5em}-\text{CH}_2\text{C}\equiv\text{CCl}$$

$$\qquad\qquad\qquad 57 \qquad : \qquad 7$$

$$+ \quad \overset{\text{Cl}}{\bigcirc}\hspace{-0.3em}-n\text{-C}_5\text{H}_{11} \quad + \quad \bigcirc\hspace{-0.3em}\bigcirc\hspace{-0.3em}-\text{Cl} \quad (248)$$

$$: \qquad 32 \qquad : \qquad 4$$

$$63\%$$

$$n\text{-C}_8\text{H}_{17}\diagdown \overset{+}{\underset{\diagup}{\text{C}}}=\overset{\text{Br}}{\underset{\diagdown}{\text{C}}} \overset{\overset{+}{\text{IPh, Cl}^-}}{\diagup}$$

$$\xrightarrow{\text{NaHCO}_3} n\text{-C}_8\text{H}_{17}\text{C}\equiv\text{CBr} + n\text{-C}_8\text{H}_{17}\text{C}\equiv\text{CCl} +$$

$$\qquad\qquad\qquad 78 \qquad : \qquad 13 \qquad :$$

$$\overset{\text{Cl}}{\bigcirc}\hspace{-0.3em}-n\text{-C}_5\text{H}_{11} \quad + \quad \overset{\text{Br}}{\bigcirc}\hspace{-0.3em}-n\text{-C}_5\text{H}_{11}$$

$$\qquad 9 \qquad : \qquad 0$$

$$(93\%)$$

or　　　　　　　　　　　　　　　　　　　　　(249)

$$n\text{-C}_8\text{H}_{17}\diagdown \overset{+}{\underset{\diagup}{\text{C}}}=\overset{\text{Cl}}{\underset{\diagdown}{\text{C}}} \overset{\overset{+}{\text{IPh, Br}^-}}{\diagup}$$

$$11 \quad : \quad 52 \quad : \quad 37 \quad : \quad 0$$

$$(72\%)$$

The production of 1-chloro-1-decyne and 1-chloro-3-*n*-pentylcyclopentene from the former iodonium compound and 1-bromo-1-decyne from the latter cannot be accounted for by the α-elimination mechanism.

Given the high electronegativity of the fluorine atom, (β-fluorovinyl)phenyliodonium salts might be expected to undergo efficient α-elimination reactions. However, sodium bicarbonate treatment of (Z)-2-fluoro-1-hexadecenyl(phenyl)iodonium chloride gives only a low yield of 1-fluoro-3-undecylcyclopentene (equation 250)[104]. Although 1-fluoro-1-hexadecyne was not found in this study, it was noted that fluoroalkynes are highly reactive and that the fluorohexadecyne might have escaped detection at 0 °C.

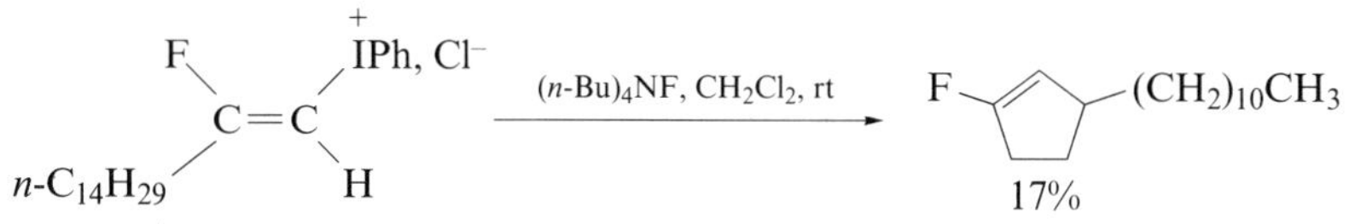

$$n\text{-C}_{14}\text{H}_{29}\diagdown \overset{+}{\underset{\diagup}{\text{C}}}=\overset{\text{F}}{\underset{\diagdown}{\text{C}}} \overset{\overset{+}{\text{IPh, Cl}^-}}{\diagup}$$

$$\xrightarrow{(n\text{-Bu})_4\text{NF, CH}_2\text{Cl}_2,\ \text{rt}} \overset{\text{F}}{\bigcirc}\hspace{-0.3em}-(\text{CH}_2)_{10}\text{CH}_3 \quad (250)$$

$$17\%$$

F. Reactions Catalyzed by Transition Metals

Two examples of Pd(II)-catalyzed carbomethoxylations of vinyl(phenyl)iodonium salts have been reported (equations 251 and 252)[125,126]. The mild reaction conditions and stereospecificity of carbonylation recommend further applications of vinyliodonium compounds for the synthesis of α,β-unsaturated carboxylate esters. By way of comparison, similar carbobutoxylations of vinyl halides (Br, I) typically require higher temperatures (60–100 °C) and longer reaction times, and they sometimes proceed with low stereospecificity[151].

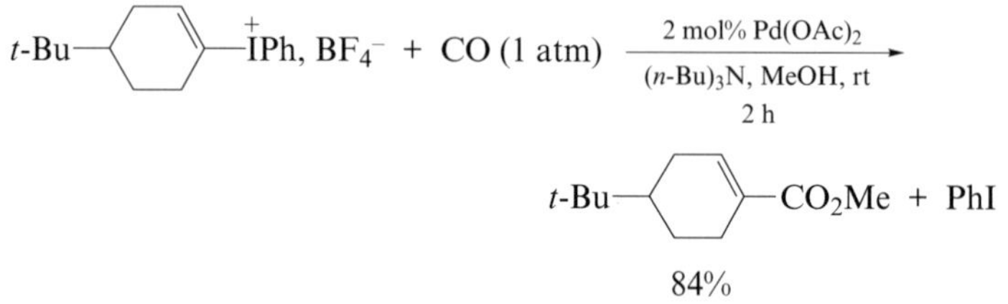

$$t\text{-Bu}-\!\!\!\bigcirc\!\!\!-\overset{+}{\text{IPh}}, \text{BF}_4^- + \text{CO (1 atm)} \xrightarrow[\substack{(n\text{-Bu})_3\text{N, MeOH, rt} \\ 2\text{ h}}]{2 \text{ mol\% Pd(OAc)}_2}$$

$$t\text{-Bu}-\!\!\!\bigcirc\!\!\!-\text{CO}_2\text{Me} + \text{PhI} \qquad (251)$$

84%

$$\begin{array}{c} \text{PhCH}_2\text{CH}_2 \\ \diagdown \\ \text{Me} \diagup \end{array}\!\!\! \text{C}=\text{CH}-\overset{+}{\text{IPh}}, \text{BF}_4^- + \text{CO (1 atm)} \xrightarrow[\substack{(n\text{-Bu})_3\text{N, MeOH, rt} \\ 2\text{ h}}]{10 \text{ mol\% Pd(OAc)}_2}$$

$(Z:E = 91:9)$

$$\begin{array}{c} \text{PhCH}_2\text{CH}_2 \\ \diagdown \\ \text{Me} \diagup \end{array}\!\!\! \text{C}=\text{CHCO}_2\text{Me} + \text{PhI} \qquad (252)$$

66% $(Z:E = 89:11)$

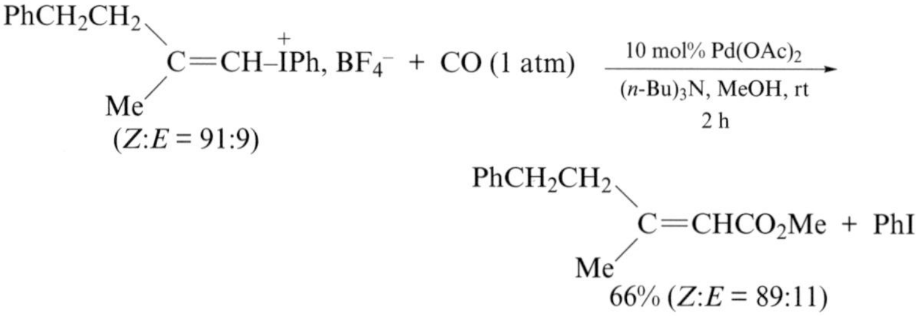

Palladium(II)-catalyzed couplings of vinyl(phenyl)iodonium salts with mono-substituted alkenes to give conjugated dienes have also been explored[132]. Such reactions proceed readily at room temperature and generally occur with retention of configuration in the vinyliodonium component and *trans*-stereoselectivity in the olefinic substrate (equations 253–256).

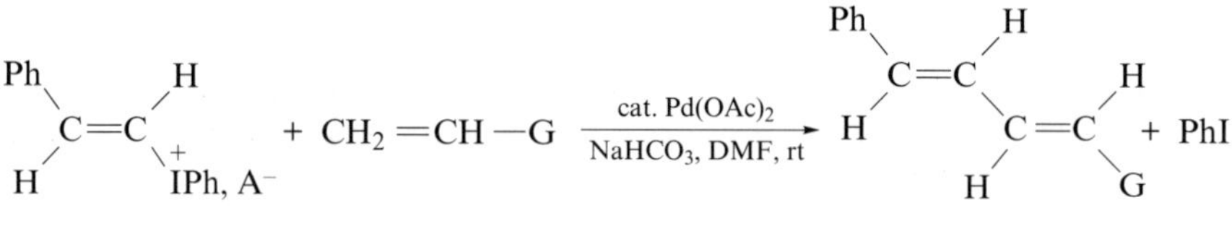

$$\begin{array}{c} \text{Ph} \quad\quad \text{H} \\ \diagdown \quad\quad \diagup \\ \text{C}=\text{C} \\ \diagup \quad\quad \diagdown \\ \text{H} \quad\quad \overset{+}{\text{IPh}}, \text{A}^- \end{array} + \text{CH}_2=\text{CH}-\text{G} \xrightarrow[\text{NaHCO}_3, \text{DMF, rt}]{\text{cat. Pd(OAc)}_2} \begin{array}{c} \text{Ph} \quad\quad \text{H} \\ \diagdown \quad\quad \diagup \\ \text{C}=\text{C} \quad\quad\quad \text{H} \\ \diagup \quad\quad \diagdown \diagup \\ \text{H} \quad\quad \text{C}=\text{C} \\ \diagup \quad\quad \diagdown \\ \text{H} \quad\quad \text{G} \end{array} + \text{PhI}$$

A$^-$, G (yield): TsO$^-$, COMe (72%); BF$_4^-$, COMe (73%); TsO$^-$, CO$_2$Me (75%); BF$_4^-$, CO$_2$Me (68%); BF$_4^-$, CHO (75%); TsO$^-$, Ph (71%)

$$(253)$$

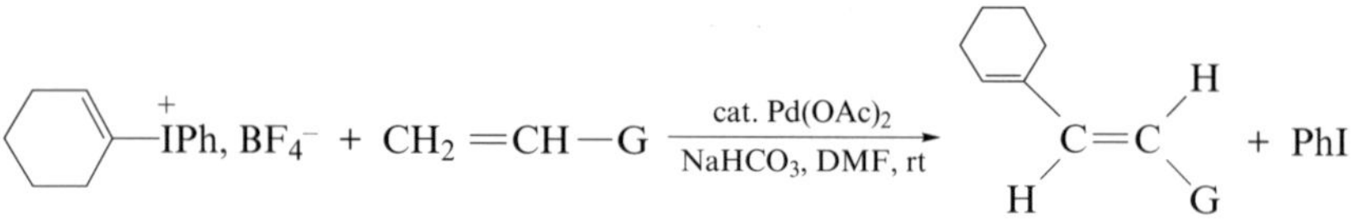

$$\bigcirc\!\!\!=\!\!\!\overset{+}{\text{IPh}}, \text{BF}_4^- + \text{CH}_2=\text{CH}-\text{G} \xrightarrow[\text{NaHCO}_3, \text{DMF, rt}]{\text{cat. Pd(OAc)}_2} \begin{array}{c} \bigcirc \\ \diagdown \quad\quad \text{H} \\ \text{C}=\text{C} \diagup \\ \diagup \quad\quad \diagdown \\ \text{H} \quad\quad \text{G} \end{array} + \text{PhI}$$

G (yield): COMe (85%), CO$_2$Me (80%), Ph (74%)

$$(254)$$

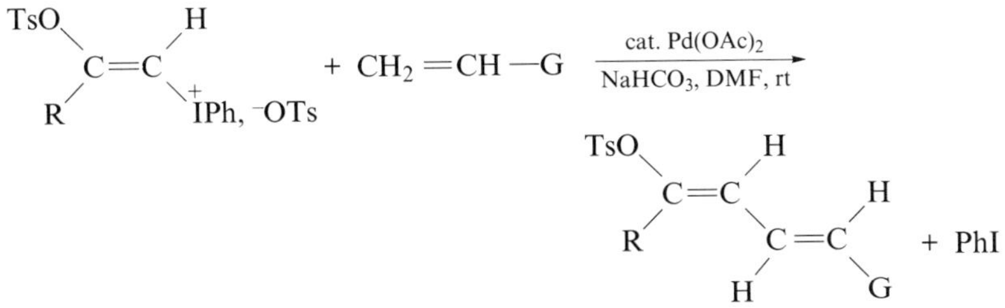

$$TsO\diagdown \atop R \diagdown C{=}C \diagup H \atop \overset{+}{I}Ph, ~^-OTs ~+~ CH_2{=}CH{-}G ~\xrightarrow[NaHCO_3, DMF, rt]{cat.~Pd(OAc)_2}$$

$$TsO\diagdown \atop R \diagdown C{=}C \diagup H \atop C{=}C \diagup H \atop H \diagdown C{=}C \diagup H \atop G ~+~ PhI \quad (255)$$

R, G (yield): H, COMe (68%); H, CO$_2$Me (60%); n-Bu, COMe (83%);
n-Bu, CO$_2$Me (75%)

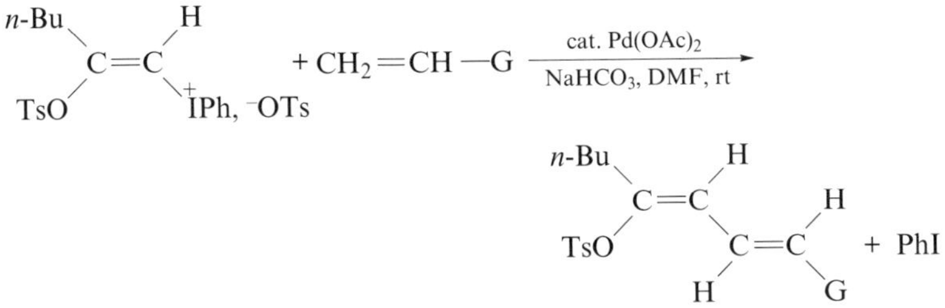

$$n\text{-}Bu\diagdown \atop TsO \diagdown C{=}C \diagup H \atop \overset{+}{I}Ph, ~^-OTs ~+~ CH_2{=}CH{-}G ~\xrightarrow[NaHCO_3, DMF, rt]{cat.~Pd(OAc)_2}$$

$$n\text{-}Bu\diagdown \atop TsO \diagdown C{=}C \diagup H \atop C{=}C \diagup H \atop H \diagdown \diagup G ~+~ PhI \quad (256)$$

G, yield (Z:E/E:E) : COMe, 71% (3.6/1); CO$_2$Me, 64% (4/1)

Although only a limited number of vinyliodonium structures have been tested, this study suggests that vinyl(phenyl)iodonium salts may offer some advantage over vinyl halides in Heck reactions with olefinic compounds[152-155]. For example, palladium(II)-catalyzed couplings of alkenyl halides with methyl acrylate, conducted in the presence of triarylphosphines and Et$_3$N, typically require extended reaction times at temperatures $\geq 100\,°$C and sometimes proceed with geometrical isomerization[152,153]. Furthermore, dienes produced under standard Heck conditions may undergo double bond shifts or give Diels–Alder adducts with the starting alkene[152,153]. The coupling of 1-bromocyclohexene with methyl acrylate (equation 257)[153] is a case in point and may be compared with the cyclohexenyliodonium counterpart shown in equation 254. The successful coupling of acrolein with (E)-(β-phenylvinyl)phenyliodonium tetrafluoroborate (equation 253) is also noteworthy, since acrolein is prone to polymerization under normal Heck conditions[155]. An interesting test of the vinyliodonium-Heck reaction would be the Pd(II)-mediated couplings of ethenyl(phenyl)- and of 2-propenyl(phenyl)iodonium salts with methyl acrylate. Thus, when either iodoethene or 2-bromopropene is treated with methyl acrylate in the presence of Pd(OAc)$_2$/Ph$_3$P/Et$_3$N at 100 °C, only Diels–Alder adducts are obtained[152].

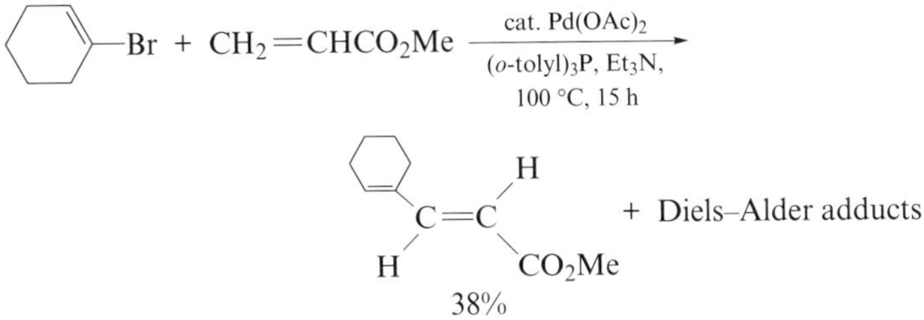

$$\langle ~ \rangle{-}Br ~+~ CH_2{=}CHCO_2Me ~\xrightarrow[\substack{(o\text{-tolyl})_3P,~Et_3N,\\100~°C,~15~h}]{cat.~Pd(OAc)_2}$$

$$\langle ~ \rangle{-}\overset{H}{\underset{H}{C{=}C}}{-}CO_2Me ~+~ \text{Diels–Alder adducts} \quad (257)$$

38%

Palladium (II)-catalyzed couplings of benzyl, aryl and vinyl halides with olefinic substrates are thought to involve the initial reduction of Pd(II) to Pd(0) by the olefin[154–156]. Oxidative addition of the organic halide to Pd(0) gives an organopalladium(II) halide, **67** (typically stabilized by coordinating ligands such as Ph$_3$P), which then adds to the alkene[154–156]. It has been suggested that the coupling reactions shown in equations 253–256 may proceed by the oxidative attachment of Pd(0) to the vinyliodonium component to give intermediates, **68**, possessing an iodine(III)-palladium(II) bond[132]. Reductive elimination of iodobenzene from **68** would give a vinylpalladium(II) tetrafluoroborate or tosylate, **69**, expected to play the same role as RPdX in normal Heck reactions. However, the oxidative addition of vinyliodonium ions to Pd(0) to give Pd(II) species, **70**, with vinyl and iodobenzene ligands also seems plausible. That oxidative additions of vinyliodonium compounds with transition metal complexes can at least occur has been demonstrated by the treatment of Vaska's complex and its rhodium(I) analog with ethenyl(phenyl)iodonium triflate (equation 258)[131]. Iodobenzene is lost during these reactions, while the ethenyl and triflate ligands are introduced with *trans*-stereochemistry.

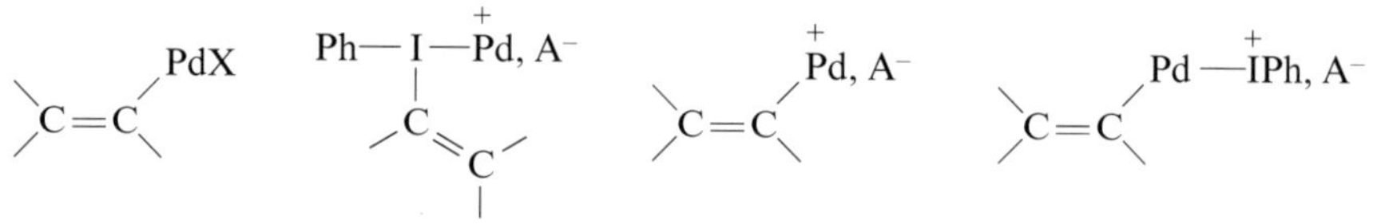

(67) X = Br, I   **(68)** A$^-$ = BF$_4^-$, TsO$^-$ **(69)** A$^-$ = BF$_4^-$, TsO$^-$ **(70)** A$^-$ = BF$_4^-$, TsO$^-$

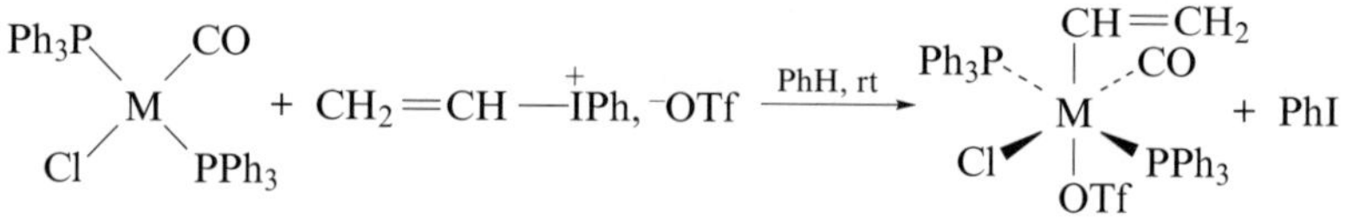

$$\text{PhH, rt} \qquad + \text{PhI} \qquad (258)$$

M (yield): Ir (75%), Rh (84%)

Palladium(II)–copper(I) cocatalyzed couplings of several (β,β-dialkylvinyl)phenyliodonium triflates with 4-isopropoxy-3-tri-n-butylstannyl-3-cyclobutene-1,2-dione (**71**) in DMF at room temperature have recently been described (equation 259)[157]. As observed in earlier studies of Cu(I)- or Pd(II)-promoted reactions of vinyliodonium salts, the vinyl ligands are introduced with retention of configuration. Since iodobenzene is a byproduct of ligand coupling, the production of 3-phenyl-4-isopropoxy-3-cyclobutene-1,2-dione might also be expected. However, the (β,β-dialkylvinyl)iodonium ions are much more reactive

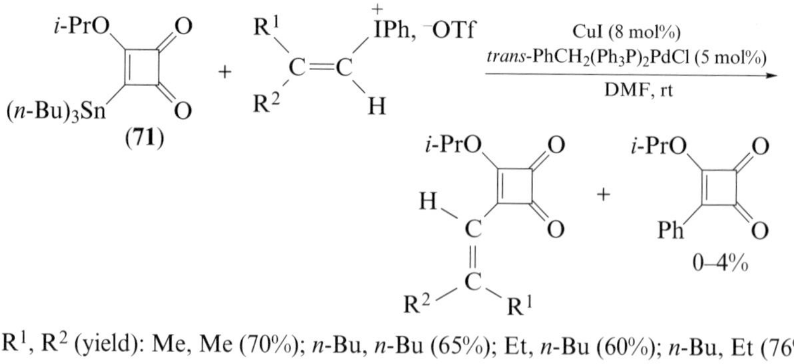

$$\begin{array}{c}\text{CuI (8 mol\%)}\\ \textit{trans}\text{-PhCH}_2(\text{Ph}_3\text{P})_2\text{PdCl (5 mol\%)}\\ \hline \text{DMF, rt}\end{array}$$

(259)

0–4%

R^1, R^2 (yield): Me, Me (70%); n-Bu, n-Bu (65%); Et, n-Bu (60%); n-Bu, Et (76%)

than iodobenzene with the catalyst system, and the yields of the 3-phenylcyclobutenedione are low.

Similar Pd(II)/Cu(I) cocatalyzed couplings of (β,β-dialkylvinyl)iodonium triflates with alkynylstannanes afford conjugated enynes, again with retention of configuration in the vinyl ligands (equation 260)[157].

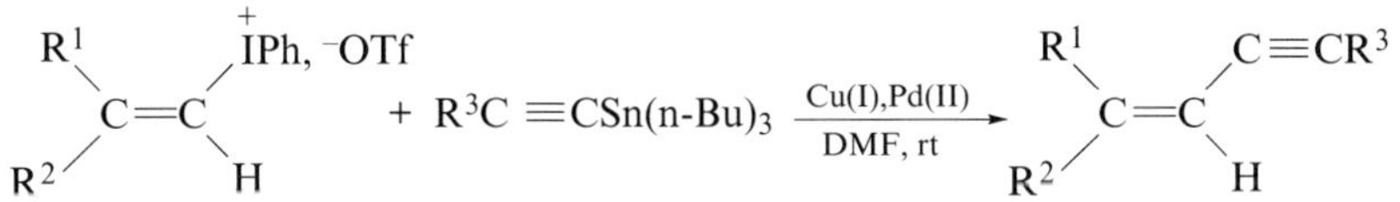

(260)

R^1, R^2, R^3 (yield): n-Bu, n-Bu, Ph (77%); Et, Me, Ph (64%); Me, Et, Ph (66%); n-Bu, n-Bu, t-BuC=O (66%); Et, n-Bu, ⌐N—C=O (67%)

G. 'Friedel-Crafts' Reactions

When (E)-[(β-arylethyl)vinyl]phenyliodonium tetrafluoroborates are heated in various solvents (at 40 °C or 60 °C), 'Friedel-Crafts' cyclizations occur and dihydronaphthalenes are obtained (equation 261)[127]. 6-Bromo-$2H$-chromene can be prepared in the same way (equation 262)[127]. The alkenylation of benzene with 4-$tert$-butyl-1-cyclohexenyl-(phenyl)iodonium tetrafluoroborate, an intermolecular version of these reactions, has also been demonstrated (equation 263)[127]. From a stereochemical standpoint, the cyclization reactions are constrained to occur with inversion of configuration at vinyl carbon, while the cyclohexenylation of benzene must proceed with retention.

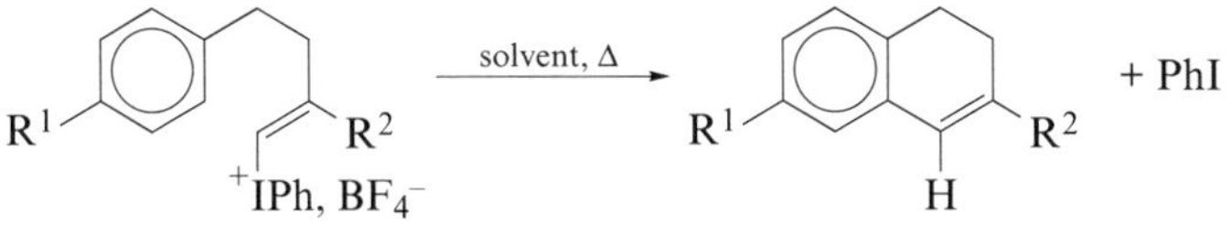

(261)

R^1, R^2, solvent (yield): H, H, CD$_3$OD (64%,GC); Cl, H, CHCl$_3$ (68%, GC); H, Me, CH$_2$Cl$_2$ (63%, isolated; starting iodonium salt 97:3–E:Z)

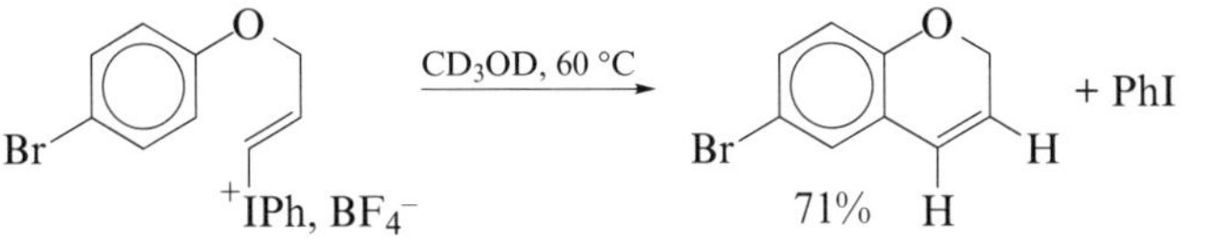

(262)

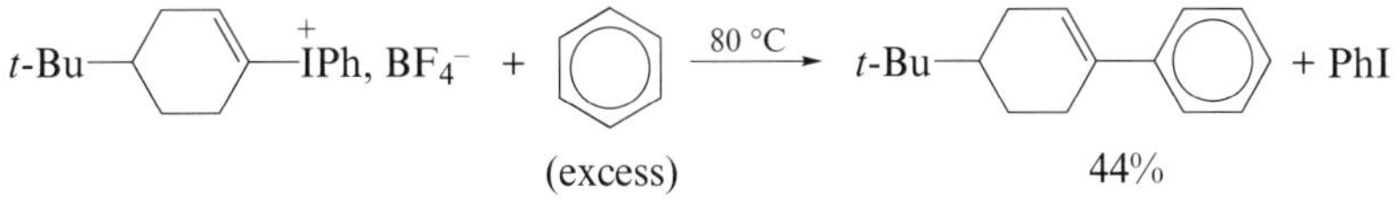

(263)

The intervention of intermediate vinyl cations in these reactions is thought to be unlikely[127]. Thus, the polarity of the solvent [i.e. CHCl$_3$, d_6-acetone, CD$_3$CN, CD$_3$OD] has little effect on the efficiency of dihydronaphthalene production from the (E)-[β-phenylethyl)ethenyl] iodonium salt. Furthermore, a Z:E (88:12) mixture of 2-methyl-4-

phenyl-1-butenyl(phenyl)iodonium tetrafluoroborate (R^1 = H, R^2 = Me) gives only a 7% yield of the expected dihydronaphthalene under conditions in which a 3:97 $Z{:}E$ mixture gives a 63% yield. An addition–elimination mechanism for cyclization of the (E)-isomer has been proposed (equation 264)[127].

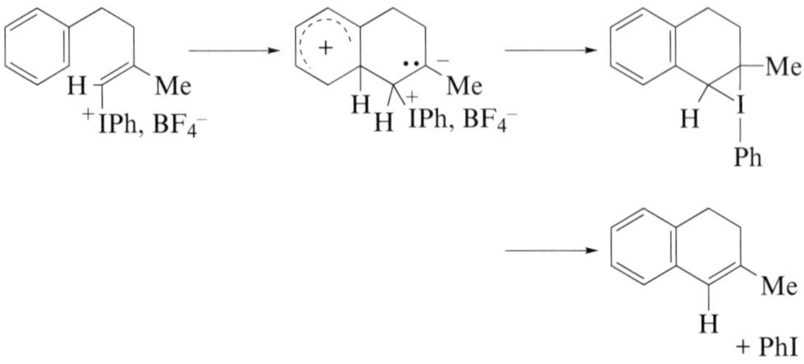

$$(264)$$

IV. ACKNOWLEDGMENTS

The author wishes to thank Cheryl A. O'Dear and Rita M. Sabo for typing the manuscript and Michael W. Justik for providing computer drawings of equations and structures for the manuscript.

V. REFERENCES

1. G. F. Koser, 'Halonium ions', in *The Chemistry of Functional Groups, Supplement D* (Eds. S. Patai and Z. Rappoport), Chap. 25, Wiley, Chichester, 1983, p. 1265.
2. M. Ochiai, *Reviews on Heteroatom Chemistry*, **2**, 92 (1989).
3. P. J. Stang, *Angew. Chem., Int. Ed. Engl.*, **31**, 274 (1992).
4. A. Varvoglis. *The Organic Chemistry of Polycoordinated Iodine*, Chap. 5, VCH Publishers, Inc. New York, 1992.
5. P. J. Stang and B. W. Surber, *J. Am. Chem. Soc.*, **107**, 1452 (1985).
6. P. J. Stang, B. W. Surber, Z-C. Chen, K. A. Roberts and A. G. Anderson, *J. Am. Chem. Soc.*, **109**, 228 (1987).
7. T. Kitamura and P. J. Stang, *J. Org. Chem.*, **53**, 4105 (1988).
8. L. Rebrovic and G. F. Koser, *J. Org. Chem.*, **49**, 4700 (1984).
9. A. J. Margida and G. F. Koser, *J. Org. Chem.*, **49**, 4703 (1984).
10. J. S. Lodaya and G. F. Koser, *J. Org. Chem.*, **55**, 1513 (1990).
11. G. F. Koser, L. Rebrovic and R. H. Wettach, *J. Org. Chem.*, **46**, 4324 (1981).
12. E. Kotali, A. Varvoglis and A. Bozopoulos, *J. Chem., Soc., Perkin Trans. 1*, 827 (1989).
13. P. J. Stang, A. M. Arif and C. M. Crittell, *Angew. Chem., Int. Ed. Engl.*, **29**, 287 (1990).
14. D. R. Fischer, B. L. Williamson and P. J. Stang, *Synlett*, 535 (1992).
15. V. V. Zhdankin, M. C. Scheuller and P. J. Stang, *Tetrahedron Lett.*, **34**, 6853 (1993).
16. M. D. Bachi, N. Bar-Ner, C. M. Crittell, P. J. Stang and B. L. Williamson, *J. Org. Chem.*, **56**, 3912 (1991).
17. B. L. Williamson, P. J. Stang and A. M. Arif, *J. Am. Chem. Soc.*, **115**, 2590 (1993).
18. P. J. Stang, B. L. Williamson and V. V. Zhdankin, *J. Am. Chem. Soc.*, **113**, 5870 (1991).
19. R. R. Tykwinski, B. L. Williamson, D. R. Fischer, P. J. Stang and A. M. Arif, *J. Org. Chem.*, **58**, 5235 (1993).
20. R. R. Tykwinski, J. A. Whiteford and P. J. Stang, *J. Chem. Soc., Chem. Commun.*, 1800 (1993).
21. P. J. Stang and C. M. Crittell, *Organometallics*, **9**, 3191 (1990).
22. G. F. Koser, X. Chen, K. Chen and G. Sun, *Tetrahedron Lett.*, **34**, 779 (1993).
23. P. J. Stang, T. Kitamura, M. Boehshar and H. Wingert, *J. Am. Chem. Soc.*, **111**, 2225 (1989).

24. M. Ochiai, M. Kunishima, K. Fuji, Y. Nagao and M. Shiro, *Chem. Pharm. Bull.*, **37**, 1948 (1989).
25. M. Kunishima, 'Synthesis of Hypervalent Alkynyl Iodine Compounds and Studies on Their Utility as Electrophilic Reagents', Ph.D. Dissertation (Japanese), Institute for Chemical Research, Kyoto University, 1989.
26. F. M. Beringer and S. A. Galton, *J. Org. Chem.*, **30**, 1930 (1965).
27. M. Ochiai, T. Ito, Y. Takaoka, Y. Masaki, M. Kunishima, S. Tani and Y. Nagao, *J. Chem. Soc., Chem. Commun.*, 118 (1990).
28. M. Ochiai, M. Kunishima, Y. Nagao, K. Fuji, M. Shiro and E. Fujita, *J. Am. Chem. Soc.*, **108**, 8281 (1986).
29. V. V. Zhdankin, R. Tykwinski, R. Caple, B. Berglund, A. S. Koz'min and N. S. Zefirov, *Tetrahedron Lett.*, **29**, 3717 (1988).
30. M. Ochiai, M. Kunishima, K. Fuji and Y. Nagao, *J. Org. Chem.*, **54**, 4038 (1989).
31. M. Ochiai, M. Kunishima, K. Sumi, Y. Nagao, E. Fujita, M. Arimoto and H. Yamaguchi, *Tetrahedron Lett.*, **26**, 4501 (1985).
32. M. Ochiai, M. Kunishima, S. Tani and Y. Nagao, *J. Am. Chem. Soc.*, **113**, 3135 (1991).
33. B. L. Williamson, R. R. Tykwinski and P. J. Stang, *J. Am. Chem. Soc.*, **116**, 93 (1994).
34. P. J. Stang and J. Ullmann, *Synthesis*, 1073 (1991).
35. P. J. Stang, V. V. Zhdankin and A. M. Arif, *J. Am. Chem. Soc.*, **113**, 8997 (1991).
36. M. Ochiai, Y. Masaki and M. Shiro, *J. Org. Chem.*, **56**, 5511 (1991).
37. G. F. Koser, G. Sun, C. W. Porter and W. J. Youngs, *J. Org. Chem.*, **58**, 7310 (1993).
38. T. Kitamura, R. Furuki, L. Zheng, T. Fujimoto and H. Taniguchi, *Chem. Lett.*, 2241 (1992).
39. T. Kitamura, L. Zheng, H. Taniguchi, M. Sakurai and R. Tanaka, *Tetrahedron Lett.*, **34**, 4055 (1993).
40. P. J. Stang and V. V. Zhdankin, *J. Am. Chem. Soc.*, **112**, 6437 (1990).
41. P. J. Stang and V. V. Zhdankin, *J. Am. Chem. Soc.*, **113**, 4571 (1991).
42. P. J. Stang, R. Tykwinski and V. V. Zhdankin, *J. Org. Chem.*, **57**, 1861 (1992).
43. R. R. Tykwinski and P. J. Stang, *Tetrahedron*, **49**, 3043 (1993).
44. E. B. Merkushev, L. G. Karpitskaya and G. I. Novosel'tseva, *Dokl. Akad. Nauk SSSR (Engl. Transl.)*, **245**, 140 (1979).
45. P. J. Stang and R. Tykwinski, *J. Am. Chem. Soc.*, **114**, 4411 (1992).
46. R. M. Moriarty, R. K. Vaid and G. F. Koser, *Synlett*, 365 (1990).
47. H. J. Lucas and E. R. Kennedy, *Org. Synth., Coll. Vol. III*, 482 (1955).
48. H. J. Lucas, E. R. Kennedy and M. W. Formo, *Org. Synth., Coll. Vol. III*, 483 (1955).
49. H. Saltzman and J. G. Sharefkin, *Org. Synth., Coll. Vol. V*, 658 (1973).
50. C. Willgerodt, *Chem. Ber.*, **25**, 3494 (1892); see p. 3498.
51. K. H. Pausacker, *J. Chem. Soc.*, 107 (1953).
52. J. G. Sharefkin and H. Saltzman, *Org. Synth., Coll. Vol. V*, 660 (1973).
53. M. R. Almond, J. B. Stimmel, E. A. Thompson and G. M. Loudon, *Org. Synth., Coll. Vol. VIII*, 132 (1993); see p. 135.
54. S. Spyroudis and A. Varvoglis, *Synthesis*, 445 (1975); see p. 446.
55. O. Neiland and B. Karele, *J. Org. Chem. USSR (Engl. Transl.)*, **6**, 889 (1970).
56. G. F. Koser and R. H. Wettach, *J. Org. Chem.*, **42**, 1476 (1977); see p. 1477.
57. N. S. Zefirov, V. V. Zhdankin, Yu. V. Dan'kov, A. S. Koz'min and O. S. Chizov, *J. Org. Chem. USSR (Engl. Transl.)*, **21**, 2252 (1985); see p. 2252.
58. R. T. Hembre, C. P. Scott and J. R. Norton, *J. Org. Chem.*, **52**, 3650 (1987); see p. 3653.
59. N. S. Zefirov, V. V. Zhdankin, Yu. V. Dan'kov and A. S. Koz'min, *J. Org. Chem. USSR (Engl. Transl.)*, **20**, 401 (1984); see p. 401 (first reported preparation).
60. V. V. Zhdankin, C. M. Crittell, P. J. Stang and N. S. Zefirov, *Tetrahedron Lett.*, **31**, 4821 (1990); see p. 4822.
61. J. R. Dalziel, H. A. Carter and F. Aubke, *Inorg. Chem.*, **15**, 1247 (1976), see p. 1247.
62. P. J. Stang, V. V. Zhdankin, R. Tykwinski and N. S. Zefirov, *Tetrahedron Lett.*, **33**, 1419 (1992); see pp. 1419 and 1422.
63. T. Kitamura, R. Furuki, K. Nagata, L. Zheng and H. Taniguchi, *Synlett*, 193 (1993); see p. 194.
64. G. P. Baker, F. G. Mann, N. Sheppard and A. J. Tetlow, *J. Chem. Soc.*, 3721 (1965); see p. 3726.
65. J. C. Martin, *Science*, **221**, 509 (1983).
66. G. F. Koser, 'Hypervalent halogen compounds', in *The Chemistry of Functional Groups, Supplement D* (Eds. S. Patai and Z. Rappoport), Wiley, Chichester, Chap. 18, 1983, p. 721.
67. C-K. Lee, T. C. W. Mak, W-K. Li and J. F. Kirner, *Acta Crystallogr.*, **B33**, 1620 (1977).

68. N. W. Alcock, R. M. Countryman, S. Esperas and J. F. Sawyer, *J. Chem. Soc., Dalton Trans.*, 854 (1979).
69. G. F. Koser, R. H. Wettach, J. M. Troup and B. A. Frenz, *J. Org. Chem.*, **41**, 3609 (1976).
70. B. C. Schardt and C. L. Hill, *Inorg. Chem.*, **22**, 1563 (1983).
71. H. Siebert and M. Handrich, *Z. Anorg. Allg. Chem.*, **426**, 173 (1976).
72. N. W. Alcock and R. M. Countryman, *J. Chem. Soc., Dalton Trans.*, 851 (1979).
73. J. Gallos, A. Varvoglis and N. W. Alcock *J. Chem. Soc., Perkin Trans. 1*, 757 (1985).
74. G. W. Bushnell, A. Fischer and P. N. Ibrahim, *J. Chem. Soc., Perkin Trans. 2*, 1281 (1988).
75. H. J. Lucas and E. R. Kennedy, *Org. Synth., Coll. Vol. III*, 485 (1955).
76. J. B. Lambert, G-t. Wang, R. B. Finzel and D. H. Teramura, *J. Am. Chem. Soc.*, **109**, 7838 (1987).
77. J. B. Lambert, G-t. Wang and D. H. Teramura, *J. Org. Chem.*, **53**, 5422 (1988).
78. T. Kitamura, R. Furuki and H. Taniguchi and P. J. Stang, *Tetrahedron*, **48**, 7149 (1992).
79. M. Ochiai, M. Kunishima, K. Fuji, M. Shiro and Y. Nagao, *J. Chem. Soc., Chem. Commun.*, 1076 (1988).
80. M. Ochiai, M. Kunishima, Y. Nagao, K. Fuji and E. Fujita, *J. Chem. Soc., Chem. Commun.*, 1708 (1987).
81. P. J. Stang, M. Boehshar, H. Wingert and T. Kitamura, *J. Am. Chem. Soc.*, **110**, 3272 (1988).
82. T. T. Nguyen and J. C. Martin, 'Heterocyclic Rings containing Halogens', in *Comprehensive Heterocyclic Chemistry*, Vol.1, Pergamon Press, New York, 1984, p. 563.
83. R. L. Amey and J. C. Martin, *J. Org. Chem.*, **44**, 1779 (1979).
84. M. Ochiai, K. Oshima and Y. Masaki, *Tetrahedron Lett.*, **32**, 7711 (1991).
85. Z-D. Liu and Z-C. Chen, *Synth. Commun.*, **22**, 1997 (1992).
86. R. R. Tykwinski, P. J. Stang and N. E. Persky, *Tetrahedron Lett.*, **35**, 23 (1994).
87. B. L. Williamson, P. Murch, D. R. Fischer and P. J. Stang, *Synlett*, 858 (1993).
88. Z-D. Liu and Z-C. Chen, *J. Org. Chem.*, **58**, 1924 (1993).
89. T. Kitamura, T. Tanaka and H. Taniguchi, *Chem. Lett.*, 2245 (1992).
90. P. J. Stang and C. M. Crittell, *J. Org. Chem.*, **57**, 4305 (1992).
91. K. K. Laali, M. Regitz, M. Birkel, P. J. Stang and C. M. Crittell, *J. Org. Chem.*, **58**, 4105 (1993).
92. A. Schmidpeter, P. Mayer, J. Stocker, K. A. Roberts and P. J. Stang, *Heteroatom Chem.*, **2**, 569 (1991).
93. K. M. Lancer and G. H. Wiegand, *J. Org. Chem.*, **41**, 3360 (1976).
94. P. J. Stang, M. Boehshar and J. Lin, *J. Am. Chem. Soc.*, **108**, 7832 (1986).
95. P. J. Stang, H. Wingert and A. M. Arif, *J. Am. Chem. Soc.*, **109**, 7235 (1987).
96. Y. Tamura, T. Yakura, J. Haruta and Y. Kita, *Tetrahedron Lett.*, **26**, 3837 (1985).
97. Y. Kita, T. Yakura, H. Terashi, J. Haruta and Y. Tamura, *Chem. Pharm. Bull.*, **37**, 891 (1989).
98. Y. Kita, H. Tohma and T. Yakura, *Trends in Organic Chemistry*, **3**, 113 (1992).
99. T. Kitamura and P. J. Stang, *Tetrahedron Lett.*, **29**, 1887 (1988).
100. P. J. Stang, *Chem. Rev.*, **78**, 383 (1978).
101. P. J. Stang, D. P. Fox, C. J. Collins and C. R. Watson, Jr., *J. Org. Chem.*, **43**, 364 (1978).
102. M. Ochiai, M. Kunishima, K. Fuji and Y. Nagao, *J. Org. Chem.*, **53**, 6144 (1988).
103. M. Ochiai, K. Uemura, K. Oshima, Y. Masaki, M. Kunishima and S. Tani, *Tetrahedron Lett.*, **32**, 4753 (1991).
104. M. Ochiai, K. Uemura and Y. Masaki, *J. Am. Chem. Soc.*, **115**, 2528 (1993).
105. M. D. Bachi, N. Bar-Ner, P. J. Stang and B. L. Williamson, *J. Org. Chem.*, **58**, 7923 (1993).
106. F. M. Beringer and L. L. Chang, *J. Org. Chem.*, **37**, 1516 (1972).
107. P. J. Stang and T. Kitamura, *J. Am. Chem. Soc.*, **109**, 7561 (1987).
108. P. J. Stang and T. Kitamura, *Org. Synth.*, **70**, 215 (1991).
109. T. Kitamura, T. Tanaka, H. Taniguchi and P. J. Stang, *J. Chem. Soc., Perkin Trans. 1*, 2892 (1991).
110. T. Kitamura, I. Mihara, H. Taniguchi and P. J. Stang, *J. Chem. Soc., Chem. Commun.*, 614 (1990).
111. G. Maas, M. Regitz, U. Moll, R. Rahm, F. Krebs. R. Hector, P. J. Stang, C. M. Critell and B. L. Williamson, *Tetrahedron*, **48**, 3527 (1992).
112. J. Burgess, M. E. Howden, R. D. W. Kemmitt and N. S. Sridhara, *J. Chem. Soc., Dalton Trans.*, 1577 (1978).
113. P. J. Stang, C. M. Crittell and A. M. Arif, *Organometallics*, **12**, 4799 (1993).
114. P. M. Gallop, M. A. Paz, R. Flückiger, P. J. Stang, V. V. Zhdankin and R. R. Tykwinski, *J. Am. Chem. Soc.*, 115, 11702 (1993).

115. C. Willgerodt, *Die Organischen Verbindungen mit Mehrwertigem Jod*, F. Enke, Stuttgart, 1914.
116. R. Kh. Freidlina, E. M. Brainina and A. N. Nesmeyanov, *Bull. Acad. Sci. U.R.S.S., Classe Sci. Chim.*, 647 (1945); *Chem. Abstr.*, **40**, 4686–4687 (1946).
117. E. M. Brainina and R. Kh. Freidlina, *Bull. Acad. Sci. U.R.S.S., Classe Sci. Chim.*, 623 (1947); *Chem. Abstr.*, **42**, 5863a (1948).
118. E. L. Colichman and J. T. Matschiner, *J. Org. Chem.*, **18**, 1124 (1953).
119. A. N. Nesmeyanov, T. P. Tolstaya, N. F. Sokolova, V. N. Varfolomeeva and A. V. Petrakov, *Dokl. Akad. Nauk SSSR (Engl. Transl.)*, **198**, 386 (1971).
120. A. N. Nesmeyanov, T. P. Tolstaya and A. V. Petrakov, *Dokl. Akad. Nauk SSSR (Engl. Transl.)*, **197**, 343 (1971).
121. A. N. Nesmeyanov, T. P. Tolstaya, A. V. Petrakov and A. N. Gol'tsev, *Dokl. Akad. Nauk SSSR (Engl. Transl.)*, **235**, 425 (1977).
122. A. N. Nesmeyanov, T. P. Tolstaya, A. V. Petrakov and I. F. Leshcheva, *Dokl. Akad. Nauk. SSSR (Engl. Transl.)*, **238**, 70 (1978).
123. F. M. Beringer, P. Ganis, G. Avitabile and H. Jaffe, *J. Org. Chem.*, **37**, 879 (1972).
124. V. R. Sandel, G. R. Buske, S. G. Maroldo, D. K. Bates, D. Whitman and G. Sypniewski, *J. Org. Chem.*, **46**, 4069 (1981).
125. M. Ochiai, K. Sumi, Y. Nagao and E. Fujita, *Tetrahedron Lett.*, **26**, 2351 (1985).
126. M. Ochiai, K. Sumi, Y. Takaoka, M. Kunishima, Y. Nagao, M. Shiro and E. Fujita, *Tetrahedron*, **44**, 4095 (1988).
127. M. Ochiai, Y. Takaoka, K. Sumi and Y. Nagao, *J. Chem. Soc., Chem. Commun.*, 1382 (1986).
128. M. Ochiai, Y. Takaoka and Y. Nagao, *J. Am. Chem. Soc.*, **110**, 6565 (1988).
129. M. Ochiai, K. Oshima and Y. Masaki, *J. Chem. Soc., Chem. Commun.*, 869 (1991).
130. R. J. Hinkle and P. J. Stang, *Synthesis*, 313 (1994).
131. P. Stang and J. Ullmann, *Angew. Chem., Int. Ed. Engl.*, **30**, 1469 (1991).
132. R. M. Moriarty, W. R. Epa and A. K. Awasthi, *J. Am. Chem. Soc.*, **113**, 6315 (1991).
133. T. Kitamura, R. Furuki, H. Taniguchi and P. J. Stang, *Tetrahedron Lett.*, **31**, 703 (1990).
134. T. Kitamura, R. Furuki, H. Taniguchi and P. J. Stang, *Mendeleev Commun.*, 148 (1991).
135. M. Ochiai, Y. Takaoka, Y. Masaki, M. Inenaga and Y. Nagao, *Tetrahedron Lett.*, **30**, 6701 (1989).
136. T. Umemoto and Y. Gotoh, *Bull. Chem. Soc. Jpn.*, **60**, 3307 (1987).
137. M. Ochiai, T. Ito, Y. Takaoka and Y. Masaki, *J. Am. Chem. Soc.*, **113**, 1319 (1991).
138. L. Rebrovic and G. F. Koser, *J. Org. Chem.*, **49**, 2462 (1984).
139. Ya. N. Kreitsberga and O. Ya. Neiland, *Khim. Geterotsikl. Soedin. (Engl. Transl.)*, 746 (1989).
140. N. S. Zefirov, A. S. Koz'min, T. Kasumov, K. A. Potekhin, V. D. Sorokin, V. K. Brel, E. V. Abramkin, Yu. T. Struchkov, V. V. Zhdankin and P. J. Stang, *J. Org. Chem.*, **57**, 2433 (1992).
141. F. M. Beringer and E. M. Gindler, *Iodine Abstr. Rev.*, **3**, 1956.
142. A. N. Chekhlov, R. G. Gafurov, I. A. Pomytkin, I. V. Martynov, N. N. Aleinikov, S. A. Kashtanov and F. I. Dubovitskii, *Dokl. Akad. Nauk SSSR (Engl. Transl.)*, **291**, 1040 (1986).
143. G. Modena, *Acc. Chem. Res.*, **4**, 73 (1971).
144. Y. Apeloig and Z. Rappoport, *J. Am. Chem. Soc.*, **101**, 5095 (1979).
145. Z. Rappoport, *Acc. Chem. Res.*, **14**, 7 (1981).
146. M. Ochiai, K. Oshima and Y. Masaki, *J. Am. Chem. Soc.*, **113**, 7059 (1991).
147. M. Ochiai, Y. Takaoka, Y. Masaki, Y. Nagao and M. Shiro, *J. Am. Chem. Soc.*, **112**, 5677 (1990).
148. P. J. Stang, T. Blume and V. V. Zhdankin, *Synthesis*, 35 (1993).
149. F. M. Beringer and R. A. Nathan, *J. Org. Chem.*, **34**, 685 (1969).
150. F. M. Beringer and R. A. Nathan, *J. Org. Chem.*, **35**, 2095 (1970).
151. A. Schoenberg, I. Bartoletti and R. F. Heck, *J. Org. Chem.*, **39**, 3318 (1974).
152. H. A. Dieck and R. F. Heck, *J. Org. Chem.*, **40**, 1083 (1975).
153. J. I. Kim, B. A. Patel and R. F. Heck, *J. Org. Chem.*, **46**, 1067 (1981).
154. R. F. Heck, *Acc. Chem. Res.*, **12**, 146 (1979).
155. R. F. Heck, *Org. React.*, **27**, 345 (1982).
156. R. F. Heck, and J. P. Nolley, Jr., *J. Org. Chem.*, **37**, 2320 (1972).
157. R. J. Hinkle, G. T. Poulter and P. J. Stang, *J. Am. Chem. Soc.*, **115**, 11626 (1993).

NOTE ADDED IN PROOF

Professor Masahito Ochiai of The University of Tokushima in Japan has informed the author of the following unpublished results[158].

(1) Phenyl(propynyl)iodonium tetrafluoroborate reacts with potassium 2-phenyl-1,3-indandionate in tetrahydrofuran to give the alkynyl and spiroannulated indandiones shown in equation 265. With sodium 2-ethyl-1,3-indandionate in *tert*-butyl alcohol, a vinyl ether is produced (equation 266). These observations are relevant to the general comments in section II.D.7 concerning the fate of **26** and related alkylidenecarbenes in reactions of [5 + 0]enolates with alkyliodonium ions containing short (≤ three carbons) alkyl chains.

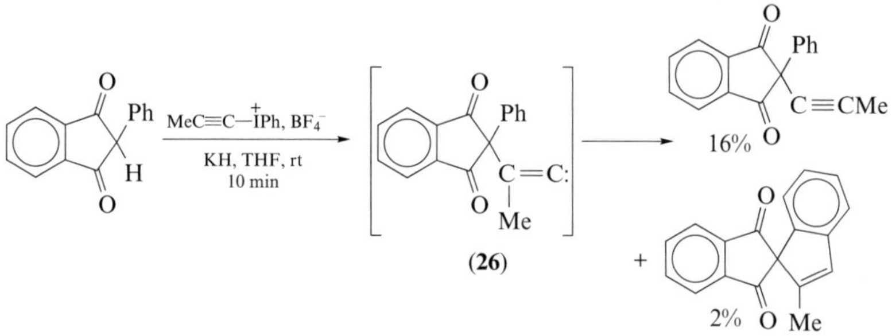

$$(265)$$

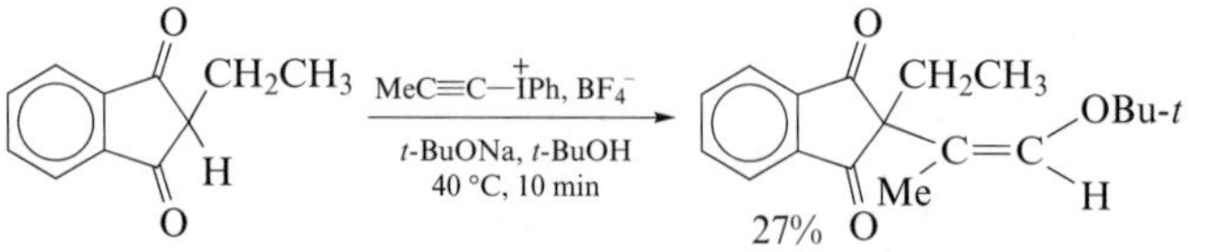

$$(266)$$

(2) The treatment of the thermodynamic enolate of 2-methylcyclohexanone with ethynyl(phenyl)iodonium tetrafluoroborate in tetrahydrofuran affords 2-ethynyl-2-methylcyclohexanone (equation 267). This is the only example known to this author of the alkynylation of an unactivated monocarbonyl compound with an alkynyliodonium salt. However, the earlier conclusion that simple enolates do not undergo alkynylation with alkynyliodonium salts (section II.D.7 and II.G) needs to be revvised.

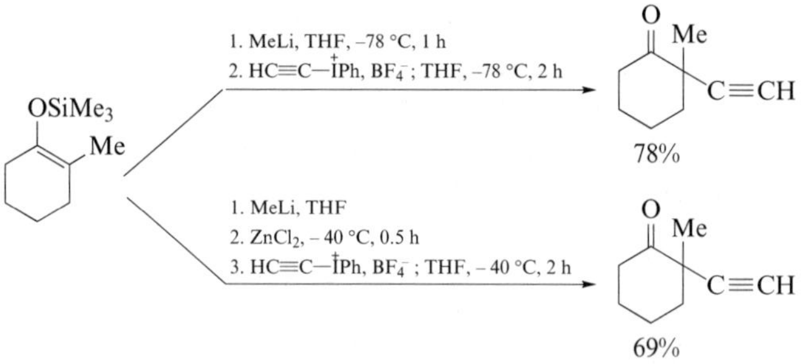

$$(267)$$

(3) (Z)-2-Fluoro-1-hexadecenyl(phenyl)iodonium chloride has been prepared in 32% yield by the treatment of 1-hexadecynyl(phenyl)iodonium chloride with cesium fluoride in aqueous acetone (see section III.B.4).

REFERENCE

158. Personal communication from M. Ochiai to G. F. Koser; letter dated August 19, 1994.

CHAPTER **22**

Reactions of organic halides mediated by transition metal compounds

JAMES R. GREEN

Department of Chemistry and Biochemistry, University of Windsor, Windsor, Ontario N9B 3P4, Canada

Supplement D2: The chemistry of halides, pseudo-halides and azides
Edited by S. Patai and Z. Rappoport   © 1995 John Wiley & Sons Ltd

I. INTRODUCTION

The scope of this chapter is based on the analogous one by Naso and Marchese published in 1983[1]. Since the time of their review, the focus of work in this area has shifted somewhat, and sections have been added or omitted to reflect that focus. New sections covering chromium(II)-mediated reactions of halides and covering cobalt-mediated radical reactions of halides have been added. In view of the relatively mature nature of the areas, sections dealing with π-allylnickel complexes, iron oxyallyl cations and cyanation reactions have been omitted.

In order to keep the chapter to a manageable size, rather strict definitions of what constitutes a halide and a transition metal have been chosen; as a result, neither f-block chemistry nor organic sulfonate chemistry (with a few asides as exceptions) are included.

Sections II and III of this chapter deal with cross-coupling reactions between organometallics and alkyl halides, mediated by organocopper compounds and transition metal complexes, respectively, Section IV deals with the nickel-induced reductive coupling between two halide compounds. The reactions of alkenes and alkynes with organic halides mediated by palladium compounds form Section V. This is followed by two new sections dealing with the reaction of chromium(II) with organic halides and the reactions of the subsequently produced organochromium(III) compounds (Section VI), and with the reactions of cobalt(I) complexes with organic halides and the radical reactions of the resulting organocobalt(III) complexes (Section VII). A section dealing with carbonylation reactions (Section VIII) concludes the chapter.

II. REACTIONS OF ORGANOCOPPER(I) REAGENTS

A. Introduction

The focus of research in copper-based carbon–carbon bond-forming reactions has shifted in the past decade. Higher-order (h.o.) cuprates have been studied in some detail, whereas progress in lower-order (Gilman) cuprates has slowed. Other areas which have been particularly active include the generation of organocopper species by the reaction of highly active copper powder ('Rieke copper') and the use of zinc cyanocuprates, which are generated from organic halides and zinc powder followed by cuprous cyanide addition.

B. Cross-coupling Reactions with Alkyl Halides

The addition of two equivalents of alkyllithium reagents to CuCN generates h.o. cuprates of the type $RR'Cu(CN)Li_2$, which have proved to be highly suitable reagents for reactions with a variety of organic halides[2,3]. Like lower-order (Gilman) cuprates, h.o. cyanocuprates give straightforward nucleophilic substitution of primary alkyl halides; it is clear, however, that they are more reactive than their lower-order cuprate counterparts (equation 1). This manifests itself in much greater success in reactions with secondary halides; iodides generally react well, and bromides are usually successfully coupled. With primary halide substrates, the bromides are the superior substrates, as the iodides sometimes give competing reductions. While primary chlorides are also often suitable partners for higher-order cuprates, their reactivity is clearly less than that of the corresponding bromides or iodides. As a result, clean monosubstitution of the bromide is possible of a mixed dihaloalkane. Many functional groups (nitrile, ester, acetal, ether, primary or secondary amine) may be present in the organic halide. Also noteworthy is the path of nucleophilic attack on secondary halides. For both lower- and higher-order cuprates, attack on secondary bromides results in clean inversion of configuration, whereas the iodide counterparts are attacked with total or near-total racemization[4].

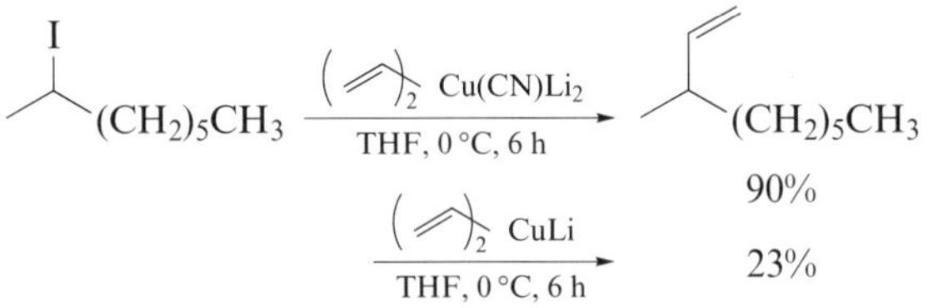

$$(1)$$

There is fairly good versatility in the nature of the ligand transferred from h.o. cuprates. Alkyl and vinyl groups may all be readily transferred from the h.o. cuprate in nearly all cases, whereas allyl, aryl or heteroaryl groups may be transferred to primary halides in good yield[5]. In substituted allyl h.o. cuprates, allyl transfer occurs predominantly with allylic transposition (equation 2)[6]. Although the 2-thienyl group is among the heteroaryl groups transferred, its reactivity is sufficiently moderated that it functions well as a non-transferred 'dummy' ligand in $R(2\text{-thienyl}) Cu(CN)Li$ cuprates (equation 3)[7].

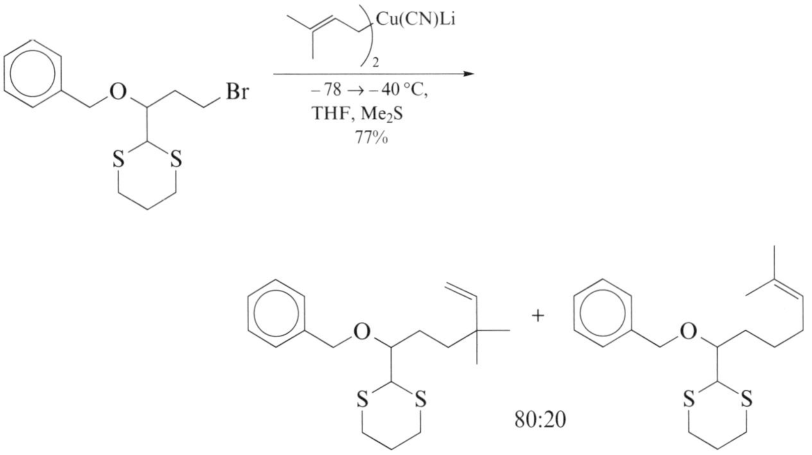

$$(2)$$

80:20

$$\text{Br}\diagup\diagdown\diagup\text{CN} \xrightarrow[\text{THF}, -78 \to 0\,°\text{C}]{s\text{-Bu}(2\text{-Thi})\text{Cu}(\text{CN})\text{Li}_2} \diagup\diagdown\diagup\diagdown\text{CN} \qquad (3)$$

$$82\%$$

$$2\text{-Thi} = \underset{S}{\diagdown\!\!\diagup}$$

An alternative to the use of the 2-thienyl group as a nontransferable ligand in h.o. cyanocuprates is the anion of DMSO (1); efficient alkylation of primary iodides at $-78\,°\text{C}$

$$\text{Li}_2(\text{CH}_3\text{SOCH}_2\text{Cu}(\text{CN})\text{R})$$

$$(1)$$

has been reported using these reagents, but comparisons with other higher-order cuprates on more difficult substrates has not been reported[8]. N-Lithiopyrrole and N-lithioimidazole can also function as nontransferable ligands, at least in selected cases[8].

Higher-order cuprates have also been made from Grignard reagents and from organozinc species; the resulting cuprate reactivity is less for mixed Li/Mg-based systems and mixed Mg/Zn-based ones, although alkylation of primary alkyl iodides may still be accomplished in these cases[2,9].

Alkyl iodides and benzylic bromides may be alkylated efficiently by another h.o. cuprate-mediated method[9]. The addition of dialkylzincs to $\text{Me}_2\text{Cu}(\text{CN})(\text{MgCl})_2$ provides an organometallic intermediate which has tentatively been assigned as $\text{R}_2\text{Cu}(\text{CN})(\text{MgX})_2{\cdot}\text{Me}_2\text{Zn}$ (equation 4); regardless of the actual solution stucture of this reagent, it is nevertheless excellent for alkylation, and tolerates significant levels of functionality in both the organometallic species (esters, nitriles, triflamide) and in the iodoalkane (ester, nitrile, triflamide, nitro, terminal alkyne). The procedure may also be altered to incorporate secondary alkylcopper nucleophiles. The amount of competing methyl transfer to the iodoalkane substrate is negligible.

$$\left(\text{AcO}\diagup\diagdown\diagup\diagdown\right)_2 \text{Zn} + \text{Me}_2\text{Cu}(\text{CN})(\text{MgCl})_2$$

$$\xrightarrow[-50 \to 0\,°\text{C}]{\text{THF}} \left(\text{AcO}\diagup\diagdown\diagup\diagdown\right)_2 \text{Cu}(\text{CN})(\text{MgX})_2 \cdot \text{Me}_2\text{Zn}$$

$$\xrightarrow[\underset{\text{Ph}}{\diagup\!\diagdown\text{NO}_2}^{\text{I}}]{\text{DMPU}, 0\,°\text{C}} \underset{\underset{\text{Ph}}{\diagup\text{NO}_2}}{\diagup\diagdown\diagup\diagdown\diagup\text{OAc}} \qquad 83\% \qquad (4)$$

DMPU = N,N′–Dimethylpropyleneurea

Alkyl halides (particularly bromides) undergo oxidative addition with activated copper powder, prepared from Cu(I) salts with lithium naphthalenide, to give alkylcopper species[10]. The alkyl halides may be functionalized with ester, nitrile and chloro functions; ketone and epoxide functions may also be tolerated in some cases[11]. The resulting alkylcopper species have been shown to react efficiently with acid chlorides, enones (conjugate addition) and (less efficiently) with primary alkyl iodides and allylic and benzylic bromides (equations 5 and 6). If a suitable ring size can be made, intramolecular reactions with epoxides and ketones are realized.

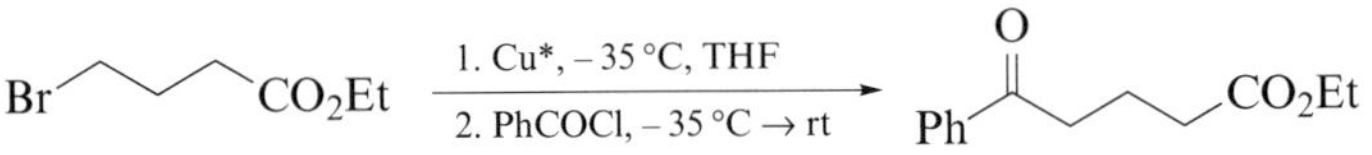

$$\text{Br}\diagup\diagdown\diagup\text{CO}_2\text{Et} \quad \xrightarrow[\text{2. PhCOCl, } -35\,°C \to rt]{\text{1. Cu*, } -35\,°C, \text{ THF}} \quad \underset{\text{Ph}}{\overset{O}{\|}}\diagup\diagdown\diagup\text{CO}_2\text{Et} \quad (5)$$

Cu* Prepared from CuI/PPh$_3$ + Li Naphthalenide   81% isolated

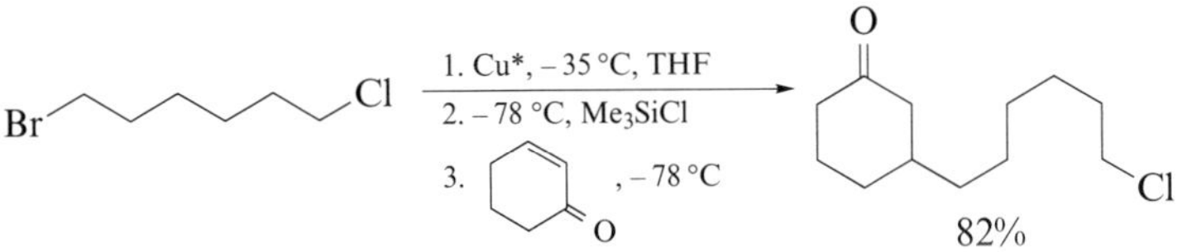

$$\text{Br}\diagup\diagdown\diagup\diagdown\diagup\text{Cl} \quad \xrightarrow[\substack{\text{2. } -78\,°C, \text{ Me}_3\text{SiCl} \\ 3. \text{ (cyclohexenone)}, -78\,°C}]{\text{1. Cu*, } -35\,°C, \text{ THF}} \quad (6)$$

82%

Surprising results have been reported in the chemistry of perfluoroalkylcopper species. Trifluoromethylcopper, prepared from metathesis of the corresponding zinc or cadmium compounds with Cu(I) salts, undergoes dimerization to give pentafluoroethylcopper[12]. This CF$_2$ insertion reaction also occurs with at least some other perfluororganocopper species, as pentafluorophenylcopper undergoes double CF$_2$ insertion in the presence of CF$_3$Cu (equations 7 and 8)[13]. The resultant perfluoroalkylcoppers have been shown to undergo coupling in respectable yields with aryl, vinyl and allyl halides (equation 9).

$$\text{CF}_3\text{MX} + (\text{CF}_3)_2\text{M} \quad \xrightarrow[\text{DMF}]{\text{CuY, } -80\,°C \to rt} \quad \underset{90\text{--}100\%}{\text{F}_3\text{CCu}} \quad \longrightarrow \quad \text{F}_3\text{C}{-}\text{CF}_2{-}\text{Cu} \quad (7)$$

$$\text{CF}_3\text{Cu} \quad \xrightarrow[\text{DMF}]{\text{F}_5\text{C}_6{-}\text{Cu}} \quad \underset{70\text{--}80\%}{\text{F}_5\text{C}_6{-}\text{CF}_2{-}\text{CF}_2{-}\text{Cu}} \quad (8)$$

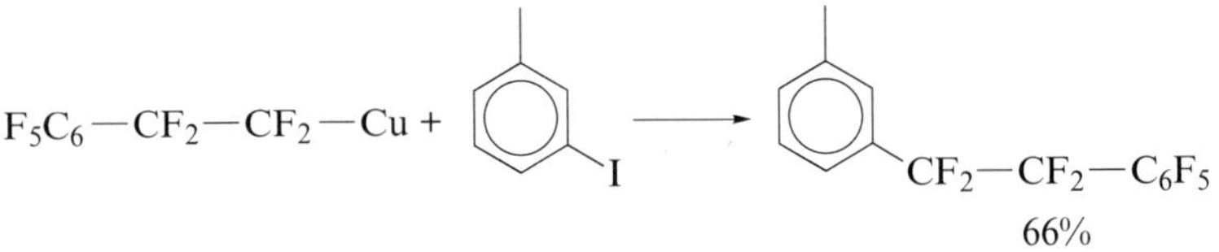

$$\text{F}_5\text{C}_6{-}\text{CF}_2{-}\text{CF}_2{-}\text{Cu} + \text{(aryl iodide)} \quad \longrightarrow \quad \underset{66\%}{\text{(aryl–CF}_2{-}\text{CF}_2{-}\text{C}_6\text{F}_5)} \quad (9)$$

Work in the reactions of alkyl halides with the more conventional Gilman cuprates has slackened considerably. Synthetically, alkylations have been performed employing alkynes as the second, disposable ligand in cuprates. In the transfer of vinyl groups, this allows aminomethylation or thiomethylation without the addition of HMPA or triethyl phosphite and without loss of alkene geometric integrity (equation 10)[14]. The addition of pentynylcopper to α-lithiated formamidines allows alkylation with primary alkyl iodides (equation 11)[15].

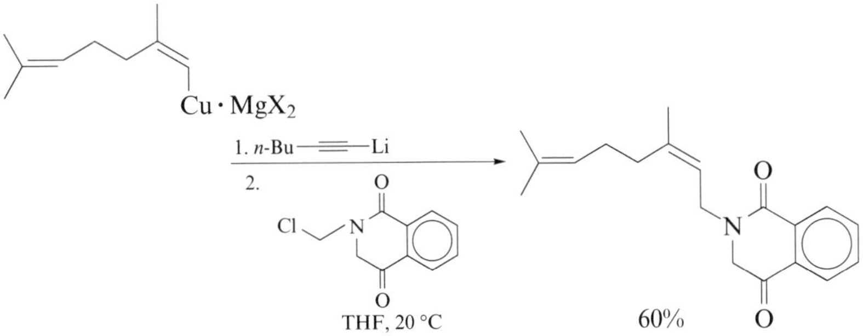

$$\xrightarrow[\substack{\text{1. } n\text{-Bu}{-}{\equiv}{-}\text{Li} \\ 2. \\ \text{THF, 20 }°C}]{\text{Cu} \cdot \text{MgX}_2} \qquad (10)$$

60%

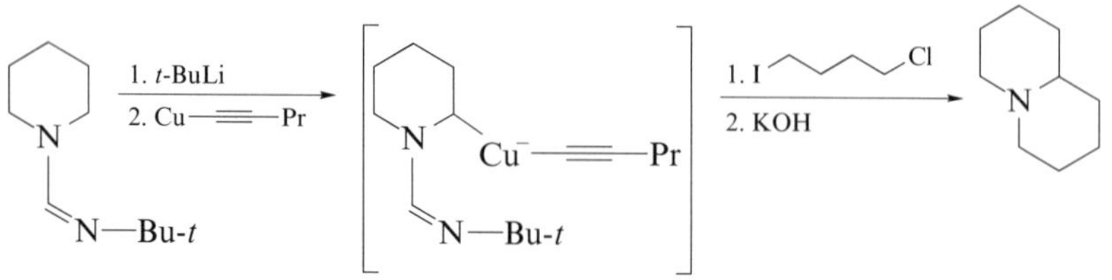

$$\text{(11)}$$

In mechanistic matters, it has been demonstrated that ω-alkenyl iodides undergo cyclization onto the vinyl function upon treatment with Me$_2$CuLi, in competition with direct substitution. This, as well as the generation of trityl radical in the reaction of Me$_2$CuLi with trityl chloride, constitutes evidence for single electron transfer in reactions of cuprates with iodides (and, to a lesser extent, bromides)[16]. The intermediacy of alkyl radicals in the substitution process (equation 12) is likely the source of the aforementioned racemization in reactions of secondary iodides[4].

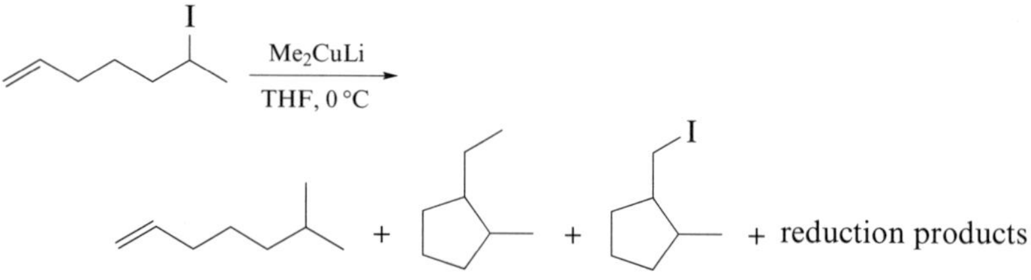

$$\text{+ reduction products} \quad \text{(12)}$$

New heterocuprates (2–4), with a diphenylphosphino, di(*tert*)butylphosphino or dicyclohexylamino nontransferable ligands, are reported to alkylate primary alkyl iodides.

$$\text{RCu(PPh}_2\text{)Li} \qquad \text{RCu(Hex-}c\text{)}_2\text{Li} \qquad \text{RCuP(Bu-}t\text{)}_2\text{Li}$$
$$\textbf{(2)} \qquad\qquad \textbf{(3)} \qquad\qquad \textbf{(4)}$$

These are clearly less reactive that the corresponding higher-order cyanocuprates; nevertheless, excellent thermal stability at 0 °C, and at least passable stability at 25 °C is reported for these reagents[17,18].

Electron transfer from copper or copper salts to alkyl halides has been used to initiate atom transfer radical additions. One modification of this process involves catalytic amounts of copper powder and fluorinated alkyl iodides; the radicals so generated may react in either inter- or intramolecular fashion with alkenes (equation 13)[19]. Alternatively, α-chloroesters with remote alkene functions undergo cyclization in the presence of cat-

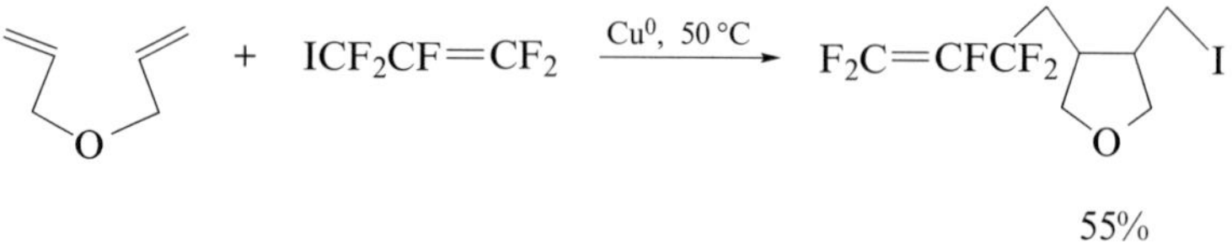

$$+ \quad \text{ICF}_2\text{CF}=\text{CF}_2 \quad \xrightarrow{\text{Cu}^0,\ 50\,°\text{C}} \quad \text{F}_2\text{C}=\text{CFCF}_2 \qquad \text{(13)}$$

$$55\%$$

$$\xrightarrow[\text{benzene, }\Delta]{\text{Cu(bpy)Cl}} \qquad 92\% \qquad \text{(14)}$$

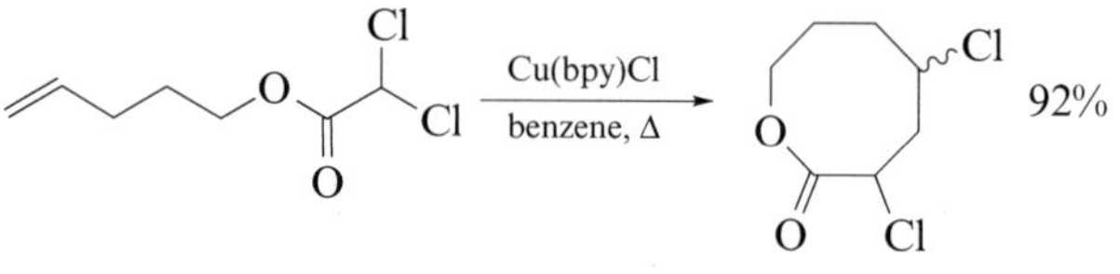

alytic amounts of CuCl–bipyridine complex (equation 14)[20]. Rings of both normal and medium size may be closed by the latter method; in both types of reaction, hydrogen atoms β to the halogen have been absent in all substrates.

C. Cross-coupling Reactions with Alkenyl Halides

The use of h.o. cyanocuprates in the substitution of vinyl halides is not as widespread as with alkyl halides, or as common as with lower-order cuprates. Nevertheless, some useful cases exist. For example, the transfer of an allyl function from h.o. allylcuprates to vinyl halides occurs quite rapidly. Unfortunately, vinyl iodides often give significant amounts of competing reduction to the alkene. This side product is nearly completely eliminated for the corresponding bromides, although in these cases double-bond stereochemical integrity is lost (equation 15)[2,5]. While vinyl iodides have also been shown to successfully couple with higher-order trialkylsilyl cuprates (equation 16)[21], other h.o. cuprate partners are rare, and triflates are generally the substrate of choice for vinylation in these cases[2,22].

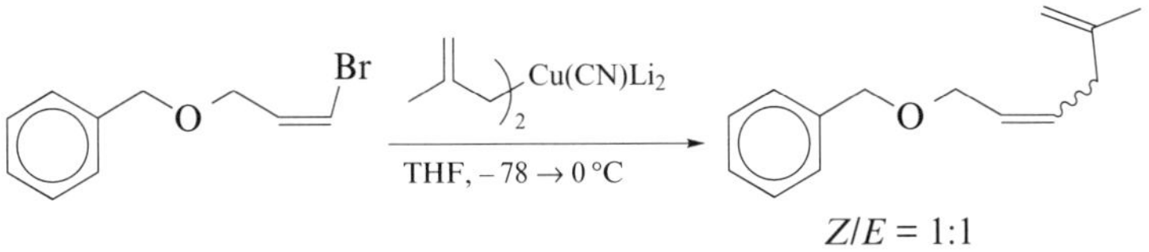

$$Z/E = 1:1 \tag{15}$$

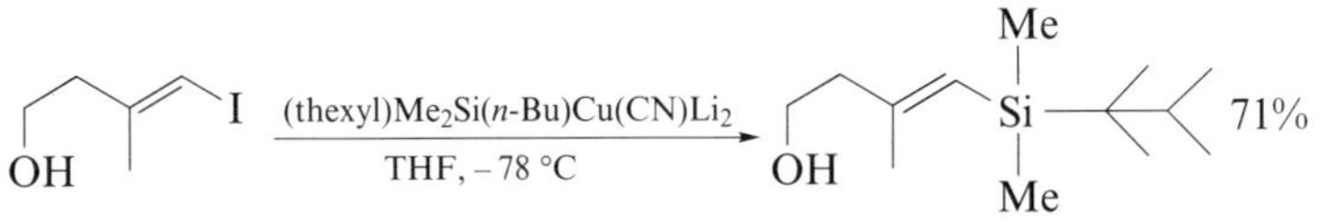

$$71\% \tag{16}$$

Like their alkyl halide counterparts, vinyl iodides and bromides undergo metal–halogen exchange with highly activated ('Rieke') copper, giving reasonable yields of vinylcoppers (equation 17)[10]. Further heating of the vinylcoppers gives modest yields of homocoupling products from the vinyl iodide precursors. Reactions of the so-generated vinylcopper species with other electrophiles has not seen much investigation.

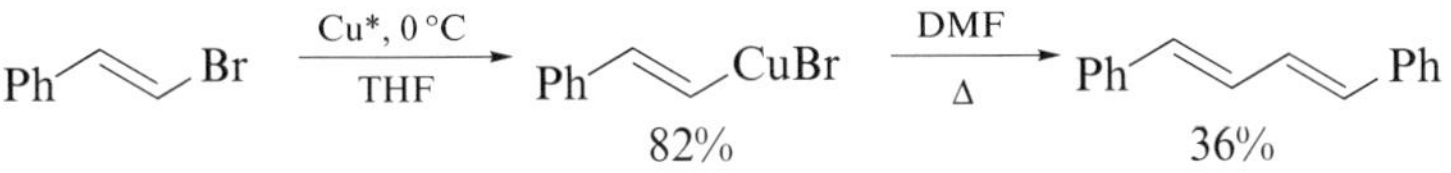

$$82\% \qquad 36\% \tag{17}$$

Perfluoroalkylation of vinyl halides has seen some recent attention. The use of trifluoro-methyl- and pentafluoroethylcopper has been mentioned previously[12,13]. Furthermore, vinyl iodides and bromides have been trifluoromethylated and pentafluoroethylated by the reaction of these substrates with the appropriate trialkylsilanes in the presence of stoichiometric amounts of potassium fluoride and CuI[23]. Only 1,2-disubstituted vinyl halides have been used as substrates, and iodides are apparently somewhat more suitable substrates. Preservation of alkene stereochemistry appears to be good in these transformations (equation 18).

$$77\% \tag{18}$$

Most of the recent developments in the vinyl halide reaction chemistry of lower-order cuprates are of zinc cyanocuprates (RCu(CN)ZnX)[24]. While only reacting with unfunctionalized iodoalkenes under forcing conditions[24], haloalkenes capable of ready addition–elimination reactions are coupled with these cuprates efficiently. These include β-iodocycloalkenones[25–27] (equation 19) and chloro-substituted cyclobutenediones (equation 20)[28]. The range of substituents which may be present in the haloalkene have not been addressed in print, whereas those for the zinc cyanocuprate include alkynes, esters, nitriles, ketones and tertiary amines.

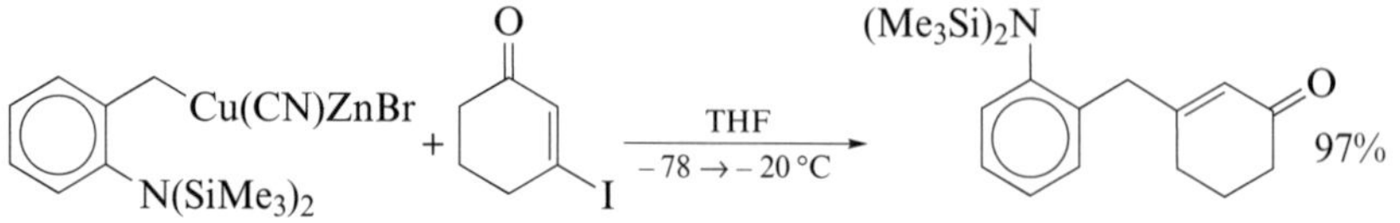

$$\text{(19)} \qquad 97\%$$

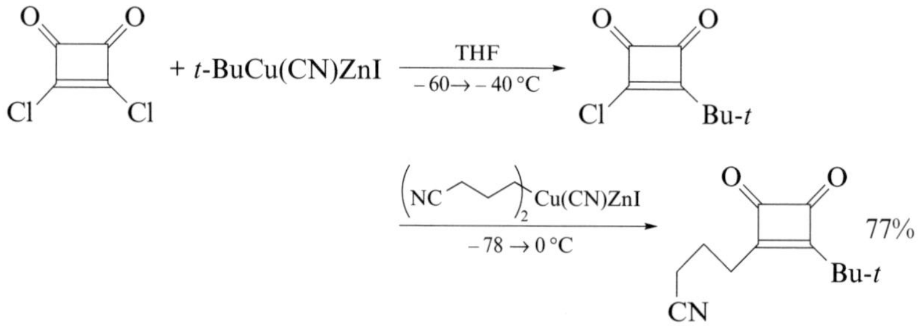

$$\text{(20)} \qquad 77\%$$

The coupling of vinyl halides and alkynes has been found to be induced by either stoichiometric amounts of CuI[29], or catalytic amounts of CuI in the presence of PPh$_3$ and K$_2$CO$_3$[30]. Disubstituted iodides and bromides are both suitable substrates, and alkene stereochemistry is once again preserved (equation 21).

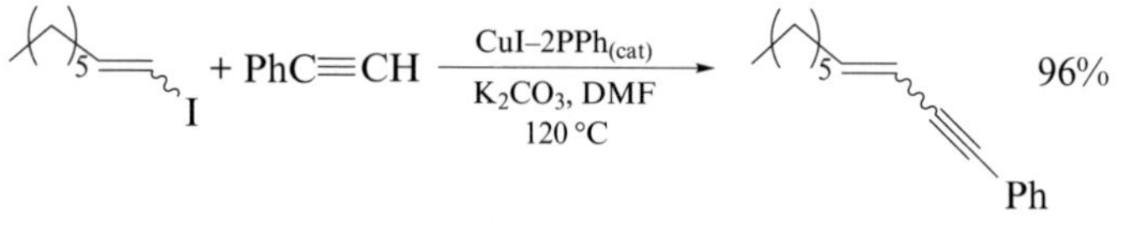

$$\text{(21)} \qquad 96\%$$

$$E/Z = 84/16 \qquad\qquad\qquad E/Z = 84/16$$

The use of halogenocycloalkenones in coupling reactions with lower-order cuprates is a viable and well established process[31]. In the presence of organobis(cuprates), the addition–elimination product undergoes a second conjugate addition, the overall result being an efficient spiroannulation (equations 22 and 23)[32]. Both chlorides and bromides have been employed as substrates in this spiroannulation reaction, and bis(cuprates) containing alkyl, aryl and vinyl anionic units have given successful results.

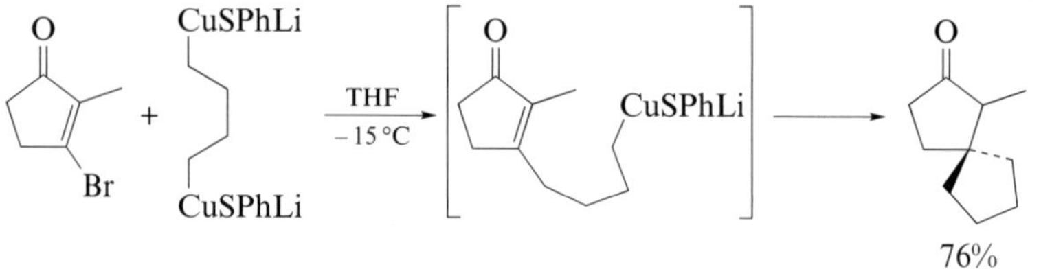

$$\text{(22)}$$

$$76\%$$

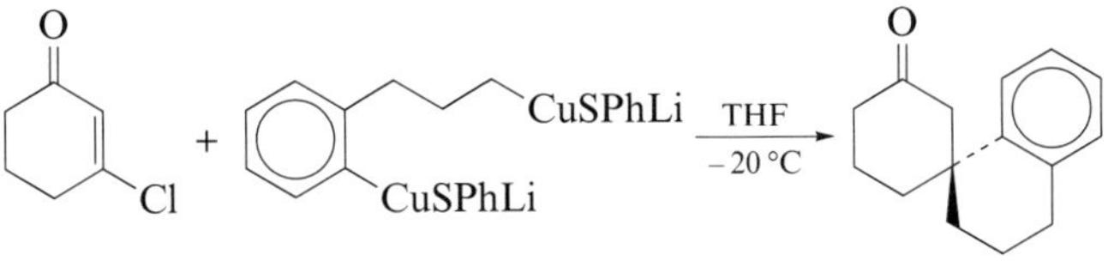

$$ \text{(structures)} \qquad 88\% \qquad (23) $$

D. Cross-coupling Reactions with Allenyl Halides

Work on the regiochemical disposition of organocopper substitutions of bromoallenes has made the overall reactivity picture clearer. For 1,3-disubstituted bromoallenes, cuprate attack is S_N2' except for very bulky cuprates (equations 24 and 25)[33]. With 3,3-disubstituted 1-bromoallenes, direct substitution is the normal pathway (equation 26), with a greater tendency for direct substitution for Gilman cuprates (R_2CuLi), a greater tendency for S_N2' substitution with the complex organocopper reagent ($RCuBr$)$MgBr$–$LiBr$ and intermediate behavior for cyanocuprates $RCu(CN)Li$[34,35]. Finally, 3-alkyl-1-bromoallenes tend to give S_N2' substitution with ($RCuBr$)$MgBr$–$LiBr$ organocoppers or cyanocuprates, and direct substitution with Gilman cuprates[35]. In many cases, the size of allene substituents and of the organocopper R group has a significant effect on the regiochemical pathway.

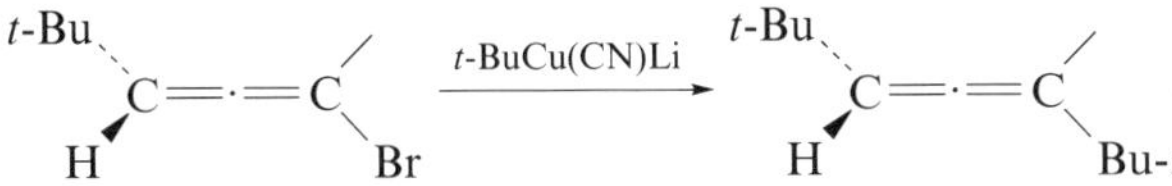

$$ \text{(structure)} \xrightarrow{\textit{t}\text{-BuCu(CN)Li}} \text{(structure)} \qquad 66\% \qquad (24) $$

$$ \text{(structure)} + Ph(CN)\,CuLi \xrightarrow{\text{THF, 0 °C}} \text{(structure)} \qquad 96\% \qquad (25) $$

$$ \textit{anti} : \textit{syn} > 99:1 $$

The stereochemical aspects of the reactions of allenic halides with cuprates have also been investigated. In cases where S_N2' reactivity is observed, addition is predominantly *anti* (equation 25)[33]. If vinyl substitution is the reaction pathway, the result is again dependent upon the organocopper reagent, with predominant retention of configuration with cyanocuprates and ($RCuBr$)$MgBr$–$LiBr$ organocoppers, and more modest levels of inversion realized with Gilman cuprates (equation 26)[35,36].

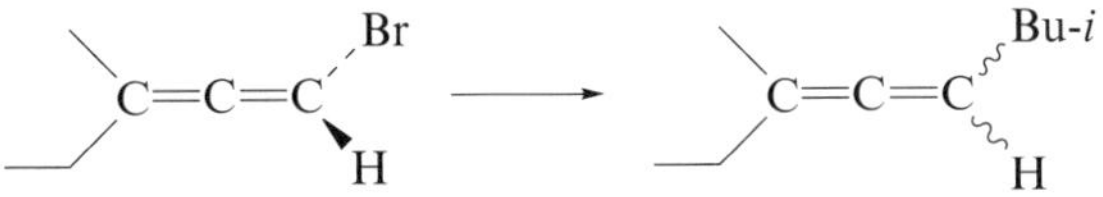

$$ \text{(structure)} \longrightarrow \text{(structure)} \qquad (26) $$

CONDITIONS

i-Bu_2CuLi, – 60 °C, Et_2O	70%, 68% ee (inversion)
(i-$BuCuBr$)$MgBr$·$LiBr$, – 60 °C, Et_2O	67%, 95% ee (retention)

E. Cross-coupling Reactions with Allyl and Propargyl Halides

The reaction of allyl halides with organocopper species usually gives regiochemical mixture of direct and S_N2' attack. Exceptions to this trend are 3,3-disubstituted 1-bromoalkenes, which give reliable direct substitution with lower-order cuprates or RCu organocopper reagents. This fact has been put to use in the synthesis of several prenelated natural products, including ubiquinone-10 (**5**)[37] and collectochlorins B and D (**6** and **7**)[38]. This property has also been employed, in conjunction with a trimethylsilyl substituent, in

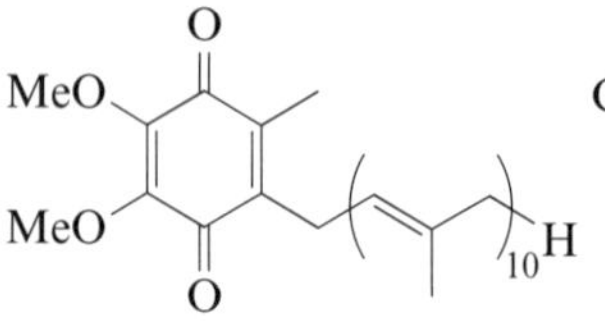

(**5**) 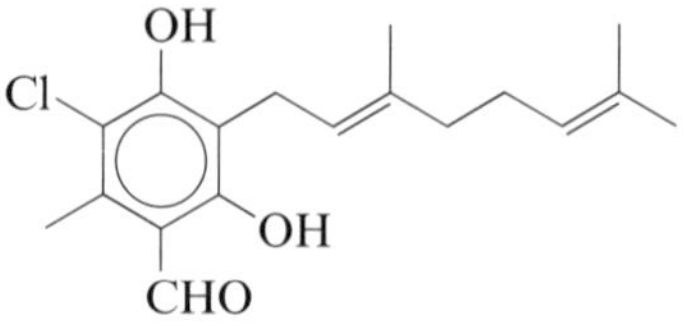   (**6**) 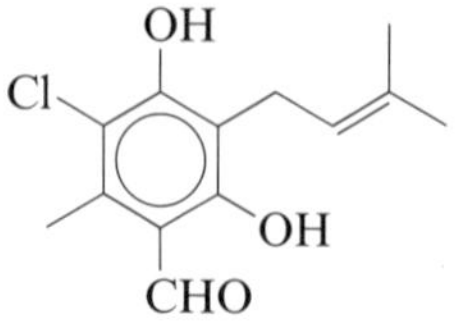   (**7**)

the regioselective and stereospecific preparation of trisubstituted vinylsilanes (equation 27)[39]. These substrates also present one of the rare cases where h.o. acetylenic cyanocuprates couple successfully.

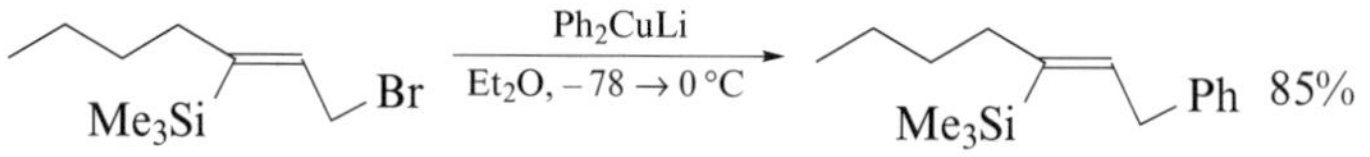

The addition of a full equivalent of CuCN to certain *ortho*-lithiated aromatics allows the introduction of allylic and propargylic side chains. Due to the nature of the organolithium species originally involved, the exact nature of the organocopper intermediate is unknown (equation 28)[40].

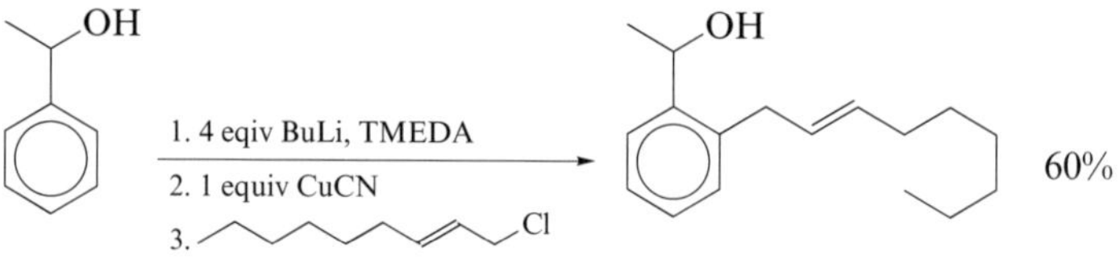

Allyl halides are also alkylated by zinc cyanocuprates. Allylic bromides, iodides and chlorides react successfully with these reagents, and a number of functional groups in the allylic halide (ester, sulfide, sulfoxide, ether, alkyl halides, acetals) stand up well to the conditions of reaction (equations 29 and 30)[41-43]. The regiochemistry of attack is predominantly S_N2'. 1,3-Dichloroalkenes can be made to undergo two successive coupling reactions; to this point, only two identical R group incorporations have been reported (equation 31)[44].

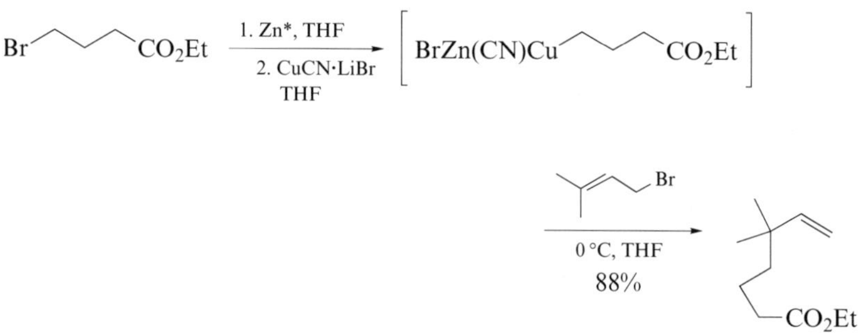

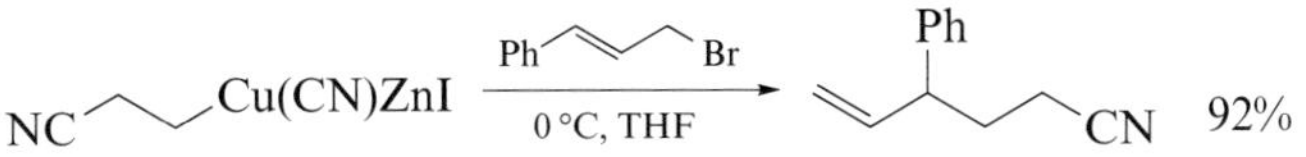

$$NC\diagdown\diagup^{Cu(CN)ZnI} + Ph\diagdown\diagup Br \xrightarrow[0\,°C,\ THF]{} Ph\diagdown\diagup\diagdown CN \quad 92\% \quad (30)$$

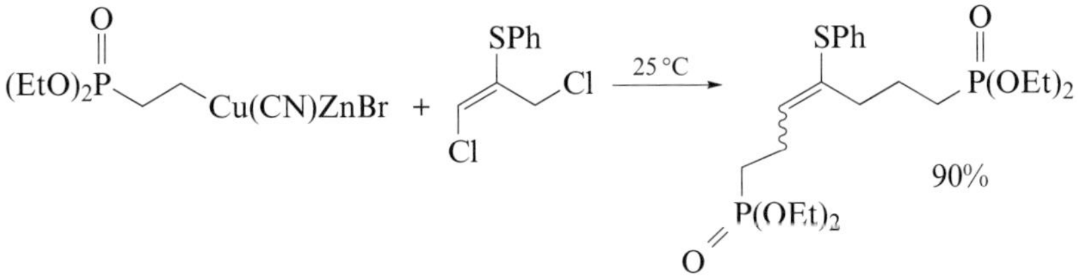

$$(EtO)_2\overset{O}{\underset{\|}{P}}\diagdown\diagup Cu(CN)ZnBr + \diagup\overset{SPh}{\underset{Cl}{}}Cl \xrightarrow{25\,°C} \diagup\overset{SPh}{}\diagdown\diagup\overset{O}{\underset{\|}{P}}(OEt)_2 \quad (31)$$

90%

Reaction of allyl bromides with ICH_2ZnI in the presence of CuI results in the iodomethylation of the allyl halide (equation 32)[45]. Unsymmetrical allyl halides have been restricted to 3,3-disubstituted cases, where predominant S_N2 substitution is observed.

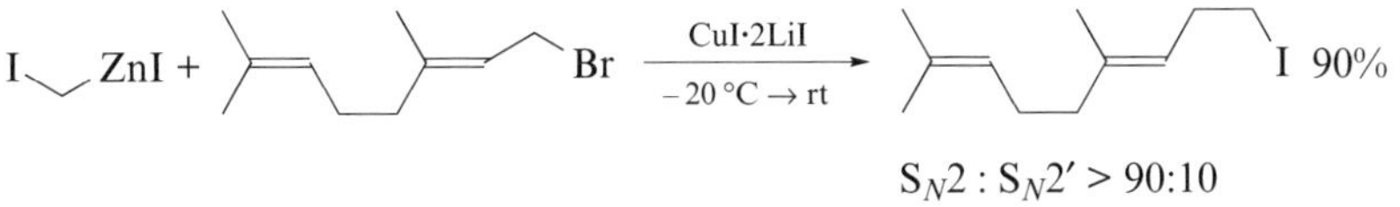

$$I\diagdown ZnI + \diagup\!\!=\!\!\diagdown\!\!\diagup\!\!=\!\!\diagup\!\diagdown Br \xrightarrow[-20\,°C\to rt]{CuI\cdot2LiI} \diagup\!\!=\!\!\diagdown\diagup\!\!=\!\!\diagup\diagdown I \quad 90\% \quad (32)$$

$$S_N2 : S_N2' > 90{:}10$$

Higher-order zinc cyanocuprates react in a similar manner with propargyl chlorides and bromides[46]. Allenes are the result (S_N2' substitution), except in the case of 1,4-dihalo-2-butynes, which undergo two successive S_N2' substitutions to afford 2,3-disubstituted buta-dienes (equations 33 and 34)[47].

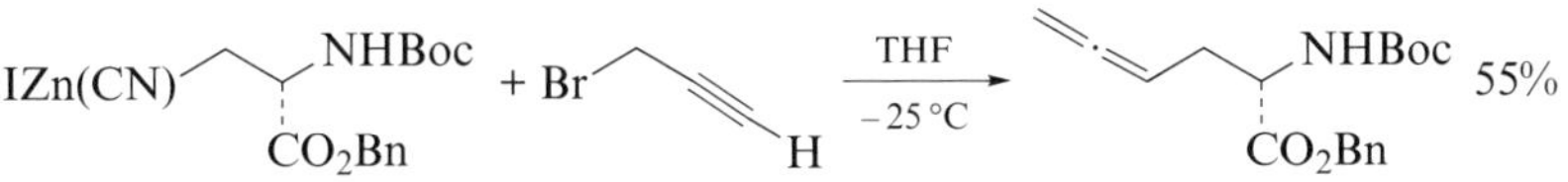

$$IZn(CN)\diagdown\diagup\overset{NHBoc}{\underset{CO_2Bn}{}} + Br\diagdown\diagup\!\!\equiv\!\!\diagdown H \xrightarrow[-25\,°C]{THF} \diagdown\!\!\cdot\!\!=\!\!\diagdown\diagup\overset{NHBoc}{\underset{CO_2Bn}{}} \quad 55\% \quad (33)$$

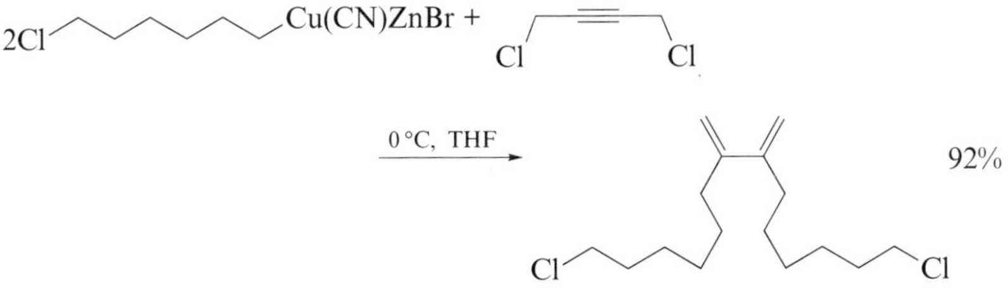

$$2Cl\diagdown\diagdown\diagdown\diagdown\diagdown Cu(CN)ZnBr + \overset{Cl}{\diagup}\!\!\equiv\!\!\overset{Cl}{\diagdown}$$

$$\xrightarrow{0\,°C,\ THF} \quad\quad 92\% \quad (34)$$

$$Cl\diagdown\diagdown\diagdown\diagdown\diagdown Cl$$

In many cases, the functionalized organozincs may also be coupled with allylic halides with catalytic amounts of copper (CuCN or $CuBr\cdot Me_2S$). Both diorganozincs (R_2Zn) and organozinc halides (RZnX) have been employed (equations 35 and 36) in this role, as homoenolate equivalents. Once again, predominant S_N2' regiochemistry is observed[48,49].

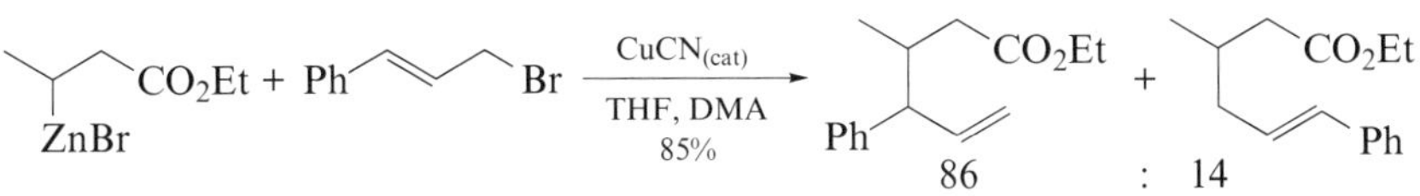

$$\diagup\!\diagdown\overset{CO_2Et}{} + Ph\diagdown\diagup Br \xrightarrow[THF,\ DMA]{CuCN_{(cat)}} Ph\diagup\diagdown\overset{CO_2Et}{}\diagdown\!\!=\!\! + \diagup\diagdown\overset{CO_2Et}{}\diagdown\!\!=\!\!\diagup Ph$$

ZnBr   85%   86   :   14

$$(35)$$

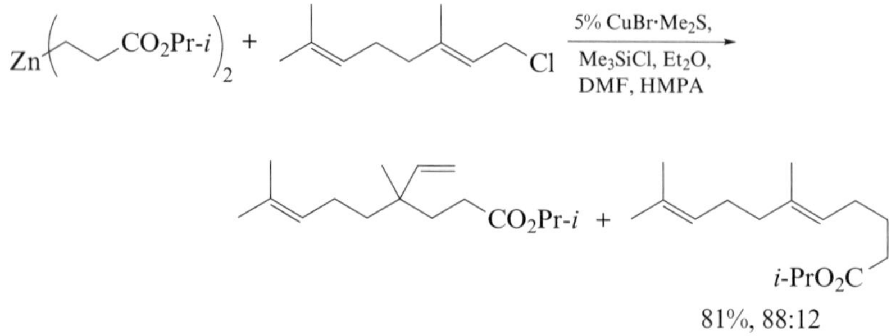

$$Zn\left(\diagdown\diagdown CO_2Pr\text{-}i\right)_2 + \diagdown\diagdown\diagdown\diagdown Cl \xrightarrow[\substack{Me_3SiCl,\ Et_2O,\\ DMF,\ HMPA}]{5\%\ CuBr\cdot Me_2S,}$$

$$\diagdown\diagdown\diagdown\diagdown CO_2Pr\text{-}i + \diagdown\diagdown\diagdown\diagdown\diagdown \quad (36)$$

$$i\text{-}PrO_2C$$

81%, 88:12

The copper-catalyzed stereospecific allylation of alkenylboranes with allyl bromides occurs in a facile manner for both disubstituted[1] and trisubstituted double bonds (equation 37)[50]. Only very small amounts of incorporation of the borane alkyl group were observed, and apparently none of the vinyl homocoupling products competed.

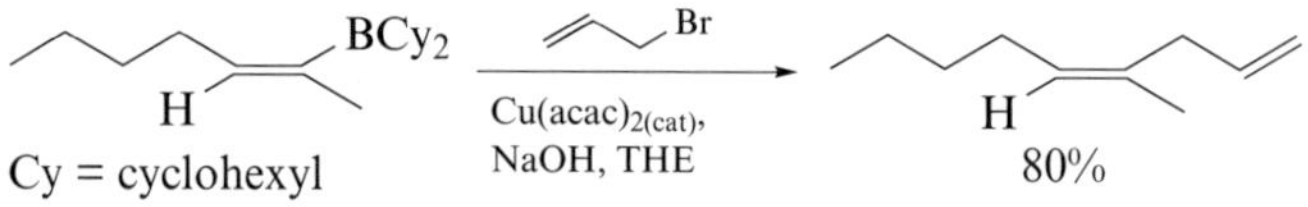

$$\diagdown\diagdown\diagdown\overset{BCy_2}{\underset{H}{\diagdown}} + \diagdown\diagdown^{Br} \xrightarrow[\substack{Cu(acac)_{2(cat)},\\ NaOH,\ THE}]{} \diagdown\diagdown\diagdown\overset{}{\underset{H}{\diagdown}}\diagdown\diagdown \quad (37)$$

Cy = cyclohexyl   80%

Burton has investigated the alkylation of allyl halides with perfluoroorganocoppers. Both perfluoroalkylcoppers and perfluoroalkenylcoppers couple with allyl bromide in high yield; the former also couples effectively with allyl chloride (equation 38)[51,13]. Double-bond stereochemistry is preserved with alkenylcopper reagents, while issues of α-/γ-regio-selectivity of attack on the allyl unit have yet to be addressed.

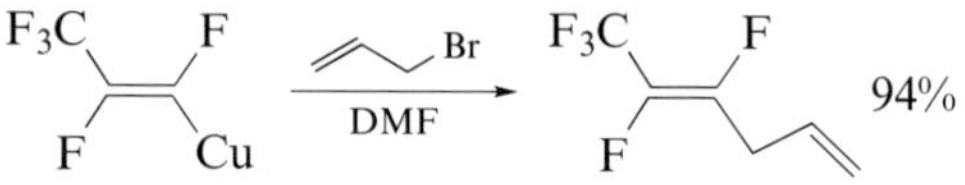

$$\underset{F}{\overset{F_3C}{\diagup}}\!\!=\!\!\underset{Cu}{\overset{F}{\diagdown}} + \diagdown^{Br} \xrightarrow[DMF]{} \underset{F}{\overset{F_3C}{\diagup}}\!\!=\!\!\underset{}{\overset{F}{\diagdown}}\diagdown\diagdown \quad 94\% \quad (38)$$

Coupling reactions between allylic halides and alkynylcoppers have been rather heavily investigated as routes to stereochemically defined polyenyl natural products. The CuCl-catalyzed reaction of acetylenic Grignard reagents with chlorine substituted allylic chlorides has been employed in the synthesis of stereochemically pure 1,3,5-undecatrienes

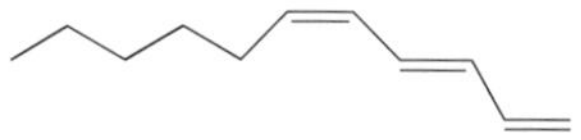

$$\diagdown\diagdown\diagdown\diagdown\diagdown\diagdown\diagdown\diagdown$$

(8)

(8)[52], whereas the CuI catalyzed allylation of terminal acetylenic alcohols by allyl chlorides has both been accomplished under phase transfer conditions and has been applied to terpenoid synthesis (equation 39)[53,54]. In each of these cases, alkynylation of the less substituted end of the allyl unit was preferred (ratios ca 90:10). The corresponding copper-catalyzed reactions of acetylenic Grignard reagents with propargyl bromides give clean direct substitution with no apparent allenic by-products (equation 40)[55].

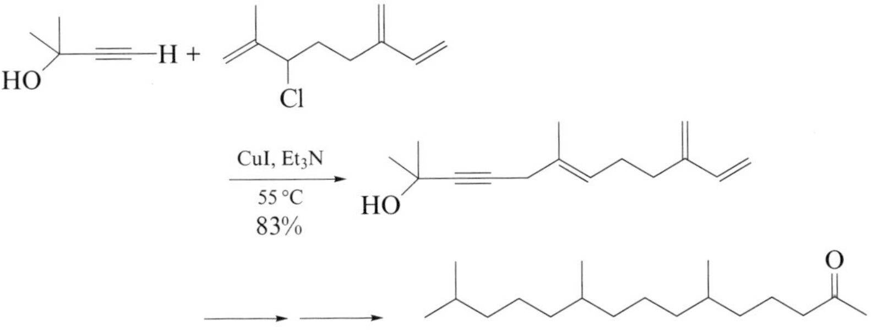

(39)

phytone

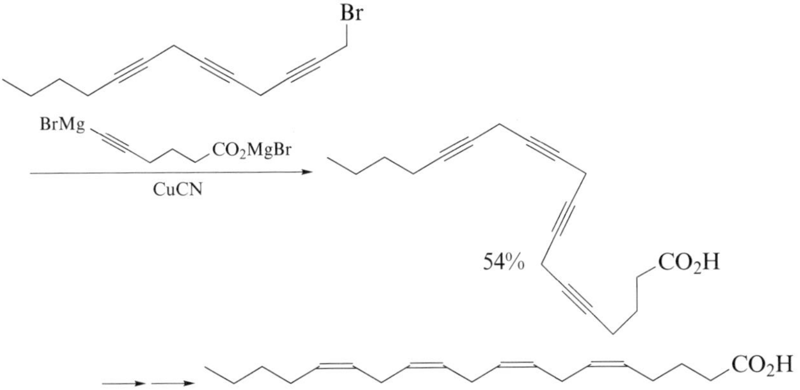

(40)

An important recent advance in the area is the demonstration that allyl chlorides readily form allylcopper species upon exposure to highly activated copper ('Rieke copper'). The resultant allylcoppers have been shown to react with a fairly wide range of nucleophiles (ketones, aldehydes, acid chlorides, enones, epoxides [after conversion to the cuprate], imines, and allyl bromides)[56]. Regiochemical considerations in such substitutions depend upon the cases studied, and are fairly complex. The allylcoppers formed have, in all cases, the least substituted C—Cu bond; allyl transfer from these reagents is selective from the γ-position of the allyl (enone electrophiles excepted)(equations 41 and 42). The allyl chloride

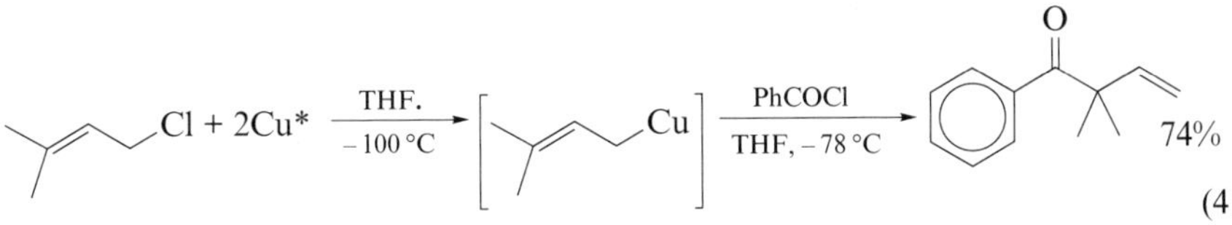

(41)

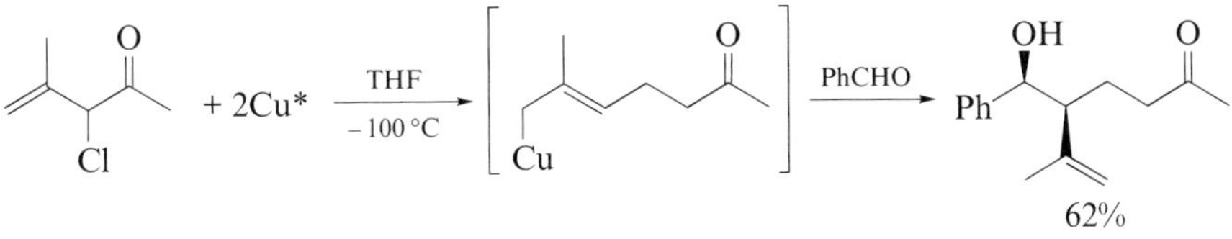

(42)

substrates may be quite highly functionalized, with even ketone-containing allyl chlorides forming reasonably stable allylcoppers at $-78\,^{\circ}$C. Allyl chlorides containing a vinylic chloride residue, however, form dicoppers (equation 43), which may be reacted selectively first at the allyl terminus and subsequently at the vinylic site[57].

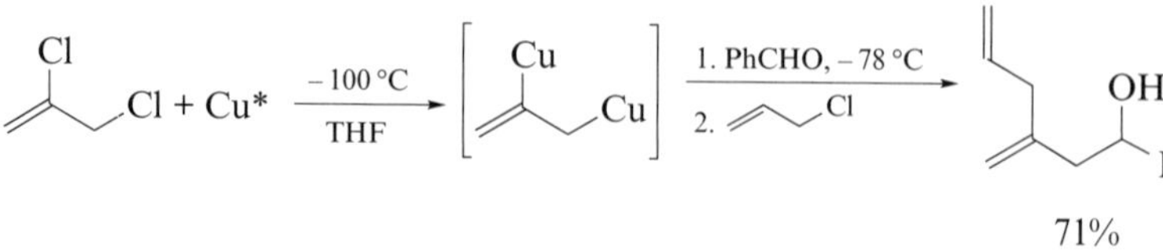

$$ (43) $$

$$ 71\% $$

F. Cross-coupling Reactions with Alkynyl Halides

Much of the new chemistry on the coupling of alkynyl halides has dealt mostly with applications of known methods. The Cadiot–Chodkiewicz-type coupling reactions (i.e. terminal acetylene coupling with alkynyl halides) have formed the cornerstone of synthesis of acetylenic retinoids (i.e. **9**)[58].

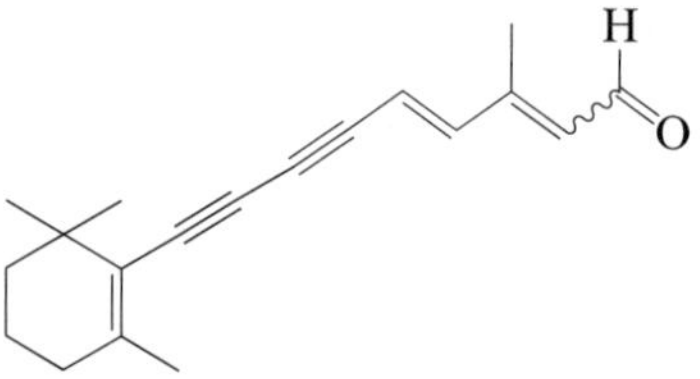

$$ (9) $$

It has proven possible to add oxidatively highly activated copper metal ('Rieke copper') to alkynyl bromides in moderate yields, to furnish acetylenic copper compounds (equation 44)[10]. Bubbling oxygen through the system resulted in the formation of homocoupling products.

$$ PhC\equiv CBr + Cu^{*} \xrightarrow{25\,^{\circ}C,\ DME} \left[PhC\equiv CCu\right] \xrightarrow{O_2} PhC\equiv C-C\equiv CPh \quad (44) $$

$$ 42\% \qquad\qquad 30\% $$

based on bromide

The coupling of vinyl organometallics to an alkyne unit has been extended to that involving the reaction of acetylenic bromides with trisubstituted vinylboranes; this reaction is catalyzed by $Cu(acac)_2$ and base, and is stereoselective with respect to the alkeny function (equation 45)[50].

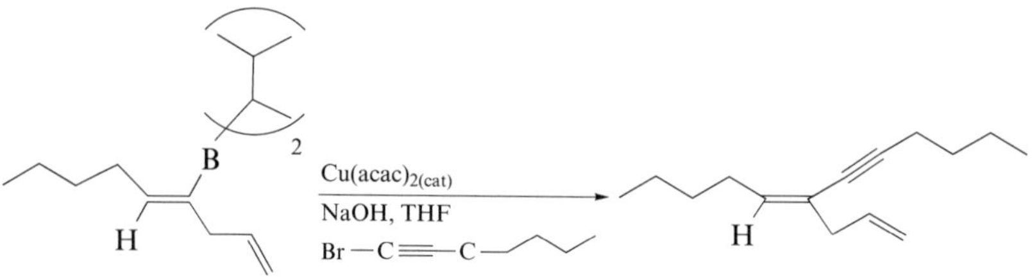

$$ (45) $$

$$ 70\% $$

It has proven possible to couple alkynyl iodides and bromides with a wide variety of zinc cyanocuprates. Coupling tolerates a number of functional groups on both the alkyne and the cuprate, including alkynyl ethers and propargyl ethers (equations 46 and 47). The cuprate partners have been primary alkyl, vinyl and aryl based[59–61], whereas secondary alkyl organometallics couple in some cases and reduce the alkynyl halide in others.

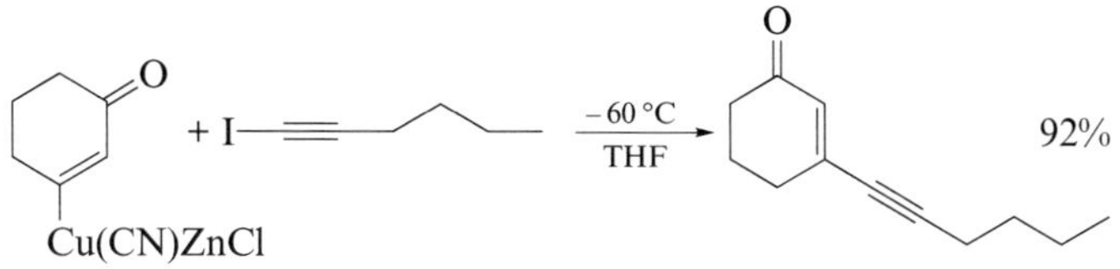

$$\text{(46)} \quad 92\%$$

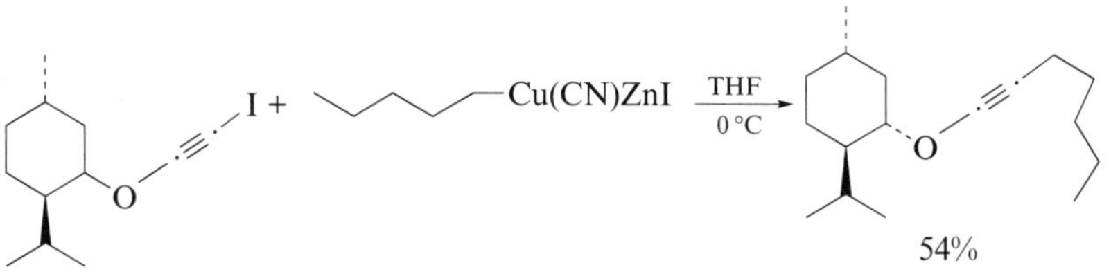

$$\text{(47)} \quad 54\%$$

G. Cross-coupling Reactions with Aryl and Heteroaryl Halides

Many examples have been reported of the copper-catalyzed coupling of aryl halides with highly stabilized carbanions, mediated by stoichiometric or catalytic amounts of copper salts at elevated temperatures[62]. A variety of anion stabilizing groups are viable participants in this reaction, such as esters, ketones, nitriles, sulfonyl groups and phosphonates[63–70]. The influence of steric effects on these couplings is highly dependent on the case in question, as *ortho*-substituted aryl halides fail to couple in some cases, whereas persubstituted arenes work well in other instances. Alternatively, *ortho*-substituents may participate in further reactions of the coupled products; OH substituents may cyclize onto ester functions, while allyloxy substituents may participate in radical cyclization of an intermediate benzylcopper (equations 48 and 49).

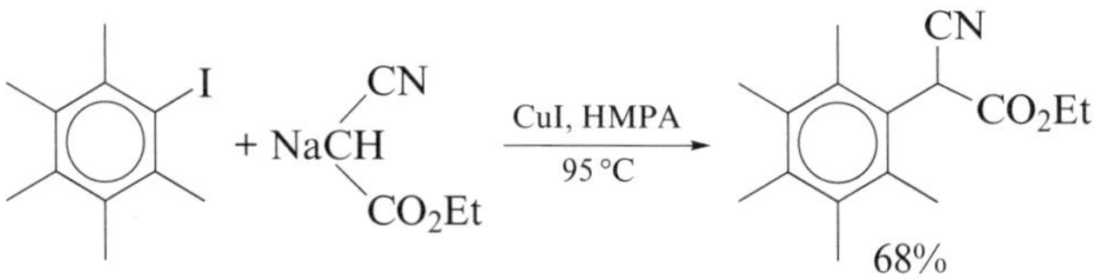

$$\text{(48)} \quad 68\%$$

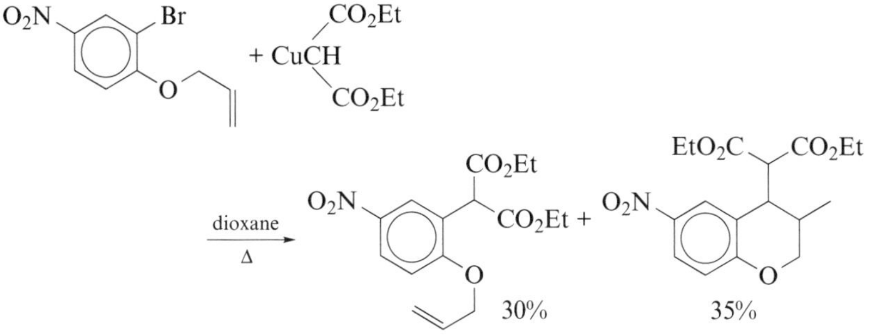

$$\text{(49)} \quad 30\% \qquad 35\%$$

Coupling of aryl iodides has also been effected with perfluoroalkylcoppers and perfluo-rovinylcopper reagents[51,13]. The former reaction may also be accomplished by employing the perfluoroalkylsilane, the aryl iodide and stoichiometric amounts of KF^{23}. Trifluoromethylation may also be accomplished, in more modest yields, by decarboxyla-tion of sodium trifluoroacetate in the presence of CuI and an aryl iodide[71]. Stannylation has been accomplished of aryl halides by the use of h.o. stannylcyanocuprates (equation 50)[72].

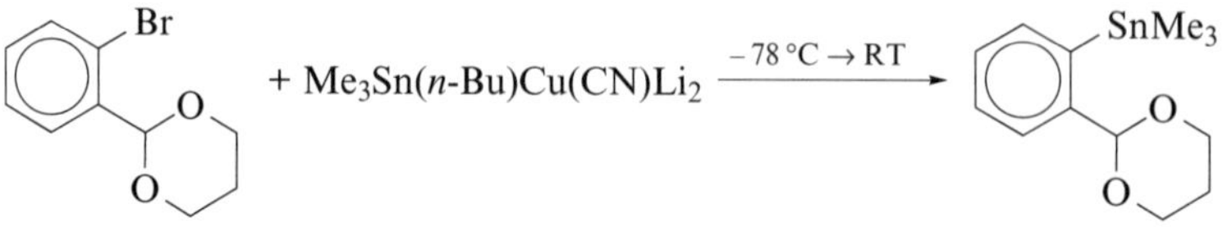

$$\text{(structure with Br, O, O)} + \text{Me}_3\text{Sn}(n\text{-Bu})\text{Cu(CN)Li}_2 \xrightarrow{-78\,^\circ\text{C} \to \text{RT}} \text{(structure with SnMe}_3\text{, O, O)} \quad (50)$$

Although aryl chlorides are not usually reactive enough to couple efficiently with alkyl-coppers, heteroaryl chlorides often couple quite well. For example, the most efficient way of alkylating chlorinated phenanthrolines and quinolines, and both 2- and 3-substitued pyridines involves the use of Grignard reagents and copper salts (equation 51)[73].

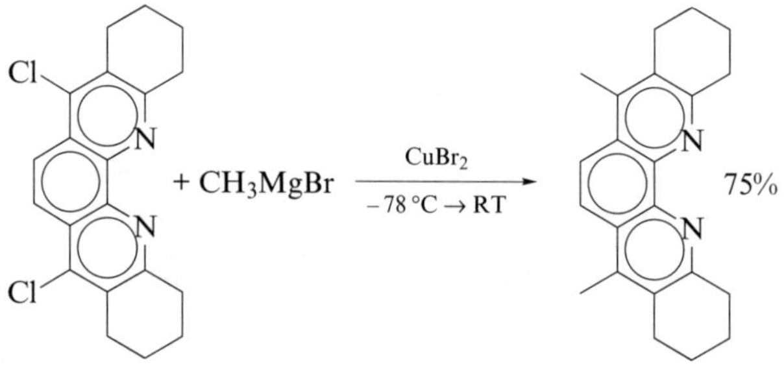

$$\text{(dichloro heteroaryl structure)} + \text{CH}_3\text{MgBr} \xrightarrow[-78\,^\circ\text{C} \to \text{RT}]{\text{CuBr}_2} \text{(dimethyl heteroaryl structure)} \quad 75\% \quad (51)$$

It has been proposed that copper-mediated substitutions of aryl halides proceed by for-mation of intermediate Cu(III) species, which reductively eliminate to the cross-coupling products (equation 52)[74].

$$\text{ArX} + \text{Cu}^{\text{I}}\text{Nu} \longrightarrow \text{Ar}-\underset{\underset{\text{X}}{|}}{\text{Cu}^{\text{III}}}-\text{Nu} \longrightarrow \text{ArNu} + \text{CuX} \quad (52)$$

The Ullman-type coupling of 3-pyridylcopper with aryl iodides has been reported; yields are improved when the arylcopper is stabilized by the addition of an equivalent of triphenylphosphine (equation 53)[75].

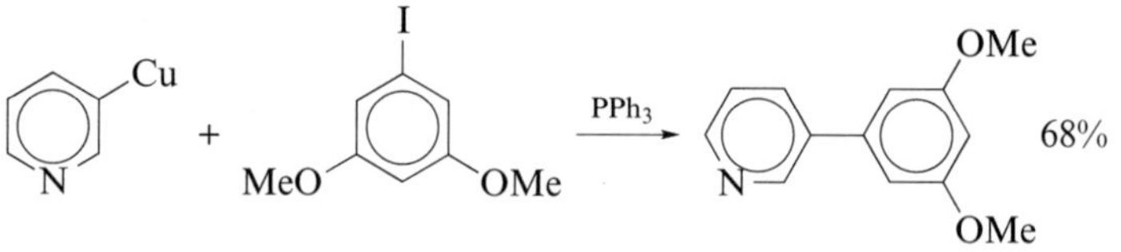

$$\text{(3-pyridylcopper)} + \text{(iodo dimethoxybenzene)} \xrightarrow{\text{PPh}_3} \text{(coupled product)} \quad 68\% \quad (53)$$

Aryl halides are excellent substrates for metal–halogen exchange by highly activated 'Rieke' copper, affording aryl coppers at or below room temperature in fair to excellent yields. Functional groups such as nitriles, ketones and fluoride substituents are tolerated, although *p*-nitrophenylcopper is rather unstable. Heating of several of these organo-coppers results in biaryl formation through Ullman coupling (equation 54)[10,11].

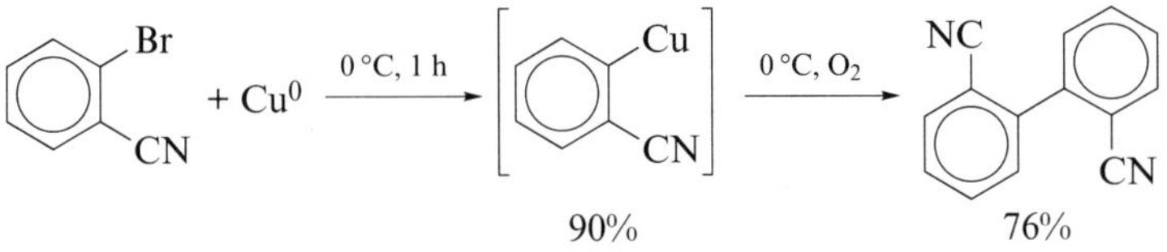

$$
\text{(54)}
$$

90%   76%

H. Cross-coupling Reactions with Acyl Halides

Acid chlorides are very effectively coupled with organocoppers to give ketones, to the point where acylation of acid chlorides is a standard method for testing the formation of organocopper species. A variety of organocoppers containing esters, nitriles, ketones, other halides and epoxides all couple efficiently. Allylcoppers and perfluorinated organocopper species[76] also couple without incident (equations 55–57)[10,11,51,56]. The

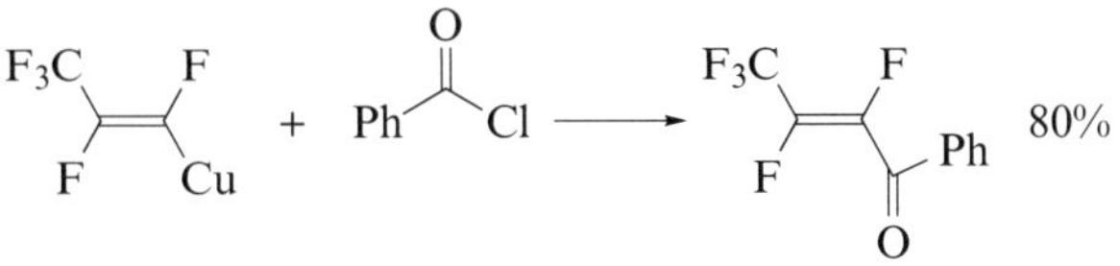

$$
\text{(55)} \quad 80\%
$$

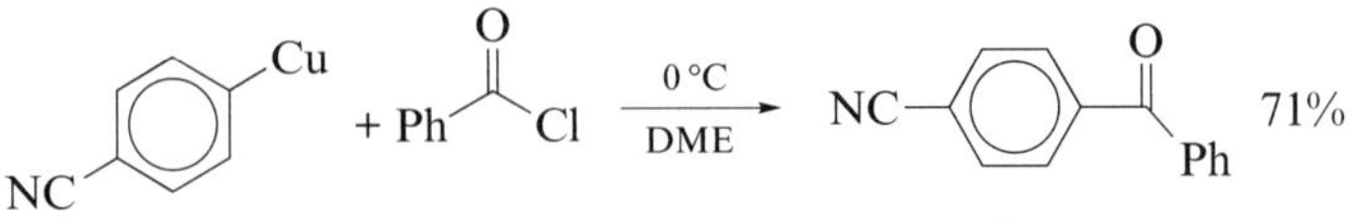

$$
\text{(56)} \quad 71\%
$$

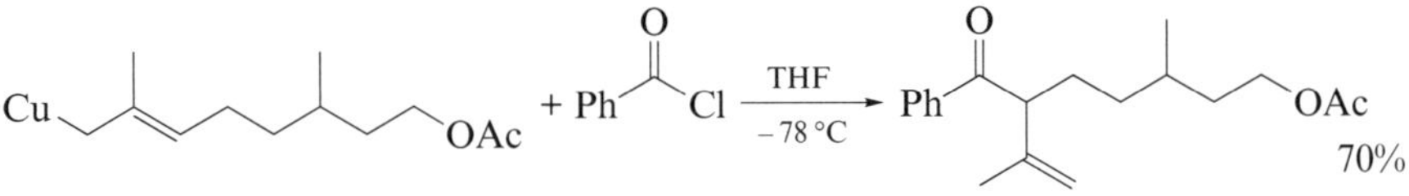

$$
\text{(57)} \quad 70\%
$$

heterocuprates **2** and **3** also couple very effectively[17], as do both higher- and particularly lower-order cuprates (**1,10**) containing DMSO as the nontransferable ligand[7].

$$\text{Li(CH}_3\text{SOCH}_2\text{CuR)}$$

(10)

Silyl cuprates give high yields of acylsilanes with acid chlorides (equation 58)[77]. Both lower-order Gilman trimethylsilylcuprates and h.o. cyanocuprates accomplish the transformation.

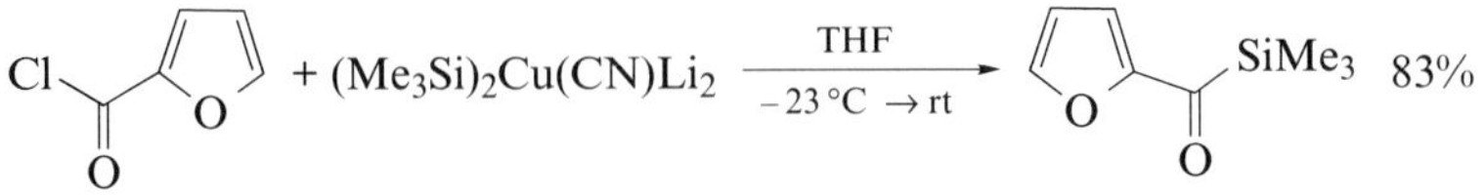

$$
\text{(58)} \quad 83\%
$$

Acid chlorides also couple smoothly with lower-order zinc cyanocuprates (equation 59) to give ketones in high yields. Although most acid chloride substrates employed have been simple ones, alkenes, benzylic and alkyl chloride functional groups have been present in the acyl halide without incident[25].

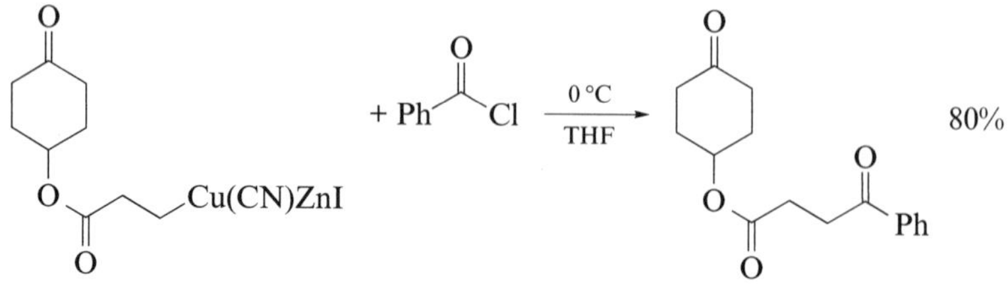

80% (59)

III. REACTIONS OF ORGANOMETALLIC REAGENTS IN THE PRESENCE OF TRANSITION METAL CATALYSTS

A. Introduction

Few areas in organic synthetic methods have received as much attention as the transition metal mediated cross-coupling reactions of organometallics and organic halides (or triflates). The emphasis within this area has also changed; the organometallic of choice is most often based on either zinc or tin, with boron-based organometallics a strong third. This is largely based on the desire for the cross-coupling reaction to tolerate as many other functions as possible, or ideally for the organometallic to be stable to other chemical transformations.

B. Cross-coupling Reactions of Grignard Reagents in the Presence of Nickel Complexes

A large number of reports on nickel-catalyzed cross-coupling reactions of Grignard reagents with aryl and vinyl halides have appeared. Many of these are applications of reactions previously developed; nevertheless, useful preparations of dienes[78-80], allylsilanes[81] or other functionalized alkenes[82], and functionalized arenes or heteroarenes[83-86] have been reported. It has proven possible in several cases to couple selectively once in a polyhaloarene[86-88] or to couple a chloroarene in preference to an alkylthioarene[89]. Allenyl bromides undergo direct coupling with secondary alkyl Grignards by a number of nickel catalysts, the most efficient being Ni(N-methylsalicylaldimine)$_2$[90]; the reaction proceeds with apparently incomplete inversion of configuration at the allenic carbon center (equation 60). At least one thiochloroformate has been shown to couple efficiently with alkyl Grignards and Ni(dppe)Cl$_2$ (equation 61)[91] [dppe = bis(1,2-diphenylphosphino)ethane]. Finally, acetylenic bromides (as well as vinyl and aryl halides) have been coupled with bridgehead Grignard reagents with NiCl$_2$(dppe)[92] (equation 62).

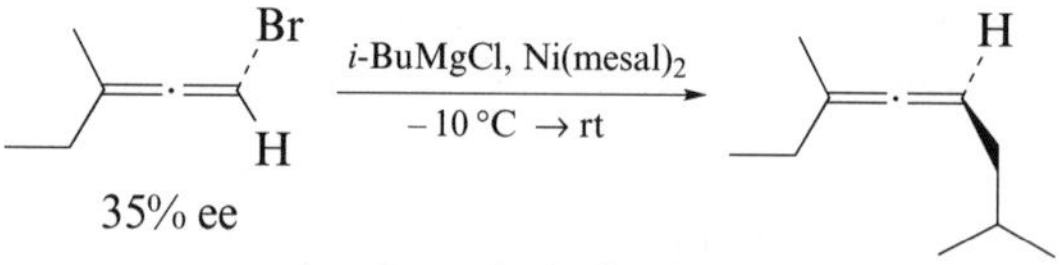

(60)

mesal = N-methylsalicylaldimine

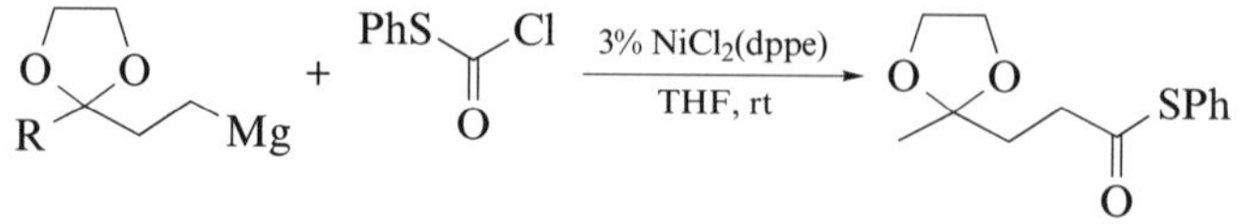

(61)

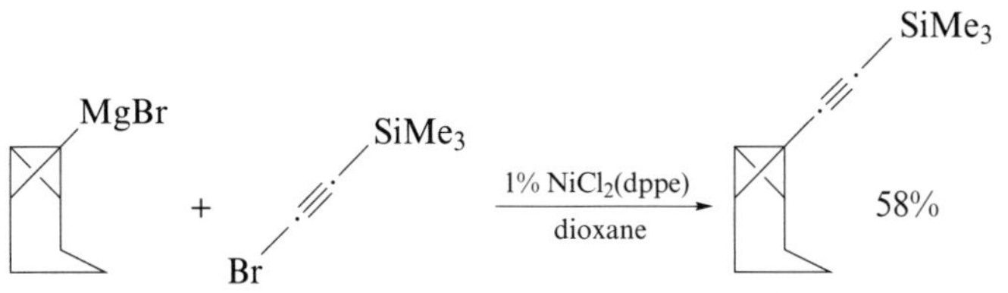

$$(62)$$

The first group of unactivated alkyl halides, namely neopentyl iodides, has been reported to couple efficiently with aryl Grignard reagents and Ni(dppf)Cl$_2$ [dppf = 1,1'-bis(diphenylphosphino)ferrocene, **17**] (equation 63)[93]. Although homocoupling of the iodide and reduction become competing reactions with alkyl Grignards, synthetically useful yields of cross-coupling are still realized in some of these cases. In two of the reported cases, β-hydrogens were present in the Grignard; even in these cases moderate yields of cross-coupling were realized.

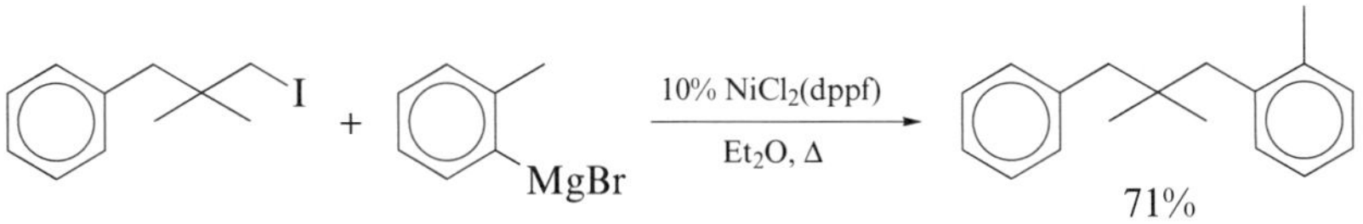

$$(63)$$

Mechanistic work on reductive elimination of arylmethylnickel(II) complexes has resulted in the isolation of the *cis*-isomers (**11**) bearing chelating phosphine ligands. Addition of phosphine to intermediates accelerated the elimination of toluene at rates which depended on the size of the phosphine and not its basicity, suggesting the intermediacy of 5-coordinate, trigonal bipyramidal species[94]. As reductive elimination of *trans*-diaxial and *cis*-equatorial organic groups is symmetry forbidden, the probable candidates for the pentacoordinate species are **12** or **13** (equation 64). An in-depth EHMO analysis of this associative mechanism is consistent with this interpretation, but could not rule out the possibility of **14** as the reactive pentacoordinate intermediate[95].

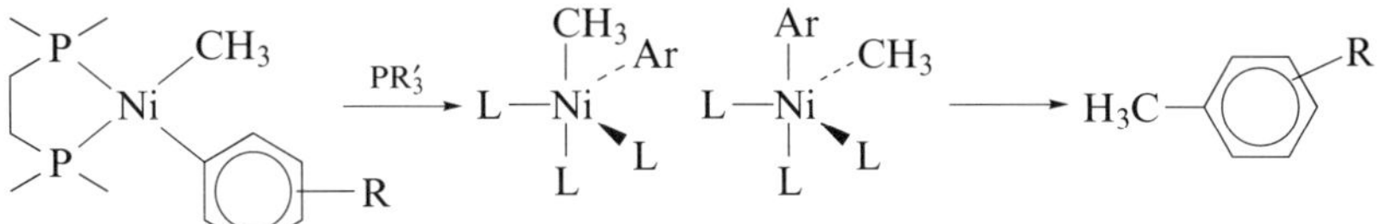

$$(64)$$

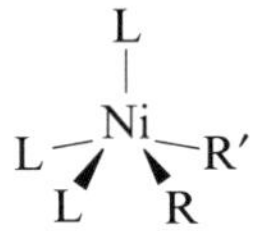

Despite a fair amount of work on asymmetric cross-coupling reactions of secondary alkyl Grignards and aryl halides, only a limited amount of progress has been made in improving coupling ee values. The highest recorded ee for the coupling of phenethylmagnesium chloride remains 94% (based on a < 100% ee ligand) for **15**; however, this class of ligands did not effectively couple 2-octyl magnesium chloride and bromobenzene[96].

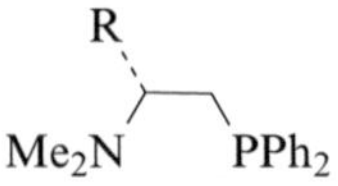

$$R$$

$$Me_2N \quad PPh_2$$

$$R = t\text{-Bu}$$

(15)

Neverthless, a number of 2-arylpropionic acids have been prepared in 79–83% ee using this methodology (equation 65).

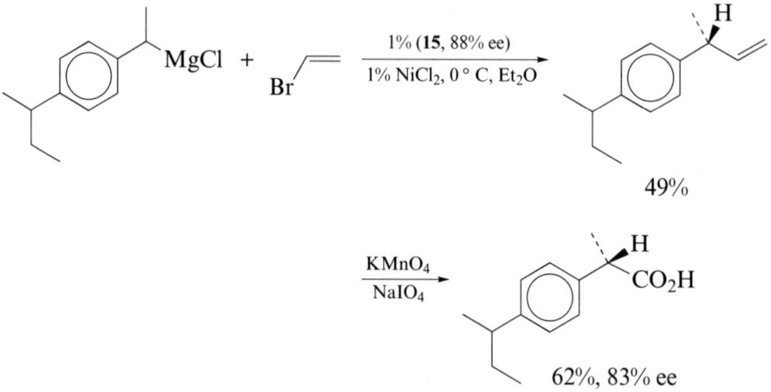

49%

$$\xrightarrow[\text{NaIO}_4]{\text{KMnO}_4}$$

62%, 83% ee   (65)

A related approach, employing catalysts with phosphorus, nitrogen and sulfur ligands, has resulted in a case where phenethyl magnesium chloride and vinyl chloride couple with NiCl$_2$ catalysis in 88% ee (equation 66)[97–99]. Furthermore, the addition of excess ZnBr$_2$ allows the formation of the opposite enantiomeric coupling product in 70% ee; this effect is general, but less pronounced, for the majority of the N-/P- and N-/P-/S- ligands tested (equation 67)[100].

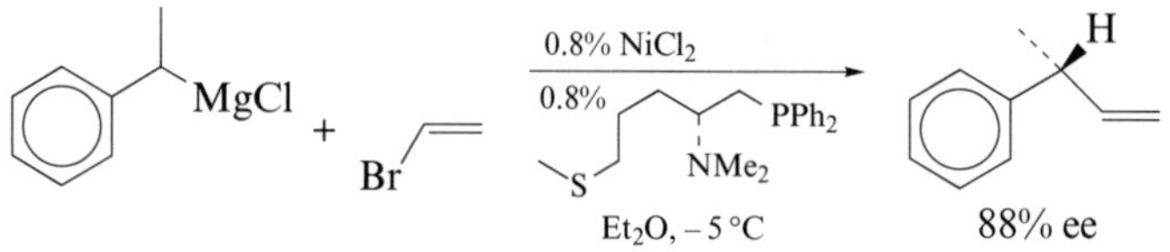

88% ee   (66)

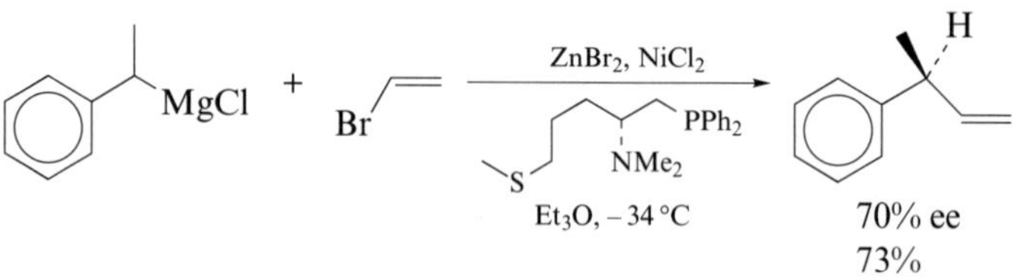

70% ee   (67)
73%

For aryl–aryl coupling reactions, monodentate phosphine ligands are normally more catalytically active than bidentate ones. Excellent asymmetric induction has consequently been realized by using ferrocene-derived **16** with NiBr$_2$, based on the assumed oxophilicity of magnesium. 2,2′-Binaphthyls may be prepared in up to 95% ee (*o,o′*-disubstituted) or 83% ee (*o*-substituted) in this manner (equation 68)[101]. Two coupling reactions of 1,5- and 1,4-dibromonaphthalene with *o*-substituted 1-bromomagnesionaphthalene have also been accomplished with the same catalysts to afford predominantly *dl* coupling products with over 95% ee (equation 69)[102].

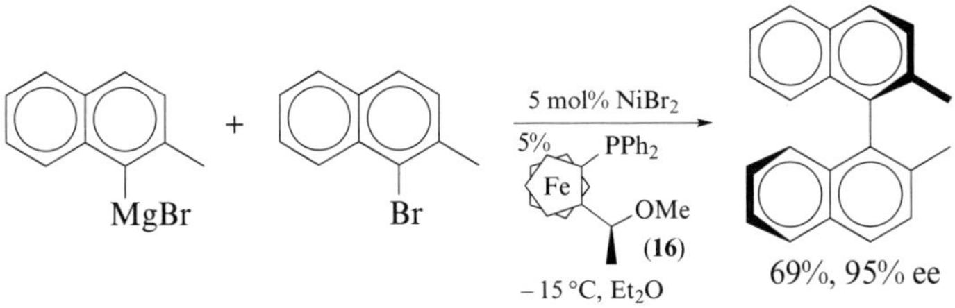

(68)

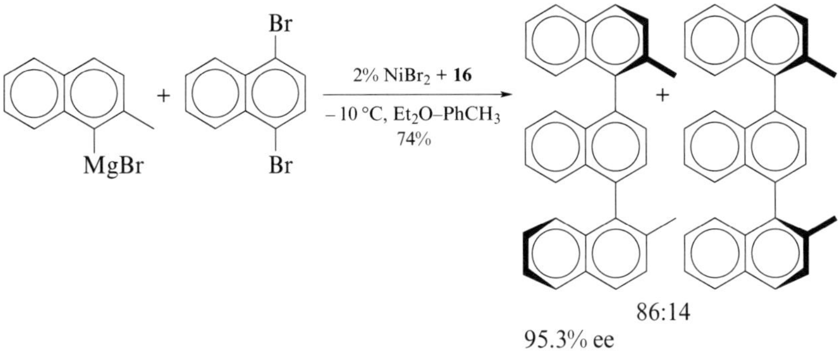

69%, 95% ee

(69)

86:14
95.3% ee

C. Cross-coupling Reactions of Grignard Reagents in the Presence of Palladium Complexes

A significant advance in the chemistry of palladium-catalyzed cross-coupling reactions has been the development of dppf (17) and its complex of $PdCl_2$, namely 18 [$PdCl_2$(dppf)].

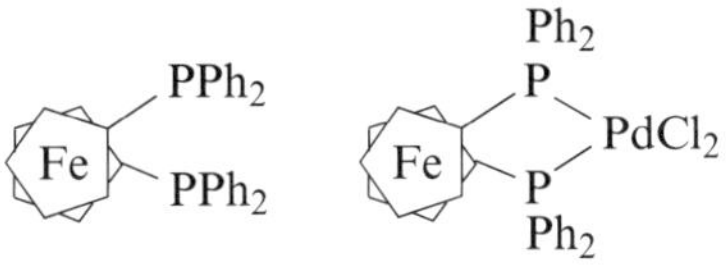

(17)   (18)

Use of the catalyst minimizes the amount of competing β-elimination and isomerization in cross-coupling involving alkyl groups, such that aryl and vinyl halides may be coupled with n- and sec-alkyl Grignard reagents (equation 70)[103]. This may be extended to aryl chlorides if additional 17 is present in the reaction mixture[104]. Aryl Grignard–alkyl iodide coupling may be accomplished with prior reduction of 18 with DIBALH (diisobutylaluminum hydride)[105]. Alkyl Grignard–alkyl iodide coupling reactions may also be effected with this catalyst, although reactions involving one or two secondary alkyl participants give suc-

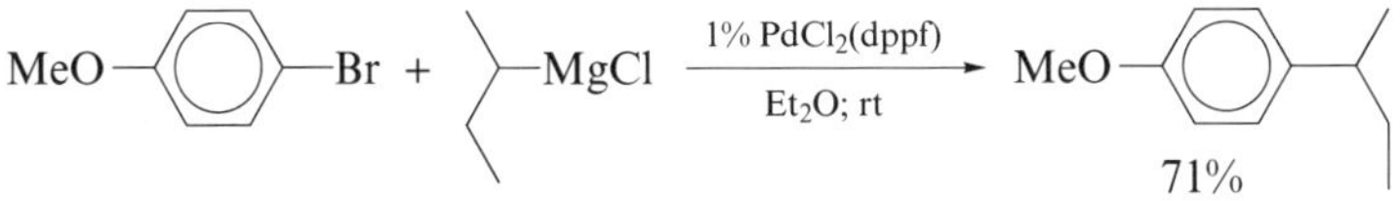

(70)

71%

cessively lower yields (equation 71)[105]. In some cases, excess ligand added to the reaction mixture further suppresses side reactions.

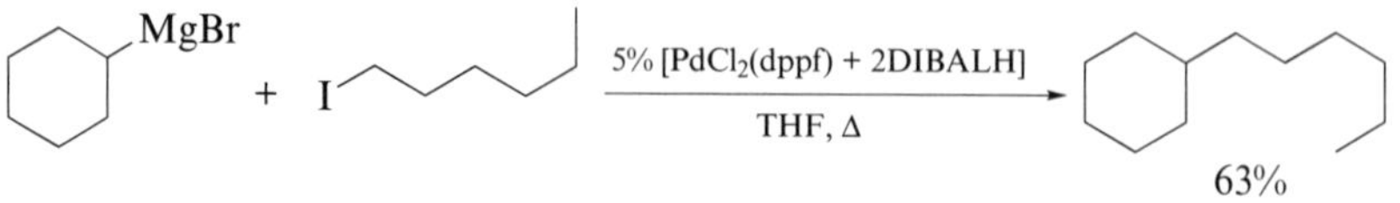

$$\text{(cyclohexyl)MgBr} + \text{I}\diagdown\diagup\diagdown\diagup \xrightarrow[\text{THF, }\Delta]{5\% \, [\text{PdCl}_2(\text{dppf}) + 2\text{DIBALH}]} \text{product} \quad 63\% \qquad (71)$$

Coupling reactions involving PdCl$_2$(dppf) are very selective. It has proven possible with catalyst to mono-alkylate or arylate dichlorobenzenes (equation 72) [although PdCl$_2$(dppb), where dppb = bis(1,2-diphenylphosphino)butane, also succeeds for the arylations] or dibromothiophenes[104,106], and to alkylate or arylate the *E*-isomer of isomeric mixtures of 1-bromo-1alkenes with excellent selectivity (equation 73)[107].

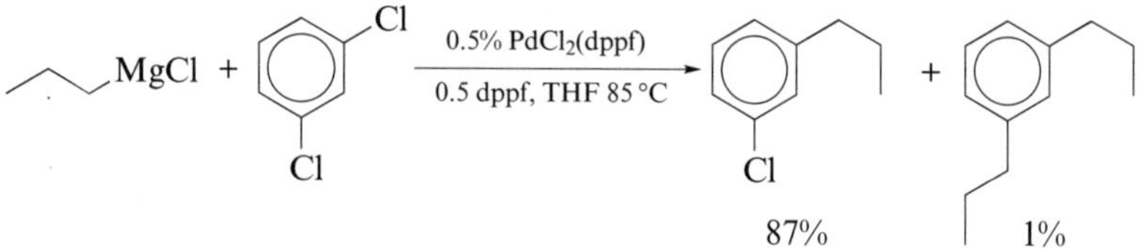

$$\diagup\diagdown\text{MgCl} + \text{(3,5-dichlorophenyl)} \xrightarrow[\text{0.5 dppf, THF 85 °C}]{0.5\% \, \text{PdCl}_2(\text{dppf})} \text{product} \, 87\% + \text{product} \, 1\% \qquad (72)$$

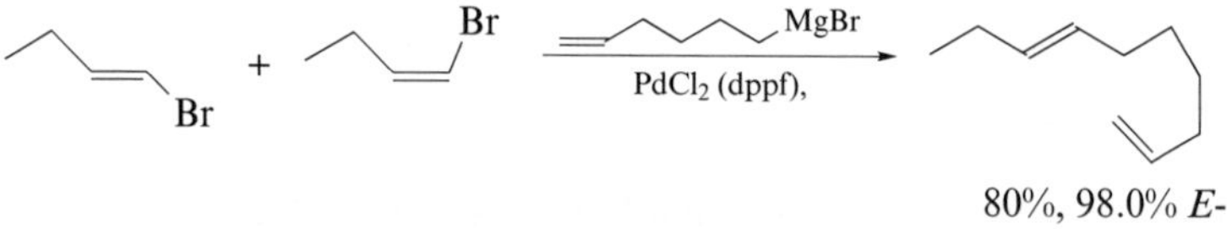

$$\diagdown\diagup\text{Br} + \diagup\diagdown\text{Br} \xleftrightarrow{\quad} \equiv\diagdown\diagdown\text{MgBr} \xrightarrow{\text{PdCl}_2 \, (\text{dppf}),} \text{product} \qquad (73)$$

$$80\%, \, 98.0\% \, E\text{-}$$

Other examples of selective coupling of polyhalogenated alkenes or heteroarenes have been reported using Pd(PPh$_3$)$_4$ as catalyst (equation 74)[108,109].

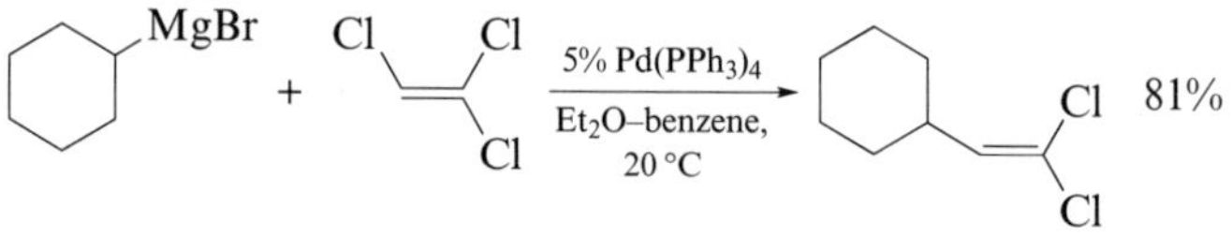

$$\text{(cyclohexyl)MgBr} + \text{Cl}_2\text{C=CCl} \xrightarrow[\substack{\text{Et}_2\text{O–benzene,} \\ 20 \, °\text{C}}]{5\% \, \text{Pd(PPh}_3)_4} \text{product} \, 81\% \qquad (74)$$

Reactions of secondary alkyl Grignard reagents with organic halides and palladium catalysts bearing chiral phosphine ligands have been shown to give useful levels of enantioselection. Although reactions of benzylic Grignard reagents and 1-bromoalkenes give ee values comparable to nickel catalysts with (aminoalkylferrocenyl)phosphine (19) ligands and ee values inferior to nickel catalysts with β-dimethylaminophosphine catalysts (15, R = *i*-Pr), much greater success is observed with 19 and α-(trimethylsilyl)magnesium bromide[110]. With this silyl-bearing Grignard and unsubstituted or *trans*-substituted 1-bromo-1-alkenes, excellent enantioselection (85–95% ee) is realized (equation 75). With

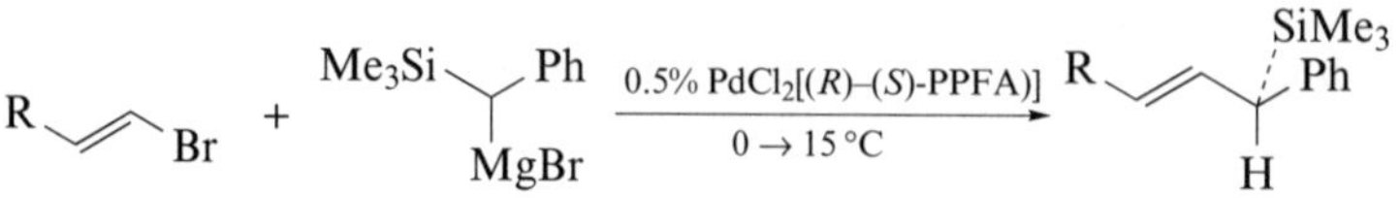

$$\text{R}\diagup\diagdown\text{Br} + \underset{\text{MgBr}}{\overset{\text{Me}_3\text{Si}\diagdown\diagup\text{Ph}}{}} \xrightarrow[0 \rightarrow 15 \, °\text{C}]{0.5\% \, \text{PdCl}_2[(R)\text{–}(S)\text{-PPFA}]} \text{R}\diagdown\diagup\overset{\text{SiMe}_3}{\underset{\text{H}}{\text{Ph}}} \qquad (75)$$

$$\text{R} = \text{H, 95\% ee (42\%)}$$
$$\text{R} = \text{Me, 85\% ee (77\%)}$$
$$\text{R} = \text{Ph, 95\% ee (93\%)}$$

cis-substituted alkene cases, much lower levels of enantioselection (15,24%) are observed, as is the case with alkynyl bromides (18% ee)[111]. Other attempts at developing aminophosphine catalysts for palladium-catalyzed couplings of Grignard reagents have produced only moderate ee values, although **20** couples phenethyl magnesium chloride and *β*-bromostyrene with ee values (40% ee) approaching that of **19** (46% ee)[112].

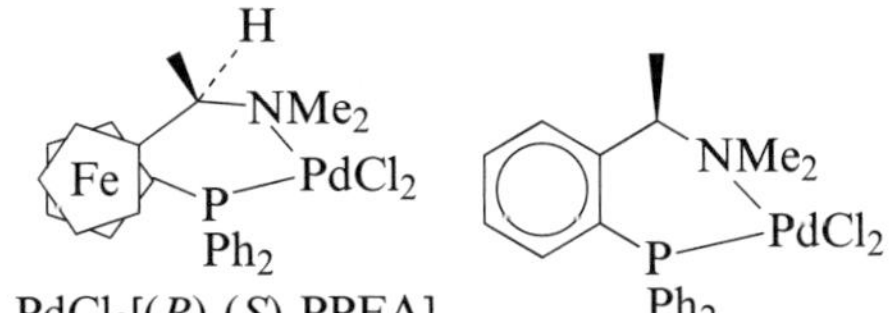

PdCl$_2$[(R)-(S)-PPFA]

(19)   **(20)**

Mechanistically, some light has been shed on the finer points of these cross-coupling reactions, particularly those dealing with the reductive elimination process. Although a *trans*-iodoarylpalladium(II) species forms initially upon oxidative addition of an aryl iodide (when monodentate phosphine ligands are present), the reductive elimination in cross-coupling occurs much more readily from a *cis*-diorganopalladium(II) species. Yamamoto and coworkers have found that in cross-coupling reactions with MeMgI, a

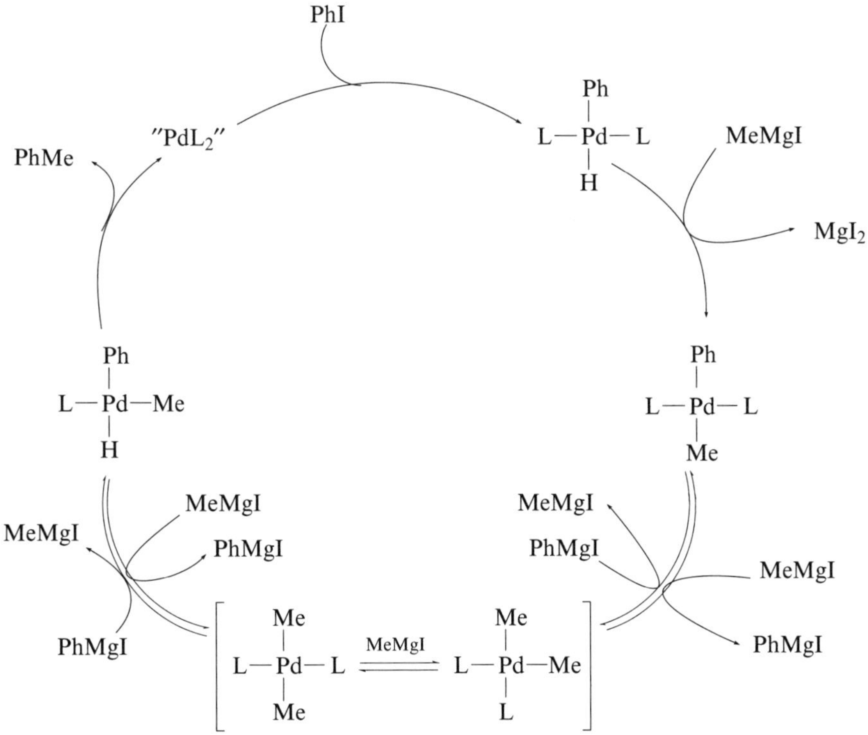

SCHEME 1

trans ⇌ *cis*-isomerization is promoted by excess Grignard reagent, probably through the involvement of *trans*- and *cis*-dimethylpalladium complexes[113]. The catalytic cycle in Scheme 1 has been proposed.

The reductive elimination itself has been studied for vinyl–alkyl coupling, in order to distinguish between a concerted reductive elimination process and a process involving migration of the alkyl group to the vinyl α-carbon, with the vinyl carbon remaining bonded to Pd (equation 76). Extended Hückel MO calculations favor the latter process, particularly where accompanying phosphine ligands allow relaxation of the P—Pd—P angle[114]. A Pd–alkene π-complex is the ultimate product of this reductive elimination, and intermediates of this type have been detected in Pd and Pt systems with dppf ligands[115]. It is apparent in alkyl–alkyl cross-coupling that the reductive elimination is more difficult and takes place by a different mechanism, which may involve either T-shaped 14-electron trivalent complex[116] or a five-coordinate Pd(IV) species[117,118].

$$
\begin{array}{ccc}
\underset{L}{\overset{L}{\diagdown}}\text{Pd}\underset{\substack{CH\\||\\CH_2}}{\overset{CH_3}{\diagup}} & \longrightarrow & \left[\underset{L}{\overset{L}{\diagdown}}\text{Pd}\underset{\substack{CH\\H\,||\\CH_2}}{\overset{CH_3}{\diagup}}\right] & \longrightarrow & \underset{L}{\overset{L}{\diagdown}}\text{Pd}-\text{C}\underset{CH_2}{\overset{H\diagdown\diagup CH_3}{}}
\end{array} \qquad (76)
$$

D. Cross-coupling Reactions of Organozinc Compounds in the Presence of Nickel or Palladium Catalysts

The past decade has seen extensive development of cross-coupling reactions of organozinc compounds and organic halides catalyzed by nickel or palladium catalysts. Although nickel-based catalysts are more reactive with respect to the organic halide partner, the number of failures with these catalysts and the greater selectivity realized with palladium-based catalysts have resulted in the almost exclusive use of the latter group of catalysts for these reactions.

The range of functional groups present in both partners is impressive. While most often employed for the coupling of sp^2–sp^2 centers, many cases exist of the use of functionalized alkyl zinc and alkynyl zinc reagents[119]. With respect to the halide partner, vinyl and aryl iodides and bromides are by far the most common candidates. Although reports of the successful use of 1-alkynyl iodides exist[119], they have not seen wide use, and there are also cases where these halides did not prove to be suitable substrates[120]. Reports of successful coupling with alkyl halides are limited to the more reactive members of that group, such as allyl[121], propargyl[122], and benzyl halides[123], α-halonitriles[124] or α-bromoesters[125]. Even among these alkyl halide partners, reported substrates which bear a β-hydrogen atom[122] are exceedingly rare. Allyl and propargyl halides provide an interesting reactivity contrast; propargyl bromides are attacked by aryl, vinyl or alkynyl zinc halides with clean S_N2' regiochemistry, affording the allenes[122], whereas allyl chlorides, with methyl- or allyl zinc halides and $NiCl_2(PPh_3)_2$ catalysis, give selective direct (S_N2-like) regiochemistry (equations 77 and 78)[121]. Acylation of the organozincs by acid chlorides may also be accomplished successfully in many cases[126].

The organozinc partner in these coupling reactions may also be the diorganozinc, R_2Zn[122]. While not an issue with simpler organic residues, it has been found with zinc bis(homoenolates) that only one of the organic units is transferred in cross-coupling reactions with the majority of halides tested (equation 79)[49].

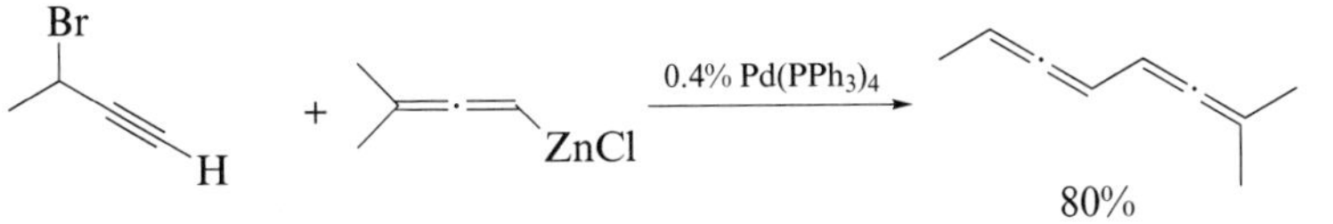

(77)

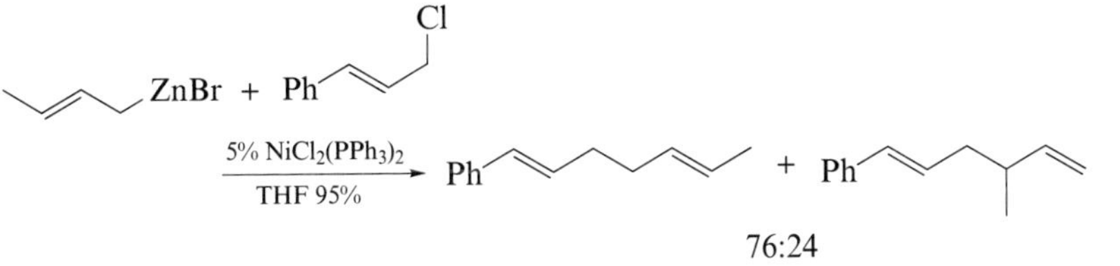

80%

(78)

76:24

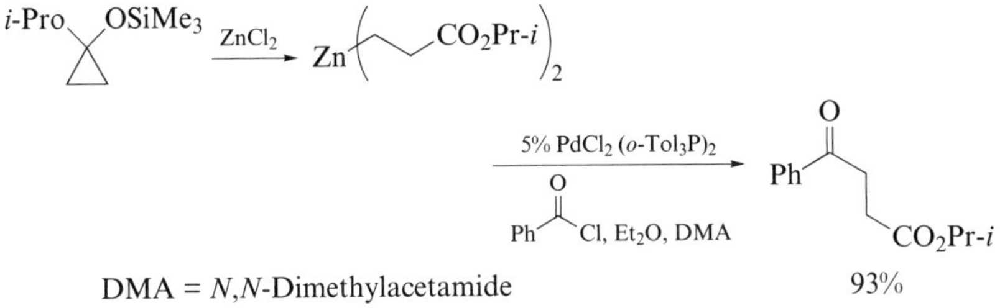

(79)

DMA = N,N-Dimethylacetamide   93%

Stereochemical issues have relatively well defined answers in palladium-mediated organozinc coupling reactions. In reactions involving vinylzincs or alkenyl halides, retention of double-bond stereochemistry is realized to a very high degree in almost all cases reported (equation 80)[127,128]. In the case of 1,2-dihaloalkenes or 2-substituted-1-haloalkenes[129,130], the E-isomer has a sufficiently greater reactivity than the Z-isomer such that a former will couple essentially to the exclusion of the latter in competition tests (equation 81). The logical assumption that the *trans*-halide of a 2-alkyl-1,1-dihaloalkene reacts preferentially over the *cis*-halide has been borne out in experimental work for the chlorides (equation 82)[131].

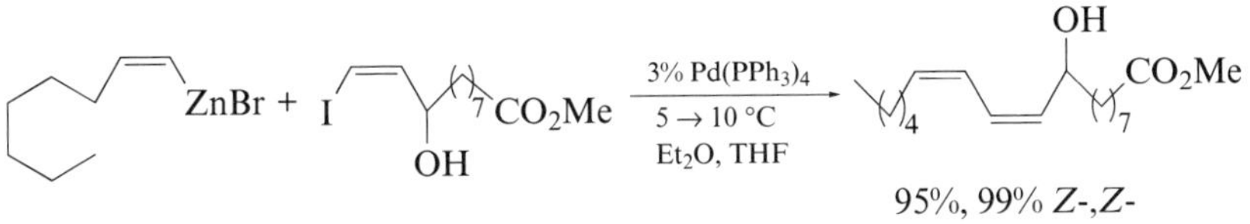

(80)

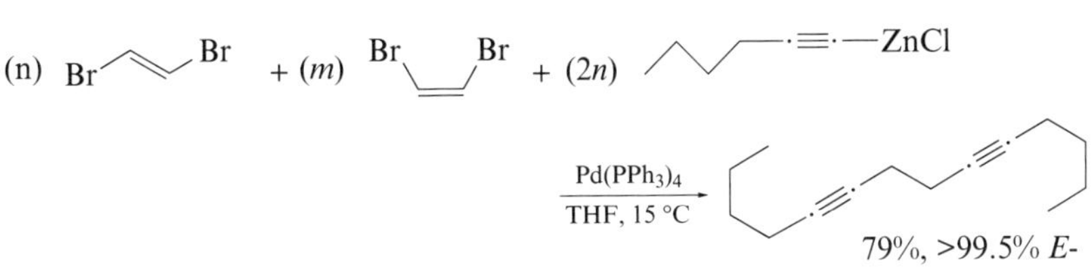

95%, 99% Z-,Z-

(81)

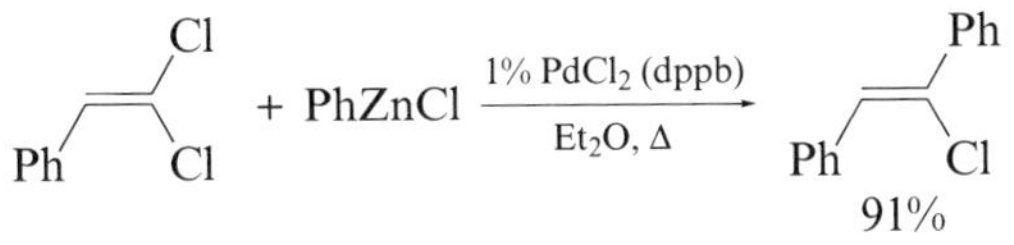

79%, >99.5% E-

(82)

91%

The situation is somewhat more complex for 1-haloallenes. With chlorides or bromides, Pd(PPh$_3$)$_4$ and PhZnCl or Ph$_2$Zn, predominant inversion of configuration is observed; with the corresponding iodides, variable levels of retention are realized[132].

Good levels of enantioselection in the reaction of α-phenethylzinc halides with bromoethene have been realized by the use of (aminoalkylferrocenyl)phosphine ligand 19 PdCl$_2$[(R)-(S)-PPFA] (85–86%% ee) (equation 83)[133]. These are superior to the identical

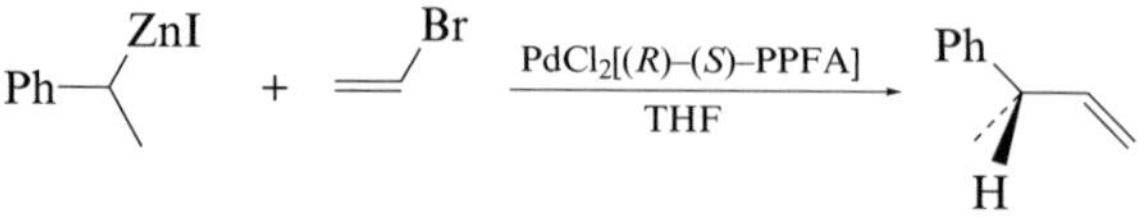

$$Ph\!-\!\!\!\underset{ZnI}{\overset{}{\diagup}} \quad + \quad \underset{}{\overset{Br}{\diagup}}\!\!=\!\! \quad \xrightarrow[\text{THF}]{\text{PdCl}_2[(R)\text{-}(S)\text{-PPFA}]} \quad \underset{H}{\overset{Ph}{\diagup}}\!\!\!\diagdown\!\!\!\diagup \quad \begin{array}{l}>95\%, \\ 80\% \text{ ee}\end{array} \qquad (83)$$

reactions involving either Grignard or aluminum-based reagents. Enantioselection with related partners (21,22) is more modest, but still significant. More recently, a C$_2$ (aminoalkylferrocenyl)phosphine (23) has catalyzed the α-phenethylation of β-bromo-

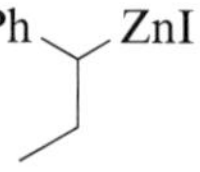

(21) 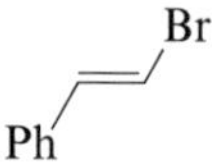   (22) 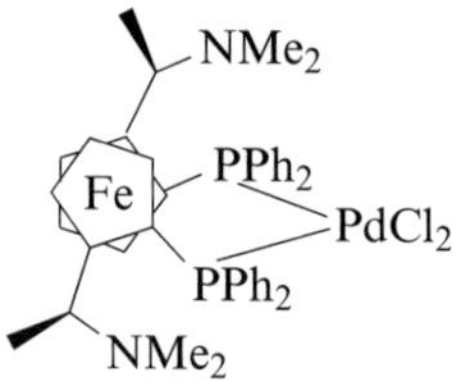   (23)

styrene in 93% ee (equation 84)[134]. Attempts to effect similar reactions with allenic zinc reagents have led to only modest ee values[135].

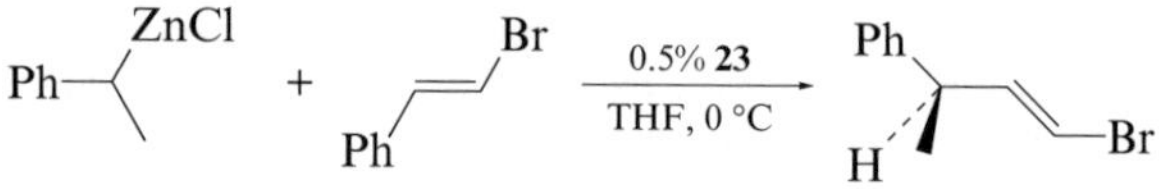

$$Ph\!-\!\!\!\underset{ZnCl}{\overset{}{\diagup}} \quad + \quad \underset{Ph}{\overset{Br}{\diagup}}\!\!=\!\! \quad \xrightarrow[\text{THF, 0 °C}]{0.5\%\ 23} \quad \underset{H}{\overset{Ph}{\diagup}}\!\!\!\diagdown\!\!\!=\!\!\!\diagdown\!\!Br \qquad (84)$$

'quantitative'
93% ee

The organic moieties of both reaction partners may be quite extensively substituted. Aromatic cases may be either π-excessive or π-deficient heterocycles of many types, or may bear either electron withdrawing or donating groups. A sample of heterocyclic aromatic halides are listed below (24–35)[136–144]. As expected, aryl halides bearing electron-

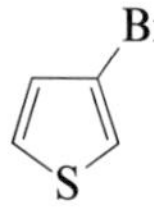

(24)   (25)   (26)   (27)   (28)
             X = Br, Cl, I   X = Br, I

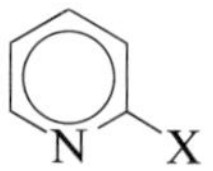

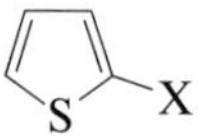

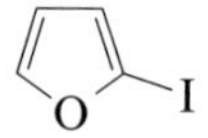

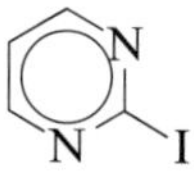

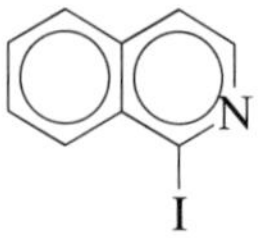

(29) 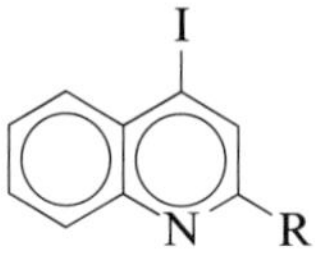   (30) 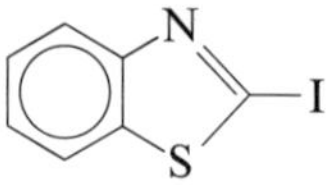   (31)   (32)

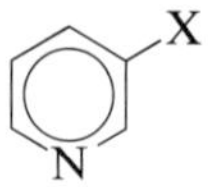

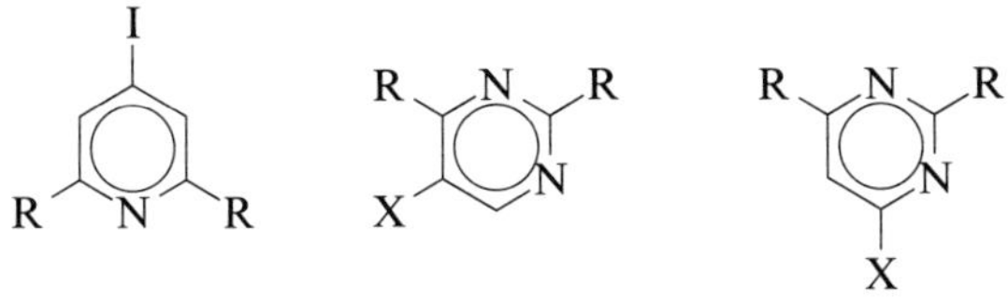

X = Cl, I   X = Br, I

(33)   **(34)**   **(35)**

withdrawing groups are more reactive than those bearing electron-donating groups, due to their greater facility for oxidative addition[145]. Since arylzinc halides are readily prepared by transmetallation from aryllithiums, cross-coupling reactions to form biaryls have been widely exploited[138]. One noteworthy application has seen the coupling of 2-thienyl- or 2-furylzinc halides with *m*- and *p*-dibromoarenes used in the preparation of benzenoid-heteroaryl polymers (equation 85)[140,146].

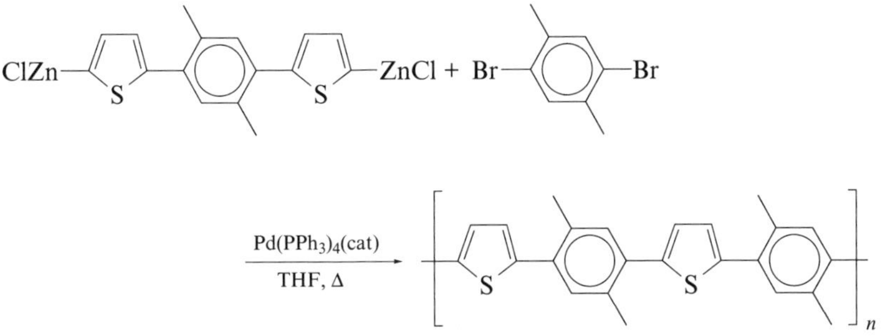

$$(85)$$

Of the couplings involving vinyl partners, a particularly popular application has been in the synthesis of highly fluorinated olefins. This has been accomplished in both ways, by attack of perfluorovinyl zinc reagents (equation 86)[145,147,148] and by attack upon perfluorovinyl iodides (equation 87)[148,149]. Organic halide–vinylzinc couplings have seen several other applications, including those of acyl anion equivalents (equaion 88)[150,151] or enolate equivalents for the α-vinylation or α-arylation of ketones (equation 89)[152,153].

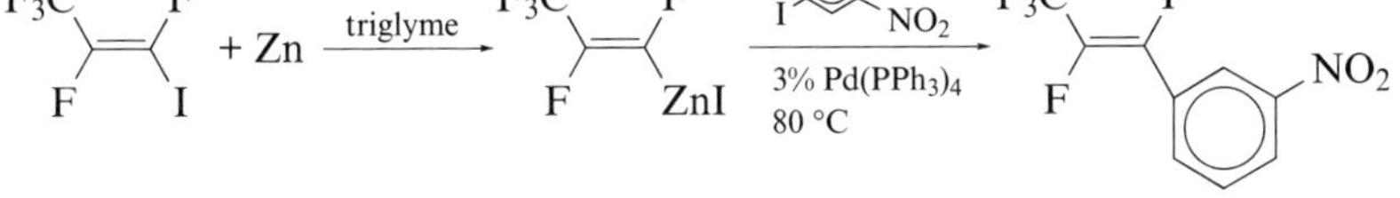

$$(86)$$

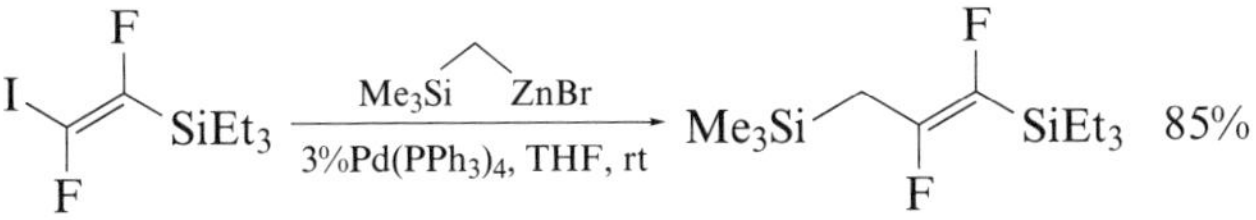

80%

$$(87)$$

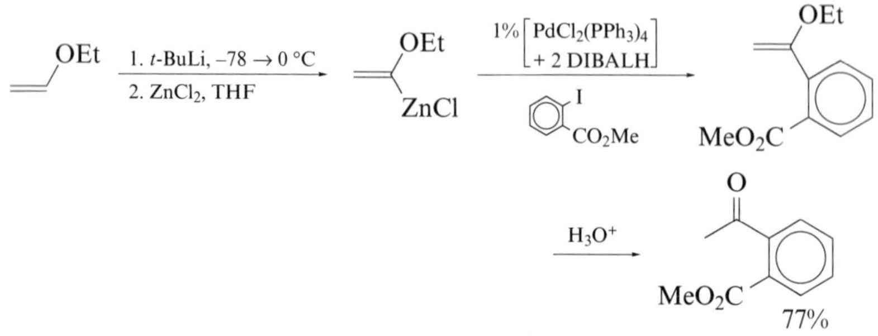

$$(88)$$

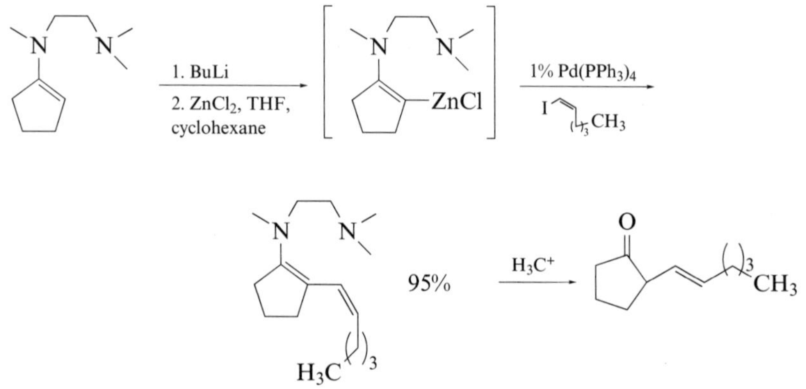

$$(89)$$

Furthermore, coupling to form C-aryl glucals, as precursors to C-aryl glycosides, was carried out both from the C-metallated glucal[154] and onto the C-iodo glucal (equations 90 and 91)[155]. A synthesis of vineomycinone B$_2$ methyl ester has resulted from this chemistry. A vinyl–vinyl coupling reaction has also formed the cornerstone of aryl steroid synthesis (equation 92)[156].

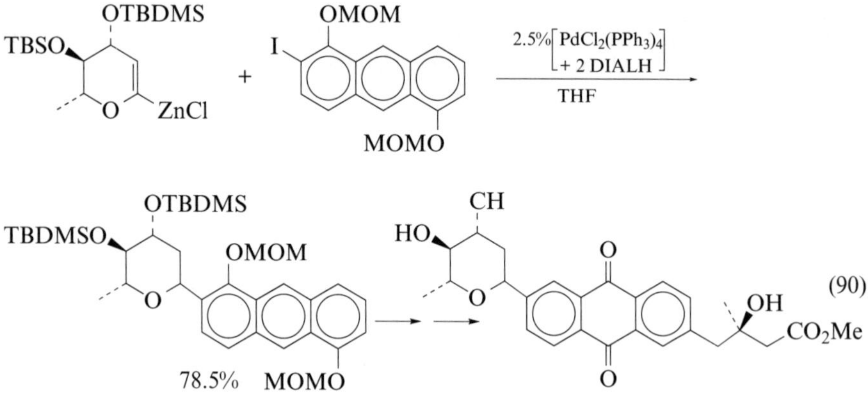

$$(90)$$

OMOM = O-methoxymethyl
OTBDMS = –OSi(t-Bu)Me$_2$

Vineomycinone B$_2$ methyl ester

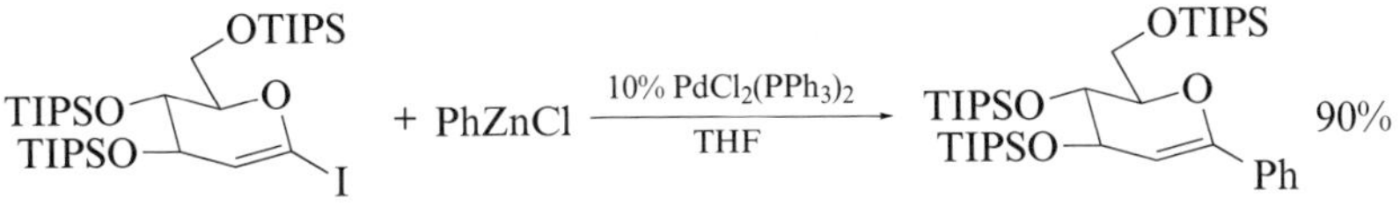

$$\text{(91)} \quad 90\%$$

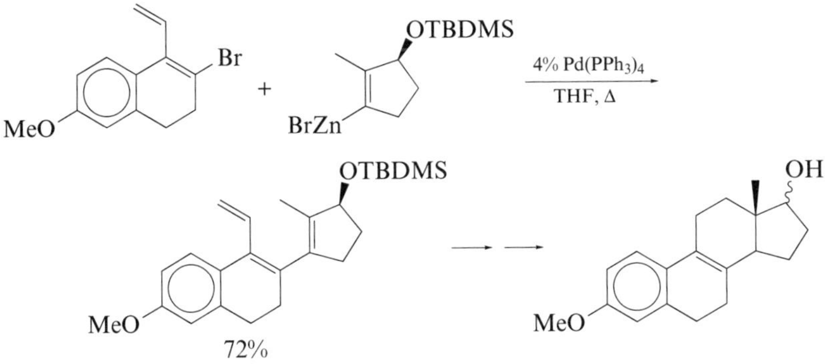

$$\text{(92)}$$

Due to the ability of generating organozinc species directly from organic halides by zinc metal, and due to the very mild reactivity of organozincs, alkylzinc halides can be generated and coupled in the presence of many functional groups, such as esters[49,157], chlorides[158], secondary amines[159] and ketones[157]. As a result, these species may be successfully employed as homoenolate and perhomoenolate equivalents (equation 93). The iodozincs generated

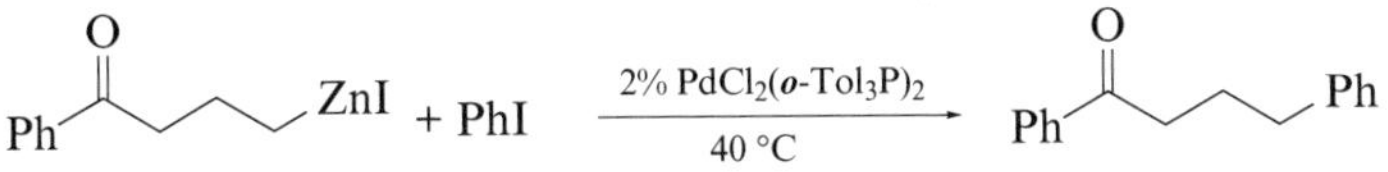

$$\text{(93)}$$

from β-iodoalanine derivatives have been coupled with aryl and vinyl iodides without compromising the enantiomeric purity of the amino acid[159]. A further application of halide–alkylzinc coupling has featured the use of a cyclopropylzinc halide and a vinyl iodide in the synthesis of the tricyclic sesquiterpenoids (±)prezizaene and (±)prezizanol (equation 94)[160].

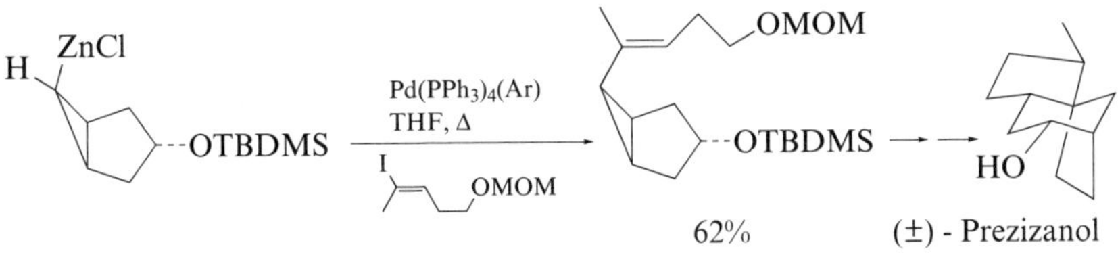

$$\text{(94)}$$

Acetylenic zinc reagents are readily accessible from 1-alkynes by a deprotonation–transmetallation sequence, and consequently have been successfully used to couple with vinyl halides in many cases (see equation 81), including butadiynyl ones[161]. Among these, coupling reactions with vinyl bromides have been used in the stereoselective synthesis of enediynes and applied to the synthesis of trienic natural products[162].

E. Cross-coupling Reactions of Organoaluminum or Organozirconium Compounds in the Presence of Nickel or Palladium Catalysts

Although the development of cross-coupling reactions in this class have been much slower than those of other organometallics, some noteworthy advances have been reported. The original work on cross-coupling reactions of vinyl alanes and vinyl zirconocene halides with vinyl and aryl halides has been elaborated more completely[163]. A comparison of metals in Pd(PPh$_3$)$_4$-catalyzed vinylmetal–vinyl iodide coupling reactions has resulted in the following relative abilites to effect the coupling: M = ZnCl ≈ ZrCp$_2$Cl > CdCl > AlBu-i_2 > B(siamyl)$_2$ > others. In cases where vinylzinc–vinyl halide coupling reactions have been described, the use of vinyl alanes is often also quite satisfactory, such as in the vinylation of α-iodocycloalkenones[153].

Organoaluminum–arene coupling reactions[119] have been extended to the point where they are feasible with heteroarenes (equation 95)[164,165]. Allyl chlorides are also suitable coupling partners with vinyl alanes under palladium catalysis[166]. High regioselectivity and stereoselectivity is only observed, however, in the cases of γ,γ-disubstituted allyl chlorides (equation 96). In the presence of stoichiometric amounts of Pd(PPh$_3$)$_4$, the incorporated nucleophile may even be H (from i-Bu$_2$AlH)[167].

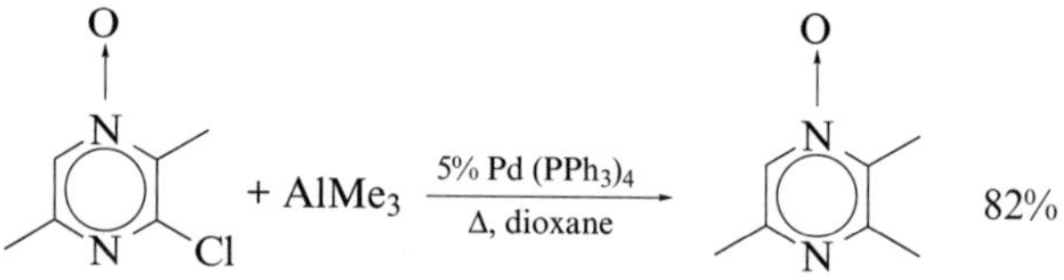

$$82\% \qquad (95)$$

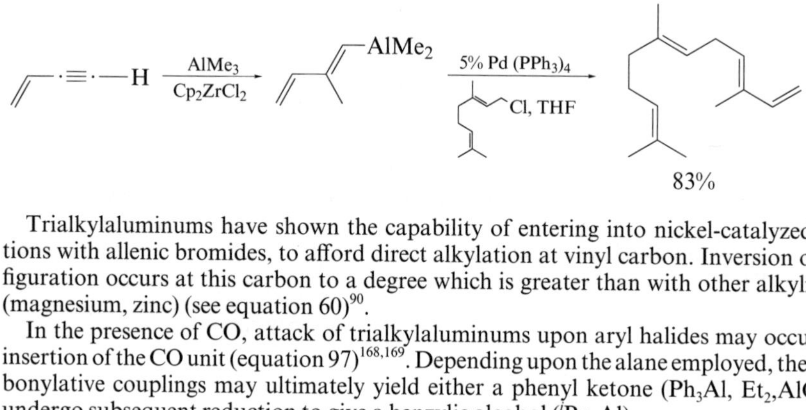

$$83\% \qquad (96)$$

Trialkylaluminums have shown the capability of entering into nickel-catalyzed reactions with allenic bromides, to afford direct alkylation at vinyl carbon. Inversion of configuration occurs at this carbon to a degree which is greater than with other alkylmetals (magnesium, zinc) (see equation 60)[90].

In the presence of CO, attack of trialkylaluminums upon aryl halides may occur with insertion of the CO unit (equation 97)[168,169]. Depending upon the alane employed, these carbonylative couplings may ultimately yield either a phenyl ketone (Ph$_3$Al, Et$_2$AlCl), or undergo subsequent reduction to give a benzylic alcohol (^{t}Bu$_3$Al).

$$99\% \qquad (97)$$

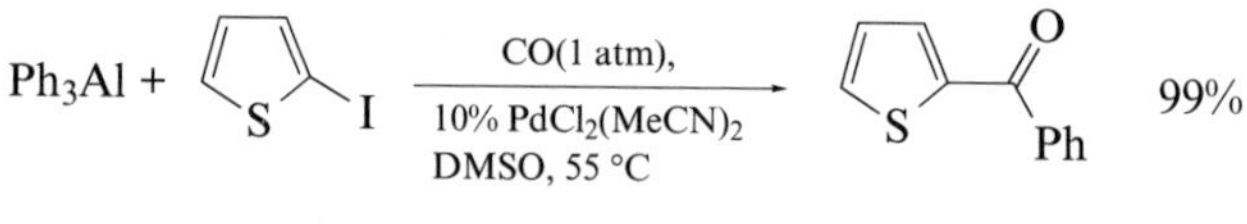

New reports of organozirconium–organic halide cross-coupling reactions have almost completely stopped. Nevertheless, the palladium-catalyzed cross-coupling of vinyl zirconocenes (from alkyne hydrozirconation) with vinyl halides has been employed in the synthesis of the lipid isobutylamide natural product anacyclin (equation 98)[170].

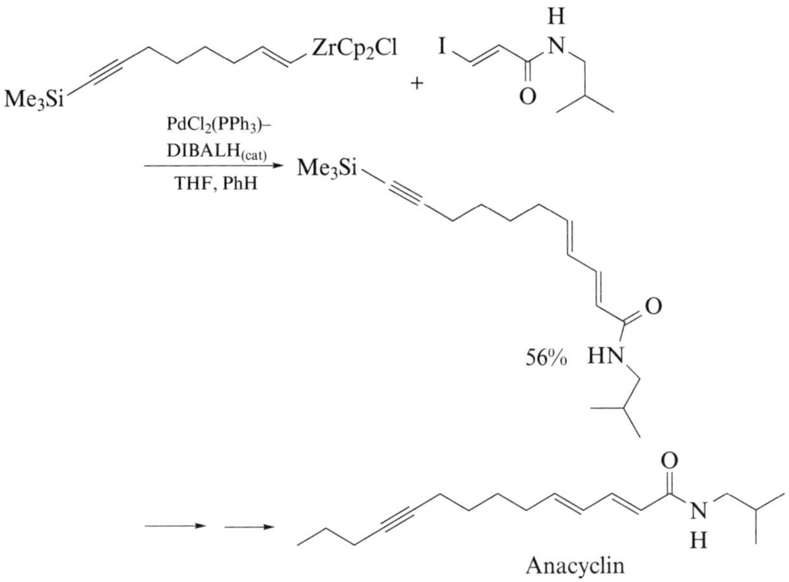

(98)

Anacyclin

F. Cross-coupling Reactions of Tetraorganotin Compounds in the Presence of Palladium Catalysts

The palladium-catalyzed cross-coupling reactions of organostannanes and organic halides has become extremely popular in the past decade. This is likely due to the greater stability of the organotin precursors relative to other organometallics which are capable of entering into analogous reactions. In theory, a problem exists with the use of tetraorganotins in such reactions in that competitive transfer of the four R groups on tin may occur. In practice, however, alkyl groups are the least readily transferred R groups $(RC \equiv C - > RCH = CH - > RCH = CHCH_2 - \approx ArCH_2 - > ROCH_2 - > alkyl)$[171], and since the most widely available organostannanes are $R'SnBu_3$, or $R'SnMe_3$, clean transfer of the desired R' group normally occurs without incident, If alkylation of a certain organic halide is required, tetraalkyltins with four identical organic groups (R_4Sn) are normally employed. The transfer of an organic group from an allylstannane normally occurs with allylic transposition[172], although the presence of an electron-withdrawing group can override this tendency (equations 99 and 100)[173]. Secondary benzylic stannanes couple with predominant inversion of configuration at the alkyltin[174].

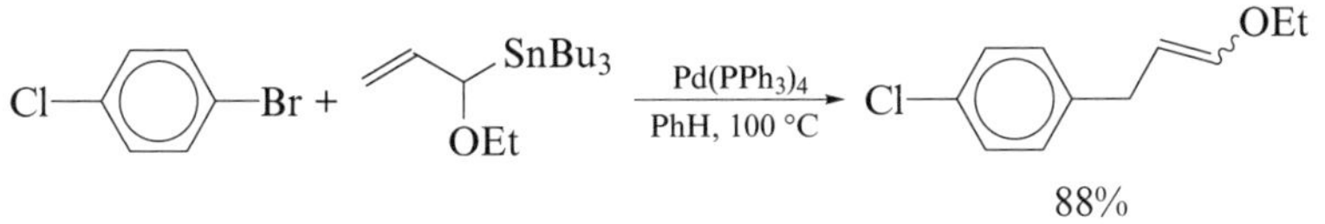

(99)

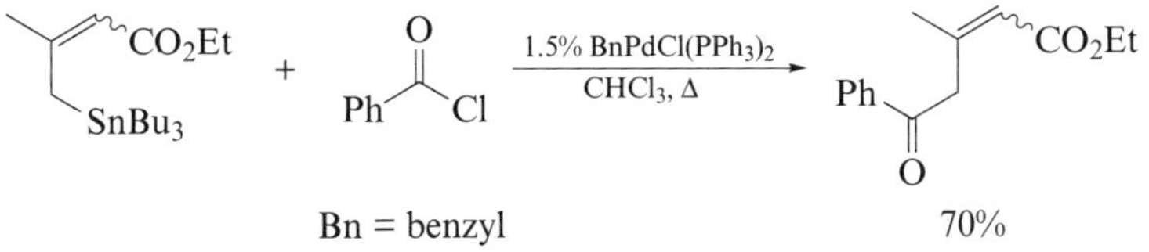

88%

(100)

Bn = benzyl   70%

While the choice of catalyst has often been a matter of trial and error, Pd(PPh$_3$)$_4$ and BnPdCl(PPh$_3$)$_2$ have been employed in the majority of cases. In several instances, a group of nonphosphine-containing 'ligandless' catalysts, [(η_3-allyl)PdCl]$_2$, (CH$_3$CN)$_2$PdCl$_2$ and LiPdCl$_3$, give unusually rapid coupling reactions under mild conditions for aryl halides in solvents such as HMPA, DMF or acetone[175]. The combinations of Pd$_2$(dba)$_3$-(2-furyl)$_3$P or Pd$_2$(dba)$_3$-AsPh$_3$ (where (dba) = dibenzylideneacetone) have been advanced recently as ideal catalysts for the couplings of vinyl-, aryl- or allylstannanes with a variety of organic halides (or triflates)[176]. These ligands facilitate coupling by being of intermediate 'donicity', being sufficiently electron rich to stabilize Pd(0) intermediates, but dissociating easily enough from Pd(II) to enhance the rate of the (rate-determining) transmetallation step. Several excellent reviews in this area have been published[177,178].

Acid chlorides have been coupled successfully with a wide range of organotin species. Vinyltins often undergo Z to E double-bond isomerization in the course of this coupling, in contrast to most other vinyltin couplings. This isomerization has been minimized in some cases by the use of 1:4 Pd$_2$(dba)$_3$:(2-furyl)$_3$P combination[176]. Many functional groups are tolerated, including tricarbonyliron functions complexed to 2-stannylated butadienes (equation 101)[179]. In the case of L-proline derived acid chloride substrates, coupling without recemization occurs with the use of PdCl$_2$(dppf) (equation 102)[180]. Intramolecular cyclization reactions via coupling with acyl halides is also fasible, forming both normal and large ring sizes in reasonable yields (equations 103 and 104)[181,182]. In many cases, a CO atmosphere is necessary to avoid decarbonylation.

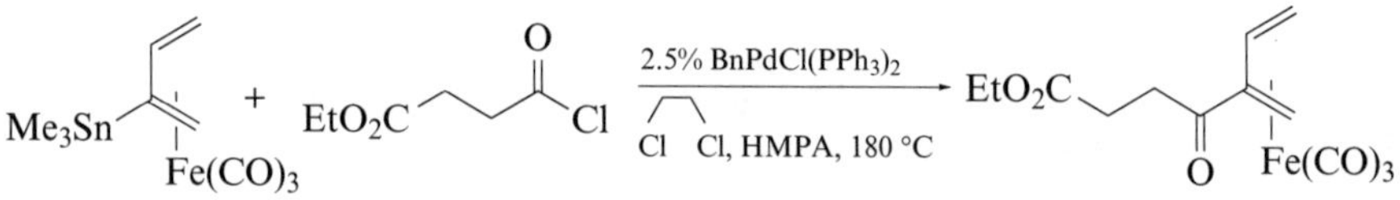

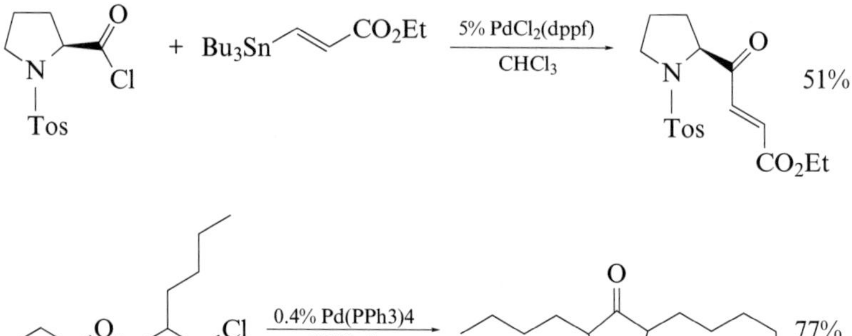

$$\text{(101)}$$

$$\text{(102)}$$

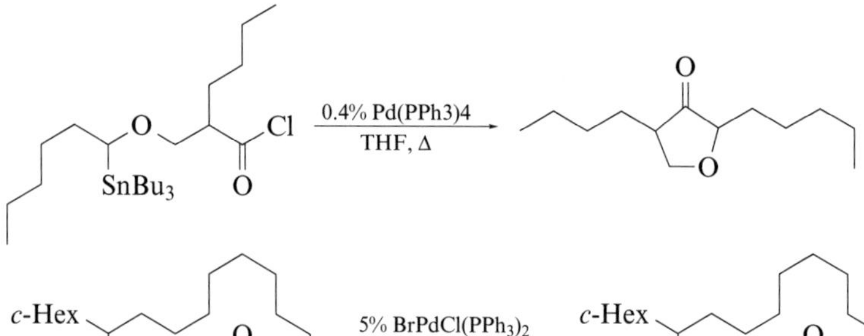

$$\text{(103)}$$

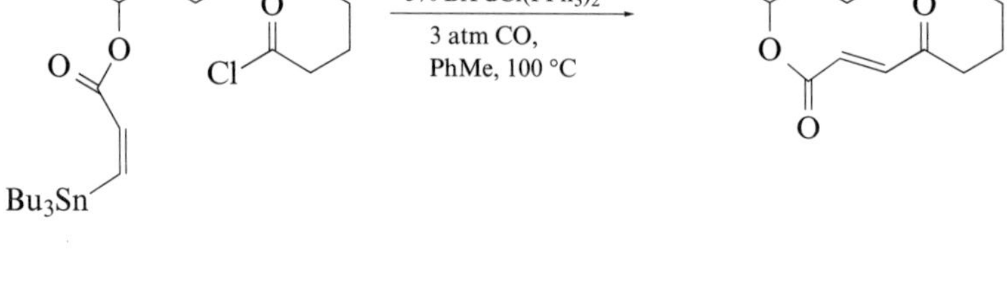

$$\text{(104)}$$

Other related species, such as alkyl chloroformates[183], dialkyl carbamyl chlorides[1], N-aryl iminoyl chlorides[184] or N-substituted isocyanide dichloride (with alkynyl stannanes only)[185] behave in an analogous manner. Acylation of acid chlorides has even proven to be possible with acyltin reagents, or by dimerization of acid chlorides with hexabutylditin and high CO pressure[186].

There are scattered reports of cross--coupling reactions between functionalized alkyl halides, such as α-bromo ketones[187], α-bromo esters[188] or α-halo ethers[189], with various organotin compounds (equations 105 and 106). Successful coupling has been observed even when the halides possess β-hydrogen atoms, and in at least one case where the organotin residue contains a hydrogen atom β to the coupling site[187]. Perfluorinated alkyl halides, which do not posses β-hydrogen atoms, also successfully couple with a wide variety of organotin species[190].

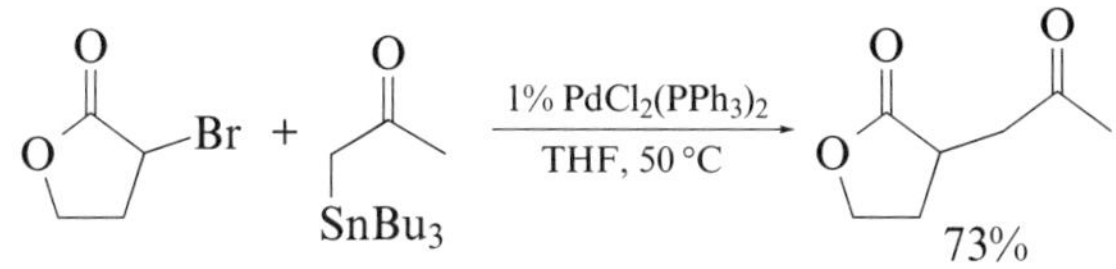

$$(105)$$

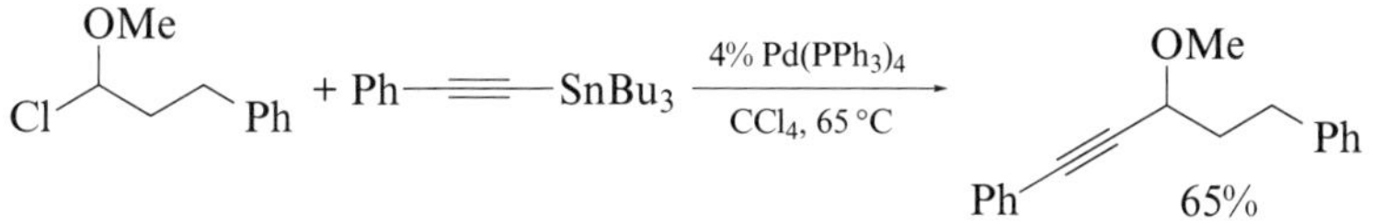

$$(106)$$

Other activated alkyl halides, namely allylic and benzylic chlorides and bromides, are suitable substrates for coupling. The regiochemistry of attack upon allyl halides depends upon the substrate; attack is predominantly at the least substituted end of the allyl unit, or at the terminus furthest removed from an electron-withdrawing group (equation 107)[191]. Double-bond stereochemistry is not reliably retained, with the most stable geometry of the double bond being realized in most cases[192]. In the case of cyclic allyl halides, nucleophilic attack occurs on the side of the ring *anti* to the halide leaving group (equation 108). This result is consistent with a substitution mechanism involving S$_N$2-like oxidative addition to form a π-allylpalladium(II) species, which undergoes subsequent nucleophilic substitution at palladium and reductive elimination with retention of configuration (Scheme 2). The incorporation of an acyl anion equivalent is possible by running the couplings under CO pressure; neither the regiochemical nor stereochemical characteristics of substitution are substantially altered, however (equation 109)[191,192].

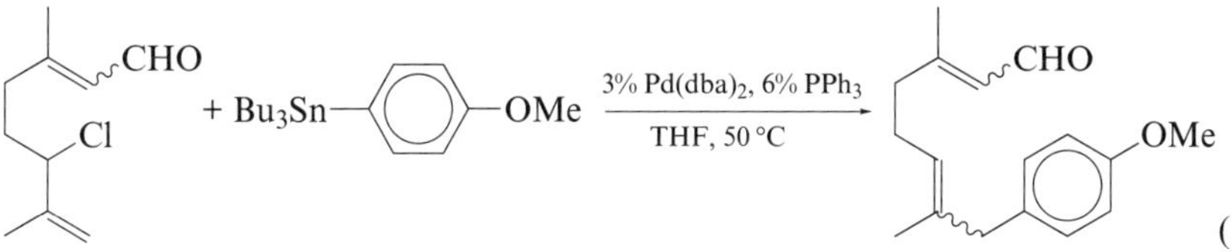

$$(107)$$

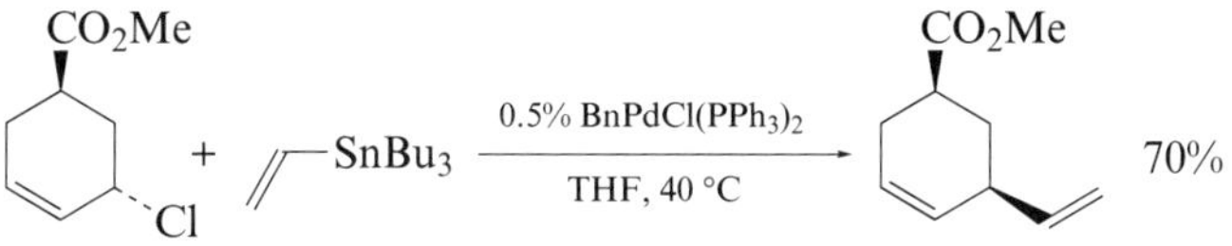

$$(108)$$

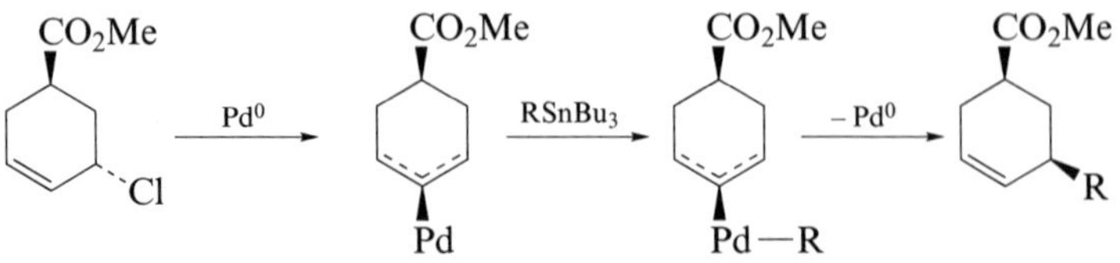

SCHEME 2

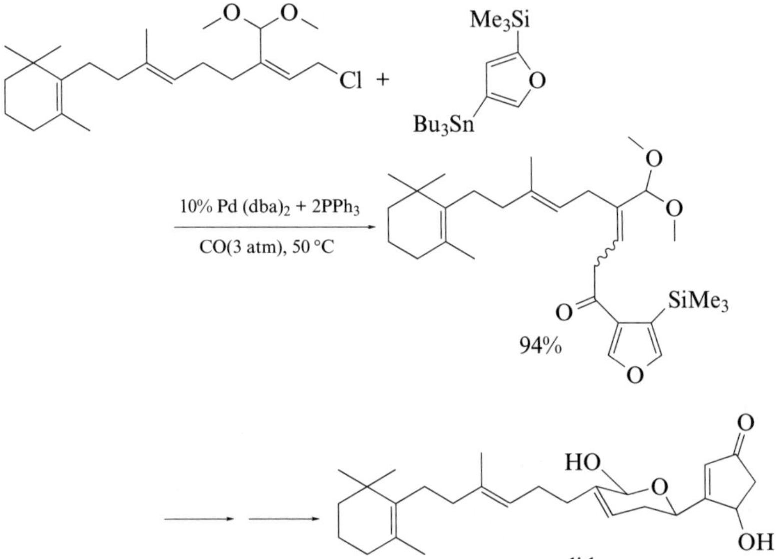

10% Pd (dba)$_2$ + 2PPh$_3$
────────────────
CO(3 atm), 50 °C

94%

manoalide

(109)

Substitution of benzyl bromides has also been accomplished, but the presence of β-hydrogen atoms in the halide gives coupling in reduced yields[177,193]. The use of deuterated benzylic substrates has revealed inversion of configuration in the overall process.

Vinyl halides couple stereospecifically with a wide range of organostannanes[194,174]. Iodides and bromides couple successfully in unactivated cases; chlorides may also be used in the cases of β-halo enones (equations 110 and 111)[195,196]. The iodides are sufficiently reactive in many cases for coupling to occur at room temperature. Hypervalent vinyl iodonium salts have been shown to be very reactive in couplings with vinyl or allyltins (equation 112)[197]. The use of CuI as a co-catalyst has enhanced yields in the case of less reactive vinyl halides[196,198]. The recent interest in enyne antibiotics has led to the use of intramolecular alkynylstannane–vinyl halide coupling reactions for the formation of large rings (equation 113)[199].

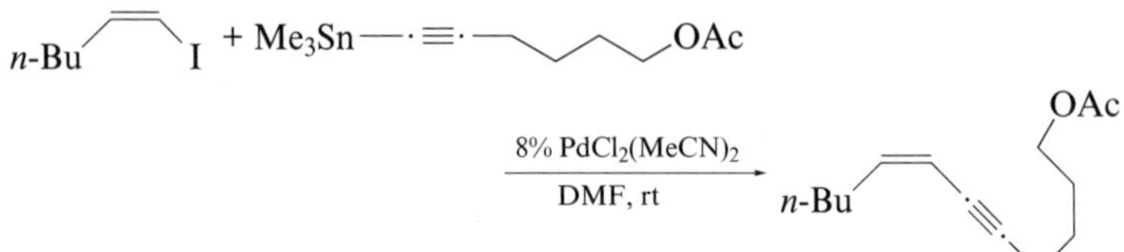

8% PdCl$_2$(MeCN)$_2$
────────────────
DMF, rt

91%   (110)

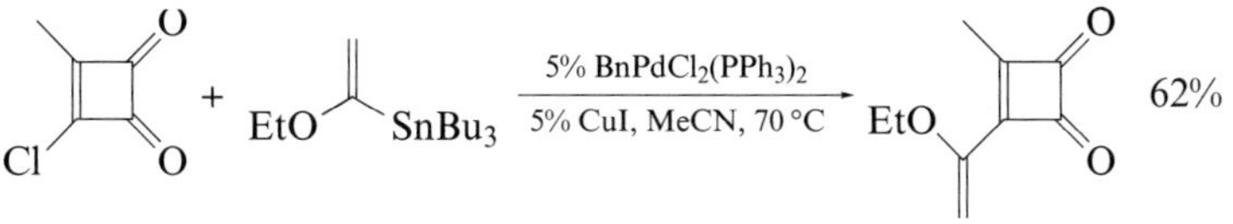

$$62\% \quad (111)$$

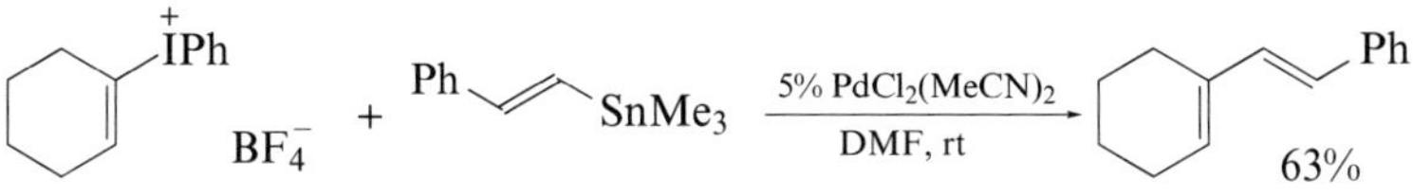

$$(112)$$

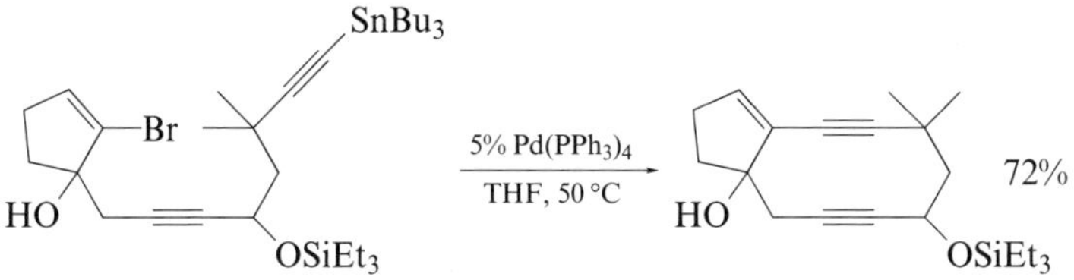

$$63\%$$

$$72\% \quad (113)$$

Carbonylative couplings are also possible with vinyl iodides when reactions are conducted under CO pressure[200]. In these cases, double-bond isomerization occurs when vinylstannanes are used; this is due to product isomerization rather than lack of stereospecificity of the coupling itself (equation 114).

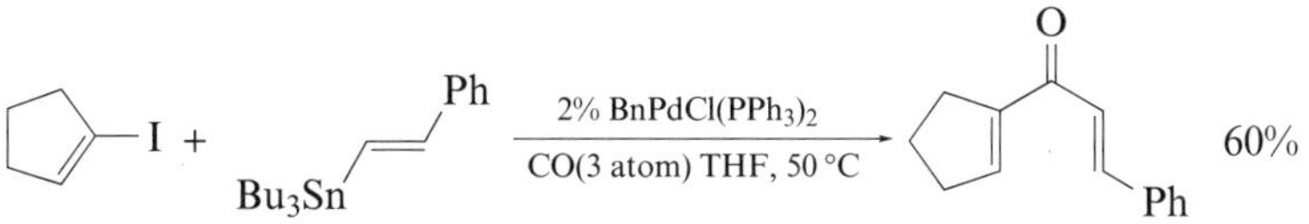

$$60\% \quad (114)$$

Many successful examples of aryl halide coupling have been reported. Again bromides and iodides are most suitable substrates, while chlorides have reacted successfully when complexed to a chromium tricarbonyl unit[201,202]. Both electron-withdrawing and electron-donating groups are tolerated on the arene, although the reaction is slowed in the latter case. *Ortho*-substituents do not seriously affect coupling, and the reactions are often successful with 2,6-disubstituted substrates[203–205]. The most common organotin partners are vinylstannanes (equations 115 and 116)[204–206], while acetylenic tins[201,203,207], heterocyclic tins[208], allyltin (equation 99)[173,174] and functionalized alkyltin (lacking β-hydrogens) reagents (equation 117)[209] have also been employed regularly. Arylstannane–aryl halide coupling reactions are known, but have been used rather infrequently[207,210]. The vinylstannane–aryl halide couplings have been conducted intramolecularly, and small[211] and large ring systems have been constructed in this manner (equation 116)[212].

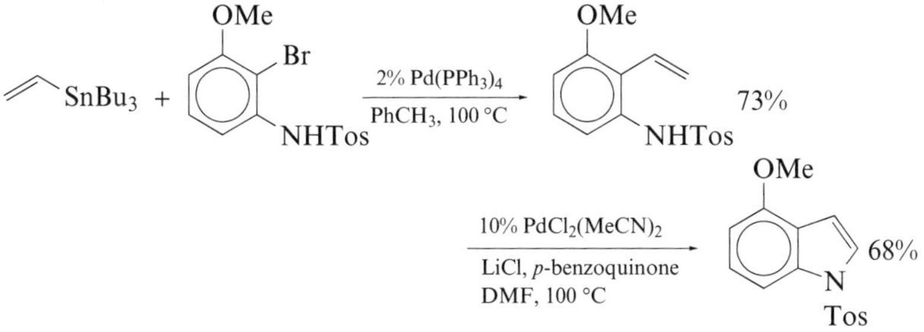

$$73\%$$

$$68\% \quad (115)$$

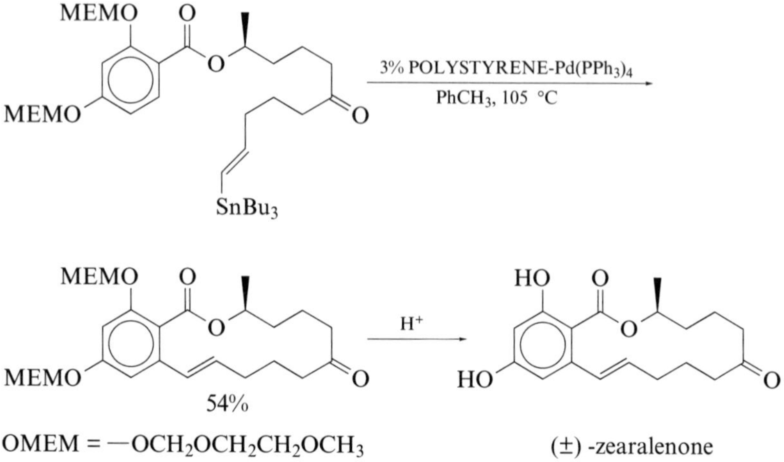

$$(116)$$

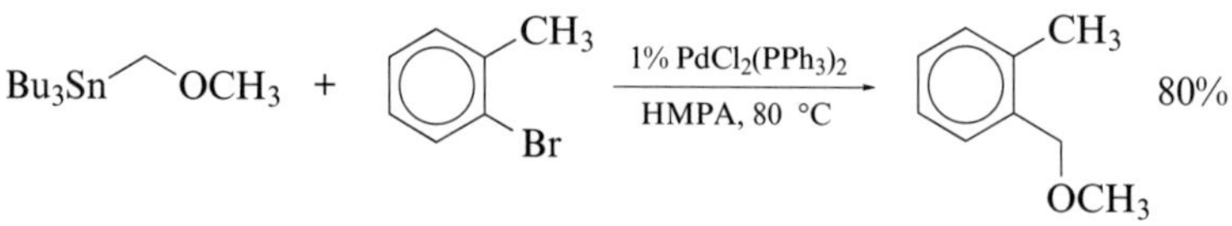

OMEM = $-OCH_2OCH_2CH_2OCH_3$ $(\pm)$ -zearalenone

$$(117)$$

The relative ability of aryl halides and aryl triflates to participate in these reactions (triflates have also been heavily explored in cross-coupling reactions with organostannanes[213]) depends upon the catalyst employed. Whereas the reactivity order is I > Br > OTf with Pd(PPh$_3$)$_4$ or in the absence of LiCl, PdCl$_2$(PPh$_3$)$_2$ + LiCl gives the order I > OTf > Br[213]. This reversal with the latter catalyst has been explained as stemming from a rate enhancement of oxidative addition to the triflate, driven by oxygen complexation to the coordinatively unsaturated catalyst. Stille and coworkers have used the greater reactivity of aryl bromides to perform two sucessive coupligs in bromoaryl triflates to ultimately prepare functionalized indoles[205].

A number of aromatic heterocyclic halides have also proven to be suitable substrates for coupling with organotins[208]. Both π-deficient and π-excessive heterocycles may participate, including halofurans[208], thiophenes[208,214], pyridines[208,215] and more highly nitrogenated heterocycles[216,217]. Finally, 1-haloalkynes (bromides, iodides) recently have been shown to be capable of entering into cross-coupling reactions with vinyltins[218].

It is possible to accomplish organostannane–organic halide cross-coupling reactions via *in situ* formation of the organostannane from another organic halide, or an allylic acetate or carbonate. Allylic acetates for example, undergo inter- or intramolecular couplings with aryl, vinyl or heteroaryl bromides upon heating with a palladium catalyst and hexaalkylditin or (tributylstannyl)diethylalane (equations 118 and 119)[219,220]. The most likely route of these transformations involves conversion of the allylic acetate to an allyl tin, followed by a conventional coupling. Carbon–carbon bond formation occurs at the least substituted end of the allyl moiety in intermolecular cases, and to form five-membered rings in preference to seven-membered rings in the intramolecular cases. Intramolecular aryl halide–aryl halide (bromide, iodide) couplings have been accomplished in an analogous manner; both six- and seven-membered rings are formed efficiently in this process (equation 120)[221,222]. In one case, an intramolecular benzyl–benzyl coupling has also been effected[222].

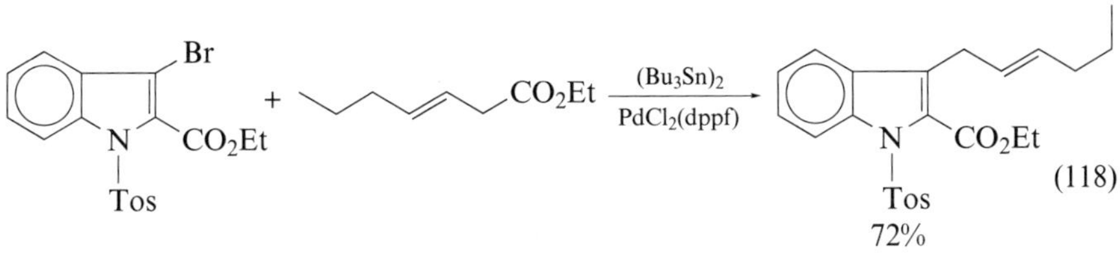

(118)

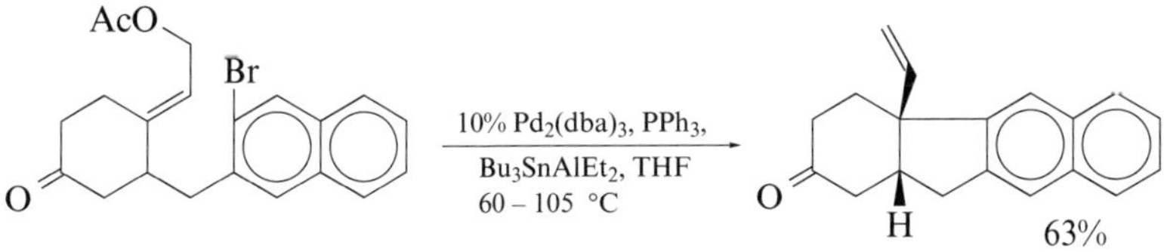

(119)

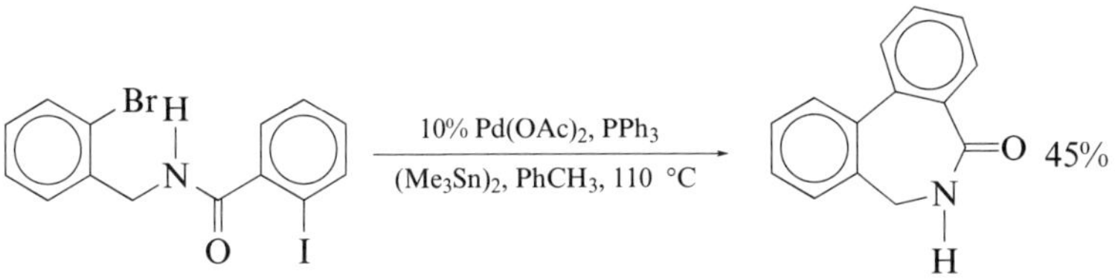

(120)

G. Cross-coupling Reactions of Organoboron Compounds in the Presence of Palladium Complexes

Palladium-mediated coupling reactions of organoboron compounds with organic halides have been developed extensively in the last decade[223]. The organoboron compounds in question have fallen into two classes generally: organoboranes and organoboronic acids or esters. Soderquist and coworkers have recently reported reactions of partially oxidized organoboranes $(R^tBR^d(OR))$[224] (R^t = transferred ligand, R^d = nontransferred ligand), but these have not yet found the level of use of the other two classes. A wide range of functional groups tolerate these couplings, including OH (including phenols), NH_2 and nitro groups. While other catalysts have been employed in several instances, $Pd(PPh_3)_4$ has been the workhorse catalyst. The most noteworthy difference from cross-coupling reactions of other organometallics is the requirement for the presence of a base, commonly $NaOH_{(aq)}$, $Na_2CO_{3(aq)}$, $NaHCO_{3(aq)}$ or NaOEt–EtOH. More recently, TlOH has been advanced as a base which causes rate enhancements in the transmetallation step of the coupling and therefore the coupling itself[225]. Other bases such as Tl_2CO_3, Cs_2CO_3 and K_3PO_4 have been able to induce coupling in difficult cases under anhydrous conditions[226,227]. The role of these bases has been ascribed to conversion of organopalladium halide (36) to organopalladium alkoxide (37) in the catalytic cycle of the coupling, and to the greater reactivity of 37 with the organoboron compounds in the transmetallation step (Scheme 3)[223,226]. The formation of a borate $[R^2BY_2OR^1]^-$ (38) and an enhanced rate of R^2 transfer relative to the trivalent organoboron compound (R^2BY_2) is not expressly required by this mechanistic proposal. Nevertheless, the presence of 38 has been detected under the conditions of coupling, and its importance has been emphasized by at least some workers[224,228]. A further consequence of the necessity for this base is one of selectivity. With the exclusion of any of the aforementioned bases, both organostannanes and organozincs undergo palladium-catalyzed cross-coupling reactions highly selectively over organoboronates[229].

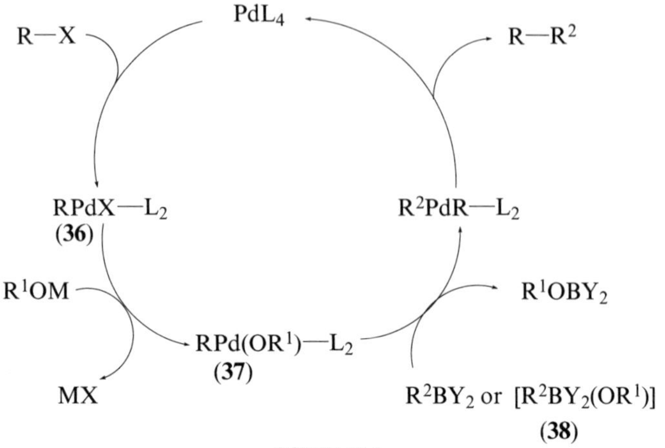

SCHEME 3

Alkene–alkene couplings have been investigated extensively using this chemistry, due partially to the ready availability of stereochemically defined vinylboron compounds by alkyne hydroboration. High levels of stereospecificity are observed ($\geq 98\%$) in nearly all cases in both the vinyl halide and vinylboron halves. With E-disubstituted vinylboron compounds, high yields are realized with both organoboranes and organoboronic esters or acids[226]. For the corresponding Z-isomers, vinyl(disiamyl)boranes and vinyl(dicyclohexyl)boranes give seriously compromised coupling yields, whereas organoboronates couple in excellent yields (equation 121)[230]. Both iodides and bromides prove to be acceptable

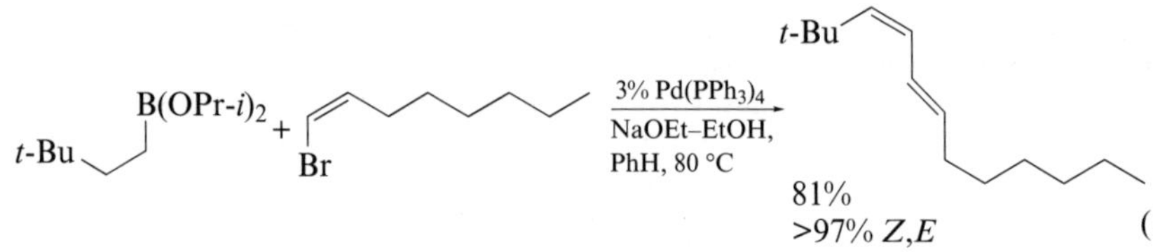

$$81\%$$
$$>97\% \; Z,E \qquad (121)$$

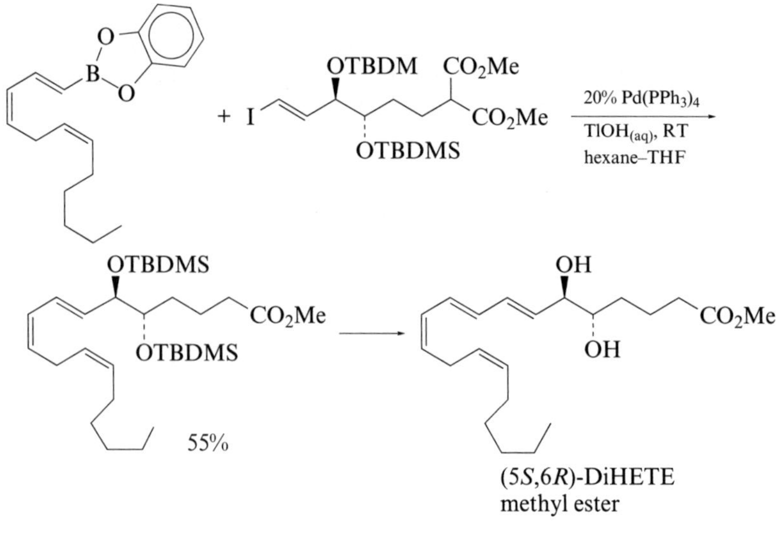

(122)

(5S,6R)-DiHETE
methyl ester

substrates, with β-chloro enones or enals also sufficiently reactive[231]. In the case of 1,1-dibromoalkenes, the *trans*-bromide reacts in preference to the *cis*-bromide[232]. With bases such as TlOH, the iodides may couple at room temperature or even 0 °C.

There are several cases of applications of vinylboron–vinyl halide coupling reactions in synthesis of targeted natural products or related molecules[225,233,234]; the synthesis of (5S,6R)-5,6-DiHETE methyl ester (where DiHETE = Dihydroxy-7,9,11,14-eicosatetra-enoic acid) is based on such a reaction for the formation of its trienic portion (equation 122)[233].

Other organic halides also have been substituted by vinylboron compounds. Acetylenic bromides couple with the vinyl(disimyl)boranes, affording enynes stereospecifically[224,226]. Di- and trisubstituted boronic esters have been reported to couple stereospecifically with both aryl and 2-thienyl iodides[223,230]. Boranes and partially oxidized boranes have also been reported to give this transformation[224,235]. A vinylborane–allyl bromide cross-coupling has been reported to be central to the synthesis of humulene (equation 123)[236].

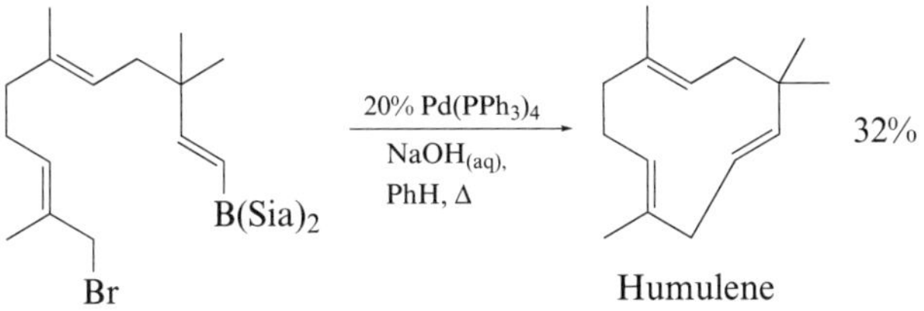

$$20\% \text{ Pd(PPh}_3)_4$$
$$\text{NaOH}_{(aq)},$$
$$\text{PhH, } \Delta$$

B(Sia)$_2$

Br

Humulene

32%   (123)

Aryl–aryl bond-forming reactions involving organoboron compounds also have received extensive attention. Aryl boronic acids or their esters are used most frequently here, due to their ready preparation from aryl Grignards[237], aryllithiums[238] or arylsilanes. Extensive substitution may be present in both the boronate and the aryl halide, including *ortho*-substitution and even *o,o*-disubsitution in one (but not both) partner (equations 124 and 125)[227,237]. Heteroaryl boronic acids and esters such as those derived from thiophenes[239] and furans[240] are also suitable for coupling. In some cases, boronate hydrolysis has proved to be a competing process, and alternate base–solvent combinations, or even anhydrous conditions, have been developed to obviate this problem[227,239]. Heterocyclic halides derived from thiophene[239], pyridine[238,241], pyrimidines[241] and related systems pose no particular problem, and vinyl halides have coupled without incident in a limited number of cases[241,242].

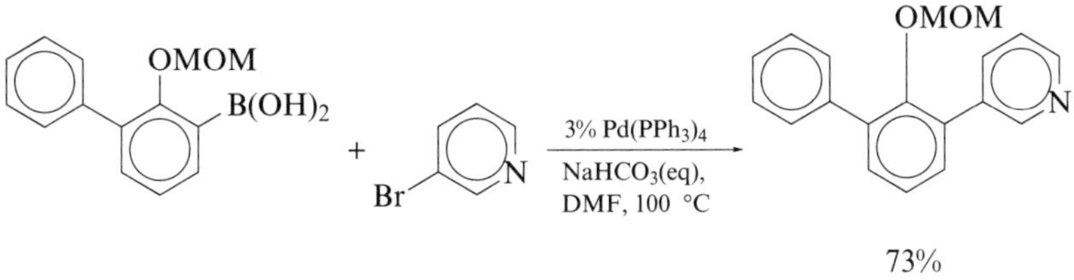

OMOM
B(OH)$_2$
+
Br

$$3\% \text{ Pd(PPh}_3)_4$$
$$\text{NaHCO}_3\text{(eq)},$$
$$\text{DMF, 100 °C}$$

OMOM

73%

(124)

OMe
B
O
O
+ I

$$2\% \text{ Pd(PPh}_3)_4$$
$$\text{K}_3\text{PO}_4, \text{DMF, 100 °C}$$

OMe

95%

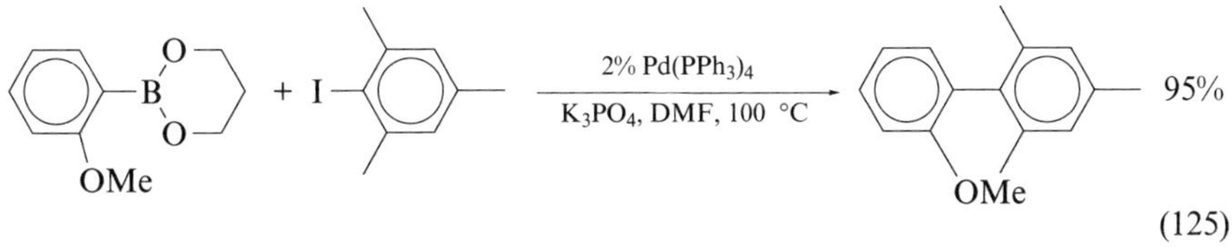

(125)

Aryl–aryl cross-coupling reactions have been exploited extensively, particularly by Snieckus and coworkers, in the synthesis of polyaryls and other aromatic nuclei[243].

Alkylation using organoboron reagents has been accomplished by the use of alkyl 9-BBN (9-BBN = 9-borabicyclo[3.3.1]nonyl); these couple stereoselectively with bromoalkenes and with haloarenes using $PdCl_2(dppf)$ as catalyst (equation 126)[244,228]. Many functional groups survive the transformation, and intramolecular cases forming five- and six-membered rings are allowed (equation 127). Under anhydrous conditions, even primary alkyl–primary alkyl coupling (including with a neopentyl iodide) may be accomplished in synthetically useful yields[245]. Under a CO atmosphere, carbonylative versions of these couplings occur[246]. Alternatively, alkyl group incorporation onto vinyl and aryl bromides may be accomplished using alkyl boronic esters in conjunction with thallium bases[247].

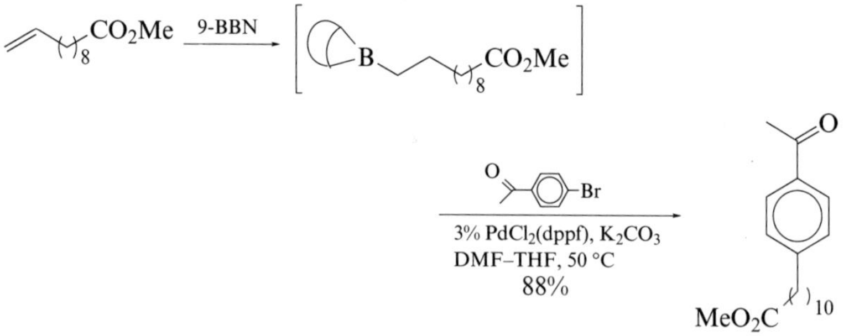

(126)

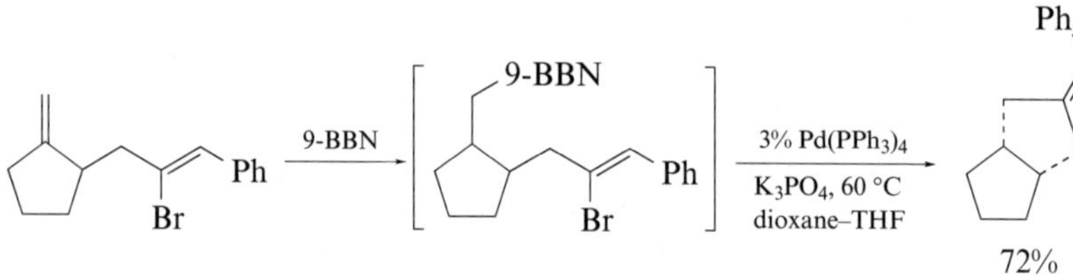

(127)

H. Cross-coupling Reactions of Organosilanes in the Presence of Palladium Catalysts

The ideal organometallics for cross-coupling reactions would probably be organotrimethylsilanes, by virtue of the stability of the majority of these compounds to handling and many other reactions. To a degree, this has been accomplished by the use of fluoride ion to render the organic groups on silicon more nucleophilic[248]. More often, organotrimethylsilanes are simply not reactive enough and additional electronegative groups on silicon, such as fluoride or alkoxy groups, are necessary; this enhances the organosilane–fluoride ion combination in its formation of pentacoordinate silicate anion and the transmetallation of any organopalladium intermediate.

A limited number of vinyltrimethylsilanes and acetylenic trimethylsilanes couple with aryl and vinyl iodides in the presence of TASF (tris(diethylamino)sulfonium difluoromethylsilicate) or TBAF (tetrabutylammonium fluoride) and $(\eta^3\text{-}C_3H_5PdCl)_2$ catalyst[249]. Acetylenic trimethylsilanes couple with vinyl bromide under even milder conditions. In the absence of an organic trimethylsilane, TASF itself transfers a methyl group in reasonable yields to the aryl halide[250].

Couplings of vinylsilanes are more general for $(vinyl)SiMe_2F$ and $(vinyl)SiMe_n(OR)_m$ ($n = 1, m = 2$, and $m = 1, n = 2$)[251] compounds. Vinyl and aryl iodides react easily, while the corresponding bromides are less reliable (equation 128). Double-bond stereochemistry is retained in the vinyl halide partner, whereas Z-vinylsilanes isomerize to a small degree in some cases. Discrete pentavalent siliconate anions bearing catecholate ligands also give vinylation of iodoarenes[252].

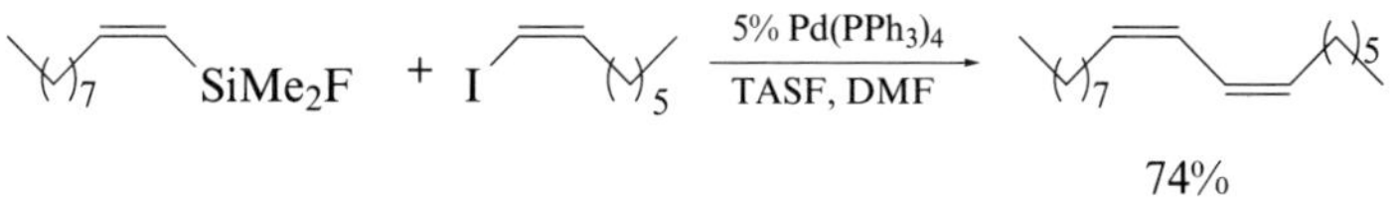

$$\text{(128)}$$

74%

Transfer of aryl groups from silicon may be performed in an analogous manner, save that the difluorides, (aryl)SiEtF$_2$ and (aryl)Si(n-Pr)F$_2$, give the best results (equation 129)[253]. Unsymmetrical biaryls are available cleanly by way of this protocol. If a CO atmosphere is also present with these reagents, diaryl ketones are realized, providing the arene portion of the arylsilane does not bear an electron-withdrawing group (equation 130)[254].

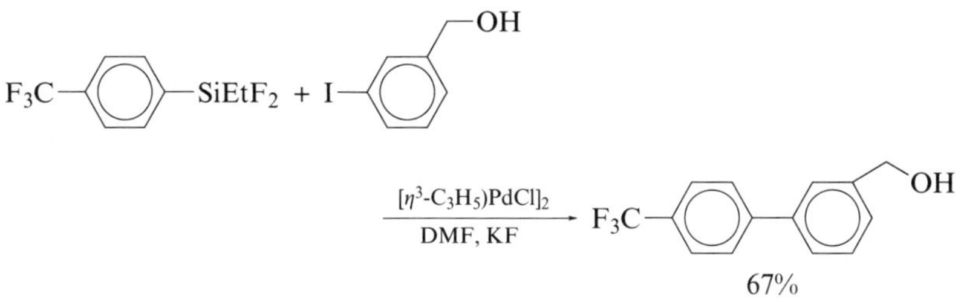

$$\text{(129)}$$

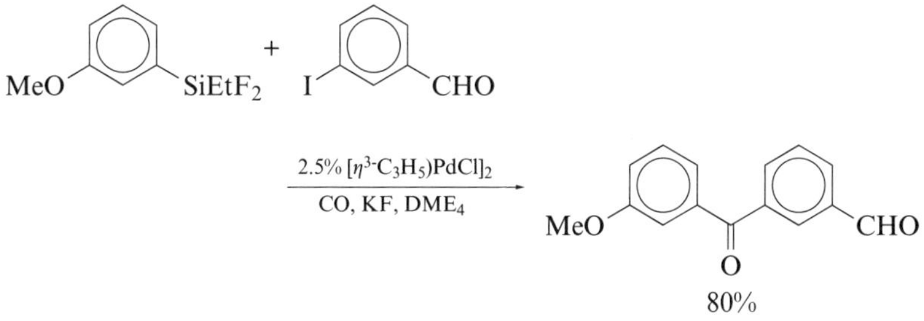

67%

$$\text{(130)}$$

80%

Allylation of aryl halides may also be accomplished by the use of allyltrifluorosilanes (equation 131). Clean reaction at the γ-position of the allylsilane occurs when the iodoarenes are employed[255].

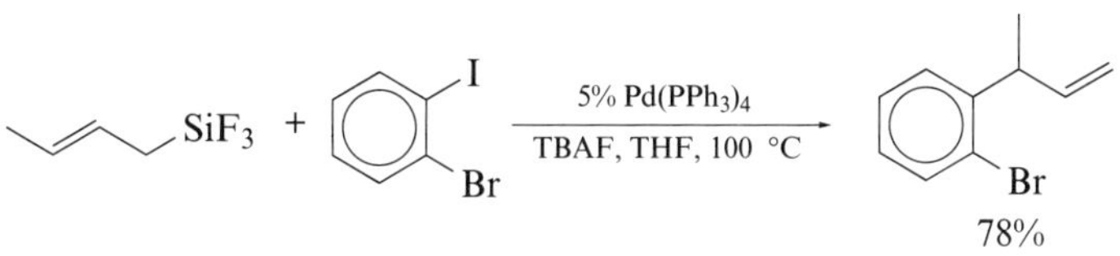

$$\text{(131)}$$

78%

IV. REDUCTIVE COUPLING OF HALIDES BY MEANS OF NICKEL(0) COMPLEXES

Detailed electrochemical studies on the synthesis of biaryls from the electrogenerated Ni^0(dppe)-catalyzed dimerization of bromobenzene has implicated the steps in Scheme 4, involving Ni(I) and Ni(III) as well as Ni(II) and Ni(0) as being critical to the coupling[256].

$$Ni^{II}Cl_2L_2 \xrightarrow{+ e^-, -Cl^-} Ni^{I}ClL_2 \xrightarrow{+ e^-, -Cl^-} Ni^0L_2$$

$$PhBr + NiL_2 \longrightarrow PhNi^{II}BrL_2$$

$$PhNi^{II}BrL_2 + e^- \longrightarrow PhNi^{I}L_2 + Br^-$$

$$PhBr + PhNi^{I}L_2 \longrightarrow Ph_2Ni^{III}BrL_2$$

$$Ph_2Ni^{III}BrL_2 \longrightarrow Ph\!-\!Ph + Ni^{I}BrL_2$$

$$Ni^{I}BrL_2 + e^- \longrightarrow Ni^0L_2 + Br^-$$

$$L_2 = dppe$$

SCHEME 4

The reduction of Ni(II) to Ni(I) is likely the rate-limiting step. Additional steps may also operate in the presence of large amounts of a chemical reductant (i.e. Zn metal) or in the presence of stoichiometric amounts of Ni(0)[257].

In synthetic aspects of Ni(0) reductive coupling reactions, three alternatives to the now-standard $NiX_{2(cat)}$–PPh_3–Zn combination have been offered. In one case, electrochemical reduction replaces the chemical reductant[258]. Alternatively, the use of complex reducing agents (CRA), containing nickel(NiCRA-bipy), prepared from varying amounts of NaH, t-AmONa, $Ni(OAc)_2$ and bipyridine (bipy) has been reported to allow homocoupling with a wide range of aryl halides[259]. Finally, Rieke and coworkers have reported several couplings using stoichiometric amounts of highly active nickel metal[260]. Other minor changes from the standard combination, such as the use of other phosphine ligands, or no ligand, have been reported[261]. The addition of iodide salts, usually either KI or n-Bu$_4$NI, enhances the rate of coupling[261,262].

It has been found that, in addition to the more commonly used bromides and iodides, aryl chlorides may participate in reductive homocoupling reactions in synthetically useful yields, by any of these three types of procedures[257–259]. Additionally, pyridine-based aryl halides also couple without incident (equations 132 and 133)[257,262]. Although vinyl halides dimerize in excellent yields, some loss in stereochemical integrity occurs, particularly with the Z-isomers (equation 134)[261]. The reaction of 1,1-dibromoalkenes takes a different course depending upon the substrate and source of nickel. With $NiBr_2(PPh_3)_2$/PPh_3/Zn/Et$_4$NI, the expected cumulene (39) is produced[263]; the NiCRA-bipy conditions, on the other hand, give conjugated diynes with dibromostyrene (equations 135 and 136)[264].

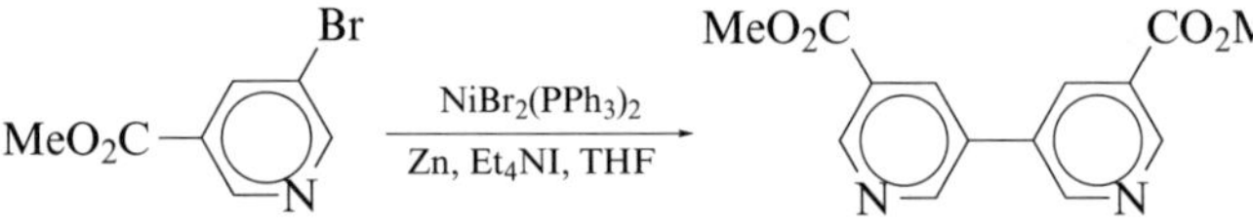

$$\text{MeO}_2\text{C}\!-\!\!\!\overset{\displaystyle Br}{\underset{N}{\bigcirc}} \xrightarrow[\text{Zn, Et}_4\text{NI, THF}]{\text{NiBr}_2(\text{PPh}_3)_2} \quad \text{MeO}_2\text{C}\!-\!\!\!\underset{N}{\bigcirc}\!\!-\!\!\!\underset{N}{\bigcirc}\!\!-\!\text{CO}_2\text{Me} \qquad 69\% \quad (132)$$

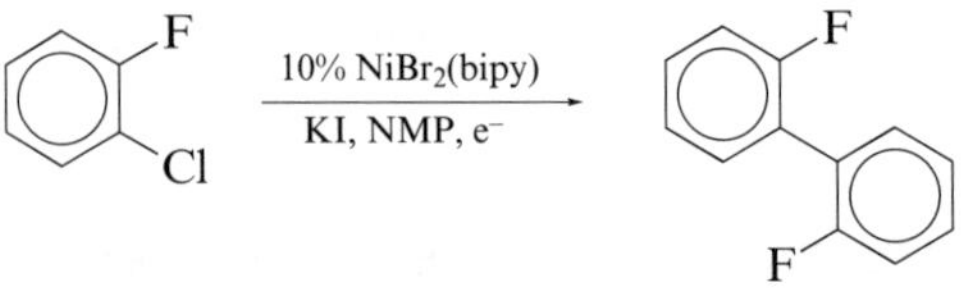

$$\overset{\displaystyle F}{\underset{\displaystyle Cl}{\bigcirc\!\!\!\bigcirc}} \xrightarrow[\text{KI, NMP, e}^-]{10\% \text{ NiBr}_2(\text{bipy})} \quad \underset{F}{\overset{F}{\bigcirc\!\!\!\bigcirc}}\!\!-\!\!\bigcirc\!\!\!\bigcirc \qquad 68\% \quad (133)$$

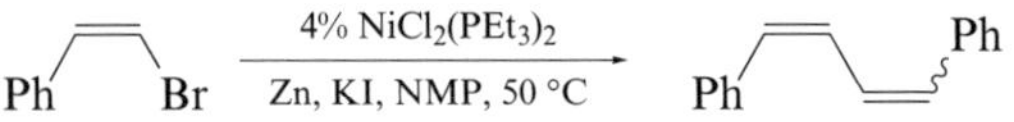

$$\underset{Ph}{\diagup}\!\!\!\diagdown\!\!\!Br \xrightarrow[\text{Zn, KI, NMP, 50 °C}]{4\% \text{ NiCl}_2(\text{PEt}_3)_2} \quad Ph\diagup\!\!\!\diagdown\diagup\!\!\!\diagdown Ph \qquad 76\% \quad (134)$$

$$E,E : E,Z : Z,Z = 3{:}20{:}77$$
$$NMP = N\text{-methylpyrrolidinone}$$

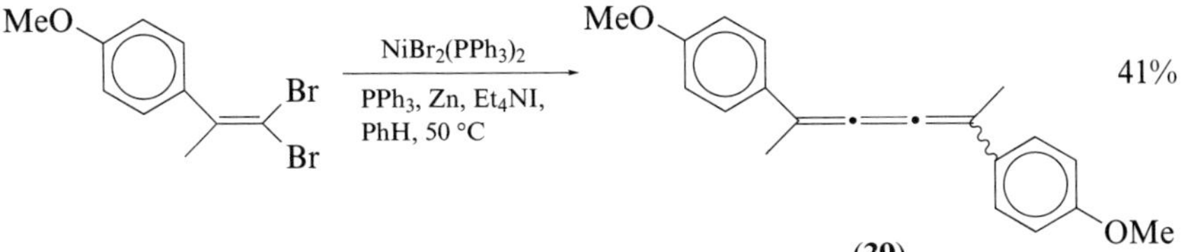

(39)

(135)

41%

$$\text{PhCH}{=}\text{CBr}_2 \xrightarrow[\text{Ni(OAc)}_2, \text{bipy, THF}]{\text{NaH, t-AmONa}} \text{Ph}{-}{\equiv}{-}{-}{\equiv}{-}\text{Ph} \quad 75\% \quad (136)$$

In certain cases, cross-couplings between different organic halides have been observed. Most commonly, this occurs successfully when one of the halides is especially reactive to oxidative addition, i.e. benzylic or α- to a carbonyl function[260]. The other halide partners in these cases have been acid chlorides, or vinyl or aryl halides (equation 137). Greater than statistical amounts of cross-coupling are also realized in cases pairing aryl bromides with electron-donating groups with aryl chlorides bearing electron-donating groups, although all possible products are formed to some degree.

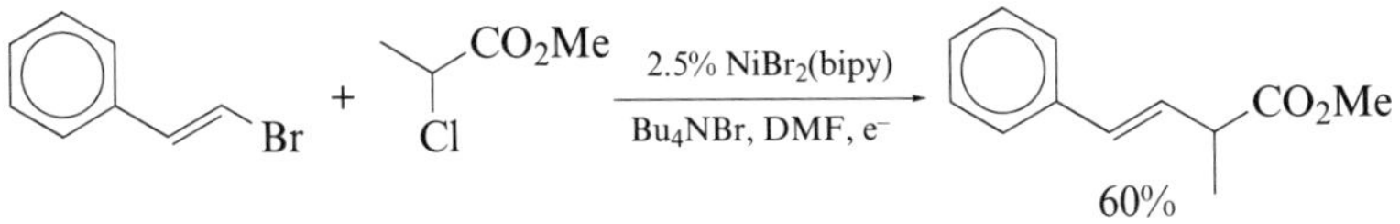

In some cases, Ni(0) compounds give controlled oligomerization of dihalides to afford cyclized products (equations 138 and 139). o-Dihalobenzene has been reported to cyclotrimerize, giving a new arene ring[265]. 2,3-Dibromothiophene, on the other hand, cyclotetramerizes relatively efficiently to form an eight-membered ring. To this point, these cyclizations have been carried out only in the presence of stoichiometric amounts of Ni(0).

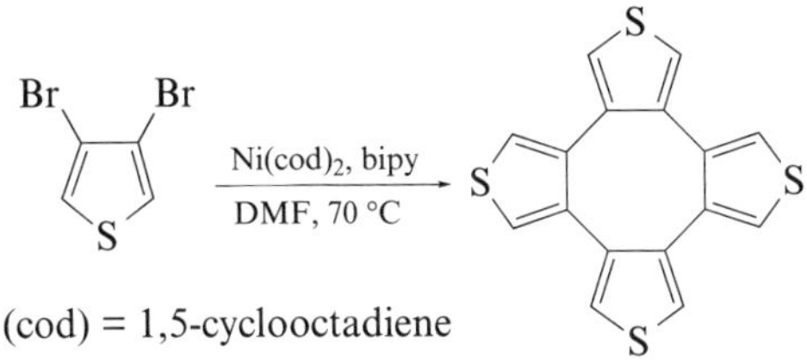

(cod) = 1,5-cyclooctadiene

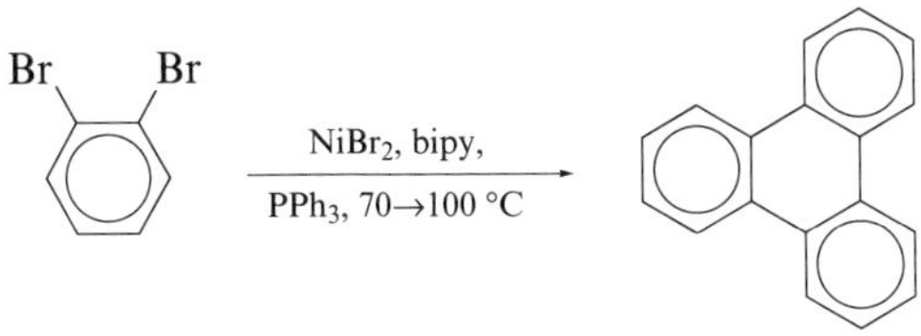

A slightly different course of reaction is available to α,α′-dibromo-o-xylene. In the presence of highly reactive nickel powder, the synthetic equivalent of o-xylylene is generated, which undergoes cycloaddition with normal Diels–Alder dienophiles in a nonstereospecific manner (equation 140)[266]. The exact nature of 40 has not been deduced.

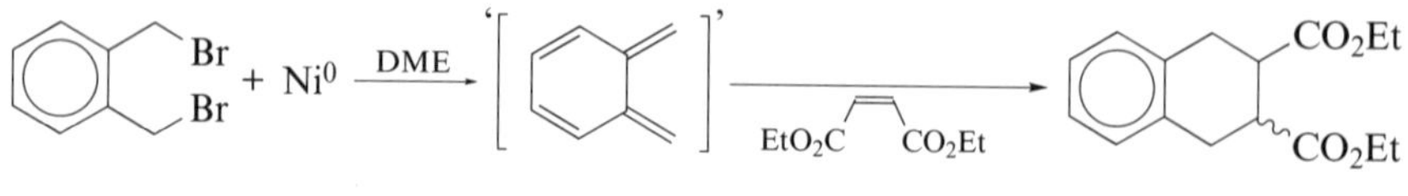

(40)   90%   (140)

V. REACTIONS OF HALIDES WITH ALKENES OR ACETYLENES IN THE PRESENCE OF PALLADIUM CATALYSTS

A. Reactions of Acetylenes

The palladium-catalyzed reaction of terminal alkynes with organic halides to give over-all substitution of the organic halide for the acetylenic hydrogen is a reliable and well estab-lished procedure[1]. The analogous polyalkynylation of perhaloarenes has also proven to be relatively straightforward. It is even possible to incorporate six alkynyl or diynyl groups onto hexabromobenzene by coupling under relatively standard conditions (equation 141)[267].

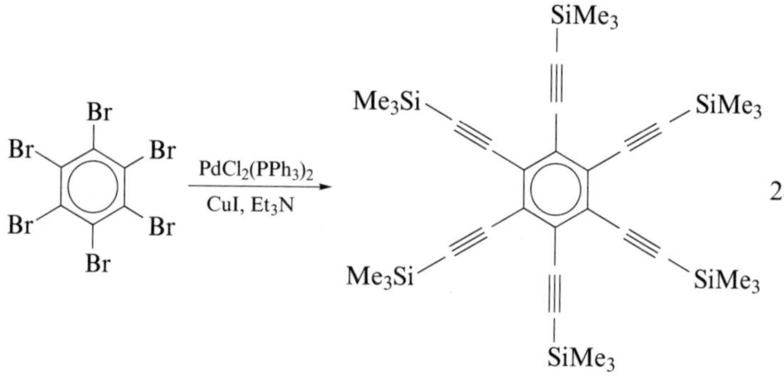

28% (141)

Selective alkynylation of organic dihalides may often be accomplished. Each of 1,2 dichloroethene[268], 1,1-dichloroethene[269] and 6,6'-dibromo-2,2'-bipyridine monocouple selectively under standard conditions [amine, $PdCl_2(PPh_3)_2$, CuI][268,270]. 1,2-Dibromobenzene performs in a similar manner only if the CuI co-catalyst is omitted[271].

Substituent patterns play the expected role in the regioselective monocoupling of dihaloarenes. In o-dibromo cases, the bromine substituent ortho- or para- to an electron-withdrawing substituent reacts more rapidly, while an electron-donating substituent in the analogous positions causes selective reaction at the alternate bromine substituent (equation 142)[272]. The relative rates of coupling of para-substituted aryl bromides ($NO_2 > CN > CHO > CO_2Me > CO_2H \approx Cl > Me > H > OMe > NH_2$) have been quantified[272]. In addition, the $PdCl_2(PPh_3)_2:Cu_2Br_2$ ratio of 2:3 has been found to be optimal for these couplings. In dibromopyridines, coupling occurs selectively at the 2-position over the 5- (3-) position[273]. Several examples of the preferential coupling at iodine relative to bromine sub-stituents also have been reported[274].

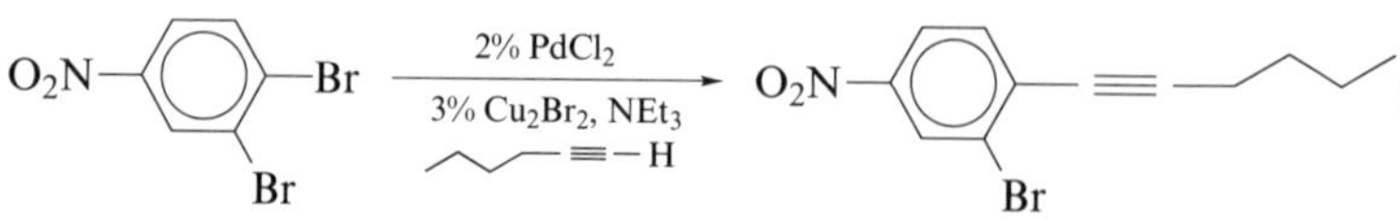

92% (142)

Although heterocyclic chlorides and dichloroalkenes participate relatively well in alkyne coupling reactions, aryl chlorides and simple vinyl chlorides normally have not been sufficiently reactive. Aryl chlorides can be induced into coupling by substitution by work-ing with their chromium tricarbonyl complexes[275]. The choice of nonphosphine-contain-ing catalysts [$PdCl_2(RCN)_2$, R = Me, Ph], CuI and piperidine as base has allowed simple vinyl chlorides to couple stereospecifically with alkynes at room temperature[276], with the E-isomer measurably faster than the Z-isomer.

Alkyne coupling reactions under Pd/Cu conditions have been extended to other func-tional groups. Allenes couple with alkynes without incident (equation 143)[277], as do dithio-carbonyl chlorides and thiocarbamoyl chlorides[278]. Alkynyl chloride cross-couple with alkynes in at least some cases, to form diyncs[279].

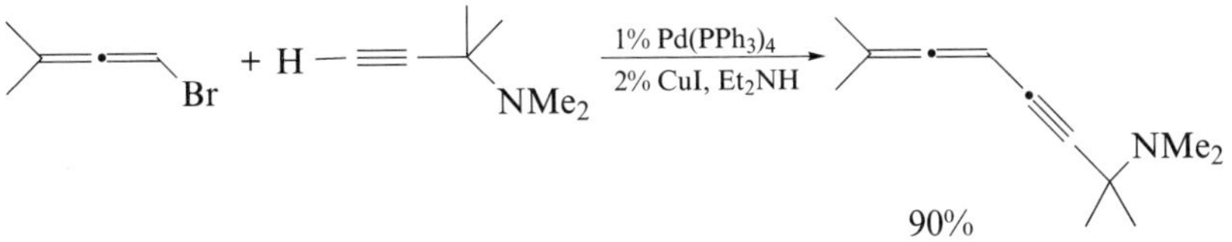

$$\begin{array}{c} \text{(143)} \end{array}$$

90%

Alkyne to vinyl halide coupling reactions have been used extensively in synthesis, due to the popularity of enediyne antitumour antibiotics[280]. An example is given in equation 144.

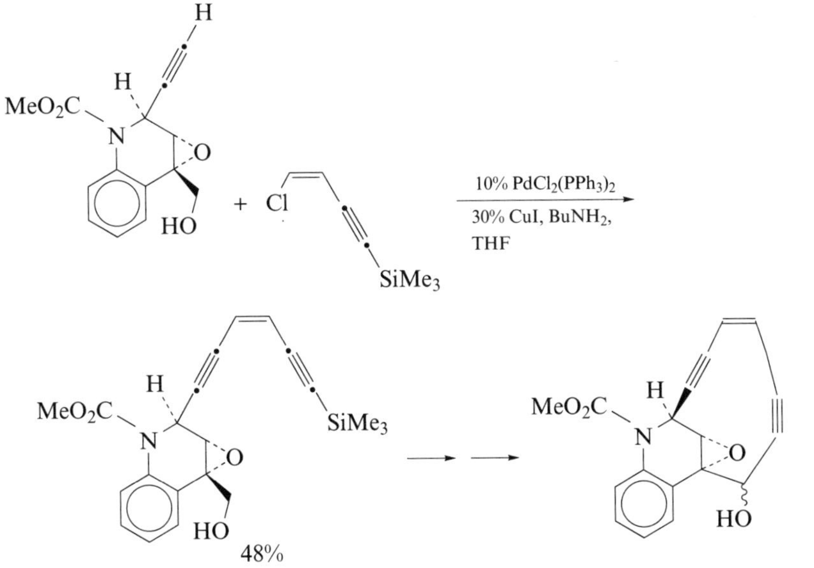

48%

(144)

B. Heck and Related Reactions

Significant developments have been reported in Heck-type alkylations of alkenes or alkynes. Several reports have addressed the lack of reactivity of aryl chlorides in their reac-tions with alkenes. Two approaches to this problem have met with success. In the first case, aroyl chlorides are used in place of aryl chlorides; under the conditions of reaction, oxida-tive addition followed by decarbonylation occurs. With activated alkenes, good yields of arylation are realized, while coupling occurs in moderate yields with unactivated cases (equation 145)[281]. An analogous approach for coupling alkenoyl chlorides has met with some success, but yields are quite variable and substrate-dependent (equation 146)[282]. Conceptually related to this is the use of arene- or alkenesulfonyl chlorides, which lose SO_2 under the conditions of their reaction with alkenes[283]. Based on a limited number of examples, their usefulness roughly parallels that of acid chlorides.

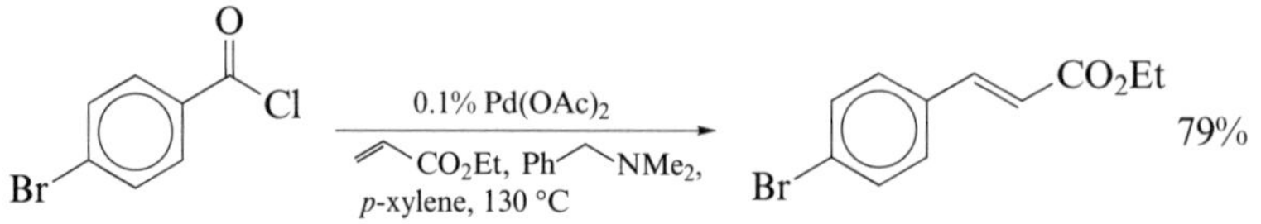

$$
\text{(145)} \qquad 79\%
$$

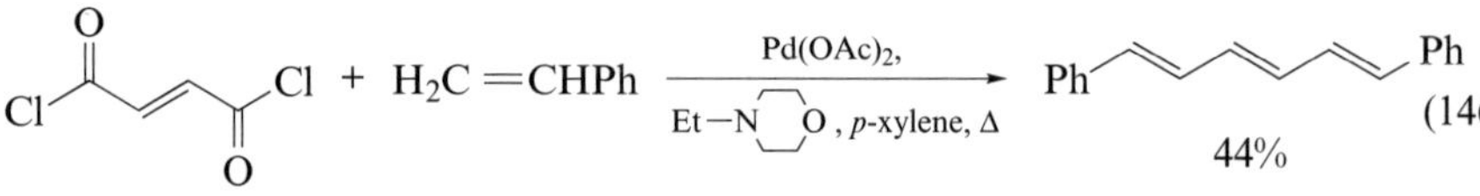

$$
\text{(146)} \qquad 44\%
$$

The second approach involves preheating the aryl chloride with NaI and catalytic amounts of $NiBr_2$ and subjecting the mixture to fairly conventional coupling conditions (equation 147)[284]. It is proposed that equilibrium quantities of aryl iodide are generated *in situ*, which is sufficiently reactive to undergo alkene arylation with activated alkenes. Both electron donating and withdrawing groups are tolerated on the arene.

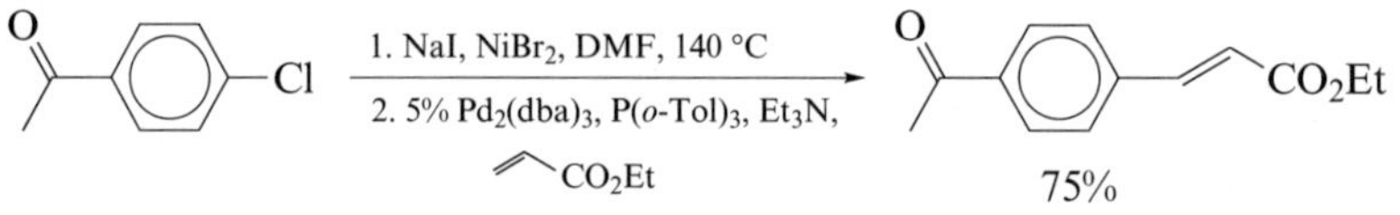

$$
\text{(147)} \qquad 75\%
$$

Conditions of Heck reactions with alkenes have been made considerably more mild through the addition of tetrabutylammonium halides and weak bases such as KOAc or K_2CO_3. The reagents have been employed in polar aprotic solvents such as DME, or under phase transfer conditions; in their presence, yields are often increased and, in the case of iodides, reactions may be accomplished even at room temperature[285-288].

Much work has gone into the optimization of results with functionalized alkenes. In the reaction of cyclic alkenes, *cis*-decomposition of organopalladium halide–alkene complex gives σ-palladium complex **41**. This subsequently undergoes *syn*-β-hydride elimination; since only one such hydrogen is available, deconjugation to 3-substituted alkenes should

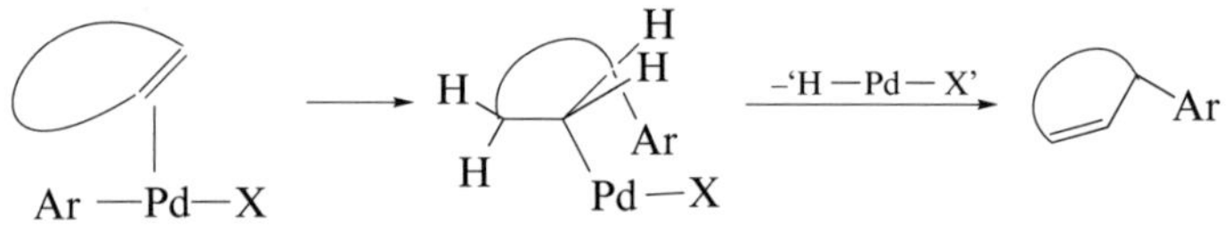

$$
\text{(148)}
$$

(41)

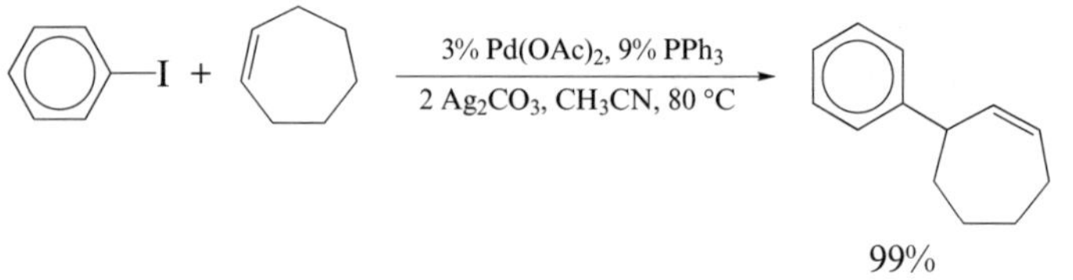

$$
\text{(149)} \qquad 99\%
$$

TABLE 1. Factors influencing regioselectivity in arylation of acyclic vinyl ethers

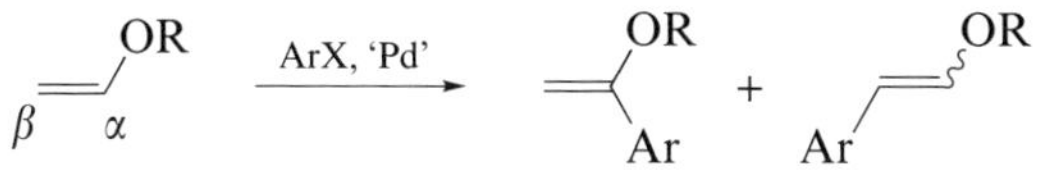

α-product   β-product

Parameter	Effect
Substituent on Ar	Electron-donating favors α
	Electron-withdrawing favors β
Nature of X	β-Favored by —Cl (—COCl) > —Br > —I
Solvent	Poorly coordinating solvent favors β
	Coordinating solvent favors α
	Added ligand favors α
β-Substitution	Favors α
Coordinating group in —OR	Favors β
Added Tl salts	Favors α

be realized (equation 148). Under conventional Heck conditions, however, cyclic alkenes give mixture of double-bond regioisomers with both aryl and vinyl halides. With the addition of silver salts and catalytic amounts of PPh_3 ('Larock conditions B'), 3-substituted cycloalkenes are formed cleanly (equation 149)[288,289]. Smaller catalytic amounts of PPh_3 ('Larock conditions C') give better yields with many functionalized aryl halides, but do not suppress double-bond isomerization.

Heteroatom substitutions have well-understood effects on the path of reaction[290]. Cyclic enols and enamides[291] substitute exclusively at the α-position, by virtue of the relative electron deficiency at this position and at the palladium center (equations 150 and 151). In acyclic enol ethers, mixtures of α- and β-substitution products are observed, the relative proportion of which are dependent upon the substituents and the identity of the halogen in the aryl halide, the coordinating ability of solvent (or added ligand) and (to a lesser degree) substituents in the enol ether (Table 1)[292]. By appropriate choice of conditions, high β-/α-ratios[293] or high α-/β-ratios[294] of substitution may be achieved (equations 152 and 153). Although the use of chiral phosphine ligands [i.e. (R)-BINAP] (where BINAP = 2,2'-bis(diphenylphosphino)-1,1'-binaphthyl) in the arylation of oxygen heterocycles (4H-1,3-dioxin, 2,3-dihydrofuran) by aryl iodides have resulted in enantiomerically enriched products, triflates have given much higher ee values in most cases[295]. A few other heteroatom-substituted alkenes have been studied with respect to their Heck reactions. Vinylphosphonates behave in a manner analogous to carbonyl-substituted alkenes, undergoing highly selective β-substitution[296]. Vinylsilanes also undergo predominant attack at the β-position to afford substituted vinylsilanes if silver salts are added; in their absence, desilylation occurs[297].

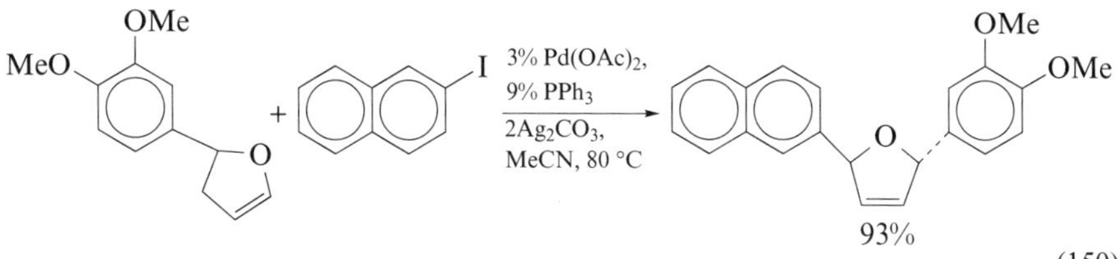

93%

(150)

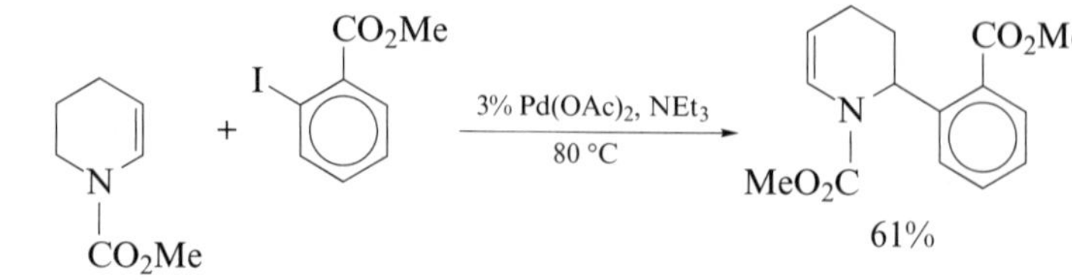

$$\text{(151)}$$

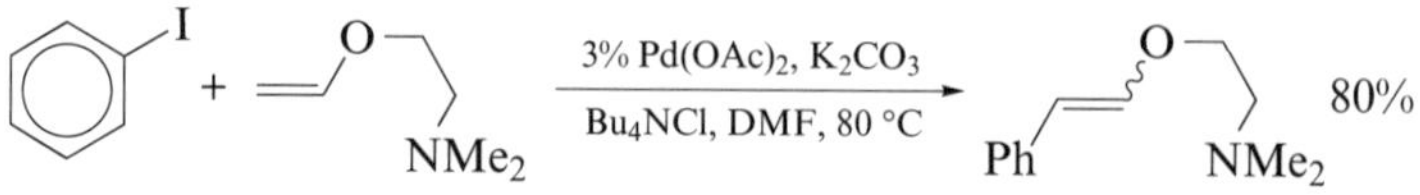

Structure: tetrahydropyridine with CO_2Me on N, plus aryl iodide with CO_2Me

3% Pd(OAc)$_2$, NEt$_3$
80 °C

product 61%

$$\text{(152)}$$

aryl iodide + vinyl ether with NMe_2

3% Pd(OAc)$_2$, K$_2$CO$_3$
Bu$_4$NCl, DMF, 80 °C

product 80%

$\beta\text{-/}\alpha\text{- = 50/ 1}$

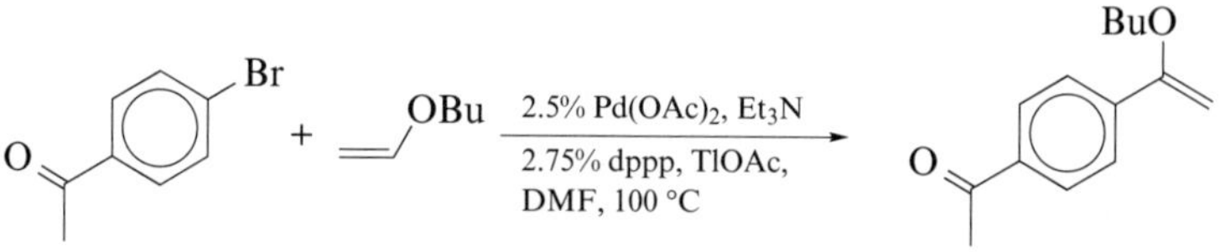

$$\text{(153)}$$

aryl bromide ketone + $\equiv$—OBu

2.5% Pd(OAc)$_2$, Et$_3$N
2.75% dppp, TlOAc,
DMF, 100 °C

dppp = bis(1,3-diphenylphophine)propane

$\alpha\text{-/}\beta\text{- = >99/1}$
82% after hydrolysis

Intramolecular Heck reactions are possible. Heterocycle synthesis has been known for some time[1], while preparation of carbocycles[298], fused bicyclics[299–302] and spiro bicyclic ring systems[289,302] are more recent developments, but may also be accomplished (equation 154). Preference in ring sizes is 5- > 6- >> others. In the synthesis of fused [6,6]- and [5,6]-bicyclics from vinyl iodides (or better, from triflates), the use of palladium catalyst PdCl$_2$[(R)-BINAP] gives high enantiomeric excesses (equation 155)[303].

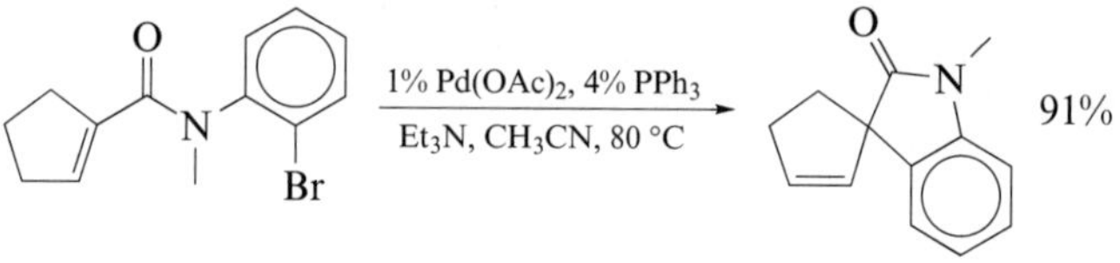

$$\text{(154)}$$

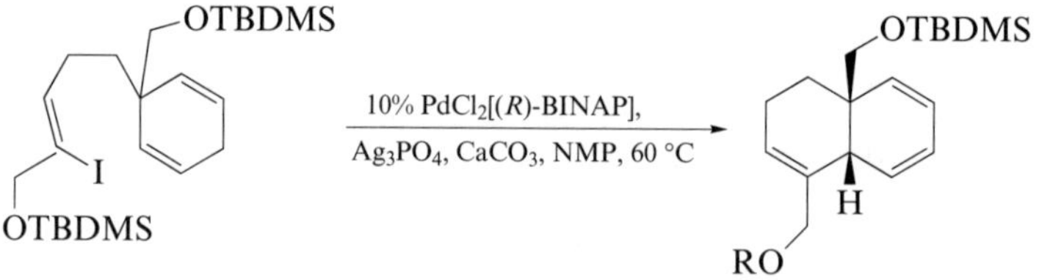

1% Pd(OAc)$_2$, 4% PPh$_3$
Et$_3$N, CH$_3$CN, 80 °C

product 91%

$$\text{(155)}$$

10% PdCl$_2$[(R)-BINAP],
Ag$_3$PO$_4$, CaCO$_3$, NMP, 60 °C

R = TBDMS, 63%, 83% ee
+ R = H, 35%, 92% ee

There are several potential pathways open to the σ-palladium intermediate, of the type **41**, other than loss of hydride to form an alkene. The most straightforward of these is reduction by formic acid or formate salts to give the alkane (equation 156)[304]. Alkynes[305] and dienes[306] react in a similar manner under these types of condition, to give alkenes. Alternatively, many examples exist of the trapping of the σ-alkylpalladium intermediate by highly stabilized carbanionic nucleophiles such as malonate[307–310], by cyanide ion[311] or by carbonylation[312]. Alkynes[307], conjugated dienes[308], cumulated dienes[309] and even nonconjugated dienes[312] are most often used in this process (equation 157). Both inter- and intramolecular versions of these reactions have been reported. In the case of the alkyne and monoene reactions, it is likely that the mechanism is different from a conventional Heck alkylation; initial anionic attack on the alkyne (or alkene)–palladium π-complex is followed by reductive elimination from the resulting diorganopalladium (equation 158)[307]. A consequence of this mechanism is that the incorporated nucleophile and organic halide are stereoselectively *anti*. This process is efficient only for the formation of five-membered rings. Oxygen and nitrogen nucleophiles have been incorporated in an analogous manner[313].

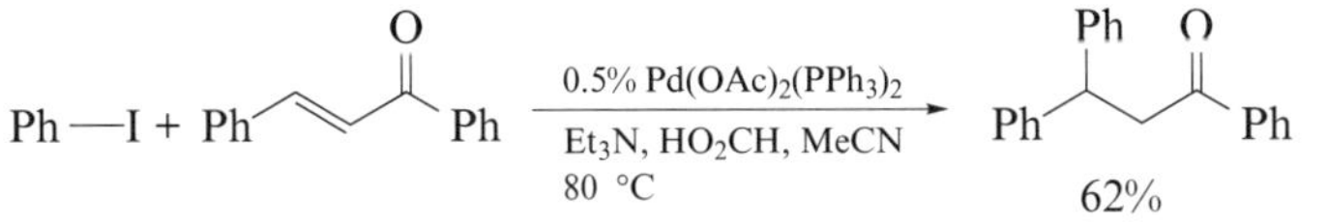

$$Ph\!-\!I + Ph \xrightarrow[\substack{Et_3N, HO_2CH, MeCN \\ 80\ °C}]{0.5\%\ Pd(OAc)_2(PPh_3)_2} \quad (156)$$

62%

$$\xrightarrow[Na_2CO_3, DMF, 80\ °C]{5\%\ Pd\,(OAc)_2,\ Bu_4NCl} \quad (157)$$

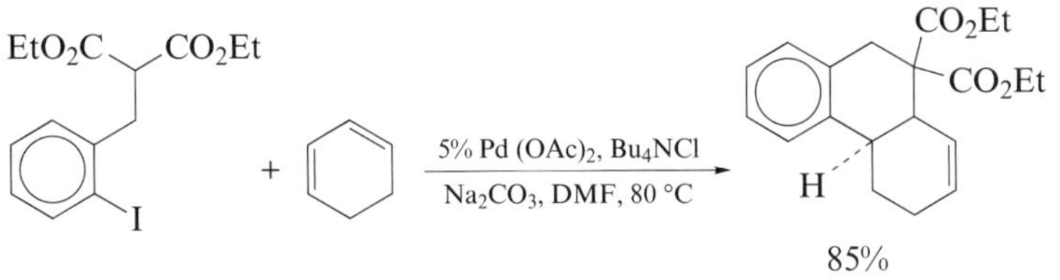

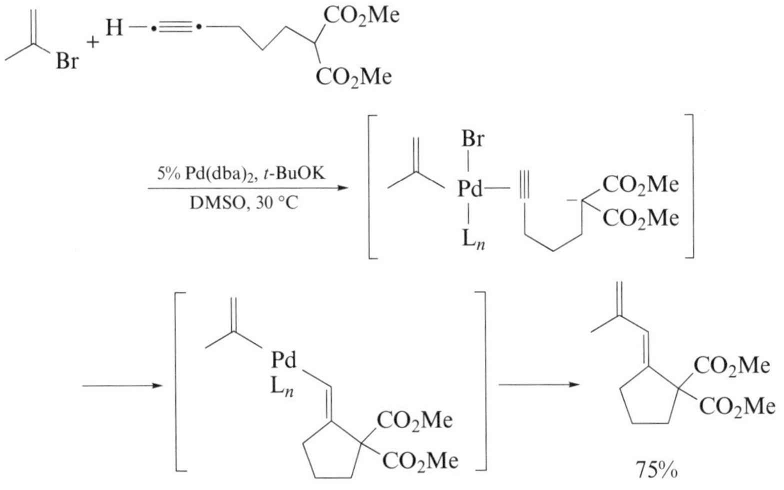

85%

$$\xrightarrow[DMSO, 30\ °C]{5\%\ Pd(dba)_2,\ t\text{-BuOK}} \quad (158)$$

75%

In still other cases, the σ-palladium intermediate of the type **41** may be combined in cross-coupling reactions with organotin[314], organoboron[315] or organozinc[316] reagents. Success in these processes is most often realized when the initial Heck-type attack is intramolecular, and onto an alkyne or diene (equation 159). This is likely due to a more rapid Heck reaction in the intramolecular cases relative to intermolecular coupling, and the greater lifetime of the vinyl- or allylpalladium species, relative to a σ- alkylpalladium intermediate. For alkynes, the incorporated nucleophile (organozinc, tin or boron) and the original organic halides are added stereoselectively *syn*. Nevertheless, intramolecular cases and cases where the initial Heck attack is on a bicycloalkene are known (equation 160)[317]. In these latter cases, the competing β-elimination process is hampered for the intermediate alkylpalladium.

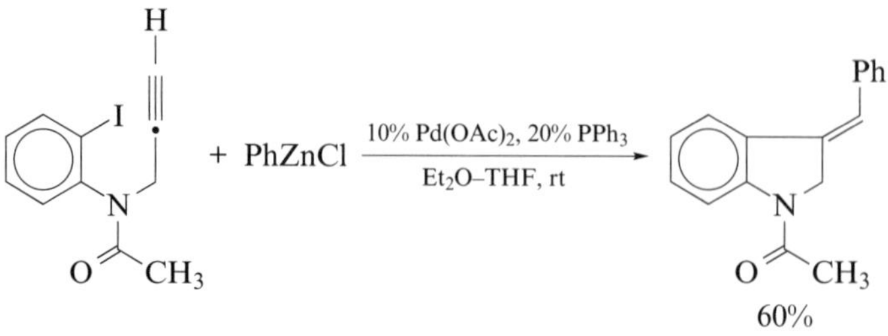

(159)

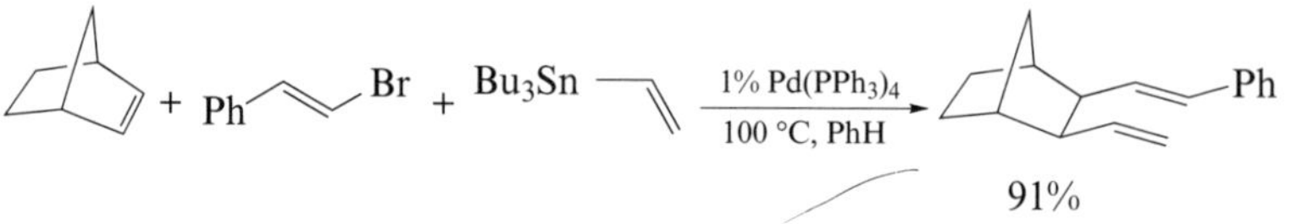

(160)

The final option available to a σ-alkylpalladium intermediate from Heck alkylation occurs if another alkene or alkyne function is situated properly to participate in a further Heck-type carbopalladation (equation 161)[318,319]. In properly constructed systems, more than one further carbopalladation is feasible, and many examples of these 'cascade carbopalladations' have been reported. Several have been quite spectacular (equation 162)[320]. Fused, spirocyclic and bridged bicyclic ring systems have been prepared in this manner. The process may also create as many as five rings in one step, with five-, six- and three-membered rings[321] being the most suitable for preparation (equation 163). Alternatively, the proper orientation of double and triple bonds allows cyclotrimerization to highly functionalized arenes or fulvenes (equation 164)[322,279].

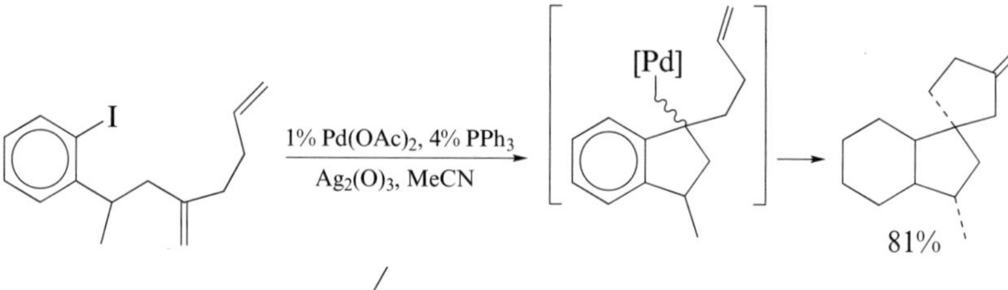

(161)

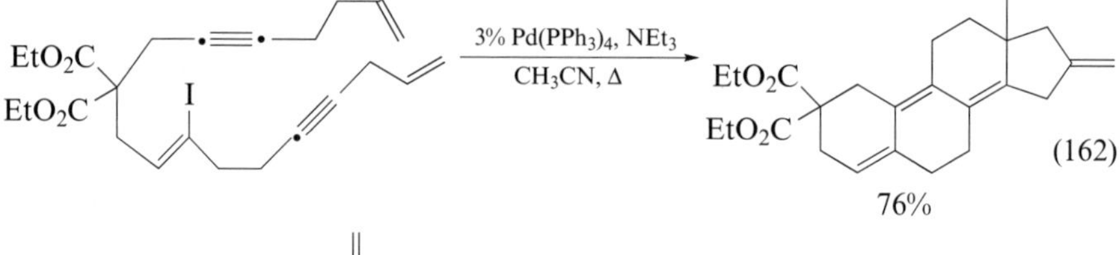

(162)

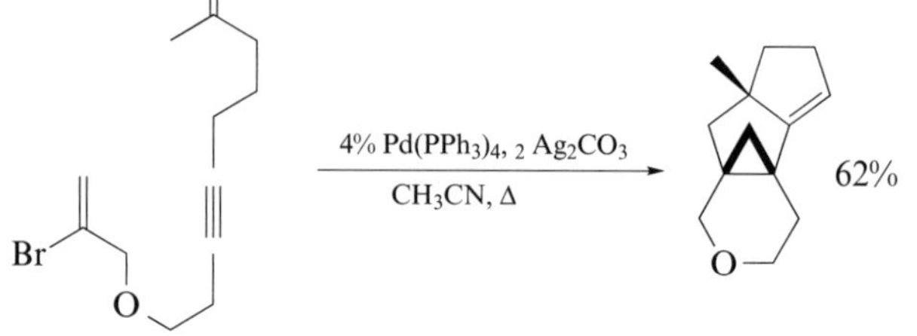

(163)

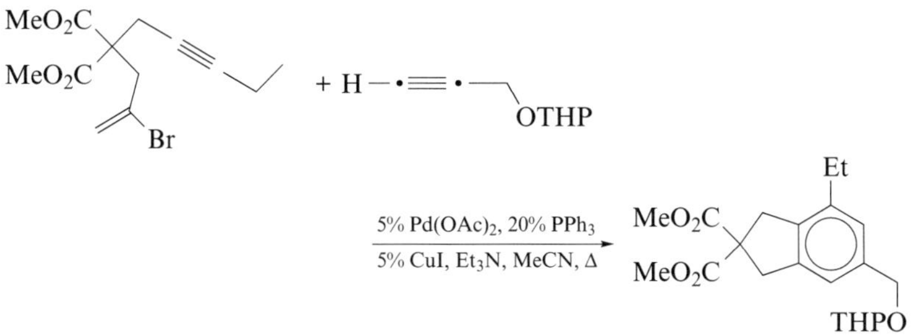

(164)

THP = tetrahydropyronyl   84%

With a combination of organic halide, alkene or alkyne, and palladium catalyst, the reaction may take a superficially similar, but different pathway other than a Heck-type reaction. In these, the halogen is retained in the products, as an alkyl or vinyl halide. These results occur when the starting organic halide is allyl, α- to a carbonyl or similar group, or attached to a perhalogenated carbon atom. In much of the early work, mixtures with Heck-like elimination products were obtained, due to elimination caused by the presence of amine bases in the reaction mixture (equation 165)[323].

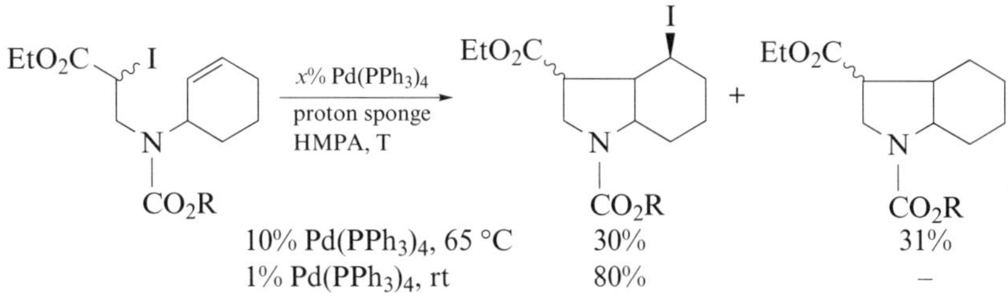

(165)

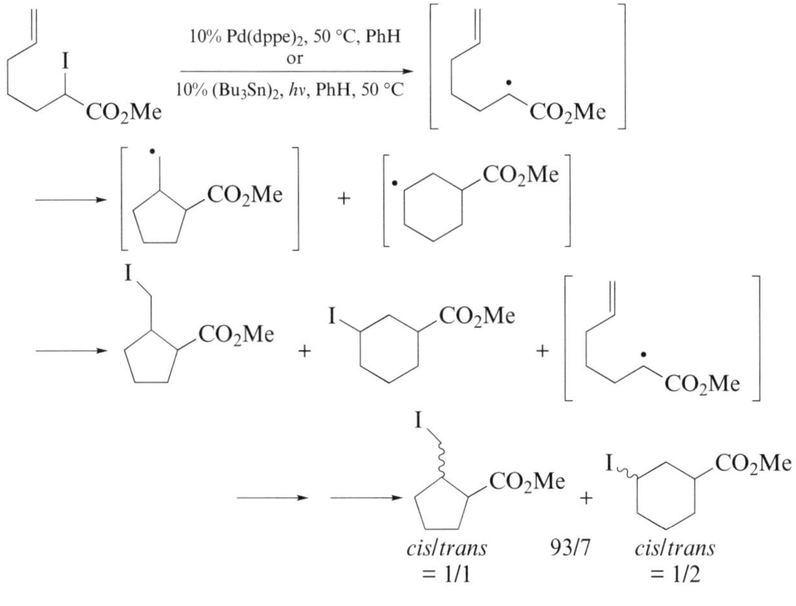

10% Pd(PPh$_3$)$_4$, 65 °C	30%	31%
1% Pd(PPh$_3$)$_4$, rt	80%	—

(166)

cis/trans   93/7   cis/trans
= 1/1   = 1/2

Evidence based on product mixtures now suggests, at least in the cases of α-halocarbonyl and perhaloalkyl starting marterials, that these reactions are in fact atom transfer radical cyclizations (equation 166)[324,325]. In them, the palladium catalyst is proposed to have roles both as the radical initiator and as a trap for iodine, similar to the more commonly used hexabutylditin. Intramolecular allyl halide–alkyne cyclizations proceed with *trans*-addition to the triple bond; this is evidence that a still different mechanism may be operating in these cases (equation 167)[1,326].

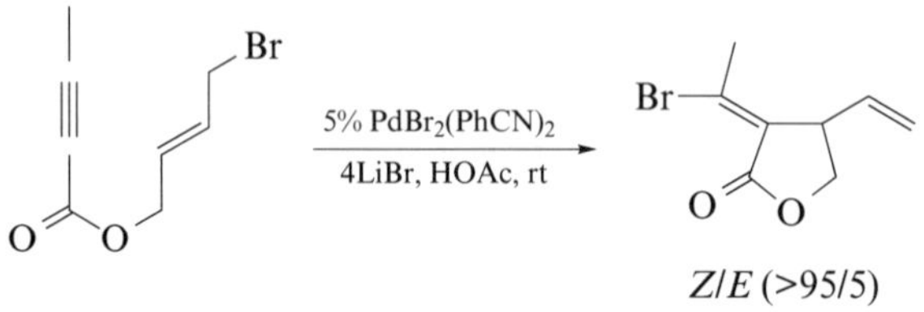

$$\text{5\% PdBr}_2\text{(PhCN)}_2 \quad / \quad \text{4LiBr, HOAc, rt} \qquad 82\% \qquad (167)$$

$$Z/E \ (>95/5)$$

The method has been used most often for cyclization reactions. Five- and six-membered rings may be formed efficiently in the common cases and, when fused to an azetidinone, seven-membered rings are formed in synthetically useful yields (equation 168)[327]. Perhaloalkane–olefin additions have focused on intermolecular cases[328,329]; in some cases, yields are enhanced by the additional use of trimethylaluminum as a radical initiator[328]. Several other transition metal complexes are known to initiate the same processes, particularly $RuCl_2(PPh_3)_2$[330].

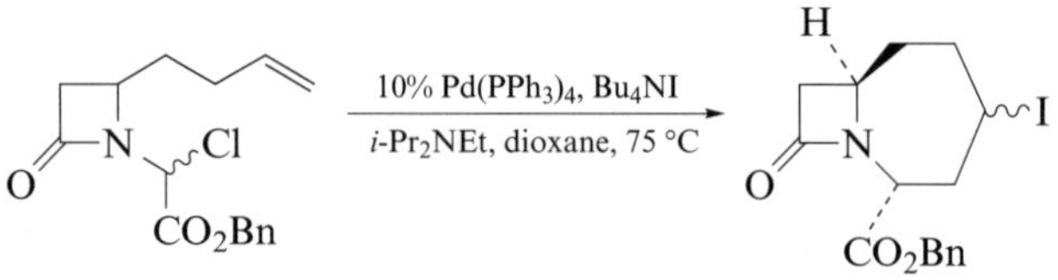

$$\text{10\% Pd(PPh}_3)_4\text{, Bu}_4\text{NI} \quad / \quad i\text{-Pr}_2\text{NEt, dioxane, 75 °C} \qquad 46\% \qquad (168)$$

VI. COUPLING REACTIONS OF ORGANIC HALIDES MEDIATED BY CHROMIUM(II)

In the presence of anhydrous Cr(II) halide salts, organic halides undergo one-electron reduction to form radical species, which in the presence of a second equivalent of Cr(II) undergo a second one-electron reduction to form organochromium(III) species (Scheme 5)[331]. This process works well for allyl, vinyl and acetylenic halides, but not for simple alkyl cases. Alkylchromium(III) species can be formed readily, however, in the presence of catalytic amounts of vitamin B_{12} or cobalt phthalocyanine (CoPc)[332].

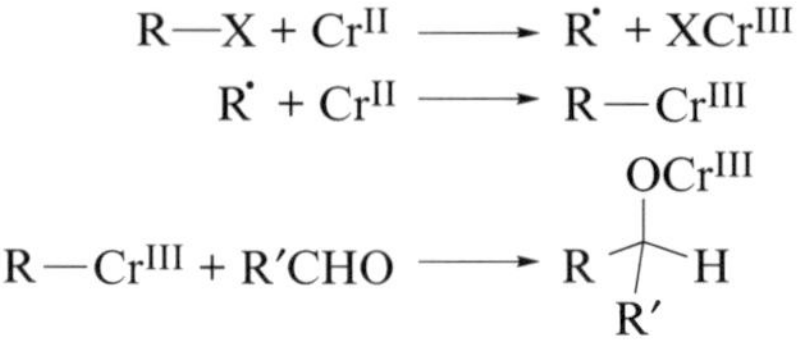

$$R\!-\!X + Cr^{II} \longrightarrow R^{\bullet} + XCr^{III}$$

$$R^{\bullet} + Cr^{II} \longrightarrow R\!-\!Cr^{III}$$

$$R\!-\!Cr^{III} + R'CHO \longrightarrow R\!-\!\underset{R'}{\overset{OCr^{III}}{\underset{\big|}{\overset{\big|}{C}}}}\!-\!H$$

SCHEME 5

Alkyl-[332], vinyl-[333], allyl-[334] and alkynylorganochromium compounds[335] so generated react with carbonyl compounds, particularly aldehydes and ketones, by nucleophilic attack, to afford alcohols in high yields. In practice, $CrCl_2$ or $CrCl_3$ and a reductant ($LiAlH_4$ or Na/Hg amalgam) are added to a mixture of the organic halide and the carbonyl compound; in many cases, catalytic amounts of other transition metal salts, particularly $NiCl_2$, facilitate the condensation.

The allyl halide–carbonyl combinations have been particularly well developed, due to the useful stereochemical outcome of the condensations. The addition occurs with allylic migration, to give highly stereoselective *anti*-addition products when alkyl or silyl[336] groups are substituted on the vinyl carbon (equation 169). Heteroatom substitution at C-3 results in lowered levels of both stereoselection and regioselection[337]. Propargyl halides also normally react by S_E2' regiochemistry to give allenes, although it appears that C-3 substitution is necessary on the alkyne for clean allylic migration to occur[338]. For both chiral aldehydes and chiral allyl halides, predominant Felkin–Ahn control is realized, with selectivities which may range from negligible to quite high under carefully selected conditions (equations 170 and 171)[339]. Intramolecular versions of this reaction are known, and in them the allylchromium–aldehyde condensations are fairly good at forming medium or macrocyclic rings. For example, such a Cr(II)-mediated allyl bromide–aldehyde condensation forms the cornerstone of a synthesis of (±)-costunolide (equation 172)[340].

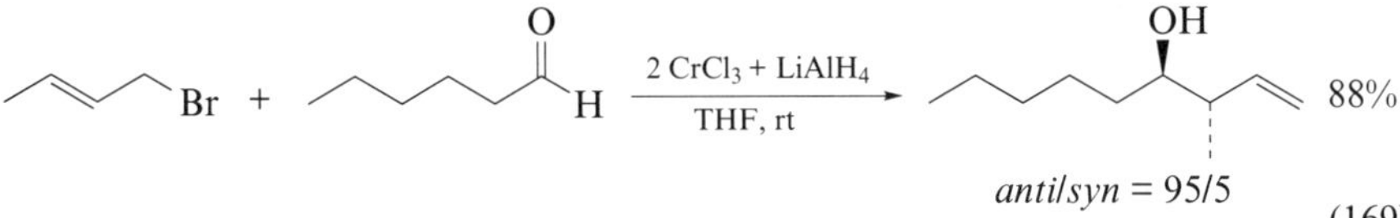

$$anti/syn = 95/5 \qquad (169)$$

R = THP, OTBDMS R^2 = Me, Et, Bu, Ph   89:11 – > 99:1

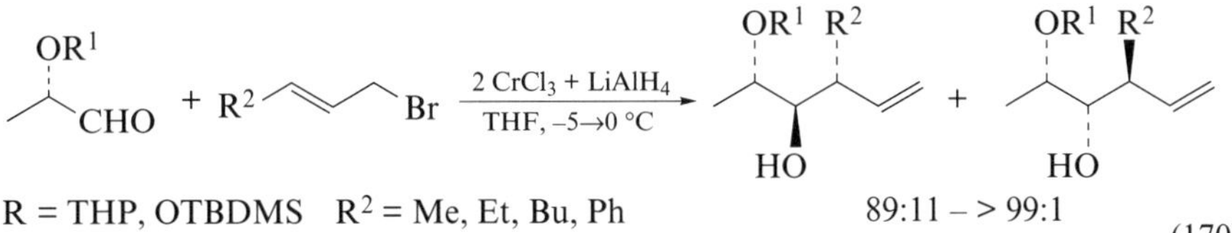

$$(170)$$

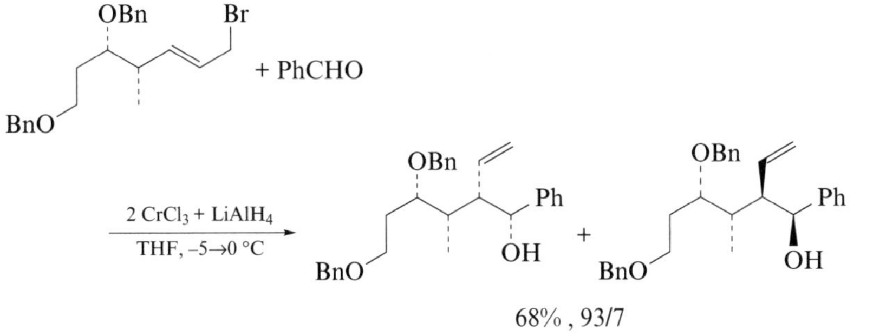

68%, 93/7

$$(171)$$

42%   (±) – costunolide

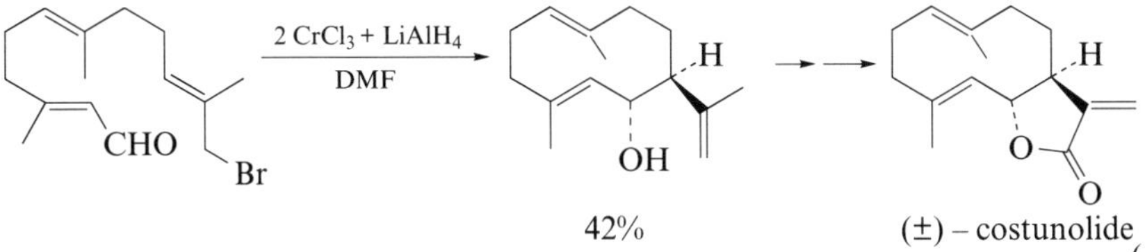

$$(172)$$

The Cr(II)-mediated addition of vinyl halides to aldehydes has been used increasingly frequently. Both E- and Z-vinyl halides normally give the more stable (E-) olefinic products, due to the geometrical instability of the intermediate vinyl radical, although there are reports of retention of double-bond stereochemistry for at least some disubstituted vinyl iodides[341]. A wide range of functional groups are tolerated (ester, amide, nitrile, conventional protecting groups and even ketones [which react more slowly]), and both intra- and intermolecular reactions have been successful (equation 173)[342].

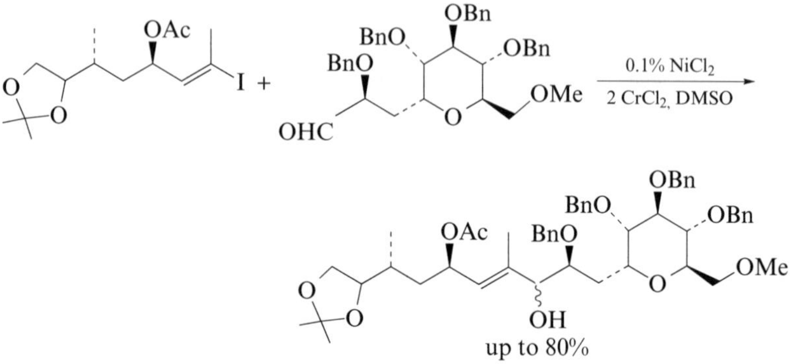

(173)

up to 80%

The previously mentioned lack of reactivity of alkyl halides with Cr(II) reagents does not extend to *gem*-dihaloalkanes. In the presence of sufficient Cr(II), these halides form geminal dimetallic alkanes, which react with aldehydes, and more slowly with ketones, to give products of olefination [343]. In most cases, *gem*-diiodides give much more ready reaction than the bromides or chlorides. *Gem*-dihalides bearing a wide range of alkyl groups give this reaction, as do silyl- and sulfur-substituted alkyl groups (equation 174)[344]. Even iodoform and haloform participate, giving vinyl halides as products (equation 175)[345]. *E*-Alkenes always predominate over their *Z*-counterparts, usually with > 90:10 selectivity.

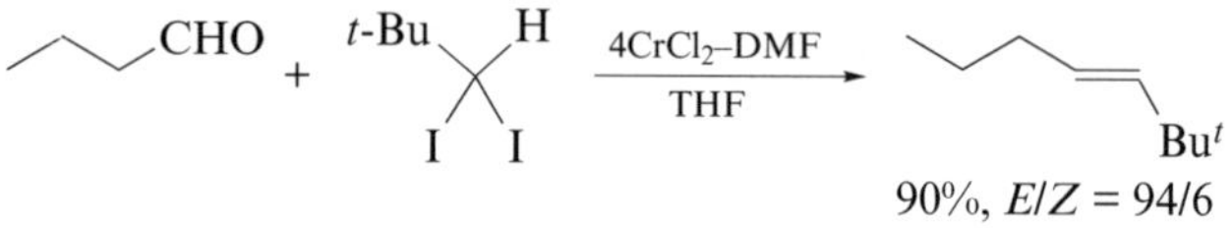

90%, $E/Z = 94/6$ (174)

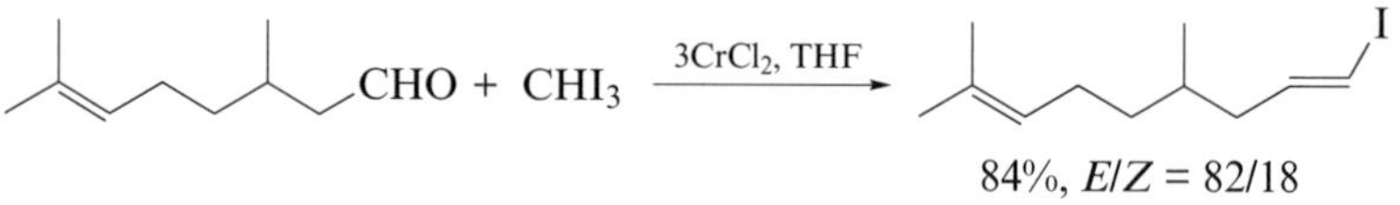

84%, $E/Z = 82/18$ (175)

α-Chloroalkyl sulfides also react with aldehydes, the reaction stopping at the alcohol stage. Good *anti-/syn*- ratios are realized in the presence of added ligands such as TMEDA; while the ratios become excellent with added diphos, the overall conversion drops off significantly[346].

The radical intermediates from Cr(II) reduction of alkyl halides can in principle be used synthetically, but have only seen limited attention to this point. ω-Haloalkynes (bromides, iodides), in the presence of excess Cr(ClO$_4$)$_2$, undergo cyclization reactions to form *exo*-alkylidene cycloalkanes (equation 176)[347]. These reactions proceed by the radical cyclization of intermediate **42** onto the alkyne unit, which undergoes subsequent reduction by Cr(II) to give a hydrolytically unstable vinylchromium(III). Rings of four, five and six members can be formed. Alternatively, α-iodo esters undergo intramolecular atom transfer radical cyclizations onto alkynes or alkenes with catalytic or stoichiometric amounts of

Cr(OAc)$_2$ (equation 177). There is even some evidence that tandem atom transfer cyclizations may be possible by this method[348].

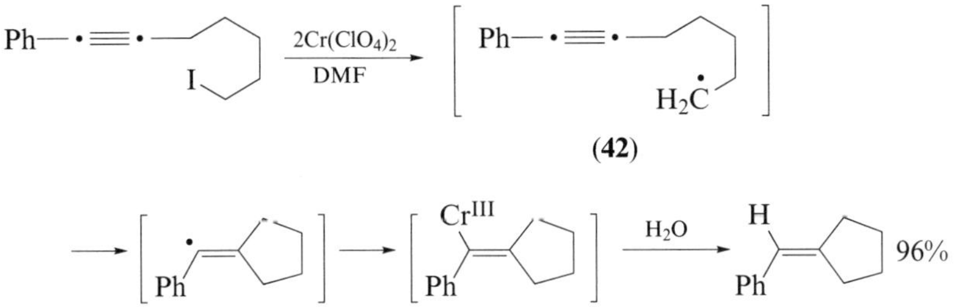

(42)

96%   (176)

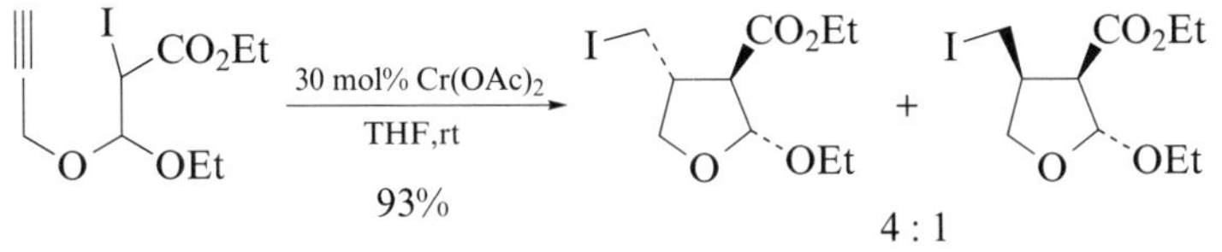

(177)

93%

4 : 1

VII. COBALT-MEDIATED RADICAL REACTIONS OF ORGANIC HALIDES

Among the most useful radical reactions in synthesis are atom transfer radical reactions
initiated by the interaction of stoichiometric amounts of organocobalt(I) complexes with
organic halides[349]. The resulting organocobalt(III) intermediates have weak cobalt–
carbon bonds (84–125 kJ mol^{-1} for primary alkylcobalts[350]) and are as such readily susceptible to homolysis by the addition of light or heat to form radicals[351]. In tertiary or aryl/vinyl
cases, heat is not even required for this to occur. Most commonly, the organocobalt(I)
complexes stem from Co(salen) complexes, vitamin B$_{12}$ or its analogs, or cobaloximes
(**43–45**), which are reduced in the presence of the halide to Co(I) by electrochemical

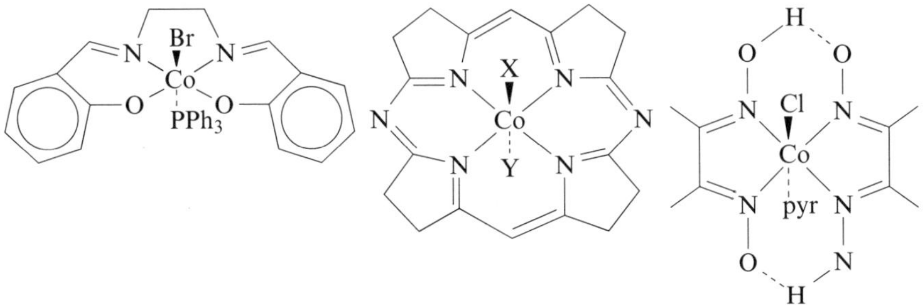

(43)   **(44)**   **(45)**

methods or by chemical reduction (by NaBH$_4$, Na/Hg, Zn). Radicals so generated add in
either an intramolecular or intermolecular sense to carbon–carbon multiple bonds. The
radical so generated is nucleophilic, and electron-poor double bonds are therefore better
substrates in both intra- and intermolecular cases (equation 178)[352,353].

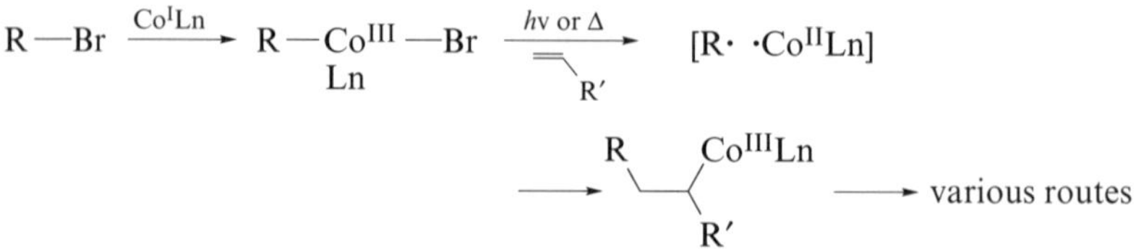

$$R-Br \xrightarrow{Co^ILn} \underset{Ln}{R-Co^{III}-Br} \xrightarrow[R']{hv \text{ or } \Delta} [R\cdot \;\cdot Co^{II}Ln]$$

$$\longrightarrow \underset{R'}{\overset{R\quad Co^{III}Ln}{\diagdown\diagup}} \longrightarrow \text{various routes} \quad (178)$$

Unlike many other type of radical addition reactions, the product is most often an alkyl-cobalt(III) species capable of further manipulation. These product Co—C bonds have been converted in good yields to carbon–oxygen (alcohol, acetate), carbon–nitrogen (oxime, amine), carbon–halogen, carbon–sulfur (sulfide, sulfinic acid) and carbon–selenium bonds (equations 179 and 180)[354]. Exceptions to this rule are the intermolecular additions to electron-deficient olefins, in which the putative organocobalt(III) species eliminates to form an α,β-unsaturated carbonyl compound or styrene[353] or is reduced (under electrochemical conditions) to the alkane (equation 181)[355].

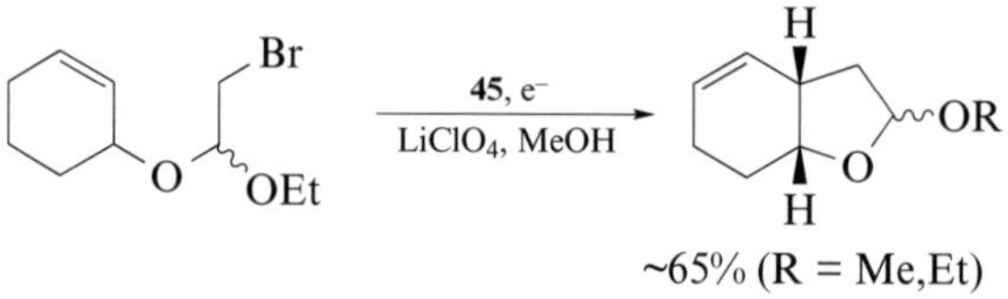

$$\text{(179)} \qquad \sim 65\% \;(R = Me, Et)$$

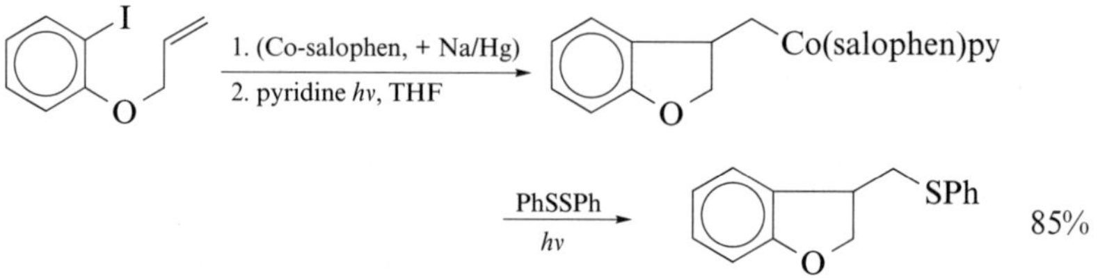

$$\text{(180)} \qquad 85\%$$

$$\text{(181)} \qquad 55\%$$

dmg = dimethylglyoximate

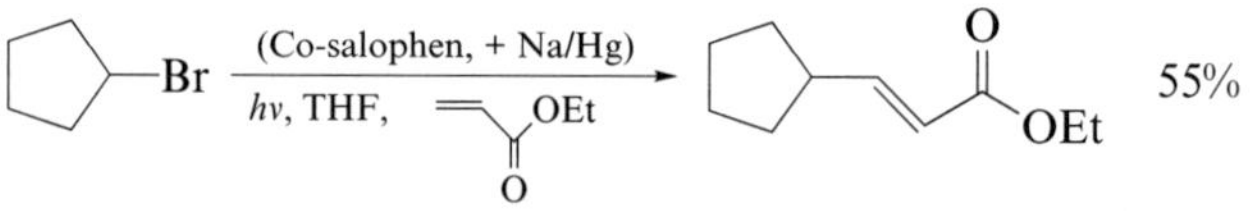

$$\text{(182)} \qquad \sim 40\%$$

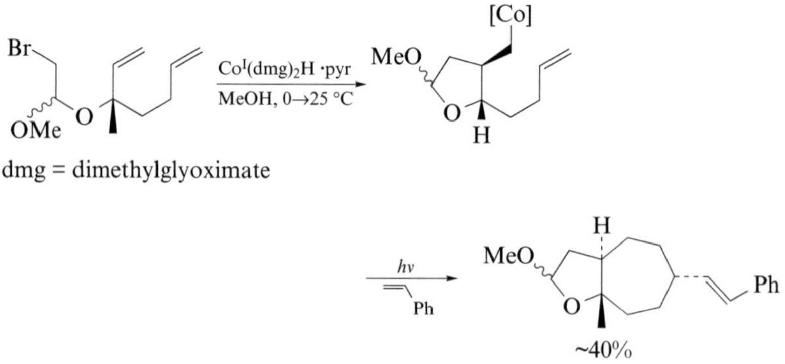

By pairing an addition to an enone in the final step with an initial intramolecular radical cyclization, it is possible to accomplish tandem radical reactions of cobalamine(III) alkyls[356]. Tandem cyclizations are also most often done, with the final termination step being elimination to the alkene or trapping with styrene (equation 182)[357]. It is also possible to enter into ring-expansion reactions by retrograde radical cyclization–recyclization processes in modest yields[358].

The nucleophilic nature of these radicals allows addition to carbon–nitrogen multiple bonds, and in selected cases this has been demonstrated. Suitable substrates for such additions include protonated pyridines, thiazoles and imidazoles (equation 183), and nitroalkyl anions[359]. Yamamoto and coworkers have also described cobalt-mediated multiple additions of carbon tetrahalides, benzyl bromides or α-bromo ketones to isocyanides, which are postulated to occur through the corresponding alkylcobalt(III) complexes[360].

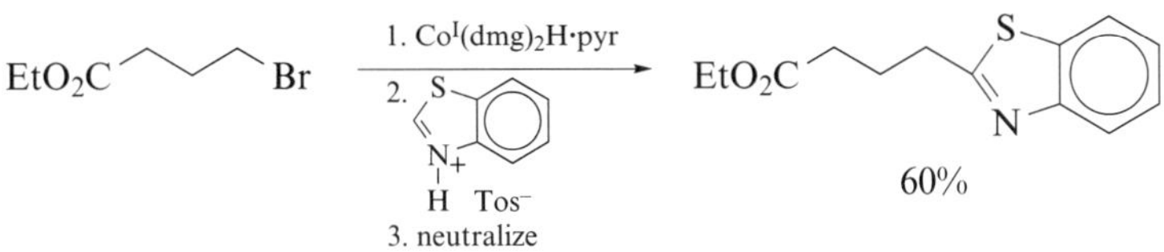

$$EtO_2C\diagdown\diagdown\diagdown Br \xrightarrow[\substack{2.\ \text{S}\\ \\ 3.\ \text{neutralize}}]{1.\ Co^I(dmg)_2H\cdot pyr} EtO_2C\diagdown\diagdown\diagdown \qquad (183)$$

60%

In addition to alkyl, aryl and vinylcobalt/radical species, acid chlorides may be converted into acylcobalt species by reaction with the cobalt(I) salens. These generate acyl radicals under photolysis, which participate in similar reactions to the alkyl radicals in most cases (equations 184 and 185). Acylcobalt(III) species bearing an α-aryl or vinyl substituent, on the other hand, undergo concomitant decarbonylation to afford a benzylic or allyl radical, which then may undergo a number of bond-forming processes, including homocoupling[361].

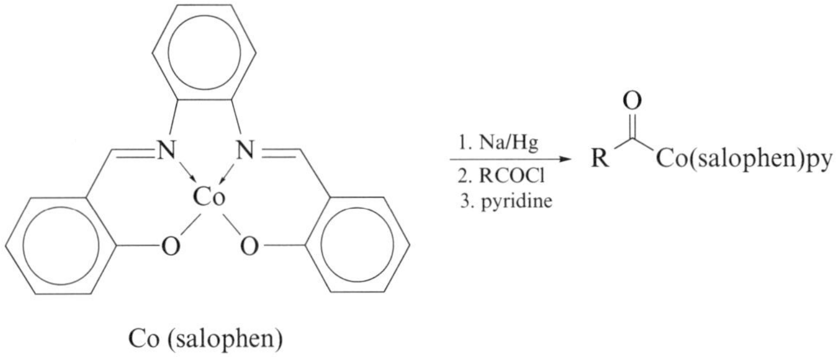

$$\xrightarrow[\substack{2.\ RCOCl\\3.\ pyridine}]{1.\ Na/Hg} R\overset{O}{\diagdown}Co(salophen)py \qquad (184)$$

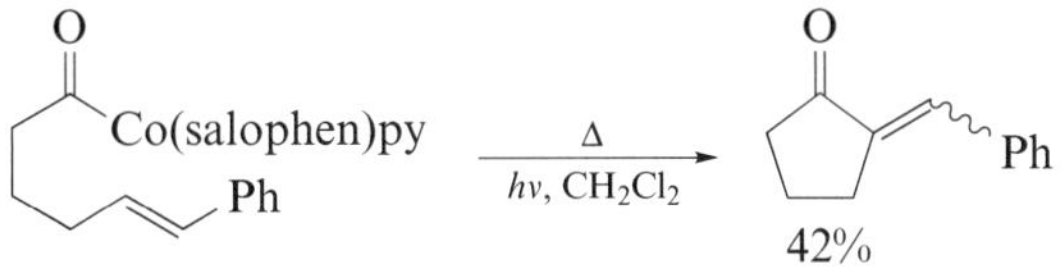

Co (salophen)

$$\xrightarrow[hv,\ CH_2Cl_2]{\Delta} \qquad (185)$$

42%

Several synthetic targets have been attacked by exploitation of this methodology; examples include an enantiospecific synthesis of acromelic acid A and a formal synthesis of physovenine (equations 186 and 187)[362]. Recent work has focused on rendering the radical generation and addition processes catalytic in cobalt; successes have been achieved by using more readily reduced cobaloxime complexes and carefully controlled conditions[363].

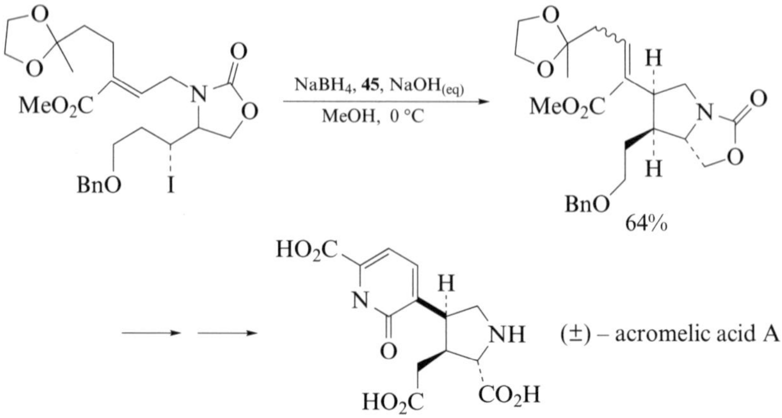

$(\pm)$ – acromelic acid A   (186)

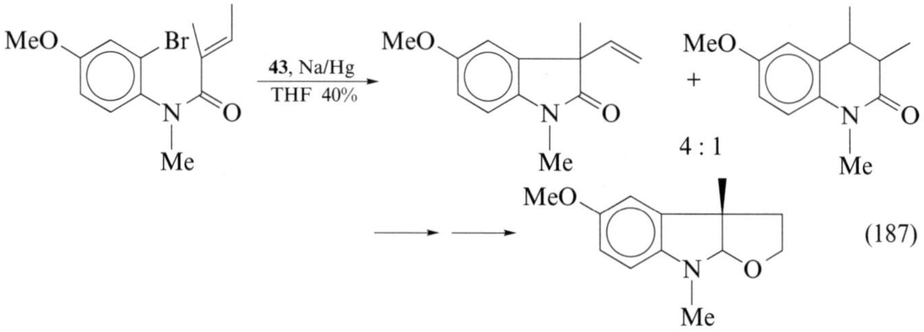

physovenine intermediate

VIII. CARBONYLATION REACTIONS

A. Carbonylation with CO in the Presence of Palladium Catalysts

Under appropriate conditions, the reaction of aryl halides with palladium catalysts, under CO pressure, leads to double carbonylations[364]. These have been performed in the presence of amines, alcohols, water or a water–calcium hydroxide–isopropanol mixture to afford α-ketoamides[365], α-ketoesters[366], α-ketoacids[367] and α-hydroxyacids[368], respectively (equations 188–190). While the reactions in all these cases are run under elevated pressures of CO, the pressures required for double carbonylation are relatively less for ketoamide synthesis than for the acid derivatives. The most successful catalysts for this process are $PdCl_2L_2$, where the ligands (L) are phosphines with basicity greater than PPh_3. For amide derivatives, dppb and $P(Me_2Ph)_2$ give the best results, while for acids, trialkylphosphine ligands are the choice. For esters, the ideal cone angle of the ligand (140–170°) appears to be more important than the ligand basicity. Finally, amines or alcohols of moderate size are preferred. Amines such as diethylamine or dipropylamine give the best bouble-/mono-carbonylation ratios in α-ketoamide preparation, whereas secondary alcohols are preferred for α-ketoesters. Results for amide preparation are generally more satisfactory than for any of the acid derivatives. The preparation of α-ketoamides from vinyl halides may also occur, and follows roughly the same pattern as for aryl halides[369].

$$\text{PhI} + \text{Et}_2\text{NH} \xrightarrow[\text{CO (20 atm), 60 °C}]{2\% \text{ PdCl}_2 \text{ (PMePh}_2)_2} \text{PhCOCONEt}_2 + \text{PhCONEt}_2 \qquad (188)$$
$$100\% \qquad 93:7$$

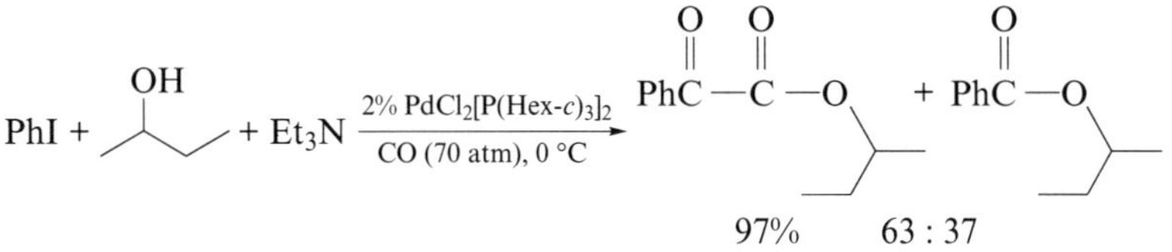

$$\text{PhI} + \underset{\text{OH}}{\diagup\!\!\!\diagup} + \text{Et}_3\text{N} \xrightarrow[\text{CO (70 atm), 0 °C}]{2\% \text{ PdCl}_2[\text{P(Hex-}c)_3]_2} \text{PhC}\overset{\text{O O}}{\underset{\|\ \|}{-\text{C}-\text{O}}} + \text{PhC}\overset{\text{O}}{\underset{\|}{-\text{O}}} \qquad (189)$$
$$97\% \qquad 63:37$$

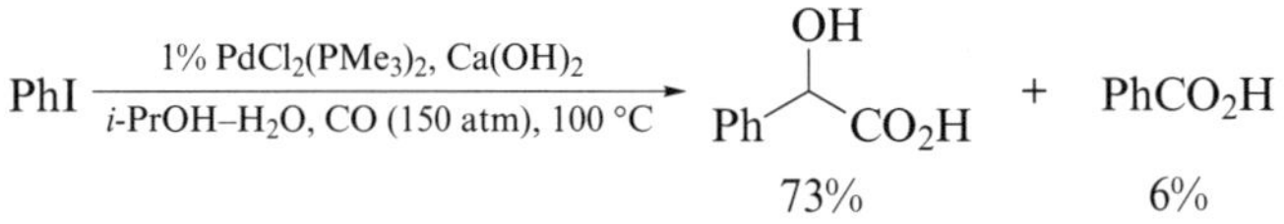

$$\text{PhI} \xrightarrow[\text{\textit{i}-PrOH–H}_2\text{O, CO (150 atm), 100 °C}]{1\% \text{ PdCl}_2(\text{PMe}_3)_2, \text{ Ca(OH)}_2} \underset{\text{Ph}\quad\text{CO}_2\text{H}}{\overset{\text{OH}}{\diagup\!\!\!\diagup}} + \text{PhCO}_2\text{H} \qquad (190)$$
$$73\% \qquad 6\%$$

Mechanistically, the formation of the double carbonylation products results from CO insertion into the arylmetal species, followed by amine or alkoxide attack on a coordinated CO ligand to form **46**, which then undergoes reductive elimination to form the observed products (Scheme 6)[366,370,371].

$$\text{R}\!-\!\text{X} + \text{PdL}_n \longrightarrow \text{R}\!-\!\text{Pd(X)L}_n$$

$$\text{R}\!-\!\text{Pd(X)L}_n + \text{CO} \longrightarrow \left[\text{R}\!-\!\overset{\text{O}}{\underset{\|}{\text{C}}}\!-\!\text{Pd(CO)L}_n\right]^+ \text{X}^- \text{ or } \text{R}\!-\!\overset{\text{O}}{\underset{\|}{\text{C}}}\!-\!\text{Pd(CO)XL}_n$$
$$(48)$$

$$\left[\text{R}\!-\!\overset{\text{O}}{\underset{\|}{\text{C}}}\!-\!\text{Pd(CO)L}_n\right]^+ \text{X}^- \text{ or } \text{R}\!-\!\overset{\text{O}}{\underset{\|}{\text{C}}}\!-\!\text{Pd(CO)XL}_n + \text{NuH} + \text{R}'_3\text{N}$$

$$\longrightarrow \text{R}\!-\!\overset{\text{O}}{\underset{\|}{\text{C}}}\!-\!\text{Pd}(\overset{\text{O}}{\underset{\|}{\text{C}}}\text{Nu})\text{L}_2 + \text{R}_3\text{NH}^+\text{X}^-$$
$$(46)$$

$$\text{R}\!-\!\overset{\text{O}}{\underset{\|}{\text{C}}}\!-\!\text{Pd}(\overset{\text{O}}{\underset{\|}{\text{C}}}\text{Nu})\text{L}_2 + \text{CO} \longrightarrow \text{R}\!-\!\overset{\text{O}}{\underset{\|}{\text{C}}}\!-\!\overset{\text{O}}{\underset{\|}{\text{C}}}\!-\!\text{Nu} + \text{Pd(CO)}_m\text{L}_n$$

SCHEME 6

Several new aspects of single carbonylation reactions have been uncovered. Aside from simple alkoxycarbonylation, the intermediate acylpalladium complex is also capable of reacting in an intramolecular fashion with the double bond of an alkene to give ketones. This process in turn generates an alkylpalladium,, which may be further carbonylated to the ester in the presence of methanol (equations 191 and 192). Vinyl[372], aryl[373] and allyl halides[374] all are capable of being the initiating organic halide in this process.

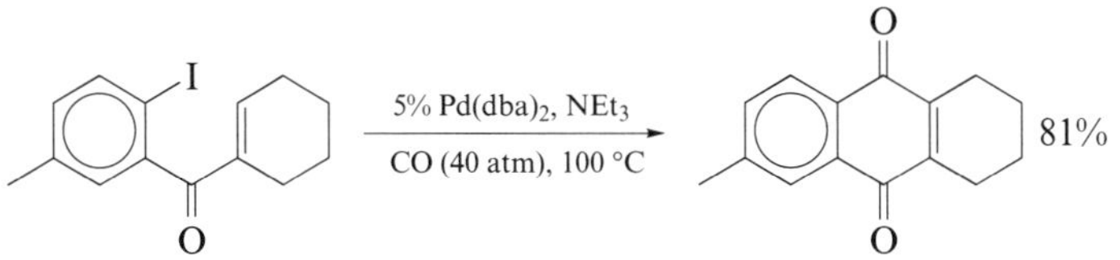

$$\xrightarrow[\text{CO (40 atm), 100 °C}]{5\% \text{ Pd(dba)}_2, \text{ NEt}_3} \qquad 81\% \qquad (191)$$

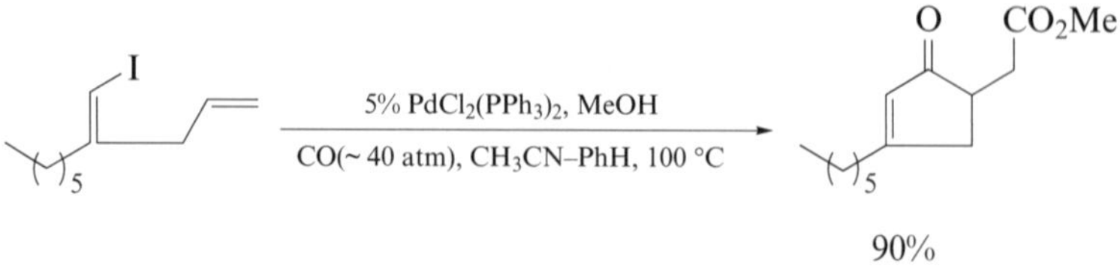

$$\text{(192)}$$

$$\xrightarrow[\text{CO(~ 40 atm), CH}_3\text{CN–PhH, 100 °C}]{\text{5\% PdCl}_2(\text{PPh}_3)_2,\ \text{MeOH}}$$

90%

A second potential fate of the palladium acyl intermediate is to enter into reaction with nucleophilic species other than alcohols or amines. Organoaluminum, organoboron, organotin or selected organosilicon reagents may attack to give ketones ultimately; these have been discussed earlier, in Sections III.E. to III.H. Highly stabilized carbanions (i.e. malonate) may react either inter- or intramolecularly, giving ketones with palladium-based catalysts or other catalysts (equation 193)[375]. Terminal alkynes also may attack to give alkynyl ketones, although further reaction of the alkyne function in the product is the norm (equation 194)[376–378]. In conjunction with appropriately substituted aryl- or vinyl-lamine or phenolic functions, useful isoquinolone and flavone syntheses may be accomplished in this manner[376,377]. Fluoride ion, particularly CsF, will attack the acylpalladium intermediate to afford acyl fluorides[379].

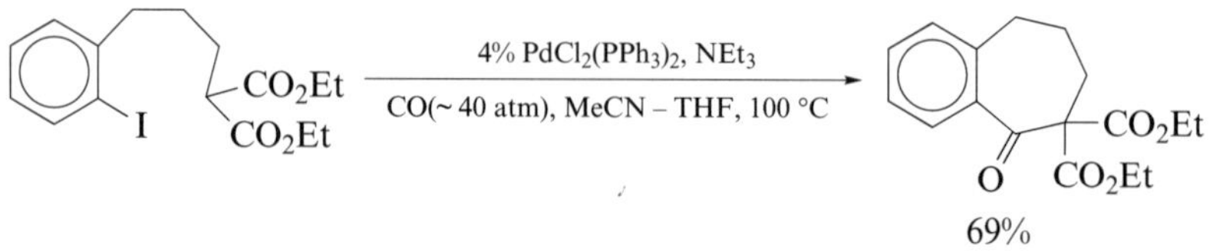

$$\xrightarrow[\text{CO(~ 40 atm), MeCN – THF, 100 °C}]{\text{4\% PdCl}_2(\text{PPh}_3)_2,\ \text{NEt}_3}$$

$$\text{(193)}$$

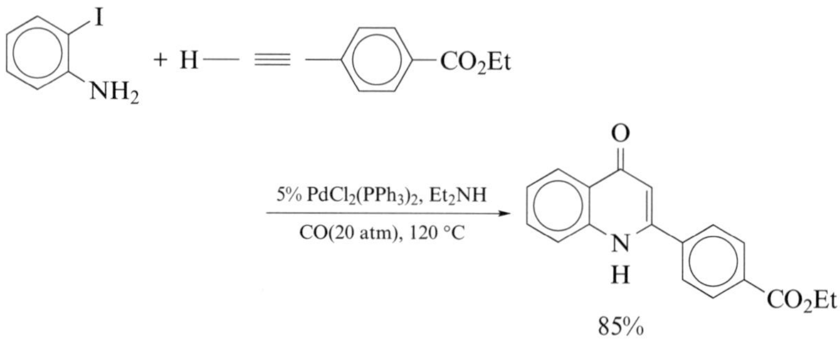

69%

$$\text{+ H} \equiv \text{—} \langle \rangle \text{—CO}_2\text{Et}$$

$$\xrightarrow[\text{CO(20 atm), 120 °C}]{\text{5\% PdCl}_2(\text{PPh}_3)_2,\ \text{Et}_2\text{NH}}$$

$$\text{(194)}$$

85%

Although aldehydes may be prepared by formylations of organic halides with CO, palladium cataysts and hydrogen gas, the conditions required are drastic. Several substitutes for hydrogen gas, however, allow formylation to be accomplished under much milder conditions. Poly(methylhydrosiloxane) (PMHS), sodium formate and Bu$_3$SnH all accomplish this in conjunction with Pd(PPh$_3$)$_4$ for aryl iodides, and in the case of the latter two reagents, for some aryl bromides[380, 381]. Extremely mild conditions (1–3 atm CO) are required with Bu$_3$SnH; vinyl iodides also give formylation with this hydride equivalent[381].

Carbonylation reactions of aryl chlorides may be accomplished by two different approaches. Coordination of the arene ring to a tricarbonylchromium unit facilitates oxidative addition, and carbonylation may consequently be accomplished[382]. Alternatively, Pd(dippp)$_2$ (47) has proven to be a catalyst capable of converting chloroarenes to aldehydes, amides, esters or acids in good yields. This catalyst may be generated *in situ*, by the addition of Pd(OAc)$_2$ and dippp. Both electron-withdrawing and electron-donating

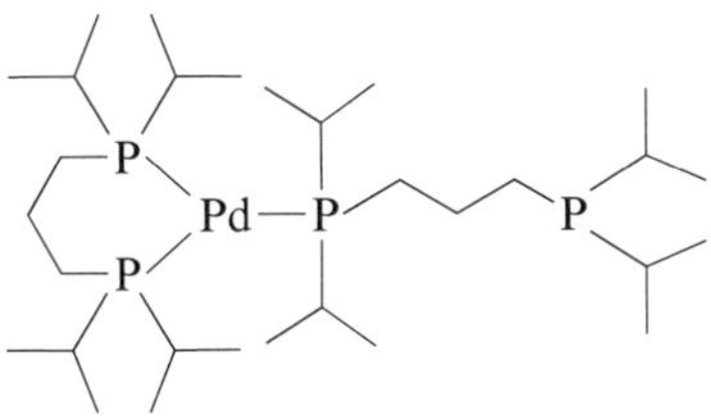

(47) Pd (dippp)$_2$
dippp = bis(diisopropylphosphino)propane

groups are permissible on the arene ring (equation 195)[383]. Bis(tricyclohexylphos-phino)palladium dichloride is less effective in the same role, but gives useful amounts of carbonylation to carboxylic acids in basic aqueous solutions[384].

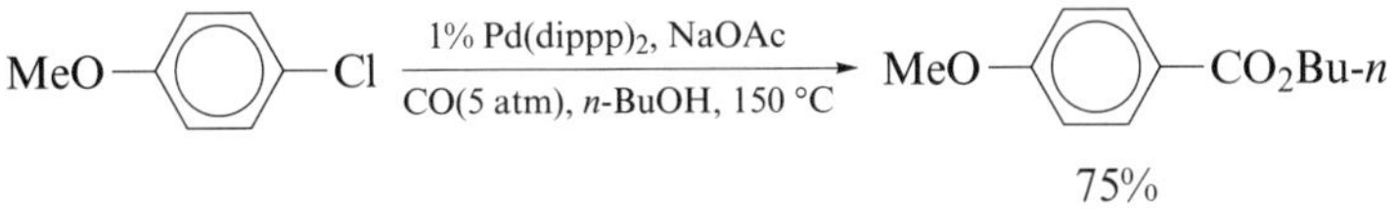

$$MeO\text{—}\langle\bigcirc\rangle\text{—}Cl \xrightarrow[\text{CO(5 atm), }n\text{-BuOH, 150 °C}]{1\% \text{ Pd(dippp)}_2,\text{ NaOAc}} MeO\text{—}\langle\bigcirc\rangle\text{—}CO_2Bu\text{-}n \quad (195)$$

75%

While the vast majority of carbonylation reactions of organic halides are on aryl or vinyl substrates, primary and secondary alkyl iodides may be carbonylated in good yields by the use of PtCl$_2$(PPh$_3$)$_2$, under high CO pressure (equation 196). This is likely due to the rela-tive stability of σ-alkyl platinum species to β-elimination. Either esters or aldehydes may be the product, if the carbonylation is carried out in the presence of alcohols or hydrogen gas, respectively[385]. In addition, the same catalyst alkoxycarbonylates 1-iodoalkynes clean-ly under similar conditions. Under the conditions of reaction, the platinum catalyst is reduced to Pt(CO)$_2$(PPh$_3$)$_2$; not coincidentally, this platinum complex carbonylates alkyl iodides well, particularly under photolytic conditions[386].

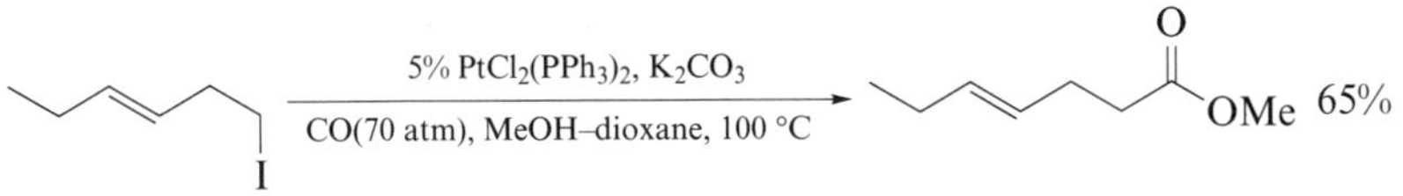

$$\xrightarrow[\text{CO(70 atm), MeOH–dioxane, 100 °C}]{5\% \text{ PtCl}_2(\text{PPh}_3)_2,\text{ K}_2\text{CO}_3} \quad 65\% \quad (196)$$

In some instances, alkyl halide carbonylation can occur using other catalysts, as well. Mixed palladium–rhodium catalysts may afford carboxylic esters from alkyl halides in the presence of borate esters or titanium or aluminum alkoxides[387]. Curiously, double carbonylation of perfluoroalkyl iodides containing β-hydrogen atoms, including one secondary iodide, undergoes competitive double and single carbonylation with unexcep-tional catalysts [PdCl$_2$(PPh$_3$)$_2$ or PdCl$_2$[P(Hex-c)$_3$]$_2$][388].

Although not as popular as with other carbonylation catalysts, biphasic conditions (par-ticularly with phase transfer catalysts) may be employed to accomplish carbonylation under very mild conditions for benzylic and vinyl bromides[389].

Mechanistic work has revealed some details of the carbonylation process. For benzyl halides and iodoarenes, the rate-determining step for alkoxycarbonylation has been found to be the alcoholysis of acylpalladium 48, and in particular the reductive elimination of the acylpalladium alkoxide[371,390,391]. In difficult cases, such as with chloroarenes, the oxidative addition of haloarene is the slowest step[383].

Two approaches to asymmetric carbonylation reactions have met with some success. In the carbonylation reactions of α-methylbenzyl bromide under phase transfer catalysis, the use of Pd(dba)$_2$ and oxaphospholane ligand **49** resulted in the preparation of 2-arylpropionic acids with up to 64% ee (42% ee at reasonable conversion) (equation 197)[392]. More recently, prochiral alkenyl iodide **50** underwent alkoxycarbonylation with up to 57% ee in the presence of (R)-binap ligands and a silver or thallium salt (equation 198)[393].

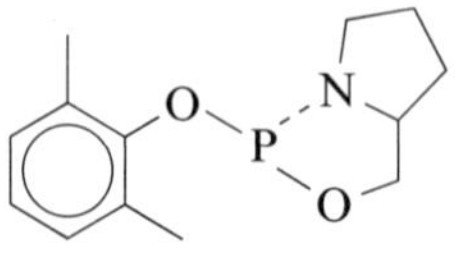

(49)

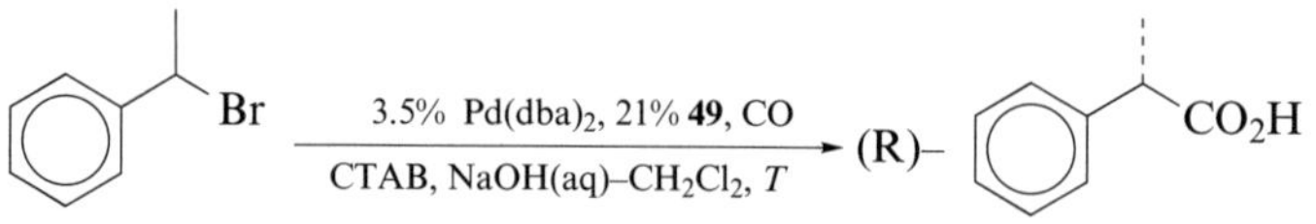

(197)

$$T = \quad \begin{array}{ll} 10\ °C & 9\%,\ 64\%\ ee \\ 18\ °C & 65\%,\ 42\%\ ee \end{array}$$

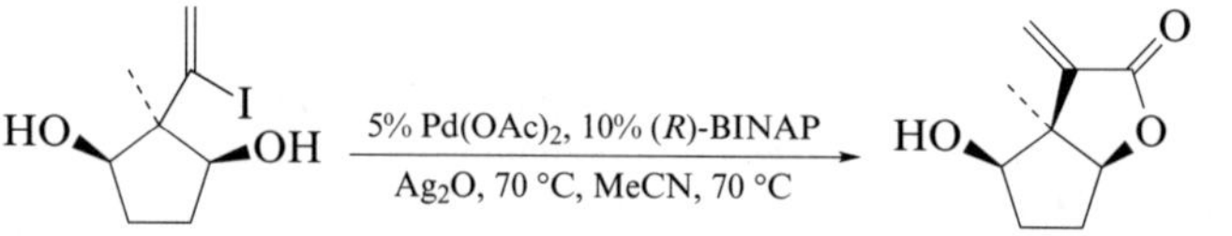

(198)

(50)   44%, 57% ee

B. Carbonylation with CO in the Presence of Nickel Catalysts

The effectiveness in carbonylations of Ni(CO)$_4$ is well documented, as well as its toxicity. Substitutes for this catalyst are therefore of much interest, and [Ni(CN)(CO)$_3^-$], generated *in situ* from Ni(CN)$_2$, CO and aqueous NaOH under phase transfer conditions, fulfills this role in many cases[394]. Under these conditions (1 atm CO), several types of organic halides are carbonylated, including allyl halides[394], benzyl chlorides (with lanthanide salts)[395], aryl iodides[396], vinyl bromides[397] and dibromocyclopropanes (equation 199)[398].

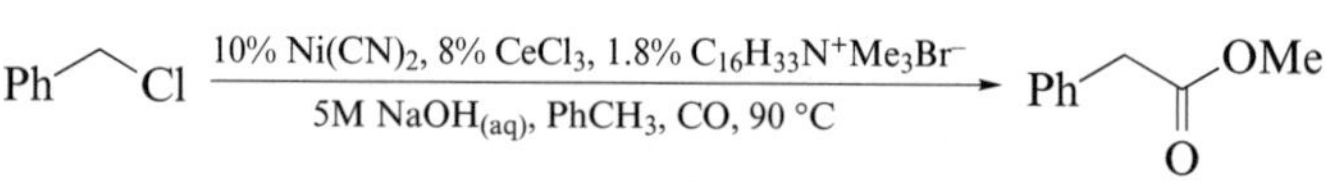

Two other Ni(CO)$_4$ substitutes, Ni(CO)$_3$PPh$_3$ and Ni(COD)$_2$/dppe, prove to be appropriate for the catalysis of tandem metallo-ene/carbonylation reactions of allylic iodides (Scheme 7)[399]. This process features initial oxidative addition to the alkyl iodide, followed by a metallo-ene reaction with an appropriately substituted double or triple bond, affording an alkyl or vinyl nickel species. This organonickel species may then either alkoxycarbonylate or carbonylate and undergo a second cyclization on the pendant alkene to give **51**, which then alkoxycarbonylates. The choice of nickel catalyst and use of diene versus enyne influences whether mono- or biscyclization predominates (equations 200 and 201).

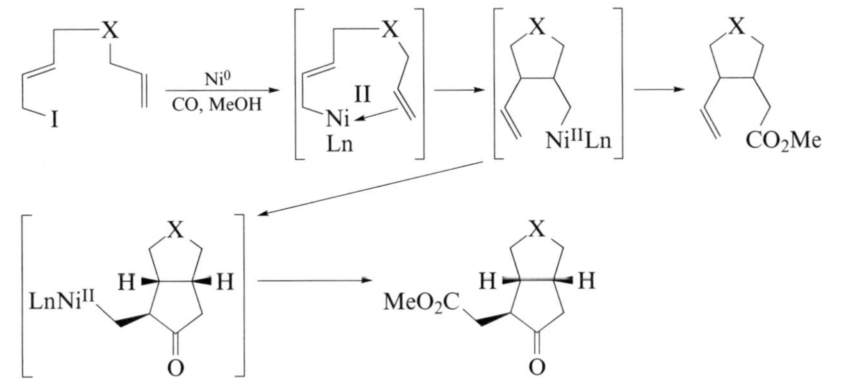

(51)

SCHEME 7

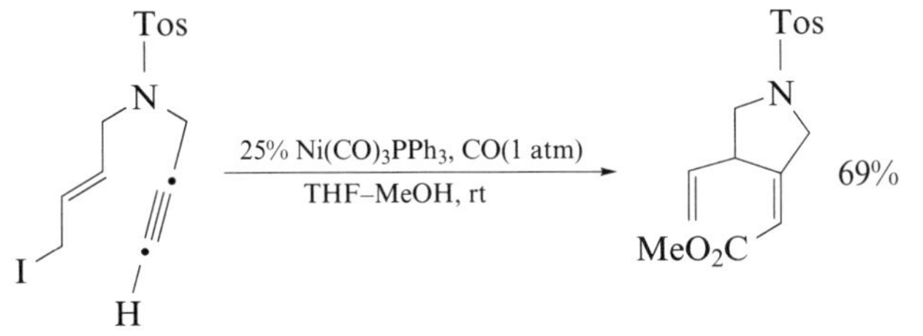

$$25\% \text{ Ni(CO)}_3\text{PPh}_3, \text{CO(1 atm)} \quad \text{THF–MeOH, rt}$$

69%   (200)

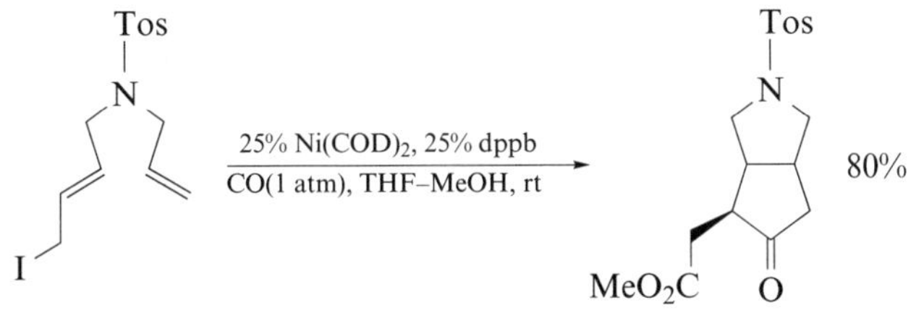

$$25\% \text{ Ni(COD)}_2, 25\% \text{ dppb} \quad \text{CO(1 atm), THF–MeOH, rt}$$

80%   (201)

C. Carbonylation with Metal Carbonyls

Despite the toxicity of volatile metal carbonyls, particularly $Ni(CO)_4$, several useful transformations have been developed employing these reagents. Monocarbonylation of *gem*-dibromocyclopropanes may be accomplished with $Ni(CO)_4$ in the presence of alcohols, amines or (less successfully) thiols, to afford cyclopropane carboxylic esters, amides or thioesters, respectively (equation 202)[400]. Silylamine or silylsulfide reagents may take the place of amines or thiols[401]. The intermediacy of a nickel enolate in the carbonylations is evidenced by its trapping with some electrophiles or ring opening with suitably placed leaving groups[402].

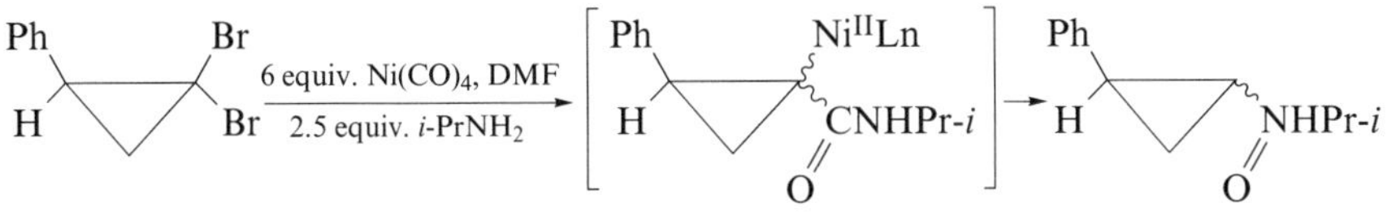

78%   (202)

Moretó and coworkers have made improvements to the Chiusoli reaction, the $Ni(CO)_4$-mediated carbonylation cyclization of allyl halides and alkynes, by conducting it in methanol[403]. It has subsequently been applied in the synthesis of methylenomycin B, in an intramolecular sense to provide bicyclo[3.3.0]octenones, and in intermolecular cases to form both fused bicyclic cyclopentenones and spirocyclopentenones (equations 203 and 204)[403-405].

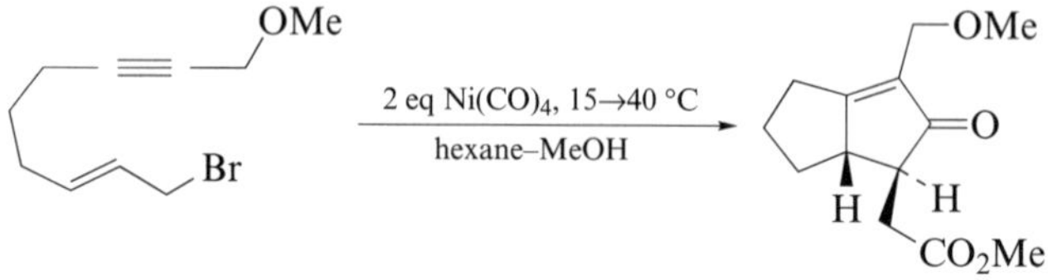

$$\text{(203)} \quad 50\%$$

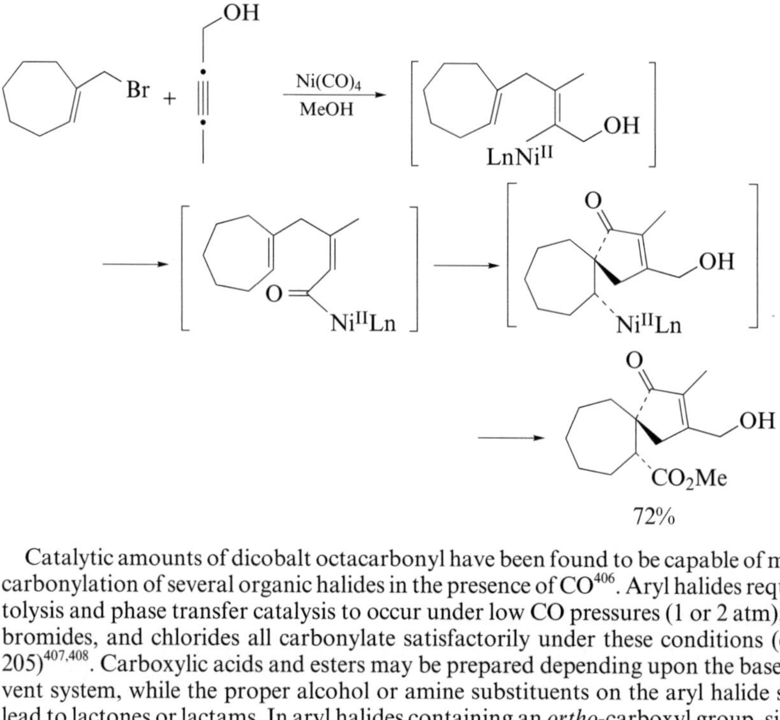

$$\text{(204)} \quad 72\%$$

Catalytic amounts of dicobalt octacarbonyl have been found to be capable of mediating carbonylation of several organic halides in the presence of CO[406]. Aryl halides require photolysis and phase transfer catalysis to occur under low CO pressures (1 or 2 atm); iodides, bromides, and chlorides all carbonylate satisfactorily under these conditions (equation 205)[407,408]. Carboxylic acids and esters may be prepared depending upon the base and solvent system, while the proper alcohol or amine substituents on the aryl halide substrate lead to lactones or lactams. In aryl halides containing an *ortho*-carboxyl group, significant amounts of double carbonylation are seen[409]. The photostimulated reactions have been proposed to occur by an $S_{RN}1$ condensation of *in situ* formed $Co(CO)_4^-$.

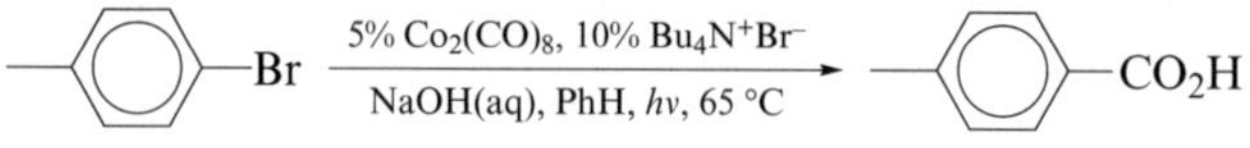

$$\text{(205)} \quad 97\%$$

Nonphotolytic carbonylations catalyzed by $Co_2(CO)_8$ may also occur. With molecular sieves replacing a base, iodoalkanes containing β-hydrogen atoms carbonylate to esters (equation 206)[410]. In addition, either $Co_2(CO)_8$ or preformed $NaCo(CO)_4$ are capable of carbonylating or doubly carbonylating benzyl halides under low CO pressures, depending upon the solvent and base employed[406,411].

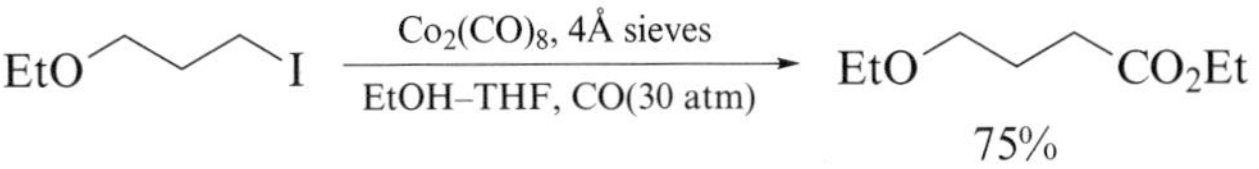

$$\text{EtO} \diagdown\diagup\diagdown \text{I} \quad \xrightarrow[\text{EtOH–THF, CO(30 atm)}]{\text{Co}_2(\text{CO})_8, \text{4Å sieves}} \quad \text{EtO} \diagdown\diagup\diagdown \text{CO}_2\text{Et} \qquad (206)$$

75%

In the presence of alkyl halides and base, alkyltetracarbonylcobalt complexes are formed with $Co_2(CO)_8$; these species $[RCo(CO)_4]$ carbonylate a wide range of aryl halides or heterocyclic halides to various products, which depend upon the specific conditions. In the presence of alcohols, carboxylic esters are formed. Under phase transfer conditions and with iodomethane, mixtures of methyl ketone and carboxylic acid formation are realized (equation 207). In the presence of sodium sulfide or $NaBH_4$ in water–$Ca(OH)_2$ (equation 208) good amounts of double carbonylation are realized under very mild conditions[412–414].

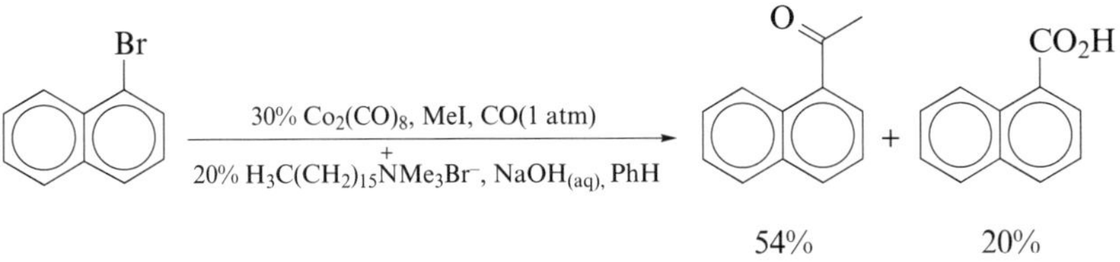

$$(207)$$

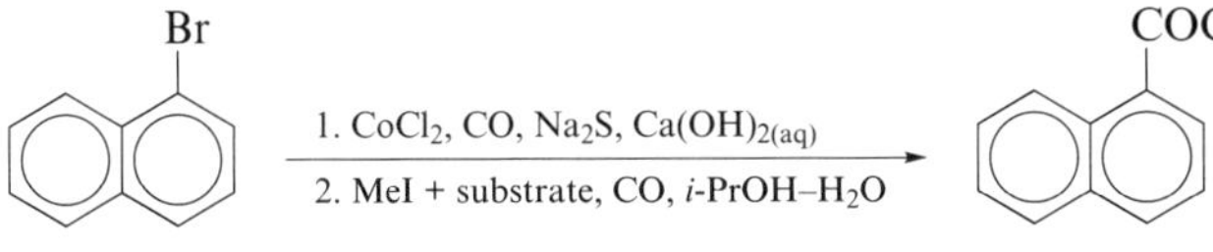

54%   20%

$$80\% \quad (208)$$

Carbonylation with iron carbonyls parallels that of cobalt carbonyls. Benzylic chlorides and bromides are carbonylated with $Fe(CO)_5$ in the presence of base. Esters are realized when carbonylation is performed in alcohols under 1 atm of CO with catalytic amounts of iron pentacarbonyl[415]. Under phase transfer conditions, two predominant routes are available. With catalytic amounts of iron under a CO atmosphere and strongly basic conditions, the carboxylic acids are realized in reasonable yields[415,416], whereas mild bases [$Ca(OH)_2$], stoichiometric amounts of iron carbonyl and the omission of CO give dibenzyl ketones[417]. In at least a few cases, it is possible to prepare unsymmetrical methyl benzyl ketones[418]. des Abbayes and coworkers have observed the formation of acyltetracarbonyl anion (**52**) under the reaction conditions, and have proposed the catalytic cycle in Scheme 8 for the ketone formation[418].

Work on other halides is much more limited. *Gem*-dibromocyclopropanes give modest yields of alkoxycarbonylation with alkoxide bases[419]. Aryl iodides may be carbonylated by catalytic mixtures of $Fe(CO)_5$–$Co_2(CO)_8$ predominantly to the carboxylic acids or predominantly to the benzophenones, depending upon conditions[420].

Electrochemical reductive methods have also been used for carbonylative formation of acyltetracarbonyl anion formation with 1-halopentanes (bromides, iodides and chlorides) under anhydrous conditions with one equivalent of $Fe(CO)_5$; the aldehyde is formed in respectable yields upon subsequent acidic workup (equation 209)[421].

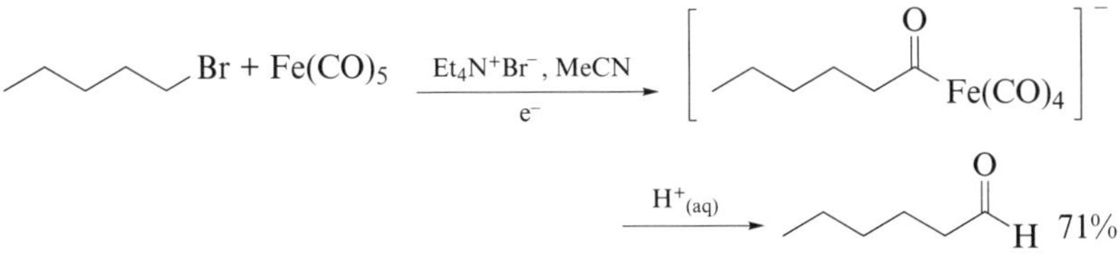

$$71\% \quad (209)$$

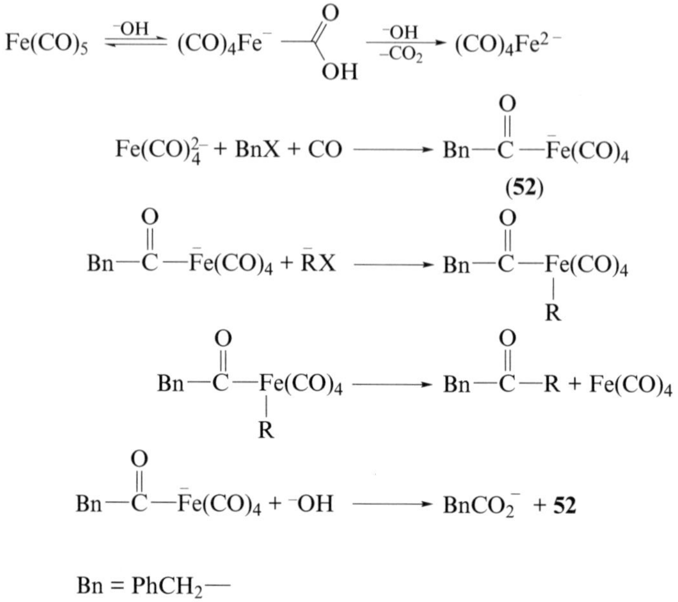

$$Fe(CO)_5 \underset{}{\overset{^{-}OH}{\rightleftharpoons}} (CO)_4Fe^- \quad \overset{O}{\underset{OH}{\diagdown C \diagup}} \quad \underset{-CO_2}{\overset{^{-}OH}{\longrightarrow}} (CO)_4Fe^{2-}$$

$$Fe(CO)_4^{2-} + BnX + CO \longrightarrow \overset{O}{\overset{\|}{Bn-C}}-\bar{Fe}(CO)_4$$

$$(52)$$

$$\overset{O}{\overset{\|}{Bn-C}}-\bar{Fe}(CO)_4 + \bar{R}X \longrightarrow \overset{O}{\overset{\|}{Bn-C}}-\overset{|}{\underset{R}{Fe}}(CO)_4$$

$$\overset{O}{\overset{\|}{Bn-C}}-\overset{|}{\underset{R}{Fe}}(CO)_4 \longrightarrow \overset{O}{\overset{\|}{Bn-C}}-R + Fe(CO)_4$$

$$\overset{O}{\overset{\|}{Bn-C}}-\bar{Fe}(CO)_4 + {}^-OH \longrightarrow BnCO_2^- + 52$$

$$Bn = PhCH_2-$$

SCHEME 8

IX. REFERENCES

1. F. Naso and G. Marchese, in *The Chemistry of Halides, Pseudo-Halides and Azides* (Eds. S. Patai and Z. Rappoport), Chapter 26, Wiley, Chichester, (1983).
2. B. H. Lipshutz and S. Sengupta, *Org. React.*, **41**, 135 (1992); B. H. Lipshutz, *Synlett*, 119 (1990); B. H. Lipshutz, *Synthesis*, 325 (1987); B. H. Lipshutz, R. S. Wilhelm and J. A. Kozlowski, *Tetrahedron*, **40**, 5005 (1994).
3. For a discussion of the nature of higher-order cyanocuprates, see: B. H. Lipshutz, S. Sharma and E. L. Ellsworth, *J. Am. Chem. Soc.*, **112**, 4032 (1990); S. H. Bertz, *J. Am. Chem. Soc.*, **112**, 4031 (1990).
4. B. H. Lipshutz, R. S. Wilhelm, J. A. Kozlowski and D. Parker, *J. Org. Chem.*, **49**, 3928 (1984).
5. A. El Marini, M. L. Roumestant, L. Pappalardo and P. Viallefont, *Bull. Soc. Chim. Fr.*, 554 (1989).
6. B. H. Lipshutz, R. Crow, S. H. Dimcock, E. L. Ellsworth, R. A. J. Smith and J. R. Behling, *J. Am. Chem. Soc.*, **112**, 4063 (1990); B. H. Lipshutz, C. Ung, T. R. Elworthy and D. C. Reuter, *Tetrahedron Lett.*, **31**, 4539 (1990).
7. B. H. Lipshutz, M. Koerner and D. A. Parker, *Tetrahedron Lett.*, **28**, 945 (1987); B. H. Lipshutz, P. Fatheree, W. Hagen and K. L. Stevens, *Tetrahedron Lett.*, **33**, 1041 (1992).
8. C. R. Johnson and D. S. Dhanoa, *J. Org. Chem.*, **52**, 1885 (1987).
9. B. H. Lipshutz, D. A. Parker, S. Nguyen, K. E. McCarthy, J. C. Barton, S. E. Whitney and H. Kotsuki, *Tetrahedron*, **42**, 2873 (1986); C. E. Tucker and P. Knochel, *J. Org. Chem.*, **58**, 4781 (1993).
10. G. W. Ebert and R. D. Rieke, *J. Org. Chem.*, **53**, 4482 (1988); D. E. Stack, B. T. Dawson and R. D. Rieke, *J. Am. Chem. Soc.*, **113**, 4672 (1991).
11. R. D. Rieke, R. M. Wehmeyer, T. -C. Wu and G. W. Ebert, *Tetrahedron*, **45**, 443 (1989); G. W. Ebert and W. R. Klein, *J. Org. Chem.*, **56**, 4744 (1991).
12. D. M. Wiemers and D. J. Burton, *J. Am. Chem. Soc.*, **108**, 832 (1986).
13. Z. -Y. Yang, D. M. Wiemers and D. J. Burton, *J. Am. Chem. Soc.*, **114**, 4402 (1992).
14. C. Germon, A. Alexakis and J.-F. Normant, *Synthesis*, **40**, 43 (1984).

15. P. D. Edwards and A. I. Meyers, *Tetrahedron Lett.*, **25**, 939 (1984).
16. E. C. Ashby and D. Coleman, *J. Org. Chem.*, **52**, 4554 (1987).
17. S. H. Bertz, G. Dabbagh and G. M. Villacorta, *J. Am. Chem. Soc.*, **104**, 5824 (1982).
18. S. F. Martin, J. R. Fishpaugh, J. M. Power, D. M. Giolando, R. A. Jones, C. M. Nunn and A. H. Cowley, *J. Am. Chem. Soc.*, **110**, 7226 (1988).
19. Z.-Y. Yang, B. V. Nguyen and D. J. Burton, *Synlett*, 141 (1992); Z.-Y. Yang and D. J. Burton, *J. Org. Chem.*, **57**, 4676 (1992); Z.-Y. Yang and D. J. Burton, *J. Org. Chem.*, **56**, 5125 (1991).
20. F. O. H. Pirrung, W. J. M. Steeman, H. Hiemstra, W. N. Speckamp, B. Kaptein, W. H. J. Boesten, H. E. Schoemaker and J. Kamphuis, *Tetrahedron Lett.*, **33**, 5141 (1992) and references therein; H. Nagashima, N. Ozaki, K. Seki, M. Ishii, and K. Itoh, *J. Org. Chem.*, **54**, 4497 (1989).
21. B. H. Lipshutz, D. C. Reuter and E. L. Ellsworth, *J. Org. Chem.*, **54**, 4975 (1989).
22. B. H. Lipshutz and T. R. Elworthy, *J. Org. Chem.*, **55**, 1695 (1990).
23. H. Urata and T. Fuchikami, *Tetrahedron Lett.*, **32**, 91 (1991).
24. P. Knochel and R. D. Singer, *Chem. Rev.*, **93**, 2117 (1993); P. Knochel, M. J. Rozema, C. E. Tucker, C. Retherford and S. A. Rao, *Pure Appl. Chem.*, **64**, 361 (1992).
25. H. G. Chen, C. Hoechstetter and P. Knochel, *Tetrahedron Lett.*, **30**, 4795 (1989).
26. C. E. Tucker, T. N. Majid and P. Knochel, *J. Am. Chem. Soc.*, **114**, 3983 (1992).
27. C. Jubert and P. Knochel, *J. Org. Chem.*, **57**, 5425 (1992).
28. A. Sidduri, N. Budries, R. M. Laine and P. Knochel, *Tetrahedron Lett.*, **33**, 7515 (1992).
29. T. Ogawa, K. Kusume, M. Tanaka, K. Hayami and H. Suzuki *Synth. Commun.*, **19**, 2199 (1989).
30. K. Okuro, M. Furuune, M. Miura and M. Nomura, *Tetrahedron Lett.*, **33**, 5363 (1992).
31. E. Piers and L. J. Browne, *J. Org. Chem.*, **40**, 2694 (1975).
32. P. A. Wender and A. W. White, *J. Am. Chem. Soc.*, **110**, 2218 (1988).
33. E. J. Corey and N. W. Boaz, *Tetrahedron Lett.*, **25**, 3059 (1984).
34. A. M. Caporusso, C. Polizzi and L. Lardicci, *J. Org. Chem.*, **52**, 3920 (1987).
35. C. Polizzi, C. Consoloni, L. Lardicci and A. M. Caporusso, *J. Organomet. Chem.*, **417**, 289 (1991).
36. H. H. Mooiweer, C. J. Elsevier, P. Wijkens and P. Vermeer, *Tetrahedron Lett.*, **26**, 65 (1985).
37. Y. Fujita, M. Ishiguro, T. Onishi and T. Nishida, *Bull. Chem. Soc. Jpn.*, **55**, 1325 (1982).
38. H. Saimoto and T. Hiyama, *Tetrahedron Lett.*, **27**, 597 (1986).
39. J. Kang, W. Cho and W. K. Lee, *J. Org. Chem.*, **49**, 1838 (1984).
40. D. F. Taber, B. S. Dunn, J. F. Mack and S. A. Saleh, *J. Org. Chem.*, **50**, 1987 (1985).
41. M. C. P. Yeh and P. Knochel, *Tetrahedron Lett.*, **29**, 2395 (1988).
42. L. Zhu, R. M. Wehmeyer and R. D. Rieke, *J. Org. Chem.*, **56**, 1445 (1991).
43. S. A. Rao, T.-S. Chou, I. Schipor and P. Knochel, *Tetrahedron*, **48**, 2025 (1992).
44. H. G. Chen, J. L. Gage, S. D. Barrett and P. Knochel, *Tetrahedron Lett.*, **31**, 1829 (1990).
45. P. Knochel, T.-S. Chou, H. G. Chen, M. C. P. Yeh and M. J. Rozema, *J. Org. Chem.*, **54**, 5202 (1989).
46. M. J. Dunn and R. F. W. Jackson, *J. Chem. Soc., Chem. Commun.*, 319 (1992).
47. L. Zhu and R. D. Rieke, *Tetrahedron Lett.*, **32**, 2865 (1991).
48. H. Ochiai, Y. Tamaru, K. Tsubaki and Z.-i. Yoshida, *J. Org. Chem.*, **52**, 4418 (1987).
49. E. Nakamura, S. Aoki, K. Sekiya, H. Oshino and I. Kuwajima, *J. Am. Chem. Soc.*, **109**, 8056 (1987).
50. M. Hoshi, Y. Masuda, Y. Nunokawa and A. Arase, *Chem. Lett.*, 1029 (1984); A. Arase, M. Hoshi and Y. Masuda, *Chem. Lett.*, 2093 (1984).
51. D. J. Burton and S. W. Hansen, *J. Am. Chem. Soc.*, **108**, 4229 (1986).
52. J.-M. Gaudin and C. Morel, *Tetrahedron Lett.*, **31**, 5749 (1990).
53. T. Jeffery, *Tetrahedron, Lett.*, **30**, 2225 (1989).
54. G. Mignani, C. Chevalier, F. Grass, G. Allmang and D. Morel, *Tetrahedron Lett.*, **31**, 5161 (1990).
55. K. Inami, T. Teshima, J. Emura and T. Shiba, *Tetrahedron Lett.*, **31**, 4033 (1990).
56. D. E. Stack, B. T. Dawson and R. D. Rieke, *J. Am. Chem. Soc.*, **114**, 5110 (1992).
57. D. E. Stack and R. D. Rieke, *Tetrahedron Lett.*, **33**, 6575 (1992).
58. H. Hopf and N. Krause, *Tetrahedron Lett.*, **26**, 3323 (1985).
59. M. C. P. Yeh and P. Knochel, *Tetrahedron Lett.*, **30**, 4799 (1989).
60. C. J. Rao and P. Knochel, *J. Org. Chem.*, **56**, 4593 (1991).
61. H. Sörensen and A. E. Greene, *Tetrahedron Lett.*, **31**, 7597 (1990).
62. J. Lindley, *Tetrahedron*, **40**, 1433 (1984).

63. S. Sugai, F. Ikawa, T. Okazaki and T. Nishida, *Chem. Lett.*, 597 (1982).
64. H. Suzuki, T. Kobayashi, Y. Yoshidu and A. Osuka, *Chem. Lett.*, 193 (1983).
65. A. Osuka, T. Kobayashi and H. Suzuki, *Synthesis*, 67 (1983).
66. H. Suzuki, T. Kobayashi and A. Osuka, *Chem. Lett.*, 589 (1983).
67. J.-i. Setsune, T. Ueda, K. Matsukawa and T. Kitao, *Chem. Lett.*, 1931 (1984).
68. H. Suzuki, K. Watanabe and Q. Li, *Chem. Lett.*, 1779 (1985).
69. H. Suzuki, Q. Yi, J. Inoue, K. Kusume and T. Ogawa, *Chem. Lett.*, 887 (1987).
70. J.-i. Setsune, T. Ueda, K. Shikita, K. Matsukawa, T. Iida and T. Kitao, *Tetrahedron*, **42**, 2647 (1986).
71. H. Suzuki Y. Yoshida and A. Osuka, *Chem. Lett.*, 135 (1982).
72. B. H. Lipshutz, S. Sharma and D. C. Reuter, *Tetrahedron, Lett.*, **31**, 7253 (1990).
73. T. W. Bell, L.-Y.Hu and S. V. Patel, *J. Org. Chem.*, **52**, 3847 (1987).
74. W. R. Bowman, H. Heaney and P. H. G. Smith, *Tetrahedron Lett.*, **25**, 5821 (1984).
75. H. Malmberg and M. Nilsson, *Tetrahedron*, **42**, 3981 (1986).
76. T. D. Spawn and D. J. Burton, *Bull. Soc. Chim. Fr.*, 876 (1986).
77. A. Capperucci, A. Degl'Innocenti, C. Faggi, A. Ricci, P. Dembech and G. Seconi, *J. Org. Chem.*, **53**, 3612 (1988).
78. A. Hosomi, T. Masunari, Y. Tominaga and M. Hojo, *Bull. Chem. Soc. Jpn.*, **64**, 1051 (1991).
79. A. Hosomi, K. Otaka and H. Sakurai, *Tetrahedron Lett.*, **27**, 2881 (1986).
80. A. Hosomi, Y. Sakata and H. Sakurai, *Tetrahedron Lett.*, **26**, 5175 (1985).
81. M. Tiecco, L. Testaferri, M. Tingoli, D. Chianelli and E. Wenbert, *Tetrahedron Lett.*, **23**, 27 (1982).
82. V. Fiandanese, G. Marchese, F. Naso and L. Ronzini, *Synthesis*, 1034 (1987).
83. L. N. Pridgen, L. Snyder and J. Prol, Jr., *J. Org. Chem.*, **54**, 1523 (1989).
84. D. E. Bergstrom and P. A. Reddy, *Tetrahedron Lett.*, **23**, 4191 (1982).
85. G. Barbarella, A. Bongini and M. Zambianchi, *Tetrahedron*, **38**, 3347 (1982).
86. L. Pridgen, *J. Org. Chem.*, **47**, 4319 (1982).
87. K. C. Eapen, S. S. Dua and C. Tamborski, *J.Org. Chem.*, **49**, 478 (1984).
88. G. S. Reddy and W. Tam, *Organometallics*, **3**, 630 (1984).
89. M. R. H. Elmoghayar, P. Groth and K. Undheim, *Acta Chem. Scand.*, **B37**, 109 (1983).
90. A. M. Caporusso, F. Da Settimo and L. Lardicci, *Tetrahedron Lett.*, **26**, 5101 (1985).
91. V. Fiandanese, G. Marchese and F. Naso, *Tetrahedron Lett.*, **29**, 3587 (1988); C. Cardillicchio, V. Fiandanese, G. Marchese and L. Ronzini, *Tetrahedron Lett.*, **26**, 3595 (1985).
92. G. Kottirsch and G. Szeimies, *Chem. Ber.*, **123**, 1495 (1990).
93. K. Yuan and W. J. Scott, *Tetrahedron Lett.*, **32**, 189 (1991).
94. S. Komiya, Y. Abe, A. Yamamoto and T. Yamamoto, *Organometallics*, **2**, 1466 (1983).
95. K. Tatsumi, A. Nakamura, S. Komiya, A. Yamamoto and T. Yamamoto, *J. Am. Chem. Soc.*, **106**, 8181 (1984).
96. T. Hayashi, M. Konishi, M. Fukushima, K. Kanehira, T. Hioki and M. Kumada, *J. Org. Chem.*, **48**, 2195 (1983).
97. B. K. Vriesema and R. M. Kellogg, *Tetrahedron Lett.*, **27**, 2049 (1986).
98. J. H. Griffin and R. M. Kellogg, *J. Org. Chem.*, **50**, 3261 (1985).
99. B. K. Vriesma, M. Lemaire, J. Butler and R. M. Kellogg, *J. Org. Chem.*, **51**, 5169 (1986).
100. G. A. Cross and R. M. Kellogg, *J. Chem. Soc., Chem. Commun.*, 1746 (1988); G. Cross, B. K. Vriesema, G. Boven, R. M. Kellogg and F. van Bolhuis, *J. Organomet. Chem.*, **370**, 357 (1989).
101. T. Hayashi, K. Hayashizaki, T. Kiyoi and Y. Ito, *J. Am. Chem. Soc.*, **110**, 8153 (1988).
102. T. Hayashi, K. Hayashizaki and Y. Ito, *Tetrahedron Lett.*, **30**, 215 (1989).
103. T. Hayashi, M. Konishi, Y. Kobori, M. Kumada, T. Higuchi and K. Hirotsu, *J. Am. Chem. Soc.*, **106**, 158 (1984).
104. T. Katayama and M. Uemo, *Chem. Lett.*, 2073 (1991).
105. P. L. Castle and D. A. Widdowson, *Tetrahedron Lett.*, **27**, 6013 (1986).
106. A. Carpita and R. Rossi, *Gazz. Chim. Ital.*, **115**, 575 (1985).
107. R. Rossi and A. Carpita, *Tetrahedron Lett.*, **27**, 2529 (1986).
108. V. Ratovelomanana, G. Linstrumelle and J.-F. Normant, *Tetrahedron Lett.*, **26**, 2575 (1985).
109. A. Carpita, R. Rossi and C. A. Veracini, *Tetrahedron*, **41**. 1919 (1985).
110. T. Hayashi, M. Konishi, H. Ito and M. Kumada, *J. Am. Chem. Soc.*, **104**, 4962 (1982).
111. T. Hayashi, Y. Okamoto and M. Kumada, *Tetrahedron Lett.*, **24**, 807 (1983).
112. H.-J. Krenzfeld, C. Döbler and H.-D. Albrecht, *J. Organomet. Chem.*, **336**, 287 (1987).

113. F. Ozawa, K. Kurihara, M. Fujimori, T. Hidaka, T. Toyoshima and A. Yamamoto, *Organometallics*, **8**, 180 (1989).
114. M. J. Calhorda, J. M. Brown and N. A. Cooley, *Organometallics*, **10**, 1431 (1991).
115. J. M. Brown and N. A. Cooley, *Organometallics*, **9**, 353 (1990).
116. A. Moravskiy and J. K. Stille, *J. Am. Chem. Soc.*, **103**, 4182 (1981).
117. W. de Graaf, J. Boersma, W. J. J. Smeets, A. L. Spek and G. van Koten, *Organometallics*, **8**, 2907 (1989).
118. P. K. Byers, A. J. Canty, M. Crespo, R. J. Puddephatt and J. D. Scott, *Organometallics*, **7**, 1363 (1987).
119. E.-i. Negishi, *Acc. Chem. Res.*, **15**, 340 (1982).
120. E.-i. Negishi, H. Matsushima, M. Kobayashi and C. L. Rand, *Tetrahedron Lett.*, **24**, 3823 (1983).
121. K. Sekiya and E. Nakamura, *Tetrahedron Lett.*, **29**, 5155 (1988).
122. K. Ruitenberg, H. Kleijn, H. Westmijze, J. Meijer and P. Vermeer, *Recl. Trav. Chim. Pays-Bas*, **101**, 405 (1982).
123. D. S. Elmis and T. L. Gilchrist, *Tetrahedron*, **46**, 2623 (1990).
124. T. Frejd and T. Klingstedt, *Synthesis*, 40 (1987).
125. T. Klingstedt and T. Frejd, *Organometallics*, **2**, 598 (1983).
126. E.-i. Negishi, V. Bagheri, S. Chatterjee, F.-T. Luo, J. A. Miller and A. T. Stoll, *Tetrahedron Lett.*, **24**, 5181 (1983).
127. J.-M. Duffault, J. Einhorn, and A. Alexakis, *Tetrahedron Lett.*, **32**, 3701 (1991).
128. M. Gardette, N. Jabri, A. Alexakis and J.-F. Normant, *Tetrahedron*, **40**, 2741 (1984).
129. A. Carpita and R. Rossi, *Tetrahedron Lett.*, **27**, 4351 (1986).
130. B. P. Andreini, A. Carpita and R. Rossi, *Tetrahedron Lett.*, **27**, 5533 (1986).
131. A. Minato, K. Suzuki and K. Tamao, *J. Am. Chem. Soc.*, **109**, 1257 (1987).
132. C. J. Elsevier and P. Vermeer, *J. Org. Chem.*, **50**, 3042 (1985).
133. T. Hayashi, T. Hagihara, Y. Katsuro and M. Kumada, *Bull. Chem. Soc. Jpn.*, **56**, 363 (1983).
134. T. Hayashi, A. Yamamoto, M. Hojo and Y. Ito, *J. Chem. Soc., Chem. Commun.*, 495 (1989).
135. P. Mortinet, R. Sauvêtre and J.-F. Normant, *J. Organomet. Chem.*, **378**, 1 (1989).
136. A. Minato, K. Suzuki, K. Tamao and M. Kumada, *Tetrahedron Lett.*, **25**, 83 (1984).
137. A. S. Bell, D. A. Roberts and K. S. Ruddock, *Tetrahedron Lett.*, **29**, 5013 (1988).
138. E.-i. Negishi, T. Takahashi and A. O. King, *Org. Synth.*, **66**, 67 (1987).
139. H. Yamanaka, M. An-Naka, Y. Kondo and T. Sakamoto, *Chem. Pharm. Bull.*, **33**, 4309 (1985).
140. A. Pelter, M. Rowland and I. H. Jenkins, *Tetrahedron Lett.*, **28**, 5213 (1987).
141. T. L. Gilchrist and R. J. Summersell, *Tetrahedron Lett.*, **28**, 1469 (1987).
142. F. Tellier, R. Sauvêtre, J.-F. Normant, Y. Dromzee and Y. Jeannin, *J. Organomet. Chem.*, **331**, 281 (1987).
143. T. Sakamoto, S. Nishimura, Y. Kondo and H. Yamanaka, *Synthesis*, 485 (1988).
144. Löffler and G. Himbert, *Synthesis*, 232 (1991).
145. P. L. Heinze and D. J. Burton, *J. Org. Chem.*, **53**, 2714 (1988).
146. A. Pelter, J. M. Maud, I. Jenkins, C. Sadeka and G. Coles, *Tetrahedron Lett.*, **30**, 3461 (1989).
147. B. Jiang and Y. Xú, *J. Org. Chem.*, **56**, 7336 (1991); *Tetrahedron Lett.*, **33**, 511 (1992).
148. P. Martinet, R. Sauvêtre and J.-F. Normant, *J. Organomet. Chem.*, **367**, 1 (1989).
149. F. Tellier, R. Sauvêtre and J.-F. Normant, *J. Organomet. Chem.*, **328**, 1 (1987).
150. C. E. Russell and L. S. Hegedus, *J. Am. Chem. Soc.*, **105**, 943 (1983).
151. S. Sengupta and V. Snieckus, *J. Org. Chem.*, **55**, 5681 (1990).
152. E.-i. Negishi and K. Akiyoshi, *Chem. Lett.*, 1007 (1987).
153. E.-i. Negishi, Z. R. Owczarczyk and D. R. Swanson, *Tetrahedron Lett.*, **32**, 4453 (1991).
154. M. A. Tius, J. Gomez-Galeno, X.-q. Gu and J. H. Zaidi, *J. Am. Chem. Soc.*, **113**, 5775 (1991).
155. R. W. Friesen and R. W. Loo, *J. Org. Chem.*, **56**, 4821 (1991).
156. T. L. Gilchrist and R. J. Summersell, *J. Chem. Soc., Perkin Trans. 1*, 2603 (1988).
157. Y. Tamaru, H. Ochiai, T. Nakamura and Z.-i. Yoshida, *Angew. Chem., Int. Ed. Engl.*, **26**, 1157 (1987); *Angew. Chem.*, **99**, 1193 (1993); *Tetrahedron Lett.*, **27**, 955 (1986).
158. Y. Tamaru, H. Ochiai, F. Sanda and Z.-i. Yoshida *Tetrahedron Lett.*, **26**, 5529 (1985); Y. Tamaru, H. Ochiai, T. Nakamura, K. Tsubaki and Z.-i. Yoshida, *Tetrahedron Lett.*, **26**, 5559 (1995).
159. R. F. W. Jackson, M. J. Wythes and A. Wood, *Tetrahedron Lett.*, **30**, 5941 (1989).

160. E. Piers, M. Jean and P. S. Marrs, *Tetrahedron Lett.*, **28**, 5075 (1987).
161. E. C. Stracker and G. Zweifel, *Tetrahedron Lett.*, **32**, 3329 (1991).
162. F. Tellier and C. Descoins, *Tetrahedron Lett.*, **31**, 2295 (1990).
163. E.-i. Negishi, T. Takahashi, S. Baba, D. E. Van Horn and N. Okukado, *J. Am. Chem. Soc.*, **109**, 2393 (1987).
164. H. Ohta, A. Inoue, K. Ohtsuka and T. Watanabe, *Heterocycles*, **23**, 133 (1985).
165. E.-i. Negishi. F.-T. Luo, R. Frisbee and H. Matsushita, *Heterocycles*, **18**, 117 (1982).
166. E.-i. Negishi and H. Matsushita, *Org. Synth.*, **62**, 31 (1984); E.-i. Negishi, S. Chatterjee and H. Matsushita, *Tetrahedron Lett.*, **22**, 197 (1981); H. Matsushita and E.-i. Negishi, *J. Am. Chem. Soc.*, **103**, 2882 (1981).
167. M. W. Hutzinger and A. C. Oehlchlager, *J. Org. Chem.*, **56**, 2918 (1991).
168. N. A. Bumagin, A. B. Ponomaryov and I. P. Beletskaya, *Tetrahedron Lett.*, **26**, 4819 (1985).
169. Y. Wakita, T. Yasunaga and M. Kojima, *J. Organomet. Chem.*, **288**, 261 (1985).
170. L. Crombie, A. J. W. Hobbs and M. A. Horsham, *Tetrahedron Lett.*, **28**, 4875 (1987).
171. J. W. Labadie and J. K. Stille, *J. Am. Chem. Soc.*, **105**, 6129 (1983).
172. J. W. Labadie, D. Tueting and J. K. Stille, *J. Org. Chem.*, **48**, 4634 (1983); A Goliaszewski and J. Schwartz, *Tetrahedron*, **41**, 5779 (1985); J.-P. Quintard, G. Dumartin, B. Elissondo, A. Rahm and M. Pereyre, *Tetrahedron*, **45**, 1017 (1989).
173. Y. Yamamoto, S. Hatsuya and J.-i. Yamada, *J. Org. Chem.*, **55**, 3118 (1990); H. Takayama and T. Suzuki, *J. Chem. Soc., Chem. Commun.*, 1044 (1988).
174. J. W. Labadie and J. K. Stille, *J. Am. Chem. Soc.*, **105**, 669 (1983).
175. I. P. Beletskaya, *J. Organomet. Chem.*, **250**, 551 (1983).
176. V. Farina and B. Krishnan, *J. Am. Chem. Soc.*, **113**, 9585 (1991).
177. J. K. Stille, *Angew. Chem., Int. Ed. Engl.*, **25**, 508 (1986) *Angew. Chem.*, **98**, 504 (1986); J. K. Stille, *Pure Appl. Chem.*, **57**, 1771 (1985).
178. T. N. Mitchell, *Synthesis*, 803 (1992).
179. P. J. Colson, M. Franck–Neumann and M. Sedatri, *Tetrahedron Lett.*, **30**, 2393 (1989).
180. G. T. Crisp and T. P. Bubner, *Synth. Commun.*, **20**, 1665 (1990).
181. R. J. Linderman, D. M. Graves, W. R. Kwochka, A. F. Ghannam and T. V. Anklekar, *J. Am. Chem. Soc.*, **112**, 7438 (1990).
182. J. E. Baldwin, R. M. Adlington and S. W. Ramcharitar, *Tetrahedron*, **48**, 2957 (1992).
183. L. Balas, B. Jousseaume, H. Shin, J.-B. Verlhac and F. Wallian, *Organometallics*, **10**, 366 (1991).
184. T.-a. Kobayashi, T. Sakakura and M. Tanaka, *Tetrahedron Lett.*, **26**, 3463 (1985).
185. Y. Ito, M. Inouye, H. Yokota and M. Murakami, *J. Org. Chem.*, **55**, 2567 (1990) and references cited therein.
186. J.-B. Verlhac, E. Chanson, B. Jousseaume and J.-P. Quintard, *Tetrahedron Lett.*, **26**, 6075 (1985).
187. M. Kosugi, I. Takano, M. Sakurai, H. Sano and T. Migita, *Chem. Lett.*, 1221 (1984).
188. J. H. Simpson and J. K. Stille, *J. Org. Chem.*, **50**, 1759 (1985).
189. R. K. Bhatt, D.-S. Shin, J. R. Falck and C. Mioskowski, *Tetrahedron Lett.*, **33**, 4885 (1992).
190. S. Matsubara, M. Mitani and K. Utimoto, *Tetrahedron Lett.*, **28**, 5857 (1987).
191. F. K. Sheffy, J. P. Godschalx and J. K. Stille, *J. Am. Chem. Soc.*, **106**, 4833 (1984).
192. S. Katsumura, S. Fujiwara and S. Isoe, *Tetrahedron Lett.*, **28**, 1191 (1987); **29**, 1173 (1988).
193. R. Sustmann, J. Lau and M. Zipp, *Tetrahedron Lett.*, **27**, 5207 (1986).
194. J. K. Stille and J. H. Simpson, *J. Am. Chem. Soc.*, **109**, 2138 (1987).
195. Y. Rubin, C. B. Knobler and F. Diederich, *J. Am. Chem. Soc.*, **112**, 1607 (1990).
196. L. S. Liebeskind and J. Wang, *Tetrahedron Lett.*, **31**, 4293 (1990).
197. R. M. Moriarty and W. R. Epa, *Tetrahedron Lett.*, **33**, 4095 (1992).
198. C. R. Johnson, J. P. Adams, M. P. Braun and C. B. W. Senanayake, *Tetrahedron Lett.*, **33**, 919 (1992).
199. M. Hirama, K. Fjiwara, K. Shigematu and Y. Fukazawa, *J. Am. Chem. Soc.*, **111**, 4120 (1989); K. Fujiwara, A. Kurisaki and M. Hirama, *Tetrahedron Lett.*, **31**, 4329 (1990).
200. W. F. Goure, M. E. Wright, P. D. Davis, S. S. Labadie and J. K. Stille, *J. Am. Chem. Soc.*, **106**, 6417 (1984); see also N. A. Bumagin, I. G. Bumagina, A. N. Kashin and I. P. Beletskaya, *Zh. Org. Khim.*, **18**, 1037 (1982); *Chem. Abstr.*, **97**, 216343 (1982).
201. M. E. Wright, *J. Organomet. Chem.*, **376**, 353 (1989).
202. J. Brocard, L. Pelinski and J. Lebibi, *J. Organomet. Chem.*, **336**, C47 (1987).
203. D. E. Rudisill and J. K. Stille, *J. Org. Chem.*, **54**, 5856 (1989).

204. E. Dubois and J.-M. Beau, *Tetrahedron Lett.*, **31**, 5165 (1990); R. W. Friesen and C. F. Sturino, *J. Org. Chem.*, **55**, 5808 (1990).
205. M. Krolski, A. F. Renaldo, D. E. Rudisill and J. K. Stille, *J. Org. Chem.*, **53**, 1170 (1988).
206. A. Takle and P. Kocieński, *Tetrahedron Lett.*, **30**, 1675 (1989).
207. M. Bochmann and K. Kelly, *J. Chem. Soc., Chem. Commun.*, 532 (1989).
208. T. R. Bailey, *Tetrahedron Lett.*, **27**, 4407 (1986); A. Dondoni, G. Fantin, M. Fogagnolo, A. Medici and P. Pedrini, *Synthesis*, 693 (1987).
209. M. Kosugi, M. Ishiguro, Y. Negishi, H. Sano and T. Migita, *Chem. Lett.*, 1511 (1984); M. Kosugi, T. Sumiya, T. Ogata, H. Sano and T. Migita, *Chem. Lett.*, 1225 (1984); M. Kosugi, I. Hagiwara, T. Sumiya and T. Migita, *Bull. Chem. Soc. Jpn.*, **57**, 242 (1984).
210. M. Kosugi, T. Ishikara, T. Nogami and T. Migita, *Nippon Kagaku Kaishi*, 520 (1985); *Chem. Abstr.*, **104**, 68496 (1986).
211. J. C. Bradley and T. Durst, *J. Org. Chem.*, **56**, 5459 (1991).
212. A. Kalivretenos, J. K. Stille and L. S. Hegedus, *J. Org. Chem.*, **56**, 2883 (1991).
213. A. M. Echavarren and J. K. Stille, *J. Am. Chem. Soc.*, **109**, 5478 (1987).
214. Y. Yang, A.-B. Hörnfeldt and S. Gronowitz, *Synthesis*, 130 (1989).
215. J. A. Porco, F. J. Schoenen, T. J. Stout, J. Clardy and S. L. Schreiber, *J. Am, Chem. Soc.*, **112**, 7410 (1990); Y. Yamamoto, Y. Azuma and H. Mitoh, *Synthesis*, 564 (1986).
216. R. M. Moriarty, W. R. Epa and A. K. Awasthi, *Tetrahedron Lett.*, **31**, 5877 (1990); V. Nair, G. A. Turner, G. S. Buenger and S. D. Chamberlain, *J. Org. Chem.*, **53**, 3051 (1988).
217. O. M. Minnetian, I. K. Morris, K. M. Snow and K. M. Smith, *J. Org. Chem.*, **54**, 5567 (1989); Y. Yamamoto, T. Seko and H. Nemoto, *J. Org. Chem.*, **54**, 4734 (1989).
218. L. S. Libeskind and R. W. Fengl, *J. Org. Chem.*, **55**, 5359 (1990); I. Beaudet, J.-L. Parrain and J.-P. Quintard, *Tetrahedron Lett.*, **33**, 3647 (1992).
219. Y. Yokoyama, M. Ikeda, M. Sato, H. Suzuki and Y. Murakami, *Heterocycles*, **31**, 1505 (1990); Y. Yokoyama, S. Ito, Y. Takahashi and Y. Murakami, *Tetrahedron Lett.*, **26**, 6457 (1985).
220. B. M. Trost and R. Walchli, *J. Am. Chem. Soc.*, **109**, 3487 (1987).
221. T. R. Kelly, Q. Li and V. Bhusan, *Tetrahedron Lett.*, **31**, 161 (1990).
222. R. Grigg, A. Teasdale and V. Sridharan, *Tetrahedron Lett.*, **32**, 3859 (1991).
223. A. Suzuki, *Pure Appl. Chem.*, **63**, 419 (1991); *Pure Appl. Chem.*, **57**, 1749 (1985).
224. I. Rivera and J. A. Soderquist, *Tetrahedron Lett.*, **32**, 2311 (1991) and references cited therein.
225. J.-i. Uenishi, J.-M. Beau, R. W. Armstrong and Y. Kishi, *J. Am. Chem. Soc.*, **109**, 4756 (1987).
226. N. Miyaura, K. Yamada, H. Suginome and A. Suzuki, *J. Am. Chem Soc.*, **107**, 972 (1985).
227. T. Watanabe, N. Miyaura and A. Suzuki, *Synlett*, 207 (1992).
228. N. Miyaura, T. Ishiyama, H. Sasaki, M. Ishikawa, M. Satoh and A. Suzuki, *J. Am. Chem. Soc.*, **111**, 314 (1989).
229. M. Ogima, S. Hyuga, S. Hara and A. Suzuki, *Chem. Lett.*, 1959 (1989).
230. N. Miyaura, M. Satoh and A. Suzuki, *Tetrahedron Lett.*, **27**, 3745 (1986); M. Satoh, N. Miyaura and A. Suzuki, *Chem. Lett.*, 1329 (1986).
231. N. Satoh, T. Ishiyama, N. Miyaura and A. Suzuki, *Bull. Chem. Soc. Jpn.*, **60**, 3471 (1987).
232. W. R. Roush, K. J. Moriarty and B. B. Brown, *Tetrahedron Lett.*, **31**, 6509 (1990); W. R. Roush and R. J. Scioti, *Tetrahedron Lett.*, **33**, 4691 (1992).
233. K. C. Nicolaou, J. Y. Ramphal, J. M. Palazon and R. A. Spanevello, *Angew. Chem., Int. Ed. Engl.*, **28**, 587 (1989); *Angew. Chem.*, **101**, 621 (1989).
234. Y. Kobayashi, T. Shimazaki, H. Takaguchi and F. Sato, *J. Org. Chem.*, **55**, 5324 (1990).
235. J. A. Soderquist and G. León-Colón, *Tetrahedron Lett.*, **32**, 43 (1991).
236. N. Miyaura, H. Suginome and A. Suzuki, *Tetrahedron Lett.*, **25**, 761 (1984).
237. R. B. Miller and S. Dugar, *Organometallics*, **3**, 1261 (1984).
238. M. J. Sharp, W. Cheng and V. Snieckus, *Tetrahedron Lett.*, **28**, 5093 (1987).
239. S. Gronowitz, V. Bobošik and K. Lawitz, *Chem. Scr.*, **23**, 120 (1984); S. Gronowitz and D. Peters, *Heterocycles*, **30**, 645 (1990).
240. W. A. Cristofoli and B. A. Keay, *Tetrahedron Lett.*, **32**, 5881 (1991).
241. N. M. Ali, A. McKillop, M. B. Mitchell, R. A. Rebelo and P. J. Wallbank, *Tetrahedron*, **48**, 8117 (1992).
242. S. Abe, N. Miyaura and A. Suzuki, *Bull. Chem. Soc. Jpn.*, **65**, 2863 (1992).
243. C. M. Unrau, M. G. Campbell and V. Snieckus, *Tetrahedron Lett.*, **33**, 2773 (1992); B. I. Alo, A. Kandil, P. A. Patil, M. J. Sharp, M. A. Siddiqui, V. Snieckus and P. D. Josephy, *J. Org. Chem.*, **56**, 3763 (1991); B. Zhao and V. Snieckus, *Tetrahedron Lett.*, **32**, 5277 (1991); V. Snieckus, *Chem, Rev.*, **90**, 879 (1990).

244. Y. Nomoto, N. Miyaura and A. Suzuki, *Synlett*, 727 (1992); N. Miyaura, M. Ishiwara and A. Suzuki, *Tetrahedron Lett.*, **33**, 2571 (1992).
245. T. Ishiyama, S. Abe, N. Miyaura and A. Suzuki, *Chem. Lett.*, 691 (1992).
246. T. Ishiyama, N. Miyaura and A. Suzuki, *Bull. Chem. Soc. Jpn.*, **64**, 1999 (1991); T. Ishiyama, N. Miyaura and A. Suzuki, *Tetrahedron Lett.*, **32**, 6923 (1991); Y. Wakita, T. Yasunaga, M. Akita and M. Kojima, *J. Organomet. Chem.*, **301**, C17 (1986).
247. M. Sato, N. Miyaura and A. Suzuki, *Chem. Lett.*, 1405 (1989).
248. Y. Hatanaka and T. Hiyama, *Synlett*, 845 (1991).
249. Y. Hatanaka and T. Hiyama, *J. Org. Chem.*, **53**, 918 (1988).
250. Y. Hatanaka and T. Hiyama, *Tetrahedron Lett.*, **29**, 97 (1988).
251. Y. Hatanaka and T. Hiyama, *J. Org. Chem.*, **54**, 268 (1989); K. Tamao, K. Kobayashi and Y. Ito, *Tetrahedron Lett.*, **30**, 6051 (1989).
252. A. Hosomi, S. Kohra and Y. Tominaga, *Chem. Pharm. Bull.*, **36**, 4622 (1988).
253. Y. Hatanaka, S. Fukushima and T. Hiyama, *Heterocycles*, **30**, 303 (1990).
254. Y. Hatanaka, S. Fukushima and T. Hiyama, *Tetrahedron*, **48**, 2113 (1992).
255. Y. Hatanaka, Y. Ebina and T. Hiyama, *J. Am. Chem. Soc.*, **113**, 7075 (1991).
256. C. Amatore and A. Jutland, *Organometallics*, **7**, 2203 (1989).
257. I. Colon and D. R. Kelsey, *J. Org. Chem.*, **51**, 2627 (1986).
258. G. Meyer, Y. Rollin and J. Perichon, *J. Organomet. Chem.*, **333**, 263 (1987); M. A. Fox, D. A. Chandler and C. Lee, *J. Org. Chem.*, **56**, 3246 (1991).
259. M. Lourak, R. Vandresse, Y. Fort and P. Caubère, *J. Org. Chem.*, **54**, 4840 (1989); R. Vanderesse, M. Lourak, Y. Fort and P. Caubère, *Tetrahedron Lett.*, **27**, 5483 (1986).
260. S.-i. Inaba and R. D. Rieke, *J. Org. Chem.*, **50**, 1373 (1985) and references cited therein; A. Conan, S. Sibille, E. d'Incan and J. Périchon, *J. Chem. Soc., Chem. Commun.*, 48 (1990).
261. K. Takagi, N. Hiyama and K. Sasaki, *Bull. Chem. Soc. Jpn.*, **57**, 1887 (1984); K. Takagi, H. Mimura and S. Inokawa, *Bull. Chem. Soc. Jpn.*, **57**, 3517 (1984) and references cited therein.
262. M. Iyoda, H. Otsuka, K. Sato, N. Nisato and M. Oda, *Bull. Chem. Soc. Jpn.*, **63**, 80 (1990) and references cited therein.
263. M. Iyoda, M. Sakaitani, T. Miyazaki and M. Oda, *Chem. Lett.*, 2005 (1984).
264. Y. Vanderesse, Y. Fort, S. Becker and P. Caubère, *Tetrahedron Lett.*, **27**, 3517 (1986).
265. Z.-h. Zhou and T. Yamamoto, *J. Organomet. Chem.*, **414**, 119 (1991).
266. S.-i. Inaba, R. M. Wehmeyer, M. W. Forkner and R. D. Rieke, *J. Org. Chem.*, **53**, 339 (1988).
267. R. Diercks, J. C. Armstrong, R. Boese and K. P. C. Vollhardt, *Angew. Chem., Int. Ed. Engl.*, **25**, 268 (1986); *Angew. Chem.*, **98**, 270 (1986); R. Boese, J. R. Green, J. Mittendorf, D. L. Mohler and K. P. C. Vollhardt, *Angew. Chem., Int. Ed. Engl.*, **31**, 1643 (1992); *Angew. Chem.*, **104**, 1643 (1992).
268. Reference 1 and D. Guillerm and G. Linstrumelle, *Tetrahedron Lett.*, **26**, 3811 (1985).
269. V. Ratovelomanana, A. Hammond and G. Linstrumelle, *Tetrahedron Lett.*, **28**, 1649 (1987).
270. I. A. Butler and C. Soucy-Breau, *Can. J. Chem.*, **69**, 1117 (1991).
271. G. Just and R. Singh, *Tetrahedron Lett.*, **28**, 5981 (1987).
272. R. Singh and G. Just, *J. Org. Chem.*, **54**, 4453 (1989).
273. J. W. Tilley and S. Zawoiski, *J. Org. Chem.*, **53**, 386 (1988).
274. W. Tao, S. Nesbitt and R. F. Heck, *J. Org. Chem.*, **55**, 63 (1990).
275. D. Villemin and E. Schigeko, *J. Organomet. Chem.*, **293**, C10 (1985).
276. M. Alami and G. Linstrumelle, *Tetrahedron Lett.*, **32**, 6109 (1991).
277. T. Jeffrey-Luong and G. Linstrumelle, *Synthesis*, 32 (1983); G. Markl, P. Attenberger and   J. Kellner, *Tetrahedron Lett.*, **29**, 3651 (1988).
278. K. Hartke, H.-D. Gerber and U. Roesrath, *Tetrahedron Lett.*, **30**, 1073 (1989).
279. G. C. M. Lee, B. Tobias, J. M. Holmes, D. A. Harcourt and M. E. Garst, *J. Am. Chem. Soc.*, **112**, 9330 (1990).
280. K. C. Nicolaou and A. L. Smith, *Acc. Chem. Res.*, **25**, 497 (1992); K. C. Nicolaou and W.-M. Dai, *Angew. Chem., Int. Ed. Engl.*, **30**, 1387 (1991); *Angew. Chem.*, **103**, 1453 (1991).
281. A. Spenser, *J. Organomet. Chem.*, **265**, 323 (1984); A. Spenser, *J. Organomet. Chem.*, **247**, 117 (1983).
282. A. Kasahara, T. Izumi and N. Kudou, *Synthesis*, 704 (1988); H.-U. Blazer and A. Spencer, *J. Organomet. Chem.*, **233**, 267 (1982).
283. A. Kasahara, T. Izumi, K. Miyamoto and T. Sakai, *Chem. Ind. (London)*, 192 (1989); A. Kasahara, T. Izumi and T. Ogihara, *Chem. Ind. (London)*, 792 (1988).

284. J. J. Bozell and C. E. Vogt, *J. Am. Chem. Soc.*, **110**, 2655 (1988).
285. T. Jeffery, *J. Chem. Soc., Chem. Commun.*, 1287 (1984).
286. T. Jeffery, *Tetrahedron Lett.*, **26**, 2667 (1985).
287. T. Jeffery, *Synthesis*, 70 (1987).
288. R. C. Larock, W. H. Gong and B. E. Baker, *Tetrahedron Lett.*, **30**, 2603 (1989); R. C. Larock and W. H. Gong, *J. Org. Chem.*, **54**, 2047 (1989).
289. M. T. Abelman, T. Oh and L. E. Overman, *J. Org. Chem.*, **52**, 4130 (1987).
290. G. D. Daves, Jr. and A. Hallberg, *Chem. Rev.*, **89**, 1433 (1989); G. D. Daves, Jr., *Adv. Met.-Org. Chem.*, **2**, 59 (1991); R. C. Larock, *Pure Appl. Chem.*, **62**, 653 (1990).
291. K. Nilsson and A. Hallberg, *J. Org. Chem.*, **55**, 2464 (1990).
292. C.-M. Andersson, A. Hallberg and G. D. Daves, Jr., *J. Org. Chem.*, **52**, 3529 (1987); C.-M. Andersson and A. Hallberg, *J. Org. Chem.*, **53**, 235 (1988).
293. C.-M. Andersson, J. Larsson and A. Hallberg, *J. Org. Chem.*, **55**, 5757 (1990).
294. W. Cabri, I. Candiani, A. Bedeschi and R. Santi, *Tetrahedron Lett.*, **32**, 1753 (1991).
295. T. Sakamoto, Y. Kondo and H. Yamanaka, *Tetrahedron Lett.*, **33**, 6845 (1992); F. Ozawa, A. Kubo and T. Hayashi, *J. Am. Chem. Soc.*, **113**, 1417 (1991) and references cited therein.
296. Y. Xu, X. Jin, G. Huang and Y. Huang, *Synthesis*, 556 (1983).
297. K. Karabelas and A. Hallberg, *J. Org. Chem.*, **51**, 5286 (1986); *J. Org. Chem.*, **53**, 4909 (1988); *J. Org. Chem.*, **54**, 1773 (1989).
298. R. Grigg, P. Stevenson and T. Worakun, *J. Chem. Soc., Chem. Commun.*, 1073 (1984).
299. P. C. Amos and D. A. Whiting, *J. Chem. Soc., Chem. Commun.*, 510 (1987).
300. Y. Zhang, B. O'Connor and E.-i. Negishi, *J. Org. Chem.*, **53**, 5588 (1988).
301. R. C. Larock, H. Song, B. E. Baker and W. H. Gong, *Tetrahedron Lett.*, **29**, 2919 (1988).
302. E.-i. Negishi, Y. Zhang and B. O'Connor, *Tetrahedron Lett.*, **29**, 2915 (1988).
303. Y. Sato, S. Watanabe and M. Shibasaki, *Tetrahedron Lett.*, **33**, 2589 (1992) and references cited therein; Y. Sato, T. Honda and M. Shibasaki, *Tetrahedron Lett.*, **33**, 2593 (1992).
304. S. Cacchi, F. La Torre and G. Palmieri, *J. Organomet. Chem.*, **268**, C48 (1984); S. Cacchi and G. Palmieri, *Synthesis*, 575 (1984); S. Cacchi and G. Palmieri, *J. Organomet. Chem.*, **282**, C3 (1985).
305. E. Uhlig and B. Hipler, *Tetrahedron Lett.*, **25**, 5871 (1984); A. Arcadi, S. Cacchi and F. Marinelli, *Tetrahedron Lett.*, **27**, 6397 (1986); B. Burns, R. Grigg, V. Sridharan and T. Worakun, *Tetrahedron Lett.*, **29**, 4325 (1988).
306. B. Burns, R. Grigg, V. Sridharan and T. Worakun, *Tetrahedron Lett.*, **29**, 4329 (1988).
307. G. Fournet, G. Balme and J. Gore, *Tetrahedron*, **47**, 6293 (1991); D. Bouyssi, G. Balme, G. Fournet, N. Monteiro and J. Gore, *Tetrahedron Lett.*, **32**, 1641 (1991).
308. M. Uno, T. Takahashi and S. Takahashi, *J. Chem. Soc., Chem. Commun.*, 785 (1987); *J. Chem. Soc., Perkin Trans. 1*, 647 (1990). R. C. Larock and C. A. Fried, *J. Am. Chem. Soc.*, **112**, 5882 (1990).
309. M. Ahmar, J.-J. Barieux, B. Cazes and J. Gore, *Tetrahedron*, **43**, 513 (1987); B. Cazes, *Pure Appl. Chem.*, **62**, 1867 (1990).
310. R. C. Larock, Y.-d. Lu, A. C. Bain and C. E. Russell, *J. Org. Chem.*, **56**, 4589 (1991).
311. S. Torii, H. Okumoto, H. Ozaki, S. Nakayasu, T. Tadokoro and T. Kokani, *Tetrahedron Lett.*, **33**, 3499 (1992) and references cited therein.
312. R. Grigg, P. Kennewell and A. J. Teasdale, *Tetrahedron Lett.*, **33**, 7789 (1992).
313. A. Arcadi, A. Burini, S. Cacchi, M. Delmastro, F. Marinelli and B. R. Pietroni, *J. Org. Chem.*, **57**, 976 (1992); D. Bouyssi, J. Gore and G. Balme, *Tetrahedron Lett.*, **33**, 2811 (1992); K. Iritani, S. Matsubara and K. Utimoto, *Tetrahedron Lett.*, **29**, 1799 (1988).
314. B. Burns, R. Grigg, P. Ratanankul, V. Sridharan, P. Stevenson, S. Sukirthalingam and T. Worakun, *Tetrahedron Lett.*, **29**, 5565 (1988); F.-T. Luo and R.-T. Wang, *Tetrahedron Lett.*, **32**, 7703 (1991); J. M. Nuss, B. H. Levine, R. A. Rennels and M. M. Heravi, *Tetrahedron Lett.*, **32**, 5243 (1991).
315. B. Burns, R. Grigg, V. Sridharan, P. Stevenson, S. Sukirthalingam and T. Worakun, *Tetrahedron Lett.*, **30**, 1135 (1989).
316. R. Grigg, V. Loganathan, S. Sukirthalingham and V. Sridharan *Tetrahedron Lett.*, **31**, 6573 (1990); F.-T. Luo and R.-T. Wang, *Heterocycles*, **31**, 2181, 2365 (1991): R.-T. Wang, F.-L. Chou and F.-T. Luo, *J. Org. Chem.*, **55**, 4846 (1990).
317. M. Kosugi, H. Tamura, H. Sano and T. Migita, *Tetrahedron*, **45**, 961 (1989) and references cited therein; M. Catellani, G. P. Chiusoli and S. Concari, *Tetrahedron*, **45**, 5263 (1989).
318. M. M. Abelman and L. E. Overman, *J. Am. Chem. Soc.*, **110**, 2328 (1988).

319. Y. Zhang and E.-i. Negishi, *J. Am. Chem. Soc.*, **111**, 3454 (1989); G.-z. Wu, F. Lamaty and E.-i. Negishi, *J. Org. Chem.*, **54**, 2507 (1989).
320. Y. Yang, G.-z. Wu, G. Angel and E.-i. Negishi, *J. Am. Chem. Soc.*, **112**, 8590 (1990); B. M. Trost and Y. Shi, *J. Am. Chem. Soc.*, **113**, 701 (1991).
321. R. Grigg, M. J. Dorrity, J. F. Malone, V. Sridharan and S. Sukirthalingham, *Tetrahedron Lett.*, **31**, 1343 (1990); F. E. Meyer, P. J. Parsons and A. de Meijere, *J. Org. Chem.*, **56**, 6487 (1991).
322. L. J. Silverberg, G. Wu, A. L. Rheingold and R. F. Heck, *J. Organomet. Chem.*, **409**, 411 (1991); S. Torii, H. Okumoto and A. Nishimura, *Tetrahedron Lett.*, **32**, 4167 (1991); F. E. Meyer and A. de Meijere, *Synlett*, 777 (1991); E.-i. Negishi, L. S. Harding, Z. Obczarczyk, M. M. Mohamud and M. Ay, *Tetrahedron Lett.*, **33**, 3253 (1992).
323. M. Mori, N. Kanda, I. Oda and Y. Ban, *Tetrahedron*, **41**, 5465 (1985); M. Mori, Y. Kubo and Y. Ban, *Tetrahedron*, **44**, 4321 (1988).
324. D. P. Curran and C.-T. Chang, *Tetrahedron Lett.*, **31**, 933 (1990).
325. Q.-Y. Chen, Z.-Y. Yang, C.-X. Zhao and Z.-M. Qiu, *J. Chem. Soc., Perkin Trans. 1*, 563 (1988).
326. S. Ma and X. Lu, *J. Org. Chem.*, **56**, 5120 (1991).
327. M. Mori, N. Kanda and Y. Ban, *J. Chem. Soc., Chem. Commun.*, 1375 (1986); M. Mori, N. Kanda, Y. Ban and K. Aoe, *J. Chem. Soc., Chem. Commun.*, 12 (1988).
328. K. Maruoka, H. Sano, Y. Fukutani and H. Yamamoto, *Chem. Lett.*, 1689 (1985).
329. T. Ishihara, M. Kuroboshi and Y. Okada, *Chem. Lett.*, 1895 (1986); Z.-Y. Yang and D. J. Burton, *Tetrahedron Lett.*, **32**, 1019 (1991).
330. H. Ishibashi, N. Uemura, H. Nakatani, M. Okazaki, T. Sato, N. Nakamura and M. Ikeda, *J. Org. Chem.*, **58**, 2360 (1993); N. Kamigata, T. Fukushima and M. Yoshida, *J. Chem. Soc., Chem. Commun.*, 1559 (1989); T. K. Hayes, A. J. Freyer, M. Parvez and S. M. Weinreb, *J. Org. Chem.*, **51**, 5501 (1986).
331. P. Cintas, *Synthesis*, 248 (1992).
332. K. Takai, K. Nitta, O. Fujimura and K. Utimoto, *J. Org. Chem.*, **54**, 4732 (1989).
333. K. Takai, K. Kimura, T. Kuroda, T. Hiyama and H. Nozaki, *Tetrahedron Lett.*, **24**, 5281 (1983).
334. T. Hiyama, Y. Okude, K. Kimura and H. Nozaki, *Bull. Chem. Soc. Jpn.*, **55**, 561 (1982).
335. K. Takai, T. Kuroda, S. Nakatsukasa, K. Oshima and H. Nozaki, *Tetrahedron Lett.*, **26**, 5585 (1985).
336. D. M. Hogson and C. Wells, *Tetrahedron Lett.*, **33**, 4761 (1992).
337. J. Augé, *Tetrahedron Lett.*, **29**, 6107 (1988); P. A. Wender, J. W. Grissom, U. Hoffmann and R. Mah, *Tetrahedron Lett.*, **31**, 6605 (1990).
338. K. Belyk, M. J. Rozema and P. Knochel, *J. Org. Chem.*, **57**, 4070 (1992).
339. J. Mulzer, L. Kattner, A. R. Strecker, C. Schröder, J. Buschmann, C. Lehmann and P. Luger, *J. Am. Chem. Soc.*, **113**, 4218 (1991); S. F. Martin and W. Li, *J. Org. Chem.*, **54**, 6129 (1989).
340. H. Shibuya, K. Ohashi, K. Kawashima, K. Hori, N. Murakami and I. Kitagawa, *Chem. Lett.*, 85 (1986); see also W. C. Still and D. Mobilio, *J. Org. Chem.*, **48**, 4785 (1983).
341. H. Jin, J.-i. Uenishi, W. J. Christ and Y. Kishi, *J. Am. Chem. Soc.*, **108**, 5644 (1986).
342. S. Aoyagi, T.-C. Wang and C. Kibayashi, *J. Am. Chem. Soc.*, **114**, 10653 (1992); M. Rowley and Y. Kishi, *Tetrahedron Lett.*, **29**, 4909 (1988).
343. T. Okazoe, K. Takai and K. Utimoto, *J. Am. Chem. Soc.*, **109**, 951 (1987).
344. K. Takai, Y. Kataoka, T. Okazoe and K. Utimoto, *Tetrahedron Lett.*, **28**, 1443 (1987).
345. K. Takai, K. Nitta and K. Utimoto, *J. Am. Chem. Soc.*, **108**, 7408 (1986).
346. S. Nakatsukasa, K. Takai and K. Utimoto, *J. Org. Chem.*, **51**, 5045 (1986).
347. J. K. Crandall and W. J. Michaely, *J. Org. Chem.*, **49**, 4244 (1984).
348. T. Libbers and H. J. Schäfer, *Synlett*, 861 (1991) and references cited therein.
349. G. Pattenden, *Chem. Soc. Rev.*, **17**, 361 (1988); R. Scheffold, A. Abrecht, R. Orlinski, H.-R. Ruf, P. Stamouli, O. Tinembart, L. Walder and C. Weymuth, *Pure Appl. Chem.*, **59**, 363 (1987).
350. J. Halpern, S.-H. Kim and T. W. Leung, *J. Am. Chem. Soc.*, **106**, 8317 (1984).
351. B. Giese, J. Hartung, J. He, O. Hüter and A. Koch, *Angew. Chem., Int. Ed. Engl.*, **28**, 325 (1989); *Angew. Chem.*, **101**, 334 (1989).
352. H. Bhandal, G. Pattenden and J. J. Russell, *Tetrahedron Lett.*, **27**, 2299 (1986); V. F. Patel, G. Pattenden and J. J. Russell, *Tetrahedron Lett.*, **27**, 2303 (1986).
353. V. F. Patel and G. Pattenden, *J. Chem. Soc., Chem. Commun.*, 871 (1987).
354. V. F. Patel and G. Pattenden, *Tetrahedron Lett.*, **28**, 1451 (1987); B. P. Branchaud, M. S. Meier and M. N. Malekzadeh, *J. Org. Chem.*, **52**, 212 (1987).

355. R. Scheffold, M. Dike, S. Dike, T. Herold and L. Walder, *J. Am. Chem Soc.*, **102**, 3642 (1980); R. Scheffold, *Chimia*, **39**, 203 (1985).
356. S. Busato, O. Tinembart, Z.-d. Zhang and R. Scheffold, *Tetrahedron*, **46**, 3155 (1990).
357. A. Ali, D. C. Harrowven and G. Pattenden, *Tetrahedron Lett.*, **33**, 2851 (1992).
358. J. E. Baldwin, R. M. Adlington and T. W. Kang, *Tetrahedron Lett.*, **32**, 7093 (1991).
359. B. P. Branchaud and Y. L. Choi, *J. Org. Chem.*, **53**, 4638 (1988); B. P. Branchaud and G.-X. Yu, *Tetrahedron Lett.*, **29**, 6545 (1988).
360. K. Sugano, T. Tanase, K. Kobayashi and Y. Yamamoto, *Chem. Lett.*, 921 (1991); Y. Yamamoto and H. Yamazaki, *Organometallics*, **7**, 2411 (1988).
361. D. J. Coveney, V. F. Patel and G. Pattenden, *Tetrahedron Lett.*, **28**, 5949 (1987); V. F. Patel and G. Pattenden, *Tetrahedron Lett.*, **29**, 707 (1988).
362. J. E. Baldwin and C.-S. Li, *J. Chem. Soc., Chem. Commun.*, 261 (1988); A. J. Clark and K. Jones, *Tetrahedron*, **48**, 6875 (1992).
363. B. P. Branchaud and W. D. Detlefsen, *Tetrahedron Lett.*, **32**, 6273 (1991); B. Giese, P. Erdmann, T. Göbel and R. Springer, *Tetrahedron Lett.*, **33**, 4545 (1992).
364. J. Collin, *Bull. Soc. Chim. Fr.*, 976 (1988).
365. T. Kobayashi and M. Tanaka, *J. Organomet. Chem.*, **233**, C64 (1982); F. Ozawa, H. Soyama, T. Yamamoto and A. Yamamoto, *Tetrahedron Lett.*, **23**, 3383 (1982).
366. T. Sakakura, H. Yamashita, T.-a. Kobayashi, T. Hiyashi and M. Tanaka, *J. Org. Chem.*, **52**, 5733 (1987); F. Ozawa, N. Kawasaki, T. Yamamoto and A. Yamamoto, *Chem. Lett.*, 567 (1985).
367. T. Kobayashi, H. Yamashita, T. Sakakura and M. Tanaka, *J. Mol. Catal.*, **41**, 379 (1987).
368. T.-a. Kobayashi, T. Sakakura and M. Tanaka, *Tetrahedron Lett.*, **28**, 2721 (1987).
369. T.-i. Son, H. Tanagihara, F. Ozawa and A. Yamamoto, *Bull. Chem. Soc. Jpn.*, **61**, 1251 (1988).
370. F. Ozawa, H. Soyama, H. Yanagihara, I. Aoyama, H. Takino, K. Izawa, T. Yamamoto and A. Yamamoto, *J. Am. Chem. Soc.*, **107**, 3235 (1985).
371. K. Ozawa, N. Kawasaki, H. Okamoto, T. Yamamoto and A. Yamamoto, *Organometallics*, **6**, 1640 (1987).
372. J. M. Tour and E.-i. Negishi, *J. Am. Chem. Soc.*, **107**, 8289 (1985); E. Amari, M. Catellani and G. P. Chiusoli, *J. Organomet. Chem.*, **285**, 383 (1985).
373. E.-i. Negishi and J. M. Tour, *Tetrahedron Lett.*, **27**, 4869 (1986); S. Torii, H. Okumoto and L. H. Xu, *Tetrahedron Lett.*, **31**, 7175 (1990).
374. E.-i. Negishi, G. Wu and J. M. Tour, *Tetrahedron Lett.*, **29**, 6745 (1988); G. Wu, I. Shimoyama and E.-i. Negishi, *J. Org. Chem.*, **56**, 6506 (1991).
375. T.-a. Kobayashi and M. Tanaka, *Tetrahedron Lett.*, **27**, 4745 (1986); E.-i. Negishi, Y. Zhang, I. Shimoyama and G. Wu, *J. Am. Chem. Soc.*, **111**, 8018 (1989).
376. S. Torii, H. Okumoto and L. H. Xu, *Tetrahedron Lett.*, **32**, 237 (1991); S. Torii, L. H. Xu and H. Okumoto, *Synlett*, 695 (1991).
377. V. N. Kalinin, M. V. Shostakovsky and A. B. Ponomaryov, *Tetrahedron Lett.*, **31**, 4073 (1990).
378. Y. Huang and H. Alper, *J. Org. Chem.*, **56**, 4534 (1991).
379. T. Sakakura, M. Chaisupakitsin, T. Hayashi and M. Tanaka, *J. Organomet. Chem.*, **334**, 205 (1987).
380. I. Pri-Bar and O. Buchman, *J. Org. Chem.*, **49**, 4009 (1984).
381. V. P. Baillargeon and J. K. Stille, *J. Am. Chem. Soc.*, **108**, 452 (1986).
382. J.-F. Carpentier, Y. Castanet, J. Brocard, A. Mortreux and F. Petit, *Tetrahedron Lett.*, **33**, 2001 (1992); *Organometallics*, **10**, 4005 (1991) and references cited therein.
383. Y. Ben-David, M. Portnoy and D. Milstein, *J. Am. Chem. Soc.*, **111**, 8742 (1989); Y. Ben-David, M. Portnoy and D. Milstein, *J. Chem. Soc., Chem. Commun.*, 1816 (1989).
384. V. V. Grushin and H. Alper, *J. Chem. Soc., Chem. Commun.*, 611 (1992).
385. R. Takeuchi, Y. Tsuji, M. Fujita, T. Kondo and Y. Watanabe, *J. Org. Chem.*, **54**, 1831 (1989).
386. T. Kondo, Y. Tsuji and Y. Watanabe, *Tetrahedron Lett.*, **29**, 3833 (1988).
387. K. E. Hashem, J. B. Woell and H. Alper, *Tetrahedron Lett.*, **25**, 4879 (1984); J. B. Woell, S. B. Fergusson and H. Alper, *J. Org. Chem.*, **50**, 2134 (1985) and references cited therein.
388. H. Urata, Y. Ishii and T. Fuchikami, *Tetrahedron Lett.*, **30**, 4407 (1989).
389. H. Alper, K. Hashem and J. Heveling, *Organometallics*, **1**, 775 (1982); *Tetrahedron Lett.*, **24**, 2965 (1983).
390. D. Milstein, *Acc. Chem. Res.*, **21**, 428 (1988) and references cited therein.
391. L. Huang, F. Ozawa and A. Yamamoto, *Organometallics*, **9**, 2603 (1990).

392. H. Arzoumanian, G. Buono, M. Choukrad and J.-F. Petrignani, *Organometallics*, **7**, 59 (1988).
393. T. Suzuki, Y. Uozumi and M. Shibasaki, *J. Chem. Soc., Chem. Commun.*, 1593 (1991).
394. F. Joó and H. Alper, *Organometallics*, **4**, 1775 (1985); R. del Rosario and L. S. Stuhl, *Tetrahedron Lett.*, **33**, 3999 (1982).
395. I. Amer and H. Alper, *J. Am. Chem. Soc.*, **111**, 927 (1989).
396. I. Amer and H. Alper, *J. Org. Chem.*, **53**, 5147 (1988).
397. H. Alper, I. Amer and G. Vasapollo, *Tetrahedron Lett.*, **30**, 2615 (1989).
398. V. V. Grushin and H. Alper, *Tetrahedron Lett.*, **32**, 3349 (1991).
399. W. Oppolzer, T. H. Keller, M. Bedoya-Zurita and C. Stone, *Tetrahedron Lett.*, **30**, 5883 (1989); W. Oppolzer, T. H. Keller, D. L. Kuo and W. Pachinger, *Tetrahedron Lett.*, **31**, 1265 (1990).
400. T. Hirao, Y. Haranko, Y. Yamana, Y. Hamada, S. Nagata and T. Agawa, *Bull. Chem. Soc. Jpn.*, **59**, 1341 (1986).
401. T. Hirao, S. Nagata, Y. Yamana and T. Agawa, *Tetrahedron Lett.*, **26**, 5061 (1985).
402. T. Hirao, S. Nagata and T. Agawa, *Tetrahedron Lett.*, **26**, 5795 (1985).
403. F. Camps, J. Coll, J. M. Moretó and J. Torras, *Tetrahedron Lett.*, **26**, 6397 (1985); *J. Org. Chem.*, **54**, 1969 (1989).
404. F. Camps, J. Coll, J. M. Moretó and J. Torras, *Tetrahedron Lett.*, **28**, 4745 (1987).
405. F. Camps, A. Llebaria, J. M. Moretó and L. Pagès, *Tetrahedron Lett.*, **33**, 109, 113 (1992).
406. F. Foà and F. Francalanci, *J. Mol. Catal.*, **41**, 89 (1987); H. des Abbayes, *New. J. Chem.*, **11**, 535 (1987).
407. J.-J. Brunet, C. Sidot and P. Caubère, *J. Org. Chem.*, **48**, 1166 (1983).
408. T. Kashimura, K. Kudo, S. Mori and N. Sugita, *Chem. Lett.*, 299 (1986); *Chem. Lett.*, 851 (1986).
409. T. Kashimura, K. Kudo, S. Mori and N. Sugita, *Chem. Lett.*, 483 (1986).
410. H. Urata, N.-X. Hu, H. Maekawa and T. Fuchikami, *Tetrahedron Lett.*, **32**, 4733 (1991); A. Miyashita, T. Kawashima, S. Kaji, K. Nomura and H. Nohira, *Tetrahedron Lett.*, **32**, 781 (1991).
411. F. Francalanci, A. Gardano and M. Foà, *J. Organomet. Chem.*, **282**, 277 (1985); B. Fell, H. Chrobaczek and W. Kohl, *Chem. Ztg.*, **109**, 167 (1985).
412. M. Foà, F. Francalanci, E. Bencini and A. Gardano, *J. Organomet. Chem.*, **285**, 293 (1985).
413. M. Miura, F. Akase, M. Shinohara and M. Nomura, *J. Chem. Soc., Perkin Trans. 1*, 1021 (1987).
414. K. Itoh, M. Miura and M. Nomura, *Bull. Chem. Soc. Jpn.*, **61**, 4151 (1988); M. Miura, F. Akase and M. Nomura, *J. Org. Chem.*, **52**, 2623 (1987).
415. G. C. Tustin and R. T. Hembre, *J. Org. Chem.*, **49**, 1761 (1984).
416. G. Tanguy, B. Weinberger and H. des Abbayes, *Tetrahedron Lett.*, **24**, 4005 (1983).
417. G. Tanguy, B. Weinberger and H. des Abbayes, *Tetrahedron Lett.*, **25**, 5529 (1984).
418. H. des Abbayes, J.-C. Clément, P. Laurent, G. Tanguy and N. Thilmont, *Organometallics*, **7**, 2293 (1988); P. Laurent, G. Tanguy and H. des Abbayes, *J. Chem. Soc., Chem. Commun.*, 1754 (1986).
419. F. Reyne, P. Brun and B. Waegell, *Tetrahedron Lett.*, **31**, 4597 (1990).
420. J.-J. Brunet and M. Taillefer, *J. Organomet. Chem.*, **384**, 193 (1990) and references cited therein.
421. K. Yoshida, M. Kobayashi and S.-i. Amono, *J. Chem. Soc., Perkin Trans. 1*, 1127 (1992).

CHAPTER **23**

Eliminations involving carbon–halogen bonds and leading to highly strained rings

MARK S. BAIRD

Department of Chemistry, University of Wales, Bangor, Gwynedd LL57 2UW, Wales, UK

and

IVAN G. BOLESOV

Department of Chemistry, Moscow State University, Moscow, Russia

This chapter will cover eliminations of HX, X_2 and $XSiR_3$ (X = halogen) which lead to the formation of cyclopropenes, bicyclobutanes and propellanes and aspects of other elimi-

Supplement D2: The chemistry of halides, pseudo-halides and azides
Edited by S. Patai and Z. Rappoport   © 1995 John Wiley & Sons Ltd

nations which are relevant to this area. As such it will cover not only 1,2-elimination but also various aspects of 1,3-, 1,4- and 1,n-elimination. It will not cover detailed analysis of eliminations in unstrained systems which may be found elsewhere[1a].

I. DEHYDROHALOGENATIONS OF HALOGENATED CYCLOPROPANES

A. 1,2-Eliminations

The dehydrohalogenation of a monobromo- or monochlorocyclopropane provides one of the simplest routes to a range of cyclopropenes[1b]. For 3,3-disubstituted cyclopropanes carrying alkyl or aryl groups, the reaction is normally achieved using potassium t-butoxide or potassium hydroxide in DMSO and in some cases the cyclopropene distils directly from the reaction mixture (Table 1).

TABLE 1

R	R^1	X	Reagent	Yield (%)	Reference
Me	Me	Br	KOBu-t, DMSO, 90–100 °C	84	2
			KOH, DMSO, 18-crown-6	82–92	3
			or TBACl, 110–125 °C		
Me	Me	Cl	KOBu-t, DMSO, 90–100 °C	74	2
Me	Me	Br	KOBu-t, DMSO	50	4
Me	CH=CH$_2$	Br	KOBu-t, DMSO	50	5
Me	CMe=CH$_2$	Cl	KOBu-t, DMSO	28	5
Me	CMe=CH$_2$	Br	KOBu-t, DMSO	61	5
Me	c-C$_3$H$_5$	Br	KOBu-t, DMSO	68	5,6
Me	c-C$_3$H$_5$	Cl	KOBu-t, DMSO	22	5
Me	C≡CH	Br	KOBu-t, DMSO	60	7
Me	C≡CCMe$_3$	Br	KOBu-t, DMSO	58	7
Me	CN	Br	KOH, DMSO, 20 °C	30	8
Me	CMe(OCH$_2$CH$_2$O)	Br	KOBu-t, DMSO		9
Me	Ph	Br	KOBu-t, DMSO	59	
			KOH, DMSO, 18-crown-6	51	3,4
			or TBACl, 110–125 °C		
Me	C$_6$H$_4$-4-OMe	Br	KOBu-t, DMSO	47	4
—(CH$_2$)$_3$—		Br	KOBu-t, DMSO	43	5
—(CH$_2$)$_3$—		Br	KOH, TBACl, DMSO	70	3
c-C$_3$H$_5$	c-C$_3$H$_5$	Br	KOBu-t, DMSO		10,11
Me	C≡C—(c-C$_3$H$_5$)	Br	KOBu-t, DMSO	31	12
Me	C≡C—(c-C$_3$H$_5$)	Br	KOBu-t, DMSO		7
Ph	Ph	Br	KOBu-t, DMSO	80	4
Ph	C(Ph)=CH$_2$	Br	KOBu-t, THF	48	13
Ph	CH=CPh$_2$	Br	KOBu-t, THF, 20 h, 20 °C	38	13

This method has also been applied to the preparation of spiro-fused cyclopropenes[14–21,24] and dicyclopropenes, e.g.:

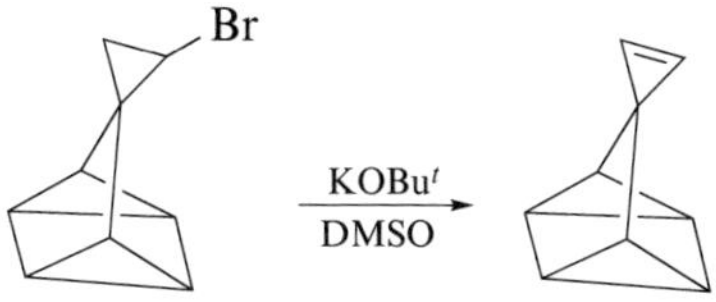

(73%)   (Refs. 15 and 16)

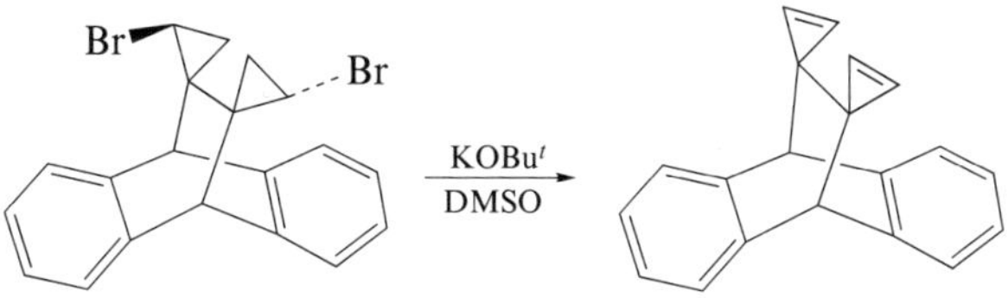

(75%)   (Ref. 14)

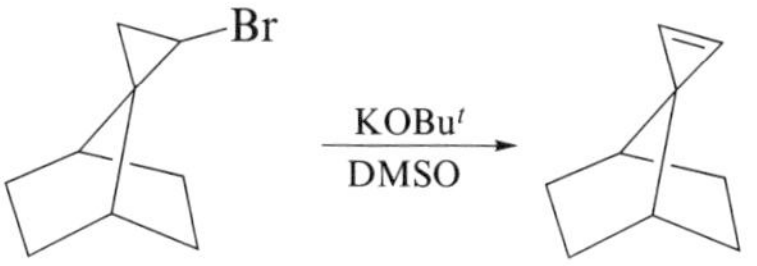

(60%)   (Ref. 15)

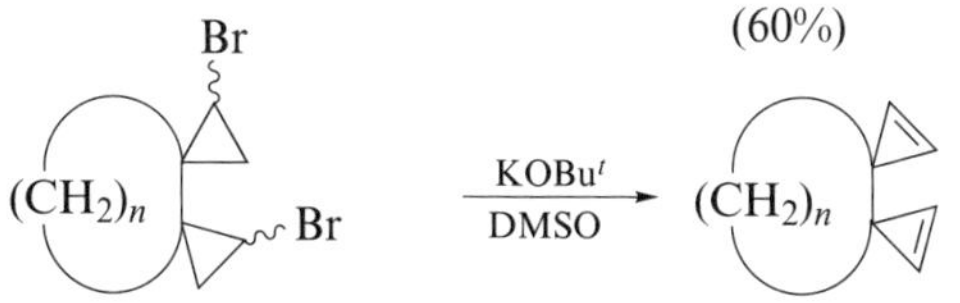

n = 2–5   (Refs. 18 and 20)

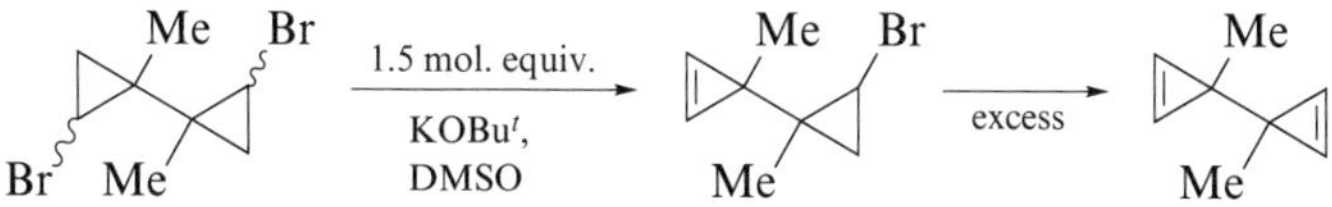

(44%)   (Ref. 21)

It is noticeable that all the above examples do not have alkyl substituents at C-1 or C-2, and that they are generally 3,3-disubstituted. The reason for this is often that the product cyclopropenes in other cases either react further, e.g. by migration of the double bond induced by base[9], or, in the case of 1-halo-1-alkylcyclopropanes, eliminate directly to produce the less strained methylenecyclopropane[22]:

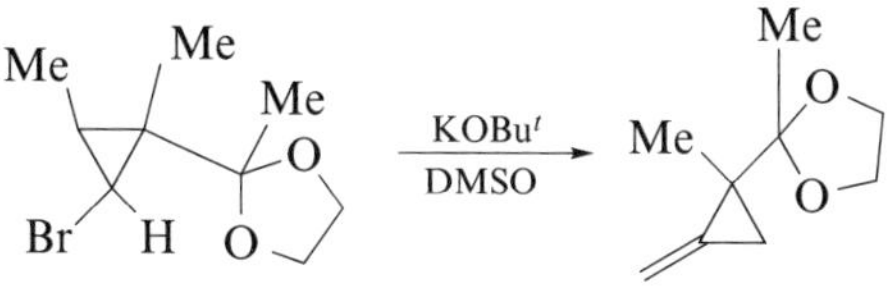

(Ref. 7)

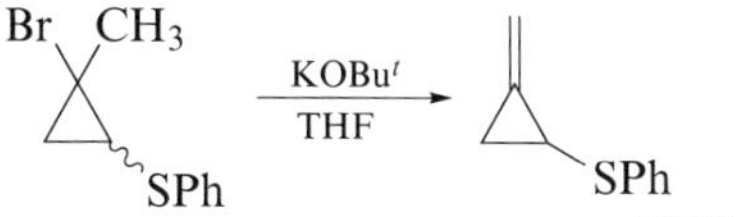

(70%)   (Ref. 22)

The reaction of either *cis-* or *trans-***1** with potassium *t*-butoxide in tetrahydrofuran at 25 °C leads to a *t*-butyl ether (**2**), apparently arising by attack of *t*-butoxide ion on an intermediate 1,4-di-*t*-butylmethylenecyclopropene. If the reaction is carried out at low temperature and the volatile materials are distilled directly into a cold trap, the cyclopropene can be trapped, albeit in low yield (10%), by added cyclopentadiene or detected directly by low-temperature NMR[23]. In a related example, a 1,1-dihalo-2-bromo-3-methylcyclopropane (**2a**) leads to products which are also apparently derived through an intermediate 1-chloro-3-methylenecyclopropene which undergoes nucleophilic addition (See Ref. 80).

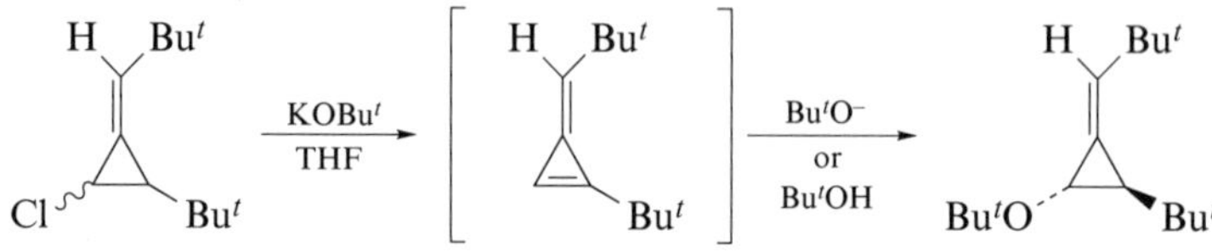

(**1**)   (**2**)

The presence of an electron-withdrawing substituent at C-2 of the cyclopropane may lead to less vigorous conditions being required for elimination to occur.

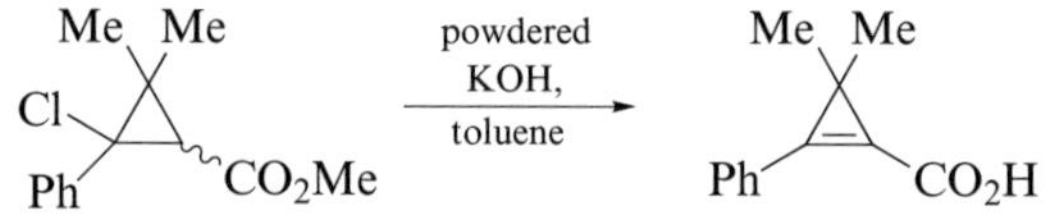

(50%, Ref. 25)

In the above example both stereoisomers apparently eliminate. A 2-aryl substituent has a similar effect, and in this case the individual *cis-* and *trans-*isomers have both been shown to undergo elimination, e.g.:

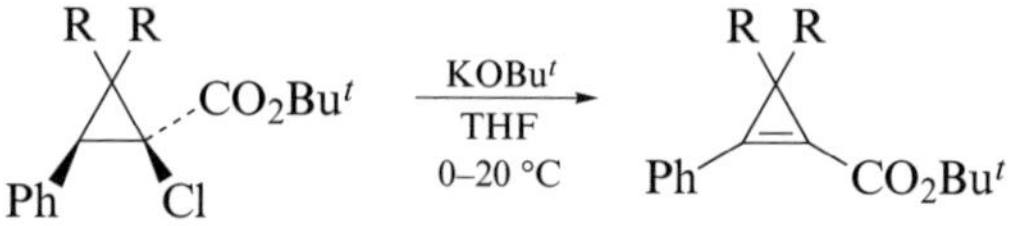

(Ref. 26–28)

R = Me 67%, R = Ph 71%, R R = —(CH$_2$)$_4$— 76%
RR = —(CH$_2$)$_2$— 77%, RR = —(CH$_2$)$_{33}$— 88%, RR = —(CH$_2$)$_5$— 66%

The *trans-*esters react much faster than the *cis-*isomers. In the case of compounds unsubstituted at C-3, stable cyclopropene esters are not formed even when lithium dialkylamides are used as bases[28].

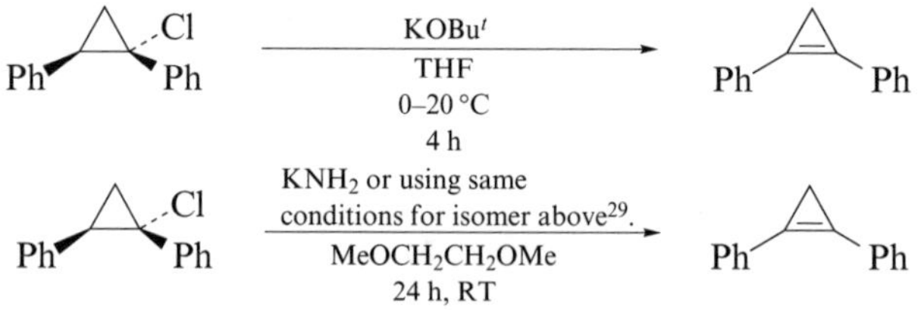

(90%, Ref. 29)

(Ref. 30)

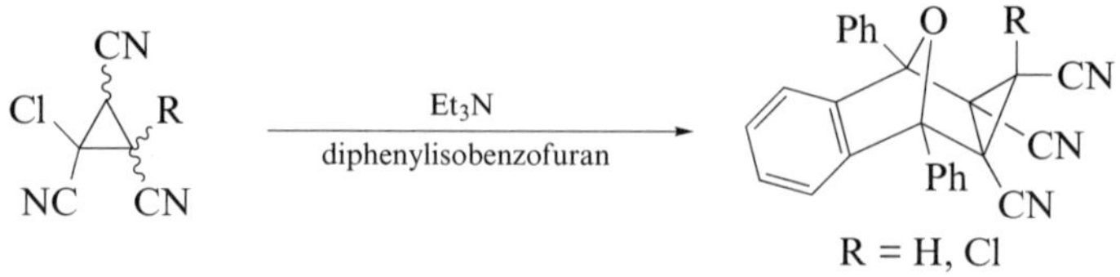

(Ref. 31)

R = H, Cl

With 2,3-disubstituted 1-halocyclopropanes there are two alternative modes of elimination, but one is generally preferred:

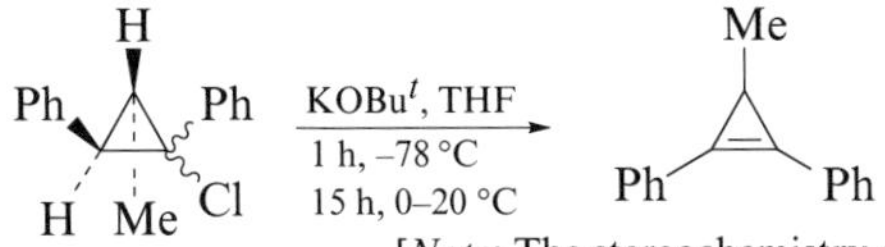

(80 – 88%, Ref. 32)

[*Note*: The stereochemistry of the starting material does not appear to be correct in the paper.]

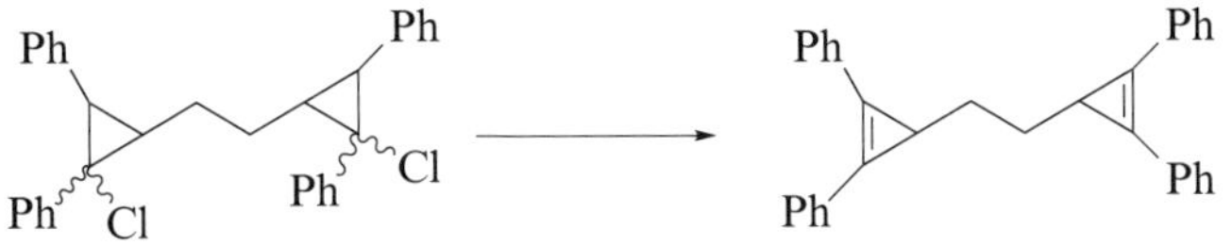

(Ref. 24)

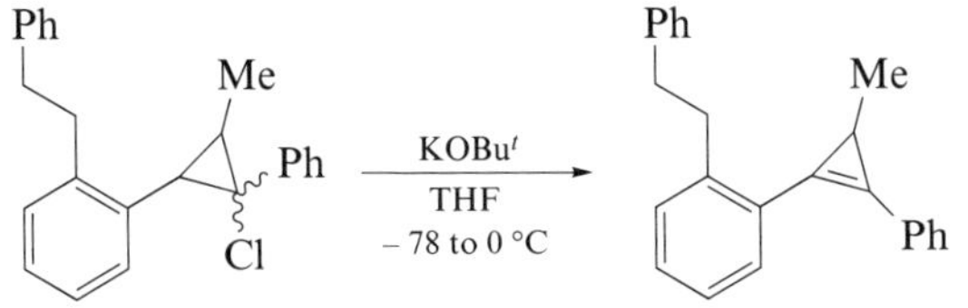

(Ref. 33)

(60%)

The *exo*-monochloride **3** eliminates at 25 °C to give largely cyclooctatetraene, apparently derived by a 1,4-elimination as in **4**; at higher temperatures in tetraglyme as solvent, the tetraene **5**, X = H, may be distilled directly from the reaction mixture. In contrast, the *endo*-monochloride (**3**) is very unreactive and, under forcing conditions, leads only to intractible materials. The corresponding dibromocyclopropane does eliminate at low temperature to give the bromocyclopropene **6**, which may be trapped by addition to diphenylisobenzofuran in moderate yield; once again, at higher temperatures the product is **5**, X = Br[34,35]. The monochloride **7**, R = H is also unreactive to KOBu-*t* in THF but, despite an earlier report[36], the addition of DMSO to the solvent mixture does lead to a low yield of a product of addition to the derived cyclopropene[37].

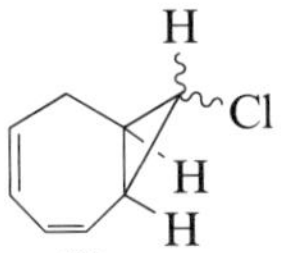

(**3**) 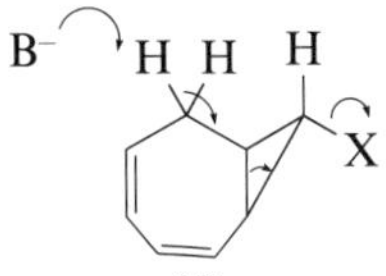   (**4**) 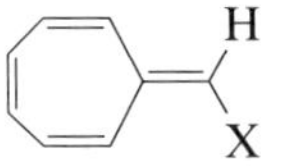   (**5**) 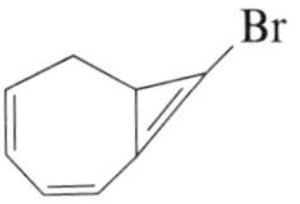   (**6**)

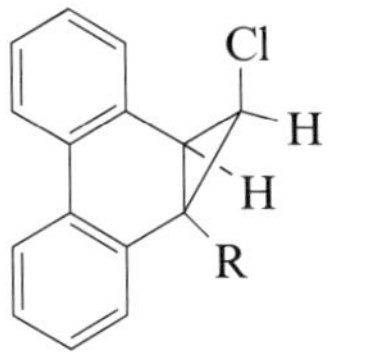

(**7**) 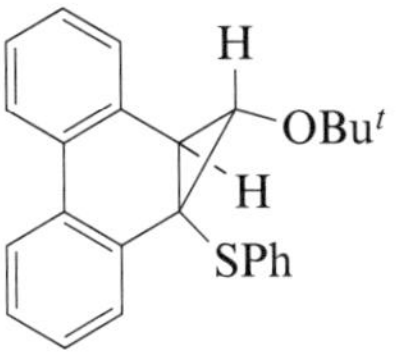   (**8**)

The corresponding thioether (**7**, R = SPh) is also very resistant to dehydrochlorination, but a low conversion into the ether **8** is observed on reaction with *t*-BuOK–THF–18-crown-6 at – 50 to 20 °C[38].

The dehydrohalogenation has also been applied to bicyclic lactones; cyclopropenes can be isolated for seven-membered ring lactones but for smaller rings undergo addition of *t*-butanol[39–41]:

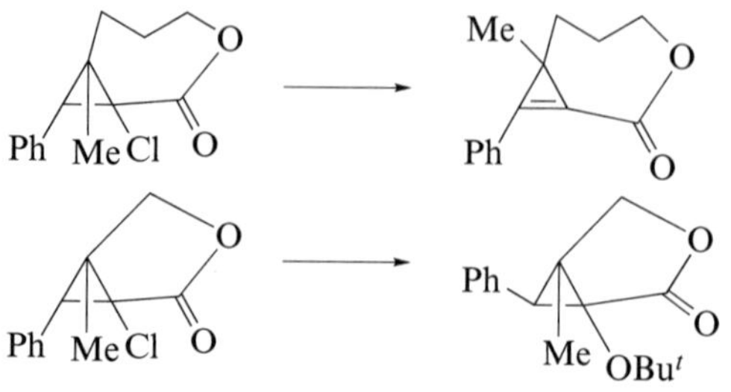

(Ref. 39 and 40)

(Ref. 39)

A valuable extension of this method is the use of base adsorbed on a solid support[42]. Thus, monobromocyclopropanes dehydrobrominate with KOBu^t on silica at 160 °C. Yields are up to 75% for cyclopropene itself on a small scale but fall considerably on scale-up (14%)[43]:

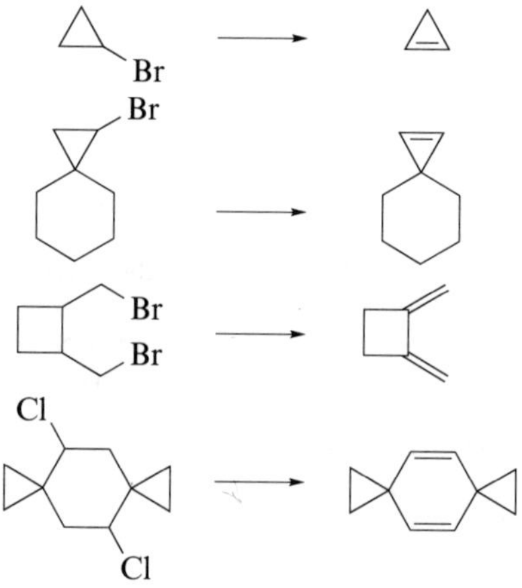

Potassium t-butoxide on Chromosorb W leads to dehydrochlorination of 1-chloro-2-methylenecyclopropane to produce the highly reactive methylenecyclopropene, which may be distilled directly into a cold trap and detected directly by NMR or trapped by addition to dienes[44-46]:

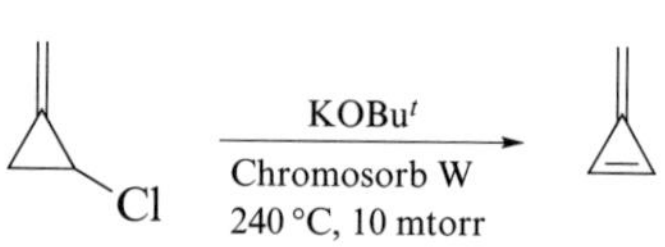

In the same way 2-vinyl-1-chlorocyclopropene is converted into the parent, highly unstable 1-vinylcyclopropene[44]:

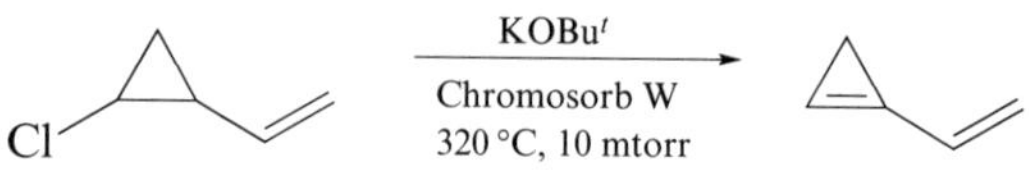

In a novel variation, the same product is obtained from a mixture of stereoisomers of dichloro-spiropentane[44]:

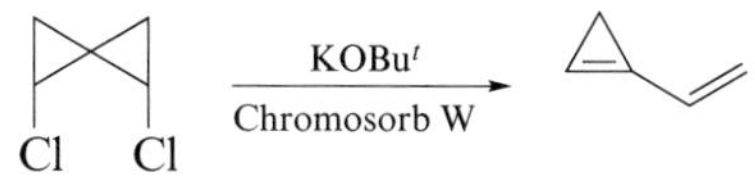

(9)   **(10)**

The mechanism proposed for this reaction is an initial 1,2-dehydrochlorination, followed by a radical fragmentation[44]. The source of the hydrogen atom could be the *t*-butanol; support for this is found in the identification of propanone among the reaction products. It would be interesting, however, to explore the direct transformation of **9** to **10** through a 1,4-dehalogenation. The dehydrochlorination of **9** in solution follows a quite different course, leading to the incorporation of *t*-butyloxy groups.

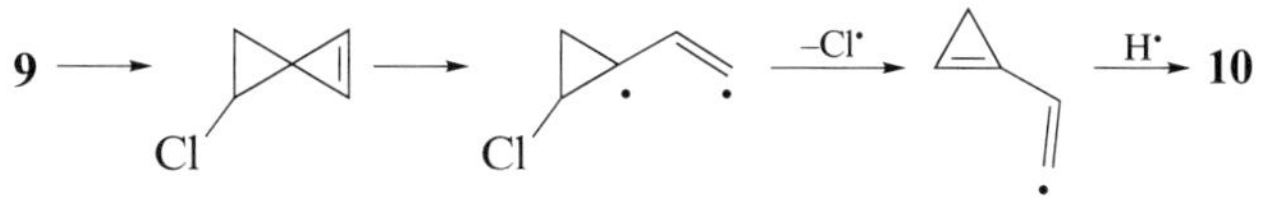

Dehydrofluorination of monofluorocyclopropanes is a much less widely studied process and, under the reaction conditions shown below, the product cyclopropene is sufficiently acidic to react further[47]:

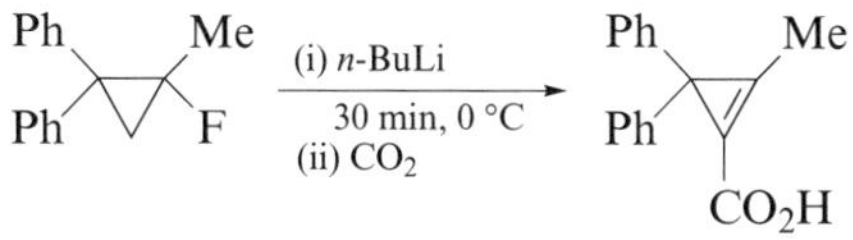

In one important series of compounds, the mono-halocyclopropane is generated *in situ* from a 1,3-dehydrohalogenation of a 1,3-dihalopropan-2-one acetal using a metal amide; with an excess of reagent, the cyclopropene produced in metallated *in situ* and can be further functionalized by trapping with electrophiles[48]:

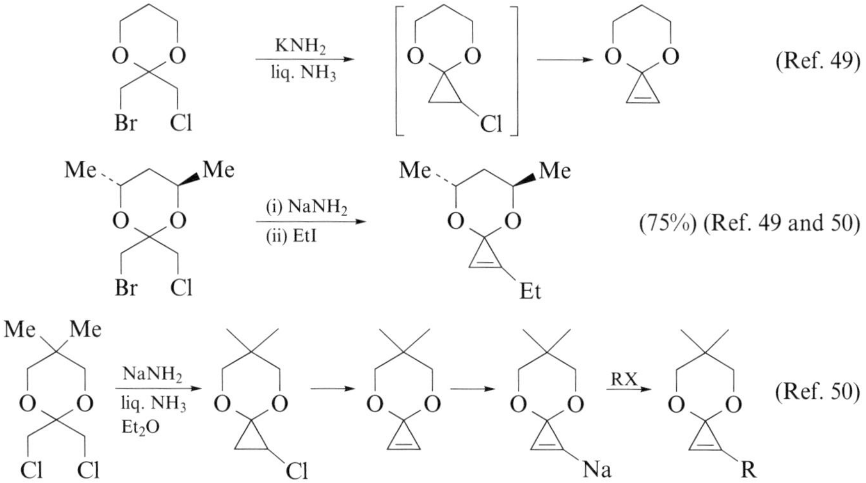

(Ref. 49)

(75%) (Ref. 49 and 50)

(Ref. 50)

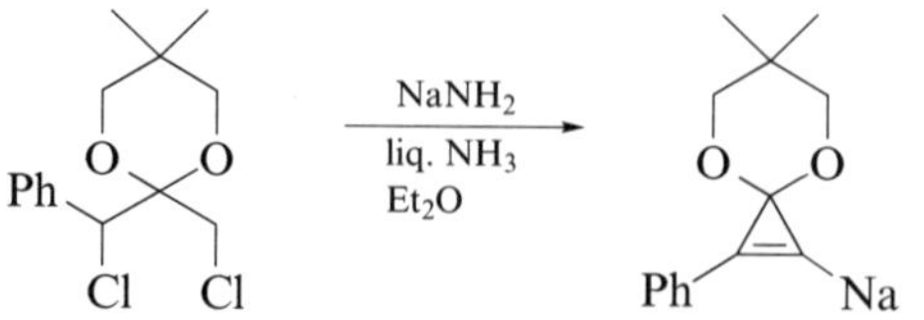

(73%) (Ref. 50)

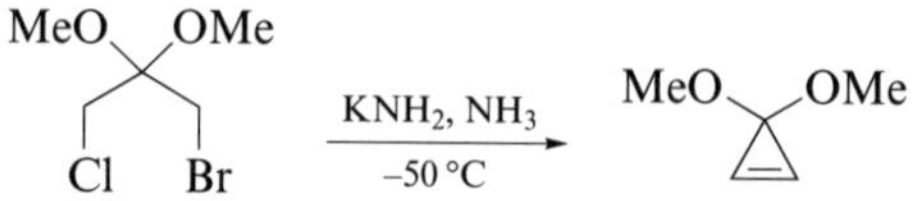

(Ref. 48, 51 and 52)

The dehydrohalogenation of 1,1-dichloro- and 1,1-dibromocyclopropanes to 1-halocyclopropenes has in some cases been of value, and has been applied in particular to the preparation of polyhalogenated compounds:

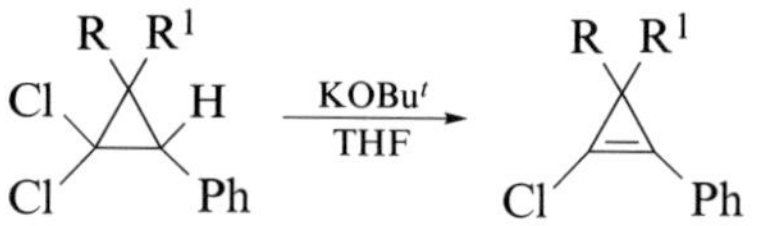

(Refs . 53 and 54

R = R^1 = H, 45%
R = R^1 = Me 87%
R = R^1 = Ph 51%
R = H, R^1 = Me 53%

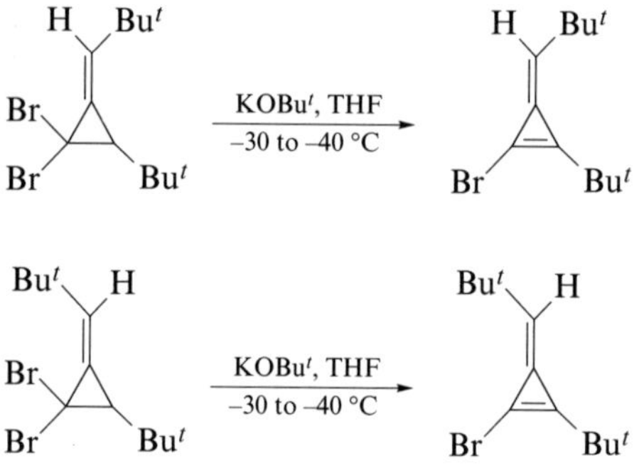

(Ref. 55)

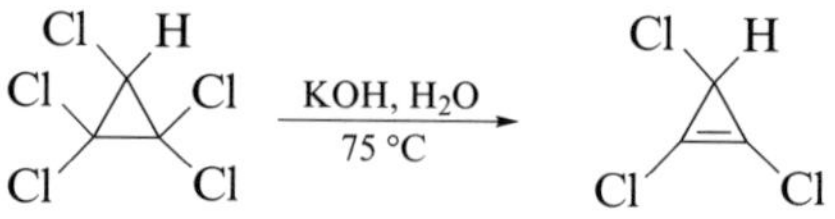

(77%, Refs. 56–58)

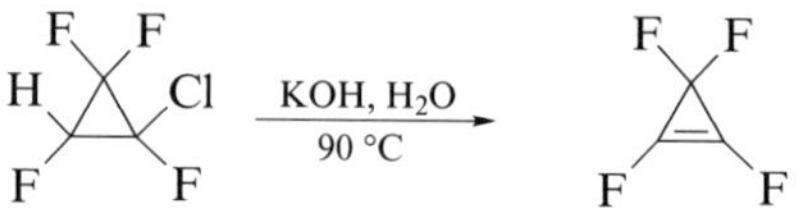

(26%, Ref. 59)

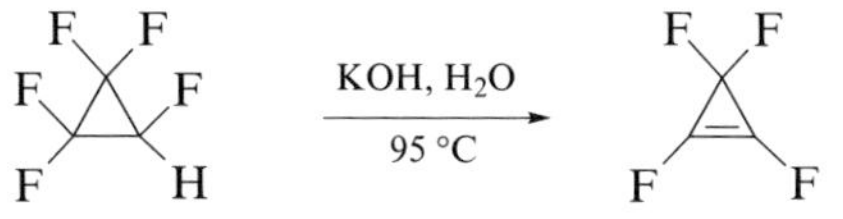

(36%, Ref. 59)

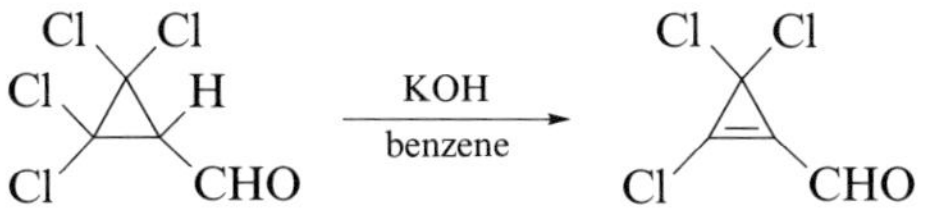

(Ref. 60)

In the last case yields are low because of further reaction of the cyclopropene with nucleophiles[61]. In some cases this latter reaction can be of preparative value;[61] thus 2-phenyl-1,1-dichlorocyclopropane may be converted into a highly reactive alkoxycyclopropene by reaction with alkoxide ion[62].

In other cases the method is not successful, because the elimination in followed by a further reaction through prototropic shifts [63–75,36], such as:

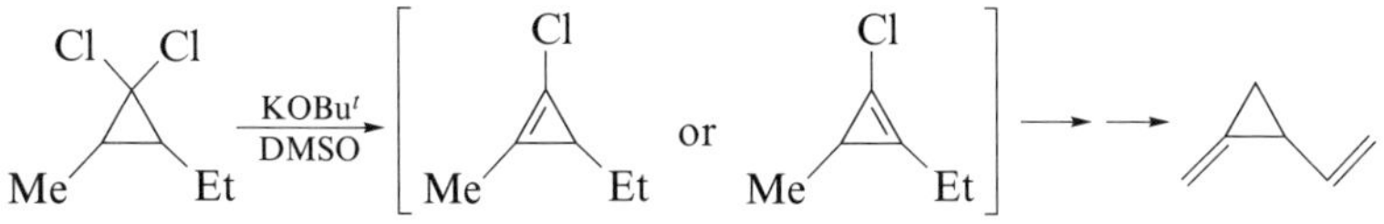

(Ref. 67)

The sequence of dehydrohalogenation, prototropic shifts, second dehydrohalogenation and further prototropic shifts has been put to extensive use in routes to cyclopropabenzenes; this is described more fully in Section IV.

A third possibility is the fragmentation of the ring. Thus, in the case of 2,2-dimethyl-1,1-dihalocyclopropanes ring-opening occurs to give enynes and, in some cases, to produce allenic carbenes:[64,76–78,69]

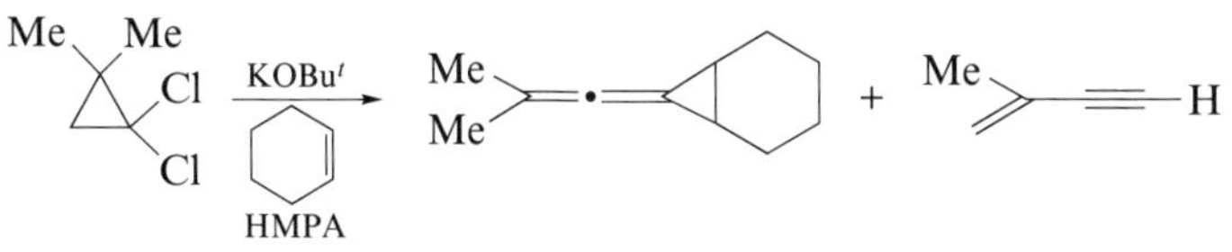

(Ref. 76)

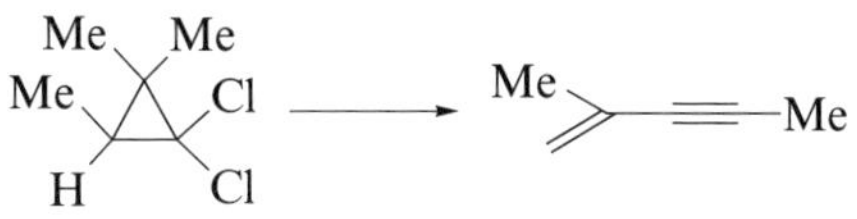

(Ref. 77)

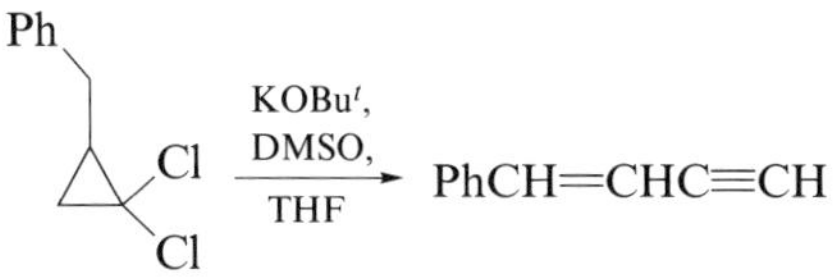

PhCH=CHC≡CH   (30%, Ref. 78)

Both types of product are thought to arise through an initially formed cyclopropene. The formation of enynes has been rationalized in terms of a 1,4-dehydrohalogenation coupled to ring-opening as in **11**, while the allene is apparently derived by trapping of the allenic carbene derived by 1,2-dehydrohalogenation as in **12**. The latter is supported by a labelling study, although a second such study in a closely related system has led to a different labelling pattern (see Section IX). The formation of allenic carbenes in this and other related reactions is described in Section IX.

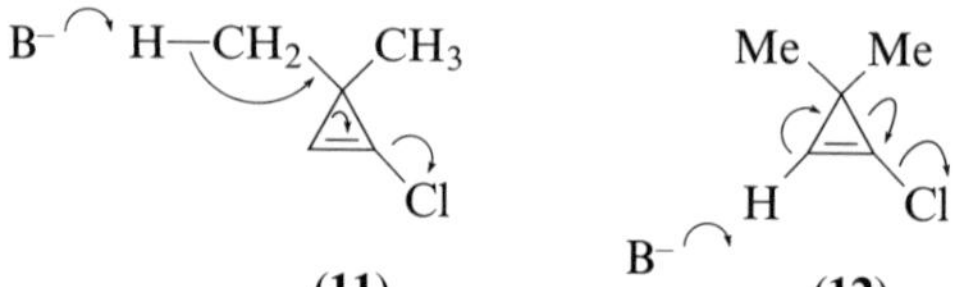

(11)   **(12)**

Ring opening also occurs in the cases of the methylenecyclopropane **13** and the alkoxy-cyclopropane **14**[75]:

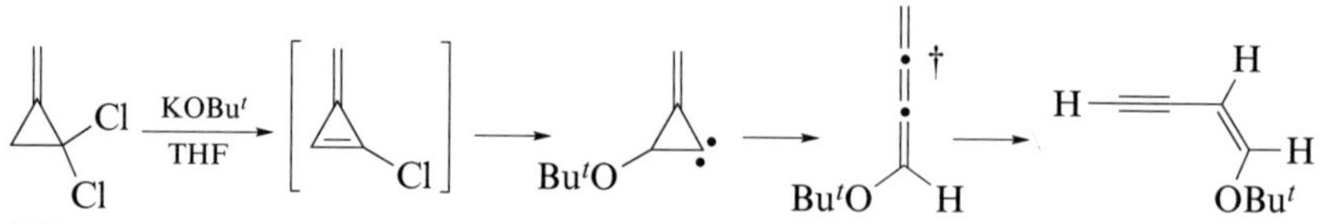

(13)   (Ref. 75)

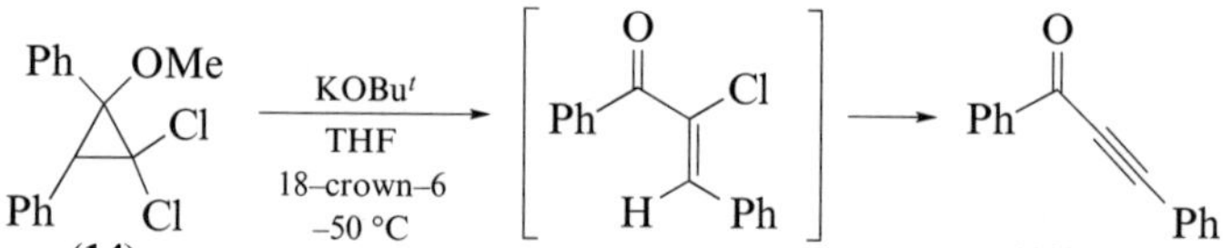

(14)   (43%)   (Ref. 79)

It is interesting to note in the first case that reaction of 2-bromomethyl-1,1-dichlorocyclopropane with KOBu-t in t-butanol, which may well also occur through 1-chloro-2-methylenecyclopropene, does not lead to ring-opening, but instead to nucleophilic addition to produce *trans*-1-chloro-2-t-butoxy-3-methylenecyclopropane[80]. In the latter case, it is noteworthy that the analogous ether **15** leads to no identifiable product under the above conditions, but in the presence of diphenylisobenzofuran the allene **16**, is isolated[79]. This could be derived by ring-opening of an intermediate 1-chlorocyclopropene to a vinyl-

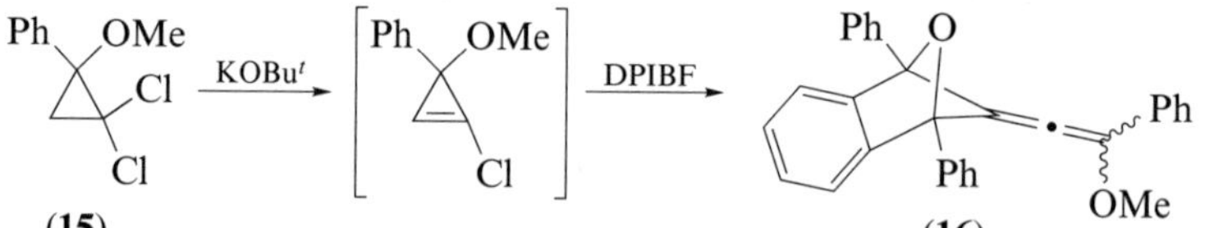

(15)   **(16)**

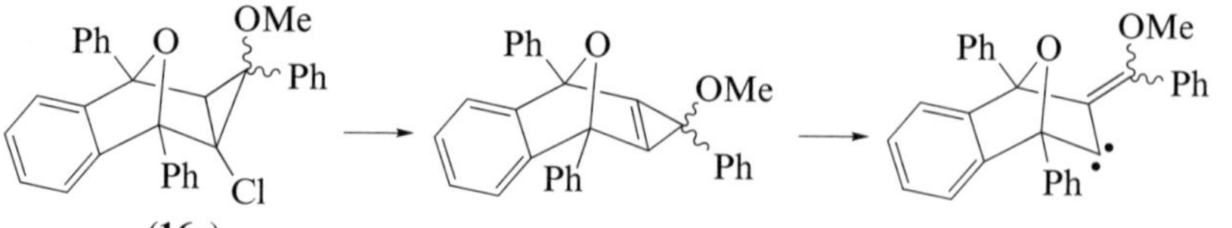

(16a)

[†] Note that this is presumably incorrectly represented in the paper.

carbene and 1,4-addition of this to the furan followed by loss of HCl, but a more likely alternative is that the cyclopropene is first trapped in a [4 + 2]-cycloaddition to give **16a**, which undergoes a second dehydrohalogenation to the corresponding cyclopropene followed by ring-opening to a vinylcarbene and rearrangement[79].

In other cases, a ring-expansion is observed, again apparently occurring through a halocyclopropene as an intermediate[81]:

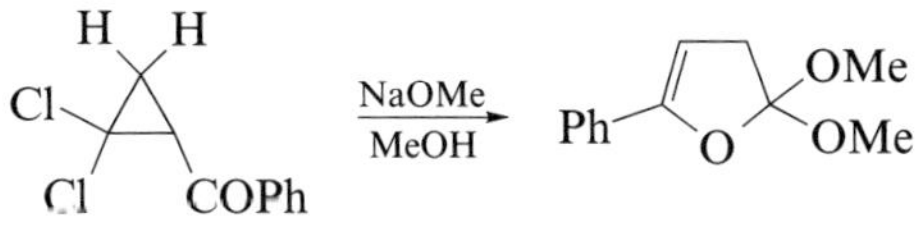

$$Cl \underset{Cl}{\overset{H\ H}{\triangle}} COPh \xrightarrow[\text{MeOH}]{\text{NaOMe}} Ph \overset{OMe}{\underset{O}{\diagdown}} OMe$$

There are isolated reports of the isolation of cyclopropenes from monoalkyl systems, although in the examples given below no yields are provided[82]:

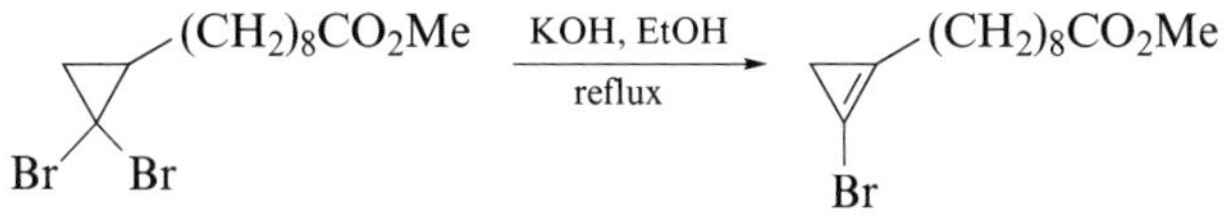

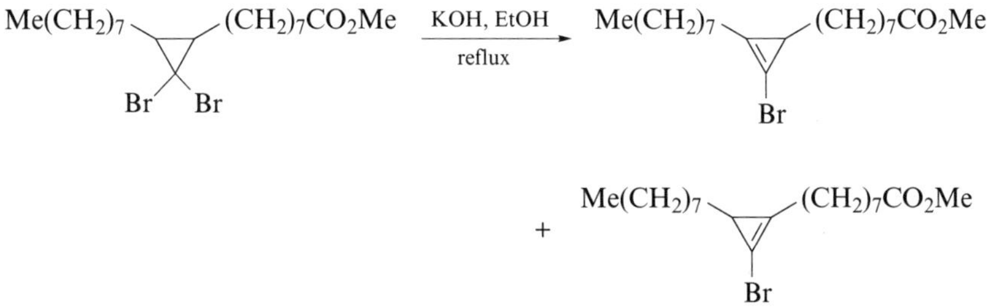

$$\underset{Br\quad Br}{\triangle}\!\!-(CH_2)_8CO_2Me \xrightarrow[\text{reflux}]{\text{KOH, EtOH}} \underset{Br}{\triangle}\!\!-(CH_2)_8CO_2Me$$

$$Me(CH_2)_7\!\!-\underset{Br\quad Br}{\triangle}\!\!-(CH_2)_7CO_2Me \xrightarrow[\text{reflux}]{\text{KOH, EtOH}} Me(CH_2)_7\!\!-\underset{Br}{\triangle}\!\!-(CH_2)_7CO_2Me$$

$$+\quad Me(CH_2)_7\!\!-\underset{Br}{\triangle}\!\!-(CH_2)_7CO_2Me$$

Elimination may also be carried out using an alkyllithium when the rate of lithium–hydrogen exchange exceeds that of lithium–halogen exchange; this is usually true only when the halogen is F or Cl[83,84]:

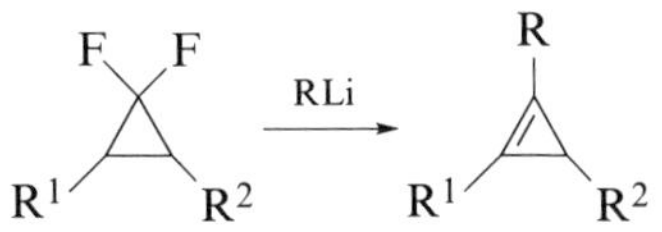

$$R^1\underset{F\ F}{\overset{}{\triangle}}R^2 \xrightarrow{\text{RLi}} R^1\underset{}{\overset{R}{\triangle}}R^2$$

R^1	R^2	R	%	
Ph	H	Bu	91	(Ref. 83)
Ph	H	Me	68	
Ph	H	Ph	63	
—(CH$_2$)$_6$—		Bu	85	
—(CH$_2$)$_6$—		Me	29	

$$\underset{Cl\quad Cl}{\overset{Me\ Me}{\underset{H\ \ \ Cl}{\triangle}}} \xrightarrow{\text{MeLi}} \underset{Cl\quad Cl}{\overset{Me\ Me}{\triangle}} \qquad \text{(Ref. 84)}$$

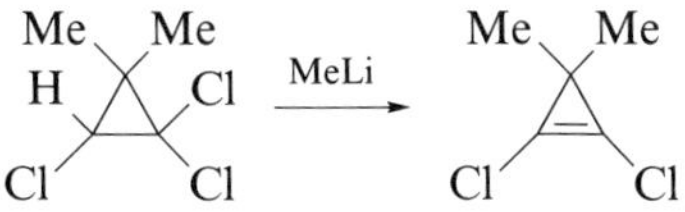

In the case of bromides, lithium-bromine exchange in normally preferred to lithium–hydrogen exchange and alternative reactions ensue:

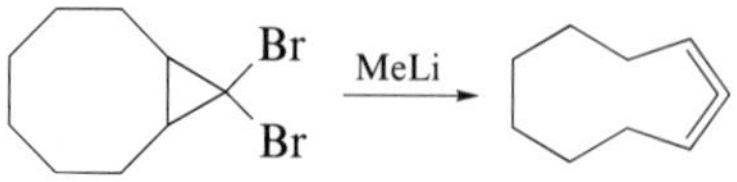

The dichloride **17**, R = H is dehydrohalogenated to a cyclopropene by addition of nucleophiles such as phenylthiolate but, in the absence of a trap, it ring-opens to a carbene which in turn is intercepted by insertion into the solvent[36,37,85]. In the case of related cyclopropanes e.g. **17**, R = CH$_2$CH$_2$Ph, the carbene may be trapped in intramolecular reactions[38]:

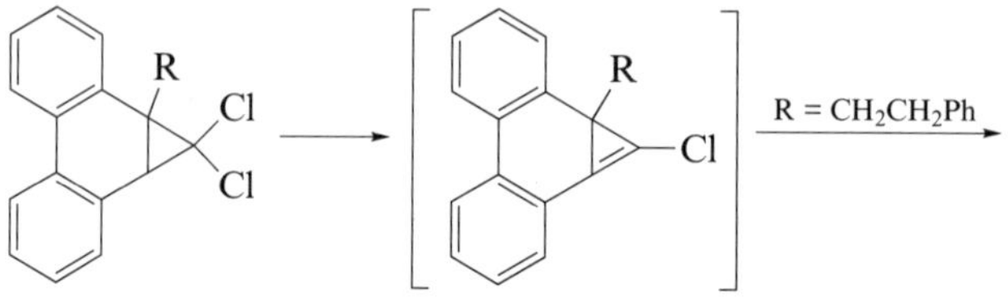

(**17**)

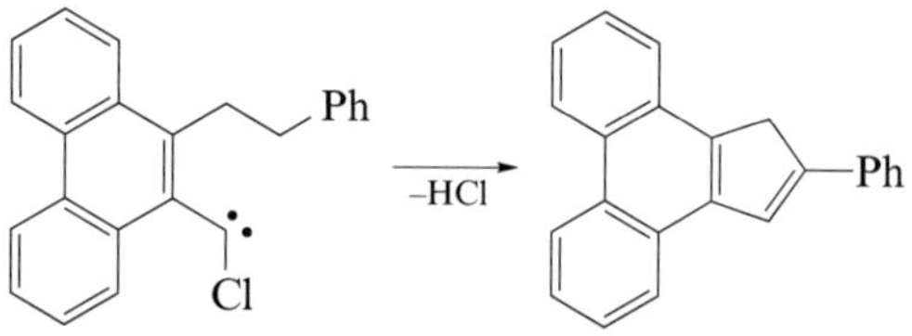

Dehydrochlorination of the bicyclic dichlorides **18** using potassium *t*-butoxide in DMSO has been known for some time to lead to the cyclopropenes **19** which, in the absence of a trap for the cyclopropene, undergo a complex series of further reactions by prototropic shifts and then a second dehydrohalogenation. In some cases, however, the cyclopropene can be intercepted if the reaction is carried out in the presence of a furan[86].

With *n* = 4 the cyclopropene is trapped in up to 40% yield by diphenylisobenzofuran, though the yield is only 12% for *n* = 6; with *n* = 5, the cyclopropene is trapped in 12% yield, while the hydrocarbon **20**, already proposed as a product of further reaction of the cyclopropene, is trapped in 28% yield. Furan only intercepts the cyclopropene derived from **18**, *n* = 4, and even then in very low yield (3%). In the case of *n* = 3, the cyclopropene is trapped in low yield by diphenylisobenzofuran (14%), but with furan the product of trapping of ring-opened carbene **21** is seen[86,88]. The same ring-opening was observed when the cyclopropene **19** *n* = 3, was generated by dehalogenation of 1,6,6-trichlorobicyclo [3.1.0] hexane using methyllithium[87], although the stereochemistry is now reassigned.

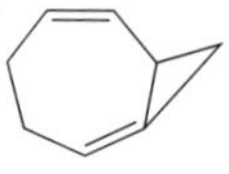

(**20**) 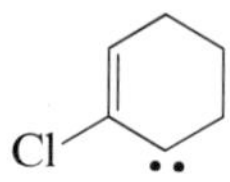   (**21**)

The hetero-substituted bicyclo[3.1.0]hexane **22** undergoes a similar dehydrohalogena-tion to **23** on reaction with KOBu^t; the cyclopropene can be intercepted by [4 + 2]-cycload-dition to diphenylisobenzofuran, but there is no evidence for double dehydrohalogena-tion[89]. Attempted trapping with furan leads, however, to the adduct **24** derived by addition of a ring-opened carbene **25** to the alkene. The sulphur analogue **26** is also dehydrohalo-genated to **27**, which is trapped either by cycloaddition to furan (46%) or, in the absence of an added trap, by addition of *t*-butanol to the cyclopropene[90].

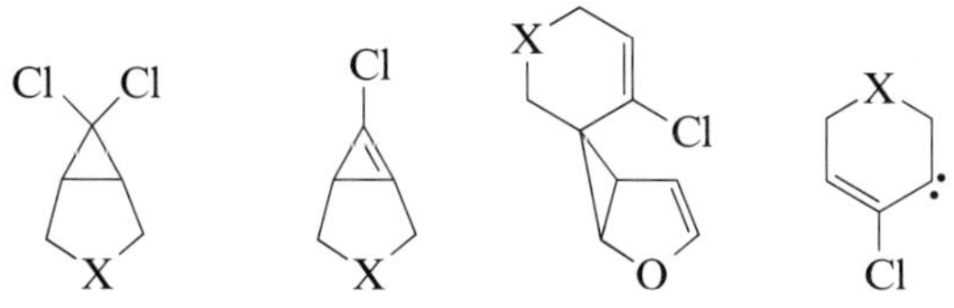

X = O (**22**)   (**23**)   (**24**)   (**25**)
X = S (**26**)   (**27**)

B. 1,4- and Related Eliminations

In addition to the very common 1,2-dehydrohalogenation of halocyclopropanes described above, there are a number of examples of a 1,4-elimination, some of which have considerable potential in the synthesis of unusual ring-systems[91–97]. Thus the tricyclic chlo-ride **28** which has no hydrogens available for a 1,2-dehydrohalogenation undergoes instead a 1,4-elimination leading to [5]metacyclophane[91]:

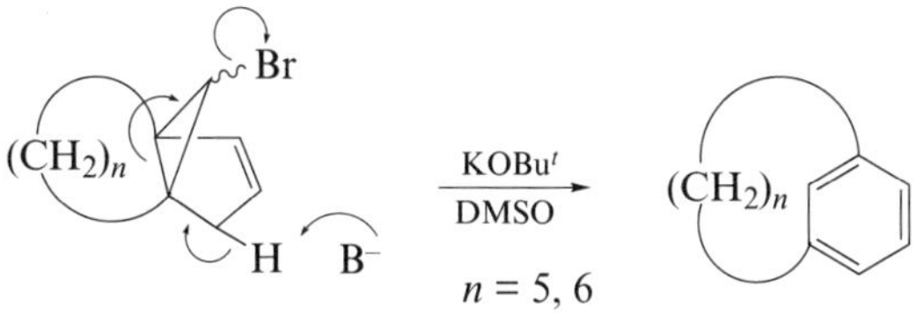

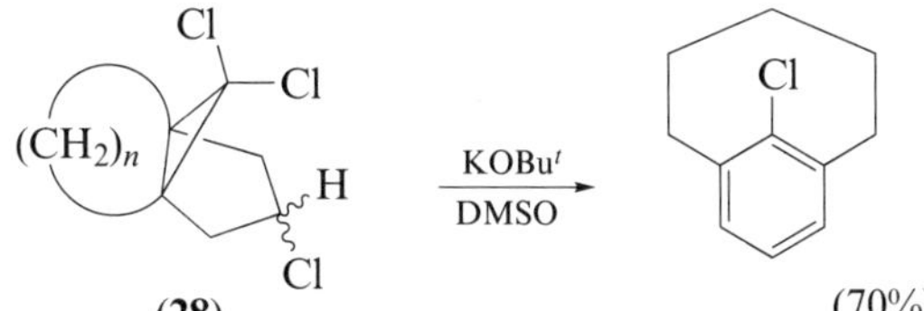

The monochloride **29**, in which the halogen is *exo* to the cyclopentene ring, is dehydro-halogenated to a mixture of Dewar benzenes thought to be derived from one primary prod-uct (**30**); this is apparently derived by deprotonation at the allylic position, ring-opening of the derived anion in a stereocontrolled electrocyclic process to give **31** and subsequent or concurrent displacement of halide ion[97]. The isomer, **32**, gives only polymers under these conditions, perhaps derived by further reaction of a metacyclophane[97]. The higher homo-logue **33** is converted into [5]metacyclophane (**34**) in near-quantitative yield, having the halogen geometry at the cyclopropane required for the 1,4-dehydrohalogenation. The iso-mer **35** also leads to some of the cyclophane on dehydrohalogenation, and both products may be derived through proton abstraction to give an allylic anion followed by direct dis-placement of the halide to form an intermediate bicyclo[1.1.0]butane, which may rearrange to both observed products[94].

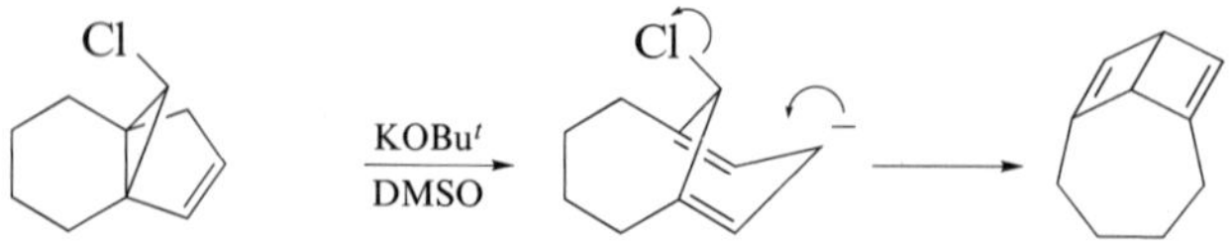

(29) (31) (30)

(Ref. 97)

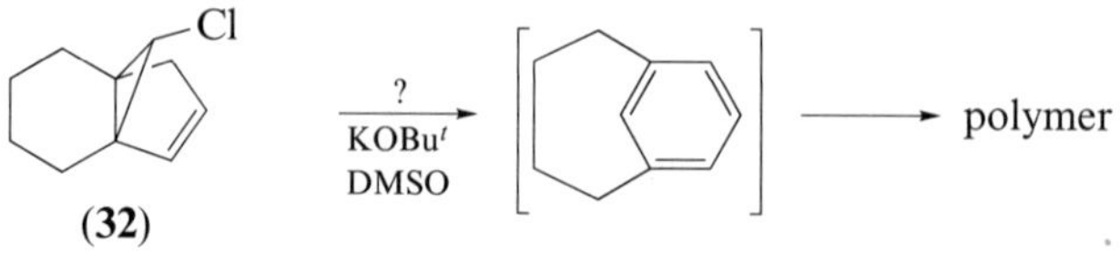

(32)

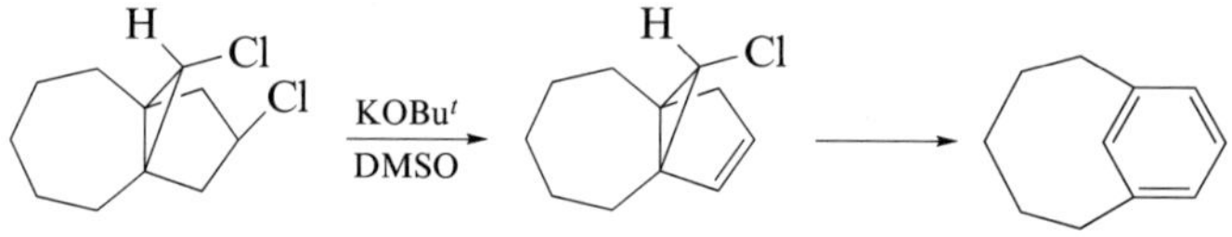

(33) (34)

(Ref. 94)

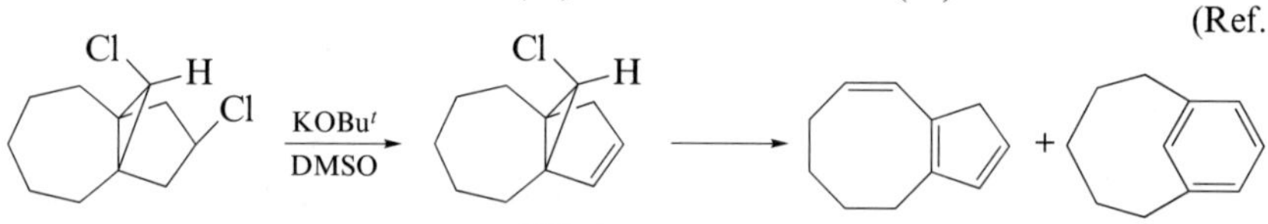

(35)

Compound **36**, X = Cl, leads to 1,6-methano-[10]annulene (**37**), 4-methylazulene (**38**) and **39** on dehydrohalogenation with strong base[99]; in the same way, **36**, X = Br, leads to **37**, a reduced amount of **38** and the monobromide **40**[99]. Each of the products arises by a distinct reaction pathway, though a 1,4-elimination to leading **41** is thought to be involved in the case of **37**.

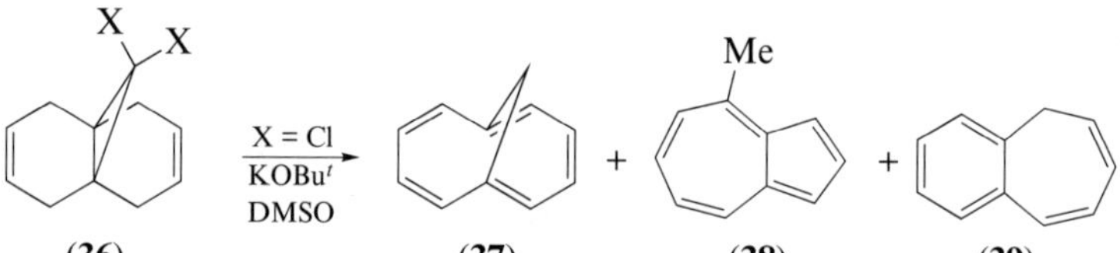

(36) (37) (38) (39)

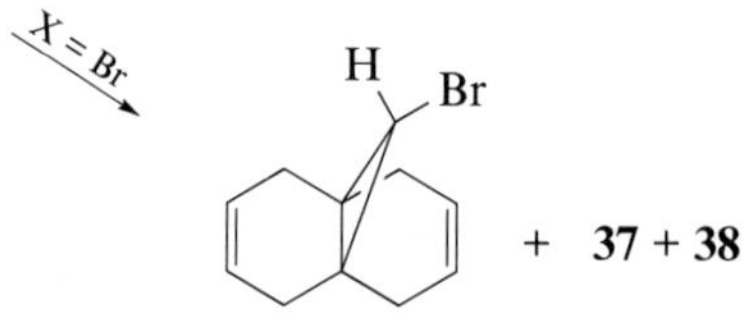

(40)

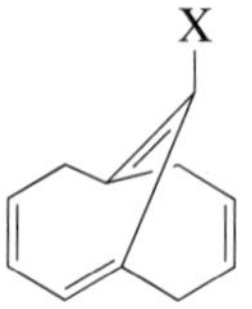

(41)

Similarly, the propellane **42** is converted into **43**, and **44** into **45**[100a, b]:

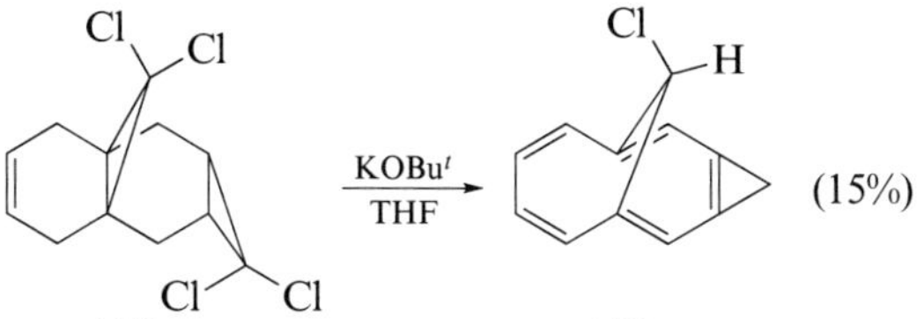

(42)   (43) (15%)

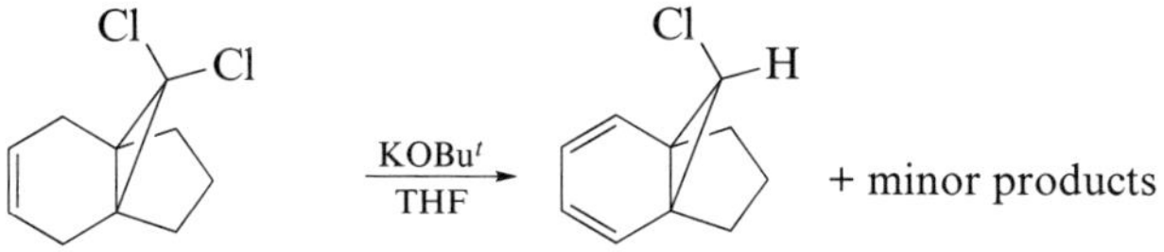

(44)   (45) + minor products

The key step in the first reaction is, once again, a 1,4-dehydrohalogenation followed by a stereocontrolled prototropic shift and electrocyclic closure of the derived triene to a bicycloheptadiene (shown below); the dichlorocyclopropane then undergoes a typical sequence of 1,2-dehydrochlorination followed by prototropic shifts to lead to the [10]annulene. The chlorodiene **45** may be produced by a similar sequence of 1,4-dehalogenation, prototropic shift and cyclization[100b]. The mechanisms of these and related reactions have been the subject of labelling studies.

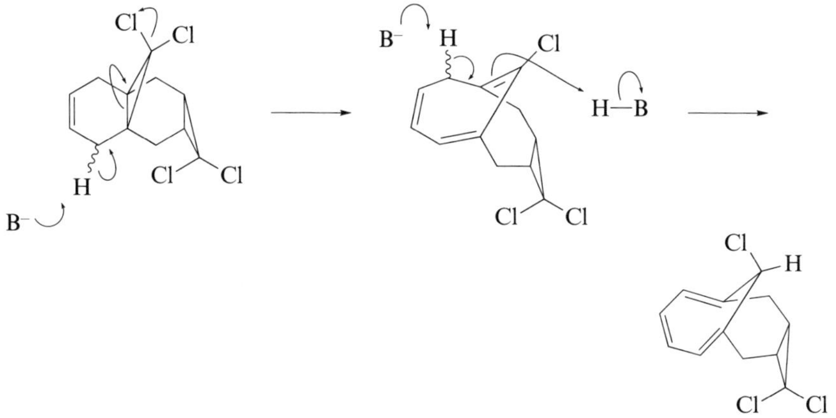

The mixture of cyclopropanes stereoisomeric at the cyclopropane group to **46** reacts with potassium *t*-butoxide in DMSO to give **47–49**[101].

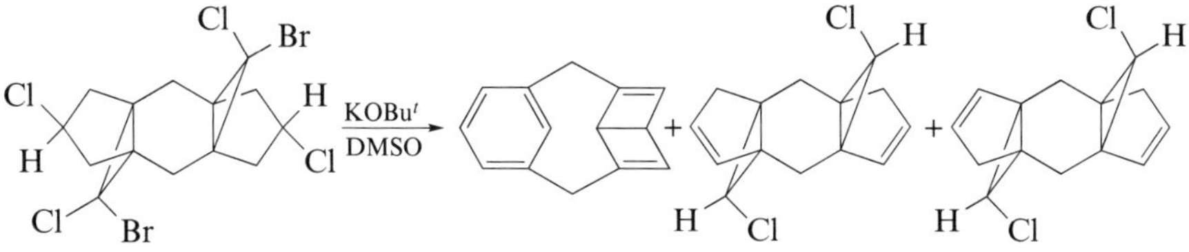

(46)   (47)   (48)   (49)

It is believed that only the isomer shown gives **47**. The first step of the sequence leading to **47** is thought to be a reductive debromination of **46** to **50**:

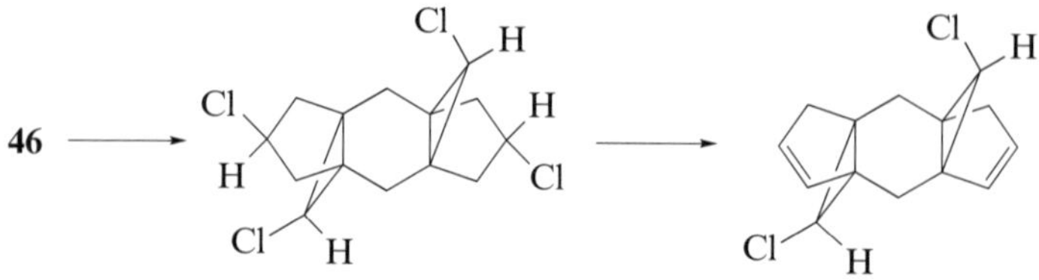

(50)   (51)

Such reductions are reported to be common in reactions of 1,1-dibromocyclopropanes with KOBu^t–DMSO (Ref. 10 in Reference 101). Double dehydrohalogenation would then lead to **51**; deprotonation at the allylic methylene group and ring-opening of the bicyclic anion would produce **52** and **53** which finally fragments to **47**.

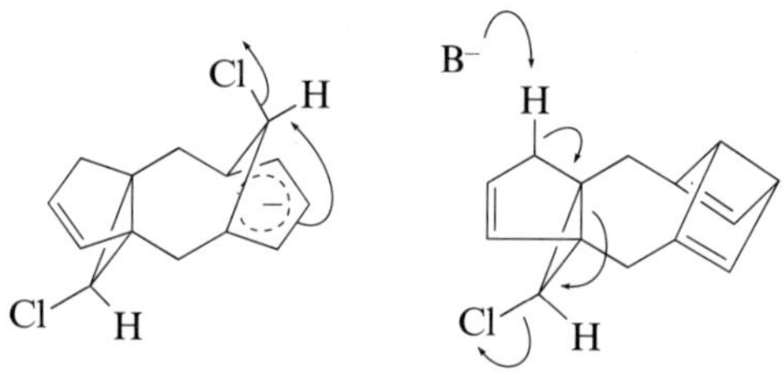

(52)   (53)

The step **52** → **53** is a direct analogy of the preparation of Dewar benzene discussed above[97]. The final fragmentation requires the *syn*-stereochemistry of the cyclopropane chlorines relative to the five-membered rings, as present in the stereoisomer of **46** shown above[101,102].

II. ELIMINATION OF HALOGEN

The reactions of either *cis*- or *trans*-1,2-dibromo-1,2-dimethylcyclopropane with *t*-butyllithium lead to 1,2-dimethylcyclopropene at low temperature[103,104]:

although it is not possible to judge from the available evidence whether *trans*- or *cis*-elimination occurs at a higher rate. Application of this method to a range of 1,2-dibromo-cyclopropanes fused to other ring systems at C$_1$ and C$_2$ leads to highly strained short-lived cyclopropenes which are trapped by reaction with excess alkyllithium followed by carboxylation[103,104]. Evidence for the intermediate cyclopropenes was provided by their trapping by [4 + 2]-cycloaddition to diphenylisobenzofuran when the reactions were carried out in the presence of the trapping agent; no adducts were formed when this was added subsequent to the *t*-butyllithium, indicating the high reactivity of the cyclopropenes. The use of methyllithium in place of *t*-butyllithium led to sluggish reactions and to the formation of products which apparently included tetramers.

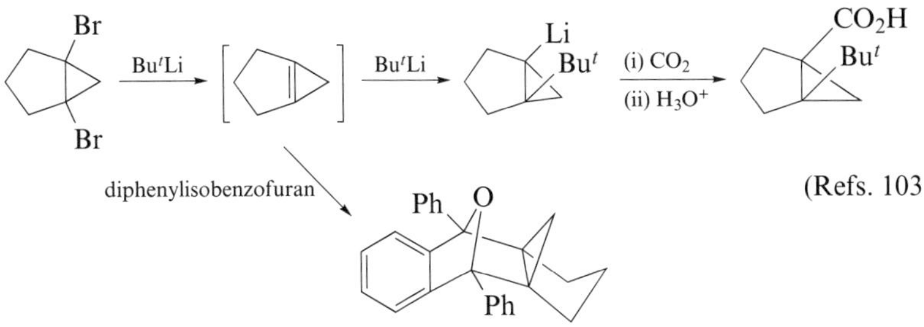

(Refs. 103 and 104)

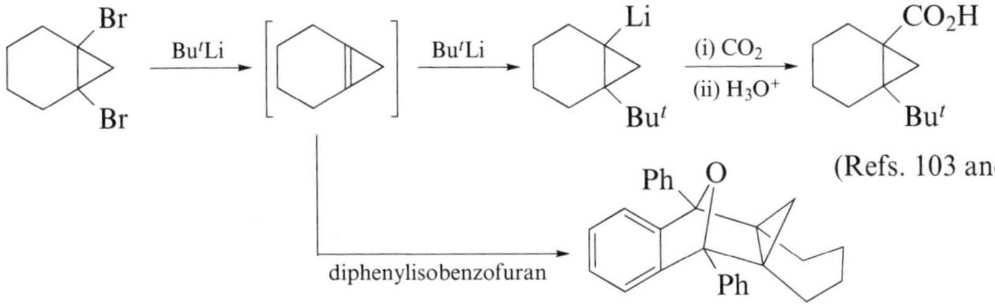

(Refs. 103 and 104)

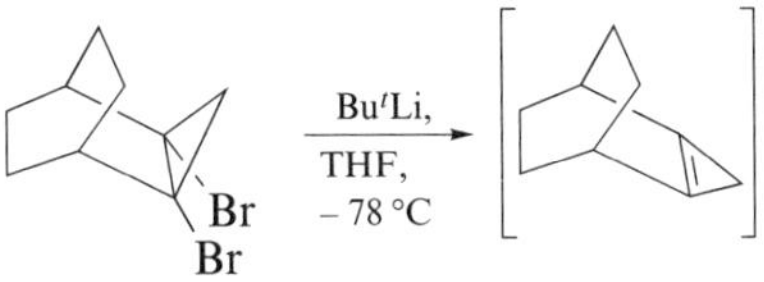

(Ref. 105)

The corresponding di-iodides also reacted with methyllithium in solution at −78 °C to give complex products apparently derived by further reaction of the initially formed cyclopropenes. However, on reaction with potassium atoms or with solid methyllithium on glass helices at −15 to 25 °C in an argon stream the only products were the corresponding methylenecycloalkenes, which were condensed at 20 K; nonetheless, in each case the cyclopropene is thought to be an intermediate[104,106]. Indeed, 1,2-dibromo-1,2-dimethylcyclopropane is converted into the corresponding cyclopropene under these conditions.

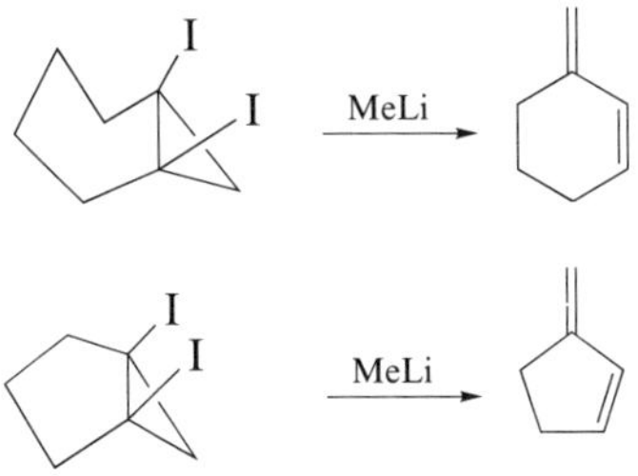

The reaction is considerably exothermic, and the real temperature at the reaction site may be rather higher than the overall zone temperature. This may explain the difference in products from those obtained in the gas-phase dehalosilylation of 7,7-dichloro-1-trimethylsilylbicyclo[4.1.0]heptane[44].

Clearly, in all these bicyclic cases the reaction must be occurring through an overall *cis*-1,2-elimination. The application of these reactions to more highly strained systems is described in Sections VII and VIII. One limitation of this route to cyclopropenes is the availability of the dihalocyclopropanes—which could often, in principle, be made most simply by addition of halogen to the product cyclopropene. However, the application of a number of variations of the Hunsdieker reaction to readily available cyclopropanecar-boxylic acids does provide a simple alternative[104]:

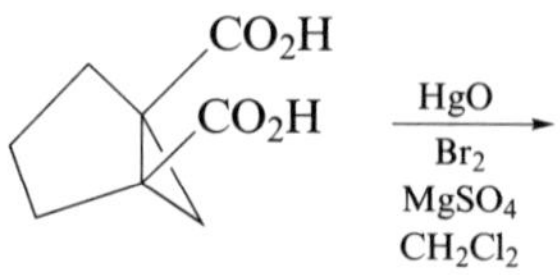

The sequence of halogen addition to a cyclopropene to form a dihalocyclopropane, followed by regeneration of the cyclopropene by 1,2-dehalogenation, does provide one possible method of protecting the highly reactive ring system. Addition of bromine often leads to a complex product mixture because ring-opening competes with direct addition; however, addition of iodine does usually lead to a mixture of *cis-* and *trans-*di-iodocyclo-propanes. In an early application of this to the protection of sterculic acid, (**54**), the natur-al product could be regenerated by reaction with hydroxide ion, albeit only in 30% yield[107]:

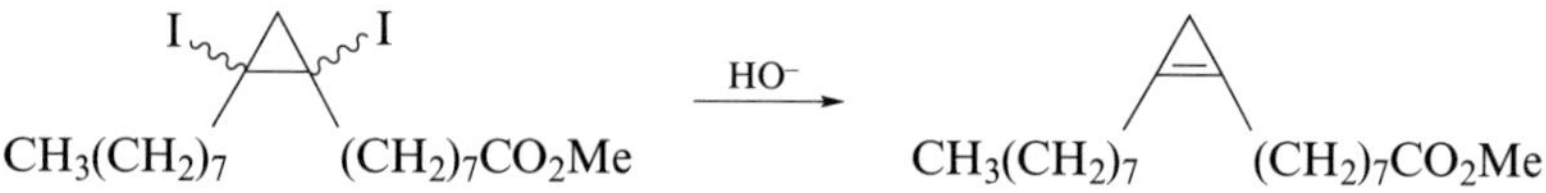

(**54**)

However, the elimination may be achieved in high yield from either *cis-* or *trans*-1,2-di-iodo-1,2-dibutylcyclopropanes using one molecular equivalent of butyllithium at low tem-perature. Since no reaction occurred at the ester group under these conditions, this has been applied in synthesis esters as of sterculic acid, α-hydroxysterculic acid (**55**)[108] and **56**, a potential inhibitor of mycolic acid synthesis[109]:

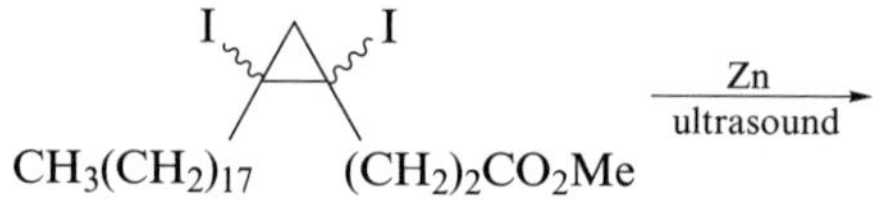

(Ref. 109)

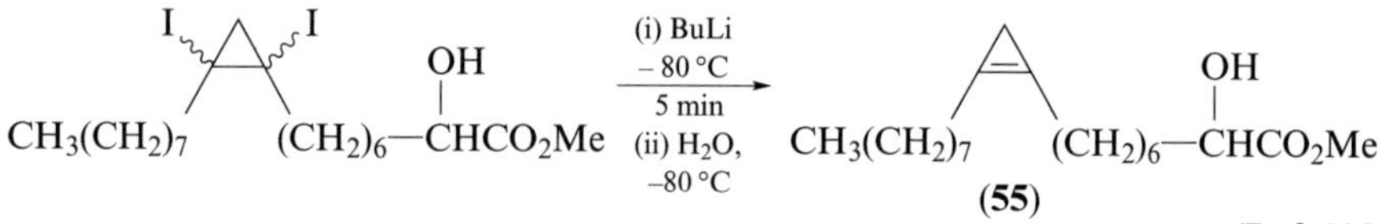

(Ref. 108)

In the first case the deiodination was brought about by the use of zinc and ultrasound[109]*; however in other cases the use of a dialkyl phosphite and sodium hydride at 20 °C also pro-vides a very mild and efficient method[110]. In the last case, both *cis-* and *trans*-di-iodides reacted at similar rates, but while the *cis*-isomer reacted also with the related reagent diethylphosphite and triethylamine to give the cyclopropene in 20 h at 20 °C, the *trans*-iso-mer remained largely unchanged under these conditions.

---

*The use of zinc in ethanol to 1,2-dehalogenate 1,2-dichlorotetraflurocyclopropane has been known for some time[59].

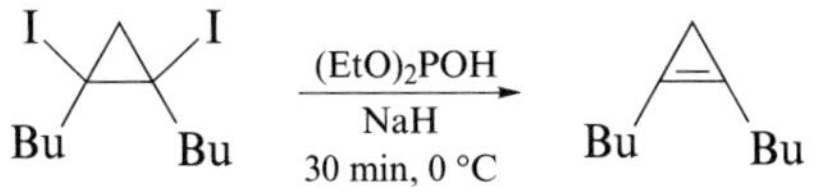

(84%, Ref. 110)

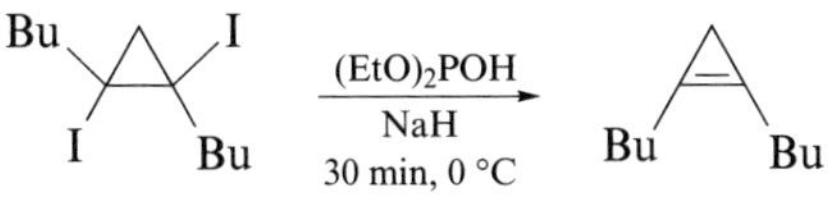

(84%, Ref. 110)

The 1,2-dehalogenation reaction is also successful with 1,1,2-trihalo- (**56**) or 1,1,2,2-tetrahalocyclopropanes (**57**) and, in the former case, provides a valuable entry to 1-halcyclopropenes which often undergo ready lithium-halogen exchange to produce synthetically valuable 1-lithiocyclopropenes. The elimination occurs extremely rapidly when X = Br even at –95 °C. When X = Cl and Y = Br, the reaction occurs in a few minutes at a temperature of –40 °C, whereas with X = Y = Cl, the reaction occurs at a reasonable rate only at ambient temperature[111,112,84,87].

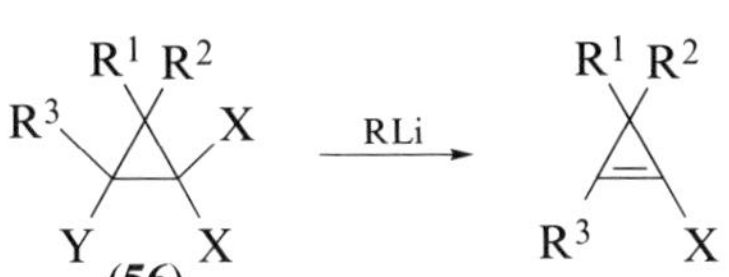

X = Y = Br	–90 °C, 2 min
X = Cl, Y = Br	–90 °C, 2 min
X = Br, Y = Cl	0 °C, 2 min
X = Y = Cl	20 °C, 10 min

The elimination is successful with a range of substituents at C-2 or C-3 of the cyclopropane, although in some cases the derived cyclopropenes ring-open even at room temprature to give vinylcarbenes, which may be trapped in inter- or intramolecular processes. It is successful when X or Y = Br even if R = H, but when X = Y = Cl and R = H, an alternative 1,2-elimination of HCl occurs. By careful control of the quantity of reagent, it is possible to carry out the elimination in the presence of functional groups which are relatively reactive to such reagents (final two examples below)[112–118,120,80].

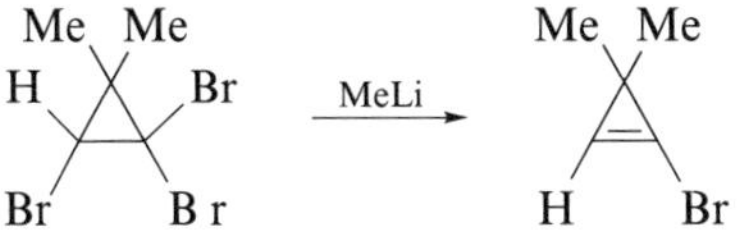

(Ref. 111)

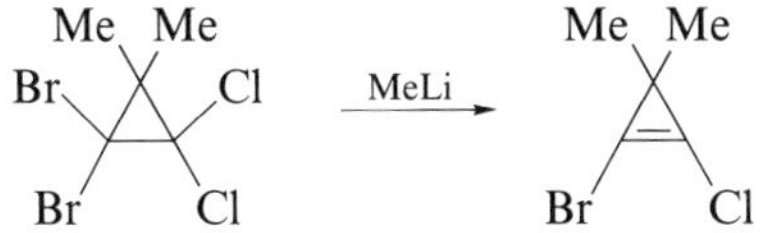

(Ref. 112)

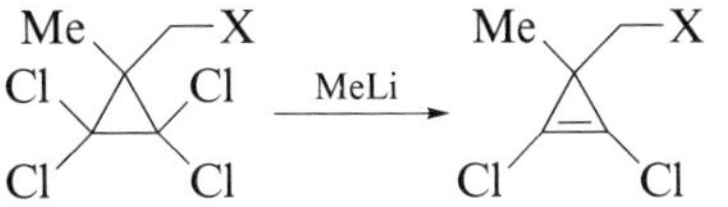

X = OMe, NPr$^i_2$, Ph, Cl   (Refs. 113 and 114)

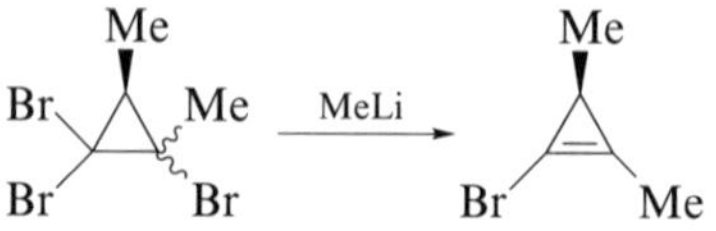

Me   Me
Br Me   MeLi →   △
Br Br   Br Me   (Ref. 115)

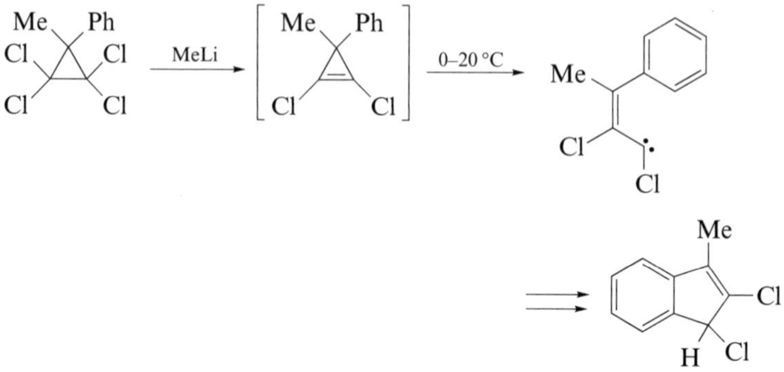

Me Ph   Me Ph
Cl Cl   MeLi → [   ]   0–20 °C → Me—C(Ph)=C(Cl)—C̈l
Cl Cl   Cl Cl

⟹ (indanyl structure) Me, Cl, H Cl   (Ref. 117)

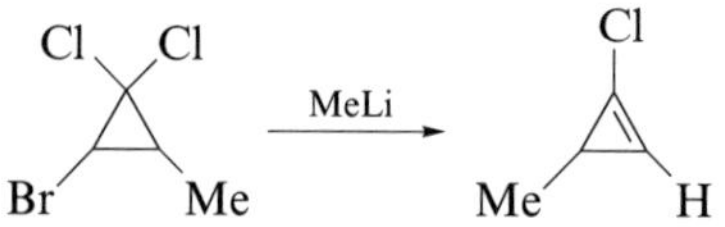

Cl Cl   Cl
△   MeLi → △   (Ref. 80) 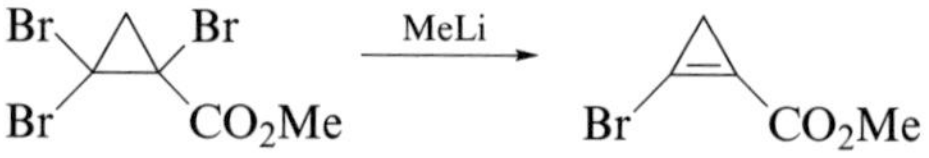
Br Me   Me H

Br Br   MeLi
△ → △   (Ref. 118) 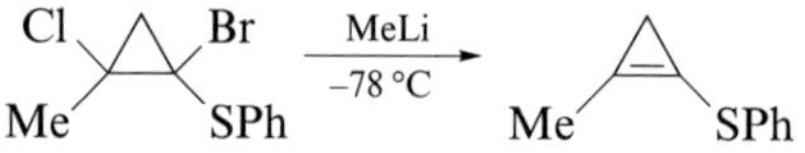
Br CO$_2$Me   Br CO$_2$Me

Cl Br   MeLi, −78 °C
△ → △   (Ref. 119) 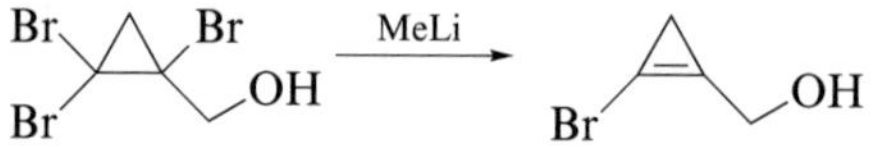
Me SPh   Me SPh

Br Br   MeLi
△ OH → △ OH   (Ref. 120)
Br   Br

The further reaction of the product halocyclopropene with a second molecular equivalent of alkyllithium by lithium–halogen exchange occurs more slowly, requiring some minutes at − 40 °C to ambient temperature for bromides and about an hour at ambient temperature for chlorides; once again the lithium–halogen exchange is successful for 1-bromocyclopropenes even when there is a hydrogen at C-2, but for 1-chlorocyclopropenes having a hydrogen at C-2 an alternative 1,2-elimination occurs (see Section IX). The lithiocyclopropenes may be trapped by a range of electrophiles, or by water[112–118,120]; in the latter case, removal of the ether solvent from the crude reaction mixture leads to solid lithiocyclopropene which, on quenching with water, produces the cyclopropene free of solvent[115]:

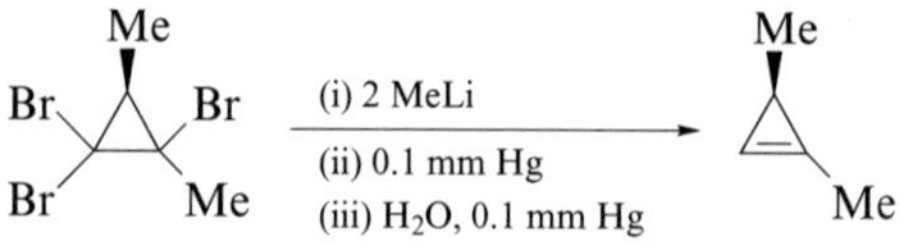

Me   Me
Br Br   (i) 2 MeLi   → △
Br Me   (ii) 0.1 mm Hg   Me
   (iii) H$_2$O, 0.1 mm Hg

The tri- and tetrahalides are in many cases readily available by dihalocarbene addition to the corresponding halo- or dihalo-alkene; in other cases they may be obtained on a synthetically useful scale from carboxylic acids[118]:

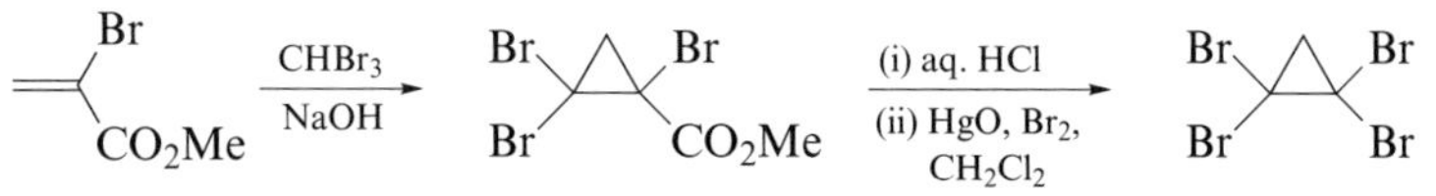

No information is available as to the stereochemistry of these eliminations. What is certain, however, is that the 1,2-elimination must be an extremely rapid process; thus no competition is seen from 1,1-dehalogenation in these processes, even though 1,1-dibromocyclopropanes are known to form allenes through the fomation of a 1-lithio-1-bromocyclopropane and a cyclopropylidene extremely rapidly even at – 90 °C. Although lithiobromides such as 58 may be trapped by electrophiles at below – 105 °C, or at higher temperatures in the presence of a stabilizing group such as an ether, there is no evidence for the formation of such intermediates in the 1,2-elimination from 56 or 57.

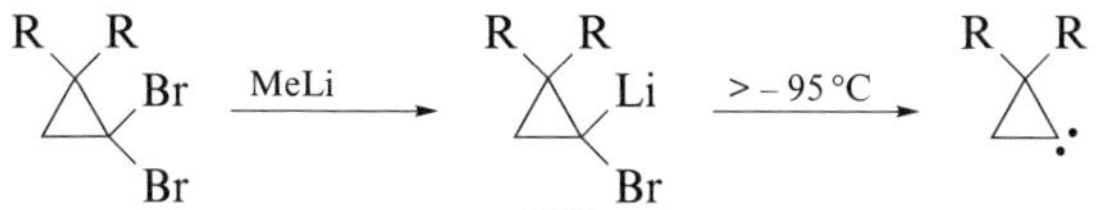

The dehalogenations may also be carried out in the gas phase over methyllithium supported on glass helices, a reaction which may be applied also to simple dihalides[44]:

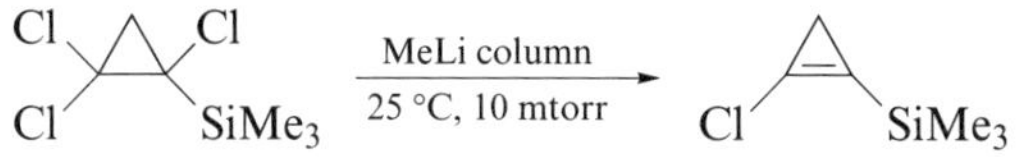

The 1,2-debromination of tri- and tetrahalocyclopropenes may also be carried out using a dialkylphosphite and either sodium hydride or a tertiary amine at ambient temperature[110]:

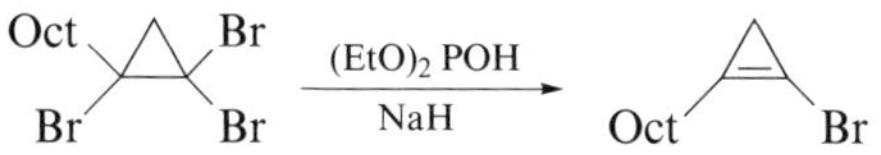

Trapped as an adduct with diphenylisobenzofuran

In the case of volatile cyclopropenes, these can be distilled directly from the reaction mixture as they form by using a less volatile dialkylphosphite and sodium hydride.

III. 1,2-ELIMINATION OF A TRIALKYLSILYLHALIDE

The fluoride-promoted dehalosilylation of 1-trimethylsilyl-2,2-dihalocyclopropanes provides one of the easiest routes to simple monohalocyclopropenes, the products being trapped by [4 + 2]-addition to furan or diphenylisobenzofuran (DPIBF)[121]. In the same way, reaction of 1-trimethylsilyl-7,7-dichlorobicyclo[4.1.0]heptane with CsF in diglyme in the presence of DPIBF leads to trapping of 7-chlorobicyclo[4.1.0]hept-(1,7)-ene[121].

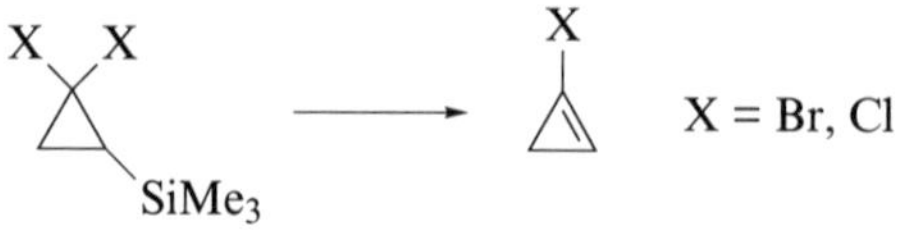

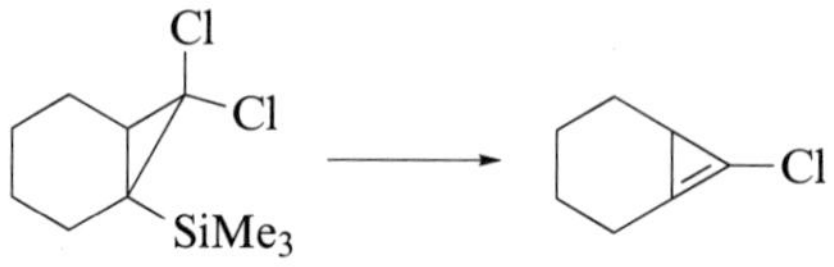

$$X = Br, Cl$$

The reaction is successful in the presence of a number of functional groups:

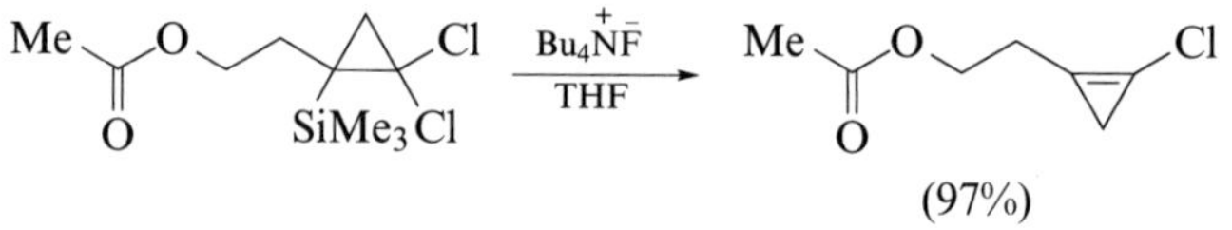

(97%)   (Ref. 122)

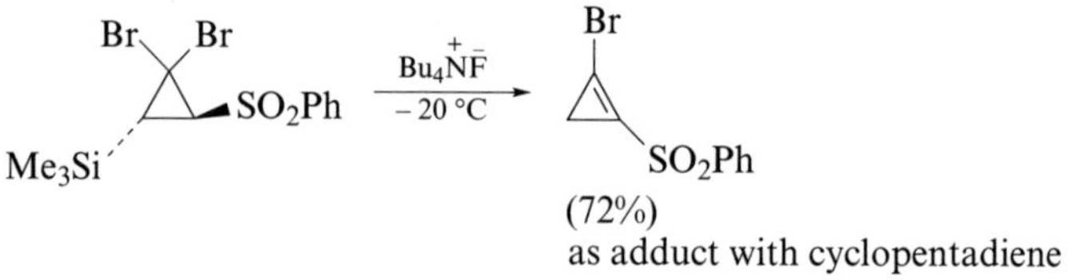

(72%)
as adduct with cyclopentadiene   (Ref. 123)

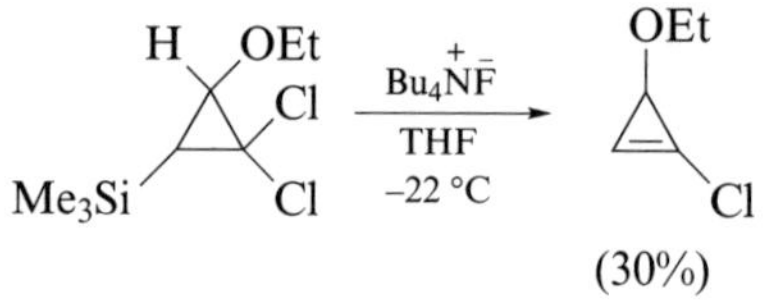

(30%)   (Ref. 79)

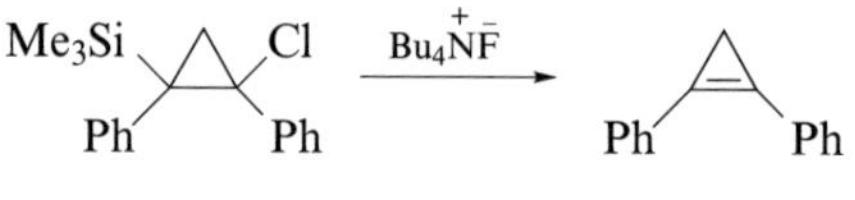

(Ref. 124)

The dehalosilylation of 1,1-dichloro-2-bromo-2-trimethylsilylcyclopropane leads to 1-bromo-2-chlorocyclopropene[125–127]:

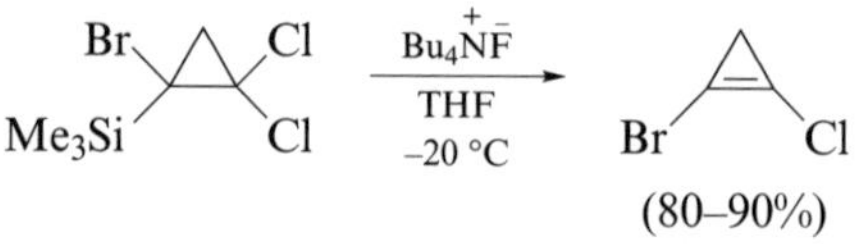

(80–90%)

In a similar reaction, 1,2-dibromocyclopropene may be prepared, characterized by NMR and trapped in high yield by [4 + 2]-cycloaddition to dienes (93% with DPIBF) while reaction of the tribromide with butyllithium leads to debromination[127]:

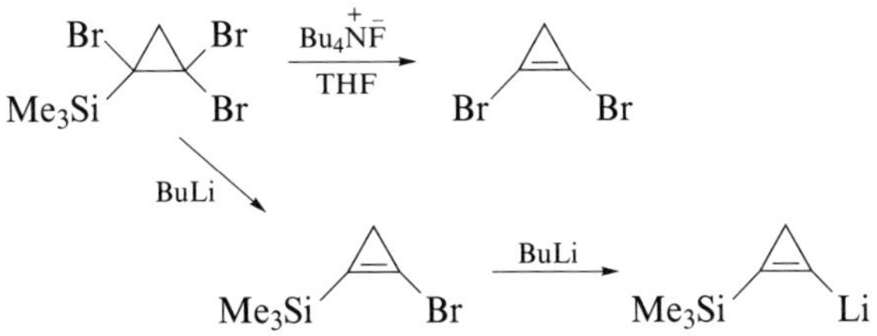

The 1,2-dihalides are readily trapped by bismethylenecycloalkanes to give products which undergo a double dehydrohalogenation to cyclopropa-aromatics, a reaction discussed in Section IV.

The dehalosilylation can be carried out in the gas phase over solid tetra-n-butylammonium fluoride deposited on glass helices, and even **59** can be transferred to a cold finger and trapped with cyclopentadiene, or as a remarkably stable ene-dimer[44,128].

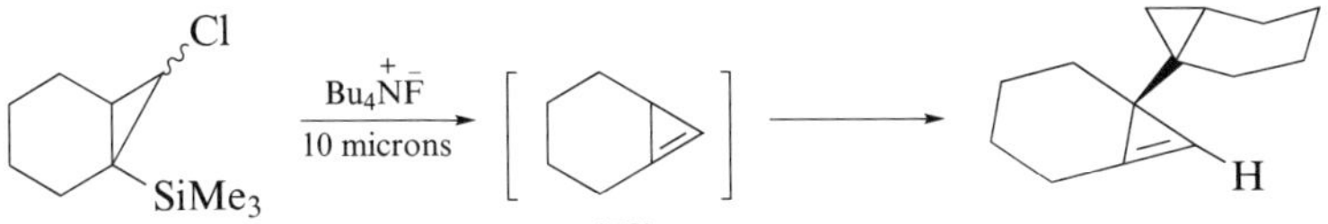

In the final case, both isomers of the monochloride apparently react, and the product cyclopropene is stable for several hours at ambient temperature. The reactions of the corresponding dichloro-analogues (**60**, n = 3–6) in solution with tetrabutylammonium fluoride in THF lead to cyclopropenes (**61**) which can either be isolated or trapped by dienes when n = 5 or 6. The bicyclohexene (**61**, n = 3) isomerises to a carbene (**62**); the same cyclopropene-to-carbene rearrangement has been observed when the cyclopropene is generated either by dehalogenation[87] or by dehydrohalogenation[86] of the corresponding 6,6-dichlorobicyclo[3.1.0]hexane. The bicycloheptene (**61**, n = 4) leads largely to **63**, apparently derived by trapping of the ring-opened carbene **64** by chloride ion, rather than products derived by trapping of the alternative vinylcarbene **65**[122]. This is in sharp contrast to the report that the cyclopropene rearranges in the gas phase to give 2-chlorocycloheptadiene[128], derived by a 1,2-hydrogen shift in **65**; the structure of the product of this reaction has, however, now been reassigned as **66** rather than the cycloheptadiene[122].

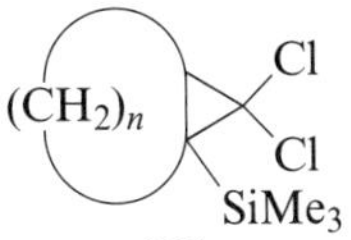

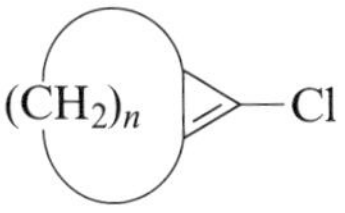

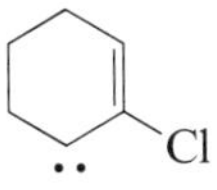

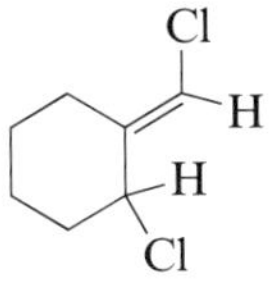

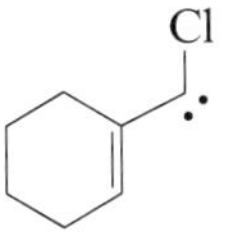

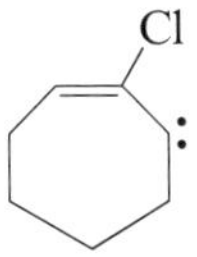

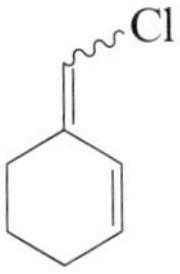

The gas-phase elimination has also been applied to the preparation of bicyclopropenes and of spiropentadiene, which may be distilled into a cold trap, characterized by NMR and trapped by dienes:

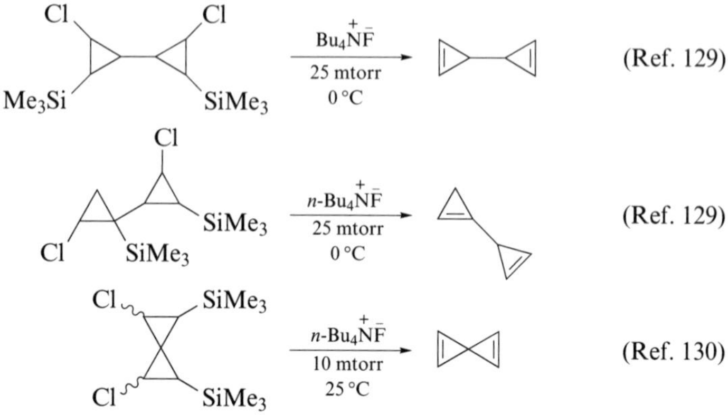

$$\text{(Ref. 129)}$$

$$\text{(Ref. 129)}$$

$$\text{(Ref. 130)}$$

The bicyclopropane **67** leads to a polymer on reaction with tetra-n-butylammonium fluoride at low pressure, although there is some evidence for the formation of bicyclopropenyl; under the same conditions **68** also leads to polymer, but on carrying out the reaction in solution in the presence of cyclopentadiene, the bicyclopropene can be trapped in reasonable yield (61%)[131]:

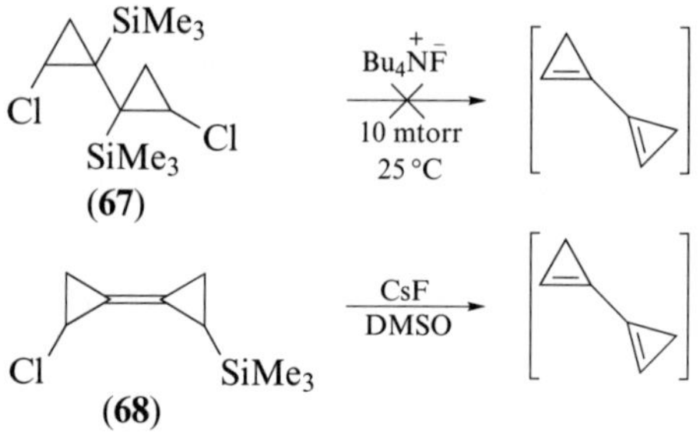

(67)

(68)

Gas-phase elimination is also of value in the preparation of simple cyclopropenes, particularly when these are needed free of solvent[42]:

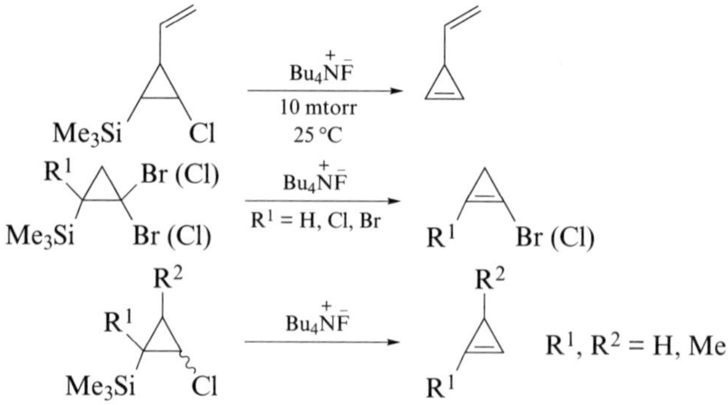

$R^1 = H, Cl, Br$

$R^1, R^2 = H, Me$

IV. ELIMINATIONS LEADING TO CYCLOPROPA-AROMATICS

The double sequence of dehydrohalogenation of a 7,7-dihalobicyclo[4.1.0]heptane derivative to give a bicyclo[4.1.0]hept-1(7)-ene, followed by base-induced prototropic shifts to
move the double bond into the six-membered ring[132–141], finds continuing applications in
routes to cycloproparenes[100,142]:

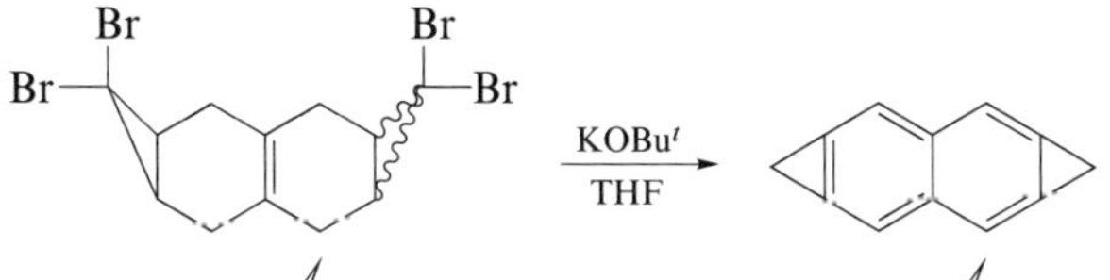

(Ref. 142)

(Ref. 100)

(49%)

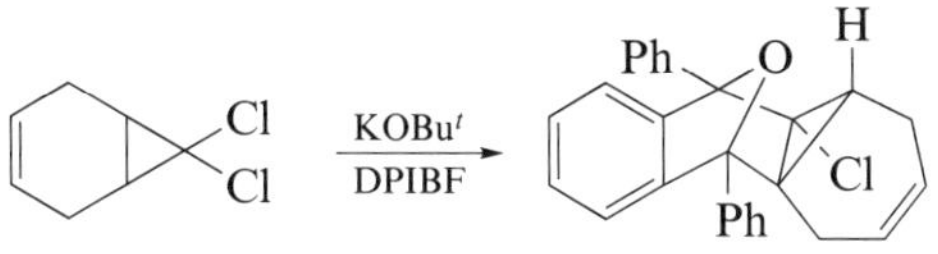

(Ref. 86)

(69)   (70)   (25%)

The mechanisms of these reactions have been discussed at some length. It is interesting to
note, however, that the reaction of 69 with KOBu-t in the presence of DPIBF has been
shown to lead to 70 in moderate yield, finally establishing its intermediacy in the well
known preparation of cyclopropabenzene itself[86]. The mechanisms are in some cases rather
more complex than might appear at first sight. Thus, the trichloride 71 does not lead to a
cyclopropabenzene labelled at entirely C-7, but rather to a mixture with ring-labelled product. Moreover, the corresponding dibromodichloro-compound 72 leads to both chloro-
and bromocyclopropabenzene. The rearranged products are explained in terms of the
removal of a proton at C-3 with the formation of a new bicyclic system, followed by the
loss of two molecules of halogen halide[141]:

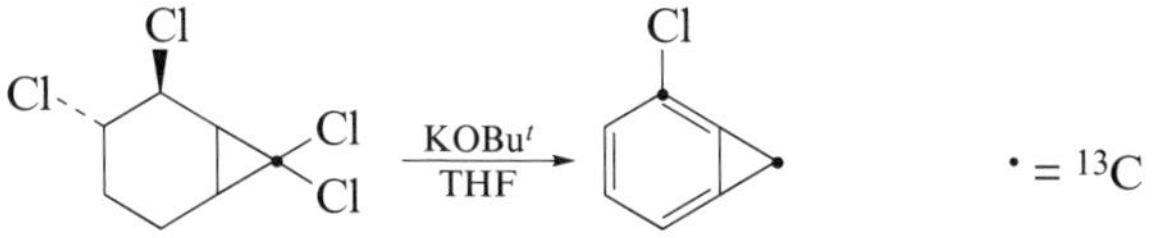

(71)

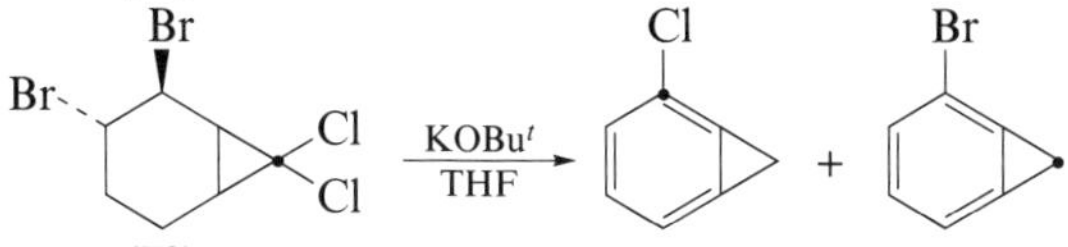

(72)

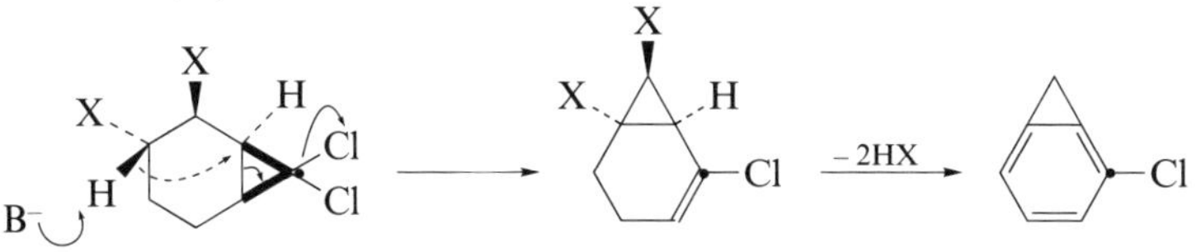

In contrast, the dehydrohalogenation of 3,4,7,7-tetrahalobicyclo[4.1.0]heptanes leads to unrearranged 3-halobicyclo[4.1.0]hepta-1,3,5-trienes[141].

The addition of a 1,2-dihalocyclopropene (see Section III) to a 1,2-bismethylenecycloalkane provides a simple route to 1,6-dihalobicyclo[4.1.0]heptanes. The elimination of two moles of HX from these has provided a very valuable extension to the use of the corresponding 7,7-dihalides[125,126,143–148]:

(Ref. 125)

(Ref. 143)

(Ref. 143)

(Ref. 144)

$n = 2$, 53%; $n = 3$, 55%; $n = 4$, 83%.

(Ref. 145)

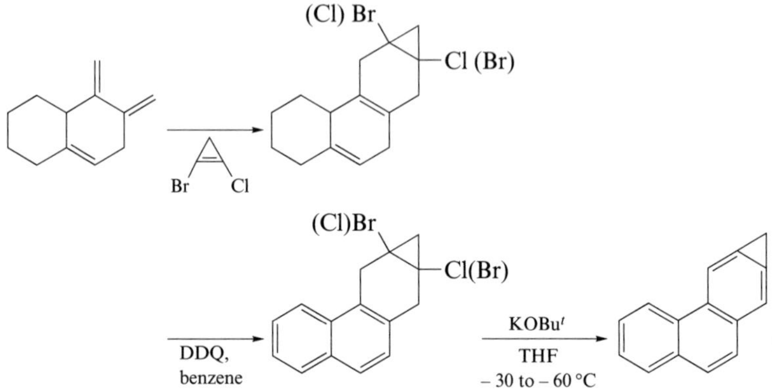

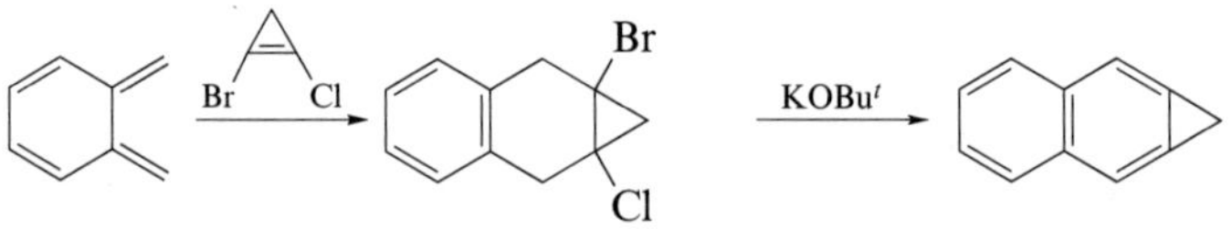

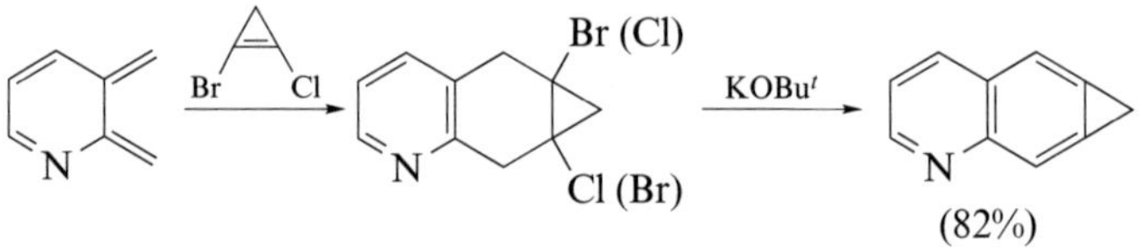

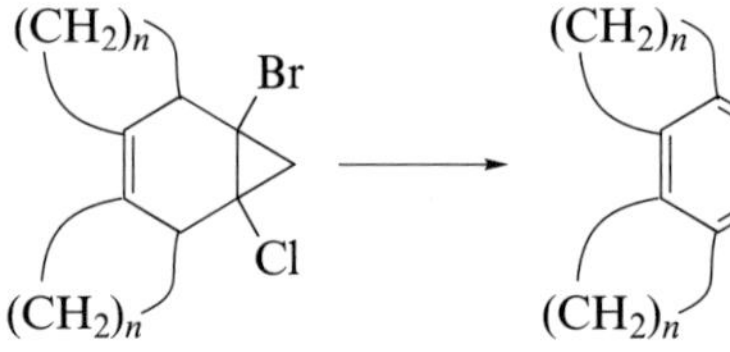

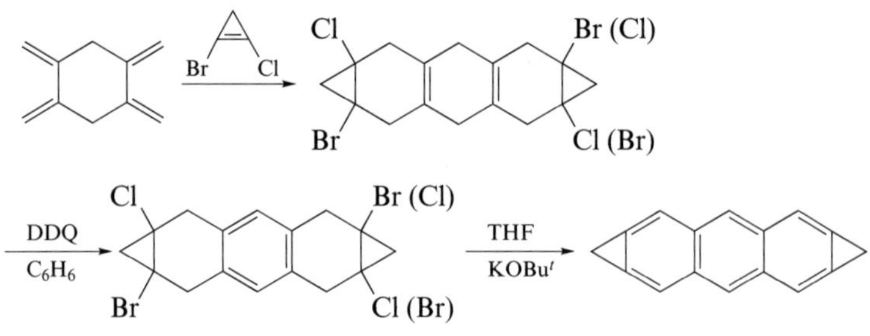

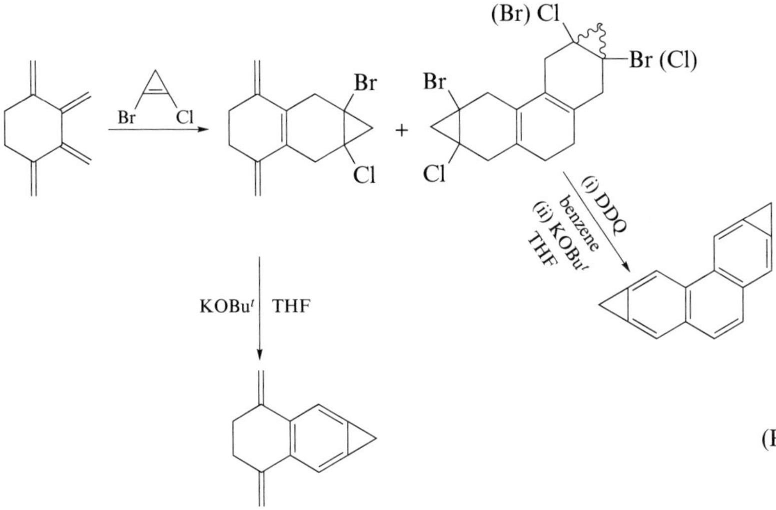

(Ref. 145)

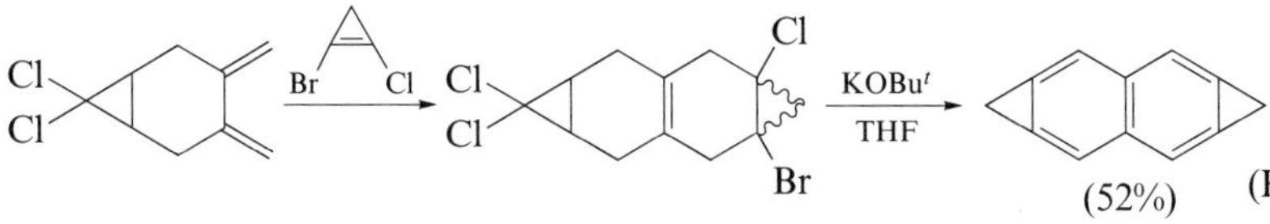

(57%)

(52%) (Ref. 145)

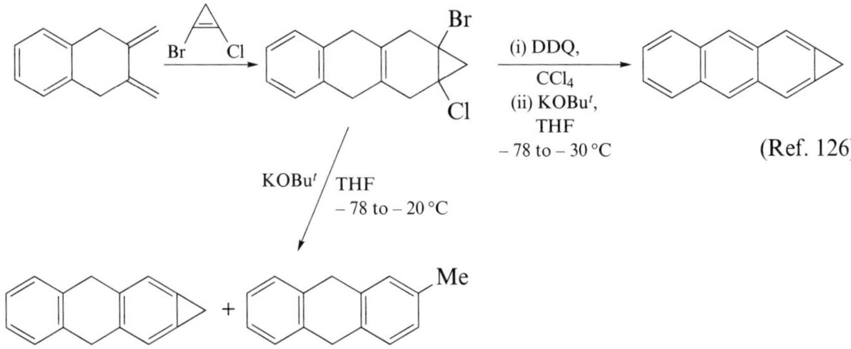

(Ref. 126)

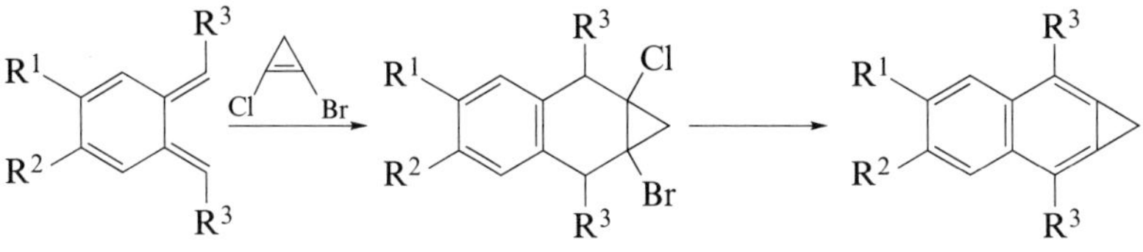

$R^1, R^3 = H$	47%
$R^1 = R^2 = Me, R^3 = H$	46%
$R^1 = Me, R^2, R^3 = H$	30%
$R^1, R^2 = H, \mathbf{R}^3 = Me$	19%

(Ref. 146)

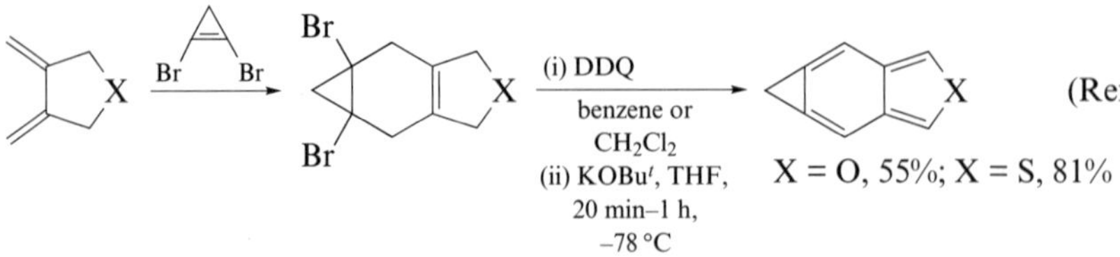

(Ref. 147)

X = O, 55%; X = S, 81%

The dibromide **73** dehydrohalogenates to give a cyclopropa(c)thiophene which is trapped in 57% yield by added isobenzofuran. A stepwise dehydrobromination-trapping has yet to be completely excluded and no products are obtained from attempted gas-phase dehydrobromination[90].

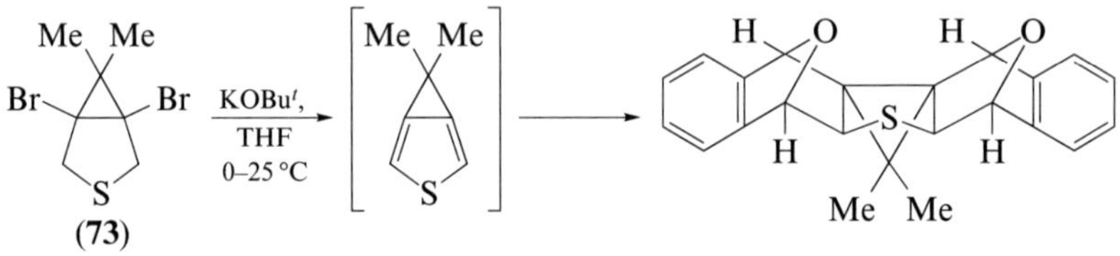

A variation of the above method, which avoids the need to prepare the 1,2-bismethyl-enecycloalkane, is to trap the 1,2-dihalocyclopropene by (4 + 2)-cycloaddition to a furan and then to effect a double debromoalkoxylation using low valent titanium[149,150]. A related debromotosylation is used in the preparation of **74**, although the yield is low [143]. Debromoalkoxylation is a potentially valuable but little used route to cyclopropenes, the parent compound being obtained from the reaction of 2-ethoxy-1-bromocyclopropane with zinc[151].

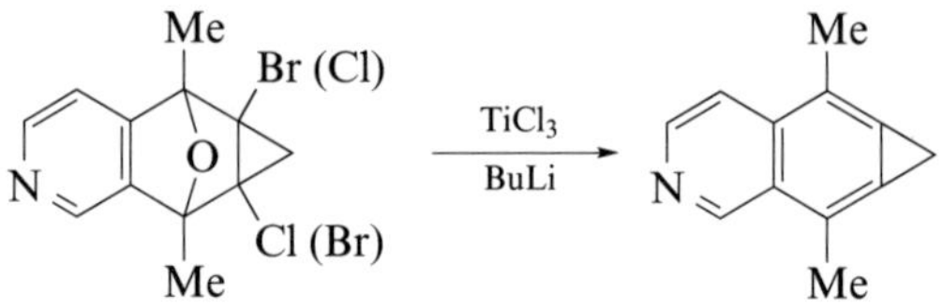

(Ref. 149)

(15%)

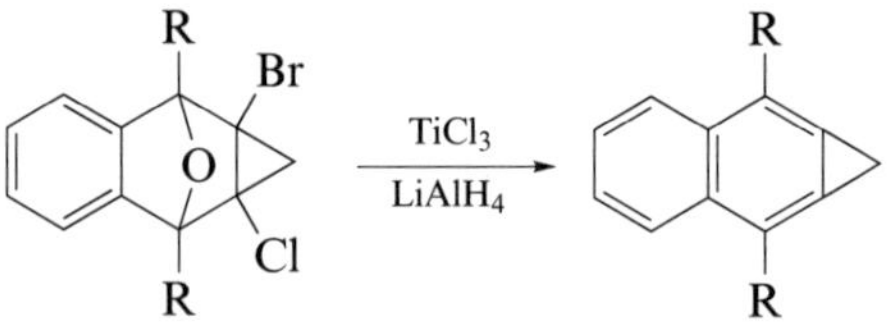

(Ref. 150)

R = H 60%, R = Ph 72%

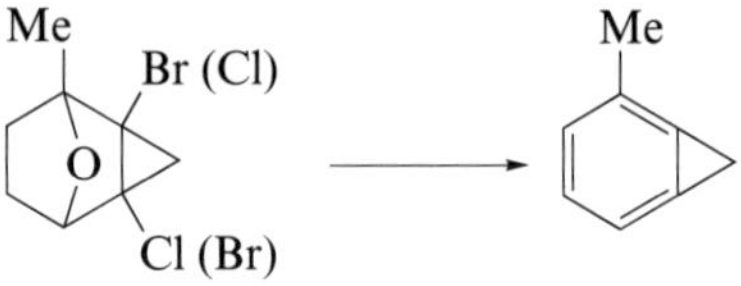

(Ref. 150)

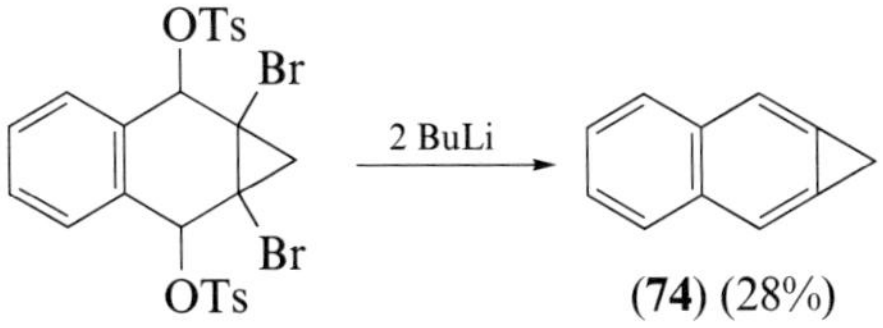

(Ref. 143)

The double dehydrohalogenation of tetrahalocyclopropene adducts of 1,2-bismethyl-enecycloalkanes leads to dihalocyclopropabenzenes:

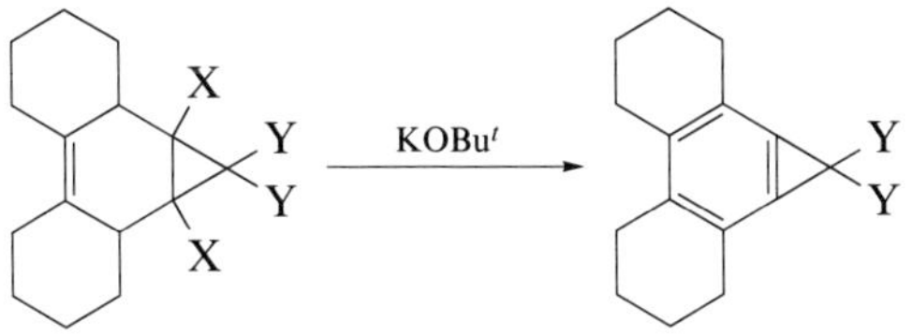

(Ref. 152)

X, Y = halogen

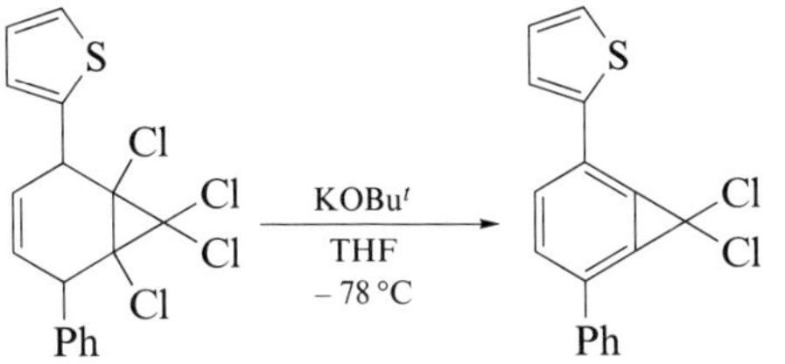

(28%) (Ref. 153)

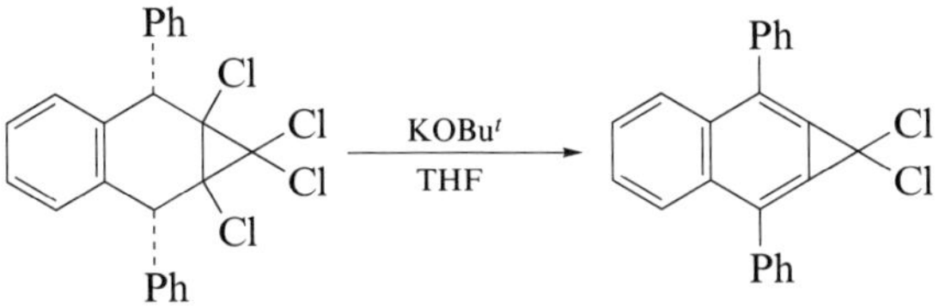

(Refs. 154 and 155)

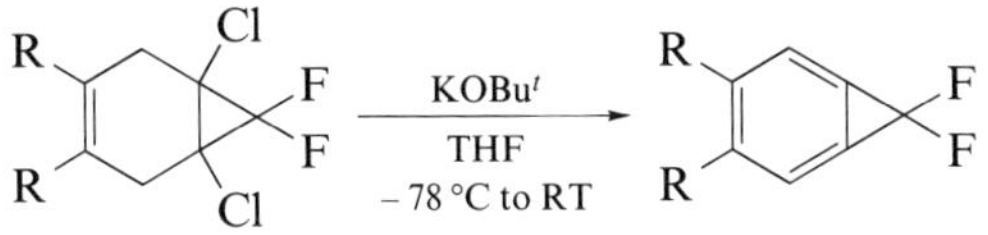

(Refs. 156 and 158)

R = H   Me   Ph   R,R = $-(CH_2)_4-$ $-(CH_2)_3-$ $-(CH_2)_2-$
%   60   78   58   65   86   60

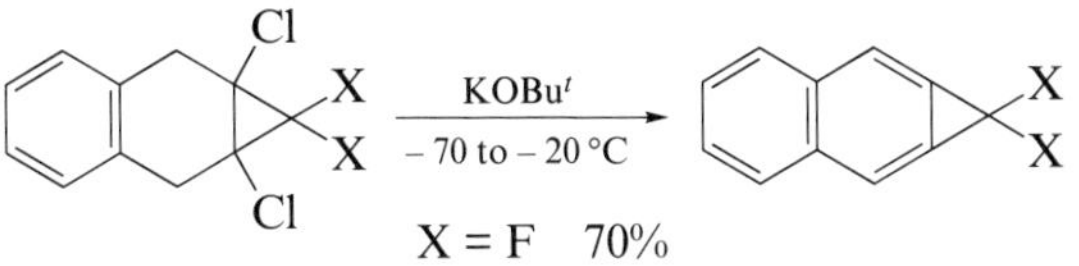

X = F   70%
X = Cl 66%

(Ref. 157)

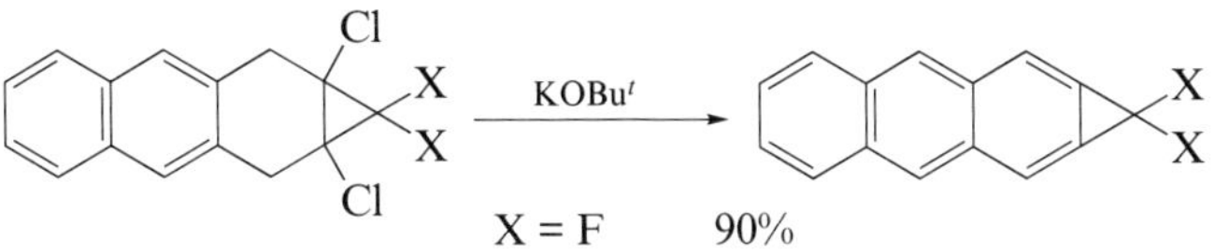

X = F   90%
X = Cl   ca. 66%

(Ref. 159)

V. 1,1- AND 1,3-ELIMINATION LEADING TO CYCLOPROPENES

The elimination of hydrogen chloride from an allylic halide to cyclopropenes and is of particular value in the preparation of the parent hydrocarbon and of simple alkyl derivatives. Although yields are often low the reagents are very readily avaiable:

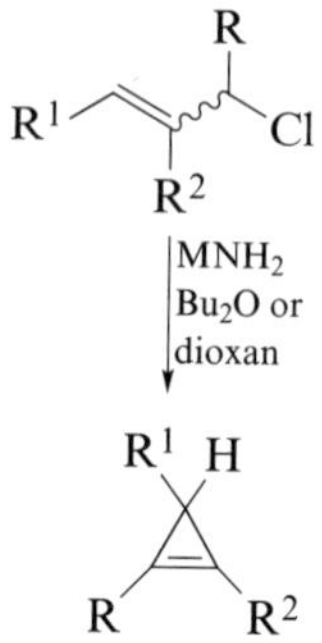

R	R¹	R²	M	%	Ref.
H	H	H	Na	10	160
H	H	Et	Na	95	161
Me	H	H	Li	30	162
H	H	Me	Li	42	162
H	Me	H	Li	35	163, 162

The choice of the metal is critical, as in other cases the initially formed cyclopropene rearranges to a methylenecyclopropane[164,161]:

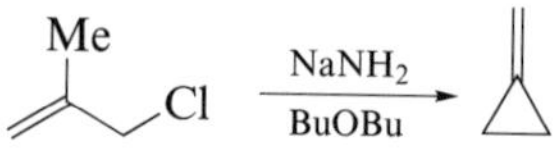

(Ref. 165)

(Ref. 166)

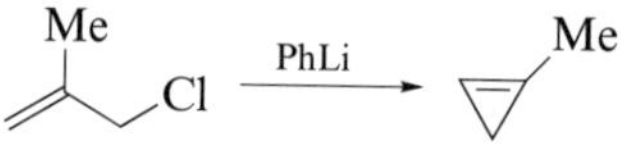

(trapped in 60–80% yield by dienes; Ref. 167)

The regioselectivity of the above reactions indicates that they involve 1,1- rather than 1,3-eliminations. Other examples show that both may occur[167,168,169]:

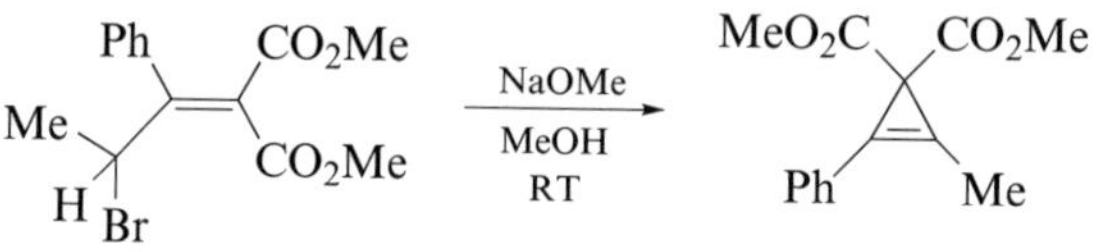

(19%)

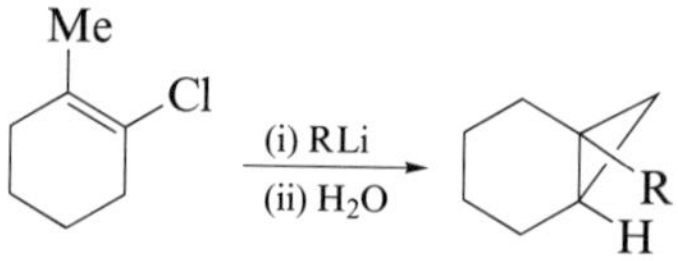

R	%
Ph	90
Bu	52
Me	47

(Labelling studies are consistent with a reaction occurring through an elimination to produce 2-methylenecyclohexylide which ring-closes to bicyclo[4.1.0]hept-1,6-ene, which undergoes addition of RLi; Ref. 168)

In a closely related reaction, 1,3-dihaloprop-1-enes undergo dehalogenation on reaction with butyllithium[170,171]:

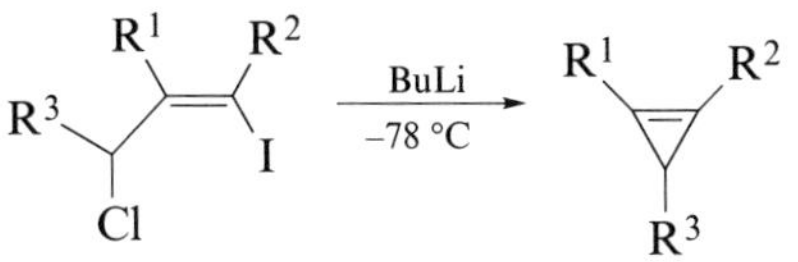

R¹	R²	R³	%
Allyl	SiMe₃	H	95
Allyl	SiMe₃	Me	78
n-Pr	SiMe₃	H	58
Ph	SiMe₃	H	90
Bu	Ph	H	83
Me	C₆H₁₃	H	84
Me	C₆H₁₃	Me	70
Allyl	C₆H₁₃	H	90
Ph	H	H	46

An apparent double 1,3-dehalogenation of 1,1,3,3-tetrahalopropanes also provides cyclopropenes in moderate to good yield[172-175]:

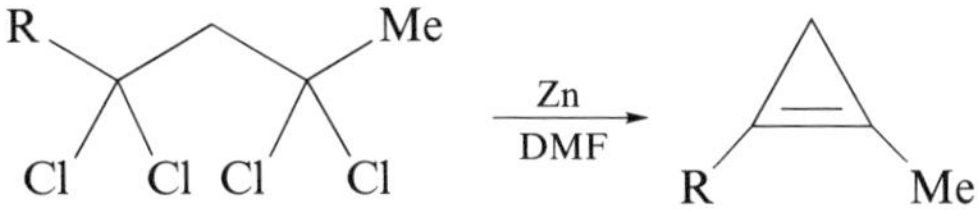

R	%
Me	73
Et	77
H	56
Cl	29
CH₂CH₂Cl	42

A novel variation of a double dehydrohalogenation leading to a cyclopropene is the reaction of ester **75** with base[176]:

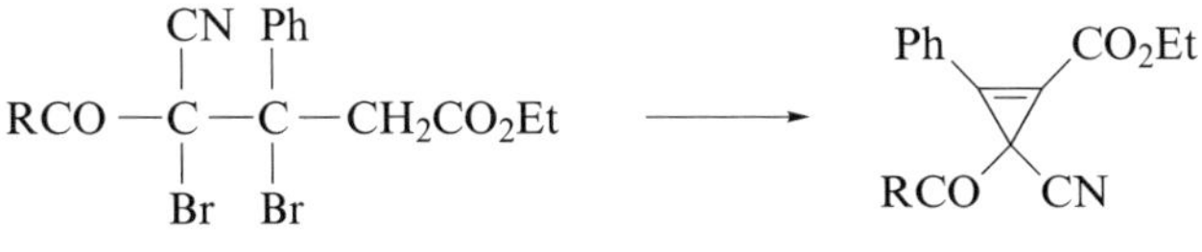

(**75**)

The sequence would appear to be initiated by proton removal adjacent to the ester, but whether this is followed first by loss of the 3-halogen and then a 1,2-dehydrohalogenation or first by loss of the 2-halogen and then a 1,3-elimination is not clear.

VI. 1,3-DEHALOGENATION OF 1-HALO-2-HALOMETHYLCYCLOPROPANES

The halomethylcyclopropanes **76–78** react with methyllithium by a 1,3-dehalogenation, apparently initiated by a lithium-halogen exchange of one of the geminal halogens[177]:

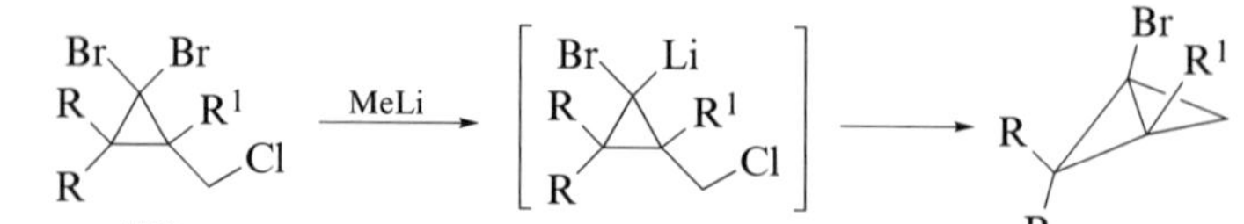

(Ref. 177)

R = H, R¹ = Me; R = Me, R¹ = H

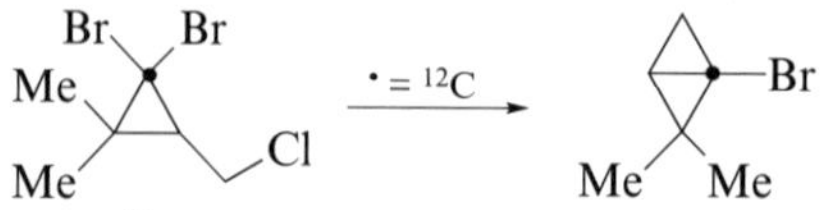

(Ref. 178)

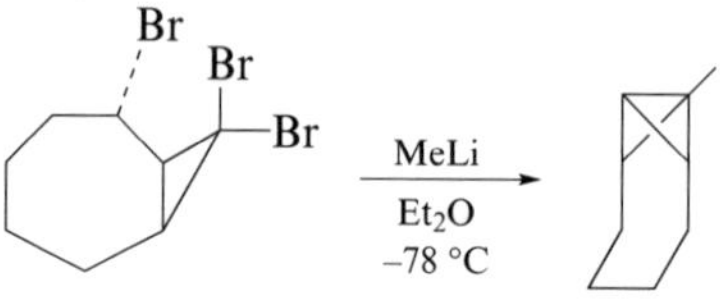

(58%)　　　　(Ref. 12 in Ref. 179)

(77)

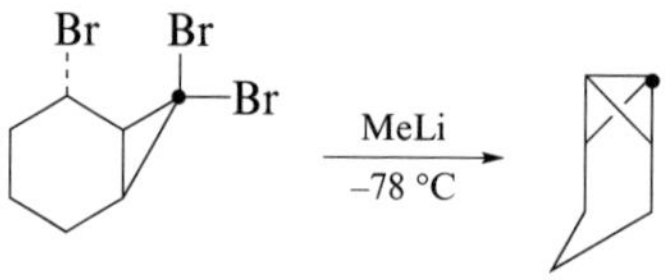

(42%)　　　　(Ref. 12 in Ref. 179)

(78)

• = ¹²C

The last reactions establish the stereochemistry of the eliminated bromide as being *anti* to the organolithium intermediate.

A similar elimination is thought to occur in the transformation of the pentahalide **79** (X = Br) into **80**, though it is interesting to note that the corresponding pentachloride **79** (X = Cl) reacts by 1,2-dehalogenation to a cyclopropene; the only major difference appears to be that the former reaction occurs at low temperature and the latter at room temperature due to the different rates of lithium–halogen exchange[114].

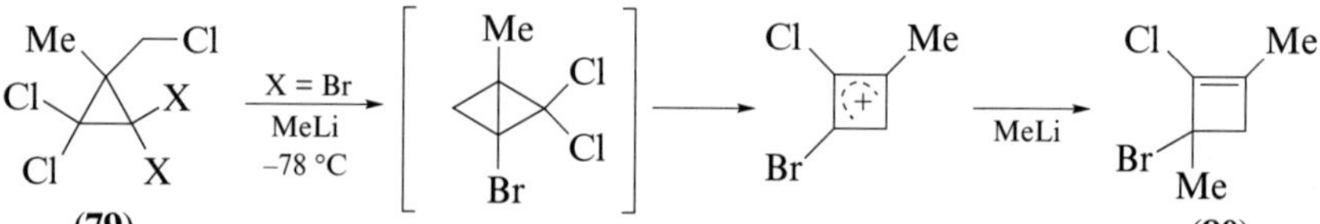

The di-iodides (**81**), formed from cyclopropenes by addition of iodine with rearrangement also eliminate halogen on reaction with methyllithium, providing a simple route to 1-vinyl-2-alkylbicyclo[1.1.0]butanes[180].

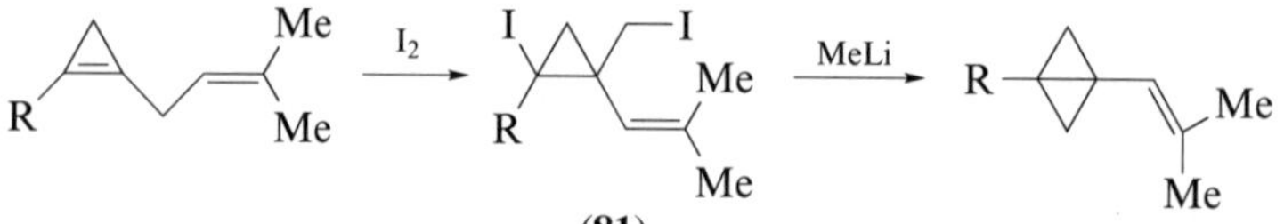

In a very elegant extension of the original 1,3-dehalogenation leading to bicyclobutanes, the parent [1.1.1]propellane is obtained as an ethereal solution in one pot from **82**[181,182]:

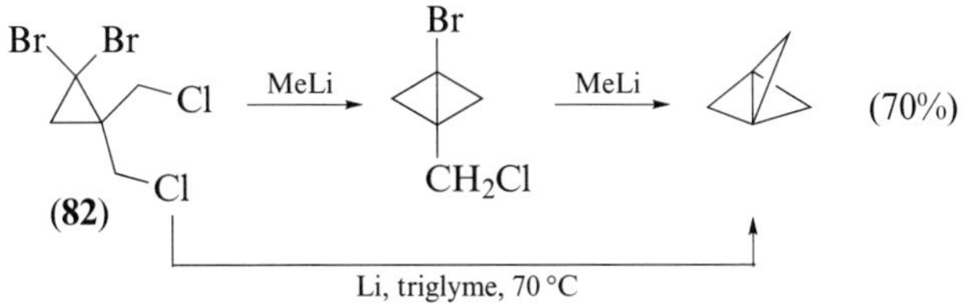

The use of lithium powder in *n*-decane/triglyme[181], or *n*-butyllithium/tetramethylenedi-amine[183], leads to the compound largely free of solvent in 25–38% (purity 70–75%) or 40%. The compound may be purified by addition of iodine in pentane under irradiation to give **83**, which can be deiodinated by NaCN–DMSO and distilled from the reaction mixture, apparently by nucleophilic attack at one of the iodines.[187] A similar 1,3-deiodination was reported to occur efficiently[184], while the corresponding 1,3-debromination of **84**, derived by a different route, was the basis of the first route to the propellane[185,186]:

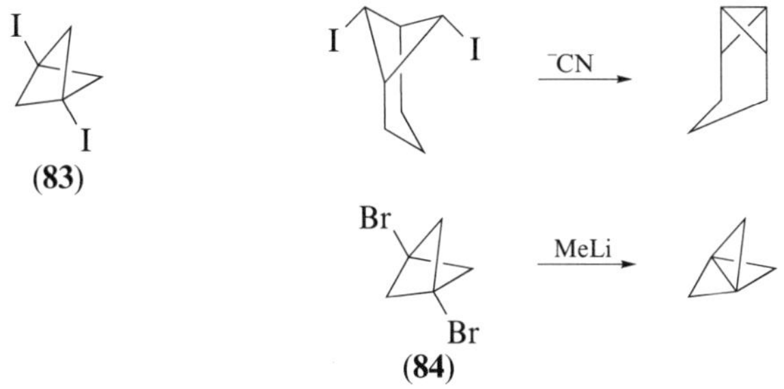

In a similar way the homologues may be cyclized[182]:

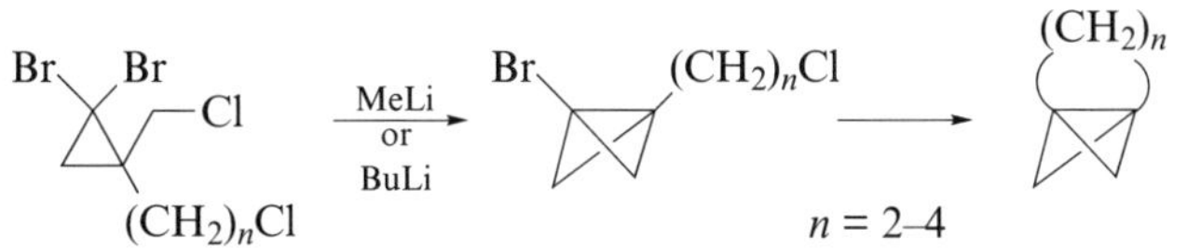

The [*n*.1.1]propellane (*n* = 2) could be isolated, but its formation was established on the basis of decomposition products.

It is interesting to note that the addition of a further ring to tricyclo[4.1.0.0²,⁷]heptanes may be achieved through 1,5- or 1,6-dehalogenations induced by reaction with butyl-lithium in ether at −40 °C, although in some cases the propellanes are not stable[188].

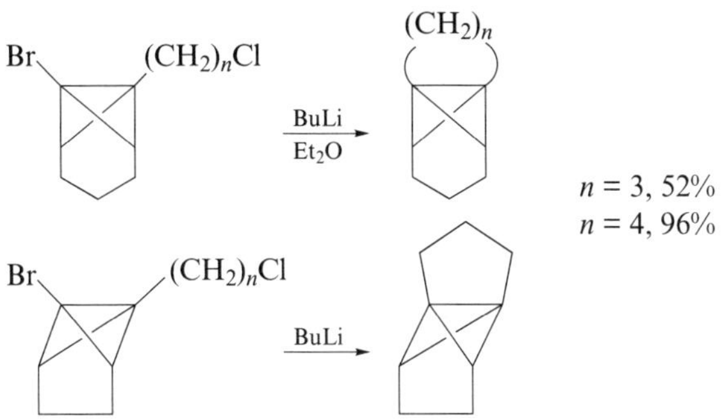

In the same way a 1,3-dehalogenation leads to [1.1.1]propellanes[189,190]:

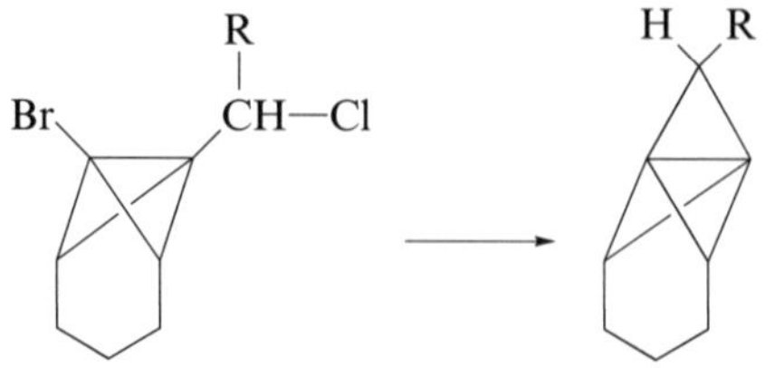

R = H   Me   Pr^i cyclohexyl
yield (%) = 71   68   60   61

VII. ELIMINATIONS LEADING TO BICYCLO[1.1.0]BUT-1(3)-ENES

Attempts to generate unbridged bicyclo[1.1.0]but-1(3)-enes led only to the trapping of ring-opened trienes and enynes derived from these, and labeling indicated that these were not derived through a symmetrical intermediate but by direct elimination of lithium bromide from **85** together with ring cleavage[191]. The labelling pattern is consistent with cleavage of bonds C_2—C_3 and C_1—C_4 rather than the opposite pair; the reason for this is not clear. An alternative mechanism for ring-opening involving single electron transfer from lithium diisopropylamide (LDA) to the bromide, followed by opening of a radical anion, is thought to be less likely because no evidence has been found for SET when similar halides are treated with lithium diisopropylamide.

$\bullet = {}^{12}C$

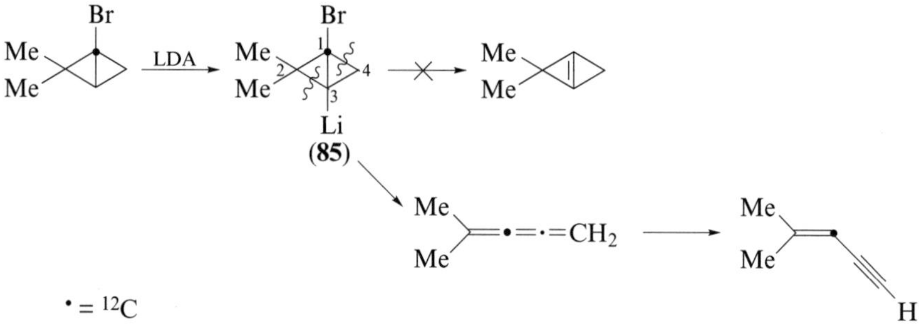

Highly strained bicyclo[1.1.0]but-1(3)-enes are obtained by dehydrohalogenation of bridged 1-halobicyclobutanes, and may be trapped in Diels-Alder reactions, although yields are variable[192–201]:

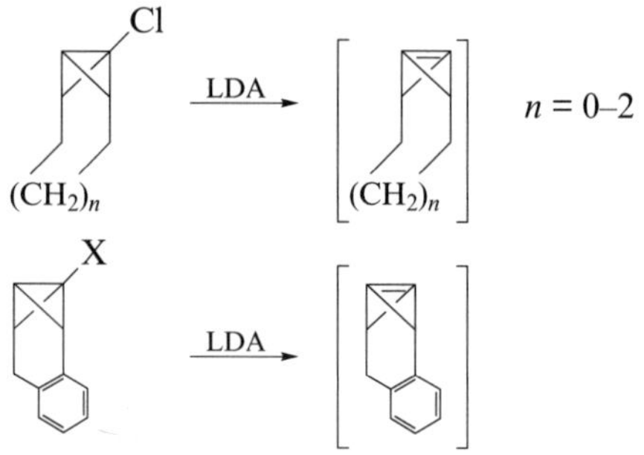

$n = 0$–2

The intermediacy of the bicyclobutene has been supported by a number of experiments including the trapping of ^{12}C-labelled compound by lithium thiophenolate. The labelled products are obtained in a 56:44 ratio; the reason the ratio in not exactly 1:1 may reflect the fact that the tricycles can adopt two alternative equilibrating chair forms[202].

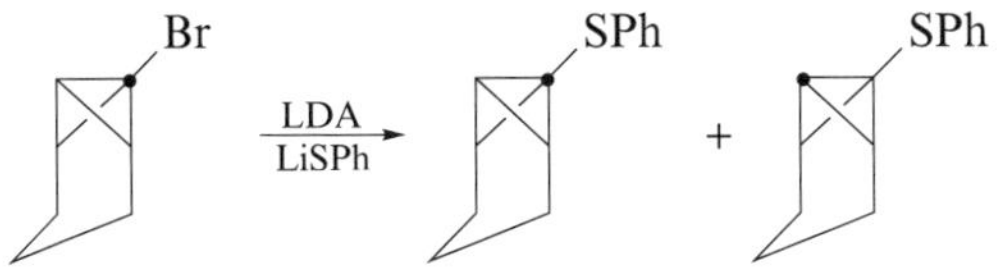

Debromosilylation has also been used in routes to bicyclo[1.1.0]but-1(3)-enes and once again trapped by dienes:

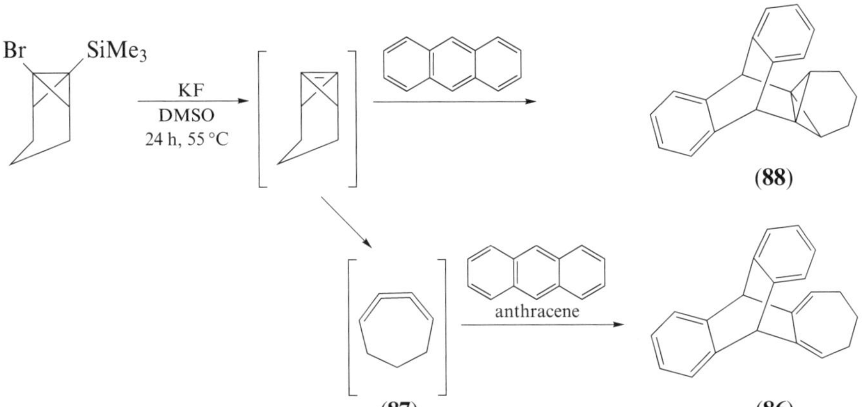

(88)

(87)

(86)

In the case of trapping with anthracene, a second product (**86**) is also isolated which apparently arises by trapping of the cycloheptatriene **87**; it does *not* arise from **88**[203]. At 80 °C the adducts of the triene with dienes is obtained exclusive of analogues **88** in 30–78% yield.

The reaction of the bromosilane with caesium fluoride in DMF at room temperature in the presence of (Ph₃P)₄Ni leads to the dimer **89**, apparently resulting from Ni(O) catalysed reaction of the cumulene **87**[204].

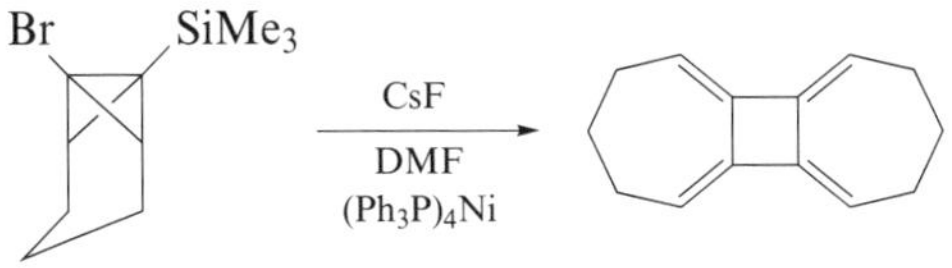

(89)

(32%)

VIII. 1,2- AND 1,3-ELIMINATIONS IN QUADRICYCLANES

The dehydrochlorination of **90** to 1,7-dehydroquadricyclane was established some time ago by trapping as a (4 + 2)-cycloadduct with anthracene is 35–45% yield, with some evidence for the alternative dehydrohalogenation to the 1(5)-isomer in reactions with butyl lithium[205-207].

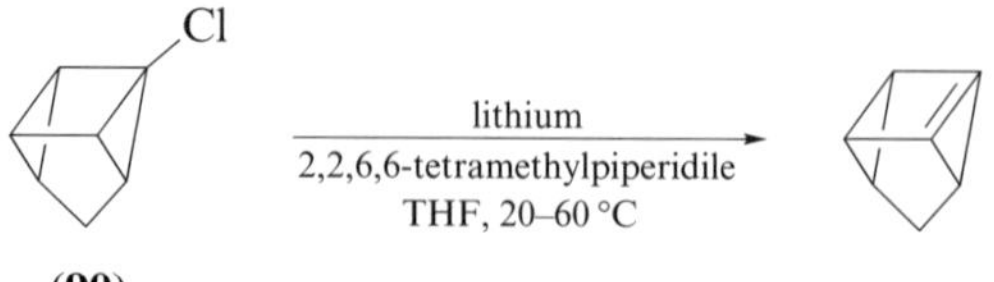

(90)

The isomeric 1,5-dehydroquadricyclane may be generated from the dihalides (**91**). The lithiation occurs at − 78 °C but elimination and trapping only on warming to room temperature; trapping leads to two stereoisomeric [4 + 2]-adducts with DPIBF in 40% yield[208]. No evidence was obtained for the intermediacy of norbornenyne in the reaction of 2-chlorobicyclo [2.2.1]hepta-2,5-diene with butyllithium in THF[208], despite the proof of the intermediacy of 2-norbornyne in related reactions[210].

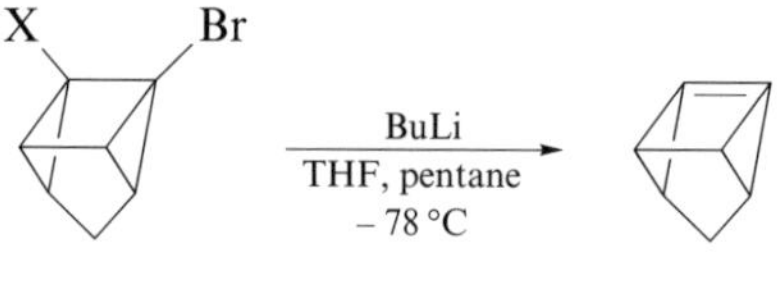

(91)

X = Br, Cl

Elimination of HCl from the quadricyclane (**92**) using either *t*-butyllithium or *n*-butyl lithium and KOBu-*t* in ether at 0 °C generates the highly strained dehydroquadricyclene as an intermediate which may by trapped by addition to DPIBF or trimethylisoindole[209].

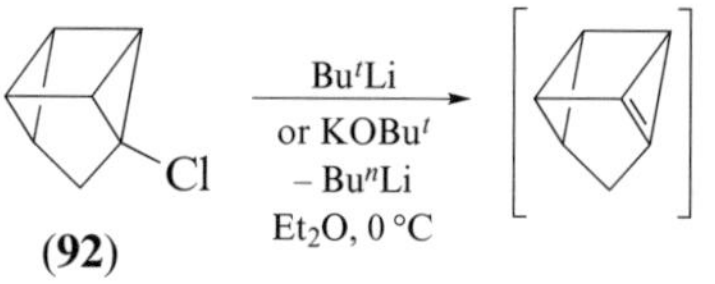

(92)

A series of dihalides (**93**) reacts with *t*-butyllithium at low temperature to give **94**. The reactions apparently proceed by lithium–halogen exchange to produce the [3.1.1]propellane derivative **95** which adds *t*-butyllithium[211].

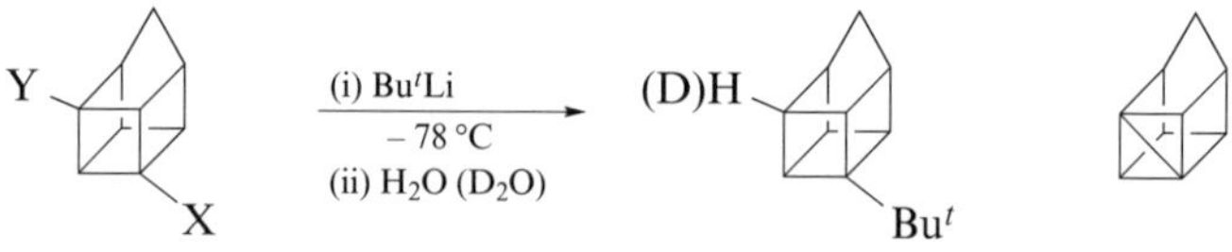

(93)   **(94)**   **(95)**

X, Y = Br, Cl, I

The isomeric dihalide **96** reacts in a similar manner with *t*-butyllithium in pentane-ether at 0 °C to give 1-*t*-butylhomocubane (39%), a result interpreted in terms of the intermediacy of 1(7)-homocubene[212]:

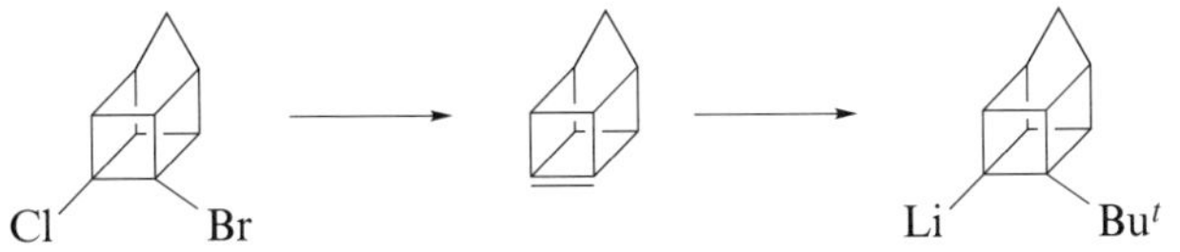

(96)

IX. DEHALOGENATION AND DEHYDROHALOGENATION OF HALOGENATED CYCLOPROPENES

The dehydrochlorination of 1,1-dichloro-2,2-dimethyl-cyclopropane has been reported to lead to 3-methylbut-3-en-1-yne and, in the presence of cyclohexane, a small amount of the cyclopropane **97** derived by trapping of 3-methyl-buta-1,2-dien-1-ylidene is isolated[76]:

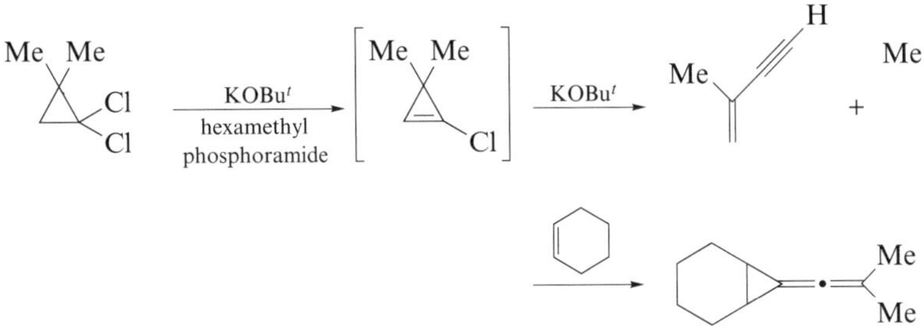

(97)

The alkyne is formally derived by a 1,4-dehydrochlorination with ring opening, although subsequent work has shown that the cyclopropene ring-opens even at ambient temperature to 3-chloro-3-methylbut-1-yne and so a 1,2-dehydrochlorination of this is an alternative. The carbene could be formed by several routes from the cyclopropene, and an early experiment with a ^{14}C-labelled compound indicated that the label at C-1 of the cyclopropane became C-1 of the carbene:

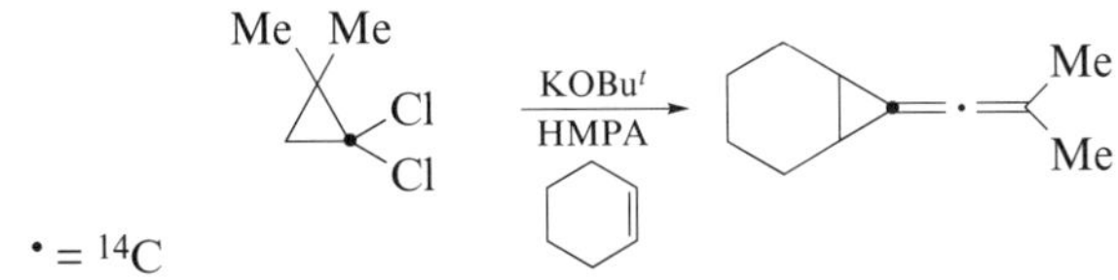

A later study in which the ^{12}C-labelled cyclopropene was treated with methyllithium led, however, to an alternative labelling pattern[84,213]:

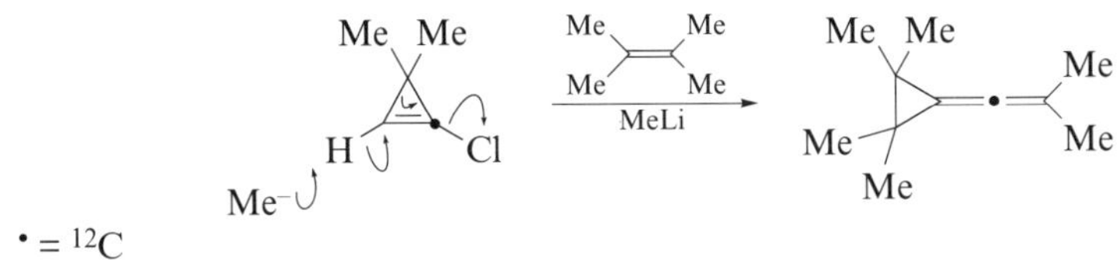

explained in terms of lithium–hydrogen exchange with loss of chloride as shown. The pinene derivative **(98)** undergoes an intramolecular reaction which leads to a similar labelling pattern[84,213]:

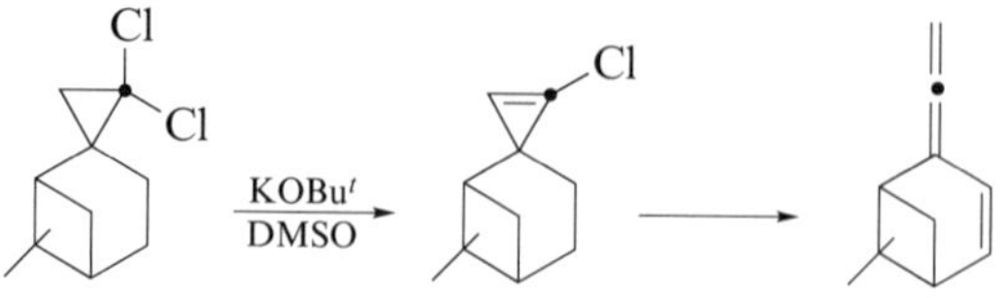

(98)

Whatever the detailed mechanism of these reactions, it is clear that a symmetrical 'cyclopropyne'-type intermediate is not involved.

In a synthetically useful variation of this reaction, allenic carbenes may also be generated from a range of 3,3-disubstituted dibromocyclopropanes by reaction with either aqueous or powdered sodium hydroxide and a phase transfer catalyst, and trapped in moderate to good yield by alkenes in a two-phase system, e.g.[214]:

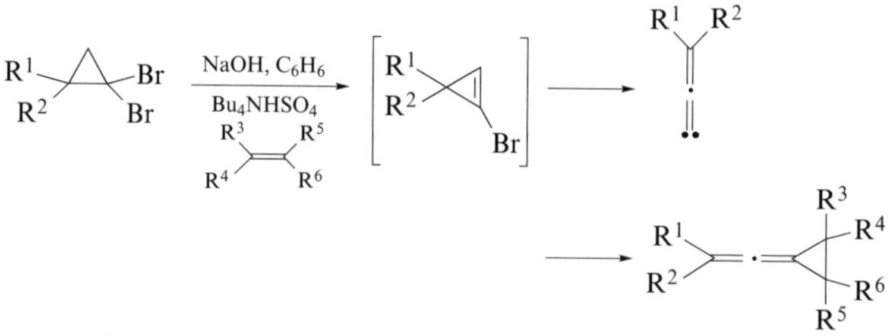

$R^1, R^2 =$ Ph or Me; $R^1 =$ naphthyl, $R^2 =$ Ph; $R^1 =$ Ph, $R^2 =$ Me

It is interesting to note that reaction of the difluorides **99** with 3 mol. equiv. of butyllithium leads to related ring openings[83]:

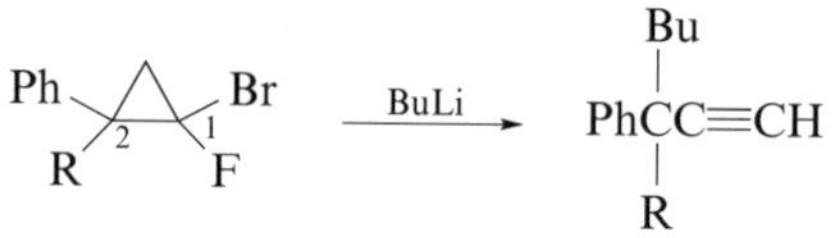

(99)

A labelling study once again rules out a cyclopropyne as an intermediate and supports a mechanism involving the cleavage of the C_2—C_3 bond, i.e. once again C-1 becomes C-2 of the product alkyne.

In a related process, tetrahalocyclopropanes react with 2 mol. equiv. of methyllithium in the presence of an alkyne to give adducts of allenic carbenes[113–115,215]:

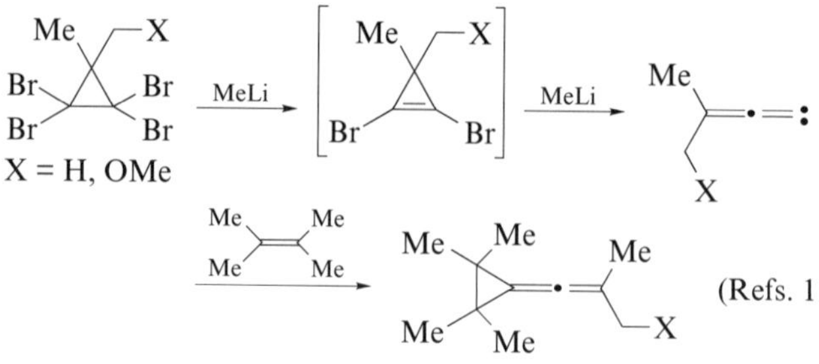

(Refs. 113 and 114)

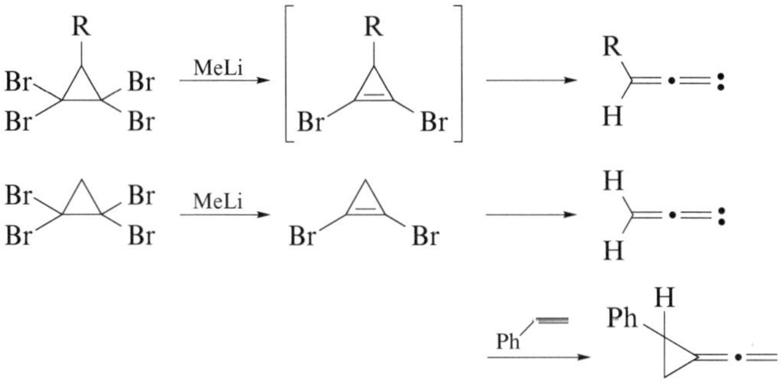

(Refs. 115)

(Refs. 215)

Similar reactions can be carried out using dialkylphosphite and sodium hydride in place of methyl lithium[261].

The advantage of this method is that the product vinylidenecyclopropanes carrying an olefinic hydrogen do not isomerize under the reaction conditions to alkynes. The final example represents the first solution trapping of the parent allenic carbene, an important component of interstellar space.

The mechanism of the elimination has once again been shown by labelling not to involve the formation of cyclopropyne, and is consistent with lithium–bromine exchange to give **100** followed by, or concurrent with, fragmentation[217]:

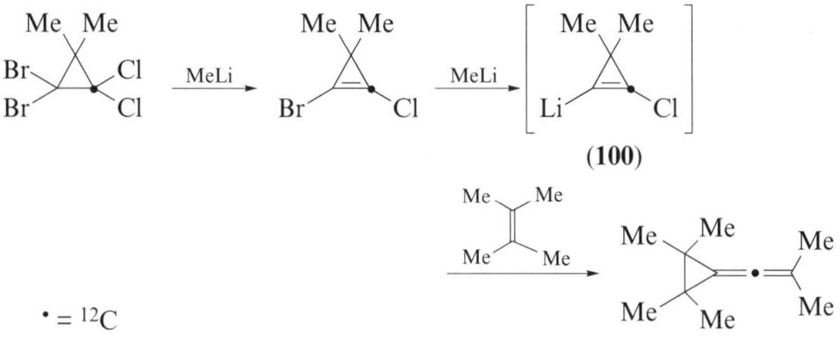

$\bullet = {}^{12}C$

X. REFERENCES

1. (a) J. R. Gandler, 'Mechanisms of base-catalysed alkene-forming 1,2-eliminations', in *The Chemistry of Double-bonded Functional Groups* (Ed. S. Patai), Chap.12, Wiley, Chichester 1989, p.733.
   (b) For a recent review of routes to cyclopropenes see B. Halton and M. G. Banwell, in *Cyclopropenes* in *The Chemistry of the Cyclopropyl Group* (Ed. Z. Rappoport), Vol. 2, Chap. 21, Wiley, Chichester, 1987, p. 1224.
2. P. Binger, *Synthesis.*, 190 (1974).
3. A. I. D'yachenko, S. A. Agre, T. Y. Rudashevskaya, R. N. Shafran and O. M. Nefedov, *Izv. Akad. Nauk SSSR, Ser. Khim.*, 2820 (1984).
4. N. V. Bovin, L. S. Surmina, N. I. Yakushkina and I. G. Bolesov, *Zh. Org. Chim.*, 9, 1888 (1973); *Chem. Abstr.*, 88, 17290n.
5. N. I. Yakushkina and I. G. Bolesov, *Zh. Org. Khim.*, 15, 954 (1979); *Chem. Abstr.*, 91, 123429h.
6. N. I. Yakushkina, G. R. Zhurina, L. S. Surmina, Y. K. Grishin, D. V. Bazhenov, V. V. Plemenkov and I. G. Bolesov, *Z. Obshch. Khim.*, 52, 1604 (1982); *Chem. Abstr.*, 97, 182567y.
7. A. L. Ivanov and I. N. Domnin, *Zh. Org. Khim.*, 24, 2547 (1988).
8. M. M. Latypova, V. V. Plemenkov, V. N. Kalinina, I. G. Bolesov, *Zh. Org. Khim.*, 20, 542 (1984).

9. M. Bertrand and H. Monti, *Compt. Rend. Acad. Sci. Paris.*, **264**, 998 (1967).
10. O. M. Nefedov, I. E. Dolgii, E. V. Bulusheva and A. Y. Shteinschneider, *Izv. Akad. Nauk SSSR, Ser. Khim.*, 1901 (1976); *Chem. Abs.*, **86**, 89247g.
11. O. M. Nefedov, I. E. Dolgii, E. V. Bulusheva and A. Y. Shteinschneider. *Izv. Akad. Nauk SSSR, Ser. Khim.*, 1535 (1979); *Chem. Abs.*, **91**, 174859.
12. O. M. Nefedov, I. E. Dolgii, I. B. Shvedova and E. A. Baidzhigitova, *Izv. Akad. Nauk SSSR, Ser. Khim.*, (1978) 1339; *Chem. Abs.*, **89**, 108235.
13. H. E. Zimmerman and D. J. Kreil, *J. Org. Chem.*, **47**, 2060 (1982).
14. G. C. Johnson and R. G. Bergman, *Tetrahedron Lett.*, 2093 (1979).
15. A. Riemann, R. W. Hoffmann, J. Spanget-Larsen and R. Gleiter, *Chem. Ber.*, **118**, 1000 (1985).
16. D. N. Butler and I. Gupta, *Can. J. Chem.*, **60**, 415 (1982).
17. D. N. Butler, I. Gupta, W. W. Ng and S. C. Nyburg, *J. Chem. Soc., Chem. Commun.*, 596 (1980).
18. I. J. Landheer, W. H. de Wolf and F. Bickelhaupt, *Tetrahedron Lett.*, 2813 (1974).
19. F. C. Peelan, G. G. A. Rietveld, I. J. Landheer, W. H. de Wolf and F. Bickelhaupt, *Tetrahedron Lett.*, 4187 (1975).
20. J. W. van Straten, I. J. Landheer, W. H. de Wolf and F. Bickelhaupt, *Tetrahedron Lett.*, 4499 (1975).
21. W. H. de Wolf, W. Stol, I. J. Landheer and F. Bickelhaupt, *Recl. Trav. Chim,. Pays-Bas*, **90**, 405 (1971).
22. A. Weber, G. Sabbioni, R. Galli, U. Stampfli and M. Neuenschwander, *Helv. Chim. Acta*, **71**, 2026 (1988).
23. W. E. Billups and L.-J. Lin, *Tetrahedron Lett.*, **24**, 1683 (1983).
24. A. Padwa and M. J. Pulwer, *J. Chem. Soc., Chem. Commun.*, 783 (1982). A. Padwa, M. J. Pulwer and R. J. Rosenthal, *J. Org. Chem.*, **49**, 856 (1984).
25. E. Schmitz, H. Sonnenschein and R. J. Kuban, *Tetrahedron Lett.*, **26**, 4911 (1985).
26. V. Sander and P. Weyerstahl, *Angew. Chem., Int. Ed. Engl.*, **15**, 244 (1976).
27. V. Sander and P. Weyerstahl, *Chem. Ber.*, **111**, 3879 (1978).
28. W. Norden, V. Sander and P. Weyerstahl, *Chem. Ber.*, **116**, 3097 (1983).
29. Z. Yoshida and H. Miyahara, *Chem. Lett.*, 335 (1972).
30. R. Breslow, P. Gal, H. W. Chang and L. J. Altman, *J. Am. Chem. Soc.*, **87**, 5139 (1965).
31. R. Breslow, D. A. Cortes, B. Jaun and R. D. Mitchell, *Tetrahedron Lett.*, **23**, 795 (1982).
32. A. Padwa and M. J. Pulwer, *Org. Synth.*, **60**, 53 (1981).
33. A. Padwa, C. S. Chou, R. J. Rosenthal and L. W. Terry, *J. Org. Chem.*, **53**, 4193 (1988).
34. R. H. Parker and W. M. Jones, *Tetrahedron Lett.*, **25**, 1245 (1984).
35. W. E. Billups, M. M. Haley and G. A. Lee, *Chem. Rev.*, **89**, 1147 (1989).
36. W. E. Billups, L. E. Reed, E. W. Casserly and L. P. Lin, *J. Org. Chem.*, **46**, 1326 (1981).
37. B. Halton and D. L. Officer, *Tetrahedron Lett.*, **22**, 3687 (1981).
38. P. Mueller and N. Pautex, *Helv. Chim. Acta*, **71**, 1630 (1988).
39. L. Hulskamper and P. Weyerstahl, *Chem. Ber.*, **114**, 746 (1981); G. Frenking, L. Hulskamper and P. Weyerstahl, *Chem. Ber.*, **115**, 2826 (1982).
40. M. A. Hashem and P. Weyerstahl, *Tetrahedron*, **40**, 2003 (1984).
41. L. Hulskamper and P. Weyerstahl, *Chem. Ber.*, **117**, 3497 (1984).
42. The use of solid supported reagents in elimination reactions has been reviewed: W. E. Billups and D. J. McCord, *Angew. Chem., Int. Ed. Engl.*, **33**, 1332 (1994).
43. J. M. Denis, R. Niamayoua, M. Vata and A. Lablache-Combier, *Tetrahedron Lett.*, **21**, 515 (1980).
44. W. E. Billups and L.-J. Lin, *Tetrahedron*, **42**, 1575 (1986).
45. S. W. Staley and T. D. Norden, *J. Am. Chem. Soc.*, **106**, 3699 (1984).
46. W. E. Billups, L.-J. Lin and E. W. Casserley, *J. Am. Chem. Soc.*, **106**, 3698 (1984).
47. H. M. Walborsky and E. J. Powers, *Isr. J. Chem.*, **21**, 210 (1981).
48. R. Breslow, J. Pecoraro and T. Sugimoto, *Org. Synth.*, **57**, 41 (1977); R. Breslow, J. Pecoraro and T. Sugimoto, *Org. Synth. Coll. Vol. VI*, 361 (1988).
49. D. Boger, C. E. Brotherton and G. I. Georg, *Org. Synth.*, **65**, 32 (1987).
50. M. Isaka, S. Ando, Y. Morinaka and E. Nakamura, *Tetrahedron Lett.*, **32**, 1339 (1991).
51. K. B. Baucom and G. B. Butler, *J. Org.Chem.*, **37**, 1730 (1972).
52. R. Breslow, J. Pecorano and T. Sugimoto, *Org. Synth.*, **57**, 41 (1977).
53. K.-O. Henseling and P. Weyerstahl, *Chem. Ber.*, **108**, 2803 (1975).
54. K.-O. Henseling, D. Quast and P. Weyerstahl, *Chem. Ber.*, **110**, 1027 (1977).

55. W. E. Billups and A. J. Blakeney, *J. Am. Chem. Soc.*, **98**, 7817 (1976).
56. J. Sepiol and R. L. Soulen, *J. Org. Chem.*, **40**, 3791 (1975).
57. S. W. Tobey and R. West, *J. Am. Chem. Soc.*, **88**, 2481 (1966); **86**, 1459 (1964).
58. S. W. Tobey and R. West, *Tetrahedron Lett.*, 1179 (1963).
59. P. B. Sargeant and C. G. Krespan, *J. Am. Chem. Soc.*, **91**, 415 (1969).
60. C. Raulet, *Compt. Rend. Acad. Sci. Paris, Ser. C*, **287**, 337 (1978).
61. J. Arct, B. Migaj and J. Zych, *Bull. Acad. Sci. Pol.*, **25**, 697 (1977); P. Weyerstahl, G. Blume and C. Muller, *Tetrahedron Lett.*, 3869 (1971); V. D. Novokreshchennykh, S. S. Molchanov and Y. S. Shabarov, *Zh. Org. Khim.*, **14**, 546 (1978); **15**, 485 (1979); J. Arct, B. Migaj and A. Leonczynski, *Tetrahedron*, **37**, 3689 (1981).
62. I. Crossland, *Acta Chem. Scand.*, **B41**, 310 (1987).
63. T. C. Shields and P. D. Gardner, *J. Am. Chem. Soc.*, **89**, 5425 (1967).
64. W. Billups and L. E. Reed, *Tetrahedron Lett.*, 2239 (1977).
65. W. Billups, B. A. Baker, W. Y. Chow, K. Leavell and E. Lewis, *J. Org. Chem.*, **40**, 1702 (1975).
66. W. Billups, W. Chow and J. H. Cross, *J. Chem. Soc. Commun.*, 252 (1974).
67. W. Billups, T. C. Shields, W. Chow and N. C. Deno, *J. Org. Chem.*, **37**, 3676 (1972).
68. W. E. Billups, W. Y. Chow, K. H. Leavell and E. S. Lewis, *J. Org. Chem.*, **39**, 274 (1974).
69. W. E. Billups, J. D. Buynak and D. Butler. *J. Org. Chem.*, **44**, 4218 (1979); W. E. Billups, J. D. Buynak and D. Butler, *J. Org. Chem.*, **45**, 4636 (1980).
70. T. C. Shields, B. A. Shoulders, J. F. Krause, C. L. Osborn and P. D. Gardner, *J. Am. Chem. Soc.*, **87**, 3026 (1965).
71. E. V. Dehmlow and G. Hofle, *Chem. Ber.*, **107**, 2760 (1974).
72. A. R. Browne, B. Halton and C. W. Spangler, *Tetrahedron*, **30**, 3289 (1974).
73. J. Arct, B. Migaj and A. Leonczynski, *Tetrahedron*, **37**, 3689 (1981).
74. A. V. Tarakanova, Y. K. Grishin, A. G. Vashakidze, E. M. Mil'vitskaya and A. F. Plate, *Zh. Org. Khim.*, **8**, 1619 (1972).
75. W. E. Billups, A. J. Blakeney, N. A.Rao and J. D. Buynak, *Tetrahedron*, **37**, 3215 (1981).
76. L. Crombie, P. J. Griffiths and B. J. Walker. *J. Chem. Soc., Chem. Commun.*, 1206 (1969).
77. T. C. Shields, B. A. Loving and P. D.Gardner, *J. Chem. Soc., Chem. Commun.*, 556 (1967) T. C. Shields and W. E. Billups, *Chem. Ind.*, 1967, 1999; T. C. Shields, W. E. Billups and A. R. Lepley, *J. Am. Chem. Soc.*, **90**, 4749 (1968).
78. L. S. Surmina, A. A. Formanovskii and I. G. Bolesov, *Zh. Org. Khim.*, **14**, 883 (1978).
79. P. Mueller and N. Pautex, *Helv. Chim. Acta*, **74**, 55 (1991).
80. U. Staempfli and M. Neuenschwander, *Chimia*, **42**, 379 (1988).
81. I. G. Tishchenko, O. G. Kulinkovich and N. V. Masalov, *Synthesis*, 268 (1982).
82. S. Jamal, I. Ahmad, J. Iqbal and M. Ahmad, *J. Chem. Res.*, S301 (1983); M2638.
83. M. Suda, *Tetrahedron Lett.*, **21**, 4355 (1980).
84. M. S. Baird, S. R. Buxton and J. S. Whitley, *Tetrahedron Lett.*, **25**, 1509 (1984).
85. W. E. Billups, L. P. Lin and W. Y. Chow, *J. Am. Chem. Soc.*, **96**, 4026 (1973).
86. B. Halton, M. D. Diggins and A. J. Kay, *J. Org. Chem.*, **57**, 4080 (1992).
87. M. S. Baird and W. Nethercott, *Tetrahedron Lett.*, **24**, 605 (1983).
88. B. Halton and E. G. Lovett, *Struct. Chem.*, **2**, 147 (1990).
89. B. Halton, J. H. Bridle and E. G. Lovett, *Tetrahedron Lett.*, **31**, 1313 (1990).
90. I. J. Anthony, Y. B. Kang and D. Wege, *Tetrahedron Lett.*, **31**, 1315 (1990).
91. J. W. van Straten, W. H. de Wolf and F. Bickelhaupt, *Tetrahedron Lett.*, 4667 (1977).
92. L. W. Jenneskens, F. J. J. De Kanter. L. A. M. Turkenburg, H. J. R. De Boer, W. H. de Wolf and F. Bickelhaupt, *Tetrahedron*, **40**, 4401 (1984).
93. P. A. Kraakman, J.-M. Valk, H. A. G. Niederlander, D. B. E. Brouwer, F. M. Bickelhaupt, W. H. de Wolf, F. Bickelhaupt and C. H. Stam, *J. Am. Chem. Soc.*, **112**, 6638 (1990).
94. L. A. M. Turkenburg, P. M. L. Blok, W. H. de Wolf and F. Bickelhaupt, *Tetrahedron Lett.*, **22**, 3317 (1981).
95. J. W. van Straten, W. H. de Wolf and F. Bickelhaupt, *Tetrahedron Lett.*, 4667 (1977).
96. F. Bickelhaupt and W. H. de Wolf, *Recl. Trav. Chim. Pays-Bas*, **107**, 459 (1988).
97. L. A. M. Turkenberg, J. W. van Straten, W. H. de Wolf and F. Bickelhaupt, *J. Am. Chem. Soc.*, **102**, 3256 (1980).
98. F. Bickelhaupt, *Pure Appl. Chem.*, **62**, 373 (1990).
99. M. G. Banwell and C. Papamihail, *J. Chem. Soc., Chem. Commun.*, 1182 (1981).
100. (a) B. Halton and S. G. G. Russell, *Aust. J. Chem.*, **45**, 911 (1992).

(b) M. G. Banwell, B. Halton, T. W. Hambley, N. K. Ireland, C. Papamihail, S. G. G. Russell and M. R. Snow, *J. Chem. Soc., Perkin Trans. 1*, 715 (1992).
101. G. W. Wijsman, D. S. van Es, W. H. de Wolf and F. Bickelhaupt, *Angew. Chem., Int. Ed. Engl.* **32**, 726 (1993).
102. F. Bickelhaupt, *Pure Appl. Chem.*, **62**, 373 (1990); G. B. M. Kostermans, P. van Dansik, W. H. de Wolf and F. Bickelhaupt, *J. Am. Chem. Soc.*, **109**, 7887 (1987).
103. K. B. Wiberg and G. Bonneville, *Tetrahedron Lett.*, **23**, 5385 (1992).
104. K. B. Wiberg, D. R. Artis and G. Bonneville, *J. Am. Chem. Soc.*, **113**, 7969 (1991).
105. P. J. Chenier and D. A. Southard, *J. Org. Chem.*, **54**, 3519 (1989).
106. K. B. Wiberg and D. R. Artis, *Strain and Its Implications in Organic Chemistry*, NATO ASI Series (Eds. A. de Meirese and S. Blechert) Kluwer, 1989, Vol. 273, p. 349.
107. D. A. Rosie and G. G. Shone, *Lipids*, **6**, 623 (1971).
108. M. S. Baird and B. Grehan, *J. Chem. Soc., Perkin Trans.1*, 1547 (1993).
109. S. Hartmann, D. E. Minnikin, H,-J. Romming, M. S. Baird, C. Ratledge and P. R. Wheeler, *Chem. Phys. Lipids*, **71**, 99 (1994).
110. A. R. Al Dulayymi and M. S. Baird, *J. Chem. Soc., Perkin Trans 1*, 1547 (1994).
111. M. S. Bairds, H. H. Hussain and W. Nethercott, *J. Chem. Soc., Perkin Trans. 1*, 1845 (1986).
112. M. S. Baird and H. H. Hussain, *Tetrahedron*, **45**, 6221 (1989).
113. J. R. Al-Dulayymi and M. S. Baird, *Tetrahedron*, **45**, 7601 (1989).
114. J. R. Al-Dulayymi and M. S. Baird, *Tetrahedron Lett.*, **29**, 6147 (1988).
115. M. S. Baird, H. L. Fitton, W. Clegg and A. McCamely, *J. Chem. Soc., Perkin Trans. 1*, 321 (1993).
116. J. R. Al Dulayymi, M. S. Baird, H. L. Fitton and L. Rajaram, *J. Chem. Soc., Perkin Trans.1*, 1633 (1994).
117. M. S. Baird, and S. Benedetti, unpublished results.
118. J. R. Al Dulayymi, M. S. Baird, E. Roberts and P. Tomasin, upublished results.
119. M. S. Baird, M. F. Shortt, H. H. Hussain and J. R. Al-Dulayymi, *J. Chem. Soc., Perkin Trans.1*. 1945 (1993).
120. M. S. Baird and S. Harkins, unpublished results.
121. T. H. Chan and D. Massuda, *Tetrahedron Lett.*, 3383 (1975).
122. M. G. Banwell, M. Corbett, J. Gulbis, M. F. Mackay, and M. E. Reum, *J. Chem. Soc., Perkin Trans. 1*, 945 (1993).
123. P. Engel, C. Lang, M. Muhlebach and M. Neuenschwander, *Chimia*, **46**, 380 (1992).
124. E. Schaumann, C. Friese and G. Adiwidjaja, *Tetrahedron*, **45**, 3163 (1989).
125. W. E. Billups, L.-J. Lin, B. E. Arney, W. A. Rodin and E. W. Casserly, *Tetrahedron Lett.*, **25**, 3935 (1984).
126. W. E. Billups, E. W. Casserly and B. E. Arney, *J. Am. Chem. Soc.*, **106**, 440 (1984).
127. B. R. Dent, B. Halton and A. M. F. Smith, *Aust. J. Chem.*, **39**, 1621 (1986).
128. W. E. Billups, G.-A. Lee, B. E. Arney and K. H. Whitmire, *J. Am. Chem. Soc.*, **113**, 7980 (1991).
129. W. E. Billups and M. M. Haley, *Angew.Chem.*, **101**, 1735 (1989).
130. W. E. Billups and M. M. Haley, *J. Am. Chem. Soc.*, **113**, 5084 (1991).
131. M. M. Haley, Ref. 125 in Reference 42.
132. W. E. Billups, A. J. Blakeney and W. Y. Chow, *J. Chem. Soc., Chem. Commun.*, 1461 (1971).
133. W. E. Billups, A. J. Blakeney and W. Y. Chow, *Org. Synth.*, **55**, 12 (1976).
134. W. E. Billups, *Acc. Chem. Res.*, **11**, 245 (1978).
135. W. E. Billups and W. J. Chow, *J. Am. Chem. Soc.*, **95**, 4099 (1973).
136. B. Halton, *Chem. Rev.*, **73**, 113 (1973).
137. A. Kumar, S. R. Tayal and D. Devaprabhakara, *Tetrahedron Lett.*, 863 (1976); P. J. Garratt and A. Koller, *Tetrahedron Lett.*, 4177 (1976); W. E. Billups, W. T. Chamberlain and M. Y. Asim, *Tetrahedron Lett.*, 571 (1977); D. Davalian and P. J. Garratt, *Tetrahedron Lett.*, 1976, 2815.
138. A. R. Browne and B. Halton, *Tetrahedron*, **33**, 345 (1977) A. R. Browne, B. Halton and C. W. Spangler, *Tetrahedron*, **30**, 3289 (1974).
139. L. K. Bee, P. J. Garratt and M. M. Mansuri, *J. Am. Chem. Soc.*, **102**, 7076 (1980).
140. B. Halton, *Chem. Rev.*, **89**, 1161 (1989).
141. B. Halton C. J. Randall, G. J. Gainsford and W. T. Robinson, *Aust. J. Chem.*, **40**, 475 (1987).
142. B. Halton, R. Boese, D. Blaser and Q. Lu, *Aust. J. Chem.*, **44**, 265 (1991).
143. P. Mueller and J.-P.Schaller, *Helv. Chim. Acta*, **73**, 1228 (1990).
144. W. E. Billups, B. E. Arney and L.-J. Lin, *J. Org. Chem.*, **49**, 3437 (1984).

145. W. E. Billups, M. M. Haley, R. C. Claussen and W. A. Rodin, *J. Am. Chem. Soc.*, **113**, 4331 (1991).
146. P. Mueller and D. Rodriguez, *Helv. Chim. Acta*, **68**, 975 (1985).
147. I. J. Anthony and D. Wege, *Tetrahedron Lett.*, **28**, 4217 (1987).
148. P. Mueller and H.-C. Nguyen-Thi, *Helv. Chim. Acta*, **67**, 467 (1984).
149. P. Mueller and J. P. Schaller, *Tetrahedron Lett.*, **30**, 1507 (1989).
150. P. Mueller and J.-P. Schaller, *Chimia*, **40**, 430 (1986).
151. A. I. D'yachenko, N. M. Abramova, T. Y. Rudashevskaya, O. A. Nesmeyanova and O. M. Nefedov, *Izv. Akad. Nauk SSSR, Ser. Khim*, 1193 (1982); *Chem. Abs.*, **97**, 109579.
152. B. Halton and D. L. Officer, *Aust. J. Chem.*, **36**, 1291 (1983); B. Haetan, P. J. Milsom and A. D. Woolhouse, *J. Chem. Soc., Perkin Trans. 1.*, 731 (19777).
153. R. Neidlein, V. Poignee, W. Kramer and C. Gluck, *Angew. Chem.Int. Ed. Engl.*, **25**, 731 (1986).
154. A. R. Browne and B. Halton, *J. Chem. Soc., Perkin Trans. 1*, 1177 (1977).
155. A. R. Browne and B. Halton, *J. Chem. Soc., Chem. Commun.*, 1341 (1972).
156. P. Mueller and D. Rodriguez, *Helv. Chim. Acta*, **69**, 1546 (1986).
157. P. Mueller and H.-C. N.Thi, *Tetrahedron Lett.*, **21**, 2145 (1980).
158. C. Gluck, V. Poignec and H. Schwager, *Synthesis*, 260 (1987).
159. P. Mueller and D. Rodriguez, *Helv. Chim. Acta*, **66**, 2542 (1983); P. Mueller and M. Rey, *Helv. Chim. Acta*, **65**, 1157 (1982); P. Mueller and M. Rey, *Helv.Chin.Acta*, **64**, 354 (1981).
160. G. L. Closs and K. D. Krantz, *J. Org. Chem.*, **31**, 638 (1966).
161. S. Arora, P. Binger and R. Koster, *Synthesis*, 146 (1973).
162. R. Koster, S. Arora and P. Binger, *Justus Liebigs Ann. Chem.*, 1219 (1973).
163. R. Koster, S. Arora and P. Binger, *Angew.Chem., Int. Ed. Engl.*, **9**, 810 (1970).
164. R. Koster, S. Arora and P. Binger, *Angew. Chem.*, **81**, 186 (1969).
165. R. Koster, S. Arora and P. Binger, *Justus Liebigs Ann. Chem.*, 1219 (1973).
166. M. A. Battiste, D. D. McRitchie, P. G. Gassman, W. F. Reus, J. N. Chasman and J. Haywood-Farmer, *Tetrahedron Lett.*, 2097 (1979).
167. R. M. Magid, T. C. Clarke and C. D. Duncan, *J. Org. Chem.*, **36**, 1320 (1971).
168. P. G. Gassman, J. J. Valcho and G. S. Proehl, *J. Am. Chem. Soc.*, **101**, 231 (1979).
169. A. S. Berg, *Acta Chem. Scand.*, **B34**, 241 (1980).
170. E. Negishi, L. D. Boardman, H. Sawada, V. Bagheri, A. T. Stoll, J. M. Tour and C. L. Rand, *J. Am. Chem. Soc.*, **110**, 5383 (1988).
171. A. T. Stoll and E. Negishi, *Tetrahedron Lett.*, **26**, 5671 (1985).
172. A. A. Kamyshova, T. A. Ryzhkova, E. T. Chukovskaya and R. K. Freidlina, *Dokl. Akad, Nauk SSSR*, **260**, 1370 (1981) *Chem. Abs.*, **96**, 85088q.
173. A. A. Kamyshova, E. T. Chukovskaya and R. K. Friedlina, *Izv. Akad, Nauk SSSR, Ser. Khim.*, 2839 (1980); *Chem. Abs.*, **87**, 67869e.
174. R. K. Freidlina, A. A. Kamyshova and E. T. Chuckovskaya, *Izv. Akad. Nauk SSSR, Ser. Khim.*, 2839 (1980).
175. R. K. Freidlina, A. A. Kamyshova and E. T. Chuckovskaya, *Izv. Akad. Nauk SSSR, Ser. Khim.*, 353 (1983); *Chem. Abs.*, **98**, 197550g.
176. H. N. Al-Jallo, A. Biaty and F. N. Al-Azawi, *J. Heterocycl. Chem.*, 1347 (1977).
177. N. O. Nilsen, L. Skattebol, M. S. Baird, S. R. Buxton and P. D. Stowey, *Tetrahedron Lett.*, **25**, 2887 (1984).
178. A. Duker and G. Szeimies, *Tetrahedron Lett.*, **26**, 3555 (1985).
179. G. Szeimies, *Strain and its Implications in Organic Chemistry*, NATO ASI Series (Eds. A de Meijere and S. Blechert), Kluwer, 1989, Vol. 273, p. 361.
180. M. S. Baird, B. Grehan and S. N. Huda, unpublished results.
181. J. Belzner, U. Bunz, K. Semmler, G. Szeimies, K. Opitz and A.-D. Schluter, *Chem.Ber.*, **122**, 397 (1989).
182. J. Fuchs and G. Szeimies, *Chem. Ber.*, **125**, 2517 (1992).
183. P. Kaszynski and J. Michl, *J. Am. Chem. Soc.*, **110**, 5225 (1988).
184. S. Mazur, A. H. Schroder and M. C. Weiss, *J. Chem. Soc.,Chem. Commun.*, 262 (1977).
185. K. B. Wiberg, W. P. Dailey. F. H. Walker, S. T. Waddell, L. S. Crocker and M. Newton, *J. Am. Chem. Soc.*, **107**, 7247 (1985).
186. K. B. Wiberg and F. H. Walker, *J. Am. Chem. Soc.*, **104**, 5239 (1982).
187. F. Alber and G. Szeimies, *Chem. Ber.*, **125**, 757 (1992).
188. J. Morf, Ref. 30 in Refernce, 179; J. Morf and G. Szeimies, *Tetrahedron Lett.*, **27**, 5363 (1986).

189. K. Semmler, G.Szeimies and J. Belzner *J. Am. Chem. Soc.*, **107**, 6410 (1985).
190. J. Belzner, B. Gareiss, K. Polborn, W. Schmid, K. Semmler and G. Szeimies, *Chem. Ber.*, **122**, 1509 (1989).
191. A. Duker and G. Szeimies, *Tetrahedron Lett.*, **26**, 3555 (1985).
192. U. Szeimies-Seebach, A. Schoffer, R. Romer and G. Szeimies, *Chem. Ber.*, **114**, 1767 (1981); Ref. 15 in Reference 179.
193. K.-D. Baumgart, H. Harnisch, U. Szeimies-Seebach and G. Szeimies, *Chem. Ber.*, **118**, 2883 (1985).
194. H.-G. Zoch, A.-D. Schluter and G. Szeimies, *Tetrahedron Lett.*, **22**, 3835 (1981).
195. A. D. Schluter, H. Harnisch, J. Harnisch, U. Szeimies-Seebach and G. Szeimies, *Chem. Ber.*, **118**, 3513 (1985).
196. P. Chakrabarti, P. Seiler, J. D. Dunitz, A. D. Schluter, G. Szeimies, *J. Am. Chem. Soc.*, **103**, 7378 (1981).
197. G. Szeimies, J. Harnisch and O. Baumgartel, *J. Am. Chem. Soc.*, **99**, 5183 (1977).
198. H.-G. Zoch, E. Kinzel and G. Szeimies, *Chem. Ber.*, **114**, 968 (1981).
199. J. Harnisch, H. Legner, U.Szeimies-Seebach and G. Szeimies, *Tetrahedron Lett.*, 3683 (1978).
200. U. Szeimies-Seebach, J. Harnisch, G. Szeimies, M. Van Meerssche, G. Germain and J.-P. Declerq, *Angew.Chem. Int. Ed. Engl.*, **17**, 848 (1978).
201. U. Szeimies-Seebach and G. Szeimies, *J. Am. Chem. Soc.*, **100**, 3966 (1978).
202. Ref. 12 in Reference 179; R. Freeman, H. W. D. Hill and R. Kaptein, *J. Magn. Reson.*, **7**, 327 (1972).
203. H.-G. Zoch, G. Szeimies, R. Romer, G. Germain and J.-P. Declerq, *Chem. Ber.*, **116**, 2285 (1983).
204. S. Hashmi, K. Polborn and G. Szeimies, *Chem. Ber.*, **122**, 2399 (1989).
205. J. Harnisch, O. Baumgartel, G. Szeimies, M. Van Meerssche, G. Germain and J.-P. Declerq, *J. Am. Chem. Soc.*, **101**, 3370 (1979).
206. O. Baumgartel and G. Szeimies, *Chem. Ber.*, **116**, 2180 (1983).
207. O. Baumgartel, J. Harnish, G. Szeimies, M. Van Meerssche, G. Germain and J.-P. Declerq, *Chem. Ber.*, **116**, 2205 (1983).
208. J. Kenndoff, K. Polborn and G. Szeimies, *J. Am. Chem.Soc.*, **112**, 6117 (1990).
209. J. Podlech, K. Polborn and G. Szeimies, *J. Org. Chem.*, **58**, 4113 (1993).
210. P. G. Gassman and I. Gennick *J. Am. Chem. Soc.*, **102**, 6864 (1980).
211. J. Schafer and G. Szeimies, *Tetrahedron Lett.*, **31**, 2263 (1990).
212. J. Schafer and G. Szeimies, *Tetrahedron Lett.*, **29**, 5253 (1988).
213. M. S. Baird, S. R. Buxton and H. H. Hussain, *J. Chem. Res.*, S310 (1986).
214. K. Isagawa, K. Mizuno, H. Sugita and Y. Otsuji, *J. Chem. Soc.. Perkin Trans. 1*, 2283 (1991).
215. J. R. Al Dulayymi and M. S. Baird, submitted for publication.
216. M. S. Baird and A. R. Al Dulayymi, unpublished results.
217. M. S. Baird, *Tetrahedron Lett.*, **25**, 4829 (1984).

CHAPTER **24**

Recent advances in the S$_{RN}$1 reaction of organic halides

ROBERTO A. ROSSI, ADRIANA B. PIERINI and ALICIA B. PEÑÉÑORY

Departamento de Química Orgánica, Facultad de Ciencias Químicas, Universidad Nacional de Córdoba, INFIQC, Suc. 16, CC 61, 5016 Córdoba, Argentina

Supplement D2: The chemistry of halides, pseudo-halides and azides
Edited by S. Patai and Z. Rappoport © 1995 John Wiley & Sons Ltd

---

I. INTRODUCTION

It is widely known that substitution reactions can be achieved on aliphatic and activated aromatic compounds by polar mechanisms. In 1966, Kornblum and coworkers[1] and Russell and Danen[2], independently, proposed a new process by which a substitution involving electron transfer (ET) reactions can be achieved on activated aliphatic halides. In 1970 Kim and Bunnett[3] proposed the same mechanism for substitution of unactivated aromatic halides. From then on, this process, termed $S_{RN}1$, has been extensively studied. Therefore its actual scope has considerably grown and nowadays it has to be recognized and taken into consideration for achieving substitution of many different types of aromatic and aliphatic substrates with different types of nucleophiles, as well as a mechanism of widely synthetic capabilities.

Among the halides that react through this process are: unactivated aromatic and heteroaromatic halides, vinyl halides, activated alkyl halides [nitroalkyl, nitroallyl, nitrobenzyl and other benzylic halides substituted with electron-withdrawing groups (EWG) as well as the heterocyclic analogues of these benzylic systems] and non-activated alkyl halides that have proved to be unreactive or poorly reactive towards polar mechanisms (bicycloalkyl, neopentyl and cycloalkyl halides and perfluoroalkyl iodides).

Besides halides, other leaving groups have also been reported, such as N_3, NO_2, SPh, $^+NMe_3$, N_2SR, HgX, etc.

A considerable number of nucleophiles have been shown to react through this mechanism: stabilized carbanions, anions derived from elements of the VI A group (S, Se, Te) and anions from elements of the V A group (P, As, Sb). More recently cyanide anion, nitrogen and oxygen (e.g. aromatic alkoxides) nucleophiles have been added to the list. These nucleophiles behave as C— rather than as N— or O— nucleophiles.

In the organometallic field the reaction of R_3Sn^- ions (R = CH_3, Ph) with aromatic and aliphatic substrates has been reported. Another example is the substitution of aryl or vinyl halides by iron(I) porphyrins under electrochemical induction[4a], and the electrochemical arylation of metal carbonyl anions to form $C_5H_5(CO)_3M$—σ—Ar (M = W, Mo; Ar = Ph, $C_6H_4NO_2$)[4b].

Several reviews have been published on the subject, such as those referring to activated alkyl halides[5], photochemically induced reactions[6], ET reactions[7], substituted aliphatic nitro compounds[8,9], alkyl mercurials[10], sp[3] non-activated alkyl halides[11,12], sp[2] carbon centres[13,14], electrochemical induction[15], synthetic aspects[15d,e,16-20] and photochemical induction of the reaction at heteroatomic carbon centres[21].

In this review we will present the distinctive features of the proposed mechanism as well as main experimental and theoretical evidence for it. The reaction of organic halides will be further discussed according to the following sequence: mechanistic features of the $S_{RN}1$ reaction in Section II, alkyl halides with electron-withdrawing groups in Section III, alkyl halides without electron-withdrawing groups in Section IV, aromatic halides in Section V and vinyl halides in Section VI.

II. MECHANISTIC FEATURES OF THE $S_{RN}1$ REACTION

In this process, represented by equations 1–8 in Scheme 1, a substitution is achieved with radicals and radical anions as intermediates. According to this mechanistic proposal the radical anion of the substrate is formed in the initiation step (equation 1) by an electron transfer (ET) from a suitable electron source such as solvated electrons from dissolved alkaline metals, sodium amalgam, reducing agents such as Fe^{+2}, SmI_2, electrons from a cathode or by a negatively charged species (nucleophile) under thermal, sonication or photochemical stimulation.

The formed radical anion fragments into a radical and the anion of the leaving group (equation 2). Coupling between the radical and the nucleophile present in the reaction media forms the radical anion of the substitution product (equation 3), which by an ET to the substrate regenerates a radical anion responsible for continuing the propagation cycle of the process (equation 4).

Reduction of the radicals by ET from the radical anion of the substitution product (or from $ArX^{-\bullet}$) are the main proposed termination steps of the mechanism (equations 5 and 6). Another is hydrogen atom abstraction of the radical from the solvent SH (equation 7).

The absence of dimerization in $S_{RN}1$ reactions indicates that it is not an important termination step (equation 8), which may be related to the low concentration of the radical R[•] in the chain reaction. However, 17% of the dimerization product 1,1'-biadamantyl was found in the reaction of 1-iodoadamantane with the less reactive carbanionic nucleophiles, such as acetone enolate ion in DMSO[22]. Likewise, under appropriate experimental conditions, the dimerization product 4,4'-dicyanobiphenyl (39%) was the principal product of the termination step in the electrochemical induced reaction of 4-chlorobenzonitrile with 2-pyridinethiolate ions in liquid ammonia[23].

$$\text{Initiation:} \quad RX + \text{Donor} \longrightarrow (RX)^{-\bullet} \tag{1}$$

$$(RX)^{-\bullet} \longrightarrow R^\bullet + X^- \tag{2}$$

$$\text{Propagation:} \quad R^\bullet + Nu^- \longrightarrow (RNu)^{-\bullet} \tag{3}$$

$$(RNu)^{-\bullet} + RX \longrightarrow RNu + (RX)^{-\bullet} \tag{4}$$

$$\text{Termination:} \quad (RNu)^{-\bullet} + R^\bullet \longrightarrow R^- + RNu \tag{5}$$

$$R^- \xrightarrow{\text{SH}} RH \tag{6}$$

$$R^\bullet \xrightarrow{\text{SH}} RH \tag{7}$$

$$2R^\bullet \longrightarrow R{-}R \tag{8}$$

SCHEME 1

For this mechanism to work efficiently, the overall reaction must be exergonic. The initiation step need not be fast but the chain propagation steps must be very fast to allow for long chains to build up. However, in general the chain length is modest, reflecting the competing termination steps. The ET of equation 4 is usually exergonic, and thus should be very fast. Fragmentation of radical anions of the substrate are also known to be in general quite fast. It follows that the crucial step of the mechanism must be the coupling between the radical and the nucleophile (equation 3). If this reaction cannot compete efficiently, the chain will be short, or even non-existent. The exergonicity of the overall process defines the exergonicity of the sum of the propagation steps.

In those cases in which the ET takes place from the nucleophile, the radical anion formed can fragment inside the solvent cage to give a radical-anion pair which can couple with the $Nu^{\bullet}$ inside the solvent cage to give the substitution product by a cage collapse mechanism (equation 9). Another possibility is that the radicals diffuse from the cage to enter the $S_{RN}1$ propagation cycle (equation 10). In this case coupling with the nucleophile will take place.

$$RX + Nu^- \xrightarrow{h\nu} \left[RX^{-\bullet}\ Nu^{\bullet}\right] \longrightarrow \left[R^{\bullet}\ X^-Nu^{\bullet}\right] \begin{array}{c} \xrightarrow{cage\ collapse} RNu + X^- \qquad (9) \\ \\ \xsearrow{S_{RN}1} \\ R^{\bullet} + X^- + Nu^{\bullet} \quad (10) \end{array}$$

ET from a nucleophile to an alkyl halide followed by coupling of the radicals through a cage collapse mechanism has been proposed by Lund[24], Bordwell[25], Ashby[26] and coworkers and by others[27].

The initiation reaction has been proposed to be dissociative for aliphatic halides, thus leading directly to a radical-anion pair[15a,28].

In the last few years it has been proposed that the dichotomy polar vs ET is not a real one and that the mechanism through which a substitution takes place can be viewed within a complete spectrum ranging from polar to ET[29]. The polar mechanism can be considered to take place by a *single electron shift*, in other words, a single ET accompanied by bond breaking and/or bond formation.

This view is opposed to the generally accepted one for polar processes in which the electrons are considered to move in pairs. Whenever there exist electronic or steric factors that favour the ET taking place without the significant geometric modifications proposed for the transition states of a polar substitution reaction, the mechanism in play will be of the $S_{RN}1$ or radical–radical anion type[11]. In these cases the radical anion of the alkyl or aryl halide is formed by an intermolecular ET and the reaction coordinate for the process is solvent and geometrical intramolecular reorganizations. Whenever the ET reaction prevails over the *single electron shift*, a $S_{RN}1$ OR cage collapse mechanism can take place.

A. The Initiation Step

Photoinduced ET from the nucleophile is one of the most popular forms of performing an $S_{RN}1$ reaction. However, the mechanism of this initiation is not fully understood.

It has been proposed for aliphatic and aromatic halides that the photochemical initiation involves a charge-transfer complex (ctc) between the nucleophile and the substrate in which the substitution takes place, and that it is photochemically energized so that complete transfer of one electron occurs (equation 11)[6,9,30].

$$RX + Nu^- \longrightarrow \left[RX, Nu^- \right]_{ctc} \xrightarrow{h\nu_{\lambda ctc}} (RX)^{-\bullet} + Nu^\bullet \qquad (11)$$

Kornblum and coworkers[31a] have determined the quantum yield for the ET substitution reactions of *p*-nitrocumyl chloride with azide ions (3.5) and quinuclidine (6000). Furthermore, by studying the wavelength dependence of the quantum yields, they have obtained evidence that photochemical initiation proceeds by means of a charge-transfer complex. Similar results have been obtained in the reaction of acetone enolate ion with PhI and PhBr in DMSO, whereas other mechanisms are in competition when PhI reacts with potassium diethyl phosphite[31b].

Haloarenes, such as 2-bromonaphthalene or *p*-bromobiphenyl, form in liquid ammonia CTC with acetonitrile or propionitrile carbanions, which are responsible for the photo-stimulated ET of the initiation step. The quantum yields of these reactions ($\lambda > 313$ nm) range from 8 to 31[32].

The photostimulated reaction of halonaphthoxides and sodium sulfite in water solution gives the substitution product (*ca.* 100%) by the $S_{RN}1$ mechanism with high quantum yields[33a]. The initiation step depends on the halogen of the halonaphthoxide. Thus, it has been suggested that the initiation step for 1-chloro-2-naphthoxide ion is determined by the ET reaction between the excited triplet of the substrate and its ground state, the length of the chain reaction being equal to 50. However, with 1-bromo-2-naphthoxide ion as sub-strate, the initiation step is the photohomolytic cleavage of the C—Br bond, and the length of the chain is equal to 20[33a].

The substitution of 1-chloro-2-naphthoxide ion by sulfite ion in water can also be initi-ated by visible light (436 nm) with the complex [Ru(bipy)$_2$]Cl as the sensitizer and the complex [Co(bipy)$_3$](ClO$_4$)$_2$ as the intermediate electron carrier[33b]. Another possibility is a dye photoinitiated reaction. In the latter example, the excited triplet of the dye (fluorescein, eosine or erythrosine) receives an electron from SO$_3^{-2}$ whose radical anion (SO$_3$)$^-$[^•] reacts with halonaphthoxides to give finally the substitution product[33c].

Acceleration by KI in the substitution reaction of aryl halides with potassium diethyl phosphite or with the 2-naphthoxide ion has also been explained on the basis of an ET through the exciplex of the charge-transfer complex formed between the aryl halide and the iodide ions[34a]. It has also been reported that iodide ions catalysed the photostimulated reaction of bromoarenes with diethyl phosphite ion[34b].

The photoinduced initiation reaction may have the disadvantage of poor quantum yields, arising from a fast backward ET which annihilates the ion pair before its cage separation. This means a poorly efficient source of radicals. However, if the photochemi-cal ET is the initiation of a very efficient radical chain process, a poor quantum yield in the reaction may turn to an advantage because small extent of production of the intermediates (Ar$^\bullet$, ArX$^{-\bullet}$, ArNu$^-$[^•]) will disfavour the proposed termination steps of the mechanism.

Thermal or spontaneous initiation is possible when the nucleophile is a powerful reduc-ing agent. Even though light is not necessary in these cases, it often provides significant rate acceleration.

Initiation by electrochemical induction may have the disadvantage of low yields of substitution due to the reduction of the radicals formed near the electrode, mainly in those cases in which the radical anion of the halide compound fragments at a considerably high rate. Redox catalysis, that is activation involving a suitable ET mediator, is an important means to avoid termination steps in electrochemically induced reactions. This approach has been extensively studied by the Savéant group[15]. A general equation has been proposed in order to predict the yield of ET-initiated $S_{RN}1$ chain reactions and related mechanisms under preparative electrochemical conditions in the presence of a redox mediator[35].

Recently, sonication has been proposed as another possibility for initiation. The reaction of *p*-nitrobenzyl bromide with 2-nitropropane anion may be thus initiated (equation 12)[36].

$$O_2N-\text{\textbenzene}-CH_2Br + Me_2CNO_2^- \xrightarrow{\;)))\;}$$

$$O_2N-\text{\textbenzene}-CH_2-C(CH_3)_2NO_2 \quad + \quad O_2N-\text{\textbenzene}-CHO \quad (12)$$

$$\underset{\text{C-alkylation}}{\qquad\qquad} \qquad\qquad \underset{\text{O-alkylation}}{\qquad\qquad}$$

Under optimal irradiation conditions, the C/O alkylation ratio is practically reversed with respect to that of the silent reaction, indicating a direct intervention of sonic waves in the ET. Sonication has also been proposed to initiate the reaction of haloarenes with Ph_2P^- ions in liquid ammonia at room temperature[37].

B. Radicals and Radical Anions as Intermediates

Inhibition by radical traps or radical anion scavengers has been extensively used in providing evidence for the mechanism with both aliphatic and aromatic substrates. The most commonly employed inhibitors are compounds that add irreversibly to radicals [di-t-butylnitroxide (DTBN), 2,2,6,6-tetramethyl-1-piperidinyloxy (TEMPO), galvinoxyl, etc.] and good reversible electron acceptors such as dinitrobenzenes (DNB) which intercept the radical anions[38].

On the other hand, difficulty has been found in detecting the reaction intermediates using ESR spectroscopy. Initial studies[39] with aryl halides using this technique at low temperature in solid matrices identified the π^* and σ^* radical-anions of PhI. Dissociation of the radical anion (PhI)$^{-\bullet}$ to phenyl radical and iodide anion was also observed. These studies therefore provided evidence for the first two steps in the aromatic $S_{RN}1$ mechanism.

According to the spectroscopic and other qualitative proposals, the first radical anion formed in the case of aryl halides is of π nature. This radical anion gives, by an intramolecular ET to the C—X bond, a σ radical anion, which dissociates into the aryl radical and the anion of the leaving group. The C—X bond strength is decreased once the radical anion is formed, as has been shown by thermodynamic cycles[40].

The aromatic system has been the subject of recent theoretical studies performed with semiempirical methods (MNDO, AM1). According to them, the most stable radical anions of PhX (X = Cl, Br) are of π nature while for X = I the most stable radical anion is of σ nature[41]. The difference in energy between both radical anions correlates with the frangibility proposed for the halides (Cl<Br<I). The transition state for the intramolecular ET has been located for (PhI)$^{-\bullet}$. The bending of the halogen atom accompanied by the C—X bond elongation are the reaction coordinates for this reaction which thus corresponds to the avoided crossing between two potential surfaces, representing each radical anion.

The extensive determination of fragmentation rates of aryl halide radical anions, due to Savéant and coworkers[15a] by electrochemical methods, indicates that they range from values of $10^{-2}\,s^{-1}$ for nitro-substituted phenyl halides up to $10^{10}\,s^{-1}$ for p-cyanophenyl halides. These values are in agreement with measurements by pulse radiolysis[42]. The fragmentation rates for unsubstituted phenyl halides are too high to be measured even by electrochemical techniques. Besides, 1-bromo- and 1-iodoanthraquinone radical anions have been shown to dissociate from their photoexcited state (Section V. D).

For non-activated haloalkanes, a dissociative electron capture[7,15,28] has been proposed mainly based on theoretical calculations and experimental findings in the gas phase, solid matrices and electrochemistry in polar solvents. The ET to the halide is accompanied by bond fragmentation and therefore radical anion intermediates are not formed. However, this is not necessarily the case for bridgehead alkyl halides. For these compounds, in which the radical formed by fragmentation of the radical anion cannot achieve planarity and thus release strain by rehybridization, the σ radical anion has been proposed to exist as an intermediate [43].

For activated aliphatic compounds, the radical anion intermediates are so unstable that generally they cannot be detected by ordinary ESR techniques[44]. However, more recently, Symons and Bowman have extensively studied the $S_{RN}1$ reactions of several halonitroalkanes by this technique at low temperature in matrices. They have unambiguously identified radical anions derived from 2-chloro- and 2-iodo-2-nitropropane $[Me_2C(X)NO_2]^{-\cdot}$, and also followed the loss of $X^{-\cdot}$ ions from them[45]. The results show that these radical anions are pyramidal at nitrogen, and the resulting stabilization inhibits the dissociation. Structure **1**, with significant delocalization into the C—X σ-bond, is strongly favoured for this radical anion. This structure leads smoothly, by reorganization of the $\pi*$ and $\sigma*$-orbitals, to the required transition state for loss of halide ion.

$$Me \cdots C \underset{Me}{\overset{X^{\cdots\cdots}}{\diagdown}} N^{+} \cdots O^{-} \quad O^{-}$$

(1)

The addition of an electron to $Me_2C(X)NO_2 (X = Br,Cl)$ in the presence of a large excess of nucleophiles has also been studied[46]. The ESR results clearly indicate that efficient addition of benzenesulphinate and nitrite ions to the $^{\cdot}C(Me_2)NO_2$ radical competes successfully with anion return of bromide or chloride.

As expected, the ESR spectrum obtained from the irradiation of the $Me_2C(Cl)NO_2$ in CD_3OD in the presence of bromide ion did not show any features due to $Me_2C(Br)NO_2^{-\cdot}$. However, the reverse reaction in which $Me_2C(Br)NO_2$ solutions in CD_3OD containing LiCl were irradiated gave clear features for $Me_2C(Cl)NO_2^{-\cdot}$ radical anions.

Electron capture to form stable radical anions has also been observed using ESR spectroscopy at low temperature for p-nitrobenzyl[47,48] and p-nitrocumyl derivatives[48], for α-substituted-2-methyl-5-nitrofurans[49] and for 4-nitroimidazole derivatives[50]. The structures of these radical anions reveal that there is a significant overlap between the aromatic nitro $\pi*$ orbitals and the C—X $\sigma*$ orbitals. As for the 2-substituted-2-nitropropanes case, it has been proposed that the fragmentation of these radical anions proceeds by smooth reorganization of molecular orbitals to the required transition state for loss of halide anions. For these compounds two distinct radical anions $\pi*$ and $\sigma*$ do not exist, but in effect, they are extreme valence bond forms of the actual structures. Similar results have been proposed from theoretical calculations[41, 51].

Not only radical scavengers, radical reduction and/or radical dimerization products but also radical probes were used in order to prove the presence of radicals as intermediates along the $S_{RN}1$ propagation cycle. Thus the formation of cyclized and uncyclized substitution products was taken as an indication of radical intermediates in the reaction of neopentyl-type halides containing a cyclizable probe of the 5-hexenyl type **2**. These reactions were performed with PhS^- and Ph_2P^- ions as nucleophiles (equation 13)[52].

$$\text{(13)}$$

(2)

$X = CH_2, O$

Similar experimental evidence was obtained in the reaction of secondary halides (6-halo-1-heptenes) with Me_3Sn^- ions[26b], of tertiary chlorides (e.g. 6-chloro-6-methyl-1-heptene)[53] and of substituted cyclohexyl bromides with Ph_2P^- ions[54]. On the other hand, ring opening or closure reactions were not observed in the $S_{RN}1$ reactions of 1-chloro-1-cyclopropyl-1-nitroethane and 2-chloro-2-nitro-6-heptene with different nucleophiles. This was ascribed to the ability of a nitro group to stabilize a radical centre[55].

Other evidence favouring the presence of radicals is the stereochemical outcome of the reaction. In the aliphatic field, complete loss of optical activity has been determined in the substitution reaction of the nitro group of optically active 2-(p-nitrophenyl)-2-nitrobutane[56] as well as in the alkylation of the anion of benzyl cyanide and of N,N-diethyl-α-aminopropionitrile by p-nitrobenzyl chloride[57]. Formation of epimeric products in the substitution reactions of nitrocyclohexanes was additional evidence[58]. The proportion of epimers was found to reflect the bulk of the incoming nucleophile PhS⁻, N_3^-, $p\text{-MeC}_6\text{H}_4\text{SO}_2^-$, O_2NCMe_2) relative to the substituent present at the reaction site.

The independence of the product distribution ratio on the concentration of the substrate and the nature of the leaving group, observed in competition experiments of a pair of nucleophiles with a series of substrates having a common aryl group but different leaving groups, has also been regarded as evidence for the existence of radicals as intermediates[59].

C. Coupling Reactions

Several methods have been employed to determine the rate constant of the addition of nucleophiles to radicals. Relative reactivities of pairs of nucleophiles toward the same radical have been obtained from the ratio of the two substitution products[14]. The absolute value of the rate constant for the coupling of aromatic radicals with nucleophiles has been determined by cyclic voltammetry. A large number of these values are close to the diffusion limit[15].

For the same nucleophile, it has been determined that the driving force of the coupling reaction increases with the standard potential of the $RNu/RNu^{-\bullet}$ couple. In other words, the lower the LUMO energy of the RNu product, the faster the addition of the nucleophile to the radical. Related concepts have been proposed for the coupling reaction of carbanionic nucleophiles with phenyl and 1-adamantyl radicals.

The fragmentation of the radical anion $(RNu)^{-\bullet}$ along the chain propagation cycle of the process has been taken as evidence of the proposed mechanism. For example, in the case of dihalobenzenes YArX, the radical anion formed upon the first substitution $YArNu^{-\bullet}$ may transfer the extra electron to the C—Y bond (intramolecular ET) or to YArX (intermolecular ET). The ratio between monosubstituted and disubstituted products formed will depend on the relative rate constants for both types of competing ET reactions.

Other radical anions that fragment are those derived from the coupling of phenyl radicals with CH_2NO_2, CH_2CN and $PhCH_2S^-$ nucleophiles. This reaction has been explained on the basis of the stability of the benzyl radicals formed upon fragmentation.

Fragmentation reactions have also been found for the radical anions formed by coupling of aliphatic or aromatic radicals with the anions Ph_2As^-, Ph_2Sb^-, $PhSe^-$ and $PhTe^-$. In these cases scrambling of moieties has been determined[60] and the product distribution has been used as mechanistic evidence (Sections IV. A.1 and V.C. 3).

$S_{RN}1$ vs $S_{RN}2$. Recently, the possibility of an $S_{RN}2$ instead of an $S_{RN}1$ mechanism has been discussed for aromatic halides[61]. In the $S_{RN}2$ process a direct coupling between the nucleophile and the radical anion of the substrate is proposed (equation 14):

$$ArX^{-\bullet} + Nu^- \longrightarrow ArNu^{-\bullet} + X^- \tag{14}$$

This possibility has been disregarded, mainly on the basis of the extremely short lifetimes of the radical anions of unsubstituted phenyl and naphthyl halides or those lacking strong electron-attracting substituents. Based on the fragmentation rates of these radical anions[15a] the $S_{RN}2$ proposal is very unlikely. The reaction of equation 14 would have to be extremely fast in order to compete with the fragmentation of the radical anions that do so with rate constants of at least $10^4 \, s^{-1}$. Other experimental evidence considered in order to exclude the $S_{RN}2$ proposal includes stereochemical evidence[57], independence of the product distribution on the leaving group ability[62,63], use of radical probes, scrambling of the substitution products, etc[63].

Moreover, several mechanisms for the suggested direct displacement of the halide ion from the radical anion by a nucleophile were examined and all were considered unacceptable because of the violation of quantum-mechanical principles or incompatibility with experimental observations[64].

On the other hand, Russell and coworkers have proposed that the substitution and enolate dimerization products, formed in the reactions of 2-substituted-2-nitropropanes ($XCMe_2NO_2$, X = Cl, NO_2, p-MePhSO$_2$) with nucleophiles that easily lose one electron, such as the mono enolate anions $ArC(OLi)=CHR$ (R = Me, Et, i-Pr, n-Bu) and t-BuC(OLi)=CH$_2$, can be rationalized on the basis of a free radical chain mechanism involving bimolecular substitution or ET reactions between the enolate anion and the intermediate nitroalkane radical anion[62]. An $S_{RN}2$-type mechanism has also been recently suggested for the reaction of pentafluoronitrobenzene with several nucleophiles in aqueous media[65].

III. ALKYL HALIDES WITH ELECTRON-WITHDRAWING GROUPS

A. α-Halonitroalkanes

Nowadays, the α-haloaliphatic nitro compounds $R^1R^2C(X)NO_2$, which are among the first substrates proposed to undergo an $S_{RN}1$ process[2], are the most extensively studied compounds in $S_{RN}1$ reactions at an sp^3 carbon. A wide range of alkyl groups has been investigated[13], from a simple methyl group to cyclic, heterocyclic and even glycocyl groups[66a]. Also they react with a wide variety of nucleophiles (for a comprehensive list see References 8 and 13).

The synthetic potential[18] of these reactions is principally due to the possibility of nitrous acid (HNO_2) elimination, making it possible to synthesize tri- and tetrasubstituted olefins, and also due to the substitution of the nitro group by a nucleophile.

B. Nitrobenzyl and Cumyl Derivatives

The o- and p-nitrobenzyl derivatives[5] and the analogous p-nitrocumyl derivatives[5,31a,66b], especially the chlorides, have been extensively studied and reviewed by Kornblum and coworkers. They react with a wide range of organic and inorganic nucleophiles, providing a novel and powerful means of synthesis.

Norris and Randles[67,68] have largely studied the regiochemistry of the coupling between p-nitrobenzyl radicals and ambient nitronate anions, in the association step of $S_{RN}1$ reactions, and showed that it depends on steric factors. Branching at the positions adjacent to the reaction sites (C_β) causes a shift in the product distribution toward O-alkylation and away from C-alkylation. In some cases no products were formed. The authors have formulated rules which predict when C-alkylation will occur in the association step involving p-nitrobenzyl radicals and nitronate anions (equation 15).

$$\begin{bmatrix} \text{ArCR}^1\text{R}^2 \\ \text{O} \\ \quad\diagdown\overset{+}{\text{N}}=\text{CR}^3\text{R}^4 \\ \quad\quad\text{O}^- \end{bmatrix}^{-\bullet} \xleftarrow[a]{\text{O}-\text{alk}} \begin{array}{c} \text{Ar}\overset{\bullet}{\text{C}}\text{R}^1\text{R}^2 \\ + \\ \text{R}^3\text{R}^4\text{CNO}_2^- \end{array} \xrightarrow[b]{\text{C}-\text{alk}} \begin{bmatrix} \text{ArCR}^1\text{R}^2 \\ | \\ \text{R}^3\text{R}^4\text{CNO}_2 \end{bmatrix}^{-\bullet} \quad (15)$$

$$\text{Ar} = p\text{-O}_2\text{NC}_6\text{H}_4$$

Based on the above results it was assumed that both C- and O-alkylation are kinetically controlled processes, i.e. that pathway a in equation 15 is effectively irreversible[68].

The possible reversibility of the C-alkylation of nitronate anions under $S_{RN}1$ conditions was studied by examining the degree of interconversion of the C-alkylation product **3** and **4** in crossed experiments with the nitronate ions **5** and **6**[69].

$$\text{ArCMe}_2\text{CMe}_2\text{NO}_2 \qquad\qquad \text{ArCMe}_2\text{CMe(Et)NO}_2$$

$$\textbf{(3)} \qquad\qquad\qquad\qquad\qquad \textbf{(4)}$$

$$\text{M}^+\text{CMe}_2\text{NO}_2^- \qquad\qquad \text{M}^+\text{CMe(Et)NO}_2^-$$

$$\textbf{(5)} \qquad\qquad\qquad\qquad\qquad \textbf{(6)}$$

The fact that the only product formed was the reduction product p-nitrocumene clearly demonstrates the irreversibility of pathway b and that the C-alkylation products do not isomerize to O-alkylation products, as was previously proposed by Kornblum and coworkers[70].

The dissociation (i.e. fragmentation of the C_α—C_β bond) of C-alkylates was proposed to take place in a process involving transfer of two electrons and not through dissociation of radical anions[69]. Thus, the p-nitrocumene found was suggested to arise from protonation of the corresponding anion, generated by fragmentation of the dianion of the C-alkylates.

Similar crossed experiments demonstrated the irreversibility of the association step between p-nitrobenzyl radicals and anions of β-keto esters[69].

Steric and electronic effects on the rate and regiochemistry of the reaction between p-nitrobenzyl substrates and tertiary carbanions were also studied[71]. Thus, increasing the size of the alkyl groups attached to the benzylic or anionic carbons of the substrates causes substantial decrease in the proportions of C-alkylation product. In contrast with the previous reaction with nitronate anions, formation of reduction products is observed instead of a significant O-alkylation.

The electronic nature of the groups attached to the anionic carbon also affects the proportion of C-alkylation. The amount of C-alkylation in the reaction of ArCH(t-Bu)Cl with a series of α-methylated carbanions, having similar steric interactions, decreases when the stabilizing substituents on the carbanion are changed from cyano to alkoxycarbonyl and to acetyl.

Although the $S_{RN}1$ reaction is shown to be prone to steric limitations, most of the C-alkylated products obtained in these studies are sterically congested molecules, whose preparation would be difficult by other methods. Indeed, the reaction of sterically hindered p-nitrobenzyl halides with nucleophiles by an $S_{RN}1$ mechanism affords a synthetic route to extremely sterically crowded molecules[72].

Whereas m-nitrobenzyl chloride reacts with nucleophiles by an S_N2 mechanism[5], m-nitrocumyl chloride[5] has been reported to react with nucleophiles by a mechanism involving radical anions, free radicals and cage collapse of radical pairs rather than by an $S_{RN}1$ process.

On the other hand, α-t-butyl-m-nitrobenzyl chloride (7) was clearly demonstrated to undergo $S_{RN}1$ substitution reactions with a variety of anions (equation 16)[73].

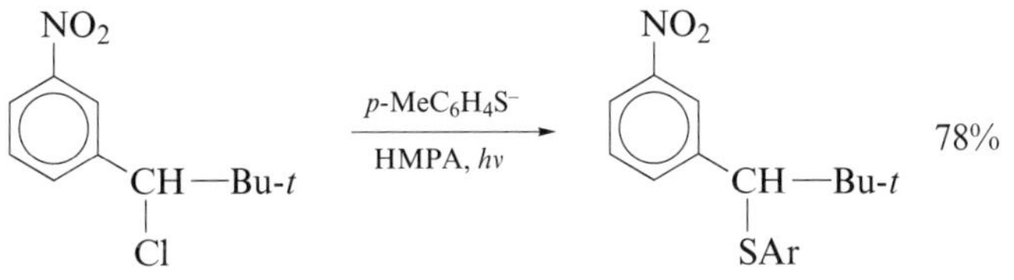

$$\text{(7)} \xrightarrow[\text{HMPA, } h\nu]{p\text{-MeC}_6\text{H}_4\text{S}^-} \qquad 78\% \qquad (16)$$

The reaction of 2-chloro-1,1-dialkyl-6-nitroacenaphthenes (8–10) with azide and p-toluenethiolate ions takes place by the $S_{RN}1$ mechanism[74]. This reaction gives the substitution products, despite the fact that the nitro group and the chlorine-bearing benzylic carbon are attached to different aromatic rings.

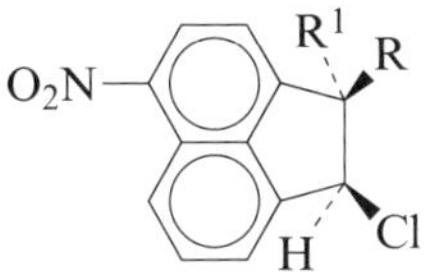

(8) R = R^1 = Me
(9) R = Me, R^1 = Et
(10) R = Et, R^1 = Me

Stereochemical evidence supports the fact that the reaction of the stereoisomers 9 and 10 takes place through an effectively planar benzylic radical, which is preferentially attacked from the face remote from the α-ethyl group. Because of the presence of geminal alkyl groups α to the reaction site, the reaction fails with nucleophiles such as the anions derived from 2-nitropropane and 2-ethylmalononitrile and with sodium p-toluenesulphinate.

C. Heterocyclic Analogues of p-Nitrobenzyl Derivatives

Different heterocyclic analogues of p-nitrobenzyl derivatives have been described to react with nucleophiles by the $S_{RN}1$ mechanism.

2-(Halomethyl)-5-nitrofurans have been reported to give a substitution product with the anion of 2-nitropropane[75a] by the $S_{RN}1$ mechanism, whereas with different thiolates ions[75b] they react by an S_N2 mechanism. Recently the reactions of 2-(bromomethyl)-5-nitrofuran with benzenethiolate and 4-chlorobenzenethiolate ions were assumed to be $S_{RN}1$ substitutions[75c].

The reaction[76a] of 2-chloromethyl-5-nitrothiophene with the 2-nitropropane anion was recently reinvestigated, and extended to various nitronate anions[76b]. These reactions, taking place by the $S_{RN}1$ mechanism, afforded good yields of new 5-nitrothiophenes bearing a trisubstituted ethylenic double bond at the 2-position. Some of these derivatives showed the same antiprotozoan activity as that of reference compounds.

When the benzylic carbon in the 5-nitrofuran derivatives is also a neopentylic one[77], the S_N2 reaction is prevented by the α-$tert$-butyl group and the substitution reaction with different nucleophiles proceeds by the $S_{RN}1$ mechanism. Similar behaviour has been found for nitrothiophene derivatives[78]. Thus, substitution reactions at neopentyl carbons bearing 4- and 5-nitrothienyl groups with various nucleophiles have been reported. The mechanisms of these substitutions are quite different, however. The substitutions in the 5-nitro series take place by the $S_{RN}1$ mechanism (equation 17), whereas those in the 4-nitro series occur by the ionic $S_N(AEAE)$ process, which involves initial attack of a nucleophile at the 5-position of the thiophene ring.

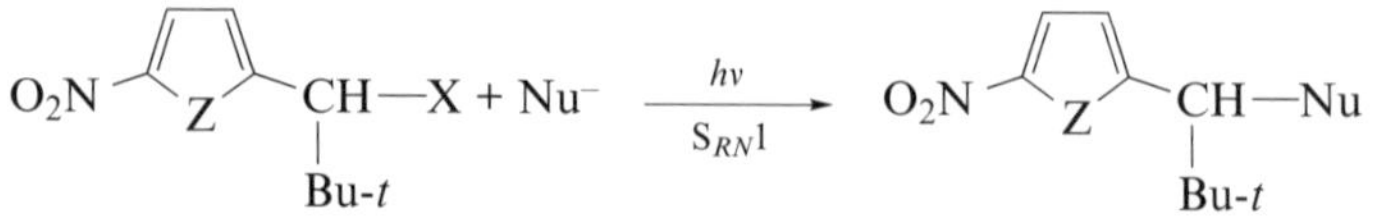

$$O_2N \underset{Z}{\diagup\diagdown} CH-X + Nu^- \xrightarrow[S_{RN}1]{h\nu} O_2N \underset{Z}{\diagup\diagdown} CH-Nu \qquad (17)$$

$$\underset{Bu\text{-}t}{} \qquad\qquad\qquad \underset{Bu\text{-}t}{}$$

Z = O, S

On the other hand, the analogues 1-methyl-2-(chloroneopentyl)-4- and 5-nitropyrroles react with azide, thiocyanate and p-toluenethiolate ions through cationic species in an S_N1-type process, rather than by the $S_N(AEAE)$ OR $S_{RN}1$ mechanism[79].

2-(Halomethyl)-1-methyl-5-nitroimidazoles have also been reported to react by the $S_{RN}1$ mechanism with tertiary nitronate anions[80,81], nitroimidazole anion[82] and 3-nitrolactam anions[83].

The m-nitrobenzyl chloride analogue 2-(chloromethyl)-1-methyl-4-nitroimidazole reacts with the 2-nitropropane anions by the $S_{RN}1$ mechanism under phase-transfer conditions. A base-promoted nitrous acid elimination from the C-alkylated product gives alkenylimidazole derivatives[84].

Recently[85], the first example of $S_{RN}1$ substitution involving a nitroimidazole compound in which the nitro group is o- to the side-chain at which the substitution occurs was described (equation 18).

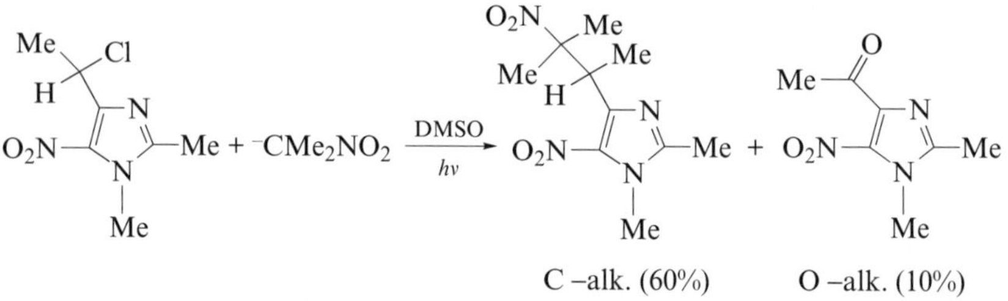

$$(18)$$

C –alk. (60%)   O –alk. (10%)

Contrary to what was found in the reactions of 2-substituted-5-nitroimidazoles[80,82], elimination of nitrous acid from the C-alkylation product in the presence of only the nitronate anion as a base was not observed. A good base is necessary to promote the elimination to give the corresponding tetrasubstituted olefin (equation 19).

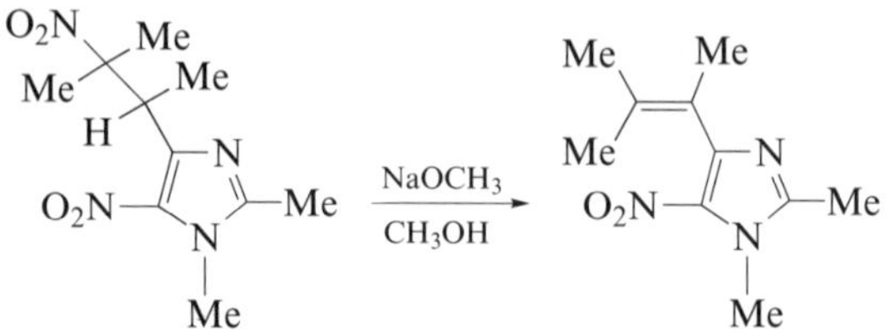

$$(19)$$

Vanelle, Crozet and coworkers[86,87] have reported the $S_{RN}1$ reaction of 2-chloromethyl-3-nitroimidazo[1,2-a]pyridine (**11**) with nitronate anions. In **11** the nitro group is in position 4 of the imidazole ring and the chloromethyl group is in position 5, *o*- to the nitro group, and there is a new feature with regard to previously studied nitroimidazoles: a fused pyridine ring. When the reaction was performed with analogue **12**[86] lacking the nitro group, no products were found, showing that the pyridine moiety is not sufficiently electron-withdrawing for an $S_{RN}1$ reaction and that only an additional strong electron-withdrawing group such as the nitro group is able to promote this type of ET reaction (equations 20 and 21).

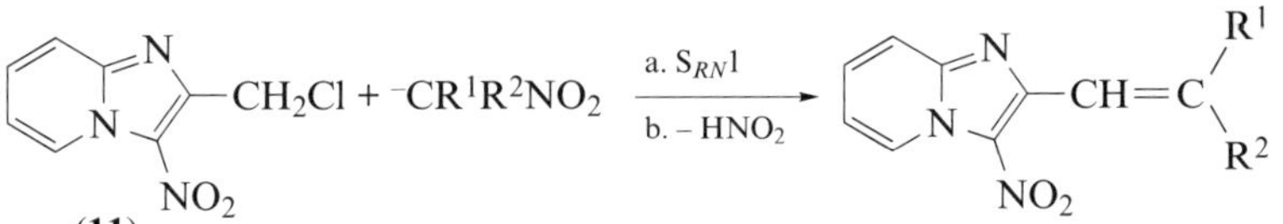

$$\text{(11)} \quad \text{—CH}_2\text{Cl} + {}^-\text{CR}^1\text{R}^2\text{NO}_2 \xrightarrow[\text{b. }-\text{HNO}_2]{\text{a. }S_{RN}1} \text{—CH}=\text{C}\begin{smallmatrix}R^1\\R^2\end{smallmatrix} \quad (20)$$

$$\text{(12)} \quad \text{—CH}_2\text{Cl} + {}^-\text{CR}^1\text{R}^2\text{NO}_2 \quad \nrightarrow \quad \text{no reaction} \quad (21)$$

Analogous nitroimidazoles with fused heterocyclic rings **13**[88a] and **14**[88b] have been also shown to react with nitronate anions (equations 22 and 23).

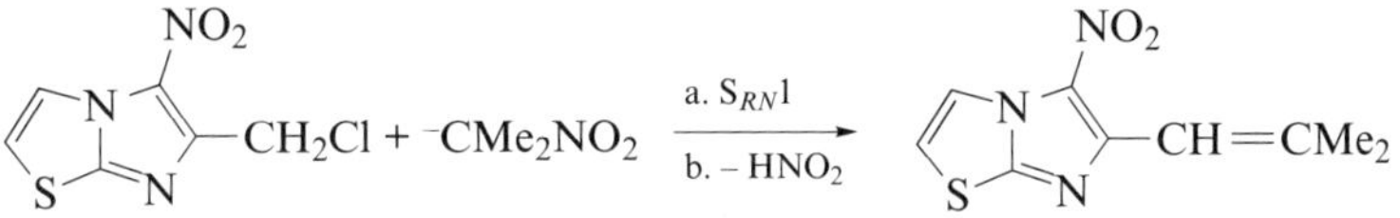

$$\text{(13)} \quad \text{—CH}_2\text{Cl} + {}^-\text{CMe}_2\text{NO}_2 \xrightarrow[\text{b. }-\text{HNO}_2]{\text{a. }S_{RN}1} \text{—CH}=\text{CMe}_2 \quad (22)$$
$$(97\%)$$

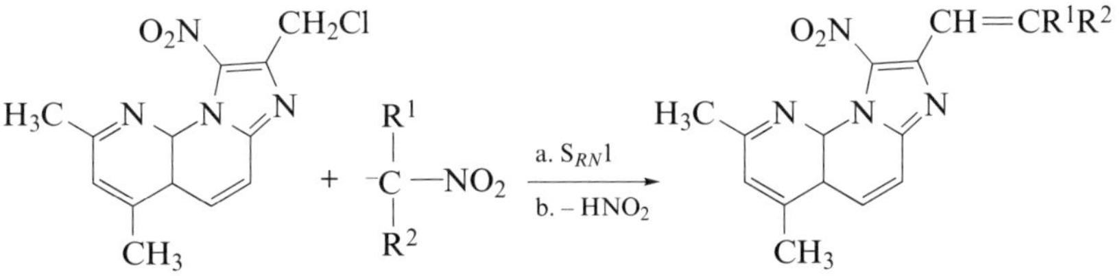

$$\text{(14)} \quad + \quad {}^-\underset{R^2}{\overset{R^1}{\text{C}}}\text{—NO}_2 \xrightarrow[\text{b. }-\text{HNO}_2]{\text{a. }S_{RN}1} \quad (23)$$
$$(59\text{–}97\%)$$

$$R^1 = R^2 = Me; \ R^1R^2 = -(CH_2)_n- \quad n = 4\text{–}7, \ -(CH_2)_{11}-;$$

The fact that no C- or O-alkylation products were formed in the reaction of analogues of **13** and **14** lacking the nitro group indicates the importance of the presence of this EWG for the formation of C-alkylation products by the $S_{RN}1$ mechanism. These reactions constitute a powerful synthetic tool for obtaining new nitroheterocycles having potential pharmacological properties.

More recently, the reaction of 1-chloromethyl-5-nitroisoquinoline with 2-nitropropane anion was the first reported example of substitution by the $S_{RN}1$ mechanism in the isoquinoline series. The reaction gave C-alkylation product, and the mechanism was confirmed by inhibitory effects of dioxygen, p-DNB, cupric chloride and TEMPO[89].

D. Other Benzylic Derivatives

Benzylic substrates with EWG like anthraquinone[90] (15) and 1,4-benzoquinone[91,92] (16) have been shown to react with the anion of 2-nitropropane by the $S_{RN}1$ mechanism (equations 24 and 25).

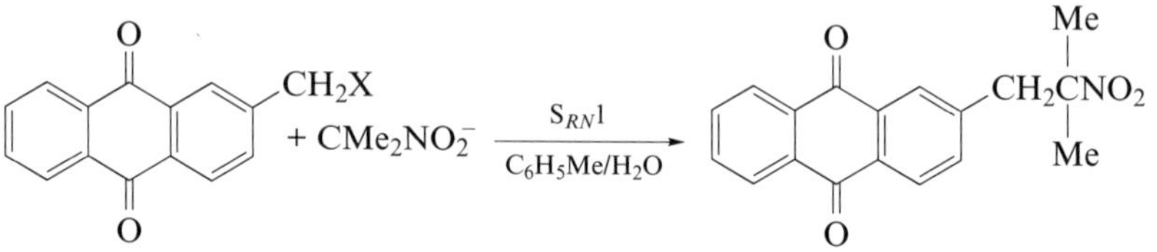

$$\text{(15)} \qquad \text{X = Cl, Br} \qquad\qquad\qquad\qquad 77\% \qquad (24)$$

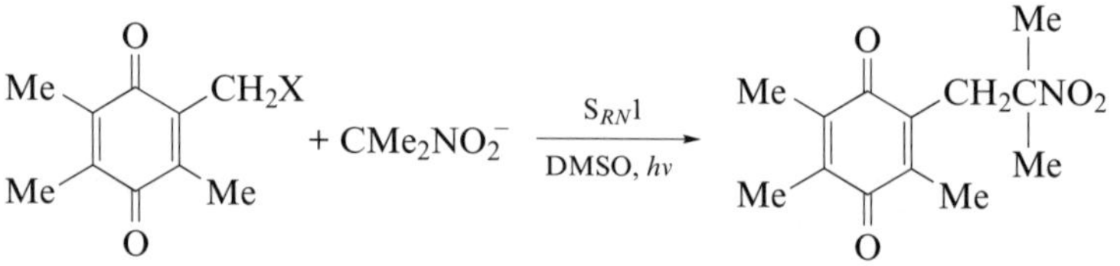

$$\text{(16)} \qquad \text{X = F, Cl} \qquad\qquad\qquad\qquad 70\text{–}83\% \qquad (25)$$

More recently the C-alkylation of 15 by 2-nitropropane anion was extended to the heterocycle nitronate anion derived from 1-methyl-3-nitropyrrolidin-2-one[93].

The importance of the quinone group is well-established for several types of bioreductive alkylating agents which exhibit anti-tumour activities and are used in treatment of tumours and leukemia.

The reaction of the fluoro compound 16 (X = F)[92] is the first example of the displacement of fluoride ion in $S_{RN}1$ reactions at an sp^3 carbon atom.

Crozet, Vanelle and coworkers[94] have also described the first example of an $S_{RN}1$ reaction involving an oxazole system activated by a non-fused tetrasubstituted p-benzoquinone 17, with 2-nitropropane anion (equation 26).

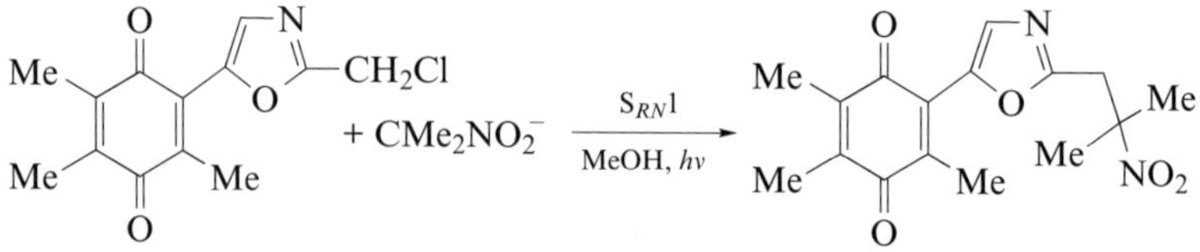

$$\text{(17)} \qquad\qquad\qquad\qquad\qquad 88\% \qquad (26)$$

On the other hand, 4-chloromethylthiazole activated by a substituted p-benzoquinone in position 2 does not react with 2-nitropropane anion, neither by C- nor by O-alkylation ($S_{RN}1$ and S_N2, respectively)[95].

The cyano group is another group that clearly facilitates substitution reactions via initial ET, although not as effectively as a nitro group[96]. Thus, p-cyanobenzyl chloride gives the aldehyde when treated with 2-nitropropane anion, whereas *p*-cyanocumyl chloride gives the substitution product (equations 27 and 28).

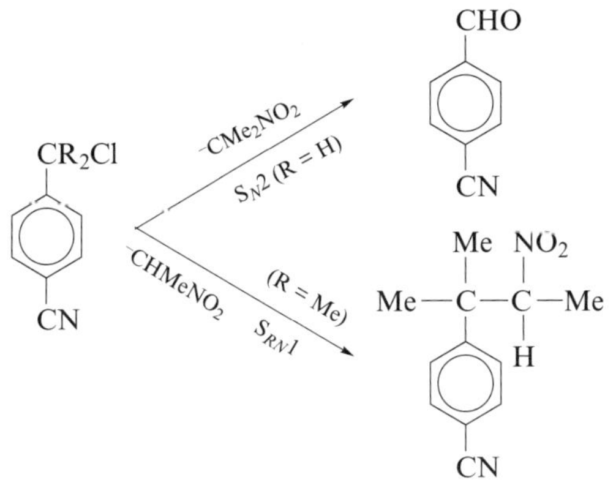

$$(27)$$

$$(28)$$

E. Other Activated Alkyl Halides

Similarly to what has been shown above[96], that an aromatic cyano group in benzylic substrates facilitates nucleophilic substitution of the $S_{RN}1$ type, it has been demonstrated that an α-cyano group is also able to activate an haloalkane. Thus, the substitution of 2-bromo-2-cyanopropane with different nitronate anions by the $S_{RN}1$ mechanism has been described[97].

Poly(α-chloroacrylonitrile) decomposes to low molecular weight compounds when treated with nucleophiles [*N,N*-diethyldithiocarbamate (Et$_2$NCS$_2^-$), PhS$^-$ and azide ions]. An$_2$ $S_{RN}1$ mechanism was suggested for this reaction, in which an ET to the polymer leads to a radical and chloride ion. Coupling with the nucleophile and decomposition are the main reactions proposed for the radical intermediates[98]. The reaction of 2-chloro-2-methylpropionitrile, as a model compound, with *N,N*-diethyldithiocarbamate (52% yield) and PhS$^-$ (61% yield) was studied[98].

Another substrate activated by a C=O group that reacts by the $S_{RN}1$ mechanism is 2-oxo-3,3-dimethylbicyclo[2.2.2]oct-1-yl chloride (Section IV.A.2).

Sterically hindered *p*-cyano- and *p*-nitro-α-haloisobutyrophenones (**18**) have been reported to react with the 2-nitropropane anion by the $S_{RN}1$ mechanism to form the C-alkylation product[99]. However, a *p*-nitro group is necessary for the reactions with diethyl malonate and diethyl methylmalonate, benzenethiolate and *p*-toluenesulphinate ions (equation 29).

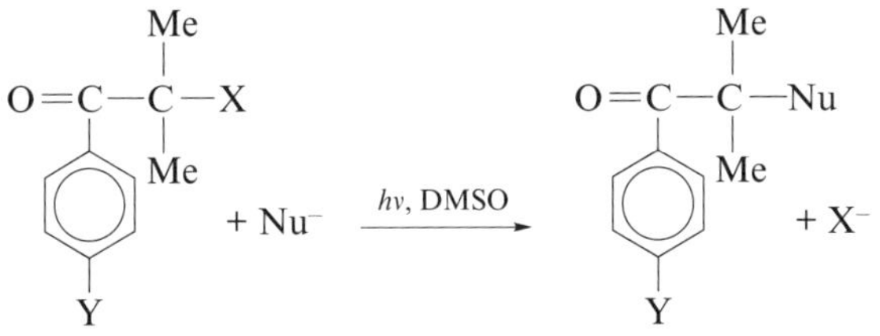

$$(29)$$

X = Cl, Br; Y = CN   Nu = $^-$CMe$_2$NO$_2$
X = Cl, Br; Y = NO$_2$ Nu = $^-$CMe$_2$NO$_2$, $^-$CR(CO$_2$Et)$_2$, PhS$^-$, ArSO$_2^-$

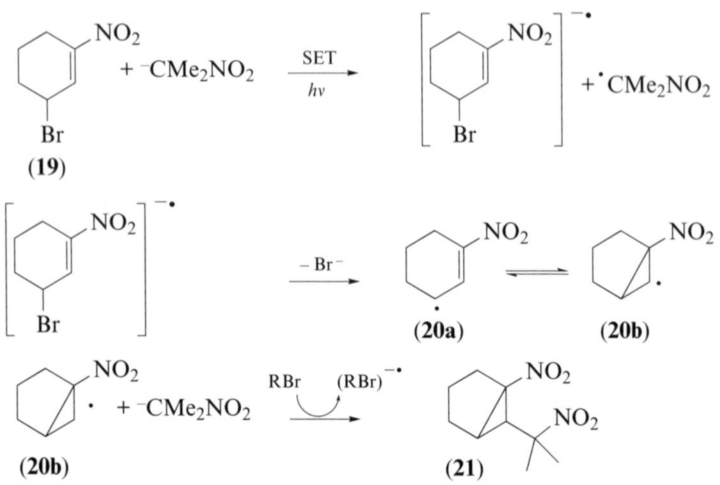

SCHEME 2

Substituted allyl halides also give substitution products by a radical chain process. Thus 3-bromo-l-nitrocyclohex-1-ene (**19**) undergoes $S_{RN}1$ reaction with the 2-nitropropane anion giving product **21** derived from valence tautomerism **20a** $\rightleftharpoons$ **20b** of the radical intermediate (scheme 2)[100].

Formation of the rearranged product was ascribed to a more rapid reaction of the low-nucleophilic anion with the unstabilized cyclopropyl radical **20b** rather than with the stabilized nitro-olefin radical **20a**[100].

In the reactions of *p*-nitrophenylallyl chlorides (**22**) with different carbanions[101], the mechanism involved depends on the alkyl substituent R. Thus, the methyl derivative (R = Me) undergoes S_N2 or S_N2' reaction, the *i*-propyl derivative (R = *i*-Pr) undergoes mixed $S_{RN}1'/S_N2$ reactions and the *t*-butyl derivative (R = *t*-Bu) undergoes $S_{RN}1'$ reaction. The $S_{RN}1'$ designation implies allylic rearrangement of the intermediate radical (equations 30 and 31).

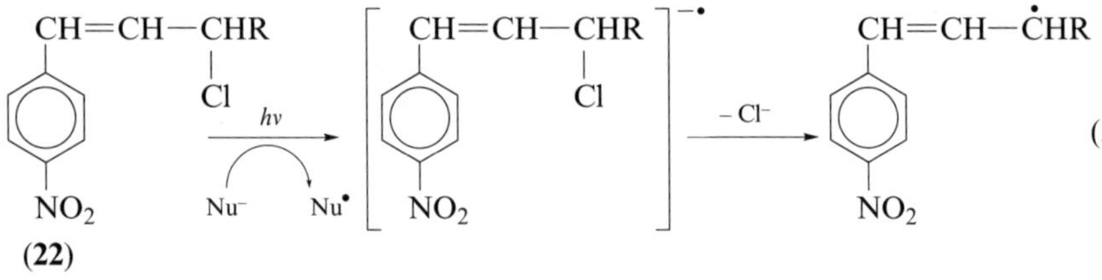

$$\text{(30)}$$

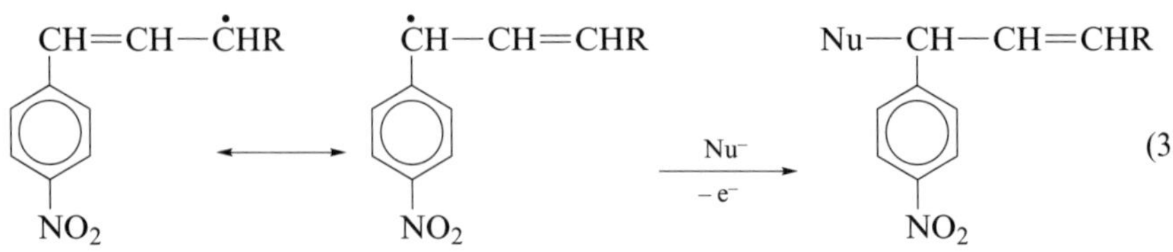

$$\text{(31)}$$

Nu$^-$ = $^-$CMe$_2$NO$_2$, $^-$CMe(CO$_2$Et)$_2$, *p*-MeC$_6$H$_4$SO$_2$$^-$
R = *t*-Bu

F. Nature of the Nucleophile

1. Carbon nucleophiles

The most commonly used nucleophiles for $S_{RN}1$ reactions of activated aliphatic compounds have been primary and secondary nitronate anions $R^1R^2CNO_2^-$ from the simplest anion of nitromethane[102] to more complex secondary nitronateanions[5,8,13] in which R^1 and R^2 are aliphatic, cyclic and heterocyclic [76b,81,83,87,88b,93,103–105].

$S_{RN}1$ substitution of different compounds by nitronate anions followed by elimination of nitrous acid is a good method for the synthesis of various substituted olefins. A clear example is the base-promoted HNO_2 elimination from the C-alkylation products formed by the $S_{RN}1$ reaction between 2-chloromethyl-3-nitroimidazo[1,2-a]pyridine (**11**) and various aliphatic, cyclic and heterocyclic nitronate anions[87], affording new potential pharmacological derivatives with a trisubstituted double bond at the 2-position. A representative example is shown in equation 32.

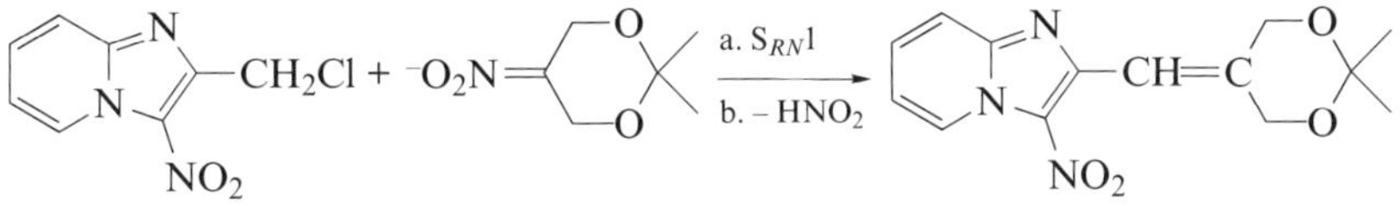

(**11**)   79%   (32)

A similar procedure can be applied to the synthesis of a series of 5-nitroimidazoles[83] with antibacterial properties.The $S_{RN}1$ reaction of various 1-alkyl-2-chloromethyl-5-nitroimidazoles (**23**) with heterocyclic nitronate anions prepared from 3-nitrolactams, under phase transfer catalysis, afforded trisubstituted olefins (equation 33).

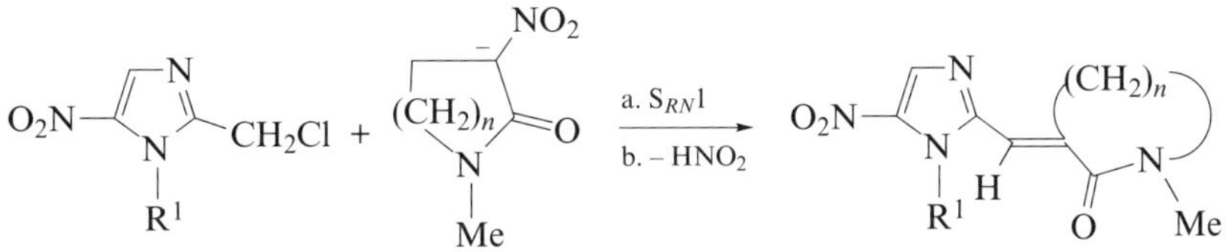

(**23**) $n = 1-4$ variable yield   (33)

Disubstituted anions of the $RCZY^-$ type react with α-halonitroalkanes by the $S_{RN}1$ mechanism. After elimination of nitro and ester groups, or nitro and keto groups, the products can be converted into α,β-unsaturated ketones and esters[106–108], nitriles[108] and sulphones[108] making this reaction a good method for synthesis of tri- and tetrasubstituted olefins (equation 34).

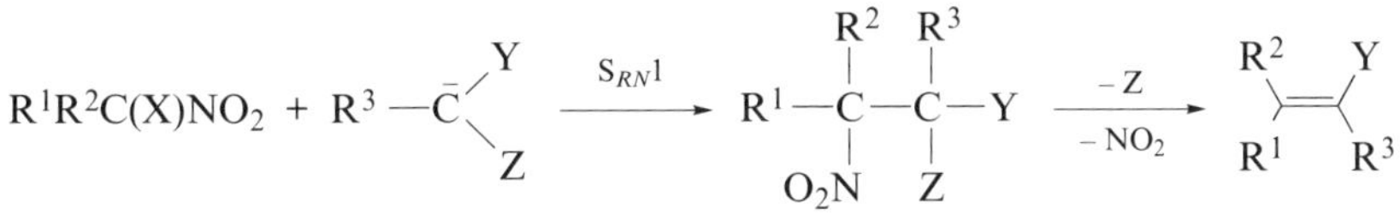

$$R^1R^2C(X)NO_2 + R^3-\overset{Y}{\underset{Z}{C}} \xrightarrow{S_{RN}1} R^1-\overset{R^2}{\underset{O_2N}{C}}-\overset{R^3}{\underset{Z}{C}}-Y \xrightarrow[-NO_2]{-Z} \overset{R^2}{\underset{R^1}{}}=\overset{Y}{\underset{R^3}{}} \quad (34)$$

X = Br, Cl

Y = CO_2Et, COR, CN, SO_2Ar;   Z = CO_2Et
Y = COR;   Z = COR
Y = R^3 = COR; Y = COR, $R^3 = CO_2R$ Z = H

Russell and coworkers[62,109,110] have shown that simple enolates undergo free radical-chain nucleophilic substitution reactions with α-chloronitroalkanes by an $S_{RN}2$ rather than an $S_{RN}1$ mechanism, and competition with a chain dimerization process was also observed. Using two equivalents of the enolate anion in the reaction allows complete elimination of HNO_2 to yield α,β-unsaturated ketones. The synthetic potential of these reactions has also been reported[110].

The reaction between cyanide ions and p-nitrocumyl chloride is the only example reported of an $S_{RN}1$ reaction involving this anion[66b].

2. Nitrogen nucleophiles

Not many N-centred anions have been reported to react via $S_{RN}1$ reactions. One of the most suitable is the azide anion which reacts with several substrates[8,74,77,78] by this mechanisme.

Kornblum and coworkers[66b] have also reported the reaction between p-nitrocumyl chloride and nitrite ion or various amines like piperidine, pyrrolidine, quinuclidine and DABCO.

A new group of N-centred anions derived from diazoles has been recently investigated. Thus, anions of imidazole and benzimidazole[111,112], pyrazole[111], triazole[111], 5(6)-nitrobenzimidazole[112], 5-nitroindazole[112], 6-nitroindazole[112] and 4(5)-nitroimidazole[82,113] have been reported to undergo $S_{RN}1$ reaction with different substrates. An example of the last anion is given in equation 35.

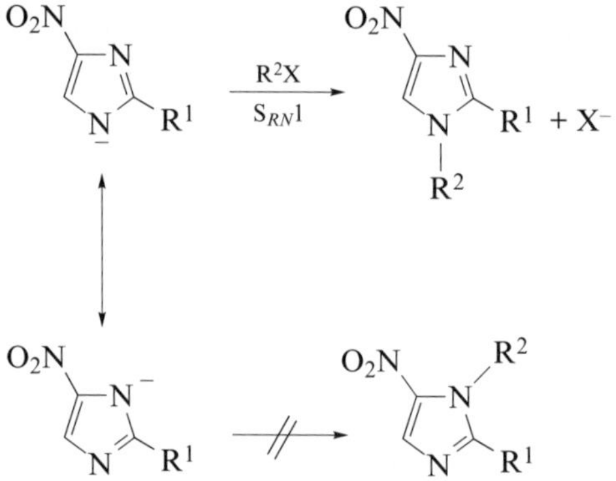

$$(35)$$

$$R^1 = H, Me; \quad R^2 = Me_2(NO_2)C, \ p\text{-}O_2NC_6H_4CH_2,$$

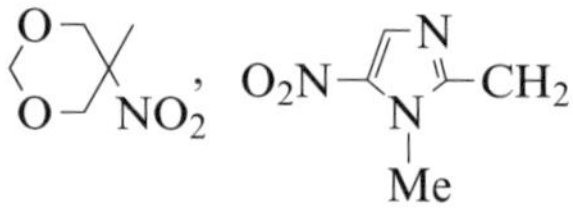

3. Sulphur and oxygen nucleophiles

Benzenethiolate and various arenethiolate (p-methylphenyl, p-chlorophenyl, p-nitrophenyl, o-nitrophenyl) ions have been shown to react by the $S_{RN}1$ mechanism with different substrates such as p-nitrobenzyl halides[5,8,13], p-nitrocumyl halides[66b], α-haloisobutyrophenone[99], nitroacenaphthenes[74] and 5-nitro-2-thienylneopentyl chloride[78]. An example of the last substrate is given in equation 17.

Thiolate ions have been reported to undergo reactions with aliphatic α-halonitro compounds[5,114,115] to yield the substitution products α-nitrosulphides or disulphides by oxidative dimerization. The substitution reaction is favoured by weakly nucleophilic thiolate ions and proceeds by an $S_{RN}1$ mechanism. On the other hand, strongly nucleophilic thiolate ions favour the redox reaction by an ionic abstraction (X-philic) mechanism.

A number of heterocyclic S—N ambident anions have also been shown to react via the sulphur atom with 2-halo 2-nitropropanes by the $S_{RN}1$ mechanism[5,116]. The authors suggest[116] that the addition of ambident anions to alkylnitro radicals (CR_2NO_2) takes place under kinetic control via the more nucleophilic centre.

Benzenesulphinate[99] and p-toluenesulphinate ions[77,78,99] have been reported to participate as nucleophiles in $S_{RN}1$ reactions with most substrates.

The thiocyanate anion[8] has been shown to undergo the $S_{RN}1$ reaction at higher temperatures (50–100 °C) in low yield (10–40%) with 2-iodo-2-nitropropane, but not with the bromo analogue.

Only one example of an oxygen-centred nucleophile, 1-methyl-2-naphthoxide, has been proposed to participate in an $S_{RN}1$ reaction with p-nitrocumyl chloride[66].

4. Phosphorous nucleophiles

Dialkylphosphite ions have been shown to undergo $S_{RN}1$ reactions with different halonitro substrates[117]. The analogous thiophosphite ions[117] react by this mechanism only with p-nitrobenzyl or p-nitrocumyl chlorides, but undergo X-philic reactions with 2-chloro-2-nitropropane.

Relative reactivities of both anions towards $\cdot CMe_2NO_2$, p-$NO_2C_6H_4CH_2\cdot$ and p-$NO_2C_6H_4C\,Me_2\cdot$ have also been reported with a change of the counterion and the solvent[117]. The results indicate that the $(EtO)_2PS^-$ ion is a better trap than the $(EtO)_2PO^-$ ion for α-nitroalkyl radicals, particularly when ion pairing is important.

IV. ALKYL HALIDES WITHOUT ELECTRON–WITHDRAWING GROUPS

Nucleophilic substitution reactions on unactivated alkyl halides have been known for a long time. The available mechanisms depend on the aliphatic moiety, the nucleophile, the leaving group and the reaction conditions[118]. Besides the polar mechanisms of nucleophilic substitution reactions (S_N1, S_N2 and related mechanisms), several alkyl halides react with nucleophiles by an ET reaction.

The alkyl halides that react by the $S_{RN}1$ mechanism have a relatively low reactivity toward polar nucleophilic substitution due to steric, electronic or strain factors[11], as shown for several bicycloalkyl, cycloalkyl and neopentyl halides and for t-butyl chloride. In the following section the reaction of these alkyl halides with different nucleophiles will be discussed.

A. Bicycloalkyl Halides

1-Substituted bicycloalkyl halides are very unreactive toward nucleophilic substitution reactions. The low reactivity in S_N1 reactions has been attributed to the fact that a planar configuration at the bridgehead carbon cannot be obtained without the introduction of considerable strain[119]. On the other hand, the S_N2 reaction is precluded because a backside approach of the nucleophile cannot occur at a bridgehead position for a steric reason. The lack of reactivity of 1-halobicycloalkanes toward nucleophiles by polar mechanisms makes them attractive substrates for the $S_{RN}1$ mechanism.

1. 1-Halo and 1,3-dihaloadamantanes

1-Haloadamantanes (1-AdX) react in liquid ammonia under photostimulation with diphenylphosphide (Ph_2P^-) [X=Cl (40%), Br, I (70–76%)[120] and Ph_2As^- (X = Br (65%))][121] ions, giving an unrearranged substitution product (isolated as the oxide in the case of Ph_2P^- ions); see equation 36.

$$1\text{-AdX} + Ph_2Z^- \ (Z = P, As) \xrightarrow{\ hv\ } 1\text{-AdZPh}_2 \qquad (36)$$

The anions PhS^-, $PhSe^-$ and $PhTe^-$ react with 1-adamantyl iodide under similar experimental conditions giving different results[122]. Thus, 1-adamantyl phenyl sulphide was formed in 45% yield in the reaction with PhS^- while with $PhSe^-$ and $PhTe^-$ ions the products in equation 37 were formed.

$$1\text{-AdI} + PhZ^- \xrightarrow{\ hv\ } Ph_2Z + 1\text{-AdZPh} + (1\text{-Ad})_2Z \qquad (37)$$

$$
\begin{array}{cccc}
Z = Se & 10\% & 74\% & 16\% \\
Z = Te & 31\% & 69\% &
\end{array}
$$

This product distribution, which can be easily rationalized by the $S_{RN}1$ mechanism, was accounted for by the rates of the fragmentation reactions of the proposed radical-anion intermediate, as sketched in Scheme 3 for $PhSe^-$[122].

The products were ascribed to the formation of a $\sigma*$ radical anion **24** which undergoes three competitive reactions, namely reversion to starting materials, ET to the substrate to give 1-adamantyl phenyl selenide (**25**) and fragmentation of the Ph—Se bond to give the 1-adamantaneselenate ion (**26**) and phenyl radical (equation 38). The intermediates **26** and phenyl radical can react to give back **24** or they can diffuse apart. In this case anion **26** can react with 1-Ad' radical to give 1-Ad$_2$Se (equation 39), whereas the Ph radical reacts with $PhSe^-$ ion to give Ph_2Se (equation 40).

However, if the nucleophiles have a low antibonding $\pi*$ MO, as in the case with the 1-naphthaleneselenate ion, the radical anion formed does not fragment because the odd electron is localized in the naphthyl moiety as a $\pi*$ radical anion. In this system only the substitution product 1-naphthyl-1-adamantyl selenide was formed in 50% yield[122].

The differences in the bond-fragmentation rates of the radical-anion intermediates of these reactions have been studied. The frangibility of the C—Z bond in the radical-anion intermediate was found to increase in the order S<Se<Te. The photostimulated reactions of 1-AdI with Se^{2-} or Te^{2-} in liquid ammonia are also known[122].

$$1\text{-Ad}^\bullet + {}^-SePh \rightleftharpoons [1\text{-AdSePh}]^{-\bullet} \rightleftharpoons 1\text{-AdSe}^- + Ph^\bullet \qquad (38)$$
$$\phantom{1\text{-Ad}^\bullet + {}^-SePh \rightleftharpoons}(\mathbf{24})\phantom{[1\text{-AdSePh}]^{-\bullet} \rightleftharpoons 1\text{-Ad}}(\mathbf{26})$$

$$ET \ \Big| \ k_{ET} \, [1\text{-AdI}]$$

$$\downarrow$$

$$1\text{-AdSePh}$$
$$(\mathbf{25})$$

$$1\text{-AdSe}^- + 1\text{-Ad}^\bullet \rightleftharpoons [1\text{-Ad}_2Se]^{-\bullet} \xrightarrow{\ ET\ } 1\text{-Ad}_2Se \qquad (39)$$
$$(\mathbf{26})$$

$$PhSe^- + Ph^\bullet \rightleftharpoons [Ph_2Se]^{-\bullet} \xrightarrow{\ ET\ } Ph_2Se \qquad (40)$$

SCHEME 3

By competition experiments the Ph_2P^- ion is 830 times more reactive than the PhS^- ion toward the 1-adamantyl radical in DMSO, but only 8.4 times more reactive toward the *p*-anisyl radical. This reactivity difference was attributed to the greater stability of the 1-adamantyl radical compared with the *p*-anisyl radical[123].

Carbanions, such as the enolate ions of acetone, acetylmorpholine and others were unreactive toward 1-AdI in liquid ammonia under photostimulation, and only the reduction product AdH and the dimer $(1\text{-Ad})_2$ were formed. It has been proposed that even though the 1-Ad˙ radicals are formed under photostimulation, they do not couple with carbanions to give the $S_{RN}1$ product, at least at a competitive rate with other reactions[121].

However, these reactions succeeded in DMSO. The results are summarized in equation 41[22,124].

$$1\text{-AdI} + {^-CH_2COR} \xrightarrow[\text{DMSO}]{hv} 1\text{-AdCH_2COR} + \text{AdH} \tag{41}$$

(27a) R = Me   (28a) R = Me (20%) 17%
(27b) R = Ph   (28b) R = Ph (65%)   9%

Good yields of substitution product **30** (75%) were also obtained in the reaction of 1-AdI with the anion of anthrone (**29**) (equation 42)[124].

$$\tag{42}$$

(29)   (30)

On the other hand, no reaction was observed under irradiation with nitronate ion (**31**) alone. It is known that in $S_{RN}1$ reactions a more reactive nucleophile can induce the reaction of a less reactive one (entrainment reaction)[5,8,13,14]. Thus nitronate ion gave the substitution product **32** in the presence of acetophenone **27b** (58%) or acetone **27a** (87%) enolate ions (equation 43).

$$1\text{-AdI} + {^-CH_2NO_2} \xrightarrow{hv} 1\text{-AdCH_2NO_2} \tag{43}$$
$$\quad\quad\quad\quad\;\; (31) \quad\quad\quad\quad\quad (32)$$

Qualitatively, **27a** is more reactive than **27b** in independent experiments, but in competition experiments **27b** is more reactive than **27a**. The results suggest that **31** is unable to initiate the $S_{RN}1$ reaction, but it does propagate the chain process very efficiently.

Also, by competition experiments, 1-Ad˙ radicals are quite selective with these carbanionic nucleophiles, the apparent reactivity being in the order: **29** (80)>**31**(32)>**27b** (11)>**27a** (1.0)[124].

The reactivity of carbanions with 1-AdI differs for different reaction steps. Thus, the photostimulated ET rate from the carbanion to 1-AdI increases with the pK_a of the conjugate acid of the carbanion (the pK_a values in DMSO are acetone: 26.5, acetophenone: 24.7 and nitromethane: 17.2[125]). The reactivity in the coupling reaction does not follow the pK_a order of the ketones.

From a frontier orbital point of view the most important interaction in this coupling reaction is between the HOMO of the nucleophile and the SOMO of the radical[126]. Thus, in the propagation cycle the reactivity of the carbanions would depend on their HOMO (pK_a) as well as on the SOMO of the radical anion formed, the HOMO–SOMO energy difference taken as an indication of the loss in π energy[124]. The HOMO energy of the carbanions decreases along with pK_a decrease in their conjugate acid, and the same tendency is followed by the SOMO energy of the corresponding radical anions. However, the HOMO–SOMO energy difference becomes smaller for more stabilized radical anions. It was proposed that the reactivity of a carbanion toward a radical increases as the pK_a decreases whenever the ΔE_π of the reaction follows a similar tendency[124] (ΔE_π **27a**>**27b**≈**31**≈ **29**).

The acetylacetone anion (**33**) is expected to be highly reactive according to its pK_a; however, an important loss in π energy occurs in the coupling reaction with this nucleophile. In the anion, both carbonyl groups stabilize the negative charge (low HOMO). However, in the radical anion **34** formed, this stabilization is prevented by the sp^3 carbon atom which separates the two carbonyl groups and only one carbonyl stabilizes the unpaired spin distribution (high SOMO) (equation 44).

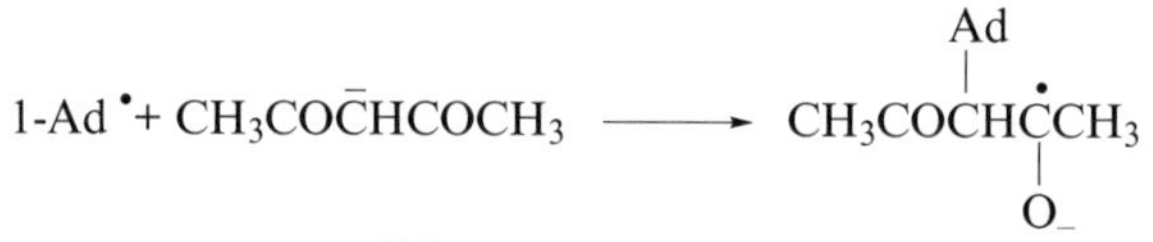

$$1\text{-Ad}^{\bullet} + CH_3CO\overset{-}{C}HCOCH_3 \longrightarrow \underset{\underset{O_-}{|}}{\overset{\overset{Ad}{|}}{CH_3COCH\overset{\bullet}{C}CH_3}} \qquad (44)$$

$$(\mathbf{33}) \qquad\qquad\qquad\qquad (\mathbf{34})$$

Carbonylation of 1-AdBr and 1-AdCl by the complex reducing agent 'NaH—RONa—FeCl$_3$—CO', which led to the dicarbonyl dianion Na$_2$Fe(CO)$_4$ in 1,2-dimethoxyethane has been performed. This reaction gives a mixture of 1-AdCO$_2$H and 1-AdCO$_2$R (80–88% overall carbonylation yield), and it was suggested to occur by the S$_{RN}$1 mechanism[127].

1,3-Dihaloadamantanes (**35**) react with Ph$_2$P$^-$ ions in liquid ammonia to afford mainly the disubstitution product **36**, together with **37** and **38** (equation 45)[128a].

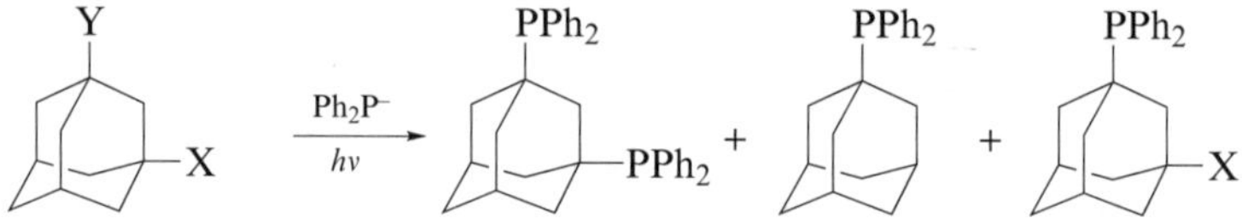

(**35**)	(**36**)	(**37**)	(**38**)
(**35a**) X = Y = Cl	74%	14%	3%
(**35b**) X = Cl, Y = Br	76%	6%	7%
(**35c**) X = Y = Br	88%	8%	<2%

These results indicate that when the intermediate 3-halo-1-adamantyl radical couples with Ph$_2$P$^-$ ions, it forms the radical anion (**38**)$^{-\bullet}$ which reacts via two competitive reactions. One is the intramolecular ET to the C—X σ^* bond, which fragments to give the radical **39** that enters the propagation cycle to give ultimately **36** or is being reduced to **37**. The other possibility is the intermolecular ET to the substrate to give the monosubstitution product **38** (equation 46).

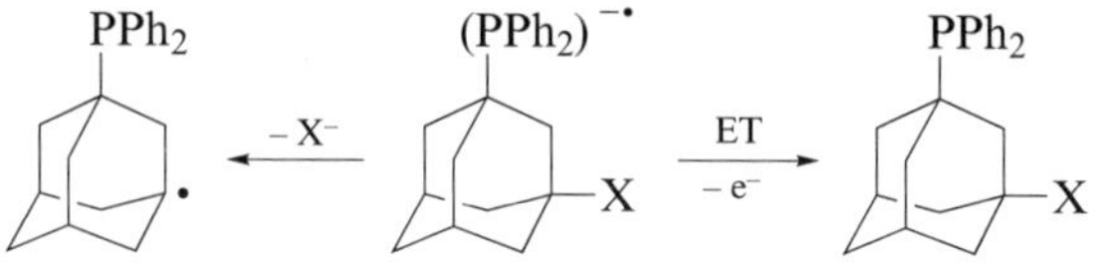

The photostimulated reaction of **35b** with p-dinitrobenzene (p-DNB) gave only 33% of **36**, and as much as 38% of **38** (X = Cl). Clearly p-DNB does not inhibit the reaction of **35b** as effectively as of the intramolecular ET of the intermediate radical anion (**38**)$^{-\bullet}$ (X = Cl).

Recently, it has been reported[128b] that substrates **35b**, **35c**, 1-chloro-3-iodo (**35d**), 1-bromo-3-iodo (**35e**) and 1-3-diiodo (**35f**) adamantanes react with Me$_3$SnM (M = Li, Na) in THF, giving a product distribution that depends on the halogens and the experimental conditions. Thus **35b**, **35c** (reactants added in an inverse order) and **35d** react with Me$_3$SnM to give high yields of the disubstitution product 1,3-bis(trimethylstannyl) adamantane by the $S_{RN}1$ mechanism. However, substrates **35d-f** react with Me$_3$SnM affording mainly 1,3-dehydroadamantane **40**.

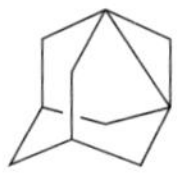

(40)

Based on trapping experiments with t-butylamine and dicyclohexylphosphine the formation of the propellane has been proposed to occur through radical intermediates[128b].

On the other hand, 1-bromo-3-fluoro- and 1-fluoro-3-iodoadamantane react with Me$_3$SnM to give the monosubstitution product 1-fluoro-3-trimethylstannyladamantane in 97% and 92% yields, respectively[128b].

2. 1-Halo and 1,4-dihalobicyclo[2.2.2]octanes

The photostimulated reaction of 1-iodobicylo[2.2.2]octane (**40a**) with Ph$_2$P$^-$ ions in liquid ammonia gave, after oxidation, the substitution product **41** (equation 47) while **40b** does not react in the dark or under photostimulation[129].

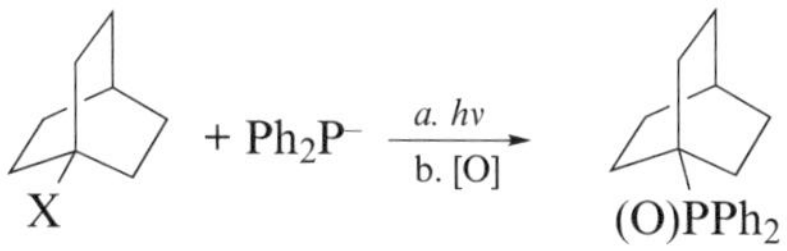

$$\text{(47)}$$

(**40a**) X = I   (**41**) 87%
(**40b**) X = Cl   0%

In the reaction of this nucleophile with **42a** the only product found was **43a**, in which the substitution occurred at the iodine (equation 48). Both halogens were substituted in the reaction of Ph$_2$P$^-$ ions with the 1,4-bromoiodo (**42b**) or 1,4-diiodo (**42c**) derivative giving only the disubstitution product **44** (equation 48).

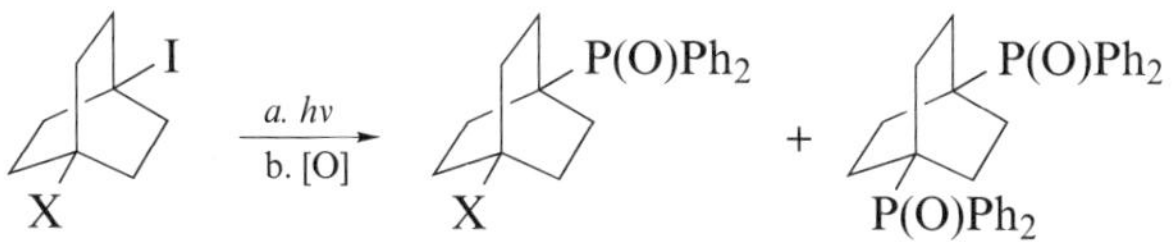

$$\text{(48)}$$

(**42a**) X = Cl   (**43a**) X = Cl   (**44**)
(**42b**) X = Br
(**42c**) X = I

The different behaviour of these substrates indicates that in the radical anion $(43)^{-\bullet}$, the fragmentation rate to give **45** is faster than the ET to **42** for X = Br and I, while for X = Cl the intermolecular ET to give **43a** (equation 49) is faster than the fragmentation reaction.

$$\underset{(\mathbf{45})}{\text{[structure PPh}_2]} \xleftarrow[- X^-]{k_{ET}\ \text{Intramolecular}} \underset{(\mathbf{43})^{-\bullet}}{\left[\text{structure PPh}_2,\ X\right]^{-\bullet}} \xrightarrow[- e^-]{k_{ET}\ \text{Intermolecular}} \underset{(\mathbf{43a})\ X = Cl}{\text{[structure PPh}_2,\ X]} \qquad (49)$$

The reaction of **42b** with Me$_3$SnLi in THF gave three stannanes **46–48** as substitution products, in a ratio 4:2:1 respectively (equation 50)[130].

$$\underset{(\mathbf{42b})}{\overset{Br}{\underset{I}{\text{[structure]}}}} \xrightarrow{\text{Me}_3\text{SnLi}} \underset{(\mathbf{46})}{\overset{I}{\underset{\text{SnMe}_3}{\text{[structure]}}}} + \underset{(\mathbf{47})}{\overset{Br}{\underset{\text{SnMe}_3}{\text{[structure]}}}} + \underset{(\mathbf{48})}{\overset{\text{SnMe}_3}{\underset{\text{SnMe}_3}{\text{[structure]}}}} \qquad (50)$$

These results suggest an unprecedented halogen nucleofugality (Br>I) for a halogen–metal exchange (HME) or ET reaction. However, competition experiments of 1-iodo and 1-bromobicyclo[2.2.2]octanes established that the iodine is *ca* twice as reactive as the bromine derivative[131].

Based on the results of deuterium labelling and radical trapping experiments with dicyclohexylphosphine (DCHP), it was proposed that the reaction occurs by the S$_{RN}$1 mechanism. A major difference, however, is an additional propagation step involving iodine atom abstraction from **42b** by the radical intermediate **49**, to give the substitution product **46** and radical intermediate **50**, that continues the chain propagation cycle of the S$_{RN}$1 mechanism (equation 51).

$$\underset{(\mathbf{49})}{\overset{\bullet}{\underset{\text{SnMe}_3}{\text{[structure]}}}} + \underset{(\mathbf{42b})}{\overset{I}{\underset{\text{Br}}{\text{[structure]}}}} \longrightarrow \underset{(\mathbf{46})}{\overset{I}{\underset{\text{SnMe}_3}{\text{[structure]}}}} + \underset{(\mathbf{50})}{\overset{\bullet}{\underset{\text{Br}}{\text{[structure]}}}} \qquad (51)$$

The reaction of Me$_3$Sn$^-$ ions with **42d** (X = F) gave only the substitution product **51d** (X = F), but with **42a** (X = Cl), products **51a** (X = Cl), **51c** (X = I) and **48** were also formed (equation 52)[132].

$$\underset{\substack{(\mathbf{42d})\ X = F \\ (\mathbf{42a})\ X = Cl}}{\overset{X}{\underset{I}{\text{[structure]}}}} + \text{Me}_3\text{SnLi} \longrightarrow \underset{\substack{(\mathbf{51a})\ X = Cl \\ (\mathbf{51c})\ X = I \\ (\mathbf{51d})\ X = F}}{\overset{X}{\underset{\text{SnMe}_3}{\text{[structure]}}}} + \underset{(\mathbf{48})}{\overset{\text{SnMe}_3}{\underset{\text{SnMe}_3}{\text{[structure]}}}} \qquad (52)$$

It is interesting to note that 1-chloro-4-methylbicyclo[2.2.2]octane does not react with Me$_3$Sn$^-$ ions but the 1-chloro-4-iodo derivative **42a** gives substitution of both halogens.

This can easily be explained by the $S_{RN}1$ chain reactions, with substitution at chlorine by an intramolecular entrainment reaction[131].

2-Oxo-3,3-dimethylbicyclo[2.2.2]oct-1-yl chloride (52), having a C=O π-electron acceptor, does not react with Ph_2P^- ions in liquid ammonia in the dark but, under irradiation, it gives, after oxidation, a 69% yield of product 53[153] (equation 53). On the other hand, 1-chloro-3,3-dimethylbicyclo[2.2.2]octane was completely unreactive under the same experimental conditions.

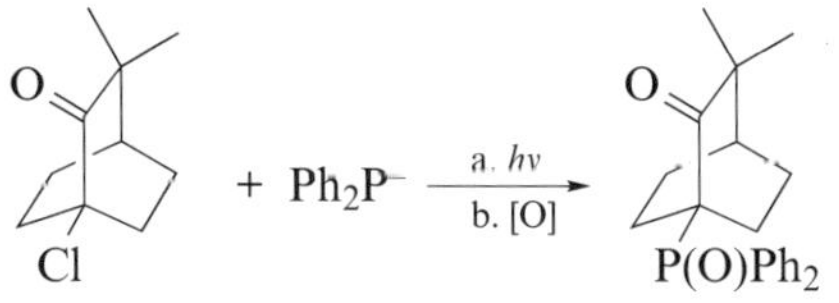

$$\text{(52)} \quad + \text{ Ph}_2\text{P}^- \xrightarrow[\text{b. [O]}]{\text{a. } h\nu} \text{(53)} \tag{53}$$

By competition experiments it was found that 52 is ≥700 times more reactive than 1-AdCl, and only 0.40 times less reactive than 1-AdBr. The LUMO of 52 belongs to the carbonyl group, and has a similar value to the LUMO of the C—Br bond in 1-AdBr. When 52 receives the electron, it is located in the antibonding π* MO of the carbonyl group to form the π* radical anion 54 which, by an intamolecular ET to the antibonding σ* MO of the C—Cl bond, forms the σ* radical anion 55, which then fragments to give chloride ions and the radical 56 to propagate the chain reaction (equation 54)[43, 133].

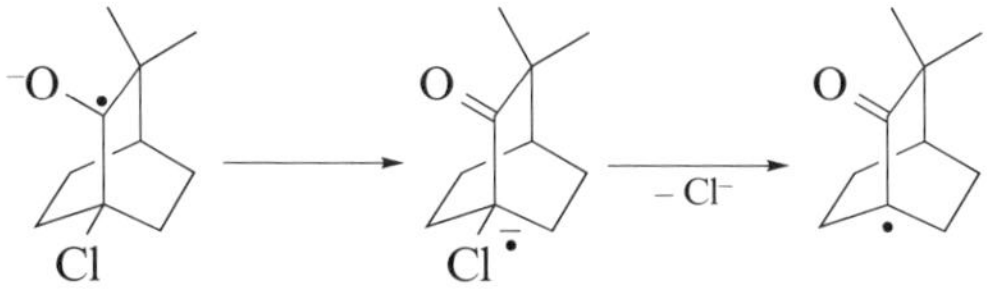

$$\text{(54)} \longrightarrow \text{(55)} \xrightarrow{-\text{Cl}^-} \text{(56)} \tag{54}$$

(54)   (55)   (56)
π* radical anion   σ* radical anion

Thus the increased reactivity of this compound vs the parent 1-chloro-3,3-dimethyl-bicyclo[2.2.2]octane can be explained through an intramolecular entrainment reaction.

3. Other bicycloalkyl halides

a. 1-Halo and 1,4-dihalo bicyclo[2.2.1]heptanes (norbornanes). The reaction of 1-iodonorbornane (57a) with $NaSnMe_3$, $LiPh_2P$ and other nucleophiles was studied[134]. Based on product distribution, deuterium labelling studies and the use of radical and radical-anion trapping reagents, it was concluded that 57a reacts to give the reduction product 58 (15%) and the substitution product 59 (83%) by an ET process, 1-norcaranyl radicals being intermediate of these reactions, and the product 59 may be formed by cage collapse or $S_{RN}1$ reaction (equation 55).

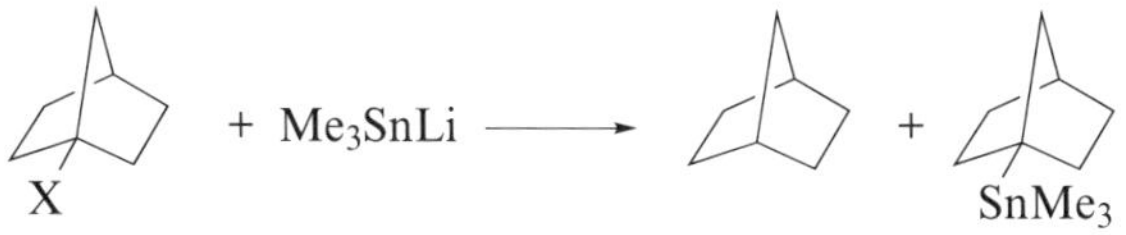

$$\text{(57a) X = I} \quad + \text{ Me}_3\text{SnLi} \longrightarrow \text{(58)} \quad + \quad \text{(59)} \tag{55}$$

(57a) X = I
(57b) X = Br

The effect of carbanion traps (t-BuNH$_2$) was determined for the reaction of **57a** and **57b** with LiSnMe$_3$ in THF[135]. Whereas **57b** appears to react exclusively by the radical pathway, **57a** was proposed to react by the radical (*ca* 60%) and HME (*ca* 40%) pathways.

The reactions of **57a** with LiPPh$_2$ and lithium diisopropylamide were sluggish in THF, **58** being the main product. No reaction was observed with the lithium salt of 2-nitro-propane nor with lithium i-propanethiolate[134].

Based on the product analysis of the reaction of Me$_3$SnLi with various 1,4-dihalo bicyclo[2.2.1]heptanes, it was proposed that in this system a polar mechanism can compete effectively with free radical chain processes. The HME pathway is predominant for the dibromo, bromoiodo and diiodo compounds, a competition between HME and ET was suggested with the chloroiodo compound and the ET reaction is important for the fluoroiodo and chlorobromo derivatives[135].

b. 4-Halotricyclanes.   4-Substituted tricyclanes are among the most unreactive substrates in solvolytic reactions[136]. However, the photostimulated reaction of 4-iodotri-cyclane (**60**) with Ph$_2$P$^-$ ions in liquid ammonia gives the substitution product **61** in 58% yield (equation 56)[137]. 1-Chlorotricyclane was completely unreactive under the same exper-imental conditions.

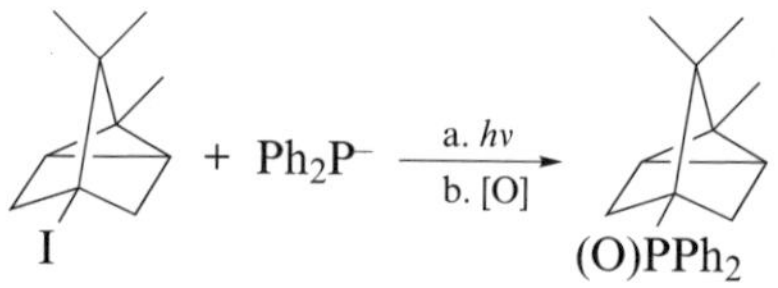

$$\text{(60)} \quad + \text{ Ph}_2\text{P}^- \xrightarrow[\text{b. [O]}]{\text{a. } h\nu} \text{(61)} \tag{56}$$

(60) **I**   **(61)** (O)PPh$_2$

c. 9-Bromotriptycene.   9-Bromotriptycene (**62**) reacts with Ph$_2$P$^-$ to give product **63** in 71% yield (equation 57), and with Ph$_2$As$^-$ ions it gives a 41% yield of the substitution product[120]. In the reaction of 9,10-dibromotriptycene with Ph$_2$P$^-$ ions, the disubstitution product was isolated in 49%[120].

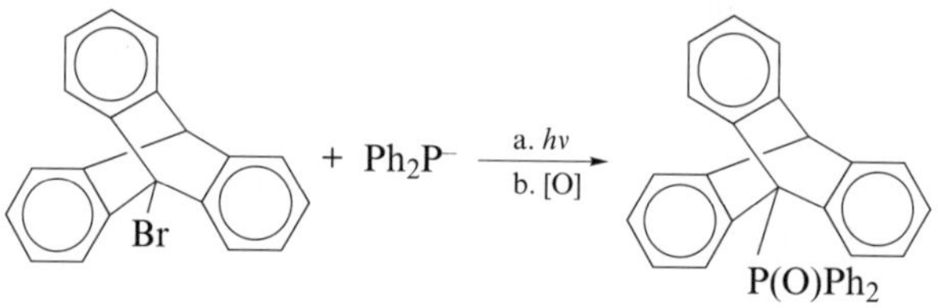

$$\text{(62)} \quad + \text{ Ph}_2\text{P}^- \xrightarrow[\text{b. [O]}]{\text{a. } h\nu} \text{(63)} \tag{57}$$

(62) Br   **(63)** P(O)Ph$_2$

d. Reactivity of bicycloalkyl halides.   The reactivity of bridgehead halides in S$_{RN}$1 can be explained through strain effects[138]. It has been shown that they react by the S$_{RN}$1 mech-anism with iodide or bromide as the leaving group, while no reaction occurred with the chloro compound (except 1-AdCl) in the photostimulated reaction with Ph$_2$P$^-$ ions in liquid ammonia[11,12].

A theoretical study was performed to determine the stability of bicycloalkyl halide radical anions as well as the dependence of the reduction potential of the corresponding neutral compounds on the angular strain energy of the parent hydrocarbons. It was proposed that an increase in strain at the bridgehead carbon is accompanied by a decrease in the photostimulated outer-sphere ET rate for the parent RX compounds[43].

t-Butyl chloride, which has a solvolytic rate[139] comparable to that of some bicycloalkyl halides with small strain energy, reacts with Ph$_2$P$^-$ ions in liquid ammonia to give the substitution product. This reaction was inhibited by p-DNB and the free radical 2,2,6, 6-tetramethyl-1-piperidinyloxy (TEMPO)[53].

When the reaction was performed with a tertiary chloride that has a radical probe, such as 6-chloro-6-methyl-1-heptene (**64**), the only product found was the rearranged phosphine **65**, isolated in 43% yield (equation 58).

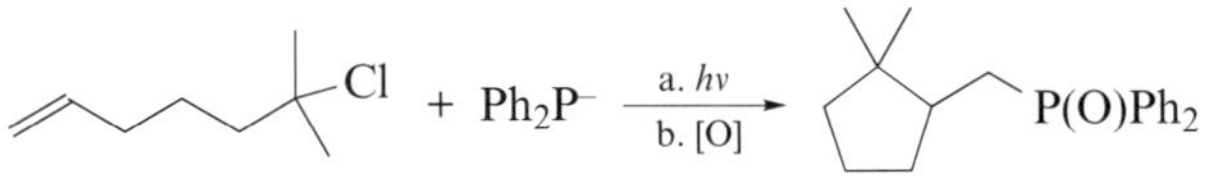

$$\text{(64)} \quad + \ Ph_2P^- \quad \xrightarrow[\text{b. [O]}]{\text{a. } h\nu} \quad \text{(65)} \quad P(O)Ph_2 \qquad (58)$$

(**64**)   (**65**)

These results indicate that t-butyl chloride and **64** react by the $S_{RN}1$ mechanism. The product distribution suggests that the radical formed as intermediate is trapped faster by the double bond to give the rearranged radical, than by the Ph_2P^- ions. Under the same experimental conditions tertiary bromides gave only elimination[53].

On the other hand, it has also been suggested that the reaction of nitrobenzene radical anion or dianion with t-butyl iodide to give p-(t-butyl)nitrobenzene (18%) occurs by an $S_{RN}1$ reaction[140].

Carbonylation by the $S_{RN}1$ mechanism of 2-methyl-2-dodecyl chloride and bromide by the carbonyl dianion $Na_2Fe(CO)_4$ has been also reported[127].

B. Cycloalkyl Halides

Studies on nucleophilic substitution reactions of cycloalkyl halides have shown their behaviour to depend on the ring size, the nucleophile, the leaving group and the reaction conditions.

1. Halo and gem-dihalo cyclopropanes

Substituted cyclopropanes react by the S_N1-type mechanism, but a disrotatory ring opening is necessary to assist the departure of the leaving group (electrocyclic process)[141]. This process has a relatively high activation energy due to the fragmentation of two bonds in a concerted step. When the cyclopropane carries the leaving group as well as a substituent at the carbon atom of the ring that stabilizes the positive charge, a substitution by the S_N1 mechanism without ring opening can take place[142]. On the other hand, halocyclopropanes do not usually undergo nucleophilic substitution by an S_N2-type reaction[143].

A similar reactivity has been found for gem-dihalocyclopropanes where nucleophilic substitution normally involves an elimination–addition sequence[143].

The photostimulated substitution reaction of 7-bromonorcarane (7-bromobicyclo-[4.1.0]heptane) (**66**) (ca 1:1 of the two isomers exo:endo) with several nucleophiles has been reported. Thus **66** gave high yields of substitution products in liquid ammonia with Ph_2P^- and Ph_2As^- ions while lower yields were obtained with PhS^- ions (equation 59)[144].

Acetone enolate ion did not react with **66**, whereas pinacolone enolate ion reacted to give 18% of **67d**[145]. However, good yields of the substitution product were obtained in the photostimulated reaction of 7-iodonorcarane (**68**) with acetophenone enolate ion in DMSO (87%). This reaction is inhibited by p-DNB and it is sluggish in the dark (equation 60)[146].

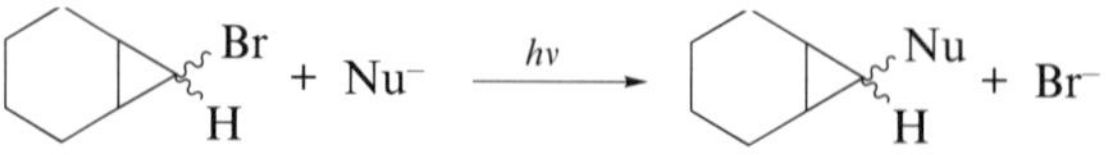

$$\text{(66)} \quad + \text{Nu}^- \xrightarrow{h\nu} \quad + \text{Br}^- \qquad (59)$$

(66)　　　　　$Ph_2P^-/[O]$　　　　(67a) Nu = $P(O)Ph_2$

　　　　　　　　Ph_2As^-　　　　　　(67b) Nu = $AsPh_2$

　　　　　　　　PhS^-　　　　　　　(67c) Nu = SPh

　　　　　　　　$^-CH_2COCMe_3$　　　(67d) Nu = CH_2COCMe_3

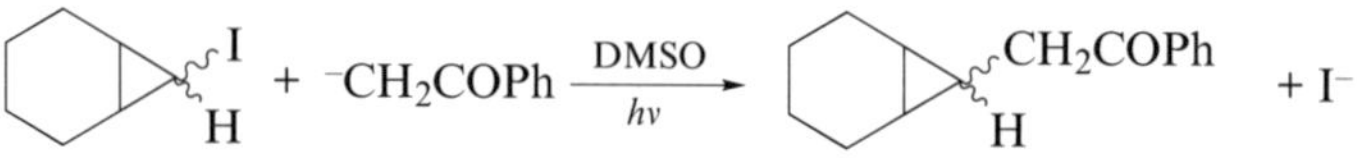

$$\text{(68)} \quad + {}^-CH_2COPh \xrightarrow[h\nu]{DMSO} \quad + I^- \qquad (60)$$

(68)　　　　　　　　　　　　　　　(69)　exo:endo

Even though the starting substrate **68** was a mixture *ca* 1:1 of the *exo:endo* isomers, product **69** was formed in a 16–22: 1 *exo:endo* ratio, suggesting that the coupling reaction of the 7-norcaranyl radical with acetophenone enolate ion is quite selective.

Disubstitution **71** and monosubstitution **72** products were obtained, although in overall low yield, in the reaction of 7,7-dibromonorcarane **70** with PhS⁻ [145,147], PhSe⁻ and PhTe⁻ ions (equation 61)[147]. The ratio of disubstituted to monosubstituted product decreases in the order PhS⁻ (4) >PhSe⁻ (0.6) > PhTe⁻ (*ca* 0.08).

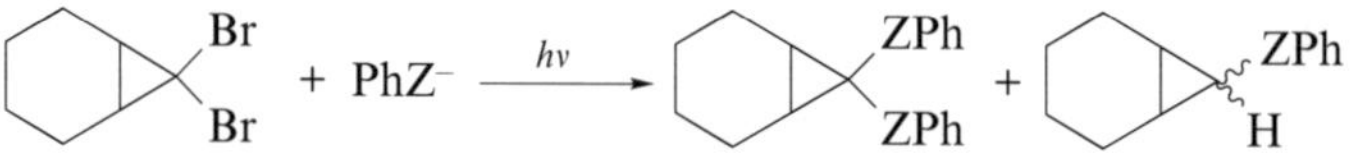

$$\text{(70)} \quad + \text{PhZ}^- \xrightarrow{h\nu} \quad + \qquad (61)$$

(70)　　　　Z = S　　　　(71a) Z = S　　　　(72a) Z = S

　　　　　　Z = Se　　　 (71b) Z = Se　　　 (72b) Z = Se

　　　　　　Z = Te　　　 (71c) Z = Te　　　 (72c) Z = Te

There was a fast dark reaction between **70** and Ph_2P^- ions to give 7-bromonorcarane (**66**) (87%). This reaction, not inhibited by DTBN, could indicate that **70** undergoes a nucleophilic attack on bromine to give the norcaranyl anion, which by protonation gives **66**[148]. On the other hand, 7,7-dichloronorcarane (**73**) reacts with Ph_2P^- ions only under irradiation to form the disubstitution (60% yield) and the reduced monosubstitution **67a** (15% yield) products (equation 62). The reaction was completely inhibited by p-DNB[147].

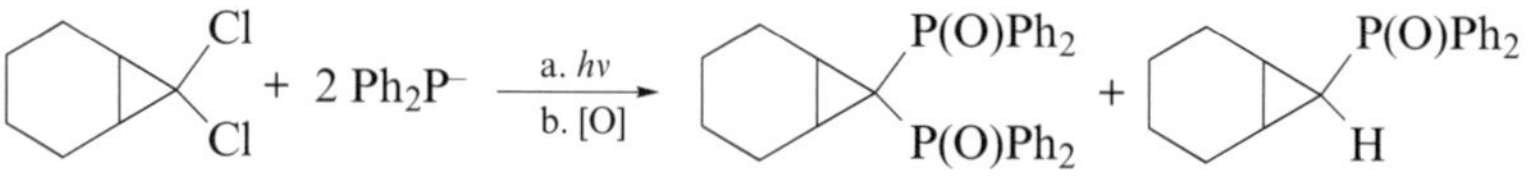

$$\text{(73)} \quad + 2\,Ph_2P^- \xrightarrow[\text{b. [O]}]{\text{a. } h\nu} \quad + \qquad (62)$$

(73)　　　　　　　　　　　　　　　　　　(67a)

Disubstitution product (46%) was the only product obtained in the photostimulated reaction of **70** with pinacolone enolate ion while n-butanethiolate ion led to the monosubstitution product in a relatively low yield (22%)[145].

Several 1,1-dibromocyclopropanes (**74**) were allowed to react under irradiation with PhS⁻ ions in DMSO. In this system the disubstitution product **75** (*ca* 10–40% yield) and the monosubstitution product **76** (*ca* 4–50% yield) were formed in variable yields (equation 63)[149]. In all these reactions no cyclopropane ring opening has been reported.

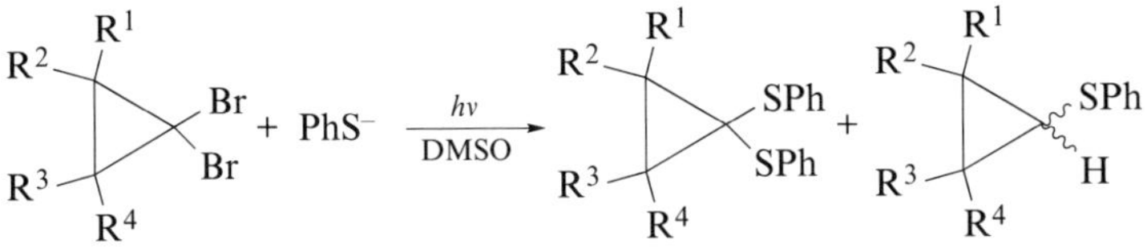

$$R^2 \overbrace{\underset{R^3}{\overset{R^1}{\diagup}}}^{} \underset{Br}{\overset{Br}{<}} + \text{PhS}^- \xrightarrow[\text{DMSO}]{hv} R^2 \overbrace{\underset{R^3}{\overset{R^1}{\diagup}}}^{} \underset{SPh}{\overset{SPh}{<}} + R^2 \overbrace{\underset{R^3}{\overset{R^1}{\diagup}}}^{} \underset{H}{\overset{SPh}{<}} \quad (63)$$

$$R^4 \qquad\qquad R^4 \qquad\qquad R^4$$

(74)   **(75)**   **(76)**

2. Cyclobutyl halides

It is known that polar substitution reactions of bromocyclobutane are accompanied by opening and rearrangement of the ring leading to cyclopropylcarbinyl and butenyl derivatives through an S_N1 mechanism. Some cases of substitution without ring opening are known[150].

Cyclobutyl chloride does not react with Ph_2P^- ions in the dark or under photostimulation in liquid ammonia. On the contrary, cyclobutyl bromide (**77**) reacts with Ph_2P^- ions in the dark (15 min, 42% of substitution product **78**) as well as under photostimulation (83% of substitution) (equation 64). *p*-DNB does not affect the dark reaction while it partially inhibits the light catalysed one. In the latter case, the amount of product obtained is similar to that formed under dark condition[151].

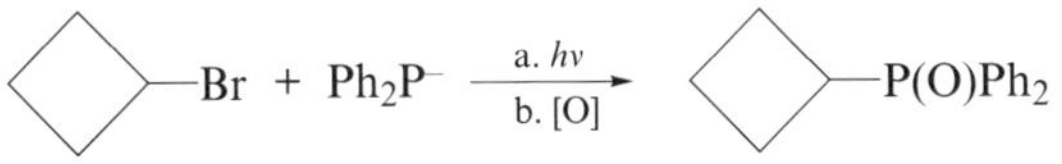

$$\text{◇}-\text{Br} + Ph_2P^- \xrightarrow[\text{b. [O]}]{\text{a. } hv} \text{◇}-\text{P(O)Ph}_2 \quad (64)$$

(77)   **(78)**

The fact that the light catalysed reaction was partially inhibited by *p*-DNB suggests that an $S_{RN}1$ process can be in competition with other mechanisms. On the other hand, even though the important dark reaction was not inhibited by *p*-DNB, no cyclopropylcarbinyl or butenyl derivatives were found, indicating that a polar nucleophilic substitution is not likely. The experimental evidence, as well as the fast rate of the reaction at low temperature (–33 °C) in dark conditions, suggest that it takes place with radicals as intermediates, probably through a cage-collapse mechanism[151].

3. Cyclopentyl halides

Cyclopentyl chloride (**79**) reacts under irradiation with Ph_2P^- ions to give product **80**. This reaction was almost completely inhibited by *p*-DNB (equation 65)[151].

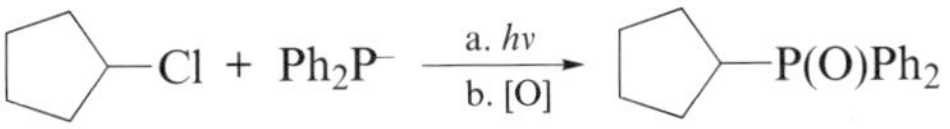

$$\text{⬠}-\text{Cl} + Ph_2P^- \xrightarrow[\text{b. [O]}]{\text{a. } hv} \text{⬠}-\text{P(O)Ph}_2 \quad (65)$$

(79)   **(80)**

On the other hand, cyclopentyl bromide reacts almost quantitatively with Ph_2P^- ions in the dark or under irradiation. Both reactions were partially inhibited by *p*-DNB. All these results suggest that **79** reacts with Ph_2P^- ions by the $S_{RN}1$ mechanism, while the bromine derivative reacts by a polar and ET reaction[151].

4. Cyclohexyl halides

Cyclohexyl chloride **81a** reacts with Ph_2P^- ions under irradiation to form 33% of the substitution product. The reaction was strongly inhibited by p-DNB, suggesting that **81a** reacts exclusively by the $S_{RN}1$ mechanism (equation 66)[54, 151].

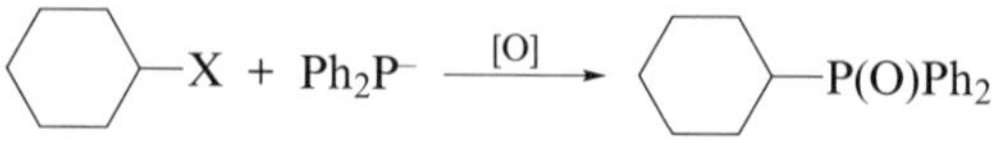

$$\langle\rangle\text{-X} + Ph_2P^- \xrightarrow{[O]} \langle\rangle\text{-P(O)Ph}_2 \tag{66}$$

(81a) X = Cl   **(82)**
(81b) X = Br

A dark reaction accelerated by light was observed with cyclohexyl bromide **(81b)**. p-DNB did not affect the dark reaction while it partially inhibited the photostimulated one. The results suggest that **81b** reacts, under irradiation, mostly by the $S_{RN}1$ mechanism. The dark reaction may be due to a polar or an ET reaction through cage collapse[54, 151].

In order to determine the degree of competition between both processes, the photo-stimulated reaction of the radical probe 3-bromo-2-tetrahydropyranyl allyl ether **(83)** with Ph_2P^- ions in liquid ammonia was studied. In this reaction both the substitution **84** (20% yield) and the cyclized **85** (69% yield) substitution products were formed (equation 67)[54, 151].

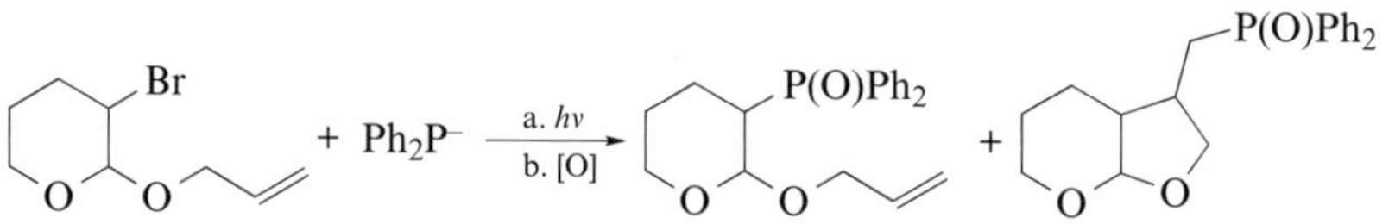

(83)   **(84)**   **(85)**

These results suggest that the major reaction pathway is the formation of free cyclohexyl radicals which cyclize, followed by reaction with Ph_2P^- ions by the $S_{RN}1$ mechanism (equation 68).

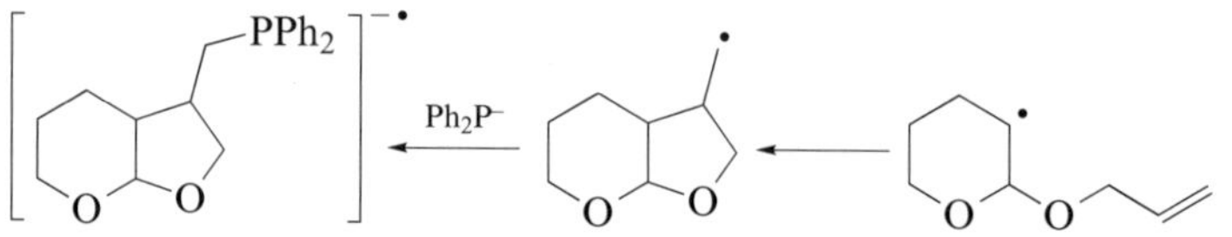

(87)   **(86)**

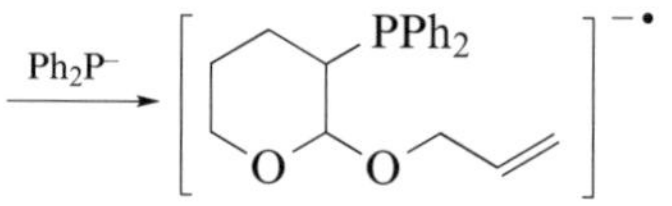

(84)$^{-\bullet}$ (68)

Under irradiation **83** gives radical **86** (equation 68), which rearranges to radical **87** or reacts with Ph_2P^- ions to give radical anion **84**$^{-\bullet}$ and ultimately the product **84**. Radical **87** also reacts with Ph_2P^- ions to give ultimately the product **85**. In the competition reactions of radical **86** only the formation of **84** depends on the concentration of Ph_2P^-, thus in dilute solutions product **85** was formed in a 93% yield.

The carbonylation reaction of cyclohexyl chloride and bromide by the dianion $Na_2Fe(CO)_4$ has been suggested to proceed by the $S_{RN}1$ mechanism[127].

2-Haloadamantanes are halocyclohexanes that have even lower reactivity toward nucleophiles than 1-haloadamantanes or the simple halocyclohexanes.

2-Bromoadamantane (**88**) reacts slowly with Ph_2P^- ions to give the substitution product **89** (equation 69), a reaction that is accelerated by light. Reactions under both conditions are inhibited by p-DNB[152] which suggests that they take place by the $S_{RN}1$ mechanism. The reaction with 2-chloroadamantane was sluggish.

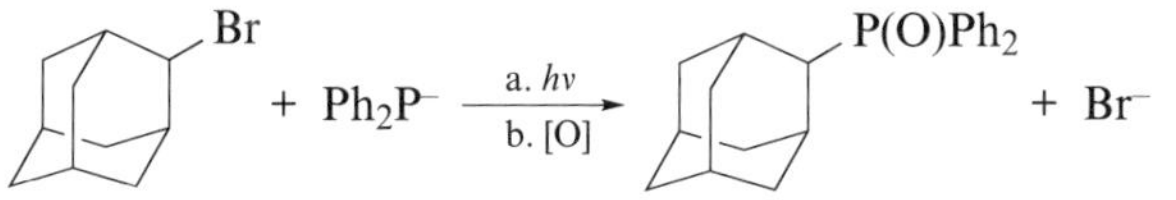

$$\text{(88)} + Ph_2P^- \xrightarrow[\text{b. [O]}]{\text{a. } hv} \text{(89)} + Br^- \qquad (69)$$

The relative reactivity of 1- and 2-bromoadamantanes toward Ph_2P^- ions was studied in liquid ammonia by competition experiments. 1-Bromoadamantane was determined to be 1.44 times more reactive than **88**[152].

5. Cycloheptyl halides

Cycloheptyl chloride (**90a**) reacts with Ph_2P^- ions to give substitution only upon irradiation. This reaction was completely inhibited by p-DNB (equation 70)[151].

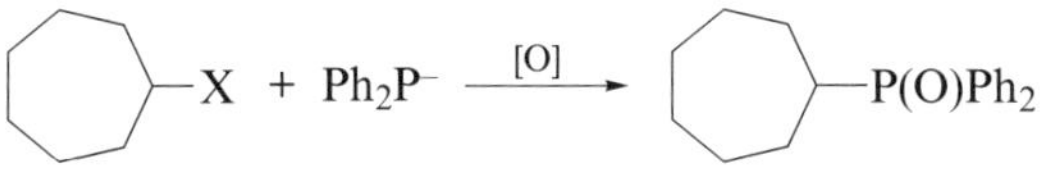

$$\text{X} + Ph_2P^- \xrightarrow{\text{[O]}} \text{P(O)Ph}_2 \qquad (70)$$

(**90a**) X = Cl
(**90b**) X = Br

(**91**)

On the other hand, **90b** reacts very rapidly in the dark to give the substitution product **91** in high yields. The reaction is partially inhibited by p-DNB, suggesting that this substrate reacts by both polar and ET mechanism, while **90a** reacts only by the $S_{RN}1$ process[151].

C. Neopentyl Halides

Neopentyl halides are among the most unreactive substrates in polar nucleophilic substitution reactions[118]. Due to the fact that the halogens are on a primary carbon atom, neopentyl halides seldom react by the S_N1 mechanism. Beacuse of the steric hindrance by the t-butyl group to backside attack, the S_N2 reaction in the neopentyl system is notoriously slow. However, with nucleophiles that are good electron donors, the ET reaction competes with the polar mechanism.

Radical probes **92a** and **92b** react with PhS^- and Ph_2P^- ions in liquid ammonia or DMSO, to give uncyclized **93** and cyclized **94** substitution products (equation 71)[52].

These results suggest an ET reaction giving the radical intermediate **95**, which reacts, for instance, with PhS^- ions to give **93**[.], or cyclizes to **96** to give finally **94** (equation 72). The rate constant for the coupling of neopentyl radicals with PhS^- ions in DMSO[52] is about $1.2 \times 10^8 \, M^{-1}$.

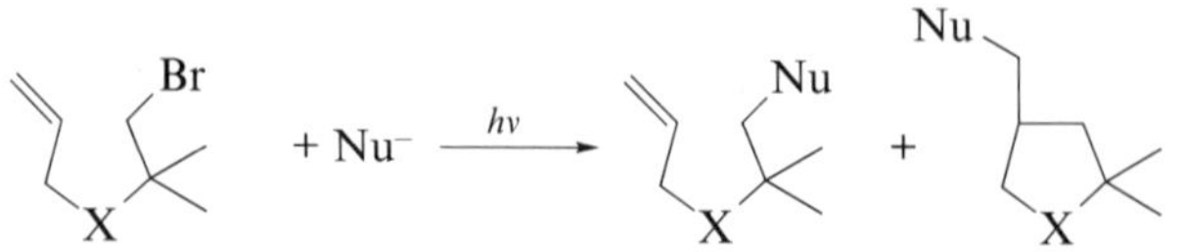

$$\text{(71)}$$

(92a) X = –CH$_2$⁻   Nu⁻ = PhS⁻   (93)   (94)
(92b) X = O   Ph$_2$P⁻

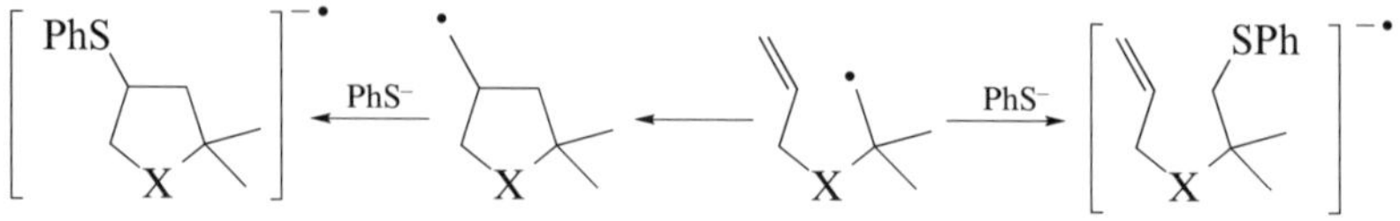

$$\text{(72)}$$

(94)⁻·   (96)   (95)   (93)⁻·

Neopentyl bromide reacts with PhS⁻ (in 90 min), Ph$_2$P⁻ (in 10 min) and Ph$_2$As⁻ (in 5 min) ions in liquid ammonia to give high yield of substitution products[153]. The reaction of arsenic metal with Na metal in liquid ammonia produces an 'As⁻³' species[154]. In the reaction of neopentyl bromide with 'As⁻³' ions a 33% yield of *tris*-neopentylarsine was formed[155].

The substitution product was obtained in modest yield when the reaction was performed with PhSe⁻ ions[153]. On the other hand Na$_2$Se, formed in liquid ammonia from Se and Na metals, reacted with neopentyl iodide (97) to give dineopentyl diselenide (100) from oxidation of 98 (equation 73b), together with a small amount of dineopentyl selenide (99) (equation 73a)[155].

$$Me_3CCH_2I + Se^{2-} \xrightarrow{h\nu} Me_3CCH_2Se^- + (Me_3CCH_2)_2Se \qquad \text{(73a)}$$
(97)   (98)   (99)

$$Me_3CCH_2Se^- \xrightarrow{[O]} (Me_3CCH_2)_2Se_2 \qquad \text{(73b)}$$
(98)   (100)

Neopentyl chloride (101a) and neophyl chloride (101b) react with Ph$_2$P⁻ ions in liquid ammonia forming high yields of the substitution product 102a and 102b (equation 74).

$$RMe_2CCH_2Cl + Ph_2P^- \xrightarrow{h\nu} RMe_2CCH_2PPh_2 \qquad \text{(74)}$$
(101a) R = Me   (102a) R = Me
(101b) R = Ph   (102b) R = Ph

By competition experiments 101b was determined to react *ca* 20 times faster that 101a. This reactivity difference can be ascribed to the LUMO differences between the compounds. In 101a, the LUMO belongs to the σ* C—Cl bond and it is higher in energy than the LUMO of 101b which corresponds to the π* phenyl moiety. Thus, the experimental and theoretical results suggest that the phenyl ring accelerates the reaction by an intramolecular entrainment process[156].

Several carbanions, such as acetone enolate or diethyl malonate ions, failed to react with neopentyl bromide in liquid ammonia under irradiation. In DMSO dehalogenation (100%) was the main reaction of 97 in the presence of the enolate ion of the acetone (27a) whereas with the enolate ion of acetophenone (27b) it gave the substitution product in 55% yield (equation 75). This reaction showed slight inhibition by *p*-DNB or TEMPO[157].

$$Me_3CCH_2I + {}^-CH_2COR \xrightarrow{h\nu,\ DMSO} Me_3CCH_2CH_2COR + I^- \qquad (75)$$

(**97**) (**27a**) R = Me; (**27b**) R = Ph → (**103a**) R = Me; (**103b**) R = Ph

Anthrone ion (**29**) was another nucleophile that reacts with **97** giving good yields of substitution product **104** (52%) (equation 76)[157].

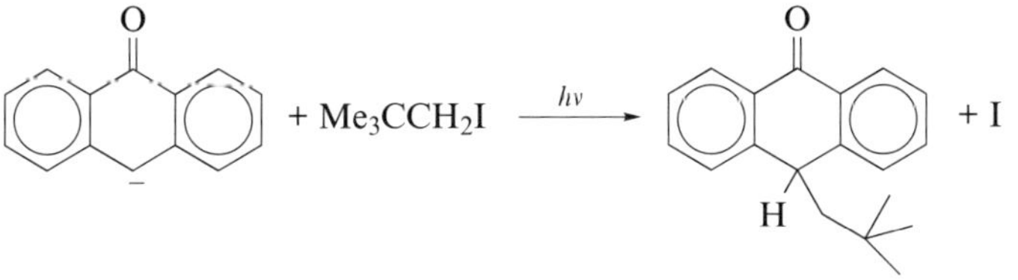

(76)

(**29**) (**97**) (**104**)

Nitromethane anion (**31**) failed to initiate the reaction with **97**. However, when the reaction was performed in the presence of the enolate ion of the acetone (**27a**) as initiator, the corresponding substitution product **105** was the only product formed in 69% yield (equation 77)[157].

$$(\mathbf{97}) + {}^-CH_2NO_2 + {}^-CH_2COCH_3 \xrightarrow{h\nu,\ DMSO} Me_3CCH_2CH_2NO_2 \qquad (77)$$

(**31**) (**27a**) (**105**)

These results can be explained by taking into account the different reactivities in the initiation and propagation steps of the carbanions in the photostimulated reaction with **97** (Section IV.A.1).

The relative reactivity of **97** and 1-AdI in competition experiments toward acetophenone enolate ion as nucleophile was studied. 1-AdI is *ca* 4.9 times more reactive than **97**. This result shows that despite the fact that 1-AdI is a strained structure compared with the flexible structure of **97**, it reacts faster due to the stabilization of the radical intermediate which is greater for tertiary than for primary carbons[157].

Disubstitution product **107** was obtained in the reaction of 1,3-diiodo-2,2-dimethylpropane (**106**) with **27b** (equation 78). The monosubstitution product is not an intermediate in this process[157]. These results are in agreement with a fast intramolecular ET to the σ C—I bond of the monosubstituted radical-anion intermediate.

$$ICH_2-\underset{CH_3}{\overset{CH_3}{\underset{|}{\overset{|}{C}}}}-CH_2I + 2\ {}^-CH_2COPh \xrightarrow{h\nu}$$

(**106**) (**27b**)

$$PhCOCH_2CH_2-\underset{CH_3}{\overset{CH_3}{\underset{|}{\overset{|}{C}}}}-CH_2CH_2COPh + 2\ I^- \qquad (78)$$

(**107**)

In the reaction of the neopentyl iodide radical probe **108** with lithium propiophenone in HMPA, the O-substituted **109** (66%) and the C-substituted **110** compounds (11%) were found as major products, together with a small amount of **111** and other minor products (equation 79)[158]. This reaction is not light catalysed or inhibited by di-*t*-butylnitroxide (DTBN).

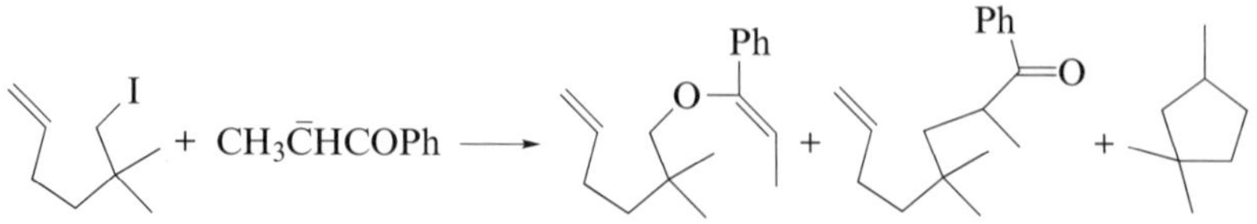

$$(108) + CH_3\bar{C}HCOPh \longrightarrow \quad (109) + (110) + (111) \quad (79)$$

However, the reaction carried out in the presence of hydrogen donors, such as 1,4-cyclohexadiene or dicyclohexylphosphine, gave **111** in 28% yield, together with **109** (36%) and **110** (6%). 1,4-Cyclohexadiene does not only reduce the radical intermediate to give the hydrocarbon **111**, but also decreases the viscosity of the HMPA solvent, thus increasing the diffusion of the radicals from the solvent cage. The radical probe can rearrange and is then reduced[158].

The reaction of **108** with $Ph_2PM(M=Li,Na,K)$ of high purity in THF gave the substitution product **112**, the cyclized product **113** and the hydrocarbon cyclized **111** (equation 80)[159].

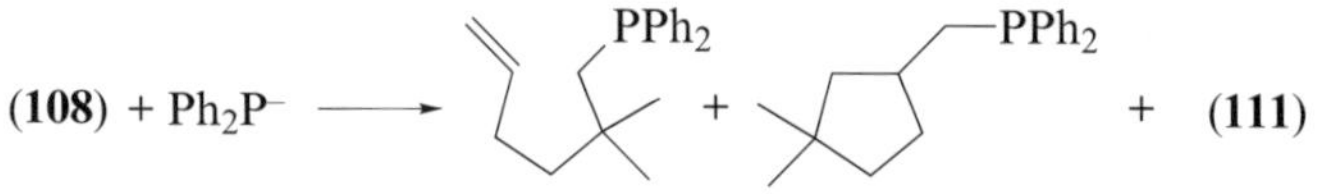

$$(108) + Ph_2P^- \longrightarrow \quad (112) + (113) \quad + (111) \quad (80)$$

With Ph_2PM (M=Na, K) the yield of **112** was 17–22%, whereas that of **113** was 67–75%. With Li as a counter ion, **112** was formed in 43% and **113** in 48% yield. The hydrocarbon **111** was obtained in 4–11% yield. The main product formed with the bromides and chlorides was **112**, and mere traces of **113** but no **111** were found.

V. AROMATIC HALIDES

A. Nucleophiles Derived from the IV Group

1. Carbon nucleophiles

By far one of the most important reactions through the $S_{RN}1$ mechanism is formation of a C—C bond by the reaction of aryl halides with carbanions derived from hydrocarbons, ketones, esters, amides, nitriles and even, with some limitations, from aldehydes. The reactions of cyanide ions and carbonyl complexes of Co and Fe also form a new C—C bond.

a. Carbanions derived from hydrocarbons. Indenyl and phenyl substituted indenyl anions react with PhBr to give phenylated and polyphenylated indenes in DMSO[160]. For instance, the indenyl anion afforded 1-phenylindene (25%), 1,3- (2%) and 1,1- (8%) diphenyl indenes while 1- and 2-phenyl indenyl anions gave phenylation at $C_{(1)}$. The reaction of 1,3-diphenylindenyl anion (**114**) afforded the products **115** and **116** in 50% and 9% yield, respectively (equation 81)[61].

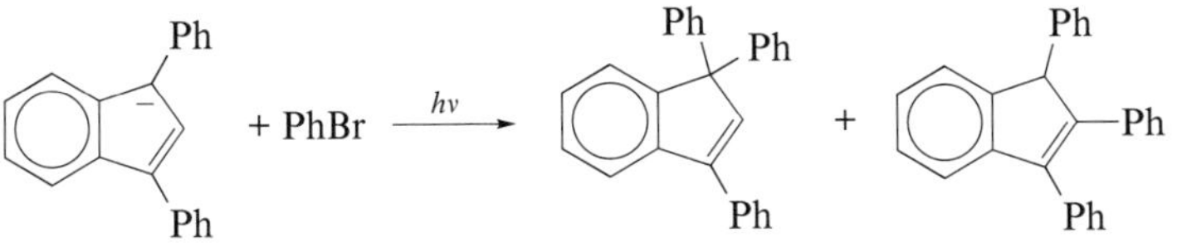

(81)

(114)   (115)   (116)

Neither **115** nor its radical anion rearranges to **116**, so that both compounds are formed by the coupling of phenyl radicals with **114**. The regioselectivity of the reaction has been proposed to be governed by the basicity of the system.

Cyclooctadienyl anion reacts spontaneously with 9-bromoanthracene and 1-bromonaphthalene to yield aryl derivatives of cyclooctadienes. With PhBr or PhCl, photostimulation is necessary for the reaction to occur[162].

The photostimulated reaction of PhX (X = Cl,Br,I) with triphenylmethyl anion (**117**) in THF leads to phenylated products **118** and **119**, although in low yield, together with phenylation of the THF solvent (equation 82)[163].

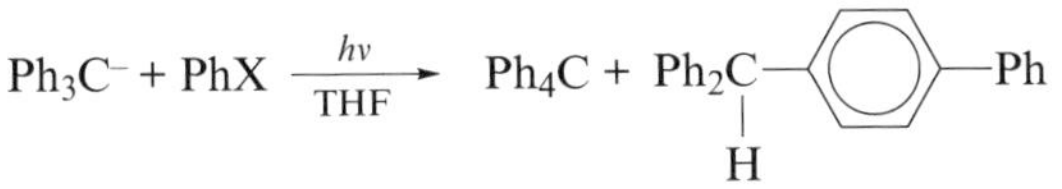

$$Ph_3C^- + PhX \xrightarrow[\text{THF}]{hv} Ph_4C + Ph_2C \text{—} \bigcirc \text{—} Ph$$

(82)

(117)   (118)   (119)

The formation of **119** is due to the attack of phenyl radical at the *p*-position of the carbanion. The **119**:**118** ratio (1.7:3.7) depends on the concentration of both reactants. Competition between $S_{RN}1$ and the cage collapse mechanism has been proposed for this system.

Polybromostryryl carbanions (**120**) are stable in THF ($-78\,^{\circ}$C) but, under irradiation, decompose rapidly (with a quantum yield of 13) through an $S_{RN}1$-type mechanism[164]. Under these conditions **120** forms a biradical **121** (equation 83) which can be reduced by **120**. The radical **122**, thus formed followed the reaction shown in equation 84.

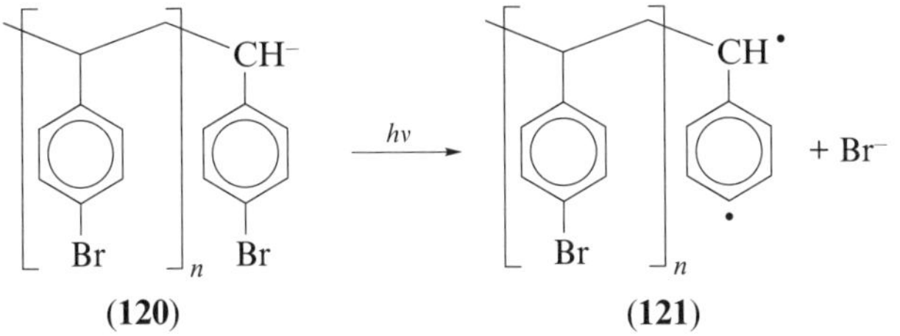

(83)

(120)   (121)

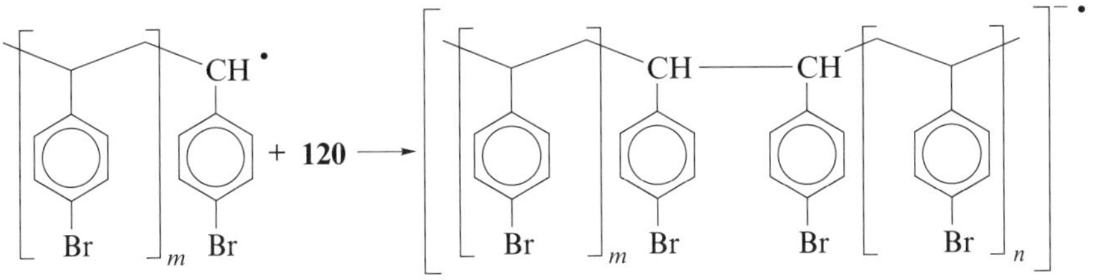

(84)

(122)   (123)

The radical anion **123** formed can fragment to give an aryl radical which is able to react again with **120**. These reactions constitute the propagation steps of the mechanism[164].

The reaction of p-dibromobenzene with t-BuLi in dioxane forms p-bromophenyl lithium, which by addition of HMPA polymerizes to polyphenylenes[165]. The same reaction in THF ($-78\,^\circ$C), after warming to room temperature, also leads to polyphenylenes[166]. Although there are no mechanistic studies, it was suggested that this polymerization occurs by an $S_{RN}1$ reaction [166].

b. Carbanions derived from enolate ions.   Many ketone enolate ions react under stimulation with solvated electrons or under irradiation in liquid ammonia with aryl and hetaryl halides, rendering the substitution product in good yield[14,17].

The dark reaction of pinacolone enolate ion and PhI in DMSO was described[167]. The same reaction is stimulated by light and inhibited by radical scavengers. This system was used to study the reactivity of different ketone enolate ions with PhI. In *competition* experiments, the following reactivity order was determined: 2-acetylcyclohexanone (unreactive) < phenylacetone (0.39) < cyclohexanone (0.67) < pinacolone (1.00) < acetone (1.09) < 2-butanone (1.10) < 3-pentanone (1.40)[168].

Under photostimulation, acetone enolate ion (**27a**) reacts with PhI in DMSO affording **124a** (88%) and **125a** (11%). The reaction with the enolate ion of acetophenone (**27b**) gives the substitution product **124b** (68%) and **125b** (13%) (equation 85)[169].

$$\text{PhI} + {}^-\text{CH}_2\text{COR} \xrightarrow{\ hv\ } \text{PhCH}_2\text{COR} + \text{Ph}_2\text{CHCOR} \qquad (85)$$

(**27a**) R = Me	(**124a**) R = Me	(**125a**) R = Me
(**27b**) R = Ph	(**124b**) R = Ph	(**125b**) R = Ph

In the photostimulated reaction of PhI with the monoanion of acetylacetone or with the diethylmalonate anion, no substitution products were found. On the other hand, in the reaction of anthrone anion (**29**) with PhI a 95% yield of the substitution product (**126**) was formed (equation 86)[169].

(**29**)	(**126**)

Anion **29** was found to be 2.2 times more reactive than **27b**, and nucleophile **27b** was 7.5 times more reactive than enolate ion **27a**. Based on these relative reactivities, carbanion **29** is 16.5 times more reactive than **27a**[169].

Moreover, nitromethane anion, one of the less reactive nucleophiles by itself in photostimulated reactions, reacts in the presence of good electron donors with similar apparent reactivity to **27b**[169].

As indicated in Section IV.A.1, the reactivity of carbanions in photostimulated reactions with haloarenes depends on the nature of the reaction step. In the initiation step the rate of photostimulated ET from the carbanion to the acceptor PhI increases with the pK_a value of the corresponding conjugate acids. In the propagation cycle the reactivity of the carbanions would depend on their pK_a values as well as on the energy of the SOMO of the radical anion intermediate, that is, on the HOMO–SOMO energy difference (loss in π energy).

On the other hand, the *relative reactivity* of different aryl bromides toward pinacolone enolate ion[170] is given in Table 1.

All the substrates have the same nucleofugal group, and there is a correlation between the relative reactivity and the standard reduction potential (E° of ArBr or $E_{1/2}$ of the unsub-

TABLE 1. Relative reactivity of different aryl bromides toward pina-colone enolate ion

	$k_{relative}$	E^0 (ArBr), V	$k_{cleavage}$ C—Br bond	$E_{1/2}$ (ArH), V
PhBr	1.0	−2.44	10^{10}	−2.38
3-Bromothiophene	1.5	−2.14		−2.10
4-Bromobiphenyl	9.7	−1.99		−2.00
2-Bromopyridine	16	−2.30	$>10^8$	−2.12
1-Bromonaphthalene	198	−2.19	10^8	−1.54
9-Bromoanthracene	810	−1.70	3×10^5	−1.09
4-Bromobenzophenone	0.4	−1.63	6×10^2	−1.68

stituted arene). As the reduction potential gets more positive, the substrate reacts more easily by the $S_{RN}1$ mechanism. The ET of the radical anion of the substitution product to the ArBr determines the relative reactivity. It is also interesting to note that as the reactivity *increases*, the rate of fragmentation of the aryl halide radical anion intermediates *decreases*. As already stated, all the propagation steps of the $S_{RN}1$ mechanism are important for determining the relative reactivities.

With the substrate 4-bromobenzophenone, whose radical anion has a very slow rate of fragmentation (6×10^2 s^{-1} in liquid ammonia)[171], the trend changes, possibly because of an equilibrium with the radical anion of bromobenzene which, although thermodynamically less stable, fragments very rapidly.

Carbanions occasionally react with aryl halides spontaneously, mostly under irradiation, or by supplying electrons either from dissolved metals or from a cathode. However, certain Fe^{+2} salts catalyse the $S_{RN}1$ reactions with carbanions. That was the case for the reaction of PhBr or PhI with acetone or pinacolone enolate ions in liquid ammonia or DMSO[172a], as well as for the reaction of the enolate ion of several carbanions with several aryl and hetaryl halides in DMSO[172b]. Since these reactions are inhibited by p-DNB and p-cymene, and the relative reactivity of nucleophiles is similar to that determined in photo-stimulated or spontaneous reactions, it seems that FeCl$_2$ initiates the $S_{RN}1$ process.

The reaction of sodium enolates of t-butyl acetate, N-acetylmorpholine and derivatives with iodoarenes catalyzed by Fe^{+2} salts in liquid ammonia gives good yields of substitution products. Bromoarenes react more sluggishly and chloroarenes almost do not react[173]. p-Diiodobenzene and p-chloroiodobenzene reacted with N-acetylmorpholine enolate and Fe^{+2} ions to give the disubstitution product in 55% and 51% yields, respectively[173].

The reaction of PhI with acetophenone enolate ion 27b in DMSO is catalysed by SmI$_2$, giving 47% of the substitution product. On the other hand, PhCl and PhBr did not react, but 1-chloro and 1-bromo naphthalenes reacted with 27b and SmI$_2$ to give 93% yield of the substitution product[174].

The reaction of aryl and hetaryl halides with ketone enolate ions can also be initiated by sodium amalgam in liquid ammonia. In these reaction conditions, neither the carbonyl group nor the aromatic moiety are reduced, as was the case in the reaction initiated with solvated electrons. Thus, the ketones below have been synthesized, mostly in good yields[175].

The reaction of carbanions with hetaryl halides has been reported. Thus, pinacolone enolate ion reacts with 2-halothiazoles (127) affording the substitution product 128 in good yields (44–67%) (equation 87)[176].

However, in the dark, or when the photostimulated reactions were inhibited with p-DNB, no ketones 128 were found, and the main products were carbinols formed by condensation reactions.

The photostimulated reaction of ketone enolate ions with 6-iodo-9-ethylpurine (129) in liquid ammonia leads to the substitution product 130 as a mixture of the keto and enol

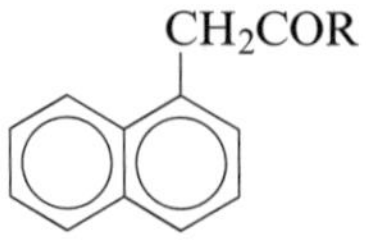

CH$_2$COR

R = Me (98%)   R = Me (49%)   H Ph
R = Ph (98%)   R = Ph (98%)   (70%)

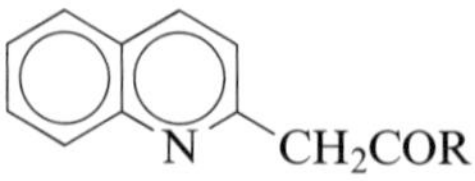

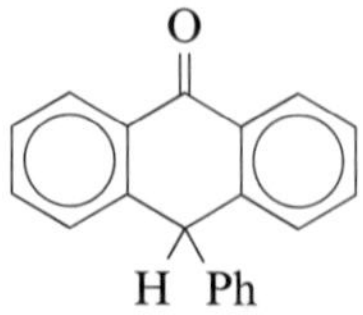

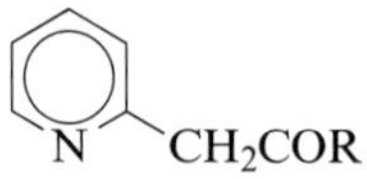

CH$_2$COMe

CH$_2$COR

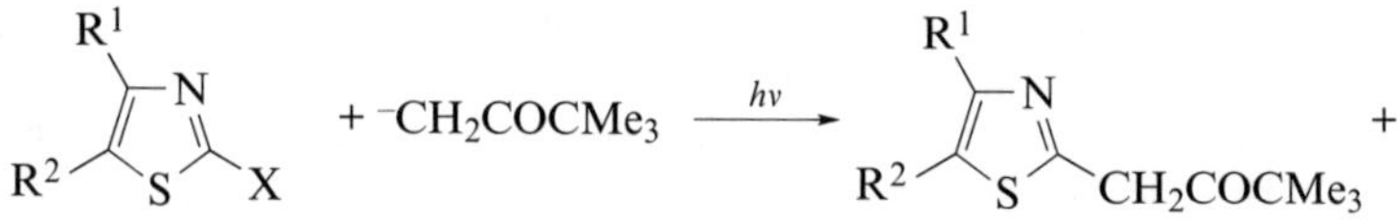

COPh

R = Me (15%)   (78%)
R = Ph (75%)

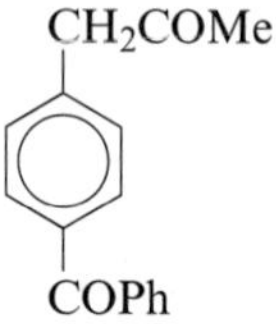

$$\text{(127)} + {}^-\text{CH}_2\text{COCMe}_3 \xrightarrow{h\nu} \text{(128)} + \text{X}^- \quad (87)$$

R^1, R^2 = H, Me; X = Cl, Br

forms in good yields (equation 88)[177]. The ketone studied (yield % of substitution product) were acetone (70%), cyclopentanone (65%), cyclohexanone (50%), 2-methyl-1-tetralone (80%), acetophenone (70%) and acetylfuran (67%)[177].

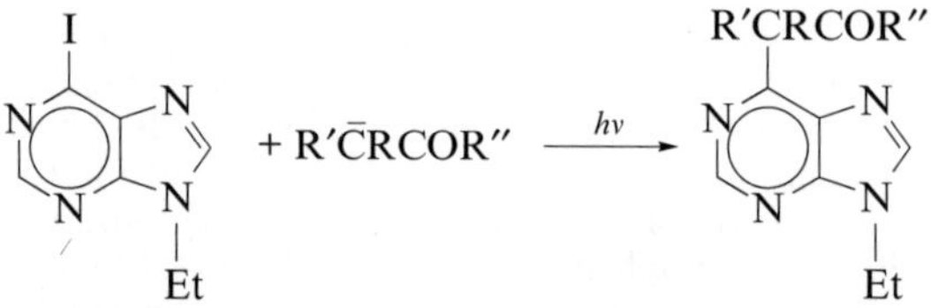

$$\text{(129)} + \text{R}'\bar{\text{C}}\text{RCOR}'' \xrightarrow{h\nu} \text{(130)} \quad (88)$$

The photostimulated reaction of these anions in DMSO or liquid ammonia with 3-halo-2-amino derivatives of benzo[b]thiophenes gave only modest yields of substitution products[178].

Disubstitution 132 [2,3 (63%), 3,5 (43%), 2,5 (85%)] is the main reaction of dihalopyridines (131) [X=Cl (2,3) and (3,5); X=Br (2,5)] with pinacolone enolate ion in liquid ammonia. The presence of monohalo substituted products as intermediates has been disregarded (equation 89)[179].

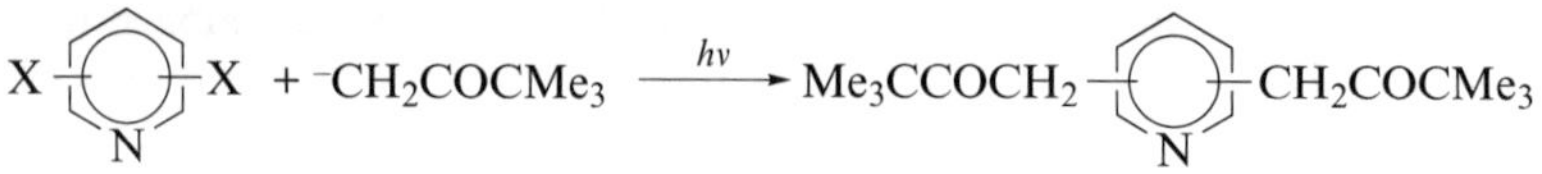

$$\text{X}\text{—}\overset{}{\underset{N}{\bigcirc}}\text{—X} + {}^-\text{CH}_2\text{COCMe}_3 \xrightarrow{h\nu} \text{Me}_3\text{CCOCH}_2\text{—}\overset{}{\underset{N}{\bigcirc}}\text{—CH}_2\text{COCMe}_3 \quad (89)$$

(131)   (132)

Ethyl phenylacetate and even the *tertiary* carbanion methyl diphenylacetate gave 84 and 42%, respectively, of the disubstitution product in their reaction with 2,6-dibromopyridine[179]. However, only monosubstitution product at position 4, **134** was obtained in 70% yield in the reaction of 4,7-dichloroquinoline (**133**) with pinacolone enolate ion (equation 90)[179].

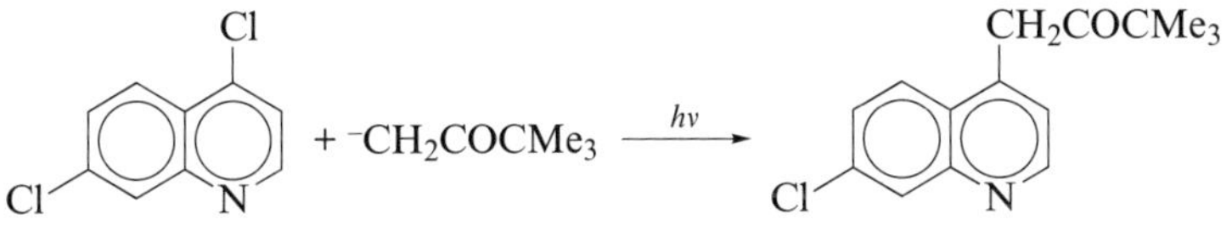

$$(90)$$

This result indicates that radical anion (**133**)$^{-\cdot}$ fragments selectively at the C—Cl bond at position 4 and that the intermolecular ET of the radical anion (**134**)$^{-\cdot}$ is faster than the fragmentation of the C—Cl bond of position 7.

Aryl halides bearing nitro groups seldom react by the $S_{RN}1$ mechanism, due to the slow fragmentation rate of their radical anion which does not allow the chain to be propagated. However, an exception is *o*-iodonitrobenzene radical anion, which fragments at a relatively fast rate $(8 \times 10^4 \mathrm{s}^{-1})$[180], compared with the *m*- and *p*-isomers (fragmentation rates 0.31 and 0.9 s^{-1}, respectively). This fragmentation rate enhancement was ascribed to a steric inhibition of coplanarity of the C—I bond with the aromatic ring by the NO_2 group.

The substitution reaction of *o*-iodonitrobenzene (**135**) with pinacolone enolate ion in liquid ammonia has been reported to afford **136** (66%) (equation 91).

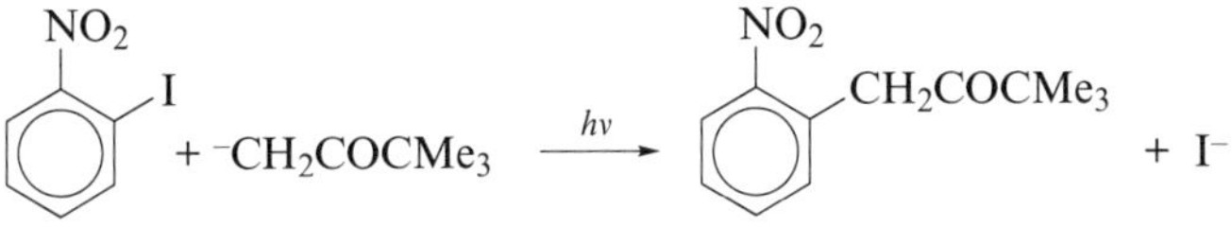

$$(91)$$

This reaction was suggested to occur by an $S_{RN}1$ and a cage collapse process[181]. On the other hand, other halonitrobenzenes, such as *m*- and *p*-iodonitrobenzenes and *o*-chloro or *o*-bromo nitrobenzenes, did not react under the same conditions.

The reaction of 2,4-dinitrohalobenzenes (F,Cl,Br,I) with the anions of heterocyclic ketene aminals was reported to give arylated products by the $S_{RN}1$ mechanism. This proposal was based on ESR spectroscopy, ESR spin trapping with nitroso-*t*-butane and the inhibition of the reaction by FeCl$_3$[182].

The reaction of *o*- and *p*-halonitrobenzenes (Cl, Br, F) with the sodium salt of ethyl cyanoacetate in DMSO gave almost quantitatively the substitution products[183]. These reactions were found to be markedly diminished by adding small amounts of *p*-, *m*- and *o*-DNBs, but were not influenced by addition of radical scavengers[184a]. Based on these results and kinetic studies it was suggested that they proceed via a non-chain radical nucleophilic substitution[184].

As previously indicated, monoanions of β-dicarbonyl compounds fail to react with phenyl halides; however, the 1,3-dianions are suitable nucleophiles and react quite well through the terminal carbanion site[185].

However, β-dicarbonyl carbanions such as **138** (R^1 = H, Et; $R^2 = R^3 = CO_2R$, COMe; R^1 = CN, $R^2 = CO_2Et$, COMe) react with more electrophilic substrates, such as *o*-, *m*- and *p*-bromobenzonitriles (**137**), to give the substitution products **139** in good yields (63–90%) (equation 92)[186].

$$\underset{(137)}{\overset{\text{CN}}{\bigcirc}}\text{-Br} + \underset{\underset{R^3}{|}}{R^1-\bar{C}-R^2} \xrightarrow{h\nu} \underset{(139)}{\overset{\text{CN}}{\bigcirc}}\underset{R^3}{\overset{R^1}{\underset{|}{C}}}_{R^2} \qquad (92)$$

The loss of π energy that hinders the reaction of these carbanions with the phenyl radical diminishes with electrophilic radicals, favouring the coupling.

2-Bromo-3-cyanopyridine (140) reacts with the anions 138 even faster than 137 giving excellent yields of the substitution product 141 (80–92%) (equation 93).

$$\underset{(140)}{\overset{\text{CN}}{\bigcirc}}\text{+ 138} \xrightarrow{h\nu} \underset{(141)}{\overset{\text{CN}}{\bigcirc}}\underset{R^3 \quad R^2}{\overset{R^1}{\underset{|}{C}}} \qquad (93)$$

The reactions of monoanions of β-dicarbonyl and β-cyanocarbonyl compounds with aryl halides have been induced electrochemically[187]. Thus 4-bromobenzophenone and p-chlorobenzonitrile give, with 138 in DMSO, good yields of substitution products.

Several halopyridines bearing not only CN^{186}, but also CF_3 substituents, react with malonate ions in liquid ammonia. The results with 2-chloropyridines (142) depend on the position of the CF_3 group as shown in equation 94. With 3-CF_3 there was no reaction[188].

$$\underset{\overset{\quad}{\underset{N}{\bigcirc}}\text{Cl}}{\overset{\text{CF}_3}{\bigcirc}} + \text{MeC}(CO_2Et)_2 \xrightarrow{h\nu} \underset{\overset{\quad}{\underset{N}{\bigcirc}}\underset{\underset{Me}{|}}{C(CO_2Et)_2}}{\overset{\text{CF}_3}{\bigcirc}} \qquad (94)$$

(142a) 5–CF_3 100%
(142b) 6–CF_3 44%

N,N-Dialkylacetamide enolate ions (143) were proved to react with aryl halides in liquid ammonia giving N,N-dimethyl-α-arylacetamides together with some N,N-dimethyl-α, α-diarylacetamides[189]. In order to synthesize some herbicides of the amide family, a series of N,N-dimethyl α-aryl and α,α-diaryl acetamides were prepared by the reaction of aryl halides (PhI, p-iodoanisole, p-iodotoluene, 1-iodonaphthalene, 9-bromophenan-threne, 9-bromoanthracene and p-iodobenzoic acid) with 143 (equation 95)[190]. The product distribution was shown to be dependent on the 143/aryl halide ratio. With a ratio of 5, approximately 50% of 144 and 20% of 145 were obtained.

$$\underset{(143)}{\text{CH}_2\text{CON(CH}_3)_2} + \text{ArX} \xrightarrow{h\nu} \underset{(144)}{\text{ArCH}_2\text{CON(CH}_3)_2} + \underset{(145)}{\text{Ar}_2\text{CHCON(CH}_3)_2} \qquad (95)$$

Similar studies were performed with the nucleophile 146 (Ar$'$ = phenyl, 9-phenanthryl and 9-anthryl) and aryl halides (PhI, 1-iodonaphthalene, 9-bromophenanthrene and 9-bromoanthracene) (equation 96).

When electron-withdrawing groups were attached to the aryl halide as in p-chloro, p-bromo and p-iodobenzonitriles and 4-bromobenzophenone, most of the reaction with 143 follows the benzyne mechanism, yielding anilines. When the aryl halides were added to a

$$\text{ArI} + {}^-\text{CHAr'CON(CH}_3)_2 \xrightarrow{hv} \text{ArAr'CHCON(CH}_3)_2 \qquad (96)$$

(146)

solution of K metal and **143**, a very good yield of substitution products was obtained, uncontaminated by the products derived from benzyne[190].

Enolate ions from lactams, such as N-methylpyrrolidinone (**147**) did not react with haloarenes in the dark[191], but **147** reacted with PhX (Cl,Br,I) and p-iodoanisole under irradiation to give the substitution product **148** (50–60% yield) (equation 97)[192].

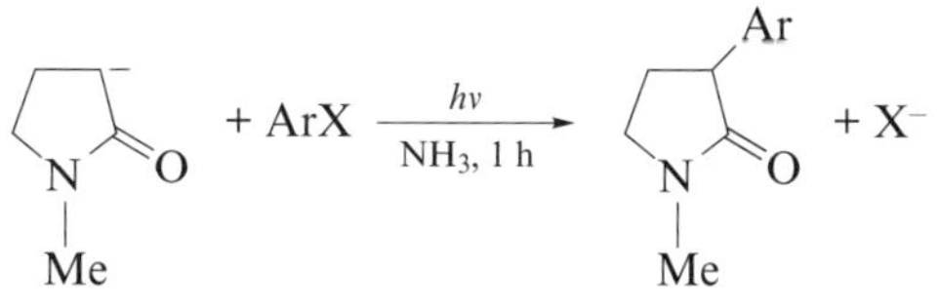

$$(97)$$

(147)   **(148)**

The enolate ion of (+)-camphor (**149**) formed in liquid ammonia reacts with PhBr, PhCl, 1-chloronaphthalene, 4-bromobiphenyl and p-chloroanisole to give mainly the *endo*-isomer **150a** (71–100% yield), with an *endo*- **150a** : *exo*-**150b** ratio of 99:1 in all cases (equation 98)[93].

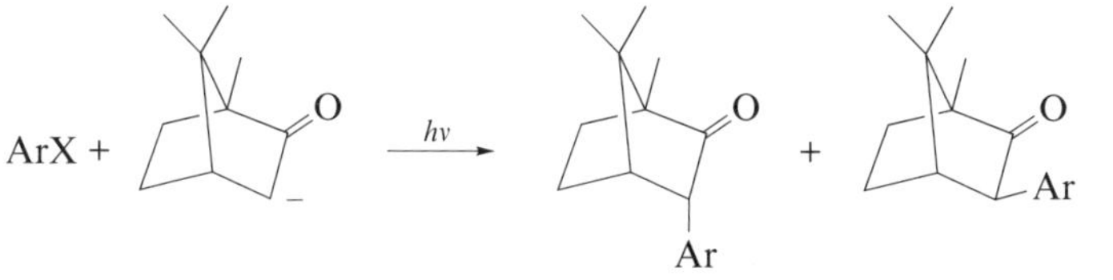

$$(98)$$

(149)   *endo*-(**150a**)   *exo*-(**150b**)

The reaction of the chiral auxiliary amide enolate ion (**151**) (M = Li) and 1-iodonaphthalene in liquid ammonia afforded the substitution products **152** (46%) and **153** (5.7%) in a ratio **152/153** of 8.2 (equation 99)[194].

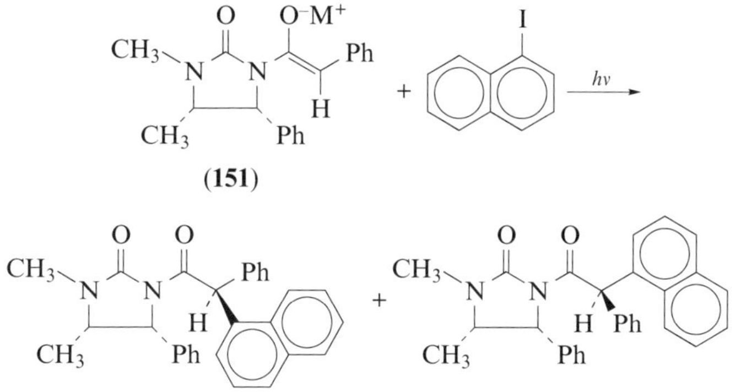

(151)

$$(99)$$

(152)   **(153)**

The reaction is highly dependent on the metal counter ion. All the ions studied display selection: with M = Na (ratio **152/153** = 2.2), M = K (ratio **152/153** = 1.8) and M = Cs (ratio **152/153** = 1.7), but highest selectivity was reached with Li^+ at low temperature ($-78\,°C$) and with Ti(IV) where the **152/153** ratios were >99. The selection that takes place in the coupling reaction between 1-naphthyl radical and **151** was suggested to occur by a preferential approach to one face of the planar enolate double bond due to the steric hindrance by the imidazolidyl-2-one group that has a certain stereoisomeric disposition.

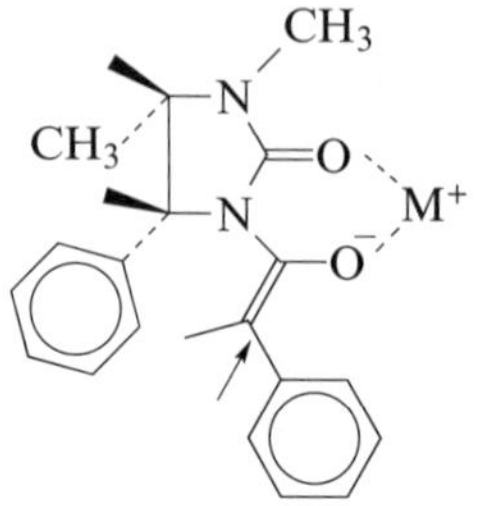

The remarkable selectivity could be explained in terms of the better complexation capacity of the Ti(IV) and Li$^+$, which by coordination with both carbonyl groups avoid the rotation around the N—CO bond and consequent formation of the other conformer in which the two carbonyl groups are in a *trans* conformation.

c. Anions of nitroalkanes.   The anions from nitroalkanes have been shown by different techniques (ESR[195] and optical absorption spectra[196]) to be very good traps for aryl and alkyl radicals. Furthermore, they are typical nucleophiles in the $S_{RN}1$ reaction with aliphatic halides. However, these nucleophiles have a low reactivity with aryl halides because the radical anion formed in the coupling with an aryl radical fragments very rapidly (equation 100) to the stable benzyl radical.

$$Ar^{\bullet} + {}^-CH_2NO_2 \longrightarrow (ArCH_2NO_2)^{-\bullet} \longrightarrow ArCH_2^{\bullet} + NO_2^- \qquad (100)$$

Nevertheless, the electrolysis of 4-bromobenzophenone (**154**) in DMSO with nitropropanide anion (**155**) in the presence of dibenzo-18-crown-6 gave products **156** and **157** (equation 101).

$$4\text{-}(C_6H_5CO)C_6H_4Br + {}^-CMe_2NO_2 \xrightarrow[\text{Dibenzo-18-crown-6}]{e^-/\text{electrode}}$$

$$\textbf{(154)} \qquad\qquad\qquad \textbf{(155)}$$

$$4\text{-}(C_6H_5CO)C_6H_4CHMe_2 + 4\text{-}(C_6H_5CO)C_6H_4C(Me)=CH_2 \qquad (101)$$

$$\textbf{(156)} \qquad\qquad\qquad\qquad \textbf{(157)}$$

These results were interpreted by considering that the radical anion of 4-bromobenzophenone **154**$^{\bullet}$ fragments to give the radical **158**, which couples with **155** to form radical anion **159** (equation 102). **159** fragments to give radical **160** and nitrite anion (equation 103). Radical **160** is further reduced to ultimately give product **156**, or gives ultimately by H atom transfer the olefin **157**[197].

$$[4\text{-}(C_6H_5CO)C_6H_4Br]^{-\bullet} \xrightarrow{-Br^-} 4\text{-}(C_6H_5CO)C_6H_4^{\bullet} \xrightarrow{155}$$

$$\textbf{(154)}^{-\bullet} \qquad\qquad\qquad \textbf{(158)}$$

$$[4\text{-}(C_6H_5CO)C_6H_4CMe_2NO_2]^{-\bullet} \qquad (102)$$

$$\textbf{(159)}$$

$$159 \quad \longrightarrow \quad 4\text{-}(C_6H_5CO)C_6H_4\overset{\bullet}{C}Me_2 + NO_2^- \qquad (103)$$
$$(160)$$

By competition experiments between nucleophile **155**, $(EtO)_2PO^-$ and PhS^- toward radical **158**, it was concluded that **155** is more reactive than $(EtO)_2PO^-$ and as reactive as PhS^- ions[197].

The anion of nitromethane (**31**) failed to react with PhI under irradiation in DMSO, as was the case with 1-AdI. However, when the reaction was carried out in the presence of acetone (**27a**) or pinacolone (**27b**) enolate ions, 1-phenyl-2-nitroethane (**161**) was formed as major product, together with small amounts of benzene and toluene (equation 104). No products from the coupling with the ketone enolate ions were found[169].

$$PhI + {}^-CH_2NO_2 \; (+ \; \textbf{27a} \; or \; \textbf{27b}) \quad \xrightarrow{\;h\nu\;} \quad PhCH_2CH_2NO_2 \qquad (104)$$
$$(\textbf{31}) \qquad\qquad\qquad\qquad\qquad\qquad\qquad (\textbf{161})$$

It has been suggested that the ketone enolate ions **27** initiate the photostimulated cycle. The coupling reaction of the phenyl radical with **31** is faster than with **27**, forming a radical anion that fragments to give benzyl radicals, which are reduced to toluene or react with **31** to afford a new radical anion $(\textbf{161})^{-\bullet}$, responsible for forming the substitution product **161** (equation 105).

$$PhCH_2^{\bullet} + \textbf{31} \quad \longrightarrow \quad (PhCH_2\,CH_2NO_2)^{-\bullet} \quad \xrightarrow{\;-e^-\;} \quad \textbf{161} \qquad (105)$$
$$(\textbf{161})^{-\bullet}$$

The radical anion $(\textbf{161})^{-\bullet}$ does not fragment, because in this case a less stable non-benzylic primary alkyl radical would be formed[169].

d. Nitrile stabilized carbanions. Acetonitrile anions react with halobenzenes under stimulation by solvated electrons or light to give the product $PhCH_2CN$ together with benzene, $PhCH_3$, 1,2-diphenylethane and other minor products[198]. When the phenyl radical couples with $^-CH_2CN$, the radical anion $(PhCH_2CN)^{-\bullet}$ is formed which, by ET, gives $PhCH_2CN$, or fragments into benzyl radicals which are reduced to toluene or dimerize to 1,2-diphenylethane[60]. However, with haloarenes that have low-lying π^* MOs, such as 2-chloropyridine, 1- and 2-chloronaphthalenes, 4-chlorobiphenyl and 4-bromobenzophenone, only the substitution products were formed[199]. It was reported that 3-bromoquinoline-1-oxide reacts with PhCH(Et)CN and NaOH at room temperature in HMPT to give the substitution product in 91% yield[200].

The photostimulated reaction of propionitrile anions with haloarenes in liquid ammonia has been studied as a synthetic route to α-arylpropionic acids[32]. Thus, 2-bromonaphthalene gave 76% yield of the product, whereas 2-bromo-6-methoxynaphthalene, *p*-bromobiphenyl and 1-chloro-2-metoxynaphthalene yielded 81%, 52% and 52% of the substitution products, respectively.

2-Bromopyridine (**162**) reacts with phenylacetonitrile anion (**163**) in liquid ammonia to afford the substitution product **164** in high yields (88%) (equation 106)[201].

$$\text{(106)}$$
$$(\textbf{162}) \qquad\qquad (\textbf{163}) \qquad\qquad\qquad\qquad (\textbf{164})$$

However, the photostimulated reaction of **162** and KCH_2CN gave (2-pyridyl) acetonitrile (75%) and 2-aminopyridine (16%). This last compound was suggested to arise from an S_NAr pathway of amination. On the other hand, in the photostimulated reaction of 2-

chloroquinoline, in addition to the substitution product (2-quinolyl)acetonitrile (**165**) (50%), minor products such as **166** and **167** have been reported (equation 107). In the dark, the main product **167** has been suggested to be formed through an S_N(ANRORC) pathway[201].

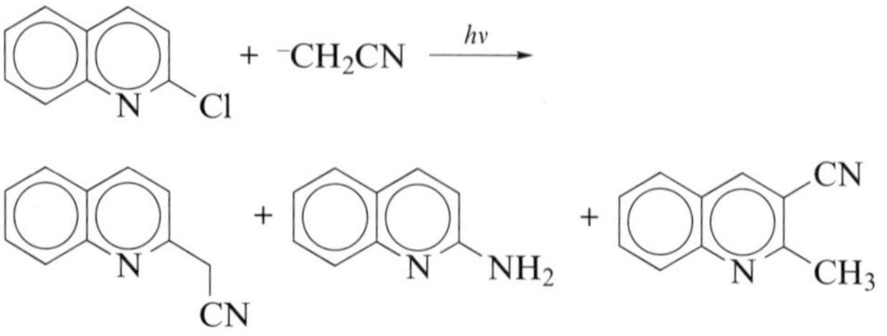

$$(107)$$

(165) **(166)** **(167)**

Carbanion **163** reacts with 2,6-dibromopyridine to afford the products **168** and **169** (equation 108)[179]. With carbanion **163b** the reduction product was formed in 52% yield.

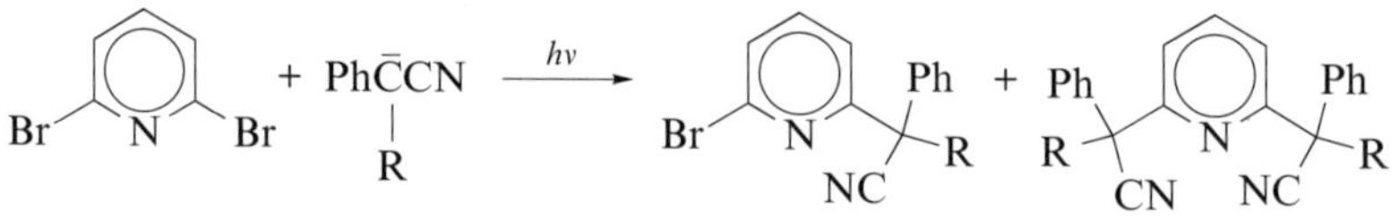

$$(108)$$

(163a) R = H   **(168a)** R = H (34%)   **(169a)** R = H (36%)
(163b) R = Et   **(168b)** R = Et (17%)   **(169b)** R = Et (25%)

The photostimulated reaction of carbanion **163a** with 2,4-dichloropyrimidine afforded exclusively the 4-substituted product[179].

Halogenated heterocycles (2-, 3-, 4-bromopyridines, 2-chloro-3-bromoquinolines, 2-chloropyrimidine) react with anion **163a**. The substitution products then undergo oxidative decyanation under phase-transfer catalytic conditions to afford phenyl hetaryl ketones in good yield (33–77%) (equation 109)[202].

$$\text{Hetaryl-X} + {}^-\text{CH(Ph)CN} \xrightarrow[\text{b. PTC [O]}]{\text{a. } h\nu \ (S_{RN}1)} \text{Hetaryl}-\text{COPh} \qquad (109)$$

(163a)

Carbanion **170** derived from cyclohexenylidene acetonitrile reacts with *p*-bromo or *p*-iodo anisole to afford a mixure of **171a** and **171b**, the two isomers of the monosubstitution product (equation 110)[203].

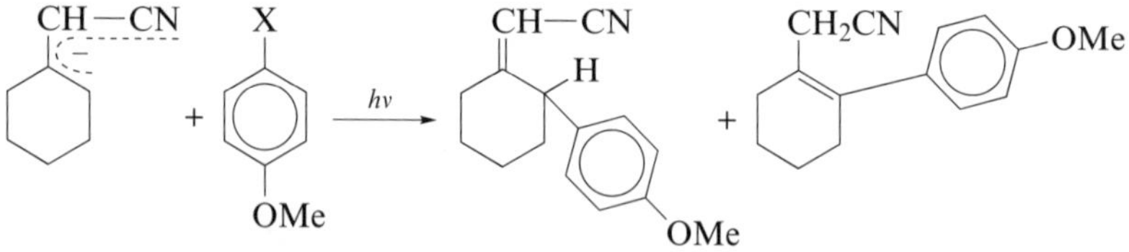

$$(110)$$

(170)   X = Br,I   **(171a)**   **(171b)**

The ratio of **171a** and **171b** depends on the method used to neutralize the reaction, though both isomers were always formed. Only arylation at γ was observed. Coupling of the radical in the γ-position gave the most stable radical anion in which the double bond and the nitrile group are conjugated.

The electrochemically induced reaction of *p*-chlorobenzonitrile with malononitrile anion in liquid ammonia, using 4,4′-bipyridine as a redox mediator, gave the substitution product **172** in 85% yield (equation 111)[204].

$$\underset{\text{CN}}{\underset{|}{\overset{\text{Cl}}{\overset{|}{\bigcirc}}}} + \ ^-\text{CH(CN)}_2 \xrightarrow{\ e^-\ \text{cathode}\ } \underset{\text{CN}}{\underset{|}{\overset{\text{NC—CH—CN}}{\overset{|}{\bigcirc}}}} + \ \text{Cl}^- \qquad (111)$$

$$\textbf{(172)}$$

e. Cyanide ions as nucleophiles. Most of the nucleophiles studied react with aromatic radicals with a rate constant close to the diffusion limit as determined by electrochemical techniques[205,206]. An exception is CN^- ion, one of the less reactive nucleophiles toward aromatic radicals. Thus, it has been determined by electrochemical methods that the rate constant of the coupling of phenyl radicals with CN^- ions is $\leq 4 \times 10^5 \ M^{-1}s^{-1}$, and the highest rate constant reported corresponds to the coupling with 2-cyanophenyl radical $(9.5 \times 10^8 \ M^{-1} \ s^{-1})[207]$.

When an aryl radical couples with a nucleophile, it forms a radical anion $(ArNu)^{-\bullet}$, and if $E^\circ [ArNu/(ArNu)^{-\bullet}] \ll E^0[ArX/(ArX)^{-\bullet}]$, the ET reaction of equation 112 is practically irreversible from left to right, with a rate constant close to the diffusion limit (catalytic system),

$$(ArNu)^{-\bullet} + ArX \longrightarrow ArNu + (ArX)^{-\bullet} \qquad (112)$$

However, if $E^\circ [ArNu/(ArNu)^{-\bullet}] \gg E^0 [ArX/(ArX)^{-\bullet}]$, the ET reaction is close to the diffusion limit from right to left (non-catalytic system), and the radical anion $(ArNu)^{\bullet}$ has to be oxidized by the anode[208]. A detailed study of the electrochemically induced reaction of *p*-bromobenzophenone with CN^- ions in liquid ammonia showed that this is a non-catalytic system[208].

The photostimulated reaction between 1-halo-2-naphthoxide ions and CN^- ions gave the substitution product, which was suggested to be formed by the $S_{RN}1$ mechanism. The quantum yield of the reaction with 1-chloro-2-naphthoxide ions is greater than unity, and depends on the pH, being *ca* 15 at pH greater than 10. These reactions are inhibited by O_2, nitroxyl radicals and electron acceptors[209].

f. Carbonylation reactions. Carbonylation of organic halides constitute interesting reactions because they allow a C—C bond formation with simultaneous introduction of a functional group. In recent years carbonylation of aryl and vinyl halides has been achieved by means of different reagents by the $S_{RN}1$ mechanism.

Thus, it has been shown that $Co(CO)_4^-$ ion is a good nucleophile for the catalytic carbonylation of aryl halides in the presence of the electron donor 'NaH—RONa' associate in THF and under photostimulation (equation 113)[210].

$$ArX \xrightarrow[\text{CO (1 atm), THF, } hv \text{ (350 nm)}]{\text{NaH-}t\text{-C}_5\text{H}_{11}\text{ONa-Co(OAc)}_2} ArCO_2Na \qquad (113)$$

The photostimulated carbonylation of aryl and vinyl halides by $Co(CO)_4^-$ ions under phase-transfer catalysis conditions $[Co_2(CO)_8, C_6H_6/NaOH_{aq}, CO, Bu_4N^+Br^-, 65\,^\circ C]$ has been reported[211,212]. Thus PhBr, *o*- and *p*-bromo toluenes, *p*-bromoanisole, *p*-bromofluorobenzene, 4-bromoacetophenone and 1- and 2-bromonaphthalenes gave $ArCO_2Na$ in 90–98% yields. Under the conditions used, carbonylation did not occur with unsubstituted PhCl, but when it was substituted, as in *p*-chlorobenzoic acid, or *p*-chlorophenylacetic acid, carbonylation was quantitative (98% yield).

These photostimulated carbonylations seem to occur by the $S_{RN}1$ condensation of $Co_2(CO)_4^-$ ions with aryl halides. The propagation cycle is outlined in Scheme 4.

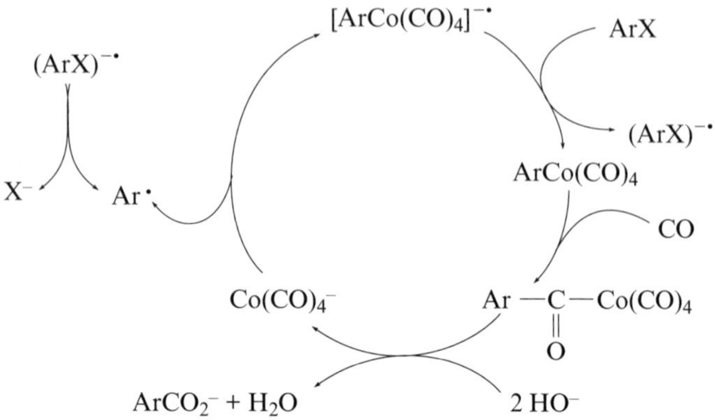

$$[ArCo(CO)_4]^{-\bullet} \qquad ArX$$

$$(ArX)^{-\bullet}$$

$$(ArX)^{-\bullet}$$

$$ArCo(CO)_4$$

$$X^- \qquad Ar^\bullet$$

$$CO$$

$$Co(CO)_4^- \qquad Ar-\overset{\overset{\displaystyle O}{\parallel}}{C}-Co(CO)_4$$

$$ArCO_2^- + H_2O \qquad 2\ HO^-$$

SCHEME 4

Carbonylations of *o*-dihalobenzenes and *o*-halobenzoic acids catalysed by $Co_2(CO)_8$ in aqueous NaOH and under photostimulation were also described[213] to give moderate yields of the diacids.

A novel method was reported for the carbonylation of aryl halides by cobalt salt catalysts, such as $Co(OAc)_2$, $CoCl_2$, $CoSO_4$, $Co(OH)_2$, $Co(OH)_3$, CoO and Co_2O_3 in an aqueous alkaline solution and under irradiation[214].

This reaction proceeds not only for halides bearing electron-withdrawing groups, but also for simple aryl halides such as PhCl, which were previously reported to be unreactive[212]. This reaction did not occur without irradiation, even at higher temperature; it seems to be photoinitiated, but the reaction mechanism and the actual active species are not known.

The reduction of aryl iodides to arenes by $KHFe(CO)_4$ under mild conditions was performed[215]. This reaction involves an ET from $HFe(CO)_4^-$ ions to the aryl iodide. The resulting aryl radicals either abstract hydrogen atom or combine with the $HFe(CO)_4^-$ radical species to form $ArHFe(CO)_4$ which yields ArH by reductive elimination.

The catalytic carbonylation of PhI to benzoic acid can be achieved under mild conditions by a bimetallic system: $(HFe(CO)_4^- - Co(CO)_4^-$[216]. ET from $HFe(CO)_4^-$ ions to PhI initiate the $S_{RN}1$ reaction; the phenyl radical reacts with the $Co(CO)_4^-$ ion which acts as the actual carbonylation catalyst.

2. Tin nucleophiles

Although the reaction of triorganostannyl ions (R_3Sn^-) with ArX has long been known, contradictory results have been found according to the solvent or the reaction conditions. The reaction of PhBr with Me_3SnNa gave benzene (96%), together with some $PhSnMe_3$ (4%)[217].

The reaction of *o*-, *m*- and *p*-bromotoluenes with Bu_3SnLi in THF gave the substitution product. However, *cine*-substitution products were found with *p*-chloro- and *p*-fluorotoluenes, indicating that a benzyne mechanism operates. In the presence of Li metal, the yield of the *ipso* substitution products increased. These results suggest that the reaction, at least partially, occurs by a radical mechanism[218].

Recently, the reaction of R_3Sn^- ions with ArX has been studied in liquid ammonia. The mechanism in play was shown to depend on the structure of the tin nucleophile, the aromatic substrates (whether they are mono- or dihalo-substituted), the halogens (Cl, Br, I) and the experimental conditions. Depending on these variables, it is possible for the reactions to follow exclusively either an $S_{RN}1$ mechanism to give substitution products, or an HME mechanism to give dehalogenation products. Competition between both mechanisms is also possible[219].

Thus, whereas Ph_3Sn^- ions do not react in the dark with p-chloro- and p-bromotoluenes, p-dichlorobenzene, 1-chloronaphthalene and 2-chloroquinoline, showing that there is no HME reaction, good yields of substitution products (62–80%) were obtained under irradiation (equation 114).

$$ArX + Ph_3Sn^- \xrightarrow{\ h\nu\ } ArSnPh_3 + X^- \qquad (114)$$

Although 1-bromonaphthalene reacts in the dark through an HME mechanism, this reaction is relatively slow and, under irradiation, the $S_{RN}1$ mechanism is the main reaction.

When iodide in p-iodotoluene and p-iodoanisole and bromide in p-dibromobenzene serve as the leaving group, there is a fast HME reaction giving only dehalogenation products. It is noteworthy that in the case of p-dibromobenzene, the HME reaction leads to PhBr which does not undergo an HME process, but reacts under irradiation by the $S_{RN}1$ mechanism as shown by the formation of Ph_4Sn (40%) together with the disubstitution product (22%).

With Me_3Sn^- ion, only compounds in which chloride is the leaving group, like p-chloroanisole, 1-chloronaphthalene, p-dichlorobenzene and 2-chloroquinoline, do not react by an HME mechanism. They react by the $S_{RN}1$ mechanism to give good yields (70–100%) of substitution products (equation 115). 2-chloroquinoline reacts in the dark in a reaction which is partially inhibited by p-DNB. p-Bromoanisole and presumably all the aryl bromides and iodides react with this nucleophile by an HME mechanism.

$$ArX + Me_3Sn^- \xrightarrow{\ h\nu\ } ArSnMe_3 + X^- \qquad (115)$$

With o-dichlorobenzene, 58% of the disubstitution product **173** together with 17% of the monosubstituted product **174** were formed (equation 116)[220].

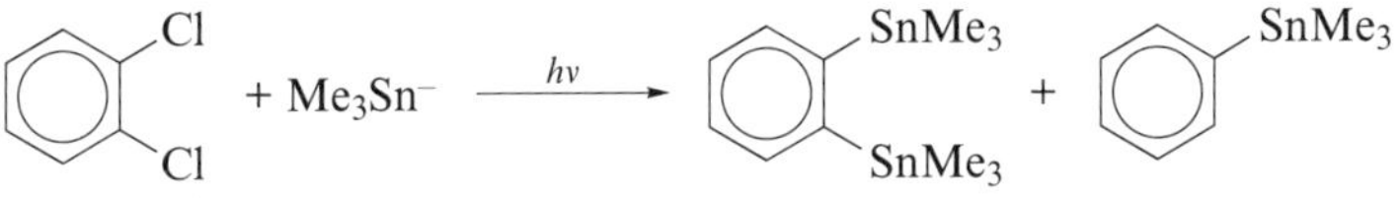

$$(116)$$

(173)   **(174)**

By *competition experiments*, it was found that Ph_3Sn^- and Me_3Sn^- react at the same rate as Ph_2P^- with p-anisyl radicals in liquid ammonia, and this is probably the diffusion controlled rate. However, Me_3Sn^- ion is more reactive than Ph_3Sn^- ion in its overall reactivity, since it is probably more reactive in the initiation step of the $S_{RN}1$ mechanism[219].

The reaction of an organostannyl compound bearing alkyl–Sn and aryl–Sn bonds with sodium metal in liquid ammonia gave selectively the alkyl–Sn bond fragmentation. Thus the reaction of p-chloroanisole with Me_3Sn^- ions gave p-MeOC$_6$H$_4$SnMe$_3$. Treatment of the latter with sodium metal gave p-MeOC$_6$H$_4$SnMe$_2^-$ ions (equation 117), which react with p-chlorotoluene under irradiation to give the diaryl dimethylstannane in 89% yield in a one-pot reaction (equation 118)[220].

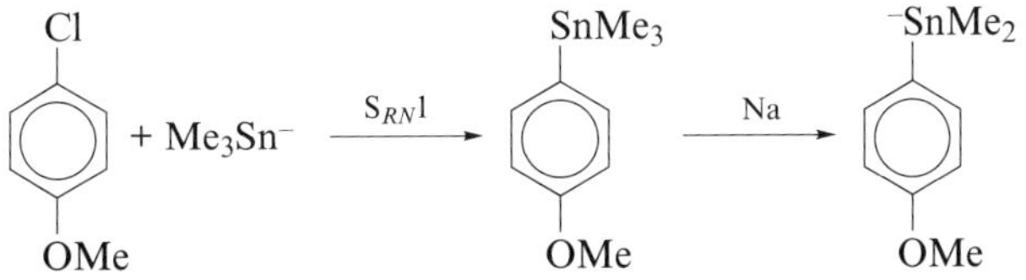

$$(117)$$

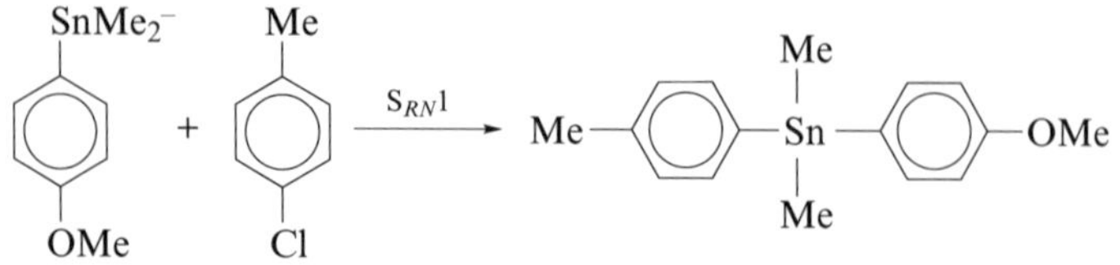

$$\text{(118)}$$

B. Nucleophiles Derived from the V Group

1. Nitrogen nucleophiles

Although the aromatic $S_{RN}1$ reaction was discovered during the reaction of aryl iodides with NH_2^- ions in spontaneous or K metal catalysed reactions[3], few examples have been studied with this nucleophile.

The photostimulated reaction of 1-bromo-2,4,6-trimethylbenzene (a substrate with no *o*-hydrogen atoms in order to avoid the benzyne mechanism) with NH_2^- ions gave 1-amino-2,4,6-trimethylbenzene (70%) and the reduction product 1,3,5-trimethylbenzene (6%). This reaction did not occur in the dark[221]. By competition experiments of NH_2^- ions with Ph_2P^- ions toward 2,4,6-trimethylphenyl radicals, it was found that Ph_2P^- ions are 6.4 times more reactive than NH_2^- ions[221], whereas NH_2^- ions are twice more reactive than acetone enolate ions toward the same radical in liquid ammonia[222].

The reaction of phenylamide ions with PhI stimulated by K metal gave diphenylamine (19%), *o*- and *p*-phenylanilines (11% each)[3]. However, the photostimulated reaction of phenylamide ions with PhI, *p*-iodoanisole and *p*-iodobenzonitrile was quite sluggish (*ca* 8% in 3 h)[223].

On the other hand, the photostimulated reaction of 2-naphthylamide ions with iodobenzene, *p*-iodoanisole or 1-iodonaphthalene gave mainly substitution at C_1 of the naphthyl moiety, affording amines **175** (45–63%) and traces of *N*-aryl-substitution product **176** (1–6%) (equation 119)[224].

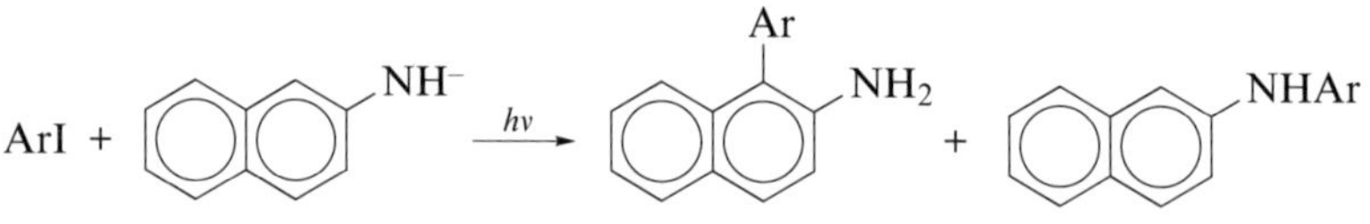

$$\text{(119)}$$

(175)   **(176)**

Pyrrole anion is unreactive in liquid ammonia under irradiation with PhBr or 1-chloronaphthalene. However, the reactions of aryl chlorides (*p*-chlorobenzonitrile, 3- and 4-chloropyridines and 4-chlorodiphenyl sulphone) with 2,5-dimethylpyrrole anion under electrochemical inducement in the presence of a redox mediator gave the C_3-substituted product in moderate yields (35–40%) (equation 120)[225]. The rate constant of the coupling reaction between this nucleophile and aryl radicals is about $5–8 \times 10^9 \text{ M}^{-1} \text{ s}^{-1}$ determined by electrochemical methods[225].

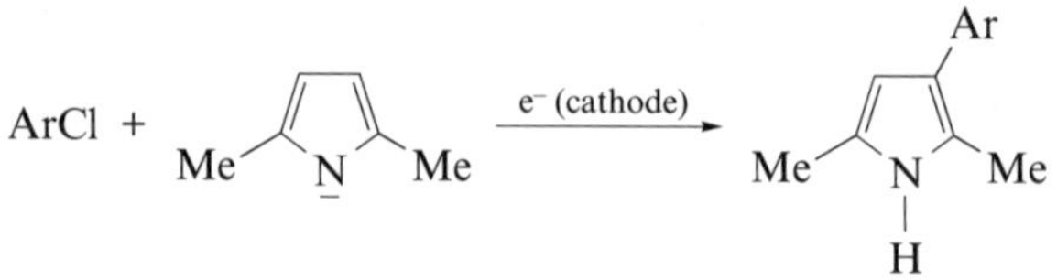

$$\text{(120)}$$

The electrochemically induced $S_{RN}1$ reaction of aryl chlorides with redox mediators and pyrrole anion gave mainly 2-aryl pyrroles (52–67%) and, in a lower amount, 3-aryl pyrroles

(3–14%) (equation 121). Disubstitution products were also observed in small yields. No products derived from N-substitution were reported[226].

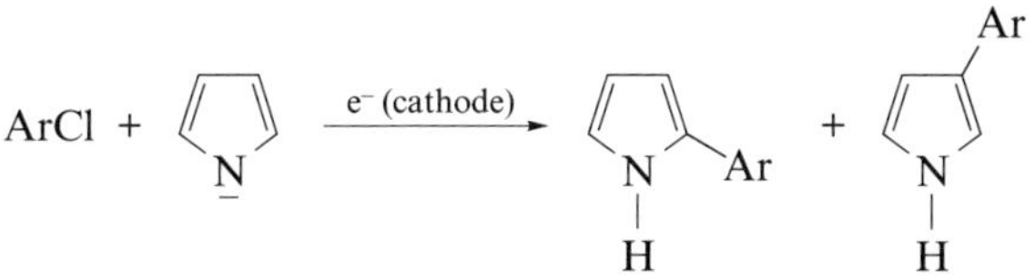

$$\text{ArCl} + \underset{\overset{|}{N}}{\boxed{}} \xrightarrow{e^- \text{ (cathode)}} \underset{\underset{H}{\overset{|}{N}}}{\boxed{}}\text{Ar} + \underset{\underset{H}{\overset{|}{N}}}{\boxed{}}^{\text{Ar}} \qquad (121)$$

In the reaction of p-chlorobenzonitrile and 4-chloropyridine with the indolyl anion under the same experimental conditions as with pyrrole anion, only the substitution product in position 3 was formed in 60% yield (equation 122)[226].

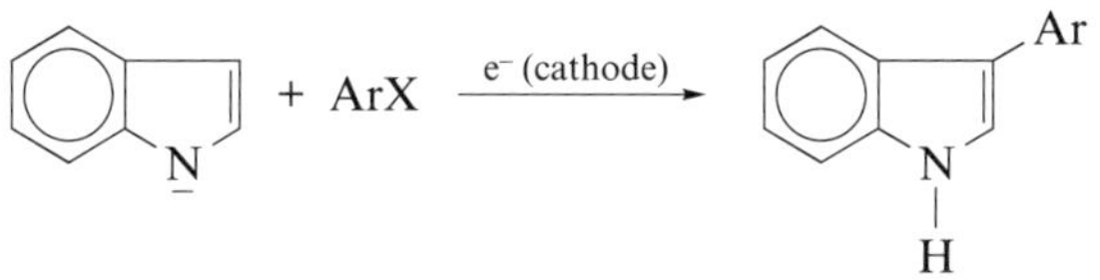

$$\underset{\overset{|}{N}}{\boxed{}} + \text{ArX} \xrightarrow{e^- \text{ (cathode)}} \underset{\underset{H}{\overset{|}{N}}}{\boxed{}}^{\text{Ar}} \qquad (122)$$

The electrochemically induced $S_{RN}1$ reaction of haloarenes with uracil anion (**177**) in DMSO gave 5-aryl uracils **178** in 30–55% yields (equation 123)[227]. Thus, good yields of the 5-aryl uracils were obtained in the reaction with 4-bromobenzophenone (55%), 4-chlorobenzonitrile (50%) and 1-iodo-4-nitrobenzene (55%). With 1-iodo-2-trifluoromethylbenzene (and with phthalonitrile as a redox catalysis), 30% of the 5-aryl uracil was formed.

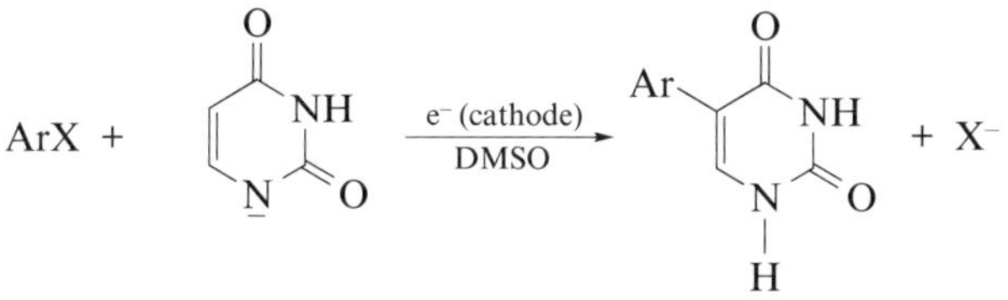

$$\text{ArX} + \underset{\textbf{(177)}}{\boxed{}} \xrightarrow[\text{DMSO}]{e^- \text{ (cathode)}} \underset{\textbf{(178)}}{\boxed{}} + \text{X}^- \qquad (123)$$

The reaction of hexafluorobenzene (**179**) with pyrrole anion in THF for 24 h leads to the p-disubstitution product (38%) on nitrogen, with successively lesser amounts of the higher polysubstitution products[228]. The reaction of **179** with pyrazole anion (**180**) (6 h) gave quantitatively the N-hexasubstituted product **181**. The reaction was partially inhibited by m-DNB and 75% of **181** and 17% of the p-disubstitution product **182** were formed in 24 h (equation 124)[228].

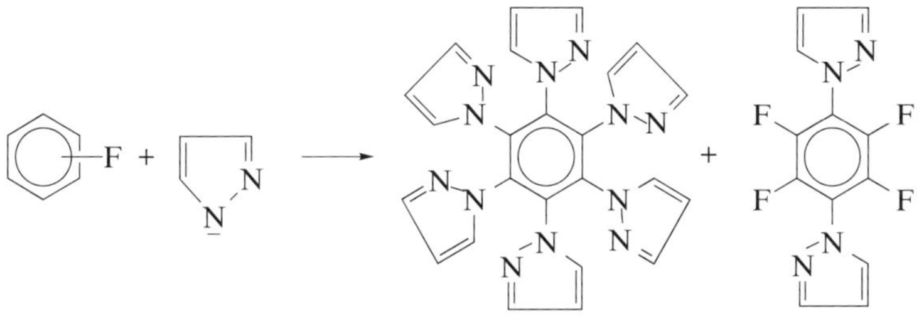

$$\boxed{}\!\!-\!\text{F} + \underset{\overset{|}{N}}{\boxed{}} \longrightarrow \quad \text{(181)} \quad + \quad \text{(182)} \qquad (124)$$

(**179**)   (**180**)   (**181**)   (**182**)

The preference for the hexasubstitution product is demonstrated by the fact that with a ratio of C_6F_6: pyrazole anion of 1:2, **181** is formed in 88%, whereas **182** is formed in 11% yield. Other substitution products are formed in low yield or not detected[228]. It was proposed that this reaction proceeds by the $S_{RN}1$ mechanism based on the fact that hexasubstitution was the main product obtained and that inhibition by m-DNB occurred.

In the reaction of **179** with 3,5-dimethyl pyrazole anion, the hexasubstitution product is formed in 72%, with 8–9% of disubstitution and 18–22% of the tetrasubstitution product. With 3,5-dimethyl-4-nitropyrazole anion and imidazole anion, the hexasubstitution products are formed quantitatively[228].

It has been proposed that the reaction of 2-chloro-5-nitropyridine (**183**) with the anions of benzotriazole, imidazole, pyrrole and phthalimide in DMF occurs by the $S_{RN}1$ mechanism. Thus, the reaction of **183** with the anion of benzotriazole gives 95% of the substitution product **184** (equation 125)[229].

$$\text{(183)} \qquad\qquad\qquad\qquad \text{(184)} \qquad\qquad (125)$$

In the reaction with imidazole anion (71% of product) and with pyrrole anion (67% of product), the substitution occurs only on nitrogen. The reaction of benzotriazole anion with 2-chloro-3-nitropyridine is slower compared with that of the 5-nitro isomer. This reaction was ESR active, and it was completely inhibited by $FeCl_3$[229].

The photostimulated reaction of aryl halides with carbazole[230], phenothiazine[231] and benzimidazole[231] anions in DMSO proceeds by the $S_{RN}1$ mechanism, giving N-arylated products, although in very low yields.

2. Phosphorous nucleophiles

Several types of phosphanion nucleophiles react with haloarenes by the $S_{RN}1$ mechanism. Examples are P^{-3} ions formed by the reaction of phosphorous with Na metal in liquid ammonia[154], Ph_2P^- ions[232], butylphenylphosphonite[233], O,O-diethylthiophosphonite[233] and N,N,N',N'-tetramethylphosphonamide ions[233].

One of the most studied nucleophiles has been the Ph_2P^- ions, which gave good yields of substitution in thermal (iodoarenes) or photostimulated reactions. However, the photostimulated reaction of 1-bromo-2,4,6-trimethylbenzene with Ph_2P^- ions gave 45% of the substitution product and 50% of the reduction product 1,3,5-trimethylbenzene. Probably because of the steric hindrance, the rate of the coupling reaction between 2,4,6-trimethylphenyl radical and Ph_2P^- ions is rather slow, increasing the percentage of reduction product[221].

p-Iodoanisole reacts in the dark with Ph_2P^- ions in liquid ammonia at room temperature to give 26% of the substitution product after 45 min, but this reaction is accelerated by ultrasound (87% in 45 min). Similar behaviour was observed with 1-bromonaphthalene (10% and 94% reaction, respectively)[37].

Sodium amalgam Na(Hg) reacts with haloarenes in liquid ammonia to afford the dehalogenation products in a selective reaction; thus PhCl did not react, but PhBr and 1-chloronaphthalene were quantitatively dehalogenated[234]. The reaction of 1-chloronaphthalene and 2-chloroquinoline with Ph_2P^- ions catalysed by Na(Hg) gave good yields of substitution products (71% and 94%, respectively) (equation 126)[235].

$$\text{ArX} + \text{Ph}_2\text{P}^- \xrightarrow{\text{Na(Hg)}} \text{ArPPh}_2 + \text{X}^- \tag{126}$$

When the haloarene radical anion fragments rapidly, and close to the Na(Hg) surface, only dehalogenation was obtained. However, by using benzonitrile as a redox catalyst, good yields of substitution were found. Thus, the reaction of p-bromoanisole with Ph_2P^- ions and benzonitrile gave 85% of the substitution product.

It has been shown that the diphenylphosphinite ion (Ph_2PO^-) reacts under irradiation with PhI in liquid ammonia to give good yields of Ph_3PO[233]. This nucleophile was prepared in liquid ammonia by the reaction of Ph_2PHO with t-BuOK or Ph_3PO with an excess of Na metal[236].

With $(\text{PhCH}_2)_3\text{PO}$ as substrate, $(\text{PhCH}_2)_2\text{PO}^-$ ions were formed by reaction with Na metal, and the photostimulated reaction with PhBr or PhI gave high yields (80–85%) of the substitution product (equation 127)

$$\text{PhX} + (\text{PhCH}_2)_2\text{PO}^- \xrightarrow{h\nu} (\text{PhCH}_2)_2(\text{Ph})\text{PO} + \text{X}^- \tag{127}$$

In the reaction of $\text{Ph}_2(\text{PhCH}_2)\text{PO}$ (where two Ph—P bonds compete with one PhCH_2—P bond) with Na metal (equation 128) and 1-iodonaphthalene in liquid ammonia, the substitution product 1-naphthyldiphenylphosphine oxide (79%) was obtained after irradiation (equation 129). This result, in agreement with electrochemical determinations[237], shows that the PhCH_2—P bond fragments faster than the Ph—P bond. No products derived from the fragmentation of the Ph—P bond were found.

$$\text{Ph}_2(\text{PhCH}_2)\text{PO} \xrightarrow[\text{b. } t\text{-BuOH}]{\text{a. Na}} \text{Ph}_2\text{PO}^- + \text{PhCH}_3 \tag{128}$$

$$\tag{129}$$

A good yield of substitution product was obtained when Ph_2PO^- ions, prepared as in equation 128, reacted under irradiation with p-bromo- and p-iodoanisole.

Potassium dialkyl phosphite ions react rapidly with aryl iodides in liquid ammonia under irradiation to form diethyl esters of arylphosphonic acid in almost quantitative yields[238]. The photostimulated reaction of $(\text{EtO})_2\text{PO}^-$ ions with 5-chloro-7-iodo-8-isopropoxyquinoline gave 70% of the substitution product with retention of the 5-chloro substituent[239]. The same nucleophile was formed in THF, and in a mixture of solvents (1:4 THF: MeCN) reacts under irradiation with ArX (o-, m-, p-iodoanilines, 2-iodo, 3-iodo, 2-bromo and 3-bromopyridines and 3-bromoquinoline) rendering the substitution product in high yields (70–98%) (equation 130)[240].

$$(\text{EtO})_2\text{PO}^- + \text{ArX} \xrightarrow{h\nu} (\text{EtO})_2\text{P(O)Ar} \tag{130}$$

These reactions, especially those of the less reactive bromoarenes, are accelerated in the presence of I^- ions[34]. Thus, the photostimulated reaction of bromotoluenes with $(\text{EtO})_2\text{PO}^-$ ions gave better yields (84–99%) in the presence of NaI than in its absence (40–49%).

A mechanism with similar propagation steps to the usual $S_{RN}1$ has been proposed in the reaction of diaryliodonium salts with triphenylphosphine; in this case, both the nucleophile and leaving group are neutral. This reaction is catalysed by light or by the thermal decomposition of di-t-butyl peroxalate. The propagation steps that have been proposed in the reaction of p-tolyliodonium hexafluorophosphate with triphenylphosphine are given in equations 131–133[241].

$$\overset{\bullet}{\text{TolPPh}_3} + \text{Tol}_2\text{I}^+ \quad \xrightarrow{\text{ET}} \quad \text{TolPPh}_3^+ + \text{Tol}_2\text{I}^{\bullet} \qquad (131)$$

$$\text{Tol}_2\text{I}^{\bullet} \quad \longrightarrow \quad \text{TolI} + \text{Tol}^{\bullet} \qquad (132)$$

$$\text{Tol}^{\bullet} + \text{Ph}_3\text{P} \quad \longrightarrow \quad \overset{\bullet}{\text{TolPPh}_3} \qquad (133)$$
$$\text{Tol} = p\text{-Tolyl}$$

The key propagation step of the chain is the ET reaction in which the phosphoranyl radical reduces the iodonium salt (equation 131).

C. Nucleophiles Derived from the VI Group

1. Oxygen nucleophiles

Aliphatic alkoxide ions are known to be unreactive as nucleophiles in the $S_{RN}1$ mechanism. Primary and secondary alkoxides reduce aromatic halides to arenes[242] by an ET chain process, which at the same time leads to the oxidation of the alkoxide into the corresponding carbonyl group. Tertiary alkoxides (e.g. t-butoxide) can be used as bases in the reaction media as well as to initiate the photostimulated process.

On the other hand, aromatic alkoxide ions were believed to be unreactive since substitution was not observed in the reaction of PhO⁻ ions with halobenzenes stimulated electrochemically[243], by solvated electrons from alkali metals[244] or sodium amalgam[245] or under light stimulation[14,246].

More recent studies, however, have proved that these anions, mainly di-t-butyl substituted phenoxides and 1- and 2-naphthoxide ions, are excellent nucleophiles under electrochemical or photostimulated conditions. These anions behave as bidentate nucleophiles and couple with radicals through the carbons of their aromatic ring. This has been proved to be a powerful route to biaryls unsymmetrically substituted by EWG and electron-acceptor groups, which are of interest in non-linear optics, as well as in the synthesis of cyclic compound (Section V.E.2).

a. Electrochemically induced reactions. PhO⁻ ions were found to be reactive under electrochemical induction in liquid ammonia with 4-bromobenzophenone[247,248], 4-chlorobenzonitrile[248] and 2-chloroquinoline[248] (equation 134).

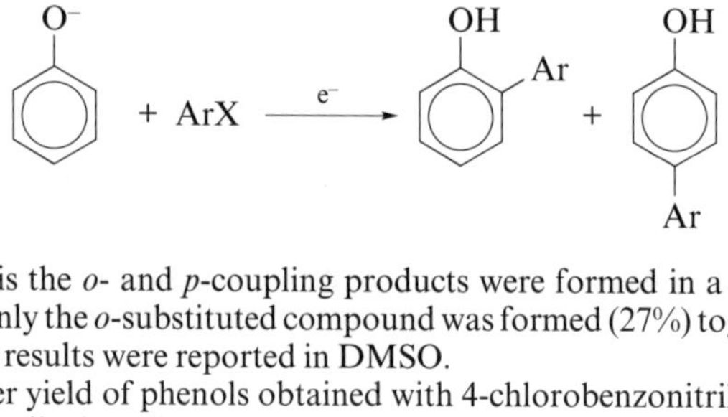

$$(134)$$

Upon electrolysis the o- and p-coupling products were formed in a 2:1 ratio. With 2-chloroquinoline, only the o-substituted compound was formed (27%) together with quinoline (68%). Similar results were reported in DMSO.

The overall lower yield of phenols obtained with 4-chlorobenzonitrile vs 4-bromobenzophenone was ascribed to the faster fragmentation rate of the former radical anion (9.3×10^8 s⁻¹ vs 590 s⁻¹ for 4-bromobenzophenone). This difficulty was overcome by redox catalysis. The requirements of the mediator, the reaction conditions and the general equations predicting the yields were established for this system[35].

The 2,6- and 2,4-di-t-butyl phenoxide ions were shown to be more reactive than PhO⁻ ions. The reaction of these nucleophiles, mainly of 2,6-di-t-butylphenoxide ion (185), was studied with a considerable variety of compounds in order to achieve the synthesis of biaryls of importance in non-linear optics[249,250]. In these nucleophiles, the t-butyl groups substitute two of the possible coupling positions in order to perform a selective synthesis of either the o- or the p-isomer. The substituents can be easily removed later[251,252].

Among the compounds shown to react with 185 are 2-, 3- and 4-chlorobenzoni-trile[35,249,253], 2-,3- and 4-chloropyridines[35], 2-chloro-5-cyanopyridine[35,249] and chloro(trifluoromethyl)pyridines (equation 135)[232].

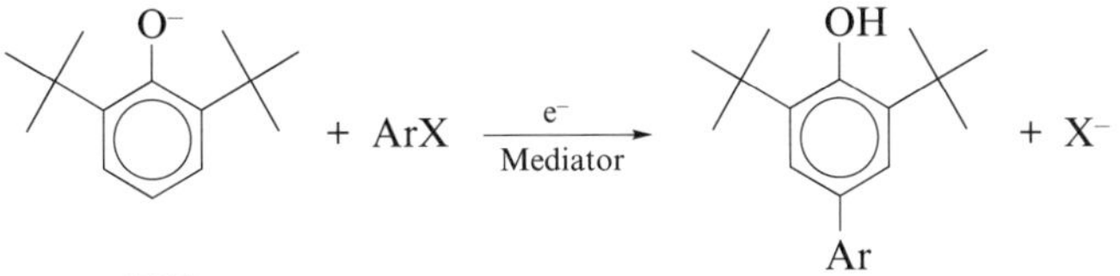

(185) $\quad$ (135)

With chloro(trifluoromethyl)pyridines, the yield of the isolated product was about 50%, and with 2-chloro-5-cyanopyridine, the yield was 75% under the same experimental conditions[35]. With 2-chloropyridine no coupling product was observed.

The electroinduced substitution of 4-chloropyridine with 2,6-dialkylphenoxide (alkyl = pentyl, isopropyl, methyl) ions was performed in a mixture of ammonia with THF[254]. The yields of products were lower than that obtained with 185 and they decreased when the steric hindrance of the substituents decreased and when the EWG character increased. Quaternarization of the substitution product by linear alkyl halides, followed by deprotonation, gives pyridiniophenoxide zwitterions[254–255]. N-alkylation of 4-(3,5-di-t-butyl-4-hydroxyphenyl)pyridine with 3-bromopropyl methacrylate or 6-bromohexyl methacrylate followed by treatment with base to give the final zwitterionic methacrylate has been reported[256].

Unsymmetrical donor–acceptor polyaryls were synthesized following this approach through the reaction of 185 with cyano-substituted polyaryl bromides[257] 186 to give the products 187 (equation 136). Thus, 186a (Ar = p-NCC$_6$H$_4$) gave 25% of 187, 186b (Ar = p-(p-NCC$_6$H$_4$)C$_6$H$_4$ gave <10% of 187) and 186c (Ar = 4-pyridyl) gave 35% of 187. The yields reported correspond to reactions interrupted before total conversion of the substrates, in order to avoid possible secondary reactions.

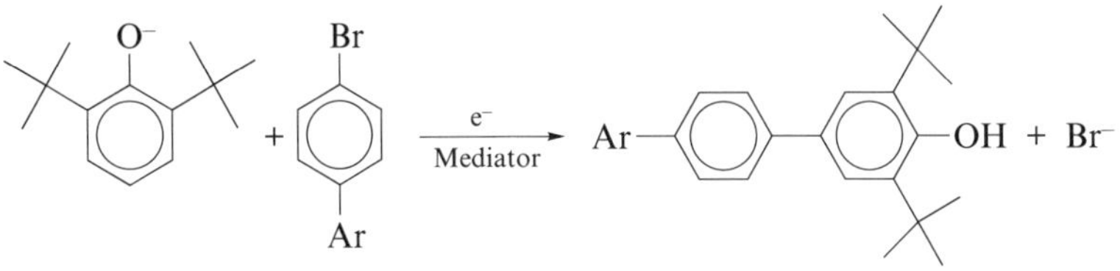

(185) $\quad$ (186) $\quad$ (187) $\quad$ (136)

Substitution of p-dichlorobenzene by 185 was carried out in liquid ammonia/benzonitrile in the presence of a mediator whose role is also to oxidize the anion radical of the monosubstitution product in order to decrease the degree of disubstitution[258]. The monosubstitution product (67%) was then further substituted to give a phosphine and (phosphoniophenyl)phenoxide zwitterions or was carboxylated to obtain 3′,5′-di-t-butyl-4′-hydroxy-1,1′–biphenyl-4-carboxylic acid[258].

The electroinduced reaction of 2-t-butyl-1-naphthoxide ions in liquid ammonia with aryl chlorides has been reported to give 60–85% yield of 4-aryl-2-t-butyl-1-naphthols[259].

Methyl or aryl sulphonyl hydroxybiphenyls **188** could be electrosynthesized in liquid ammonia from chlorophenyl sulphones and **185** (equation 137)[260].

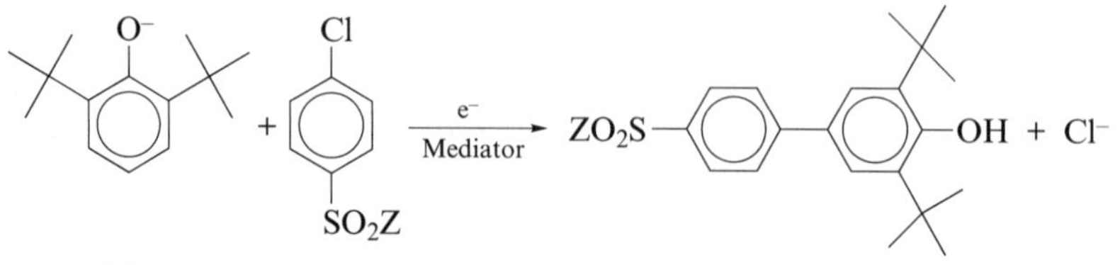

(**185**)   Z = Me, Ph, 4-ClC$_6$H$_4$ (**188**) 90%

b. Photostimulated reactions. Under photostimulation p-methylphenoxide ion (**189a**) reacted sluggishly with PhI[246,261] or p-iodobenzonitrile (20% of **190a**) but gave good yields of substitution in its reaction with p-bromobenzonitrile (65% of **190a**)[261]. The reaction of PhBr and p-methoxyphenoxide ion **189b** gave 40% of **190b** and the disubstitution product **191** (12%). It was determined that **189b** reacted much faster than **189a** (equation 138)[261].

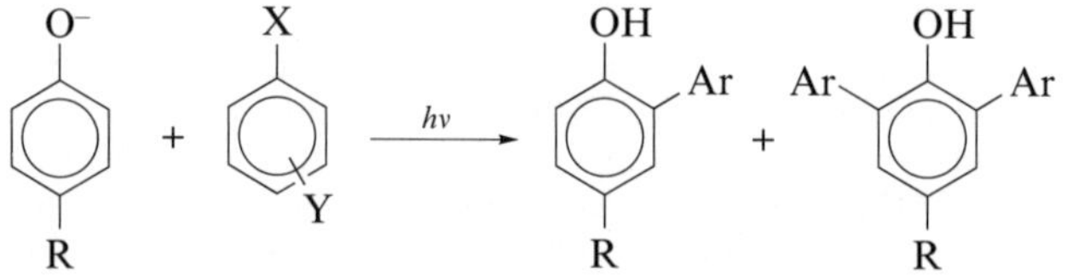

(**189a**) R = Me   ArX   (**190a**) R = Me   (**191**) R = OMe
(**189b**) R = OMe   (**190b**) R = OMe

The nucleophile **185** appeared also to be very reactive under photostimulation. Good percentage of substitution was obtained in its reaction with p-CN, o-CN, o-CONH$_2$, o-COCH$_3$ and o-OCH$_3$ substituted bromobenzenes[261].

The 1- and 2-naphthoxide ions also reacted with p-bromobenzonitrile[261], p-iodoanisole[262] and 1- and 2-iodo- substituted naphthalenes[262,263]. In the reaction of 1-naphthoxide ions a mixture of 2- and 4-monosubstituted naphthyl derivates (**192** and **193**, respectively) together with the 2,4-disubstituted compounds **194** were observed (equation 139). Thus, p-bromobenzonitrile, 1-iodo- and 2-iodonaphthalenes gave **192** (25%, 22%, <1%), **193** (40%, 36%, 37%) and **194** (<1%, 20%, 15%), respectively.

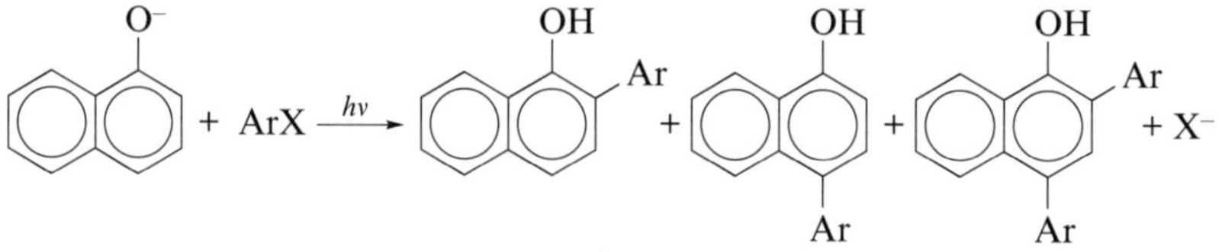

(**192**)   (**193**)   (**194**)

When the reaction of the 2-methyl-1-naphthoxide ions was performed with 1-iodo-, 2-iodo-, 6-(i-Pr)-2-iodo- , or 3,5-di-(OMe)-2-iodo naphthalenes only the 4-substitution product was formed in 50–70% yield. In the photostimulated reaction of 4-substituted

1-naphthoxide ion **195** with 1-iodo- and 2-iodonaphthalene in liquid ammonia, the 2-substitution product **196** (35%) and the 4-addition compound **197** (23–34%) were formed (equation 140). The addition compound was not observed in the reaction of 1-iodonaphthalene in DMSO or MeCN, where the yields of **196** were 32–47%, respectively.

$$\text{(195)} \qquad \text{(196)} \qquad \text{(197)}$$

2-Naphthoxide ions (**198**) react to give only the 1-Ar substitution product **199**. Thus, the reaction of **198** with p-bromobenzonitrile (85%), p-iodoanisole (48%), 1-iodo- and 2-iodonaphalene and derivatives (44–65%) has been reported to give the yields indicated above of the substitution products **199** (equation 141)[261–263].

$$\text{(198)} \qquad\qquad \text{(199)}$$

Fluorinated biaryl derivatives were obtained by substitution of p-F, p-CF$_3$ and p-CF$_3$O bromobenzenes with 2,4- and 2,6-di-t-butylphenoxide ion and p-MeO and p-CF$_3$O phenoxide ions as well as with **198** under photostimulation[264]. Depending on the nucleophile and the substrate, o-disubstitution and 4-*addition* products were obtained.

2-Chloropyridines bearing a CF$_3$ group on position 3,4,5 or 6 were found to be suitable substrates for photostimulated $S_{RN}1$ reactions with **185**, **198** and 2,4-di-t-butylphenoxide ions. Similar reactions were performed under electrochemical induction[252]. The results obtained indicate that synthetically useful yields are obtained in the reaction of 2-chloro-5-trifluoromethylpyridine with these nucleophiles. When the CF$_3$ group is in position 3, a lower percentage of substitution product is obtained. These differences were ascribed to steric interactions between the CF$_3$ group and the OH of the incoming nucleophile[188].

It is possible to obtain reasonable yield of substitution product when the CF$_3$ group is in position 6, even though this substrate generates an unhindered but poor electrophilic radical, whereas the substrate in which the CF$_3$ group is in position 4 is not anticipated to be a useful substrate for synthesizing trifluoromethyl heterobiaryl derivatives[188], 2-Amino-3-chloro-5-trifluoromethyl pyridine was reported to give high yields of substitution with **198** and 2,4-di-t-butylphenoxide ions[188].

The regiochemistry of the reaction of phenoxide and naphthoxide ions with radicals was explained on the basis of the perturbation theory of the frontier MO involved in the coupling reaction, which are the HOMO of the nucleophile and the SOMO of the radical[262b]. The position of the coupling depends on the charge distribution in the HOMO of the nucleophile. Perturbation theory and thermodynamic stability of the formed radical anions follow the same tendency.

The reactivity, through the $S_{RN}1$ mechanism, of **198** allows the synthesis of naphthylquinolines and naphthylisoquinolines **201** via its coupling reaction with halobenzopyridine **200** derivatives, followed by protection of the OH group (equation 142)[265].

Another synthesis is by the reaction of iodonaphthalenes with anions from hydroxyquinoline **202** to give **203** (equation 143)[265].

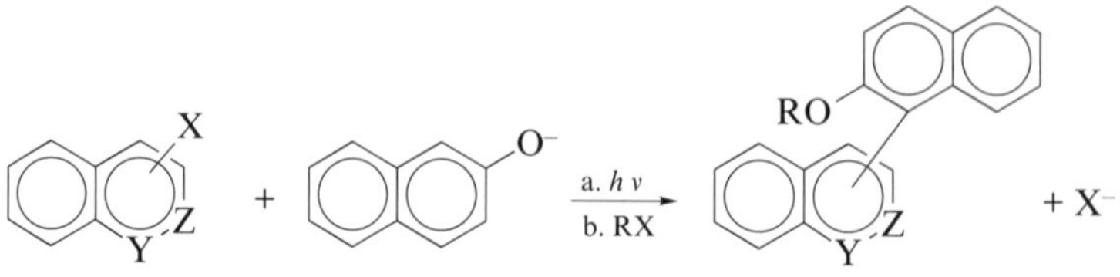

$$\text{(200a) Y = N, Z = CH (198)} \quad \text{(201)}$$
$$\text{(200b) Y = CH, Z = N}$$

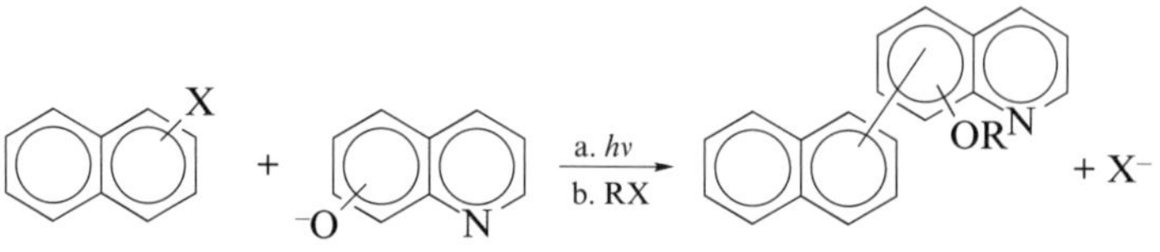

$$\text{(202)} \quad \text{(203)}$$

This strategy of synthesis can be compared with the reaction of either acetylchloropyridine, in which the acetyl and chloro groups are o- to each other, or o-bromobenzamide treated with anions from acetonaphthone[265].

Similar types of compounds were synthesized through the substitution reaction of the iodine in position 7 of clioquinol derivative (5-chloro-7-iodo-8-hydroxyquinoline) with o- and p-substituted phenoxide ions[266].

c. Polymerization reactions. Aromatic poly(ether sulphone)s and poly(ether ketone)s are frequently synthesized by an aromatic nucleophilic substitution reaction in which a bis(aryl halide), which is activated by a sulphone or carbonyl group, is condensed with bisphenoxide ions[267].

In this system polycondensations of 4,4'-dihalodiphenyl sulphones with t-butylhydroquinone resulted in high molecular weight polymers when the halogen was F, Cl or Br while low molecular weight polymers were obtained for the I derivative[268]. It has been proposed that substitution of the halide takes place by polar mechanism while the reductive elimination of the halide takes place by an ET mechanism, the ArO$^-$ ion acting both as the nucleophile and the electron donor.

An $S_{RN}1$-type ($S_{RN}1'$) mechanism has been proposed in the synthesis of poly(2,6-dimethyl-1,4-phenylene ether) through the anion-radical polymerization of 4-bromo-2,6-dimethylphenoxide ions (204) under phase-transfer catalysed conditions[269]. Ions 204 are oxidized to give an oxygen radical 205. The propagation consists of the radical nucleophilic substitution by 205 at the *ipso* position of the bromine in 204 (equation 144). The anion-radical 206 thus formed eliminates a bromide ion to form a dimer phenoxy radical 207 (equation 145). A polymeric phenoxy radical results by continuation of this radical nucleophilic substitution.

This mechanism resembles the $S_{RN}1$ process. The main differences are that the initiation requires the oxidation of the ArO$^-$ ions, then a phenoxy radical is the propagating radical

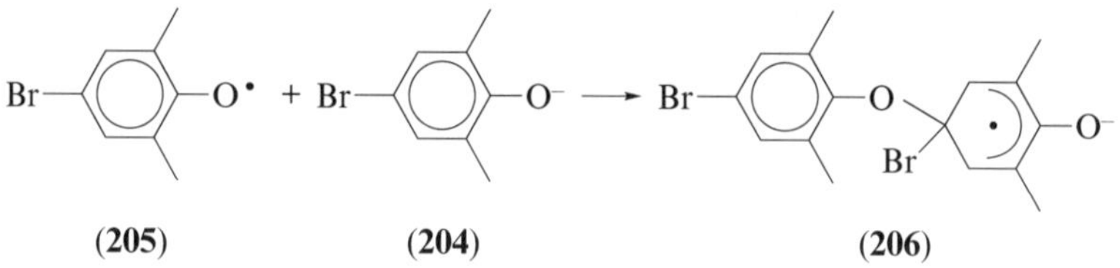

$$\text{(205)} \quad \text{(204)} \quad \text{(206)}$$

$$\textbf{206} \xrightarrow[-Br^-]{} \quad Br\text{—}\langle\bigcirc\rangle\text{—O—}\langle\bigcirc\rangle\text{—O}^\bullet \qquad (145)$$

$$(\textbf{207})$$

and, in the coupling reaction, a radical anion is formed which eliminates halide anion to give a radical while in the $S_{RN}1$ process the radical anion transfers the extra electron to the halide substrate to re-initiate the propagation cycle.

Copolymerization of 4-bromo-2,6-dimethylphenol with 2,4,6-trimethylphenol, 4-t-butyl-2,6-dimethylphenol, 2,2-bis(4-hydroxy-3,5-dimethylphenyl)propane, 4-hydroxy-3,5-dimethylbenzyl alcohol and a 4-substituted-2,6-di-t-butylphenol are discussed[269].

2. Sulphur nucleophiles

One of the first nucleophiles recognized to react under irradiation with ArI by the $S_{RN}1$ mechanism was PhS$^-$ ion, which has led to the synthesis of many aryl phenyl sulphides in liquid ammonia. Bromoarenes also react, but much more slowly[270].

The reaction of PhS$^-$ ion with 4-bromobenzophenone (**154a**) and 4-haloacetophenones (**154b**) was studied in MeCN, DMSO and DMF under thermal (Δ, 60 °C) or photochemical activation to yield 80–97% yield of products **208** (equation 146)[271].

$$4\text{-}XC_6H_4COR + PhS^- \longrightarrow (4\text{-}PhS)C_6H_4COR \qquad (146)$$

(**154a**) X = Br, R = Ph   (**208a**) R = Ph
(**154b**) X = Br, Cl, F; R = Me   (**208b**) R = Me

The $S_{RN}1$ mechanism was confirmed by the observed quantum yield (Φ) values (3 to 6), which decreased to less than unity with the additives 1,4-benzoquinone, p-DNB or tetracyanoquinodimethane (even at low concentration, ca 0.25 mol%). The thermal reaction was also inhibited with these additives.

Under electrochemical conditions **154a** (and other haloarenes) react with PhS$^-$ ions to give **208a**. The catalytic character of this reaction was clearly shown[272].

The coupling rate of PhS$^-$ ions with aryl radicals has been determined in different solvents. Thus 9-anthryl radical has a rate constant of 4.5×10^8 M^{-1}s^{-1} in DMSO, 2-quinolyl radical a rate constant of 1.4×10^7 M^{-1}s^{-1} in liquid ammonia and p-cyanophenyl radical a rate constant of 2×10^{10} M^{-1}s^{-1} in MeCN[273].

In a study which compared the solvents in an electrochemically induced $S_{RN}1$ reaction, it was found that although MeCN is a better hydrogen donor than DMSO, PhS$^-$ ions react faster with aryl radicals in MeCN[274].

The photostimulated reaction of 3-iodopyridine, 4-bromoisoquinoline and 3-bromoquinoline with PhS$^-$ ions in liquid ammonia, in the presence of MeO$^-$ ions, gave both the substitution and the reduction products, the latter derived from the reaction of hetaryl radicals with MeO$^-$ ions and MeOH. The ratio of reduction to substitution products suggests a similar rate constant for the coupling of PhS$^-$ ions with the three hetaryl radicals[275].

Good yields of products were also obtained in DMF or HMPA at 80 °C[276]. Thus, 2-bromo- or 2-iodopyridine gave 52% and 58% yield with PhS$^-$ ions in DMF. A better yield was obtained in HMPA (65% with 2-bromopyridine).

In the reaction of 2-chloro, 2-bromo-, 2-iodo- and 3-bromothiophenes **209** with PhS$^-$ ion in MeCN, after quenching with MeI, the products **210** (10–45%), **211** (14–30%) and Ph$_2$S were obtained (equation 147)[277].

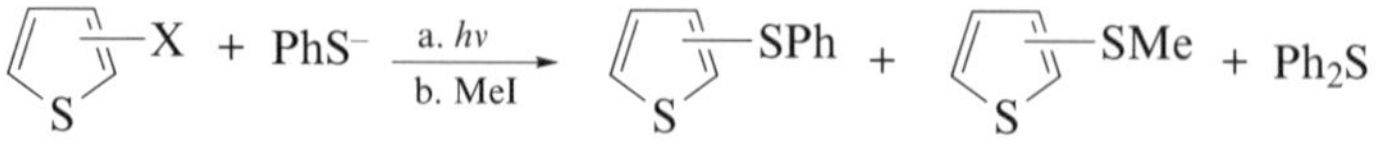

$$\text{(209)} \quad + \text{ PhS}^- \xrightarrow[\text{b. MeI}]{\text{a. } hv} \quad \text{(210)} \quad + \quad \text{(211)} \quad + \text{ Ph}_2\text{S} \quad (147)$$

When the thienyl radical **212** couples with PhS⁻ ions, the radical anion **210**⁻˙ formed has three competitive reactions: bond fragmentation to give the starting materials, ET to the substrate to give product **210**, or a C—S bond fragmentation to give the anion **213** and Ph˙ radical (equation 148).

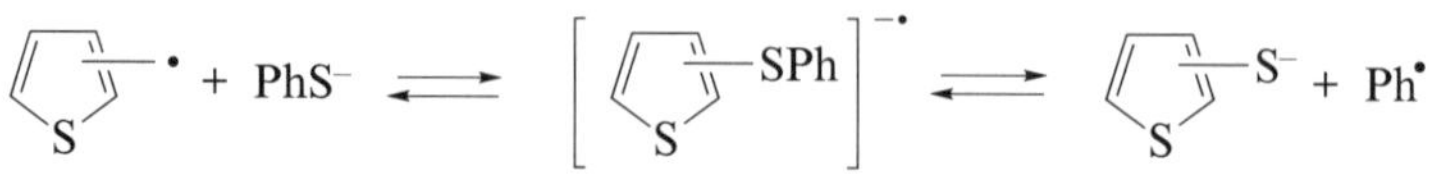

$$\text{(212)} \quad + \text{ PhS}^- \rightleftharpoons \left[\text{(210)}^{-\cdot} \right] \rightleftharpoons \text{(213)} \quad + \text{ Ph}^\bullet \quad (148)$$

Anion **213** is trapped by MeI to give **211**. Phenyl radicals can react with PhS⁻ ions to give finally the product Ph₂S. This step has been proved to be irreversible under these experimental conditions[278], although the equilibrium constant ($ca\ 3 \times 10^5$) has been measured[279].

5-Bromo- and 5-iodo-1,2-dimethylimidazole did not react under irradiation with PhS⁻ ions in liquid ammonia or DMF[280].

In order to find a synthetic route to diphenyl thioether derivatives of peptides and amino acids, the reaction of iodinated aryl amino acid, such as monoiodo (O-methyl) Boc-tyrosine methyl ester (**214**) with Boc-protected p-mercaptophenylalanine methyl ester (**215**) (R=Me), or the free acid (**215**, R = H) was studied in liquid ammonia, 90% yield of **216** from **215**, R = Me were obtained (equation 149)[281].

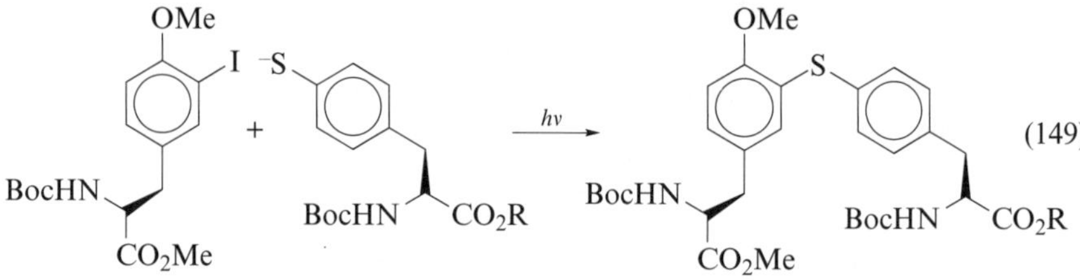

$$\text{(214} \quad + \quad \text{(215)} \xrightarrow{hv} \quad \text{(216)} \quad (149)$$

It has been reported that the acid or the amine functionality could be left unprotected without affecting the yield, and without giving racemization. Other products synthesized by reaction of **215** with monoaryl iodides and diiodides are compounds **217** and **218**.

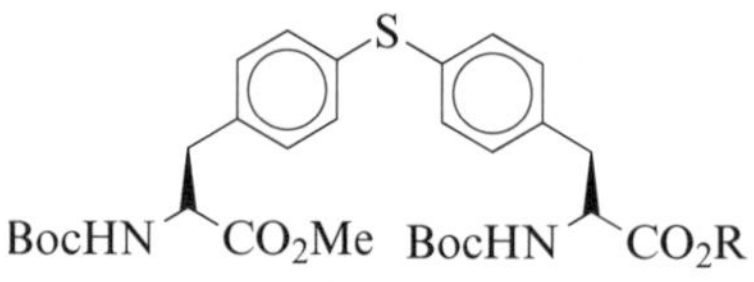

(**217**) R = Me, 84%

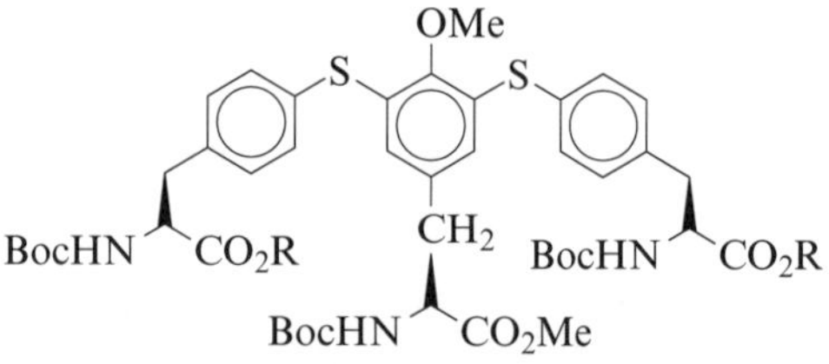

(**218**) R = Me, >95%

When diiodohydroxyphenylglycine derivative **219** (with the amino group protected with Boc) was irradiated with the anion **215** in liquid ammonia, the completely racemized product was obtained, but when unprotected **219** was used, no racemization was observed; the disubstitution product **220** was obtained in >95% yield (equation 150).

(**219**) (**215**)

(**220**)

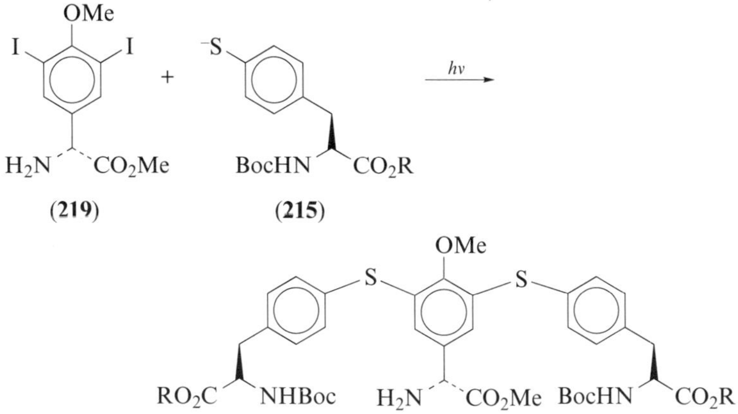

(150)

Many dihalobenzenes afford disubstitution products when they react with PhS^- ions and little, if any, monosubstitution products are formed. For instance, *m*-chloroiodo, *p*-chloroiodo, *m*- and *p*-dibromo benzenes gave the disubstitution products[282a], whereas *m*-fluoroiodobenzene gave 96% of the monosubstitution product[270]. However, with other sulphanions, the ratio of mono- to disubstitution varies, depending on the nature of the nucleophile and the haloarenes[279]. Thus, the sulphanions **221**–**225** were studied in the photostimulated reaction with chloroiodobenzenes.

(**221**) (**222**) (**223**) (**224a**) Z = O (**224b**) Z = S (**225**)

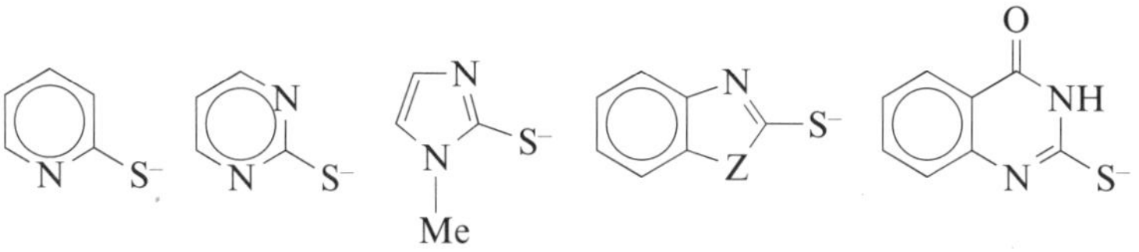

The reaction of *o*- (**226a**), *m*- (**226b**) and *p*- (**226c**) chloroiodobenzenes with these sulphanions gave the monosubstituted product **227** and the disubstitution product **228** (equation 151).

(**226**) (**227**) (**228**)

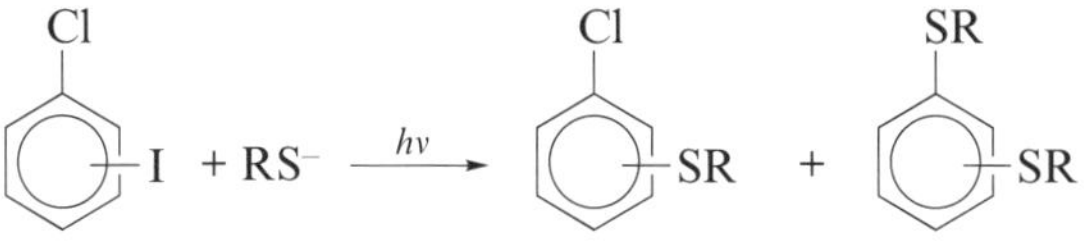

(151)

Thus, **226a–c** react with PhS^- ions giving **228** (77–91%) and no **227**, but with **221** they gave **227** (70–87%) and **228** (10–12%). With **222**, only **227** (100%) was formed.

Similar rate constants ($1.2–3.2 \times 10^9$) for the reactions of 2-quinolyl radical with sulphanions **221–225** have been determined in liquid ammonia. The partition between the monosubstitution **227** and the disubstitution products **228** depends not only on the stability of the radical-anion intermediate, but also on the concentration of the substrate. With a concentration of 50 mM of **226b**, the yield of **227** is 90% and of **228** is 3%; with 5 mM of **226b** the product **227** is not formed, and 100% yield of **228** is obtained. Another complication is that the products **227** can also enter in the propagation cycle yielding the products **228**[279].

The reaction of 5-chloro-6-nitroquinoxaline with *p*-methylbenzenethiolate ions, suggested to occur by the $S_{RN}1$ mechanism, has been reported to give disubstitution (both the nitro and chloro as leaving groups) without formation of any monosubstitution compound[282b]. On the other hand, the photostimulated reaction of ethers **229** derived from 5-chloro-7-iodo-8-hydroxy quinolines with different sulphanions has been studied and, in these cases, only monosubstitution products were obtained (equation 152)[239].

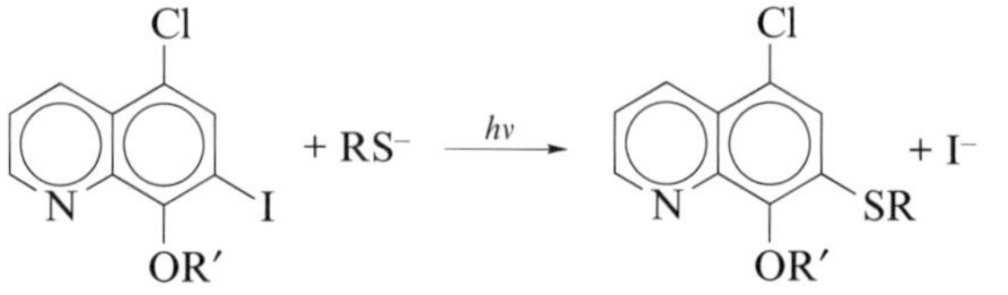

$$+ \text{RS}^- \xrightarrow{h\nu} \qquad + \text{I}^- \qquad (152)$$

(229)   **(230)**

The percentage yield of the monosubstituted products **230** obtained with the following RS⁻ ions is indicated:

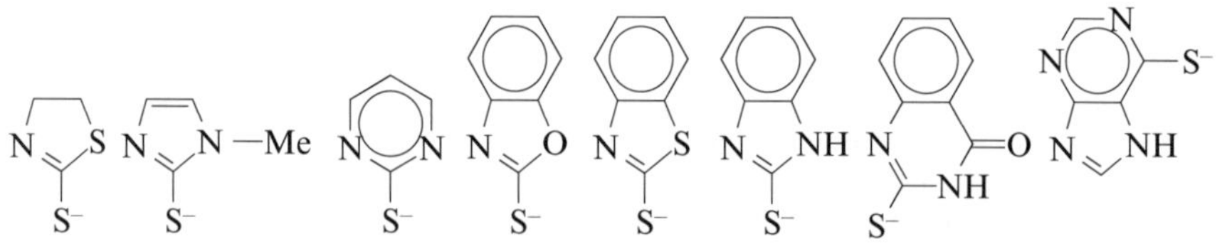

77%   75%   90%   80%   100%   100%   100%   88%

Only with PhS⁻ ions and *N*-methylimidazole-2-thiolate ions was the disubstitution product (8–10%) detected. These compounds, among others with the quinoline moiety, have amoebicide activity[266].

The reaction of haloarenes with 2- and 4-pyridinethiolate ions gave the products **232** in liquid ammonia. The yields are increased by the presence of Bu₄NOH (equation 153)[283]. Thus, PhI and **231a** or **231b** gave **232** (68% and 47%, respectively). *m*-Nitroiodobenzene gave 32% and 47%, respectively, of **232**. With PhCl, PhBr, *o*- and *p*-nitrochlorobenzene the yields were very low.

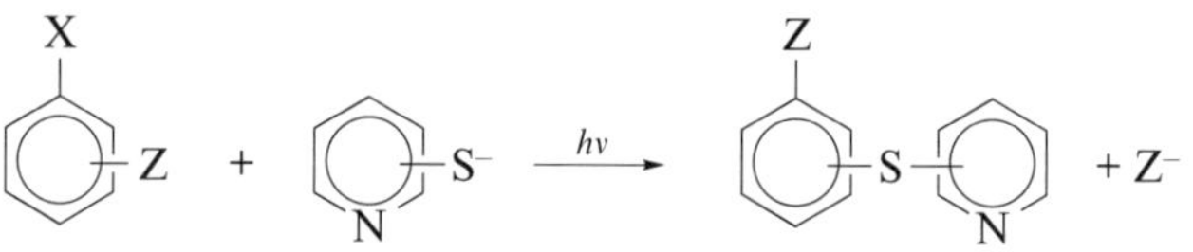

$$+ \qquad \xrightarrow{h\nu} \qquad + \text{Z}^- \qquad (153)$$

(231a) 2-S⁻   **(232)**
(231b) 4-S⁻

Arenediazonium tetrafluoroborates react with PhS⁻ ions in DMSO to give diaryl sulphides by the $S_{RN}1$ mechanism[284]. However, when the aromatic ring has halogen as a substituent, as in **233**, the monosubstitution product **234** and the disubstitution product **235** were obtained depending on the halogen and its position (equation 154).

Thus, when X = 2-F and 3-F only **234** is formed (80%), and when X = 4-F mainly **234** (75%) is formed with <3% yield of **235**. When X = 4-Cl, Br and I and 3-Cl, **234** is formed

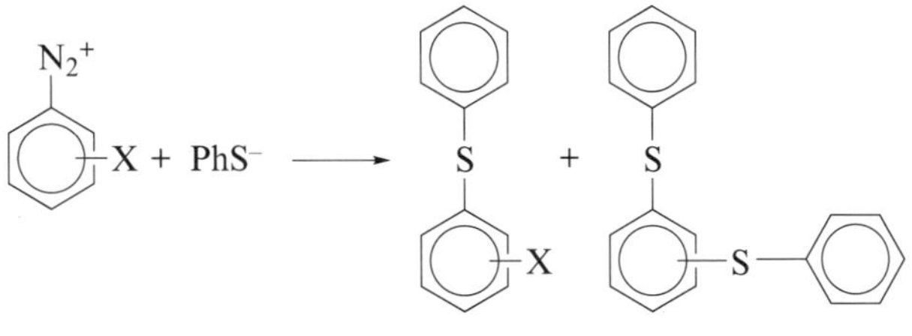

(154)

(233)   (234)   (235)

in *ca* 8–13% yield whereas **235** is obtained in 63–69%. When X = F, the intermolecular ET rate is faster than the bond fragmentation rate, but when X = Cl, Br or I, the bond fragmentation is faster than the intermolecular ET rates.

The reaction between $PhN_2^+ BF_4^-$ (**236**) and *p*-chlorobenzenethiolate ion (**237**) gave the monosubstitution product **238a**, (34%), the disubstitution product **238b** (11%) and trisubstitution product **238c** (9%) (equation 155)[284].

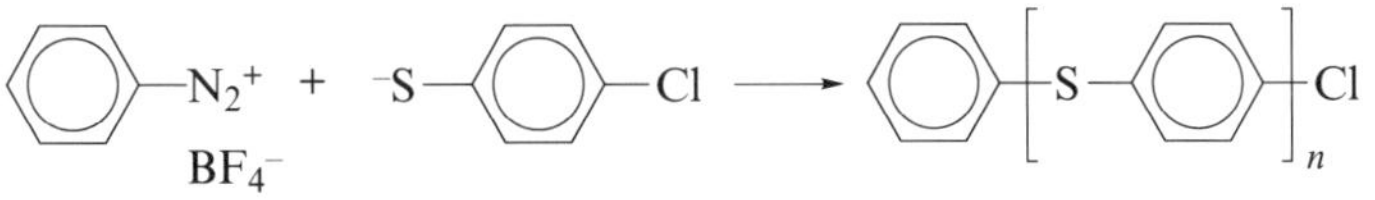

(155)

(236)   (237)   (238) (a) $n = 1$
                                                 (b) $n = 2$
                                                 (c) $n = 3$

Based on these findings, a polymerization was attempted with *p*-bromobenzenethiolate ions and catalytic amounts of $PhN_2^+ BF_4^-$ as initiator. When the reaction was performed in DMSO, a poly(*p*-phenylene sulphide) (PPS) was obtained with $n = 6$–7[285]. When the diazonium salt was $p\text{-}BrC_6H_4N_2^+ BF_4^-$ (catalyst that has another leaving group), the PPS obtained was with $n = 9$–10[285].

Based on ESR spectroscopy, the yield and molar mass of the polymers, it has been suggested that the formation of PPS by polymerization of copper(I) 4-bromobenzenethiolate at 200 °C in quinoline probably occurs by an $S_{RN}1$-type mechanism. A free radical mechanism may also be involved[286].

The products formed in the photostimulated reaction of PhI with EtS⁻ ions, after quenching with $PhCH_2Cl$, are shown in equation 156[278].

$$PhI + EtS^- \xrightarrow[\text{b. } PhCH_2Cl]{\text{a. } h\nu} PhSEt + PhSCH_2Ph \qquad (156)$$

These results can be interpreted considering that, in the coupling of phenyl radical with EtS⁻ ion, the radical anion formed either transfers its odd electron to give the product PhSEt, or undergoes an S—Et bond fragmentation to give PhS⁻ ion, that is trapped in turn by $PhCH_2Cl$, rendering the observed product[278].

Quite different results were obtained for haloarenes with lower π* MOs, such as 1-iodonaphthalene, with which only the substitution product was found. This was the case even with *t*-BuS⁻ ion as nucleophile[287]. On the other hand, fragmentation was observed in the reaction of 1-iodonaphthalene with $PhCH_2S^-$ ion, but 2-chloroquinoline, that has a lower π* MO, only gave the substitution product (69%)[287].

These results show that substrates with low π* MO may react with RS⁻ ions without bond fragmentation, but this reaction also depends on the structure of the sulphanions. Thus, the photostimulated reaction of *o*-bromobenzamide, *o*- and *m*-bromobenzonitriles, *o*-bromoacetophenone, *o*-bromobenzaldehyde and *p*-bromobenzophenone with EtS⁻ ions gave the substitution products (70–90%)[288].

o-Bromobenzonitrile reacts with sulphanion **239a** to give the substitution product **240** (85%). With **239b**, **240** (70%) and products derived from S—C bond fragmentation **241** (15%) were obtained. With **239c** and **239d**, product **241** is formed in 85–100% yield (equation 157)[288].

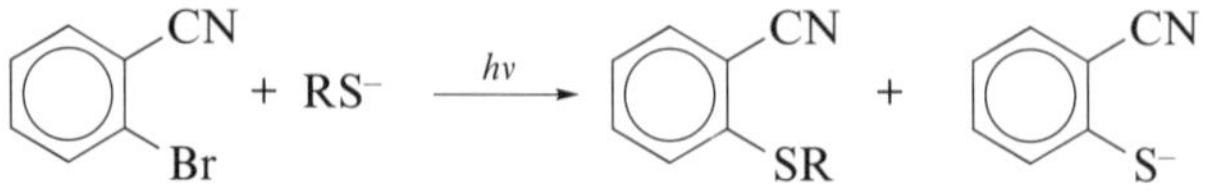

$$(239)\ \textbf{(a)}\ R = CH_2CH_2OH \qquad \textbf{(240)} \qquad \textbf{(241)}$$
$$\textbf{(b)}\ R = CH_2CH_2CO_2Et$$
$$\textbf{(c)}\ R = PhCH_2$$
$$\textbf{(d)}\ R = CH_2CO_2Et$$

However, with substrates of low-lying π^* MO such as p-bromobenzophenone, 2-chloroquinoline, or cyanobromopyridines, only the substitution products are obtained in very good yields. With the last substrates, the substitution products lead to a ring-closure reaction (Section V.E.1).

4-Bromoisoquinoline did not react with MeS$^-$ ions in liquid ammonia, but 80% yield of the substitution product 4-isoquinolyl methyl sulphide was formed in the presence of amide ions. Isoquinoline and many other nitrogen-containing heteroaromatic compounds are known to react with amide ions to give anionic adducts **242**[289], which have been proposed to initiate the $S_{RN}1$ reaction by an ET to the substrate, which leads to the hetaryl radical. The radical couples only with MeS$^-$ ions [290].

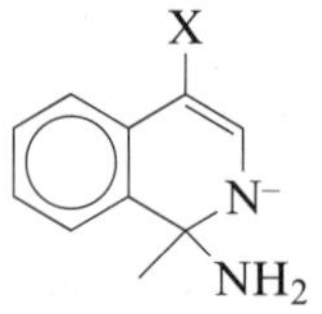

$$(242)\ X = SMe, Br$$

In the photostimulated reaction of 1-iodonaphthalene with a bidentate $^-S(CH_2)_nS^-$ ion, mainly the disubstitution product **243a** ($n = 2 = 3$: 50%, $n = 4$: 25%) and the sulphide **244a** (9–17%) from fragmentation of the radical-anion intermediate were formed (equation 158)[291]. With 2-iodonaphthalene the yields of **243b** were lower (27–40%) and 10–14% of **244b** were formed. With the bromonaphthalenes the yields were even lower.

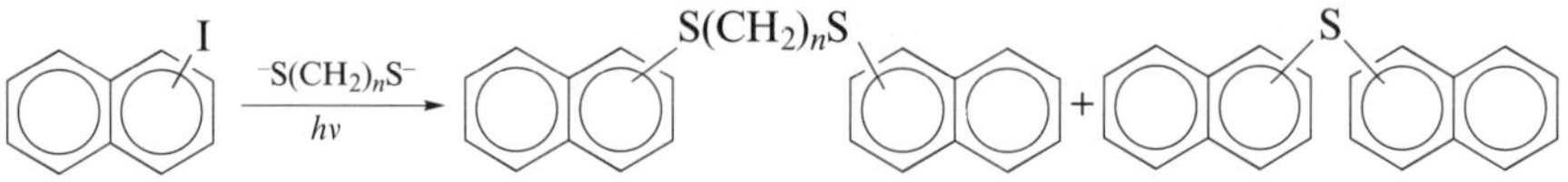

$$(243a)\ 1\text{-naphthyl} \qquad (244a)\ 1\text{-naphthyl}$$
$$(243b)\ 2\text{-naphthyl} \qquad (244b)\ 2\text{-naphthyl}$$

$$(158)$$

It has been proposed that reaction of p-nitrochlorobenzene with dithiols (such as HSCH$_2$CH$_2$SH, formed $in\ situ$ from thiouronium salts) under phase transfer conditions to give 4,4'-dinitrodiphenyl sulphide (10%), 4,4'-dinitrodiphenyl disulphide (20%) and 1,2-bis(4-nitrophenylthio)ethane (20%) occurs by the $S_{RN}1$ mechanism in competition with S_NAr. These reactions are partially inhibited by p-benzoquinone and gave ESR signals[292].

The thermal (170–195 °C) reaction in tetraglyme between 1-bromonaphthalene, 9-chloroanthracene, 9-bromoanthracene and 9,10-dichloroanthracene with RS⁻ and PhS⁻ ions has been suggested to occur by a competition between S_NAr and $S_{RN}1$ mechanisms[293].

Although there are no mechanistic studies, by comparison with the results obtained in the reaction of other dichalcogenides and *peri*-disubstituted arenes, the thermal reaction (150 °C) of 1,8-dichloronaphthalene with Na_2S_2 followed by reduction with $NaBH_4$ to give 46% yield of 1,8-naphthalene dithiol[294a] can be proposed to occur by the $S_{RN}1$ mechanism. The polycondensation of *p*-dichlorobenzene or 4,4′-dichlorodiphenyl sulfone with Na_2S has also been ascribed to the $S_{RN}1$ mechanism[294b].

3. Selenium and tellurium nucleophiles

a. Selenide and telluride ions. Na_2Se can be formed in liquid ammonia from the metal, and it reacts under irradiation with PhI to give $(PhSe)_2$ (78%), after oxidation of the substitution product PhSe⁻ formed, and Ph_2Se (12%)[295]. The reaction between the substitution product PhSe⁻ and phenyl radicals explains the formation of Ph_2Se. The reaction of Ph_2Se with Na metal in liquid ammonia to form PhSe⁻ has been used as a method to achieve the synthesis of several organoselenium compounds.

The synthesis of poly(*p*-phenyleneselenide) in 80% yield and with a molecular weight of *ca* 10,000 has been reported from the reaction of *p*-dibromobenzene and Na_2Se prepared *in situ* directly from the elements in DMF at 120–140 °C[296].

The reaction of aryl halides with Na_2Te_2 in HMPA or DMF at 110–170 °C gave the diarylditellurides Ar_2Te_2 in moderate yields. Thus, 9-bromoanthracene gave 40%, 2-chloronaphthalene gave 20% and 2-bromonaphthalene gave 30% of the ditelluride[297]. With PhBr, PhI and 1-chloronaphthalene the yields were less than 10%. With Na_2Te as nucleophile, the reaction with 2-bromonaphthalene gave 30% yield of bis(2-naphthyl) telluride and PhI gave Ph_2Te in 43% yield[297].

In the same experimental conditions (DMF, 110 °C), the reaction of 1,4,5,8-tetrachloronaphthalene with Na_2Se_2 gave the substitution products **245** (55%) and **246** (10% yield) equation 159)[298].

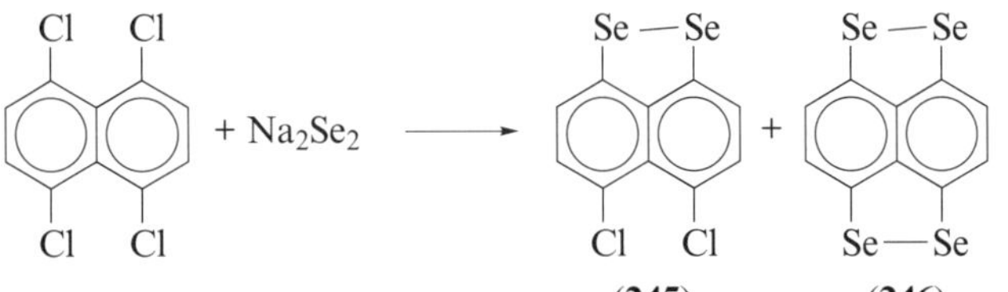

$$(159)$$

$$(245) \qquad (246)$$

1,8-Dichloronaphthalene reacts with Na_2Z_2 in HMPA to give the substitution products **247** (equation 160)[299].

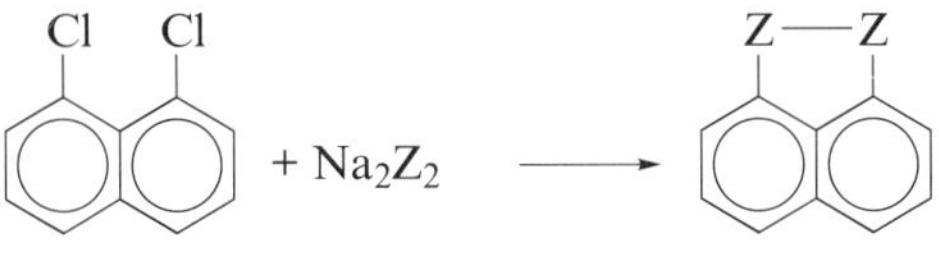

$$(160)$$

Z = S   (**247a**) Z = S (46%)
Z = Se   (**247b**) Z = Se (69%)

The reaction of ArI with tellurium and Rongalite ($HOCH_2SO_2Na$) as tellurating reagent, gives good yield of symmetrical Ar_2Te[300]. Thus, the following compounds were prepared[301]:

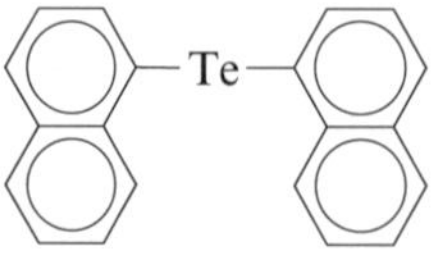

70% 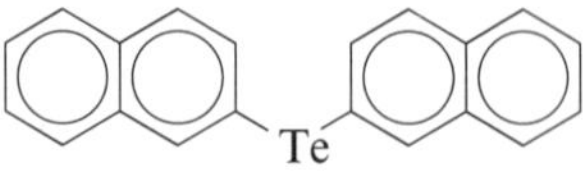   61%

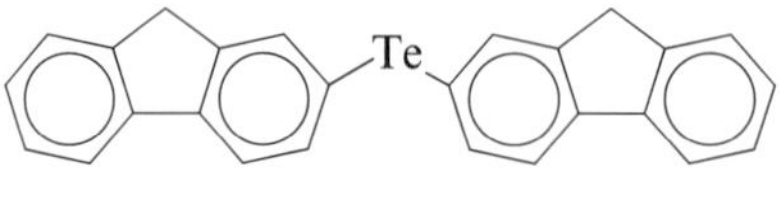

48% 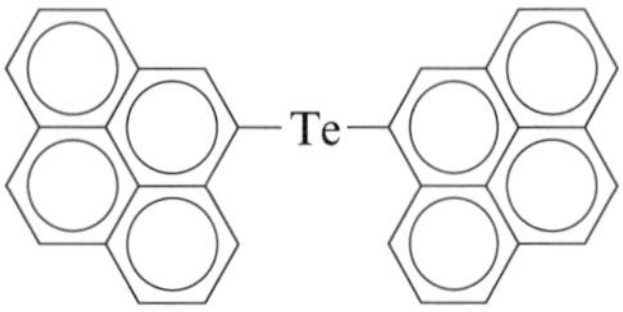   60%

Sodium telluride prepared *in situ* from tellurium and NaH in *N*-methyl-2-pyrrolidone reacts with iodoarenes to give diaryl tellurides[302].

The nucleophiles Z_2^{-2} (Z = Se, Te) can be electrogenerated from the elements using sacrificial Z electrodes in MeCN. The synthesis of ArZZAr is possible when the electrolysis is performed in the presence of haloarenes (equation 161)[303].

$$\text{ArX} + Z_2^{-2} \xrightarrow{\;e^-\,/\,electrode\;} \text{ArZZAr} \tag{161}$$

2-Chloroquinoline	Z = Se	(248a) (70%)
	Z = Te	(248b) (50%)
9-Bromoanthracene	Z = Se	(248c) (54%)

An improvement in this technique of sacrificial electrodes (Se and Mg) for the formation of the Z_2^{-2} ions was achieved by using an undivided cell and by addition of fluoride ions to avoid the precipitation of the Se_2Mg[304]. Thus, the reaction of 2-chloroquinoline with Se_2^{-2} gave 79% yield of 248a. Following this methodology the diselenides 250 were prepared by reaction with aryl halides 249 (equation 162)[303].

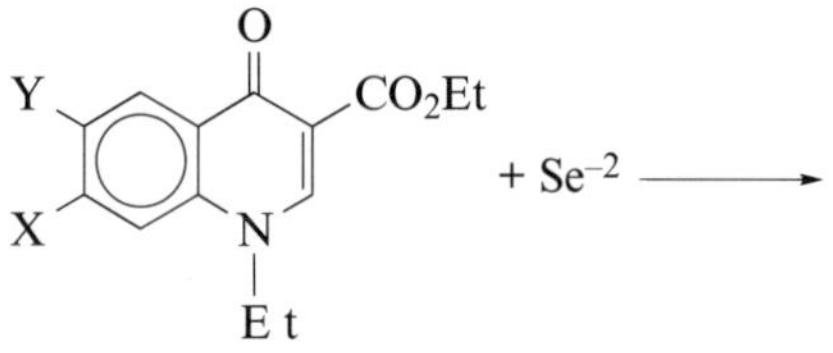

(249a) X = Br, Y = H
(249b) X = Cl, Y = F

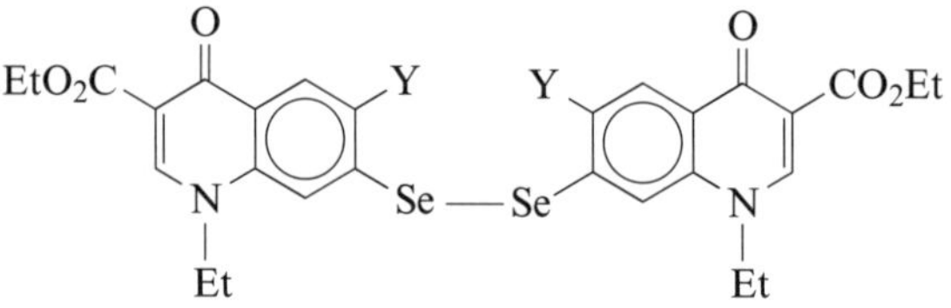

(162)

(250a) Y = H (70%)
(250b) Y = F (82%)

The yields with Te_2^{-2} are lower. For instance, **249a** gave only 35% of the substitution product.

b. Areneselenate and arenetellurate ions. The photostimulated reaction of *p*-iodoanisole or 2-bromopyridine (2-BrPyr) with PhSe⁻ ions gave substitution product ArSePh **251**, together with the symmetricals Ph₂Se **252** and Ar₂Se **253** (equation 163)[305].

$$ArX + PhSe^- \xrightarrow{\ hv\ } ArSePh \quad + \quad Ph_2Se \quad + \quad Ar_2Se \qquad (163)$$

ArX = *p*-IC₆H₄OMe	**(251a)** Ar = *p*-An	**(252)**	**(253a)** Ar = *p*-An
= 2-BrPyr	**(251b)** Ar = 2-Pyr		**(253b)** Ar = 2-Pyr

This scrambling of aromatic rings has been ascribed to a reversible coupling and fragmentation of the radical anion intermediates during the $S_{RN}1$ chain process (Section IV. A. 1).

One possibility to avoid the fragmentation of radical anions lies in the lowering of their antibonding π^* MO. For instance, the photostimulated reactions of polycyclic or hetero-cyclic halides, such as 1-chloronaphthalene, 2-chloroquinoline, 4-chlorobiphenyl and 9-bromophenanthrene with PhSe⁻ ions give good yields of substitution products ArSePh $(50-72\%)$[306]. In this case a stable π^* radical anion is formed.

Considering that there is competition between the rate of fragmentation (k_f) and the rate of ET $[k_t(ArX)]$, the scrambling reaction could be avoided favouring the rate of ET by increasing the concentration of ArX. The photostimulated reaction of *p*-iodoanisole with PhSe⁻ ions in MeCN gave PhSeAn (17%), Ph₂Se (8%) and An₂Se (7%). When the reaction was performed with a 2.5 M concentration of the aryl iodide, PhSeAn was formed as the only product (67%)[307]; the same happened in the reaction with *p*-bromoanisole (7.8 M). There was no photostimulated reaction of *p*-chloroanisole (15.6 M) with PhSe⁻ ions in the absence of *p*-iodoanisole. The fact that in the reaction with 15.6 M *p*-chloroanisole and 0.025 M *p*-iodoanisole only PhSeAn was formed, indicates that the latter initiates the reaction, and the radical-anion intermediate formed in the coupling of *p*-anisyl radical with PhSe⁻ ion transfers its odd electron to the *p*-chloroanisole (entrainment reaction).

Assuming that the rate constant of the ET from (PhSeAn)⁻˙ to *p*-iodoanisole (k_t difference in $E_{1/2}$ *ca* 1.2 V) is *ca* 10^{10} M⁻¹ s⁻¹ in MeCN[308], the rate constant k_f of the radical anion can be estimated as $\leq 10^8$ s⁻¹, and the k_t from (PhSeAn)⁻˙ to *p*-Br and *p*-Cl anisole as *ca* 2.6×10^9 M⁻¹ s⁻¹ and 1.3×10^9 M⁻¹ s⁻¹, respectively.

The ions PhSe⁻ and PhTe⁻ can be prepared electrochemically by reducing the corresponding (PhZ)₂ (Z = Se,Te) in MeCN and then, by addition of bromobenzonitriles, the electrolysis produced the substitution products **254**. With *p*-(Z = Se, 58%, Z = Te, 42%), *m*-(Z = Se, 42%) and *o*-(Z = Se, 36%), substitution products in the given yields were obtained, as indicated in equation 164[309].

$$BrC_6H_4CN + PhZ^- \xrightarrow{\ e^-/\ \text{elctrode}\ } NCC_6H_4ZPh \qquad (164)$$
$$(\mathbf{254})$$

With PhTe⁻ ions the symmetrical telluride (*p*-NCC₆H₄)₂Te was formed. With *p*-chlorobenzonitrile, whose radical anion has a slow rate of fragmentation, the yield of **254** was 70%[309a]. However, with *o*- and *m*-chlorobenzonitriles the yields of the products were low, but in the presence of 1,2-di(4-pyridyl)ethylene as redox catalyst, the yields could be improved[310].

In the reaction of 2-, 3- and 4-bromobenzophenones with electrogenerated PhZ⁻ ions (Z = Se,Te) in MeCN, with azobenzene as a redox catalyst and with the addition of carbon acids such as fluorene or malononitrile, good yields of substitution products **255a** $(62-86\%)$ were obtained in the reaction with PhSe⁻ ions, and no symmetrical selenide **256a** was found. With PhTe⁻ ions, the yields of **255b** were lower (45–48%) and the symmetrical tellurides **256b** were formed (31–36%) (equation 165)[311].

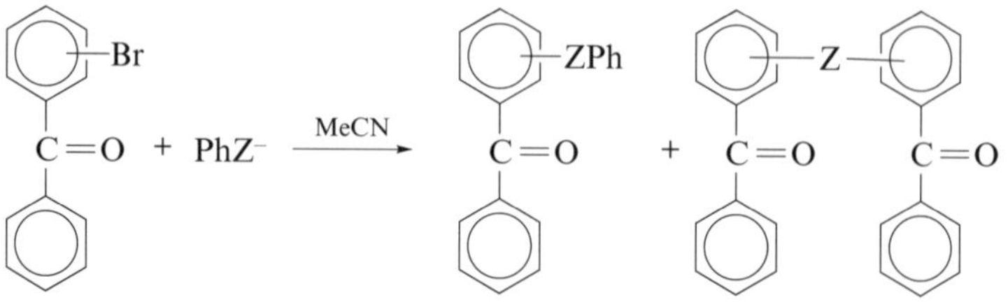

$$(255a) \ Z = Se \qquad (256a) \ Z = Se$$
$$(255b) \ Z = Te \qquad (256b) \ Z = Te$$

The electrosynthesis in MeCN of the following derivatives of Se and Te has been carried out with redox catalysts[312].

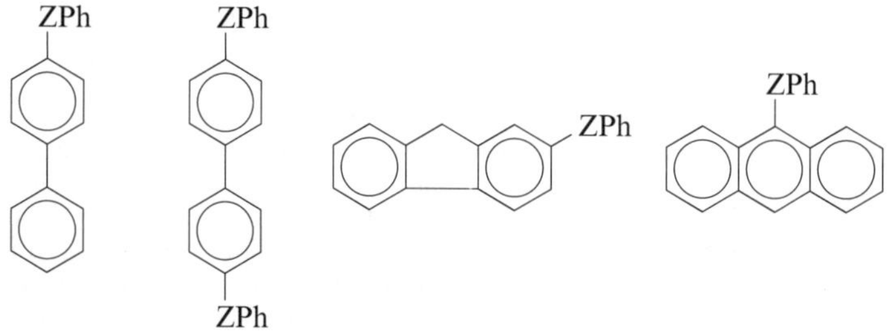

Z = Se (65%)   Z = Se (46%)   Z = Se (53%)   Z = Se (74%)
            Z = Te (16%)   Z = Te (68%)

D. Nucleophiles Derived from the VII Group

The radical anions of 1-bromo and 1-iodoanthraquinones (AQI), electrochemically generated in MeCN containing tetrabutyl ammonium (TBA⁺) perchlorate, undergo reduction to anthraquinone (AQH) on irradiation. TBA⁺ was determined to be the source of the hydrogen[313,314]. However, this photoreduction process is inhibited by added iodide ions, and this could be explained by partitioning of the 1-anthraquinolyl radical (AQ˙) intermediate between reaction with TBA⁺ and iodide ions, reforming, in the latter case, the ground state radical anion (equation 166).

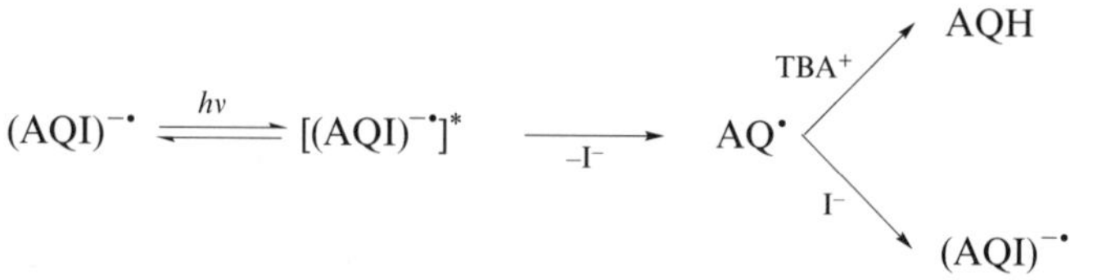

The reaction of [(AQI)⁻˙]* with TBA⁺ or iodide ions in a bimolecular process was rejected based on kinetic analysis[315]. The reaction of aryl radicals substituted by strong electron-withdrawing substituent with iodide ions[316] and of aryl radicals with I₃⁻ ions[317] are known.

E. Ring Closure Reactions

1. $S_{RN}1$ and ring closure reactions of aryl halides bearing an ortho substituent

There are different approaches to ring closure reactions by the $S_{RN}1$ mechanism. By far the most studied system is when the aromatic moiety has an appropriate substituent in *ortho* position to the leaving group[16,19].

The first examples of these ring closure reactions were reported by Bard and Bunnett[318] and Beugelmans and Roussi[319], who studied the photostimulated reactions of *o*-amino-halobenzenes (**257**, X = Br,I) with carbanions in liquid ammonia to perform the synthesis of indoles **259** (equation 167).

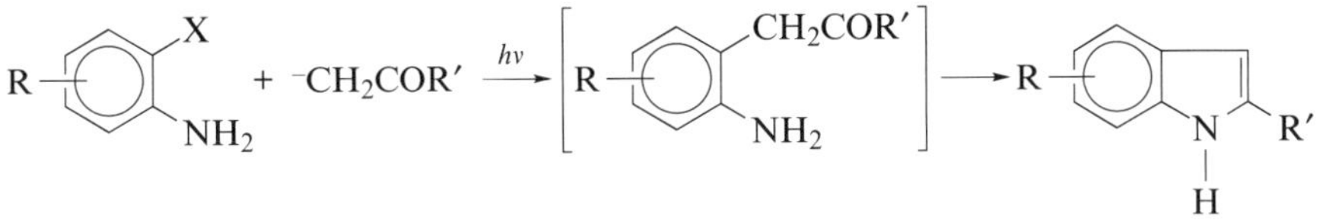

These reactions give as intermediates the substitution products **258** that undergo spontaneous cyclization in good yields. Most of the examples studied involve carbanions derived from ketones. With cyclic ketones, such as cyclohexanone[320], the yields are lower.

Carbanions derived from aldehydes or α-dicarbonyl compounds, with one of the carbonyl groups protected as a dimethyl acetal, also react with *o*-iodoanilines to give indoles[319].

Ring closure reactions can be induced electrochemically in liquid ammonia. For instance, the reaction of *o*-iodoaniline with acetaldehyde enolate ion gives 95% yield of indole, and with acetone enolate ion 87% of 2-methylindole[321].

Pyridine rings, such as 2-chloro-3-aminopyridine (**260**), gave 4-azaindoles (**261**) in photostimulated reactions with aldehyde and ketone enolate ions in good yields [R = R' = H (62%)[322]; R = H, R' = Me (45%[318], 54%[322]), R' = *i*-Pr (61%)[323], R' = *t*-Bu (100%)[323], R' = 2-pyridyl (21%)[322]; R = Me, R' = H (59%)[322], R' = Ph (30%)[322]; RR' = – (CH$_2$)$_4$ – (52%)[322]] (equation 168).

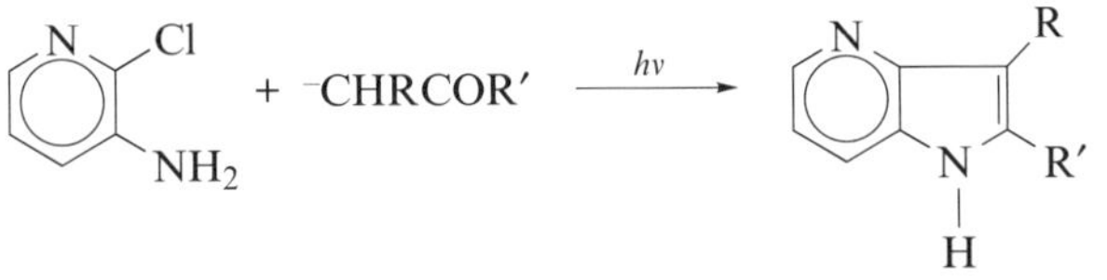

An interesting approach to the synthesis of indoles is the combination of directed *ortho*-lithiation followed by an $S_{RN}1$ reaction[324]. Thus, the selective *ortho*-lithiation of 2-fluoropyridine (**262**) by LDA followed by iodination afforded 2-fluoro-3-iodopyridine (**263**) in high yield (75%). Substitution of the 2-fluorine atom under S_NAr conditions is a convenient synthesis of 2-substituted 3-iodopyridines (**264**) (equation 169), which can react with acetone or pinacolone enolate ions in liquid ammonia, followed by acidic treatment, to give the corresponding substituted 7-azaindoles (**265**) [R = H, R' = Me (75%), R' = *t*-Bu (78%); R = Me, R' = *t*-Bu (70%), R' = Me (95%)] (equation 170).

Sequential *ortho*-lithiation, iodination and $S_{RN}1$ substitution was extended to the three aminopyridines (protected as pivaloylamino derivatives) providing a straightforward access to azaindoles. Thus, lithiation of the 2-,3- and 4-isomers occurred respectively at

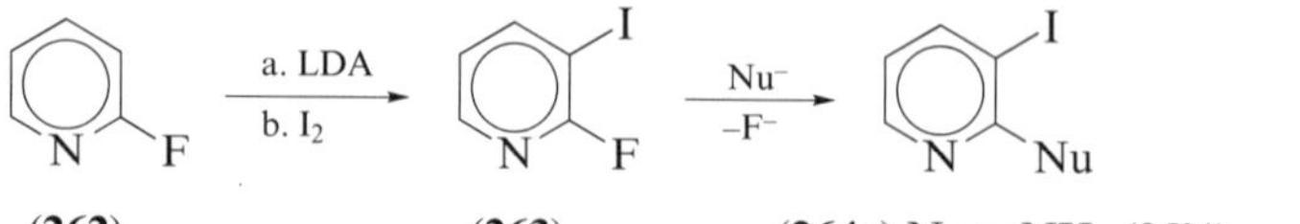

(169)

(262) (263) (264a) Nu = NH$_2$ (95%)
(264b) Nu = NHMe (96%)

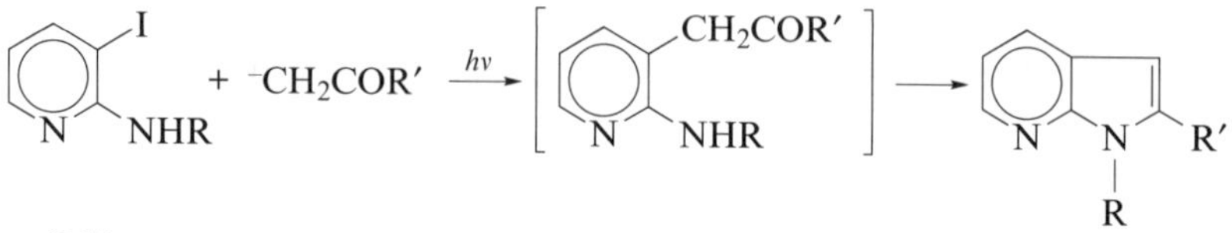

(170)

(264) (265)

position 3, 4 and 3, giving the iodo derivatives **266–268**, respectively, when treated with iodine at −75 °C.

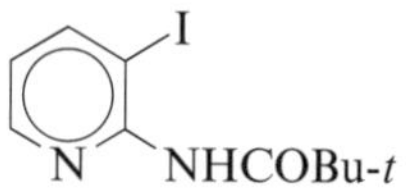

(266) 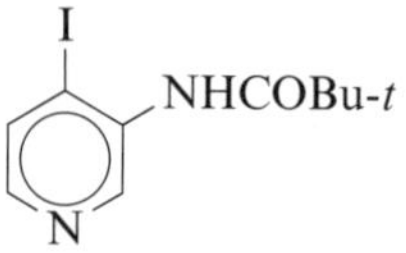 (267) 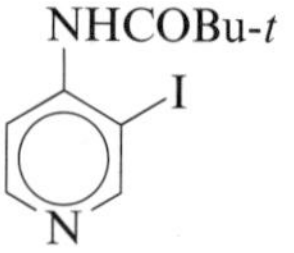 (268)

These iodo(pivaloylamino)pyridines underwent photostimulated S$_{RN}$1 reaction with acetone or pinacolone enolate ions in liquid ammonia to give, after acidic treatment, the azaindoles **269–271** in good yields[324].

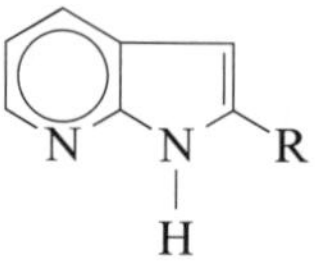

(269a) R = Me (80%)   (270a) R = Me (98%)   (271a) R = Me (98%)
(269b) R = t-Bu (90%)   (270b) R = t-Bu (99%)   (271b) R = t-Bu (98%)

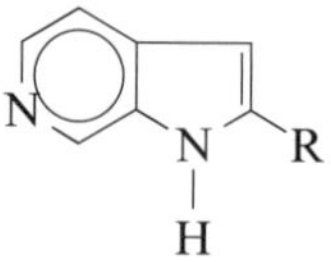

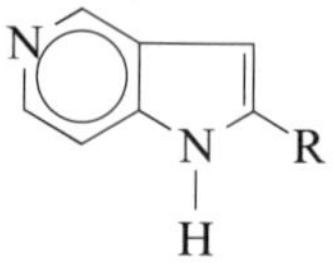

In all these systems the amino group reacts spontaneously with the keto group to give the ring closure product. Alkoxy groups also give ring closure reactions when deprotected after the S$_{RN}$1 reaction. Thus, in the reaction of *o*-iodoanisole (**272**) with acetaldehyde or ketone enolate ions the substitution product **273** is obtained. After deblocking of the phenolic group it gives the expected benzo[*b*]furan derivatives **275** (equation 171)[325].

This approach was used for the synthesis of furo[3,2-*h*]quinolines (**277**). The reaction of 5-chloro-7-iodo-8-methoxyquinoline (**229**) with several ketone enolate ions gave the substitution product **276** (70–80 % yield, R = Me, *t*-Bu, 2-furyl, *p*-anisyl) which, when treated with HBr at 100 °C, quantitatively led to **277** (equation 172)[326].

It is interesting to note that the chlorine atom in position 5 did not react under these conditions. Also, 5,7-dichloro-8-*i*-propoxyquinoline gave only **276** (R = *t*-Bu) when irradiated with pinacolone enolate ion. However, 5,7-dibromo-8-methoxyquinoline gave the disubstitution product (60%)[326].

The photostimulated reaction of 2-bromo-3-*i*-propoxypyridine (**278**) and ketone enolate ions in liquid ammonia afforded the substitution products **279**, which gave quantitatively the furo[3,2-*b*]pyridines (**280**) (equation 173). With 2-pentanone enolate ion, **278**

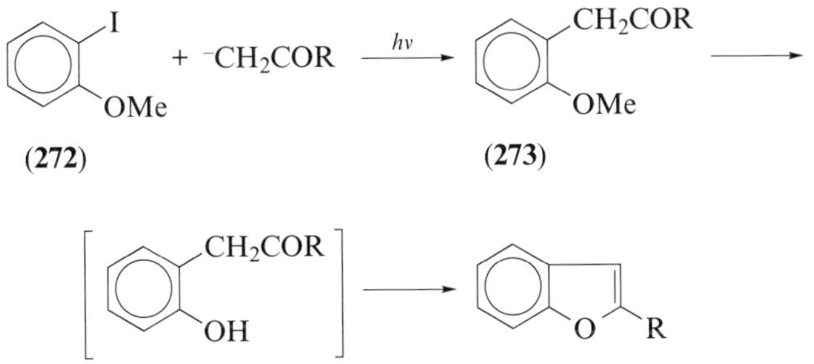

(272)   (273)   (171)

(274)   (275) R = H (40%); Me (67%);
                    i-Pr (66%); t-Bu (100%)

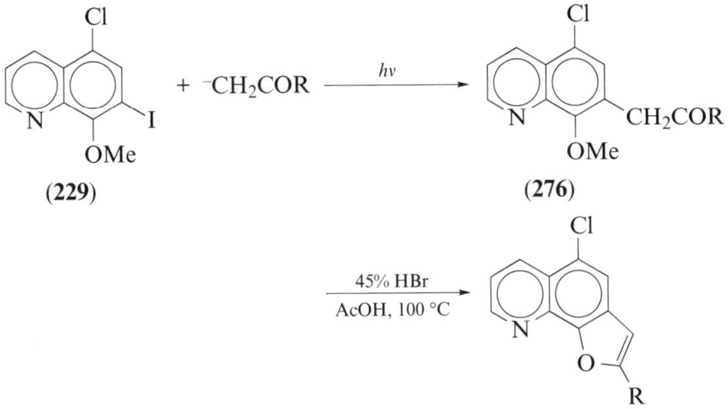

(229)   (276)

(172)

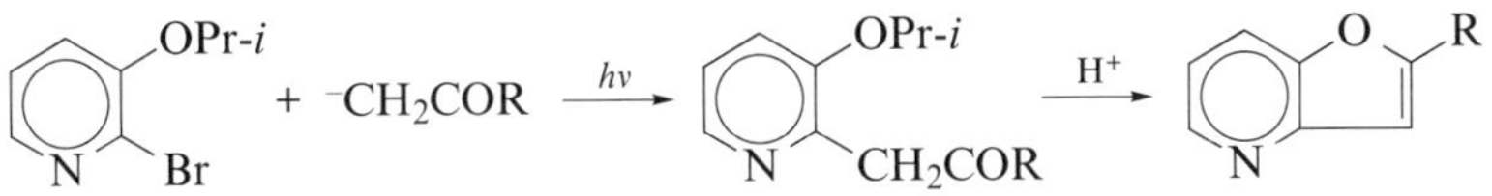

(277)

(278)   (279a) R = t-Bu (70%)   (280)   (173)
               (279b) R = p-anisyl (30%)

gave the substitution product, which cyclized to 2-ethyl-3-methylfuro[3,2-b]pyridine in 86% yield[326].

The percentage of product 279b was increased to 98% when the reaction was performed in DMSO. The product 2-phenylfuro[3,2-b]pyridine was obtained in 70% yield with acetophenone enolate ion in DMSO[326].

The photostimulated reaction of benzylamines with an o-iodo (bromo) substituent (281) and ketone enolate ions gave the product 282 that cyclized spontaneously to give 283 (equation 174)[327].

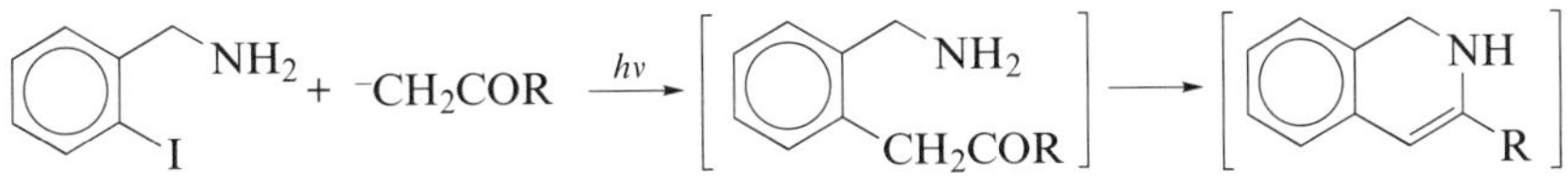

(281)   (282)   (283)
                                            (174)

The intermediate products **283** are not isolated, but they are easily oxidized by air to give hydroxyisoquinolines, or dehydrogenated by Pd/C to give substituted isoquinolines, or reduced with NaBH$_4$ to give tetrahydroisoquinolines in very good yields (equation 175).

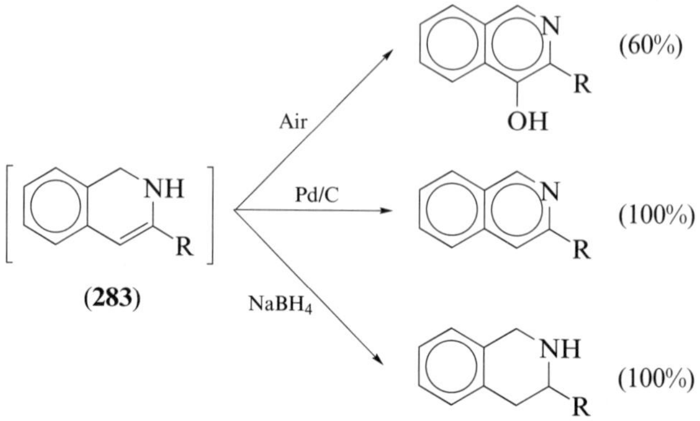

Benzo[c]phenanthridines are an important class of the isoquinoline alkaloid family. It was possible to synthesize a variety of these compounds based on the $S_{RN}1$ reactions of appropriate derivatives of o-iodobenzylamines with nucleophiles[328]. For instance, the photostimulated reaction of **281** (R^1 = R^2 = H, OMe; R^1 = OPr–i, R^2 = OMe; R^{1}R^2 = OCH$_2$O) with the appropriate derivative of tetralone enolate ions **284** (R^3 = R^4 = H, OMe; R^3 = OMe, R^4 = H; R^3 = OPr–i, R^4 = OMe; R^{3}R^4 = OCH$_2$O) gave the substitution compound, which spontaneously cyclizes to **285** (equation 176).

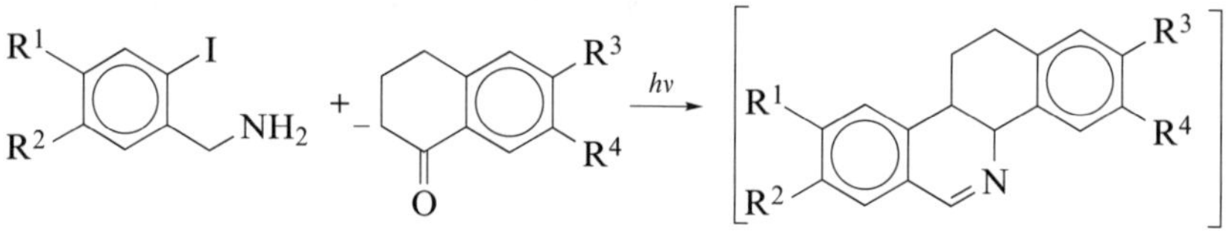

Compounds **285** are spontaneously oxidized by air to **286**, which are further dehydrogenated by Pd/C to **287** (equation 177).

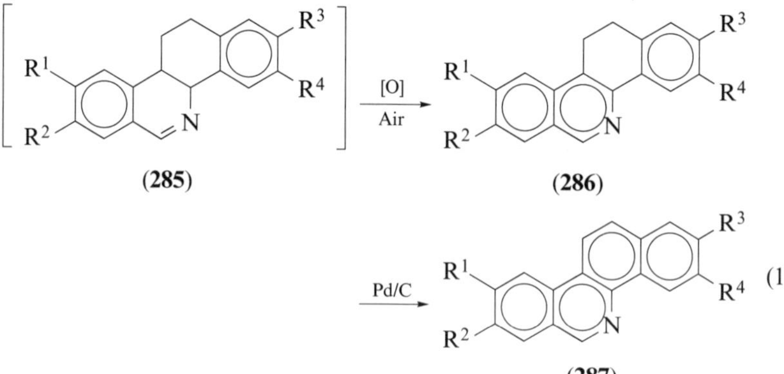

Another system which leads to ring closure reaction is *o*-halobenzoic acids. For instance, the reaction of *o*-iodobenzoate ion (288) with acetone enolate ion gives the substitution product 289, which leads to the isocoumarin 290 in acidic conditions, in high overall yield (80%) (equation 178)[329].

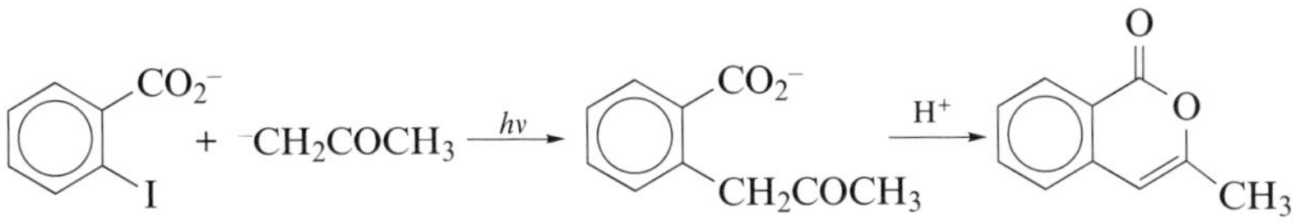

(288)   (289)   (290)   (178)

The reaction of 2-iodobenzoic acids derivatives (288) with tetralone enolate ions (284) gives the products 291 which, after acid treatment, afford the ring closure compounds 292 (60–75% yield) (equation 179)[328].

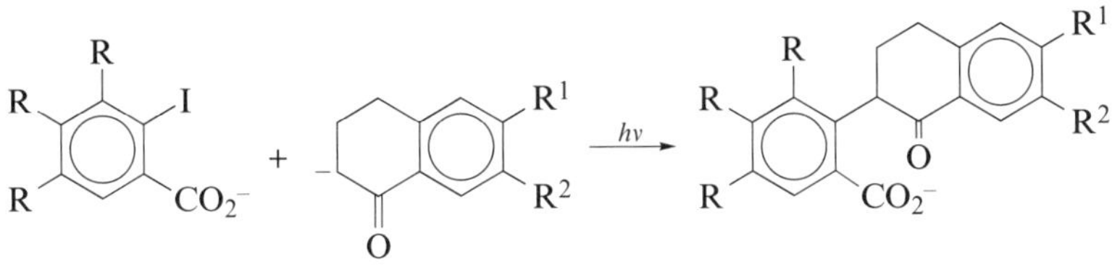

(288)   (284)   (291)

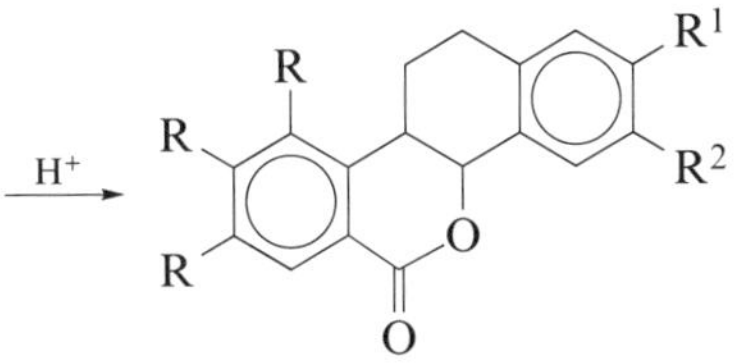

(292)   (179)

One of the most interesting applications of the photostimulated carbonylation of aryl halides by $Co(CO)_4^-$ ions under phase-transfer catalysis conditions is the synthesis of benzolactams and lactones 293 starting from aryl halides bearing amino or hydroxy groups on a side chain *ortho* to the halogen (equation 180)[212].

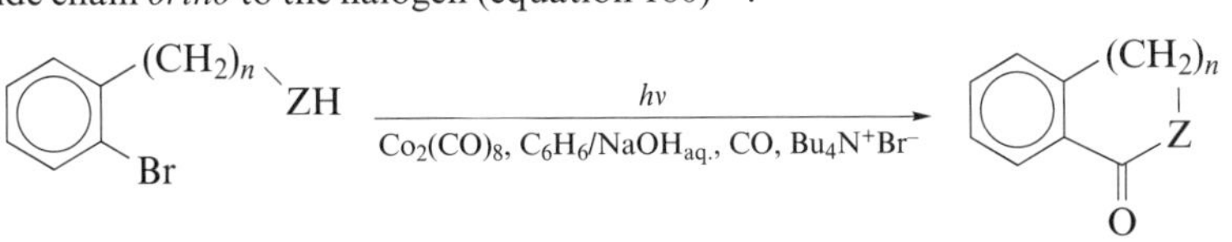

Z = O, NH, NCH$_2$Ph; n = 1,2   (293) 60–95%

It has been suggested that photostimulated carbonylation occurs by the $S_{RN}1$ mechanism, and the $ArCO_2^-$ intermediate cyclizes to give 293.

Another system is the reaction of *o*-halobenzamides (294), which leads to isocarbostyrils 295 [R^1 = R^2 = R^3 = OMe (80%); R^{1}R^2 = OCH$_2$O, R^3 = H (70%); R^1 = R^2 = OMe, R^3 = H (75%)] (equation 181). With *N*-methylbenzamides the yields of substitution products are lower[329,330].

This reaction has been used to synthesize the precursors of berberine alkaloids [benzo(*c*)phenanthridine alkaloids][331]. Thus, the photostimulated reaction of *o*-iodoben-

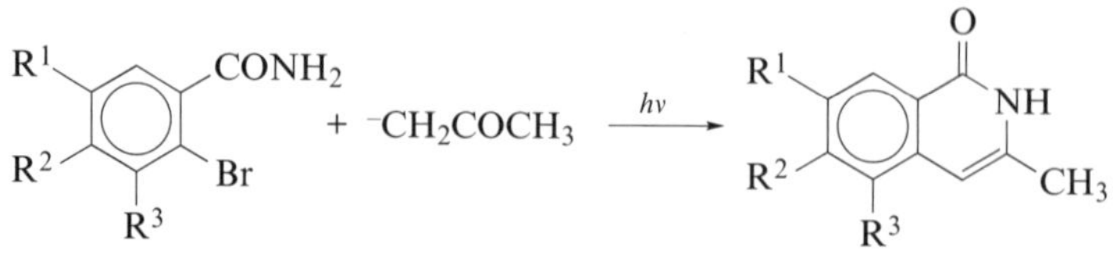

$$(181)$$

(294)   **(295)**

zamide derivatives with the ketone enolate ion **296** gives the ring closure products **297** in
ca 80% yield (equation 182).

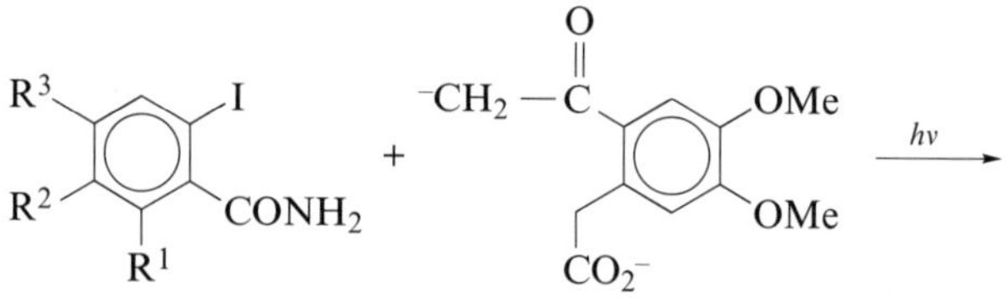

$R^1 = R^2 = R^3 = H$
$R^1 = H, R^2 = R^3 = OMe$
$R^1 = R^2 = OMe, R^3 = H$

(296)

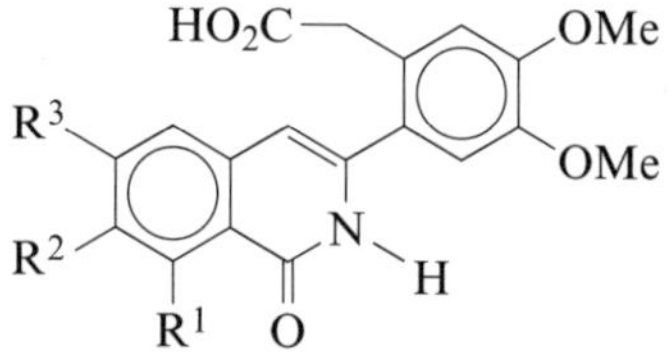

$$(182)$$

(297)

Another $S_{RN}1$ reaction that leads to a substitution product that cyclized in a subsequent
step is the reaction of (o-iodoaryl)acetic acid derivatives with ketone enolate ions. Thus,
substrate **298** reacts under irradiation with ketone enolate ions to give **299** [R = Me
(75–80%), R = i-Pr (75–80%), R = t-Bu (85–90%)] (equation 183)[332]. The substitution
products **299** are the key intermediates for the synthesis of 3-benzazepines and 3-
benzoxepines[332].

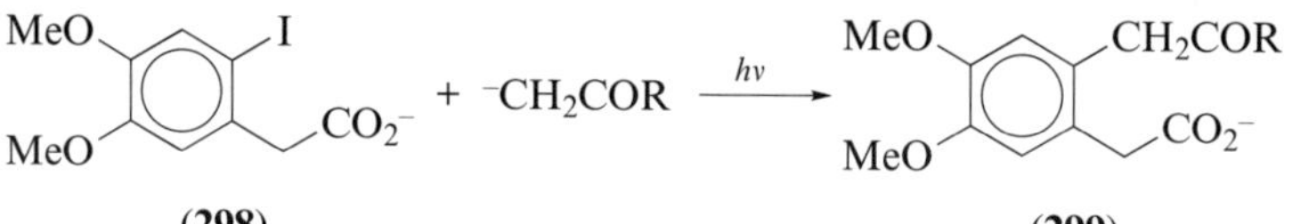

$$(183)$$

(298)   **(299)**

Another system studied is the photostimulated reaction of o-bromobenzaldehyde or 2-
bromoacetophenone (**300**) with $^-SCH_2CO_2Et$ as a nucleophile. The substitution product
301 in the basic condition forms the anion **302** which leads to benzothiophenes **303** (equation 184)[288]. The yields are only moderate due to the fragmentation of the radical-anion
intermediate.

In the photostimulated reaction of $^-SCH_2CO_2Et$ ions with 2-bromo-3-cyano or 3-
bromo-4-cyano pyridines, the substitution product formed is deprotonated, and the anion
formed reacts with the cyano group to give ultimately the ring closure substitution
products **304** (90%) and **305** (98%), respectively. The yields are high, because in these cases
fragmentation of the radical-anion intermediates does not take place[288].

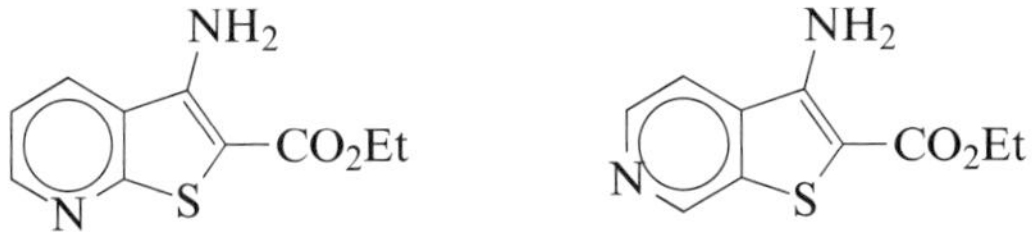

(304)   (305)

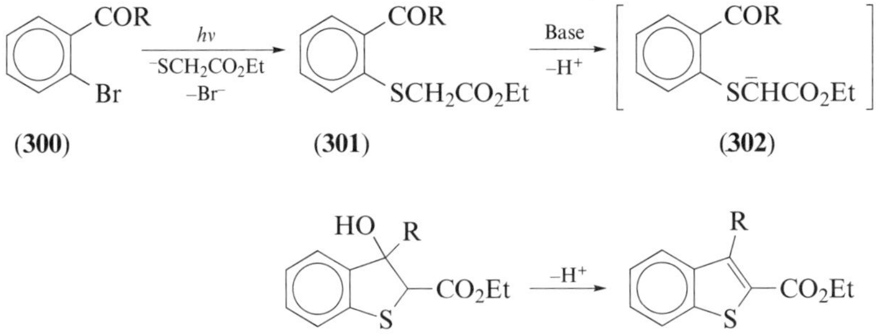

(300)   (301)   (302)

(184)

(303a) R = Me (40%)
(303b) R = H (55%)

The photostimulated reaction of *o*-bromoacetophenone **306a** or propiophenone **306b** with enolate ions of acetophenones (R^1, R^2 = H, Me, OMe) in DMSO gives the substitution products **307**, that spontaneously cyclize to **308** (isolated as the *i*-propyl ethers) (equation 185). With acetophenone **306a** the yields are higher than with **306b**, due to the fact that the former has no β-hydrogen[333].

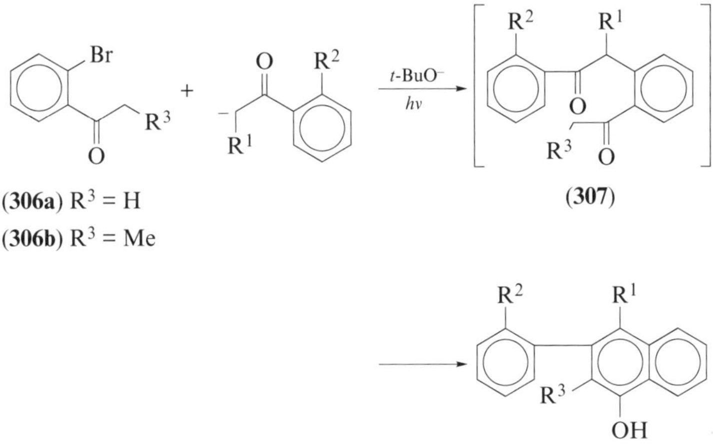

(306a) R^3 = H
(306b) R^3 = Me   (307)

(185)

(308)

In the photostimulated reaction of **306** carrying different ring substituents (H, OMe, Me) with the enolate ions of substituted 2-aceto or 2-propionaphthophenones, it was possible to synthesize substituted 2,2'-binaphthalenes **309** (25–80%) in a one-pot reaction[333].

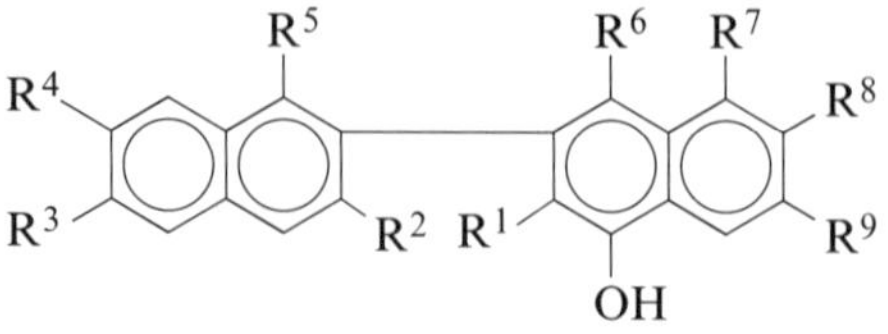

(309)

The substitution products, formed in the reaction of acetone or pinacolone enolate ions with aromatic substrates bearing a F_3C group either *ortho* or *para* to the halogen, undergo reactions in which fluoride ions are eliminated. In the case of the *ortho*-isomer **310**, a ring closure product **311** is formed by intramolecular reaction (equation 186)[334].

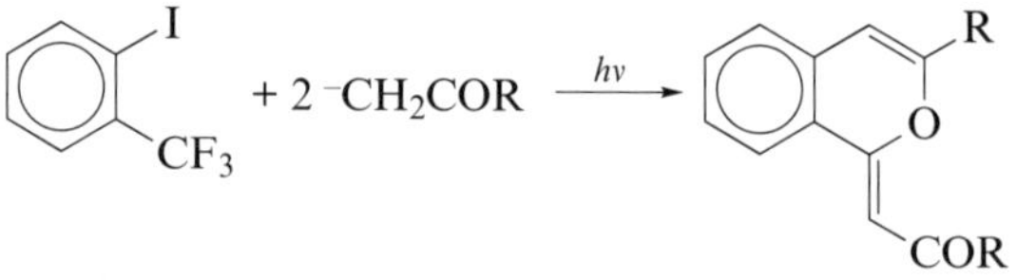

$$+ 2\ {}^-CH_2COR \xrightarrow{hv} \qquad (186)$$

(310)   (311)

2. Intramolecular ring closure reactions

An interesting type of ring closure reaction is the intramolecular $S_{RN}1$ reaction of substrates that have both a nucleophilic centre and a nucleofugal group. The first recognition of this reaction was the intramolecular photostimulated cyclization of ketone enolate ions, such as **312** to give **313** [$n = 1$ (99%), $n = 3$ (73%), $n = 5$ (25–35%)] (equation 187)[246, 335]. In these cases, only one enolate ion is formed. When the ketone has two possible carbanionic centres, both isomers of the ring closure products are formed, although in low yield[335].

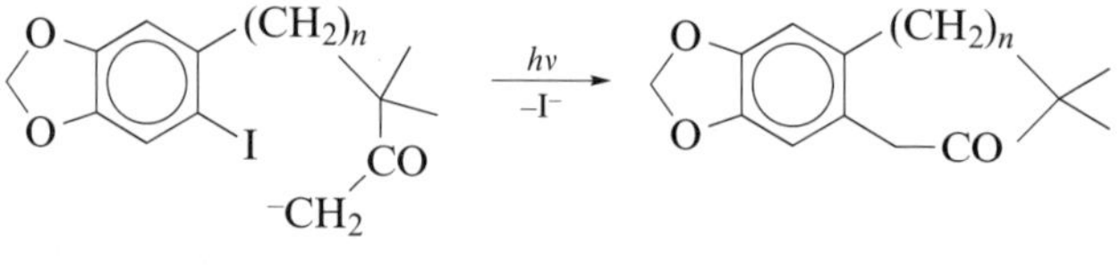

$$\xrightarrow[-I^-]{hv} \qquad (187)$$

(312)   (313)

When *N*-acyl-*o*-chloroaniline (**314**) is treated with LDA in THF–hexane solution, it forms the enolate ions which undergo cyclization to afford oxindoles **315**. When R, R′ = Me, Ph, *n*-Bu, the yield of **315** are 63–82%. When R = PhCH₂, R′ = H the yield is 32% (equation 188)[336].

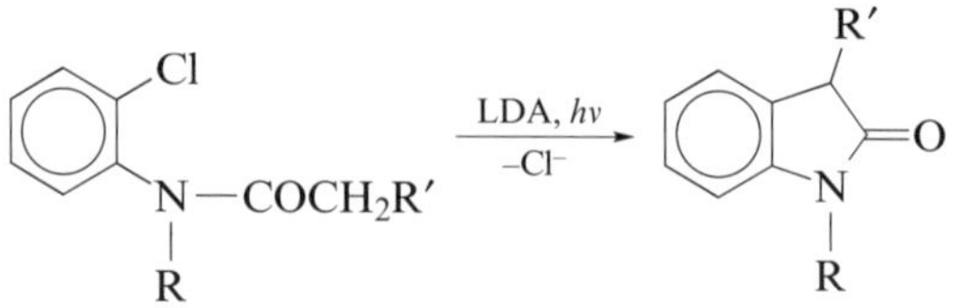

$$\xrightarrow[-Cl^-]{LDA,\ hv} \qquad (188)$$

(314)   (315)

It has been reported that when R = n-C_9H_{19} and R′ = Me, 68% of the ring closure product is formed, but when R = n-$C_{18}H_{37}$ and R′ = Me, only 8% yield of the cyclized product was obtained[337].

When the photostimulated reaction of equation 188 was carried out in the presence of NH_2^- ions in liquid ammonia, besides oxindoles **315**, products derived from the benzyne mechanism were formed[338]. However, when the substrate has no o-hydrogens, such as **316** good yields of oxindoles were obtained (equation 189).

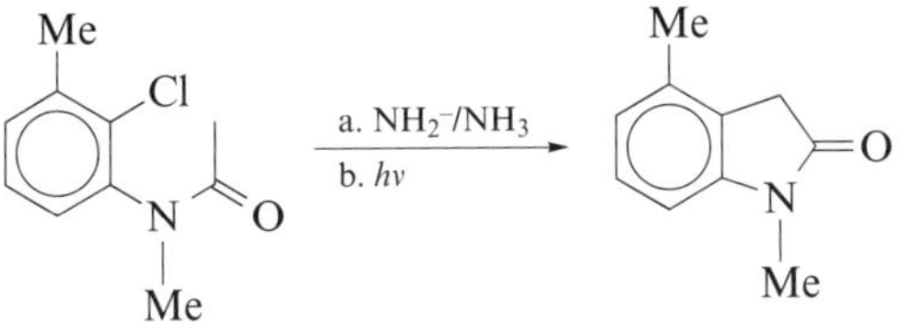

$$(189)$$

(316)   (80%)

When the substrate has no alkyl groups on N, the dianion formed by LDA in THF also reacts under irradiation to afford oxindoles in good yields. Photocyclization of 2-chloro-3-(N-methylacetamido)pyridine by means of KNH_2 in liquid ammonia or LDA in THF gave azaoxindoles in good yields (62 and 83%, respectively)[338].

N-methyl α,β-unsaturated anilides (**317**) underwent intramolecular arylation exclusively at the α-position to afford 3-alkylideneoxindoles (**318**) in good yields. Although they react more slowly than **314** with LDA in THF, they react quite rapidly (15 min irradiation) in liquid ammonia to give excellent yields of **318** (equation 190)[338].

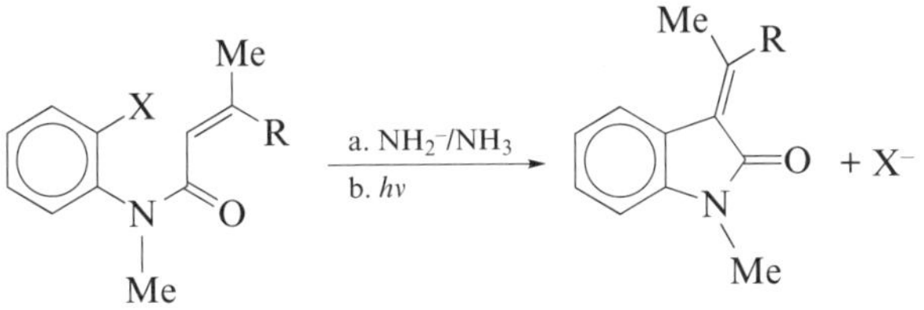

$$(190)$$

(317a) R = H, X = Cl   **(318a)** R = H, (63%, *trans:cis* 3:1)
(317b) R = Me, X = Cl   **(318b)** R = Me, (100%)
(317c) R = Me, X = I   **(318b)** R = Me, (90%)

The reaction of N-methyl-N-acetyl-2-chloro-3-aminopyridine with LDA and irradiation in THF–hexane (–78 °C) gave azaindole in 83% yield[336]. This type of reaction has been used to prepare the key intermediate for the synthesis of Eupolauramine[339]. Thus, o-metallation of 3-bromopyridine and treatment with MeNCO gave the anion **319**, which reacts with $PhCOCH_2Br$ forming the product **320** (equation 191).

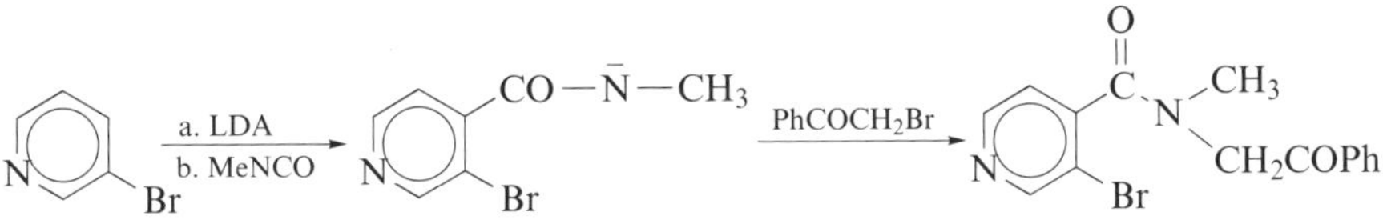

(319)   **(320)**

$$(191)$$

The product **320** when treated with LDA in THF (0 °C) forms the carbanion **321**, that by irradiation gives the ring closure product **322** in 87% yield (equation 192)[339].

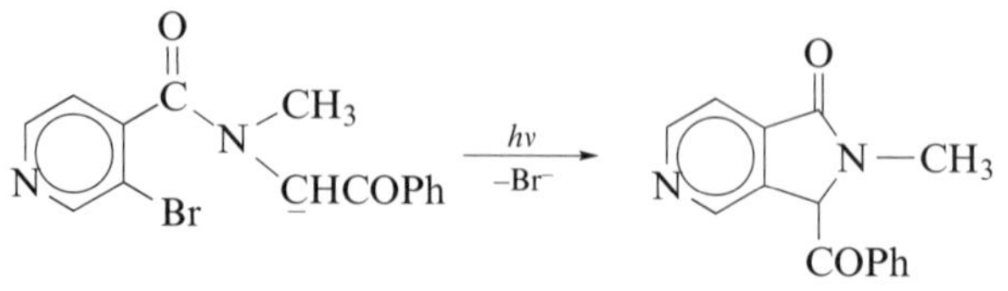

(192)

(321)   **(322)**

The photostimulated reaction of the carbanion formed by the reaction of N-acyl-N-methyl-o-chlorobenzylamines (**323**) (R^1 = Me, R^2 = H; R^1 = R^2 = Me; R^1 = Me, R^2 = Et) with NH_2^- ions in liquid ammonia gave 1,4-dihydro-3 ($2H$)-isoquinolones (**324**) (R^1 = Me, R^2 = H; R^1 = R^2 = Me; R^1 = Me, R^2 = Et) in good yields (equation 193)[338].

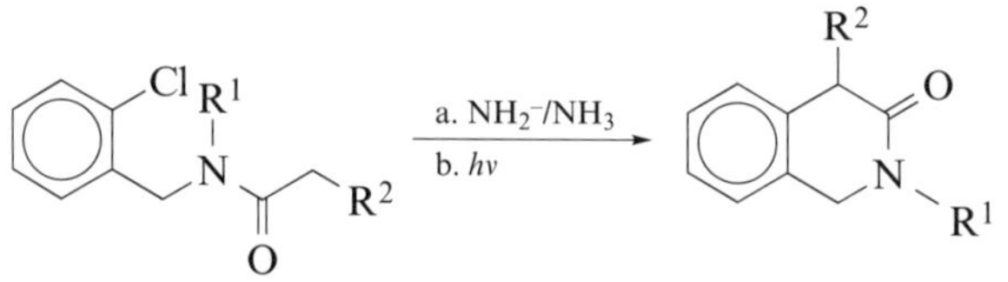

(193)

(323)   **(324)**

When the nucleophilic centre and the nucleofugal group are in m-position, the photostimulated reaction gives ultimately the ring closure product by an intermolecular followed by an intramolecular $S_{RN}1$ reaction. Thus, the reaction of the enolate ion of ω-(m-bromophenyl)-3,3-dimethylalkan-2-one (**325**) in liquid ammonia gives [m-m] cyclophadienones (**326**) (equation 194)[340].

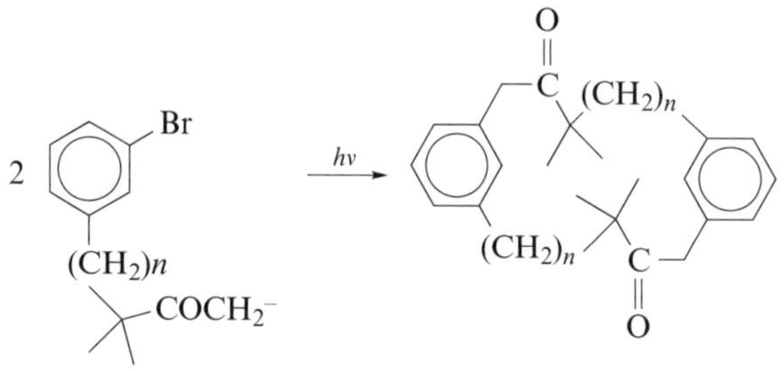

(194)

(325) n = 0–5   **(326)** 20–33%

There is no photostimulated reaction of o-bromo-(**327a**) or o-iodothioanilides (**327b**) in DMSO with t-BuOK or NaH, but in the presence of acetone enolate ion (**27a**), quantitative yields of benzothiazoles (**329**) were obtained in the reaction of **327b** (equation 195)[341].

These results suggest that anion **328** is unable to initiate the process, a reaction that is performed by the enolate ion of acetone. However, the rate of the coupling reaction between the anion of acetone and the radical intermediate formed is slower than the rate of intramolecular ring closure reaction of the radical.

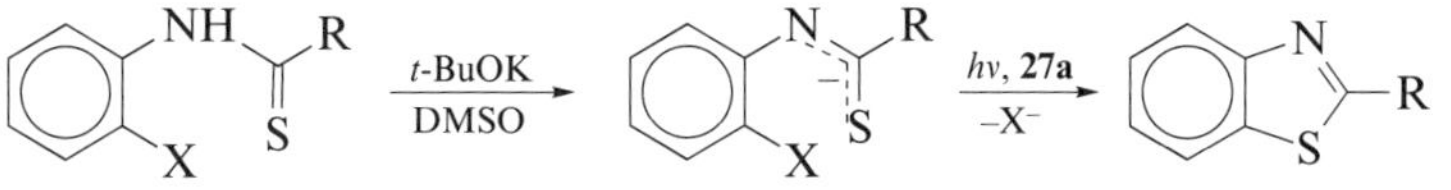

$$\text{(327)(a) } R = Ph; X = Br \quad \text{(328)} \quad \text{(329)}$$
$$\text{(b) } R = Ph, Me; X = I$$

In the photostimulated reaction of substrates with two leaving groups in *o*-position, such as *o*-dibromobenzene and acetone enolate ion, a disubstitution product (330) is formed. In the basic reaction conditions it leads, by an aldol condensation, to a final mixture of two acetylmethyl indenes 331a and 331b in 64% yield (equation 196)[342].

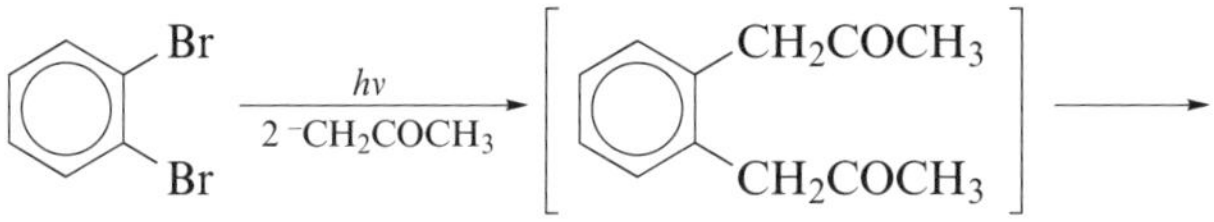

(330)

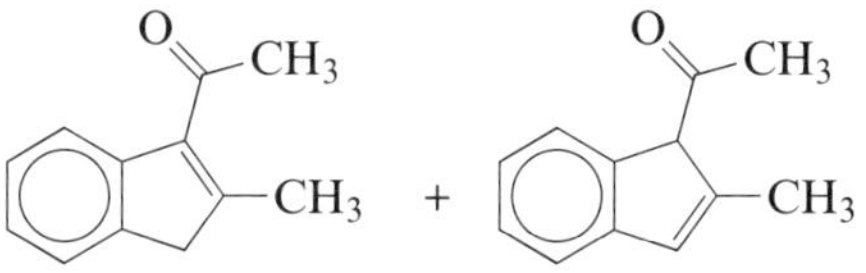

(331a)   (331b)   (196)

Another approach to ring closure is the reaction of a substrate having two leaving groups in *o*-position and a bidentate nucleophile. For instance, in the photostimulated reaction of *o*-bromochlorobenzene with 3,4-toluenedithiolate ions (332), 3-methylthianthrene (333) is formed (55%). When the substrate, such as *o*-diiodobenzene, has a better leaving group, an increase in the yield of 333 to 64% is observed (equation 197)[343].

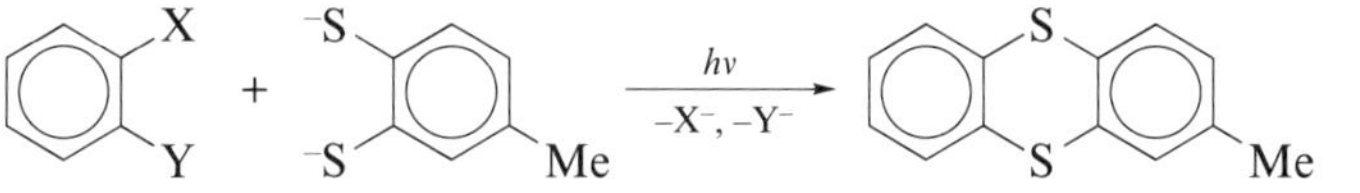

(197)

(332)   (333)

The reaction of 1,2-diiodoacenaphthene (334) with Na_2Te in DMF at room temperature gave the cyclic bis-telluride (335) in 35% yield (equation 198)[301].

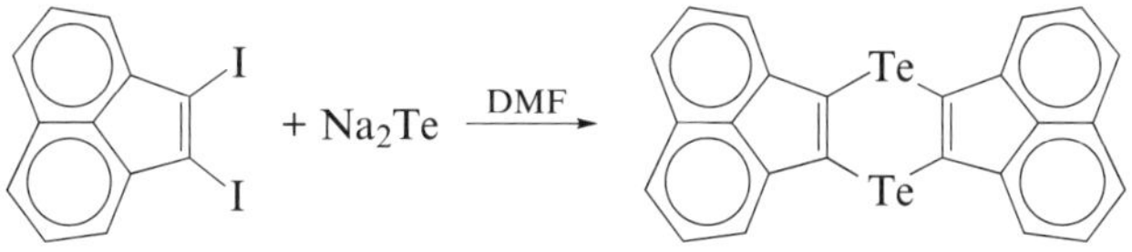

(198)

(334)   (335)

Another possibility of ring closure reaction in the $S_{RN}1$ mechanism is when the haloarene is *o*-substituted with an appropriate substituent bearing a double bond which is able to trap the aryl radical intermediate (a clock reaction). The resulting *alkyl* radical can react with the nucleophile by the $S_{RN}1$ mechanism. Thus, when o-(but-3-enyloxy)iodobenzene (**336**) was treated with nucleophiles under irradiation, the substitution product **337**, resulting from direct substitution, and **338**, arising via cyclization followed by substitution, were formed (equation 199)[52].

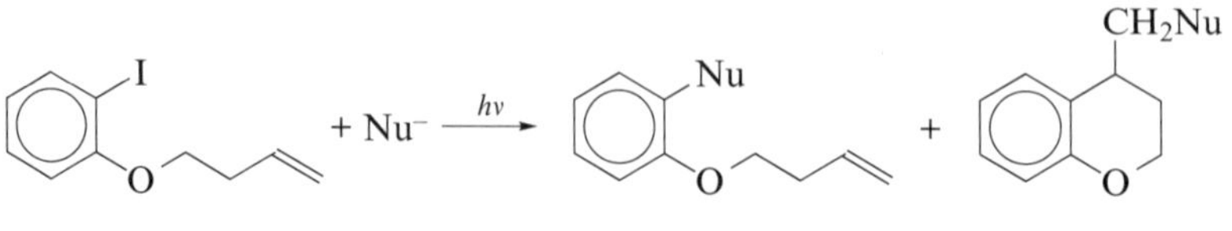

$$ \text{(336)} \qquad Nu^- = Ph_2P^- \qquad \text{(337a)} \qquad \text{(338a)} \tag{199} $$
$$ \qquad\qquad\qquad PhS^- \qquad \text{(337b)} \qquad \text{(338b)} $$

In this system the intermediate aryl radical **339** undergoes two competing reactions: association with the Nu⁻ to give the radical anion **337**⁻·, which finally gives **337**, or cyclization to give radical **340**, which reacts with the Nu⁻ to ultimately give **338** (equation 200).

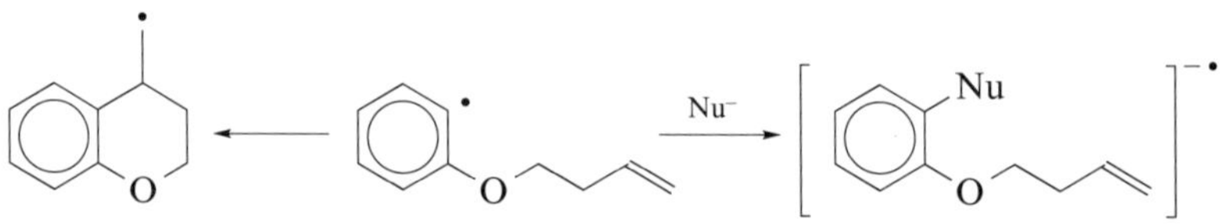

$$ \text{(340)} \qquad\qquad \text{(339)} \qquad\qquad \text{(337)}^{-\bullet} \tag{200} $$

This competition depends on the Nu⁻, its concentration and the solvent. Thus, with Ph_2P^- ions, one of the most reactive nucleophiles, radical **339** is trapped rapidly to give **337a** (67%) and **338a** (33%) in liquid ammonia (but 24% and 75%, respectively, in MeCN). With the less reactive PhS⁻ ions, **339** cyclizes in liquid ammonia to give mainly **338b** (75%) and **337b** in only 5.8%[52].

Aromatic rings can also trap aryl radicals in the propagation cycle of the $S_{RN}1$ mechanism to give ring closure product. The reaction of o-dihalobenzenes **341** with 2-naphthalenethiolate ion (**342**) in liquid ammonia under photostimulation gives the ring closure product **343** as well as the substitution product **344** (equation 201)[344].

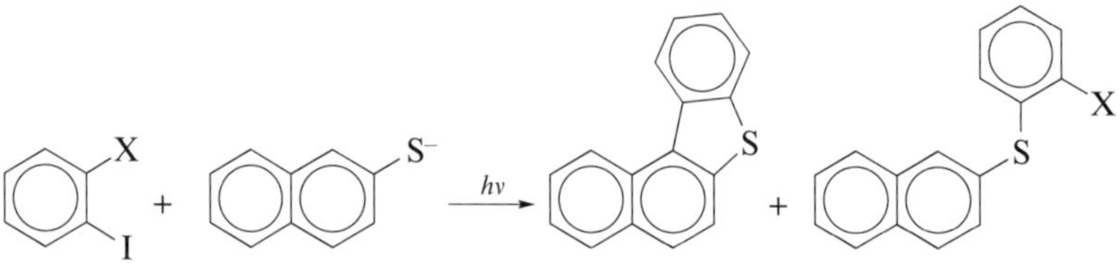

$$ \text{(341)} \qquad \text{(342)} \qquad \text{(343)} \qquad \text{(344)} \tag{201} $$

In the reaction of **341** (X = I), the only product obtained was **343** (62% yield). With **341** (X = Br), products **343** (46%) and **344** (X = Br, 56%) were obtained in similar yields. With **341** (X = Cl) the only product obtained was **344** (X = Cl). In all these reactions the reduction product **344** (X = H) was not observed. All these results can be explained according to the propagation cycle of the $S_{RN}1$ mechanism, but with some additional steps. When substrate **341** receives an electron it fragments into I⁻ ions (the better leaving group) and radical **345**, which reacts with **342** to give radical anion **344**⁻˙ (equation 202).

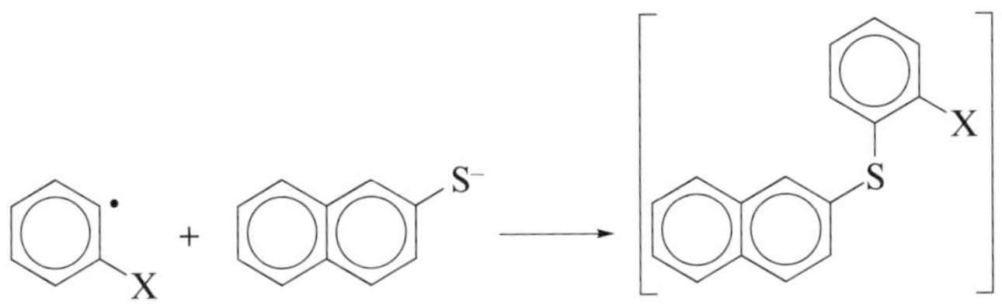

$$(202)$$

(**345**)   (**342**)   (**344**)⁻˙

The intermediate **344**⁻˙ undergoes two competing reaction: an intermolecular ET to the substrate to give substitution product **344**, or intramolecular ET to the C–X σ* MO, which fragments to give radical **346**. The latter is trapped intramolecularly by the α-position of the naphthalene ring to give radical **347** which, in the basic reaction conditions, is deprotonated to give radical anion **343**⁻˙ that finally gives **343** (equation 203).

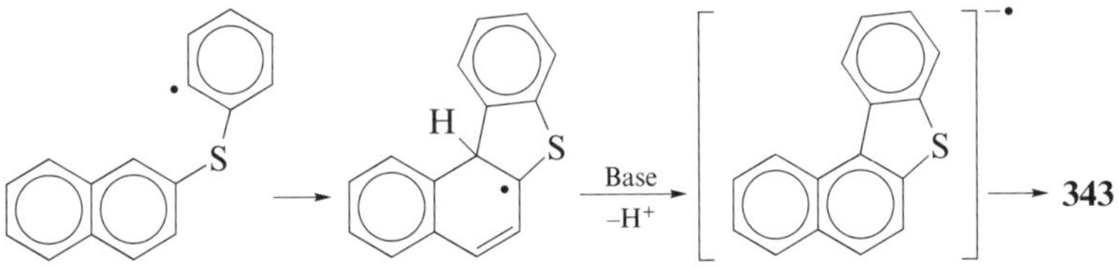

(**346**)   (**347**)   (**343**)⁻˙

With radical anion **344**⁻˙ (X = I), the intramolecular ET is much faster than the intermolecular ET, and thus only **343** is obtained. In **344**⁻˙ (X = Br) the rates of both ET reactions are similar, and both products are obtained in similar yields. However, in **344**⁻˙ (X = Cl), only the intermolecular ET is observed. MO calculations are in agreement with this interpretation [344].

The reaction of pentachloropyridine (**348**) with an excess of benzenethiolate ions in 1,3-dimethyl-2-imidazolidinone as solvent leads to complete halogen replacement: pentakis(phenylthio)pyridine (**349**) and 1,3,4-tris(phenylthio)[1]benzothieno[3,2-c]pyridine (**350**) were thus formed (equation 204)[345].

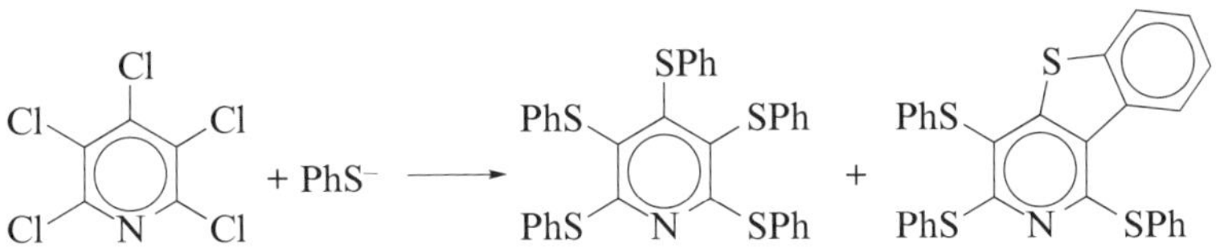

$$(204)$$

(**348**)   (**349**)   (**350**)

When the reaction was performed in the dark and with p-DNB, almost no ring closure product **350** was obtained, but under irradiation with a tungsten lamp 30% yield was reported. These result suggest that **350** is formed via radical intermediates.

In the photostimulated reaction of 2-naphthoxide ion (**198**) with an *o*-dihalobenzene, an aromatic σ radical may be formed very close to the oxygen functionality along the chain propagation cycle of the $S_{RN}1$ mechanism. This spatial proximity and the fact that the intramolecular coupling between the two moieties will form a relatively stable radical anion will favour the reaction between both reactive centres. Thus in the photostimulated reaction of *o*-dihalobenzenes with **198** in liquid ammonia, the formation of the monosubstitution **351** and of the cyclization product **352** were reported in yields that depend on the substrate and on the reaction conditions (equation 205)[346].

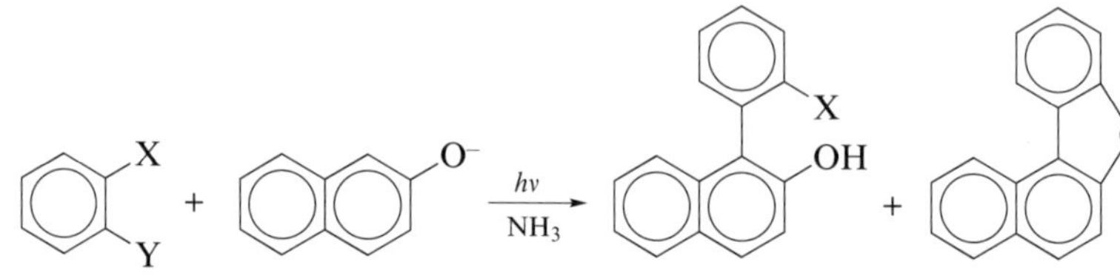

(205)

(198) **(351)** **(352)**

All the results indicate that **198** reacts with *o*-dihalobenzenes by the $S_{RN}1$ mechanism as a bidentate nucleophile through a stepwise process in which the monosubstitution product **351** is an intermediate, and that the oxygen functionality is able to react intramolecularly quite efficiently with the aromatic σ radical to give the cyclized product. The proposed mechanistic steps are summarized in equations 206 and 207 for *o*-diiodobenzene.

(206)

(353) **(198)** **(354)** **(355)**

(355)$^{-\bullet}$ **(356)** **(352)**$^{-\bullet}$

(207)

(352)

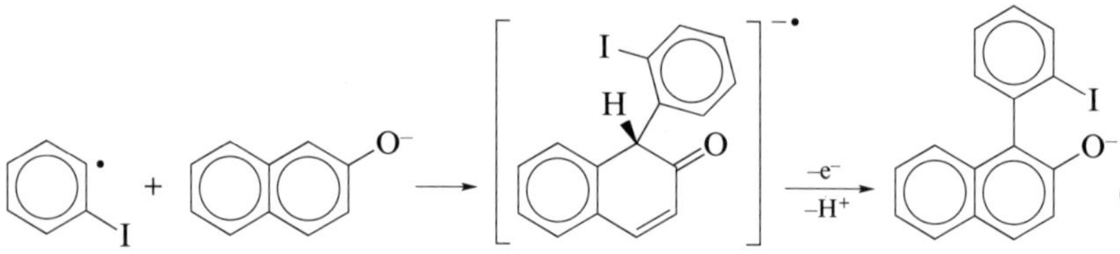

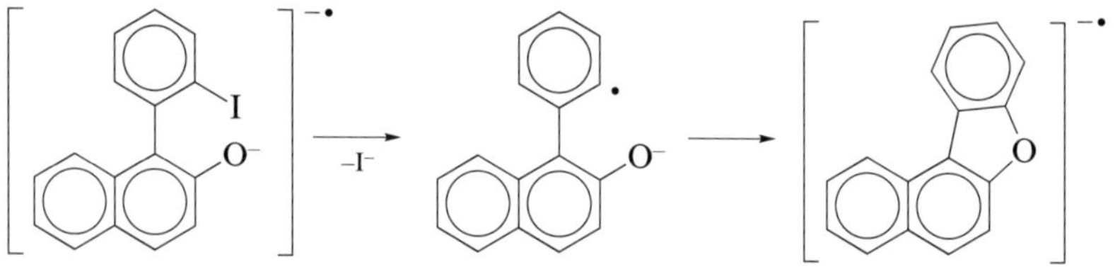

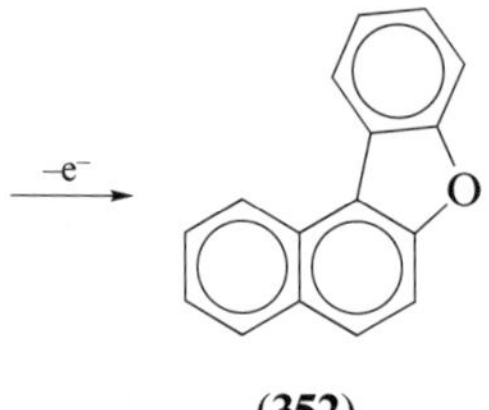

The o-iodophenyl radical **353** formed couples with the nucleophile at the C_1 position to give the radical anion **354**, which by ET and deprotonation under the basic reaction conditions, gives **355** (equation 206). The radical dianion of **355**, formed by photostimulated ET fragments to give the radical and anion **356** which cyclizes to give compound **352** (equation 207).

Similar cyclization reactions to polychlorodibenzodioxins have been reported under irradiation of the sodium salt of the conjugate base of perchloro-o-phenoxyphenol in methanol in the presence of a sensitizer and excess triethylamine. Photodecomposition of the conjugate base of perchloro-o-phenoxyphenol in methanol reveals a small amount of cyclization, while irradiation in methanol in the presence of a 10-fold excess of triethylamine increase the quantum yield of cyclization by 17-fold. These results were interpreted in terms of an ET to the substrate to from a radical dianion, which reacts through an intramolecular $S_{RN}1$ mechanism[347].

An interesting approach to the synthesis of the dibenzo[b,d]pyran-6-one skeleton of benzoocumarins has been recently reported[348]. It involves the $S_{RN}1$ o-arylation of phenoxide ions by o-bromobenzonitrile followed by SiO_2-catalysed lactonization (equation 208). Good yields were observed for phenols bearing either electron-donating (t-Bu, OMe) or electron-withdrawing (OCF$_3$, F, CN) groups at the p-position. The exception was p-nitrophenol, which gave no substitution. The coupling takes place at the less hindered o-position of the nucleophile.

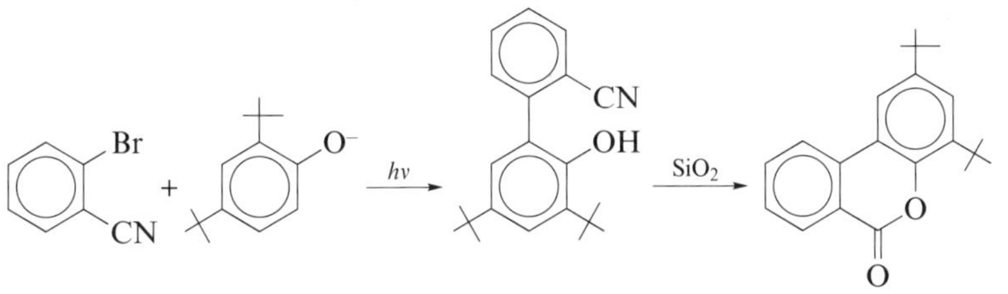

$$(208)$$

As a route to the synthesis of benzonaphthopyranones, the reaction of o-bromobenzonitrile derivatives **357** with **198**, was investigated. In this case a mixture of two products was obtained. The minor product was that resulting from coupling on C_3 (5%), and high yields of substitution products **358** from coupling at C_1 were obtained (R = H, 76%; R = OMe, 84%). These products can be lactonized by refluxing their solution in CHCl$_3$ in the presence of SiO$_2$ (equation 209). The reactions were also performed with chiral phenoxide ions derived from hydroxyphenyl amino acids, namely tyrosine and p-hydroxyphenylglycine. High percentage yields of optically pure dibenzo(b,d)pyran-6-ones were obtained with nucleophiles that do not racemize under $S_{RN}1$ conditions[348].

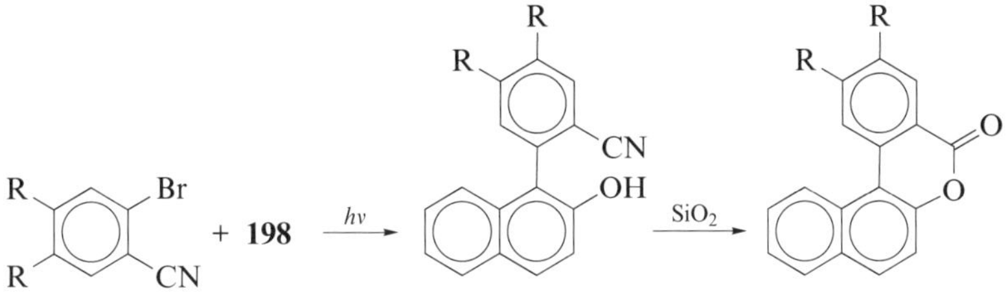

(357)   (358)

$$(209)$$

The reaction of (2-cyanophenyl)azo phenyl sulphide or t-butyl sulphide with PhO$^-$ ions in DMSO followed by lactonization is another route to the synthesis of these compounds[349].

The bidentate behaviour of phenoxide ions appears to be a synthetic route to the precursor **359** of dibenz[d, f]azonine alkaloids such as bractazonine (equation 210)[350].

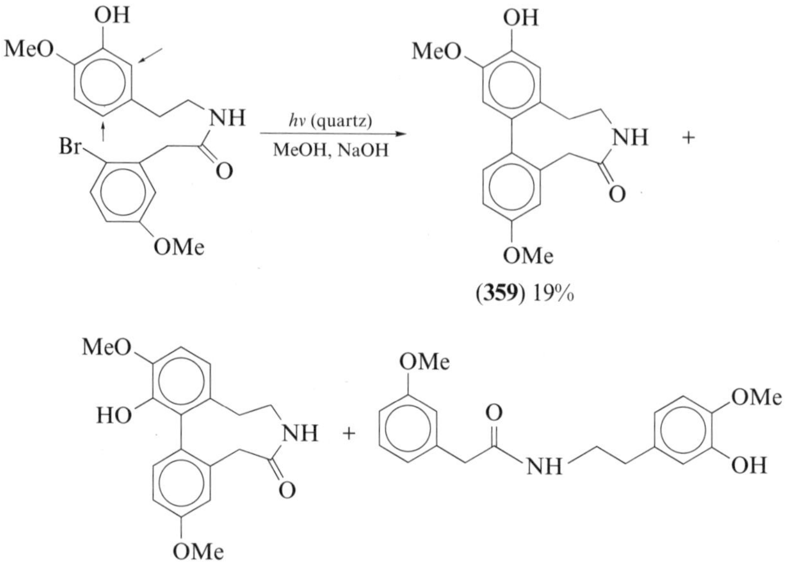

(**359**) 19%

38%   9%

VI. VINYL HALIDES

The $S_{RN}1$ mechanism has been proposed for the carbonylation of vinyl bromides and chlorides with NaCo(CO)$_4$ under photostimulation[211,212] (section V.A.1), and for the vinylation of iron porphyrines under electrochemical induction[4a].

More recently, the reaction of the dye **360** with different nucleophiles has been reported to yield the corresponding derivatives **361** (**361a**, Nu = OMe; **361b**, Nu = NHMe; **361c**, Nu = OPh; **361d**, Nu = SPh; **361e**, p-H$_2$NC$_6$H$_4$S) (equation 211)[351].

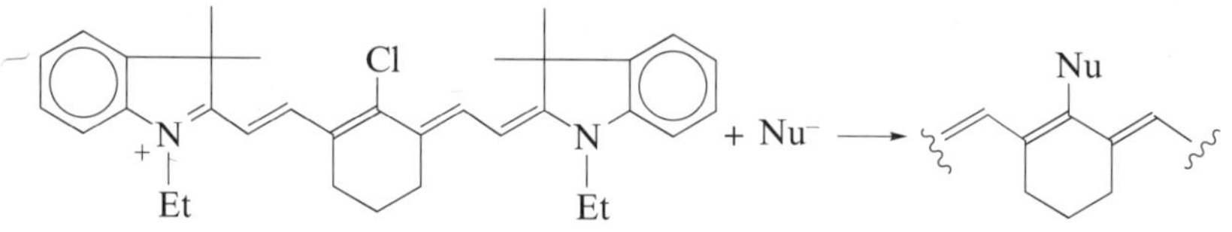

(**360**)   (**361**)

The yields of **361b–e** were reduced in the presence of oxygen and PhNO$_2$, while the reaction with MeOH was unaffected. It was suggested that **360** reacts by an $S_{RN}1$-type mechanism leading to products **361b–c** with radical cations as intermediates.

The presence of radical intermediates was supported by observation of the ESR signal for a mixture of **360** and PhSH in DMF. On the other hand, the reaction with MeONa to give **361a** may involve an addition of the nucleophile followed by elimination of chloride ions.

Recently, evidence for the vinylic nucleophilic substitutions suggests the occurrence of an ionic elimination–addition along with the originally proposed $S_{RN}1$ route[352]. Thus, the reaction of β-bromostyrene with pinacolone enolate ions and $FeCl_2$ as catalyst in DMSO gave a mixture of products **362–366** in yields that depend on the reaction time, 10 min or (3 h).

$PhCH=CHCH_2COCMe_3$ (**362**), 21% (45%); $PhCH_2CH=CHCOCMe_3$ (**363**), 7% (14%);

$$\begin{array}{c} Me \\ | \\ PhC\equiv CH\ (\textbf{364}),\quad 21\%\ (18\%);\quad PhC\equiv CCCMe_3\ (\textbf{365}),\ 42\%\ (9\%); \\ | \\ OH \end{array}$$

$$PhCH=CH—C\equiv CPh\ (\textbf{366}),$$

The amount of the substitution products **362** and **363** was reduced in the presence of p-DNB, while the yield of **365** was unaffected. The formation of **365** was ascribed to the reversible addition of the conjugated base of the elimination product phenylacetylene **364** to the ketone carbonyl. This step is likely to be reversible and will gradually allow the build-up of the substitution products by addition of the nucleophile to **364**, as an alternative route to the $S_{RN}1$ process[353].

VII. REFERENCES

1. N. Kornblum, R. E. Michel and R. C. Kerber, *J. Am. Chem. Soc.*, **88**, 5662 (1966).
2. G. A. Russell and W. C. Danen, *J. Am. Chem. Soc.*, **88**, 5663 (1966).
3. J. K. Kim and J. F. Bunnett, *J. Am. Chem. Soc.*, **92**, 7463, 7464 (1970).
4. (a) D. Lexa and J. M. Savéant, *J. Am. Chem. Soc.*, **104**, 3503 (1982).
    (b) L. I. Denisovich, N. A. Ustymyuk, M. G. Peterleitner, V. N. Vinogradova and D. N. Kravtsov, *Izv. Akad. Nauk SSSR. Ser. Khim.*, 2635 (1987); *Chem. Abstr.*, **108**, 45786a (1989).
5. N. Kornblum in *Supplement F: The Chemistry of Amino, Nitroso and Nitro Compounds and Their Derivatives* (Ed. S. Patai), Chapter 10, Wiley, Chichester, 1982.
6. W. R. Bowman, in *Photoinduced Electron Transfer* (Eds. M. A. Fox and M. Chanon), Part C, Elsevier, The Hague, 1988, p. 487.
7. L. Eberson, *Electron Transfer Reactions in Organic Chemistry*, Springer-Verlag, New York, 1987.
8. W. R. Bowman, *Chem. Soc. Rev.*, **17**, 283 (1988).
9. J. Prousek, *Chem. Listy*, **78**, 284 (1988); *Advances in Chem.*, **19**, 132 (1984).
10. G. A. Russell, *Acc. Chem. Res.*, **22**, 1 (1989).
11. R. A. Rossi, A. B. Pierini and S. M. Palacios, *Adv. Free-Radical Chem.*, Vol. 1 (Ed. D. D. Tanner), Jai Press, London (1990), p. 193; *J. Chem. Educ.*, **66**, 720 (1989).
12. R. A. Rossi and A. N. Santiago, *Trends Org. Chem.*, **3**, 193 (1992).
13. R. K. Norris, in *The Chemistry of Functional Groups, Supplement D: The Chemistry of Halides, Pseudohalides and Azides* (Eds. S. Patai and Z. Rappoport), Chapter 16, Wiley, Chichester, p. 681, 1983.
14. R. A. Rossi and R. H. de Rossi, *Aromatic Substitution by the $S_{RN}1$ Mechanism*, ACS Monograph 178, Washington, D. C., 1983.
15. (a) J. M. Savéant, *Adv. Phys. Org. Chem.*, **26**, 1 (1990).
    (b) J. M. Savéant, *Nouv. J. Chem.*, **16**, 131 (1992).
    (c) C. P. Andrieux, P. Hapiot and J. M. Savéant, *Chem. Rev.*, **90**, 723 (1990).
    (d) J. Pinson and J. M. Savéant, in *Electroorganic Synthesis* (Eds. R. D. Little, and N. L. Weinberg), Marcel Dekker, New York, 1991, p. 29.
    (e) C. Degrand, R. Prest and P. L. Compagnon, in *Electroorganic Synthesis* (Eds. R. D. Little, and N. L. Weinberg), Marcel Dekker, New York, p. 45, 1991.
16. R. Beugelmans, *Bull. Soc. Chim. Belg.*, **93**, 547 (1984).
17. R. K. Norris, in *Comprehensive Organic Synthesis* (Ed. B. M. Trost), Vol. 4, Pergamon, New York, 1991, p. 451.

1478   R. A. Rossi, A. B. Pierini and A. B. Peñéñory

18. N. Kornblum, *Aldrichimica Acta*, **23**, 71 (1990).
19. J. F. Bunnett, E. Mitchel and C. Galli, *Tetrahedron*, **41**, 4119 (1985).
20. Z. B. Chen, Z. Zhang and C. Z. Xia, *Youji Huaxue*, **11**, 113 (1991); *Chem Abstr.*, **114**, 246466u (1991).
21. A. Lablache-Combier, in *Photoinduced Electron Transfer* (Eds. M. A. Fox and M. Chanon), Part C, Elsevier, The Hague, 1988, p.134.
22. G. L. Borosky, A. B. Pierini and R. A. Rossi, *J. Org. Chem.*, **55**, 3705 (1990).
23. R. Ettayeb, J. M. Savéant and A. Thiébault, *J. Am. Chem. Soc.*, **114**, 10990 (1992).
24. (a) K. Daasbjerg, T. Lund and H. Lund, *Tetrahedron Lett.*, **30**, 493 (1989).
     (b) J. S. Kristensen and H. Lund, *Acta Chem. Scand.*, **44**, 524 (1990).
     (c) K. Daasbjerg, S. U. Pedersen and H. Lund, *Acta Chem. Scand.*, **45**, 424 (1991).
     (d) L. V. Jorgensen and H. Lund, *Acta Chem. Scand.*, **47**, 577 (1993).
25. (a) F. G. Bordwell and D. L. Hughes, *J. Org. Chem.*, **48**, 2206 (1983).
     (b) F. G. Bordwell and C. A. Wilson, *J. Am. Chem. Soc.*, **109**, 5470 (1987).
     (c) F. G. Bordwell, M. J. Baush and C. A. Wilson, *J. Am. Chem. Soc.*, **109**, 5465 (1987).
     (d) F. G. Bordwell and J. A. Harrelson, Jr., *J. Am. Chem. Soc.*, **111**, 1052 (1989); *J. Org. Chem.*, **54**, 4893 (1989).
26. (a) E. C. Ashby, R. N. DePriest and W.-Y. Su, *Organometallics*, **3**, 1718 (1984).
     (b) E. C. Ashby, W. Y. Su and T. N. Pham, *Organometallics*, **4**, 1493 (1985).
     (c) E. C. Ashby and T. N. Pham, *Tetrahedron Lett.*, **28** 3183 (1987).
     (d) E. C. Ashby and D. Coleman, *J. Org. Chem.*, **52**, 4554 (1987).
     (e) E. C. Ashby, W. S. Park, A. B. Goel and W. Y.-Su, *J. Org. Chem.*, **50**, 5184 (1985).
     (f) E. C. Ashby, *Acc. Chem. Res.*, **21**, 414 (1988).
27. (a) J. Tanaka, H. Morishita, M. Nojima and S. Kusabayashi, *J. Chem. Soc., Perkin Trans. 2*, 1009 (1989).
     (b) J. Tanaka, M. Nojima and S. Kusabayashi, *J. Am. Chem. Soc.*, **109**, 3391 (1987).
28. (a) J. Bertran, I. Gallardo, M. Moreno and J. M. Savéant, *J. Am. Chem. Soc.*, **114**, 9576 (1992).
     (b) J. M. Savéant, *Acc. Chem. Res.*, **26**, 455 (1993) and references cited therein.
29. (a) S. S. Shaik, *J. Am. Chem. Soc.*, **103**, 3692 (1981).
     (b) S. S. Shaik and A. Pross, *J. Am. Chem. Soc.*, **104**, 2708 (1982).
     (c) A. Pross and S. S. Shaik, *Acc. Chem. Res.*, **16**, 363 (1983).
     (d) A. Pross, *Acc. Chem. Res.*, **18**, 212 (1985).
30. S. Hoz and J. F. Bunnett, *J. Am. Chem. Soc.*, **99**, 4690 (1977).
31. (a) P. A. Wade, H. A. Morrison and N. Kornblum, *J. Org. Chem.*, **52**, 3102 (1987).
     (b) M. A. Fox, J. Younathan and G. E. Fryxell, *J. Org. Chem.*, **48**, 3109 (1983).
32. B. Q. Wu, F-W. Zeng, M-J. Ge, X-Z, Cheng and G-S. Wu, *Science in China*, **34B**, 777 (1991).
33. (a) V. L. Ivanov, L. Eggert and M. G. Kuz'min, *Khim. Vys. Energii*, **21**, 337 (1987), *Chem. Abstr.*, **109**, 169594b (1988)) and references cited therein.
     (b) V. S. Savvina and V. L. Ivanov, *High Energy Chem.*, **24**, 205 (1990).
     (c) V. L. Ivanov, J. Aurich, L. Eggert and M. G. Kuz'min, *J. Photochem. Photobiol.*, *A*, **50**, 275 (1989).
34. (a) R. Beugelmans and M. Chbani, *Nouv. J. Chim.*, **18**, 949 (1994).
     (b) A. Boumekouez, E. About-Jaudet, N. Collignon and P. Savignac, *J. Organometal. Chem.*, **440**, 297 (1992).
35. N. Alam, C. Amatore, C. Combellas, A. Thiébault and J. N. Verpeaux, *J. Org. Chem.*, **55**, 6347 (1990).
36. M. J. Dickens and J. L. Luche, *Tetrahedron Lett.*, **32**, 4709 (1991).
37. P. G. Manzo, R. A. Alonso and S. M. Palacios, *Tetrahedron Lett.*, **35**, 677 (1994).
38. J. F. Bunnett, *Acc. Chem. Res.*, **11**, 413 (1978).
39. M. C. R. Symons, *J Chem. Soc.*, *Chem. Commun.*, 408 (1977).
40. (a) E. M. Arnett and R. A. Flowers, II, *Chem. Soc. Rev.*, 9 (1993).
     (b) X. M. Zhang, *J. Chem. Soc., Perkin Trans. 2*, 2275 (1993).
     (c) K. Daasbjerg, *J. Chem. Soc. Perkin Trans 2*, 1275 (1994).
41. (a) A. B. Pierini, J. S. Duca, Jr. and M. T. Baumgartner, *Theochem*, **311**, 343 (1994).
     (b) A. B. Pierini and J. S. Duca, Jr., submitted.
42. (a) P. Neta and D. Behar, *J. Am. Chem. Soc.*, **103**, 103 (1981).
     (b) M. Meot-Ner, P. Neta, R. K. Norris and K. Wilson, *J. Phys. Chem.*, **90**, 168 (1986).

43. A. B. Pierini, A. N. Santiago and R. A. Rossi, *Tetrahedron*, **47**, 941 (1991).
44. G. A. Russell, R. K. Norris and E. J. Panek, *J. Am. Chem. Soc.*, **93**, 5839 (1971); G. A. Russell and A. R. Metcalfe, *J.Am. Chem. Soc.*, **101**, 2539 (1979); N. H. Anderson, M. McMillan and R. O. C. Norman, *J. Chem. Soc. (B)*, 1075 (1970).
45. (a) M. C. R. Symons and W. R. Bowman, *Tetrahedron Lett.*, **22**, 4549 (1981).
    (b) M. C. R. Symons and W. R. Bowman, *J. Chem. Soc., Perkin Trans. 2*, 25 (1983).
46. W. R. Bowman and M. C. R. Symons, *J. Chem. Res. (S)*, 162 (1984).
47. M. C. R. Symons and W. R. Bowman, *J. Chem. Soc., Chem. Commun.*, 1445 (1984).
48. M. C. R. Symons and W. R. Bowman, *J. Chem. Soc., Perkin Trans. 2*, 583 (1988).
49. M. C. R. Symons and W. R. Bowman, *J. Chem. Soc., Perkin Trans. 2*, 1133 (1987).
50. M. C. R. Symons and W. R. Bowman, *J. Chem. Soc., Perkin Trans. 2*, 1077 (1988).
51. K. E. Miller and J. J. Kozak, *J. Phys. Chem.*, **89**, 401 (1985).
52. A. L. Beckwith and S. M. Palacios, *J. Phys. Org. Chem.*, **4**, 404 (1991).
53. A. N. Santiago and R. A. Rossi, *J. Chem. Soc., Chem. Commun.*, 206 (1990).
54. S. M. Palacios and R. A. Rossi, *J. Phys. Org. Chem.*, **3**, 812 (1990).
55. G. A. Russell and D. F. Dedolf, *J. Org. Chem.*, **50**, 2498 (1985).
56. N. Kornblum and O. A. Wade, *J. Org. Chem.*, **52**, 5301 (1987).
57. D. Cabaret, N. Maigrot and Z. Welvart, *Tetrahedron*, **41**, 5357 (1985).
58. (a) R. K. Norris and R. J. Smyth-King, *Tetrahedron*, **38**, 1051 (1982).
    (b) R. K. Norris and R. J. Smyth-King, *J. Chem. Soc., Chem. Commun.*, 79 (1981).
59. C. Galli and J. F. Bunnett, *J. Am. Chem. Soc.*, **103**, 7140 (1981).
60. R. A. Rossi, *Acc. Chem.Res.*, **15**, 164 (1982).
61. D. B. Denney and D. Z. Denney, *Tetrahedron*, **47**, 6577 (1991).
62. G. A. Russell, B. Mudryk and M. Jawdosiuk *J. Am. Chem. Soc.*, **103**, 4610 (1981).
63. R. A. Rossi and S. M. Palacios, *Tetrahedron*, **49**, 4485 (1993).
64. J. F. Bunnett, *Tetrahedron*, **49**, 4477 (1993).
65. J. Marquet, Z. Jiang, I. Gallardo, A. Batle and E. Cayón, *Tetrahedron Lett.*, **34**, 2801 (1993).
66. (a) B. Aebisher, R. Meuwly and A. Vasella, *Helv. Chim. Acta*, **67**, 12236 (1984).
    (b) N. Kornblum, L. Cheng, T. M. Davies, G. W. Earl, N. L. Holy, R. C. Kerber, N. M. Kestner, J. W. Manthey, M. T. Musser, H. W. Pinnick, D. H. Snow, F. W. Stuchal and R. T. Swiger, *J. Org. Chem.*, **52**, 196 (1987).
67. R. K. Norris and D. Randles, *J. Org. Chem.*, **47**, 1047 (1982).
68. R. K. Norris and D. Randles, *Aust. J. Chem.*, **35**, 1621 (1982).
69. R. K. Norris and T. A. Wright, *Aust. J. Chem.*, **38**, 1107 (1985).
70. N. Kornblum, P. Ackermann and R. T. Swiger *J. Org. Chem.*, **45**, 5294 (1980).
71. B. D. Jacobs, S-J. Kwon, L. D. Field, R. K. Norris, D. Randles, K. Wilson and T. A. Wright, *Tetrahedron Lett.*, **26**, 3495 (1985).
72. L. D. Field, T. W. Hambley, B. D. Jacobs, K. Wilson and R. K. Norris, *Aust. J. Chem.*, **41**, 443 (1988).
73. S. D. Barker and R. K. Norris, *Aust. J. Chem.*, **36**, 81 (1983).
74. F. I. McLure and R. K. Norris, *Aust. J. Chem.*, **40**, 523 (1987).
75. (a) C. D. Beadle and W. R. Bowman, *J. Chem. Res. (S)*, 150 (1985).
    (b) C. D. Beadle, W. R. Bowman and J. Prousek, *Tetrahedron Lett.*, **25**, 4979 (1984).
    (c) J. Prousek, *Collect. Czech. Chem. Commun.*, **53**, 851 (1988).
76. (a) P. J. Newcombe and R. K.Norris, *Aust. J. Chem.*, **32**, 2647 (1979).
    (b) P. Vanelle, S. Ghezali, J. Maldonado, M. P. Crozet, F. Delmas, M. Gasquet and P. Timon-David, *Eur J. Med. Chem.*, in press.
77. M. S. K. Lee, P. J. Newcombe, R. K. Norris and K. Wilson, *J. Org. Chem.*, **52**, 2796 (1987).
78. F. I. Flower, P. J. Newcombe and R. K. Norris, *J. Org. Chem.*, **48**, 4202 (1983).
79. M. C. Harsányi and R. K. Norris, *Aust. J. Chem.*, **40**, 2063 (1987).
80. M. P. Crozet and J. M. Surzur, *Tetrahedron Lett.*, **26**, 1023 (1985).
81. (a) M. P. Crozet and P. Vanelle, *Tetrahedron*, **43**, 5477 (1989).
    (b) P. Vanelle, M. P. Crozet, J. Maldonado and M. Barreau, *Eur. J. Med. Chem.*, **26**, 167 (1991).
82. A. T. O. M. Adebayo, W. R. Bowman and W. G. Salt, *J. Chem. Soc., Perkin Trans. 1*, 2819 (1987).
83. O. Jentzer, P. Vanelle, M. P. Crozet, J. Maldonado and M. Barreau, *Eur. J. Med. Chem.*, **26**, 687 (1991).

1480   R. A. Rossi, A. B. Pierini and A. B. Peñéñory

84. M. P. Crozet, P. Vanelle, O. Jentzer and J. Maldonado, *Comp. Rend. Acad. Sci. Paris, Ser. 2*, **306**, 967 (1988).
85. M. P. Crozet, P. Vanelle, O. Jentzer and J. Maldonado, *Tetrahedron Lett.*, **31**, 1269 (1990).
86. P. Vanelle, J. Maldonado, N. Madadi, A. Gueiffier, J-C. Teulade, J. P. Chopal and M. P. Crozet, *Tetrahedron Lett.*, **31**, 3013 (1990).
87. P. Vanelle, N. Madadi, C. Raubaud, J. Maldonado and M. P. Crozet, *Tetrahedron*, **47**, 5173 (1991).
88. (a) P. Vanelle, N. Madadi, J. Maldonado, L. Giraud, J. F. Sabuco and M. P. Crozet, *Heterocycles*, **32**, 2083 (1991).
    (b) P. Vanelle, S. Ghezali, J. Maldonado, O. Chavignon, A. Gueiffier, J. C. Teulade and M. P. Crozet, *Heterocycles*, **36**, 1541 (1993).
89. P. Vanelle, P. Rathelot, J. Maldonado and M. P. Crozet, personal communication.
90. M. P. Crozet, O. Jentzer and P. Vanelle, *Tetrahedron Lett.*, **28**, 5531 (1987).
91. M. P. Crozet, L. Giraud, J. F. Sabuco, P. Vanelle and M. Barreau, *Tetrahedron Lett.*, **32**, 4125 (1991).
92. M. P. Crozet, L. Giraud, J. F. Sabuco and P. Vanelle, *Tetrahedron Lett.*, **33**, 1063 (1992).
93. M. P. Crozet, P. Vanelle, O, Jentzer, S. Donini and J. Maldonado, *Tetrahedron*, **49**, 11253 (1993).
94. M. P. Crozet, J. F. Sabuco, I. Tamburlin, M. Barreau, L. Giraud and P. Vanelle, *Heterocycles*, **36**, 45 (1993).
95. M. P. Crozet, J. F. Sabuco, M. Barreau and P. Vanelle, *Comp. Rend, Acad. Sci. Paris, Ser. 2*, in press.
96. (a) N. Kornblum and M. Fifolt, *J. Org. Chem.*, **45**, 360 (1980).
    (b) N. Kornblum and M.Fifolt, *Tetrahedron*, **45**, 1301 (1989).
97. F. Ros and J. de la Rosa, *J. Org. Chem.*, **53**, 2868 (1988).
98. M. Takeishi,T. Yoshita, I. Kuroda, N. Takahashi, S. Utsumi, N. Shiozawa and R. Sato, *Reactive Polymers*, **17**, 297 (1992).
99. (a) G. A. Russell and F. Ros, *J. Am. Chem. Soc.*, **107**, 2506 (1985).
    (b) G. A. Russell and F. Ros, *J. Am. Chem. Soc.*, **104**, 7349 (1982).
100. W. R. Bowman, D. S. Brown, C. T. W. Leung and A. P. Stutchbury, *Tetrahedron Lett.*, **26**, 539 (1985).
101. S. D. Barker and R. K. Norris, *Aust. J. Chem.*, **36**, 527 (1983).
102. N. Kornblun and A. S. Erickson, *J. Org. Chem.*, **46**, 1037 (1981).
103. M. P. Crozet and P. Vanelle, *Tetrahedron Lett.*, **26**, 323 (1985).
104. M. P. Crozet, G. Archaimbault, P. Vanelle and R. Nouguier, *Tetrahedron Lett.*, **26**, 5133 (1985).
105. P. Vanelle, J. Maldonado, M. P. Crozet, K. Senouk, F. Dalmas, M. Gasquet and P. Timon-David, *Eur. J. Med. Chem.*, **26**, 709 (1991).
106. G. A. Russell, B. Mudryk and M. Jawdosiuk, *Synthesis*, 62 (1981).
107. R. Beugelmans, A. Lechevallier and H. Rousseau, *Tetrahedron Lett.*, **24**, 1787 (1983).
108. N. Ono, R. Tamura, H. Eto, I. Hamamoto, T. Nakatsuka, J. Hayami and A. Kaji, *J. Org. Chem.*, **48**, 3678 (1983).
109. G. A. Russell, B. Mudryk, F. Ros and M. Jawdosiuk, *Tetrahedron*, **38**, 1059 (1982).
110. G. A. Russell, B. Mudryk, M. Jawdosiuk and Z. Wrobel, *J. Org. Chem.*, **47**, 1879 (1982).
111. R. Beugelmans, A. Lechevallier, D. Kiffer and P. Maillos, *Tetrahedron Lett.*, **27**, 6209 (1986).
112. A. T. O. M. Adebayo, W. R. Bowman and W. G. Salt, *J. Chem. Soc., Perkin Trans. 1*, 1415 (1989).
113. A. T. O. M. Adebayo, W. R. Bowman and W. G. Salt, *Tetrahedron Lett.*, **27**, 1943 (1986).
114. W. R. Bowman and G. D. Richardson, *Tetrahedron Lett.*, **22**, 1551 (1981); S. I. Al Khalil and W. R. Bowman, *Tetrahedron Lett.*, **25**, 461 (1984).
115. A. Amrollah-Madjdabadi, R. Beugelmans and A. Lechevallier, *Tetrahedron Lett.*, **28**, 4525 (1987).
116. W. R. Bowman, D. Rakshit and M. D. Valmas, *J. Chem. Soc., Perkin Trans. 1*, 2327 (1984).
117. G. A. Russell, F. Ros, J. Hershberger and H. Tashtoush, *J. Org. Chem.*, **47**, 1480 (1982).
118. (a) J. March, *Advanced Organic Chemistry*, 4th ed., Wiley, New York, 1992.
    (b) F. A. Carey and R. J. Sundberg, *Advanced Organic Chemistry*, 2nd ed., Part A, Plenum Press, New York, 1984.
119. (a) P. Muller, J. J. Mareda and P. v. R. Schleyer, *Helv. Chim. Acta*, **70**, 1017 (1987).
    (b) E. W. Della, P. M. W. Gill and C. H. Schiesser, *J. Org. Chem.*, **53**, 4354 (1988).

120. S. M. Palacios, A. N. Santiago and R. A. Rossi, *J. Org. Chem.*, **49**, 4609 (1984).
121. R. A. Rossi, S. M. Palacios and A. N. Santiago, *J. Org. Chem.*, **47**, 4654 (1982).
122. S. M. Palacios, R. A. Alonso and R. A. Rossi, *Tetrahedron*, **41**, 4147 (1985).
123. E. R. N. Bornancini, R. A. Alonso and R. A. Rossi, *J. Org. Chem.*, **52**, 2166 (1987).
124. R. A Rossi, A. B. Pierini and G. L. Borosky, *J. Chem. Soc., Perkin Trans. 2*, in press.
125. F. Bordwell, *Acc. Chem. Res.*, **21**, 456 (1988).
126. R. A. Rossi, *J. Chem. Educ.*, **59**, 310 (1982).
127. J. J. Brunet, C. Sidot and P. Caubere, *J. Org. Chem.*, **46**, 3147 (1981).
128. (a) A. E. Lukach, A. N. Santiago and R. A. Rossi, *J. Phys. Org. Chem.*, in press.
     (b) W. Adcock and C. I. Clark, *J. Org. Chem.*, **58**, 7341 (1993).
129. A. N. Santiago, V. S. Iyer, W. Adcock and R. A. Rossi, *J. Org. Chem.*, **53**, 3016 (1988).
130. W. Adcock, V. S. Iyer, G. B. Kok and W. Kitching, *Tetrahedron Lett.*, **24**, 5901 (1983).
131. W. Adcock, V. S. Iyer, W. Kitching and G. Young, *J. Org. Chem.*, **50**, 3706 (1985).
132. W. Adcock and A. N. Abeywickrema, *J. Org. Chem.*, **47**, 2951 (1982).
133. A. N. Santiago, K. Takeuchi, Y. Ohga, M. Nishida and R. A. Rossi, *J. Org. Chem.*, **56**, 1581 (1991).
134. E. C. Ashby, X. Sun and J. L. Duff, *J. Org. Chem.*, **59**, 1270 (1984).
135. W. Adcock and H. Gangodawila, *J. Org. Chem.*, **54**, 6064 (1989).
136. R. C. Bingham and P. v. R. Schleyer, *J. Am. Chem. Soc.*, **93**, 3189 (1971).
137. A. N. Santiago, D. G. Morris and R. A. Rossi, *J. Chem. Soc., Chem. Commun.*, 220 (1988).
138. K. B. Wiberg, *Angew. Chem., Int. Ed. Engl.*, **25**, 312 (1986).
139. P.v. R. Schleyer and R. D. Nicholas, *J. Am. Chem. Soc.*, **83**, 2700 (1961).
140. N. K. Danilova and V. D. Shteingarts, *J. Org. Chem. USSR*, **22**, 701 (1986).
141. R. B. Woodward and R. Hoffmann, *The Conservation of Orbital Symmetry*, Verlag Chemie, Weinheim, 1970.
142. R. Jorristsma, H. Steinberg and T. J. de Boer, *Recl. Trav. Chim. Pays-Bas*, **100**, 184 (1981).
143. J. Arct. B. Migaj and A. Leoncynski, *Tetrahedron*, **37**, 3689 (1981).
144. R. A. Rossi, A. N. Santiago and S. M. Palacios, *J. Org. Chem.*, **49**, 3387 (1984).
145. G. F. Meijs, *J. Org. Chem.*, **49**, 3863 (1984).
146. M. A. Nazareno and R. A. Rossi, *Tetrahedron*, **50**, 9267 (1994).
147. A. N. Santiago and R. A. Rossi, *J. Chem. Res. (S)*, 172 (1988).
148. G. F. Meijs, *Tetrahedron Lett.*, **26**, 105 (1985).
149. G. F. Meijs, *J. Org. Chem.*, **51**, 606 (1986).
150. R. H. Mazur, W. N. White, D. A. Semenov, C. C. Lee, M. S. Silver and J. D. Roberts, *J. Am. Chem. Soc.*, **81**, 4390 (1959).
151. M. A. Nazareno, S. M. Palacios and R. A. Rossi, *J. Phys. Org. Chem.*, **6**, 421 (1993).
152. A. N. Santiago and R. A. Rossi, unpublished results.
153. A. B. Pierini, A. B. Peñéñory and R. A. Rossi, *J. Org. Chem.*, **50**, 2739 (1985).
154. E. R. N. Bornancini, R. A. Alonso and R. A. Rossi, *J. Organometal. Chem.*, **270**, 177 (1984).
155. E. R. N. Bornancini, S. M. Palacios, A. B. Peñéñory and R. A. Rossi, *J. Phys. Org. Chem.*, **2**, 255 (1989).
156. J. S. Duca, Jr., A. B. Pierini and R. A. Rossi, unpublished results.
157. A. B. Peñéñory and R. A. Rossi, unpublished results.
158. E. C. Ashby and J. N. Argyropoulos, *J. Org. Chem.*, **50**, 3274 (1985).
159. E. C. Ashby, R. Gurumurthy and R. W. Ridlehuber, *J. Org. Chem.*, **58**, 5832 (1993).
160. L. M. Tolbert and S. Siddiqui, *J. Org. Chem.*, **49**, 1744 (1984).
161. L. M. Tolbert and S. Siddiqui, *Tetrahedron*, **38**, 1079 (1982).
162. M. A. Fox and N. J. Singletary, *J. Org. Chem.*, **47**, 3412 (1982).
163. L. M. Tolbert and D. P. Martone, *J. Org. Chem.*, **48**, 1185 (1983).
164. I. Königsberg and J. Jagur-Grodzinski, *J. Polym. Sci., Polym. Chem.*, **22**, 2713 (1984).
165. J. M. Tour and E. B. Stephens, *J. Am. Chem. Soc.*, **113**, 2309 (1991).
166. J. M. Tour, E. B. Stephens and J. F. Davis, *Macromolecules*, **25b**, 499 (1992).
167. R. G. Scamehorn and J. F. Bunnett, *J. Org. Chem.*, **42**, 1449 (1977).
168. R. G. Scamehorn, J. M. Hardacre, J. M. Lukanich and L. R. Sharpe, *J. Org. Chem.*, **49**, 4881 (1984).
169. G. L. Borosky, A. B. Pierini and R. A. Rossi, *J. Org. Chem.*, **57**, 247 (1992).
170. C. Galli, *Gazz. Chim. Ital.*, **118**, 365 (1988).
171. J. M. Savéant and A. Thiébault, *J. Electroanal. Chem.*, **89**, 335 (1978).

172. (a) C. Galli and J. F. Bunnett, *J. Org. Chem.*, **49**, 3041 (1984).
    (b) C. Galli and P. Gentili, *J. Chem Soc.*, *Perkin Trans 2*, 1135 (1993).
173. M. v. Leevween and A. McKillop, *J. Chem. Soc., Perkin Trans. 1*, 2433 (1993).
174. M. A. Nazareno and R. A. Rossi, *Tetrahedron Lett.*, **35**, 5185 (1994).
175. E. Austin, C. G. Ferrayoli, R. A. Alonso and R. A. Rossi, *Tetrahedron*, **49**, 4495 (1993).
176. S. C. Dillender, Jr., T. D. Greenwood, M. S. Hendi and J. F. Wolfe, *J. Org. Chem.*, **51**, 1184 (1986).
177. V. Nair and S. D. Chamberlain, *J. Am. Chem. Soc.*, **107**, 2183 (1985).
178. L. Beltran, C. Gálves, M. Prats and J. Salgado *J. Heterocycl. Chem.*, **29**, 905 (1992).
179. D. R. Carber, T. D. Greenwood, J. S. Hubbard, A. P. Komin, Y. P. Sachdeva and J. W. Wolfe, *J. Org. Chem.*, **48**, 1180 (1983).
180. W. C. Danen, T. T. Kensler, J. G. Lawless, M. F. Marcus and M. D. Hawley, *J. Phys. Chem.*, **73**, 4389 (1969); T. Teherani and A. J. Bard, *Acta Chem. Scand.*, **B37**, 413 (1983).
181. C. Galli, *Tetrahedron*, **44**, 5205 (1988).
182. W.-Y. Zhao and Z.-T. Huang, *J. Chem. Soc., Perkin Trans. 2*, 1967 (1991).
183. X.-M. Zhang, D.-L. Yang, Y.-C. Liu, W. Chen and J.-L. Cheng, *Res. Chem. Interm.*, **11**, 281 (1989).
184. (a) X.-M. Zhang, D.-L. Yang and Y.-C. Liu, *J. Org. Chem.*, **58**, 224 (1993).
    (b) X.-M. Zhang, D.-L. Yang, X.-Q. Jia and Y.-C. Liu, *J. Org. Chem.*, **58**, 7350 (1993).
185. J. F. Bunnett and J. C. Sundberg, *J. Org. Chem.*, **41**, 1702 (1976).
186. R. Beugelmans, M. Bois-Choussy and B. Boudet, *Tetrahedron*, **38**, 3479 (1982).
187. M. A. Oturan, J. Pinson, J. M. Savéant and A. Thiébault, *Tetrahedron Lett.*, **30**, 1373 (1989).
188. R. Beugelmans and J. Chastanet, *Tetrahedron*, **49**, 7883 (1993).
189. R. A. Rossi and R. A. Alonso, *J. Org. Chem.*, **45**, 1239 (1980).
190. S. M. Asis, S. M. Palacios and R. A. Rossi, *Bull. Soc. Chim. Fr.*, **130**, 111 (1993).
191. J. D. Stewart, S. C. Fields, S. K. Kochar and H. W. Pinnick, *J. Org. Chem.*, **52**, 2110 (1987).
192. R. A. Alonso, C. H. Rodriguez and R. A. Rossi, *J. Org. Chem.*, **54**, 5983 (1989).
193. B.-Q. Wu, F.-W. Zeng, Y. Zhao and G.-S. Wu, *Chinese J. Chem.*, **10**, 253 (1992).
194. G. A. Lotz, S. M. Palacios and R. A. Rossi, *Tetrahedron Lett.*, **35**, 7711 (1994).
195. G. A. Russell and A. R. Metcalfe, *J. Am. Chem. Soc.*, **101**, 2359 (1979).
196. D. Veltwisch and K. D. Asmus, *J. Chem. Soc.*, *Perkin Trans. 2*, 1143 (1982).
197. C. Amatore, M. Gareil, M. A. Oturan, J. Pinson, J. M. Savéant and A. Thiébault, *J. Org. Chem.*, **51**, 3757 (1986).
198. (a) J. F. Bunnett and B. F. Gloor, *J. Org. Chem.*, **38**, 4156 (1973).
    (b) R. A. Rossi, R. H. de Rossi and A. B. Pierini, *J. Org. Chem.*, **44**, 2662 (1979).
199. R. A. Rossi, R. H. de Rossi and A. F. López, *J. Org. Chem.*, **41**, 3367, 3371 (1976).
200. M. Hamana, G. Iwasaki and S. Saeki, *Heterocycles*, **17**, 177 (1982).
201. M. P. Moon, A. P. Komin, J. F. Wolfe and G. F. Morris, *J. Org. Chem.*, **48**, 2392 (1983).
202. C. K. F. Hermann, Y. P. Sachdeva and J. F. Wolfe, *J. Heterocycl. Chem.*, **24**, 1061 (1987).
203. R. A. Alonso, E. Austin and R. A. Rossi, *J. Org. Chem.*, **53**, 6065 (1988).
204. C. Combellas, M. Lequan, R. M. Lequan, J. Simon and A. Thiébault, *J. Chem. Soc., Chem. Commun.*, 542 (1990).
205. C. Amatore, C. Combellas, J. Pinson, M. A. Oturan, S. Robveille, J. M. Savéant and A. Thiébault, *J. Am. Chem. Soc.*, **107**, 4846 (1985).
206. C. Amatore, M. A. Oturan, J. Pinson, J. M. Savéant and A. Thiébault, *J. Am. Chem. Soc.*, **107**, 3451 (1985).
207. C. Amatore, C. Combellas, S. Robveille, J. M. Savéant and A. Thiébault, *J. Am. Chem. Soc.*, **108**, 4754 (1986).
208. C. Amatore, J. M. Savéant, C. Combellas, S. Robveille and A. Thiébault, *J. Electroanal. Chem.*, **184**, 25 (1985).
209. V. L. Ivanov and A. Kherbst, *J. Org. Chem. USSR*, **25**, 1542 (1989).
210. J. J. Brunet, C. Sidot and P. Caubere, *J. Organometal. Chem.*, **204**, 229 (1980).
211. J. J. Brunet, C. Sidot and P. Caubere, *Tetrahedron Lett.*, **22**, 1013 (1981).
212. J. J. Brunet, C. Sidot and P. Caubere, *J. Org. Chem.*, **48**, 1166 (1983).
213. T. Kashimura, K. Kudo, S. Mori and N. Sugita, *Chem. Lett.*, 483 (1986).
214. K. Kudo, T. Shibata, T. Kashimura, S. Mori and N. Sugita, *Chem. Lett.*, 577 (1987).
215. J. J. Brunet and M. Taillefer, *J. Organometal. Chem.*, **348**, C5 (1988).
216. (a) J. J. Brunet and M. Taillefer, *J. Organometal. Chem.*, **361**, C9 (1989).
    (b) J. J. Brunet, A. Montauzon and M. Taillefer, *Organometallics*, **10**, 341 (1991).

217. K. R. Wursthorn, H. G. Kuivila and G. F. Smith, *J. Am. Chem. Soc.*, **100**, 2789 (1978).
218. J. P. Quintard, S. Hauvette-Frey and M. Pereyre, *J. Organomet. Chem.*, (a) **112**, C11 (1976). (b) **159**, 147 (1978).
219. C. C. Yammal, J. C. Podestá and R. A. Rossi, *J. Org. Chem.*, **57**, 5720 (1992).
220. C. C. Yammal and R. A. Rossi, unpublished results.
221. R. A. Alonso, A. Bardon and R. A. Rossi, *J. Org. Chem.*, **49**, 3584 (1984).
222. M. J. Tremelling and J. F. Bunnett, *J. Am. Chem. Soc.*, **102**, 7375 (1980).
223. M. T. Baumgartner, Ph.D. Thesis, National University of Córdoba (1990).
224. A. B. Pierini, M. T. Baumgartner and R. A. Rossi, *Tetrahedron Lett.*, **28**, 4653 (1987).
225. M. Chahma, C. Combellas, H. Marzouk and A. Thiébault, *Tetrahedron Lett.*, **32**, 6121 (1991).
226. M. Chahma, C. Combellas and A. Thiébault, *Synthetic*, 366 (1994).
227. M. Medicbélle, M. A. Oturan, J. Pinson and J. M. Savéant, *Tetrahedron Lett.*, **34**, 3412 (1993).
228. R. N. Henrie II and W. H. Yeager, *Heterocycles*, **35**, 415 (1993).
229. W.-Y. Zhao, Y. Liu and Z.-T. Huang, *Synthesis Commun.*, **23**, 591 (1993).
230. C. Z. Xia, Z. B. Chen and Z. Zhang, *Chinese Chem. Lett.*, **2**, 131 (1991).
231. C. Z. Xia, Z. B. Chen and Z. Zhang, *Chinese Chem. Lett.*, **2**, 429 (1991).
232. J. E. Swartz and J. F. Bunnett, *J. Org. Chem.*, **44**, 340 (1979).
233. J. E. Swartz and J. F. Bunnett, *J. Org. Chem.*, **44**, 4673 (1979).
234. E. Austin, R. A. Alonso and R. A. Rossi, *J. Chem. Res.*, 190 (1990).
235. E. Austin, R. A. Alonso and R. A. Rossi, *J. Org. Chem.*, **56**, 4486 (1991).
236. E. R. N. Bornancini and R. A. Rossi, *J. Org. Chem.*, **55**, 2332 (1990).
237. v. H. Matschiner, A. Tzschach and A. Steinert, *Z. Anorg. Allg. Chem.*, **373**, 237 (1970).
238. J. F. Bunnett and X. Creary, *J. Org. Chem.*, **39**, 3612 (1974); J. F. Bunnett and R. H. Weiss, *Org. Synth.*, **58**, 134 (1978).
239. R. Beugelmans and M. Bois-Choussy, *Tetrahedron*, **42**, 1381 (1986).
240. J. J. Bulot, E. E. Aboujaonfr, N. Collignon and R. Savignac, *Phosphorus and Sulfur*, **21**, 197 (1984).
241. J. A. Kampmeier and T. W. Nalli, *J. Org. Chem.*, **58**, 943 (1993).
242. G. A. Tomaselli and J. F. Bunnett, *J. Org. Chem.*, **57**, 2710 (1992); G. A. Tomaselli, J. Cui, Q. Chen and J. F. Bunnett, *J. Chem. Soc., Perkin Trans. 2*, 9 (1992).
243. C. Amatore, J. Chaussard, J. Pinson, J. M. Savéant and A. Thiébault, *J. Am. Chem. Soc.*, **101**, 6012 (1979).
244. R. A. Rossi and J. F. Bunnett, *J. Org. Chem.*, **39**, 3020 (1973).
245. R. A. Rossi and A. B. Pierini, *J. Org. Chem.*, **45**, 2914 (1980).
246. M. F. Semmelhack and T. Bargar, *J. Am. Chem. Soc.*, **102**, 7765 (1980).
247. C. Amatore, C. Combellas, J. Pinson, J. M. Savéant and A. Thiébault *J. Chem. Soc., Chem. Commun.*, 7 (1988).
248. N. Alam, C. Amatore, C. Combellas, J. Pinson, J. M. Savéant, A. Thiébault and J. N. Verpeaux *J. Org. Chem.*, **53**, 1496 (1988).
249. C. Combellas, H. Gautier, J. Simon, A. Thiébault, F. Tournilhac, M. Barzoukas, D. Josse, I. Ledoux, C. Amatore and J. N. Verpeaux, *J. Chem. Soc., Chem. Commun.*, 203 (1988).
250. G. Bacquet, P. Bassoul, C. Combellas, J. Simon, A. Thiébault and F. Tournilhac, *Adv. Materials*, **2**, 311 (1990).
251. M. Tashiro, *Synthesis*, 921 (1979).
252. P. Boy, C. Combellas and A. Thiébault, *Synlett*, **12**, 923 (1991).
253. N. Alam, G. Amatore, C. Combellas, A. Thiébault and J. N. Verpeaux, *Tetrahedron Lett.*, **28**, 6171 (1987).
254. C. Combellas, C. Suba and A. Thiébault, *Tetrahedron Lett.*, **33**, 4923 (1992).
255. C. Combellas, C. Suba and A. Thiébault, *Tetrahedron Lett.*, **33**, 5741 (1992).
256. C. Combellas, M. A. Petit, A. Thiébault, G. Froyer and D. Bosc, *Makromol. Chem.*, **193**, 2445 (1992).
257. P. Boy, C. Combellas, A. Thiébault, C. Amatore and A. Jutand, *Tetrahedron Lett.*, **33**, 491 (1992).
258. C. Combellas, H. Marzouk, C. Suba and A. Thiébault, *Synthesis*, 788 (1993).
259. C. Combellas, C. Suba and A. Thiébault, *Tetrahedron Lett.*, **35**, 5217 (1994).
260. P. Boy, C. Combellas, S. Fielding and A. Thiébault, *Tetrahedron Lett.*, **32**, 6705 (1991).
261. R. Beugelmans and M. Bois-Choussy, *Tetrahedron Lett.*, **29**, 1289 (1988).
262. (a) A. B. Pierini, M. T. Baumgartner and R. A. Rossi, *Tetrahedron Lett.*, **29**, 3451 (1988). (b) A. B. Pierini, M. T. Baumgartner and R. A. Rossi, *J. Org. Chem.*, **56**, 580 (1991).

263. R. Beugelmans, M. Bois-Choussy and O. Tang, *Tetrahedron Lett.*, **29**, 1705 (1988).
264. R. Beugelmans and J. Chastanet, *Tetrahedron Lett.*, **32**, 3487 (1991).
265. R. Beugelmans and M. Bois-Choussy, *J. Org. Chem.*, **56**, 2518 (1991).
266. R. Beugelmans, M. Bois-Choussy, P. Gayral and M. C. Rigothier, *Eur. J. Med. Chem.*, **23**, 539 (1988).
267. V. Percec, R. S. Clough, P. L. Rinaldi and V. E. Liman, *Macromolecules*, **24**, 5889 (1991).
268. V. Percec, R. S. Clough, M. Grigoras, P. L. Rinaldi and V. E. Litman, *Macromolecules*, **26**, 3650 (1993).
269. V. Percec, J. H. Wang and R. S. Clough, *Makromol. Chem. Macromol. Symp.*, **54/55**, 275 (1992).
270. J. F. Bunnett and X. Creary, *J. Org. Chem.*, **39**, 3173 (1974).
271. M. Julliard and M. Chanon, *J. Photochem.*, **34**, 231 (1986).
272. J. Pinson and J. M. Savéant, *J. Am. Chem. Soc.*, **100**, 1506 (1978).
273. C. Amatore, J. Pinson, J. M. Savéant and A. Thiébault, *J. Am. Chem. Soc.*, **104**, 817 (1982).
274. C. Thobie-Gautier, M. Genesty and C. Degrand, *J. Org. Chem.*, **56**, 3452 (1991).
275. J. A. Zoltewicz and G. A. Locko, *J. Org. Chem.*, **48**, 4214 (1983).
276. S. Kondo, M.Nakanishi and K. Tsuda, *J. Heterocycl. Chem.*, **21**, 1243 (1984).
277. M. Novi, G. Garbarino, G. Petrillo and C. Dell'Erba, *J. Org. Chem.*, **52**, 5382 (1987).
278. J. F. Bunnett and X. Creary, *J. Org. Chem.*, **40**, 3740 (1975).
279. C. Amatore, R. Beugelmans, M. Bois-Choussy, C. Combellas and A. Thiébault, *J. Org. Chem.*, **54**, 5683 (1989).
280. W. R. Bowman and P. F. Taylor, *J. Chem. Soc., Perkin Trans. 1*, 919 (1990).
281. D. W. Hobbs and W. C. Still, *Tetrahedron Lett.*, **28**, 2805 (1987).
282. (a) J. F. Bunnett and X. Creary, *J. Org. Chem.*, **39**, 3611 (1974).
    (b) J. N. Nasielski, C. Moucheron and R. Nasielski-Hinkens, *Bull. Soc. Chem. Belg.*, **101**, 491 (1992).
283. I. A. Rybakova and R. I. Shektman, *Izv. Akad. Nauk SSSR, Ser. Khim.*, 833 (1987); *Chem. Abstr.*, **108**, 221560h (1988).
284. G. Petrillo, M. Novi, G. Garbarino and C. Dell'Erba, *Tetrahedron*, **42**, 4007 (1986).
285. M. Novi, G. Petrillo and M. L. Sartirana, *Tetrahedron Lett.*, **27**, 6129 (1986).
286. A. C. Archer and P. A. Lovell, *Makromol. Chem. Macromol. Symp.*, **54/55**, 257 (1992).
287. R. A. Rossi and S. M. Palacios, *J. Org. Chem.*, **46**, 5300 (1981).
288. R. Beugelmans, M. Bois-Choussy and B. Boudet, *Tetrahedron*, **39**, 4153 (1983).
289. G. Illuminati and F. Stegel, *Adv. Heterocycl. Chem.*, **34**, 306 (1983).
290. J. A. Zoltewicz and T. M. Oestreich, *J. Org. Chem.*, **56**, 2805 (1991).
291. R. Beugelmans and H. Ginsburg, *Tetrahedron Lett.*, **28**, 413 (1987).
292. P. Sing and G. Arora, *Tetrahedron*, **44**, 2625 (1988).
293. S. D. Pastor, *Helv. Chim. Acta*, **71**, 859 (1988).
294. (a) K. Yui, Y. Aso, T. Otsubo and F. Ogura, *Chem. Lett.*, 551 (1986).
    (b) V. Z. Annenkova, L. M. Antonik, I. V. Shafeeva, T. L. Vakul'skaya, U. Y. Vitkorskii and M. G. Voronkov, *Vysokomd. Soedin, Ser. B*, **28**, 137 (1986); *Chem. Abstr.*, **105**, 6848r (1986).
295. R. A. Rossi and A. B. Peñéñory, *J. Org. Chem.*, **46**, 4580 (1981).
296. D. J. Sandman, M. Rubner and L. Samuelson, *J. Chem. Soc., Chem. Commun.*, 1133 (1982).
297. D. J. Sandman, J. C. Stark, L. A. Acampora and P. Gagne, *Organometallics*, **2**, 549 (1983).
298. J. C. Stark, R. Reed, L. A. Acampora, D. J. Sandman, S. Jansen, M. T. Jones and B. M. Foxman, *Organometallics*, **3**, 732 (1984).
299. K. Yui, Y. Aso, T. Otsubo and F. Ogura, *Chem. Lett.*, 551 (1986).
300. H. Suzuki and M. Inouyu, *Chem. Lett.*, 389 (1985).
301. H. Suzuki, P. Seetharamaier, I. Masahiko and T. Ogawa, *Synthesis*, 468 (1989).
302. H. Suzuki and T. Nakamura, *Synthesis*, 549 (1992).
303. C. Thobie-Gautier and C. Degrand, *J. Org. Chem.*, **56**, 5703 (1991).
304. (a) C. Thobie-Gautier, C. Degrand, M. Nour and P.-L.Compagnon, *J. Electroanal. Chem.*, **344**, 383 (1993).
    (b) M. Genesty, C. Thobie, A. Gautier and C. Degrand, *J. Appl. Electrochem.*, **23**, 1125 (1993).
305. A. B. Pierini, A. B. Péñénory and R. A. Rossi, *J. Org. Chem.*, **49**, 486 (1984).
306. A. B. Pierini and R. A. Rossi, *J. Org. Chem.*, **44**, 4667 (1979).
307. A. B. Peñéñory and R. A. Rossi, *J. Phys. Org. Chem.*, **3**, 266 (1990).
308. See Reference 7, p. 34.

309. (a) C. Degrand, *J. Chem. Soc., Chem. Commun.*, 1114 (1986).
     (b) C. Degrand, *J. Org. Chem.*, **52**, 1421 (1987).
310. (a) C. Degrand, *J. Electroanal, Chem.*, **238**, 239 (1987).
     (b) C. Degrand, R. Prest and M. Nour, *Phosphorus and Sulfur*, **38**, 201 (1988).
311. C. Degrand, R. Prest and P. L. Compagnon, *J. Org. Chem.*, **52**, 5229 (1987).
312. C. Degrand, *Tetrahedron*, **46**, 5237 (1990).
313. R. G. Compton, B. A. Coles, M. B. G. Pilkington and D. Bethell, *J. Chem. Soc., Faraday Trans.*, **86**, 663 (1990).
314. R. G. Compton, A. C. Fisher, R. G. Wellington, D. Bethell and P. Lederer, *J. Phys. Chem.*, **45**, 4749 (1991).
315. D. Bethell, R. G. Compton and R. G. Wellington, *J. Chem. Soc., Perkin Trans. 2*, 147 (1992).
316. V. D. Parker, *Acta Chem. Scand.*, **B 35**, 533 (1981); Q. Y. Chen and M. F. Chen, *J. Chem. Soc., Perkin Trans. 2*, 1071 (1991).
317. A. N. Abeywickrema and A. L. J. Beckwith, *J. Org. Chem.*, **52**, 2568 (1987).
318. R. R. Bard and J. F. Bunnett, *J. Org. Chem.*, **45**, 1546 (1980).
319. R. Beugelmans and G. Roussi, *J. Chem. Soc., Chem. Commun.*, 950 (1979).
320. R. Beugelmans and G. Roussi, *Tetrahedron*, **37**, 393 (1981).
321. K. Boujlel, J. Simonet, G. Roussi and R. Beugelmans, *Tetrahedron Lett.*, **23**, 173 (1982).
322. R. Fontan, C. Galvez and P. Viladoms, *Heterocycles*, **16**, 1473 (1981).
323. R. Beugelmans, B. Boudet and L. Quintero, *Tetrahedron Lett.*, **21**, 1943 (1980).
324. L. Estel, F. Marsais and G. Quéguiner, *J. Org. Chem.*, **53**, 2740 (1988).
325. R. Beugelmans and H. Ginsburg, *J. Chem. Soc., Chem. Commun.*, 508 (1980).
326. R. Beugelmans and M. Bois-Choussy, *Heterocycles*, **26**, 1863 (1987).
327. R. Beugelmans, J. Chastanet and G. Roussi, *Tetrahedron Lett.*, **23**, 2313 (1982); *Tetrahedron*, **40**, 311 (1984).
328. R. Beugelmans, J. Chastanet, H. Ginsburg, L. Quintero-Cortes and G. Roussi, *J. Org. Chem.*, **50**, 4933 (1985).
329. R. Beugelmans, H. Ginsburg and M. Bois-Choussy, *J. Chem. Soc., Perkin Trans. 1*, 1149 (1982).
330. R. Beugelmans and M. Bois-Choussy, *Synthesis*, 729 (1981).
331. R. Beugelmans and M. Bois-Choussy, *Tetrahedron*, **48**, 8285 (1992).
332. R. Beugelmans and H. Ginsburg, *Heterocycles*, **23**, 1197 (1985).
333. R. Beugelmans, M. Bois-Choussy and Q. Tang, *J. Org. Chem.*, **52**, 3880 (1987); *Tetrahedron*, **45**, 4203 (1989).
334. J. F. Bunnett and C. Galli, *J. Chem. Soc., Perkin Trans. 1*, 2515 (1985).
335. M. F. Semmelhack and T. M. Bargar, *J. Org. Chem.*, **42**, 1481 (1977).
336. J. F. Wolfe, M. C. Sleevi and R. R. Goehring, *J. Am. Chem. Soc.*, **102**, 3646 (1980).
337. G. S. Wu, T. C. Tao, J. J. Cao and X. L. Wei, *Acta Chem. Sinica*, **50**, 614 (1992).
338. R. R. Goehring, Y. P. Sachdeva, J. S. Pisipati, M. C. Sleevi and J. F. Wolfe, *J. Am. Chem. Soc.*, **107**, 435 (1985).
339. R. R. Goehring, *Tetrahedron Lett.*, **33**, 6045 (1992).
340. (a) S. Usui and Y. Fukazawa, *Tetrahedron Lett.*, **28**, 91 (1987).
     (b) Y. Fukasawa, Y. Takeda, S. Usui and M. Kodama, *J. Am. Chem. Soc.*, **110**, 7842 (1988).
341. W. R. Bowman, H. Heaney and P. H. G. Smith, *Tetrahedron Lett.*, **23**, 5993 (1982).
342. J. F. Bunnett and P. Sing, *J. Org. Chem.*, **46**, 5022 (1981).
343. A. B. Pierini, M. T. Baumgartner and R. A. Rossi, *J. Org. Chem.*, **52**, 1089 (1987).
344. M. T. Baumgartner, A. B. Pierini and R. A. Rossi, *J. Org. Chem.*, **58**, 2593 (1993).
345. C. J. Gilmore, D. D. MacNicol, A. Murphy and M. A. Russell, *Tetrahedron Lett.*, **25**, 4303 (1984).
346. M. T. Baumgartner, A. B. Pierini and R. A. Rossi, *Tetrahedron Lett.*, **33**, 2323 (1992).
347. P. K. Freeman and R. Srinivasa, *J. Org. Chem.*, **51**, 3939 (1986).
348. R. Beugelmans, M. Bois-Choussy, J. Chastanet, M. Legleuher and J. Zhu, *Heterocycles*, **36**, 2723 (1993)
349. G. Petrillo, M. Novi, C. Dell'Erba and C. Tavani, *Tetrahedron*, **47**, 9297 (1991).
350. H. G. Theuns, H. B. M. Lenting, C. A. Salemink, H. Tanaka, M. Shibata, K. Ito and R. J. J. Ch. Lousberg, *Heterocycles*, **22**, 2007 (1984).
351. L. Strekowski, M. Lipowska and G. Patonay, *J. Org. Chem.*, **57**, 4578 (1992).
352. J. F. Bunnett, X. Creary and J. E. Sundberg, *J. Org. Chem.*, **41**, 1707 (1976).
353. C. Galli and P. Gentili, *J. Chem. Soc., Chem. Commun.*, 570 (1993).

CHAPTER **25**

Biochemistry of halogenated organic compounds

KENNETH L. KIRK*

Department of Health & Human Services, National Institutes of Health, Bethesda, Maryland 20892, USA

---

*This chapter was written by the author in his private capacity. No official support nor endorsement is intended nor should be inferred.

Supplement D2: The chemistry of halides, pseudo-halides and azides
Edited by S. Patai and Z. Rappoport © 1995 John Wiley & Sons Ltd.

I. INTRODUCTION

The replacement of a carbon–hydrogen or carbon–hydroxyl bond with a carbon–halogen bond has proven to be a very valuable strategy in the development of analogues of biologically active compounds as potential medicinal and pharmacological agents. The biochemistry of every class of organic compound—carboxylic acids, alcohols, aldehydes and ketones, amino acids, amines, steroids and other terpenoids, nucleosides and nucleotides, carbohydrates, etc.—is well represented by the study of halogenated analogues, often as a central theme. The classic research on fluoroacetate, fluorocorticosteroids and 5-fluorouracil are obvious examples wherein halogenated analogues have been the keys to important biochemical discoveries. Accordingly, there is a massive volume of important literature related directly to the biochemistry of synthetic halogenated biomolecules. However, as pervasive as these analogue studies have been in biochemistry and pharmacology, they represent only one aspect of the biochemistry of halogenated compounds. The effectiveness of halogenated compounds as pesticides and the many industrial applications of polyhalogenated compounds have been very beneficial to society. However, toxic properties and the persistence of many of these compounds in the biosphere have led to serious environmental problems. An active and productive area of biochemical and pharmacological research has been directed towards defining mechanisms of action of pesticides and mechanisms of toxicity of polyhalogenated compounds.Likewise, medical applications of halogenated compounds as anesthetics and as artificial oxygen carriers have been accompanied by important biochemical implications. Finally, while the chemical synthesis of halogenated organic compounds has produced an enormous catalogue of new materials that have had major impact on biochemistry, the versatile halogenating mechanisms of nature likewise are responsible for a wide variety of halogenated organic compounds. The role of iodine in thyroid hormones, both as determinants of receptor interactions and in regulatory functions, is becoming increasingly better understood. In addition, we now recognize that, in early work, both the number of haloperoxidases in nature and the number of metabolites produced by these enzymes had been grossly underestimated.

In this chapter, representative examples of these different themes encompassing the biochemistry of halogenated organic compounds will be described. Where appropriate, the topics will be placed in perspective with brief historical development. Biological halogenation will be discussed first, and representative metabolites described. Continuing the topic of naturally occurring halogenated organic compounds, recent research on the biochemistry of thyroid hormones will be considered next. Halogenated examples from various classes of compounds will then be given, and current trends in this research will be developed, particularly in the cases where halogenated molecules have been the keys to the design of new medicinal agents. Finally, the biochemistry of halogenated pesticides, perhalogenated industrial compounds and biological dehalogenation mechanisms wil be discussed. Material will be taken primarily from the literature published after the previous review that appeared in an earlier part of this Series[1]. Certain of these topics are discussed in considerably greater detail in a recently published two-volume monograph[2,3].

II. BIOHALOGENATION

The belief had been common until recently that the occurrence of covalently bound halogen in living organisms was an infrequent event[1]. The thyroid hormones (Section III) represent an obvious and important example. Now it is recognized that naturally occurring halogen-containing compounds are widespread in nature, particularly as secondary metabolites of bacteria, fungi and marine organisms. Likewise, increasing numbers of haloperoxidases, once thought to be rare, are being identified in nature. As a reflection of the potent biological activity of many of these halogenated metabolites, they often play important roles as toxic chemical deterrents to predators. The recognition that many of these secondary metabolites—both halogenated and non-halogenated—possess potent anti-fungal, anti-bacterial and/or anti-tumor activity has prompted extensive research into isolation and identification of compounds from these sources as potential chemotherapeutic agents.

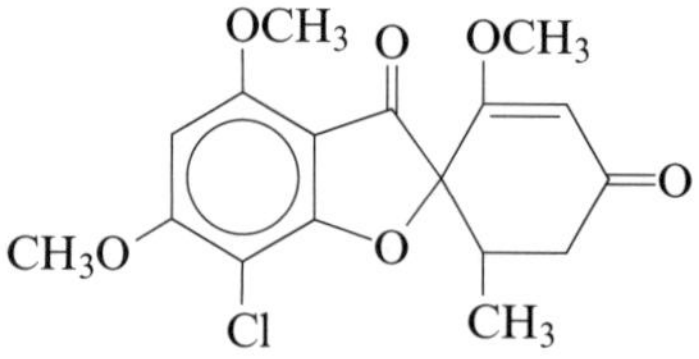

Griseofulvin

(1)

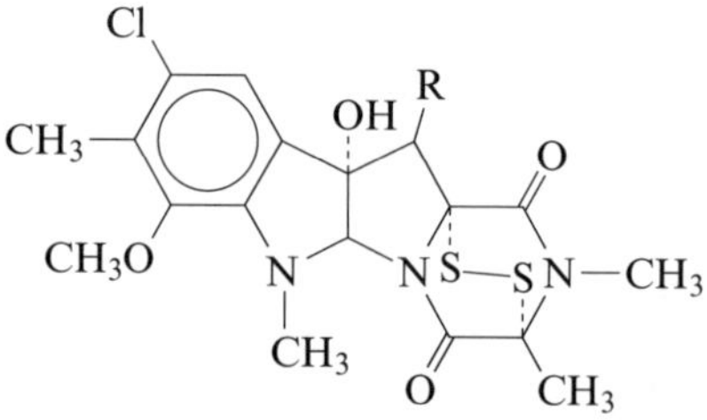

Sporidesmin: R = OH
Sporidesmin b:R = H

(2)

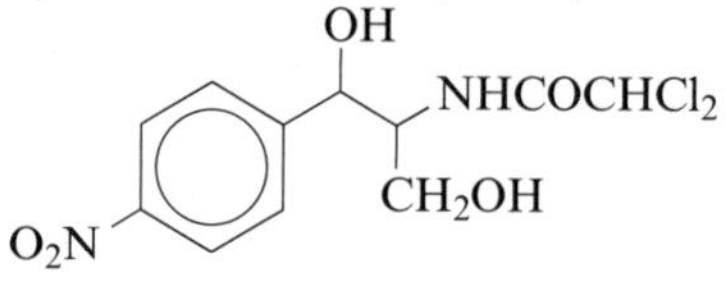

Chloramphenicol

(3)

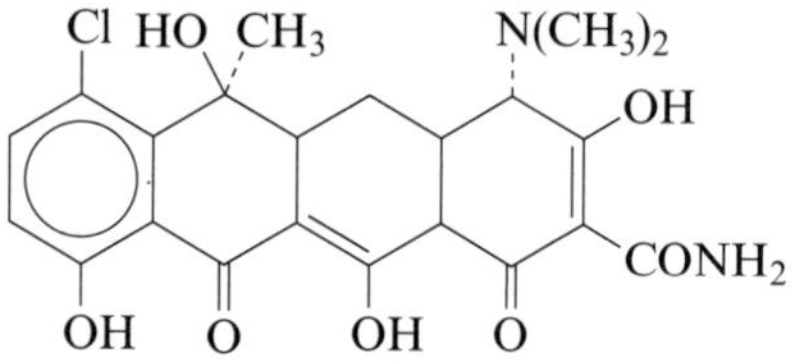

Chlortetracycli ne

(4)

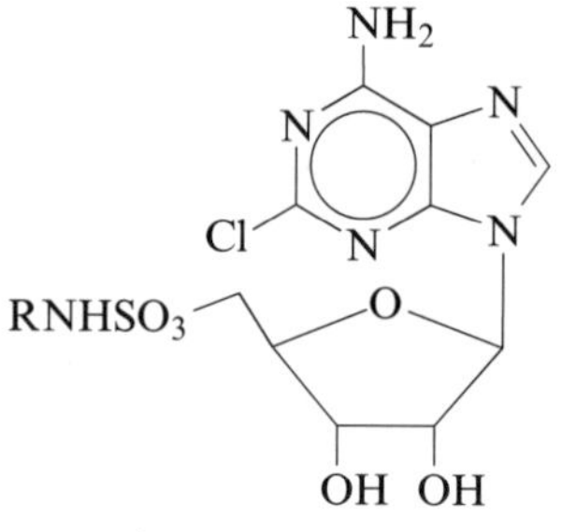

(5) Ascamycin; R = NH₂CH(CH₃)CO —

A. Examples of Halometabolites

Chlorometabolites are widely distributed in fungi, early examples of which chlorometabolites are the antifungal agent griseofulvin (**1**) and the toxic sporidesmins (e.g. **2**). Chloramphenicol (**3**), chlortetracycline (**4**) and ascamycin (**5**) are examples of chlorinated antibiotics produced by bacteria.

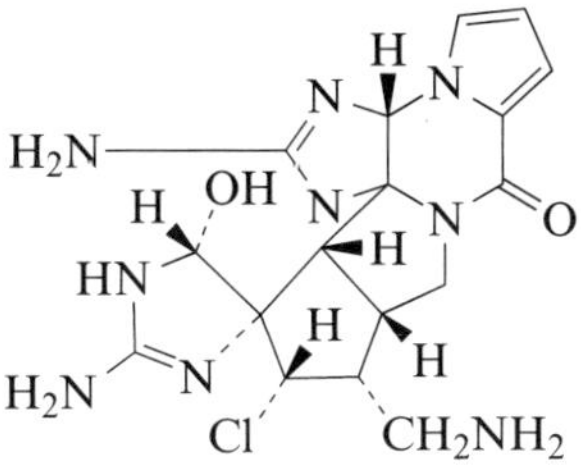

(**6**)

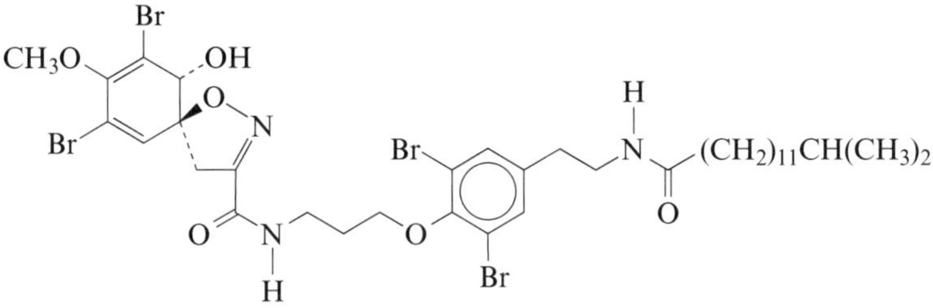

(**7**)

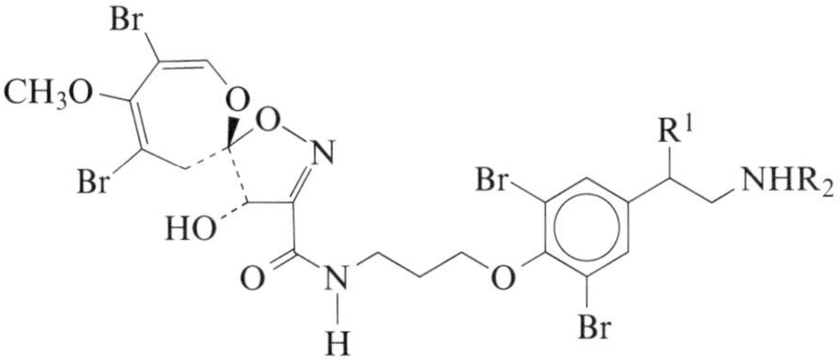

(**8**)

(a) $R^1 = R^2 = H$
(b) $R^1 = OH, R^2 = H$
(c) $R^1 = OH, R^2 = CH_3$
(d) $R^1 = OH, R^2 = CO(CH_2)_{11}CH(CH_3)_2$

(e) $R^1 = H, R^2 = CH=$

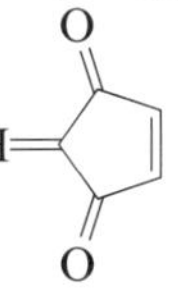

Sea water contains high concentrations of halides (Cl⁻, 19,000 mg/liter; Br⁻, 65 mg/liter; I⁻/IO₃⁻, 5×10^{-4} mg/liter)[4]. It thus is not surprising that marine plants and animals are particularly rich sources of halometabolites, including halogenated terpenes, acetogenins and alkaloids. Examples of complex biologically active, halogen-containing compounds include the guanidine-derived palau'amine **6** isolated from a sponge[5], and the bromotyrosine derivatives, araplysillin-II **(7)**[6] and psammaplysins **(8)**[7], likewise produced by marine sponges. Volatile halogenated compounds, including the simple halomethanes **9–13**, are produced by tropical seaweeds as well as by macroalgae from temperate zones[8,9]. Volatile halocarbons produced by algae are released into the atmosphere at rates of nanograms to micrograms per gram of dry algae per day. Based on this and an estimate of the global macroalgal biomass, the estimate has been made that as much as 10^{10} g of volatile organobromine may be released into the atmosphere per year from this source, an amount that would constitute a major contribution to the atmospheric accumulation of this type of compound[9].

CH_3Br	$CHBr_2I$	$CHBr_2Cl$	$CHCl_3$	CCl_4
(9)	**(10)**	**(11)**	**(12)**	**(13)**

Participation of Br⁻ in marine terpene biosynthesis (Section II.C) is reflected in the impressive number (several hundred) of halogenated terpenes, some with useful biological activities, that have been identified. For example, the polyhalogenated sesquiterpenes isoobtusol **(14)** and isoobtusol acetate **(15)**, isolated from the red alga *Laurencia obtussa*, showed potent activity against HeLa 229 human carcinoma cells[10,11].

Halogenated marine prostaglandins, including (7E)-punaglandin-4 **(16)** (named after *puna*, Hawaiian for coral), and chloro-, bromo- and iodovulones **(17)**, recently isolated from coral, show potent anti-tumor activity[12,13]. Examples of synthetic halogenated prostaglandins will be discussed in Section IV.D.

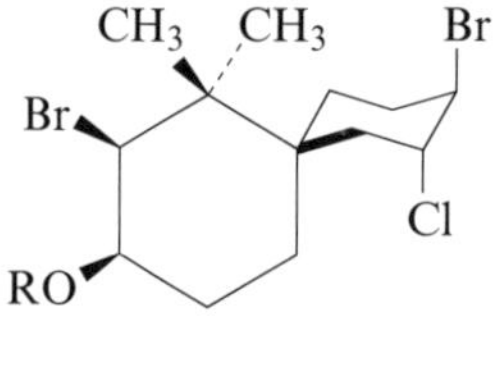

(14) R = H
(15) R = Ac

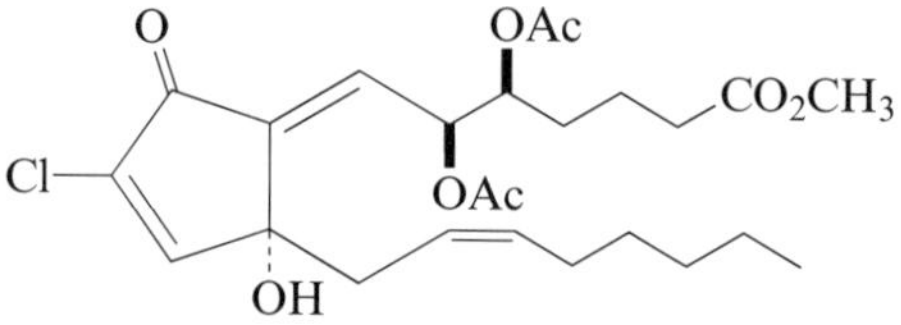

(16) (7E)-Punaglandin

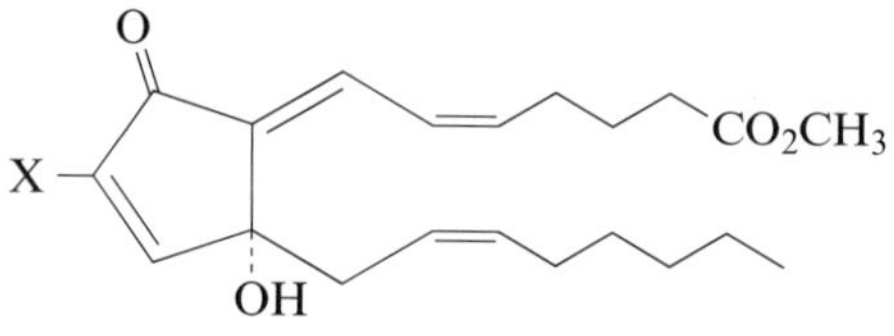

(17) X = Cl, Chlorovulone I
X = Br, Bromovulone I
X = I, Iodovulone I

B. Haloperoxidases

1. Function

Haloperoxidases catalyze the hydroperoxide-dependent oxidation of Cl^-, Br^- and/or I^- to electrophilic halogenating species that halogenate organic substrates (equation 1). As predicted by the relative ease of oxidation of halide ions ($I^- > Br^- > Cl^- >> F^-$), chloroperoxidases oxidize Cl^-, Br^- and I^-, bromoperoxidases oxidize Br^- and I^-, iodoperoxidase oxidize only I^- while no peroxidase can oxidize F^- [14].

$$H_2O_2 + AH + H^+ + X^- \xrightarrow{\text{Haloperoxidase}} AX + 2H_2O \qquad (1)$$

2. Occurrence

Horseradish peroxidase (HPO) (an iodoperoxidase), lactoperoxidase (a bromoperoxidase) and myeloperoxidase (a chloroperoxidase) were the only peroxidases known in the early 1950s. Thyroid peroxidase (TPO), chloroperoxidase (CPO) from the fungus *Caldarimyces fumago* and eosinophil peroxidase (EPO) from white blood cells became known a decade later. Research over the past several years has identified a large number of different haloperoxidases from several sources, with marine organisms being a particularly rich source. TPO plays an obvious and important role in mammalian metabolism (Section III). Likewise, myloperoxidase (MPO), present in neutrophils, and EPO both catalyze the formation of highly reactive hypohalite species, and thus have important functions in mammalian defense mechanisms against invading microorganisms. The recent report that MPO, though the intermediate formation of HOBr, can catalyze the formation of the highly microbicidal singlet oxygen, provided the first identified mammalian physiological role for Br^- (equation 2) [14,15].

$$H_2O_2 + Br^- + H^+ \longrightarrow HOBr + H_2O$$

$$H_2O_2 + HOBr \longrightarrow H_2O + H^+ Br^- + \boxed{O_2(^1\Delta_g)} \qquad (2)$$

3. Structure

Most haloperoxidases, including, for example, EPO, CPO, LPO, TPO and HPO, consist of a glycoprotein and ferriprotoporphyrin IX (**18**) as a heme component. The metal present carries out the heterolytic cleavage of H_2O_2 and stores one oxidizing equivalent, the

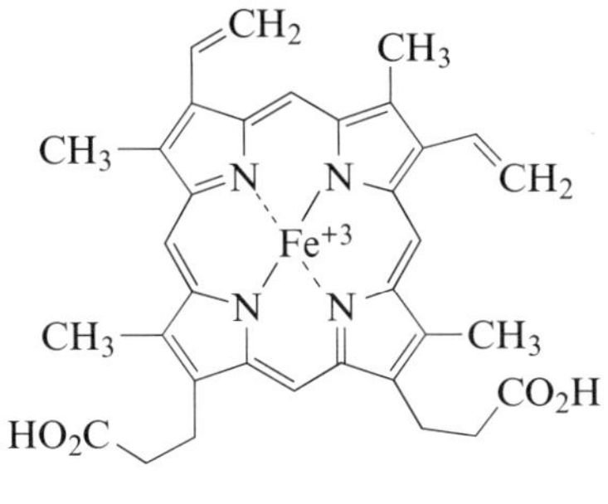

(**18**)

Ferriprotoporphyrin IX

K. L. Kirk

Mechanism A: Hypohalous acid (HOX) Intermediate

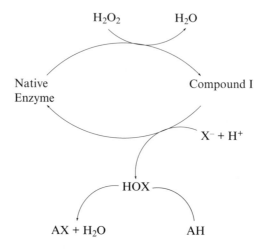

Mechanism B: Enzyme-bound (Compound EOX) Intermediate

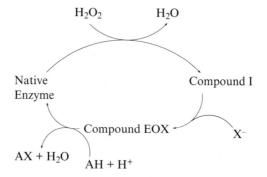

FIGURE 1. Two proposed mechanisms for haloperoxidase-catalyzed halogenations. Reproduced from Reference 14 by permisson of Ellis Horwood Ltd

porphyrin controls the oxidation potential and stores one oxidizing potential while the protein environment of the heme controls the reactivity of the metal and stabilizes reactive intermediates. As glycoprotein structures become elucidated (the complete amino acid sequences of several haloperoxidases are known, for example HPO, CPO and TPO), the effect of glycoprotein structure on reactivity has become an area of extensive scrutiny[16].

There are several examples of haloperoxidases containing metals other than iron, and many of these are non-heme enzymes. For example, a series of bromoperoxidases have been isolated recently from seaweed[17,18] and an actinomycete[19] that require vanadium for halogenating activity.

4. Mechanism of halogenation

Reaction of a heme-containing haloperoxidase with peroxides leads to an unstable intermediate ($t_{1/2}$ < 30 s) called Compound I. Reaction of Compound I with halide ion produces a native enzyme and a halogenating intermediate. This intermediate carries out halogenation reactions that resemble normal electrophilic halogenations with respect to substrate specificity and product distributions. However, the structure of this halogenating intermediate has been a matter of considerable research and controversy. For example, detailed studies by several groups on the mechanism of halogenation by CPO from *C. fumago* have attempted to determine if the active halogenating species is free hypohalous acid (HOX) or is an enzyme-bound hypohalite species (EOX) (Figure 1). The absence of stereoselectivity in CPO halogenations appears to favor the formation of free HOX, while several kinetic studies gave support for the existence of EOX as the reactive species. The active halogenating species produced by other haloperoxidases have also been investigated. For more thorough discussions of these issues see reviews by Neidleman and Geigert[14] and Kirk[15].

C. Biosynthesis of Halometabolites by Marine Microorganisms

To illustrate mechanisms by which complex halogenated secondary metabolites are elaborated, the haloperoxidase-mediated biosynthesis of selected halometabolites elaborated by marine microorganisms will be described. The discussion given by Neidleman[20] of chloramphenicol biosynthesis provides an example of bacterial halometabolite biosynthesis, while fungal halometabolite biosynthetic schemes are found in the monograph by Turner[21].

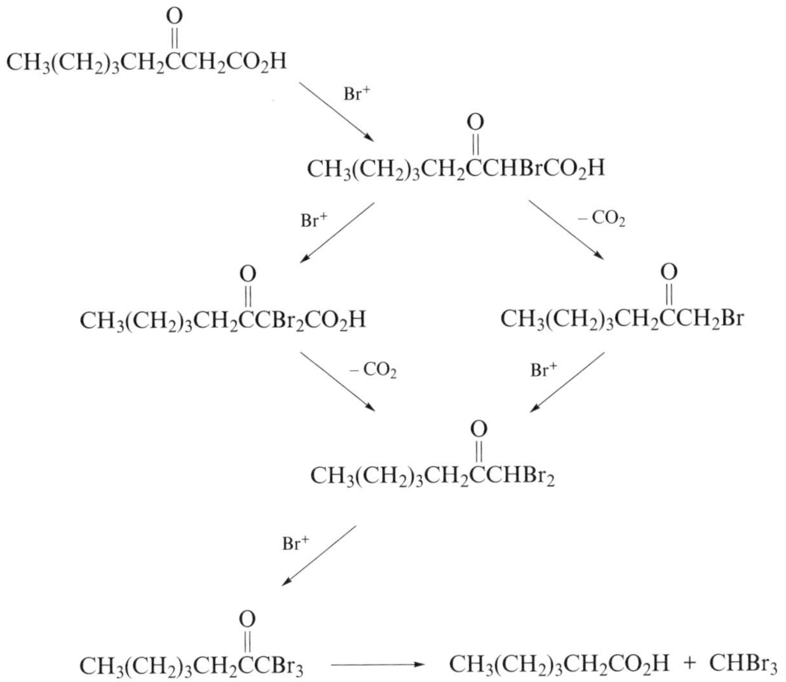

$$\text{CH}_3(\text{CH}_2)_3\text{CH}_2\overset{\overset{\text{O}}{\|}}{\text{C}}\text{CH}_2\text{CO}_2\text{H}$$

$\xrightarrow{\text{Br}^+}$

$$\text{CH}_3(\text{CH}_2)_3\text{CH}_2\overset{\overset{\text{O}}{\|}}{\text{C}}\text{CHBrCO}_2\text{H}$$

$\overset{\text{Br}^+}{\swarrow} \qquad \overset{-\text{CO}_2}{\searrow}$

$$\text{CH}_3(\text{CH}_2)_3\text{CH}_2\overset{\overset{\text{O}}{\|}}{\text{C}}\text{CBr}_2\text{CO}_2\text{H} \qquad \text{CH}_3(\text{CH}_2)_3\text{CH}_2\overset{\overset{\text{O}}{\|}}{\text{C}}\text{CH}_2\text{Br}$$

$\overset{-\text{CO}_2}{\searrow} \qquad \overset{\text{Br}^+}{\swarrow}$

$$\text{CH}_3(\text{CH}_2)_3\text{CH}_2\overset{\overset{\text{O}}{\|}}{\text{C}}\text{CHBr}_2$$

$\overset{\text{Br}^+}{\swarrow}$

$$\text{CH}_3(\text{CH}_2)_3\text{CH}_2\overset{\overset{\text{O}}{\|}}{\text{C}}\text{CBr}_3 \longrightarrow \text{CH}_3(\text{CH}_2)_3\text{CH}_2\text{CO}_2\text{H} + \text{CHBr}_3$$

(3)

1. Acetogenin biosynthesis

Despite the abundance of Cl^- in seawater, most of the biohalogenation reactions of marine microorganisms are carried out by bromoperoxidases, and chlorine-containing metabolites are thought to be produced by nucleophilic attack of Cl^- on bromonium or iodonium intermediates[22]. Enzyme-catalyzed brominations of typical fatty acids, such as β-keto acids, lead to a wide variety of acetate-derived secondary metabolites. For example, Beissner and coworkers[23] demonstrated the mechanism shown in equation 3 for the bromoperoxidase catalyzed formation of halogenated ketones and bromoform from 3-oxooctanoic acid.

The presence of bromine in biosynthetic intermediates can trigger additional reactions that produce a variety of complex metabolites. For example, McConnell and Fenical[24] have demonstrated that Favorskii rearrangement of 1,1,3,3-tetrabromo-2-heptanone (**19**) produces the unsaturated acid **20** (equation 4). These and other transformations are reviewed by Fenical[25].

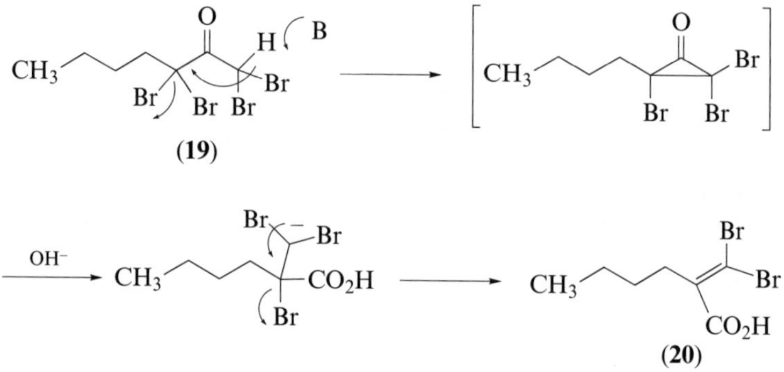

(4)

2. Terpene biosynthesis

Proton-catalyzed cyclization of an acyclic terpene precursor is often a key step in the formation of more complex cyclic terpenes in terrestrial organisms. Bromonium-ion catalyzed cyclizations play a comparable role in the marine environment with concomitant incorporation of bromine into the product. These parallel pathways are illustrated in equation 5. In addition, subsequent rearrangements of halogen-containing terpenoids can lead to more complex metabolites, including non-halogenated products. For example, the frequent deviation of the 'normal' head-to-tail arrangement of isoprene units in marine terpenes has been explained by the rearrangement of bromoterpenoid intermediates, as illustrated in equation 6 by the formation of the rearranged sesquiterpene **21**[26].

III. BIOCHEMISTRY OF THE THYROID HORMONES

Biohalogenation is essential for the survival of vertebrates. The thyroid hormones, 3,3',5,5'-tetraiodothyronine (thyroxine, T_4, **22**) and 3,5,3'-triiodothyronine (T_3, **23**), products of iodination of thyroglobulin in the thyroid gland, play key roles in development and metabolic regulation. Examples of biological processes affected by thyroid hormones include growth and differentiation, regulation of basal metabolism, oxygen consumption, intermediary anabolic and catabolic metabolism, and receptor protein synthesis. The myriad medical problems associated with either over-production or under-production of thyroid hormones attest to their importance in human physiology.

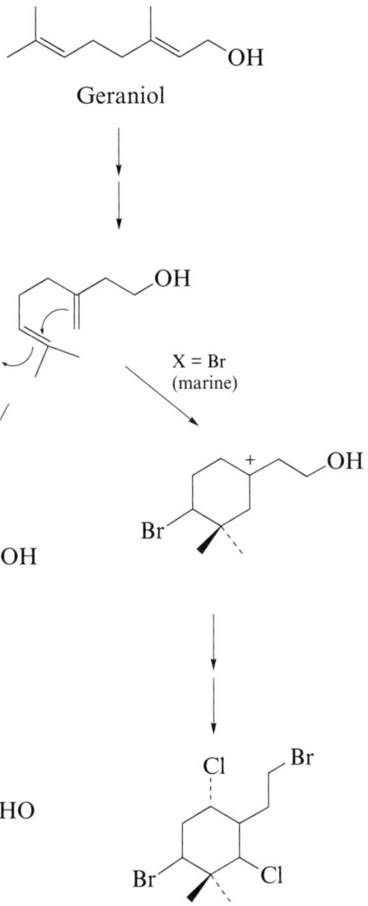
OH
Geraniol
OH
X⁺
X = Br
(marine)
X = H
(terrestrial)
OH
Br
OH
H
Cl
Br
CHO
Br
Cl

(5)

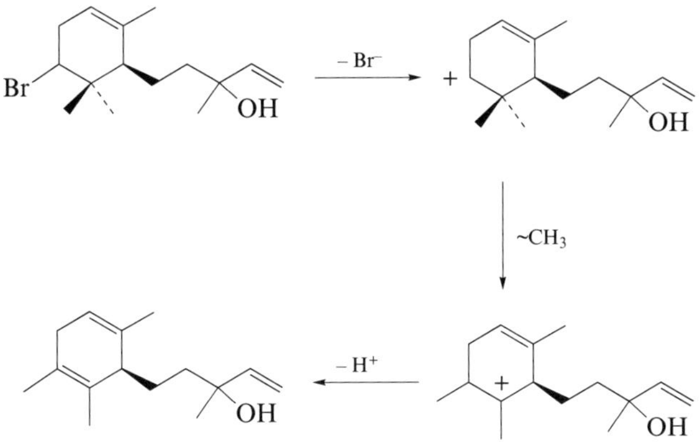
Br
OH
– Br⁻
+
OH
~CH3
OH
– H⁺
+
OH

(6)

(**21**)

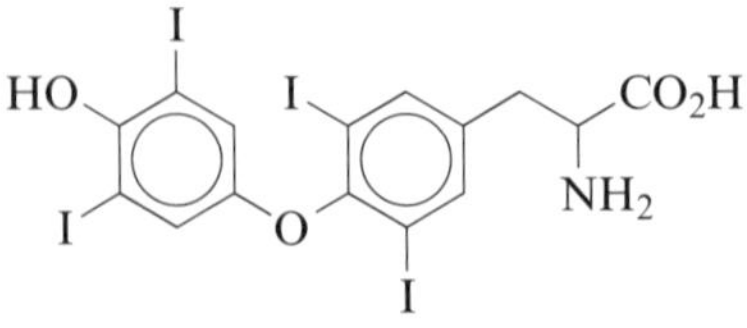

T$_4$

(22)

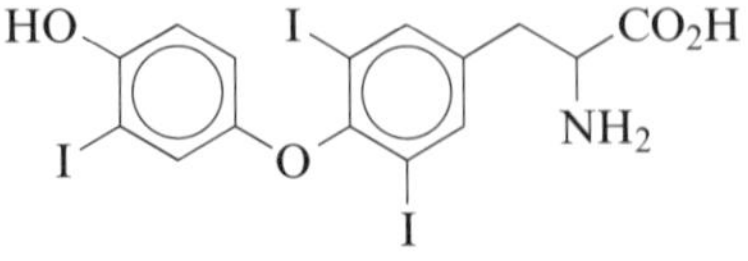

T$_3$

(23)

There are many reviews available on the biochemistry of thyroid hormones[27–30] includ-ing the excellent review by Doonan in this Series[1]. Biosynthesis and metabolism of thyroid hormones will be reviewed briefly in this chapter. Attention will be given to recent developments relating to the mechanisms responsible for the physiological responses to thyroid hormones.

A. Historical

The presence of organically bound iodine in a biologically active constituent of the thy-roid glands was established by Baumann in 1896. Crystals of thyroxine were isolated by Kendall in 1914 and the compound was given the erroneous structure of the indolepropi-onic acid (24). Because of the presumed indolic nature of the compound, this was given the name thyroxindole, later shortened to thyroxin. In 1926, Harrington identified the product of catalytic reduction of thyroxin as the new amino acid thyronine. From this he correctly established the structure of thyroxin, and confirmed this by total synthesis. The isolation by Gross and Pitt-Rivers and by Rosche and coworkers of 3,5,3'-triiodothyro-nine was of major importance in that this amino acid proved to be more potent than T$_4$. Braveman and coworkers subsequently demonstrated the conversion of T$_4$ to T$_3$ *in vivo* and T$_3$ is now accepted as the more important physiologically active hormone. 3,3',5'-Triiodothyronine (reverse T$_3$, rT$_3$) as well as virtually all of the other possible deiodination products of T$_4$ have now been isolated. More detailed discussions of the very interesting history of thyroid hormone research can be found in specialized reviews[27–29].

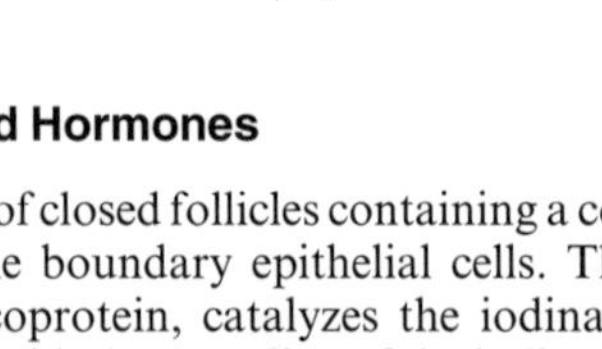

(24)

B. Biosynthesis of Thyroid Hormones

The thyroid gland consists of closed follicles containing a colloid consisting primarily of thyroglobulin secreted by the boundary epithelial cells. Thyroid peroxidase (TPO), a membrane-bound heme glycoprotein, catalyzes the iodination of tyrosine residues in thyroglobulin, as well as the oxidative coupling of the iodinated residues to give T$_4$ and T$_3$ residues. Proteolytic cleavage of the thyroglobulin protein backbone releases the hor-mones, which are then secreted and transported by specific transport proteins ultimately to cellular receptors. Although both T$_4$ and T$_3$ are secreted, most of the T$_3$ is produced by deiodination of T$_4$ subsequent to secretion.[29,31,32]

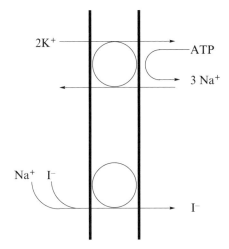

FIGURE 2. A schematic diagram of Na^+/K^+-ATPase-dependent I^- transport

1. Iodide transport

The total body iodine pool, amounting to 15–20 mg iodine, consists of plasma inorganic I^- (PII), protein-bound plasma iodine (PBI) and intrathyroidal iodine. While the thyroid gland makes up only about 0.03% of the weight of the adult human body, it contains 70–80% of the total body iodine, a fact that demonstrates the efficiency with which the thyroid concentrates I^-. This concentration process consists of (1) transport of I^- across the basolateral membrane of the thyroid gland (the iodide pump), (2) accumulation within the cell, (3) diffusion across the apical membrane into the follicular lumen following the electrochemical gradient and (4) storage. The initial transport step is carried out against a concentration gradient that maintains a 30-fold higher concentration in the thyroid than in the circulation under normal conditions. This is accomplished by I^-/Na^+ cotransport whereby I^- is driven by the Na^+ gradient, this gradient being established by extrusion of intracellular Na^+ by a Na^+/K^+-ATPase pump (Figure 2)[33,34].

2. Thyroglobulin iodination and the coupling reaction

TPO, peroxide and I^- form an enzyme-associated iodinating species [TPO-I_{oxid}(I^-)] that carries out the iodination of Tyr residues in thyroglobulin to give monoiodotyrosine (MIT) and diiodotyrosine (DIT) residues[29,31]. In the formation of T_4, TPO-catalyzed coupling of two proximal DIT residues produces the T_4 residue. This occurs by either a one-electron oxidation of two proximal DIT residues or a two-electron oxidation of one DIT residue positioned near another DIT residue, with formation of dehydroalanine as the 'lost side chain' (Figure 3)[29,32,35]. In thyroid hormone biosynthesis, the tertiary structure of thyroglobulin plays an important role, in that the hormonogenic Tyr residues must not only be in close proximity, but also held in the proper orientation for coupling subsequent to iodination. Thyroglobulins from several species, including human, have been cloned and sequenced, and from this, donor and acceptor Tyr residues have been identified[36].

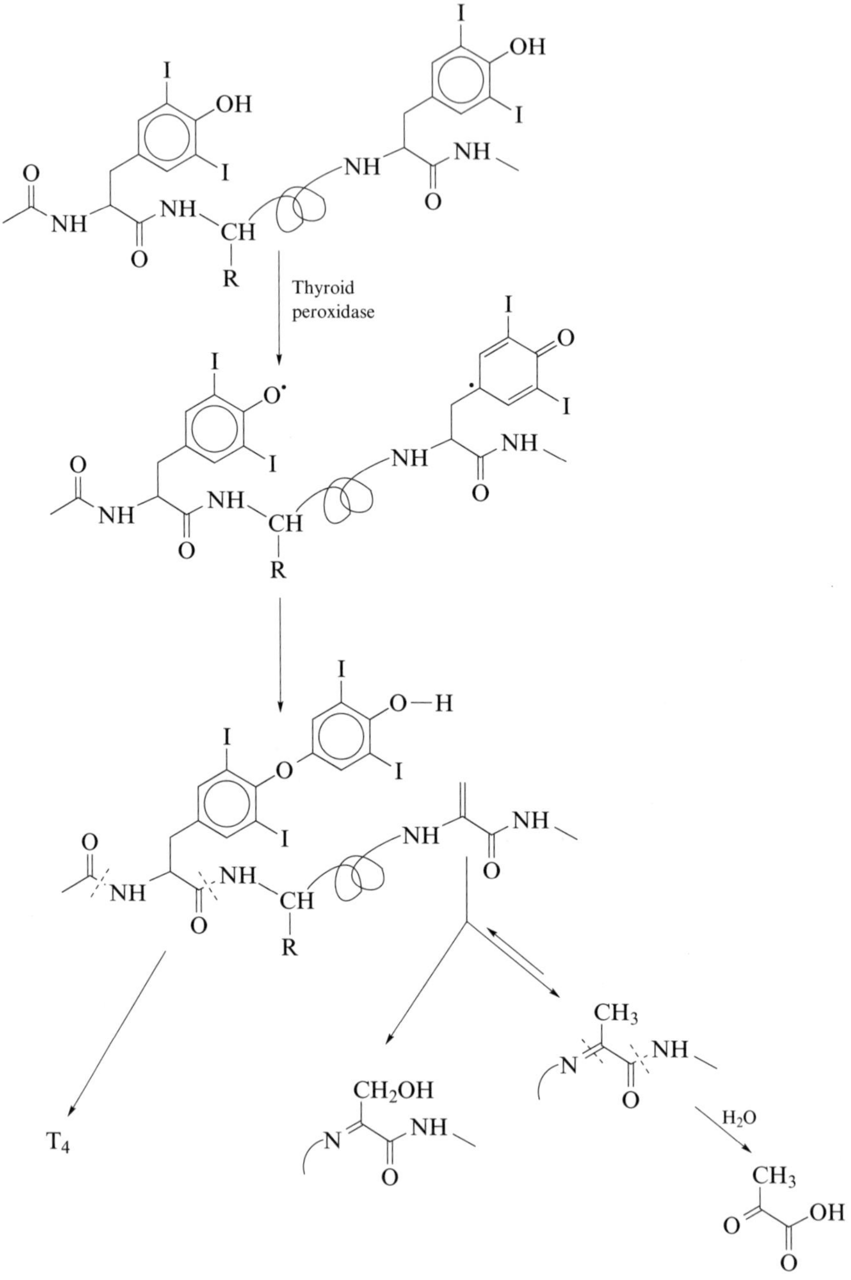

FIGURE 3. Schematic diagram of coupling of 3,5-diiodotyrosine residues of thyroglobulin to form thyroxine. Subsequent to coupling, proteolytic and hydrolytic reactions produce free thyroxine and pyruvic acid. Reproduced from Reference 29 by permission of John Wiley & Sons, Inc.

C. Release and Transport of Thyroid Hormones

The hormonogenic iodination and coupling reactions of thyroglobulin are extracellular processes. The poorly diffusible iodinated precursor protein serves as a storage depot for hormone and for iodine. Secretion involves endocytosis and lysosomal proteolysis. In this process, released free MIT and DIT are deiodinated and the resulting I^-, in a conservation step, is recycled, while T_3 and T_4 are released into the circulation. Here, the poorly soluble hormones quickly associate with two plasma proteins, thyroxine-binding globulin (TBG) and thyroxine-binding prealbumin (TBPA), that carry the hormones to the target cells[29].

Most of the hormonal activities of thyroid hormones result from the binding of T_3 to chromatin-associated nuclear receptors. Recent evidence suggests that, in several cells, translocation of T_3 and T_4 from plasma to the intracellular compartment is accomplished by a receptor-mediated, limited capacity, active transport system[37]. In hepatocytes, it has also been shown that an energy-dependent stereospecific active transport system is involved in the translocation of T_3 from the cytosol into the nucleus[38,39].

D. Thyroid Hormone Receptors

Over the decades, several theories have been proposed and discarded to explain the diverse biological responses to thyroid hormones. For example, early researchers attempted to look for direct interaction of the hormones with energy-producing reactions. Functional roles for iodine and/or the diphenyl ether structural unit were considered. For example, in 1950, Nieman proposed that a reversible oxidation of T_4 to a quinoid form was coupled with electron transfer to an energy-generating system, whereas, in 1957, Szent-Györgi proposed that energy transfer involved the excited triplet state of iodine (reviewed by Jorgensen)[29]. However, evidence, such as a lag period before the onset of hormonal response, and the ability of inhibitors of protein synthesis to block the response, gradually made it clear that stimulation of protein synthesis plays an important role in the action of thyroid hormones. In 1972, Oppenheimer reported binding of T_3 to the nuclei of rat liver and kidney cells and isolated a 60,000–79,000 molecular weight protein that served as a high-affinity, low-capacity binding site. Oppenheimer has reviewed this and other work that established a fundamental role of a T_3 nuclear receptor in thyroid hormone action[40]. According to the nuclear receptor hypothesis, activation of the nuclear receptor stimulates the synthesis of specific mRNA sequences that code for T_3-inducible proteins. Observed biological responses reflect the involvement of these proteins in diverse biological functions, including such processes as energy metabolism.

The molecular events associated with the binding of T_3 to the nuclear receptor have been the focus of much research. A mechanism for the control of gene transcription involves association of DNA-binding proteins with specific DNA sequences. There is evidence that the T_3 nuclear receptor may function as such a regulatory protein. The nuclear T_3 receptor has been located on promoter regions of human growth hormone genes and lactogen genes from human lymphoblastoid IM-9 cells. The presence of T_3 was shown to be necessary for the binding of the T_3 receptor to the growth hormone gene promoter, suggesting that this binding leads to recognition of specific gene regions by the receptor protein.[41,42] Supporting the regulatory function of the nuclear T_3 receptor is the fact that it has been shown to belong to a family of DNA-binding proteins that have sequence homology to a cellular oncogene[43,44].

Structural requirements for binding to the nuclear receptor and for biological activity have been studied extensively. The 'mystique of iodine' was shown to be illusionary, first by the activity of other halogenothyronines, and secondly by the activity shown by a series of 3'-alkyl-3,5-diiodothyronines[27]. A schematic formulation of ligand–receptor interaction is shown in Figure 4.

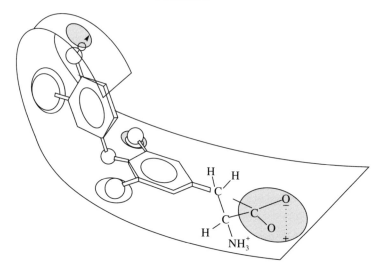

FIGURE 4. A schematic representation of the receptor–T_3 complex. Reproduced from reference 27 by permission of Elsevier Science Publishers BV

Based on NMR data showing that thyroid hormone analogues function as electron acceptors in molecular complexes with aromatic donors, Chae and McKinney proposed a 'stacking complexation' receptor binding mechanism for thyroid hormones and analogues thereof, as well as for the binding of halogenated aromatic compounds to the Ah (dioxin) receptor. From these and other studies, they suggested a possible role of thyroid hormone-binding proteins in mediating carcinogenicity and toxicity of dioxin and certain other aromatic compounds (see Section V.B)[45].

Although the importance of the nuclear T_3 receptor is well established, there also is evidence for the existence of non-nuclear T_3 receptors. For example, Segal and Ingbar[46] demonstrated an immediate T_3-mediated increase in calcium accumulation in rat thymocytes, an increase that was insensitive to inhibitors of protein biosynthesis. A direct effect on oxidative phosphorylation has been suggested to result from T_3 binding to saturable receptors of mitochondrial membranes. The possible roles of non-nuclear T_3 receptors have been reviewed by Sterling[47].

E. Regulation and Metabolism of Thyroid Hormones

1. TRH and TSH

The thyroid–pituitary–hypothalamus axis controls thyroid hormone homeostasis. Thyrotropin-releasing hormone (TRH), released from the hypothalamus, stimulates the synthesis and release of thyroid-stimulating hormone (thyrotropin, TSH) from the anterior pituitary. TSH increases the release of thyroid hormones by several mechanisms, including stimulation of the I⁻ pump. While lower than normal levels of T_3 and T_4 cause an exaggerated response of the pituitary to TRH, released thyroid hormones, in feedback control, blunt the stimulating action of TRH on the pituitary. For further discussion of TSH and TRH biochemistry, see, for example, the review by Kannan[48].

2. Iodothyronine deiodinases and regulation

Iodothyronine deiodinases have been grouped into three classes—Types I, II and III—based on such parameters as substrate specificities, kinetics and sites of deiodination. The conversion of T_4 to the more active T_3 demonstrates the importance of deiodinases in thyroid function. The further deiodination of T_3 to inactive iodothyronines provides a further mechanism for attenuating the action of the thyroid hormones (see below).

In addition to the obvious deactivating role of deiodinases, there has been recent evidence that a relationship exists between regulation of deiodination of thyroid hormones in target cells and the intracellular effects of T_4 and T_3 on pituitary and hypothalamus function. In the rat pituitary, and probably the human, type-II deiodinase-catalyzed conversion of T_4 to T_3 is a prerequisite for inhibition of TRH release. rT_3, produced from T_4 by type-III deiodinase, is a potent inhibitor of type-II deiodinase. In a postulated regulatory circuit, rT_3 formed from T_4 by type-III deiodinase in surrounding CNS (Central Nervous System) tissue enters the pituitary and inhibits type-II enzyme. The resulting decrease in T_3 concentration, in turn, causes an increase in TSH secretion[49].

3. Metabolism of thyroid hormones

Approximately 80% of T_4 disposal is accomplished by reductive deiodination to T_3 and rT_3. Further monodeiodination, producing a cascade of lesser iodothyronines, degrades one-third to one-half of the T_3 and one-third of the rT_3. Approximately 20% of radio-labeled T_4 is excreted as thyronine (T_0). Other metabolic pathways combine side-chain modification and/or O-sulfate or O-glucuronide formation with deiodination. Tri- and tetraiodothyroacetic acids, **25** and **26**, are examples of products of side-chain modification.

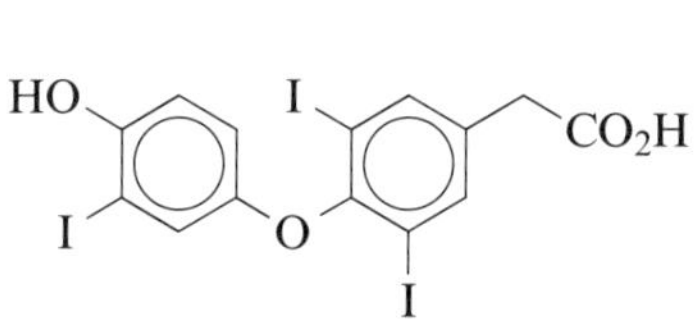

(25) 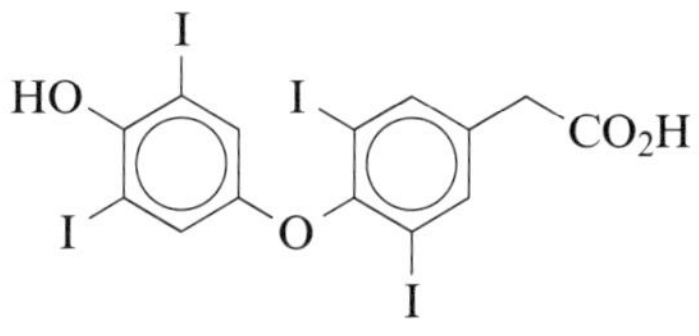   (26)

Oxidative deiodination of T_4 in a process that includes ether link cleavage produces DIT, I^-, and iodinated protein. Since this mechanism accounts for 40–50% of all T_4 degradation during human leukocyte phagocytosis, the suggestion has been made that this process may contribute to the bactericidal capacity of the leukocyte[50].

F. Summary

The thyroid hormones, T_3 and T_4, are involved in a host of biological processes in all vertebrates. Many of these are mediated through the T_3 nuclear receptor, and involve regulation of gene transcription. In an important feat of evolution, mechanisms were established to concentrate the trace element iodine in the thyroid gland. This element not only provides the requisite bulky, non-polar substituent required for receptor binding, but also provides, through deiodination, mechanisms for modulating hormone levels that are interactive with TRH–TSH control levels.

IV. BIOCHEMISTRY OF SELECTED HALOGENATED ORGANIC COMPOUNDS

In this section, the biochemistry of different classes of halogenated organic compounds will be discussed by examination of representative examples. Mechanistic aspects of biological properties will be stressed in many of the examples given. Recent comprehensive reviews have been published[2,51].

A. Carboxylic Acids

As a reflection of the important role of carboxylic acids in intermediary metabolism, compounds that serve as carboxylic acid mimics or as inhibitors of carboxylic acid processing enzymes have been extremely useful biochemical tools. A host of halogenated carboxylic acids have been synthesized and studied in this regard. Selected examples are given in this section. More extensive reviews are available[52,53].

1. Haloacetates as enzyme inhibitors

In 1874, Steinmauer observed the several toxic manifestations of bromoacetate that had been injected into frogs, including weakness and paralysis of muscles and respiratory and cardiac depression. In 1930, Lundsgaard found that iodoacetate was even more toxic, and made the critical discovery that both compounds inhibited lactate formation in muscle. As these initial observations were extended, iodoacetate, along with fluoride, became a very important tool for the study of intermediary metabolism. As the biochemical reactions that make up the process of glycolysis became known, the site of action of iodo- and bromoacetate were delineated. In the oxidation of glyceraldehyde-3-phosphate to phosphoglyceric acid by glyceraldehyde-3-phosphate dehydrogenase (GDP), the NADPH required for the reduction of pyruvate to lactate is produced. Research over several years demonstrated that GDP is extremely sensitive to haloacetate, and the inhibition of this enzyme thus accounts for the inhibition of lactate formation. The molecular mechanism of this inhibition, alkylation of a critical SH residue, is illustrative of the general functioning of these α-haloacetates as alkylating agents of nucleophilic centers on macromolecules. Based on this property, these analogues have been used in a host of studies as enzyme inhibitors and to label enzyme active sites. Early work is discussed in an extensive review by Webb[54]. In a related application, haloacetate-mediated carboxymethylation of proteins is used routinely prior to degradation.

Dichloroacetic acid (DCA) is an inhibitor of pyruvate dehydrogenase kinase, the enzyme that, through phosphorylation, deactivates the pyruvate dehydrogenase (PDH) complex. This enzyme complex carries out the irreversible oxidative decarboxylation of pyruvate to acetyl CoA. Inhibition of PDH kinase funnels the metabolism of pyruvate to Krebs cycle oxidation or incorporation into lipids at the expense of gluconeogenesis. Because of the importance of pyruvate in the biosynthesis and metabolism of several classes of compounds—carbohydrates, amino acids, lipids and ketone bodies, for example—striking and diverse effects on mammalian metabolism result, including an increase in respiratory quotient and a lowering of blood glucose. Potential medical applications of DCA include treatment of hyperlactatemia, a usually fatal disorder of acid–base metabolism caused by such conditions as hypotension and shock[55].

2. Fluoroacetic acid, fluorocitric acid, and 'lethal synthesis'

Perhaps no sequence of events has had greater impact on strategic approaches to medicinal chemistry and biochemistry than the research in the early 1950s on the toxic properties of fluoroacetic acid. As the topic of several reviews, this important story will be discussed only briefly here.

$$^-O_2C \diagdown \diagup \overset{\displaystyle O}{\underset{\displaystyle O}{\parallel}}{\diagup}^{CO_2^-} + FCH_2 \overset{\displaystyle O}{\diagup \parallel} SCoA \longrightarrow \quad ^-O_2C \diagdown \diagup \overset{^-O_2C \quad OH}{\underset{H \quad F}{\diagup}} CO_2^- \qquad (7)$$

(27)

2R,3R-erythro-Fluorocitrate

The classic studies by Sir Rudolf Peters demonstrated that the stereospecific biosynthesis ('lethal synthesis') of 2R,3R-erythro-fluorocitric acid (fluorocitrate) (27) (equation 7) is the underlying mechanism of toxicity of fluoroacetate. The inhibition by fluorocitrate of aconitase, the enzyme that catalyzes the dehydration of citrate to cis-aconitate in the Krebs cycle, was established, and widely accepted, as the explanation of the potent toxicity of fluoroacetate. In later work, however, Kun and coworkers noted that, although fluorocitrate is a reversible inhibitor of aconitase, the concentrations required (Ki ≈ 60–200 μM) seemed inconsistent with the toxic properties of fluorocitrate. For example, a lethal dose of 0.2–0.5 μg injected into the brain of an adult rat resulted in an estimated cerebral concentration of the active isomer of only 10 nM[56]. Subsequent research showed that fluorocitrate is a potent inhibitor of citrate mitochondrial transport. Identification of a mitochondrial enzyme that catalyzes the formation of a citrate–glutathione thioester, together with the fact that fluorocitrate irreversibly inhibits this enzyme, suggested that this might be a molecular site of fluorocitrate toxicity[57]. In addition, it is known that acetylcholine synthesis in mitochondria is dependent on citrate as a source of acetyl-CoA. Fluorocitrate-mediated blockade of acetylcholine biosynthesis through inhibition of citrate transport would be consistent with the apparent action of fluorocitrate as a neurotoxin[58].

Other mechanisms have been considered. Clarke has proposed that the acute toxicity of fluorocitrate may be caused by irreversible inhibition of aconitase by (2E,4R)-4-fluoroaconitate, formed by the dehydration of fluorocitrate by aconitase[59]. Hornfeldt and Larson reported evidence that seizures accompanying fluorocitrate toxicity may result from Ca^{+2} chelation in the spinal cord[60].

3. Fluorinated carboxylic acids as mechanistic probes

The 'lethal synthesis' of fluorocitrate from fluoroacetate is an important example of how a fluorinated analogue can mimic its parent in biological systems. The special advantages of substitution of a C—H bond by C—F bond in designing analogues of biologically active compounds are now well recognized, and have been discussed in several reviews. Fluorinated carboxylic acids have been used extremely effectively by several groups as enzyme inhibitors and as mechanistic and stereochemical probes. This topic has been reviewed by Walsh[52]. For example, fluoroacetate has been used as a prochiral substrate analogue to show that malate synthase catalyzed formation of malate from acetyl-CoA is non-stereoselective[61]. Likewise, fluoropyruvate has been used to study the stereochemistry of several enzyme-catalyzed reactions of pyruvate. Kun and coworkers made effective use of fluoromalate in their studies on fluorocitrate toxicity, since this analogue stimulates the tricarboxylic acid carrier but is not appreciably metabolized[57]. Fluorophosphoenolpyruvate has been used in many mechanistic studies[52]. In a recent example, (Z)-3-fluorophosphoenolpyruvate (28) acts as a pseudosubstrate and is converted to a stable analogue 29 of the catalytic intermediate by EPSP synthase (5-enolpyruvoylshikimate-3-

phosphate synthase), an important enzyme in the shikimate pathway[62]. Many additional examples of the use of fluorinated carboxylic acids as mechanistic probes are given in recent reviews[52,53,56].

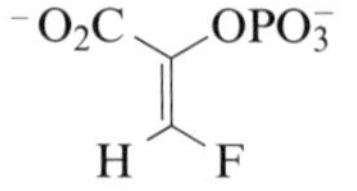

(28) 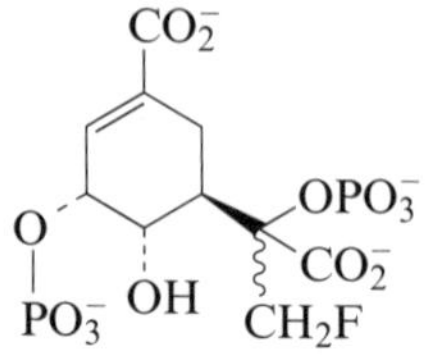 **(29)**

B. Halogenated Aldehydes and Ketones

1. Haloacetaldehydes and etheno adduct formation

Bromo- and chloroacetaldehyde react readily with adenine and cytosine, either free or as constituents of nucleosides and nucleotides (Figure 5)[63]. The etheno adducts so formed are highly fluorescent, a discovery that has made this reaction a very valuable research

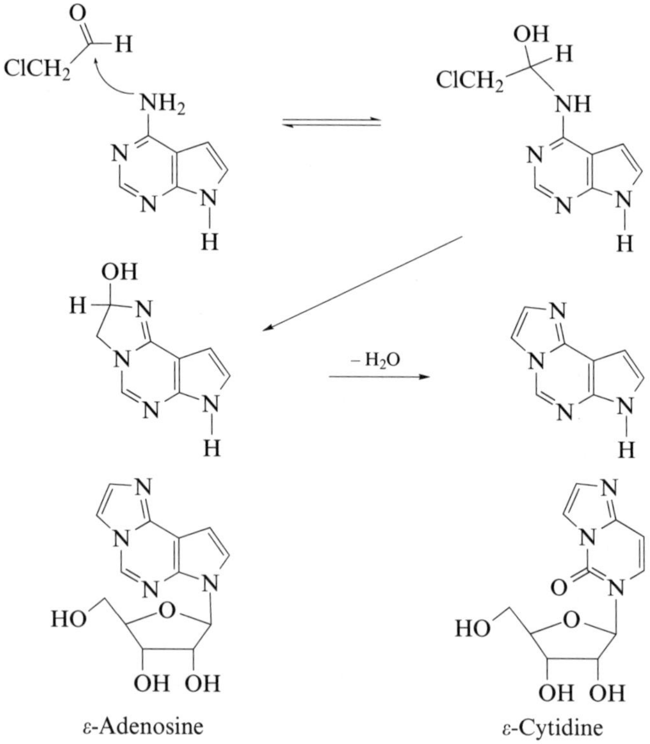

ε-Adenosine ε-Cytidine

FIGURE 5. The mechanism for the formation of etheno adducts by the reaction of chloroacetaldehyde and base components of nucleic acids. Structures of ε-adenosine and ε-cytidine

tool[64,65]. For example, the highly fluorescent ε-adenosine-containing analogues of 3'- and 5'-AMP, ADP, ATP and 3',5'-cyclic-AMP functioned effectively in various roles as phosphoryl, pyrophosphoryl and adenyl donors, and as allosteric effectors[65]. Of particular utility is the fact that facile reaction of CAA (chloroacetaldehyde) and BAA (bromoacetaldehyde) with polynucleotides occurs only at exposed bases[66]. This selectivity has been used to study DNA and RNA structure, as well as to probe for sites in genes that become exposed only when the gene is transcriptionally active[67].

2. Halogenated aldehydes and ketones as enzyme inhibitors

The facile reaction of CAA and BAA with nucleosides and nucleotides is one example of many of the applications of the bifunctional reactivity of halogenated aldehydes and ketones in modification of biomolecules. In an early example of the extensive use of halogenated ketones as protease substrate analogues, 1-*N*-tosylamido-2-phenylethyl chloromethyl ketone (TPCK) **30** was synthesized as a chymotrypsin substrate analogue. Stoichiometric inhibition was accompanied by loss of one histidine residue as a result of alkylation by the chloromethyl moiety[68]. A host of similar analogues were subsequently prepared and used as selective enzyme inhibitors, in particular for the identification of amino acid residues located at enzyme active sites[69].

(30)

Important refinements of this strategy have involved preparation of halomethyl ketones as transition-state analogue inhibitors of protease and other hydrolytic enzymes. In this approach, compounds are designed to resemble the tetrahedral intermediate formed during the enzymatic cleavage of a peptide bond. Carboxaldehyde-containing peptide and amino acid analogues are inhibitors of serine, cysteine and metallo proteases through formation of a tetrahedral structure that mimics the transition state associated with enzymatic bond cleavage. This structure is formed by reaction of the aldehyde with an active site serine hydroxyl group to form a hemiacetal, or by formation of the hydrated aldehyde[70,71]. Extension of this strategy to less-reactive ketones, whereby recognition elements could be situated on both sides of the carbonyl group, was accomplished by increasing the electrophilicity of the ketone by α-fluorine substitution. In applications of this strategy, extensive synthetic efforts have produced a number of potent inhibitors of several proteases. Examples of synthetic targets include potent pepsin inhibitors, for example **31**[72], the renin inhibitors **32**[73] and **33**[74], HIV protease inhibitors, for example, **34**[75], and several elastase inhibitors, exemplified by **35**[76].

Fluoroketone analogues are also inhibitors of esterases. For example, the difluoroketone analogue **36** of sn-glycerol phospholipids is a potent competitive inhibitor of phospholipase A_2[77] while the trifluoromethyl ketone **37** analogue of an insect juvenile hormone **38** is a potent and selective inhibitor of juvenile hormone esterase[78].

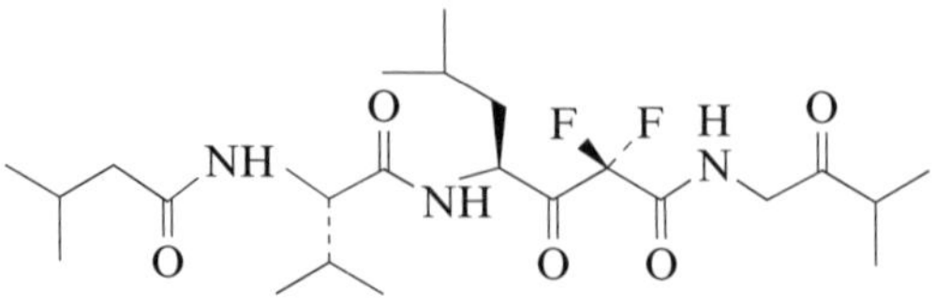

(31)

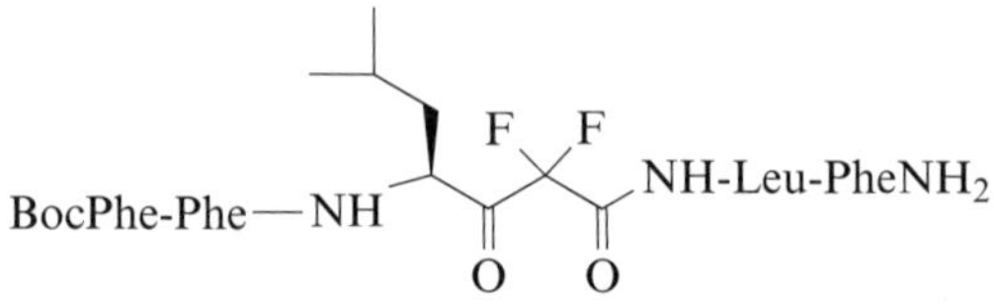

BocPhe-Phe—NH ... NH-Leu-PheNH₂

(32)

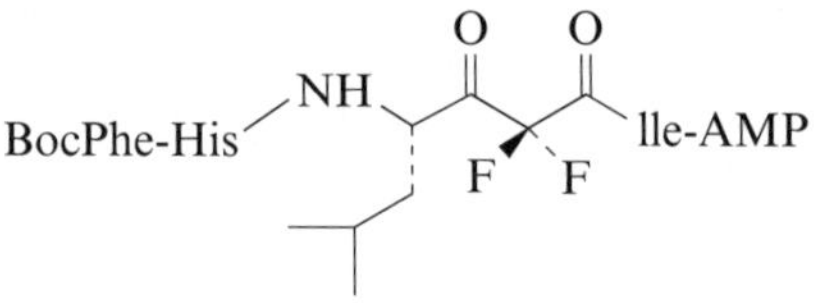

BocPhe-His ... Ile-AMP

(33)

Z-Val-NH ... NH-Val-Z

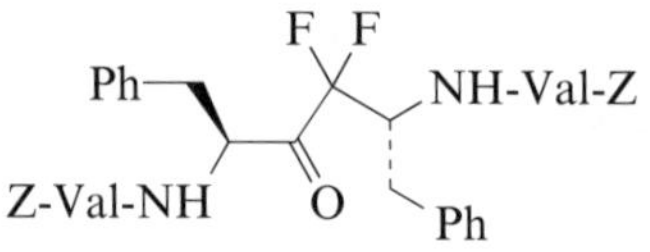

(34)

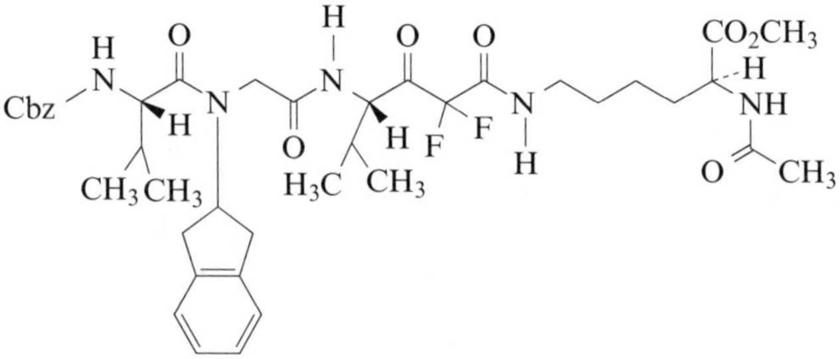

(35)

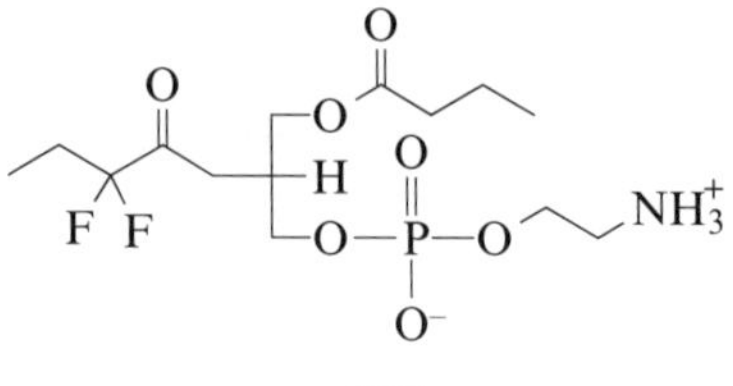

(36)

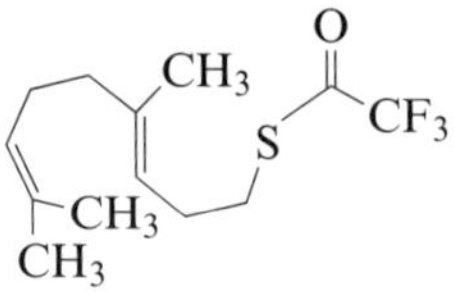

(37)

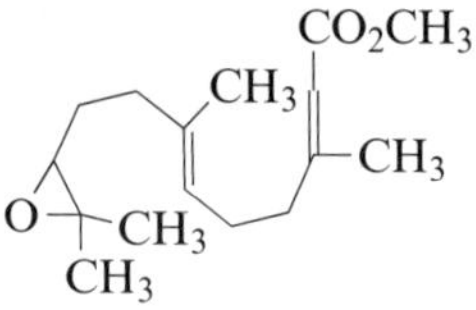

(38)

C. Halogenated Terpenoids

1. Fluorinated analogues of terpene biosynthetic intermediates

The effective use of fluorinated analogues as mechanistic probes, taking particular advantage of the high electronegativity of this element, is well represented by a series of detailed studies of the reactions by which the terpene biosynthetic precursors, mevalonic acid **39**, isopentenyl pyrophosphate (IPP) **40** and dimethylallyl pyrophosphate (DMAPP) **41** are processed to geranyl pyrophosphate (**42**) (equation 8). For example, 6-fluoro-mevalonic acid lactone (6F-Mev) (**43**) is converted enzymatically to 6-Mev pyrophosphate (6-Mev-PP). The latter is not a substrate but is an extremely potent inhibitor of Mev-PP decarboxylase, the enzyme that catalyzes the conversion of Mev-PP to IPP[79,80]. In the next steps of terpene biosynthesis, fluorinated analogues were used to provide evidence for carbonium ion intermediates for both the isomerization of IPP to DMAPP, catalyzed by isomerase, and for the coupling of IPP and DMAPP to give geranylpyrophosphate, catalyzed by prenyl transferase. In each reaction, if trifluoromethyl- or fluoro-substituents were introduced adjacent to a position at which a positive charge would be required in a carbonium ion mechanism, much slower reaction rates were observed[81–83].

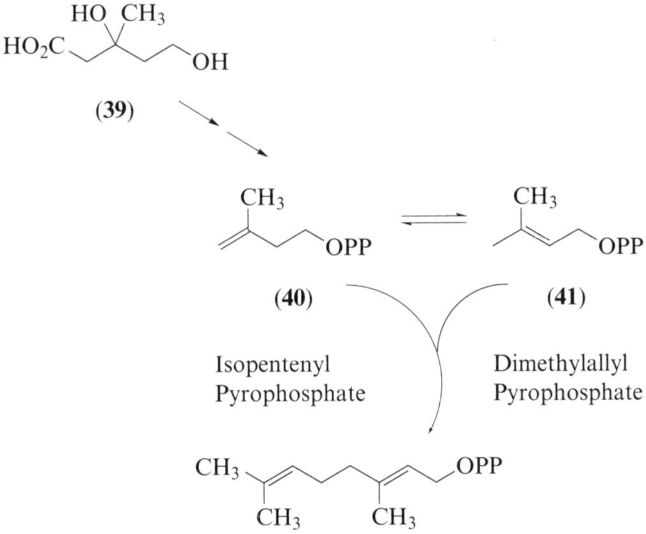

(8)

(40)

Isopentenyl
Pyrophosphate

(41)

Dimethylallyl
Pyrophosphate

(42)

Geranyl Pyrophosphate

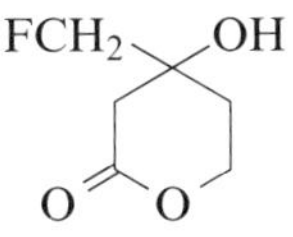

(43)

2. Halogenated steroids

Extensive synthetic and biochemical studies of halogenated steroids have produced a wealth of new medicinal and pharmacological agents. The discovery of the increased anti-inflammatory activity of 9α-halocorticosteroids (**44**) by Fried and Sabo in the early 1950s[84] was a very important event in the field of steroid biochemistry, in that previous wisdom held that any modifications of a steroid molecule would likely lead to lower activity. The subsequent research that led to the development of such important anti-inflammatory agents as dexamethasone, as well as useful halogenated analogues of other classes of steroids, such as steroid sex hormones, has been reviewed extensively and will not be discussed here[85–87].

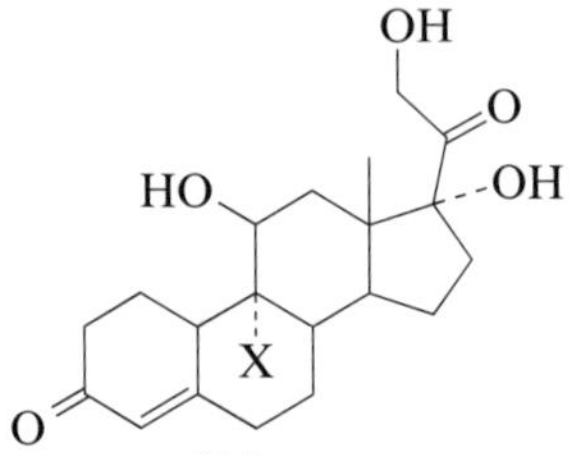

(44)
9α-Halo-17α-hydroxycorticosterone
(X = F, Cl, Br, I)

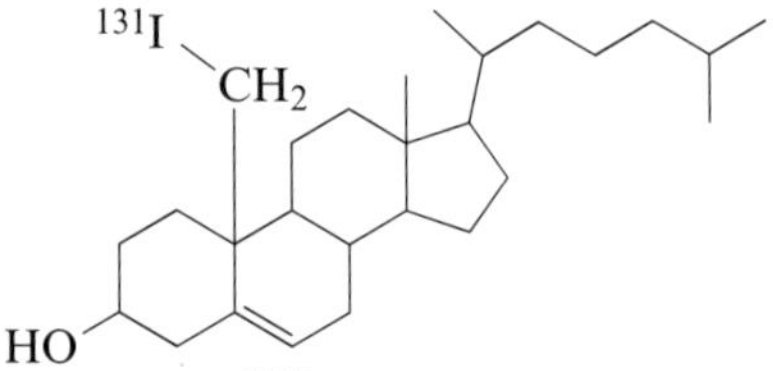

(45)
19-[131I]Iodocholesterol

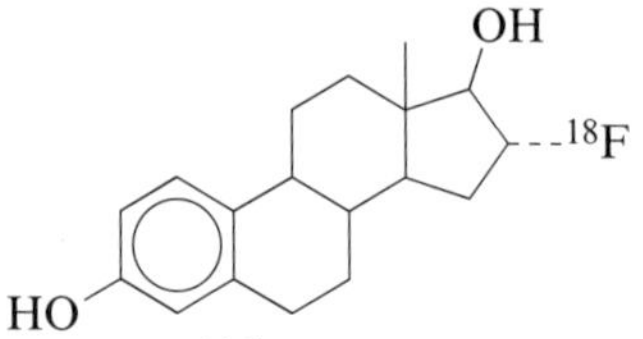

(46)
16α-[18F]Fluoroestradiol

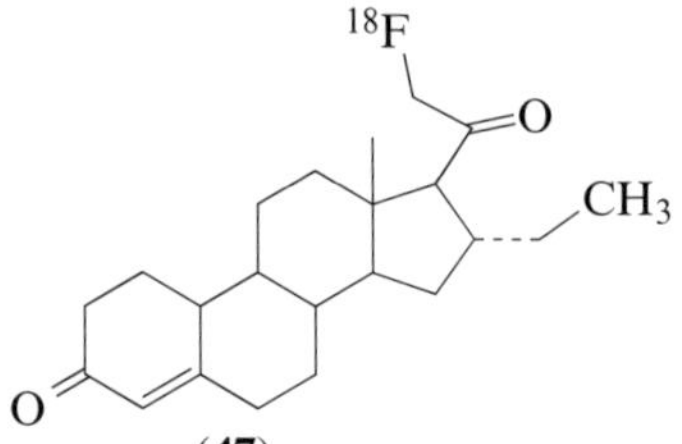

(47)
12-[18F]Fluoro-16α-ethyl-
19-norprogesterone

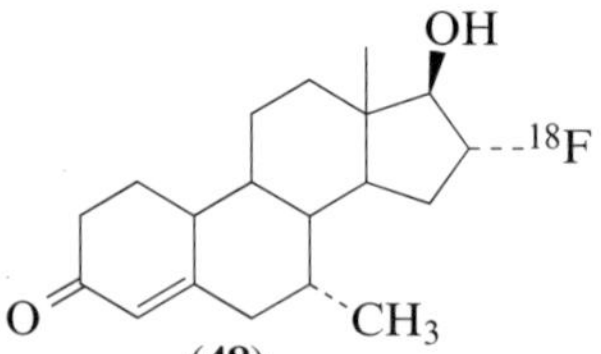

(48)
16α-[18F]Fluoro-7α-methyl-19-norprogesterone

An area of recent major interest has been in the development of radiohalogenated steroid derivatives for use as biological tracers. For example, 19-[31I]iodocholesterol (**45**) was used for the first imaging of the human adrenal gland *in vivo* by scintillation scanning[88]. More recently, radiohalogenated analogues of steroid sex hormones have been studied extensively as steroid receptor-based scanning agents for the detection of cancer of the breast, ovary and prostate. 16α-[^{18}F]-fluoroestradiol-17β (**46**) showed an excellent correlation between uptake at sites of primary carcinomas and loci of metastases, as measured by PET imaging, and tumor estrogen receptor concentration determined *in vitro* after excision[89]. In a comprehensive program designed to develop progesterone-receptor imaging agents—a more reliable *in vivo* marker for estrogen receptor concentrations than are estrogen receptor-based probes—several radiohalogenated nortestosterones and progestins were examined. For example, selective uptake of 21-[^{18}F]fluoro-16α-ethyl-19-norprogesterone (**47**) gives this analogue promise as a potentially clinically used PET imaging agent for progesterone receptors[90]. Similarly, several [^{18}F]fluoro-labeled androgens, for example 16α [^{18}F]fluoro-7α-methyl-19-nortestosterone (**48**), have been prepared as potential imaging agents for prostatic cancer[91].

The above examples are only a few of an extensive and growing list of halogenated steroids that show promise as *in vivo* scanning agents for a variety of clinical purposes. More complete discussions are found in recent reviews[87,92].

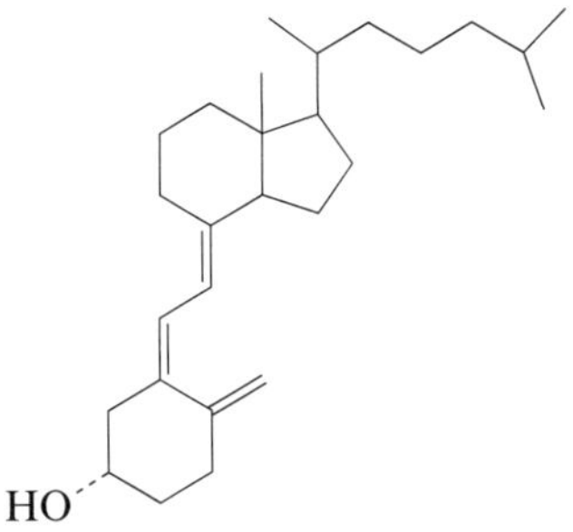

Vitamin D$_3$ (D$_3$)
(**49**)

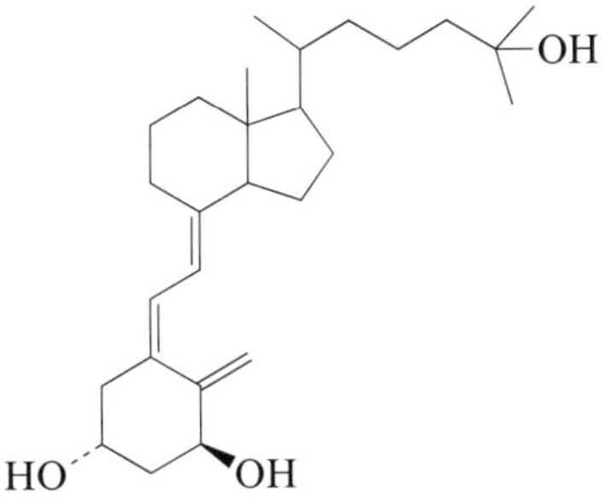

1,25-(OH)$_2$-D$_3$
(**50**)

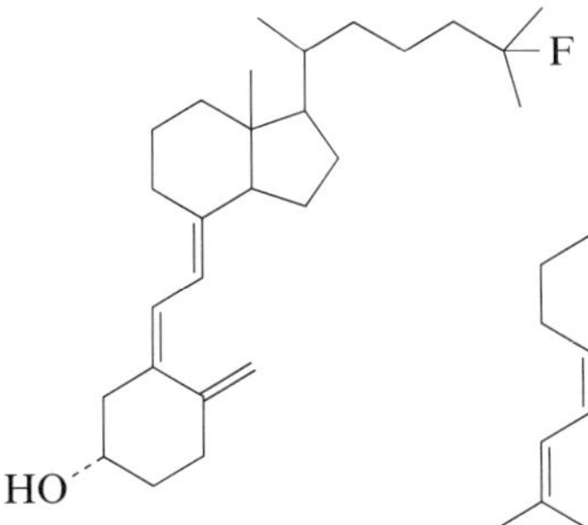

25-F-D$_3$
(**51**)

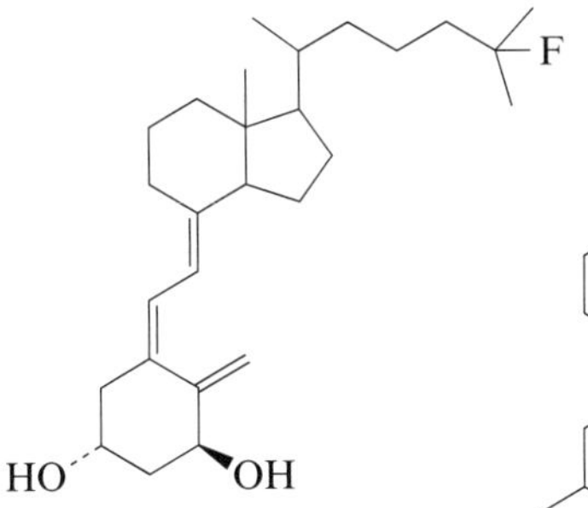

1α-OH-25-F-D$_3$
(**52**)

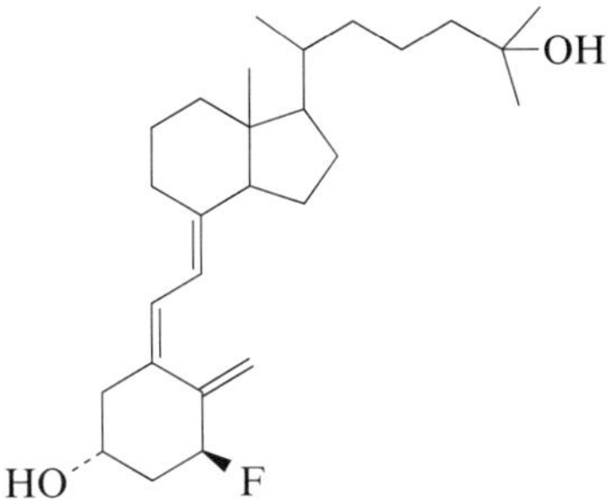

1α-F-25-OH-D$_3$
(**53**)

3. Fluorinated analogues of vitamin D₃(D₃)

Successive hydroxylation of vitamin D_3 (D_3) (49) at C-25 and C-1 produces 1,25-dihydroxyvitamin D_3 [1,25-$(OH)_2$-D_3, (50)], the active hormone. Several strategies have been used to antagonize, increase or modulate D_3 activity by strategic incorporation of fluorine. For example, 25-fluoro-D_3, 51, 1-α-OH-25-fluoro-D_3, 52, and l-α-fluoro-25-OH-D_3, 53, are examples of analogues prepared as potential D_3 antagonists wherein fluorine has been introduced at a position at which hydroxylation is required for production of the active hormone[93]. In an example of a strategy used to attempt to produce D_3 analogues with increased activity, 2-β-fluoro-1α-OH-D_3, 54, was prepared. This analogue, which incorporates the *trans*-fluorohydrin moiety, a structural feature known to increase the activity of corticosteroids, was found to be more active than D_3 itself[94]. Strategies to increase metabolic stability of D_3 have focused on incorporation of fluorine at positions known to be targets for hydroxylative deactivation. A five- to ten-fold increase of antirachitic activity of 24,24-difluoro-1α,25-$(OH)_2$-D_3, 55, over 1,25-$(OH)_2$-D_3 apparently results from increased biological stability because of blockade of C-24 hydroxylation[95]. Likewise, 24(R)-fluoro- 1α,25-$(OH)_2$-D_3, 56, an analogue having significant anti-rachitogenic activity, had significantly longer biological half-life than 1, 25-$(OH)_2$-D_3, presumably through specific blockade of 24(R)-hydroxylation[96]. In an example unrelated to hydroxylation, 6-fluoro-D_3, 57, has no D_3-activity, but binds strongly to the D_3 receptor, suggesting an

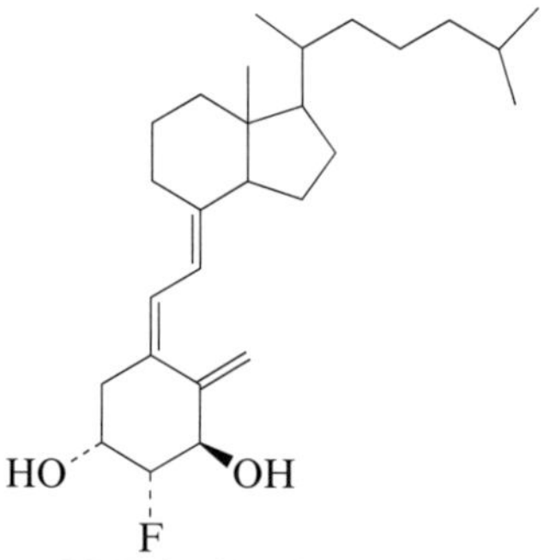

2β-F-1α-OH-D₃

(54)

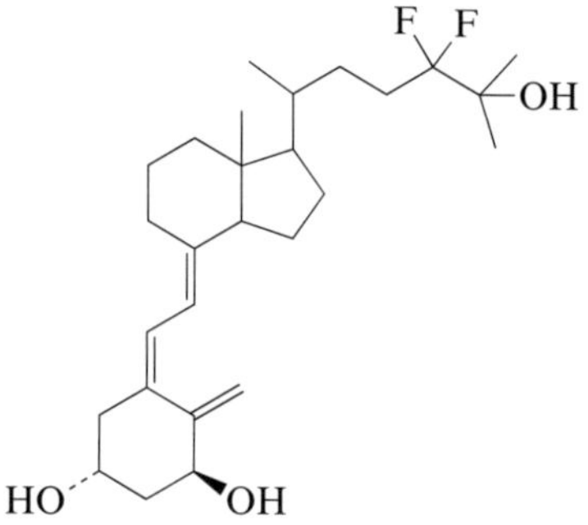

$24,24$-F₂-1α,25-$(OH)_2$-D₃

(55)

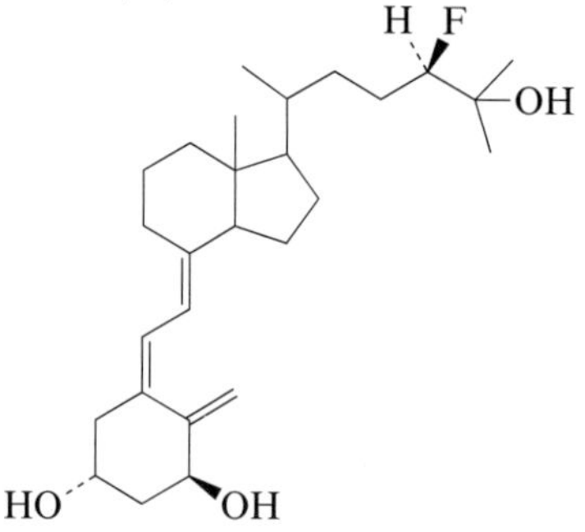

$24(R)$-F-1α, 25-$(OH)_2$-D₃

(56)

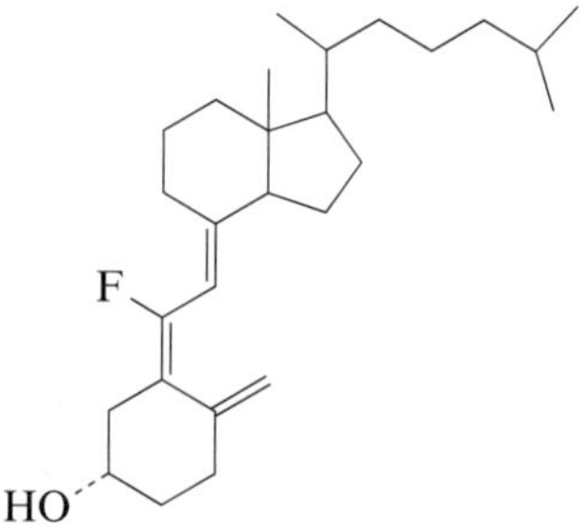

6-F-D₃

(57)

important contribution of the triene moiety to receptor activation[97]. More detailed discussions of these and other fluorinated D_3 analogues can be found in recent reviews[87,93].

4. Fluorinated retinoids

The photochemical isomerization of 11-*cis*-rhodopsin to all-*trans* bathorhodopsin is the primary photochemical process of vision. The effects of selective fluorination of retinals and derived rhodopsins on photoisomerization processes related to vision have been studied extensively. For example, 10-fluoro- (58), 12-fluoro- (59), and 14-fluororetinal (60) and the corresponding fluororhodopsins were prepared, one goal being the development of ^{19}F-NMR probes for studying photochemical double bond isomerizations and subsequent dark reactions[98]. Altered photochemical behavior of 9-*cis*-10-fluororhodopsin and batho-10-fluororhodopsin was attributed to interaction of fluorine situated at configurationally critical double bonds with amino acid residues of opsin[98–100]. Phenylretinals with fluorine labels at both *ortho* positions have been used to provide direct evidence for restricted ring/chain rotational equilibration in the protein-bound chromophores[101].

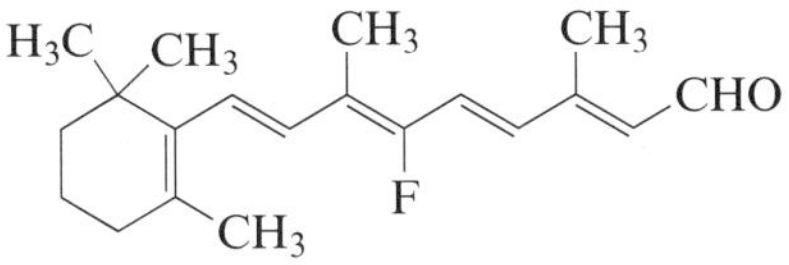

(58)

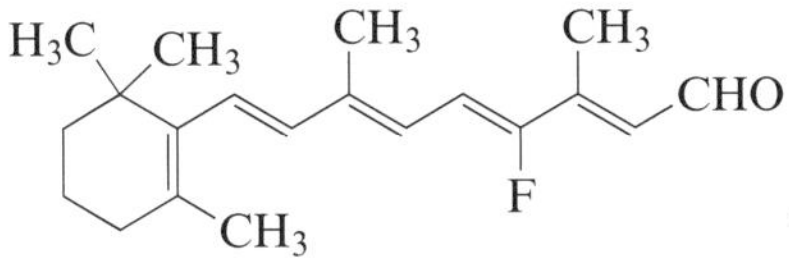

(59)

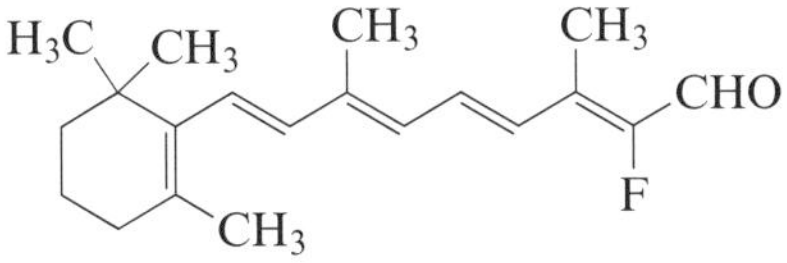

(60)

D. Halogenated Prostaglandins

Arachidonic acid-derived eicosanoic acids, including the primary prostaglandins, prostacyclins, throboxanes and leukotrienes (unsaturated eicosanoic acids having hydroxy and/or hydroperoxy substitution), have important and diverse physiological roles. Drug development based on prostaglandin analogues has included synthesis of many halogenated derivatives of each class of eicosanoid. Introduction of fluorine has been used in several strategies of drug design, including attempts to increase activity and selectivity, to block metabolism and to increase chemical stability, using similar rationales as used with steroids, vitamin D_3, and other classes. Although a large number of fluorinated compounds have been produced, the range of useful analogues comparable to fluorinated steroids has not been achieved. Selected examples will be given below. Several reviews are available for more thorough treatment[102–105]. Naturally occurring halogenated marine prostaglandins were discussed in Section II.

1. Halogenated primary prostaglandins

Approaches to the improvement of prostaglandin activity have included introduction of a carbon–fluorine bond at positions expected to impede metabolic inactivation. For example, $PGF_{2\alpha}$ (61) has smooth muscle stimulatory activity as well as potent luteolytic activity. (+)-12-Fluoro-$PGF_{2\alpha}$ (62) has a 10-fold increase in luteolytic activity and a significant decrease in effect on smooth muscles, properties attributed to the fact that this analogue is not a substrate for placental 15-hydroxyprostaglandin dehydrogenase[106]. In another example, 2,2-difluoro-$PGF_{2\alpha}$ (63) and 2,2-difluoro-PGE_2 (64) have longer durations of action, presumably because of blockade of β-oxidation, a major metabolic pathway of prostaglandin metabolism[103].

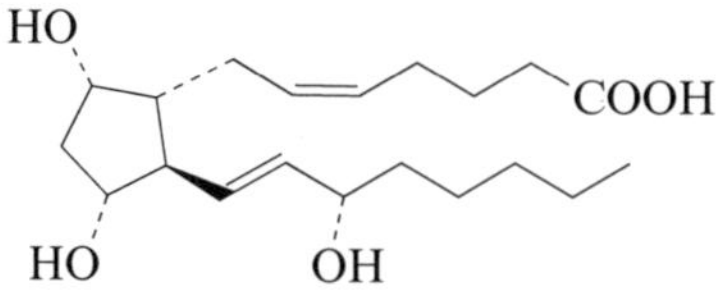

PGF$_{2\alpha}$
(61)

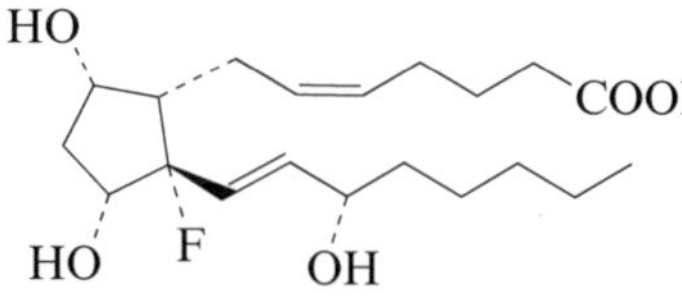

12-Fluoro-PGF$_{2\alpha}$
(62)

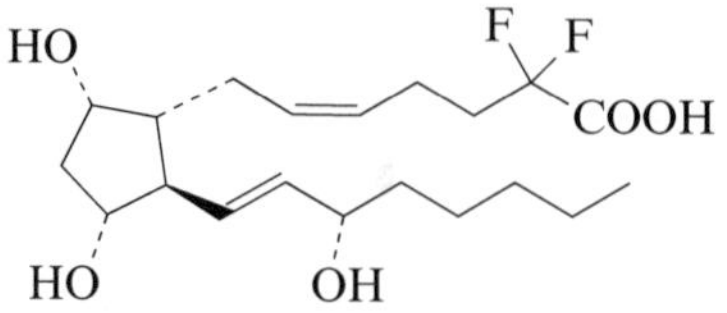

2,2-Difluoro-PGF$_{2\alpha}$
(63)

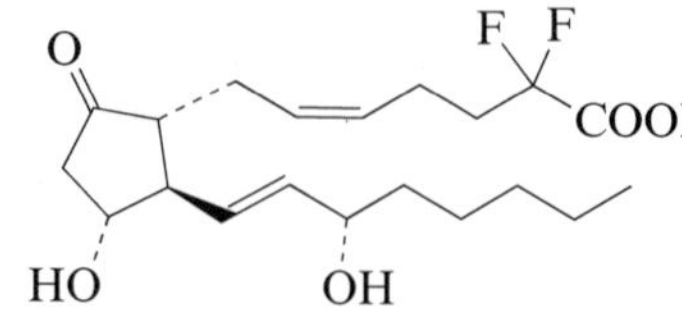

2,2-Difluoro-PGE$_2$
(64)

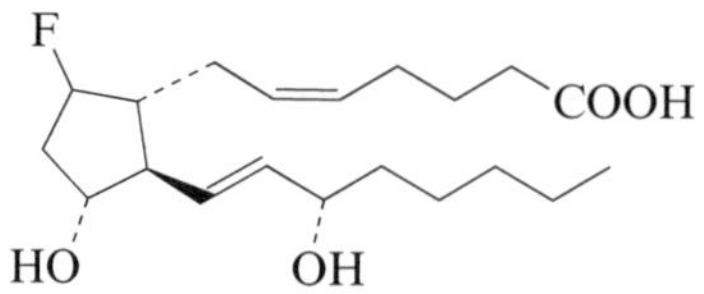

(65)

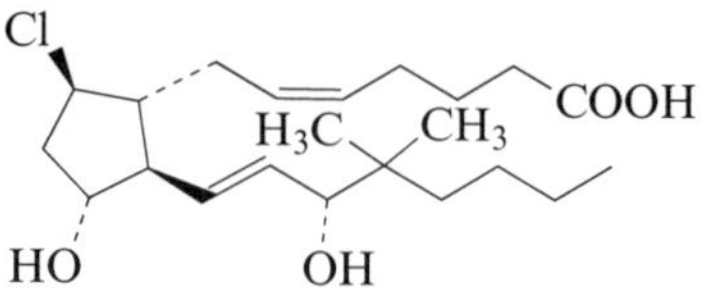

(66)

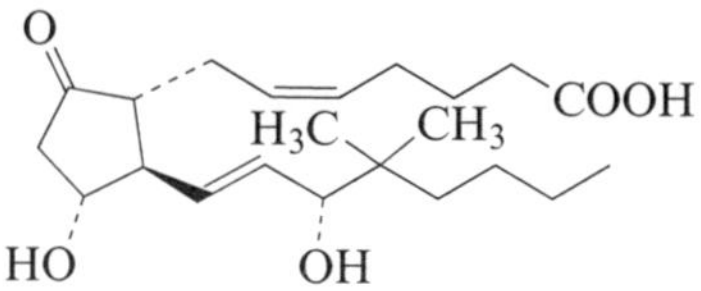

(67)

Several analogues of primary prostaglandins have been made wherein a halogen has been substituted for a carbonyl or hydroxyl group. In certain cases increased activities are found. For example, the 9-deoxy-9β-fluoro-11α-hydroxy analogue, **65** (replacement of the C-9 hydroxyl with fluorine), was three to four times more potent than PGE$_2$ as a bronchodilator[107]. PGE$_2$ inhibits ulcer formation and promotes healing of ulcers. Chemically stable PGE$_2$ analogues, wherein the C-9 hydroxyl group has been replaced with halogen, have been explored as anti-ulcer agents. For example, 9-deoxy-9β-chloro-16,16-dimethyl-PGF$_{2\alpha}$ (**66**) is comparable to 16,16-dimethyl-PGE$_2$ (**67**) in preventing gastric erosion but less potent in causing diarrhea[108].

2. Halogenated analogues of prostacyclin and thromboxane

Prostacyclin (PGI$_2$, **68**) and thromboxane (TXA$_2$, **69**) have potent and opposing biological properties that are important in vasculature regulation. The extremely, unstable TXA$_2$ ($t_{1/2}$ = 30 s at 37 °C) strongly contracts the aorta and induces platelet aggregation while PGI$_2$ ($t_{1/2}$ in blood at 37 °C of two-three min) is a potent inhibitor of platelet

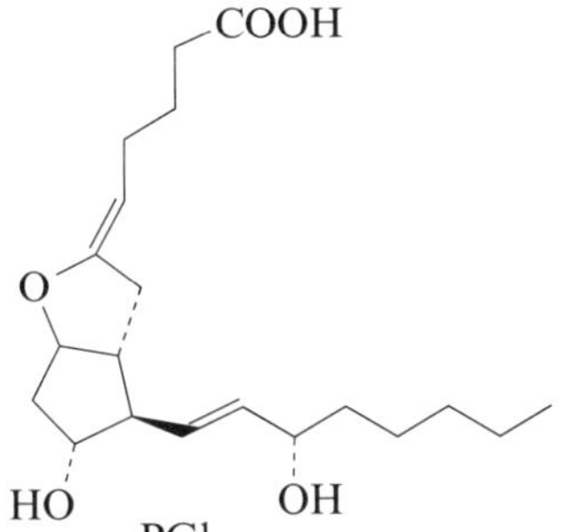

PGI$_2$
(**68**)

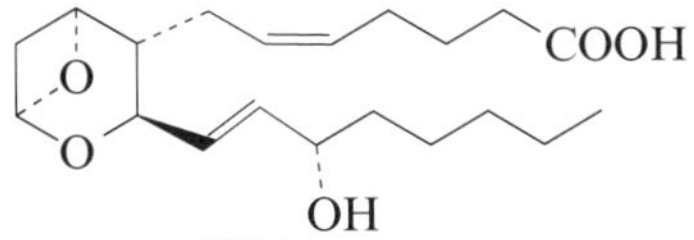

TXA$_2$
(**69**)

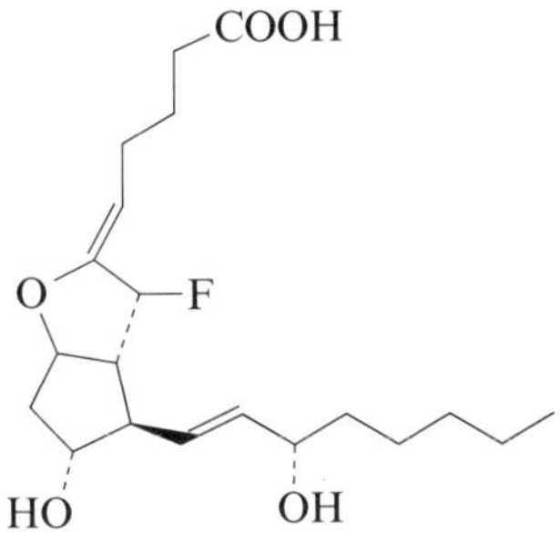

7-Fluoro-PGI$_2$
(**70**)

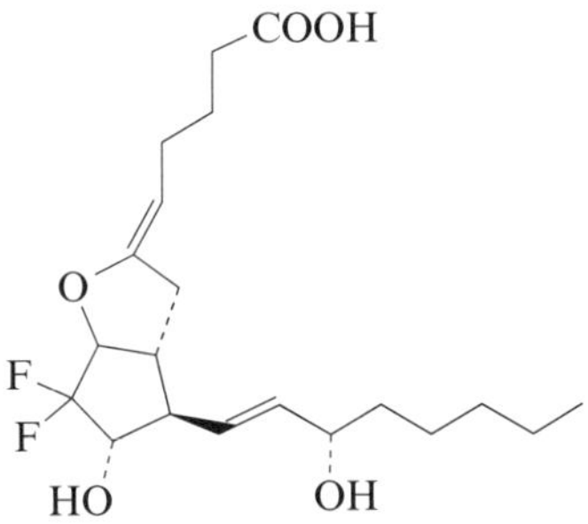

10,10-Difluoro-13-dehydro-PGI$_2$
(**71**)

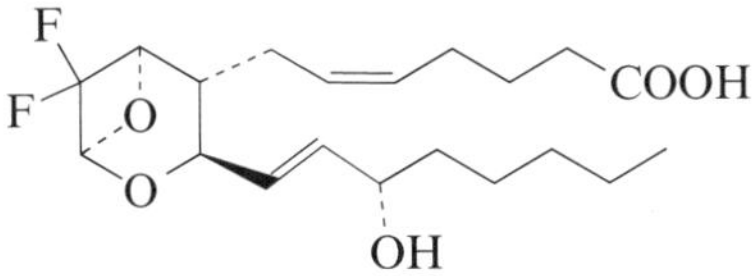

10,10-Difluoro-TXA$_2$
(**72**)

aggregation. Chemically more stable analogues of PGI$_2$ and TXA$_2$ would have obvious clinical potential, and also would facilitate biochemical and pharmacological investigations of processes mediated by these important eicosanoids. Strategies based on selective fluorination have had noteworthy success in the search for such analogues. For example, the presence of the highly electronegative fluorine in 7-fluoro-PGI$_2$ (**70**) decreases the electron density of the hydrolytically labile enol ether to the extent that this analogue has a half-life of one month in pH 7.4 buffer. 7-Fluoro-PGI$_2$ is a potent inhibitor of platelet aggregation with an initial activity 10% that of PGI$_2$. Unlike PGI$_2$, which rapidly loses activity, 7-fluoro-PGI$_2$ retains complete activity for at least four hours[109]. The potent and stable analogue, 10,10-difluoro-14-dehydro-PGI$_2$ (**71**), synthesized by Fried and coworkers, had prolonged *in vitro* activity, but *in vivo* activity was not prolonged[110]. In a similar approach to stable thromboxane, the hydrolytically stable 10,10-difluoro-TXA$_2$ (**72**) was found to be four to five times more potent than TXA$_2$ with respect to stimulation of platelet aggregation[111,112]

3. Halogenated analogues of arachidonic acid

Geminal difluoro-substituted analogues of arachidonic acid have been used as substrates for PGH synthase as an alternative approach to difluoro-analogues of TXA$_2$ and PGI$_2$, with unexpected but useful outcomes. For example, although 10,10-difluo-

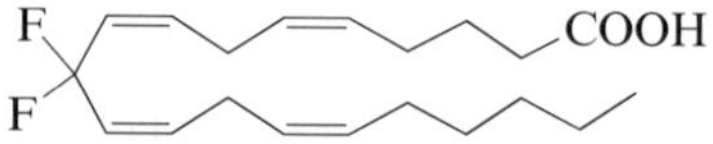

10,10-Difluoroarachidonic acid
(**73**)

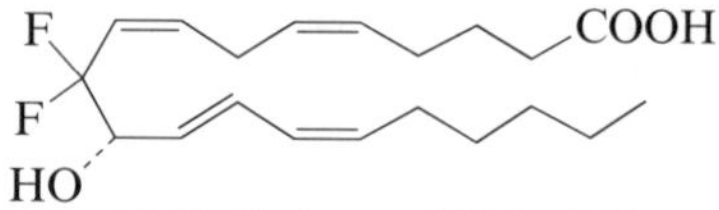

10,10-Difluoro-11S-HETE
(**74**)

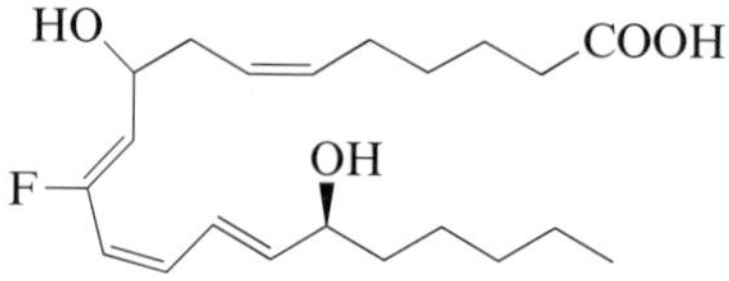

10-Fluoro-8,15-di-HETE
(**75**)

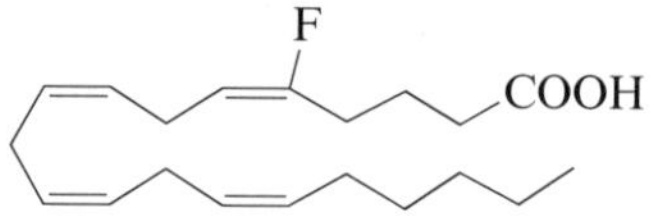

5-Fluoroarachidonic acid
(**76**)

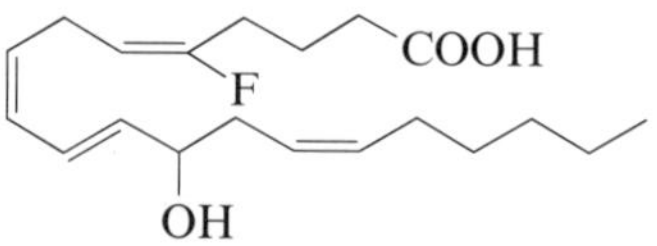

5-Fluoro-12-HETE
(**77**)

20,20,20-Trifluoro-AA
(**78**)

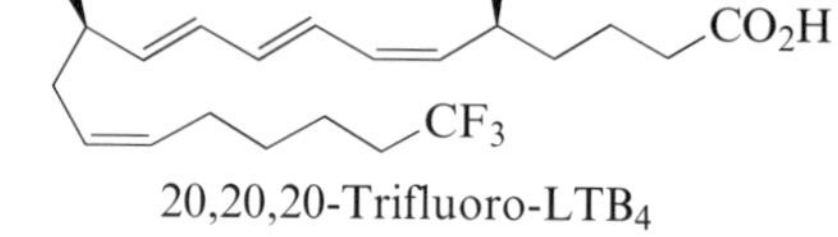

20,20,20-Trifluoro-LTB$_4$
(**79**)

roarachidonic acid (**73**) was a substrate for PGH synthase, no cyclization to prostaglandins was observed. Instead, fluorinated analogues of the biologically important hydroxye-icosanoic acids (HETEs) were formed, including 10,10-difluoro- 11S-HETE (**74**) and 10-fluoro-8,15-di-HETE (**75**)[113]. In related work, 5-fluoroarachidonic acid (**76**) was convert-ed enzymatically to the 5-fluoro analogue (**77**) of 12-HETE, a major product of arachidonic acid found in platelets[114]. 20,20,20-Trifluoroarachidonic acid (**78**) has been converted enzymatically to 20,20,20-trifluoro-LTB$_4$ (Leukotriene B$_4$, **79**). These analogues will be important in biochemical studies of the parent LTB$_4$, a leukotriene that has impor-tant functions in neutrophils[115,116].

E. Halogenated Nucleosides and Nucleotides

A large number of halogenated analogues are among the multitudinous synthetic nucle-osides and nucleotides that have been prepared in the search for effective antiviral and anti-tumor agents. A comprehensive survey of this field is beyond the scope of this chapter. Selected examples will be given with an emphasis on the biochemical rationales involved.

1. Halogenated pyrimidines

The folic acid-dependent conversion of deoxyuridine monophosphate (dUMP) to deoxythymidine monophosphate (dTMP) carried out by thymidylate synthase is an absolute requirement for DNA synthesis. An unusually high demand for uracil (ura) by certain tumor cells suggested that such an inhibitor of this process could have tumor cell selectivity. 5-Fluorouracil (fl^{5}ura) (**80**), along with 5-fluorocytosine (fl^{5}cyt) (**81**) and 5-fluoroorotic acid (fl^{5}oro) (**82**), were synthesized by Heidelberger in 1957 as part of a

| fl^{5}ura | fl^{5}cyt | fl^{5}oro |
| (**80**) | (**81**) | (**82**) |

program designed to develop such inhibitors. Additional activity of fl^{5}ura was considered possible through ribosylation, phosphorylation and incorporation of 5-fluorouridine triphosphate (fl^{5}UTP) into RNA. These expectations proved valid, and fl^{5}ura became an important antitumor agent, especially for treatment of solid tumors of the colon, breast and prostate. fl^{5}dUMP, (5-fluorodeoxyuridine 5-monophosphate) formed enzymatically *in vivo* from fl^{5}ura, forms a tight ternary complex with thymidylate synthase and tetrahy-drofolate, thus making the enzyme unavailable for thymidine synthesis (Figure 6)[117]. Toxic mechanisms are also manifested by participation of fl^{5}UTP in the synthesis of several RNA fractions, including rRNA, mRNA, tRNA and snRNA. To a lesser extent, fl^{5}dUTP is a substrate for DNA synthesis, a fact that may be implicated in cytotoxicity. There has been considerable controversy and research focused on the precise mechanism(s) of activity of fl^{5}ura against different tumor strains, and it appears that different mechanisms may be operating, depending on the conditions and/or tumor type. These issues are covered more thoroughly in reviews[118,119].

Although 5-fluorodeoxycytidine (fl^{5}dC) (**83**) is more toxic than fl^{5}dU (5-fluoro-deoxyuridine), and is less potent as an inhibitor of thymidylate synthase, there has recently been renewed interest in the antitumor activity of this analogue. Tumor cells have

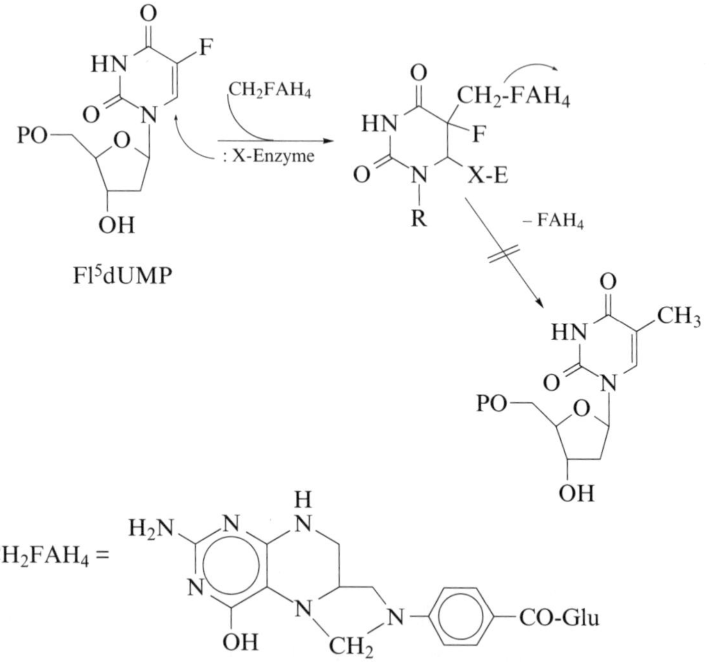

FIGURE 6. In the stable covalent ternary complex formed between thymidylate synthase, CH_2THF_4 and fl^5dUMP, the presence of fluorine at C-5 blocks deprotonation and elimination of FAH_4

high levels of dC kinase and dC deaminase, enzymes involved in the *in vivo* conversion of dC to dUMP. Coadministration of optimal doses of tetrahydrouridine, an inhibitor of dC deaminase, protects normal cells from fl^5dC, and leads to selective accumulation of fl^5dUMP in tumor[120].

In a process similar to the inhibition of thymidylate synthase by fl^5dUMP, fl^5dC blocks (cytosine-5)-methyltransferase- (DCMtase)-catalyzed methylation of dC residues in DNA. In this case, a catalytic SH group on the enzyme adds to the 6-position of the pyrimidine. After transfer of a methyl group from AdoMet ((S)-adenosyl metionine) to the 5-position of this -enzyme–inhibitor complex, the presence of fluorine blocks the elimination of enzyme-SH (Figure 7)[121].

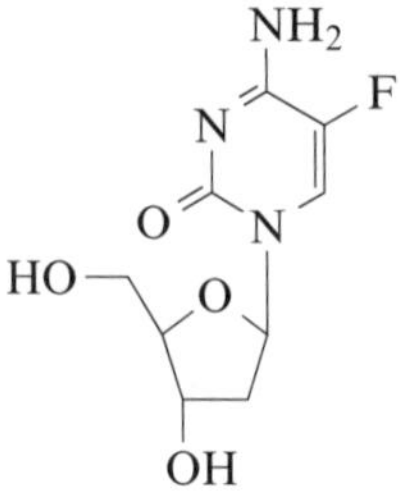

fl^5dC
(83)   (84) 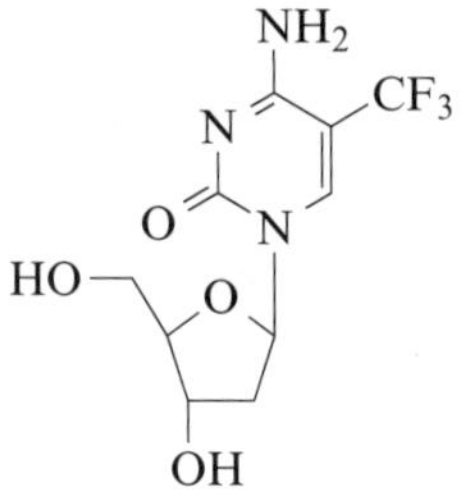   (85)

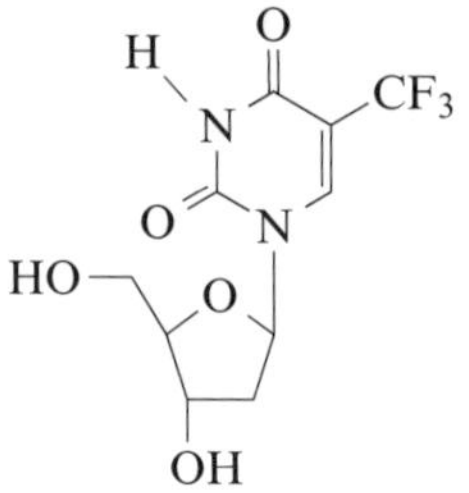

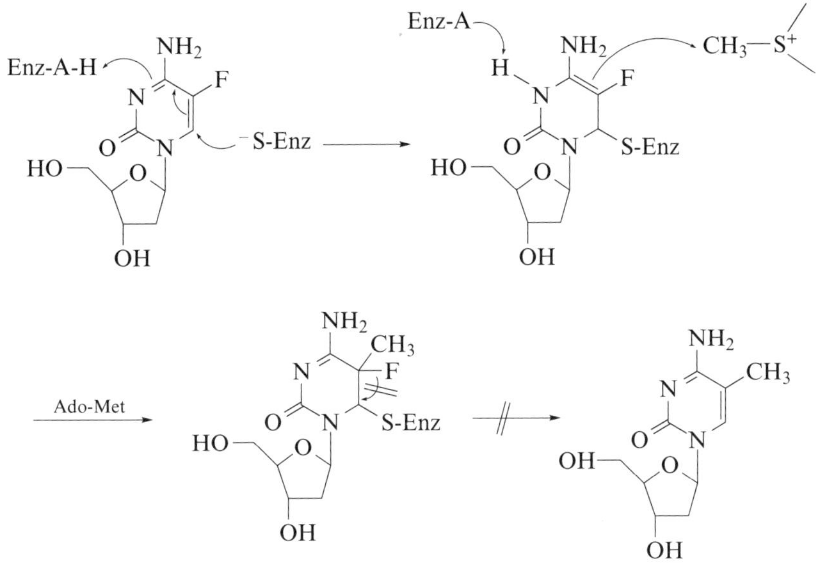

FIGURE 7. Schematic representation of the proposed mechanism of inhibition of (cytosine-5)-methyltransferase by fl^{5}dC

5-Trifluoromethyldeoxyuridine (CF$_3$dU) (**84**) has potent antiviral and antitumor activities, but its use has been complicated by rapid metabolism and by toxicity. In an approach similar to that used with fl^{5}dC, 5-trifluoromethyldeoxycytidine (CF$_3$dC) (**85**) has been used as a prodrug for CF$_3$dU, with tumor-cell selectivity achieved by coadministration of tetrahydrouridine. Antitumor effects are DNA-directed[122].

Whereas the effectiveness of 5-fluoropyrimidines as antimetabolites derives in part from the isosteric relationship between hydrogen and fluorine, the DNA-directed biological properties of 5-chloro-, 5-bromo- and 5-iodouridine (cl^{5}dU, br^{5}dU, io^{5}dU; **86**) reflect the ability of these halogen-substituents to mimic the presence of the 5-methyl group of thymidine. Br^{5}dU and io^{5}dU received early attention as antitumor and antiviral agents, and io^{5}dU, used for topical treatment of hepatic keratitis, was the first antiviral agent to be licensed by the US FDA[123]. Toxicity related to non-selective incorporation of the 5-halouridines into DNA has limited the *in vivo* use of these analogues. On the other hand,

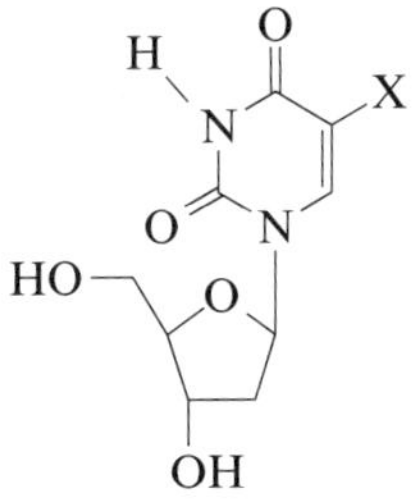

X = Br, Cl, I
(**86**)

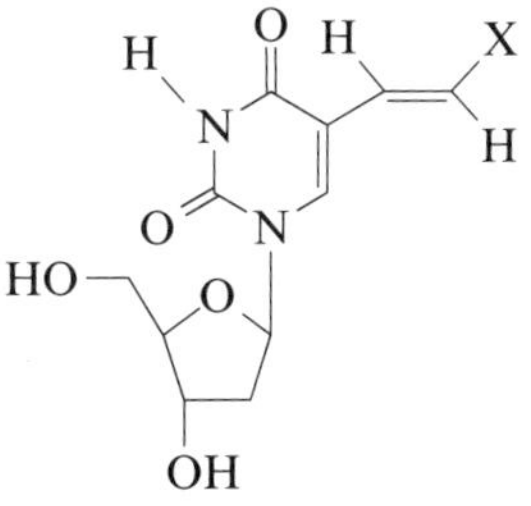

X = Br, BVdU
X = I, IVdU
(**87**)

these agents have been extremely valuable for the study of many aspects of DNA structure and function, such as cell cycle kinetics, cell differentiation and mutagenesis[124–126].

Development of new, more selective antiviral agents related to io^{5}dU and br^{5}dU have produced promising new candidates, including, for example, (E)-5-bromo- and (E)-5-iodovinyldeoxyuridine (BVdU and IVdU) (**87**). Both BVdU and IVdU are effective in the treatment of varicella-zoster virus (VZV), responsible for shingles, and herpes simplex virus 1 (HSV-1)[127].

Many other halogenated pyrimidine nucleosides have been prepared as potential anti-tumor and antiviral agents, including those incorporating altered sugar moieties. Prolonged *in vivo* activity through pro-drug strategies, greater resistance to metabolic enzymes, or more specific action on viral or tumor target enzymes are among guiding principals in many of these approaches. These strategies are discussed more thoroughly in recent reviews[104,117,119,128]. Examples of pyrimidine nucleosides having halogenated sugar residues will be discussed below.

2. Halogenated purines

The first fluorinated purine prepared, 2-fluoroadenosine (fl^{2}A, **88**), is highly cytotoxic, but has no therapeutic potential because of a lack of tumor-cell selectivity. In contrast, the corresponding arabinosyl analogue, ara-2-fluoroadenosine (ara-fl^{2}A) (**89**), has good tumor-cell selectivity and greater metabolic stability than fl^{2}A. Activity arises from inhibition of DNA polymerase α and ribonucleotide reductase, resulting in a 'self-potentiating' inhibition of DNA synthesis[129]. Phase I clinical trials revealed unexpected toxicity[130] perhaps related to *in vivo* formation of 2-fluoroadenine (fl^{2}ade)[131]. Selectivity appears to be related to preferential uptake and a higher rate of metabolism in tumor cells[132].

2′-Deoxyribosides, but not ribosides, of 2-fluoro-, 2-bromo- and 2-chloroadenine are cytotoxic and effective against a leukemia L110 system[133].

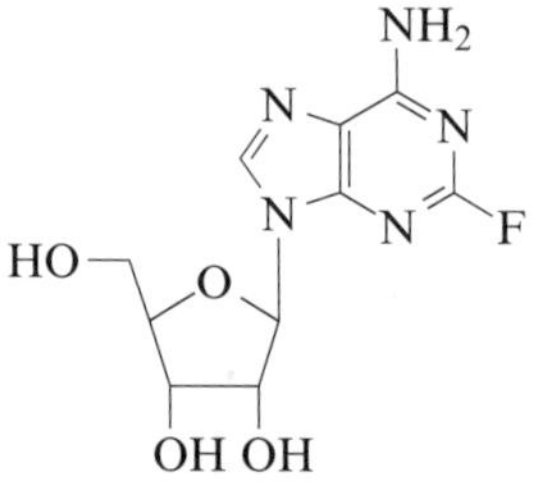

(**88**) 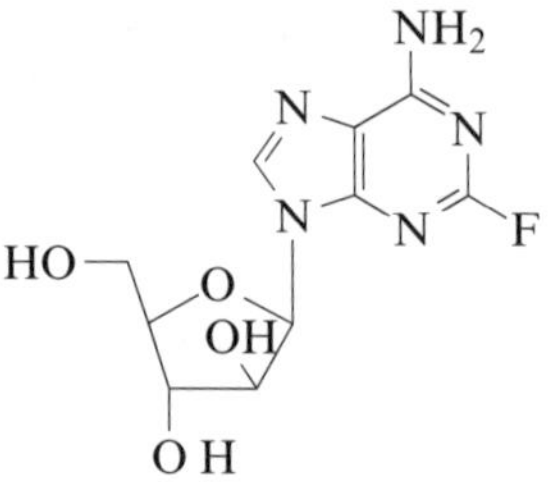   (**89**)

3. Sugar-halogenated nucleoside analogues

The presence of halogen on the sugar portion of a nucleoside can influence the biological behavior of the analogue in several ways. Altered intermolecular interactions can result from the substitution of halogen for an OH group that takes part in hydrogen bonding. Fluorine at the 2′-position has been found to increase the hydrolytic stability of the nucleoside linkage, increasing biological half-life. Replacement of a hydroxyl group required for phosphate ester formation can produce nucleic acid chain terminating analogues. A more subtle but nonetheless important factor involves alterations in furanose configuration that result from halogen substitution. These are among the considerations that have guided the design of sugar-halogenated nucleoside analogues as potential anti-tumor and anti-viral agents, and for use in other biochemical studies. A host of analogues have been prepared,

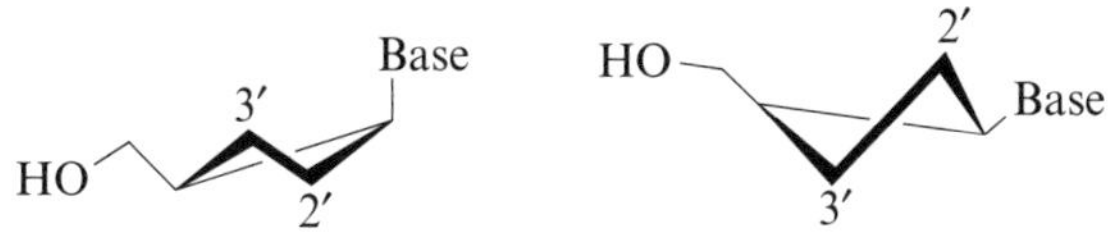

C (3') *endo*   C (2') *endo*
(N-Form)   (S-Form)

FIGURE 8. The N and S forms of nucleosides

and many show significant clinical promise. Illustrative but limited examples are given in this section.

2'-Deoxy-2-haloribonucleosides and oligonucleotides have received much attention, particularly with respect to altered ribose conformation and biological activity. Nucleosides can exist in either a C-3' *endo* (N) or C-2' *endo* (S) configuration (Figure 8). RNA is confined to the 3'-*endo* conformation while the more flexible DNA is predominantly in the C-2' *endo* conformation, but can assume a C-3' *endo* configuration. A linear relationship has been found between the electronegativity of a 2'-substituent and population of C-3' (N) configuration in adenine nucleotides, with 2'-deoxy-2' fluoroadenosine having a high population (67%) of the N-conformation[134]. Several studies have demonstrated that a 2'-halogen, and especially a 2'-fluorine, can impose a stabilized ribose configuration on polynucleotides containing this substitution. For example, a double-helical RNA configuration is a structural requirement for polynucleotide interferon inducers. Among a series of $(I)_n \cdot (C)_n$ and $(A)_n \cdot (U)_n$ analogues containing 2'-deoxy-2'-halogenated ribonucleotides, $(2'\text{-FdI})_n \cdot (C)_n$ and $(2'\text{-CldI})_n \cdot (C)_n$ showed unusually high interferon inducing activities. Increased thermal stability and resistance to nucleases of $(2'\text{-FdI})_n \cdot (C)_n$ suggested that this analogue should have an increased biological half-life[135].

Several 5-substituted 1-(2'-deoxy-2'-halogeno-β-D-arabinosyl)cytosines and uracils show potent anti-viral and anti-tumor activities. These analogues were designed to have fluorine in the arabinose configuration, based on the activities of *ara*-A and *ara*-C, and a substituent in the 5-position, based on the known anti-viral and anti-tumor activities of certain 5-substituted pyrimidines. Examples of potent antiviral analogues include 1-(2'-deoxy-2'-fluoroarabinosyl)-5-methyl uracil (FMAU, **90**) and 1-(2'-deoxy-2'-fluoroarabinosyl)-5-iodo cytosine (FIAC, **91**)[136]. In these analogues, fluorine in the arabino ('up') configuration is a requirement for activity. A fluorine-induced restricted rotation of the base about the sugar–base bond has been proposed to account for this. Selectivity results from higher activity of the analogues toward viral-encoded enzymes relative to normal enzymes.

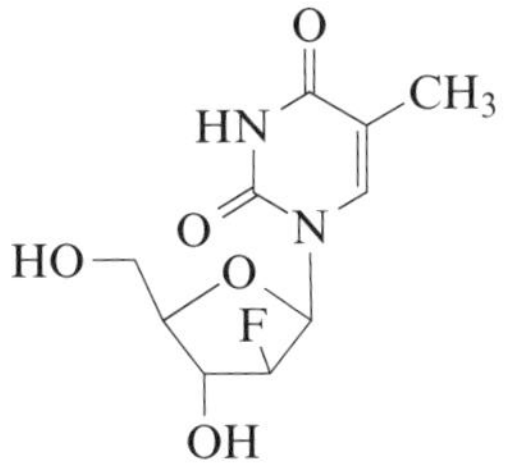

FMAU
(**90**)

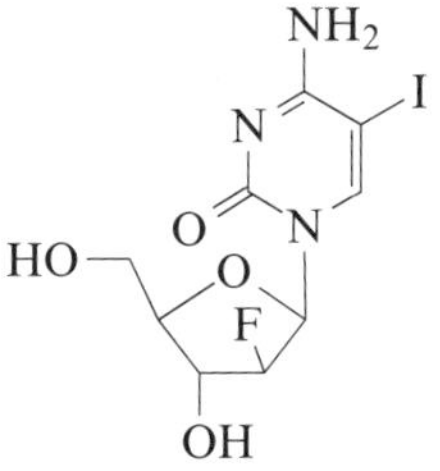

FIAC
(**91**)

Many other pyrimidine and purine nucleosides having halogenated ribose moieties have also been prepared. Halogen has been substituted on every available position of the ribose ring. This substitution has been combined with other structural variations in the sugar as well as in the base portion of the molecule.

F. Halogenated Carbohydrates

1. Halogenated carbohydrates as mechanistic probes

Halogenated analogues of carbohydrates have received much attention as tools for the study of glycolysis, glyconeogenesis and sugar transport mechanisms. Fluorinated analogues have been particularly useful in defining hydrogen-bonding interactions at macromolecular recognition sites for carbohydrates. Examples of these applications are given in this section.

In a process critical for maintaining proper blood glucose levels, glycogen phosphory-lase cleaves successive 1,4-glycosidic bonds from the non-reducing end of glycogen, the principal storage depot of glucose, to produce glucose-1-phosphate. A series of fluorinat-ed hexoses were important in defining the binding sites of glucose in the glycogen phos-phorylase–glucose complex, the complex that regulates the activity of glycogen phospho-rylase. By replacement of hydroxyl groups with either fluorine or hydrogen, hydroxyl groups were identified that functioned as hydrogen-bond donors and acceptors in interac-tions of glucose with the phosphorylase enzyme[137].

A series of 1-fluoroglycosides have been shown to be excellent substrates for glycosi-dases. For example, α- and β-maltosyl fluoride (**92** and **93**) are both good substrates for β-amylase, the enzyme that cleaves the 1,4-glycosidic bond of starch to produce β-maltose residues. These, and other fluorinated analogues, proved to be powerful tools to help delin-eate a general mechanism for the enzymatic cleavage of glycosidic bonds[138].

α-Maltosyl fluoride   β-Maltosyl fluoride
(**92**)   (**93**)

Fluorinated sugar analogues have been used extensively to probe mechanisms of sugar transport. The classic work of Barnett and coworkers on the mechanism of glucose trans-port made extensive use of fluorinated analogues of the substrate. By skillful replacement of hydroxyl groups with either fluorine or hydrogen, hydroxyl groups were identified that functioned as hydrogen-bond donors and acceptors in interactions with transporter proteins, both for the intestine active Na^+/glucose co-transport system and the erythrocyte facilitated glucose transporter[139,140].

More recently, the use of fluorine to probe hydrogen-bonding interactions has been demonstrated in the study of antigen–antibody interactions. A series of deoxyfluorogly-cosides were used as probes to define binding sites of antisachharide immunoglobulins[141].

Chlorinated sugars have been used to study the mechanism of the sweetness of organic compounds, and certain of these have been found to be several orders of magnitude sweet-er than sucrose. Preparation of several chlorinated analogues of sucrose revealed that halo-gen substitution on C-4, C-1′ and C-6′ positions greatly increased the sweetness of sucrose.

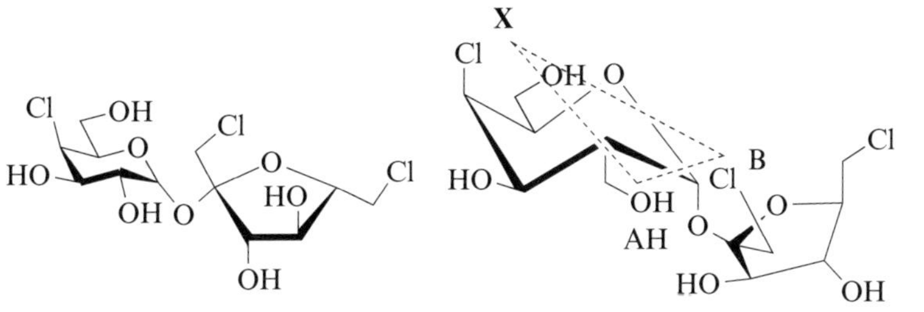

(94)

FIGURE 9. The extreme sweetness of trichlorogalactosucrose is explained by optimal interaction with an AH,B,X tripartite glucophore

For example, the trichloro-*galacto*-sucrose **94** is 2000 times sweeter than sucrose. Hough and Khan have rationalized the extreme sweetness of these trichloro analogues on the basis of an optimal interaction with an AH,B,X tripartite glucophore (Figure 9)[142]. Recent studies have extended examples of these intensely sweet sugars[143].

Protein glycosylation is another important area of carbohydrate biochemistry wherein halogenated analogues have been important mechanistic probes. Many proteins of eucaryotes have covalently linked oligosachharides as an important structural feature. Glycosylation of proteins is an integral part of such important factors as protection of the protein from its environment, transport, compartmentalization, conformational control and intermolecular association. Cell-surface glycoproteins play important roles in the immune response and in other cellular recognition phenomena. Alteration of protein glycosylation using sugar analogues, including halogenated analogues, has been used extensively to study glycosylation mechanisms.

The N-linked glycoproteins, an important class of membrane-bound glycoproteins, contain oligosaccharides linked to asparagine through an N–glycosidic bond. Although there is much diversity of structure of these N-linked glycoproteins, they all contain a core asparagine-linked pentasaccharide [Man$_3$(GlcNAc)$_2$ → Asn], indicative of their common synthetic pathway. In this biosynthetic pathway, a common oligosaccharide precursor is synthesized by stepwise addition of activated monosaccharides (derivatives of UDP (uridine 5′-diphosphate), GDP (guanosine 5′-diphosphate), and the polyisoprenoid, dolichol) to a carrier molecule, dolichol phosphate. The dolichol-linked oligosaccharide is then transferred *en bloc* to an asparagine residue on the protein, with release of dolichol phosphate (Dol-P) (Figure 10). Trimming and addition of new sugars produces the different classes of N-linked glycoproteins[144]. Intervention of the biosynthesis of N-linked glycoproteins, either at the dolichol pathway stage or at later processing, has been one approach to the study of mechanisms of glycosylation, and also as an approach to medicinal agents.

Examples of sugar derivatives that have been used in attempts to inhibit glycosylation at early stages of biosynthesis include deoxyglucose, 2-fluorodeoxyglucose (2-FDG), 2- and 4-fluorodeoxymannose (2- and 4-FDM) and the UDP and GDP derivatives thereof (Figure 11). For example, 4-FDM inhibits the glycosylation of viral G protein in vesicular stomatitis virus in infected BHK-2 1 cells through inhibition of the addition of GDP-Man to Man(GlcNAc)$_2$-PP-Dol. GDP-4-FDM was shown to be the inhibitory species[145,146].

Halogenated sugar analogues also have been used in attempts to block the processing of precursor oligosaccharide–protein conjugates to the glycoprotein products. Impetus for this work comes in part from the recognition that tumor cells often have altered membrane properties that are attributed to altered cell-surface glycoproteins. Several strategies were used to develop analogues that would be selectively toxic to tumor cells (see reviews by Sharma and coworkers[147] and by Bernacki and Korytnyk[148] for concise discussions of this

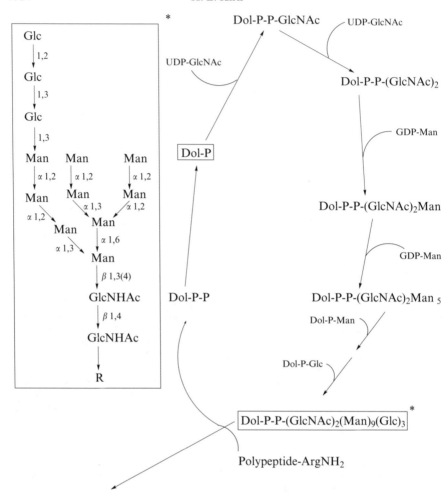

FIGURE 10. A schematic representation of the dolichol (Dol) pathway for the glycosylation of proteins. Reproduced from K. L. Kirk, *Biochemistry of the Elements, Volume 9B: Biochemistry of Halogenated Compounds*, by permission of Plenum Publishing Corp.

rationale.) For example, haloacetamido analogues of 2-amino-2-deoxyglucose and 2-amino-2-deoxygalactose, and O-acetylated derivatives thereof were synthesized, in recognition of the fact that glucosamines are important constituents of cell-surface glycoproteins[149]. Tetra-*O*-acetyl-2-*N*-bromoacetamido-2-deoxy-D-glucose (**95**) and -galactose (**96**) were found to be effective in Ehrlich ascites tumor-bearing mice[150]. Several other halogenated analogues of amino hexoses have been studied, with promising results in some cases.

Many other examples of the use of halogenated sugar analogues in glycoprotein research could be cited, including additional rational approaches to chemotherapeutic agents. These are discussed in more detail in recent reviews[147,151].

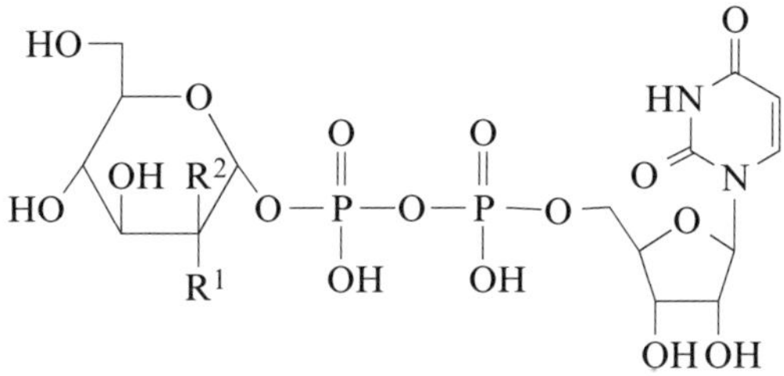

R¹ = R² = H (UDP-DG)
R¹ = F; R² = H (UDP-2FDG)
R¹ = H; R² = F(UDP-2FDM)

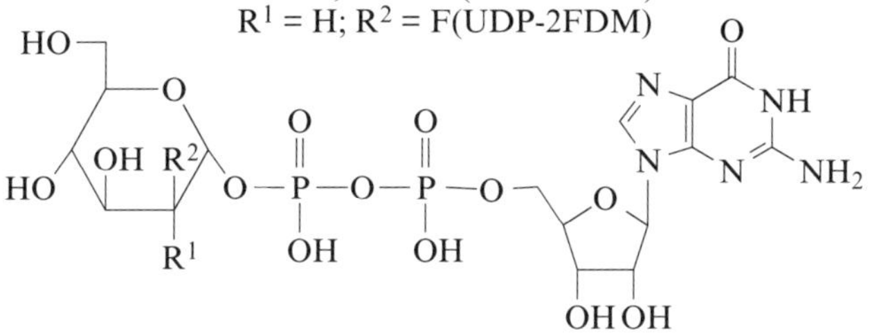

R¹ = R² = H (GDP-DG)
R¹ = F; R² = H (GDP-2FDG)
R¹ = H; R² = F(GDP-2FDM)

FIGURE 11. UDP and GDP conjugates of 2-deoxyglucose (2-DG), 2-deoxy-2-fluoroglucose (2-FDG) and 2-deoxy-2-fluoromannose(2-FDM)

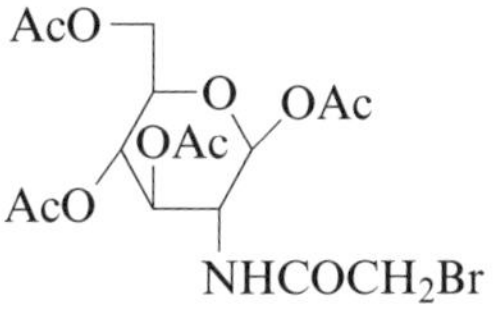

(95)

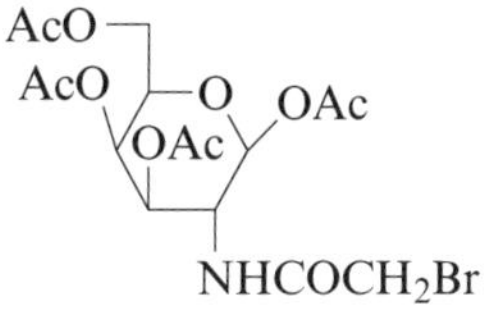

(96)

2. Halogenated analogues and carbohydrate metabolism

Halogen at C-1 of a hexose is at the locus of reactions involved with formation and cleavage of glycosidic bonds. Enzymatic processing of hexoses by enzymes in the glycolytic pathway, the pentose phosphate shunt and the aldose reductase sorbitol pathway can be affected by halogen substituted on internal carbons. Much early research with fluorinated sugars was related to the behavior of these analogues with metabolic enzymes. See reviews for discussions of this work[151,152].

An example of practical applications of this research is found in the use of ^{18}F-labelled 2-deoxy-2-fluoro glucose ([18F]2-FDG) (97) as a PET scanning agents. Glucose is the sole source of metabolic energy in the brain. 2-FDG efficiently crosses the blood–brain barrier and is phosphorylated to 2-FDG-6-phosphate by hexokinase. 2-FDG-6-phosphate is cleared slowly from brain tissue because of low membrane solubility, the low activity of glucose 6-phosphatase in the brain and resistance to further metabolism. Regional distributions of 'trapped' radiolabel, as determined by PET, is then related to regional glucose consumption. $[^{18}F]$-2-FDG is now used routinely in PET scanning as a valuable research and medical diagnostic tool[153].

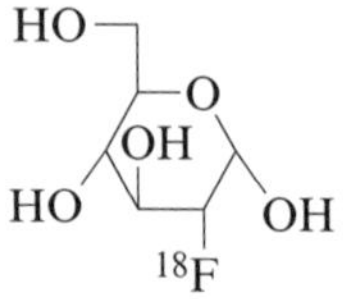

(97)

G. Halogenated Amino Acids

Analogues of amino acids can serve as inhibitors of specific enzymes, regulators of amino acid biosynthesis, substrates for biological peptide bond formation and precursors for other biologically active molecules such as aminergic neurotransmitters. Research on the biological behavior of analogue-containing peptides has provided a wealth of information regarding structure and function of individual amino acids in the native peptides. A large number of halogenated amino acid analogues have been made, and studies with these in all aspects of amino acid biochemistry over the past several decades have generated an enormous amount of literature. In this section, examples of research will be described that will illustrate various aspects of this biochemistry.

1. Halogenated analogues and amino acid biosynthesis

The 10 amino acids essential in the human diet (Arg, His, Ile, Leu, Lys, Met, Phe, Thr, Trp, Val) are synthesized by non-human organisms by multistep pathways starting from simple metabolic precursors. Amino acid biosynthesis is controlled by feedback inhibition and suppression of synthesis of biosynthetic enzymes. The ability of an amino acid analogue to block biosynthesis of the parent amino acid often contributes to the toxicity of the analogue. Mutants resistant to the toxic effects of the analogue can be valuable tools for studying various aspects of cellular mechanism (examples to be given below).

Beginning with early studies on the effects of 2-, 3- and 4-fluorophenylalanine (2-, 3-, 4-FPhe) on growth of microorganism[154], halogenated amino acids have been particularly useful probes of amino acid biosynthetic mechanisms. For example, halogenated analogues of Phe, Tyr and Trp have been used effectively in studying details of the biosynthesis of chorismate and of the branching of this key intermediate leading to the formation of Phe, Tyr and Trp. Several fluorinated aromatic amino acids function as false-feedback inhibitors and/or suppressors of biosynthetic enzymes involved in aromatic amino acid biosynthesis.

Toxic affects of 2-fluoro-L-histidine (2-FHis) (98) on bacteria have been attributed in part to the action of this analogue as a potent feed back inhibitor of ATP:phosphoribosyl transferase, the first committed enzyme in histidine biosynthesis. No other halogenated analogues of histidine were active in this regard[155].

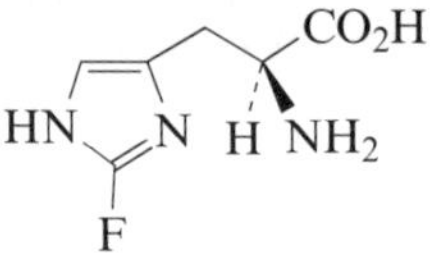

(98)

2. Effects of halogenation on amino acid metabolism

Halogenated aromatic amino acids have been important tools in studies of the mechanism of hydroxylation and dehalogenation of haloaromatic compounds, studies that uncovered the rearrangement known as the 'NIH' shift. Early studies by Weissman and Koe[156] and by Kaufmann[157] demonstrated that the monooxygenase-catalyzed hydroxylation of 4-FPhe produced tyrosine with release of an equivalent of fluoride. Later work by Guroff, Daly and coworkers at NIH revealed that facile halogen migration (the 'NIH' shift) occurred during similar hydroxylation of 4-ClPhe and 4-BrPhe (equation 9)[158]. The arene oxide mechanism that was established for this hydroxylation and halogen migration has been implicated in other processes, including the carcinogenicity of polycyclic aromatic hydrocarbons. 3-FPhe is hydroxylated, without rearrangement, to 3-FTyr. The latter is acutely toxic to mammals by virtue of being metabolized to fluoroacetate in the liver. The absence of toxicity of 4-FPhe is one consequence of the absence of the NIH shift during microsomal hydroxylative defluorination of this isomer to give Tyr.

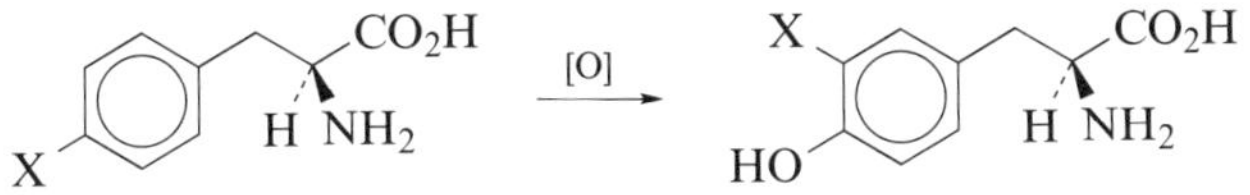

$$ (9) $$

X = Cl, Br

4-ClPhe is a competitive inhibitor of both phenylalanine and tryptophan hydroxylases, and produces selective irreversible inhibition *in vivo* of both enzymes[159]. As a tryptophan hydroxylase inhibitor, 4-ClPhe has been used extensively in a wide range of psychopharmacology experiments based on *in vivo* serotonin depletion. The ability of 4-ClPhe to produce increased levels of Phe *in vivo* has made it useful in the development of models for phenylketonuria[160].

3,4-Dihydroxyphenylalanine (DOPA) (99) is produced by tyrosine hydroxylase-catalyzed hydroxylation of Tyr. Recent interest in the use of [18F]-6-F-DOPA (100) as a PET scanning agent for regional dopaminergic brain function is based on its conversion, in the brain, to [18F]-6-F-dopamine (101)[161]. The fact that fluorine in the 6-position of DOPA, dopamine and other catecholamines retards methylation by catechol-*O*-methyl transferase presumably increases the biological half-life of the tracer. In contrast, fluorine in the 5-position increases the rate of methylation[162].

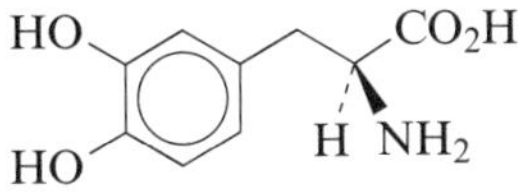

(99)

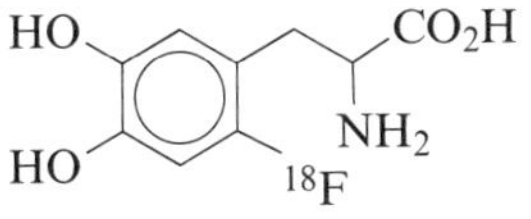

(100)

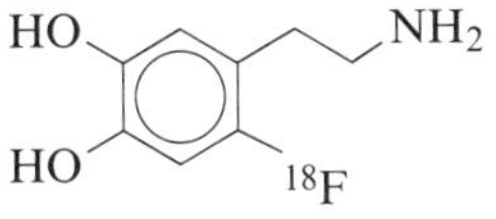

(101)

3. Halogenated amino acids as mechanism-based enzyme inhibitors

Introduction of a reactive group adjacent to the reaction locus has proven to be a very effective approach to irreversible inhibitors of many pyridoxal phosphate (PLP) dependent

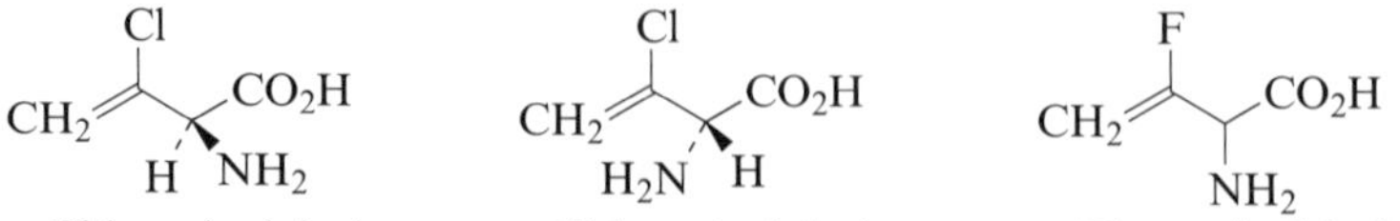

L-Chlorovinylglycine   D-Chlorovinylglycine   L-Fluorovinylglycine

FIGURE 12. 3-Halovinylglycines, mechanism-based inhibitors of alanine racemase

FIGURE 13. (1-Aminoethyl)phosphonic acid (Ala-P) and β-chloro-, β,β-dichloro-, and β,β,β-trichloro analogues of Ala-P, inhibitors of alanine racemase

enzymes. Included are enzymes, such as deaminases and decarboxylases, that are important for the processing of amino acids. Antibacterial agents and other chemotherapeutic agents have resulted from this work.

The synthesis of β-fluoro-D-alanine (β-F-D-Ala) (102) by Kollonitch and coworkers, and the demonstration that this analogue is a potent antibacterial agent was a seminal development[163]. Irreversible inhibition of bacterial alanine racemase, resulting in blockade of bacterial cell-wall biosynthesis, was shown to be the underlying mechanism of this activity. Many other halogenated alanine derivatives have been studied. Recent advances include the synthesis of 3-halovinyl glycines (Figure 12)[164] and halogenated phosphonate analogues of alanine (Figure 13)[165].

In PLP-dependent enzymatic reactions, the Schiff base formed by reaction of the substrate with PLP provides an electron sink for stabilization of the negative charge that results from the bond-breaking process required in the reaction (racemization, decarboxylation, aldol reaction, elimination, etc.). The elegant work of Walsh and coworkers provided evidence that, subsequent to Schiff base formation, a common intermediate is formed from several different alanine analogues that are alanine racemase inhibitors. From this they proposed the elimination–Michael addition sequence shown in Figure 14 as the mechanism for inhibition[166].

Although this mechanism was widely accepted, later work by Metzler and coworkers[167] and by Walsh and coworkers[168,169] revealed that, whereas the eneamino aldimine complex is formed in the inhibition process, subsequent elimination produces amino acrylate. Nucleophilic addition of amino acrylate to the Schiff base formed between the enzyme lysine residue and PLP results in irreversible inhibition (Figure 15). Support for this mechanism includes the isolation of the pyruvate derivative 103 (the Schnackerz adduct) upon denaturization. The significance of the reversed polarity pathway as an alternative mechanism is discussed effectively in a review by Walsh[170].

Much of the potential of the PLP-dependent amino acid decarboxylase inhibitors as medicinal and pharmacological agents stems from the crucial role of many amino acids as precursors of bioactive amines, including amine neurotransmitters. Early work by Kollonitch demonstrated the effectiveness of α-halomethyl and dihalomethyl analogues of amino acids as selective irreversible decarboxylase inhibitors. Among effective inhibitors were α-fluoromethyl-DOPA (104) (inhibits dopamine formation), α-fluoromethylhistidine (105) (inhibits histamine formation) and α-fluoromethylglutamate (106) (inhibits GABA formation)[171]. Enantioselective syntheses have produced $\alpha(S)$- (fluoromethyl)- tryptophan (107) and $\alpha(S)$-(fluoromethyl)-5-hydroxytryptophan (108). The former is a substrate for tryptophan hydroxylase, and the latter, the product of this hydroxylation, is an inhibitor of aromatic amino acid decarboxylase[172].

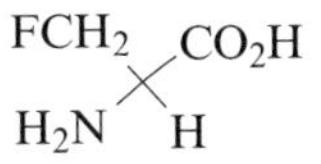

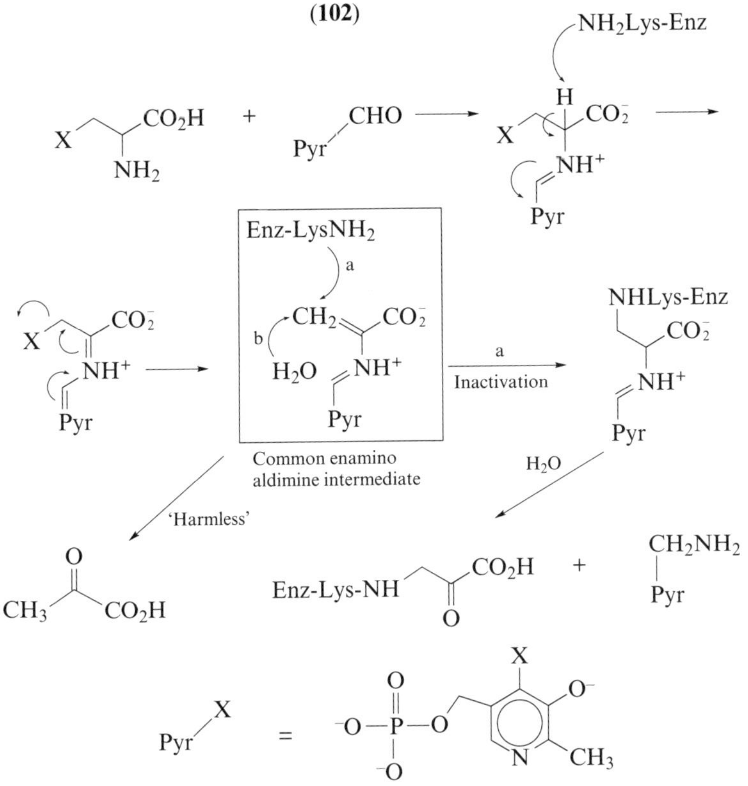

FIGURE 14. The initial elimination–Mechael addition mechanism proposed for the inhibition of alanine recemase by β-haloalanines. Reproduced from K. L. Kirk, *Biochemistry of the Elements, Volume 9B: Bichemistry of Halogenated Compounds*, by permission of plenum Publishing Corp.

The polyamines putrescine, spermidine and spermine have important roles in cellular replication and differentiation. In mammalian cells, ornithine decarboxylase (ODC) catalyzed decarboxylation of ornithine is the sole source of putrescine, the precursor of spermine and spermidine (equation 10). Based on this, several irreversible ODC inhibitors have been studied as potential chemotherapeutic agents. α-Difluoromethylornithine (DFMO) (109), one example of a series of mechanism-based ODC inhibitors, has proven to be very useful in the study the pharmacological consequences of ODC inhibition[173]. Although initial clinical anti-tumor activity was disappointing, DFMO has shown promise in treatment of protozoal infections, including African trypanosomiases (sleeping sickness).

Arginine decarboxylase (ADC) catalyzed decarboxylation of arginine is the initial step in an alternative pathway to putrescine in bacteria and higher plants. α-Difluoromethylarginine (DFMA) (110)[174] is an effective mechanism-based inhibitor of ADC and has been used to study compensatory processes involving ADC and ODC[175].

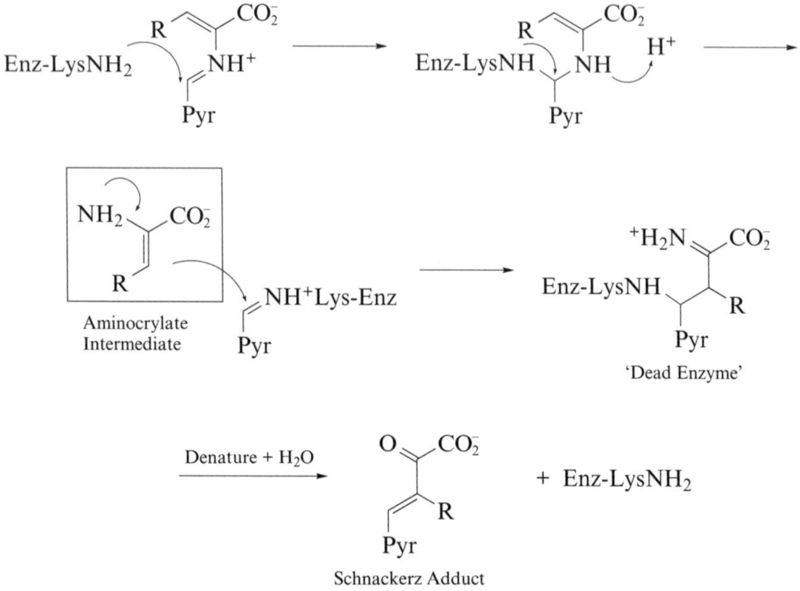

(103)

FIGURE 15. A revised mechanism for the inhibition of alanine recemase based on the release of α-aminoacrylate from the eneamino–PLP complex. Reproduced from K. L. Kirk, *Biochemistry of the Elements, Volume 9B: Biochemistry of Halogenated Compounds*, by permission of Plenum Publishing Corp.

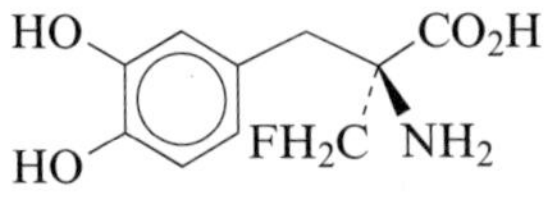

(104) 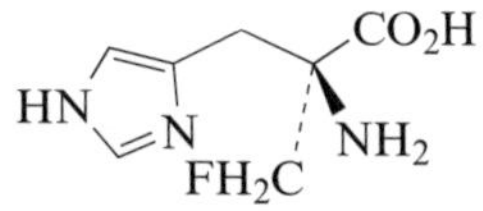   **(105)**

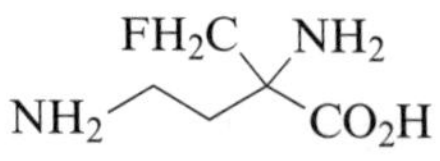

(106) 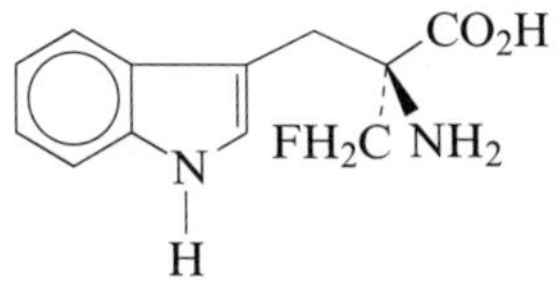   **(107)**

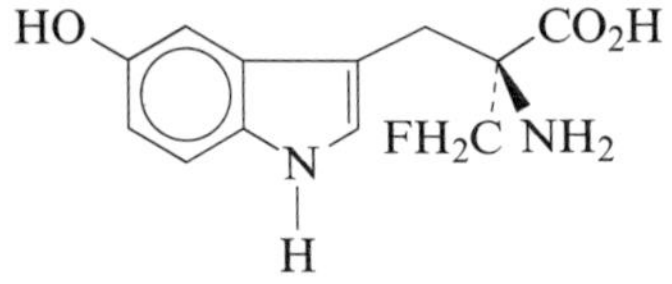

(108)

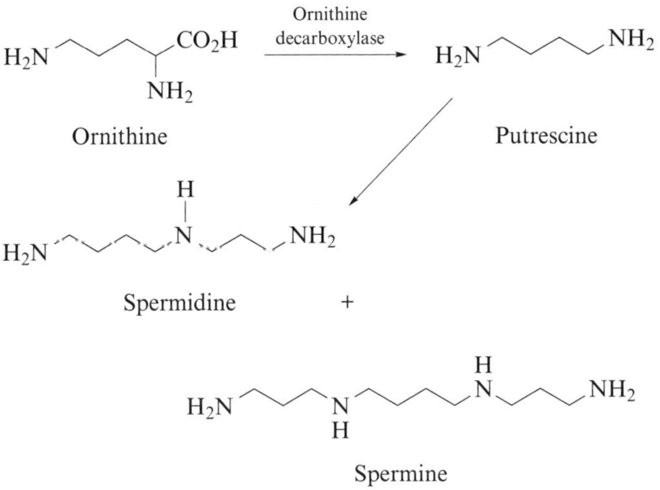

Ornithine decarboxylase

Ornithine → Putrescine (10)

Spermidine +

Spermine

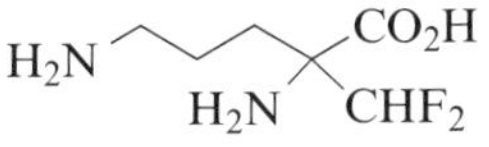

DFMO (109)   DFMA (110)

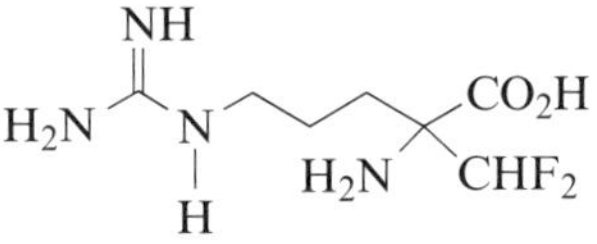

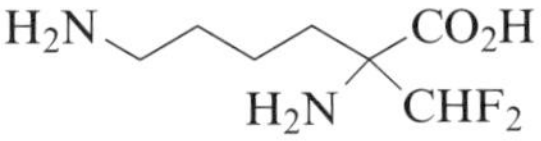

DFML (111)

In related work, α–difluoromethyllysine (DFML) (111) was synthesized and shown to be a potent irreversible inhibitor of lysine decarboxylase, responsible for the biosynthesis of cadaverine. This analogue has been examined as a potential drug for the treatment of certain mycoplasmic infections[176].

The neurotransmitter, γ-aminobutyrate (GABA), functions as an inhibitory amino acid by increasing the post-synaptic membrane permeability to K^+. GABA is formed by decarboxylation of glutamate and is inactivated by GABA-T catalyzed transamination to give succinate semialdehyde. Inhibitors of GABA-T that can cross the blood–brain barrier can function as anti-convulsants by causing an increase in GABA levels. Several fluorinated analogues of GABA have been studied as GABA-T inhibitors. Examples are (S)-4-amino-

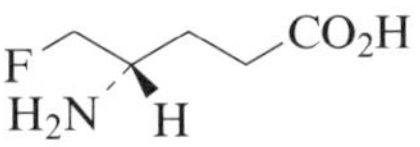

(112)

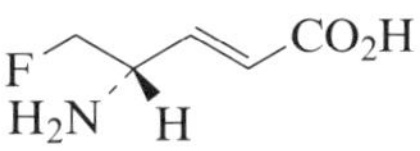

(113) 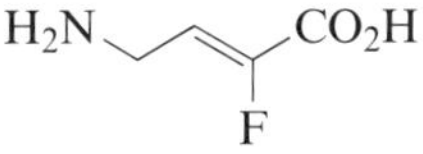   (114)

5-fluoropentanoic acid (**112**), (*S*)(*E*)-4-amino-5-fluoropent-2-enoic acid (**113**) and (*Z*)-4-amino-2-fluorobut-2-enoic acid (**114**)[177,178].

4. Incorporation of halogenated amino acids into proteins

An area of research of obvious importance and immense scope involves the incorporation of amino acid analogues into proteins. Incorporation is accomplished either by chemical synthesis or by biochemical synthesis using the protein assembly mechanisms of nature. Halogenated, especially fluorinated, amino acid analogues have been particularly useful tools in this research. In this section, selected examples of research using such analogues will be described to illustrate strategies and principles employed.

Biosynthetic techniques have been used to incorporate fluorinated analogues into protein sequences in many cell types for a variety of research purposes. Included are studies on the specificities of amino acid tRNA synthases, the study of mechanisms of the individual steps of protein biosynthesis, the use of the incorporated analogues as biological tracers for ^{19}F-NMR studies and investigations of functional and structural consequences of analogue incorporation. Fluorinated analogues are particularly useful in that the minimal steric perturbations introduced with fluorine frequently permit a fluorinated analogue to pass the strict editing processes that serve to insure fidelity in translation of genetic information to peptide bonds.

In the course of these studies several strategies have been developed to increase the level of analogue incorporation into microorganisms. In one approach, an auxotrophic bacterium unable to synthesize a particular amino acid is starved for that amino acid. Addition of the corresponding analogue can result in high levels of incorporation, often, however, with low protein yield due to loss of cell viability[179]. Induction of target protein biosynthesis combined with the use of auxotrophs can lead to efficient incorporation and greatly improved yields. In one example, a Tyr auxotroph was produced that overproduces aspartate transcarbamoylase (the enzyme that catalyzes the first committed step in pyrimidine biosynthesis) when starved for uracil. Addition of 3-FTyr in uracil-depleted cells resulted in 85% replacement of Tyr in the enzyme with retention of high enzyme activity. ^{19}F-NMR was used to study the communication between the catalytic and regulatory subunits of the enzyme[180,181]. In an impressive example of the adaption of an auxotroph to an amino acid analogue, Wong developed a Trp auxotroph of *Bacillus subtilis* that grew much better on 4-FTrp than on Trp. Wong suggested that this may be the first free-living organism in the past two billion years to learn to thrive on an altered genetic code. Since this organism can be supported on 4 FTrp, it is a convenient source of 4-FTrp-labelled proteins endogenous or clonable to *B. subtilis*, including proteins vital to cell growth[182,183].

3-FTyr has been incorporated into the *lac* repressor, a macromolecular component of the well-studied lactose operon. In this case, induction and replication of a prophage in the cell line greatly increased production of labelled *lac* repressor protein. ^{19}F-NMR was used to study the number and environment of Tyr residues in the protein[184]. Chain terminating (non-sense) mutations of the *lac* repressor gene at positions corresponding to the eight Tyr and two Trp residues in the *lac* repressor were subsequently used to systematically study the site of incorporation of the fluorinated analogues[185].

There are many other recent studies wherein fluorinated amino acids have been incorporated into bacterial protein, and into mammalian protein expressed in bacteria cells, for ^{19}F-NMR experiments. For example, 5-FTrp and 3-FPhe were incorporated into the D-galactose chemosensory receptor of *E. coli*. Site-directed mutagenesis was use to assign ^{19}F-signals[186].

The above examples are meant to illustrate some of the strategies used to incorporate fluorinated amino acids efficiently into microbial cellular protein. The review by Gerig[187] includes a discussion of the theory and practice of ^{19}F-NMR as applied to biological macromolecules, and contains many additional examples of fluoroamino acid labeled proteins. The review by Sykes and Weinen[179] cites many additional studies and contains a discussion of NMR spectra. The review by Wheatly[188] tabulates several examples of 4-FPhe incorporation. Incorporation of fluorinated amino acids into proteins and peptides is also covered more thoroughly in a recent chapter[189].

Much research also has been directed toward the isolation of fluoroamino acid-labeled proteins from mammalian sources. Analogue-containing proteins have been used for ^{19}F-NMR studies, as well as for examination of the effects of analogue incorporation on protein structure and function. Attention also has been paid to the effects of fluoroamino acid analogues on completion of protein chains, and to the behavior of analogue-containing protein toward protein processing. An early example of structural effects is seen in the effect of 4-FPhe on cell division. Wheatly and Henderson[190] found that the ability of HeLa cells to enter into the G_2 phase of the cell cycle was almost completely inhibited by 0.2 mM D,L-4FPhe, suggesting that division-related proteins containing 4-FPhe may malfunction.

threo-2-FAsn

(115)

threo-β-Fluoroasparagine (*threo*-FAsn) **(115)** is an example of an analogue that can alter post-translational processing of proteins. This analogue is highly toxic to certain mammalian cells in culture, an observation that has increased significance in light of the fact that some lymphoma and leukemia cells require asparagine due to a deficiency of asparagine synthase. As discussed above, asparagine-linked glycosides are an important structural feature of many proteins. Inhibition of protein glycosylation due to incorporation of *threo*-FAsn was thus considered to be a likely contributing factor to toxicity. Indeed, in a cell-free translation system from Krebs II ascite tumor cells, *threo*-FAsn was an effective inhibitor of protein glycosylation. Inhibition was blocked by Asn, further supporting incorporation of *threo*-FAsn as a toxic mechanism[191].

A His residue is often found at the active site of enzymes where it functions as a catalyst in acid–base and nucleophilic processes. Substitution of fluorine on His reduces the pK_a by about 5 pH units, and this dramatic drop in basicity is reflected in altered biological properties of FHis-containing proteins. The presence of 2-FHis in cell cultures inhibits the stimulation of several enzymes, for example, the stimulation of pineal gland N-acetyltransferase activity, in cell culture and *in vivo*. This stimulation is accompanied by His and cycloheximide-sensitive incorporation of 2-FHis into cellular protein[192,193]. A direct comparison of His and 2-F-His in mouse L cells showed that the analogue is incorporated at about 17% the efficiency of the parent[194]. 4-FHis showed none of the above biological activity.

There are many additional studies of *in vitro* incorporation of fluorinated amino acids into mammalian protein that could be cited. In addition, there are many examples of *in vivo* incorporation. The first report appears to be that of Westhead and Boyer[195] who showed that 4-FPhe could be incorporated to a significant degree into rabbit protein by maintaining the animal on a diet that contained the analogue. Gerig and coworkers have isolated, purified and studied by ^{19}F-NMR, proteins, including hemoglobin and carbonic anhydrase, from animals maintained on 4-FPhe-containing diet[196–198]. The toxicity of certain amino acids may be associated with protein biosynthesis, although other mechanisms can also operate. For example, 2-FHis (but not 4-FHis) is toxic at *ca* 250 mg/kg in the mouse, and becomes incorporated into all tissues examined[199].

Chemical synthesis has provided an additional route to peptides containing halogenated amino acids. Early ^{19}F-NMR studies of proteins were performed on semi-synthetic polypeptides prepared by attachment of fluorinated probes to the polypeptide. For example, Heustis and Raftery modified ribonuclease by trifluoroacetylation of Lys residues 1 and 7. They then used ^{19}F-NMR to study conformational changes brought about by the presence of inhibitors[200]. In his review, Gerig provides several other examples of this strategy[187].

More recently, total synthesis has been used increasingly to incorporate amino acid analogues, including halogenated analogues, into polypeptides. For example, Val in angiotensin II (Asp-Arg-Val-Tyr-Ile-His-Pro-Phe) was replaced with $\gamma,\gamma,\gamma,\gamma,\gamma,\gamma$-hexafluorovaline (HFVal) in order to develop a probe for hormone–receptor interactions. This HFVal analogue had 133% activity compared to angiotensin II and, furthermore, was resistant to proteolytic digestion[201]. In a subsequent study, introduction of HFVal into sarconsin^1-containing analogues of angiotensin II produced long-acting inhibitors of the hormone[202]. In another example, the amino-terminal decapeptide of ribonuclease was synthesized containing 4-FHis as a replacement for His at position 12. Although the synthetic polypeptide formed a stable non-covalent complex with native ribonuclease (21–24)[203], the complex was devoid of catalytic activity, presumably due to the lowered basicity of 4-FHis now present at the active site[204].

In a related study with halogenated His analogues, 4-FHis- and 2-CF$_3$-His-containing analogues of thyrotropin releasing factor (pyro-Glu-His-ProNH$_2$) did not bind to pituitary GH$_4$ cells *in vitro*, nor stimulated prolactin release, but the analogues did elicit *in vivo* cardiovascular effects of the native tripeptide. *In vivo* prolactin release of the analogues was also two to three times that of native TRH[205,206].

H. Halogenated Amines

Included in this section will be a discussion of the biochemistry of halogenated amine neurotransmitters, including dopamine (DA), norepinephrine (NE), epinephrin (EPI) (Figure 16) and serotonin (5HT) (**130**). DA is an important central neurotransmitter, in particular with respect to control of motor functions. NE is the primary neurotransmitter of the sympathetic nervous system, eliciting responses through interactions with adrenergic receptors (consisting of several subtypes of α- and β-adrenergic receptors). Effects of halogen substitution on agonist properties and metabolism will be considered. The use of halogenated compounds designed as antagonists and inhibitors of amine-processing enzymes also will be discussed.

1. Structure–activity relationships of halogenated sympathomimetic amines

β-Phenethylamine (**116**) can be considered the parent compound of sympathomimetic amines, including the catecholamines. The presence of phenolic hydroxyl groups on positions 3 and 4 of the aromatic ring of phenethylamine is important for maximum direct adrenergic activity. Compounds having halogen as a replacement for a ring hydroxyl group often have greatly reduced agonist activity, and many function as antagonists at β-adrenergic receptors[207]. For example, 3,4-dichloroisoproterenol (DCI) (**117**) is a potent and selective β-adrenergic antagonist that has been used extensively in pharmacological studies. Halogenation of amphetamine (**118**) and methamphetamin (**119**), amines that exert most of their CNS effects by inhibition of uptake of NE and EPI into noradrenergic neurons, can have profound effects on biological activity. 4-Chloroamphetamine (PCA, **120a**) is a very potent inhibitor of NE and EPI uptake into noradrenergic neurons. 4-Bromo and 4-chloro analogues of amphetamine and methamphetamine (**120**, **121**) have severe and long-lasting neurotoxic effects selective to serotonergic neurons[208]. Fuller and

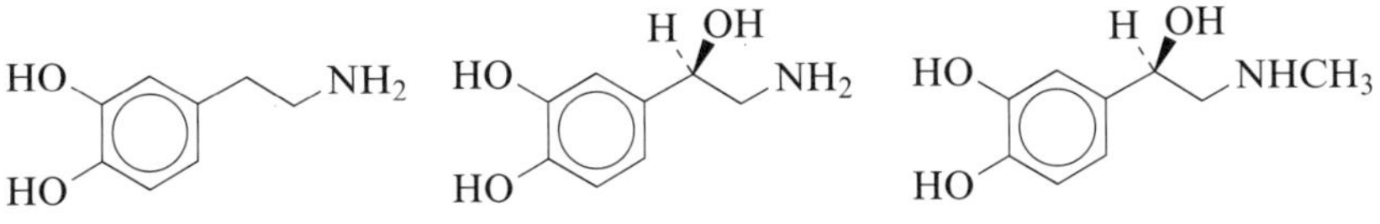

Dopamine
(DA)

Norepinephrine
(NE)

Epinerphrine
(EPI)

FIGURE 16. Structures of the naturally occurring catecholamines

Molloy[209] have studied the biological properties of a series of β-fluoro- and β,β- difluo-rophenethylamines with respect to the effects of the drastically lowered amine pK_a on biodistribution, interactions with catabolic enzymes and modulation of endogenous amine levels.

Ring-halogenated catecholamines (3- and 4-hydroxyl groups retained) have proven to be quite interesting with respect to altered selectivities towards α- and β-adrenergic

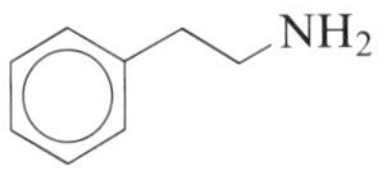

Phenethylamine
(116)

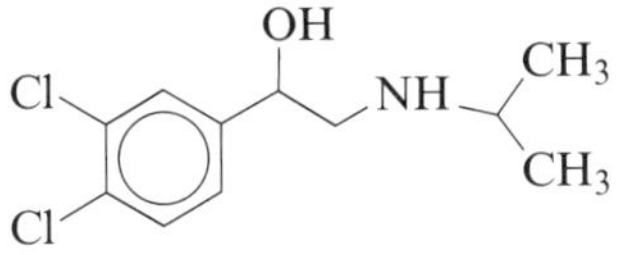

Dichloroisoproterenol
(117)

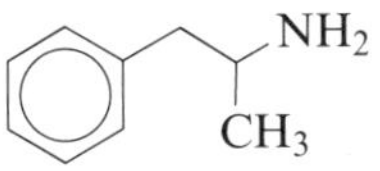

Amphetamine
(118)

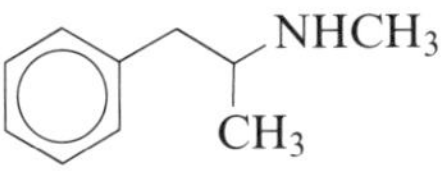

Methamphetamine
(119)

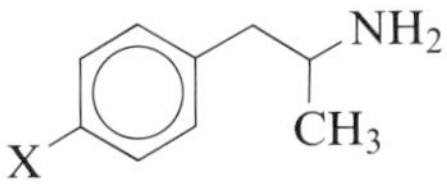

(120a) X = Cl, PCA
(120b) X = Br

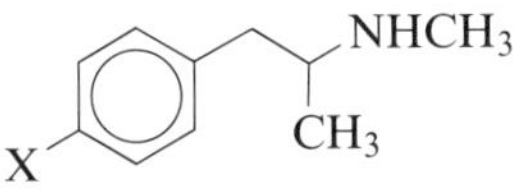

(121) X = Cl, Br

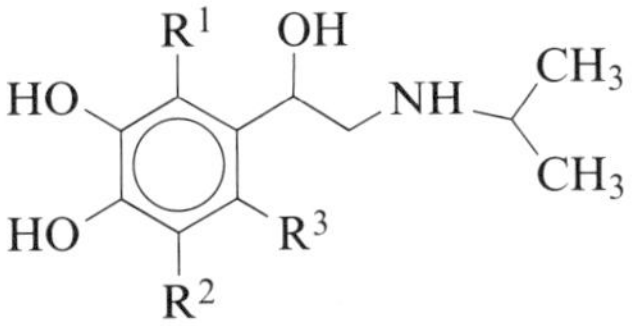

(122a) R^1 = R^2 = R^3 = H (Isoproterenol)
(122b) R^1 = Cl, R^2 = R^3 = H
(122c) R^2 = Cl, R^1 = R^3 = H
(122d) R^3 = Cl, R^1 = R^2 = H

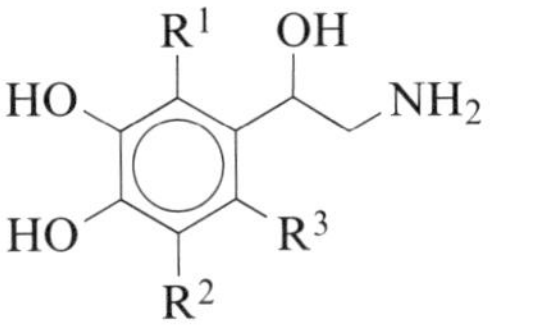

(123a) R^1 = F, R^2 = R^3 = H
(123b) R^2 = F, R^1 = R^3 = H
(123c) R^3 = F, R^1 = R^2 = H

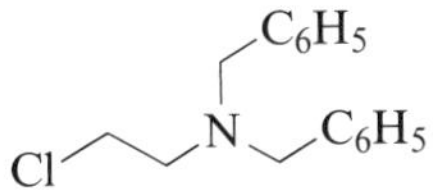

Dibenamine
(124)

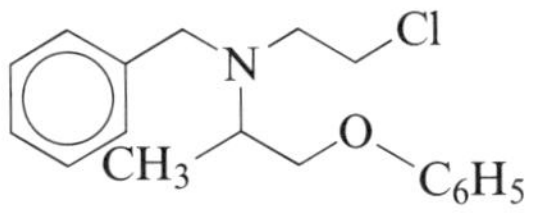

Phenoxybenzamine
(125)

receptors. Kaiser and coworkers[210] showed that, whereas 5-chloro- and 6-chloroisoproterenol (122c,d) had reduced, β-adrenergic activity, 2-chloroisoproterenol (122b) was some 5 times more potent than isoproterenol (122a). Examination of the adrenergic activities of 2-, 5- and 6-fluoro-NE (FNE) (123a–c) revealed rather dramatic effects on α- and β-adrenergic selectivities. 2-FNE had potency comparable to NE at β-adrenergic receptors but was essentially inactive at α-adrenergic receptors. In contrast, 6-FNE was comparable to NE at α-adrenergic receptors, but was much less potent at β-adrenergic receptors. 5-FNE was comparable to NE as an α-adrenergic agonist and, depending on the system, was equal to or somewhat more potent than NE as a β-adrenergic agonist[211]. Similar results were obtained with fluorinated analogues of ISO, EPI and phenylephrine[212].

Replacement of the β-hydroxyl group of β-hydroxyphenylethylamines with halogen produces irreversible α-adrenergic blocking agents. Examples of these mustard-type analogues, of which a huge number are known, are dibenamine (124) and phenoxybenzamine (125). There is much evidence to support the concept that these compounds are converted *in vivo* to ethyleneimminium ions that react with and inactivate α-adrenergic receptors[207,213].

2. Halogenated amines and dopaminergic receptors

Whereas ring fluorination of NE results in marked effects on adrenergic receptor selectivities, 2-, 5- and 6-fluorodopamine (FDA, 126a–c) were comparable to DA at both adrenergic and dopaminergic (D_1, D_2 and D_3) receptors. This was somewhat unexpected in view of the strict structural requirements of dopamine receptors[214]. The similar biochemical behavior of 6-FDA to DA is advantageous in the use of [^{18}F]-L-DOPA-derived [^{18}F]-DA for PET studies of dopaminergic neurons.

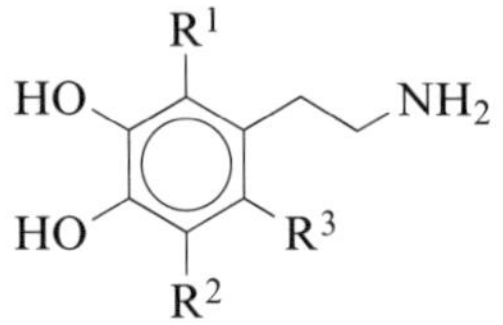

(126a) $R^1 = $ F, $R^2 = R^3 = $ H

(126b) $R^2 = $ F, $R^1 = R^3 = $ H

(126c) $R^3 = $ F, $R^1 = R^2 = $ H

The 'dopamine theory' of schizophrenia is based in part on the correlation of the effectiveness of anti-psychotic drugs with their affinities for dopaminergic receptors[215,216]. Examples of widely used antipsychotic drugs are phenothiazines (127), thioxanthines (128) and butyrophenones (129). The discovery that the *para*-fluoro substituent in butyrophenones decreases morphine-like effects but increases tranquilizing properties led to the development of such potent neuroleptics as spiroperidol (129a) and haloperidol (129b). A 4-fluorobenzoyl group appears to be an absolute structural requirement for maximum potency[217].

3. Halogenated amines and serotonergic receptors

Several ring-halogenated analogues of 5-hydroxytryptamine (serotonin, 5HT, 130) have been prepared. Preliminary results suggest little effect of ring fluorination on receptor binding. Halogenated analogues of the 5HT-derived pineal hormone, melatonin (131a),

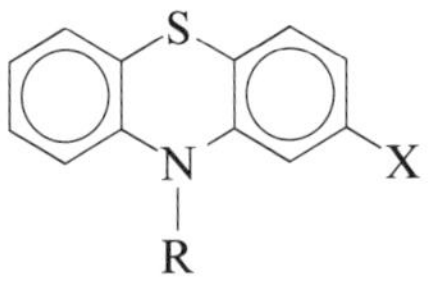

(127)

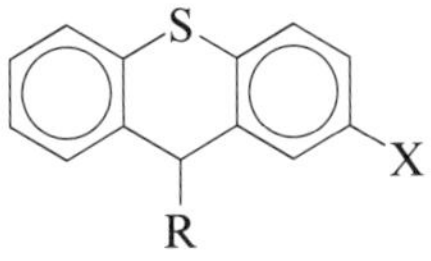

(128)

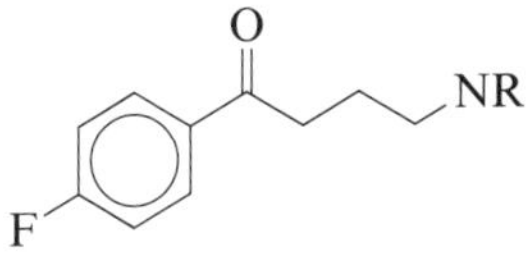

(129)

(129a) NR=N $\overset{O}{\underset{C_6H_5}{\diagup}}$NH   Spiroperidol

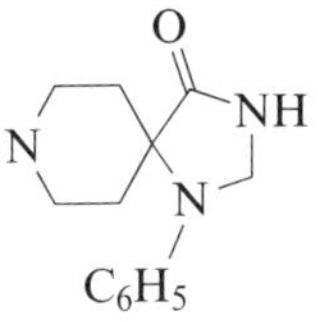

(129b) NR=N $\overset{OH}{\underset{C_6H_4Cl-p}{\diagdown}}$ Haloperidol

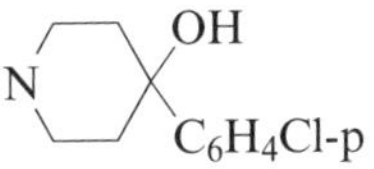

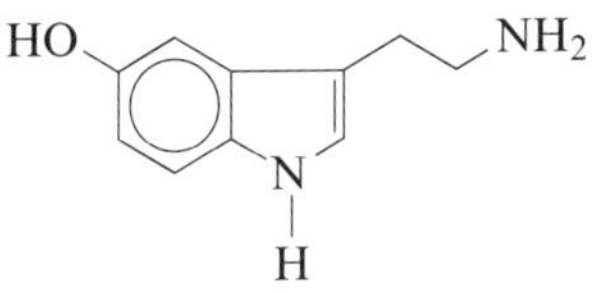

(130)

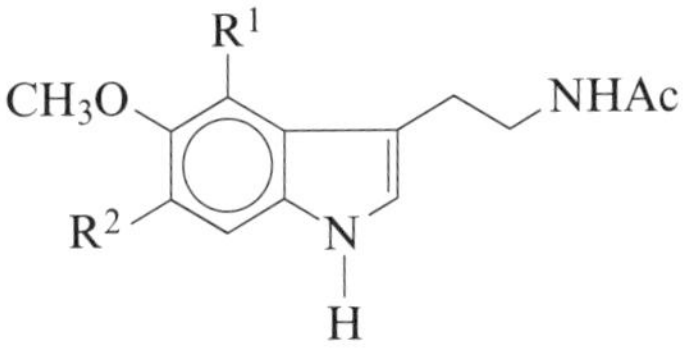

(131a) $R^1 = R^2 = H$
(131b) $R^1 = H, R^2 = F$
(131c) $R^1 = R^2 = F$

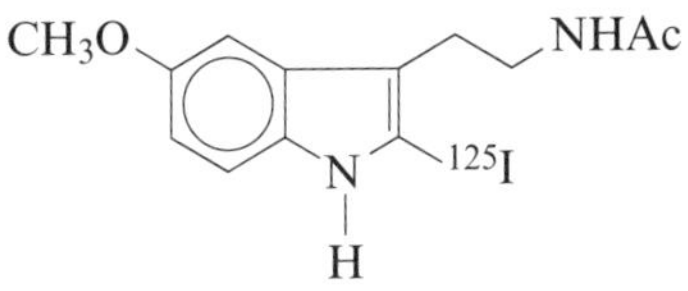

(132)

have been studied in efforts to develop analogues resistant to ring-hydroxylative metabolism. 6-Fluoro- and 4,6-difluoromelatonin (131b,c) were found to mimic melatonin with respect to pituitary LH response to LHRH *in vitro*[218] but biological half-life was not prolonged[219]. A major advance in melatonin pharmacology has come with the discovery that

2-[^{125}I]iodo-melatonin (**132**) is a selective, high-affinity ligand suitable for autoradiographic visualization of melatonin binding sites[220].

A mechanism of irreversible toxicity of *para*-chloroamphetamine (PCA, **120a**) to serotonergic neurons has recently been proposed. According to this proposal, PCA induces 5HT release from serotonergic neurons. The ability of PCA also to function as a monoamine oxidase inhibitor leads to an accumulation of extraneuronal 5HT. Nonenzymatic oxidation produces the neurotoxin, 6-hydroxy-5HT, which is taken up into the neuron where cross-linking to macromolecular structures leads to neuronal destruction[221,222].

4. Halogenated monoamine oxidase (MAO) inhibitors

The search for selective inhibitors for the two forms of MAO (MAO A and MAO B) remains an important area of medicinal chemistry[223]. Mood-elevating effects of MAO inhibitors make them useful drugs for the treatment of depressive illness. Complications of MAO treatment include elevation of blood pressure due to blockade of the oxidative metabolism of tyramine, a pressor amine present in many foods (the 'cheese effect'). Since this inhibition has been attributed to the action of intestinal MAO A, the search for selective MAO B inhibitors has been particularly intense.

Although many reversible inhibitors of MAO A have been prepared, largely based on α-methylamines, MAO B-selective reversible inhibitors are rare. A recent example is found in 4-chloro-β-methylphenethylamine (**133**)[224].

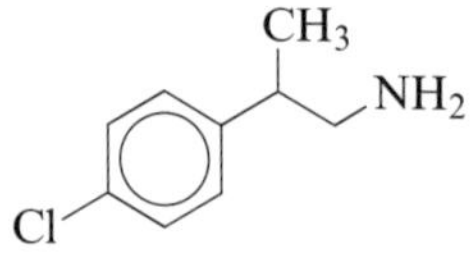

(133)

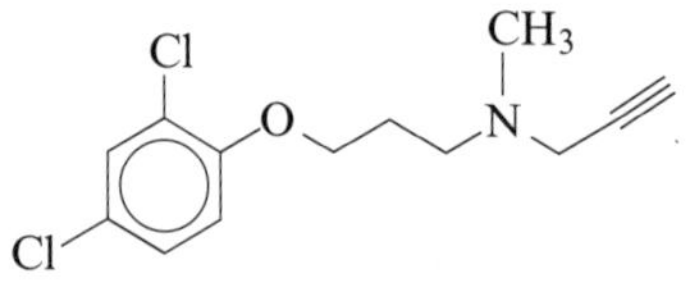

Clorgyline   Deprenyl
(134)   (135)

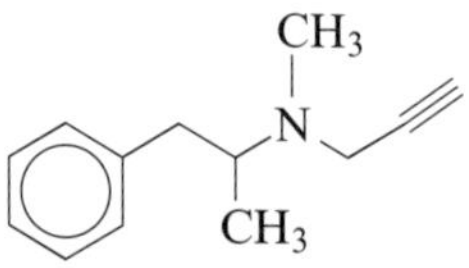

Several hundreds of irreversible MAO inhibitors have been prepared, exemplified by the MAO A-selective clorgyline (**134**) and the MAO-B selective deprenyl (**135**). The search for selective MAO B inhibitors has produced a series of mechanism-based inhibitors containing a 3-fluoroallylamine as a critical structural unit (figure 17). The proposed mechanism of inactivation involves nucleophilic addition of the cofactor or of an active site nucleophile to the double bond. Activation of the double bond occurs through MAO-mediated oxidation of the allylic amine to the electron-deficient iminium species[225-227].

5. Fluoroquinolones

Beginning with the report of nalidixic acid (**136**) in 1962, the quinolone antibiotics have developed into an important, clinically useful class of antibacterial agents, active against both gram-negative and gram-positive bacteria[228]. These drugs exert their action through

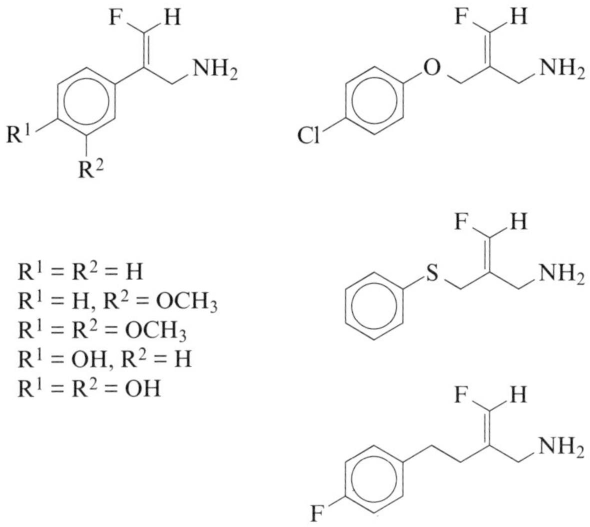

FIGURE 17. Examples of 3-fluoroallylamines that function as selective MAO B inhibitors

inhibition of bacterial DNA gyrase. A significant development in structure–activity relationships in this series was the report in 1980 that substitution of fluorine in the 6-position, as in norfloxacin (137), produced a significant increase in activity[229]. Subsequently, a host of 6-fluoroquinolone antibiotics have been produced, many clinically relevant, in the search for more selective, less toxic, drugs[230,231]. Several recent reviews describe the history and structure–activity studies of these compounds[232].

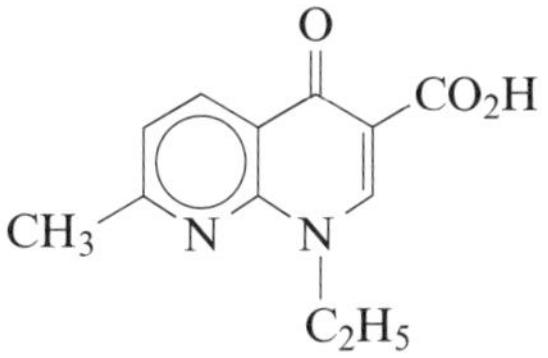

(136)

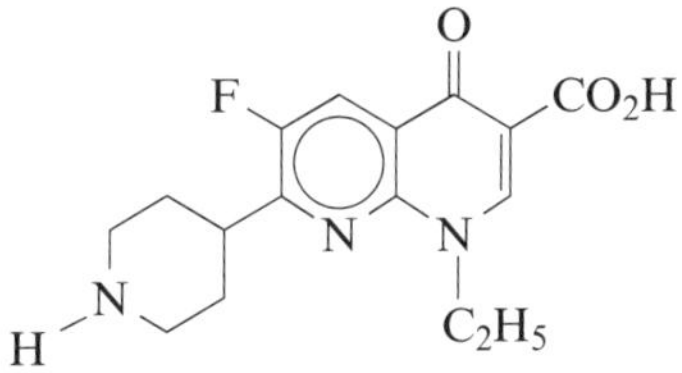

(137)

V. BIOCHEMISTRY OF POLYHALOGENATED ORGANIC MOLECULES

Several classes of polyhalogenated compounds have had widespread practical applications in industry and commerce. Examples are halogenated insecticides and polyhalogenated aromatic hydrocarbons, and related phenols, dioxins and dibenzofurans. Some of these compounds have proven to be quite toxic, and environmental concerns have been increased by the recognition that bioaccumulation concentrates these lipid soluble materials in the food chain. Biochemical mechanisms related to the biological activities and toxic properties of polyhalogenated compounds will be reviewed briefly in this section. Halogenated volatile anesthetics and artificial oxygen carriers are examples of medicinal applications of polyhalogenated compounds. These topics will also be discussed.

A. Biochemistry and Toxicology of Chlorinated Insecticides

1. 2,2-Bis(p-chlorophenyl)-1,1,1-trichloroethane (DDT)

DDT (**138**), first synthesized by Othmar Zeidler in 1873, is a broad-spectrum insecticide that is toxic to invertebrates and vertebrates. Although topical application and prolonged exposure to DDT dust have been shown to have minimal harmful effects on humans, DDT is acutely toxic when injected. Despite the initial optimism that accompanied the first impressive successes of DDT in insect and insect-borne disease control (malaria and typhus, for example), the appearance of DDT resistance in insects and the growing recognition of environmental problems have largely curtailed its use. Nonetheless, DDT is thought to have saved more lives than penicillin. The mechanism of action of DDT has also made this insecticide a useful pharmacological tool[233-235].

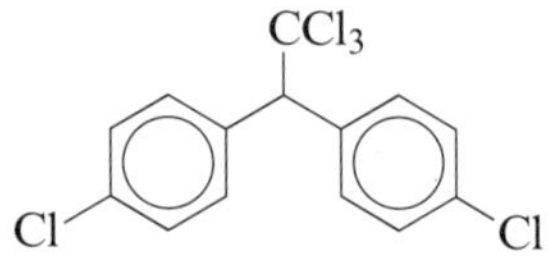

DDT
(138)

Symptoms of DDT poisoning include hyper-responsiveness, tremor, loss of coordination, convulsions and death, symptoms indicative of neurotoxicity. Extensive electrophysiological studies have shown that DDT administration causes a continued firing of a neuron following an initial depolarizing potential. This has been shown to result from an inhibition of the closing of an opened Na^+ channel. Structure–activity relationships and consideration of molecular mechanisms have led to a model wherein DDT can penetrate the lipid membrane of the open Na^+ channel to reach, and block, the gating mechanism[235].

The impressive insecticidal properties of DDT have prompted extensive research designed to develop DDT analogues that would overcome the problems of environmental persistence and insect resistance. Ethoxychlor (**139**) and EDO (**140**) are two examples of readily degradable DDT analogues that are resistant to the action of DDT dehydrochlorinase, the insect enzyme largely responsible for resistance[236].

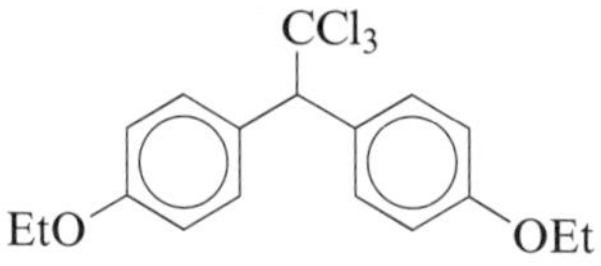

Ethoxychlor
(139)

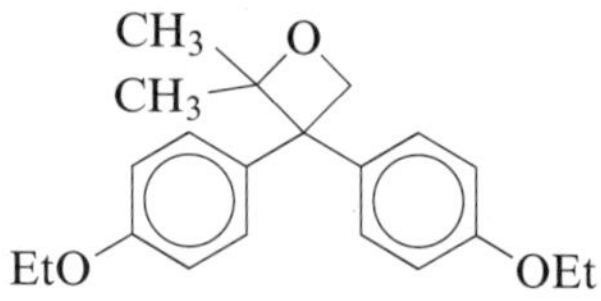

EDO
(140)

1-(ortho-Chlorophenyl)-1-(para'-chlorophenyl)-2,2-dichloroethane [(o,p')-DDD, mitotane, 141], present as a contaminant in technical grade DDT, has been found to cause adrenal atrophy in dogs. Mitotane is now an FDA-approved adrenocorticolytic used in the treatment of hyperadrenocorticism (Cushing's syndrome) caused by adrenal tumors or hyperplasia. Equally effective, but metabolically more stable and less toxic, is the structurally related mitometh (**142**)[237].

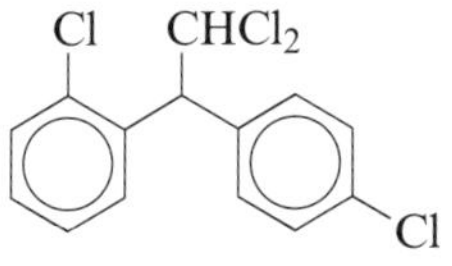

o,p-DDD (Mitotane)
(141)

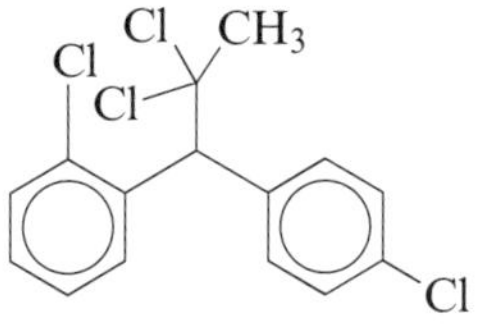

Mitometh
(142)

2. Chlorinated cyclodiene, hexachlorocyclohexane and polychloronorbornane insecticides

A group of highly chlorinated polycyclic insecticides, exemplified by heptachlor (143) and aldrin (144), is prepared from hexachlorocyclopentadiene and cyclopentadiene. The γ-isomer (lindane, 145) of hexachlorocyclohexane (HCH) is the toxic component of HCH, first prepared by Michael Faraday. Toxaphene consists of a complex mixture of chlorinated camphenes, of which 146 represents a particularly toxic isomer.

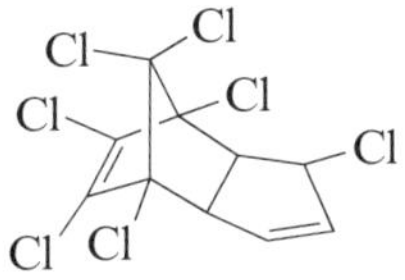

Heptachlor
(143)

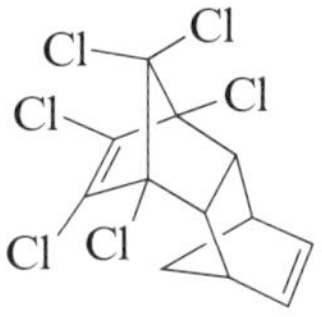

Aldrin
(144)

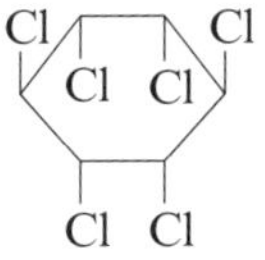

Lindane (γ-isomer)
(145)

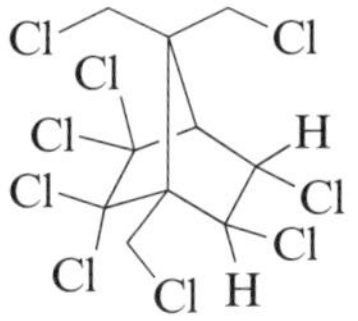

Toxaphene (β-component)
(146)

As with DDT, much research has been carried out to determine the mechanisms of action of these polychlorinated insecticides. Toxicity of the cyclodiene and related insecticides is characterized by a sudden and violent onset of convulsions. The primary site of action appears to be GABAergic neurons. Functioning similarly to the naturally occurring neuroexitant picrotoxinin (PTX), these insecticides appear to block the GABA-gated chloride channel, blocking influx of Cl⁻ and preventing hyperpolarization of the neuron, thus thwarting the inhibitory action of GABA[238].

3. Chlordecone (Kepone)

The contamination of a chlordecone (147) production plant in Hopewell, Virginia, in 1975, the accompanying contamination of the nearby James river and the severe illness of exposed plant workers focused attention on the toxic properties of this insecticide.

Included in the symptoms of exposed workers were disabling neurological disorders, as well as testicular damage, sterility and production of abnormal sperm. Experimental evidence has accumulated which indicates that chlordecone does not act by inactivation of the Na^+ channel[239,240], or by binding to the GABA-gated Cl^- channel[241]. Chlordecone appears to be a non-selective neurotoxic agent, with possibly several modes of action. Inhibition of membrane-bound ATPases, alteration of neurotransmitter and neuropeptide levels, and disruption of calcium regulation have all been implicated[242]. For example, Tilson and coworkers have found a possible link between calcium mobilization, polyamine synthesis and chlordecone neurotoxicity[243]. Chlordecone also has potent estrogenic effects, both *in vitro* and *in vivo*, possibly related to direct action on the estrogen receptor[244].

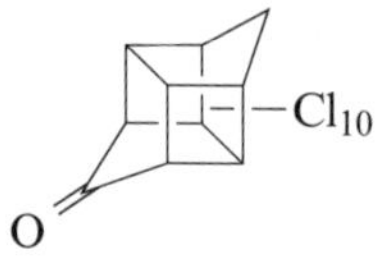

Chlordecone (Kepone)
(**147**)

B. Biochemistry and Toxicology of Halogenated Biphenyls, Dioxins and Related Compounds

Polychlorinated biphenyls, terphenyls, polybrominated biphenyls, polychlorinated phenols and polychlorinated naphthalenes are important members of a series of halogenated aromatic compounds that have had important industrial applications. While thermal and chemical stabilities of these compounds increase their industrial utility, these same properties increase their resistance to degradation, with a corresponding increase in environmental pollution potential. Recognition that many of these compounds are toxic, or are contaminated with toxic impurities, has led to a curtailment of usage of many of these materials. In addition, polyhalogenated dioxins and furans, including the particularly toxic 2,3,7,8-tetrachloro-*p*-dioxin (TCDD, dioxin, **148**), have been found as contaminants formed during the industrial preparation of polyhalogenated biphenyls and phenols.

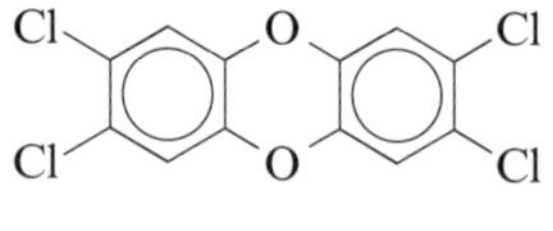

2,3,7,8-Tetrachloro-*p*-dioxin
(TCDD)
(**148**)

The toxicology and biochemical properties of TCDD have received particular scrutiny, research spurred, in part, by such events as the 'Saveso episode' in 1976, and the spraying of Agent Orange in Vietnam during the period 1965–1971. Since many polyhalogenated aromatic compounds share common toxic manifestations that appear to be mediated by a common mechanism, the brief summary of recent research given in this section will focus on TCDD toxicity.

1. TCDD (Dioxin) toxicity and the aromatic hydrocarbon receptor

There is a wide variation in sensitivity of animal species to the toxic effects of halogenated aromatic hydrocarbons, a fact that has complicated extrapolation of data from animal studies to humans. Symptoms of TCDD toxicity include wasting, hepatotoxicity and porphyria, dermal lesions (chloracne), loss of lymphoid tissue and suppression of thymus-dependent immunity, teratogenicity and reproductive illness, carcinogenicity and the induction of several enzymes[245]. The broad spectrum of symptoms implied toxicity at a fundamental cellular level. Indeed, TCDD-induced increase in aryl hydrocarbon hydroxylase (AHH) activity was shown to be initiated by reversible binding of TCDD to a cytosolic protein (the aromatic hydrocarbon, or aH, receptor), followed by translocation of the receptor–ligand complex to the nucleus[246]. Among evidence that related induction of AHH to toxicity is the fact that, for a large number of aromatic compounds, there was correlation between toxicity and affinity for the aH receptor. Nonetheless, this correlation *per se* does not explain TCDD toxicity and, in fact, raises important questions. For example, many compounds that show high affinity for the aH receptor, and induce AHH, are *not* toxic. Binding thus appears to be a necessary, but not sufficient, condition for toxicity.

Research in several disciplines has focused on events subsequent to binding of TCDD to the aH receptor. For example, translocation of the receptor–TCDD complex to the nucleus was proposed to activate genes for expression of enzymes related to drug response (such as AHH). In the toxic response, an additional battery of genes would be activated that are involved with cell division and differentiation[245,247]. Based on two separate SARs, (structure activity relationship) McKinney and coworkers have proposed the existence of two receptors, one for binding and one for toxicity[248]. According to this two-receptor model, the receptor–TCDD complex, after translocation to the nucleus, associates with a second receptor protein associated with chromatin. Subsequent nuclear events leading to the biological response would be activated by the complex formed by binding of the two receptor proteins to TCDD[249].

2. Toxicity of halogenated aromatic compounds and thyroid function

The potential for certain halogenated aromatic hydrocarbons to function as agonists or antagonists for thyroid hormones has long been recognized. SAR studies related to the biochemical behavior of aromatic hydrocarbons and binding to the aH receptor have been used to support a model for toxicity of halogenated aromatic hydrocarbons that directly involves interference with proper thyroid hormone function. McKinney and coworkers[250,251] developed a model wherein structural requirements for binding and toxicity include an essential accessible planar face capable of undergoing a dispersive interaction with receptor protein. Lateral hydrogens would facilitate lateral polarization. The presence of these structural features in thyroxine, TCDD, and a polychlorinated biphenyl (PCB), are illustrated in Figure 18. This model suggests that TCDD toxicity may be mediated through the thyroid hormone receptor, and could be caused, in part, by the expression of persistent thyroid hormone activity.

C. Halogenated Volatile Anesthetics

Research during World War II produced many new fluorinated compounds and, in the early 1950s, a concerted effort was made to find among them nonflammable volatile compounds suitable for general anesthesia. Of hundreds tested, most, with notable exceptions, were found unsuitable for a variety of reasons, for example, toxicity or low potency. Examples of clinically promising halogenated anesthetics are discussed briefly in this section[252].

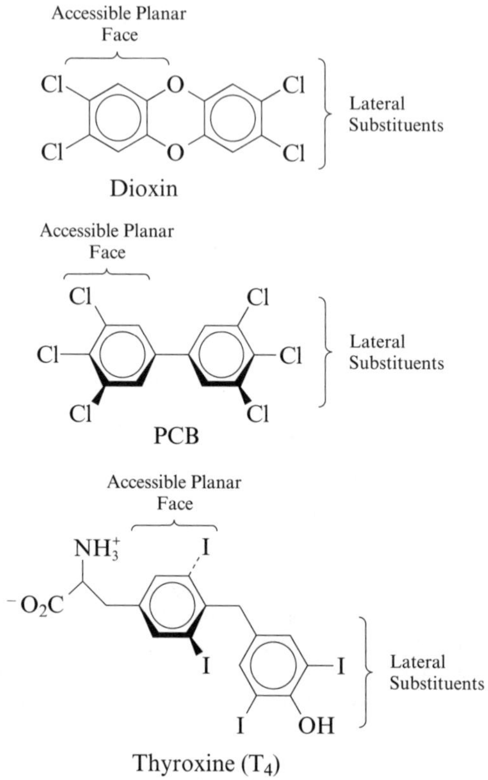

Dioxin

PCB

Thyroxine (T₄)

FIGURE 18. Schematic comparison of dioxin, a polychlorobiphenyl (PCB), and thyroxine, emphasizing the common features of an accessible planar face for binding to the Ah receptor, and lateral substituents to facilitate binding to a nuclear receptor

Halothane (**149**), first used clinically in the 1950s, is some three times more potent than ether and onset of anesthesia is rapid. Mild, transient hepatotoxicity occurs in about 20% of patients. A much rarer but more severe toxicity has been attributed to a drug-induced hypersensitivity reaction. Under aerobic conditions, halothane is oxidatively metabolized to trifluoroacetyl halide that apparently acylates tissue molecules. The bound trifluoroacetyl moiety functions as a hapten in sensitive individuals triggering the immune response[253].

Enflurane (**150**) and, to a greater extent, isoflurane (**151**) are metabolically much more stable than halothane. Isoflurane was introduced into clinical use in 1981. The metabolism of isoflurane and enflurane have been the subjects of several studies[254]. Halogenated anes-

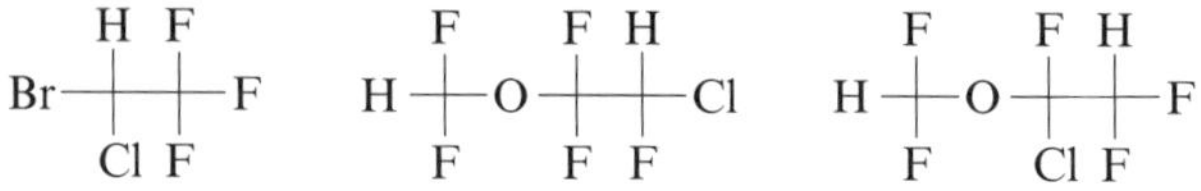

Halothane	Enflurane	Isoflurane
(**149**)	(**150**)	(**151**)

thetics have also been used as *in vivo* and *in vitro* biological probes to study mechanisms of anaesthetic action using ^{19}F-NMR[255,256].

D. Perfluorocarbons as Artificial Oxygen Transporters

Perfluorocarbons (PFCs), formed by complete substitution of fluorine for hydrogen in aliphatic hydrocarbons, have many unique properties. Of increasing medical significance is the dramatic solubility of gases in PFCs—50 %vol of oxygen and 400 %vol of carbon dioxide[2,57]. The demonstration by Clark and Gollan that mice submerged in oxygenated perfluorobutyltetrahydrofurans (**152**) survived unharmed for prolonged periods impressively demonstrated the physiological availability of gases dissolved in PFCs[258]. Subsequent research on possible medical applications of oxygenated PFCs has focused on such areas as liquid ventilation (directly related to the observations of Clark and Gollan), delivery of oxygen to ischemic tissues, oxygen perfusion and preservation of organs prior to transplantation, oxygen enhancement of tumors to increase sensitivity to radiation and cytotoxic agents, use as contrast media and for NMR imaging, and treatment of decompression sickness. However, the use of emulsified PFCs as a replacement for red blood cells to transport oxygen and carbon dioxide has received the most attention[759,260].

$$R^1 = F, R^2 = CF_2CF_2CF_2CF_3$$
$$R^1 = CF_2CF_2CF_2CF_3, R^2 = F$$

Perfluorobutyltetrahydrofuran

(**152**)

Intravenous infusion of the plasma-insoluble PFCs into animals results in fatal embolisms. However, following extensive research with a large number of PFCs and emulsifiers, the development of several PFC emulsions dispersed in isotonic solution has produced promising results. The first commercial product, Fluosol-DA 20% (F-DA), available in 1978, consists of perfluorodecalin (14%) and perfluorotripropylamine (6%) in emulsifiers[260].

Replacement of blood with PFC emulsions remains an attractive, but elusive, goal. However, despite problems that remain to be solved, impressive results have been achieved with these first-generation PFC emulsions. A major advantage comes from the low viscosity and low particle size of these emulsions that permits higher tissue blood flow. There are many medical situations where improved microcirculation can be of immense benefit. For example, following myocardial infarction, hemodilution with PFC emulsions may be expected to improve circulation through collateral arteries, and to penetrate deeply into hypoxic tissue beds, re-oxygenating sludged cells[259]. Improvements in both PFC structure and emulsifiers should produce even more dramatic results.

VI. METABOLISM OF HALOGENATED ORGANIC COMPOUNDS—BIODEHALOGENATION

The metabolism of halogenated organic compounds is related directly to their environmental impact and toxicology. Accordingly, there has been extensive research into metabolic mechanisms, some of which detoxify and remove environmental pollutants, others which are responsible for *in vivo* toxicity. A brief overview in this subject will be given in this section.

The metabolic dehalogenation of halogenated compounds *in vivo* is initiated primarily by two classes of enzymes, the cytochrome P-450-dependent monooxygenases and glutathione S-transferases. The relative importance of these pathways and the ultimate product of metabolic dehalogenation are influenced by substrate structure, including the identity of the halogen. Several reviews are available[261-263].

A. Metabolism of Halogenated Alkanes

1. Monooxygenase-catalyzed oxidation of halogenated alkanes

Oxidative metabolism of alkyl chlorides, bromides and iodides is initiated primarily by insertion of an oxygen atom into an α-C—H bond. The geminal halohydrin product rapidly loses hydrogen halide to form a carbonyl compound (an aldehyde, ketone, acid chloride or carbonyl chloride, depending on substrate structure).

Because of the toxicity and possible carcinogenicity of trihalomethanes, as well as their widespread industrial and laboratory use, their metabolism has been studied extensively. Phosgene, the major initial product of chloroform metabolism, reacts with cellular protein and is implicated in liver-directed toxicity. The rate of metabolism of trihalomethanes follows the expected order, $CHI_3 > CHBr_3 > CHCl_3$, whereas the order of hepatotoxicity is $CHBr_3 > CHCl_3 >> CHI_3$. The lack of hepatotoxicity of iodoform may be related to the instability of carbonyl diiodide, with rapid hydrolysis to carbon dioxide or decomposition to carbon monoxide preventing reaction with cellular macromolecules[262]. The hepatotoxicity of halothane is also related to formation of reactive acylating agents by α-hydroxylation.

The mechanism of hydroxylation of alkanes catalyzed by cytochrome P-450 has been studied extensively. The process involves activation of molecular oxygen to a perferryl species, and transfer of perferryl oxygen to the substrate. With respect to the transfer of oxygen to heteroatom (including halogen) substituted alkanes, Guengerich and Macdonald have proposed a mechanism that involves a single-electron oxidation to form an initial heteroatom-centered radical cation[264].

Cytochrome P-450 also can metabolize perhalogenated compounds having no C—H bonds. For example, carbon tetrachloride and bromotrichloromethane are metabolized by rat microsomal cytochrome P-450 to phosgene and 'electrophilic chlorine' (possibly a chlorine radical or iron hypochlorite[261]), in a process that occurs maximally at a dioxygen concentration of 5% (equation 11). Irreversible toxicity of carbon tetrachloride may be associated with lipid peroxidation mediated by trichloromethylperoxy radicals, since carbon tetrachloride-induced lipid peroxidation is also maximal at a dioxygen content of 5%[265]. However, other cytotoxic mechanisms have received experimental support, including an inhibition of Ca^{++} transport[266].

$$CCl_4 \xrightarrow[1\,e^-]{P\text{-}450} {}^{\bullet}CCl_3 + X^{\bullet}$$
$$\diagdown O_2$$
$$CCl_3\text{-}O\text{-}O^{\bullet} \longrightarrow COCl_2 + \text{'Electrophilic Chlorine'}$$

$$(11)$$

2. GSH-dependent metabolism of haloalkanes

Many halogenated hydrocarbons are substrates for GSH-transferase-catalyzed nucleophilic substitution reactions that produce S-substituted glutathione (GSH) derivatives. These are normal S_N2 displacements of halide with thiolate anion that occur with inversion of configuration with chiral substrates. Iodomethane and iodopropane form the expected thioethers. Detoxification of fluoroacetate in mammals is catalyzed by a specific GSH-S-transferase activity, and fluoride and carboxymethylcysteine have been identified as metabolites[267,268].

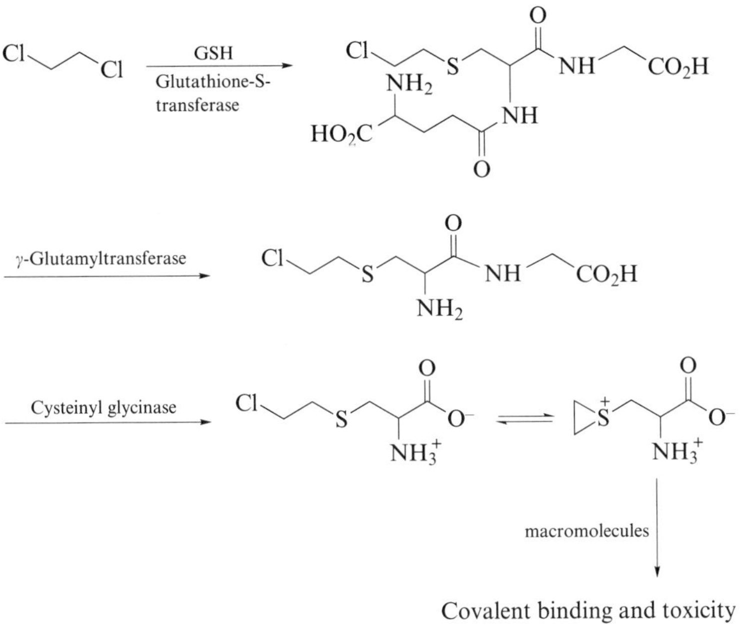

(12)

Covalent binding and toxicity

Although GSH conjugation is normally associated with detoxification, certain halogenated compounds are among substrates that are activated to toxic metabolites by this process. For example, 1,2-dichloroethane is converted to a reactive episulfonium intermediate during GSH-dependent metabolism (equation 12). Reaction of this intermediate with macromolecules *in vivo* has been linked to the carcinogenicity and hepato- and nephrotoxicity of 1,2-dichloroethane[269,270], now classified as a carcinogen by the National Cancer Institute. The carcinogenic activity of 1,2-dibromoethane also appears to be related to GSH-linked activation. In this case, a product of DNA alkylation, S-[2-(N^7-guanyl)ethyl]glutathione (**153**), is formed both *in vivo* and *in vitro*[271].

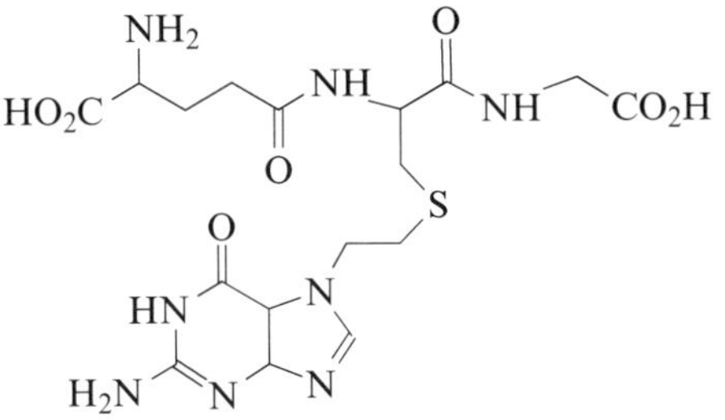

S-[2-(N^7-Guanyl)ethyl]glutathione
(**153**)

B. Metabolism of Halogenated Alkenes

1. Monooxygenase-catalyzed formation of oxirane and its reactions

To a large extent, metabolism of halogenated alkenes involves cytochrome P-450-catalyzed formation and subsequent reactions of halogenated oxiranes. Ease of oxirane formation increases with increasing halogen substitution. Subsequent reactions include alkylation of cellular molecules, conjugation with low-molecular-weight nucleophiles such as GSH, rearrangement or hydrolysis to vicinal diols.

As with cytochrome P-450 oxidations of alkanes, the mechanism of P-450-catalyzed formation of oxiranes has been extensively studied, but certain details remain in dispute[272]. Examples of mechanistic schemes include formation of a transient $[Fe^{IV}—O]^{+2}$-olefin complex followed by collapse to an [Fe—O]–olefin adduct that collapses further to the epoxide and other products[264]. An alternative mechanism involves a 2a + 2s addition of olefin to the Fe=O double bond to give an intermediate metallocene, followed by rearrangement to products[273-275].

Intense scrutiny of the biochemical behavior of vinyl chloride, the simplest member of this class of compound, has been prompted by health and environmental problems associated with its industrial use. An association between vinyl chloride and cancer in vinyl chloride polymerization factories was discovered in 1974. Target organs include liver, brain and lung[276]. It is now well established that vinyl chloride is bioactivated by cytochrome P-450 monooxygenase to chlorooxirane that spontaneously rearranges to chloroacetaldehyde (equation 13). Conjugation of the latter with glutathione and subsequent enzymatic processing provides the primary pathway for deactivation. Toxic, mutagenic and carcinogenic activity is associated with electrophilic reactions of chlorooxirane and chloroacetaldehyde with biological nucleophiles. In the process linked more closely with mutagenic and carcinogenic properties, the more electrophilic chlorooxirane reacts mainly with DNA bases while chloroacetaldehyde primarily alkylates proteins. DNA-etheno adducts formed by cyclization reaction of the halooxirane, or the corresponding haloacetaldehyde, with exocyclic and endocyclic nitrogens of the nucleotide bases, have been implicated in the carcinogenesis of vinyl halides (see Section IV.B)[277,278].

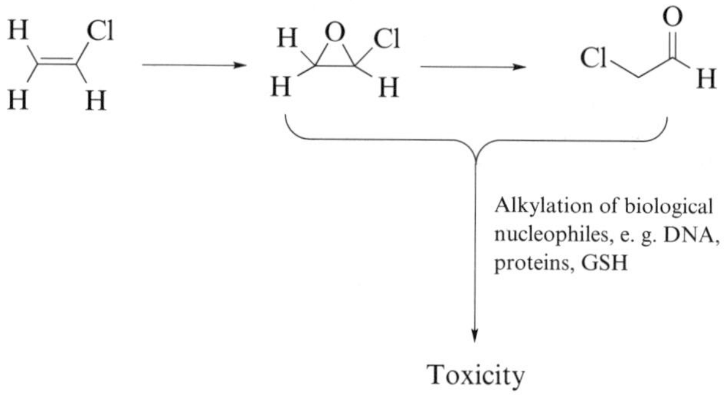

(13)

The action of cytochrome P-450 monooxygenases on 1,1-dichloroethylene, cis- and trans- 1,2-dichloroethylene, trichloroethylene and tetrachloroethylene produces products that are derived from rearrangements of the corresponding halooxirane. For example, the principal products from each isomer of 1,2-dichloroethylene are dichloroacetic acid and dichloroethanol, derived enzymatically from dichloroacetaldehyde, the product of rearrangement of 1,2-dichlorooxirane. Neither isomer of dichloroethylene is mutagenic[263].

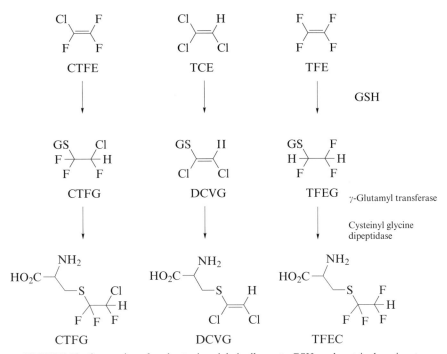

FIGURE 19. Conversion of nephrotoxic polyhaloalkenes to GSH- and cysteinyl-conjugates

2. GSH-dependent metabolism and toxicity of halogenated alkenes

Many polyhalogenated alkenes are potent nephrotoxins. 2-Chloro-1, 1,2-trifluoroethylene (CTFE), trichloroethylene (TCE) and tetrafluoroethylene (TFE) are notable examples. The initial step in bioactivation of such compounds is the GSH-transferase-catalyzed addition or addition–elimination of GSH. For example, in the liver, the action of GSH- transferase converts CTFE to S-(2-chloro- 1,1,2-trifluoroethyl)glutathione (CTFG), TCE to S-(1,2-dichlorovinylglutathione (DCVG) and TFE to S-

(1,1,2,2-tetrafluoroethyl)glutathione (TFEG) (Figure 19). These adducts are transported to the kidney and metabolized by γ-glutamyltransferase and cysteinyl glycine dipeptidase to the corresponding S-cystienyl conjugates. These can be acylated and excreted as non-toxic mercapturic acids, or processed further by cysteine conjugate β-lyase (β-lyase). Much data demonstrate that β-lyase-catalyzed extrusion of reactive thiol intermediates from the cysteine adduct is the ultimate activating step in the nephrotoxicity of these olefins. For example, in a model study, the β-lyase-catalyzed formation of chlorodifluorothionoacetyl chloride from CTFC was demonstrated by formation of the corresponding diethyl thioamide in the presence of diethyl amine (equation 14)[279]. After adminstration of CTFE and TFE to rats, direct *in vivo* detection of protein adducts derived from CTFC and TFEC was achieved using ^{19}F-NMR and stable lysine adducts were isolated[280].

C. Metabolism of Halogenated Aromatic Compounds

Aromatic hydrocarbons are subject to cytochrome P-450-catalyzed hydroxylation in a process that is similar to olefin epoxidation. As discussed in Section IV.G, halogen migration observed during the hydroxylation of 4-ClPhe and similar substrates, led to the discovery of a general mechanism of oxidation that invokes arene oxide intermediates and the 'NIH shift.' Arene oxides and their oxepin tautomers have not been isolated as products of metabolism of benzenoid compounds, but their presence has been inferred by the isolation of phenols, dihydrodiols and dihydrophenolic GSH conjugates derived therefrom[262].

The metabolism of more complex halogenated aromatic compounds can also be initiated by monooxygenase-catalyzed hydroxylation. For example, hydroxylative metabolism of TCDD is accompanied by dehalogenation and the NIH shift[281]. The degree of metabolic stability of PCBs increases with increasing chlorine substitution. For this reason, environmental PCBs that pass through the food chain become more and more highly chlorinated. Metabolism is greatly facilitated if there is at least one pair of adjacent, unsubstituted carbon atoms. 2,4,5,2',4',5'-hexachlorobiphenyl is the most abundant isomer found in human adipose tissue[282].

Of considerable importance with respect to removal of environmental pollutants is the recent discovery that anaerobic microorganisms can reductively dehalogenate polyhalogenated aromatic compounds, including PCBs[283]. Neidleman and Geigert have tabulated a series of naturally occurring microorganisms that are capable of dehalogenating potential environmental pollutants[284].

VII. REFERENCES

1. S. Doonan, in *The Chemistry of the Carbon–Halogen Bond* (Ed. S. Patai), Wiley, London, 1973, pp. 865–915.
2. K. L. Kirk, *Biochemistry of Halogenated Organic Compounds*, Plenum, New York, 1991.
3. K. L. Kirk, *Biochemistry of Elemental Halogens and Inorganic Halides*, Plenum, New York, 1991.
4. W. Fenical, *Science*, **215**, 923 (1982).
5. R. B. Kinnel, H.-P. Gehrken and P. J. Scheuer, *J. Am. Chem. Soc.*, **115**, 3376 (1993).
6. A. Longeon, M. Guyot and J. Vacelet, *Experientia*, **46**, 548 (1990).
7. T. Ichiba, P. J. Scheuer and M. Kelly-Boggs, *J. Org. Chem.*, **58**, 4149 (1993) and references cited therein.
8. O. J. McConnell and W. Fenical, *Phytochemistry*, **16**, 367 (1977).
9. P. M. Gschwend, J. K. MacFarland and K. A. Newman, *Science*, **277**, 1033 (1985).
10. A. G. Gonzalez, J. M. Arteaga, J. D. Martin, M. L Rodriguez, J. Fayos and M. Martinez-Ripolls, *Phytochemistry*, **17**, 947 (1978).
11. Reviewed by M. H. G. Munro, R. T. Luibrand and J. W. Blunt, in *Bioorganic Marine Chemistry*, Vol. I (Ed. P. J. Scheuer), Springer-Verlag, Berlin, 1987, pp. 93–176.

12. M. Suzuki, Y. Morita, A. Yanagisawa. B. J. Baker, P. J. Scheuer and R. Noyori, *J. Org. Chem.*, **53**, 286 (1988).
13. K. Iguchi, S. Kaneta, K. Mori, Y. Yamada, A. Honda and Y. Mori, *Tetrahedron Lett.*, **26**, 5787 (1985).
14. Reviewed in S. L. Neidleman and J. Geigert, *Biohalogenation: Principles, Basic Roles, and Applications*, Ellis Horwood, Chichester, 1986.
15. Reviewed in Reference 3, pp. 155–189.
16. See, for example, J. H. Dawson, *Science*, **240**, 433 (1988).
17. E. de Boer, K. Boon and R. Wever, *Biochemistry*, **27**, 1629 (1988).
18. R. Wever, E. de Boer, H. Plat and B. E. Krenn, *FEBS Lett.*, **216**, 1 (1987).
19. H. Plat, B. E. Krenn and R. Wever, *Biochem. J.*, **248**, 277 (1987).
20. S. L. Neidleman, *CRC Crit Rev. Microbiol.*, **5**, 333 (1975).
21. W. B. Turner, *Fungal Metabolites*, Academic Press, New York, 1971.
22. S. L. Neidleman and J, Geigert, *Trends Biotechnol*, **1**, 21 (1983).
23. R. S. Beissner, W. J. Guilford, R. M. Coates and L. P. Hager, *Biochemistry*, **20**, 3724 (1981).
24. O. J. McConnell and W. Fenical, *Tetrahedron Lett.*, 4159 (1977).
25. W. Fenical, *Recent Adv. Phytochem.*, **13**, 219 (1979).
26. W. Fenical, *Science*, **215**, 923 (1982).
27. E. Frieden, *Trends Biochem. Sci.*, **6**, 50 (1981).
28. N. M. Alexander, in *Biochemistry of the Essential Ultratrace Elements* (Ed. E. Frieden), Plenum, New York, 1984, pp. 33–53.
29. E. C. Jörgensen, in *Burger's Medicinal Chemistry*, Part III, 4th ed. (Ed. M. E. Wolf), Wiley, New York, 1981, pp. 103–145.
30. Reference 3, pp. 135–153.
31. J. T. Neary, M. Soodak and F. Maloof. *Methods Enzymol.*, **107**, 445 (1984).
32. J. Nunez, *Methods Enzymol.*, **107**, 476 (1984).
33. J. Wolff and N. S. Halmi, *J. Biol. Chem.*, **238**, 847 (1963).
34. S. J. Weiss, N. J. Philip and E. F. Grollman, *Endocrinology*, **114**, 1090 (1984).
35. J.-M. Gavaret, H. J. Cahnmann and J. Nunez, *J. Biol. Chem.*, **256**, 9167 (1981).
36. L. Lamas, P. C. Anderson, J. W. Fox and J. T. Dunn, *J. Biol. Chem.*, **264**, 13541 (1989).
37. J. -P. Blondeau, J. Osty and J. Francon, *J. Biol. Chem.*, **263**, 2685 (1988).
38. J. H. Oppenheimer and H. L. Schwartz, *J. Clin. Invest.*, **75**, 147 (1985).
39. A. D. Mooradian, H. L. Schwartz, C. N. Mariash and J. H. Oppenheimer, *Endocrinology*, **117**, 2449 (1985).
40. J. H. Oppenheimer, *Science*, **203**, 971 (1979).
41. J. W. Barlow, M. L. J. Voz, P. H. Eliard, M. Mathy-Hartert and P. De Nayer, *Proc. Natl. Acad. Sci. USA*, **83**, 9021 (1986).
42. J. W. Barlow and P. De Nayer, *Acta Endocrinol.*, **117**, 327 (1988).
43. J. Sap, A. Muñoz, K. Damm, Y. Goldberg, J. Ghysdael, A. Leutz, H. Beug and B. Vennström, *Nature*, **324**, 635 (1986).
44. C. Weinberger, C. C. Thompson, E. S. Ong, R. Lebo, D. J. Gruol and R. M. Evans, *Nature*, **324**, 641 (1986).
45. K. Chae and J. D. McKinney, *J. Med. Chem.*, **31**, 357 (1988).
46. J. Segal and S. H. Ingbar, *Endocrinology*, **115**, 160 (1984).
47. K. Sterling, in *The Thyroid Gland* (Ed. L. van Middlesworth), Year Book Medical Publishers, Chicago, 1986, pp 203–229.
48. C. R. Kannan, *The Pituitary Gland*, Plenum Medical Book Co., New York, 1987, pp. 145–169.
49. J. Köhrle, G. Grabant and R.-D. Hesch, *Horm. Res.*, **26**, 58 (1987).
50. A. Balsam, F. Sexton, M. Borges and S. H. Ingbar, *J. Clin. Invest.*, **72**, 1234 (1983).
51. J. T. Welch and S. Eswarakrishnan, *Fluorine in Bioorganic Chemistry*, Wiley, New York, 1991.
52. C. Walsh, *Adv. Enzymol.*, **55**, 197 (1983).
53. Reference 2, pp 1–39.
54. J. L. Webb, *Enzyme and Metabolic Inhibitors*, Vol. III, Academic Press, New York, 1966, pp 1–283.
55. Reviewed by D. W. Crabb, E. A. Yount and R. A. Harris, *Metabolism*, **30**, 1024 (1981).
56. E. Kun, in *Biochemistry Involving Carbon–Fluorine Bonds* (Ed. R. Filler), ACS Symposium Series, No. 28, American Chemical Society, Washington, D.C., 1976, pp. 1–22.
57. E. Kun, E. Kirsten and M. L. Sharma, *Proc. Natl. Acad. Sci. USA*, **74**, 4942 (1977).

58. E. Kirsten, M. L. Sharma and E. Kun, *Mol. Pharmacol.*, **14**, 172 (1978).
59. D. D. Clarke, *Neurochemical Research*, **16**, 1055 (1991).
60. C. S. Hornfeldt and A. A. Larson, *Eur. J. Pharmacol.*, **179**, 307 (1990).
61. M. A. Marletta, P. A. Srere, and C. Walsh, *Biochemistry*, **21**, 3719 (1981).
62. M. C. Walker, C. R. Jones, R. L. Somerville and J. A. Sikorski, *J. Am. Chem. Soc.*, **114**, 7601 (1992).
63. N. K. Kochetkov, V. N. Shibaev and A. A. Kost, *Tetrahedron Lett.*, 1993 (1971).
64. J. R. Barrio, J. A. Secrist, III and N. J. Leonard, *Proc. Natl. Acad. Sci USA*, **69**, 2039 (1972).
65. N. J. Leonard, *CRC Crit. Rev. Biochem.*, **15**, 125 (1984).
66. K. Kimura, M. Nakanishi, T. Yamamoto and M. Tsuboi, *Biochem. J.*, **81**, 1699 (1977).
67. T. Kohwi-Shigamatsu, R. Gelinas and H. Weintraub, *Proc. Natl. Acad. Sci. USA*, **80**, 4389 (1983).
68. E. Shaw in *The Enzymes* (Ed. P. D. Boyer), Academic Press, New York, 1970, pp. 91–146.
69. J. C. Powers, *Methods Enzymol*, **46**,197 (1977).
70. D. O. Shah, K. Lai and D. G. Gorenstein, *J. Am. Chem. Soc.*, **106**, 4272 (1984).
71. R. E. Galardy and Z. P. Kortylewicz, *Biochemistry*, **23**, 2083, (1984).
72. M. H. Gelb, J. P. Svaren and R. H Abeles, *Biochemistry*, **24**, 1813 (1985).
73. K. Fearon, A. Spaltenstein, P. B. Hopkins and M. H. Gelb, *J. Med. Chem.*, **30**, 1617 (1987).
74. S. Thaisrivongs, D. T. Pals, W. M. Kati, S. R. Turner, L. M. Thomasco and W. Watt, *J. Med. Chem.*, **29**, 2080 (1986).
75. H. L. Sham, N. E. Wideburg, S. G. Spanton, W. E. Kohlbrenner, D. A. Betebenner, D. J. Norbeck, J. J. Plattner and J. W. Erickson, *J. Chem. Soc., Chem. Commun.*, 110 (1991).
76. J. W. Skiles, C. Miao, R. Sorcek, S. Jacober, P. W. Mui, G. Chow, S. M. Weldon, G. Possanza, M. Skoog, J. Keirns, G. Letts and A. S. Rosenthal, *J. Med. Chem.*, **35**, 4795 (1992).
77. M. H. Gelb, *J. Am. Chem. Soc.*, **108**, 3146 (1986).
78. G. D. Prestwich, W.-S. Eng, R. M. Roe and B. D. Hammock, *Arch. Biochem. Biophys.*, **228**, 639 (1984).
79. J.-F. Nave, J. d'Orchymont, J.-B. Ducep, F. Piriou, and M. J. Jung, *Biochem. J.*, **227**, 247 (1985).
80. J. E. Reardon and R. H. Abeles, *Biochemistry*, **26**, 4717 (1987).
81. J. E. Reardon and R. H. Abeles, *Biochemistry*, **25**, 5609 (1986).
82. C. D. Poulter and H. Rilling, *Acc. Chem. Res.*, **11**, 307 (1978).
83. C. D. Poulter, E. A. Mash, J. C. Argyle, O. J. Muscio and H. Rilling, *J. Am. Chem. Soc.*, **101**, 6761 (1979).
84. J. Fried and E. F. Sabo, *J. Am. Chem. Soc.*, **75**, 2273, (1953).
85. R. Filler, in *Organofluorine Chemicals and their Industrial Applications* (Ed. E. Banks), Ellis Horwood, Chichester, 1979, pp. 123–153.
86. P. S. Chen, Jr. and P. Borrevang, in *Handbook of Experimental Pharmacology*, Vol. XX/2, *Pharmacology of Fluorides* (Eds. O. Eicher, A. Farah, H. Herken and A. D. Welch), Springer-Verlag, Berlin, 1970, pp. 193–252.
87. Reference 2 , pp. 65–103.
88. J. H. Thrall, J. E. Freitas and W. H. Beierwaltes, *Semin. Nucl. Med.*, **VII**, 23 (1978).
89. M. A. Mintum, M. J. Welsh, B. A. Siegel, C. J. Mathias, J. W. Brodack, A. H. McGuire and J. A. Katzenellenbogen, *Radiology*, **169**, 45 (1988).
90. M. G. Pomper, J. A. Katzenellenbogen, M. J. Welch, J. W. Brodack and C. J. Mathias, *J. Med. Chem.*, **31**, 1360 (1988).
91. A. Liu, C. S. Dence, M. J. Welch and J. A. Katzenellenbogen, *J. Nucl. Med.*, **33**, 724 (1992).
92. S. J. Brandes and J. A. Katzenellenbogen, *Nucl. Med. Biol.*, **15**, 53 (1988).
93. For a review, see Y. Kobayashi and T. Taguchi, in *Biomedicinal Aspects of Fluorine Chemistry* (Eds. R. Filler and Y. Kobayashi), Kodansha Ltd., Tokyo; Elsevier Biomedical Press, Amsterdam, 1982, pp. 33–53.
94. J.-I. Oshida, M. Morisaki and N. Ikekawa, *Tetrahedron Lett.*, **21**, 1755 (1980).
95. S. Okamoto, Y. Tanaka, H. F. DeLuca, Y. Kobayashi and N. Ikekawa, *Am. J. Physiol.*, **244**, E159 (1983).
96. S.-J. Shiuey, J. J. Partridge and M. R. Uskokovic, *J. Org. Chem.*, **53**, 1040 (1988).
97. F. Wilhelm, W. G. Dauben, B. Kohler, A. Roesle and A. W. Norman, *Arch Biochem. Biophys.*, **233**, 127 (1984).
98. Y. Schichida, T. Ono, T. Yoshizawa, H. Matsumoto, A. Asato, J. P. Zingoni and R. S. H. Liu, *Biochemistry*, **26**, 4422 (1987).

99. R. S. H. Liu, F. Crescitelli, M. Denny, H. Matsumoto and A. E. Asato, *Biochemistry*, **25**, 7026 (1986).
100. T. Mirzadegan, C. Humblet, W. C. Ripka, L. U. Colmenares and R. S. H. Liu, *Photochem. Photobiol.*, **56**, 883 (1992).
101. L. U. Colmenares and R. S. H. Liu, *J. Am. Chem. Soc.*, **116**, 6933 (1992).
102. Reference 2, pp. 105–126.
103. W. E. Barnette, *Crit. Rev. Biochem.*, **15**, 201(1984).
104. R. Filler and S. M. Naqvi, in *Biomedicinal Aspects of Fluorine Chemistry* (Eds. R. Filler and Y. Kobayashi), Kodansha Ltd., Tokyo; Elsevier Biomedical Press, Amsterdam, 1982, pp, 1–32.
105. A. Yasuda, in *Organofluorine Compounds in Medicinal Chemistry and Biomedical Applications* (Eds. R. Filler, Y. Kobayashi and L. M. Yagupolskii), Elsevier Science Publishers, 1993, pp. 275–307.
106. P. A. Greico, W. Owens, C. -L. J. Wang, E. Williams and W. J. Schillinger, *J. Med. Chem.*, **23**, 1072 (1980).
107. C. E. Arroniz, J. Gallina, E. Martinez, J. M. Muchowski, E. Velarde and W. H. Rooks, *Prostaglandins*, **16**, 47 (1978).
108. O. Loge and B. Radüchel, *Naunyn-Schmiedeberg's Arch. Pharmacol., Suppl.*, **325**, R33 (1984).
109. Y. Mizuno, A. Ichikawa and K. Tomita, *Prostaglandins*, **26**, 785 (1983).
110. Y. Hatano, J. D. Kohli, L. I. Goldberg, J. Fried and M. M. Mehrotra, *Proc. Natl. Acad. Sci. USA*, **77**, 6846 (1987).
111. T. A. Morinelli, A. K. Okwu, D. E. Mais, P. V. Halushka, V. John, C.-K. Chen and J. Fried, *Proc. Natl. Acad. Sci. USA*, **86**, 5600 (1989).
112. S. Witkowski, Y. K. Rao, R. H. Premchandran, P. V. Haluska and J. Fried, *J. Am. Chem. Soc.*, **114**, 8464 (1992).
113. P. -Y. Kwok, F. W. Muellner and J. Fried, *J. Am. Chem. Soc.*, **109**, 3692 (1987).
114. T. Taguchi, T. Takigawa, A. Igarashi, Y. Kobayashi, Y. Tanaka, W. Jubiz and R. G. Briggs, *Chem. Pharm. Bull.*, **35**, 1666 (1987).
115. Y. Tanaka, T. M. Klauck, W. Jubiz, T. Taguchi, Y. Hanzawa, A. Igarashi, K. Inazawa, Y. Kobayashi and R. G. Biggs, *Arch. Biochem. Biophys.*, **263**, 178 (1988).
116. B. S. Tsai, R. H. Kieth, D. Villani-Price, R. A. Haack, R. F. Bauer, R. Leonard, Y. Abe and K. C. Nicolaou, *Prostaglandins*, **37**, 287 (1989).
117. Reviewed by D. V. Santi, A. L. Pogoloti, Jr., E. M. Newman and Y. Wataya, in *Biomedicinal Aspects of Fluorine Chemistry*, (Eds. R. Filler and Y. Kobayashi), Kodansha Ltd., Tokyo; Elsevier Biomedical Press, Amsterdam, 1982, pp 123–142.
118. C. E. Myers, *Pharmacol. Rev.*, **33**, 1 (1981).
119. Reference 2, pp 127–192.
120. D. A. Boothman, T. V. Briggle and S. Greer, *Cancer Res.*, **47**, 2354 (1987).
121. L. Chen, A. M. MacMillan and G. L. Verdine, *J. Am. Chem. Soc.*, **115**, 5318 (1993) and references cited therein.
122. J. A. Mekras, D. A. Boothman and S. Greer, *Cancer Res.*, **45**, 5270 (1985).
123. Reviewed by W. H. Prusoff, M. Zucker, W. R. Mancini, M. J. Otto, T.-S. Lin and J.-J. Lee, in *Proceedings of the 1st International TNO Conference on Antiviral Research, Antiviral Research, Suppl. 1* (Eds. A. Billau, E. De Clercq and H. Schellekens), Elsevier Science Publishers, Rotterdam, 1985, pp. 1–10.
124. See, for example, S. G. Paison, J. A. Hartigan, V. Kumar and D. K. Biswas, *DNA*, **6**, 419 (1987).
125. J. W. Gray and B. H. Mayall (Eds.), *Cytometry*, **6**, 499 (1985) and references cited therein.
126. E. R. Kaufman, *Mutat. Res.*, **176**, 133 (1987).
127. E. De Clercq and R. T. Walker, *Pharmacol. Ther.*, **26**, 1 (1984).
128. L. W. Hertel and R. J. Ternansky, in *Organofluorine Compounds in Medicinal Chemistry and Biomedical Applications* (Eds. R. Filler, Y. Kobayashi and L. M. Yagupolskii), Elsevier Science Publishers, Amsterdam, 1993, pp. 23–71.
129. W.-C. Tseng, D. Derse, Y.-C. Cheng, R. W. Brockman and L. L. Bennett, *Mol. Pharmacol.*, **21**, 474 (1981).
130. J. J. Hutton, D. D. Von Hoff, J. Kuhn, J. Philips, M. Hersh and G. Clark, *Cancer Res.*, **44**, 4183 (1984).
131. P. Huang and W. Plunkett, *Biochem. Pharmacol.*, **36**, 2945 (1987).
132. J. R. Barrueco, D. M. Jacobsen, C.-H. Chang, R. W. Brockman and F. M. Sirotnak, *Cancer Res.*, **47**, 700 (1987).

133. Reviewed by J. A. Montgomery, *Cancer Res.*, **42**, 3911 (1982).
134. S. Uesigi, H. Miki, M. Ikehara, H. Iwahashi and Y. Kyogoku, *Tetrahedron Lett.*, 4073 (1979).
135. E. De Clercq, B. D. Stollar, J. Hobbs, T. Fukui, N. Kakiuchi and M. Ikehara, *Eur. J. Biochem.*, **107**, 279 (1980).
136. Reviewed by J. J. Fox, K. A. Watanabe, T. C. Chou, R. F. Schinazi, K. F. Soike, I. Fourel, G. Gantz and C. Trepo, in *Fluorinated Carbohydrates, Chemical and Biochemical Aspects* (Ed. N. F. Taylor), ACS Symposium Series, No. 374, American Chemical Society, Washington, D.C., 1988, pp. 176–190.
137. S. G. Withers, I. P. Street and M. D. Percival, in *Fluorinated Carbohydrates, Chemical and Biochemical Aspects* (Ed. N. F. Taylor), ACS Symposium Series, No. 374, American Chemical Society, Washington, D.C., 1988. pp. 59–77.
138. S. Chiba, C. F. Brewer, G. Okada, H. Matsui and E. J. Hehre, *Biochemsitry*, **27**, 1564 (1988).
139. J. E. G. Barnett, in *CIBA Foundation Symposium: Carbon–Fluorine Compounds, Chemistry, Biochemistry, and Biological Activities*, Associated Scientific Publishers, New York, 1972, pp. 95–115.
140. J. E. G. Barnett, G. D. Holman, R. A. Chalkley and K. A. Munday, *Biochem. J.*, **145**, 417 (1975).
141. C. P. J. Glaudemans and P. Kovac, in *Fluorinated Carbohydrates, Chemical and Biochemical Aspects* (Ed. N. F. Taylor), ACS Symposium Series, No. 374, American Chemical Society, Washington, D.C., 1988, pp 78–108.
142. L. Hough and R. Khan, *Trends Biochem. Sci.*, **3**, 61 (1978).
143. C.-K. Lee, *Carbohydr. Res.*, **162**, 53 (1987).
144. Reviewed by R. Kornfeld and S. Kornfeld, *Annu. Rev. Biochem.*, **54**, 631 (1985).
145. T. J. Grier and J. R. Rasmussen, *J. Biol. Chem.*, **259**, 1027 (1984).
146. W. McDowell, T. J. Grier, J. R. Rasmussen and R. T. Schwarz, *Biochem. J.*, **248**, 523 (1987).
147. M. Sharma, R. J. Bemacki and W. Korytnyk, in *Fluorinated Carbohydrates, Chemical and Biochemical Aspects* (Ed. N. F. Taylor), ACS Symposium Series, No. 374, American Chemical Society, Washington, D.C., 1988, pp. 191–206.
148. R. J. Bernacki and W. Korytnyk, in *The Glycoproteins, Vol IV, Glycoproteins, Glycolipids, and Proteoglycans, Part B* (Ed. M. I. Horowitz), Academic Press, New York, 1982, pp, 245–263.
149. T. P. Fondy, S. B. Roberts, A. S. Tsiftsoglou and A. C. Sartorelli, *J. Med. Chem.*, **21**, 1222 (1978).
150. P. Simon, W. J. Burlingham, R. Conklin and T. P. Fondy, *Cancer Res.*, **39**, 3897 (1979).
151. Reference 2, pp 193–252.
152. N. F. Taylor, A. Romaschin and D. Smith, in *Biochemistry Involving Carbon–Fluorine Bonds* (Ed. R. Filler), American Chemical Society, Washington, D.C., 1976, pp. 99–116.
153. Reviewed by M. E. Phelps and J. C. Maziotta, *Science*, **228**, 799 (1985).
154. Reviewed by R. E. Marquis, in Handbook of Experimental Pharmacology, Vol XX/ 11 (Eds. O. Eicher, A. Farah, H. Herken and A. D. Welch), Springer-Verlag, New York, 1970, pp. 166–192.
155. K. L. Kirk and L. A. Cohen, in *Biochemistry Involving Carbon–Fluorine Bonds* (Ed. R. Filler), American Chemical Society, Washington, D.C., 1976, pp 23–36.
156. A. Weisman and B. K. Koe, *J. Pharmacol. Exp. Ther.*, **155**, 135 (1967) and references cited therein.
157. S. Kaufmann, *Biochim. Biophys. Acta*, **51**, 619 (1961).
158. G. Guroff, J. W. Daly, D. M. Jerina, J. Renson, B. Witkop and S. Udenfriend, *Science*, **157**, 1524 (1967).
159. E. M. Gál and D. H. Whiteacre, *Neurochem. Res.*, **7**, 13 (1982) and references cited therein.
160. C. V. Vorhees, R. E. Butcher and H. K. Berry, *Neurosci. Biobehav. Rev.*, **5**, 177 (1981) and references cited therein.
161. E. S. Garnett, G. Firnau and C. Nahmias, *Nature*, **305**, 137 (1983).
162. C. R. Creveling and K. L. Kirk, *Biochem. Biophys. Res. Commun.*, **130**, 1123 (1985).
163. J. Kollonitch, L. Barash, F. M. Kahan and H. Kropp, *Nature*, **243**, 346 (1973).
164. N. A. Thornberr, H. G. Bull, D. Taub, W. J. Greenlee, A. A. Patchett and E. H. Cordes, *J. Am. Chem. Soc.*, **109**, 7543 (1987).
165. Y. Vo-Quang, D. Carniato, L. VoQuang, A.-M. Lacoste, E. Neuzil and F. Le Goffic, *J. Med. Chem.*, **29**, 148 (1986).
166. E. Wang and C. Walsh, *Biochemistry*, **17**, 1313 (1978).
167. J. J. Likos, H. Ueno, R. W. Feldhaus and D. E. Metzler, *Biochemistry*, **21**, 4377 (1982).

168. D. Roise, K. Soda, T. Yagi and C. T. Walsh, *Biochemistry*, **23**, 5195 (1984).
169. B. Badet, D. Roise and C. T. Walsh, *Biochemistry*, **23**, 5188 (1984).
170. C. T. Walsh, *Annu. Rev. Biochem.*, **53**, 493 (1984).
171. J. Kollonitch, in *Biomedicinal Aspects of Fluorine Chemistry* (Eds. R. Filler and Y. Kobayashi), Kodansha Ltd., Tokyo; Elsevier Biomedical Press, Amsterdam, 1982, pp. 93–122.
172. D. E. Zembower, J. A. Gilbert and M. M. Ames, *J. Med. Chem.*, **36**, 305 (1993).
173. B. W. Metcalf, P. Bey, C. Danzin, M. J. Jung, P. Casara and J. P. Vevert, *J. Am. Chem. Soc.*, **100**, 2551(1978).
174. A. Kallio, P. P. McCann and P. Bey, *Biochemistry*, **20**, 3163 (1981).
175. A. J. Bitonti, P. J. Casara, P. P. McCann and P. Bey, *Biochem. J.*, **242**, 69 (1988).
176. H. Pösö, P. P. McCann, R. Tanskanen, P. Bey and A. Sjoerdsma, *Biochem. Biophys. Res. Commun.*, **125**, 205 (1984).
177. R. B. Silverman and C. George, *Biochemistry*, **27**, 3285 (1988).
178. R. B. Silverman and M. A. Levy, *Biochemistry*, **20**, 1197 (1981).
179. B. D. Sykes and J. H. Weiner, *Magn. Res. Biol.*, **1**, 171 (1980) and references cited therein.
180. D. B. Wacks and H. K. Schachman, *J. Biol. Chem.*, **260**, 11651 (1985).
181. D. B. Wacks and H. K. Schachman, *J. Biol. Chem.*, **260**, 11659 (1985).
182. J. T.-F. Wong, *Proc. Natl. Acad. Sci. USA*, **80**, 6303 (1983).
183. P. M. Bronskill and J. T. -F. Wong, *Biochem. J.*, **249**, 305 (1988).
184. P. Lu, M. A. C. Jarema, K. Mosser and W. E. Daniel, Jr., *Proc. Natl. Acad. Sci. USA*, **73**, 3471 (1976).
185. M. A. C. Jarema, P. Lu and J. H. Miller, *Proc. Natl. Acad. Sci. USA*, **78**, 2702 (1981).
186. L. A. Luck and J. J. Falk, *Biochemistry*, **30**, 4248, 4257, 6484 (1991).
187. J. T. Gerig, in *Biological Magnetic Resonance* (Eds. L. S. Berliner and J. Reuben), Plenum Press, New York, 1978, pp. 139–203.
188. D. N. Wheatly, *Int. Rev. Cytol.*, **55**, 109 (1978).
189. K. L. Kirk, in *Fluorinated Amino Acids: Synthesis and Perspectives* (Eds. V. P. Kukhar' and V. A. Soloshonok), Wiley, John Wiley & Sons Ltd., 1995, pp. 343-401.
190. D. N. Wheatly and J. Y. Henderson, *Nature*, **247**, 281 (1974).
191. G. Hortin, A. M. Stem, B. Miller, R. H. Abeles and I. Boime, *J. Biol. Chem.*, **258**, 4047 (1983).
192. D. C. Klein, J. L. Weller, K. L. Kirk and R. W. Hartley, *Mol. Pharmacol.*, **13**, 1105 (1977).
193. D. C. Klein and K. L. Kirk, in *Biochemistry involving Carbon–Fluorine Bonds* (Ed. R. Filler), American Chemical Society, Washington, D.C., 1976, pp. 37–56 and references cited therein.
194. P. F. Torrence, R. M. Friedman, K. L. Kirk, L. A. Cohen and C. R. Creveling, *Biochem. Pharmacol.*, **28**, 1565 (1979).
195. E. W. Westhead and P. D. Boyer, *Biochim. Biophys. Acta*, **54**, 145 (1961).
196. J. T. Gerig, J. C. Klinkenborg and R.A. Nieman, *Biochemistry*, **22**, 2076 (1983).
197. M. P. Gamcsik, J. T. Gerig and D. H. Gregory, *Biochim. Biophys. Acta*, **912**, 303 (1987).
198. M. P. Gamcsik and J. T. Gerig, *FEBS Lett.*, **196**, 71 (1986).
199. C. R. Creveling, W. L. Padgett, E. T. McNeal, L. A. Cohen and K. L. Kirk, *Life Sci.*, **19**, 1197 (1992).
200. W. H. Heustis and M. A. Raftery, *Biochemistry*, **10**, 1181(1971).
201. W. H. Vine, K. Hsieh and G. R. Matrshall, *J. Med. Chem.*, **24**, 1043 (1981).
202. K. Hsieh, P. Needleman and G. R. Marshall, *J. Med. Chem.*, **30**, 1097 (1987).
203. B. M. Dunn, C. DiBello, K. L. Kirk, L. A. Cohen and I. M. Chaiken, *J. Biol. Chem.*, **249**, 6295 (1974).
204. H. C. Taylor and I. M. Chaiken, *Fed. Proc.*, **36**, 864 (1977).
205. V. M. Labroo, L. A. Cohen, D. Lozovsky, A. -L. Siren and G. Feuerstein, *Neuropeptides*, **10**, 29 (1987).
206. G. Feuerstein, D. Lozovsky, L. A. Cohen, V. M. Labroo, K. L. Kirk, I. J. Kopin and A. I. Faden, *Neuropeptides*, **4**, 303 (1984) and references cited therein.
207. For a review of early work, see F. A. Smith, in *Handbook of Experimental Pharmacology*, Vol XX, *Pharmacology of Fluorides, Part 2* (Ed. F. A. Smith), Springer-Verlag, Heidelberg, 1970, pp 252–408.
208. Reviewed by R. W. Fuller, *Ann. N. Y. Acad. Sci.*, **305**, 147 (1978).
209. R. W. Fuller and B. B. Molloy, in *Biochemistry Involving Carbon–Fluorine Bonds* (Ed. R. Filler), American Chemical Society, Washington, D.C., 1976, pp 77–98.
210. C. Kaiser, D. F. Colella, A. M. Pavloff and J. R. Wardell, Jr., *J. Med. Chem.*, **17**, 1071 (1974).

211. K. L. Kirk, D. Cantacuzene, Y. Nimitkitpaisan, D. McCulloh, W. L. Padgett, J. W. Daly and C. R. Creveling, *J. Med. Chem.*, **22**, 1493 (1979).
212. Reviewed by K. L. Kirk, in *Selective Fluorination in Organic and Bioorganic Chemistry* (Ed. J. T. Welch), American Chemical Society, Washington, D.C., 1991, pp. 136–155.
213. Reviewed by N. Weiner, in *Goodman and Gilman's The Pharmacological Basis of Therapeutics* (Eds. A. G. Gilman, L. S. Goodman, T. W. Rall and F. Murad), Macmillan, New York, 1985, pp. 181–214.
214. L. I. Goldberg, J. D. Kohli, D. Cantacuzene, K. L. Kirk and C. R. Creveling, *J. Pharmacol. Exp. Ther.*, **213**, 509 (1980).
215. A. Carlsson, *Am. J. Psychiat.*, **135**, 164 (1978).
216. D. M. Barnes, *Science*, **235**, 430 (1987).
217. Reviewed by C. Kaiser and P. E. Setler, in *Burger's Medicinal Chemistry*, Part III (Ed. M. E. Wolff), Wiley, New York, 1981, pp. 859–980.
218. J. E. Martin, K. L. Kirk and D. C. Klein, *Endocrinology*, **106**, 398 (1980).
219. J. E. Martin, unpublished results.
220. M. L. Dubocovich and J. S. Takahashi, *Proc. Natl. Acad. Sci. USA*, **84**, 3916 (1987).
221. D. L. Commins, K. J. Axt, G. Vosmer and L. S. Seiden, *Brain Res.*, **403**, 7 (1987).
222. D. L. Commins, K. J. Axt, G. Vosmer and L. S. Seiden, *Brain Res.*, **419**, 253 (1987).
223. Reviewed by C. J. Fowler and S. B. Ross, *Med. Res. Rev.*, **4**, 323 (1984).
224. H. Kinemuchi, Y. Arai, Y. Toyoshima, T. Tadona and K. Kisara, *Jpn. J. Pharmacol.*, **46**, 197 (1987).
225. M. G. Palfreyman, P. Bey and A. Sjoerdsma, *Essays Biochem.*, **23**, 28 (1987).
226. I. A. McDonald, J. M. Lacoste, P. Bey, M. G. Palfreyman and M. Zreika, *J. Med. Chem.*, **28**, 186 (1985).
227. P. Bey, J. Fozard, J. M. Lacoste, I. A. McDonald, M. Zreika and M. G. Palfreyman, *J. Med. Chem.*, **27**, 9 (1984).
228. H. C. Neu, in *Annual Review of Medicine: Selected Topics in Clinical Science*, Vol. **43** (Eds. W. P. Creger, C. H. Coggins and E. W. Hancock), Annual Reviews, Inc., Palo Alto, 1992, pp. 465–486.
229. H. Koga, A. Itoh, S. Murayama, S. Suzue and T. Irikura, *J. Med. Chem.*, **23**, 1358 (1980).
230. For example, J. M. Domagala, A. J. Bridges, T. P. Culbertson, L. Gambino, S. E. Hagen, G. Karrick, K. Porter, J. P. Sanchez, J. A. Sesnie, F. G. Spense, D. Szotek and J. Wemple, *J. Med. Chem.*, **34**, 1142 (1991).
231. P. Claireford, D. Bouzard, B. Ledousal, E. Coroneos, S. Bazile and N. Mareau, *Bioorganic & Medicinal Chemistry Letters*, **2**, 643 (1992).
232. For example, D. T. W. Chu, in *Organofluorine Compounds in Medicinal Chemistry and Biomedical Applications* (Eds. R. Filler, Y. Kobayashi and L. M. Yagupolskii), Elsevier Science Publishers B. V., Amsterdam 1993, pp. 165–207; M. Q. Zhang and A. Haemers, *Pharmazie*, **46**, 687 (1991).
233. L. G. Costa in *Toxicology of Pesticides: Experimental, Clinical and Regulatory Perspectives* (Eds. L. G. Costa, C. L. Galli and S. D. Murphy), Springer-Verlag, Berlin, 1987, pp. 1–10.
234. D. E. Woolley, in *Mechanisms of Action of Neurotoxic Substances* (Eds. K. N. Prasad and A. Vernadakis), Raven Press, New York, 1982, pp. 95–141.
235. M. S. Quraishi, *Biochemical Insect Control, Its Impact on Economy, Environment, and Natural Selection*, Wiley, New York, 1977, pp. 98–123.
236. G. T. Brooks, *Xenobiotica*, **16**, 989 (1986) and references cited therein.
237. B. L. Jensen, M. W. Caldwell, L. G. French and D. G. Briggs, *Toxicol. Appl. Pharmacol.*, **87**, 1 (1987).
238. Reviewed by L. G. Costa, in *Toxicology of Pesticides: Experimental, Clinical and Regulatory Perspectives* (Eds. L. G. Costa, C. L. Galli and S. D. Murphy), Springer-Verlag, Berlin, 1987, pp. 77–91.
239. See, for example, H. A. Tilson, P. M. Hudson and J. S. Hong, *J. Neurochem.*, **47**, 1870 (1986).
240. D. W. Herr, J. A. Gallus and H. A. Tilson, *Psychopharmacology (Berlin)*, **91**, 320 (1987).
241. J. R. Bloomquist, P. M. Adams and D. M. Soderlund, *Neurotoxicology*, **7**, 11 (1986).
242. Reviewed by D. Desaiah, *Neurotoxicology*, **3**, 103 (1982).
243. H. A. Tilson, D. Emerich and S. C. Bondy, *Brain Res.*, **379**, 147 (1986).
244. B. Hammond, B. S. Katzenellenbogen, N. Krauthammer and J. McConnell, *Proc. Natl. Acad. Sci. USA*, **76**, 6641(1979).
245. Reviewed by A. Poland and J. C. Knutson, *Annu. Rev. Pharrnacol. Toxicol.*, **22**, 517 (1982).

246. A. Poland, E. Glover and A. S. Kende, *J. Biol. Chem.*, **251**, 4936 (1976).
247. J. C. Knutson and A. Poland, *Cell*, **30**, 225 (1982).
248. J. D. McKinney, K. Chae, E. E. McConnell and L. S. Birnbaum, *Environ. Health Perspect.*, **60**, 57 (1985).
249. J. D. McKinney, J. Fawkes, S. Jordan, K. Chae, S. Oatley, R. E. Coleman and W. Briner, *Environ. Health Perspect.*, **61**, 41 (1985).
250. J. D. McKinney, T. Darden, M. A. Lyerly and L. G. Pedersen, *Quant Struct. -Act. Relat.*, **4**, 166 (1985).
251. J. D. McKinney, R. Fannin, S. Jordon, K. Chae, U. Rickenbacher and L. Pedersen, *J. Med. Chem.*, **30**, 79 (1987).
252. Reviewed by D. F. Halpern, in *Organofluorine Compounds in Medicinal Chemistry and Biomedical Applications* (Eds. R. Filler, Y. Kobayashi and L. M. Yagupolskii), Elsevier Science Publishers, Amsterdam, 1993, pp. 101–133.
253. H. Satoh, H. W. Davies, T. Takemura, J. R. Gillette, K. Maeda and L. R. Pohl, *Prog. Drug Metab.*, **10**, 187 (1987).
254. R. K. Stoelting, *Pharmacology and Physiology in Anesthetic Practice*, J. B. Lippincott Co., Philadelphia, 1987, pp. 35–68 and references cited therein.
255. A. M. Wyrwicz, C. B. Conboy, K. R. Ryback, B. G. Nichols and P. Eisele, *Biochim. Biophys. Acta*, **927**, 86 (1987).
256. A. S. Evers, B. A. Berkowitz, and D. A. d'Avignon, *Nature*, **328**, 157 (1987).
257. J. G. Riess and M. Le Blanc, *Pure Appl. Chem.*, **54**, 2382 (1982).
258. L. C. Clark, Jr. and F. Gollan, *Science*, **152**, 1755 (1966).
259. N. S. Faithfull, *Anaesthesia*, **42**, 234 (1987) and references cited therein.
260. K. C. Lowe, *Comp. Biochem. Physiol.*, **87A**, 825 (1987) and references cited therein.
261. M. W. Anders and L. R. Pohl, in *Bioactivation of Foreign Compounds* (Ed. M. W.Anders), Academic Press, Orlando, 1985, pp. 283–315.
262. T. L. Macdonald, *Crit. Rev. Toxicol.*, **11**, 85 (1984).
263. D. Henschler, in *Bioactivation of Foreign Compounds* (Ed. M. W. Anders), Academic Press, Orlando, 1985, pp. 317–437.
264. F. P. Guengerich and T. L. Macdonald, *Acc. Chem. Res*, **17**, 9 (1984).
265. H. Kieczka and H. Kappus, *Toxicol, Lett.*, **5**, 191 (1980).
266. M. Younes and C. -P. Siegers, *Biochem. Pharmacol.*, **33**, 3001(1984).
267. R. J. Mead, D. L. Moulden and L. E. Twigg, *Aust. J. Biol. Sci.*, **38**, 139 (1985).
268. A. I. Soiefer and P. J. Kostyniak, *J. Biol. Chem.*, 10787 (1984).
269. M. W. Anders, L. Lash, W. Dekant A. A. Elfarra and D. R. Dohn, *Crit. Rev. Toxicol.*, **18**, 311 (1988).
270. W. W. Webb, A. A. Elfarra, K. D. Webster, R. E. Thom and M. W. Anders, *Biochemistry*, **26**, 3017 (1987).
271. N. Koga, P. B. Inskeep, T. M. Harris and F. P. Guengerich, *Biochemistry*, **25**, 2192 (1986).
272. Reference 3, pp. 257–261.
273. J. P. Collman, T. Kodadek and J. I. Brauman, *J. Am. Chem. Soc.*, **108**, 2588 (1986).
274. J. T. Groves, G. E. Avaria-Niesser, K. M. Fish, M. Imachi and R. L. Kuczkowski, *J. Am. Chem. Soc.*, **108**, 3837 (1986).
275. See, however, A. J. Castellino and T. C. Bruice, *J. Am. Chem. Soc.*, **110**, 158 (1988).
276. H. Vainio and R. Saracci, in *The Role of Cyclic Nucleic Acid Adducts in Carcinogenesis and Mutagenesis* (Eds. B. Singer and H. Bartch), IARC Scientific Publications, No. 70, Oxford University Press, New York, 1984, pp. 15–29.
277. H. M. Bolt, *CRC Crit. Rev. Toxicol*, **18**, 299 (1988) and references cited therein.
278. H. Bartch, in *The Role of Cyclic Nucleic Acid Adducts in Carcinogenesis and Mutagenesis* (Eds. B. Singer and H. Bartch), IARC Scientific Publications, No. 70, Oxford University Press, New York, 1984, pp. 3–14.
279. W. Dekant, L. H. Lash and M. W. Anders, *Proc. Natl. Acad. Sci. USA*, **84**, 7443 (1987).
280. J. W. Harris, W. DeKant and M. W. Anders, *Chem. Res. Toxicol.*, **5**, 34 (1992).
281. H. Poiger and H. -R. Buser, in *Banbury Report 18. Biological Mechanisms of Dioxin Action* (Eds. A. Poland and R. D. Kimbrough), Cold Spring Harbor Laboratory, 1984, pp. 39–47.
282. M. H. Bickel and S. Muehlbach, *Drug Metab. Rev.*, **11**, 149 (1980).
283. J. F. Quensen III, J. M. Tiedje and S. A. Boyd, *Science*, **242**, 752 (1988).
284. Reference 14, pp. 156–175.

CHAPTER **26**

Atmospheric chemistry of organic halides

JOSEPH S. FRANCISCO

Department of Chemistry and Department of Earth and Atmospheric Sciences, Purdue University, West Lafayette, Indiana, USA

and

IAN H. WILLIAMS

School of Chemistry, University of Bath, Bath BA2 7AY, UK

I. INTRODUCTION

A. Features of the Lower Atmosphere

Organic halides are among the very great number of trace constituents of the atmosphere whose presence, nonetheless, is of profound significance for our environment. To appre-

Supplement D2: The chemistry of halides, pseudo-halides and azides
Edited by S. Patai and Z. Rappoport   © 1995 John Wiley & Sons Ltd

ciate the nature of the chemical processes which organic halides undergo and initiate in the atmosphere it is first necessary to outline briefly some important features of the physics and chemistry of the lower atmosphere where these species may become distributed[1].

Near to the earth's surface the temperature of the atmosphere falls with increasing altitude, in consequence of adiabatic cooling as the air expands under reduced pressure. In this lower region, known as the troposphere, cold air lies above warm air and thus convection currents are established which not only drive the familiar patterns of wind and weather but also are responsible for the uptake of materials from the surface. There is rapid vertical mixing of atmospheric constituents in this region, which extends for roughly the first 10–17 km of the atmosphere, and over which the temperature decreases from ~290 K to ~210 K. At some altitude, however, which varies with season and latitude, the temperature begins to rise again as the result of absorption of solar radiation. In this region, known as the stratosphere, warm air lies above cold air and there are no convection currents to cause mixing. The stratosphere extends to an altitude of ~50 km at which height the temperature rises again to ~270 K. The region where the temperature gradient reverses is the tropopause, which forms a diffuse boundary between the turbulent troposphere and the static stratosphere. The tropopause acts like a cold trap to prevent condensable materials, especially water vapour, passing from the troposphere into the stratosphere. Trace compounds introduced from the earth's surface may be rapidly transported both vertically and horizontally throughout the troposphere, but transport across the troposphere and within the stratosphere is slow.

The troposphere contains ~ 87% of the total air mass, while the stratosphere accounts for essentially all the remainder. The higher regions of the atmosphere, including the mesosphere and the thermosphere, do not need to be considered for the purposes of this review. The concentrations of the chief atmospheric constituents, N_2 (78.1%), O_2 (20.9%), Ar (0.93%), CO_2 (0.33%), Ne (0.0018%), He (0.00052%) and CH_4 (0.000165%), expressed here by volume, remain remarkably fixed over time; water vapour, however, has a variable concentration, typically ~1%.

Solar radiation of wavelength < 190 nm is absorbed before it reaches the stratosphere, which absorbs ultraviolet (UV) light in the wavelength range ~190–330 nm. In the upper stratosphere, absorption of radiation between 190 and 240 nm causes photo-dissociation of molecular oxygen into atomic oxygen (equation 1). These atoms rapidly combine with molecular oxygen in the presence of a third molecule M to form ozone (equation 2), which in turn almost completely absorbs radiation between 240 and 290 nm. The absorption by ozone of UV radiation, harmful to life, is of crucial importance. Consequently, essentially no radiation of wavelength <290 nm reaches the earth's surface.

$$O_2 + h\nu \longrightarrow 2\,O \tag{1}$$

$$O + O_2 + M \longrightarrow O_3 + M \tag{2}$$

In the troposphere ozone is produced by photodissociation (equation 3) not of O_2 but of NO_2, initiated with wavelengths <410 nm, followed by recombination (equation 2) with O_2. In the upper troposphere and lower stratosphere the photolysis of ozone (equation 4) yields excited-state oxygen atoms which react with water to produce hydroxyl radicals (equation 5). These are crucial to the removal of organic compounds from the troposphere, since they readily abstract hydrogen atoms (equation 6) to yield organic radicals which subsequently undergo further oxidative degradation.

$$NO_2 + h\nu \ (\lambda < 410 \text{ nm}) \longrightarrow NO + O \tag{3}$$

$$O_3 + h\nu \ (\lambda < 310 \text{ nm}) \longrightarrow O^* + O_2 \tag{4}$$

$$O^* + H_2O \longrightarrow 2 OH \tag{5}$$

$$RH + OH \longrightarrow R + H_2O \tag{6}$$

There are several mechanisms whereby organic compounds released into the atmosphere may be removed: (i) physical removal by precipitation ('rain-out'); (ii) chemical reaction in the troposphere; (iii) transport into the stratosphere; (iv) chemical reaction in the stratosphere. The physical and dynamic conditions of the different atmospheric regions will usually dictate the type of mechanism that occurs[2,3].

B. Sources of Atmospheric Halides

Organic halides admitted into the atmosphere have both natural and anthropogenic sources; concentrations of the principal sources are listed in Table 1. The most abundant chlorine-containing molecule in the troposphere is chloromethane, 70% of which originates from biological processes in the oceans with the remainder being produced microbiologically and by biomass burning and other anthropogenic sources[4]. About 80% of the total organic chlorine loading of the troposphere is due to compounds of anthropogenic origin, with the chlorofluorocarbons CCl_2F_2 (CFC-12), CCl_3F (CFC-11), CCl_2FCClF_2 (CFC-113) and tetrachloromethane CCl_4 (used in the manufacture of CFCs) accounting for about 70% of this. As is well known, the CFCs have been widely used as refrigerants, as aerosol spray propellants and as inflating agents in the manufacture of foam materials, for which applications their main advantages are chemical stability, volatility, nonflammability and nontoxicity. These very properties, however, confer resistance to degradation within the troposphere and give these compounds long atmospheric lifetimes. The only known process for their removal is photolytic decomposition in the stratosphere. Since their introduction in 1930, the annual usage of these materials rose nearly exponentially to $\sim 10^9 \text{ kg yr}^{-1}$. Total tropospheric chlorine is currently increasing by ~ 0.1 parts per billion by volume (ppbv) per year, with CFCs contributing about 75% of this increase[5]. However, international agreement, through the Montreal Protocol which came into force at the start of 1989, has called for the phase-out of CFCs and has promoted efforts to find replacement compounds.

TABLE 1. Tropospheric concentrations (parts per trillion by volume) of principal organohalide source gases

CH_3Cl	600	$CClF_2CClF_2$	~ 20
CCl_2F_2	453	CH_3Br	~ 15
CCl_3F	~ 260	$CHCl_3$	~ 10
CH_3CCl_3	135	$CHClCCl_2$	~ 10
$CHClF_2$	110	CF_3CCl_2F	~ 5
CCl_4	107	$CClF_3$	~ 5
CCl_2FCClF_2	64	CH_2Br_2	~ 3
CH_2Cl_2	~ 35	$CHBr_3$	~ 3
CH_2ClCH_2Cl	~ 35	$CBrClF_2$	~ 2
CCl_2CCl_2	~ 30	$CBrF_3$	~ 2

Hydrochlorofluorocarbons (HCFCs) and hydrofluorocarbons (HFCs) have been proposed as suitable alternatives to CFCs for a large variety of industrial applications. These compounds each contain one or more hydrogen atoms, in order to make them susceptible to tropospheric degradation initiated by OH radical (reaction 6). 1,1,1-trichloroethane (methyl chloroform, CH_3CCl_3), used as an industrial solvent, and $CHClF_2$ (HCFC-22), increasingly used as a substitute for CFCs, contribute about 14% and 3% of the tropospheric anthropogenic organochlorine loading.

Bromomethane CH_3Br is the most abundant organobromide present in the atmosphere. It is produced by biological processes in the oceans, along with CH_2Br_2, $CHBr_3$, CH_2BrCl and $CHBrCl_2$[6-8], although there may also be an anthropogenic contribution[9]. One significant anthropogenic source is $CBrF_3$, used as a flame retardant. Brominated CFCs are known as halons and find uses in anaesthesia. Iodomethane CH_3I is produced by marine algae and phytoplankton, but has only a short atmospheric lifetime owing to its rapid photodecomposition.

In all there are about seventy different haloalkane species which have been detected in the atmosphere, together with more than twenty haloalkenes, nearly one-hundred halogenated cycloalkane and aromatic compounds and about forty halogenated pesticides[10]. In view of their greater abundances, this review will focus upon the atmospheric chemistry of CFCs, HCFCs and HFCs.

C. Impact of Halogen Atoms upon Atmospheric Ozone Concentrations

In 1974 Rowland and Molina[11] pointed out the potential impact upon atmospheric chemistry of chlorine from anthropogenic sources. Atomic chlorine reacts with ozone to form molecular oxygen and chlorine monoxide, which then may react with atomic oxygen to form another molecule of oxygen and to regenerate atomic chlorine. The net effect is the conversion of two 'odd' oxygen species, O and O_3, into 'even' oxygen at a rate much faster than that of the direct reaction (equation 9). Thus ozone is destroyed by the ClO_x catalytic cycle of reactions 7 and 8.

$$Cl + O_3 \longrightarrow ClO + O_2 \qquad (7)$$

$$ClO + O \longrightarrow Cl + O_2 \qquad (8)$$

$$\text{net:} \quad O_3 + O \longrightarrow 2O_2 \qquad (9)$$

Only a small percentage of the chlorine released by photolysis of CFCs is present in the 'active' forms as Cl or ClO, however. Most of it is bound up in reservoir compounds such as hydrogen chloride and chlorine nitrate, formed respectively by hydrogen abstraction (equation 10) from methane and addition (equation 11) to nitrogen dioxide. Slow transport of these reservoir species across the tropopause, followed by dissolution in tropospheric water and subsequent rain-out, provide sink processes for stratospheric chlorine.

$$Cl + CH_4 \longrightarrow HCl + CH_3 \qquad (10)$$

$$ClO + NO_2 + M \longrightarrow ClONO_2 + M \qquad (11)$$

The atmospheric chemistry of the organobromides is similar to that of the organochlorides; degradation ultimately produces bromine atoms which may participate in catalytic destruction of ozone through a BrO_x catalytic cycle (reactions 12 and 13).

$$Br + O_3 \longrightarrow BrO + O_2 \qquad (12)$$

$$BrO + O \longrightarrow Br + O_2 \qquad (13)$$

$$\text{net:} \quad O_3 + O \longrightarrow 2O_2 \qquad (9)$$

The BrO_x cycle is considered to be more efficient than the ClO_x cycle for ozone destruction because bromine atoms are not removed effectively by hydrogen-atom abstraction reactions[12,13]. Moreover, the two cycles can interact synergistically with each other to enhance O_3 destruction in the lower stratosphere[14,15]. The importance of reaction 14 is that, in contrast with reactions 8 and 13, it releases halogen atoms without the involvement of atomic oxygen.

$$Cl + O_3 \longrightarrow ClO + O_2 \tag{7}$$

$$Br + O_3 \longrightarrow BrO + O_2 \tag{12}$$

$$ClO + BrO \longrightarrow Cl + Br + O_2 \tag{14}$$

$$\text{net:} \quad 2 O_3 \longrightarrow 3 O_2 \tag{15}$$

The extreme temperatures (~190 K) of the winter Antarctic stratosphere cause polar stratospheric clouds (PSCs) to form, comprising crystals of nitric acid trihydrate. The reservoir species HCl and $ClONO_2$ may undergo a heterogeneous reaction (equation 16) on the surface of a PSC particle liberating molecular chlorine[16,17]. The latter accumulates during the long dark night of the polar winter and, owing to the stable meteorological conditions of the polar vortex, is not dispersed. As soon as the sun rises in the springtime, the chlorine is photolyzed (equation 17), reaction 7 then occurs readily to destroy ozone, and the resulting dramatic decrease in ozone concentration is mirrored by the corresponding increase in concentration of chlorine monoxide.

$$HCl + ClONO_2 \longrightarrow Cl_2 + HNO_3 \tag{16}$$

$$Cl_2 + h\nu \longrightarrow 2 Cl \tag{17}$$

Regeneration of atomic chlorine, essential for catalysis, does not take place by means of reaction 8 because of the very low concentration of atomic oxygen (reaction 1 does not happen in the dark!). However, reaction 14 with bromine monoxide does occur instead. Furthermore, dimerization (equation 18) of chlorine monoxide[18] is followed by photolysis (equation 19) and dissociation (equation 20) of the chloroperoxy radicals which serve to release chlorine atoms[19]. Thus arises the now famous Antarctic ozone hole, a very significant depletion in ozone concentrations at altitudes of ~ 15–20 km during the austral spring[12,20,21].

$$2 ClO \longrightarrow ClOOCl \tag{18}$$

$$ClOOCl + h\nu \longrightarrow ClOO + Cl \tag{19}$$

$$ClOO + M \longrightarrow Cl + O_2 + M \tag{20}$$

Although atomic fluorine could in principle catalyse ozone destruction by means of an analogous FO_x cycle, this process is not very effective because fluorine atoms are readily removed by reactions 21 and 22 with methane or water, each of which forms hydrogen fluoride as a stable product.

$$F + CH_4 \longrightarrow HF + CH_3 \tag{21}$$

$$F + H_2O \longrightarrow HF + OH \tag{22}$$

The most likely fate of atomic iodine released into the troposphere by photodissociation of iodomethane is reaction 23 with ozone, but the subsequent reactions of iodine oxide are as yet unknown [22].

$$I + O_3 \longrightarrow IO + O_2 \tag{23}$$

The chemistry of organohalides in the atmosphere is dictated by both physical and chemical factors, but the nature of their substituents also serves to determine the type of reactions that occur. Organohalides containing a hydrogen are susceptible to reaction with OH radicals in the troposphere, whereas those not containing hydrogen are inert to tropospheric oxidation and are transported into the stratosphere where their degradation is initiated by UV photolysis. In this chapter we will review studies of the atmospheric degradation of the organic halides and the relevance of the gas-phase chemistry to both the troposphere and stratosphere. The main emphasis is on chemistry and the mechanistic details of the oxidation process rather than on kinetic and thermochemical data.

II. DEGRADATION MECHANISMS FOR ATMOSPHERIC ORGANOHALIDES

A. Halomethane Degradation

The atmospheric fate of a halocarbon molecule depends upon whether or not it contains a hydrogen atom. Hydrohalomethanes are oxidized by a series of reactions with radicals prominant in the troposphere, predominantly hydroxyl OH. Fully halogenated methanes are unreactive towards these radicals and consequently are transported up through the troposphere into the stratosphere, where their oxidation is initiated by UV photolysis of a carbon–halogen bond.

Oxidation of hydrohalomethanes is initiated by hydroxyl-radical attack[23-25] causing hydrogen-atom abstraction (equation 24) to yield a halomethyl radical CX_3 (where each X atom may be either a halogen or hydrogen) which may then add (equation 25) to molecular oxygen. In view of the abundance of O_2 in the troposphere, this is the only reaction of significance for the CX_3 radical. The newly formed haloalkylperoxy radical CX_3O_2 is stabilized by a collision partner M, which removes the excess energy from bond formation. Despite the presence of a third body, these reactions become effectively second order in the high pressure limit which applies within the troposphere. The rate constants vary depending on the nature of the substituents X[26-34], but rapid reaction rates mean that CX_3 radicals have short lifetimes in the troposphere.

$$CX_3H + OH \longrightarrow CX_3 + H_2O \tag{24}$$

$$CX_3 + O_2 + M \longrightarrow CX_3O_2 + M \tag{25}$$

$$CX_3O_2 + NO \longrightarrow CX_3O + NO_2 \tag{26}$$

Haloalkylperoxy radicals are in turn converted to haloalkoxy radicals CX_3O by means of reaction 26 in which nitric oxide is oxidized to nitrogen dioxide. These reactions are fast[35,36], with rates ranging from 10^{-12} to 10^{-11} cm^3 molecule^{-1} s^{-1}. In a competing process, an initially formed vibrationally-excited peroxynitrite isomerizes to a nitrate which may either dissociate (equation 27a) or else be stabilized by collision with another molecule (equation 27b); with ethylperoxy this occurs to the extent of $\leq 1\%$, but with other alkyl or haloalkyl groups this channel can become more significant[37-39].

$$CX_3O_2 + NO \longrightarrow (CX_3O_2NO)^* \longrightarrow (CX_3ONO_2)^* \longrightarrow CX_3O + NO_2 \tag{27a}$$

$$\downarrow M$$

$$CX_3ONO_2 \tag{27b}$$

In regions of the atmosphere where the NO concentrations are low, reaction 28 of alkylperoxy radicals with HO_2 could be important[40]. If the halogenated compounds behave in the same manner as alkylperoxy, the resulting hydroperoxides may be expected to undergo photolysis (equation 29) and hydrogen-atom abstraction (equation 30) by OH radical. The result of these various reactions of alkylperoxy radicals and their derivative reservoir species (nitrates and hydroperoxides) is formation of alkoxy radicals.

$$CX_3O_2 + HO_2 \longrightarrow CX_3OOH + O_2 \tag{28}$$

$$CX_3OOH + h\nu \longrightarrow CX_3O + OH \tag{29}$$

$$CX_3OOH + OH \longrightarrow CX_3O_2 + H_2O \tag{30}$$

If the haloalkoxy radical contains a hydrogen atom, this may be abstracted in a bimolecular reaction (equation 31) with O_2. The analogous reaction of CH_3O occurs with a rate constant of $10^{-15} cm^3 molecule^{-1} s^{-1}$ at 298 K[41]. Alkoxy radicals of the form $CHClXO$ (X=H, F or Cl) have recently been shown to undergo unimolecular elimination (equation 32) of HCl, although this is of only minor significance in the atmosphere[42].

$$CHX_2O + O_2 \longrightarrow CX_2O + HO_2 \tag{31}$$

$$CHClXO \longrightarrow XCO + HCl \tag{32}$$

For fully halogenated alkoxy radicals, halogen-atom abstraction by reaction with O_2 is generally endothermic and therefore improbable under atmospheric conditions. The favoured mode of reaction is unimolecular decomposition. For species of the type CX_2ClO (X=F or Cl), chlorine-atom elimination is the dominant process. Measured rate constants[43,44] show that elimination of Cl proceeds faster from $CFCl_2O$ than from CF_2ClO; *ab initio* MO calculations[45] for CF_2ClO and $CFCl_2O$ respectively show this to be exothermic by -16 and $-24\,kcal\,mol^{-1}$, with very low activation energies of only 2.4 and $-3.2\,kcal\,mol^{-1}$. The reactions of CX_2BrO and $CXBr_2O$ radicals are analogous to those of the corresponding chlorine-containing species. F-atom elimination from trifluoromethoxy CF_3O is endothermic and is predicted to have an appreciable activation energy[46]; this species may react instead with nitric oxide (equation 33)[47,48] or with nitrogen dioxide (equation 34) yielding carbonyl difluoride. A similar reaction (equation 35) may occur with trifluoromethylperoxy[49,50].

$$CF_3O + NO \longrightarrow CF_2O + FNO \tag{33}$$

$$CF_3O + NO_2 \longrightarrow CF_2O + FNO_2 \tag{34}$$

$$CF_3O_2 + NO \longrightarrow CF_2O + FNO_2 \tag{35}$$

The tropospheric fate of hydrohalomethanes, following reaction with OH radicals and leading ultimately to formation of carbonyl halides, is summarized by the mechanism shown in Figure 1.

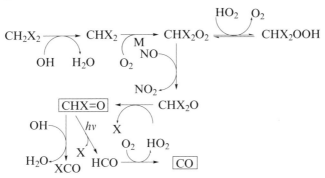

FIGURE 1. Degradation pathways for hydrohalomethanes

The oxidation scheme for halomethanes not containing a hydrogen atom is similar to that for those which do, except that it is not initiated by tropospheric reaction with hydroxyl radicals, since the fully halogenated methanes are unreactive. Consequently, substantial amounts of CFCs and halons are transported intact up into the stratosphere, where they absorb UV radiation of short wavelength and undergo photodissociation (equation 36) to a halogen atom and a trihalomethyl radical. The halogen atom Y may enter into catalytic cycles for ozone destruction, as discussed in the introduction.

$$CX_3Y + h\nu \longrightarrow CX_3 + Y \tag{36}$$

The oxidation scheme for CX_3 radicals (where each X is any of F, Cl or Br) is similar to that of CHX_2 radicals. However, abstraction of halogen from a CX_3O radical is usually endothermic and is therefore an improbable process under atmospheric conditions. The result of oxidation is the formation of carbonyl halides CX_2O.

Tetrafluoromethane and other perfluoroalkanes require radiation of wavelength < 100 nm to cause photodissociation of the very strong C—F bond. These species must therefore diffuse upwards to altitudes beyond 100 km before photolysis occurs, and consequently they have atmospheric lifetimes of thousands of years[51].

Degradation pathways for perhalomethanes are summarized in Figure 2.

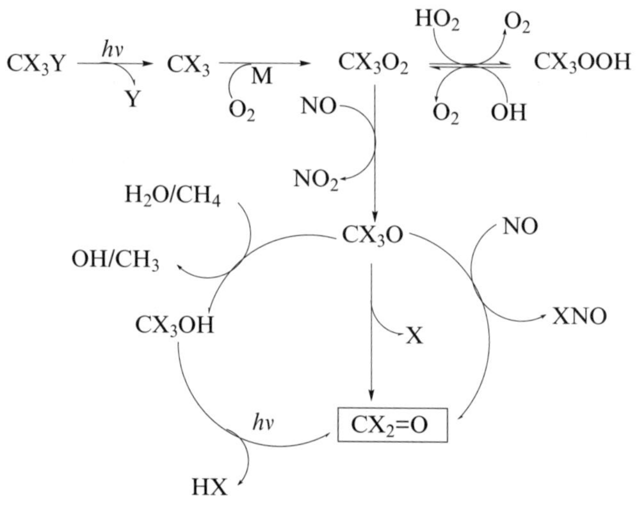

FIGURE 2. Degradation pathways for perhalomethanes

B. Carbonyl Halide Chemistry

Carbonyl halides are not directly emitted into the atmosphere, but rather are generated *in situ* as the by-product of halomethane degradation. There are two classes to be considered: those containing hydrogen (CHXO) and those that do not (CX_2O). In general, degradation may be initiated by reaction with hydroxyl radical, by photodissociation, by homogeneous hydrolysis, or by heterogeneous reaction.

Carbonyl halides containing a hydrogen atom are likely to undergo reaction in the troposphere with hydroxyl radicals, which dominate the day-time chemistry. Reaction 37 of carbonyl halides CHXO (where X = F or Cl) may occur by two mechanisms: (i) direct hydrogen abstraction or (ii) radical addition to carbonyl.

$$CHXO + OH \longrightarrow XCO + H_2O \tag{37}$$

It has been shown computationally that direct hydrogen abstraction is the favoured route[52]: the reactions are exothermic by -17.4 and -27.7 kcal mol^{-1} for CHFO and CHClO, respectively, with correspondingly low barriers of 6.0 and 2.3 kcal mol^{-1}. Experimental results for these reactions are consistent with the theoretical predictions[53]. The barrier height for CHFO is greater than that for CHClO because fluorine strengthens the CH bond; the CH bond strength in CHClO is the same as that in CH_2O.

$$CHXO + Y \longrightarrow XCO + HY \tag{38}$$

Hydrogen abstraction from CHXO may also occur by reaction with another halogen atom (equation 38). The barrier for abstraction of hydrogen from CHFO by fluorine ($Y = F$) is 1.6 kcal mol^{-1}, whereas that for abstraction of chlorine is 3.2 kcal mol^{-1}. Addition (equation 39) of halogen Y may also occur as a competing process, followed by either α-elimination of HX (equation 40) or dissciation (equation 41) of a weaker CX bond[54,55].

$$CHXO + Y \longrightarrow CHXYO \tag{39}$$

$$CHXYO \longrightarrow HX + YCO \tag{40}$$

$$CHXYO \longrightarrow X + CHYO \tag{41}$$

Carbonyl halides not containing hydrogen are not susceptible to reactions involving halogen abstraction by OH, F or Cl species, but their degradation is initiated by photo-dissociation. The $n \rightarrow \pi^*$ absorption of fluoro- and chloro-carbonyls occurs at wavelengths < 300 nm; the bromides absorb at longer wavelengths than chlorides, and the iodides at even longer wavelength still. However, the atmospheric photochemistry of carbonyl halides is not completely understood, and the photodissociation products have not been completely identified. Analysis of the UV absorption spectrum for phosgene indicates a planar ground state but a non-planar excited state; no notable fine structure is observed, and the absence of fluorescence and the diffuseness of the spectrum suggests that CCl_2O predissociates[56]. There are three dissociation reactions for phosgene.

$$CCl_2O \longrightarrow Cl + ClCO \tag{42}$$

$$CCl_2O \longrightarrow Cl_2 + CO \tag{43}$$

$$CCl_2O \longrightarrow CCl_2 + O^* \tag{44}$$

The primary reaction upon irradiation of phosgene is C—Cl bond cleavage[57], and the quantum yield for photodissociation is 1.1 ± 0.1 at 248 nm[58]. The C–Cl bond dissociation energy for reaction 42 is 67 ± 10 kcal mol^{-1} (cf. References 58 and 59) but that of the C—Cl bond in ClCO is only ~ 6 kcal mol^{-1}. Photolysis (equation 45) of CCl_2O by UV wavelengths in the range 210–315 nm should leave sufficient excess energy in the resulting ClCO radical to effect its further dissociation (equation 46).

$$CCl_2O + h\nu \longrightarrow Cl + ClCO^* \tag{45}$$

$$ClCO^* \longrightarrow Cl + CO \tag{46}$$

The gas-phase UV absorption spectrum of CF_2O between 122 and 238 nm shows three electronic transitions[60]. The quantum yield for dissociation (equation 47) is only 0.47 ± 0.03, as measured[61] by following the loss of CF_2O; if recombination (equation 48) of the photolysis products were considered, the true quantum yield would be nearly unity. However, these results only provide indirect evidence for photolysis of the C—F bond. The first direct experimental evidence for the C—F cleavage has recently been provided by Maricq and coworkers, who photolysed CF_2O with 193-nm UV light and monitored the production of FCO radical by following its $\tilde{X} \leftarrow \tilde{B}$ and $\tilde{X} \leftarrow \tilde{C}$ transitions[62.]

$$CF_2O + h\nu \longrightarrow FCO + F \tag{47}$$

$$FCO + FCO \longrightarrow CF_2O + CO \tag{48}$$

The UV spectrum of CFClO has been observed between 190 and 260 nm, and shows an absorption maximum near 203 nm[63], but no measurements of the quantum yields or dissociation products have been reported. On the basis of bond dissociation energies, it may be assumed that the primary process for photodissociation of this radical is C—Cl bond cleavage (equation 49). If the alternative process of C—F bond cleavage were to occur, it would be expected that the resulting ClCO radical would further dissociate to Cl and CO.

$$CFClO \longrightarrow FCO + Cl \tag{49}$$

Infrared photolysis[64] of CFBrO results in C—Br bond cleavage (equation 50). UV quantum yields and photodissociation products have not been studied for this molecule.

$$FBrCO \longrightarrow FCO + Br \tag{50}$$

The bromocarbonyl species CHBrO and CBr$_2$O result from oxidation of CH$_3$Br, CH$_2$Br$_2$ and CHBr$_3$ initiated by hydroxy radical attack[65]. The UV spectrum of CHBrO has a maximum at 269 nm; the absorption band starts at ~ 320 nm and extends to 240 nm: bromine-atom catalysed conversion (equation 51) to carbon monoxide and hydrogen bromide has been studied by following the loss of CHBrO and the growth of CO[66]. Examination of the computed[67] potential energy surface for CHBrO decomposition suggests C—Br bond cleavage (equation 52) as the lowest energy pathway, with α-elimination of HBr as a competing reaction (equation 53). However, since hydrogen-atom abstraction (equation 54) from HCO by Br occurs readily, these two mechanisms are indistinguishable if only the loss of CHBrO or the growth of CO is monitored. The photolysis rate calculated from the measured absorption cross sections (for actinic fluxes corresponding to the conditions 40° N, noon time, July 1) enables the tropospheric lifetime to be estimated as ~ 4 days[66].

$$CHBrO + h\nu \ (\lambda > 300 \text{ nm}) \longrightarrow CO + HBr \tag{51}$$

$$CHBrO \longrightarrow Br + HCO \tag{52}$$

$$CHBrO \longrightarrow HBr + CO \tag{53}$$

$$Br + HCO \longrightarrow HBr + CO \tag{54}$$

The UV spectrum of CBr$_2$O exhibits no fine structure. Absorption starts at 300 nm and the spectrum is shifted to higher wavelengths by ~ 40 nm as compared with CCl$_2$O. An assumed quantum yield of 1 for photolysis leads to an estimated tropospheric lifetime of ~ 3 days[66].

Another mechanism for initiating the atmospheric degradation of carbonyl halides is hydrolysis. Rapid reaction may ensue upon contact with water in clouds or rain. Although there have been measurements of hydrolysis rates, the mechanism is not well understood. The overall process may be represented as reaction 55. On this basis it has been speculated that carbon–halogen balances may be measured from monitoring CX$_2$O and HX concentrations in the atmosphere; however, it has been pointed out that this could be misleading, because reaction 55 lacks a detailed understanding of the chemical mechanism for hydrolysis[68]. A detailed study of the gas-phase hydrolysis mechanism of CF$_2$O shows that the first step is the addition (equation 56) of water. The resulting hydrate may then eliminate HF (equation 57) thereby producing fluoroformic acid, which in turn may dissociate (equation 58) to HF and CO$_2$.

$$CX_2O + H_2O \longrightarrow 2HX + CO_2 \tag{55}$$

$$CF_2O + H_2O \longrightarrow CF_2(OH)_2 \tag{56}$$

$$CF_2(OH)_2 \longrightarrow HF + FC(O)OH \tag{57}$$

$$FC(O)OH \longrightarrow HF + CO_2 \tag{58}$$

From this it is clear that to use hydrolysis of carbonyl halides as a monitor of carbon–halogen species in the atmosphere, it is necessary also to characterize the species $CF(OH)_2$ and $FC(O)OH$. A general mechanism for carbonyl halide hydrolysis can be written as shown in equations 59–62.

$$CXYO + H_2O \longrightarrow CXY(OH)_2 \tag{59}$$

$$CXY(OH)_2 \longrightarrow HX + YC(O)OH \tag{60}$$

$$\underline{YC(O)OH \longrightarrow HY + CO_2} \tag{61}$$

$$\text{net: } CXYO + H_2O \longrightarrow HX + HY + CO_2 \tag{62}$$

The activation barrier for the initial hydration depends on the halogen substituents but is $\sim 50\,\text{kcal mol}^{-1}$ for addition of a single water molecule in the gas phase[69]. However, it has been shown that this step may be very effectively catalysed by the participation of a second water molecule[70]; it is thus likely that the carbonyl halide hydrolysis may occur rapidly in contact with a water droplet.

Degradation pathways for carbonyl halides are summarised in Figure 3.

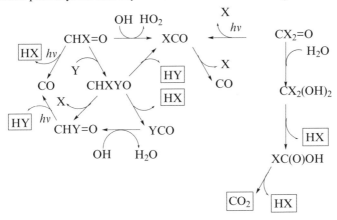

FIGURE 3. Degradation pathways for carbonyl halides

C. Formation and Destruction of $XC(O)O_x$ Radicals

A consequence of photolysis and of atom abstraction from atmospheric carbonyl halides is the production of haloformyl radicals, XCO. These may react (equation 63) with molecular oxygen to form haloformylperoxy radicals $XC(O)O_2$, which in turn may be reduced (equation 64) by nitric oxide to yield haloformyloxy radicals $XC(O)O$.

$$XCO + O_2 + M \longrightarrow XC(O)O_2 + M \tag{63}$$

$$XC(O)O_2 + NO \longrightarrow XC(O)O + NO_2 \tag{64}$$

The group of radicals $XC(O)O_x$ is likely to be important for X = halogen but not for X = H. This is because the reaction of HCO with molecular oxygen favours hydrogen abstraction (equation 65) rather than addition (equation 66). In contrast, the strength of the CF bond in FCO renders fluorine abstraction very unlikely under atmospheric conditions[71]. Consequently, addition (equation 67) of molecular oxygen to form the fluoroformylperoxy radical is more favourable[72]. The species $FC(O)O_x$ and $ClC(O)O_x$ are stable and are known to exist, but the bromine and iodine analogues have not been isolated to date.

$$HCO + O_2 \longrightarrow HO_2 + CO \qquad (65)$$

$$HCO + O_2 + M \longrightarrow HC(O)O_2 + M \qquad (66)$$

$$FCO + O_2 + M \longrightarrow FC(O)O_2 + M \qquad (67)$$

Indirect evidence for the existence of FCO was first obtained by Heras and coworkers[73] who added F_2 to CO in the presence of O_2 and found bis(fluoroformyl) peroxide as a product; the mechanism comprising equations 68–70, involving the intermediacy of FCO radicals, was postulated to account for its formation.

$$F_2 + CO \longrightarrow FCO + F \qquad (68)$$

$$FCO + O_2 \longrightarrow FC(O)O_2 \qquad (69)$$

$$2 FC(O)O_2 \longrightarrow FC(O)OOC(O)F + O_2 \qquad (70)$$

Direct proof for the existence of FCO has come from spectroscopic studies. The three infrared peaks observed in matrix isolation studies[74] have been refined by use of isotopic substitution[75], and infrared diode laser spectroscopy has been employed to obtain a high-resolution spectrum for the CO and CF stretching modes[76]. Maricq and collaborators[77] reported a low-resolution gas-phase spectrum showing a vibrational progression in the region between 220–340 nm with a spacing of about 650 cm^{-1}, and recently obtained a high-resolution UV spectrum of FCO which better defines the origin of the absorption bands in this region[78].

The major source for atmospheric FCO is expected to be photolysis of CF_2O and CFClO, but the advent of hydrofluorocarbons, such as CF_3CFH_2 (HFC-134a), as CFC-alternatives now provides an additional source by means of hydrogen-atom abstraction from CHFO by OH, F or Cl. Yet another source is addition of atomic fluorine to carbon monoxide.

Perhaps the most important process for removal of FCO radicals is reaction 69, for formation of $FC(O)O_2$, which shows little or no temperature dependence over a wide range of temperatures and pressures[72,79], Heydtmann[80] has studied reaction 71 of FCO with ozone to produce the radical FC(O)O, which is also formed as an intermediate in the fast reaction 72 of FCO with atomic oxygen leading to production of carbon dioxide, a process studied experimentally[81,82] and computationally[83].

$$FCO + O_3 \longrightarrow FC(O)O + O_2 \qquad (71)$$

$$FCO + O \longrightarrow [FC(O)O]^* \longrightarrow F + CO_2 \qquad (72)$$

The reactions of FCO radicals with O_2 and O_3 serve to interconvert $FC(O)O_x$ species. Two other potentially important interconversion reactions involve FC(O)O with ozone (equation 73) and $FC(O)O_2$ with NO (equation 74). The latter is very efficient at oxidizing nitric oxide to nitrogen dioxide[79].

$$FC(O)O + O_3 \longrightarrow FC(O)O_2 + O_2 \qquad (73)$$

$$FC(O)O_2 + NO \longrightarrow FC(O)O + NO_2 \qquad (74)$$

$$FC(O)O + NO \longrightarrow FCO + NO_2 \tag{75}$$

$$FC(O)O + CH_4 \longrightarrow FC(O)OH \tag{76}$$

Among the potential reactions for removal of FC(O)O, a key process, is reaction 75 of FC(O)O radicals with NO. Another possibility is hydrogen abstraction (equation 76) from methane, for which an upper limit of the rate constant has been measured as $\leq 2 \times 10^{-16}$ cm^3 molecule^{-1} s^{-1}, suggesting that this reaction is slow; the reaction products have not yet been positively identified.

It is crucial to assess the potential atmospheric activity of the FC(O)O$_x$ family of radicals. It is possible to devise a cycle for ozone regeneration in which these species participate as follows:

$$FCO + O_2 + M \longrightarrow FC(O)O_2 + M \tag{67}$$

$$FC(O)O_2 + NO \longrightarrow FC(O)O + NO_2 \tag{74}$$

$$FC(O)O + NO + M \longrightarrow FC(O)O\,NO \tag{77}$$

$$FC(O)ONO \longrightarrow FCO + NO_2 \tag{78}$$

$$2 \times (NO_2 + h\nu \longrightarrow NO + O) \tag{3}$$

$$2 \times (O + O_2 + M \longrightarrow O_3 + M) \tag{2}$$

$$\overline{\hspace{3cm} 3\,O_2 \longrightarrow 2\,O_3 \hspace{3cm}} \tag{79}$$

However, calculations predict[85] that the exothermicity of formation (equation 77) of the nitrite adduct FC(O)ONO is more than twice the activation energy for its dissociation into FNO and CO$_2$ (Figure 4); the measured rate of this reaction is fast, and the products are indeed as predicted[84]. No spectroscopic evidence has been found for FC(O)ONO as an intermediate, and the experimental and theoretical studies both suggest that reaction 80 terminates the CFX$_3$ photo-oxidation process. A summary of the oxidation steps of FC(O)O$_x$ radicals is shown in Figure 5.

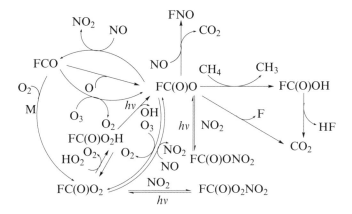

FIGURE 4. Overview of FC(O)O$_x$ radical chemistry

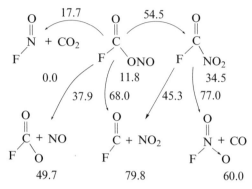

FIGURE 5. Calculated relative energies (kcal mol^{-1}) for species on the potential energy surface for FC(O)O + NO, together with barrier heights (*italics*) for their interconversion

$$FC(O)O + NO \longrightarrow FNO + CO_2 \qquad (80)$$

The chloroformyl radical ClCO is less stable than FCO but nonetheless is of atmospheric significance due to its ability to participate in the oxidation of CO to CO_2. The first step of this process is the formation of ClCO either by the addition (equation 81) of chlorine to carbon monoxide or by photolysis of phosgene (equation 82).

$$Cl + CO + M \longrightarrow ClCO + M \qquad (81)$$

$$CCl_2O + h\nu \longrightarrow Cl + ClCO \qquad (82)$$

The next step, addition (equation 83) of molecular oxygen to ClCO, involves the intermediacy of the unstable chloroformylperoxy radical, which may readily dissociate (equation 84) to chlorine monoxide and carbon dioxide.

$$ClCO + O_2 + M \longrightarrow ClC(O)O_2 + M \qquad (83)$$

$$ClC(O)O_2 \longrightarrow ClO + CO_2 \qquad (84)$$

Formation of CO_2 from $ClC(O)O_2$ may also occur by reaction with atomic oxygen (equation 85) or chlorine (equation 86). Furthermore, the chloroformylperoxy radical is reduced by nitric oxide (equation 87), like its fluorine analogue. However, the resulting chloroformyloxy radical ClC(O)O is very unstable, and the exothermicity of reaction 87 would cause dissociation (equation 88) into CO_2 and atomic Cl.

$$ClC(O)O_2 + O \longrightarrow CO_2 + O_2 + Cl \qquad (85)$$

$$ClC(O)O_2 + Cl \longrightarrow CO_2 + Cl_2O \qquad (86)$$

$$ClC(O)O_2 + NO \longrightarrow ClC(O)O + NO_2 \qquad (87)$$

$$ClC(O)O \longrightarrow CO_2 + Cl \qquad (88)$$

Bromoformyl radical BrCO may be formed either by reaction 85 of CHBrO with OH or by photolysis (equation 89) of CBr_2O, a degradation product of $CHBr_3$. No direct spectroscopic detection of BrCO has yet been reported, neither is it known that $BrC(O)O_x$ species exist nor if these species have any atmospheric chemical role. It is expected that BrCO would very readily dissociate (equation 90).

$$CBr_2O + h\nu \longrightarrow Br + BrCO \tag{89}$$

$$BrCO \longrightarrow Br + CO \tag{90}$$

D. Haloethane Degradation

Many of HCFC or HFC alternatives to CFCs are two-carbon compounds, and are therefore haloethanes of general formula CX_3CX_2H, where each X may be any of H, F or Cl. These compounds should be susceptible to tropospheric degradation by hydroxyl-mediated abstraction of the hydrogen atom which each one contains, and should not reach the stratosphere. This strategy is effective if the tropospheric lifetime of the HFC or HCFC is short relative to its rate of transport into the stratosphere. Obviously it is important to consider whether tropospheric oxidation of these compounds may lead to the production of any other halogenated species which might themselves be transported into the stratosphere.

The initial steps of degradation are very similar to those described above for halomethanes, and may conveniently be described in generic fashion using $CX_3CX_2 = R$. Abstraction (equation 6) of hydrogen from a haloethane results in formation of water and a haloethyl radical which rapidly combines with molecular oxygen to produce a halo-ethylperoxy radical (equation 91). These may react with HO_2, NO or NO_2; self-reactions are possible but unlikely, owing to the low concentrations present.

$$RH + OH \longrightarrow R + H_2O \tag{6}$$

$$R + O_2 + M \longrightarrow RO_2 + M \tag{91}$$

$$RO_2 + HO_2 \longrightarrow ROOH + O_2 \tag{92}$$

Reaction 92 with HO_2 results in formation of a hydroperoxide. Since the nature of the haloalkyl group is not likely to affect greatly the reactivity for this process, rate constants similar to that of $\sim 5 \times 10^{-12} cm^3$ molecule^{-1} s^{-1} for methylperoxy are expected. In view of the low concentrations of HO_2 radicals in the troposphere, this reaction is expected to be of only minor significance, except under conditions where the NO and NO_2 concentrations are also low.

$$RO_2 + NO_2 + M \longrightarrow RO_2NO_2 + M \tag{93}$$

$$RO_2 + NO \longrightarrow RO + NO_2 \tag{94}$$

Reaction 93 of RO_2 with nitrogen dioxide is fast (rate constants $\sim 7.5 \times 10^{-12}$ cm^3 molecule^{-1} s^{-1} and yields a haloalkylperoxynitrate, but the reverse process is also thermal-ly accessible[86,87]. For $R = CF_2ClCH_2$ and $CFCl_2CH_2$, the lifetimes with respect to thermal decomposition in the upper troposphere are ~ 2 days, which suggest that transport to the stratosphere is unlikely to occur. The major process for removal of atmospheric haloalkylperoxy radicals is likely to be the generally fast reaction 94 with nitric oxide which produces haloalkoxy radicals RO. These species may undergo one of three processes: (i) reaction with O_2 if there is an α-hydrogen atom; (ii) fission of either a carbon–halogen or carbon–carbon bond; (iii) elimination of HX, if the radical is of the form RCHClO. For CF_3CX_2O radicals lacking an α-hydrogen but containing an α-chlorine (i.e. at least one X = Cl) then C—Cl bond fission is the preferred route[88,89]. If both α-halogens are fluorine (i.e. both X = F) then C—C bond cleavage is preferred over C—F bong fission. The activation energy for C—C bond dissociation is estimated[90] as ~ 10 kcal mol^{-1}, while for C—F bond cleavage it is 25 kcal mol^{-1}.

The radical CF_3CHFO may be removed by two channels. The first is C—C bond fission (equation 95) for which the activation energy (~ 13 kcal mol^{-1}) is less than that for cleavage of either the C—H or C—F bond. The rate of dissociation at 277 K is estimated at $\sim 10^3$ s^{-1}, which has been shown[91] to be competitive under atmospheric conditions with the alternative channel, namely hydrogen abstraction (equation 96) by O_2; as expected, the yield of $CF_3C(O)F$ rises with increasing O_2 concentration.

$$CF_3CHFO \longrightarrow CF_3 + HFCO \tag{95}$$

$$CF_3CHFO + O_2 \longrightarrow HO_2 + CF_3C(O)F \tag{96}$$

The CH_3CHClO radical may undergo either C—C bond fission (equation 97) or α-elimination (equation 98). The real time observation[92] of HCl formation indicates that HCl elimination proceeds rapidly at 298 K. The resulting acetyl radical CH_3CO may add to molecular oxygen (equation 99) yielding acetylperoxy radical which is reduced by nitric oxide (equation 100). The acetyloxy radical $CH_3C(O)O$ rapidly dissociates to methyl radical and carbon dioxide. Elimination of HCl from the trifluoro-analogue CF_3CHClO has not been reported.

$$CH_3CHClO \longrightarrow CH_3C(O)H + Cl \tag{97}$$

$$CH_3CHClO \longrightarrow CH_3CO + HCl \tag{98}$$

$$CH_3CO + O_2 + M \longrightarrow CH_3C(O)O_2 + M \tag{99}$$

$$CH_3C(O)O_2 + NO \longrightarrow CH_3C(O)O + NO_2 \tag{100}$$

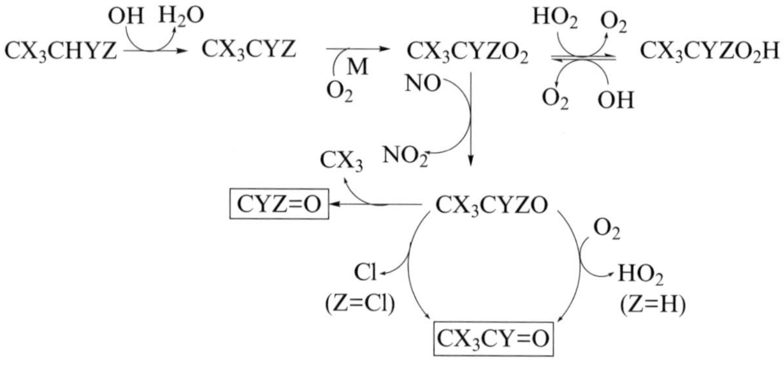

FIGURE 6. Degradation pathways for hydrohaloethanes

The atmospheric degradation of haloethanes is summarized in Figure 6. The net result is production of carbonyl halides, whose further degradation has been considered in Section II.B, and acetyl halides, whose atmospheric fate will be considered in the following section. Finally, mention may be made here of haloalkenes, such as tetrachloroethene and trichloroethene which appear in Table 1: these compounds are very reactive towards addition of OH, yielding a halogenated hydroxyethyl radical which may undergo further degradation according to the pathways outlined above.

TABLE 2. Atmospheric lifetime, ozone depletion potential (ODP) and global warming potential (GWP) of CFCs, HCFCs, HFCs and halons

Species[a]	Formula	Lifetime (yr) [b]	ODP[c]	GWP[d]
CFC-11	$CFCl_3$	55	1.0	3500
CFC-12	CF_2Cl_2	116	~1.0	7300
CFC-113	$CFCl_2CF_2Cl$	110	1.07	4200
CFC-114	CF_2ClCF_2Cl	220	~0.8	6900
CFC-115	CF_3CF_2Cl	550	~0.5	6900
HCFC-22	CHF_2Cl	15.8	0.05	1500
HCFC-123	CF_3CHCl_2	1.7	0.02	85
HCFC-124	CF_3CHFCl	6.9	0.022	430
HCFC-141b	$CFCl_2CH_3$	10.8	0.11	440
HCFC-142b	CF_2ClCH_3	22.4	0.065	1600
HCFC-225ca	$CF_3CF_2CHCl_2$	2.8, 1.5[e]	0.025	162
HCFC-225cb	CF_2ClCF_2CHFCl	8.0	0.033	680
Tetrachloromethane	CCl_4	47	1.08	1300
1,1,1-Trichloroethane	CH_3CCl_3	6.1	0.12	100
HFC-125	CF_3CHF_2	28	0	2500
HFC-134a	CF_3CH_2F	16	0	1200
HFC-143a	CF_3CH_3	41	0	2900
HFC-152a	CHF_2CH_3	1.7	0	140
H-1301	CF_3Br	~67	~16	
H-1211	CF_2ClBr	~19	~4	
H-1202	CF_2Br_2	3.9	~1.25	
H-2402	CF_2BrCF_2Br	~26	~7	
H-1201	CHF_2Br	5.6, 4.7[e]	~1.4	
H-2401	CF_3CHFBr	2.0, 3.9[e]	~0.25	
H-2311	$CF_3CHClBr$	0.8	~0.14	
Bromomethane	CH_3Br	~1.5	~0.6	

[a] The digital code for identification of these species may be interpreted as follows: the units digit is the number of fluorine atoms in the molecule, the tens digit is the number of hydrogen atoms plus one, the hundreds digit is the number of carbon atoms minus one (and is dropped if equal to zero); the residue of atoms required to saturate the carbons is assumed to be chlorine.
[b] Reference 105.
[c] Relative to CFC-11.
[d] 100 year integration time horizon; Reference 112.
[e] A. C. Brown, C. E. Canosa-Mas, A. D. Parr, K. Rothwell and R. P. Wayne, *Nature*, **347**, 541 (1990).

E. Formation and Destruction of CF₃C(O)Oₓ Radicals

The acetyl halides that result from the HFCs and HCFCs listed in Table 2 include $CF_3C(O)X$, where $X = H$, F or Cl. In the case of the aldehyde ($X = H$), the major pathway for further degradation is slow abstraction (equation 101) of the hydrogen atom by hydroxyl radical[93,94], although abstraction of hydrogen by either a fluorine (equation 102) or a chlorine (equation 103) atom is intrinsically faster[95] by a factor of 10. Any hydrogen halides produced by the latter reactions would be rained out of the atmosphere.

$$CF_3C(O)H + OH \longrightarrow CF_3CO + H_2O \qquad (101)$$

$$CF_3C(O)H + F \longrightarrow CF_3CO + HF \qquad (102)$$

$$CF_3C(O)H + Cl \longrightarrow CF_3CO + HCl \tag{103}$$

Since $CF_3C(O)F$ and $CF_3C(O)Cl$ are not known to react with OH radicals at any significant rate, the major loss process for these species is through photolysis or hydrolysis. The mechanism for hydrolysis of acetyl halides $CF_3C(O)X$, where $X = H$, F and Cl, has been examined theoretically[96]. The first and rate-determining step is formation (equation 104) of the hydrate, which may then decompose (equation 105) to trifluoroacetic acid and a hydrogen halide; these products may be removed from the atmosphere by uptake into cloud droplets followed by rain-out. Concern has been raised as to the environmental impact of trifluoroacetic acid: the suggestion is that it may be metabolised by microorganisms in soils to monofluoroacetic acid, which is toxic to mammals.

$$CF_3C(O)X + H_2O \longrightarrow CF_3CX(OH)_2 \tag{104}$$

$$CF_3XC(OH)_2 \longrightarrow CF_3C(O)OH + HX \tag{105}$$

Photolysis of trifluoroacetyl fluoride and chloride results in cleavage of the carbon–halogen bond. The UV absorption band of $CF_3C(O)Cl$ starts at 315 nm, with a maximum at 260 nm, while that for $CF_3C(O)F$ is blue shifted[97]; this may suggest that photolysis of the chloride is more important in the troposphere, while photolysis of the fluoride occurs in the stratosphere.

$$CF_3C(O)X + hv \longrightarrow CF_3CO + X \tag{106}$$

The atmospheric chemistry of trifluoroacetyl radical CF_3CO is not well known. It was implicated to explain the products observed[98] in photolysis of hexafluoroacetone in the presence of Br_2 and Cl_2, but its first direct observation came from rapid-scan infrared spectroscopic studies in a matrix[99], and more recently its laser-induced fluorescence spectrum has been observed[100]; the band origin for the first excited state of the radical appears at 384 nm. A weak UV absorption band which onsets at 250 nm and continues to increase in intensity below 200 nm has been attributed to CF_3CO[101].

The most abundant reaction partner for trifluoroacetyl radical is molecular oxygen. The product of addition (equation 107) is trifluoroacetylperoxy radical $CF_3C(O)O_2$, which may be converted into trifluoroacetyloxy $CF_3C(O)O$ in two ways. The dominant process in the upper atmosphere is likely to be reaction with nitric oxide (equation 108). The alternative would be self-reaction (equation 109), resulting in disproportionation.

$$CF_3CO + O_2 + M \longrightarrow CF_3C(O)O_2 + M \tag{107}$$

$$CF_3C(O)O_2 + NO \longrightarrow CF_3C(O)O + NO_2 \tag{108}$$

$$2CF_3C(O)O_2 \longrightarrow 2CF_3C(O)O + O_2 \tag{109}$$

$$CF_3C(O)O \longrightarrow CF_3 + CO_2 \tag{110}$$

The $CF_3C(O)O$ radical is unstable (equation 110) with respect to dissociation into trifluoromethyl radicals and carbon dioxide[102]. The activation energy is estimated as 4.9 kcal mol^{-1}. Although $CF_3C(O)O_2$ has been observed spectroscopically using UV absorption, to date there is no such direct evidence for the $CF_3C(O)O$ radical[103]. A theoretical examination[102] of possible reactions of CF_3CO and $CF_3C(O)O$ radicals with atmospheric species such as O, O_2, O_3 and NO_2 suggests that these species will not participate in catalytic cycles that could perturb stratospheric ozone concentrations; this prediction has since been confirmed experimentally[103]. The degradation pathways for CX_3CO radicals are summarized in Figure 7.

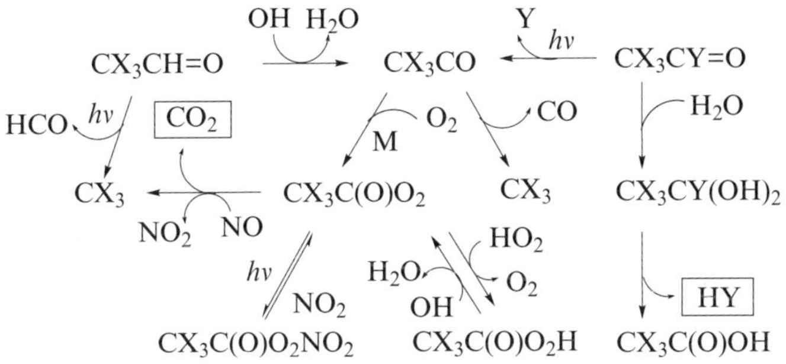

$$CX_3CH=O \xrightarrow{\text{OH } H_2O} CX_3CO \xleftarrow{\begin{array}{c}Y\\hv\end{array}} CX_3CY=O$$

FIGURE 7. Degradation pathways for trifluorocarbonyl halides

III. ENVIRONMENTAL IMPLICATIONS OF ATMOSPHERIC ORGANOHALIDES

A. Stratospheric Ozone Depletion

The total amount of ozone in the atmosphere, if compressed to 1 atm pressure, is equivalent to a skin only 3 mm thick. This is distributed in trace amounts throughout the whole atmosphere, but the largest concentrations are found in a definite layer within the stratosphere, centred at ~ 25–30 km in altitude. Above this layer, the concentration of molecular oxygen is too low for reactions 1 and 2 to generate significant concentrations of ozone; below it, there is insufficient short-wavelength radiation to effect dissociation (equation 1) of O_2. The rise in temperature with increasing altitude in the stratosphere is due to the conversion of solar radiation into heat by means of the chemical processes for formation and destruction of ozone, and the UV absorption of ozone itself. A delicate balance exists whereby the ozone which protects living organisms on the earth from harmful radiation in turn depends upon biological systems both for its natural production (from oxygen of largely biological origin) and for its natural destruction (by means of various chemical cycles involving, for example, oxides of nitrogen which are also at least partly of biological origin). An appreciation for the complexities of ozone chemistry in the unpolluted atmosphere is essential for an understanding of the consequences of anthropogenic introduction of substances, such as organohalides, which have potential to perturb the natural balance.

The mechanisms by which atomic chlorine and bromine are introduced to the stratosphere have been discussed above in Section II, and their ability to catalyse ozone destruction by means of the ClO_x and BrO_x cycles has been reviewed in Section I.C. The possibly deleterious effect of an organohalide depends upon the altitude at which the halogen atoms are released: destruction of a large fraction of the ozone concentration at high altitude has much less impact on an absolute scale than destruction of a smaller fraction of the larger ozone concentrations at about 15–20 km. CFC-12 has more impact upon stratospheric ozone than CFC-114, which contains the same number of chlorine atoms (two) but twice the number of fluorine atoms (four), because it is photolysed at a lower altitude; the more fluorine atoms a compound contains, the more its absorption maximum is shifted to shorter wavelength, and the higher in altitude its maximum rate of photolysis occurs. Ozone depletion due to an organohalide depends upon several factors: (i) the number of chlorine and/or bromine atoms in the molecule, (ii) the total amount released into the atmosphere, (iii) the efficiency of any processes for its removal from the troposphere, (iv) the rate and mechanism of its breakdown within the stratosphere, responsible for liberating reactive chlorine or bromine, and (v) the subsequent stratospheric chemistry of these halogens.

A quantitative measure of these factors is provided by the Ozone Depletion Potential (ODP) of the compound, which represents the amount of ozone destroyed by emission of that substance over its entire atmospheric lifetime relative to that due to emission of the same mass of CFC-11[104]. The determination of an ODP requires a calculation based upon a particular model of atmospheric processes. The various models include rather detailed descriptions of gas-phase photochemical processes and parameterized descriptions of stratospheric transport, and have been successful in representing the chemical composition of the stratosphere above ~ 25 km. However, current ozone depletion occurs largely below 25 km and is not well simulated by existing models of gas-phase chemistry, which do not contain satisfactory descriptions of the complicated microphysical processes governing the development of PSCs, the heterogeneous reactions that occur on them, and the important processes of dehydration and denitrification which occur in polar regions, nor do they simulate in detail the transport processes that link polar and mid-latitude regions of the atmosphere[105]. Thus at present there is considerable uncertainty about estimates of ODPs; the fact that they are relative measures alleviates but does not eliminate the effects of short-comings in the existing models.

Inspection of Table 2 shows that organobromides are considerably more damaging than organochlorides: e.g. halon-1202 (CF_2Br_2) has an ODP of ~1.25 as compared with ~1.0 for CFC-12 (CF_2Cl_2), and halon-1301 (CF_3Br) is 16 times more damaging on a kilogram-for-kilogram basis than CFC-11 ($CFCl_3$). The effectiveness of the stratagem of replacing at least one halogen atom of a CFC by hydrogen, as in a HCFC or HFC, is also immediately apparent: e.g. HCFC-124 (CF_3CHFCl) has an ODP of only 0.022 as compared with ~ 0.5 for CFC-115 (CF_3CF_2Cl) which contains the same number of chlorine atoms. Of course, the overall impact of a particular compound depends upon how much of it is released. For example, although the ODP for 1,1,1-trichloroethane (0.12) is only one-ninth that of CFC-113 (1.07), which has the same number of chlorines but no hydrogens, the overall contribution it makes to ozone loss is more than one-third of that of CFC-113 in view of the greater amount of its annual release (~6 × 10^8 kg vs. ~1.7 × 10^8 kg for 1988)[106].

Estimates of atmospheric lifetimes depend upon the values adopted for key reaction rates [e.g. for hydrogen abstraction (equation 6) from HCFCs and HFCs] and particularly upon the assumed tropospheric concentration of the hydroxyl radical. Measurements of atmospheric concentrations of 1,1,1-trichloroethane have been used to deduce globally averaged tropospheric OH concentrations, since this compound is exclusively anthropogenic, its rate of release is known from industry production figures[107] and its loss from the atmosphere is primarily due to reaction with OH. A recent re-evaluation[108] of the rate constant for reaction 111 has suggested that tropospheric OH concentrations should be ~15% higher than previously accepted, with the implication that the atmospheric lifetimes and ODPs of compounds whose main loss process is reaction with OH in the troposphere should be reduced by ~15% from the values presented in Table 2.

$$CH_3CCl_3 + OH \longrightarrow CH_2CCl_3 + H_2O \qquad (111)$$

The 'Montreal Protocol on Substances that Deplete the Ozone Layer' requires each signatory nation to reduce its production and consumption of the CFCs 11, 12, 113, 114 and 115 to 80% of their 1986 levels by 1993 and to 50% by 1998. Figure 8 shows that production levels of the first three of these has indeed fallen dramatically since 1988, according to data reported in Reference 109 by the major industrial producers. However, the 'Montreal' measures will have little effect on the current levels of stratospheric CFCs, which would still continue to rise for many years, as illustrated by Figure 9 for the example of CFC-12. It would be necessary to impose an 85% reduction in order to stabilize atmospheric concentrations at their 1989 level[110]. Even with a total cessation of CFC emission atmospheric concentrations will not be restored to their pre-1960 levels until well

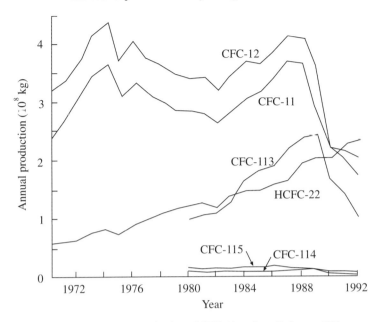

FIGURE 8. Annual production of CFCs (data from Reference 109)

into the twenty-second century. The provisions of the Montreal Review meeting have now established a deadline for phase-out of the controlled CFCs by the year 2000, except for a delay of ten years for developing countries. Some European community and other countries are adopting even stricter timetables for complete elimination of CFCs by dates between 1995 and 1997.

The necessity to avoid the use of CFCs has prompted much effort to find substitutes which possess the desirable properties of the CFCs without their undesirable and unacceptable properties. Thus HCFC-123 and HCFC-141b are prime candidates for replacement of CFC-11 in certain refrigeration systems and in foam blowing applications, and HFC-134a may replace CFC-12 in refrigeration systems, medical aerosols and certain foam blowing applications. Not only is it essential to find the right combination of physical properties, such as boiling point or heat of vaporization, chemical properties, and lack of flammability and toxicity, but it is also necessary to establish the probable mechanism for atmospheric degradation of these compounds. It may be that an HCFC or HFC undergoes breakdown in the troposphere as a result of attack by hydroxy radical, but the question arises as to the fate of the degradation products themselves: it is possible that oxidation might produce new halogenated compounds which could then either be transported into the stratosphere or else have harmful consequences for tropospheric chemistry? These matters are very much the concern of current research in this area.

B. Global Warming

The Earth's radiative and thermal balance depends on the incoming solar (UV and visible) energy and the outgoing infra-red radiation which is eventually lost to space. Infra-red energy is partially blocked at many wavelengths by naturally occurring gases, such as carbon dioxide, methane and stratospheric water vapour. Absorption of energy at the

J. S. Francisco and I. H. Williams

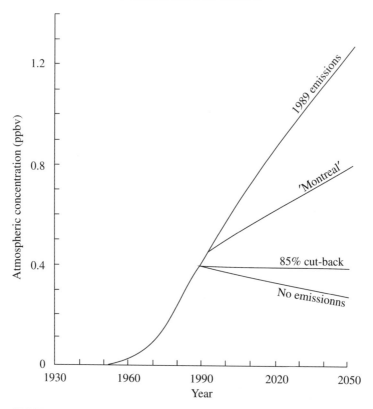

FIGURE 9. Model predictions of average atmospheric concentrations (parts per billion by volume) of CFC-12

fundamental frequencies chracterizing these molecular species causes local warming (the 'greenhouse effect') and is eventually re-radiated. Radiation may escape to space through the 'windows' between the absorption bands, but the significance of trace atmospheric gases including CFCs is that they tend to absorb infra-red radiation just in these regions of the spectrum, thus shutting up the windows. The stretching and bending vibrational modes of the C—Cl and C—F bonds of CFCs occur in the range of wavelengths between 8 and 12 μm which is otherwise virtually transparent. Thus CFCs become very effective greenhouse gases, especially in view of their long atmospheric lifetimes. Since the alternatives to CFCs may also absorb in this region of the spectrum, it is important to assess their potential impact upon the global climate in this manner.

A relative measure of the possible warming effect of the Earth's surface and the troposphere arising from each greenhouse gas is provided by the Global Warming Potential (GWP) as defined by the Intergovernmental Panel on Climate Change[111]. This depends on emissions of greenhouse gases and on their lifetimes, and the reference gas chosen is carbon dioxide. The GWP of a gas is calculated from its absorbing properties together with the amount present in the atmosphere over a specified length of time, known as the integration time horizon. The GWPs presented in Table 2 are for an integrated time horizon of one hundred years. However, carbon dioxide, CFCs, HCFCs and HFCs are purged from the atmosphere at very different rates. A century after a single emission of a typical HFC there would be none remaining in the atmosphere, but 41% of an emission of carbon dioxide would still remain. Therefore the 100-year time horizon does not provide a true comparison for many greenhouse gases since a substantial part of the global warming effect of the carbon dioxide reference is excluded. Nonetheless, these GWPs are of value in discussions of global warming[112].

The contribution of CFCs to global warming in the last decade is second only to carbon dioxide. As the GWPs in Table 2 show, the CFCs are much more potent than carbon dioxide on a kilogram-for-kilogram basis, although their total emissions are several orders of magnitude less. Many of the HCFCs and HFCs also have high GWPs, but are lower than those for CFCs in view of their shorter atmospheric lifetimes. Moreover, it is important to realize that HCFCs and HFCs may make contributions to energy efficiency that far outweigh potential warming from the compounds themselves. For a domestic refrigerator, the potential warming if the HCFC and HFC fluids in it were emitted would amount to 1% of the potential warming arising from the energy used to run the refrigerator during its life. The contribution of HCFCs and HFCs to global warming will be small compared to the contribution from other greenhouse gases. Calculations made upon the assumption of replacement of CFCs by HCFCs and HFCs suggest that the potential global warming from all fluorocarbons will be stabilized throughout the next century, in contrast to that from other greenhouse gases which will continue to rise.

IV. ACKNOWLEDGEMENTS

We are grateful to the J. S. Guggenheim Memorial Foundation for the award of a Fellowship (JSF) and to the UK Natural Environment Research Council for support from the Atmospheric Chemistry Initiative (IHW).

V. REFERENCES

1. R. P. Wayne, *Chemistry of Atmospheres*, 2nd ed., Oxford University Press, Oxford, 1991.
2. R. P. Wayne, *Sci. Prog. (Oxford)*, **74**, 379 (1990).
3. T. E. Graedel and P. J. Crutzen, *Sci. Am.*, **261**, 28 (1989).
4. P. Warneck, *Chemistry of the Natural Atmosphere*, Academic Press, San Diego, 1988.
5. *Scientific Assessment of Ozone Depletion: 1991*, World Meteorological Organization Global Ozone Research and Monitoring Project Report No. 25.
6. J. E. Lovelock, *Nature*, **256**, 193 (1975).
7. J. E. Lovelock, R. J. Maggs and R. J. Wade, *Nature*, **241**, 194 (1973).
8. H. B. Singh, L. J. Salas and R. E. Stiles, *J. Geophys. Res.*, **88**, 3684 (1983).
9. C. E. Reeves and S. A. Penkett, *Geophys. Res. Lett.*, **20**, 1563 (1993).
10. T. E. Graedel, L. D. Claxton and D. T. Hawkins (Eds.), *Atmospheric Chemical Compounds: Sources, Occurrence and Bioassay*, Academic Press, Orlando, 1986.
11. F. S. Rowland and M. J. Molina, *Rev. Geophys. Space Phys.*, **13**, 1 (1975).
12. S. Solomon, *Nature*, **347**, 347 (1990).
13. Y. L. Yung, J. P. Pinto, R. T. Watson and S. P. Sander, *J. Atmos. Sci.*, **37**, 339 (1980).
14. M. B. McElroy, R. J. Salawitch, S. C. Wofsy and J. A. Logan, *Nature*, **321**, 759 (1986).
15. M. J. Prather and R. T. Watson, *Nature*, **344**, 729 (1990).
16. M. J. Molina, T.-L. Tso, L. T. Molina and F. C.-Y. Wang, *Science*, **238**, 1253 (1987).
17. M. A. Tolbert, M. J. Rossi, R. Malhotra and D. M. Golden, *Science*, **238**, 1258 (1987).
18. L. T. Molina and M. J. Molina, *J. Phys. Chem.*, **91**, 433 (1987).
19. M. J. Molina, A. J. Colussi, L. T. Molina, R. N. Schindler and T.-L. Tso, *Chem. Phys. Lett.*, **173**, 310 (1990).
20. R. S. Stolarski, *Sci. Am.*, **258**, 20 (1988).
21. R. P. Wayne, *Proc. R. Inst.*, **61**, 13 (1990).
22. W. L. Chameides and D. D. Davies, *J. Geophys. Res.*, **85**, (C12), 7383 (1980).
23. K.-M. Jeong and F. Kaufman, *J. Phys. Chem.*, **86**, 1808 (1982).

24. R. Atkinson, *Chem. Rev.*, **86**, 69 (1986).
25. *Scientific Assessment of Stratospheric Ozone: 1989, Vol. 2, Appendix: AFEAS Report*, World Meteorological Organization Global Ozone Research and Monitoring Project Report No. 20.
26. C. J. Cobus, H. Hippler, K. Luther, A. R. Ravishankara and J. Troe, *J. Phys. Chem.*, **89**, 4332 (1985).
27. P. Borrell, C. J. Cobos, A. E. C. de Cobos, H. Hippler, K. Luther, A. R. Ravishankara and J. Troe, *Ber. Bunsenges. Phys. Chem.*, **89**, 337 (1985).
28. K. R. Ryan and I. C. Plumb, *J. Phys. Chem.*, **86**, 4678 (1982).
29. F. Caralp, R. Lesclaux and A. M. Dognon, *Chem. Phys. Lett.*, **129**, 433 (1986).
30. K. R. Ryan and I. C. Plumb, *Int. J. Chem. Kinet.*, **16**, 591 (1984).
31. J. J. Russell, J. A. Seetula, D. Gutman, F. Danis, F. Caralp., P. D. Lightfoot, R. Lesclaux, C. F. Melius and S. M. Senkan, *J. Phys. Chem.*, **94**, 3277 (1990).
32. F. Danis, F. Caralp, M. T. Rayez and R. Lesclaux, *J. Phys. Chem.*, **95**, 7300 (1991).
33. J. T. Niiranen and D. Gutman, *J. Phys. Chem.*, **97**, 4695 (1993).
34. F. F. Fenter, P. D. Lightfoot, J. T. Niiranen and D. Gutman, *J. Phys. Chem.*, **97**, 5313 (1993).
35. R. Atkinson and A. C. Lloyd, *J. Phys. Chem. Ref. Data*, **13**, 315 (1984).
36. A. M. Dognon, F. Caralp and R. Lesclaux, *J. Chim. Phys. Phys.-Chim. Biol.*, **82**, 349 (1985).
37. K. R. Darnell, W. P. L. Carter, A. M. Winer, A. C. Lloyd and J. N. Pitts, *J. Phys. Chem.*, **80**, 1948 (1976).
38. W. P. L. Carter and R. Atkinson, *J. Atmos. Chem.*, **8**, 165 (1989).
39. R. Atkinson, S. M. Aschmann, W. P. L. Carter, A. M. Winer and J. N. Pitts, *J. Phys. Chem.*, **86**, 4563 (1982).
40. M. E. Jenkin, R. A. Cox, G. D. Hayman and L. J. Whyte, *J. Chem. Soc.*, *Faraday Trans.2*, **84**, 913 (1988).
41. D. Gutman, N. Sanders and J. E. Butler, *J. Phys. Chem.*, **86**, 66 (1982).
42. E. W. Kaiser and T. J. Wallington, *J. Phys. Chem.* in press.
43. F. Wu and R. W. Carr, *J. Phys. Chem.*, **96**, 1743 (1992).
44. R. Lesclaux, A. M. Dognon and F. Caralp, *J. Photochem. Photobiol.*, *A Chem.*, **41**, 1 (1987).
45. Z. Li and J. S. Francisco, *J. Am. Chem. Soc.*, **111**, 5660 (1989).
46. J. S. Francisco, Z. Li and I. H. Williams, *Chem. Phys. Lett.*, **140**, 531 (1987).
47. J. Chen, T. Zhu and H. Niki, *J. Phys. Chem.*, **96**, 6115 (1992).
48. T. J. Bevilacqua, D. R. Hanson and C. J. Howard, *J. Phys. Chem.*, **97**, 3750 (1993).
49. K. C. Clemitshaw and J. R. Sodeau, *J. Phys. Chem.*, **91**, 3650 (1987).
50. J. Sehested and O. J. Nielsen, *Chem.Phys. Lett.*, **206**, 369 (1993).
51. A. R. Ravishankara, S. Solomon, A. A. Turnipseed and R. F. Warren, *Science*, **259**, 194 (1993).
52. J. S. Francisco, *J. Chem. Phys.*, **96**, 7597 (1992).
53. T. J. Wallington, personal communication.
54. J. S. Francisco and N. Mina-Camilde, *Can. J. Chem.*, **71**, 135 (1993).
55. J. S. Francisco and Y. Zhao, *J. Chem. Phys.*, **93**, 9203 (1990).
56. L. E. Giddings and K. K. Innes, *J. Mol. Spectrosc.*, **8**, 328 (1962).
57. C. W. Montgomery and G. K. Rollefson, *J. Am. Chem. Soc.*, **55**, 4025 (1933); **56**, 1089 (1934).
58. R. C. Hyer, S. M. Freund, A. Hartford and J. H. Atencio, *J. Appl. Phys.*, **52**, 6944 (1981).
59. D. R. Stull and H. Prophet, *JANAF Thermochemical Tables*, 2nd ed., NSRDS-NBS37, Washington, D.C., 1971.
60. G. L. Workman and A. B. F. Duncan, *J. Chem. Phys.*, **52**, 3204 (1970).
61. A. Nolle, H. Heydtmann, R. Meller, W. Schneider and G. K. Moortgat, *Geophys. Res. Lett.*, **19**, 281 (1992).
62. M. M. Maricq, J. J. Szente, Y. Su and J. S. Francisco, *J. Chem. Phys.*, **100**, 8673 (1994).
63. I. Zanon, G. Giacometti and D. Picciol, *Spectrochim. Acta*, **19**, 301 (1963).
64. Y. Zhao and J. S. Francisco, *Mol. Phys.*, **77**, 1187 (1992).
65. A. Mellouki, R. K. Talukdar, A. M. Schmoltner, T. Gierczak, M. J. Mills, S. Solomon and A. R. Ravishankar, *Geophys. Res. Lett.*, **19**, 2059 (1992).
66. H. G. Libuda, F. Zabel and K. H. Becker, *STEP-HALOCSIDE/AFEAS Workshop, Dublin, May 1991*, p. 126.
67. Y. Zhao and J. S. Francisco, *J. Phys. Chem.*, **96**, 7624 (1992).
68. J. S. Francisco, *J. Atmos, Chem.*, **13**, 285 (1993).
69. J. S. Francisco and I. H. Williams, *J. Am. Chem. Soc.*, **115**, 3746 (1993).
70. I. H. Williams, D. Spangler, D. A. Femec, G. M. Maggiora and R. L. Schowen, *J. Am. Chem. Soc.*, **105**, 31 (1983); M. R. Hand and I. H. Williams, unpublished work.

71. J. S. Francisco, A. N. Goldstein and I. H. Williams, *J. Chem. Phys.*, **89**, 3044 (1988).
72. M. M. Maricq, J. J. Szente, G. A. Khitrov and J. S. Francisco, *J. Chem. Phys.*, **98**, 9522 (1993).
73. J. M. Heras, A. J. Arvia and P. J. Aymonino, *Z. Phys. Chem.*, *N.F.*, **28**, 250 (1961); *Ann. Assoc. Quim. Argent.*, **50**, 120 (1962).
74. D. E. Milligan, M. E. Jacox, A. M. Bass, J. J. Comeford and D. E. Mann, *J. Chem. Phys.*, **42**, 3187 (1965).
75. M. E. Jacox, *J. Mol. Spectrosc.*, **80**, 257 (1980).
76. K. Nagai, G. Yamada, Y. Endo and E. Hirota, *J. Mol. Spectrosc.*, **90**, 249 (1981).
77. M. M. Maricq, J. J. Szente, G. A. Khitrov and J. S. Francisco, *Chem. Phys. Lett.*, **199**, 71 (1992).
78. M. M. Maricq, J. J. Szente, Y. Su and J. S. Francisco, *J. Chem. Phys.*, **100**, 8673 (1994).
79. T. J. Wallington, T. Ellermann, O. J. Nielsen and J. Sehested, *J. Phys. Chem.*, **98**, 2346 (1994).
80. H. Heydtmann, in AFEAS workshop on Atmospheric Degradation of HCFCs and HFCs, Boulder, Colorado, November 1993.
81. K. R. Ryan and I. C. Plumb, *Plasma Chem. Plasma Process*, **4**, 271 (1984).
82. G. Hancock and D. E. Heard, *J. Chem. Soc., Faraday Trans.*, **87**, 1039 (1991).
83. J. S. Francisco and A. Ostafin, *J. Phys. Chem.*, **94**, 6337 (1990).
84. M. M. Maricq, J. J. Szente, T. S. Dibble and J. S. Francisco, *J. Phys. Chem.*, **98**, 12294 (1994).
85. T. S. Dibble and J. S. Francisco, *J. Phys. Chem.*, **98**, 5010 (1994).
86. F. Zabel, A. Reimer, K. H. Becker and E. H. Fink, *J. Phys. Chem.*, **93**, 5500 (1989).
87. F. Kirchner, F. Zabel and K. H. Becker, *AFEAS Workshop, Dublin, May 1991*.
88. A. A. Jemi-Alade, P. D. Lightfoot and R. Lesclaux, *Chem. Phys. Lett.*, **179**, 119 (1991).
89. E. O. Edney, B. W. Gay and D. J. Driscoll, *J. Atmos. Chem.*, **12**, 105 (1991).
90. J. S. Francisco, Z. Li, A. Bradley and A. E. W. Knight, *Chem. Phys. Lett.*, **214**, 77 (1993).
91. T. J. Wallington, M. D. Hurley, J. C. Ball and E. W. Kaiser, *Environ. Sci. Technol.*, **26**, 1318 (1992).
92. A. F. Wagner, I. R. Slagle, D. Sarzynski and D. Gutman, *J. Phys. Chem.*, **94**, 1853 (1990).
93. S. Dobe, L. A. Kachatryan and T. Berces, *Ber. Bunsenges. Phys. Chem.*, **93**, 847 (1989).
94. D. J. Scollard, J. J. Treacy, H. W. Sidebottom, G. Balestra-Garcia, G. Laverdet, G. LeBrass, H. Macleod and S. Teton, *J. Phys. Chem.*, **97**, 4683 (1993).
95. T. J. Wallington and M. D. Hurley, *Int. J. Chem. Kinet.*, **25**, 819 (1993).
96. J. S. Francisco, *J. Phys. Chem.*, **96**, 4894 (1992).
97. J. S. Francisco and I. H. Williams, *Spectrochim. Acta*, **48A**, 1115 (1992).
98. B. G. Tucker and E. Whittle, *Trans. Faraday Soc.*, **63**, 80 (1967).
99. M. Kozuka and T. Isobe, *Bull. Chem. Soc. Jpn.*, **49**, 1766 (1976).
100. Z. Li and J. S. Francisco, *Mol. Phys.*, **79**, 1127 (1993).
101. M. M. Maricq, J. J. Szente, G. Khitrov and J. S. Francisco, in press.
102. J. S. Francisco, *Chem. Phys. Lett.*, **191**, 7 (1992).
103. M. M. Maricq, J. J. Szente, G. Khitrov and J. S. Francisco, in press.
104. D. J. Wuebbles, *J. Geophys. Res.*, **88**, 1433 (1983).
105. J. A. Pyle, S. Solomon, D. Wuebbles and S. Zvenigorodsky in *Scientific Assessment of Ozone Depletion: 1991*, Chap. 6, World Meteorological Organization Global Ozone Research and Monitoring Project Report No. 25, WMO, Geneva, 1992.
106. M. J. Prather, A. M. Ibrahim, T. Sasaki, F. Stordal and G. Visconti, in *Scientific Assessment of Ozone Depletion: 1991*, Chap. 8, World Meteorological Organization Global Ozone Research and Monitoring Project Report No. 25, WMO, Geneva, 1992.
107. P. Midgley, *Atmos. Environ.*, **23**, 2663 (1989).
108. R. K. Talukdar, A. Mellouki, A.-M. Schmoltner, T. Watson, S. Montzka and A. R. Ravishankara, *Science*, **257**, 227 (1992).
109. *Production, Sales and Atmospheric Release of Fluorocarbons through 1992*, Alternative Fluorocarbons Environmental Acceptability Study, Washington, D.C., 1993.
110. *Stratospheric Ozone 1988*, U.K. Stratospheric Ozone Review Group, Her Majesty's Stationery Office, London, 1988.
111. J. T. Houghton, G. J. Jenkins and J. J. Ephraums (Eds.), *Climate Change: The IPCC Scientific Assessment*, Cambridge University Press, 1990, pp. 41–64.
112. S. K. Fischer, P. J. ¡Hughes, P. D. Fairchild, C. L. Kusik, J. T. Dieckmann, E. M. McMahon and N. Hobday, *Energy and Global Warming Impacts of CFC Alternative Technologies*, Alternative Fluorocarbons Environmental Acceptability Study, Washington, D. C., 1991.

Author index

This author index is designed to enable the reader to locate an author's name and work with the aid of the reference numbers appearing in the text. The page numbers are printed in normal type in ascending numerical order, followed by the reference numbers in parentheses. The numbers in *italics* refer to the pages on which the references are actually listed.

Author index

Index compiled by K. Raven

Subject index

Index compiled by P. Raven